Devonian Climate, Sea Level and Evolutionary Events

IUGS/GSL publishing agreement

This volume is published under an agreement between the International Union of Geological Sciences and the Geological Society of London and arises from the joint efforts of IGCP 596 'Climate Change and Biodiversity Patterns in the Mid-Palaeozoic' and the International Subcommission on Devonian Stratigraphy (SDS).

GSL is the publisher of choice for books related to IUGS activities, and the IUGS receives a fee for all books published under this agreement.

Books published under this agreement are subject to the Society's standard rigorous proposal and manuscript review procedures.

It is recommended that reference to all or part of this book should be made in one of the following ways:

Becker, R. T., Königshof, P. & Brett, C. E. (eds) 2016. *Devonian Climate, Sea Level and Evolutionary Events*. Geological Society, London, Special Publications, **423**.

Kaiser, S. I., Aretz, M. & Becker, R. T. 2016. The global Hangenberg Crisis (Devonian–Carboniferous transition): review of a first-order mass extinction. *In*: Becker, R. T., Königshof, P. & Brett, C. E. (eds) *Devonian Climate, Sea Level and Evolutionary Events*. Geological Society, London, Special Publications, **423**, 387–437. First published online November 11, 2015, http://doi.org/10.1144/SP423.9

GEOLOGICAL SOCIETY SPECIAL PUBLICATION NO. 423

Devonian Climate, Sea Level and Evolutionary Events

EDITED BY

R. T. BECKER
Westfälische Wilhelms-Universität, Germany

P. KÖNIGSHOF
Forschungsinstitut und Naturmuseum Senckenberg, Germany

and

C. E. BRETT
University of Cincinnati, USA

2016
Published by
The Geological Society
London

THE GEOLOGICAL SOCIETY

The Geological Society of London (GSL) was founded in 1807. It is the oldest national geological society in the world and the largest in Europe. It was incorporated under Royal Charter in 1825 and is Registered Charity 210161.

The Society is the UK national learned and professional society for geology with a worldwide Fellowship (FGS) of over 10 000. The Society has the power to confer Chartered status on suitably qualified Fellows, and about 2000 of the Fellowship carry the title (CGeol). Chartered Geologists may also obtain the equivalent European title, European Geologist (EurGeol). One fifth of the Society's fellowship resides outside the UK. To find out more about the Society, log on to www.geolsoc.org.uk.

The Geological Society Publishing House (Bath, UK) produces the Society's international journals and books, and acts as European distributor for selected publications of the American Association of Petroleum Geologists (AAPG), the Indonesian Petroleum Association (IPA), the Geological Society of America (GSA), the Society for Sedimentary Geology (SEPM) and the Geologists' Association (GA). Joint marketing agreements ensure that GSL Fellows may purchase these societies' publications at a discount. The Society's online bookshop (accessible from www.geolsoc.org.uk) offers secure book purchasing with your credit or debit card.

To find out about joining the Society and benefiting from substantial discounts on publications of GSL and other societies worldwide, consult www.geolsoc.org.uk, or contact the Fellowship Department at: The Geological Society, Burlington House, Piccadilly, London W1J 0BG: Tel. +44 (0)20 7434 9944; Fax +44 (0)20 7439 8975; E-mail: enquiries@geolsoc.org.uk.

For information about the Society's meetings, consult *Events* on www.geolsoc.org.uk. To find out more about the Society's Corporate Affiliates Scheme, write to enquiries@geolsoc.org.uk.

Published by The Geological Society from:
The Geological Society Publishing House, Unit 7, Brassmill Enterprise Centre, Brassmill Lane, Bath BA1 3JN, UK

The Lyell Collection: www.lyellcollection.org
Online bookshop: www.geolsoc.org.uk/bookshop
Orders: Tel. +44 (0)1225 445046, Fax +44 (0)1225 442836

British Library Cataloguing in Publication Data

A catalogue record for this book is available from the British Library.
ISBN 978-1-86239-734-7
ISSN 0305-8719

Distributors
For details of international agents and distributors see:
www.geolsoc.org.uk/agentsdistributors

Typeset by Nova Techset Private Limited, Bengaluru and Chennai, India.
Printed and bound by CPI Group (UK) Ltd, Croydon CR0 4YY

Contents

Devonian climate, sea level and evolutionary events: an introduction

R. T. BECKER[1]*, P. KÖNIGSHOF[2] & C. E. BRETT[3]

[1]*Institut für Geologie und Paläontologie, Westfälische Wilhelms-Universität, Corrensstrasse 24, D-48149 Münster, Germany*

[2]*Forschungsinstitut und Naturmuseum Senckenberg, Senckenberganlage 25, D-60325 Frankfurt am Main, Germany*

[3]*Department of Geology, University of Cincinnati, 500 Geology/Physics Building, Cincinnati, OH 45221-0013, USA*

**Corresponding author (e-mail: rbecker@uni-muenster.de)*

The face of Planet Earth has changed significantly through geological time. Dynamic processes active today, such as plate tectonics and climate change, have shaped the Earth's surface and impacted biodiversity patterns from the beginning. Organisms, on the other hand, have the capacity to significantly alter Earth's hydrological and geochemical cycles, its atmosphere and climate, sediments, and even hard rocks deep down under the surface. Abiotic–biotic interactions characterize Earth's system history and, together with biotic competition and food webs, were the main trigger of evolutionary change, innovations and biodiversity fluctuations. Within the Palaeozoic, the Devonian was an especially interesting time interval as it was characterized by the 'mid-Paleozoic predator revolution' (Signor & Brett 1984; Brett 2003) and the related 'nekton revolution' (Klug *et al.* 2010), characterized by the blooms of free-swimming cephalopods, including the oldest ammonoids, and fish groups (e.g. toothed sharks and giant placoderms), the rise of more advanced vertebrates, including the oldest tetrapods (e.g. Blieck *et al.* 2007, 2010; Niedzwiedzki *et al.* 2010), the most extensive reef complexes of the Phanerozoic (e.g. Kiessling 2008), and the 'greening of land' by the diversification and spread of land plants, including the oldest forests (e.g. Stein *et al.* 2012; Giesen & Berry 2013), which resulted in new soil types and changing weathering.

These major evolutionary trends did not unfold in a long interval of environmental stability, but in times of numerous and repeated, geologically brief, global events that punctuated prolonged periods, up to several million years in duration, of relative stability, termed ecological-evolutionary subunits (EE subunits: Boucot 1990; Brett & Baird 1995; Brett *et al.* 2009). The bounding events, even those of lesser intensity, produced major re-structuring in local to global ecosystems and are seen as critical drivers of long-term evolutionary patterns (Brett 2012). These linked abiotic and biotic events and extinctions of different magnitude have been summarized by House (1983, 1985, 2002), Walliser (1984, 1996) and, more recently, by Becker *et al.* (2012). The Devonian event succession is summarized in Figure 1. Two first-order mass extinctions at the Frasnian–Famennian boundary (Kellwasser Crisis) and at the end of the Devonian (Hangenberg Crisis), characterized by the loss of major fossil groups (classes and orders) and complete ecosystems (e.g. metazoan reefs, early forests), have to be viewed in the context of a complex global event sequence. There are important similarities between discrete pulses/phases of the major biotic crises and individual smaller-scale events. In our understanding, second-order global events are characterized by sudden extinctions in many groups and ecosystems, including the complete disappearance of several widespread and diverse organism groups (orders and families). Examples are the basal Emsian *atopus* Event, where the planktonic graptolites finally died out, the Taghanic Crisis, Frasnes events and Lower Kellwasser Event. Third-order global events show globally elevated extinction rates, often at lower taxonomic level (genera and species), but within many clades and in several ecosystems. Examples are the Silurian–Devonian boundary Klonk Event, and the Daleje, Chotěc, Kačak, Condroz and *Annulata* events. Fourth-order global extinctions refer to the sudden disappearance of relatively fewer but widespread groups, which implies a global, not regional, trigger. This category may include the Lochkovian–Pragian boundary

From: BECKER, R. T., KÖNIGSHOF, P. & BRETT, C. E. (eds) 2016. *Devonian Climate, Sea Level and Evolutionary Events*. Geological Society, London, Special Publications, **423**, 1–10.
First published online August 5, 2016, http://doi.org/10.1144/SP423.15

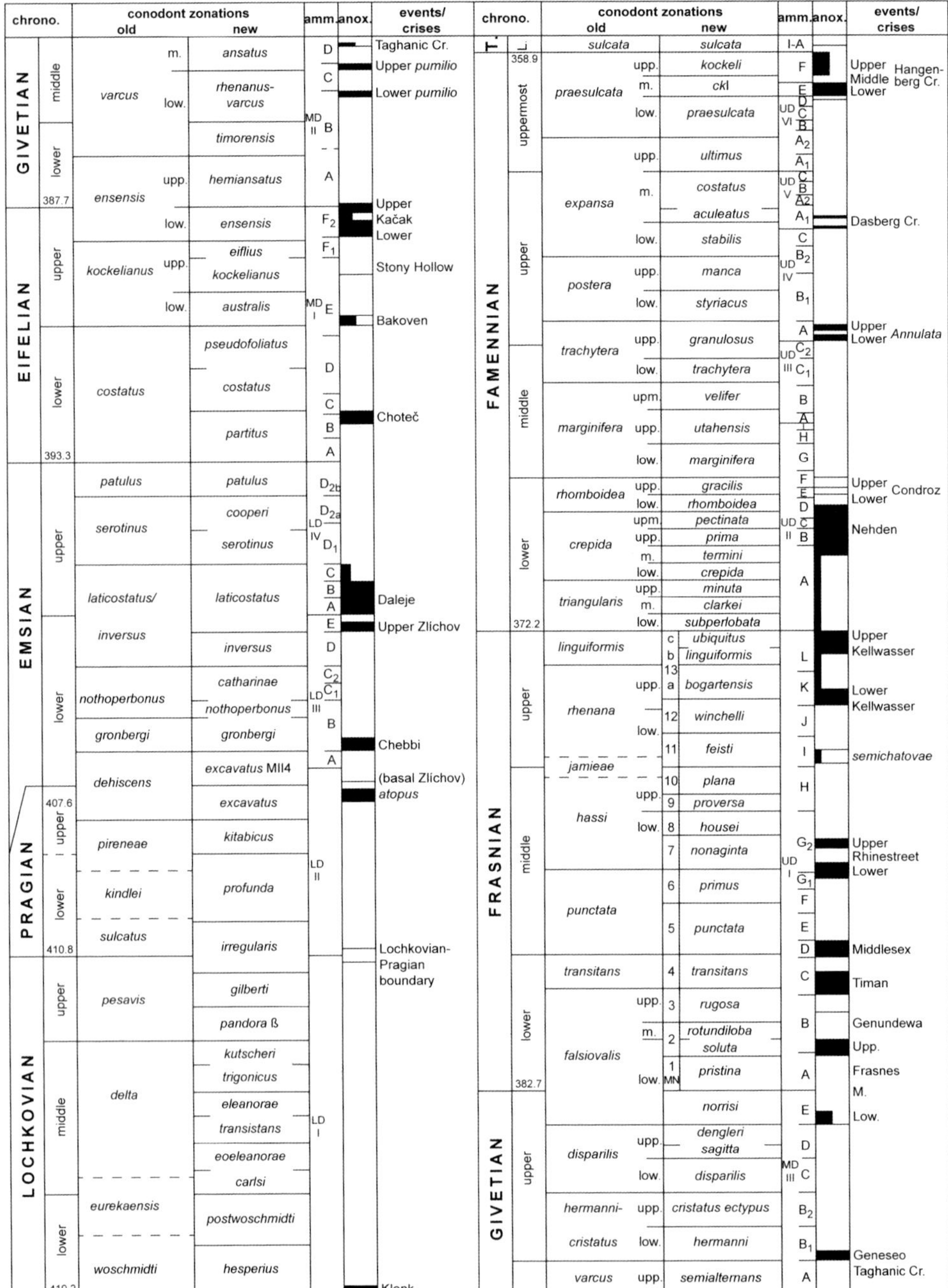

Fig. 1. The succession of Devonian global events and crises (well-known examples in bold), including maxima of anoxic sedimentation (e.g. black shales), plotted against the chronostratigraphic scale, old and new conodont zonations (e.g. Klapper 1989; Yolkin *et al.* 1994; Klapper & Becker 1999; Aboussalam 2003; Slavík & Hladil 2004; Girard *et al.* 2005; Murphy 2005; Aboussalam & Becker 2007; Slavík *et al.* 2007, 2012; Kaiser *et al.* 2009; Hartenfels 2011; Corradini & Corriga 2012; Aboussalam *et al.* 2015; Spalletta *et al.* 2015; Valenzuela-Ríos *et al.* 2015), and the zonal key for global ammonoid zones (Becker & House 2000). The substage levels shown here follow proposals to SDS but have only partly been ratified (Givetian, Frasnian) and not yet formally approved by the International Stratigraphic Commission. The given absolute age number for the base of each stage is adopted from the calibration and interpolation in Becker *et al.* (2012).

event, and the Chebbi, Upper Zlíchov, Middlesex and Dasberg events. Other intervals, such as the *pumilio*, Timan, Rhinestreet, *semichatovae* and Nehden events, are better characterized as faunal blooms, radiations and sudden organism spread with transgression rather than as extinctions (Fig. 2). The Bakoven and Stony Hollow events are strong in the Appalachian Basin, where their impact is locally far greater than the Chotěc Event (DeSantis & Brett 2011), and represent major abrupt faunal turnovers. They have been recognized elsewhere (e.g. Spain) but may not be as strong, making them third- or fourth-order events. The ranking of the named individual events may change with continuing research and additional small-scale global extinctions will probably be recognized in the future.

The widely known Devonian black shale events (Fig. 2) are linked with hypoxia/anoxia, rapid

Fig. 2. Example of thin black shales interrupting a strongly cyclic middle–upper Famennian pelagic nodular limestone succession: the two *Annulata* Event beds (top lower third) and the Dasberg Black Shale (base of the upper third) at Effenberg Quarry, northern Rhenish Massif, Germany (photograph by S. Hartenfels).

eustatic fluctuations, faunal blooms, geochemical anomalies, and stepped extinctions and survival. But not all global events were associated with the spread of anoxia in times of climatic warming and eustatic rise, others correlate with sharp regressions and climate cooling. Clearly, multiple oceanographic factors were involved. But sudden climate change appears to have been the most important common trigger, possibly linked with episodes of massive volcanism and times of significant drawdown of atmospheric CO_2. Currently, there is no unequivocal example of bolide impact events as global extinction triggers in the Devonian, although some impacts are now well known, such as the lower Frasnian Flynn Creek crater of Tennessee (Schieber & Over 2005) and the middle Frasnian Alamo impact of Nevada (e.g. Warme & Sandberg 1996; Poole & Sandberg 2015).

Despite the summarized general patterns, knowledge of many of the individual events and even a deeper understanding of the two major crises is still at an early stage, with more open questions than answers (e.g. Racki 2005). For many of the second- or third-order global events there are no summaries of their recognition in different regions or of event patterns on different continents and in different facies realms. The correlation with the listed major evolutionary developments is partly arbitrary, and the integration of biotic, sedimentological and geochemical data in sound event models is still missing or preliminary. An understanding of the major abiotic and biotic factors that shaped our planet and its ecosystems in the past is essential to the understanding of natural processes and the effects of human-induced global change at present and for forecasting possible future developments. For the Devonian period, two volumes in this book series (Becker & Kirchgasser 2007; Königshof 2009) have previously introduced interdisciplinary and international approaches to achieve progress in Devonian event research. This resulted from the fruitful cooperation of IGCP 499 on 'Devonian Land–Sea Interaction – Evolution of Ecosystems and Climate' (DEVEC) and the IUGS International Subcommission on Devonian Stratigraphy (SDS). Since then, a follow-up project, IGCP 596 on 'Climate Change and Biodiversity Patterns in the Mid-Palaeozoic' picked up open and new research directions, which were the subject of numerous international symposia and during joint IGCP 596–SDS field trips: for example, the 2013 field symposium on the Devonian and Lower Carboniferous of southern Morocco (Fig. 3), an area world famous for its mid-Palaeozoic sediments and fossils. As research on mid-Palaeozoic anoxic events was mainly concentrated around the former Rheic Ocean (e.g. Europe, USA and Morocco), some satellite projects were established in order to get a better understanding of mid-Palaeozoic events around the eastern Proto- or Palaeotethys Ocean: for example, in Vietnam and Thailand. Combined biostratigraphy, sedimentology, physical stratigraphy and isotope geochemistry in different facies settings will be one of the future challenges in order to better understand the evolution of ecosystems and climate change. In this respect, and as an outlook, research on varied and regionally distinctive shallow-water successions should be a focus in the near future. As mentioned earlier, the fruitful cooperation between IGCP and SDS produced a very large number of scientific papers and greatly increased knowledge in many respects. But there is still much work to be done in order to fully understand the complexity of abiotic–biotic interactions. There will be key roles for the integration of new and different techniques, for an expansion into currently poorly known territory (new and neglected sections/regions), and, equally important, for a thorough integration of the wealth of strongly dispersed data on different fossil groups and regions, both in datasets and in models.

Some contributions of this volume were first presented at the 2013 meeting or subsequently, for example at the IGCP 596–SDS Symposium in Brussels in September 2015 (Mottequin *et al.* 2015).

The primary goal of IGCP 596 was to assess the intensity of mid-Palaeozoic climate change and its impact on biodiversity both in marine and terrestrial successions. Key questions included:

- Was CO_2 the dominant driver of climate and how did it affect global or regional diversity?
- What other factors ruled short- and long-term biodiversity at that time?
- What caused the cooling and sudden glaciation episodes, and were there similarities between these events (e.g. CO_2 thresholds)?
- How did environmental change stimulate or influence evolutionary innovations?
- Were major marine and terrestrial biotic changes linked with Milankovitch cyclicity?
- Were there different extinction–innovation patterns in different facies and distant basins?
- How did ocean chemistry change in the Mid-Palaeozoic (e.g. ocean acidification), and what was its impact on marine organisms and the fossil record
- Is climate modelling a viable tool for understanding mid-Palaeozoic environmental change?
- Do current climate models fit the available geological and palaeontological evidence?

From the SDS side, investigations were added that were aimed at achieving a much greater time resolution for global correlation based on an increasingly refined geological timescale. Stratigraphic progress provides a more precise range of organisms

Fig. 3. Participants of the joint IGCP 596–SDS Field Symposium in the eastern Anti-Atlas of SE Morocco, March 2013, including the volume editors and Moroccan hosts, at the famous Merzouga sand dunes south of Rissani and Erfoud.

in time and space, and also of their ecosystem distributions, and, therefore, a more precise dating of abiotic and evolutionary events, including geochemical spikes or isotope-based palaeotemperature reconstructions. The focus of SDS is on a revision of the eustatic sea-level curve, which can be linked with climate pulses, and the integration of biostratigraphy with modern stratigraphic techniques, such as isotope, sequence, magneto-, element geochemistry, gamma-ray and quantitative stratigraphy. This approach is followed in the present volume, which combines recent efforts of both IGCP 596 and SDS members. This has resulted in 14 chapters that are sorted roughly according to stratigraphic age. Contributions cover all of the Devonian and successions from 10 different basins in Europe, North Africa, North and South America, Australia, and the isolated Falkland Islands in the South Atlantic. The 37 co-authors are from Europe, North America and Australia, but other areas of Africa, South America and Asia, especially of China, were treated in the previously mentioned two earlier volumes, which should be considered for further reading.

The first chapter by **Suttner & Kido (2015)** deals with a widely neglected (fourth-order) extinction and positive carbon isotope signal near the Lochkovian–Pragian boundary that was first noted by Walliser (1996). In the Carnic Alps, it correlates with a regressive phase that culminated in a peculiar megaclast unit. This study stands out as an example of the interdisciplinary approach by combining conodont biostratigraphy, sedimentology, carbon isotope analysis, magnetic susceptibility data and sediment colour analysis. Interpretations include a sedimentary model and, to address the main IGCP 591 topics, a discussion of the regional conodont diversity curve.

The paper by **Marshall (2016)** summarizes the, as yet, poorly known stratigraphy of the Devonian of the Falkland Island, mostly based on palynomorphs. As is typical for the Malvinokaffric cold-water realm, there is a thick siliciclastic succession that ranges from the Lochkovian to the upper Famennian. It is correlated with the originally adjacent Cape Basin of South Africa and with the, today, closer South American Devonian. This comparison enabled the recognition of transgressive and regressive events, which are tentatively correlated with the established Devonian event succession. This research forms a significant contribution because, for the first time, it traces some of the (sub)tropical event intervals, notably the Chotěc and Kačák events, into the high latitudes, where the restricted invertebrate record has, so far, only occasionally been tested for extinction and survival patterns.

The extensive review of the Lower Devonian brachiopod succession of the Rhenish Massif by **Jansen (2016)** provides a modern synthesis of regional litho-, bio- and event stratigraphy. It provides a new and revised synopsis of a wealth of data from older literature. Because the western Rhenish Massif is one of the classic regions for the study of brachiopod faunas, this review will become a new standard for the understanding of Lower Devonian nearshore clastic successions, their faunas and biofacies. There are important implications for the correct correlation of well-known regional chronostratigraphic terms (Gedinnian, Siegenium, type Emsian), which have been used outside their type regions. It is also an important case study that traces established outer-shelf events into the more complex inner-shelf and marginal-marine settings, where biotic turnovers appear to have been much more numerous.

Brocke *et al.* (2015) attempt successfully to correlate Basal Chotěc Event successions of the Bohemian type area into the distant Appalachian foreland basin of eastern North America. This chapter provides many new records of palynomorphs and dacryoconarids, and a welcome brief summary of the global distribution of the third-order event, which is marked by moderate extinctions in selected organism groups. In the studied sections, there are sudden blooms of prasinophyte algae and specific spores. The associated eustatic rise permitted important migrations, providing a case study of the biogeographical implications of Devonian short-term hypoxic and transgressive events. As known from other events, the episodic breakdown of biogeographical barriers can have a profound effect on global biodiversity.

Königshof *et al.* (2015) apply a multidisciplinary method set, including conodont biostratigraphy, microfacies analysis, stable isotopes, magnetic susceptibility, total organic carbon content and whole-rock geochemistry, in order to find signatures of the third-order global Kačák Event in a typical shallow-water carbonate succession of the Eifelian type region of Germany, where physical event manifestations are much less obvious than in contemporaneous pelagic limestones. This proved to be a challenge, but rare conodonts, a positive carbon isotope excursion and a cerium anomaly enabled the authors to pinpoint the beginning of the larger event interval. If lateral correlation can be achieved, it will help to refine the extinctions, innovations and palaeoecological changes of the rich Eifel macrofaunas in the course of the Kačák Event.

Two successive chapters deal with the second-order global Taghanic Crisis at the middle–upper Givetian boundary. **Narkiewicz *et al.* (2015)** reviewed the conodont record of SE Poland in order to elaborate the existing Givetian conodont biofacies model and to document onshore–offshore biofacies shifts with the major, eustatic Taghanic Onlap. Results were transferred to interpret the conodont records of widely separated basins of North America, southern Europe and North Africa. This encountered some difficulties that have to be pursued in future studies. The significance of the contributions lies in the development of a refined biofacies tool that can trace Givetian sea-level movements by quantitative faunal analysis. **Zambito *et al.* (2015)**, on the other hand, further develop carbon isotope stratigraphy as a tool to define pulses of the Taghanic Crisis. This paper provides the first detailed isotope study of the Taghanic type region in New York State, USA. Results suggest that in particular the upper crisis interval can be found easily, possibly even in facies, where other dating methods are not at hand.

Two chapters focus on the Kellwasser Crisis at the Frasnian–Famennian boundary. **Mottequin & Poty (2015)** summarize the brachiopod and coral evidence of the Ardennes, which in the middle and upper Frasnian can be assigned to a succession of clearly defined third-order sequences. The end of regionally famous reef episodes (mostly mud-mounds) was linked with regressions and sequence boundaries, whilst the Lower and Upper Kellwasser levels represent maximum transgressions. The upper Frasnian brachiopod and coral extinctions occurred in discrete steps, and only a few taxa were left at the time of the basinwide final spread of dysoxic/anoxic Matagne facies. The authors propose that the regional Upper Kellwasser beds were triggered by tsunamis, an idea that will probably be challenged and tested. It has long been known that the Kellwasser Crisis was a first-order extinction for trilobites, which terminated long-lived orders and families, including several that had existed since the Ordovician. Based on the faunal record of the Canning Basin of NW Australia, which is currently the best on a global scale, **McNamara & Feist (2015)** summarize not only the regional trilobite succession, with new taxa, but focus on the astonishing parallel morphological and evolutionary trends in associated families. This evolutionary-ecological approach gives new insights into end-Frasnian shallow-marine ecosystems, where nutrient (and food) availability seems to have been a major factor controlling adaptation, radiation and extinctions. A third paper, by **Hairapetian *et al.* (2015)**, deals indirectly with aspects of the Frasnian–Famennian global biotic crisis, and includes descriptions of the newly discovered and globally youngest thelodonts from Iran and Australia. These findings prove a survival of this fish group into the middle Famennian, which is remarkable given that armoured agnathans had been thought to have died out during the Upper Kellwasser Event. Old and new records are also placed into a palaeobiogeographical context.

Hartenfels & Becker (2016) summarize the global record of the third-order basal upper Famennian *Annulata* events, which can be recognized in more than 40 (sub)tropical basins in all continents belonging to that climatic realm. Records are assigned to 10 different tectonosedimentary event settings, which is important for distinguishing extinction v. survival and palaeoecological effects in specific environments. Based on the first event records of the Rheris Basin in southern Morocco, which yielded some regionally unique ammonoids and conodonts, the authors also establish distinctive event biofacies types. These reflect local differences of palaeobathymetry, nutrient availability, organic productivity and ventilation. Such an approach should be extended in the future to other Devonian hypoxic/anoxic events.

The final three chapters deal with the first-order global Hangenberg Crisis and short-lived end-Famennian–Lower Carboniferous glaciations. **Becker *et al.* (2016)** summarize the wealth of past and new information to provide a detailed review of chrono-, litho- and biostratigraphy around the Devonian–Carboniferous boundary. Based on these data, the authors refine the precise time framework for the stepped mass extinction, and contribute to the current revision of the Devonian–Carboniferous chronostratigraphic boundary. The Rhenish Massif, as the type region of the Hangenberg Crisis, is used as a standard to establish clearly defined lower, middle and upper crisis intervals. Building on this review, **Kaiser *et al.* (2015)** summarize the variably well-known or still insufficiently studied extinction and survival patterns for all relevant organism groups around the Devonian–Carboniferous boundary. Based on a detailed literature review, they compile the current knowledge of abiotic event aspects from impact signatures to the timing of glacial deposition, isotope excursions and the question of nutrient sources. Considering parallels with better understood Cretaceous events, all available data are synthesized in a new crisis model, which is open to future testing. A list of knowledge gaps and research directions shows how far we still are from a full understanding of this undoubted first-order mass extinction and global ecosystem overturn. **Lakin *et al.* (2016)** review the precise ages of uppermost Famennian–Visean glacial deposits, especially of South America, and provide evidence for three 'precursor glaciations' before the onset of long-term (Upper Carboniferous–Permian) icehouse climates. They stress the current uncertainties of the mostly palynomorph-based dating, especially for the onset of glaciations, when still few diamictites were formed. As with the review by **Kaiser *et al.* (2015)**, this contribution ends with a plea for more data that can better constrain the extent of glaciation in time and space.

We wish to dedicate this volume to the memory of members of IGCP 596, SDS and the Devonian–Carboniferous Boundary Task Group who, sadly and unexpectedly, passed away since our 2013 Morocco Meeting, which was the starting point for this book. These esteemed colleagues are: Kolya Bakharev (Russia), Mena Schemm-Gregory (Portugal/Germany), Vladimir Nikolaevich Pazukhin (Russia), Paul Sartenaer (Belgium), Chen Xiuqin (China) and R. Lane (USA).

The editors express thanks to all contributing authors and to the many reviewers, who invested their knowledge and time to ensure a high-quality publication. We are convinced this volume will form an enduring source of valuable information concerning Devonian biotas and stratigraphy, and will, in particular, form the foundation for a deeper understanding of the complex global climatic and environmental changes that repeatedly shaped mid-Palaeozoic ecosystems and evolutionary patterns. The editorial team of the Geological Society, notably Angharad Hills and Jo Armstrong, was most helpful in getting this volume completed and published.

References

Aboussalam, Z.S. 2003. Das 'Taghanic Event' im höheren Mittel-Devon von West-Europa und Marokko. *Münstersche Forschungen zur Geologie und Paläontologie*, **97**, 1–332.

Aboussalam, Z.S. & Becker, R.T. 2007. New upper Givetian to basal Frasnian conodont faunas from the Tafilalt (Anti-Atlas, Southern Morocco). *Geological Quarterly*, **51**, 345–374.

Aboussalam, Z.S., Becker, R.T. & Bultynck, P. 2015. Emsian (Lower Devonian) conodont stratigraphy and correlation of the Anti-Atlas (Southern Morocco). *Bulletin of Geosciences*, **90**, 893–980.

Becker, R.T. & House, M.R. 2000. Devonian ammonoid zones and their correlation with established series and stage boundaries. *Courier Forschungsinstitut Senckenberg*, **169**, 79–135.

Becker, R.T. & Kirchgasser, W.T. (eds) 2007. *Devonian Events and Correlations*. Geological Society, London, Special Publications, **278**, http://sp.lyellcollection.org/content/278/1

Becker, R.T., Gradstein, F.M. & Hammer, O. 2012. The Devonian Period. *In*: Gradstein, F.M., Ogg, J.G., Schmitz, M. & Ogg, G. (eds) *The Geologic Time Scale 2012, Volume 2*. Elsevier, Amsterdam, 559–601.

Becker, R.T., Kaiser, S.I. & Aretz, M. 2016. Review of chrono-, litho- and biostratigraphy across the global Hangenberg Crisis and Devonian–Carboniferous Boundary. *In*: Becker, R.T., Königshof, P. & Brett, C.E. (eds) *Devonian Climate, Sea Level and*

Evolutionary Events. Geological Society, London, Special Publications, **423**. First published online January 6, 2016, http://doi.org/10.1144/SP423.10

Blieck, A., Clément, G. *et al.* 2007. The biostratigraphical and palaeogeographical framework of the earliest diversification of tetrapods (Late Devonian). *In*: Becker, R.T. & Kirchgasser, W.T. (eds) *Devonian Events and Correlations*. Geological Society, London, Special Publications, **278**, 219–235, http://doi.org/10.1144/SP278.10

Blieck, A., Clément, G. & Streel, M. 2010. The biostratigraphical distributrion of earliest tetrapods (Late Devonian) – a revised version with comments on biodiversification. *In*: Vecoli, M., Clément, G. & Meyer-Berthaud, B. (eds) *The Terrestrialization Process: Modelling Complex Interactions at the Biosphere–Geosphere Interface*. Geological Society, London, Special Publications, **339**, 129–138, http://doi.org/10.1144/SP339.11

Boucot, A.J. 1990. Silurian and pre-Upper Devonian bioevents. *In*: Kauffman, E.G. & Walliser, O.H. (eds) *Extinction Events in Earth History*. Lecture Notes in Earth Sciences, **30**. Springer, Berlin, 125–132.

Brett, C.E. 2003. Durophagous predation in Paleozoic marine benthic assemblages. *In*: Kelley, P., Kowalewsky, M. & Hansen, T. (eds) *Predator–Prey Interactions in the Fossil Record*. Kluwer Academic–Plenum, Dordrecht, 401–432.

Brett, C.E. 2012. Coordinated stasis reconsidered: a perspective at fifteen years. *In*: Talent, J. (ed.) *Global Biodiversity, Extinction Intervals and Biogeographic Perturbations through Time*. Springer, Heidelberg, 23–36.

Brett, C.E. & Baird, G.C. 1995. Coordinated stasis and evolutionary ecology of Silurian to Middle Devonian marine biotas in the Appalachian basin. *In*: Erwin, D. & Anstey, R. (eds) *New Approaches to Speciation in the Fossil Record*. Columbia University Press, New York, 285–315.

Brett, C.E., Ivany, L.C., Bartholomew, A.J., DeSantis, M. & Baird, G.C. 2009. Devonian ecological-evolutionary subunits in the Appalachian Basin: a revision and a test of persistence and discreteness. *In*: Königshof, P. (ed.) *Devonian Change: Case Studies in Palaeogeography and Palaeoecology*. Geological Society, London, Special Publications, **314**, 7–36, http://doi.org/10.1144/SP314.2

Brocke, R., Fatka, O., Lindemann, R.H., Schindler, E. & Ver Straeten, C.A. 2015. Palynology, dacryoconarids and the lower Eifelian (Middle Devonian) Basal Choteč Event: case studies from the Prague and Appalachian basins. *In*: Becker, R.T., Königshof, P. & Brett, C.E. (eds) *Devonian Climate, Sea Level and Evolutionary Events*. Geological Society, London, Special Publications, **423**. First published online September 14, 2015, http://doi.org/10.1144/SP423.8

Corradini, C. & Corriga, M.G. 2012. A Přídolí–Lochkovian conodont zonation in Sardinia and the Carnic Alps: implications for a global zonation scheme. *Bulletin of Geosciences*, **87**, 635–650.

DeSantis, M. & Brett, C.E. 2011. Late Eifelian to early Givetian bioevents: timing and signature of the pre-Kačák Bakoven and Stony Hollow events. *Palaeogeography, Palaeoclimatology, Palaeoecology*, **304**, 113–135.

Giesen, P. & Berry, C.M. 2013. Reconstruction and growth of the early tree *Calamophyton* (Pseudosporochnales, Cladoxylopsida) based on exceptionally complete specimens from Lindlar, Germany (Mid-Devonian): organic connection of *Calamophyton* branches and *Duisbergia* trunks. *International Journal of Plant Sciences*, **174**, 665–686.

Girard, C., Klapper, G. & Feist, R. 2005. Subdivision of the terminal Frasnian *linguiformis* conodont Zone, revision of the correlative interval of Montagne Noire Zone 13, and discussion of stratigraphically significant trilobites. *In*: Over, D.J., Morrow, J.R. & Wignall, P.B. (eds) *Understanding Late Devonian and Permian–Triassic Biotic and Climatic Events. Towards an Integrated Approach*. Developments in Palaeontology & Stratigraphy, **20**. Elsevier, Amsterdam, 181–198.

Hairapetian, V., Roelofs, B.P.A., Trinajstic, K.M. & Turner, S. 2015. Famennian survivor turiniid thelodonts of North and East Gondwana. *In*: Becker, R.T., Königshof, P. & Brett, C.E. (eds) *Devonian Climate, Sea Level and Evolutionary Events*. Geological Society, London, Special Publications, **423**. First published online June 10, 2015, http://doi.org/10.1144/SP423.3

Hartenfels, S. 2011. Die globalen *Annulata*-Events und die Dasberg-Krise (Famennium, Oberdevon) in Europa und Nord-Afrika – hochauflösende Conodonten-Stratigraphie, Karbonat-Mikrofazies, Paläoökologie und Paläodiversität. *Münstersche Forschungen zur Geologie und Paläontologie*, **105**, 17–527.

Hartenfels, S. & Becker, R.T. 2016. The global *Annulata* Events: review and new data from the Rheris Basin (northern Tafilalt) of SE Morocco. *In*: Becker, R.T., Königshof, P. & Brett, C.E. (eds) *Devonian Climate, Sea Level and Evolutionary Events*. Geological Society, London, Special Publications, **423**. First published online August 3, 2016, http://doi.org/10.1144/SP423.14

House, M.R. 1983. Devonian eustatic events. *Proceedings of the Ussher Society*, **6**, 396–405.

House, M.R. 1985. Correlation of mid-Palaeozoic ammonoid evolutionary events with global sedimentary perturbations. *Nature*, **313**, 17–22.

House, M.R. 2002. Strength, timing, setting and cause of Mid-Palaeozoic extinctions. *Palaeogeography, Palaeoclimatology, Palaeoecology*, **181**, 5–25.

Jansen, U. 2016. Brachiopod faunas, facies and biostratigraphy of the Pridolian to lower Eifelian succession in the Rhenish Massif (Rheinisches Schiefergebirge, Germany). *In*: Becker, R.T., Königshof, P. & Brett, C.E. (eds) *Devonian Climate, Sea Level and Evolutionary Events*. Geological Society, London, Special Publications, **423**. First published online March 30, 2016, http://doi.org/10.1144/SP423.11

Kaiser, S.I., Becker, R.T. & Spalletta, C. 2009. High-resolution conodont stratigraphy, biofacies and extinctions around the Hangenberg Event in pelagic successions from Austria, Italy, and France. *Palaeontographica Americana*, **63**, 97–139.

Kaiser, S.I., Aretz, M. & Becker, R.T. 2015. The global Hangenberg Crisis (Devonian–Carboniferous transition): review of a first-order mass extinction.

In: Becker, R.T., Königshof, P. & Brett, C.E. (eds) *Devonian Climate, Sea Level and Evolutionary Events*. Geological Society, London, Special Publications, **423**. First published online November 11, 2015, http://doi.org/10.1144/SP423.9

Kiessling, W. 2008. Sampling-standardized expansion and collapse of reef building in the Phanerozoic. *Fossil Record*, **11**, 7–18.

Klapper, G. 1989. The Montagne Noire Frasnian (Upper Devonian) conodont succession. *In*: McMillen, N.J., Embry, A.F. & Glass, D.J. (eds) *Devonian of the World*. Canadian Society of Petroleum Geologists, Memoirs, **14**(III), 449–468.

Klapper, G. & Becker, R.T. 1999. Comparison of Frasnian (Upper Devonian) conodont zonations. *Bolletino della Società Paleontologica Italiana*, **37**, 339–347.

Klug, C., Kröger, B. *et al*. 2010. The Devonian nekton revolution. *Lethaia*, **43**, 465–477.

Königshof, P. (ed.) 2009. *Devonian Change: Case Studies in Palaeogeography and Palaeoecology*. Geological Society, London, Special Publications, **314**, http://sp.lyellcollection.org/content/314/1

Königshof, P., Da Silva, A.C. *et al*. 2015. Shallow-water facies setting around the Kačák Event: a multidisciplinary approach. *In*: Becker, R.T., Königshof, P. & Brett, C.E. (eds) *Devonian Climate, Sea Level and Evolutionary Events*. Geological Society, London, Special Publications, **423**. First published online June 10, 2015, http://doi.org/10.1144/SP423.4

Lakin, J.A., Marshall, J.E.A., Troth, I. & Harding, I.C. 2016. Greenhouse to icehouse: a biostratigraphic review of latest Devonian–Mississippian glaciations and their global effects. *In*: Becker, R.T., Königshof, P. & Brett, C.E. (eds) *Devonian Climate, Sea Level and Evolutionary Events*. Geological Society, London, Special Publications, **423**. First published online April 15, 2016, http://doi.org/10.1144/SP423.12

Marshall, J.E.A. 2016. Palynological calibration of Devonian events at near-polar palaeolatitudes in the Falkland Islands, South Atlantic. *In*: Becker, R.T., Königshof, P. & Brett, C.E. (eds) *Devonian Climate, Sea Level and Evolutionary Events*. Geological Society, London, Special Publications, **423**. First published online June 7, 2016, http://doi.org/10.1144/SP423.13

McNamara, K.J. & Feist, R. 2015. The effect of environmental changes on the evolution and extinction of Late Devonian trilobites from the northern Canning Basin, Western Australia. *In*: Becker, R.T., Königshof, P. & Brett, C.E. (eds) *Devonian Climate, Sea Level and Evolutionary Events*. Geological Society, London, Special Publications, **423**. First published online June 10, 2015, http://doi.org/10.1144/SP423.5

Mottequin, B. & Poty, E. 2015. Kellwasser horizons, sea-level changes and brachiopod–coral crises during the late Frasnian in the Namur–Dinant Basin (southern Belgium): a synopsis. *In*: Becker, R.T., Königshof, P. & Brett, C.E. (eds) *Devonian Climate, Sea Level and Evolutionary Events*. Geological Society, London, Special Publications, **423**. First published online June 26, 2015, http://doi.org/10.1144/SP423.6

Mottequin, B., Denayer, J., Königshof, P., Prestianni, C. & Olive, S. (eds). 2015. *IGCP 596 – SDS Symposium. Climate change and Biodiversity patterns in the Mid-Palaeozoic Abstracts*. STRATA, Travaux de Géologie sédimentaire et Paléontologie, Série 1: Communications, **16**. Association STRATA, Gaillac, France, 1–157.

Murphy, M.A. 2005. Pragian conodont zonal classification in Nevada, Western North America. *Revista Española de Paleontologia*, **20**, 177–206.

Narkiewicz, K., Narkiewicz, M. & Bultynck, P. 2015. Conodont biofacies of the Taghanic transgressive interval (middle Givetian): Polish record and global comparisons. *In*: Becker, R.T., Königshof, P. & Brett, C.E. (eds) *Devonian Climate, Sea Level and Evolutionary Events*. Geological Society, London, Special Publications, **423**. First published online June 10, 2015, http://doi.org/10.1144/SP423.2

Niedzwiedzki, G., Szrek, P., Narkiewicz, M. & Ahlberg, P.E. 2010. Tetrapod trackways from the early Middle Devonian period of Poland. *Nature*, **463**, 43–48.

Poole, F.G. & Sandberg, C.A. 2015. Alamo impact olistoliths in Antler orogenic foreland, Warm Springs-Milk Spring area, Hot Creek Range, central Nevada. *In*: Poole, F.G. & Sandberg, C.A. (eds) *Unusual Central Nevada Geologic Terranes Produced by Late Devonian Antler Orogeny and Alamo Impact*. Geological Society of America, Special Papers, **517**, 39–104, http://doi.org/10.1130/2015.2517(02)

Racki, G. 2005. Towards understanding Late Devonian global events: few answers, many questions. *In*: Over, D.J., Morrow, J.R. & Wignall, P.B. (eds) *Understanding Late Devonian and Permian-Triassic Biotic and Climatic Events. Towards an Integrated Approach*. Development in Palaeontology & Stratigraphy, **20**, 5–36.

Schieber, J. & Over, D.J. 2005. Sedimentary fill of the Late Devonian Flynn Creek crater. *In*: Over, D.J., Morrow, J.R. & Wignall, P.B. (eds) *Understanding Late Devonian and Permian-Triassic Biotic and Climatic Events. Towards an Integrated Approach*. Developments in Palaeontology & Stratigraphy, **20**. Elsevier, Amsterdam, 51–69.

Signor, P.W. & Brett, C.E. 1984. The mid-Paleozoic precursor to the Mesozoic marine revolution. *Paleobiology*, **10**, 222–236.

Slavík, L. & Hladil, J. 2004. Lochkovian/Pragian GSSP revisited: evidence about conodont taxa and their stratigraphic distribution. *Newsletters on Stratigraphy*, **40**, 137–153.

Slavík, L., Valenzuela-Ríos, J.I., Hladil, J. & Carls, P. 2007. Early Pragian conodont-based correlations between the Barrandian area and the Spanish Central Pyrenees. *Geological Journal*, **42**, 499–512.

Slavík, L., Carls, P., Hladil, J. & Koptíková, L. 2012. Subdivision of the Lochkovian Stage based on conodont faunas from the stratotype area (Prague Synform, Czech Republic). *Geological Journal*, **47**, 616–631.

Spalletta, C., Perri, M.C., Corradini, C. & Over, J.D. 2015. Proposed revision of the Famennian (Upper Devonian) standard conodont zonation. *In*: Mottequin, B., Denayer, J., Königshof, P., Prestianni, C. & Olive, S. (eds) *IGCP 596 – SDS Symposium. Climate Change and Biodiversity Patterns in the Mid-Palaeozoic*. STRATA, Travaux de Géologie sédimentaire et Paléontologie, Série 1: Communications, **16**, 135–136.

Stein, W.E., Berry, C.M., VanAller Hernick, L. & Mannolini, F. 2012. Surprisingly complex community discovered in the mid-Devonian fossil forest at Gilboa. *Nature*, **483**, 78–81.

Suttner, T.J. & Kido, E. 2015. Distinct sea-level fluctuations and deposition of a megaclast horizon in the neritic Rauchkofel Limestone (Wolayer area, Carnic Alps) correlate with the Lochkov–Prag Event. *In*: Becker, R.T., Königshof, P. & Brett, C.E. (eds) *Devonian Climate, Sea Level and Evolutionary Events*. Geological Society, London, Special Publications, **423**. First published online June 10, 2015, http://doi.org/10.1144/SP423.1

Valenzuela-Ríos, J.I., Slavík, L., Liao, J.-C., Calvo, H., Hušková, A. & Chadimová, L. 2015. The middle and upper Lochkovian (Lower Devonian) conodont successions in key peri-Gondwana localities (Spanish Central Pyrenees and Prague Synform) and their relevance for global correlations. *Terra Nova*, **27**, 409–415.

Walliser, O.H. 1984. Geologic processes and global events. *Terra Cognita*, **4**, 17–20.

Walliser, O.H. 1996. Global events in the Devonian and Carboniferous. *In*: Walliser, O.H. (ed.) *Global Events and Event Stratigraphy in the Phanerozoic*. Springer, Berlin, 225–250.

Warme, J.E. & Sandberg, C.A. 1996. Alamo megabreccia: record of a Late Devonian impact in southern Nevada. *GSA Today*, **6**, 1–7.

Yolkin, E.A., Weddige, K., Izokh, N.G. & Erina, M.V. 1994. New Emsian conodont zonation (Lower Devonian). *Courier Forschungsinstitut Senckenberg*, **168**, 139–157.

Zambito, J.J., IV, Joachimski, M.M., Brett, C.E., Baird, G.C. & Aboussalam, Z.S. 2015. A carbonate carbon isotope record for the late Givetian (Middle Devonian) Global Taghanic Biocrisis in the type region (northern Appalachian Basin). *In*: Becker, R.T., Königshof, P. & Brett, C.E. (eds) *Devonian Climate, Sea Level and Evolutionary Events*. Geological Society, London, Special Publications, **423**. First published online June 26, 2015, http://doi.org/10.1144/SP423.7

Distinct sea-level fluctuations and deposition of a megaclast horizon in the neritic Rauchkofel Limestone (Wolayer area, Carnic Alps) correlate with the Lochkov–Prag Event

T. J. SUTTNER* & E. KIDO

University of Graz, Institute of Earth Sciences, Heinrichstraße 26, A-8010 Graz, Austria

**Corresponding author (e-mail: suttner.thomas@gmail.com)*

Abstract: Distinctive facies change and magnetic susceptibility across the neritic Rauchkofel Limestone (Central Carnic Alps) document specific sea-level fluctuations that are related to the Lochkov–Prag Event. This is supported by a positive shift in the carbon isotope signal of about 1.5‰ and a distinct decline in conodont biodiversity during the late Lochkovian. A transgressive interval during the late Lochkovian is followed by a regressive phase indicated by an eye-catching megaclast horizon of early Pragian age. Some of the boulders within that horizon reach up to 10 m in diameter. A model explaining regional depositional patterns is provided.

During extensive fieldwork in Early Devonian neritic deposits at Lake Wolayer (Central Carnic Alps, Austria) we aimed to document the lateral continuity of a pronounced megaclast horizon that crops out at the NW wall of Mount Seewarte. We wanted to find out more on the origin of deposition and the source area of the large limestone boulders, some of which reach up to 10 m in diameter. Already sketched by Bandel (1969), the megaclast horizon was described as non-dolomitized boulders that may be rotated in dolomite matrix within unit ‘0c’ of the neritic Rauchkofel Limestone. It is assumed that the lower part of this formation consisting of allodapic (lithoclastic limestone and mega-conglomerate levels) and pelagic limestones and shales formed on a slope or over-steepened shelf margin (Vai 1980; Hubmann *et al.* 2003). Among others, based on the sequence between Seewarte and Hohe Warte, a model for the carbonate platform development during the Devonian in the Central Carnic Alps was proposed by Kreutzer (1990, 1992*a*, *b*), who concluded that a facies division into crinoid meadows in the south and slope to pelagic deposits in the north existed already during the Lochkovian. A true reef belt was not established at that time, although single patch reefs (stromatoporoid biostromes and microbial–coral bioherms) grew on a slightly northwards inclined carbonate platform (Kreutzer 1990; Schönlaub & Kreutzer 1997; Schönlaub & Histon 2000). However, the tectonic development of the entire area resulted in a complex nappe system that brought tectonic units of different depositional environments into close proximity (Hubmann *et al.* 2003).

Further detailed study of the Mount Seewarte section (Suttner 2005, 2007; Suttner & Kido 2011) resulted in a biostratigraphic update and the recognition of specific trans- and regressive cycles that match the Lochkovian/Pragian Boundary Event observed for pelagic settings in the Carnic Alps by Schönlaub (1986). According to Schönlaub (1986), a colour change from grey to red in flaser limestones contemporaneous with distinct changes in the conodont fauna, the disappearance of graptolites and the appearance of new dacryoconarid taxa demarcate the event interval. The model proposed here shall help to clarify depositional patterns within the neritic Rauchkofel Limestone and explain its link with the so-called Lochkov–Prag Event documented from pelagic deposits of the Carnic Alps and elsewhere.

The Lochkov–Prag Event

Event history

The Lochkov–Prag Event (in the sense of a biotic event) was defined by Walliser (1996). Earlier sedimentological descriptions of Chlupáč & Kukal (1986) identified the nature of this event as being expressed by a lithological change from dark Lochkovian to lighter Pragian deposits. That specific sedimentological change accompanied by a change in benthic and planktonic communities is interpreted as a regressive event by Chlupáč & Kukal (1986, 1988). Ziegler & Lane (1987) documented a conspicuous reduction in conodont diversity in the latest Lochkovian *Pedavis pesavis* Biozone (= *Masaraella pandora* beta-*Pedavis gilberti* biozones, see Valenzuela-Ríos 1994). A few years later, Talent *et al.* (1993) observed a similar sharp drop in conodont diversity in several sections from

From: Becker, R. T., Königshof, P. & Brett, C. E. (eds) 2016. *Devonian Climate, Sea Level and Evolutionary Events*. Geological Society, London, Special Publications, **423**, 11–23.
First published online June 10, 2015, http://doi.org/10.1144/SP423.1

east-central New South Wales (Australia), wherefore the term end-*pesavis* event was introduced. There, a gradual sedimentological change characterizes the event interval, which was followed by a significant regional regression just before the end of the biozone. Additionally, Talent *et al.* (1993) pointed out that no obvious drop in brachiopod diversity is recognized. A further brief summary on the end-*pesavis* event in Australia is provided by Young (1995). In contrast to Chlupáč & Kukal (1986), House (2002) concluded that the event (renamed the Basal Pragian Event *sensu* Chlupáč 1998) did not result in a regression but in a transgression, which follows the interpretation of Johnson *et al.* (1985, T–R cycle Ia). Regarding carbon isotope chemostratigraphy across the Lochkov–Prag Event: here a rapid increase in $\delta^{13}C$ is documented from the late Lochkovian until the earliest Pragian, which exactly spans the event interval and is observed in several European localities (Buggisch & Mann 2004; Buggisch & Joachimski 2006). Based on results from geophysical methods (magnetic susceptibility (MS) and gamma ray spectrometry) applied to sedimentary successions in the Czech Republic, Vacek (2011) concluded that the Lochkov–Prag Event was related to climate warming accompanied by enhanced carbonate productivity, while eustatic sea-level change might have played a subordinate role.

Global recognition

Localities with corresponding stratigraphical levels are already mentioned by Chlupáč & Kukal (1986) and include sections in the Czech Republic (Chlupáč & Kukal 1986; Chlupáč 1998, 2000; Slavík *et al.* 2008), Germany (Alberti 1981, 1983), Austria (Schönlaub 1980, 1985), Sardinia (Jaeger 1976; Alberti 1983), the Armorican Massif (Paris 1981) and western and eastern USA (Johnson & Murphy 1984; Johnson *et al.* 1985, 1996). The Lochkov–Prag Event is also documented from the Tafilalt Platform in Morocco (Alberti 1981, 1983; discussion in Kaufmann 1998; Kröger 2008; Klug *et al.* 2013) and from the Garra Limestone at Wellington in eastern Australia (Talent *et al.* 1993).

Regional geology and fossil content

In the Central Carnic Alps the Devonian shallow-marine sequence is best documented at Mount Seewarte in the Wolayer area (Fig. 1). There it starts with a few lithoclastic horizons and crinoidal limestones (organisms: brachiopods, gastropods, echinoderms, conodonts) followed by a short interval of dark grey nodular limestone bearing phosphatic brachiopod shells of *Opsiconidion* and a few conodonts during the Lochkovian (neritic Rauchkofel Limestone; indicated as interval 1 in Fig. 2). These are overlain by well-bedded crinoidal grainstones (neritic Rauchkofel Limestone; indicated as interval 2 in Fig. 2) and a thick unit of massive, bright grey frame- and rudstones (dominating skeletal components: calcareous algae, stromatoporoids, tabulate and rugose corals) of Pragian age (Hohe Warte Limestone; indicated as interval 3 in Fig. 2). The massive limestone is succeeded by laminated dark grey to black Seewarte Limestone, which is rich in algae, bryozoans, ostracodes and large specimens of gastropods, and interpreted as lagoonal deposits. Deposited above is the Lambertenghi Limestone, a unit that consists of a series of loferite cycles, some levels of birds-eye limestone, rimmed-grain grainstones, oncoidal limestone intervals and a few palaeosol horizons. Within the grainstone beds, abundant trochospiral gastropods, brachiopods, bivalves, ostracodes, small solitary rugose corals and some stromatoporoid colonies are observed. Overlying these cyclically deposited sediments another unit of massive, bright grey limestones (Spinotti Limestone) exposes mainly small bioherms of tabulate and rugose corals at its base, a relatively thick horizon dominated by stromatoporoids in the middle and peritidal sediments, such as birds-eye limestone and microbial laminites, that are followed by intervals of *Amphipora* biostroms towards the top of the formation. At Forcella Monumenz (near the Marinelli hut on the Italian side), close to the top of the Spinotti Limestone, a brachiopod horizon mainly consisting of specimens assigned to *Stringocephalus burtini* occurs. The interval dominated by *Amphipora* is overlain by deposits of the Kellergrat Reef Limestone that is rich in solitary rugose and tabulate corals, single, up to 0.5 m-wide stromatoporoid colonies, echinoderms and brachiopods. Conodonts derived from crinoidal debris layers that are intercalated between 10 and 20 cm-thick frame and rudstone layers of this unit exposed at the abandoned quarry on trail 149 between the Plöcken Pass and Collina, indicate a Givetian age.

Lochkovian deposits continue in slope and pelagic facies towards the north (bounded by a large fault) and east. They are named pelagic Rauchkofel Limestone and La Valute Limestone (syn. Boden Limestone), respectively. These are succeeded by the pelagic Findenig Limestone, which is late Lochkovian (*pandora* beta Biozone) at La Valute in the Mount Zermula area (Corriga *et al.* 2011) and Pragian in age at Rauchkofel Boden north of the Valentin Valley (Schönlaub 1980; Schönlaub *et al.* 1997).

Intertidal and shallow-marine limestone sequences are exposed at the Biegengebirge in the western continuity and north of the Valentin Valley

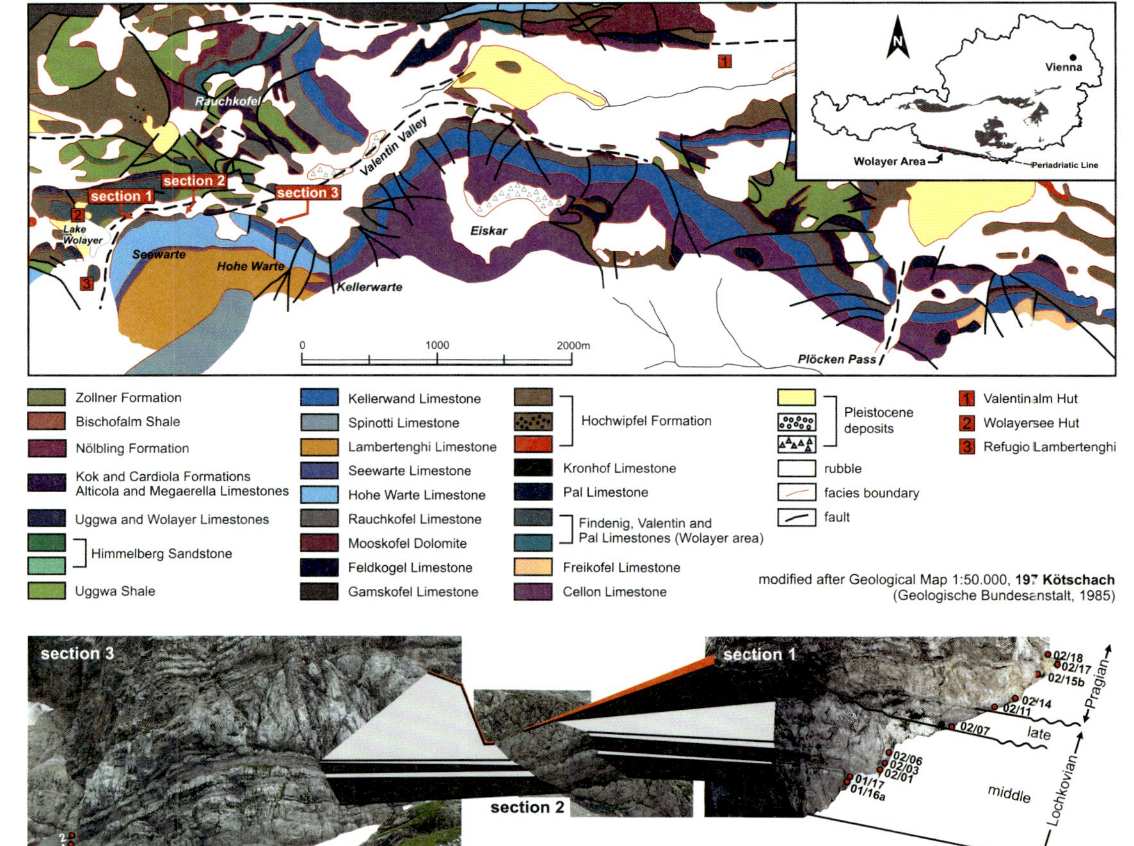

Fig. 1. Geological map of the Central Carnic Alps with three sections of neritic Rauchkofel Limestone indicated and correlated. Conodont samples S, 1 and 2 from section 3 equate to a few decimetres below Se/01/16a, Se/01/17 and Se/02/03 in section 1, respectively.

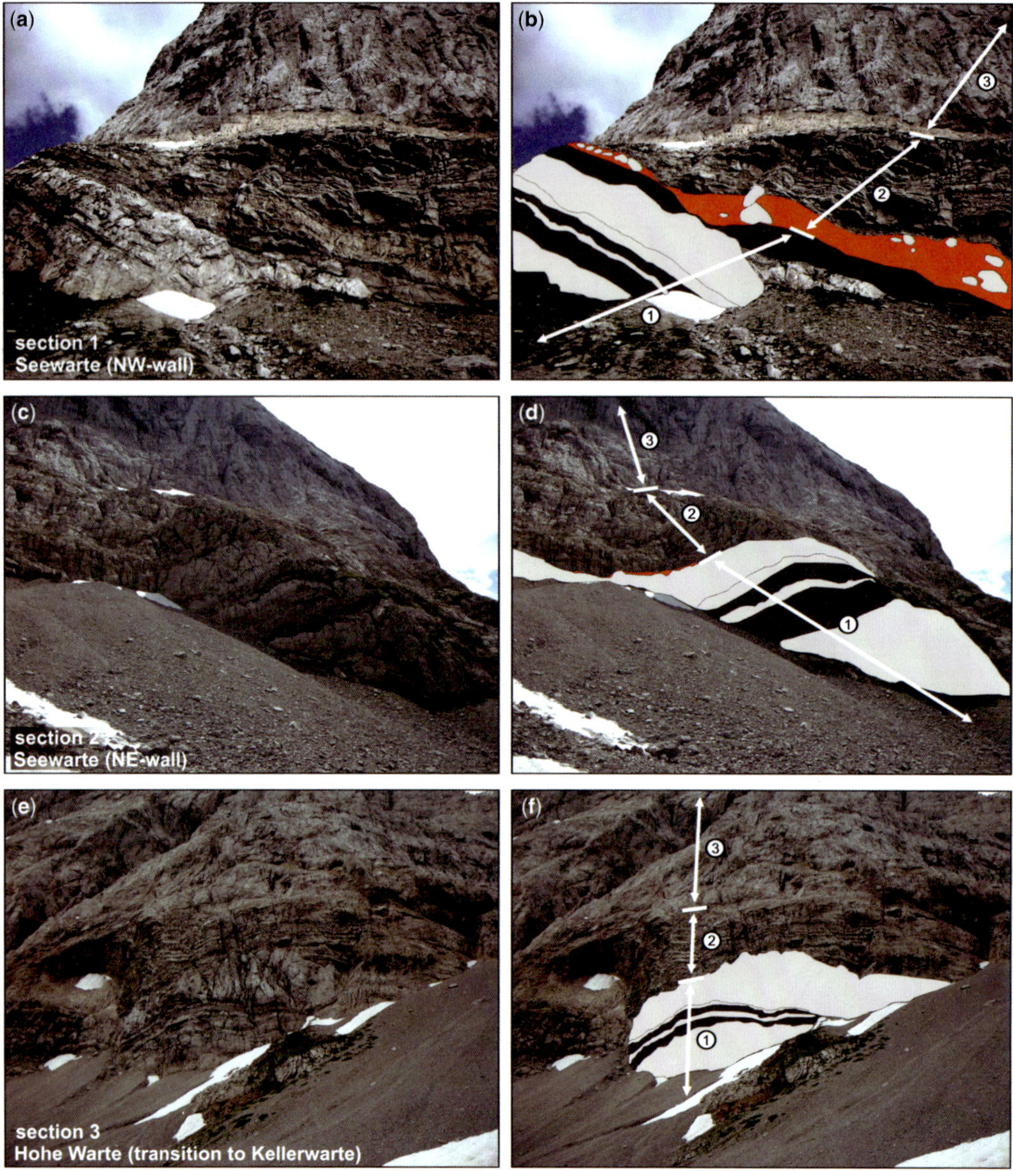

Fig. 2. Detailed photos of laterally traced sections. (**a**, **b**) Section 1 at the NW wall of Mount Seewarte; (**c, d**) section 2 at the NE wall of Mount Seewarte; (**e**, **f**) section 3 at the transition of Mount Hohe Warte to the Kellerwarte. Numbers 1–3 in (b), (d) and (f) indicate the (1) Lochkovian and (2) Pragian interval of the neritic Rauchkofel Limestone, and (3) the Pragian interval of the Hohe Warte Limestone.

at the Gamskofel to Mooskofel Massif (Kreutzer 1990, 1992*a*). Commonly, facies transitions are interrupted by complex fault systems and, therefore, not well preserved (especially at Lake Wolayer and the Hohe Warte–Kellerwarte transition). A summary on the characteristics of formations in the Carnic Alps is provided in Hubmann *et al.* (2003, 2014).

Methods

For conodont biostratigraphy limestone samples of 2–5 kg were dissolved in formic acid (HCOOH, 85% + tap water at a proportion of 1:5), sieved (fractions: 63, 125, 250 and 500 μm) and separated by heavy liquid (non-toxic sodium polytungstate;

liquid density: 2.78). For the biostratigraphic results and conodont taxonomy of section 1, please refer to Suttner (2007). Additional samples from section 1 (sample numbers: Se/02x, Se/02y, Se/02/15b, Se/02/15c, Se/02/17a) and section 3 (sample numbers: S, 1 and 2; see Fig. 1 for correlation of sampled levels) were dissolved following the same procedure (methods of Anderson *et al.* 1995; Jeppsson & Anehus 1995).

An investigation of carbon stable isotopes was carried out by Michael Joachimski and Werner Buggisch (Friedrich-Alexander Universität Erlangen-Nürnberg, Germany). For details on the method see Buggisch & Mann (2004) and for original values of section 1 see Suttner (2007), where the results of carbon isotope measurements were published.

For MS measurements of rocks, we used a field device (KT-6; resolution of 10^{-3} SI units), which resulted in about 1150 measured points along the 125 m-thick Rauchkofel Limestone. For proper measurement in the field we applied the hand-held device exclusively to dry and smooth rock surfaces.

Stratigraphic framework

Conodont faunas

Limestone samples from section 1 (Figs 1, 3 & 4) yield a diverse Lochkovian fauna that allows one to distinguish two conodont zones: the *Ancyrodelloides delta* and *pesavis* biozones (Suttner 2007). Here we update the older biostratigraphic scheme used by Suttner (2007) according to the latest conodont zonation proposed for Lochkovian strata of Sardinia and the Carnic Alps by Corradini & Corriga (2012) and the Prague Synform by Slavík *et al.* (2012). Early Pragian strata are documented with the first occurrence of *Icriodus steinachensis* beta morph (alternative biozonation according to Slavík 2004; Slavík & Hladil 2004; Slavík *et al.* 2007) in the dolomitic matrix of the megaclast horizon by Suttner (2007). In general, conodont specimens count between 50 and 1000 elements per sample during the middle Lochkovian, with distinctively decreased abundance above the megaclast horizon to an amount of less than 25 specimens per sample during the Pragian. A drop in conodont taxa from 24 to 14 species is recognized within the *Lanea eleanorae* to *Ancyrodelloides trigonicus* Biozone, which is followed by a further drop down to 7 species during the *pandora* beta-*gilberti* biozones (Fig. 4). Samples from the early Pragian *Icriodus steinachensis* Biozone yield between 2 and 13 species and are dominated mainly by coniform elements. Owing to the specific nature of the deposition, any interpretation regarding the conodont distribution across the megaclast horizon has to be considered carefully. As samples Se/02/16 and Se/02/17 were taken from two of the larger boulders that are stratigraphically misplaced (originally being deposited during the middle to late Lochkovian), the waxing and waning of conodont taxa in this interval is artificial. The only interpretation we can give for this interval is that a general trend points to a further decline in conodont species.

Conodont samples obtained from section 3 yield an abundant and diverse coniform but less diverse ozarkodiniform assemblage (*Belodella resima, Pseudooneotodus beckmanni, Decoriconus fragilis, Panderodus unicostatus, Oulodus* sp., *Wurmiella excavata* and *Zieglerodina remscheidensis*).

Chemostratigraphy

The carbon isotope plot (Fig. 3) shows a nearly continuously increasing positive trend from the *Ancyrodelloides transitans* to the lower part of the *steinachensis* Biozone with lowest values of 0.86‰ (Se/01/15) up to 3.29‰ (Se/02/20). A sudden shift of about 1.5‰ is recorded at the boundary between the *pandora* beta-*gilberti* to *steinachensis* Biozone.

Magnetic susceptibility

Trends in the MS-log show steadily increasing values during the *eleanorae* Biozone (Fig. 3, indicated by ‘A’). The above-following gap of measurements is due to an inaccessible section-part. At the top of the thick limestone bed (sample Se/02/07) a significant decline in values is recorded, which continues until shortly before the top of the *pandora* beta-*gilberti* interval (Fig. 3, indicated by ‘B’). Minor shifting, in- and decreasing values follow, which reflect a slightly increasing trend during the lowest part of the *steinachensis* Biozone. The MS-log culminates in a markedly positive excursion at the top of the megaclast horizon (Fig. 3, indicated by ‘C’). A short interval of decreasing values follows (Fig. 3, indicated by ‘D’).

Microfacies and sedimentary development

Thick, bright grey limestone beds between samples Se/02/01 and Se/02/07 consist of crinoidal pack- to grainstone (Fig. 5a). The uppermost part of that interval yields a few rugose and tabulate corals and is rich in large stromatoporoids of flat growth-type (for detailed log with fossil distribution chart, see Bandel 1969). The above-deposited late Lochkovian dark grey limestones (Fig. 5b) are bituminous, fine-grained, organodetritic carbonates with pyrite, small amounts of siliciclastics, peloids and small bioclasts. Insoluble residues from conodont samples of that interval include small phosphatic

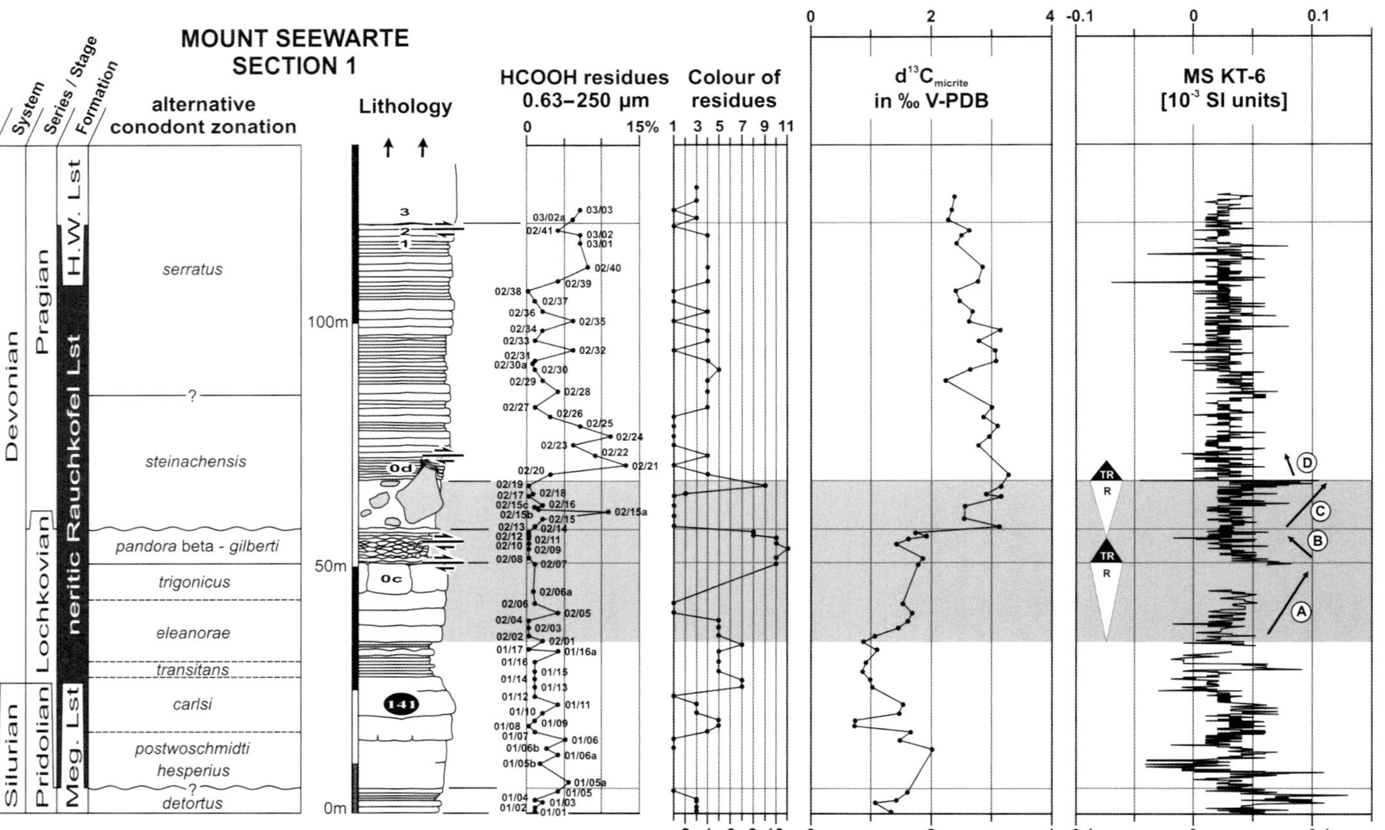

Fig. 3. Section log with plots displaying the amount of non-calcareous residues (in percent), the colour of residues (1, very light grey (N8); 2, pinkish grey (5YR8/1); 3, yellowish grey (5Y8/1); 4, light grey (N7); 5, medium light grey (N6); 6, light brownish grey (5YR6/1); 7, medium grey (N5); 8, medium dark grey (N4); 9, brownish grey (5YR4/1); 10, dark grey (N3); 11, greyish black (N2)), carbon stable isotopes (slightly adapted from Suttner 2007) and magnetic susceptibility (A to D correspond to the numbers in the facies model, Fig. 6). Lst, limestone; V-PDM, Vienna Pee Dee Belemnite; MS, magnetic susceptibility; TR, transgressive, R, regressive.

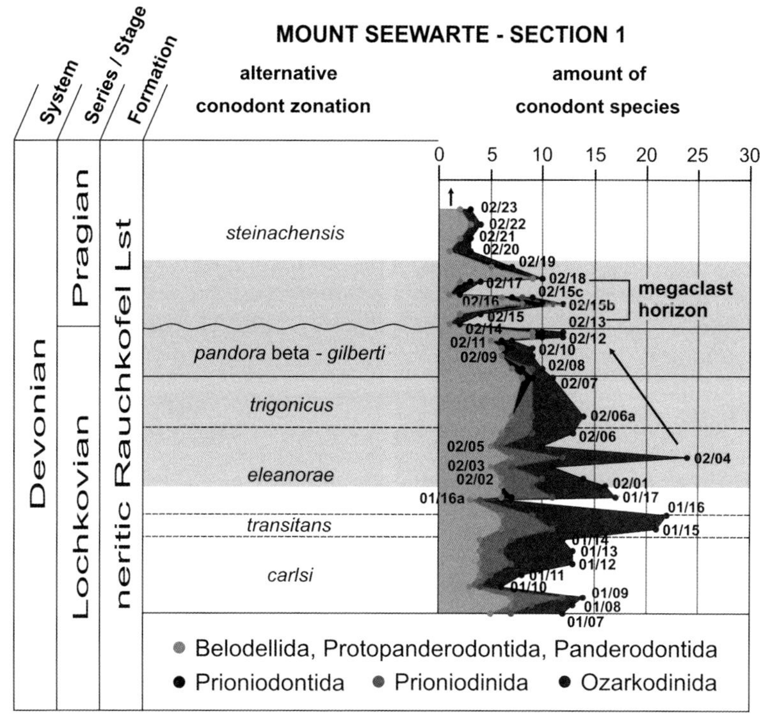

Fig. 4. Distribution of conodont taxa (rank: order) across the middle to late Lochkovian and early Pragian of section 1.

brachiopods of the genus *Opsiconidion* and conodonts. The sediment is partly dolomitized. The somewhat nodular consistency of these beds is supposedly a diagenetic effect after compaction of the primary soft to firm sediment. This interval is succeeded by two to three bright grey limestone beds that form the base of the megaclast horizon, which consists of bright grey limestone boulders floating in an orange dolomitic matrix (Fig. 5c, e, f). These boulders consist mainly of rudstone (Fig. 5d). Some of the clasts yield rugose and tabulate corals (about 20 cm-large colonies of *Favosites* sp.) and broken stromatoporoid colonies. More than half of the largest boulder protrudes from the orange matrix and is covered stepwise by overlying limestone beds. The steeply inclined onlap of orange weathering crinoidal pack- to grainstone is illustrated in Figure 5g. A few metres higher in the section, bedding is nearly horizontal again and the bed colour changes to dark grey/black. Based on how crinoidal limestone beds are onlapping around the boulder, a direction of marine currents is supposed, which give a N-NE to S-SW orientation. Similar results are provided by Bogolepova (1997) based on the orientation of mid-Lochkovian cephalopods from the pelagic La Valute Limestone (syn. Boden Limestone) measured at the Rauchkofel Boden section.

The megaclast horizon is observed only at the NW wall of the Seewarte (Fig. 2, section 1; 46°36′41″N,12°52′22″E) where it reaches a maximum thickness of 8–10 m. It can be traced for approximately 70 m, decreasing in thickness and boulder size towards the east. Similarly restricted in its lateral extension is the underlying interval of dark grey limestone, which is evident only in section 1, disappearing towards the NE wall of the Seewarte (Fig. 2, section 2; 46°36′42″N, 12°52′41″E). There, only a thin, orange dolomitic bed is observed, which seems to continue close to the first large fault between Hohe Warte and Kellerwarte (Fig. 2, section 3; 46°36′41″N, 12°53′04″E), directly overlying a thick, bright grey limestone bed (*c.* 13 m). In all sections (Fig. 2), the sequence below the megaclast horizon exposes two very distinctive dark grey intervals (2–3 m in thickness), each of which is succeeded by a thin, bright grey limestone bed (0.5 m and 2 m consecutively) slightly varying in thickness laterally. The second of these intervals is overlain by the aforementioned relatively thick, bright grey limestone bed that consists of massive limestone with strong relief on its upper surface at Mount Hohe Warte (=unit '15' of Schönlaub & Flajs 1975). According to the apparent lens-shape in section 3, this part of the unit

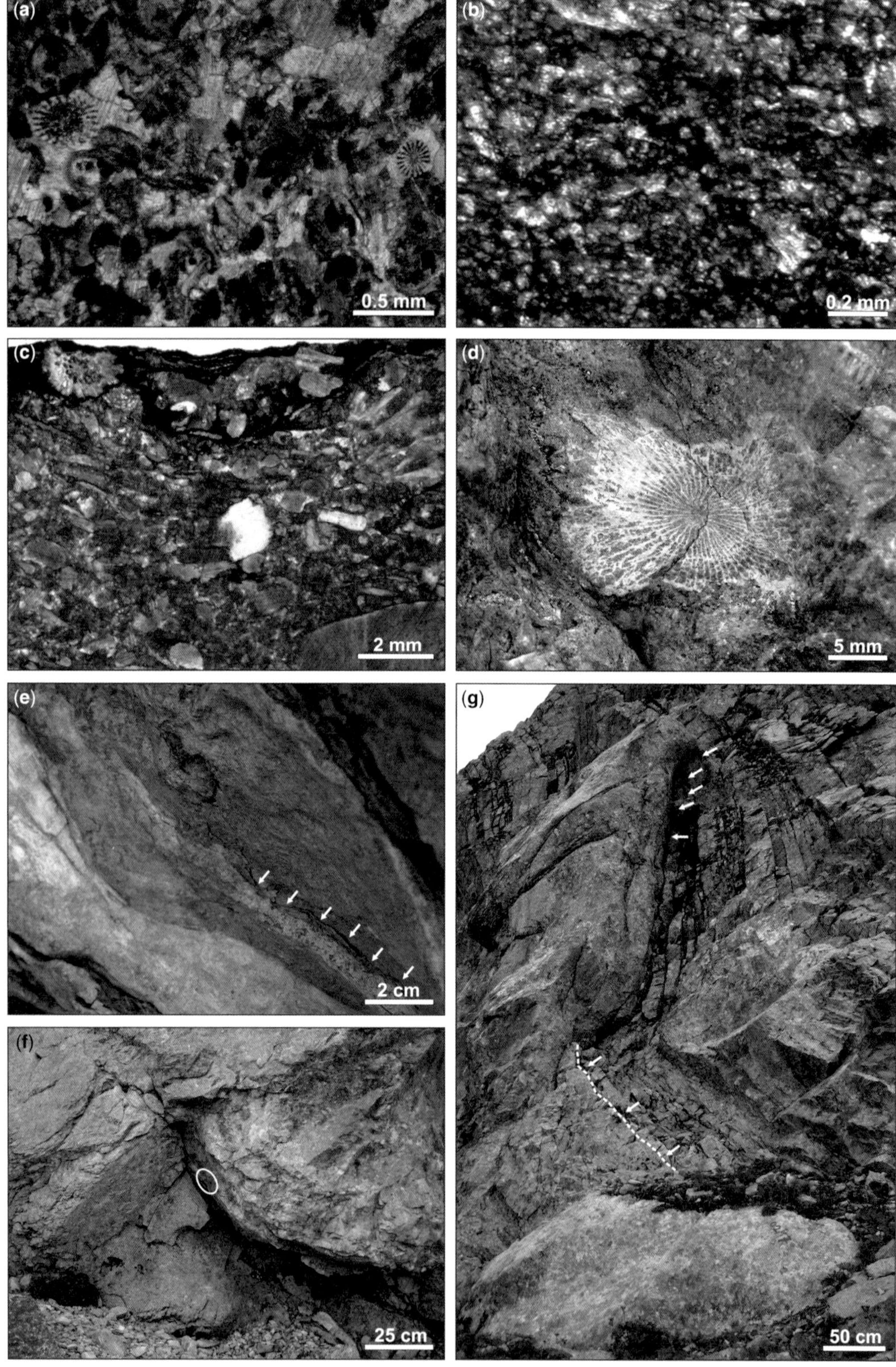
(a)
0.5 mm
(b)
0.2 mm
(c)
2 mm
(d)
5 mm
(e)
2 cm
(f)
25 cm
(g)
50 cm

was ascribed to a 'Lochkovian mound' (Hubmann & Suttner 2007). However, recent investigations have shown that the lateral restriction is the result of regional faulting, wherefore the 'mound' hypothesis has to be declined.

Discussion and conclusion

Our results confirm a 'conspicuous reduction in conodont diversity in the latest Lochkovian *pesavis* Biozone' recognized by Ziegler & Lane (1987). According to the diversity pattern of conodont taxa observed in section 1, a certain drop as well as a change in the composition of the assemblage from an ozarkodinid- to an icriodontid-dominated community is recognized. However, it is important to note that coniform taxa are more conservative and did not undergo a significant change at species level, but showed a decline in species from the *trigonicus* Biozone into the early Pragian *steinachensis* Biozone. The peak in conodont diversity in the latest Lochkovian (samples Se/02/12 and Se/02/13) is coherent with the onset of a regressive phase and most probably related to a transport-related accumulation of coniform taxa. We assume that the generally stepwise decline of conodont taxa since the *eleanorae–trigonicus* biozones until the *steinachensis* Biozone, with two peaks in diversity during the *transitans–eleanorae* biozones, is related to changing environmental conditions dependent on regional deposition patterns and predominating facies. The conodont diversity pattern from the Seewarte section does not seem to be affected by a sudden devastating decline as would be expected for a biotic extinction event. It rather seems that the conodont community is generally in decline, with taxa belonging to the Prioniodinida and Ozarkodinida becoming substituted by representatives of the Prioniodontida. Apart from conodonts, other faunas documented across the event interval (sections 1 and 3), but not studied at species level, are stromatoporoids, corals, brachiopods and crinoids (Bandel 1969; Schönlaub & Flajs 1975; Kreutzer 1990; Suttner 2007).

Results from microfacies, the rock colour, carbon isotopes and MS show a strong affinity with the change in lithology documented, for example, by Chlupáč & Kukal (1986). A clear change from very dark, clay-rich limestone to light grey limestone, succeeded by the megaclast horizon, is observed. We think that the boulders of the megaclast horizon accord with re-deposits that derived from the Early Devonian platform that developed farther west between Biegengebirge and Giramondo Pass, which probably eroded during a regressive phase in the latest Lochkovian/early Pragian (Fig. 6, scenario 1). Based on MS and facies analyses this regression was preceded by a transgressive trend that spanned almost the entire interval of the *pandora* beta-*gilberti* biozones and lasted until shortly before the biostratigraphic boundary where the regressive phase had its onset (Fig. 3). Earlier we assumed that the boulders could have derived from section 3 at the Hohe Warte–Kellerwarte transition (Fig. 6, scenario 2), but as the evidence shows that the largest boulders are allocated in the western-most continuity in section 1 (declining in size towards the east), the source area is suspected to be within the carbonate platform west of Mount Seewarte. What is still unclear is whether the megaclast horizon was produced solely by erosion during the regressive phase, or if deposition was coupled with a synsedimentary fault movement that resulted in a fault scarp and steepened palaeoslope. As other Early Devonian transgressive–regressive cycles did not produce large boulders within lithoclastic levels, such a regional tectonic event would support the depositional model. However, confirmation in the field is difficult, as the area suffered both the Variscan and Alpine Orogeny.

Unexpectedly, high-resolution $\delta^{13}C$ and geophysical logs from the neritic Seewarte section resulted in similar trends to those observed in other sections from Austria, the Czech Republic, France, Germany, Spain and Morocco (e.g. Ellwood *et al.* 2001; Buggisch & Mann 2004; Hladil *et al.* 2010; Koptíková *et al.* 2010; Vacek 2011). MS values across the Lochkov and Praha formations show a strong positive shift from below 1 up to *c.* 20 $m^3\ kg^{-1} \times 10^{-9}$ shortly after the Lochkovian/Pragian boundary, with a second maximum of *c.* 30 $m^3\ kg^{-1} \times 10^{-9}$ in the upper third of the Pragian (Koptíková *et al.* 2010). According to Koptíková *et al.* (2010), the shift at the Lochkovian/Pragian boundary is associated with the maximum regression level at the base of the Pragian, which is also recognized at the top of the megaclast horizon in the Seewarte

Fig. 5. Microfacies and outcrop details around the megaclast horizon. (**a**) Crinoidal pack- to grainstone, Se/02/02; (**b**) dolomitized, dark, clay-rich limestone, Se/02/10; (**c**) largely dolomitized matrix of the megaclast horizon yielding crinoid stem plates, coral fragments and bryozoans, Se/02/15c; (**d**) rudstone with rugose coral, Se/02/17; (**e**) detail of (**f**) thin layer around the large boulder (attachment marked by arrow heads), rich in strongly fragmented bioclasts, Se/02/15b; (**g**) two boulders floating in the dolomitic matrix of which the larger one is about 10 m (Se/02/16) and the smaller one in front about 2 m in diameter (Se/02/17). Arrow heads point to onlapping single layers covering the large boulder, and to the boundary of the matrix of the megaclast horizon with above-deposited beds.

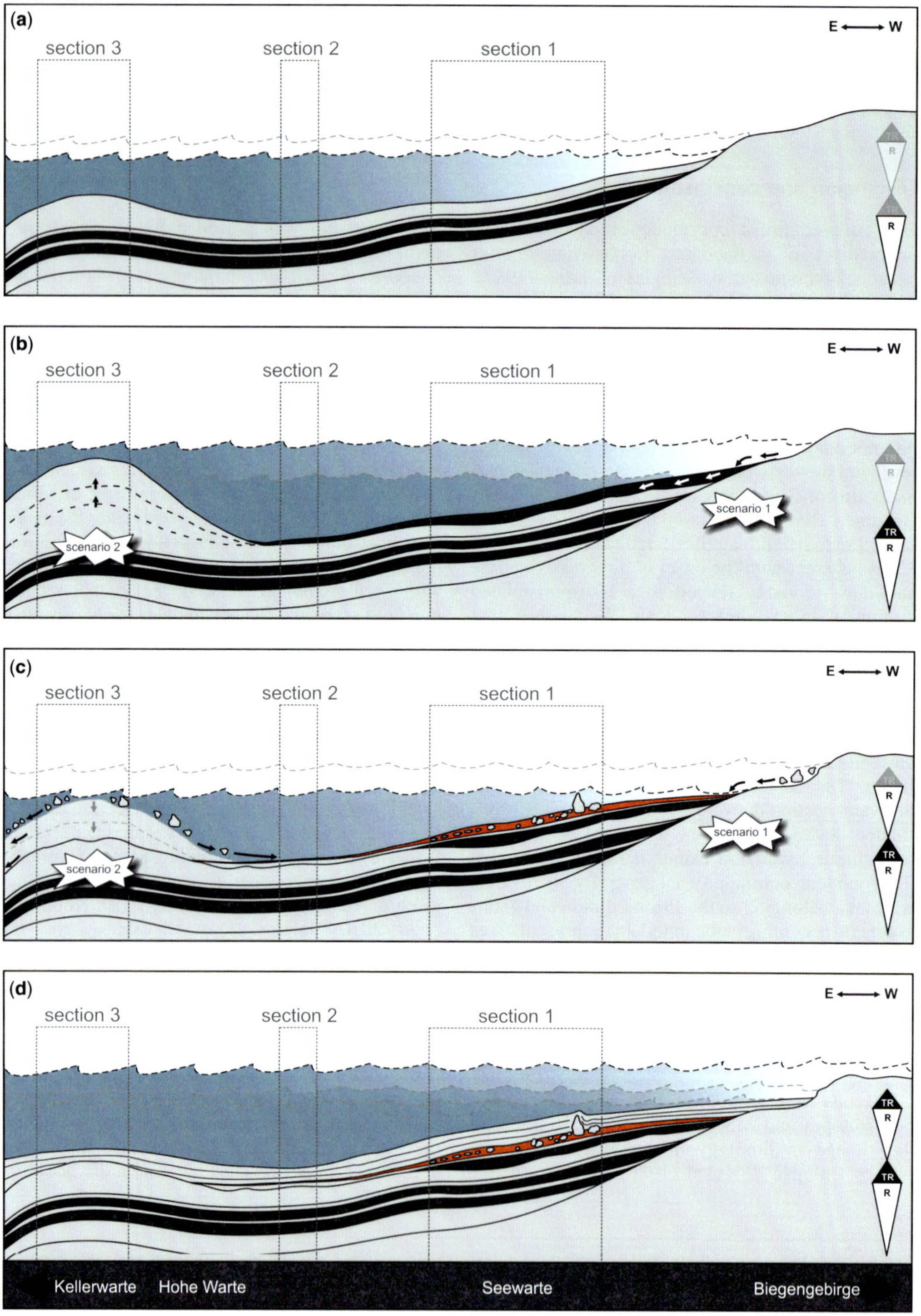

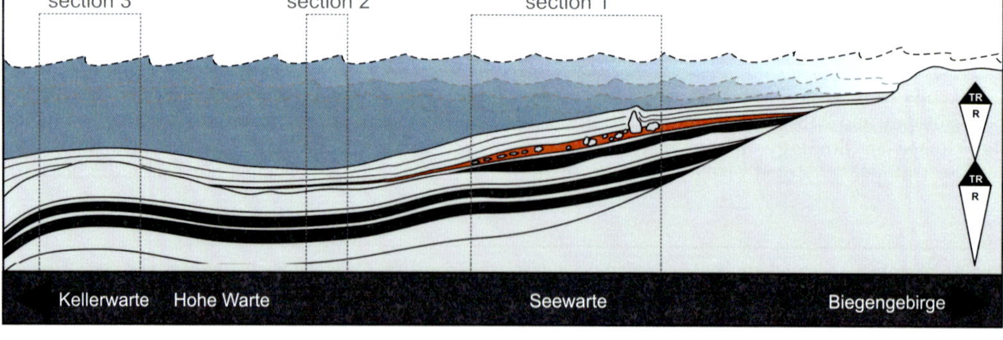

Fig. 6. Facies model reconstructing the possible mechanism of deposition of the megaclast horizon. (**a**) to (**d**) show two possible scenarios of the sedimentary development during mid-Lochkovian to early Pragian re- and transgressive cycles.

section. However, only the relative trend of the MS-log of both sections can be compared here, as MS-intensities result in the application of two different methods (Prague Synform: KLY-2 v. Carnic Alps: KT-6). Interestingly, the relative MS-trend of both sequences, from the Prague Synform and the Carnic Alps, are generally similar. What accords with the *c.* 42 m-thick interval from the 'False' Koněprusy Limestone to the base of the Zlíchov Formation in the Prague Synform correlates to a more than 400 m-thick shallow-marine development from the middle of the neritic Rauchkofel Limestone to the base of Spinotti Limestone. Another important point that complicates direct comparison of both MS-trends is the different depositional environment. A major part of the neritic sequence in the Carnic Alps consists of thick, massive, nearly pure limestone (Hohe Warte Limestone and Spinotti Limestone), wherefore MS results in relatively low values in the Seewarte section during the Pragian, with a distinctly decreased intensity of the MS signal in both areas during the early Emsian.

Although the Lochkov–Prag Event does not conform to a major biotic extinction event, at least it seems to represent an interval of geochemical and geophysical traceable perturbations in marine ecosystems of Europe and North Africa.

The authors are grateful for NAP 0001 and NAP 0017 for financial support. For help in the field we would like to thank Markus Reuter (University of Graz), Stanislava Vodrážková and Radek Vodrážka (both of Czech Geological Survey). Carlo Corradini (Università di Cagliari), a second anonymous reviewer and R. Thomas Becker (WWU Muenster) are gratefully acknowledged for their constructive comments on a former version of the typescript, especially regarding the biostratigraphic zonation and the tectonic-related interpretation of the depositional model. This is a contribution to IGCP 580 and IGCP 596.

References

Alberti, G. K. B. 1981. Daten zur stratigraphischen Verbreitung der Nowakiidae (Dacryoconarida) im Devon von NW-Afrika (Marokko, Algerien). *Senckenbergiana lethaea*, **62**, 205–216.

Alberti, G. K. B. 1983. Trilobiten des jüngeren Siluriums sowie des Unter- und Mitteldevons. *I. Abhandlungen der Senckenbergischen Naturforschenden Gesellschaft*, **520**, 1–692.

Anderson, M., Dargan, G., Brock, G. A., Talent, J. A. & Mawson, R. 1995. Maximising efficiency of conodont separations using sodium polytungstate solution. *Courier Forschungsinstitut Senckenberg*, **182**, 515–521.

Bandel, K. 1969. Feinstratigraphische und biofazielle Untersuchungen unterdevonischer Kalke am Fuß der Seewarte (Wolayer See, zentrale Karnische Alpen). *Jahrbuch der Geologischen Bundesanstalt*, **112**, 197–234.

Bogolepova, O. 1997. Orientation of Cephalopods. *In*: Schönlaub, H. P. (ed.) *IGCP-421 Inaugural Meeting Vienna, Guidebook*. Berichte der Geologischen Bundesanstalt, **40**, 117–119.

Buggisch, W. & Joachimski, M. M. 2006. Carbon isotope stratigraphy of the Devonian of Central and Southern Europe. *Palaeogeography, Palaeoclimatology, Palaeoecology*, **240**, 68–88.

Buggisch, W. & Mann, U. 2004. Carbon isotope stratigraphy of Lochkovian to Eifelian limestones from the Devonian of central and southern Europe. *International Journal of Earth Sciences*, **93**, 521–541.

Chlupáč, I. 1998. Devonian. *In*: Chlupáč, I., Havlíček, V., Kříž, J., Kukal, Z. & Štorch, P. (eds) *Palaeozoic of the Barrandian (Cambrian to Devonian)*. CGS, Prague, 101–133.

Chlupáč, I. 2000. Cyclicity and duration of Lower Devonian stages: observations from the Barrandian area, Czech Republic. *Neues Jahrbuch für Geologie und Paläotologie, Abhandlungen*, **215**, 97–124.

Chlupáč, I. & Kukal, Z. 1986. Reflection of possible Global Devonian events in the Barrandian area, C.S.S.R. *In*: Walliser, O. H. (ed.) *Global Bio-Events*. Lecture Notes in Earth History **8**. Springer-Verlag, Berlin, 169–179.

Chlupáč, I. & Kukal, Z. 1988. Possible global events and the stratigraphy of the Palaeozoic of the Barrandian (Cambrian–Middle Devonian, Czechoslovakia). *Sborník geologických věd, Geologie*, **43**, 83–146.

Corradini, C. & Corriga, M. G. 2012. A Přídolí–Lochkovian conodont zonation in Sardinia and the Carnic Alps: implications for a global zonation scheme. *Bulletin of Geosciences*, **87**, 635–650.

Corriga, M. G., Suttner, T. J., Corradini, C., Kido, E., Pondrelli, M. & Simonetto, L. 2011. The age of the La Valute Limestone-Findenig Limestone transition in the La Valute Section (Mount Zermula area, Carnic Alps). *Gortania*, **32**, 5–12.

Ellwood, B. B., Crick, R. E., García-Alcade Fernandez, J. L., Soto, F. M., Truyóls-Massoni, M., El Hassani, A. & Kovas, E. J. 2001. Global correlation using magnetic susceptibility data from Lower Devonian rocks. *Geology*, **29**, 583–586.

Hladil, J., Vondra, M., Čejchan, P., Vich, R., Koptíková, L. & Slavík, L. 2010. The dynamic time-warping approach to comparison of magnetic-susceptibility logs and application to Lower Devonian calciturbidites (Prague Synform, Bohemian Massif). *Geologica Belgica*, **13**, 385–406.

House, M. R. 2002. Strength, timing, setting and cause of mid-Paleozoic extinctions. *Palaeogeography, Palaeoclimatology, Palaeoecology*, **181**, 5–25.

Hubmann, B. & Suttner, T. 2007. Siluro-Devonian Alpine reefs and pavements. *In*: Alvaro, J. J., Aretz, M., Boulvain, F., Munnecke, A., Vachard, D. & Vennin, E. (eds) *Palaeozoic Reefs and Bioaccumulations: Climatic and Evolutionary Controls*. Geological Society, London, Special Publications, **275**, 95–107, http://doi.org/10.1144/GSL.SP.2007.275.01.07

Hubmann, B., Pohler, S., Schönlaub, H. P. & Messner, F. 2003. Paleozoic coral-sponge bearing successions in Austria. *Berichte der Geologischen Bundesanstalt*, **61**, 1–91.

HUBMANN, B., EBNER, F. ET AL. 2014. The Paleozoic Era(them). *In*: PILLER, W. E. (ed.) *The Lithostratigraphic Units of the Austrian Stratigraphic Chart 2004 (Sedimentary Successions) – Vol. I*. Abhandlungen der Geologischen Bundesanstalt, **66**, 9–133.

JAEGER, H. 1976. Das Silur und Unterdevon vom thüringischen Typ in Sardinien und seine regionalgeologische Bedeutung. *Nova Acta Leopoldina*, **45**, 263–299.

JEPPSSON, L. & ANEHUS, R. 1995. A buffered formic acid technique for conodont extraction. *Journal of Paleontology*, **69**, 790–794.

JOHNSON, J. G. & MURPHY, M. A. 1984. Time-rock model for Siluro-Devonian continental shelf, western United States. *Geological Society of America Bulletin*, **95**, 1349–1359.

JOHNSON, J. G., KLAPPER, G. & SANDBERG, C. A. 1985. Devonian eustatic fluctuations in Euramerica. *Geological Society of America Bulletin*, **96**, 567–587.

JOHNSON, J. G., KLAPPER, G. & ELRICK, M. 1996. Devonian transgressive–regressive cycles and biostratigraphy, northern antelope range, Nevada: establishment of reference horizons for global cycles. *Palaios*, **11**, 3–14.

KAUFMANN, B. 1998. Facies, stratigraphy and diagenesis of Middle Devonian reef- and mud-mounds in the Mader (eastern Anti-Atlas, Morocco). *Acta Geologica Polonica*, **48**, 43–106.

KLUG, C., KORN, D., NAGLIK, C., FREY, L. & DE BAETS, K. 2013. The Lochkovian to Eifelian succession of the Amessoui Syncline (Southern Tafilalt). *In*: *International Field Symposium 'The Devonian and Lower Carboniferous of Northern Gondwana' – Morocco 2013*. Document de l'Institut Scientifique, Rabat, **26**, 51–59.

KOPTÍKOVÁ, L., HLADIL, J., SLAVÍK, L., ČEJCHAN, P. & BÁBEK, O. 2010. Fine-grained non-carbonate particles embedded in neritic to pelagic limestones (Lochkovian to Emsian, Prague Synform, Czech Republic): composition, provenance and links to magnetic susceptibility and gamma-ray logs. *Geologica Belgica*, **13**, 407–430.

KREUTZER, L. H. 1990. Mikrofazies, Stratigraphie und Paläogeographie des Zentralkarnischen Hauptkammes. *Jahrbuch der Geologischen Bundesanstalt*, **133**, 275–343.

KREUTZER, L. H. 1992*a*. Palinspastische Entzerrung und Neugliederung des Devons in den Zentralkarnischen Alpen aufgrund von neuen Untersuchungen. *Jahrbuch der Geologischen Bundesanstalt*, **135**, 261–272.

KREUTZER, L. H. 1992*b*. Photoatlas zu den Variszischen Karbonat-Gesteinen der Karnischen Alpen (Österreich/Italien). *Abhandlungen der Geologischen Bundesanstalt*, **47**, 1–129.

KRÖGER, B. 2008. Nautiloids before and during the origin of ammonoids in a Siluro-Devonian section in the Tafilalt, Anti-Atlas, Morocco. *Special Papers in Palaeontology*, **79**, 1–110.

PARIS, F. 1981. Les Chitinozoaires dans le Paleozoique du sud-ouest de l'Europe. *Memoires de la Societe geologique et mineralogique de Bretagne*, **26**, 1–412.

SCHÖNLAUB, H. P. 1980. Carnic Alps. Field trip A. With contributions from Jaeger, H., House, M. R., Price, J. D., Göddertz, B., Priewalder, H., Walliser, O. H., Kříž, J., Haas, W. & Vai, G. B. *In*: SCHÖNLAUB, H. P. (ed.) *Second European Conodont Symposium, ECOS II, Wien, Guidebook, Abstracts*. Abhandlungen der Geologischen Bundesanstalt, **35**, 5–57.

SCHÖNLAUB, H. P. 1985. Devonian conodonts from section Oberbuchach II in the Carnic Alps (Austria). *Courier Forschungsinstitut Senckenberg*, **75**, 353–374.

SCHÖNLAUB, H. P. 1986. Significant geological events in the Paleozoic record of the Southern Alps (Austrian part). *In*: WALLISER, O. H. (ed.) *Global Bio-events*. Lecture Notes in Earth History, Springer-Verlag, Berlin, **8**, 163–167.

SCHÖNLAUB, H. P. & FLAJS, G. 1975. Die Schichtfolge der Nordwand der Hohen Warte (Mt. Coglians) in den Karnischen Alpen (Österreich). *Carinthia II*, **165/85**, 83–96.

SCHÖNLAUB, H. P. & HISTON, K. 2000. The Palaeozoic evolution of the Southern Alps. *Mitteilungen der Österreichischen Geologischen Gesellschaft*, **92**, 15–34.

SCHÖNLAUB, H. P. & KREUTZER, L. H. 1997. Stop 3: Seewarte Section. *In*: SCHÖNLAUB, H. P. (ed.) *IGCP-421 Inaugural Meeting, Vienna, Guidebook*. Berichte der Geologischen Bundesanstalt, **40**, 121–126.

SCHÖNLAUB, H. P., HISTON, K., FERRETTI, A., BOGOLEPOVA, O. & WENZEL, B. 1997. Stop 2: Rauchkofel Boden Section. *In*: SCHÖNLAUB, H. P. (ed.) *IGCP-421 Inaugural Meeting, Vienna, Guidebook*. Berichte der Geologischen Bundesanstalt, **40**, 107–120.

SLAVÍK, L. 2004. A new conodont zonation of the Pragian Stage (Lower Devonian) in the stratotype area (Barrandian, central Bohemia). *Newsletters on Stratigraphy*, **40**, 39–71.

SLAVÍK, L. & HLADIL, J. 2004. Lochkovian/Pragian GSSP revisited: evidence about conodont taxa and their stratigraphic distribution. *Newsletters on Stratigraphy*, **40**, 137–153.

SLAVÍK, L., VALENZUELA-RÍOS, J. I., HLADIL, J. & CARLS, P. 2007. Early Pragian conodont-based correlations between the Barrandian area and the Spanish Central Pyrenees. *Geological Journal*, **42**, 499–512.

SLAVÍK, L., HLADIL, J., KOPTÍKOVÁ, L., CARLS, P. & VALENZUELA-RÍOS, J. I. 2008. Integrated stratigraphy of the Lower Devonian in the Barrandian area, Czech Republic: an introduction of the project, preliminary data. *Subcommission on Devonian Stratigraphy, Newsletter*, **23**, 71–74.

SLAVÍK, L., CARLS, P., HLADIL, J. & KOPTÍKOVÁ, L. 2012. Subdivision of the Lochkovian Stage based on conodont faunas from the stratotype area (Prague Synform, Czech Republic). *Geological Journal*, **47**, 616–631.

SUTTNER, T. 2005. Bericht 2004 über paläontologische Aufnahmen im Paläozoikum der Karnischen Alpen auf Blatt 197 Kötschach. *Jahrbuch der Geologischen Bundesanstalt*, **145**, 382–383.

SUTTNER, T. 2007. Conodont stratigraphy, facies-related distribution patterns and stable isotopes (carbon and oxygen) of the uppermost Silurian to Lower Devonian Seewarte section (Carnic Alps, Carinthia, Austria). *Abhandlungen der Geologischen Bundesanstalt*, **59**, 111.

SUTTNER, T. & KIDO, E. 2011. Bericht 2007–2010 über geologische und paläontologische Aufnahmen im Unterdevon der Karnischen Alpen auf Blatt 197

Kötschach. *Jahrbuch der Geologischen Bundesanstalt*, **151**, 160–161.

Talent, J. A., Mawson, R., Andrew, A. S., Hamilton, P. J. & Whitford, D. J. 1993. Middle Palaeozoic extinction events: faunal and isotope data. *Palaeogeography, Palaeoclimatology, Palaeoecology*, **104**, 139–152.

Vacek, F. 2011. Palaeoclimatic event at the Lochkovian-Pragian boundary recorded in magnetic susceptibility and gamma-ray spectrometry (Prague Synclinorium, Czech Republic). *Bulletin of Geosciences*, **86**, 259–268.

Vai, G. B. 1980. Sedimentary environment of Devonian pelagic limestones in the Southern Alps. *Lethaia*, **13**, 79–91.

Valenzuela-Ríos, J. I. 1994. The Lower Devonian conodont *Pedavis pesavis* and the *pesavis* Zone. *Lethaia*, **27**, 199–207.

Walliser, O. H. 1996. Global events in the Devonian and Carboniferous. *In*: Walliser, O. H. (ed.) *Global Events and Event Stratigraphy in the Phanerozoic*. Springer-Verlag, Berlin, 225–250.

Young, G. C. 1995. *Timescales. 4. Devonian. Australian Phanerozoic Timescales*. Biostratigraphic Charts and Explanatory Notes Series 2. Australian Geological Survey Organisation, Australia, 1–47.

Ziegler, W. & Lane, H. R. 1987. Cycles in conodont evolution from Devonian to mid-Carboniferous. *In*: Aldridge, R. J. (ed.) *Palaeobiology of Conodonts*. Horwood Press, Chichester, 147–164.

Palynological calibration of Devonian events at near-polar palaeolatitudes in the Falkland Islands, South Atlantic

J. E. A. MARSHALL

Ocean and Earth Science, University of Southampton, National Oceanography Centre, European Way, Southampton SO14 3ZH, UK (e-mail: jeam@noc.soton.ac.uk)

Abstract: In the Devonian, the Falkland Islands were part of the high-latitude Cape Basin. West Falkland palynological assemblages are at a low thermal maturity level but also very low in diversity and dominated by simple spores with rare chitinozoans and acritarchs. The South Harbour Member contains rare verrucate and sculptured trilete spores, and is early Lochkovian. The Fish Creek Member palynofloras are late Lochkovian. The best correlative datum is the transgressive Fox Bay Formation where the palynological assemblage includes *Ramochitina magnifica* and is equivalent to the Sequence B transgression in Brazil (dated as late Pragian–earliest Emsian). The early Eifelian upper Fox Bay Formation spore assemblage is comparable to the Sequence C transgression in Brazil and the ?Choteč Event. The facies and position of the base Port Philomel Formation prasinophyte-rich black shale suggests the late Eifelian transgression (Kačák Event). Above this level is the inception of *Geminospora lemurata* (base Givetian). The upper Port Stanley Formation contains a late (but not latest) Famennian assemblage. Comparisons with both South Africa and South America indicate a number of correlative transgressive and regressive events that match to Euramerica. These correlations are at a continental scale and confirm eustatic control of these Devonian events.

There is now an established pattern of Devonian events that link sea-level change, palaeoclimate and extinctions. However, much of this information is only available from sections in Euramerica and northern Gondwana (i.e. in the Devonian palaeoequatorial to mid-latitude zones). In order to more fully understand the Devonian Earth system, we need to determine the palaeolatitudinal limits of these events. There has been significant progress in documenting and correlating these Devonian events at intermediate high palaeolatitudes (*c.* 60° S) in both Bolivia (Troth *et al.* 2011) and Brazil (Horodyski *et al.* 2013). At higher palaeolatitudes, the best-known and most complete Devonian successions are in the Cape Fold Belt of South Africa. Unfortunately, these sequences have only sporadic, poorly preserved palynofloras that are at a high level of thermal maturation. But there is another preserved fragment of the Cape Fold Belt present in the Falkland Islands in southernmost South America. For, although now an integral part of the South American Plate, the immediate Palaeozoic geological affinities of the Falkland Islands (Fig. 1) are with southern Africa and the dispersed terranes of West Gondwana (e.g. Marshall 1994). The Falkland Islands were one of a group of these microcontinents that were produced by the early Mesozoic fragmentation of Gondwana, the islands themselves undergoing significant rotation away from their original position on the eastern side of the Cape Fold Belt. They thus became attached to southern South America and then separated with it, rather than with Africa and Antarctica, during the subsequent opening of the South Atlantic Ocean.

The Palaeozoic sequences in southern Africa are both thick and laterally very extensive (e.g. Shone & Booth 2005). However, although they are well known and well defined through the results of a comprehensive geological mapping programme (e.g. South African Committee for Stratigraphy 1980), the succession, particularly in the Eastern Cape (i.e. closest to the pre-drift position of the Falkland Islands), is suspect owing to the presence of largely undetected thrust faults (e.g. Booth & Shone 2002). The sequences are also very poorly constrained chronostratigraphically. This is because the fauna, although locally rich and well described (e.g. Hiller & Theron 1989), is entirely lacking in the conodonts and goniatites that are the basis for correlation in the marine Devonian. In addition, the brachiopods and trilobites that do dominate these faunas have very few elements in common with lower-latitude sequences and are a strongly endemic representative of the Malvinokaffric Realm. In common with South America, cosmopolitan brachiopods do occur – for example, *Tropidoleptus* (Boucot *et al.* 1983) and *Rhipidothyris* (Boucot & Theron 2001) – but are very rare and have yet to be found in the Falkland Islands. As regards palynological recovery, the Palaeozoic sequences in both South Africa and the Falkland Islands are part of the Cape Fold Belt and, in

From: Becker, R. T., Königshof, P. & Brett, C. E. (eds) 2016. *Devonian Climate, Sea Level and Evolutionary Events*. Geological Society, London, Special Publications, **423**, 25–44.
First published online June 7, 2016, http://doi.org/10.1144/SP423.13

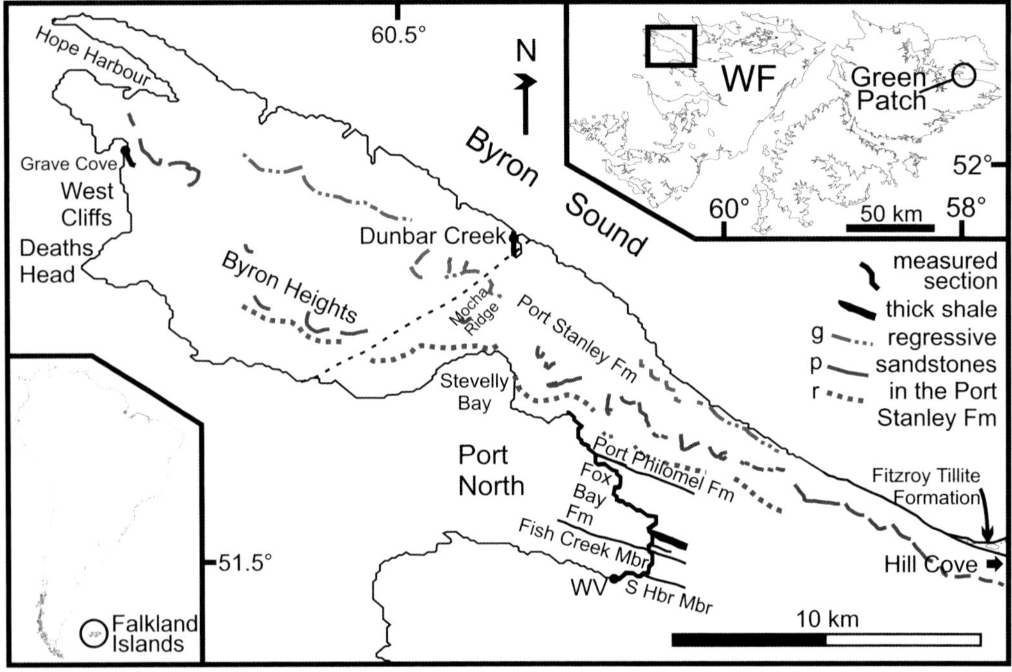

Fig. 1. Location of the Port North and Dunbar Creek sections in West Falkland. Inset maps show the location of the Falkland Islands in South America and Port North–Dunbar Creek on the island of West Falkland (WF) with Green Patch in East Falkland. The main Port North section is near-continuous from the South Harbour Member (S Hbr Mbr) at Wine Valley (WV) to the top of the Port Philomel Formation in Stevelly Bay. The section can then be correlated via a series of three regressive sandstone (r, p and g) to Dunbar Creek. The thick shale interval in Stevelly Bay re-occurs west along-strike in Grave Cove. Compiled from OSI air photographs, Google Earth and field observations.

consequence, the rocks are meta-sediments with a pervasive deformation. This has made it difficult to routinely study the palynomorphs that are both graphitized to very high levels of thermal maturity and also generally damaged by metamorphic mineral growth (e.g. Stapleton 1977).

It was the Palaeozoic sequence on East Falkland (Fig. 1) that, prior to the 180° Mesozoic rotation, was positioned immediately adjacent to the present-day coast of Natal and, hence, has a succession that is most similar to that of the Eastern Cape. It contains appreciable intervals of pelitic mudstones that similarly have an elevated thermal maturity and significant internal deformation. In contrast, the succession in West Falkland, although clearly correlative, is quite different and typically present as long gently-dipping sequences (15–25°) that are largely undeformed (Marshall 1994; Hyam *et al.* 2000). In addition, the West Falkland succession is thinner, relatively much more proximal in character, being generally arenitic, and has only thin intervals and intercalations of mudstones present. The two sequences are separated by major and long-lived syndepositional fault systems (Falkland Sound Fault: Marshall 1994; Hornby Mountain Fault: Hyam *et al.* 2000) that run through the Falkland Sound and along the east coast of West Falkland, respectively. It was on these structures that the differential deformation between East and West Falkland was accommodated (Marshall 1994; Curtis & Hyam 1998; Hyam *et al.* 2000).

There is minimal published palynological work on the Falkland Islands. These are restricted to a short illustrated report on a single sample by Cramer & Diez de Cramer (1972) from a fossil plant specimen collected during the 1907–08 Swedish Expedition (Halle 1911) and brief notes in more general accounts of Falkland Island geology (e.g. Marshall 1994; Aldiss & Edwards 1999; Meadows 1999).

Falkland Island geology and the palynological samples

The pre-Carboniferous West Falkland Group is subdivided (Fig. 2) into the Port Stephens, Fox Bay,

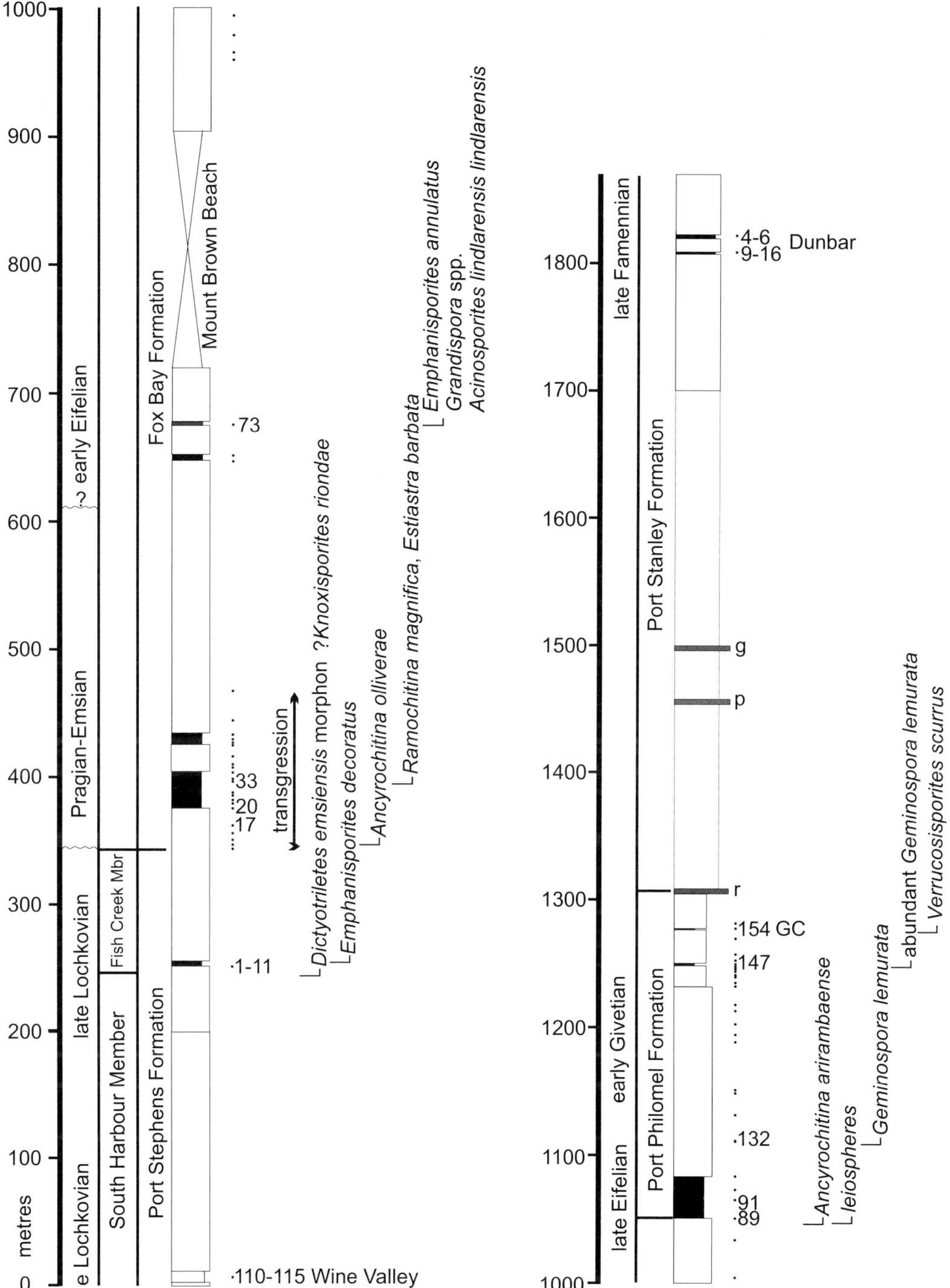

Fig. 2. Measured section from the Port North and Dunbar Creek sections. The figure shows the main stratigraphic units, the position of the palynological samples (•) and related spore inceptions, with ages estimated from the palynological assemblages. Three regressive sandstones (r, p and g) in the Port Stanley Formation are shown. The missing section of Port Stanley Formation along Mocha Ridge was estimated using field measurements, field photographs, OSI air photographs and Google Earth from the compilation in Figure 1. GC, Grave Cove.

Port Philomel and Port Stanley formations (Aldiss & Edwards 1998, 1999; Hunter & Lomas 2003; Stone *et al.* 2005). All these units are dominantly arenitic, apart from the Fox Bay Formation, which contains both an appreciable content of fine-grained sediment and a locally abundant marine fauna. Two sections (Fig. 1), one from the west side of West Falkland (Port North–Dunbar) and the other from the eastern side of East Falkland (Green Patch: Marshall in Hyam *et al.* 2000), were measured in detail and sampled for palynomorphs. The two sections are contrasting, with that from Green Patch in East Falkland containing appreciable thicknesses of dark-coloured, organic-rich pelitic mudstone. Although the 49 palynological samples collected from the 1.3 km-thick section at Green Patch yielded organic residues, these were almost entirely composed of phytoclasts with a reflectivity of 3.3% (Hyam *et al.* 2000). The spores were also rare, graphitized and poorly preserved. There was also a lack of marine palynomorphs, even at the Fox Bay Formation transgression level. Hence, this section will not be considered further here.

In contrast to Green Patch, the section from Port North to Dunbar (Figs 1 & 2) contained palynomorphs at a low level of thermal maturity (Figs 3–6) throughout the interval from the upper Port Stephens to the uppermost Port Stanley formations and forms the basis of this study. The section starts in the South Harbour Member of the Port Stephens Formation with near-continuous exposure through the overlying Fish Creek Member (Port Stephens Formation), Fox Bay Formation and Port Philomel Formation, apart from some intervals of limited exposure beneath beaches (e.g. Mount Brown Beach: Fig. 2). The continuous exposure then terminates in the bay east of Stevelly Bay (Fig. 1) at a prominent sandstone unit that marks the base of the Port Stanley Formation. This is just above a prominent shale unit that can also be traced westwards to Grave Cove (Aldiss & Edwards 1999, p. 30) and is regionally downcut at the base of the Port Stanley Formation.

The Port Stanley Formation is a sand-rich sequence, which also contains a number of prominent orthoquartzites that can be traced as near-continuous beds from Hope Harbour to, at least, Hill Cove. These more resistant beds also enable the sequence to be traced along Mocha Ridge down into the west side of Dunbar Creek, where there is a second continuous section into the youngest preserved section of the Port Stanley Formation. The distinctive orthoquartzites are intensely cemented pure sandstone with abundant evidence of a shallow-water origin with herringbone cross-stratification and low-angle bedding. They represent regressive intervals within the generally regressive Port Stanley Formation. Mapping these orthoquartzites eastwards towards Hill Cove reveals them to be truncated by the downcutting Fitzroy Tillite Formation, which is Permian in age and the local equivalent of the Dywka Group in South Africa (Trewin *et al.* 2002).

The lithology of the palynomorph samples varied from major intervals of bedded shale and siltstone to thin silt partings within flaser–lenticular bedding. The latter were not uncommon within thick intervals of sandstone and provided the palynological samples (Fig. 2) throughout much of the succession.

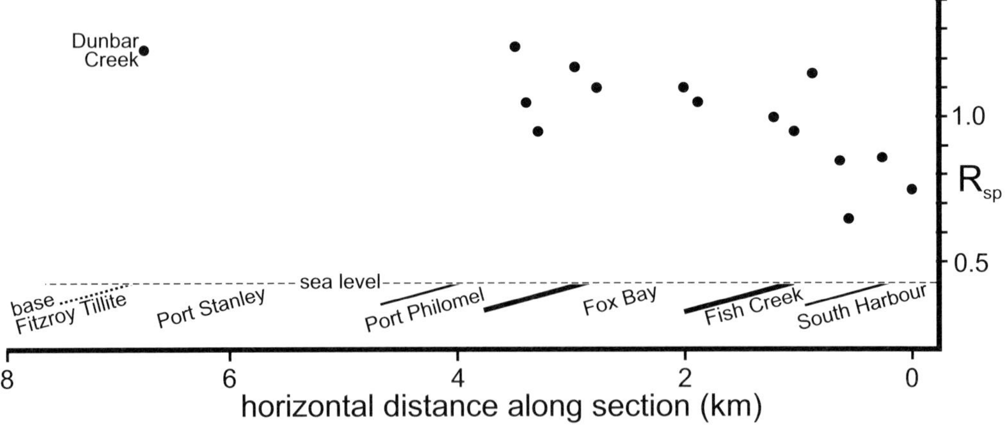

Fig. 3. Cross-plot of thermal maturity (measured as spore reflectivity, R_{sp}) v. location through the Port North section and, hence, stratigraphic position. The thermal maturity is lowest in the oldest rocks and increases as the age becomes younger, up to a maximum in Stevelly Bay. This shows that thermal maturity increases towards the axis of the Cape Fold Belt. Reflectivities were measured using the methods described in Marshall (1998).

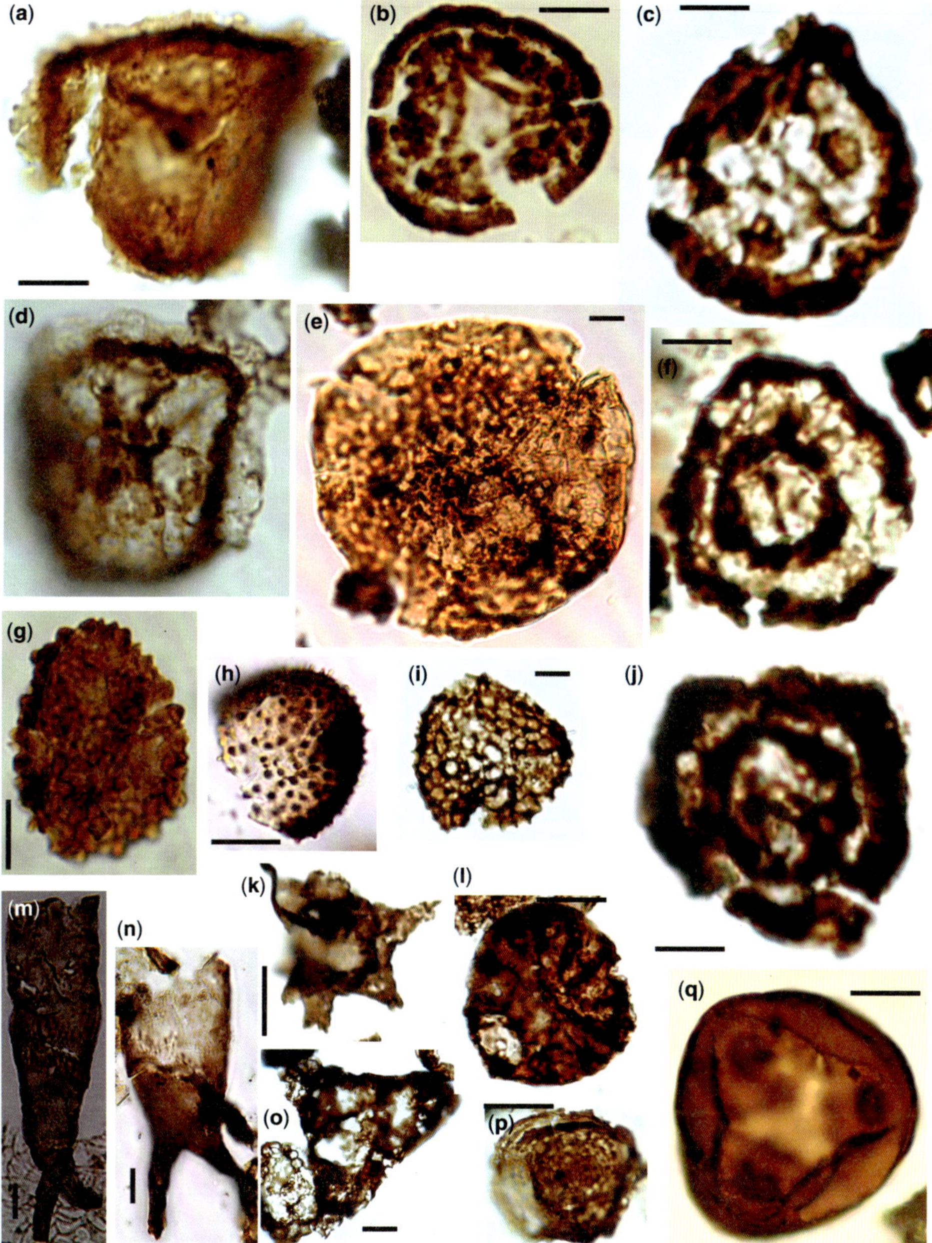

Fig. 4. Early Devonian miospores and acritarchs from the South Harbour and Fish Creek members of the Port Stephens Formation and the Fox Bay Formation, West Falkland. The scale bar is 10 μm. (**a**) & (**d**) *Zonotriletes* sp., cf. sp. 3 of Jardiné & Yapaudjian 1968: (a) PN112, long c' slip 115.0, 8.0, U14/1; South Harbour Member; (d) PN1, 139, 9, T39; Fish Creek Member. (**b**) *Coronaspora* sp., PN11, 121.3, 13.3, O21/3. (**c**) *Synorisporites tripapillatus*, PN33.1 ox, 119.4, 13.6, O19/1. (**e**) *Retusotriletes goensis*, PN6 long, 116.8, 8.6, T16. (**f**) & (**j**) ?*Knoxisporites riondae*: (f) PN20, 116, 17.1, K15/4; (j) is PN33.1 ox, 139.2, 23.2, D39/4. (**g**) Verrucate spore, PN114, 123, 4.1, Y22/2. (**h**) Simple apiculate spore, PN114, <50, 119.7, 11.1, Q19/3. (**i**) *Dictyotriletes emsiensis* morphon, PN33.1, ox, 138.6, 10.2, R38/4. (**k**) *Estiastra barbata*, PN33.2 ox, 143.5, 16, L44/3. (**l**) *Emphanisporites decoratus* PN6, 128, 24.8, N27/2. (**m**) & (**n**) *Bimerga paulae* (m) PN20, 124.2, 4.8, X23/2; (n) PN17, 124.2, 4.8, P27/3. (**o**) *Tyligmasoma alagarda*, PN33, 134.6, 17, K34/4. (**p**) *Chomotriletes vedugensis*, PN17, 127.2, 12.2, P27/3. (**q**) Tripapillate retusoid spore preserved in three dimensions from a phosphatic nodule, PN11, 3d 118.4, 11.3, Q18/3.

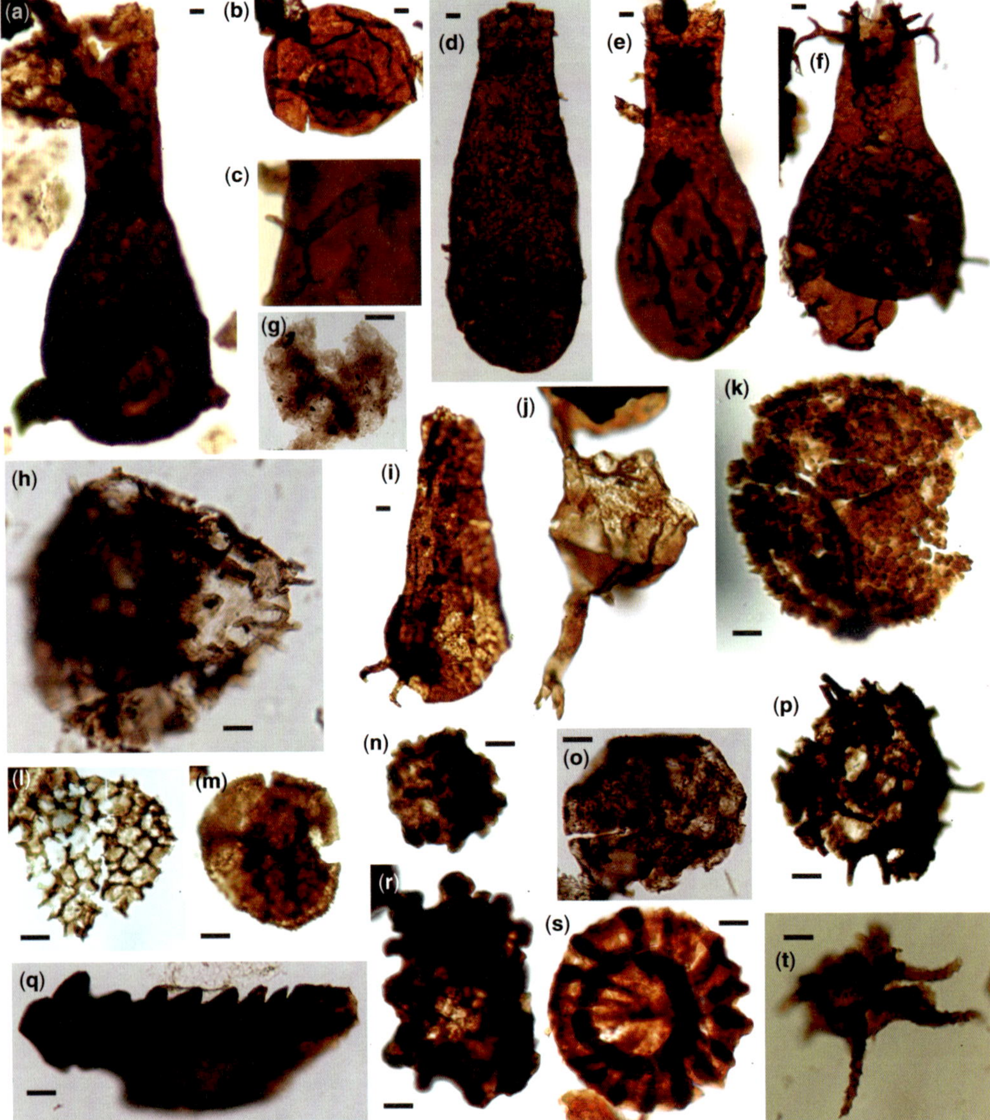

Fig. 5. Miospores, chitinozoans, acritarchs and other palynomorphs, Fox Bay and Port Philomel formations, West Falkland. The scale bar is 10 μm. (**a**) *Ancyrochitina olliverae*, PN17.2, 129.6, 10.6, R29/2. (**b**) *Hoegisphaera* cf. *glabra*, PN33.2, 142, 8.6, T42. (**c**) & (**d**) Chitinozoan comparable to '*Gotlandochitina*' *marettensis* Paris, 1981 in Vavrdová *et al.* (1996) and ?*R. magnifica* Lange, 1967 in Vavrdová *et al.* (1996). PN20 pick 1, 138.2, 16, L38/4. (**e**) & (**f**) *Ramochitina magnifica* (e) PN33.2, 133.9, 7.1, V34/1; (f) PN33.2, 145.2, 4.4, Y45/2. (**g**) Leiosphere, a poorly preserved prasinophyte from PN91, 133, 18.1, J33/3. (**h**) *Grandispora naumovae*, PN74.2, 127.3, 16.5, L27/1. (**i**) *Ancyrochitina arirambaense*, PN73.1, >88, 132.8, 8.9, T32/2. (**j**) *Exochoderma* sp. with incomplete processes, either *E. arca* (if more than three processes) or *E. triangulata* (three processes), PN33.2, 129.9, 7.6, U29/4. (**k**) *Acinosporites lindlarensis lindlarensis*, PN73.2, 128, 13.1, O27/4. (**l**) *Dictyotriletes emsiensis* morphon, PN73.2, 132.5, 9.4, S32/4. (**m**) & (**o**) *Geminospora lemurata* (m) GC6 ox 127.4, 14, N27/3; (o) PN139, 133, 12.6, Q33/1. (**n**) *Verrucosisporites scurrus*, GC6 ox 135.1, 7.4, U35/3. This specimen has variably developed sculpture that shows affinity to the variably sculptured patinate spores, such as *Archaeozonotriletes variabilis*. (**p**) Spore with long broken spines, a late Emsian innovation in Euramerica, PN73.1, >88, 137.7, 18.8, J38/1. (**q**) Scolecodont, the organic-walled tooth of an annelid worm, PN73.1, >88, 137.1, 17, K37/3. (**r**) *Verrucosisporites premnus*, GC6 ox, 132.1, 9, T32/1. (**s**) *Emphanisporites annulatus*, PN73.2, 115, 17.1, K14/4. (**t**) *Diexallophasis denticulata*, PN62, 128.3, 8.2, U28/3.

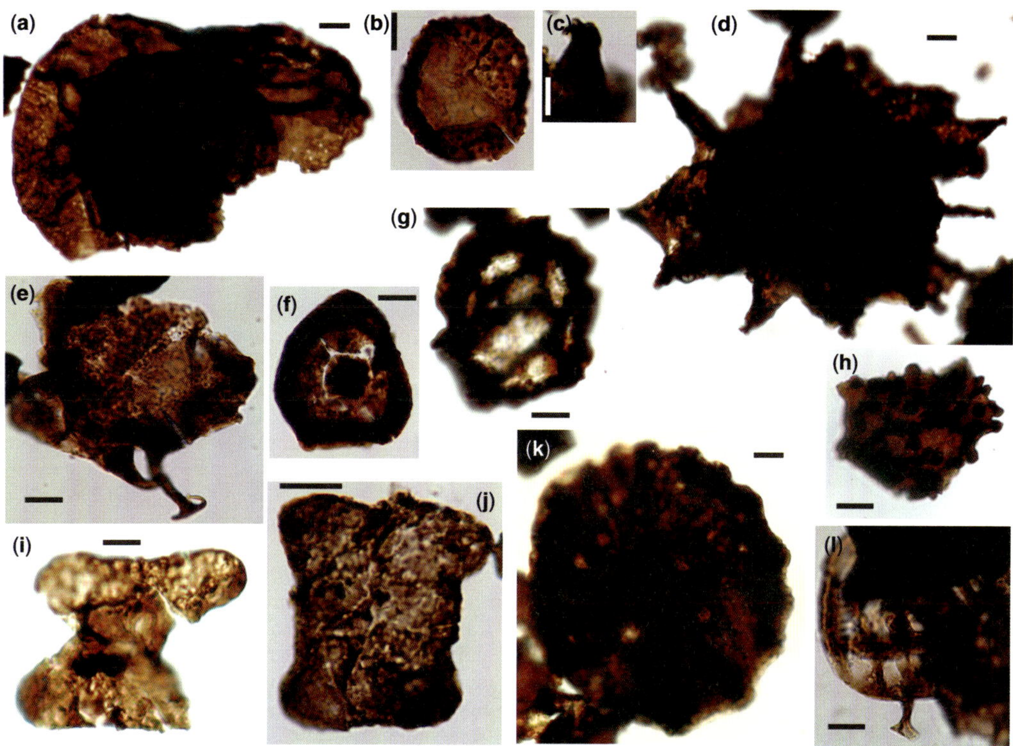

Fig. 6. Miospores and acritarchs from the uppermost part of the Port Stanley Formation, Dunbar Creek, West Falkland. The scale bar is 10 μm, except where indicated. (**a**) *Diducites versabilis* Dunbar 8, >50 < 100, 147.1, 9.6, S47/4. (**b**) *Verrucosisporites famenensis*, Dunbar 14.1, 143.5, 7.5, U43/4. (**c**) & (**d**) *Ancyrospora* with multifurcate tips, Dunbar 8 >50 < 100, 136, 11.9, Q36; (c) the scale bar is 5 μm. (**e**) Isolated bifurcate tip from either *Ancyrospora* or *Hystricosporites*, Dunbar 10.1, 129, 6.9, V29/1. (**f**) *Tumulispora ordinaria*, Dunbar 1, 112.8, 15.8, M12/1. (**g**) *Knoxisporites hederatus*, Dunbar 8, >50 < 100, 142.5, 20.7, G43/1. (**h**) *Verrucosisporites scurrus*, Dunbar 13 ox 124.3, 11.4, Q24/3. (**i**) & (**j**) *Horologinella horologia* (i) Dunbar 5 ox 131.9, 14.9, N32/1; (j) Dunbar 14, 131.1 15.3, M31/3. (**k**) *Maranhites mosesii*, Dunbar 8.2, 133.1, 11.6, Q33/1. (**l**) *Umbellasphaeridium saharicum*, Dunbar 8.3, <50, 135.1 20.1, G35/3.

The 181 samples collected from the Port North–Dunbar section were processed using standard methods, including HCl and HF, followed by decant washing to neutral and sieving at 20 μm, followed by a brief short treatment in hot HCl to solubilize neoformed fluorides. The samples were then rapidly diluted into 500 ml of water and re-sieved before storing in a vial. Palynomorph colours varied from yellow to dark brown and black, the latter through contact alteration by Jurassic dolerite dykes. The trend in thermal maturity (Fig. 3) as shown both by colour and by reflectance (Marshall 1994; Hyam *et al.* 2000) is inverted against that expected for normal sedimentary burial. The oldest palynomorphs have the palest colours. This shows the overprint of Cape Fold Belt deformation on normal burial. Some of the kerogen residues were oxidized with fuming nitric acid or concentrated nitric acid with, or without, the addition of potassium chlorate. These oxidations were not always successful, with the palynomorphs subsequently redarkening and/or fragmenting. The samples were mounted in Elvacite 2044 or Aquamount™, with all slides being stored in the School of Ocean and Earth Science, University of Southampton. Microscope co-ordinates (for Olympus BHS-313 210685) and England Finder references are given in the relevant figure captions. All taxonomic citations are listed on Table 1. Selected kerogen residues were handpicked for chitinozoans using a finely drawn pipette connected to an oral suction tube.

Remarkably, all the samples contained organic residues. However, the residues from the sections in both Port North and Green Patch were dominated by phytoclasts, with spores and spore-rich residues being relatively rare. In an attempt to increase palynomorph yield, many of the samples were sieved at a variety of mesh sizes (50, 88, 100, 150 μm) to

Table 1. *Taxonomic citations*

Acinosporites lindlarensis Riegel, 1968 var. *lindlarensis* McGregor & Camfield, 1976
Ancyrospora (Eisenack) Richardson, 1962
Auroraspora Hoffmeister, Staplin & Malloy, 1955
Cirratriradites diaphanus Steemans, 1989
Coronaspora (Rodriguez) Richardson *et al.*, 2001
Dictyotriletes emsiensis morphon of Rubinstein *et al.*, 2005
Diducites versabilis Van Veen, 1981
Emphanisporites annulatus McGregor, 1961
Emphanisporites decoratus Allen, 1965
Emphanisporites micrornatus Richardson & Lister, 1969
Geminospora lemurata (Balme) Playford, 1983
Grandispora naumovae (Kedo) McGregor, 1973
Iberospora cantabrica Cramer & Diez, 1975
Knoxisporites hederatus (Ishchenko) Playford, 1963
Knoxisporites riondae Cramer & Diez, 1975
Retusotriletes goensis Lele & Streel, 1969
Synorisporites tripapillatus Richardson & Lister, 1969
Tumulispora ordinaria Staplin & Jansonius, 1964
Verrucosisporites famenensis (Kedo) Massa *et al.*, 1979
Verrucosisporites premnus Richardson, 1965
Verrucosisporites scurrus (Naumova) McGregor & Camfield, 1982
Zonotriletes sp. cf. sp. 3 of Jardiné & Yapaudjian, 1968
Bimerga paulae Le Hérissé, 2011
Chomotriletes vedugensis Naumova, 1953
Diexallophasis denticulata (Stockmans & Willière) Loeblich, 1970
Estiastra barbata Barreda, 1986
Exochoderma arca Wicander & Wood, 1981
Exchoderma triangulata Wicander & Wood, 1981
Horologinella horologia (Staplin) Jardiné *et al.*, 1972
Maranhites mosesii (Sommer) González, 2009
Tyligmasoma alagarda (Cramer) Playford, 1977
Umbellasphaeridium deflandreii (Moreau-Benoit) Jardiné *et al.*, 1972
Ancyrochitina arirambaense Grahn & Melo, 2005
Ancyrochitina olliverae Boumendjel, 2002
Angochitina daemonii Grahn, 2000
Hoegisphaera cf. *glabra* Staplin, 1961
Ramochitina magnifica Lange, 1967

separate the phytoclasts from the smaller spores or larger chitinozoans. In addition to palynomorph abundance problems, their preservation (as distinct from thermal maturation) was also quite variable and generally poor, with pervasive exine degradation being a characteristic of most samples. This is atypical of many contemporary palynological assemblages from lower palaeolatitudes, and one possible explanation may be that the colder water environments with their lower organic matter decomposition rates meant that the palynomorphs had much longer residence times, with many showing progressive degradation. In warmer waters, palynomorphs have much faster degradation and, hence, preservation may be a consequence of rapid burial (cf. Goldring *et al.* 1998). The relative dominance of phytoclasts in the residues increases through the sequence and is indicative of the source vegetation being dominated by vegetative growth rather than reproduction. This emphasis on propagation by clonal growth rather than reproduction is a feature of modern high-latitude ecosystems (Archibold 1995).

In addition to the combination of poor preservation and phytoclast abundance, the assemblages are dominated by simple spores with a smooth and apiculate sculpture. More complex spore types are relatively uncommon throughout the sequence. Even rarer in the samples were chitinozoans and, more so, acritarchs. This scarcity of marine palynomorphs had already been noted, on the basis of a single sample, by Cramer & Diez de Cramer (1972). The anomaly being that, in intervals that were undoubtedly marine in origin (i.e. containing brachiopods, crinoids and trilobites), the palynological assemblages contained only spores and rare chitinozoans, with the normally ubiquitous acritarchs absent. This much larger study through the entire Devonian succession confirms this general observation. However, acritarchs (Fig. 5j & t) have been found together with prasinophytes (Fig. 6k) and scolecodonts (Fig. 5q) in a few samples, although at a very low abundance. Significantly, acritarchs are no more common in the much thicker shale intervals in, for example, the more distal Green Patch sequence. In these respects, the Falkland Island palynological assemblages are somewhat different from those known from northerly 'high-palaeolatitude' Gondwana localities (Fig. 5), such as Bolivia (e.g. Ottone 1996), Brazil (e.g. Melo & Loboziak 2003; Grahn & Melo 2004), northern Argentina (e.g. di Pasquo 2007*a*, *b*; Rubinstein & García Muro 2013), Uruguay (Pöthe de Baldis 1978) and Paraguay (Pöthe de Baldis 1974), which contain a diverse and abundant acritarch assemblage. However, it must be noted that these other 'high-palaeolatitude' localities are some 2000 km distant from the present-day location of the Falkland Islands and more distant in pre-Gondwana fragmentation models. This relative paucity of acritarchs, and relative abundance of both spores and acritarchs, has been previously reported (e.g. Wellman *et al.* 2000). As the sections are clearly at normal marine salinity, explanations based on a relatively proximal location are unsatisfactory. Instead, it may be that acritarchs and, perhaps, encystment are not common at high palaeolatitudes.

This account documents some of the more important palynomorphs that can be used in external correlation to demonstrate links between the Falkland Island sequence and related sections in South America and Euramerica. It provides an important record of spores at what may currently be their highest known Devonian palaeolatitudes.

The palynological assemblages

Port Stephens Formation, South Harbour Member

A closely spaced group of samples (PN110–PN115) containing palynomorphs was found in Wine Valley (Figs 1 & 2) within an intercalation that included finer-grained sediments. In assigning an age to these palynomorphs, there has to be an element of considerable caution, given that it is very different from typical Siluro-Devonian assemblages from both northern Gondwana and Euramerica. These samples contained an assemblage dominated by small simple smooth spores. Also present are rare specimens of simple apiculate (Fig. 4h) and tripapillate spores, together with a single specimen of a trilete spore that has an undivided proximal cingular feature (Fig. 4a), very similar to *Cirratriradites diaphanus* but lacking the inter-radial papillae. Another comparable form is *Zonotriletes* sp. 3 of Jardiné & Yapaudjian (1968) that was also figured by Steemans *et al.* (2008) and with an early Lochkovian first appearance in Gondwana (Steemans *et al.* 2012). There is also a single densely sculptured 'verrucate' *Cymbosporites* type spore (Fig. 4g), a morphological innovation with a first appearance in the Ludlow (Steemans *et al.* 2012). The assemblage lacks microconate spores with proximal radial ribs (emphanoid), which are present upsection, and shows that the age is pre-mid-Lochkovian (Steemans 1989). Also absent are spores with distal reticula (the *Dictyotriletes emsiensis* morphon of Rubinstein *et al.* 2005), which had a first appearance in the mid- to late Lochkovian interval. Cryptospores are also absent, which was a group that showed a significant decline in the latest Silurian (Wellman *et al.* 2013). Taken together, these palynomorphs indicate an age that is probably in the early Lochkovian.

Port Stephens Formation, Fish Creek Member

There are a number of productive palynological samples from within the lower part of the Fish Creek Member (Fig. 2) and which are an estimated 250 m above the lowest samples from Wine Valley (South Harbour Member). The samples from the Fish Creek Member include a group from close to the base of the continuously logged section, including several from a 4 m-thick interval of shale within a sequence that is otherwise dominated by sandstone of tidal channel origin. The lowest sample (PN1) has an assemblage that is entirely composed of sporomorphs, which include examples of the *Dictyotriletes emsiensis* morphon with both fine (Fig. 4i) and coarse reticula, together with a conspicuous group (Fig. 4f & j) of trilete spores with an undivided equatorial cingulum and subconcentric distal ring variously connected to the equator. These were named ?*Knoxisporites riondae* by Cramer & Diez (1975), and then recognized variously as *Knoxisporites*? *riondae* and *K.*? cf. *riondae* (Le Hérissé 1983), and *Knoxisporites riondae* and *Knoxisporites*? *riondae* (Grahn *et al.* 2013). Other forms have tripapillae and are similar to ?*Aneurospora* sp. of Kemp (1972, pl. 55, fig. 10) from Antarctica. Also present are a tripapillate spore with a thick equatorial cingulum (Fig. 4c) and, again, the zonate spore (Fig. 4d) similar to *Zonotriletes* sp. 3. The assemblages that are most similar are the NsZ (non-spinose zonate) interval Zone of Melo & Loboziak (2003) described from the Amazon Basin and that from the Solimões Basin (Rubinstein *et al.* 2005), both from Brazil. These comparisons and the sequence of overlying palynological assemblage suggest a broadly late Lochkovian age for this part of the Fish Creek Member.

The spore assemblage from the slightly higher 4 m-thick shale interval (e.g. PN6) contains many common elements with, in addition, very rare prasinophytes and acanthomorph acritarchs. It also includes the inception of the chitinozoan *Hoegisphaera* cf. *glabra* (PN9: Fig. 5b). However, the dominant palynomorph, which can reach abundances in excess of 90%, is a retusoid spore with a wide size range that can be attributed to *Retusotriletes goensis* (Fig. 4e). Also present (Fig. 4l) is a species of *Emphanisporites* with a prominent sculpture of coni that is similar to *E. decoratus* rather than the more minutely sculptured *E. micrornatus*. Such sculptured forms of *Emphanisporites* (as *E. micrornatus*) have a mid-Lochkovian first appearance (Wellman & Richardson 1996).

Fox Bay Formation

The Fox Bay Formation transgression. The Fox Bay Formation is characterized by a progressive and sustained transgression to a locally thick accumulation of shale and siltstone, with accompanying biota that included trilobites, brachiopods, bivalves, orthocones and crinoids (Cocks in Aldiss & Edwards 1999), together with acanthodian fish spines (Maisey *et al.* 2002), making it the level with the most diverse macrofauna in the West Falkland Group. This macrofauna has many elements in common with other Malvinokaffric faunas, and has been used to correlate (e.g. Edgecombe 1994) the Fox

Bay with the Gydo Formation (Bokkeveld Group) in South Africa and the Ponta Grossa Formation in South America (Melo 1989; Clarke 1913). This fauna was traditionally regarded as Emsian (Hiller & Theron 1989), mid-Emsian (Barrett & Isaacson 1989) or even late Emsian in age (Cooper 1986). The base of the Fox Bay Formation is placed at the first occurrence of persistent shales (Aldiss & Edwards 1999), which is approximately 50 m below the level of the main transgression.

Uniquely in sample PN11, the spores were recovered from a reworked phosphatic nodule with 3D preservation. Most are simple spores (Fig. 4q: a tripapillate retusoid spore) but include more complex forms (Fig. 4b) with prominent krytomes and a distal ring of verrucae, somewhat similar to specimens figured by Cramer & Diez (1975) as *Iberospora cantabrica* (their pl. 2, fig. 27), but clearly different from typical *Iberospora* and better accommodated in *Coronaspora*.

In the lower part of the Fox Bay Formation, chitinozoans re-occur in a somewhat greater diversity and abundance. In PN17, the chitinozoan fauna includes *Hoegisphaera* cf. *glabra* and a species of *Ancyrochitina* that has incomplete, but apparently solid, processes (Fig. 5a) that can be attributed to *A. ollivierae* (Boumendjel 2002). The sample in PN17 also has an 'increase' in acritarch abundance that now includes well-known forms (e.g. *Tyligmasoma alagarda* (Fig. 4o) and the ?euglenoid *Chomotriletes vedugensis* (Fig. 4p)), and the inception of the leiofusid *Bimerga paulae* with bifurcate terminations (Fig. 4m & n) that has sometimes (Le Hérissé 2011) been referred to *Bimerga bensonii*. Some 4.5 m above the level with abundant *A. olliverae*, there is a further sample (PN20) that is rich in chitinozoans but with the assemblage now dominated by *Ramochitina*. This chitinozoan has a very distinctive sculpture (Fig. 4c & d), which, when fully developed, has continuously tapering vertical processes that bifurcate equally to two equal horizontal processes that then divide again vertically. Similar sculpture is known from *Ramochitina* n. sp. cf. *R. devonica* of Azevedo-Soares & Grahn (2005) and Grahn (2005*a*) from the Amazonas Basin. In Grahn (2005*b*), this is attributed an earliest Lochkovian age. The species from the Falkland Islands differs in possessing a much less pronounced neck to body differentiation with a tapering transition between them. Other forms found in the Tequeje Formation of northernmost Bolivia with similar sculpture are '*Gotlandochitina*' *marettensis* Paris, 1981 in Vavrdová *et al.* (1996, plate V, fig. 3) and ?*R. magnifica* Lange, 1967, also in Vavrdová *et al.* (1996, pl. VI, fig. 2). These also have a body shape that is more similar to the Falkland Island specimens. It is too large to be *Ramochitina praemagnifica* (Grahn in Mendlowicz Mauller *et al.* 2009), as this is generally less than half the size of the Falkland Island specimens.

However, it is the maximum transgression shales (PN33), some 34 m higher in the section, that contain the most diverse palynomorphs, which included spores, chitinozoans and acritarchs. This assemblage provides the best age-tie for the West Falkland Group, both in terms of chronostratigraphy and in relation to a South American-based lithostratigraphy and sequence stratigraphy. The most conspicuous element in the palynological assemblage is the inception of the chitinozoan *Ramochitina magnifica* (Fig. 5e & f), which is locally very abundant, occurring in near-monotypic chitinozoan assemblages (cf. Dino 1999) together with *Hoegisphaera* cf. *glabra* and which can be attributed to the *Ramochitina magnifica* Zone of Grahn (2005*a*).

Also present with the chitinozoans is an 'acme' of the acritarch referred here to *Estiastra barbata* (Fig. 4k). A similar acme has been recognized (pers. obs.) in subsurface sections in Bolivia from a stratigraphic level in the lowest Icla Formation.

The spores are generally simple in construction and either lack sculpture or possess simple coni. However, there are rare specimens that include biform elements within a range of sculpture, which has a mid-Lochkovian first appearance (Steemans *et al.* 2012). Also present are examples of the *Dictyotriletes emsiensis* morphon (Fig. 5l) and *Emphanisporites rotatus*, together with further specimens of the robustly ribbed and conspicuously sculptured *Emphanisporites* (*E. decoratus*).

Ramochitina magnifica is a distinctive chitinozoan that has been widely recognized from many sections in South America (e.g. Lange 1967; Grahn *et al.* 2000), together with Antarctica (*Angochitina* sp. A in Kemp 1972). In the Paraná Basin of Brazil and Paraguay, *R. magnifica* has a first appearance (Grahn *et al.* 2000) 15 m above the base of the Ponta Grossa Formation. The shales of the Ponta Grossa Formation (Jaguariaiva Member) mark a major marine transgression over the 'coastal' sands of the Furnas Formation (Gerrienne *et al.* 2001; Candido & Rostirolla 2007).

In Bolivia, *R. magnifica* has been reported from the lower part of the Icla Formation (Grahn 2002; Troth *et al.* 2011) of the Tarija Basin. In sections on the Altiplano of Bolivia, it occurs at the basal part of the Formation (Ayo Ayo section: Troth *et al.* 2011).

In South America, *Ramochitina magnifica* is used to define the *R. magnifica* chitinozoan Zone (Lange 1967; Grahn 2005*b*). However, there has been some debate as to the exact age of the base of the zone. Grahn initially, and on the basis of a spore correlation (Grahn *et al.* 2000; Grahn 2002), regarded it as Pragian and, in fact, younger in age than earliest Pragian. The age of the first appearance of *R. magnifica* has subsequently been discussed in Gerrienne *et al.* (2001), where they concluded that the first appearance was definitely Lochkovian, presumably latest Lochkovian, although clearly with the main duration of the zone being of Pragian age. Subsequently, Grahn (2005*b*) regarded the age of the zone as ?latest Lochkovian–Pragian.

This palynological assemblage from the main Fox Bay transgression has a very clear parallel to those in South America. In addition, there is a very strong correlation both in terms of lithological change and macrofauna to a transgressive event that is very widespread across South America and South Africa. So, although the first appearance of *R. magnifica* is latest Lochkovian, it should be recognized that the occurrence in the Falkland Islands is coincident with the maximum transgression of the Fox Bay Formation and represents a facies-restricted occurrence as opposed to a total range. Subsequently, the Furnas–Ponta Grossa sequence in Brazil has been the subject of a number of integrated palynological and sequence stratigraphic studies described by Mendlowicz Mauller *et al.* (2009) and Grahn *et al.* (2010*a*, *b*, 2013). The same sequence of palynological inceptions and acme associated with the Ponta Grossa Formation can be recognized in the Fox Bay transgression. These include the inception of *Bimerga paulae*, and the associated assemblage of *Ramochitina magnifica*, *Hoegisphaera* cf. *glabra*, *Tyligmasoma alagarda*, *Knoxisporites riondae* and the *Dictyotriletes emsiensis* morphon. This shows a clear correlation with the Fox Bay transgression and Sequence B of the Ponta Grossa Formation that has been dated using spores as lying close to the Pragian–Emsian boundary (Grahn *et al.* 2013). This age date is based largely on the first appearance of *Dictyotriletes subgranifer*, the nominate species for the Su Zone as defined and tied to Devonian stages in Belgium (Steemans 1989). Accepting this age assignment shows that in the Paraná Basin there would be a significant '4 myr' gap above the top of the underlying Furnas Formation. Aldiss & Edwards (1999) discussed the Fish Creek Member–Fox Bay Formation boundary and described it variously as, 'conformable although variable, being sometimes abrupt, and sometimes transitional across the Falkland Islands'. Comparing these palynological results to Brazil indicates the presence of a similar stratigraphic gap at the Fish Creek–Fox Bay boundary.

It should also be emphasized that, although there are many common elements when compared with the palynological assemblages from southern South America, there are, in fact, many conspicuous Early Devonian palynomorphs that are missing from the Falkland Islands. For example, the distinctive chitinozoans (e.g. *Urochitina*, *Cingulochitina* and *Pterochitina*) that are widely known from Gondwana have yet to be found in the Falkland Island sequence.

The mid part of the Fox Bay Formation. Above the main level of the Fox Bay transgression, the succession reverts to the tidal channel facies that characterizes much of the West Falkland Group. This is very proximal in character, lacking even flaser–lenticular mud drapes. This gives a sample gap of about 175 m (and some 400 m above the Fox Bay transgression) before the next group of palynological samples. These samples (e.g. PN73) contain a significant and diagnostic assemblage from two intervals of shale interbedded with more thinly bedded sandstones, representing episodes of deepening. This new assemblage has a number of 'inceptions', including the first occurrence of the sculptured camerate spore *Grandispora naumovae* (Fig. 5h), *Emphanisporites annulatus* (Fig. 5s) and *Acinosporites lindlarensis* (Fig. 5k). Other characteristic taxa remain, including the *Dictyotriletes emsiensis* morphon.

In the São Domingoes Formation of the Paraná Basin in Brazil, there is a pair of closely spaced organic-matter-rich, hot shales (i.e. uranium-rich on account of their organic content) at the base of the formation and interpreted by Grahn *et al.* (2008, 2013) as a sequence boundary. The lower black shale has the inception of *Acinosporites lindlarensis*, whilst the upper shale has the inception of *Grandispora* (as *G. protea*) and *Emphanisporites annulatus*. In the Paraná Basin, this has been dated using spores as late Emsian in age. However, this is the same general assemblage and group of inceptions as recorded from the Icla–Huamampampa transition in Bolivia (Troth *et al.* 2011), which contains rare goniatites formerly attributed to *Tornoceras*. A reinterpretation of new and existing goniatite specimens has shown this to be earliest Eifelian in age (Becker in Troth *et al.* 2011) and, as such, the only independently dated tie-point within the South American Devonian succession. The difference in age results from a different sequence of spore inceptions in Bolivia and Brazil compared to Europe. These differences in spore inceptions are similarly found in Arabia (Breuer *et al.* 2015). Hence, this level is dated here as early Eifelian in age and, like Bolivia, represents the Choteč Event.

Above this group of assemblages, there is an estimated 200 m gap in the continuous exposure beneath the sands of Mount Brown Beach.

Port Philomel Formation

The lower boundary of the Port Philomel Formation is placed, following Aldiss & Edwards (1998,

1999), at the base of a conspicuous 32 m-thick shale bed. This shale is the thickest coherent shale interval within the West Falkland Group of West Falkland. It is completely atypical of the overlying Port Philomel Formation, which is characterized by somewhat immature sandstones with siltstones and shales in which the remains of fossil plants, particularly lycopsids, have become more common.

Some 2 m below the main development of shale, but within the transition zone, there is a single sample (PN89) that contains many specimens of a single chitinozoan species. This chitinozoan (Fig. 5i) has a smooth conical body with a long unornamented neck and ?six solid processes that are ?simple but never completely preserved. They are most similar to *Ancyrochitina arirambaense* from the late Eifelian–Givetian of Brazil, Bolivia and Paraguay (Lange 1967; Grahn 2002; Grahn *et al.* 2002; Grahn & Melo 2005). In South America, it has its inception below, and then a concurrent range with, that of *Alpenachitina eisenackii* (Grahn & Melo 2004; Grahn 2005*a*, *b*). However, *A. eisenackii*, the distinctive and very widely distributed (Paris *et al.* 2000) chitinozoan, is absent despite being present in most South and North American sections.

Some 15 palynological samples were investigated from the 32 m-thick shale. Although complicated by the contact metamorphism from a number of Jurassic dykes that cut the section, the palynological assemblage was uniquely impoverished. The samples with the most palynomorphs were from the nodular levels (PN91 and PN92: i.e. with the lowest sedimentation rate and the maximum flooding surface) and these consisted almost entirely of simple leiospheres (Fig. 5g) without acritarchs and very rare, poorly preserved spores and chitinozoans. This was confirmed both by top sieving at 88 μm to concentrate the larger chitinozoans and spores/acritarchs, and by handpicking from these wet residues. Hence, it is not possible to date the level palynologically. But it clearly represents a significant transgression that must be Eifelian, given the ages determined for the overlying and underlying intervals.

Therefore, although this thick shale bed cannot be directly dated, it is inferred here to represent the transgression at the base of the Los Monos Formation in Bolivia, which is accompanied by an epibole of the acritarch *Evittia sommeri*. This epibole has enabled this horizon to be widely recognized from Bolivia, Argentina and Brazil (Troth *et al.* 2011). Furthermore, Troth *et al.* (2011) have identified this level as a high-latitude expression of the late Eifelian Kačák Event, which is a well-known transgressive black shale level in both Euramerica and northern Gondwana (e.g. House 1996). However, it must be emphasized that this correlation to the Falkland Islands is based only on matching the sequence of transgressions within a broad framework of palynological ages.

Troth *et al.* (2011) have also shown that when the scattered records from Bolivia of the brachiopod *Tropidoleptus* are palynologically correlated, they are from within the *Evittia sommeri* Event of the basal Los Monos Formation and are late Eifelian in age. Therefore, the occurrence of this brachiopod was regarded as representing the temporary incursion of more northerly warmer water rather than the breakdown of the Malvinokaffric Realm (contrast Boucot 1999). *Tropidoleptus* is known from several occurrences (Boucot *et al.* 1983) in the Wagen Drift Formation (base Witteberg Group) in South Africa (but not the Falkland Islands). Its position at the base of the Witteberg Group parallels that of the Fox Bay–Philomel transition in occurring at a distinct change in sedimentological character. Compilations of chronostratigraphic ages for the South African sequence generally attribute a Givetian–Frasnian age to the Wagen Drift Formation. However, it is emphasized here that this age assignment has been largely based on the co-occurrence of *Tropidoleptus* in both Bolivia and South Africa, and the previous Givetian–Frasnian age attribution for the Huamamapampa Formation based on a goniatite from Bolivia that was incorrectly identified as a *Tornoceras* and, as already noted here (Becker in Troth *et al.* 2011), is now known to be earliest Eifelian in age.

It was in the Port Philomel Formation that Cramer & Diez de Cramer (1972) reported and figured an assemblage of spores and chitinozoans from the matrix of a plant specimen collected at Dunose Head by Halle (1911). The only chitinozoan formally identified by Cramer & Diez de Cramer (1972) was *Angochitina* cf. *A. capillata* Eisenack, 1937 of Lange (1967), which was used together with unnamed spores to infer an early Givetian age. *Angochitina* cf. *A. capillata* in Lange (1967) is now referred to *Angochitina daemonii* Grahn and attributed a late Emsian–early Givetian range in Brazil (Grahn 2005*b*).

Uppermost Port Philomel Formation

It is in the uppermost part of the Port Philomel Formation that a further significant change in spore assemblage occurs. This is the inception of *Geminospora lemurata* (Fig. 5o) in PN137, and marks a significant change in assemblage composition from one dominated by simple pteridophytic and lycopod spores to one dominated by progymnosperms. The abundance of *Geminospora lemurata* (Fig. 5m) increases upsection to become dominant by PN147, and in PN154 it occurs with the *Verrucosisporites scurrus* morphon (Fig. 5n).

The inception of *Geminospora lemurata* is regarded as marking the base of the Givetian (Loboziak *et al.* 1991) from its occurrence in conodont-dated sections in Euramerica. It occurs widely in Gondwana and is accepted as indicating a Givetian age. However, it becomes most abundant in the late Givetian (Marshall *et al.* 2011) following the Taghanic Extinction Event. Therefore, the base of the Givetian is placed at PN137 and the late Givetian at PN147. The association of *G. lemurata* and *V. scurrus* is characteristic of the Iquiri Formation in Bolivia.

Port Stanley Formation, Dunbar Creek

The base of the Port Stanley Formation is placed at the abrupt change to a coherent succession of sandstones that is rarely exposed due to it being covered by debris. One locality where it can be seen is in Carcass Bay (West Falkland), where it is locally erosive (Aldiss & Edwards 1999). But, in all probability, this does not represent a significant downcutting as the distinctive interval of mudstone in the uppermost Port Philomel Formation can be traced across the Falkland Islands (Aldiss & Edwards 1999).

The Port Stanley Formation in West Falkland comprises over 500 m of mature sandstone, probably half the thickness of the sequence in East Falkland (Aldiss & Edwards 1999). The formation is very sand-rich and includes three prominent orthoquartzite layers. Mudstones are difficult to find within the Port Stanley Formation, but there is a thin interval in Dunbar Creek that included rare beds of dark-coloured mudstone from which 16 samples were collected. In common with the lower parts of the sequence, these are dominated by phytoclasts, but rare palynomorphs are present. This discovery is very significant as it enables the top of the West Falkland Group to be dated and, hence, its age relationship determined relative to the glacigene sediments of the overlying Fitzroy Tillite Formation. The main element in the assemblage is *Verrucosisporites famenensis* (Fig. 6b), together with rare species of the *Auroraspora/Diducites* morphon (Fig. 6a). Also present are bifurcate (Fig. 6e) and multifurcate tipped spores (e.g. *Ancyrospora*: Fig. 6c & d), *Knoxisporites hederatus* (Fig. 6g), *Tumulispora ordinaria* (Fig. 6f), and a single specimen of *Grandispora* sp. The acritarchs include *Hurologinella horologia* (Fig. 6i & j) and a single specimen of *Umbellasphaeridium saharicum*, which has a single preserved process (Fig. 6l). Very rare prasinophytes are present, such as *Gorgonisphaeridium* sp., leiospheres and *Maranhites mosesi* (Fig. 6k). No chitinozoans were found. This assemblage is clearly still of Devonian age as it contains both spores (*Ancyrospora*) and acritarchs (*Umbellasphaeridium*) that become extinct at the Devonian–Carboniferous boundary. The age of the assemblage is also clearly Famennian from the presence of *Verrucosisporites famenensis* (Massa *et al.* 1979), the *Auroraspora* morphon (Richardson & McGregor 1986) and *Knoxisporites hederatus*. The absence of *Retispora lepidophyta* is also significant. This distinctive spore is very abundant, restricted to the latest Famennian and has a global distribution (Streel & Marshall 2006). It is, for example, well known in southern South America and Australia. It has not been found in South Africa, but occurs in the Ellsworth Whitmore Mountains in Antarctica (pers. obs.). In Devonian times, the Ellsworth Whitmore Mountains (Fig. 7) were adjacent and east of the Falkland Islands. However, during the fragmentation of Gondwana, this block became isolated as a separate microcontinent that became part of Antarctica. Hence, *Retispora lepidophyta* is present regionally and its absence in the Falkland Island sequence strongly suggests that the Dunbar Creek section has been eroded down to a level below the latest Famennian. The minimum age, based on this absence, is therefore late, but not latest, Famennian. Another index species is *Rugospora radiata*, which had a late, but not latest, Famennian inception and is well known from Gondwana, but this was not found in the Falkland Islands. If this absence is real, then the age of the youngest preserved sediments is pre-VCo spore zone. This absence of the latest Famennian is significant as the Devonian glacial diamictites in South America occur within the range of *R. lepidophyta* (Wicander *et al.* 2011).

Correlations

Figure 8 is a preliminary correlation between a composite section from the sub-Andean zone of Bolivia, the Port North–Dunbar section in West Falkland and the Cape Supergroup of South Africa. The succession from South Africa is a composite using the Table Mountain Group from the Eastern Cape, which was geographically closest to the Falkland Islands at the time of deposition. However, the Bokkeveld, Witteberg and Lake Mentz groups are from the Western Cape, where the succession is better established in ?largely unthrusted contiguous sequences. The main correlative tie at the base of the sections is the macrofauna, which links all three sequences together, along with the identification of *Ramochitina magnifica* Zone palynomorphs in both Bolivia and West Falkland. All three sequences are transgressive and were dated by Grahn *et al.* (2013) as Pragian–Emsian in age. This correlation indicates that there are strong lithological parallels between the coherent sandstones of

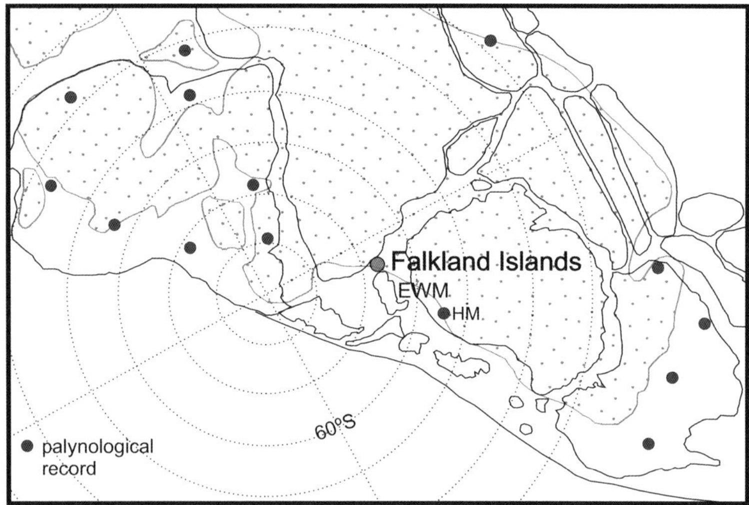

Fig. 7. Polar palaeogeographical reconstruction of Gondwana modified from Hunter & Lomas (2003). The locations of significant Devonian palynological assemblages are shown. The Ellsworth Whitmore Mountains (EWM), a distinct terrane that is now embedded within West Gondwana, is shown lying along-strike from West Falkland. Palynological assemblages from the EWM contain *Retispora lepidophyta* and show that this spore was present within the greater Cape Fold Belt system. In the Horlick Mountains (HM), a palynological assemblage has been reported (Kemp 1972) similar to that from the base of the Fox Bay Formation.

the South Harbour Member (Port Stephens Formation) and the Skurweberg Formation (Nardouw Subgroup, Table Mountain Group), with the more heterolithic Fish Creek Member (Port Stephens Formation) being equivalent to the Rietvlei Formation. This correlation at the base of the Fox Bay Formation can be extended along the margin of Gondwana into Antarctica (Fig. 7), where there is a similar palynological assemblage of spores and chitinozoa (but lacking acritarchs: Kemp 1972) in the Horlick Mountains (Bradshaw & McCartan 1983), together with associated Malvinokaffric fauna including brachiopods (Doumani *et al.* 1965).

The next tie is the inception in the uppermost Icla Formation (Bolivia) of *Emphanisporites annulatus*, *Grandispora* spp. and *Alpenachitina eisenackii* (Troth *et al.* 2011) from sediments that are goniatite dated as earliest Eifelian and represent the Choteč Event transgression. In West Falkland, the first occurrence of *E. annulatus* and *Grandispora* spp. (but not as an inception within a sequence of samples) occurs in the uppermost Fox Bay Formation and probably above a significant gap in the succession. The prominent shale at the base of the Port Philomel Formation is tied on general age and transgressive character to the basal Los Monos transgression and the *Evittia sommeri* epibole in Bolivia. The tie to South Africa is less certain, but the brachiopod *Tropidoleptus* occurs (Boucot *et al.* 1983) in the Wagen Drift and its Eastern Cape equivalent in the lower part of the Weltevrede Formation and the Los Monos Formation in Bolivia. The next tie is the inception of *Geminospora lemurata* in the lower part of the Port Philomel Formation and the lower part of the Iquiri Formation in Bolivia.

It is in the upper part of the section that the lithostratigraphic links between the Falkland Islands and South Africa become more obvious, with the many parallels between the thick sandstone sequences of the Port Stanley and Witpoort formations. Both contain distinctive regressive sandstones. In West Falkland, the palynological record shows that the very uppermost part of the sequence is late, but not latest, Famennian in age. The base of the Port Stanley Formation is no older than late Givetian in age, with some element of downcutting at the base. There is, at yet, no palynological recovery from within the middle part of the Port Stanley Formation. The three regressive orthoquartzite intervals within the Port Stanley Formation can now be dated as pre-late Famennian in age. Their lithological character as supermature sandstones with herringbone cross-stratification and low-angle bedding indicates shoreline facies, and this is the lowest sea level in the succession. It points to three significant and presumed drawdowns. Given the existence upsection of diamictites of glacial origin, these could be interpreted as representing earlier glacial drawdown episodes. But the regressive sandstones can also be interpreted as indicating major regressive events from within the later Devonian. The Taghanic Event, although generally

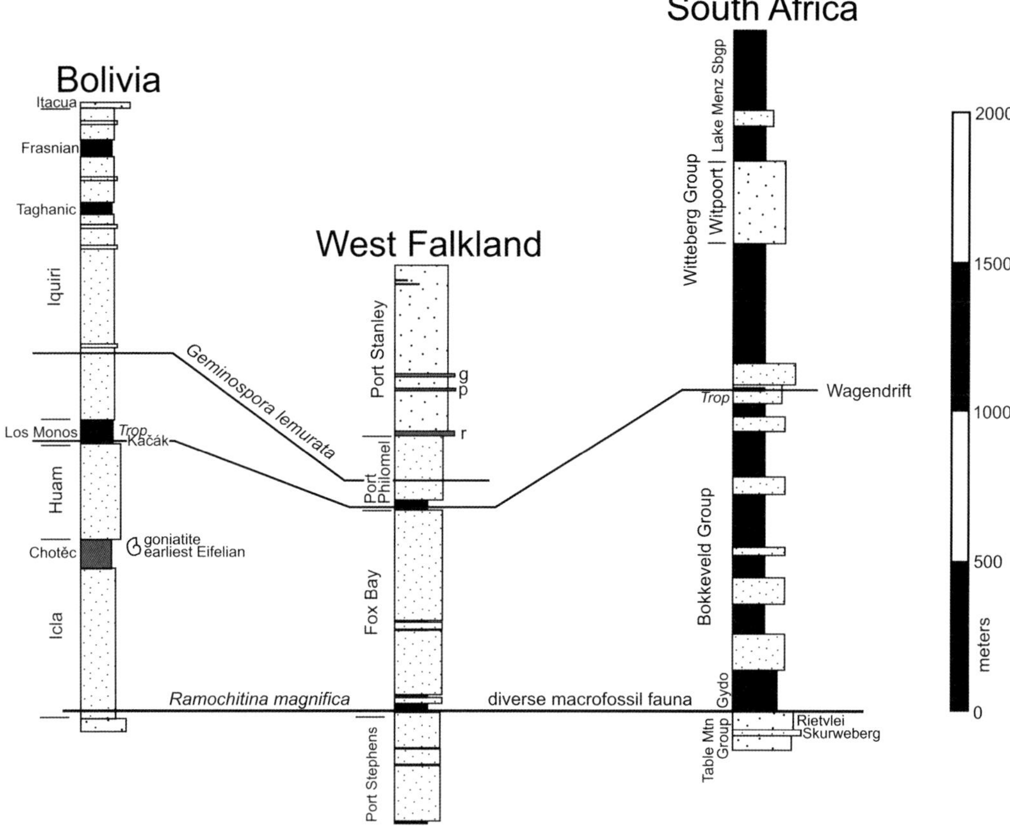

Fig. 8. Tentative correlation between Bolivia, West Falkland and South Africa. The Bolivian section is based on Campo Redondo and Bermejo in the sub-Andean zone (Troth *et al.* 2011); the Falkland Islands is based on original data presented here; and South Africa is a compilation of the Eastern (up to the Witpoort Formation) and Western Cape (above the Witpoort Formation) from South African Committee for Stratigraphy (SACS) and Theron (1970). Good correlation ties are at the level with the most diverse fauna associated with the transgression of the Fox Bay and Gydo formations tied to Bolivia with *Ramochitina magnifica*. Other ties are the brachiopod *Tropidoleptus* in both Bolivia and South Africa tied to the base Port Philomel Formation transgression in West Falkland.

transgressive, has an interval of sustained aridity within it (Marshall *et al.* 2011) accompanied by a fall in sea level. This could be represented by the lowest regressive sandstone at the base of the Port Stanley Formation, with any characteristic black shales having a low preservation potential in the nearshore marine shelf environment. Similarly, the latest Frasnian Lower and Upper Kellwasser Events, although transgressive (or the result of increased water stratification) with black shales at their onset, terminate with very significant regressive episodes that bring downcutting unconformities into most marine shelf sequences (e.g. Racki 2005). Again, these could be represented by the higher pair of regressive sandstones (Figs 1 & 2) in this more proximal environment. There are no Devonian diamictites known from the Port Stanley Formation, as the section is truncated by overlying diamictites of the Permian Fitzroy Tillite Formation. The preserved top of the Port Stanley Formation in Dunbar Creek is pre-late Famennian and is, thus, older than the latest Famennian palynologically dated glacial diamictite successions in Bolivia and Brazil (as reviewed by Lakin *et al.* 2016). There are younger diamictites in West Falkland, but these occur as sedimentary dykes (Hyam *et al.* 1997) within the Port Stephens Formation at South Harbour. They contain a palynological assemblage including *Reticulatisporites magnidictyus*, and are within the late Visean–early Serpukovian interval based on the revised zonation and age calibration of Playford & Melo (2012) and Playford (2015).

In South Africa, the age of the Witpoort Formation is very poorly constrained but clearly includes fish of Famennian age (Long *et al.* 1997; Evans 1999), with the overlying Lake Mentz Subgroup being earliest Carboniferous (Theron 1993; Streel & Theron 1999). There are isolated dropstones associated with regressive sandstones in the upper Witpoort Formation (Almond *et al.* 2002) that probably represent fallout from distant floating ice, together with thin immature diamictites at the top of the formation. This places the glaciation within the upper and top Witpoort Formation.

In Brazil, the lower unit of the Barreirinha Formation (Loboziak *et al.* 1997; Grahn & Melo 2002) is 'equivalent' to the lower part of the Witpoort Formation (late Frasnian–mid-Famennian) and is a condensed interval, and similarly represents a time of sustained regression.

As regards controls on the transgressive and regressive events, the key observation is the distance over which they can be correlated from the Falkland Islands/South Africa to Bolivia and Brazil. On palaeogeographical reconstructions (Fig. 7), this is estimated as some 90° of palaeolongitude (i.e. *c.* 3000 km). This is on a much larger scale than any of the tectonic domains present in southern South America and South Africa. Hence, the primary driver on the transgressions and regressions can be interpreted as being driven by sea level and is a true test of eustatic control of Devonian events.

Bill Braham proved an ideal field companion and remained cheerful through all the vicious woollies. The hospitality and assistance of many Falkland Islanders is acknowledged, especially Roy and Audrey McGhie at Port North, Marshall Barnes at Dunbar, and Pat and Dan Whitney at Green Patch. Adil Babiker and Shir Akbari processed the palynological samples. Danny Hyam and Morag Hunter provided valuable discussion on Falkland Island geology, and Ian Troth on all things Bolivian. Jon Lakin has revealed much about the Devonian–Carboniferous diamictites in Bolivia. Uli Jansen provided a useful insight into brachiopods. Fiona Evans and John Almond provided the invaluable opportunity to see the Devonian geology of South Africa. The project was funded by NERC GR9/91/556.

References

Aldiss, D.T. & Edwards, E.J. 1998. *Geology of the Falkland Islands. Geological Map, 1:250 000.* Falkland Islands Government.

Aldiss, D.T. & Edwards, E.J. 1999. *The Geology of the Falkland Islands.* British Geological Survey Technical Report WC/99/10.

Allen, K.C. 1965. Lower and Middle Devonian spores of North and Central Vestspitsbergen. *Palaeontology*, **8**, 687–748.

Almond, J., Marshall, J. & Evans, F. 2002. Latest Devonian and earliest Carboniferous glacial events in South Africa. *In*: *Abstracts of the 16th International Sedimentological Congress.* Rand Afrikaans University, Johannesburg, 11–12.

Archibold, O.W. 1995. *Ecology of World Vegetation.* Springer, Dordrecht.

Azevedo-Soares, H.L.C. & Grahn, Y. 2005. The Silurian–Devonian boundary in the Amazonas Basin, northern Brazil. *Neues Jahrbuch für Geologie und Paläontologie, Abhandlungen*, **236**, 79–94.

Barreda, V.D. 1986. Acritarchos Givetiano –Frasnianos de la cuence del noroeste, Provincia de Salta. Argentina. *Revista Española de Micropaleontología*, **18**, 229–245.

Barrett, S.E. & Isaacson, P.E. 1989. Devonian paleogeography of South America. *In*: McMillan, N.J., Embry, A.F. & Glass, D.J. (eds) *Devonian of the World; Proceedings of the Second International Symposium on the Devonian System.* Canadian Society of Petroleum Geologists, Memoirs, **14**(I), 655–681.

Booth, P.W.K. & Shone, R.W. 2002. A review of thrust-faulting in the Eastern Cape fold belt, South Africa and the implications for current lithostratigraphic interpretations of the Cape Supergroup. *Journal of African Earth Sciences*, **34**, 179–190.

Boucot, A.J. 1999. Southern African Phanerozoic marine invertebrates; biogeography, palaeoecology, climatology and comments on adjacent regions. *Journal of African Earth Sciences*, **28**, 129–143.

Boucot, A.J. & Theron, J.N. 2001. First *Rhipidothyris* (Brachiopoda) from Southern Africa: Biostratigraphic, paleoecological, biogeographical significance. *Journal of the Czech Geological Society*, **46**, 155–160.

Boucot, A.J., Brunton, C.H.C. & Theron, J.N. 1983. Implications for the age of South African Devonian rocks in which *Tropidoleptus* (Brachiopoda) has been found. *Geological Magazine*, **120**, 51–58.

Boumendjel, K. 2002. Nouvelles espèces de chitinozoaires du Silurien Supérieur et do Dévonien Inférieur du bassin de Timimoun (Sahara central, Algérie). *Review of Palaeobotany and Palynology*, **118**, 29–46.

Bradshaw, M.A. & McCartan, L. 1983. The depositional environment of the Lower Devonian Horlick Formation, Ohio Range. *In*: Oliver, R.L., James, P.R. & Jago, J.B. (eds) *Antarctic Earth Science, Fourth International Symposium.* Cambridge University Press, Cambridge, 238–241.

Breuer, P., Miller, M.A., Lesczyński, S. & Steemans, P. 2015. Climate-controlled palynofacies and miospore stratigraphy of the Jauf Formation, Lower Devonian, northern Saudi Arabia. *Review of Palaeobotany and Palynology*, **212**, 187–213.

Candido, A.G. & Rostirolla, S.P. 2007. Análise de fácies e revisão da estratigrafia de seqűências fa Formação Ponta Grossa, Bacia do Paraná- ênfase nos arenitos do Membro Tibagi. *Boletim de geociencias da petrobras, Rio de Janeiro*, **15**, 45–62.

Clarke, J.M. 1913. *Fosseis Devonianos do Parana.* Monographias do Servico Geologico e Mineralogico do Brasil, **1**.

Cooper, M.R. 1986. Facies shifts, sea-level changes and event stratigraphy in the Devonian of South Africa. *South African Journal of Science*, **82**, 255–258.

Cramer, F.H. & Diez, M. del C.R. 1975. Earliest Devonian miospores from the province of Leon, Spain. *Pollen et Spores*, **17**, 331–344.

CRAMER, F.H. & DIEZ DE CRAMER, M. DEL C.R. 1972. Exclusive occurrence of chitinozoans and miospores in a shale of Devonian age from the Malvinas Islands. *Ameghiniana*, **9**, 220–222.

CURTIS, M.L. & HYAM, D.M. 1998. Late Palaeozoic to Mesozoic structural evolution of the Falkland Islands: a displaced segment of the Cape Fold Belt. *Journal of the Geological Society, London*, **155**, 115–129, http://doi.org/10.1144/gsjgs.155.1.0115

DINO, R. 1999. Palynostratigraphy of the Silurian and Devonian sequence of the Paraná Basin. *In*: RODRIGUES, M.A.C. & PEREIRA, E. (eds) *Ordovician–Devonian Palynostratigraphy in Western Gondwana: Update, Problems and Perspectives*. UERJ, Faculdade de Geologia, Rio de Janeiro, 27–61.

DI PASQUO, M. 2007*a*. Asociacones palinológicas en las formaciones Los Monos (Devónico) e Itacua (Carbonífero Inferior) en Balapuca (Cuenca Tarija), sur de Bolivia. Parte 1. Formación Los Monos. *Revista Geológica de Chile*, **34**, 97–137.

DI PASQUO, M. 2007*b*. Asociacones palinológicas en las formaciones Los Monos (Devónico) e Itacua (Carbonífero Inferior) en Balapuca (Cuenca Tarija), sur de Bolivia. Parte 2. Asociaciones de la Formación Itacua e interpretación estratigráfica y cronología de las formaciónes Los Monos et Itacua. *Revista Geológica de Chile*, **34**, 163–198.

DOUMANI, G.A., BOARDMAN, R.S. *ET AL.* 1965. Lower Devonian Fauna of the Horlick Formation, Ohio Range, Antarctica. *In*: HADLEY, J.B. (ed.) *Geology and Paleontology of the Antarctic*. American Geophysical Union, Antarctic Research Series, **1299**, 241–281.

EDGECOMBE, G.D. 1994. Calmoniid trilobites from the Devonian Fox Bay Formation, Falkland Islands: systematics and biogeography. *New York State Museum Bulletin*, **481**, 55–68.

EISENACK, A. 1937. Neue Mikrofossilien des Baltischen Silurs. IV. *Paläontologische Zeitschrift*, **19**, 217–243.

EVANS, F.J. 1999. Palaeobiology of Early Carboniferous lacustrine biota of the Waaipoort Formation (Witteberg Group), South Africa. *Palaeontologia Africana*, **35**, 1–6.

GERRIENNE, P., BERGAMASCHI, S., PEREIRA, E., RODRIGUES, M.-A.C. & STEEMANS, P. 2001. An Early Devonian flora, including *Cooksonia*, from the Parana Basin (Brazil). *Review of Palaeobotany and Palynology*, **116**, 19–38.

GOLDRING, R., ASTIN, T.R., MARSHALL, J.E.A., GABBOTT, S. & JENKINS, C.D. 1998. Towards an integrated study of the depositional environment of the Bencliff Grit (Upper Jurassic) of Dorset. *In*: UNDERHILL, J.R. (ed.) *The Development, Evolution and Petroleum Geology of the Wessex Basin*. Geological Society, London, Special Publications, **133**, 355–372, http://doi.org/10.1144/GSL.SP.1998.133.01.18

GONZÁLEZ, F. 2009. Reappraisal of the organic-walled microphytoplankton genus *Maranhites*: morphology, excystment, and speciation. *Review of Palaeobotany and Palynology*, **154**, 6–21.

GRAHN, Y. 2002. Upper Silurian and Devonian Chitinozoa from central and southern Bolivia, central Andes. *Journal of South American Earth Sciences*, **15**, 315–326.

GRAHN, Y. 2005*a*. Silurian and Lower Devonian chitinozoan taxonomy and biostratigraphy of the Trombetas Group, Amazonas Basin, northern Brazil. *Bulletin of Geosciences*, **80**, 245–276.

GRAHN, Y. 2005*b*. Devonian chitinozoan biozones of western Gondwana. *Acta Geologica Polonica*, **55**, 211–227.

GRAHN, Y. & MELO, J.H.G. 2002. Chitinozoan biostratigraphy of the Late Devonian formations in well Caima PH-2, Tapajos River area, Amazonas Basin, northern Brazil. *Review of Palaeobotany and Palynology*, **118**, 115–139.

GRAHN, Y. & MELO, J.H.G. 2004. Integrated Middle Devonian chitinozoan and miospore zonation of the Amazonas Basin, northern Brazil. *Revue de Micropaléontologie*, **47**, 71–75.

GRAHN, Y. & MELO, J.H.G. 2005. *Middle and Late Devonian Chitinozoa and Biostratigraphy of the Parnaiba and Jatoba Basins, Northeastern Brazil*. Palaeontographica, Abteilung B, **272**.

GRAHN, Y., PEREIRA, E. & BERGAMASCHI, S. 2000. Silurian and Lower Devonian chitinozoan biostratigraphy of the Parana Basin in Brazil and Paraguay. *Palynology*, **24**, 147–176.

GRAHN, Y., BERGAMASCHI, S. & PEREIRA, E. 2002. Middle and Upper Devonian chitinozoan biostratigraphy of the Paraná Basin in Brazil and Paraguay. *Palynology*, **26**, 135–165.

GRAHN, Y., YOUNG, C. & BORGHI, L. 2008. Middle Devonian chitinozoan biostratigraphy and sedimentology in the eastern outcrop belt of the Parnaíba Basin, northeastern Brazil. *Revista Brasileira de Paleontologia*, **11**, 137–146.

GRAHN, Y., MENDLOWICZ MAULLER, P., BREUER, P., BOSETTI, E.P., BERGAMASCHI, S. & PEREIRA, E. 2010*a*. The Furnas/Ponta Grossa contact and the age of the lowermost Ponta Grossa Formation in the Apucarana Sub-basin (Paraná Basin, Brazil): integrated palynological age determination. *Revista Brasileira de Paleontologia*, **13**, 89–102.

GRAHN, Y., MENDLOWICZ MAULLER, P., PEREIRA, E. & LOBOZIAK, S. 2010*b*. Palynostratigraphy of the Chapada Group and its significance in the Devonian stratigraphy of the Paraná Basin, south Brazil. *Journal of South American Earth Sciences*, **29**, 354–370.

GRAHN, Y., MENDLOWICZ MAULLER, P., BERGAMASCHI, S. & BOSETTI, E.P. 2013. Palynology and sequence stratigraphy of three Devonian rock units in the Apucarana Sub-basin (Paraná Basin, south Brazil): additional data and correlation. *Review of Palaeobotany and Palynology*, **198**, 27–44.

HALLE, T.G. 1911. On the geological structure and history of the Falkland Islands. *Bulletin of the Geological Institution of the University of Uppsala*, **11**, 1–114.

HILLER, N. & THERON, J.N. 1989. Benthic communities in the South African Devonian. *In*: MCMILLAN, N.J., EMBRY, A.F. & GLASS, D.J. (eds) *Devonian of the World; Proceedings of the Second International Symposium on the Devonian System*. Canadian Society of Petroleum Geologists, Memoirs, **14**(III), 229–242.

HOFFMEISTER, W.S., STAPLIN, F.L. & MALLOY, R.E. 1955. Mississippian plant spores from the Hardinsburg formation of Illinois and Kentucky. *Journal of Paleontology*, **29**, 372–399.

HORODYSKI, R.S., HOLZ, M., GRAHN, Y. & BOSETTI, E.P. 2013. Remarks on sequence stratigraphy and

taphonomy of the Malvinokaffric shelly fauna during the KAČÁK Event in the Apucarana Sub-basin (Paraná Basin), Brazil. *International Journal of Earth Science*, **103**, 367–380.
House, M.R. 1996. The Middle Devonian Kačák Event. *Proceedings of the Ussher Society*, **9**, 79–84.
Hunter, M.A. & Lomas, S.A. 2003. Reconstructing the Siluro-Devonian coastline of Gondwana: insights from the sedimentology of the Port Stephens Formation, Falkland Islands. *Journal of the Geological Society, London*, **160**, 459–476, http://doi.org/10.1144/0016-764902-038
Hyam, D.M., Marshall, J.E.A. & Sanderson, D.J. 1997. Carboniferous diamictite dykes in the Falkland Islands. *African Journal of Earth Sciences*, **25**, 505–517.
Hyam, D.M., Marshall, J.E.A., Bull, J.M. & Sanderson, D.J. 2000. The structural boundary between East and West Falkland: new evidence for movement history and lateral extent. *Marine & Petroleum Geology*, **17**, 13–26.
Jardiné, S. & Yapaudjian, L. 1968. Lithostratigraphie et palynology du Dévonien-Gothlandian Gréseux du Bassin de Polignac (Sahara). *Revue de l'Institut Français du Pétrole*, **23**, 439–468.
Jardiné, S., Combaz, A., Magloire, G., Peniguel, G. & Vachey, G. 1972. Acritarches du Silurien terminal et du Dévonien du Sahara Algérien. *Comptes rendus 7e Congrès international de stratigraphie et de géologie du Carbonifère, Krefeld, August 1971*, **1**, 295–311.
Kemp, E.M. 1972. Lower Devonian palynomorphs from the Horlick Formation, Ohio Range, Antarctica. *Palaeontographica, Abteilung B*, **139**, 105–124.
Lange, F.W. 1967. Biostratigraphic subdivision and correlation of the Devonian in the Parana Basin, in Bigarella, J.J., Problems in Brazilian Devonian Geology. *Boletim Paranaense de Geociências*, **21/22**, 63–98.
Lakin, J.A., Marshall, J.E.A., Troth, I. & Harding, I.C. 2016. Greenhouse to icehouse: a biostratigraphic review of latest Devonian–Mississippian glaciations and their global effects. *In*: Becker, R.T., Königshof, P. & Brett, C.E. (eds) *Devonian Climate, Sea Level and Evolutionary Events*. Geological Society, London, Special Publications, **423**. First published online April 15, 2016, http://doi.org/10.1144/SP423.12
Le Hérissé, A. 1983. *Les spores du Dévonien Inférieur du Synclinorium de Laval (Massif Armoricain)*. Palaeontographica, Abteilung B, **188**, 1–81.
Le Hérissé, A. 2011. A reappraisal of F.W. Lange's 1967 algal microfossil studies. *In*: Bosetti, E.P., Grahn, Y. & Melo, J.H.G. (eds) *Essays in Honour of Frederico Waldemar Lange*. Editora Interciência Ltda, Rio de Janeiro, 151–179.
Lele, K.M. & Streel, M. 1969. Middle Devonian (Givetian) plant microfossils from Goé (Belgium). *Annales de la Société Géologique de Belgique*, **92**, 89–121.
Loboziak, S., Streel, M. & Weddige, K. 1991. Miospores, the *lemurata* and *triangulatus* levels and their fauna indices near the Eifelian/Givetian boundary in the Eifel (F.R.G.). *Annales de la Société Géologique de Belgique*, **113**, 299–313.
Loboziak, S., Melo, J.H.G., Matsuda, N.S. & Quadros, L.P. 1997. Miospore biostratigraphy of the type Barreirinha Formation (Curuá Group, Upper Devonian) in the Tapajos River area, Amazon Basin, North Brazil. *Bulletin des Centres de Recherches Exploration–Production Elf-Aquitaine*, **21**, 187–205.
Loeblich, A.R., Jr. 1970. Morphology, ultrastructure and distribution of Paleozoic acritarchs. *In*: *Proceedings of the North American Paleontological Convention, 1969, Chicago*. Allen Press, Lawrence, Kansas, Part G, 705–788.
Long, J.A., Anderson, M.E., Gess, R. & Hiller, N. 1997. New placoderm fishes from the Late Devonian of South Africa. *Journal of Vertebrate Palaeontology*, **17**, 253–268.
Maisey, J.G., Borghi, L. & De Carvalho, G.P.D. 2002. Lower Devonian fish remains from the Falkland Islands. *Journal of Vertebrate Paleontology*, **22**, 708–711.
Marshall, J.E.A. 1994. The Falkland Islands: a key element in Gondwana paleogeography. *Tectonics*, **13**, 499–515.
Marshall, J.E.A. 1998. The recognition of multiple hydrocarbon generation episodes: an example from Devonian lacustrine sedimentary rocks in the Inner Moray Firth, Scotland. *Journal of the Geological Society, London*, **155**, 335–352, http://doi.org/10.1144/gsjgs.155.2.0335
Marshall, J.E.A., Brown, J.F. & Astin, T.R. 2011. Recognising the Taghanic Crisis in the Devonian terrestrial environment; its implications for understanding land-sea interactions. *Palaeogeography, Palaeoclimatology, Palaeoecology*, **304**, 165–183.
Massa, D., Coquel, R., Loboziak, S. & Taugourdeau-Lantz, J. 1979. Essai de synthèse stratigraphique et palynologique de Carbonifère en Libye occidentale. *Annales de la Societe Géologique du Nord*, **99**, 429–442.
McGregor, D.C. 1961. Spores with proximal radial pattern from the Devonian of Canada. *Geological Survey of Canada Bulletin*, **76**, 1–11.
McGregor, D.C. 1973. Lower and Middle Devonian spores of eastern Gaspé, Canada. I. Systematics. *Palaeontographica, Abteilung B*, **142**, 1–77.
McGregor, D.C. & Camfield, M. 1976. Upper Silurian? to Middle Devonian spores of the Moose River Basin, Ontario. *Geological Survey of Canada Bulletin*, **263**, 1–63.
McGregor, D.C. & Camfield, M. 1982. Middle Devonian miospores from the Cape de Bray, Weatherall, and Hecla Bay Formations of northeastern Melville Island, Canadian Arctic. *Geological Survey of Canada Bulletin*, **348**, 1–105.
Meadows, N.S. 1999. Basin evolution and sedimentary fill in the Palaeozoic sequences of the Falkland Islands. *In*: Cameron, N.R., Bate, R.H. & Clure, V.S. (eds) *The Oil and Gas Habitats of the South Atlantic*. Geological Society, London, Special Publications, **153**, 445–464, http://doi.org/10.1144/GSL.SP.1999.153.01.27
Melo, J.H.G. 1989. The Malvinokaffric Realm in the Devonian of Brazil. *In*: McMillan, N.J., Embry, A.F. & Glass, D.J. (eds) *Devonian of the World; Proceedings of the Second International Symposium on the Devonian System*. Canadian Society of Petroleum Geologists, Memoirs, **14**(I), 669–703.
Melo, J.H.G. & Loboziak, S. 2003. Devonian-Early Carboniferous miospore biostratigraphy of the Amazon

Basin, northern Brazil. *Review of Palaeobotany and Palynology*, **124**, 131–202.

Mendlowicz Mauller, P., Grahn, Y. & Machado Cardoso, T.R. 2009. Palynostratigraphy from the Lower Devonian of the Paraná Basin, South Brazil, and a revision of the contemporary Chitinozoan biozones from Western Gondwana. *Stratigraphy*, **6**, 313–332.

Naumova, S.N. 1953. Spore–pollen complexes of the Upper Devonian of the Russian Platform and their stratigraphic significance. *Transactions of the Institute of Geological Sciences, Academy of Sciences of the USSR*, **143** (Geological Series 60), 1–204.

Ottone, E.G. 1996. Devonian palynomorphs from the Los Monos Formation, Tarija Basin, Argentina. *Palynology*, **20**, 105–155.

Paris, F. 1981. *Les Chitinozoaires dans le Paléozoïque du Sud-Ouest de l'Europe*. Mémoires de la Société géologique et minéralogique de Bretagne, **26**.

Paris, F., Winchester-Seeto, T., Boumendjel, K. & Grahn, Y. 2000. Toward a global biozonation of Devonian chitinozoans. *Courier Forschungsinstitut Senckenberg*, **220**, 39–55.

Playford, G. 1963. Lower Carboniferous microfloras of Spitsbergen. *Palaeontology*, **5**, 619–678.

Playford, G. 1977. Lower to Middle Devonian acritarchs of the Moose River basin, Ontario. *Geological Survey of Canada Bulletin*, **279**, 1–87.

Playford, G. 1983. The Devonian miospore genus *Geminospora* Balme 1962. A reappraisal based on topotypic *G. lemurata* (type species). *Memoirs of the Association of Australasian Palaeontologists*, **1**, 311–325.

Playford, G. 2015. Mississippian palynoflora from the northern Perth Basin, Western Australia: systematics and stratigraphical and palaeogeographical significance. *Journal of Systematic Palaeontology*, http://doi.org/10.1080/14772019.2015.1091792

Playford, G. & Melo, J.H. 2012. *Miospore Palynology and Stratigraphy of Mississippian Strata of the Amazonas Basin, Northern Brazil*. American Association of Stratigraphic Palynologists Foundation, Contribution Series, **47**.

Pöthe de Baldis, E.D. 1974. El microplancton del Devonico medio de Paraguay. *Revista Española de Micropaleontología*, **6**, 367–379.

Pöthe de Baldis, E.D. 1978. Paleomicroplancton adicional del Devonico Inferior de Uruguay. *Revista Española de Micropaleontología*, **9**, 235–250.

Racki, G. 2005. Toward understanding Late Devonian global events: few answers, many questions. *In*: Over, D.J., Morrow, J.R. & Wignall, P.B. (eds) *Understanding Late Devonian and Permian–Triassic Biotic and Climatic Events – Towards an Integrated Approach*. Developments in Palaeontology and Stratigraphy, **20**, 5–36.

Richardson, J.B. 1962. Spores with bifurcate processes from the Middle Old Red Sandstone of Scotland. *Palaeontology*, **5**, 171–194.

Richardson, J.B. 1965. Middle Old Red Sandstone spore assemblages from the Orcadian Basin north-east Scotland. *Palaeontology*, **7**, 559–605.

Richardson, J.B. & Lister, T.R. 1969. Upper Silurian and Lower Devonian spore assemblages from the Welsh Borderland and South Wales. *Palaeontology*, **12**, 201–252.

Richardson, J.B. & McGregor, D.C. 1986. *Silurian and Devonian Spore Zones of the Old Red Sandstone Continent and Adjacent Regions*. Geological Survey of Canada, Bulletin, **364**.

Richardson, J.B., Rodriguez, R.M. & Sutherland, S.J.E. 2001. Palynological zonation of the Mid-Palaeozoic sequences from the Cantabrian Mountains, NW Spain: implications for inter-regional and interfacies correlations of the Ludford/Přídolí and Silurian/Devonian boundaries, and plant dispersal patterns. *Bulletin of the Natural History Museum, London (Geology)*, **57**, 115–162.

Riegel, W. 1968. Die Mitteldevon flora von Lindlar (Rheinland) 2. Sporae dispersae. *Palaeontographica, Abteilung B*, **123**, 76–96.

Rubinstein, C.V. & García Muro, V.J. 2013. Silurian to Early Devonian organic-walled phytoplankton and miospores from Argentina: biostratigraphy and diversity trends. *Geological Journal*, **48**, 270–283.

Rubinstein, C.V., Melo, J.H.G. & Steemans, P. 2005. Lochkovian (earliest Devonian) miospores from the Solimões Basin, northwestern Bolivia. *Review of Palaeobotany and Palynology*, **133**, 91–133.

Shone, R.W. & Booth, P.W.K. 2005. The Cape Basin, South Africa: a review. *Journal of African Earth Sciences*, **43**, 196–210.

South African Committee for Stratigraphy 1980. Stratigraphy of South Africa. Part I (Comp. Kent, L. E.) Lithostratigraphy of the Republic of South Africa, South West Africa/Namibia, and the Republic of Bophuthatswana, Transkei and Venda. *Handbook of the Geological Survey of South Africa*, **8**, 1–690.

Stapleton, R.P. 1977. Carbonized Devonian spores from South Africa. *Pollen et Spores*, **19**, 427–440.

Staplin, F.L. 1961. Reef-controlled distribution of Devonian microplankton in Alberta. *Palaeontology*, **4**, 392–424.

Staplin, F.L. & Jansonius, J. 1964. Elucidation of some Palaeozoic Densospores. *Palaeontographica, Abteilung B*, **114**, 95–117.

Stone, P., Aldiss, D. & Edwards, E. 2005. *Rocks and Fossils of the Falkland Islands*. NERC and the British Geological Survey for Department of Mineral Resources, Falkland Islands Government. Keyworth, Nottingham.

Steemans, P. 1989. *Etude palynostratigraphique du Devonien Inferieur dans l'ouest de l'Europe*. Mémoires pour servir à l'Explication des Cartes Géologiques et Minières de la Belgique, **27**.

Steemans, P., Rubinstein, C. & Melo, J.H.G. 2008. Siluro-Devonian miospore biostratigraphy of the Urubu River area, western Amazon Basin, northern Brazil. *Geobios*, **41**, 263–282.

Steemans, P., Petus, E., Breuer, P., Mauller-Mendlowicz, P. & Gerrienne, P. 2012. Palaeozoic innovations in the micro- and megafossil plant record: from the earliest plant spores to the earliest seeds. *In*: Talent, J.A. (ed.) *Earth and Life, International Year of Planet Earth*. Springer, Dordrecht, 437–477.

Streel, M. & Marshall, J.E.A. 2006. Devonian–Carboniferous boundary global correlations and their paleogeographic implications for assembly of

Pangaea. *In*: WONG, TH. E. (ed.) *Proceedings of the XVth International Congress on Carboniferous and Permian Stratigraphy, 10–16 August 2003*, Utrecht, The Netherlands. Royal Netherlands Academy of Arts and Sciences, Amsterdam, 481–496.

STREEL, M. & THERON, J.N. 1999. The Devonian-Carboniferous boundary in South Africa and the age of the earliest episode of the Dwyka glaciation: new palynological result. *Episodes*, **22**, 41–44.

THERON, J.N. 1970. A stratigraphic study of the Bokkeveld Group (Series). *In*: *Second Gondwana Symposium (South Africa)*. Cape Town and Johannesburg, CSIR, Pretoria, 197–204.

THERON, J.N. 1993. The Devonian-Carboniferous boundary in South Africa. *Annales de la Société géologique de Belgique*, **116**, 291–300.

TREWIN, N.H., MACDONALD, D.I.M. & THOMAS, C.G.C. 2002. Stratigraphy and sedimentology of the Permian of the Falkland Islands: lithostratigraphic and palaeoenvironmental links with South Africa. *Journal of the Geological Society, London*, **159**, 5–19, http://doi.org/10.1144/0016-764900-089

TROTH, I., MARSHALL, J.E.A., RACEY, A. & BECKER, R.T. 2011. Middle Devonian sea-level change at high palaeolatitude: testing the global sea-level curve. *Palaeogeography, Palaeoclimatology, Palaeoecology*, **304**, 3–20.

VAN VEEN, P.M. 1981. Aspects of Late Devonian and Early Carboniferous palynology of Southern Ireland, IV. Morphological variation within *Diducites*, a new form-genus to accommodate camerate spores with two-layered outer wall. *Review of Palaeobotany and Palynology*, **31**, 261–287.

VAVRDOVÁ, M., BEK, J., DUFKA, P. & ISAACSON, P.E. 1996. Palynology of the Devonian (Lochkovian to Tournaisian) sequence, Madre de Dios Basin, northern Bolivia. *Vestnik Ceskeho Geologickeho Ustavu*, **71**, 333–349.

WELLMAN, C.H. & RICHARDSON, J.B. 1996. Spore assemblages from the 'Lower Old Red Sandstone' of Lorne, Scotland. *In*: CLEAL, C.J. (ed.) *Studies on Early Land Plant Spores from Britain*. Special Papers in Palaeontology, **55**, 41–101.

WELLMAN, C.H., HIGGS, K.T. & STEEMANS, P. 2000. Spore assemblages from a Silurian sequence in Borehole Hawiyah-151 from Saudi Arabia. *In*: AL-HAJRI, S. & OWENS, B. (eds) *Stratigraphic palynology of the Palaeozoic of Saudi Arabia*. Special GeoArabia Publications, **1**, 116–133.

WELLMAN, C.H., STEEMANS, P. & VECOLI, M. 2013. Palaeophytogeography of Ordovician–Silurian land plants. *In*: HARPER, D.A.T. & SERVAIS, T. (eds) *Early Palaeozoic Biogeography and Palaeogeography*. Geological Society, London, Memoirs, **38**, 461–476, http://doi.org/10.1144/M38.29

WICANDER, R. & WOOD, G.D. 1981. *Systematics and Biostratigraphy of the Organic-Walled Microphytoplankton from the Middle Devonian (Givetian) Silica Formation, Ohio, USA*. American Association of Stratigraphic Palynologists Foundation, Contribution Series, **8**.

WICANDER, R., CLAYTON, G., MARSHALL, J.E.A., TROTH, I. & RACEY, A. 2011. Was the latest Devonian glaciation a multiple event? New palynological evidence from Bolivia. *Palaeogeography, Palaeoclimatology, Palaeoecology*, **305**, 75–83.

Brachiopod faunas, facies and biostratigraphy of the Pridolian to lower Eifelian succession in the Rhenish Massif (Rheinisches Schiefergebirge, Germany)

ULRICH JANSEN

Palaeozoology III, Department of Palaeontology and Historical Geology, Senckenberg Research Institute and Museum of Natural History, D-60325 Frankfurt am Main, Germany (e-mail: Ulrich.Jansen@senckenberg.de)

Abstract: The succession of Pridolian to lower Eifelian (uppermost Silurian to lowermost Middle Devonian) rhynchonelliformean brachiopod faunas from the Rhenish Massif (Germany) is described and interpreted against the background of sedimentary sequences and facies development. The predominant rhenotypic ('Rhenish') facies is redefined as a neritic-siliciclastic facies type of the Devonian. It is subdivided into eurhenotypic, pararhenotypic and allorhenotypic subfacies, based on sedimentary features and specific brachiopod assemblages reflecting different shallow-marine palaeoenvironments under more or less terrigenous influence. The studied sedimentary successions are subdivided biostratigraphically on the basis of brachiopods, correlated on a supraregional scale, and calibrated in terms of the global chronostratigraphy. Stratigraphic intervals with characteristic brachiopod faunas are distinguished. Faunal turnovers between these are attributed to regional events in the context of short or more extended phases of palaeoenvironmental change presumably caused mainly by sea-level fluctuations in combination with varying crustal subsidence and sedimentation rates. These changes resulted in shelf-wide or more regional extinction or emigration of substantial parts of a brachiopod fauna and subsequent replacement by a largely new one. Finally, new taxa of brachiopods are introduced: *Sartenaerirhynchus* gen. nov., *Paraspirifer* (*Mosellospirifer*) subgen. nov., *Paraspirifer* (*Laurentispirifer*) subgen. nov., *Iridistrophia* (*Flabellistrophia*) subgen. nov. and *Iridistrophia* (*Flabellistrophia*) *musculosa* sp. nov.

Brachiopods are one of the most abundant and diverse marine invertebrate groups of the Palaeozoic Era. Pridolian (uppermost Silurian) through lower Eifelian (lower Middle Devonian) faunas from the Rhenish Massif (Rheinisches Schiefergebirge, Germany) document, with interruptions, brachiopod evolution during a time span of approximately 30 myr. The time interval under consideration starts with the earliest Gedinnian (Pridolian) transgression and ends with an early Eifelian faunal turnover introducing the interval of the Mid-Devonian 'Great Gap' in the broad sense (Struve 1982*b*). The Rhenish Massif is very suitable for examining evolutionary patterns of the brachiopods and possible causes of faunal change in a mid-Palaeozoic shallow-marine shelf sea along the southern margin of the Laurussia palaeocontinent. The succession of brachiopod faunas reflects palaeoenvironmental changes in a tropical epeiric sea, driven by local, regional and global forces.

The main objectives of the author's project 'Rhenish Lower Devonian brachiopods' are (1) to taxonomically revise these brachiopods, (2) to refine the brachiopod biostratigraphy, (3) to study faunal changes and to reconstruct their causes, (4) to elucidate the palaeobiology and palaeoecology of the Rhenish brachiopods and (5) to define the palaeobiogeographical positions of the Rhenish faunas. Pridolian and lower Eifelian brachiopods are included in order to study the faunal developments across the basal and upper boundaries of the Lower Devonian.

A solid taxonomical framework is necessary prior to any study on brachiopod biostratigraphy, palaeoecology, faunal change or palaeobiogeography. The taxonomic research on Rhenish brachiopods started in the nineteenth century with early monographs (e.g. von Schlotheim 1813, 1820; d'Archiac & de Verneuil 1842; Sowerby 1842; Roemer 1844; Schnur 1851, 1853; Steininger 1853; Sandberger & Sandberger 1854, 1856; Kayser 1889), was continued with many, often stratigraphically oriented studies during the twentieth century (e.g. Scupin 1900; Drevermann 1902, 1904; Fuchs 1915, 1919; Dahmer 1916, 1923, 1931, 1934, 1936*a*, *b*, 1951; Mauz 1935; Solle 1936, 1953, 1963, 1971; Simpson 1940; Vandercammen 1963; Struve 1970*b*; Jahnke 1971; Mittmeyer 1972, 1973*a*, *b*; and many more) and has seen a diminishing trend in recent decades. The current level of knowledge is variable and depends on the taxonomic group considered. While some taxa of the order Spiriferida can be

From: Becker, R. T., Königshof, P. & Brett, C. E. (eds) 2016. *Devonian Climate, Sea Level and Evolutionary Events*. Geological Society, London, Special Publications, **423**, 45–122.
First published online March 30, 2016, http://doi.org/10.1144/SP423.11

regarded as well known, e.g. species of *Arduspirifer*, *Brachyspirifer*, *Paraspirifer*, *Howellella*, *Euryspirifer* and *Acrospirifer*, many taxa of other groups have not been systematically revised since their first descriptions in the nineteenth century. These descriptions are often based on few specimens or even a single specimen only, so many morphological details and the intraspecific variation remain largely unknown up to the present day. Another difficulty concerns the original type localities and strata of these classic taxa, which are often not precisely known, but the experienced worker may be able to specify possible localities and strata on account of the rock matrix of the fossil. Because many of the classic Rhenish Lower Devonian brachiopod taxa are still poorly known, the idea arose to revise these, to re-examine their type collections and to study as numerous and well-preserved additional specimens as possible.

The Ardenno-Rhenish Massif is a classic area of Devonian stratigraphy. The traditional stages, the Gedinnian, Siegenian and Emsian of the Lower Devonian, were introduced here after a long and complex history (for an overview, see Ziegler 1979). In the Rhenish Massif, these stages are subdivided into regional substages, biostratigraphically defined mainly on the basis of brachiopods and, in particular, representatives of the order Spiriferida (e.g. Solle 1953, 1971, 1972; Mittmeyer 1974, 1982*a*, 2008; Jansen 2001*a*, 2014*b*; Schemm-Gregory & Jansen 2005, 2006*a*, *b*). In order to introduce a new way of subdividing and correlating the Rhenish successions, spiriferide taxon-range zones are proposed in the present work.

In a further step, it is intended to correlate the Ardenno-Rhenish subdivision with the international chronostratigraphy that is based on pelagic guide fossils. Persistent problems of stratigraphic correlation are pointed out. The available data confirm that a substantial lowermost part of the classic Gedinnian falls into the uppermost Silurian, as has previously been suggested by some workers (e.g. Carls 1971; Godefroid 1995; Godefroid & Cravatte 1999), and that the basal Emsian boundary in the present GSSP (Global Stratotype Section and Point) sense (Yolkin *et al.* 1997) is much older than the Siegenian–Emsian boundary in the traditional German or Rhenish sense (e.g. Carls *et al.* 2008). Finally, the basal Emsian boundary in the German sense is older than the basal Emsian boundary as defined by Belgian geologists in the Ardennes (Godefroid & Stainier 1982). A number of unresolved problems of correlation are discussed.

With regard to stratigraphic nomenclature, the author follows widely accepted stratigraphic international guidelines (e.g. Whittaker *et al.* 1991; Salvador 1994; Steininger & Piller 1999), in which a separation of material chronostratigraphic units (rock columns) and immaterial geochronological units (time intervals) is defined. This includes the use of 'lowermost', 'lower', 'middle', 'upper' and 'uppermost' for the subdivision of chronostratigraphic units and 'earliest', 'early', 'mid-', 'late' and 'latest' for the subdivision of geochronological units. A typical example for the usage of a geochronological term would be 'the late Emsian transgression', referring to the process in the geological past. A chronostratigraphic expression would be, for example, 'the lower Emsian in the Eifel region', referring to the rocks of this age in the Eifel region. In an adjectival usage, 'the lower Emsian strata/biozones/brachiopods' mean that the strata/biozones/brachiopods are referred to the rock column of the 'lower Emsian'. Depending on the context, i.e. if actual rock successions or processes in the geological past are the focus, the chronostratigraphic or the geochronological terminology is preferred in order to harmonize the general appearance of the text.

Litho- and biofacies must be taken into account in all parts of the project. Facies studies are essential to interpret the palaeoenvironments that triggered the spatio-temporal distribution of brachiopods and influenced their morphology. In the Devonian of central Europe (Rhenish Massif, Ardennes, Harz Mountains, Thuringia, Barrandian), traditionally two marine facies ('magnafacies') with different biota are distinguished: (1) the siliciclastic 'Rhenish' (= rhenotypic) facies with diverse neritic faunas reflecting shallow-marine, nearshore palaeoenvironments with agitated, turbid water and predominantly coarse siliciclastic sedimentation and (2) the 'Hercynian' (= hercynotypic) or 'Bohemian' facies with mainly pelagic faunas reflecting more offshore palaeoenvironments with calm water and predominantly argillaceous and calcareous sedimentation (e.g. Schmidt 1926, 1962; Kegel 1950; Rabien 1956; Erben 1962; Jansen 2001*a*). The brachiopod faunas considered in the present work belong to the Rhenish facies. This facies and some of its subfacies are described and interpreted with regard to specific palaeoenvironments.

This work reviews and summarizes many previous data, unpublished and published, sometimes from internationally little known or hardly accessible German publications, including results from recent studies of the present author based on a very large material from the Rhenish Lower Devonian (e.g. Jansen 2001*a*, *b*, 2012a, *b*, 2014*a*, *b*; Schemm-Gregory & Jansen 2005, 2007; Jansen *et al.* 2007). On this basis, a general overview of the faunal development is presented. The succession of the Rhenish faunas is described in stratigraphic order, and possible relationships to changing palaeoenvironments, regional and eustatic sea-level fluctuations, and global events are discussed. Some new taxonomic data are excerpts from the still ongoing monographic revision of the Rhenish brachiopods.

Abbreviations

Brachiopods. *Acr.*, *Acrospirifer*; *Al.*, *Alatiformia*; *Ard.*, *Arduspirifer*; *At.*, *Athyris*; *Bou.*, *Boucotstrophia*; *Br.*, *Brachyspirifer*; *Cr.*, *Crassirensselaeria*; *Crin.*, *Crinistrophia*; *Cry.*, *Cryptonella*; *Cun.*, *Cuninulus*; *Da.*, *Dayia*; *Din.*, *Dinapophysia*; *Eur.*, *Euryspirifer*; *Fasc.*, *Fascistropheodonta*; *Flab.*, *Flabellistrophia*; *Gib.*, *Gibbodouvillina*; *Gig.*, *Gigastropheodonta*; *Hip.*, *Hipparionyx*; *Hyst.*, *Hysterolites*; *In.*, *Inaequalibellirostrum*; *Int.*, *Intermedites*; *Ir.*, *Iridistrophia*; *Lap.*, *Lapinulus*; *Lau.*, *Laurentispirifer*; *Lept.*, *Leptostrophiella*; *Lor.*, *Loreleiella*; *Mau.*, *Mauispirifer*; *Meg.*, *Meganteris*; *Mer.*, *Meristella*; *Mes.*, *Mesodouvillina*; *Mos.*, *Mosellospirifer*; *Mult.*, *Multispirifer*; *Nuc.*, *Nucleospira*; *Ol.*, *Oligoptycherhynchus*; *Or.*, *Orthotetes*; *Pa.*, *Pachyschizophoria*; *Par.*, *Paraspirifer*; *Pla.*, *Platyorthis*; *Pleb.*, *Plebejochonetes*; *Pli.*, *Plicostropheodonta*; *Pro.*, *Proschizophoria*; *Prot.*, *Protocortezorthis*; *Ps.*, *Pseudoleptostrophia*; *Qu.*, *Quadrifarius*; *R.*, *Rhenothyris*; *Rh.*, *Rhenorensselaeria*; *Rhe.*, *Rhenoschizophoria*; *Rhen.*, *Rhenostropheodonta*; *Sart.*, *Sartenaerirhynchus*; *Sh.*, *Shaleria*; *Soll.*, *Sollispirifer*; *Str.*, *Straelenia*; *Sub.*, *Subcuspidella*; *Tei.*, *Teichostrophia*; *Tor.*, *Torosospirifer*; *Trop.*, *Tropidoleptus*; *Unc.*, *Uncinulus*; *Van.*, *Vandercammenina*.

Others. *Ac.*, *Acastella*; *B.*, *Belgicaspis*; *C.*, *Caudicriodus*; *Eoc.*, *Eocostapolygnathus*; *Fid.*, *Fidelites*; *Fo.*, *Foordites*; *Ling.*, *Linguipolygnathus*; *Mim.*, *Mimagoniatites*; *Now.*, *Nowakia*; *Po.*, *Polygnathus*; *Rhi.*, *Rhinopteraspis*; *War.*, *Warburgella*.

Geological setting

The Rhenish Massif represents a part of the Rhenohercynian Fold and Thrust Belt belonging to the central European Variscides (Kossmat 1927; Franke 2000; Walliser & Ziegler 2008). Devonian and Carboniferous rocks predominate at the surface, and rocks of Early Devonian age occupy about two-thirds of the surface (Fig. 1). During the Early Devonian, thick siliciclastic successions were deposited as 'Caledonian molasse' (Franke *et al.* 1978) on the subsiding Rhenish shelf. This material was transported by rivers and deltas onto the shelf and accumulated chiefly in mobile troughs but subordinately also on swells (e.g. Meyer & Stets 1980, 1996; Weddige *et al.* 2005; Stets & Schäfer 2011). This bathymetric differentiation was induced by synsedimentary tectonics under an extensional regime. Thicknesses of several kilometres have been reported from the Mosel and Lenne troughs, for example (Meyer & Stets 1980, 1996). Subsidence and sedimentation were in an approximate equilibrium for long periods, so mainly shallow-water, marine shelf palaeoenvironments persisted during the Early Devonian. The coastline was relatively stable in Siegenian–Emsian times and the water generally shallow, as indicated by the sedimentary and palaeontological characters of the rhenotypic facies. However, fluctuations in sedimentation rates, crustal subsidence and eustatic sea-level resulted in spatio-temporal changes of the palaeoenvironment from the supratidal to subtidal zones reflected today by variations in the rhenotypic facies (see below), a complex lithostratigraphy (Table 1), and, from the viewpoint of the present work, the distribution and composition of brachiopod faunas.

The mid-Palaeozoic succession was strongly affected by the Variscan orogeny during the Carboniferous, so the deposits on the Rhenish shelf were reduced in width, folded and disassembled by faults and overthrusts. On a smaller scale, the orogeny has led to deformation of the brachiopods.

As regards Early Devonian palaeogeography, the Rhenish Shelf was located at the southern margin of the Laurussia (Old Red) Palaeocontinent, which was separated from North Gondwana by the Rheic Ocean (between the Rhenohercynian Zone and the Armorican Terrain Assemblage = ATA) and possibly a second oceanic area, the Palaeotethys (between the ATA and stable Gondwana). According to the close palaeobiogeographical relationships between the Rhenish and Gondwanan shelves – within the 'Maghrebo-European Subrealm' ('Unterreich': Jansen 2012*a*) – these oceans could not have been very wide, and brachiopod-based biostratigraphic correlations between these areas are clearly possible (e.g. Carls 1987; Jansen 2001*a*; Jansen *et al.* 2007; Halamski & Baliński 2013).

Rhenotypic facies and palaeoenvironments

Introductory remarks

As the distribution of brachiopods is strongly dependent on facies, special attention must be paid to it in every biostratigraphic, palaeoecological or palaeobiogeographical study based on this fossil group. As noted above, two main facies types have been distinguished in the Devonian of central Europe: 'Rhenish (magna) facies' and 'Hercynian' or 'Bohemian (magna) facies' (e.g. Kayser 1878; Kayser & Holzapfel 1894; Schmidt 1926, 1962; Rabien 1956; Erben 1962; Carls *et al.* 1993; Jansen 2001*a*). The terms were explicitly applied to Devonian successions in Europe and North Africa (Erben 1962; Carls *et al.* 1993; Jansen 2001*a*). The studied brachiopod faunas from the Lower Devonian of the Rhenish Massif come from strata of the Rhenish

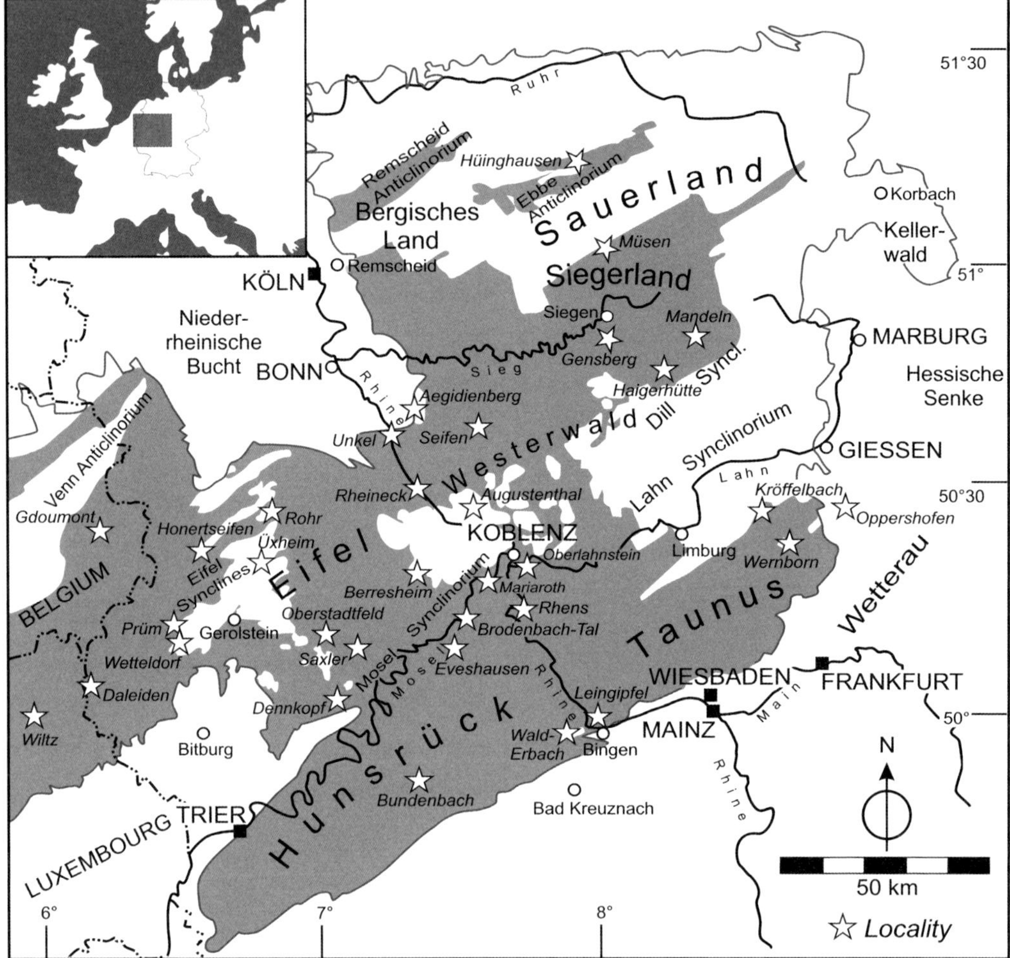

Fig. 1. Sketch map of the Rhenish Massif showing the main regions and some classic localities. The outer boundary of the Rhenish Massif is indicated with a grey line. Shaded area denotes distribution of Lower Devonian outcrops.

facies, which is the predominant facies here. Various subfacies types characterized by different faunas reflect a diversity of palaeoenvironments in space and time.

In general, the siliciclastic, largely sandy Rhenish (= rhenotypic) facies is interpreted to document shallow-marine, nearshore palaeoenvironments with agitated water, whereas the argillaceous and calcareous Hercynian (= hercynotypic) facies is interpreted to document more offshore, deeper and calmer palaeoenvironments. Palaeontologically, the Rhenish facies is characterized by neritic fossils, for example the coarsely ribbed and thick-shelled spiriferides, whereas the Hercynian facies is dominated by pelagic fossils, such as ammonoids and dacryoconaride tentaculitoids. This terminology has not been widely accepted in recent decades, because it was probably insufficient, too imprecise or partly ambiguous, as different shallow-water, calcareous deposits with benthic fossils were also included in the Hercynian facies, or different facies were separated as 'Mischfazies' ('mixed facies'; e.g. Erben 1962).

Referring to Meischner's concept (1971), Goldring & Langenstrassen (1979) preferred the distinction between 'inner shelf', 'outer shelf', 'shelf slope' and 'basinal facies', based on litho- and biofacies characters and essentially independent of the geological time or palaeogeography. In a conference abstract, more recently Becker (2008, p. 25) saw the terms 'Rhenish' and 'Hercynian' (or 'Bohemian') facies as problematic, because 'this

Table 1. *Simplified stratigraphy of the Rhenish Lower Devonian showing selected successions of the most important areas with Lower Devonian outcrops*

		Region →	Middle Rhine – Mosel		Eifel		Sauerland	Siegerland Westerwald-Dill		Taunus – Hunsrück	
EIFELIAN		lower	Wissenbach Fm.		Nohn Fm. Lauch Fm.		Schmallenberg Fm.	Wissenbach Fm.		Wissenbach Fm.	
EMSIAN	upper Emsian	upper (Kondel)	Kondel Group	Kieselgallenschiefer Fm. Flaserschiefer Fm.	Heisdorf Fm. Wetteldorf Fm.		Orthocrinus Fm.	Haigerhütte Fm.		upper Emsian sandstones and shales	
		middle (Laubach)	Laubach Group	Laubach Fm.	Wiltz Fm.		Remscheid Group	Mandeln Fm.			
		lower (Lahnstein)	Lahnstein Group	Hohenrhein Fm. Emsquarzit Fm.	Berlé Fm.		K 4				Warmsroth Fm.
	lower Emsian	upper (Vallendar)	Vallendar Group	Nellenköpfchen/Klerf Fm. Rittersturz/Gladbach Fm.	Vallendar Group	Klerf Fm. Stadtfeld Fm.	Ebbe Group: Siesel Fm.	Höllberg Fm.		Hiatus ?	Walderbach Fm.
		middle (Singhofen)	Singhofen Group	Bendorf Fm.	Singhofen Group	Neichnerberg Fm. Gefell Fm.				Oppershofen Fm. Spitznack Fm.	Wambach Fm.
		lower (Ulmen)	Wied Group	Nauort Fm. Oberbieber Fm. Deichselbach Fm.	Ulmen Group	Reudelsterz Fm. Eckfeld Fm.		Wilgersdorf Fm.		Hunsrück Slate Group	
PRAGIAN	Siegenian	upper		Aubach Fm.	Upper Siegen Group	Saxler Fm. Nitztal Fm.	Pasel Fm.	Upper Siegen Group	Unkel Fm.	Taunusquarzit Group	
		middle		Augustenthal Fm. Leutesdorf Fm.	Monschau Fm. Middle Siegen Group	Ramersbach Fm.		Middle Siegen Group	Seifen Fm. Eisernhardt Fm.		
		lower		Mayen Fm.	Lower Siegen Group		Bunte Ebbe Fm.	Lower Siegen Group		Hermeskeil Fm.	
LO	Ged.	upper	?		Kalltal Fm. (Venn)			Müsen Fm.		Bunte Schiefer Fm.	
		lower					Bredeneck Fm. Hüinghausen Fm.	Silberg Fm.			
PRI		lowermost			Grès de Gdoumont		Köbbinghausen Fm.			Kellerskopf Fm.	

terminology left a wide range of un-assigned mixed facies and shallow-water faunas [which] variably were assigned to one of both'. The Rhenish facies would 'include only the thick siliciclastic successions of the Rhenish Massif and exclude the neritic limestones of the Eifel Mountains or the thick reefal complexes of the Givetian and Frasnian'. In contrast, Struve had regarded the general character of the faunas from the shallow-marine, more calcareous or marly facies of the Middle Devonian in the Eifel region as 'rheinisch', that is 'Rhenish' (1970*a*, pp. 135–136; 1970*b*, pp. 528–532; 1982*a*) or 'rhenotypic' (1988, 1992). Becker (2008) preferred a more comprehensive term, 'neritic magnafacies', and summarized under it 'all shallow-water sediments and faunas deposited close to or at moderate distance from the shore (nearshore or inner shelf successions) and within the range of storm waves and the photic zone'. In contrast, the pelagic magnafacies would be 'characterised by sediments and faunas deposited far away from the shoreline, below the photic zone and, except major storms, gravity gliding, turbidites, and contourites, under quieter conditions'. Walliser & Ziegler (2008, p. 18) argued that one may regard the facies-related terms 'neritic' and 'Rhenish' on the one hand and 'pelagic' and 'Hercynian' on the other hand even as synonymous. According to their definition, the neritic facies would be characterized by prevailing fossils of benthic organisms and the pelagic facies by fossils of nektonic, planktonic and pseudoplanktonic organisms. Walliser *et al.* (1989) had argued in a similar way when distinguishing Devonian facies types in south China, using the term 'neritic facies' instead of the Rhenish facies. The neritic facies in this sense would include, for example, every shallow-water reef facies in the Phanerozoic, the facies of the Ordovician Armorican Quartzite or the neritic pelecypod facies of the Jurassic. Nobody would assign these rocks to the 'Rhenish facies' as this term was originally coined for a siliciclastic facies type of Devonian age. In my opinion, the Rhenish (= rhenotypic) facies should be regarded and redefined as a separate, subordinate facies of the neritic facies in the Devonian.

The Rhenish facies as redefined in this work includes sedimentary and palaeontological characters of its typical development in the interval from the Lochkovian (lower Lower Devonian) to the Frasnian (lower Upper Devonian). It shows a great preponderance of neritic fossils, mainly epibenthic suspension feeders and representatives of the shallow endobenthos, and sedimentary features of shallow-water settings. The faunas are characterized by a typical taxonomic composition, animals representing different ecomorphotypes, specific life habits and ecological needs. The focus is put on the fossil content, as it is most significant with regard to the palaeoenvironment. A redescription of this facies and its faunas follows below.

In order to keep a semantic distance from the strong regional, palaeogeographical and structural context of the term 'Rhenish', the term 'rhenotypic' is preferred and used as a pure facies term herein, following Carls *et al.* (1993). The term 'hercynotypic' is used instead of 'Hercynian'. Rhenotypic and hercynotypic facies are not restricted to palaeogeographical units but also occur outside the areas where they were introduced.

Large collections of rocks and fossils from different subtypes of the rhenotypic facies have been investigated for the present project. Previous studies were reconsidered; as an example, Fuchs (1971, 1982) established a detailed spatio-temporal facies zonation of the upper Siegenian and lower Emsian strata in the Eifel region, based on palaeontological, lithological and sedimentary characters. The collections of G. Fuchs (*c.* 1200 collection drawers) are stored in the Senckenberg Museum and were restudied by the present author.

Focusing on the composition of brachiopod assemblages, three main subfacies are distinguished: (1) eurhenotypic, (2) pararhenotypic and (3) allorhenotypic subfacies. These subfacies reflect, respectively, (1) shallow-marine palaeoenvironments with turbid water and prevailing siliciclastic-arenaceous sedimentation, characterized by typical, rich rhenotypic biota, including many ecomorphotypes of brachiopods, a diversity of pelecypods, trilobites and other fossil groups; (2) marginally marine, deltaic, coastal-lagoonal or intertidal palaeoenvironments with changeable conditions and siliciclastic sedimentation, commonly with impoverished, monospecific terebratulide faunas, pelecypods, lingulides, eurypterides, tentaculitides and early fishes; and (3) marine-shelf palaeoenvironments with clearer water and more calcareous and marly sedimentation, characterized by different rhenotypic faunas with many brachiopod species preferring these conditions, in addition tabulate and rugose corals, bryozoans, stromatoporoids and specific trilobites that are uncommon in the eurhenotypic facies. This differentiation is still of a rather general kind and should not be seen as too restrictive, as there are transitional facies with mixed characters of two or more subfacies. In addition, faunal analysis can be hampered by transport and sorting of shells that altered the taxonomic composition within the beds and may have led to the mixing of species from different biotopes. Besides, the facies may rapidly change vertically and laterally on small scales so that a detailed view is necessary. In the end, only a comprehensive facies analysis including the total fossil content, the sedimentary features and the taphonomy will allow sound reconstructions of the changing palaeoenvironments.

Sedimentary and palaeontological features

Sandstones and sandy shales are the predominant rocks of the rhenotypic facies, but siltstones, marlstones, mudstones or even limestones occur as well in different subfacies. The rocks show a wide spectrum of sedimentary features reflecting predominantly agitated and more or less turbid shallow water with currents and turbulence during deposition (e.g. Niehoff 1958; Wentzlau 1960; Struve 1961*b*, 1963*a*, 1970*b*; Wunderlich 1970; Zygojannis 1971; Winter 1977; Goldring & Langenstrassen 1979; Faber 1980; Walliser & Michels 1983; Müller 1987; Hahn 1990; Hahn & Zankl 1991; Schäfer & Stets 1995; Elkholy 1998; Kröll 2001; Stets & Schäfer 2002; Wehrmann *et al.* 2005; Elkholy & Gad 2006). For example, different types of ripples, small- and large-scale cross-stratification, high-energy horizontal bedding, slumps, tidal bedding, flaser-bedding, channels, storm deposits and reworking horizons occur, but subordinately also low-energy horizontal bedding. In outcrops, the shelly fossils are typically preserved as internal and external moulds of the valves; pelecypod and brachiopod shells are largely disarticulated. A calcareous component is present chiefly in the allorhenotypic subfacies and in some rocks of the distal eurhenotypic subfacies, where brachiopods may frequently occur as articulated shells. Specimens from the subsurface, from mines or deep drilling cores, are often preserved as shells, too, because they escaped from dissolution by humic acids.

The rhenotypic faunas include mainly neritic, benthic faunal groups, i.e. specific brachiopod, pelecypod, gastropod and trilobite taxa, a few corals, true tentaculitides (*Tentaculites*) and thick-shelled, coarsely ornamented beyrichiide ostracods (Erben 1962; Jansen 2001*a*). The taxonomic composition is further specified in the description of the subfacies types below. Ichnofossils such as *Skolithos*, *Spirophyton* and *Sabellarifex* indicate that the upper sediment was inhabited by infauna; they are most abundant in the proximal eurhenotypic or pararhenotypic subfacies; *Chondrites* is abundant in the eurhenotypic subfacies. The pararhenotypic subfacies shows an impoverished fauna, in particular with regard to the brachiopods. In the allorhenotypic subfacies, benthic faunas prevail again but have a different taxonomic composition, with multiple organisms preferring clear water; the facies shows in general a higher diversity of trilobites and corals. Remains of pelagic animals with wide geographical distribution, such as polygnathid conodonts, dacryoconaride tentaculitoids and ammonoids, are largely lacking in the rhenotypic facies but may be found as rare elements in some strata of the allorhenotypic or eurhenotypic subfacies.

The brachiopods of the rhenotypic facies, in particular those of the eurhenotypic subfacies, are typically large and thick-shelled. In many assemblages they represent the most abundant and diverse fossil group. The corresponding kind of fossil biocoenoses has been called ‘brachiopodetum’ by Struve (e.g. 1963*b*, p. 27; 1966, p. 137; 1988, pp. 97, 101). Typical Lower Devonian constituents are coarsely plicate delthyridoid and spinocyrtiid spiriferides with a wide hinge line, often large Strophomenida, some characteristic representatives of the Orthotetida, Orthida, Rhynchonellida and Athyridida, and often subglobular, ribbed Terebratulida, whereas Pentamerida and Atrypida are common only in the allorhenotypic subfacies. The diversity of the brachiopod assemblages, their taxonomic composition and the representation of ecomorphotypes are related to manifold facies variations. As well, the taxonomic composition varies in dependence on the geological age and the palaeobiogeographical position.

Interpretation

From the combination of sedimentary and palaeontological facies traits it is concluded that the rhenotypic facies reflects favourable living conditions for diverse benthic biota in intertidal and shallow to moderately deep subtidal environments of an ancient subtropical to tropical shelf sea affected by agitated water and transport of siliciclastic material. The presence of abundant and often highly diverse brachiopod faunas indicates that suitable substrates, sufficient dissolved calcium carbonate and nutrient-rich, warm and well-aerated bottom water were available, allowing the settlement and survival of these suspension feeders, at least in times of normal or fair weather. The rhenotypic brachiopods were adapted to or could at least tolerate the shallow-water conditions of the Rhenish Shelf. They show a number of morphological features that can be interpreted as adaptations to such conditions, for example thick shells, plicate shells and large muscle fields suggesting a strong musculature (Jansen 2001*a*). Carls *et al.* (1993) introduced two terms to describe their turbidity tolerance that have not been widely used so far: the bulk of the eurhenotypic brachiopods were tolerant to turbid water – they were ‘turbidicolous’, for example most of the coarsely plicate spiriferides, such as *Arduspirifer*, *Acrospirifer*, *Euryspirifer* and *Hysterolites*. In contrast, many brachiopods from the allorhenotypic subfacies lived preferably in clear water – they were ‘claricolous’, in particular species of athyridides, atrypides and pentamerides, but also some coarsely plicate spiriferides (e.g. *Intermedites*). The turbidity of the former bottom water is estimated from sedimentary and palaeontological characters of the successions. It has to be taken into account that the hydrodynamic conditions during the embedding of the fossils were often exceptional and did not represent the 'normal' fair weather situation. Possible morphological adaptations of the prevailing brachiopods, the abundance of respective ecomorphotypes (e.g. the ratio of coarsely ribbed v. smooth shells) and the presence or absence, abundance and diversity of accompanying clear-water organisms such as corals enable conclusions on the prevailing turbidity conditions.

The rhenotypic brachiopods were more or less tolerant to a certain level of hydrodynamic energy and turbidity, but whole populations were extinguished during episodic events, for example heavy tropical storms, when masses of siliciclastic material were mobilized, transported and redeposited. The brachiopod substrates were eroded, and the animals were overturned or transported and buried by sediment. The recurrent, episodic destruction of biocoenoses during sudden erosion or sedimentation events and resulting high mortality rates were probably compensated by high reproduction rates.

The palaeogeographical position of subsiding shelf areas in low tropical latitudes and along orogens with strong denudation (e.g. the Caledonides) under partly humid and hot climate conditions (Golonka *et al.* 1994; Stets & Schäfer 2002) promoted the environmental conditions of the rhenotypic facies. Fluvial deltas transported huge amounts of terrigenous material into the shelf seas, and this included also nutrients. During the Devonian, the land was colonized by plants representing an additional source of nutrients. In the Early Devonian, vegetated coastal marshlands existed (e.g. Schweitzer 1983), with still small, rootless or shallowly rooted plants (Algeo & Scheckler 1998). Organic detritus and dissolved organic material were transported into the sea, which may have led to eutrophication and plankton blooms and may have stimulated the diversification and radiation of marine organisms (e.g. Algeo & Scheckler 1998; Bambach 1999; Allmon & Martin 2014).

Subfacies of the rhenotypic facies

In this section, the subdivision of the rhenotypic facies into the three main subfacies is specified. It can be applied not only to Devonian strata in the Ardenno-Rhenish area but also to other rhenotypic successions, at least in Europe and North Africa. The characteristics outlined below are still rough guides and refer to larger facies bodies or formations.

(1) *Eurhenotypic subfacies* – (≈'Oberstadtfelder Intrafazies' *sensu* Schmidt 1926; Jux 1971; cf. 'inner' and 'outer coastal shelf environments' *sensu* Fuchs 1982). This is the typical rhenotypic subfacies encountered in many Rhenish outcrops. The most common rocks are distinctly bedded, often impure sandstones and sandy shales. Erosional surfaces, cross-stratification and ripples occur; single beds may pinch out rapidly or extend laterally over moderately long distances of some decametres. Palaeontologically, the subfacies is characterized by rich brachiopod assemblages. Their taxonomic composition changes with geological age and palaeobiogeographical position, but different taxa may often represent similar ecomorphotypes. Typical constituents are coarsely plicate delthyridoid spiriferides, for example species of *Howellella*, *Hysterolites*, *Acrospirifer*, *Euryspirifer*, *Arduspirifer*, *Sollispirifer*, *Brachyspirifer* and *Paraspirifer*. Spinocyrtiid spiriferides of the genera *Tenuicostella*, *Incertia*, *Subcuspidella* or *Alatiformia* may be common as well. Representatives of the orders Strophomenida (e.g. *Boucotstrophia*, *Gibbodouvillina*, *Gigastropheodonta*, *Plicostropheodonta*, *Rhenostropheodonta*), Orthotetida (e.g. *Iridistrophia*), Orthida (e.g. *Rhenoschizophoria*, *Platyorthis*), Athyridida (e.g. '*Athyris*', *Septathyris*) and Terebratulida (e.g. *Meganteris*, *Rhenorensselaeria*), and the enigmatic genus *Tropidoleptus* are common, whereas Pentamerida and Atrypida are uncommon. The shells are very often disarticulated and accumulated in shell beds, which probably represent mostly storm layers. Pelecypods (e.g. Eichele 2014) can even be more abundant and diverse than the brachiopods; representatives of the Palaeotaxodonta (e.g. *Palaeonucula*, *Nuculites*, *Palaeoneilo*), Pteriomorphia (e.g. *Cypricardites*, *Follmannia*, *Myalina*, *Cornellites*, *Tolmaia*, *Actinodesma*, *Leptodesma*, *Limoptera*, *Aviculopecten*, *Leiopteria*) and Heteroconchia (e.g. *Modiomorpha*, *Goniophora*, *Eoschizodus*, *Prosocoelus*, *Cypricardella*, *Grammysia*) are common. Gastropods are moderately abundant, for example species of *Platyceras*, *Bembexia*, *Murchisonia* and *Bucanella*. The tabulate coral *Pleurodictyum* or related genera and sporadic solitary rugose corals of the '*Zaphrentis*' type occur in many outcrops. Trilobites are locally common in some assemblages and at specific stratigraphic levels (Gandl 1972; Wenndorf 1990; Basse 2002, 2003; Basse & Müller 2004). They are mainly represented by species of the Homalonotinae (*Burmeisterella*, *Parahomalonotus*, *Digonus*, *Dipleura*), Asteropyginae (*Treveropyge*, *Dunopyge*, *Wiltzops*, *Rheingoldium*, *Comura*, *Kayserops*, *Rhenops*), Phacopinae (*Phacops*) and Acastidae (*Acastella*, *Acastoides*).

Orthoconic cephalopods, tentaculitides (almost exclusively *Tentaculites*), coarsely sculptured ostracods (e.g. *Zygobeyrichia*), bryozoans, crinoids, asterozoans and fish remains occur in variable abundance.

The eurhenotypic subfacies reflects marine-shelf palaeoenvironments from the lower intertidal to moderately deep subtidal zones with prevailing siliciclastic sedimentation, providing suitable living conditions for a high diversity of marine animals that are tolerant to a certain degree towards agitated and turbid water. In general, the palaeoenvironments were suitable for turbidicolous brachiopods, at least during the normal, prevailing hydrodynamic conditions.

Three subordinate subfacies types can be distinguished:

(A) Typical eurhenotypic subfacies (cf. 'Assemblage 2' and partly 'Assemblage 3' of the inner coastal-shelf environment *sensu* Fuchs 1982) with polyspecific faunas including commonly more than 10 and in some cases more than 20 articulate brachiopod species at one locality, various morphotypes including strophomenides and smooth athyridides, and in addition the whole spectrum of eurhenotypic biota; in particular, pelecypods can be diverse and abundant; trilobites, corals, bryozoans and orthoconic nautiloids occur in places. Distinctly bedded, alternating sandstones and shales with changing sand content are the common rocks. This category corresponds to nearshore, subtidal palaeoenvironments with relatively stable substrates and moderate hydrodynamic energy and turbidity largely between the fair weather wave base and the storm wave base. It can be compared with the 'heterostrate facies' in the terminology of Schäfer (1962). Examples: Hüinghausen, Seifen, Saxler, Spitznack, Stadtfeld and Hohenrhein formations.

(B) Proximal, impoverished eurhenotypic subfacies (cf. 'Assemblage 1' of the inner coastal shelf environment *sensu* Fuchs 1982) with articulate brachiopod assemblages of reduced species diversity, less than 10 species at one locality, few morphotypes of brachiopods, commonly numerous chonetoids, e.g. of the genus *Plebejochonetes*, plus winged and plicate spiriferides, e.g. of the genera *Hysterolites*, *Arduspirifer*, *Subcuspidella* or related genera. Orthides (e.g. *Platyorthis*), orthotetides (*Iridistrophia*) and subglobular, ribbed terebratulides (e.g. *Crassirensselaeria*, *Globithyris*) may occur in abundance as well; strophomenides and smooth athyridides are rare. Corals, bryozoans and orthoconic nautiloids are generally rare or absent. Pelecypods, gastropods and the tentaculitide genus *Tentaculites* are common; homalonotine trilobites and detrital plant remains may be abundant, too. Trace fossils and bioturbation indicate an active bottom life. Sedimentary structures of very shallow water, such as different kinds of ripples, signs of strong and abundant reworking, lateral facies change over short distances and storm layers are more common than in the typical eurhenotypic subfacies. The proximal eurhenotypic subfacies reflects palaeoenvironments with higher terrigenous influence than the typical eurhenotypic subfacies, more turbid water and relatively unstable substrates; these were partly hostile, high-energy palaeoenvironments ('lethal heterostrate facies' *sensu* Schäfer 1962), nearshore, lower intertidal to shallow subtidal settings, and subaqueous delta platforms or interdistributary bays. Examples: Upper Siegen Group, Klerf Formation, Wetteldorf Formation ('Wetteldorfer Sandstein'), Bredeneck Formation, Reudelsterz Formation, Nellenköpfchen Formation, Taunusquarzit Group and Emsquarzit Formation; in each of these formations parts are developed in this subfacies.

(C) Distal eurhenotypic subfacies (cf. 'outer coastal shelf environment' and partly 'Assemblage 3' of the 'inner coastal shelf environment' *sensu* Fuchs 1982) with variable brachiopod diversity; partly enriched, polyspecific brachiopod assemblages with more than 20 species at one locality. Other assemblages may consist of less than 10 species, in particular in the more muddy facies. There are essentially eurhenotypic taxa, such as species of *Iridistrophia*, *Leptostrophiella*, *Gigastropheodonta*, *Platyorthis*, *Pachyschizophoria*, *Euryspirifer*, *Arduspirifer*, *Sollispirifer*, *Alatiformia*, *Oligoptycherhynchus* and *Meganteris*, plus more or less abundant admixtures of not genuinely eurhenotypic, partly allorhenotypic taxa regarded as claricolous, e.g. *Leptaenopyxis*, *Leptaena*, *Cyrtina*, *Resserella*, *Schizophoria*, *Sieberella*, *Nucleospira*, *Meristella*, '*Athyris*', *Quadrithyris* and atrypides. Brachiopods occur often with articulated valves in shelly preservation or internal moulds of these. Corals, gastropods, orthoconic nautiloids, diverse pelecypods, bryozoans, crinoids and phacopide trilobites (mainly *Phacops*, *Kayserops*, *Treveropyge*, *Acastoides*, *Comura*) can be present. Mud-rich rocks are common, psammitic or silty, often slightly calcareous, dark grey to black shales occur with intercalations of fine-grained sandstone beds, which presumably are mostly distal storm layers or turbidites. Siliceous nodules ('Kieselgallen') are not uncommon in some formations representing the subfacies. The distal eurhenotypic subfacies is referred to open-shelf palaeoenvironments of the shallow to moderately deep subtidal, more or less above or close to the storm wave base, with low to moderate hydrodynamic energy. There were nutrient-rich settings with not too soft substrates ('vital heterostrate' to 'vital isostrate facies' *sensu* Schäfer 1962). The subfacies is transitional to the argillaceous hercynotypic facies. Examples: Haigerhütte, Kieselgallen-Schiefer, Wiltz, Mandeln and Augustenthal formations.

(2) *Pararhenotypic subfacies* – (≈'Niederrheinische Intrafazies' *sensu* Schmidt 1926; '*Modiolopsis*-Fazies' *sensu* Dahmer 1936*b*; 'globithyrid facies' *sensu* Boucot 1963; cf. 'inner shelf facies' *sensu* Goldring & Langenstrassen 1979; cf. 'coastal environment' *sensu* Fuchs 1982). Sandstones of different maturity, mudstones and intra- and extraformational conglomerates are the main sedimentary rocks of this subfacies; red and green rock colours can be common. Mud pebbles in sandstones document reworking of mud layers, for example from erosion in meandering channels. A rich inventory of very shallow-water sedimentary features occurs, changing vertically and laterally on small scales. These are oscillation and current ripples with often rapidly changing orientation, channels with longitudinal cross-stratification, high-energy horizontal bedding, erosional surfaces, ball and pillow structures and horizontal-bedded mudstones deposited under low-energy conditions. Desiccation structures, water-level marks and wind-induced striation may be present, showing episodic subaerial exposition. Palaeontologically, the facies is locally characterized by monospecific articulate brachiopod faunas of subglobular, ribbed terebratulides, e.g. *Globithyris* or *Crassirensselaeria* – different generic assignments for comparable forms from different units circulate in the literature: *Rhenorensselaeria*, *Rensselaeria*, *Mutationella*, *Paulinella* and *Retzia* (= *Trigeria*, an athyridide); these determinations are still to be checked in many cases. These brachiopods of the 'globithyrid facies' (Boucot 1963) were probably tolerant towards lowered or changing salinities (Fürsich & Hurst 1980). It is possible that they were even able to tolerate temporary subaerial exposure during low tides and could survive by closing their valves tightly and storing an amount of seawater in their voluminous mantle cavity. Finds of numerous articulated valves of *Crassirensselaeria* very close to fossil land plants *in situ* in the Odenspiel Formation near Wenden (author's unpublished data, upper Siegenian, Sauerland) support this assumption. Linguloids and pelecypods (e.g. *Modiolopsis*) are often abundant, fish remains (mainly agnathans, e.g. *Rhinopteraspis*; bonebeds), tentaculitides (*Tentaculites*), eurypterides, ostracods and early terrestrial plant remains, including large fragments, are common. Trace fossil assemblages are diverse and typically include vertical structures of the *Scolithos* type in sandstones, *Spirophyton* in muddy deposits and a diversity of arthropod tracks.

The subfacies reflects a wide spectrum of marginally marine, coastal, lagoonal, intertidal or deltaic palaeoenvironments, often with reduced or changing salinities, but always with regular, more or less marine influence; variable hydrodynamic energy, turbidity and temperature – and generally changing physical parameters. The pararhenotypic subfacies should be kept separate from the facies of purely terrestrial, freshwater palaeoenvironments of fluvial deltas and lakes, which are also present in the Rhenish Massif and sometimes not easily distinguishable (Hahn 1990; Stets & Schäfer 2002, 2009; Franke 2006). The occurrences of brachiopods and other marine animals are good indicators of marine conditions for specific strata. The overall fossil content is the crucial factor to recognize the transitional zone from the marine to the terrestrial realm. Examples: (parts of) the Lower Siegen Group and the Odenspiel, Klerf and Nellenköpfchen formations.

(3) *Allorhenotypic subfacies* – (≈'Mischfazies', partially *sensu* Erben 1962; 'Fazies-Typ C' *sensu* Winter 1977; 'Medley Facies' or 'Eifel Facies' *sensu* Struve 1982*a*). Limestones and crinoidal limestones, marly siltstones, marlstones and calcareous sandstones are the most common rocks. Often with a high diversity of articulate brachiopods (partly more than 20

species at one locality), faunas include genera regarded as rhenotypic (Struve 1982*a*, p. 423: 'in principle a typical Rhenish fauna'), for example the coarsely ribbed spiriferides *Paraspirifer*, *Quiringites*, *Intermedites*, *Alatiformia* and *Subcuspidella*. These spiriferides are associated with a substantial content of forms that are suggested to be typical clear-water, i.e. claricolous brachiopods, such as atrypides, athyridides, pentamerides, reticulariid spiriferides and the genera *Leptaena*, *Teichostrophia*, *Helaspis*, *Quadrithyris* and *Cimicinella*. The spiriferides *Intermedites* and *Quiringites* mainly occur in the marly-calcareous facies and are also interpreted as claricolous (Carls 2000, p. 74). These claricolous genera and their close relatives are uncommon, often rare or absent in strata of the eurhenotypic subfacies (with exceptions). Smooth and rounded brachiopod ecomorphotypes of different taxonomic groups can be abundant (Winter 1971). Tabulate and rugose corals, bryozoans and stromatoporoids are locally abundant and show also that the water was not too turbid, at least apart from episodic storms. In some places there occur receptaculitids, calcareous algae and a rich trilobite fauna, including phacopines (e.g. *Phacops*, *Pedinopariops*, *Geesops*), scutelluids (*Scabriscutellum*), asteropygines (*Asteropyge*), lichides (*Ceratarges*) and proetides (e.g. *Basidechenella*, *Rhenocynproetus*, *Tropidocoryphe*, *Dohmiella*). Icriodontid and rare polygnathid conodonts can be extracted from limestones. This is mainly the zone of brachiopod communities ('Brachiopoden-Fazies' *sensu* Struve 1961*b*; 'Brachiopoden-Siedlungen' *sensu* Struve 1963*a*, *b*) on the Mid-Devonian open-marine shelf of the Eifel area (cf. Winter 1977; Faber 1980). The allorhenotypic subfacies documents open-shelf palaeoenvironments, with shallow, clear-water and nutrient-rich conditions in the photic zone. The production of calcium carbonate exceeded the siliciclastic influx in many cases. It appears reasonable to exclude from the allorhenotypic subfacies biostromes and bioherms of carbonate platforms, including lagoonal limestones ('Fettkalke'), that represent a poorly fossiliferous mud facies with calcispheres or fenestral structures (Faber 1980). Examples: Heisdorf (Lower Devonian), Lauch, Nohn, Ahrdorf and Junkerberg formations (Middle Devonian, Eifel region).

Within the Rhenish Massif, the three main subfacies show core areas of distribution: The eurhenotypic subfacies is mainly represented by Lower Devonian successions of the central and southern regions (Middle Rhine and South Eifel regions, Taunus and Hunsrück), whereas the pararhenotypic subfacies has its main distribution in the northern regions (Bergisches Land, Sauerland, Siegerland). The uppermost lower Emsian of the central Rhenish Massif (Klerf and Nellenköpfchen formations) is largely developed in the pararhenotypic subfacies. The allorhenotypic subfacies is mainly represented by parts of the Eifelian beds in the Eifel region and the Sauerland.

Distribution and limits of the rhenotypic facies

The rhenotypic facies has originally been defined as a Devonian neritic facies in Europe and North Africa, based on specific neritic faunas and siliciclastic shallow-water deposits (Erben 1962; Jansen 2001*a*), but a well-founded spatio-temporal delimitation is not unproblematic. I would hesitate to assign any Ordovician or Silurian marine facies to the rhenotypic facies, as the taxonomic composition of the biota would be too different, although similar physical palaeoenvironments with brachiopods, pelecypods and trilobites existed. The Silurian brachiopod faunas from Gotland (Sweden), for example, certainly show parallels to the allorhenotypic faunas from the Eifelian of the Eifel region, but the faunas should be carefully evaluated first and compared with each other before any conclusions are drawn. The lowermost Gedinnian (Pridolian) assemblages from the lower Muno, Gdoumont, Köbbinghausen, lower Silberg and Kellerskopf formations with *Quadrifarius dumontianus* (de Koninck, 1876) still show a predominantly Silurian character, although first representatives of *Platyorthis* and *Mutationella* in the Grès de Gdoumont show the temporal proximity of the Devonian and could be interpreted, in combination with the strongly plicate spiriferide, as features of a first rhenotypic facies. In my opinion, the oldest rocks of the Ardenno-Rhenish area in doubtless rhenotypic facies are represented by the lower Gedinnian Mondrepuis, upper Muno (Parensart level), Hüinghausen and Bredeneck formations. Certainly, the evolution and radiation of the coarsely plicate, often transverse and partly large-sized delthyridoid spiriferides originating from small howellellids is an important feature of the rhenotypic facies in the Lochkovian–Givetian interval (Gourvennec 1989; Carls 2000; Schemm-Gregory 2011), the expansion of the Chonetoidea another one (e.g. Racheboeuf 1981), although some genera of this group are present in the hercynotypic facies, as well. Many brachiopod groups that characterize the rhenotypic facies in the pre-Famennian part of the Devonian are absent in the Famennian, for example most of the delthyridoid spiriferides, the leptostrophiid, strophodontid and douvillinid strophomenides, the genus *Tropidoleptus*, most of the uncinuloid rhynchonellides, the pentamerides and the atrypides, resulting in a lower brachiopod diversity in the Famennian than in the preceding Devonian stages (Brice *et al.* 2000; Curry & Brunton 2007; Hubert *et al.* 2007; Zapalski *et al.* 2007). The ribbed, subglobular terebratulide genera *Globithyris* and *Crassirensselaeria* and their relatives, typical of the pararhenotypic subfacies, went extinct before the Famennian as well. The widespread stromatoporoid–coral reefs of the Mid-Devonian and Frasnian disappeared with the Kellwasser Crisis and with these the brachiopod biotopes of the adjacent shallow-water palaeoenvironments. A number of characteristic trilobite groups also disappeared, some of them as early as with the Mid-Devonian and Frasnian events (Chlupáč *et al.* 2000). The pelecypods were reduced significantly and replaced by new genera chiefly in the late Famennian (Amler 1996). Nearshore to open-shelf palaeoenvironments with siliciclastic or mixed siliciclastic-calcareous sedimentation and rich brachiopod faunas, characteristic of the Lochkovian to Frasnian interval in many regions, were reduced in the early Famennian. The Kellwasser Crisis is a distinct break in the development of marine palaeoenvironments (e.g. Schindler 1993), extinguishing the mid-Palaeozoic tropical carbonate shelf biota, especially in the reef and peri-reefal settings but retaining the stress-resistant calcimicrobes (Copper 1998). The palaeoecological consequences of this two-phase event or the effects on the palaeoecosystems are still uncertain. From the observable faunal change it is suggested that the upper Frasnian could be defined as the upper stratigraphic limit of the rhenotypic facies. The Famennian and Carboniferous brachiopod faunas, and even more the Permian ones, are significantly different from the Lochkovian to Frasnian faunas, for example regarding the diversification of the non-chonetidine productides (Brunton *et al.* 2000); many new genera of orthotetides (Williams *et al.* 2000), rhynchonellides (Savage 1986; Sartenaer 1998; Brice 2000) and spiriferides (e.g. Spiriferoidea, Spiriferinidina; Carter & Gourvennec 2006, fig. 1102) appeared as well. The Mesozoic and Cenozoic biofacies types with neritic faunas are very different from the Devonian ones, as they are less diverse in brachiopods and dominated by the 'Modern Fauna' (e.g. Sepkoski 1990). The latter increased in the late Palaeozoic, developing all inherent features, i.e. more complex tiering and utilization of ecospace, in particular an increased number of motile, durophageous (shell-crushing) predators and the presence of a mobile, deep endobenthic fauna (e.g. Sepkoski 1990; Bush & Bambach 2011; Allmon & Martin 2014). Pelecypods, among these the endobenthic suspension feeders, the siphonate heterodonts, became more diverse and abundant during the Mesozoic; the net increase of infaunal organisms was probably controlled by a rise in predation pressure and the adoption of new strategies (Aberhan 1994). It can be stated that the palaeoecosystem 'shallow marine shelf' fundamentally changed.

The palaeogeographical distribution of the rhenotypic facies is not easy to define precisely (cf. Erben 1962). In the Lower and Middle Devonian of the Rhenish Massif and the Ardennes, the rhenotypic facies is the most common facies. In other areas of Europe and in North Africa, it is also widespread, for example it is well represented by the following units of Early Devonian age (at least substantial parts of these): Meadfoot Group (south England); St. Céneré, Montguyon and Reun ar H'Rank formations in Brittany (western France); Nogueras, Santa Cruz, Mariposas and Castellar formations in the Iberian Chains (northeastern Spain); La Vid Group and Santa Lucía Formation in León and La Ladrona Formation in Asturias (Cantabrian Mountains, north Spain); Assa, Merzâ-Akhsaï and Mdâouer-el-Kbîr formations (the 'Rich sandstones') in the Dra Plains (pre-Sahara, Morocco); Zagórze Formation in the Holy Cross Mountains (Poland); Drakov Quartzite of the Vrbno Group in the Hrubý Jeseník Mountains (Czech Republic); and Kartal Formation in the Istanbul region (Turkey). Different subtypes of the rhenotypic facies are represented by these formations. The allorhenotypic subfacies of the Middle Devonian is particularly well represented by the Skały Formation in the northern part of the Holy Cross Mountains, where brachiopod faunas with close relationships to those from the Eifel region occur (Halamski 2008). Many reports of the 'globithyrid' biofacies, for example from the Emsian of the Appalachians (Boucot 1963), the Lower Devonian of Saudi Arabia (Boucot 1984) and the Emsian–Eifelian of the central Sahara (Mergl & Massa 2004, pp. 89–90), may be assigned to the pararhenotypic subfacies. In some regions, strata in rhenotypic and hercynotypic facies follow one after another vertically or interfinger laterally in the same sections (Carls *et al.* 1972; Carls 1987, 1988: Iberian Chains, Cantabrian Mountains, Armorican Massif; Jansen *et al.* 2007: Dra Plains), so they can be correlated. Devonian successions developed in probably rhenotypic or at least similar siliciclastic or mixed siliciclastic-calcareous marine facies show a wide distribution, for example in South Africa, eastern North America, Venezuela, Bolivia, Argentina, Antarctica, southern Siberia, Mongolia, NW China, south China, Antarctica and New Zealand. Interestingly, brachiopod morphotypes of these areas are often strikingly similar to the Rhenish ones, but at least in most cases represent different, somewhat homeomorphic genera. It is suggested that the development of analogous morphologies is caused mainly by a comparable autecology. To conclude, many of these brachiopod-dominated successions could be assigned to the rhenotypic facies, but it is beyond the scope of the present work to examine this question in detail.

The rhenotypic facies is delimited in former landward direction against the terrestrial Old Red facies and in former seaward direction against the hercynotypic facies. The pararhenotypic subfacies shows manifold transitions to the Old Red facies, indicated, for example, by the disappearance of marine fossils and increasing red bed intercalations or signs of subaerial exposure within the Klerf Formation (Eifel region: Becker & Mentzel 1961; Solle 1976; Fuchs 1982; Franke 2006; Meyer 2013). On the other hand, there are gradual transitions from the distal eurhenotypic subfacies to the hercynotypic facies, for example the transition from the Haigerhütte Formation ('Kieselgallen-Schiefer') to the Wissenbach Formation at the locality 'Papiermühle Haiger' in the Dill Synclinorium (Jahnke & Michels 1982*b*; Jansen *et al.* 2001). It is expected that there are also transitions from the allorhenotypic subfacies to the hercynotypic facies or from the distal eurhenotypic subfacies to the allorhenotypic subfacies, but this is still to be verified in the field, in the Sauerland or in the Eifel region.

The hercynotypic facies

The hercynotypic facies represents a Devonian subtype of the pelagic facies; it is dominated by argillaceous rocks and pure limestones, indicating deposition in quiet water and far from the shoreline (partly 'Hercynische Fazies' *sensu* Erben 1962; 'basinal facies' *sensu* Krebs 1979). Turbidites, tempestites, gravity slides, contourites and allodapic limestones indicate episodic high-energy sedimentation events (Becker 2008). The Devonian fossil assemblages are dominated by pelagic fossils, such as ammonoids, polygnathid conodonts and dacryoconarides. Ostracods of the Thuringian ecotype, a specialized (often small-eyed to blind), diverse trilobite fauna and thin-shelled pelecypods are common. The brachiopods are generally small, often smooth or finely ribbed and thin shelled. They include, for example, smooth and biconvex forms of different taxonomic affiliation, the strophomenides *Plectodonta* and *Dalejodiscus*, the orthide *Prokopia*, or the small, enigmatic genera *Costanoplia*, *Notanoplia* and *Boucotia* (e.g. Langenstrassen 1972; Jahnke & Michels 1982*b*; Vogel *et al.* 1989; Schubert 1996). Rhenotypic brachiopods could not survive in this palaeoenvironment because the substrates were too soft and muddy, not sufficiently aerated, or both. The hercynotypic facies reflects offshore palaeoenvironments below the storm wave base, on the deeper outer-shelf and in inner-shelf basins. Typical examples are the Wissenbach Formation (Eifelian) in the Rhenish Massif, the Daleje Formation (upper Emsian) in the Barrandian area (Czech Republic) or the Khodzha-Kurgan Formation (lower Emsian in the present GSSP sense) in Uzbekistan (central Asia). Contrary to Erben's definition (1962), this concept of the hercynotypic facies excludes some Devonian neritic facies of the Barrandian area, in particular the facies of the upper Koněprusy Limestone.

Rhenish brachiopod biostratigraphy

When brachiopods are used for stratigraphic subdivision and correlation, a number of problems have to be considered. These are aspects of facies and palaeoenvironment, as discussed in the previous section, brachiopod palaeoecology, precise age assignment, differing rates of evolution in separate lineages and palaeobiogeographical differences. The foundations of any reliable biostratigraphy are detailed morphological studies and a modern taxonomy considering all morphological characters, the intraspecific variation and phylogenetic relationships. In recent years, the author has revised a number of Rhenish brachiopods, compared them with related species from other areas and used them in biostratigraphy (Jansen 2001*a*, *b*, 2014*a*; Schemm-Gregory & Jansen 2005, 2006*a*, *b*, 2007; Jansen *et al.* 2007). Rhenotypic brachiopods have turned out to be excellent index fossils (Figs 2 & 3) on a regional scale here in the Lower Devonian of the Ardenno-Rhenish area, but they are partially useful for subdivision and correlation on a more supra-regional scale as well.

Figure 3 shows a series of 22 spiriferide zones that can well be recognized in the eurhenotypic successions of the Rhenish Massif and the Ardennes. With some restrictions, they can also be used in other Devonian areas of Europe and North Africa. This biozonation is based on the total vertical and lateral ranges of well-known spiriferide taxa that are generally short-lived, abundant and widely distributed. Due to facies-dependent occurrences, the zonal boundaries can locally be rather diachronic. The spiriferide zones are taxon-range zones, but in some cases the transition of one of the guide species or subspecies to its phylogenetic successor is documented in the Rhenish Massif. Since the zones partly overlap, one and the same assemblage may belong to more than one biozone. In comparison with the pelagic biostratigraphic scales, which are

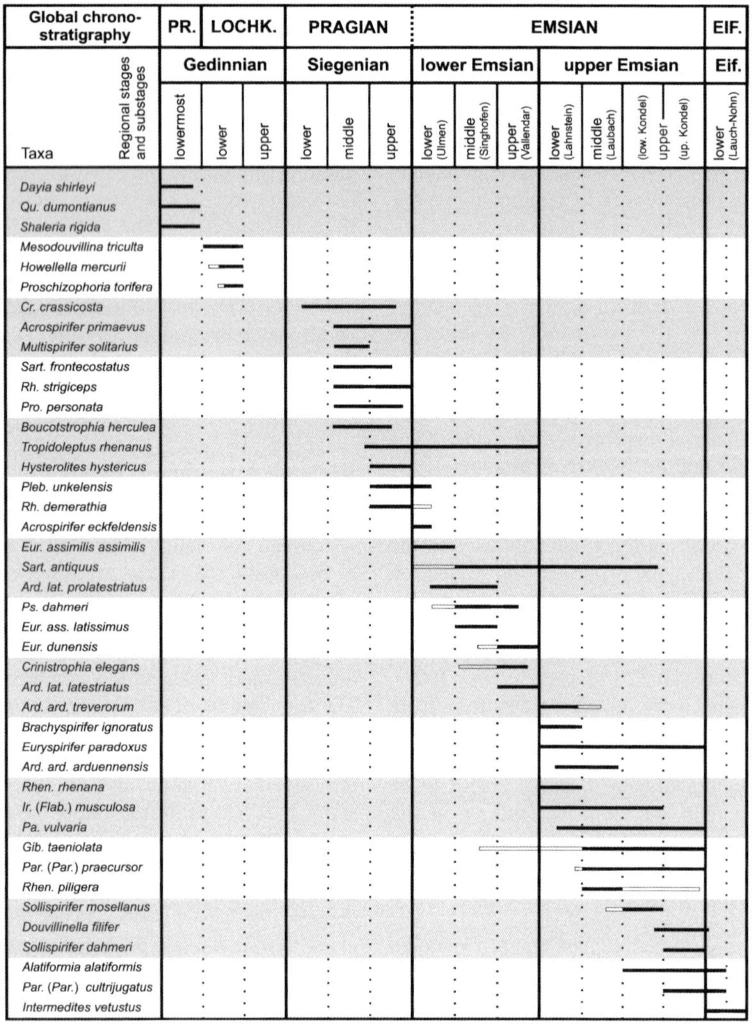

Fig. 2. Ranges of selected brachiopod taxa with stratigraphic significance in the Rhenish Massif. Black bars denote certain ranges; white bars denote uncertain ranges or cf. determinations. The Pragian–Emsian boundary is indicated provisionally (not the present GSSP level). EIF./Eif., Eifelian; LOCHK., Lochkovian; PR., Pridolian; *Ard.*, *Arduspirifer*; *ard.*, *arduennensis*; *ass.*, *assimilis*; *Cr.*, *Crassirensselaeria*; *Eur.*, *Euryspirifer*; *Flab.*, *Flabellistrophia*; *Gib.*, *Gibbodouvillina*; *Ir.*, *Iridistrophia*; *lat.*, *latestriatus*; *Pa.*, *Pachyschizophoria*; *Par.*, *Paraspirifer*; *Pleb.*, *Plebejochonetes*; *Pro.*, *Proschizophoria*; *Ps.*, *Pseudoleptostrophia*; *Qu.*, *Quadrifarius*; *Rh.*, *Rhenorensselaeria*; *Rhen.*, *Rhenostropheodonta*; *Sart.*, *Sartenaerirhynchus*.

often phylogenetic zones of abundant and widely distributed guide fossils (e.g. the Devonian conodont stratigraphy), this biostratigraphy is certainly of inferior quality but the best option available. One may draw comfort from the fact that any recorded range of a brachiopod taxon at least represents a section of a continuous evolutionary development that documents a specific time-rock unit.

Stratigraphically nearly congruent assemblages of other articulate brachiopods co-occur with the zonal index spiriferides. Accordingly, a second subdivision into 'faunal zones' is in preparation to describe a series of successive stratigraphic units on the basis of characteristic brachiopod assemblages.

The Rhenish brachiopod zonation is based on abundant and well-defined species and subspecies, partly appearing in the course of moderate to strong faunal turnovers. The latter reflect changes in the palaeoenvironment, partly major events,

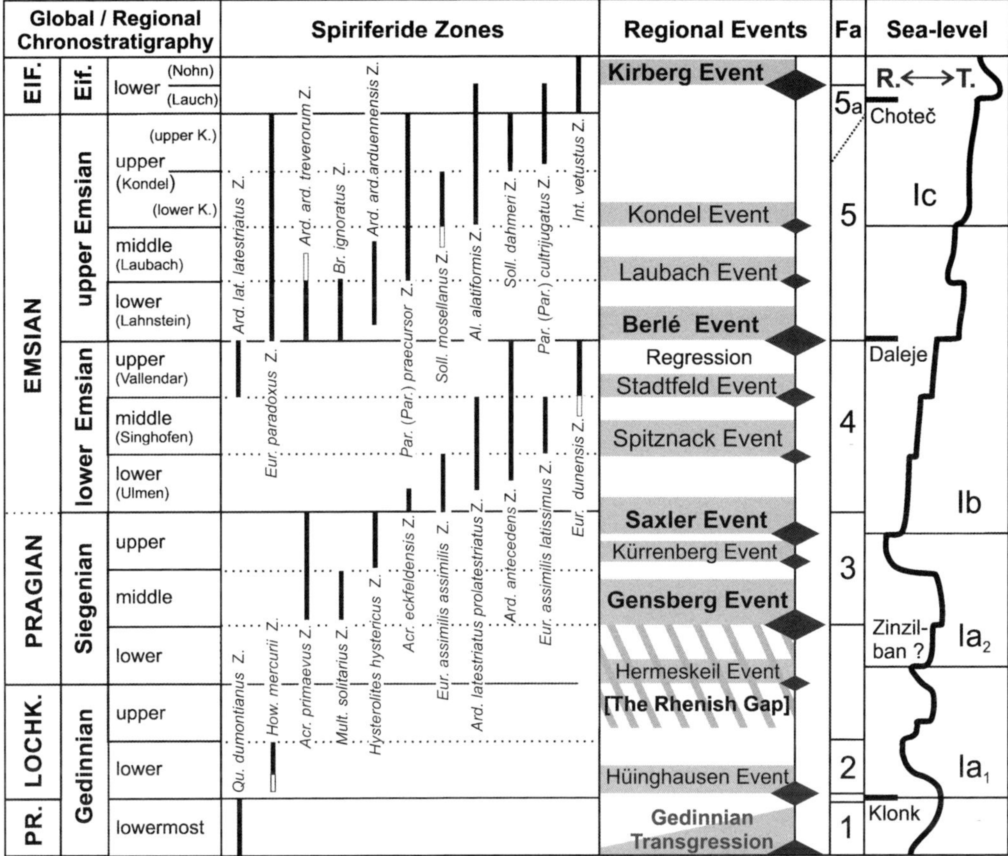

Fig. 3. Spiriferide zones, regional events (partly after Mittmeyer 2008), faunas (Fa) and sea-level fluctuations with position of global events (modified after Johnson *et al.* 1985; Johnson & Sandberg 1989; Walliser 1996, 1998) documented in the Pridolian to lower Eifelian successions of the Rhenish Massif (see text for explanation). Differentiation of T-R cycles Ia_1 and Ia_2 in the sense of Walliser (1998). The Pragian–Emsian boundary is indicated provisionally (not the present GSSP level). Diamonds symbolize the regional events, the size of the diamonds the suggested importance of the events. The extent of the 'Rhenish Gap' *sensu stricto* is indicated. *Acr.*, *Acrospirifer*; *Al.*, *Alatiformia*; *Br.*, *Brachyspirifer*; K., Kondel; R., Regression; T., Transgression; Z., Zone; for further abbreviations, see Figure 2. Faunas: 1, *Quadrifarius dumontianus* Fauna; 2, *Howellella mercurii* Fauna; 3, *Acrospirifer primaevus* Fauna; 4, *Arduspirifer antecedens* Fauna; 5, *Euryspirifer paradoxus* Fauna; 5a, *Paraspirifer cultrijugatus* Fauna.

presumably governed by fluctuations in sea-level, crustal subsidence and the input of terrigenous siliciclastic material plus nutrients. The role of changes in the palaeoclimate is difficult to estimate, but it certainly was an important factor with respect to terrestrial erosion and transport of siliciclastics and nutrients, and the development of the Devonian biota as well.

The Ardenno-Rhenish subdivision into regional Gedinnian, Siegenian and Emsian stages and their subunits (Fig. 3) is used here because the global subdivision, defined in the pelagic facies realm, cannot be applied in the area with satisfactory precision, but respective correlations are of course intended. The Ardenno-Rhenish subdivision has been modified during the last decades but remained on the whole relatively stable (e.g. Solle 1972; Mittmeyer 1974, 1982*a*, 2008; Godefroid 1982; Carls 1987; Jansen 2001*a*). It is not recommended to just replace regionally the old stage names by the current international stages, as practiced in many recent works, because there is no congruence of their boundaries (e.g. in the case of the basal Emsian boundary) or a very imprecise correlation. The lower Emsian has been subdivided into regional lower Emsian Ulmen, Singhofen and Vallendar substages (Mittmeyer

1974), and the upper Emsian into regional Lahnstein, Laubach and Kondel substages (Solle 1972, 1976). This subdivision is useful in the central Rhenish Massif but difficult to apply in other regions, where the use of informal 'lower', 'middle' and 'upper' segments of lower and upper Emsian is recommended. The Ardenno-Rhenish subdivision is largely based on the ranges of brachiopod taxa, supported mainly by trilobite, palynomorph and fish data. This biostratigraphy allows correlations with western and southwestern Europe (e.g. Armorican Massif, Cantabrian Mountains, Iberian Chains) and North Africa (e.g. Dra Plains, Ougarta Chains), where neritic–pelagic correlations are possible (e.g. Carls *et al.* 1972, 2008; Le Menn *et al.* 1976; Carls 1987, 1988; García-Alcalde & Truyóls-Massoni 1994; Jansen 2001*a*; Jansen *et al.* 2007). Resulting possibilities of correlation between the regional Rhenish stratigraphy and the global chronostratigraphy (see Becker *et al.* 2012) are discussed in the next section. The level of the still valid Pragian–Emsian GSSP boundary in Uzbekistan (Yolkin *et al.* 1997) is difficult to recognize outside the stratotype area and much different from previous concepts of both the Pragian–Zlichovian and Siegenian–Emsian boundaries in their classic type regions. The current Pragian–Emsian (GSSP) boundary is therefore not generally accepted, much debated by the scientific community and subject to a current revision (Carls *et al.* 2008; Jansen 2012*c*). Here, the classic basal Emsian boundary in the Rhenish sense is generally used. If the Emsian in the present GSSP sense is meant, it is indicated as such.

Lowermost Gedinnian (Pridolian) to lower Eifelian brachiopod faunas and their significance

Introductory remarks

In this section, the succession of brachiopod faunas from the lowermost Gedinnian (Pridolian) to the lower Eifelian of the Rhenish Massif is briefly described in stratigraphic order. Spiriferide taxon-range zones are introduced. Possibilities of stratigraphic correlation are outlined, mainly with some Ardennish, Ibero-Armorican and Pre-Saharan sections. Attempts are made to calibrate the Rhenish sections with the pelagic biostratigraphy and the global chronostratigraphy.

The faunal development in the Rhenish Massif is interpreted, taking into account the sedimentary sequences, the facies successions and, derived from these, possible changes in the palaeoenvironment. The question is addressed how facies and faunal developments correspond to eustatic sea-level fluctuations, using the global sea-level curve of Johnson *et al.* (1985) and the slightly modified ones of Johnson & Sandberg (1989) and Walliser (1996, 1998). In addition, it is taken into account that varying subsidence and sedimentation rates influenced the relative sea-level and palaeoenvironments on the Rhenish shelf. The significance of regional Rhenish events and possible relationships to known supraregional or global events are discussed.

I still hesitate to conduct a quantitative analysis of the diversity of Rhenish brachiopod faunas, because more taxonomic work is necessary, and the absolute chronological calibration is still very imprecise. Hubert *et al.* (2007) and Zapalski *et al.* (2007) reported on the Devonian species diversity per stage in the Ardennes. It has to be considered that the stages have very different durations, that the durations are strongly dependent on the individual definitions of the stage boundaries and that the state of taxonomic research (for example 'monograph effect') may have a strong influence on the numbers of species – problems that Hubert *et al.* also pointed out in their quantitative account. Irrespective of these restrictions, at least the general trends from the author's present, still-unpublished list of taxa from the Rhenish Massif may be regarded as valid: at least 18 species of articulate brachiopods are present in the lowermost Gedinnian, 16 in the lower Gedinnian, 0 in the upper Gedinnian (a total of *c.* 32 in the Gedinnian), 4 in the lower Siegenian, 42 in the middle Siegenian, 29 in the upper Siegenian (a total of *c.* 44 in the Siegenian), 46 in the lower Emsian and 84 in the upper Emsian (a total of *c.* 122 in the Emsian) – several species range through more than one substage. Some new species are expected after future revisions. Hubert *et al.* (2007) recorded 104 species in the Eifelian of the Ardennes; a decrease follows in the Givetian (55 species), a maximum in the Frasnian (137 species) and a decrease after the Kellwasser Crisis, resulting in a much lower diversity (52 species) in the relatively long Famennian.

Only some of the most important brachiopod taxa from the main stratigraphic units are mentioned in the following description, because complete faunal lists for each formation would have inflated this work enormously. Faunal lists can be obtained from the literature, although many records of taxa are in need of revision. Another restriction applies to the references, which concentrate on recent reviews, such as the articles in the 'Devonian Monograph of Germany' (Weddige, coord. 2008), which quote the vast amount of classic literature. The reference and type specimens of the publications cited below have largely been restudied by the author, so most conclusions are based on direct morphological studies and comparisons of specimens.

As integrative studies on the whole benthic faunas or assemblages, with a quantitative approach and in connection with facies and taphonomy studies, are still lacking, the description of brachiopod communities within the 'Rhenish Complex of Communities' (Boucot 1975) is still postponed. Instead, a series of successive 'faunas' is proposed, intervals with characteristic assemblages conceptually resembling 'ecological-evolutionary faunas' or 'faunal intervals' by previous workers and named by characteristic spiriferide species. These explicitly refer as possible 'natural biomeres' to the fossil faunas and must be regarded separately from 'faunal zones', which represent defined biostratigraphic units.

The geographical positions of some of the most important localities and regions of the Rhenish Massif are shown in Figure 1, and a selection of the most important lithostratigraphic units is presented in Table 1. Typical examples of brachiopod-bearing rocks are shown in Figures 4–6, some characteristic species in Figure 7.

Lowermost Gedinnian brachiopod faunas

Characteristics. The oldest brachiopod-bearing strata of the mid-Palaeozoic succession in the Ardenno-Rhenish region contain the '*Quadrifarius dumontianus* Fauna', which is moderately diverse and includes mainly brachiopods, gastropods, orthoconic cephalopods, pelecypods, a few trilobites and corals (Dahmer 1942, 1951; Richter & Richter 1942, 1954; Boucot 1960; Struve 1973; Himmler 1976). As this oldest Gedinnian interval was apparently not very long, evolutionary change within the brachiopod species has not yet been observed. Typical species are *Qu. dumontianus* (de Koninck, 1876), *Shaleria rigida* (de Koninck, 1876), *Dayia shirleyi* Alvarez & Racheboeuf, 1986, '*Chonetes*' *omalianus* de Koninck, 1876, *Bathyrhyncha sinuosa* (Fuchs, 1923), *Atrypa gedinniana* Fuchs, 1934 and *Platyorthis verneuili* (de Koninck, 1876). After a long debate on the age of this faunal complex, stimulated by increasing correlation data, an 'earliest Gedinnian' or, respectively, Pridolian (latest Silurian) age has become most probable (see discussion below).

Distribution and regional developments. Brachiopod-bearing marine strata of this interval are represented by the argillaceous, partly calcareous Köbbinghausen Formation in the Ebbe and Remscheid anticlinoria. A number of localities expose different levels within this formation (Dahmer 1951; Beyer 1952; Himmler 1976; Clasen 1988; Eiserhardt & Ribbert 2006). The Köbbinghausen Formation in the sense of this work (excluding the 'Ockrige Kalke', which are assigned to the overlying Hüinghausen Formation) contains layers with, for example, well-preserved *Da. shirleyi* or *Qu. dumontianus*. At the famous outcrop in Hüinghausen, the pelagic scyphocrinoids occur (Beyer 1952; Clasen 1988), as confirmed by Raimund Haude (pers. comm. 2014). Arkosic, strongly weathered sandstones of the Grès de Gdoumont (= 'Arkose de Weismes', 'Schichten von Weismes'; within the lower part of the Kalltal Formation of Ribbert 2006) with *Qu. dumontianus* (Fig. 4a) and *Sh. rigida* crop out along the southeastern Venn Anticlinorium, largely in Belgium (Asselberghs 1930, 1943; Dahmer 1942; Richter & Richter 1942, 1954; Boucot 1960; Carls 1971; Godefroid & Cravatte 1999), but also in Germany (Ribbert 2006, p. 36). The *Qu. dumontianus* Fauna occurs in the lower part of the Silberg Formation of the Müsen Horst, in the northern Siegerland (Fuchs 1929; Clausen 1991, 1994; Eiserhardt & Ribbert 2006). Finally, strongly deformed, but identifiable specimens of *Qu. dumontianus* and *Da. shirleyi* are known from the Kellerskopf Formation in the southern Taunus near Wiesbaden (= 'Graue Phyllite'; Dahmer 1946) and *Qu. dumontianus* from Eppenhain ('Eppenhain-Schichten') (see Wirth 1960; Struve 1973; Anderle 2006).

Brachiopod biostratigraphy. After the typical spiriferide species the considered interval is assigned to the *Qu. dumontianus* Zone. A subdivision of this unit may be possible when the precise ranges of the species co-occurring with *Qu. dumontianus* are better known. As this interval represents a stratigraphic equivalent of a lowermost part of the Gedinnian succession (Godefroid & Cravatte 1999; see below) in the classic sense in the Ardennes (Dumont 1848), it is best referred to as 'lowermost Gedinnian' (or, geochronologically, 'earliest Gedinnian') – to distinguish it from the classic lower Gedinnian represented by the younger strata bearing the guide fossil *Howellella mercurii* (Gosselet, 1880) and, possibly, early direct predecessors of this species.

Correlation and age. The age of this interval has been a matter of debate for a long time; the formations mentioned above were either assigned to the upper Silurian or to the lowermost Devonian (Richter & Richter 1937, 1954; Dahmer 1951; Schmidt 1954, 1956; Shirley 1962; Carls 1971; Struve 1973). A Silurian (mid-Ludlowian) age of the Köbbinghausen Formation was early postulated by Richter & Richter (1937, 1954) – based on the putative presence of the classic Silurian guide fossil *Dayia navicula* (Sowerby, 1839) and the trilobite fauna (e.g. *Acaste dayiana* Richter & Richter, 1954). Shirley (1962) pointed out the differences of the Rhenish faunas compared with the ones from the Ludlowian

Fig. 4. Typical examples of brachiopod-bearing rocks, I (collection SMF). Scale bar: 5 cm. (**a**) Weathered sandstone with internal moulds of *Quadrifarius dumontianus* (de Koninck, 1876) (1), SMF 98430; collected by G. Fuchs. Locality: quarry near Gdoumont, Belgium. Stratum: Grès de Gdoumont, *Qu. dumontianus* Zone, lowermost Gedinnian. (**b**) Sandstone with moulds of *Howellella mercurii* (Gosselet, 1880) (1), *Mesodouvillina triculta* (Fuchs, 1919) (2), *Protocortezorthis fornicatimcurvata* (Fuchs, 1919) (3) and other species, eurhenotypic subfacies, SMF 98431; collected by P. Carls. Locality: old railway station of Hüinghausen, Ebbe Anticlinorium, Sauerland. Stratum: 'Flaserschiefer' (= Hardt Member) of the Hüinghausen Formation, *How. mercurii* Zone, lower Gedinnian. (**c**) Sandy shale with external moulds of *Crassirensselaeria crassicosta* (Koch, 1881) (1), pararhenotypic subfacies, SMF 98435; collected by U. Jansen. Locality: abandoned quarry near Rheineck, Rhine river. Stratum: Lower Siegen Group ('untere Schwarzschiefer'), interval of the 'Rhenish Gap', lower Siegenian.

of Britain. He stressed the 'advanced' evolutionary development of the *Dayia* specimens from the Köbbinghausen Formation and the 'Graue Phyllite' (= Kellerskopf Formation) compared with the true *Da. navicula* from the Welsh Borderland. Thus, he recognized that the Köbbinghausen form represents a separate species, *Da. tenuisepta* Shirley, 1962 *nom. nud.* (= *Da. shirleyi* Alvarez & Racheboeuf,

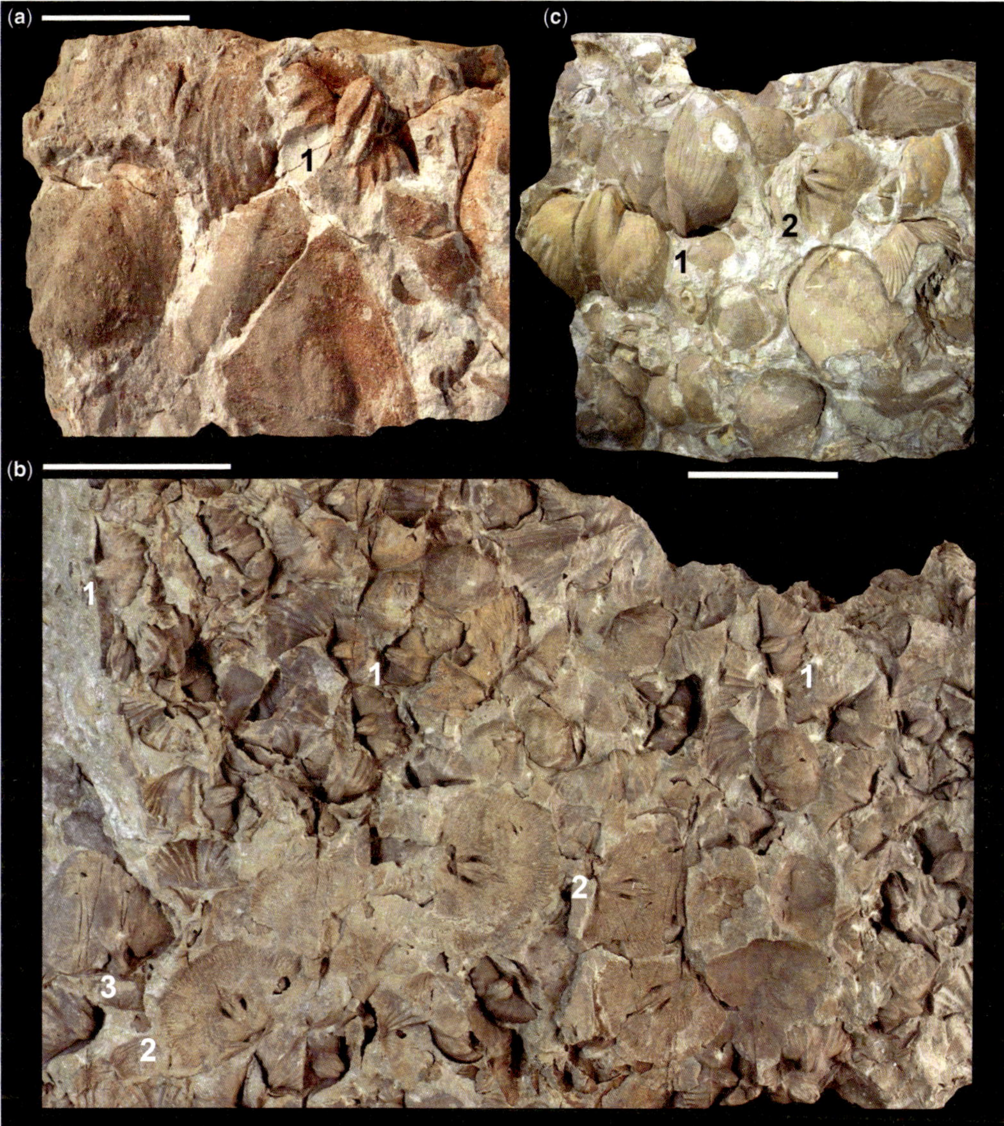

Fig. 5. Typical examples of brachiopod-bearing rocks, II (collection SMF). Scale bar: 5 cm. (**a**) Quartzitic sandstone with *Acrospirifer primaevus* (Steininger, 1853) (1) and large pelecypods, proximal eurhenotypic subfacies, SMF 98432; collected by O. Rose. Locality: Leingipfel near Rüdesheim, Rhine River. Stratum: Taunusquarzit Group, *Acr. primaevus* Zone, middle Siegenian. (**b**) Sandstone with internal moulds of numerous *Arduspirifer antecedens* (Frank, 1898) (1), *Crinistrophia* sp. (2) and *Meganteris ovata ovata* Maurer, 1879 (3), typical eurhenotypic subfacies; SMF 98190; collected by W. Zimmer. Locality: Wernborn near Usingen, Taunus. Stratum: Spitznack Formation, *Ard. antecedens* Zone, middle part of lower Emsian. (**c**) Sandstone with *Paraspirifer* (*Mosellospirifer*) *sandbergeri* Solle, 1971 (1) and *Pachyschizophoria vulvaria* (von Schlotheim, 1820) (2), typical eurhenotypic subfacies, SMF 94421. Locality: Dörrbach-Tal near Koblenz. Stratum: lowermost Laubach Group, middle part of upper Emsian.

1986) and came to the conclusion that the Rhenish *Dayia*-bearing beds had an earliest Devonian age. Based on his own evaluation of the biostratigraphic data and following Shirley (1962), Struve (1973) correlated the 'Graue Phyllite' in the Taunus, containing *Qu. dumontianus* and *Da. shirleyi* (loc. Hubertus-Hütte, Goldstein-Tal), with the Grès de Gdoumont ('Grès de Waismes') and suggested an

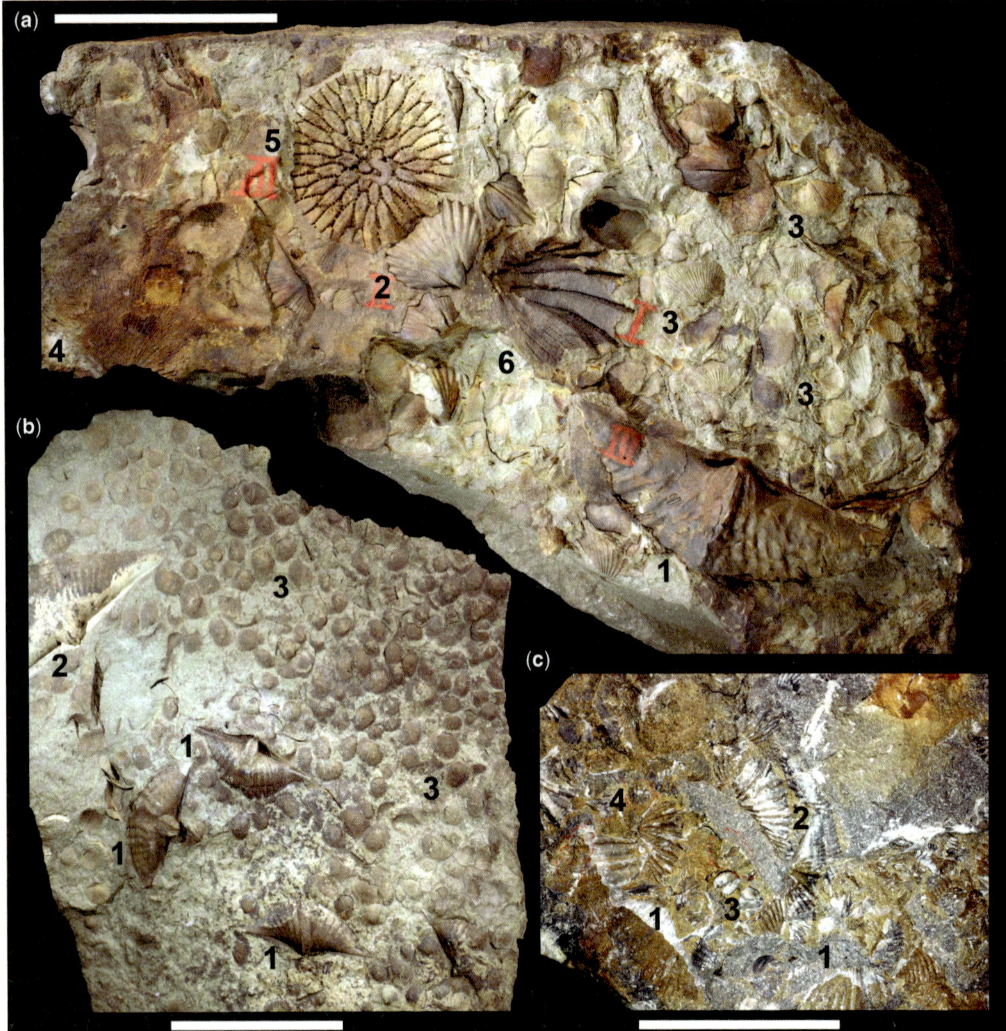

Fig. 6. Typical examples of brachiopod-bearing rocks, III (collection SMF). Scale bar: 5 cm. (**a**) Sandstone with *Euryspirifer dunensis* (Kayser, 1889) (1), *Inaequalibellirostrum inauritum* (Sandberger & Sandberger, 1856) (2), *Plebejochonetes semiradiatus* (Sowerby, 1842) (3), *Iridistrophia maior* (Fuchs, 1915) (4), the tabulate coral *Pleurodictyum problematicum* Goldfuss, 1829 (5) and the pelecypod *Cornellites* sp. (6), typical eurhenotypic subfacies, SMF 98433; collected by Rud. Richter. Locality: Oberstadtfeld, Eifel. Stratum: Stadtfeld Formation, *Eur. dunensis* Zone, upper part of lower Emsian. (**b**) Sandstone with *Arduspirifer arduennensis arduennensis* (Schnur, 1853) (1), *Euryspirifer robustiformis* Mittmeyer, 1972 (2) and masses of *Plebejochonetes semiradiatus* (Sowerby, 1842) (3), eurhenotypic subfacies, SMF 25678; collected by R. Werner 1962. Locality: 60 m east of the road from Niederprüm to Schloßheck, Prüm Syncline, Eifel region. Stratum: Wiltz Formation, *Ard. arduennensis arduennensis* Zone, lower or middle part of upper Emsian. (**c**) Sandy limestone with *Paraspirifer* (*Paraspirifer*) *cultrijugatus* (Roemer, 1844) (1), *Alatiformia alatiformis* (Drevermann, 1907) (2), athyridides (3) and a productellid (4), allorhenotypic subfacies, SMF 98434; collected by R. Werner. Locality: SE of Ellwerath, road from Oberlauch to Ellwerath, Prüm Syncline. Stratum: Lauch Formation, *Par.* (*Par.*) *cultrijugatus* and *Al. alatiformis* zones, lower part of lower Eifelian.

earliest Devonian age of both units, a late early Gedinnian or early late Gedinnian age.

In the Ardennes, strata of the lowermost Gedinnian in the sense here used are represented by two older fossiliferous horizons of the Muno Formation ('Ruisseau des Roches') on the southern flank of the Neufchâteau Synclinorium (see Godefroid 1995; Godefroid & Cravatte 1999; Bultynck *et al.* 2000).

These strata contain essentially the same species as their Rhenish correlatives, and it can be concluded that they have the same geological age. The older fossiliferous horizon of the Ruisseau des Roches contains both *Qu. dumontianus* and *Da. shirleyi*, the upper fossiliferous horizon *Qu. dumontianus* but apparently not *Da. shirleyi*.

Some more recent correlations have shed new light on the age assignment of the *Qu. dumontianus* Fauna. In principle, the age of this interval in terms of the global geochronological scale can be assessed by correlations based on three brachiopod species: *Sh. rigida*, *Qu. dumontianus* and *Da. shirleyi*. Each of these species belongs to a genus actually pointing to a Silurian age. Well-preserved specimens of *Sh. rigida* and *Da. shirleyi* occur in borehole sections of the lower Angres Member of the Noulettes Formation in the Artois region (northern France; Alvarez & Racheboeuf 1986; Brice *et al.* 1986; Jahnke 1986; Racheboeuf & Babin 1986), restricted to a succession with the chitinozoan *Urnochitina urna* (Eisenack, 1934), suggesting a Pridolian age (Paris 1986). The same age shows the presence of the conodont '*Ozarkodina*' *eosteinhornensis* (Walliser, 1964) from approximately the same level (Bultynck 1986; Racheboeuf & Babin 1986). In a Podolian section (area of Dniester River, southwestern Ukraine), the species *Dayia bohemica* Bouček, 1941, which is closely related to the Rhenish *Da. shirleyi*, occurs in nodular limestone layers of the Dzenyhorod Formation (Skala Horizon), 1.3–1.6 m below the Silurian–Devonian (S/D) boundary marked by a shale intercalation with the first occurrence of the basal Lochkovian index graptolite *Monograptus uniformis angustidens* Přibyl, 1940 and other Lochkovian fossils (Baliński 2012). '*Ozarkodina*' *eosteinhornensis* disappears just below the S/D boundary here, 40 cm below the boundary appears *Zieglerodina remscheidensis* (Ziegler, 1960) and 60 cm above the boundary the first representative of the icriodontids, *Caudicriodus hesperius* (Klapper & Murphy, 1975) (see Drygant & Szaniawski 2012; Racki *et al.* 2012). In addition, *Quadrifarius magnus* (Kozłowski, 1929), comparable to *Qu. dumontianus*, occurs little below the S/D boundary in Podolian sections (see Nikiforova 1977; Nikiforova *et al.* 1985). *Dayia bohemica* is also a typical constituent of the uppermost beds of the Pridolian Series in Bohemia (Czech Republic), where the species is associated with stratigraphically significant scyphocrinoids in the type region of the S/D GSSP boundary (Chlupáč 1977; Havlíček & Štorch 1990). In the Ebbe Anticlinorium, the Köbbinghausen Formation at Hüinghausen contains *Da. shirleyi* and again scyphocrinoids (Beyer 1952), showing the S/D boundary interval. From all these arguments it is concluded that the *Qu. dumontianus* Fauna and the lowermost Gedinnian in the Ardenno-Rhenish Massif are at least largely of late Pridolian age.

Still debated is the age of the fauna from the Grès de Gdoumont exposed at the southeastern flank of the Venn Anticlinorium, which contains *Qu. dumontianus* and *Sh. rigida* but lacks representatives of the highly significant Silurian genus *Dayia*. The unit has been assigned to the lower part of the Lower Devonian (either lower or upper Gedinnian) for a long time (e.g. Asselberghs 1930, 1943; Richter & Richter 1942, 1954; Schmidt 1954, 1956, 1959; Boucot 1960, table 1; Shirley 1962; Struve 1973; Godefroid 1982; Meyer 2013), but also to the Silurian (Ludlowian?: Dahmer 1951; Pridolian: Carls 1971; Jahnke 1986; Godefroid & Cravatte 1999). The poor trilobite fauna, in particular '*Asteropyge*' *gdoumontensis* Asselberghs, 1930, may show an evolutionary development pleading for a Devonian age (Richter & Richter 1942, p. 164; 1954, pp. 52–53), but it does not really exclude a Silurian age either. The taxon is poorly known and has to be revised (Bignon *et al.* 2014). In the age discussion of the Grès de Gdoumont, ichthyostratigraphic data were considered: Schmidt (1954, 1959) described agnathans from a similar stratigraphic position, but from non-marine beds of the northwestern flank of the Venn Anticlinorium, and dated these as late Gedinnian based on the species *Belgicaspis crouchi* (Lankester, 1868). As a result of his lithostratigraphic evaluation, Schmidt (1956, p. 52) correlated the fossiliferous part of the Grès de Gdoumont and the agnathan-bearing beds he studied and regarded them as coeval. Later studies supported Wo. Schmidt's results with respect to the Early Devonian age of the agnathans: according to Blieck *et al.* (1995), *B. crouchi* is indicative of a mid-Lochkovian age. Hance *et al.* (1992) assigned the lower part of the Marteau Formation, a stratigraphic equivalent of the agnathan-bearing succession on the northwestern flank, to the Lochkovian Interval Zone M of the spore stratigraphy. After the detailed revision of the Gedinnian lithostratigraphy around the Venn Anticlinorium by Neumann-Mahlkau (1970), however, a lower position of the Grès de Gdoumont in the SE compared with the agnathan-bearing beds in the NW appears rather possible (Neumann-Mahlkau 1970, pp. 352–353, fig. 30). Accordingly, the Gdoumont brachiopod fauna can well be distinctly older than the agnathans, in particular when considering the high possibility of the presence of hiatuses and lateral facies changes in these very shallow-marine to terrestrial sequences.

The biostratigraphic arguments based upon the brachiopods listed above are overwhelming; the arguments in favour of a Pridolian (latest Silurian) age of the Grès de Gdoumont with its marine fauna (Carls 1971; Jahnke 1986; Godefroid & Cravatte

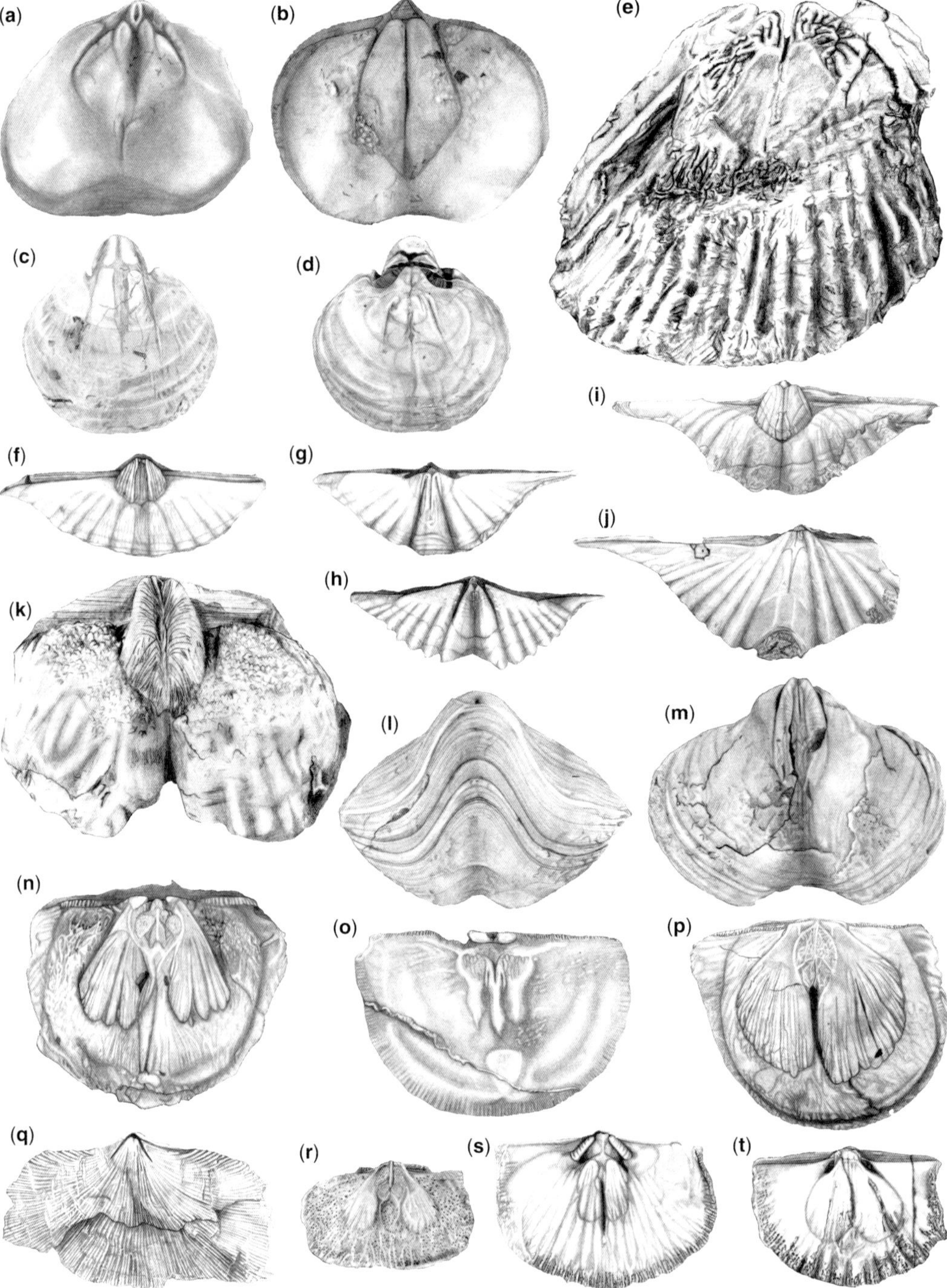

Fig. 7. Selected brachiopod species from the Gedinnian to lower part of lower Eifelian in the Rhenish Massif (Germany) to show some typical examples of the rhenotypic facies; natural size, if not indicated otherwise (drawings by Claudia Groth). (**a**) *Pachyschizophoria* sp. nov. C *sensu* Jansen, 2001*a*, internal mould of dorsal valve, SMF 59650. Locality: Winseler near Wiltz, West Eifel region. Stratum: Berlé Formation, lowermost part of upper Emsian. (**b**) The same species, internal mould of ventral valve, SMF 66976. Locality: road from Daleiden to Olmscheid, West Eifel region. Stratum: Berlé Formation, lowermost part of upper Emsian.

1999) are more convincing. This interval can best be correlated with the upper fossiliferous horizon of the 'Ruisseau des Roches' (Godefroid & Cravatte 1999). Although there is hardly any definite argument from the brachiopods to assume an earliest Devonian age, it cannot be totally excluded that upper parts of the *Qu. dumontianus* Zone reach or even just cross the S/D boundary where the facies remains suitable. An argument of a relatively young age of the Gdoumont Fauna could be the absence of *Da. shirleyi*, but this may also have facies reasons; the presence of the genus *Platyorthis* and terebratulides, such as *Mutationella barroisi* (Asselberghs, 1930) and *Podolella* sp., could be used as arguments in favour of a Devonian age. It is, however, improbable that *Qu. dumontianus* and the typical Silurian taxa associated with it extend far into the Lochkovian, because nowhere has an overlap been recorded with the range of the succeeding, lower Lochkovian species *Howellella mercurii* (Gosselet, 1880) or with the range of any other characteristic species of the *How. mercurii* Zone.

Palaeoenvironment, sea-level and events. The beds with the *Qu. dumontianus* Fauna document the first marine transgression of the middle Palaeozoic succession. Along the southeastern flank of the Venn Anticlinorium, the base of the Kalltal Formation shows a transgression on Ordovician rocks (Salm Group), starting with a basal conglomerate, which is followed by the Grès de Gdoumont or equivalents of that unit (Ribbert 2006). In general, the fossiliferous sequences of Pridolian age in the Rhenish Massif seem to overlie much older Palaeozoic rocks, but the lower contacts are often tectonically disturbed, and/or the directly underlying successions are still incompletely known or unknown in some areas (Anderle 2006, 2008; Eiserhardt & Ribbert 2006; Ribbert 2006).

The development of the Pridolian sea-level differs in various regions; a sea-level rise termed 'Basal Přídolian Event' (Chlupáč & Kukal 1988) and subsequent fall has been described from Bohemia (Kříž 1998) and a long-lasting step-by-step regression in the east Baltic (see discussion in

Fig. 7. (*Continued*) (c, d) *Cryptonella macrorhyncha* (Schnur, 1853), internal mould of articulated valves, ventral (**c**) and dorsal (**d**) views; Schnur collection, Bonn. Locality: Daleiden, West Eifel region. Stratum: Wiltz Formation, *Arduspirifer arduennensis arduennensis* Zone, lower or middle part of upper Emsian. (**e**) *Dinapophysia papilio* (Krantz, 1857), internal mould of dorsal valve, Maurer collection, Darmstadt, Mr 195. Locality: Seifen, Westerwald. Stratum: Seifen Formation, *Multispirifer solitarius* and *Acrospirifer primaevus* zones, middle Siegenian. (**f**) *Sollispirifer mosellanus* (Solle, 1953), internal mould of ventral valve, holotype SMF 93219.1a. Locality: Brodenbach-Tal, western slope of the valley, Lower Mosel region. Stratum: Flaserschiefer Formation, lower Kondel Group, *Soll. mosellanus* Zone, upper part of upper Emsian. (**g**) The same species, internal mould of dorsal valve, paratype SMF 93216a. Locality: Dennkopf, Kondelwald region, Olkenbach Syncline. Stratum: lower part of Flaserschiefer Formation, *Soll. mosellanus* Zone, lower Kondel Group, upper part of upper Emsian. (**h**) *Alatiformia janseni* Gad, 2002, internal mould of ventral valve, SMF 65331; collected by O. Vogel. Locality: field near Eveshausen, Hunsrück, Mosel Synclinorium. Stratum: Flaserschiefer Formation, lower Kondel Group, *Soll. mosellanus* Zone, upper part of upper Emsian. (**i**) *Arduspirifer arduennensis treverorum* Schemm-Gregory & Jansen, 2005, internal mould of ventral valve, holotype SMF 66025; collected by B. Gräßle. Locality: Feldberg near Oberlahnstein, field 'auf dem Mann', central Middle Rhine region. Stratum: Emsquarzit Formation, *Ard. arduennensis treverorum* Zone, lowermost part of upper Emsian. (**j**) The same subspecies, internal mould of dorsal valve, paratype SMF 93000a; collected by G. Dahmer. Locality: Feldberg near Oberlahnstein, 'Schwarzes Kreuz', central Middle Rhine region. Stratum: Hohenrhein Formation, *Ard. arduennensis treverorum* Zone, lower part of upper Emsian. (**k**) *Paraspirifer* (*Mosellospirifer*) *sandbergeri* Solle, 1971, internal mould of ventral valve (×0.8), holotype SMF 93350; collected by J. Hefter. Locality: Dörrbach-Tal near Koblenz. Stratum: Laubach Group, middle part of upper Emsian. (l, m) *Rhenothyris aequabilis tumida* Struve, 1970*b*, internal mould of articulated valves with remains of the shell in anterior (**l**) and ventral (**m**) views, holotype SMF 29336; collected by W. Struve 1949. Locality: southern slope of the Kirberg, near Üxheim, Eifel region. Stratum: Kirberg Member, lower part of Nohn Formation, lower Eifelian. (**n**) *Rhenostropheodonta piligera* (Sandberger & Sandberger, 1856), internal mould of ventral valve, lectotype Museum Wiesbaden, Sandberger collection, No. 288a. Locality: 'Lahnstein', Middle Rhine region. Stratum: most probably Laubach Group, upper Emsian. (**o**) *Rhen. rhenana* Jansen, 2014*a*, latex cast of internal mould of dorsal valve, paratype SMF 93653.2; collected by F. Drevermann. Locality: Miellen, Lahn River, central Middle Rhine region. Stratum: Hohenrhein Formation, *Brachyspirifer ignoratus* Zone, lower part of upper Emsian. (**p**) The same species, internal mould of ventral valve, holotype SMF 66712. Locality, collector and stratum same as before. (**q**) *Iridistrophia* (*Ir.*) *euzona* (Fuchs, 1919), internal mould of ventral valve, SMF 93792; collected by Rud. Richter 1937. Locality: abandoned railway station of Hüinghausen, Ebbe Anticlinorium, Sauerland. Stratum: Hüinghausen Formation, *Howellella mercurii* Zone, lower Gedinnian. (**r**) *Teichostrophia lepis subtilis* Struve, 1992, internal mould of ventral valve (×1.6), holotype SMF 50159; collected by I. & W. Struve & H.-O. Nürnberg 1968. Locality: Honigseifen, Salmerwald Syncline, Eifel region. Stratum: Lauch Formation, *Alatiformia alatiformis* Zone, lower part of lower Eifelian. (**s**) *Tropidoleptus rhenanus* Frech, 1897, internal mould of dorsal valve, SMF 66517a; collected by Rud. Richter 1914. Locality: Oberstadtfeld, Eifel region. Stratum: Stadtfeld Formation, upper part of lower Emsian. (**t**) The same species, internal mould of ventral valve, SMF 85846; collected by J. Mauz 1931. Locality: Niederstadtfeld, Eifel region. Stratum: (middle? part of) lower Emsian.

Kaljo *et al.* 2012, p. 176, and references therein). Based on a multiproxy approach, a transgression in the late Pridolian time in combination with a tendency towards eutrophication and oxygen deficiency and continued flooding with a carbonate crisis and cooling in the earliest Devonian has been suggested in a Podolian section (Racki *et al.* 2012). A transgressive interval is also assumed for the Ardenno-Rhenish area where shales with scyphocrinoids and orthoconic nautiloids occur in the Köbbinghausen Formation of the Ebbe Anticlinorium (Beyer 1952), indicating an open-marine palaeoenvironment. However, the evaluation of the boundary sequence is largely hampered by tectonics and poor outcrops there. From the occurrence of grey, finely organodetrital and micritic platy Lochkovian limestones, Walliser (1996) deduced a transgressive trend at the S/D boundary in the Barrandian, the Carnic Alps, Sardinia and the Moroccan Meseta. Because a transgressive trend had also been reported from sections of the Rhenohercynian and Saxothuringian zones of central Europe, Australia, southwestern Siberia and Podolia, he suggested a worldwide 'Silurian/Devonian Boundary Event' (S/D-Event). According to Walliser, this event was not very drastic but rather a minor, relatively gradual and globally traceable one that marked the onset of the Devonian transgressive–regressive (T-R) cycle Ia_1, as defined by Walliser (1998). It is regarded as synonymous with the 'Klonk Event' in the sense of Jeppsson (1998).

The onset of the marine succession with the *Qu. dumontianus* Fauna is probably related to a latest Silurian eustatic sea-level rise but may also be influenced by increased subsidence in particular areas where the sea could flood the continental shelf. This transgression preceded the event at the S/D boundary, and it could represent an earlier transgressive pulse of the same superordinate transgressive phase.

The basal formations of the Gedinnian in the Ardennes have regionally different ages, as evident from brachiopod and palynology data (Streel *et al.* 1987; Steemans 1989; Godefroid & Cravatte 1999; Bultynck *et al.* 2000). The diachronic onlap of marine strata is documented by the onset of the first brachiopod faunas in Gedinnian units: they belong either to the upper Pridolian *Qu. dumontianus* Zone (lower Muno Formation, Grès de Gdoumont) or the subsequent lower Lochkovian *How. mercurii* Zone (Mondrepuis Formation). In some areas, both zones are present in the same succession (southern Neufchâteau Synclinorium, Ebbe Anticlinorium, Remscheid Anticlinorium). A proceeding transgression on a bathymetrically differentiated shelf may be assumed (Godefroid & Cravatte 1999), but a regional erosion of Pridolian units and synsedimentary tectonics may have complicated the depositional history.

The facies of the rocks representing the *Qu. dumontianus* Zone reflects a shallow-marine palaeoenvironment. The coarsely ribbed spiriferide *Qu. dumontianus* may have settled preferentially in very shallow, agitated water ('*Quadrifarius* Community' *sensu* Boucot 1975), whereas *Da. shirleyi* preferred quiet water in probably more offshore, subtidal settings ('*Dayia* Community') of the deeper shelf (Timm 1981*b*, p. 156). *Dayia shirleyi* is very common in the shales of the Köbbinghausen Formation and is accompanied there by orthoconic nautiloids showing the pelagic influence. Representatives of *Quadrifarius* and *Dayia* co-occur in the lower fossiliferous horizon of the Muno Formation ('lower fauna of the Ruisseau des Roches'; Godefroid & Cravatte 1999) and in the Kellerskopf Formation of the Goldstein-Tal near Wiesbaden (Dahmer 1946; Struve 1973). The fossiliferous beds in the Goldstein-Tal have been interpreted as pro-delta deposits (Hahn 1990). It is still unclear, which factors led to the extinction of the *Qu. dumontianus* Fauna near the S/D boundary, but negative environmental effects in connection with the transgressive S/D-Event or the correlative Klonk Event may be assumed.

Lower Gedinnian brachiopod faunas

Characteristics. The lower Gedinnian (lower Lochkovian, lowermost Devonian) brachiopod assemblages of the Rhenish Massif include geologically young representatives of ancient genera that are also known from the Silurian, for example the species *Howellella mercurii* (Gosselet, 1880), *Mesodouvillina triculta* (Fuchs, 1919) and *Protocortezorthis fornicatimcurvata* (Fuchs, 1919). On the other hand, *Iridistrophia* (*Ir.*) *euzona* (Fuchs, 1919), *Proschizophoria torifera* (Fuchs, 1919), *Platyorthis verneuili* (de Koninck, 1876) and *Cyrtina utrimquesulcata* Fuchs, 1919 are early representatives of essentially Devonian genera. After the most abundant spiriferide species, the faunal complex represented by the lower Gedinnian assemblages is termed '*Howellella mercurii* Fauna'. The macrofossil content is dominated by brachiopods, which are accompanied by some pelecypods, gastropods, orthoconic cephalopods, ostracods and stratigraphically important trilobites (*Warburgella*, *Acastella*).

Distribution and regional developments. Fossiliferous successions of early Gedinnian age are represented by the Hüinghausen Formation of the Ebbe and Remscheid anticlinoria and by the overlying Bredeneck Formation in the Ebbe Anticlinorium. From here, rich and abundant marine assemblages with brachiopods have been described by Fuchs (1919), Dahmer (1951) and Richter & Richter (1954). In the Ebbe Anticlinorium, the stratigraphically important species *How. mercurii* is very common and typically developed in both formations mentioned.

Within the lower part of the Hüinghausen Formation, the 'Ockrige Kalke', masses of *Prot.* cf. *fornicatimcurvata* ('*Orbicularis*-Pflaster' *sensu* Dahmer 1951) and some specimens of *Mesodouvillina* cf. *triculta* (Fuchs, 1919) and *Atrypa westfalica* Dahmer, 1951 occur (it is proposed to replace the misleading and ambiguous term 'Ockrige Kalke' by the neutral term 'Immecke Member', following Böger (1983), who regarded this unit as a separate formation of the 'Hüinghausen Group'; the terms 'Kieselgallen-Schiefer' and 'Flaserschiefer', referring to the overlying subunits, may also be replaced; see below). It is still not clear whether *How. mercurii* or a related (sub)species is present in this unit. Particularly diverse are the faunal assemblages in the argillaceous and arenaceous, flaser-bedded shales of the 'Flaserschiefer' (= Hardt Member; cf. Böger 1983) of the higher Hüinghausen Formation, with bedding surfaces covered with abundant moulds of brachiopods (Fig. 4b), such as *Mes. triculta*, *How. mercurii*, *Ir.* (*Ir.*) *euzona* and *Prot. fornicatimcurvata*; the fauna of this subunit includes 15–20 species of articulate brachiopods. The overlying Bredeneck Formation is composed of shales and sandstone beds with slightly impoverished assemblages, including, e.g., many specimens of *How. mercurii* and *Pro. torifera*. This unit has been included in the lower Gedinnian by most workers (e.g. Richter & Richter 1954; Boucot 1960; Shirley 1962; Ziegler 1970; Eiserhardt *et al.* 1981; Koch *et al.* 1990), but a partial late Gedinnian age has also been regarded as possible (Timm *et al.* 1981), and the formation also has been assigned to the basal upper Gedinnian, based on a redefinition of the lower–upper Gedinnian boundary by trilobites (Carls 1985, fig. 4; 1987, pp. 95, 98–99). The present author would rather stick to the Rhenish tradition and regard all strata with *How. mercurii* as lower Gedinnian. A younger age of upper, poorly fossiliferous parts of the formation would still be possible. The *How. mercurii* Fauna occurs also in the Remscheid Anticlinorium (Dahmer 1951; Clasen 1988). An additional small occurrence of marine lower Gedinnian, with poorly preserved representatives of *How. mercurii*, is documented in the upper part of the Silberg Formation of the Müsen Horst, in the northern Siegerland (Clausen 1991, 1994). There are more sedimentary rocks of probable early Gedinnian age, but these are largely unfossiliferous and of terrestrial, limnic-fluvial or lacustrine origin, for example parts of the Bunte Schiefer ('Variegated Shales') Formation in the Taunus (Hahn 1990; Hahn & Zankl 1991; Anderle 2008) or parts of the Kalltal Formation along the flanks of the Venn Anticlinorium (Ribbert 2008). Locally, rocks of early or late Gedinnian age have yielded fish remains (e.g. Schmidt 1954, 1958, 1959). The precise ages of these strata are often difficult to ascertain; they introduce the 'Rhenish Gap' (see below).

Brachiopod biostratigraphy. The marine successions of the lower Gedinnian are largely represented by the *How. mercurii* Zone, as this spiriferide species is present in most of the Ardenno-Rhenish assemblages of this age. These share very few brachiopod species with lowermost Gedinnian faunas (still to be verified) and no species with the Rhenish Siegenian faunas. The upper boundary of the biozone is facies controlled. A biostratigraphic subdivision of the lower Gedinnian based on brachiopods appears possible in the future after more detailed morphological studies are undertaken.

The precise onset of *How. mercurii* is still unclear. According to Carls (1985, fig. 4), the species first occurs in the middle part of the Hüinghausen Formation, in the 'Kieselgallen-Schiefer' (= Himmelmert Member; cf. Böger 1983). As Dahmer (1951) reported '*Spirifer* (*Delthyris*) *elevatus* Dalman, 1828' (= *How. mercurii*; not Dalman's species) already from the underlying 'Ockrige Kalke' (= Immecke Member), it is possible that this species or an early predecessor occurs there, but this unit could still be unassigned in terms of the present spiriferide stratigraphy. The brachiopods from the 'Ockrige Kalke' – as far as known to the author – show differences to the congeneric ones from the 'Flaserschiefer' (= Hardt Member), which may be ecologically controlled. The species concerned are still to be studied in greater detail.

Correlation and age. Howellella mercurii and closely related species of the same genus have a good potential for supraregional correlation, because they are widespread in western Europe and North Africa (Carls 1985; Gourvennec 1985, 1989; Carls *et al.* 1993).

In the type region of the Gedinnian Stage, in the Ardennes, lower Gedinnian marine strata with the *How. mercurii* Fauna are represented by the Mondrepuis Formation exposed on the southern flank of the Dinant Synclinorium and by the upper part of the Muno Formation (Parensart Horizon) exposed on the southern flank of the Neufchâteau Synclinorium (e.g. Leriche 1912; Asselberghs 1930; Boucot 1960; Godefroid & Cravatte 1999). The localities in the Ardennes yielded essentially the same brachiopod species as the outcrops in the Rhenish lower Gedinnian. However, the marine documentation apparently does not reach the high stratigraphic level of the succession in the Ebbe Anticlinorium, because the trilobites are just represented by the ancient species *Acastella heberti* (Gosselet, 1888) (cf. Richter & Richter 1954; Gandl 1972; Carls 1987, p. 95). The Naux Limestone, intercalated at the base of the Gedinnian Levrezy Formation, a possible lateral equivalent of the Mondrepuis Formation (Asselberghs 1946, pp. 44–45) at the southeastern border of the Rocroi Massif, contains conodonts of the lower Lochkovian *Caudicriodus woschmidti* Zone (Bender 1967; Borremans & Bultynck 1986; Bultynck *et al.* 2000).

The *How. mercurii* Zone can be correlated with borehole sections in the Artois area, where the closely related and at least approximately coeval species *How. angustiplicata* (Kozłowski, 1929) as well as *Mes. triculta* have been reported from a sequence containing the chitinozoan species *Eisenackitina bohemica* (Eisenack, 1934) (see Gourvennec 1986; Jahnke 1986; Paris 1986; Racheboeuf & Babin 1986). In Bohemia and other areas, this species marks the beginning of the Lochkovian, and it is restricted to this stage (Paris *et al.* 2000).

This correlation is supported by conodonts from the 'Ockrige Kalke' of the Remscheid Anticlinorium, for example *Zieglerodina remscheidensis* (Ziegler, 1960) and *Caudicriodus woschmidti* (Ziegler, 1960) (see Ziegler 1960), indicating the lower Lochkovian (Carls *et al.* 2007), but the first species has been reported to first occur in the uppermost Silurian (Corradini & Corriga 2012; Drygant & Szaniawski 2012). The mentioned conodonts occur in the same sequence with lower Gedinnian brachiopods, such as *How.* ex gr. *mercurii* and *Mes. triculta* (see Clasen 1988), and stratigraphically important trilobites, such as *Warburgella rugulosa rhenana* Alberti, 1962 and *Acastella elsana* (Richter & Richter, 1954) (see Alberti 1962; Clasen 1988). In the Ebbe Anticlinorium, the same trilobite taxa (Richter & Richter 1954; Timm 1981*a*) and the advanced species *Ac. tiro* (Richter & Richter, 1954) and *Paracryphaeus ebbae* (Richter & Richter, 1954) occur in the succession of the Hüinghausen Formation, the latter reaching upwards into the Bredeneck Formation, while lower Lochkovian strata in the Bohemian type region have yielded *War. rugulosa rugosa* (Bouček, 1934) (see Chlupáč 1977; Chlupáč & Hladil 2000) and the Lochkovian of the Artois region *Ac. tiro* (see Morzadec 1986). According to Timm (1981*a*, table 1), the onset of *War. rugulosa rhenana* within the 'Ockrige Kalke' (= Immecke Member) could be regarded as a good approximation to the S/D boundary in the Ebbe Anticlinorium.

In the Podolian sections along the valleys of the Dniester River and its distributaries (southwestern Ukraine), the presence of *How. angustiplicata* in the Khudykivtsi (Tajna beds) and Mytkiv (= Mitkov) formations (Nikiforova *et al.* 1985) suggests an early Gedinnian age of these units, the co-occurring species *Mesodouvillina subinterstrialis* (Kozłowski, 1929) is very similar to *Mes. triculta* from the Hüinghausen Formation and probably even synonymous (Jahnke 1986), and *Iridistrophia* (*Ir.*) *praeumbracula* (Kozłowski, 1929) closely resembles the Rhenish *Ir.* (*Ir.*) *euzona*. The interval with the mentioned brachiopod species falls into the lower Lochkovian *Caudicriodus hesperius* to *C. transiens* conodont zones (Drygant & Szaniawski 2012), following above the S/D boundary, which is recognized here also by the occurrence of the basal Devonian index graptolite subspecies *Monograptus uniformis angustidens* Přibyl, 1940 (see Nikiforova 1977; Baliński 2012). The *C. hesperius* Zone marks the basal Lochkovian in the Bohemian type region (Slavík *et al.* 2012). *Iridistrophia* (*Ir.*) *umbella* (Barrande, 1848) from the Kotýs Limestone (Lochkovian) of Bohemia (Havlíček 1967; Mergl 2003) resembles the Gedinnian species *Ir.* (*Ir.*) *euzona*.

It can be concluded, that the *How. mercurii* Zone is of early Lochkovian age, and its lower boundary may serve as an approximation to the S/D boundary, although the onset of *How. mercurii* could be slightly higher.

Palaeoenvironment, sea-level and events. The aftermath of a modest sea-level rise in the context of the S/D-Event (Walliser 1985, 1996) or the Klonk Event (Jeppsson 1998) and recurrent suitable palaeoenvironments during a subsequent regressive period may have led to the onset of the rhenotypic *How. mercurii* Fauna in the Ebbe and Remscheid anticlinoria. It is proposed to describe the appearance of lower Gedinnian biota (which may still not include *How. mercurii*) in the 'Ockrige Kalke' (= Immecke Member) of the lower Hüinghausen Formation as regional 'Hüinghausen Event'. The first rich, clearly eurhenotypic assemblages of the 'Flaserschiefer' (= Hardt Member) in the higher Hüinghausen Formation document a shallow-water, open-marine shelf palaeoenvironment with strong siliciclastic input. A general regressive trend is visible in the succession from the Hüinghausen to the overlying Bredeneck and Bunte Ebbe formations. The global sea-level curve (Walliser 1996, 1998; Fig. 3) records a falling sea-level during the early Lochkovian (early Gedinnian) part of T-R cycle Ia_1, which would be consistent with the facies development in the Rhenish Massif. A slightly higher sea-level followed in the mid- to late (not latest) Lochkovian and, finally a regression occurred in the latest Lochkovian and earliest Pragian times, correlating with middle and late parts of T-R cycle Ia_1; this curve does not correspond to the facies development in the Rhenish area. Nevertheless, the overall regressive trend in the northern Rhenish Massif is probably primarily caused by an increased input of siliciclastic material by deltas from the Old Red Continent, which compensated the subsidence and may have masked effects of not very strong eustatic sea-level changes as well. Field data support this assumption: Bredeneck and Bunte Ebbe formations together, for example, reach a thickness of more than 1000 m (Ziegler 1970; Eiserhardt *et al.* 1981). The Bredeneck Formation still contains brachiopod assemblages, but with reduced diversity, including a number of specimens of *How. mercurii* and *Pro. torifera*. Interpreting the facies development within the Bredeneck Formation, a diachronous change from shallow-marine to very shallow-marine, intertidal, coastal and deltaic deposits with lagoonal and brackish conditions has been suggested (Eiserhardt *et al.* 1981; Timm 1981*b*). The regressive tendency continues with the overlying Bunte Ebbe Formation (Böger 1981; Eiserhardt *et al.* 1981; Timm 1981*b*). In the Müsen Horst, the transition from the Silberg Formation to the Müsen Formation suggests a similar development from an open-marine to a deltaic palaeoenvironment (Böger 1981; Clausen 1991, 1994; Thünker 2008). These transitions mark the beginning of the 'Rhenish Gap' *sensu stricto*, referring to a long-term interval of non-marine, restricted-marine, tidal and deltaic conditions.

Upper Gedinnian to lower Siegenian: the 'Rhenish Gap'

Characteristics. The successions between the fossiliferous levels of the lower Gedinnian and the middle Siegenian have yielded only few marine fossils in the Ardenno-Rhenish Massif. Early agnathan fishes and land plants occur at some localities (fishes: e.g. Schmidt 1959; Blieck & Jahnke 1980; Blieck *et al.* 1995; land plants: e.g. Schweitzer 1983; Meyer & Stets 1996, p. 27). The upper Gedinnian sequences are largely developed in non-marine facies, whereas the lower Siegenian beds can be developed in the marine pararhenotypic subfacies with occurrences of the terebratulide *Crassirensselaeria crassicosta* (Koch, 1881) and lingulides. Very rare, short-term, open-marine intercalations of the proximal eurhenotypic subfacies may contain species of other brachiopod orders, such as spiriferides. For the stratigraphic interval from the upper Gedinnian to the lower Siegenian in the Ardenno-Rhenish Massif, the term 'Rhenish Gap' *sensu stricto* is proposed, i.e. a gap of normal marine sedimentation. It can regionally start at an older level or end at a younger level ('Rhenish Gap' *sensu lato*), and the precise stratigraphic range may be difficult to ascertain. This gap does not represent a substantial lithic break – although hiatuses may be present – but essentially refers to a break in the normal marine record, here a lack of the eurhenotypic subfacies. Thus it is no contradiction that the 'Rhenish Gap' corresponds to very thick, poorly fossiliferous siliciclastic successions. It is comparable with Struve's 'Great Gap' ('Great Gap in the Record of Marine Devonian'; Struve 1982*b*; 1990, pp. 261–262) of the Middle Devonian, which refers to incomplete successions and more carbonatic strata documenting lagoonal or not full-marine palaeoenvironments and is supposed to have a much wider geographical distribution. The 'Rhenish Gap' is essentially present in the Ardenno-Rhenish area, but may extend to Poland in the east and south England in the west. The 'Rhenish Gap' *sensu stricto* represents a considerable time span of approximately 6–8 myr (Carls & Valenzuela-Ríos 1998: 7 myr; and estimated after the timescales of Weddige 2003; Kaufmann 2006; Becker *et al.* 2012), having the duration of a whole stage; regionally, the 'Rhenish Gap' *sensu lato* can last even longer.

Distribution and regional developments. Successions of the 'Rhenish Gap' are represented by the Bunte Schiefer Formation (Gedinnian) and the Hermeskeil Formation (lower Siegenian) in the Taunus (Hahn 1990; Hahn & Zankl 1991; Anderle 2000); the same formations in the Hunsrück (Meyer 1970); the Müsen Formation (upper Gedinnian) and the Lower Siegen Group (lower Siegenian) in the Siegerland (Pilger 1954; Pilger & Schmidt 1959; Thünker 2001, 2008); the Lower Siegen Group in the Westerwald (Meyer & Pahl 1960; Poschmann & Jansen 2003) and the East Eifel region (Meyer 1958; Meyer & Pahl 1960; Mittmeyer 1982*c*; Stets & Schäfer 2002, 2009); the Bunte Ebbe Formation (upper Gedinnian?) and Pasel Formation (Siegenian) of the Sauerland (Böger 1981; Eiserhardt *et al.* 1981; Timm 1981*b*; Langenstrassen 2008); and higher parts of the Kalltal Formation (Gedinnian, lower Siegenian?) of the North Eifel Region (Ribbert 2008).

No brachiopods or other fossils of open-marine environments have been found in the listed Gedinnian strata, and only a few have been found in some units of the lower Siegenian, where an increased marine influence is documented by pararhenotypic or, very rarely, proximal eurhenotypic subfacies. Sandstones and shales with intercalated red beds (mainly in the Gedinnian) and sedimentary features pointing to terrestrial, limnic-fluvial, lacustrine, estuarine, deltaic or marginal marine, lagoonal palaeoenvironments predominate. Remains of early land plants are common; fishes occur both in the Gedinnian (Schmidt 1954, 1958, 1959) and lower Siegenian (Blieck & Jahnke 1980). In the Siegerland and the East Eifel region, the 'Untere Siegener Schichten' (= 'Lower Siegen beds'; 'Tonschiefer-Gruppe' in early works) represent the lower Siegenian. It is here proposed to use the 'supergroup' as the lithostratigraphic category for the Siegen beds and the 'group' for the lower, middle and upper subunits, respectively. The 450–1100 m thick Lower Siegen Group (= 'Untere Siegener Schichten') is built up of shales and sandstones ('Siegen Normal Facies'). The unit has been divided into six subunits of formation rank in the central type area, the 'Siegener Schuppensattel' ('Siegen Slice Anticline') (e.g. Pilger 1954; Thünker 2001, 2008). Commonly, the pararhenotypic subfacies (Fig. 4c) is developed with occurrences of the terebratulide *Cr. crassicosta* (Koch, 1881) plus some linguloids, agnathan fish remains and pelecypods. Thick units of dark, argillaceous shales with remains of early land plants and sandstone beds occur (Meyer 1958, 2013; Meyer & Pahl 1960; Carls *et al.* 1982; Fenchel *et al.* 1985; Meyer & Stets 1996; Thünker 2001; Stets & Schäfer 2002), and there are even early terrestrial arthropods in the Hombach Section in the Westerwald (Poschmann & Jansen 2003). As very rare exceptions, poorly preserved delthyridoid spiriferides have been found in proximal eurhenotypic intercalations within the upper part of the Lower Siegen Group (Pilger & Schmidt 1959) and in the approximately coeval Hermeskeil Formation of the Hunsrück (Meyer 1970), which are too poorly preserved to provide convincing stratigraphic evidence. However, based on an examination by the author, it can be stated that they are clearly separated from the evolutionary stage of the small lower Gedinnian species of *Howellella*, as has previously been assumed by Carls (1987, p. 88). The argillaceous Mayen Formation in the Westerwald has a late Gedinnian to early Siegenian age according to spore associations (Gad 2005), but brachiopods from presumably sandy equivalents in the SE Eifel region indicate a younger age (see below).

In the southern Taunus (Anderle 1987, 2008; Hahn 1990) and at the southeastern margin of the Venn Anticlinorium (Jansen *et al.* 2004; Ribbert 2008), the 'Rhenish Gap' has a wider stratigraphic range ('Rhenish Gap' *sensu lato*) because non-marine strata start above the disappearance of the lowermost Gedinnian *Qu. dumontianus* Fauna. Due to the complicated structural situation, the reconstruction of the sedimentary succession above the Kellerskopf Formation in the southern Taunus is problematic (e.g. Anderle 1987, 2006, 2008). In the Sauerland and the Bergisches Land (e.g. Böger 1981; Eiserhardt *et al.* 1981; Timm 1981*b*; Hilden 2008; Langenstrassen 2008), the 'Rhenish Gap' extends upwards far into the Emsian.

Brachiopod biostratigraphy. The lack of significant marine faunas in strata representing the 'Rhenish Gap' inhibits any detailed biostratigraphic implications from the brachiopods. *Crassirensselaeria crassicosta* and poorly preserved, advanced delthyridoid spiriferides comparable to *Hysterolites hystericus* von Schlotheim, 1820 from the upper part of the Lower Siegen Group (Pilger & Schmidt 1959) and spiriferides comparable to *Acrospirifer primaevus* (Steininger, 1853) from the same unit (Eranil in Meyer 2013, p. 39) and the Hermeskeil Formation (Meyer 1970) suggest a post-Gedinnian (or post-Lochkovian) age – the brachiopods show similarities to middle and upper Siegenian spiriferides. By definition, the Lower Siegen Group should typify the lower Siegenian, but a biostratigraphic delimitation is hardly possible in the type region. Above, the onset of marine strata with eurhenotypic faunas ('Gensberg Event') marks the lower boundary of the middle Siegenian.

Correlation and age. The scarcity of marine faunas in the whole Ardenno-Rhenish area during the interval of the 'Rhenish Gap' resulted in many correlation problems in the past, as the extent of this break in the marine record was underestimated by many early workers (see discussions in Carls *et al.* 1982; Carls 1987, 2000; Carls & Valenzuela-Ríos 1998; Jansen 2001*a*), but Allan (1935, p. 46) and Solle (1953, p. 13; 1963, pp. 190–192) at least recognized the problem. The poorly fossiliferous Oignies and Saint-Hubert formations represent the classic upper Gedinnian along the southern flank of the Dinant Synclinorium in the Ardennes (Asselberghs 1946; Godefroid 1982); the overlying Mirwart Formation ('Grès d'Anor') – except its uppermost fossiliferous part – is regarded as a correlative of the Lower Siegen Group ('Untere Siegener Schichten'; Asselberghs 1936). As the comparable successions in the Rhenish Massif, this complex is underlain by strata of the *Howellella mercurii* Zone and overlain by strata of the *Acr. primaevus* Zone. Approximately correlating formations with red and green rock colours in the Rhenish Massif were often compared lithologically with the 'Schistes d'Oignies', for example the 'Bunte Schiefer' in the Taunus (Anderle 1987) and the 'Bunte-Ebbe-Schichten' in the Sauerland (Spriestersbach 1925, p. 393; 1942, p. 40), and a similar age was assumed. The use of palynomorphs (e.g. Streel *et al.* 1987; Steemans 1989) and fish remains (Blieck *et al.* 1995) for correlation has been attempted, but the present database is still punctual. Based on palynomorphs, the top of the Gedinnian in the Ardennes could be correlated via sections in Brittany (northwestern France) approximately with the Lochkovian–Pragian boundary in Bohemia, while the (almost) basal lower Siegenian in the Siegerland (within the Gilberg Formation) turned out to be younger, since a younger spore zone (Streel *et al.* 1987; Steemans 1989: Pragian 'Pa alpha' Zone) and the *Rhinopteraspis dunensis* Agnathan Zone (Blieck & Jahnke 1980; Blieck *et al.* 1995) could be verified there. Thus, when only regarding the type strata, there would possibly be an interval, which can be neither assigned to the Gedinnian nor to the Siegenian. However, the basal boundary of the Siegen Supergroup is not exposed at the surface in the Siegerland, and the sequence of the Gilberg Formation showed a considerable hidden thickness only visible in underground outcrops (Fenchel *et al.* 1985, p. 28). The Lower Siegen Group is apparently underlain by the Müsen Formation, which is assigned to the upper Gedinnian by lithological comparisons with strata in other areas (Fenchel *et al.* 1985) and spore data (determined by P. Steemans; in Clausen 1991, pp. 32–33). The spores from upper parts of the Martinshardt Member (= the upper part of the Müsen Formation) suggest that this unit may already reach the Siegenian. Considering the correlation data and the eustatic sea-level development (see below), it cannot be excluded that the base of the Siegenian succession in its classic type region falls into the lower Pragian.

In the Holy Cross Mountains (Poland), a similar lack of marine documentation and a hiatus in middle parts of the Lower Devonian follow above marine successions of Gedinnian age (Racki & Turnau 2000), suggesting a continuation of the 'Rhenish Gap' to the east, but a precise correlation is still not possible. In south England, the Dartmouth Group may represent an approximately coeval, poorly fossiliferous siliciclastic succession in similar facies, which has also yielded agnathans; one brachiopod assemblage (from a high level?) indicates a mid- or late Siegenian age (Evans 1981) of this unit, which is overlain by the more open-marine Meadfoot Group containing brachiopod assemblages of mid- or late Siegenian to Emsian age (House 1975; Evans 1985; Leveridge 2011, pp. 627–628, 630).

In contrast, the stratigraphic interval of the 'Rhenish Gap' is represented in western France, Spain and North Africa (e.g. Morocco) by marine successions with rich brachiopod faunas, including series of transitional forms between early Gedinnian and mid-Siegenian species, for example between *Howellella* species and representatives of *Hysterolites*, *Euryspirifer*, *Acrospirifer* and *Vandercammenina* (e.g. Carls *et al.* 1982, 1993; Carls 1987, 1988; Gourvennec 1989; García-Alcalde 1996; Jansen 2001*a*; Jansen *et al.* 2007; Schemm-Gregory 2011). It is possible to bridge the gap in Ardenno-Rhenish marine biostratigraphy in these successions.

Due to the unfavourable terrestrial to deltaic conditions, the base of the Siegenian cannot be defined biostratigraphically by marine fossils in its type region, the Siegerland mining area. Therefore, Carls (1987) proposed a redefined Gedinnian–Siegenian boundary in continuously marine successions of Spain and western France based on the first occurrence of the oldest species of the Siegenian genus *Vandercammenina*, i.e. the species *Van. sollei* Carls, 1985. Correlation with the upper Lochkovian and Pragian of the Bohemian stratigraphy is possible there by means of conodonts, chitinozoans and dacryoconarides. The boundaries Gedinnian–Siegenian (in that sense) and Lochkovian–Pragian turned out to be approximately synchronous (Carls 1987; García-Alcalde *et al.* 1990; Morzadec *et al.* 1991; Jansen *et al.* 2007). The interval of the 'Rhenish Gap' *sensu stricto* starts about in the middle of the Lochkovian and ends in wide parts of the Ardenno-Rhenish region with the base of the middle Siegenian, which is still difficult to correlate on a global scale (Jansen *et al.* 2007; Carls *et al.* 2008); this level correlates with a level in the middle Pragian in the traditional or possible future GSSP sense, and it may be close to the Pragian–Emsian boundary in the present GSSP sense.

Palaeoenvironment, sea-level and events. The successions of the 'Rhenish Gap' of normal marine documentation are interpreted as deposits of terrestrial, limnic-fluvial, lacustrine, brackish-marine, lagoonal, deltaic and intertidal palaeoenvironments (e.g. Schmidt 1956; Böger 1981; Hahn 1990; Clausen 1991; Stets & Schäfer 2002, 2009; Poschmann & Jansen 2003; Anderle 2008; Langenstrassen 2008; Ribbert 2008). In general, the sedimentary rocks assigned to the upper Gedinnian (parts of Bunte Schiefer Formation, Bunte Ebbe Formation, Müsen Formation) are interpreted as largely non-marine, while marine influences seem to regionally increase with the beginning of the lower Siegenian (Hermeskeil Formation, Lower Siegen Group). The succession of the Lower Siegen Group in the Siegerland was deposited in a limnic-fluvial to deltaic-marine palaeoenvironment (Stets & Schäfer 2002, 2009), where episodic marine influences are indicated by few respective fossil assemblages of the pararhenotypic or, very rarely, proximal eurhenotypic subfacies. The Hermeskeil Formation is referred to lagoonal, tidally influenced palaeoenvironments (Hahn 1990; Anderle 2000).

In terms of the global sea-level curve, the interval of the 'Rhenish Gap' approximately correlates with a mid- to late (not latest) Lochkovian highstand during the corresponding parts of T-R cycle Ia_1 (Walliser 1998; ≈'pre-Ia' *sensu* Johnson & Sandberg 1989), a latest Lochkovian to earliest Pragian regressive period during the latest T-R cycle Ia_1, including the 'Lochkovian–Pragian Boundary Event' (Chlupáč & Kukal 1986, 1988; Walliser 1996), and again a rise during the early Pragian or early T-R cycle Ia_2 (Walliser 1998; ≈'Ia' *sensu* Johnson *et al.* 1985; Johnson & Sandberg 1989). Compared with the later Early Devonian and Mid-Devonian eustatic sea-level, the Lochkovian to early Pragian sea-level was probably as a whole not very high and affected by relatively minor fluctuations only. High rates of sediment accumulation on the Rhenish Shelf may have compensated the higher global sea-level during the mid- and late Lochkovian. The slight transgressive tendency starting with the onsets of the Hermeskeil Formation and the Lower Siegen Group would not fit well with the latest Lochkovian–earliest Pragian regressive tendency in the global sea-level curve. However, it is possible that the correlation with the global timescale is still too imprecise, so the increasing marine influence in the Rhenish area may correlate with the following transgressive trend within the early Pragian, starting with T-R cycle Ia_2. For this increasing marine influence in the Rhenish region, Mittmeyer (2008) introduced the local term 'Hermeskeil Event'. Along the flank of the Venn Anticlinorium, the 'Grenz-Sandstein', a coarse-grained sandstone unit at the base of the Monschau Formation, could represent the same level, but this remains speculative, as biostratigraphic data are still lacking (Richter 1979; Jansen *et al.* 2004; Ribbert 2008).

Middle and upper Siegenian brachiopod faunas

Characteristics. The middle Siegenian starts in wide areas of the Rhenish Massif with the onset of marine, eurhenotypic subfacies characterized by rich benthic faunas, including brachiopods, pelecypods, trilobites, ostracods, corals, tentaculitides and echinoderms (e.g. Drevermann 1904; Quiring 1923; Dahmer 1934; Paproth 1960; Poschmann & Jansen 2003). These faunas appear after the 'Rhenish Gap' *sensu stricto*.

The following brachiopod species are particularly abundant and typical in the middle Siegenian: an early morphotype of *Leptostrophiella explanata* (Sowerby, 1842), *Boucotstrophia herculea* (Drevermann, 1904), *Gigastropheodonta gigas* (McCoy, 1852), *Fascistropheodonta sedgwicki* (d'Archiac & de Verneuil, 1842), *Plicostropheodonta* spp. ex gr. *murchisoni* (d'Archiac & de Verneuil, 1842), '*Orthotetes*' *ingens* Drevermann, 1904, *Platyorthis circularis taunica* (Fuchs, 1915), *Proschizophoria personata* (Zeiler, 1857), *Rhenoschizophoria provulvaria* (Maurer, 1886), *Dinapophysia papilio* (Krantz, 1857) (Fig. 7e), *Alatiformia affinis* (Fuchs, 1909), *Mauispirifer gosseleti* (Béclard, 1887), *Vandercammenina rhenana* (Kegel, 1913), *Acr. primaevus* (Steininger, 1853), '*Athyris*' *avirostris* (Krantz, 1857), *Nucleospira maillieuxi* Dahmer, 1931, *Septathyris aliena* (Drevermann, 1904) and *Cryptonella minor* Dahmer, 1931. The diversity is high at some localities; for example, about 40 articulate brachiopod species occur at the locality Seifen (Westerwald). The common but still poorly known species '*Uncinulus*' *frontecostatus* Drevermann, 1902 is assigned herein to the new genus *Sartenaerirhynchus* (see the later section on systematic palaeontology), which ranges with at least two species from the middle Siegenian to the upper Emsian in the Rhenish Massif and the Ardennes, where it may have a good stratigraphic potential.

Most of the middle Siegenian brachiopod species range into the upper Siegenian, but often with a change in abundance. The following taxa occur in both substages: *Lept. explanata*, *Pla. circularis taunica*, *Pro. personata*, *Rhe. provulvaria*, *Vandercammenina* cf. *trigeri* (de Verneuil, 1850), *Van. rhenana*, *Acr. primaevus*, '*At.*' *avirostris* and *Rhenorensselaeria strigiceps* (Roemer, 1844). Faunal differences between middle and upper Siegenian are largely ascribed to a distinct facies change (Poschmann & Jansen 2003).

The faunas from the middle Siegenian in its type region include strophomenides and athyridides as very common elements, whereas chonetoids and rhenorensselaeriid terebratulides are rare. In the upper Siegenian, this relationship is reversed; chonetoids and rhenorensselaeriid terebratulides are the dominant components, and chonetoids may cover bedding planes. Characteristic and very common species are *Pla. circularis taunica*, *Pro. personata*, *Rhe. provulvaria*, *Plebejochonetes unkelensis* (Dahmer, 1936*b*), *Hysterolites hystericus* von Schlotheim, 1820, *Acr. primaevus*, *Rh. strigiceps*, *Rh. demerathia* Simpson, 1940, *Crassirensselaeria crassicosta* (Koch, 1881) and *Tropidoleptus rhenanus* Frech, 1897. The species diversity is generally lower than in the middle Siegenian but can also be high at a few localities (e.g. Unkel, Westerwald; Aegidienberg, Siebengebirge; Saxler, Eifel). In particular, in the Siegerland, Westerwald and East Eifel region, the upper Siegenian is often developed in marine, proximal eurhenotypic to pararhenotypic subfacies, whereas mainly in the Sauerland it can also be developed in non-marine, limnic-fluvial or deltaic facies. Where the pararhenotypic subfacies is developed, *Cr. crassicosta* may occur in great abundance.

In spite of the general impoverishment of the fauna, some upper Siegenian assemblages still share many species with the middle Siegenian ones. It is proposed to summarize this unit as the '*Acrospirifer primaevus* Fauna'. The majority of its species disappear in highest parts of the upper Siegenian, for example *Pro. personata*, *Al. affinis*, *Acr. primaevus*, '*At.*' *avirostris* and *Cr. crassicosta*. The complex of middle and upper Siegenian faunas is readily distinguishable from the fauna of the lower Emsian.

Distribution and regional developments. Apart from the Siegerland mining area, the type region of the

Siegenian Stage (e.g. Quiring 1923; Paproth 1960; Lusznat 1968; Carls *et al.* 1982), brachiopod-bearing strata of the middle and upper Siegenian are widely distributed in the Westerwald (e.g. Drevermann 1904; Dahmer 1931, 1934, 1936*b*; Jansen 1994; Poschmann & Jansen 2003), Eifel region (Wunstorf 1932; Dahmer 1935, 1937; Simpson 1940; Mittmeyer 1982*c*, 1997, 2008), Bergisches Land (Jux 1981; Hilden 2008), southern Taunus (e.g. Rose 1936; Jentsch & Röder 1957) and southern Hunsrück (e.g. Rose 1936; Kutscher 1937, 1940, 1952; Meyer 1970). The middle Siegenian succession contains diverse brachiopod faunas, especially in the Eisernhardt Formation (Siegerland), the Seifen and Augustenthal formations (Westerwald), the Ramersbach Formation (East Eifel region) and the Monschau Formation (North Eifel region). The lithology of these formations is partly strikingly similar and characterized by flaser-bedded sandstones ('Rauhflaser' in the older German literature). Richly fossiliferous, eurhenotypic units of the upper Siegenian are, for example, parts of the Obersdorf (Siegerland), Unkel (Rhine-Westerwald), and Saxler and Nitztal (southern Central Eifel to East Eifel regions) formations.

In the Siegerland, the Middle and Upper Siegen groups ('Mittlere' and 'Obere Siegener Schichten' or, respectively, 'Rauhflaser-Schichten' and 'Herdorf-Schichten') are again developed in the 'Siegen Normal Facies', very thick sandstone and shale successions (Bauer *et al.* 1960; Lusznat 1968; Carls *et al.* 1982; Fenchel *et al.* 1985; Meyer & Stets 1996; Thünker 2001, 2008; Stets & Schäfer 2002). The Middle Siegen Group has a thickness of 1000–1200 m and the Upper Siegen Group of 1200–2000 m (Fenchel *et al.* 1985). A finer lithostratigraphic subdivision based on 'Leitschichten-Partien', smaller units of 50–150 m thickness defined by general lithofacial characters, was applied to these successions for the purpose of mapping (e.g. Pilger 1952, 1954; Bauer *et al.* 1960), but strong lateral and vertical facies changes and recurrences made it difficult to trace the units in the field (e.g. Lusznat 1968, pp. 16–17). As a result of detailed mapping, including a survey of the structural inventory, a number of formations were distinguished within the Middle and Upper Siegen groups (e.g. Bauer *et al.* 1960; Lusznat 1968; Fenchel *et al.* 1985; Thünker 2001, 2008). The eurhenotypic faunas are restricted to particular levels and scattered localities in this succession (e.g. Paproth 1960; Lusznat 1968; Thünker 2001).

The lower middle Siegenian Gensberg Fauna from the Eisernhardt Formation (Paproth 1960: 'Hauptgrauwacken-Zone'; Thünker 2001) represents a typical eurhenotypic fauna with remarkably abundant occurrences of *Rhe. provulvaria*, *Pla. circularis taunica* and '*At.*' *avirostris*. The fauna from the locality Hüttenberg near Freudenberg (Paproth 1960; Lusznat 1968) is a second rich fauna from a low level within the middle Siegenian. However, the most diverse is the classic Seifen Fauna in the Westerwald, which comes from a high level within the middle Siegenian (general accounts: Drevermann 1904; Dahmer 1934; Paproth 1960; Jahnke & Michels 1982*a*; Grigo 1993; Müller 2011). Several fossiliferous horizons within the Seifen Formation contain a high diversity of well-preserved brachiopods, altogether almost 40 species (Jansen 1994, 2014*a*; Poschmann & Jansen 2003), that are associated with many pelecypods, trilobites (Basse & Müller 2012), bryozoans, crinoids, edrioasteroids (Müller & Hahn 2010), ostracods (Grigo 1994) and corals (Grigo *et al.* 1992). These typical to distal eurhenotypic assemblages point to relatively clear-water conditions with some distance from the delta in the north. A striking feature is the abundance of large 'strophodontoid' strophomenides (Jansen 1994, 2014*a*). In the East Eifel region, the fauna from the Ramersbach Formation is similar to the Seifen Fauna (Mittmeyer 1982*c*).

The characteristic upper Siegenian assemblages from the Unkel Formation at Unkel on the Rhine river (Westerwald; Dahmer 1936*b*) are typified by bedding surfaces covered with *Pleb. unkelensis*; *Hyst. hystericus*, *Pro. personata* and *Pla. circularis taunica* are abundant as well. Sedimentary features and faunas show a proximal eurhenotypic subfacies. Similar, coeval assemblages are known from a succession with many fossiliferous levels in the Aegidienberg tunnel in the Siebengebirge (Schindler *et al.* 2004; Schemm-Gregory & Jansen 2007).

In the Siegerland and the East Eifel region north of the Siegen Main Thrust, a proximal eurhenotypic subfacies is common in the Upper Siegen Group, indicated by pelecypod–brachiopod assemblages with *Rh. strigiceps*, *Trop. rhenanus* and *Hyst. hystericus* as predominating species. With the uppermost Siegenian Saxler Formation of the Southeast Eifel region, high-diversity, typical eurhenotypic faunas recur, including approximately 30 species of articulate brachiopods (Simpson 1940), for example *Lept. explanata*, *Trop. rhenanus*, *Pleb. unkelensis*, *Pla. circularis taunica*, '*At.*' *avirostris*, *Hyst. hystericus*, an advanced morphotype of *Acr. primaevus*, *Rh. strigiceps* and *Rh. demerathia*.

In the central Rhenish Massif south of the Siegen Main Thrust (mainly SE Eifel, Mosel Valley, southwestern Westerwald; Mosel Synclinorium), the Siegenian and lowermost Emsian sequences are represented by the partly argillaceous and silty but also arenaceous 'Hunsrück Slate in the wider sense' (Mittmeyer 1980; Meyer & Stets 1996), consisting of five formations included in the Wied Group (Elkholy & Gad 2006). To reconstruct the geological development of the central Rhenish

Massif, it is important to know the age of this succession. Based on palynomorphs, its lower part, the Mayen Formation, has been dated as late Gedinnian to early Siegenian by Gad (2005), which is consistent with the opinion of Meyer & Stets (1996) that the formation laterally grades into the Lower Siegen Group in the Westerwald. However, the restudy of the brachiopod fauna from sandstones at the classic locality Berresheim (Simpson 1940; Solle 1950*a*, p. 364; Meyer 1965, p. 39) near Mayen has confirmed that the formation contains *Trop. rhenanus*, *Acr. primaevus* and *Rh. strigiceps*, suggesting a mid- or, more probably, even late Siegenian age. Mittmeyer (1980) first inferred a late Siegenian age of the Mayen Formation, then a late mid-Siegenian (Mittmeyer 1997) age for its upper part near Mayen and finally a mid-Siegenian age for the unit west of the river Rhine (Mittmeyer 2008). Possible explanations for the contradicting palynomorph v. brachiopod data could be tectonic complications or a diachronic distribution of the formation; the present age data are still too scarce.

Eurhenotypic faunas from the Wied Group are commonly restricted to the arenaceous parts of the succession. The most important of these faunas, the middle Siegenian fauna from Augustenthal near Neuwied in the Westerwald (Dahmer 1931), is remarkable because it includes about 35 articulate brachiopod species with unusual composition: abundant athyridides such as *Sept. aliena* and *Nuc. maillieuxi*; a number of specimens of the terebratulides *Cry. minor* and *Meg. ovata*; the exotic ('Bohemian') strophomenide genus *Leptaenopyxis*, abundant *Lept. explanata*, *Bou. herculea*, *Pli.* ex gr. *murchisoni* and *Gig. gigas*; the orthides *Pla. circularis taunica*, *Pro. personata* and *Rhe. provulvaria*; the spiriferides *Mau. gosseleti*, *Cyrtina latesinuata* Dahmer, 1931 and poorly known representatives of *Arduspirifer* and *Euryspirifer*. The brachiopods are associated with bryozoans, ostracods and corals, suggesting a distal, enriched eurhenotypic subfacies. The fauna closely resembles the coeval fauna from the Longlier Formation in the Ardennes (Maillieux 1936*a*: 'Quartzophyllades de Longlier').

In the northern Rhenish Massif, in the Bergisches Land and the Sauerland, strata of mid- to late Siegenian age are developed in the pararhenotypic subfacies or non-marine facies. In the southern Bergisches Land, the Wahnbach Formation contains the famous Wahnbach flora (Kräusel & Weyland 1930; Schweitzer 1983) and the Odenspiel Formation the important early vertebrate faunas from quarries near Overath (Schriel & Gross 1933) and Odenspiel (Walliser & Michels 1983). *Crassirensselaeria crassicosta* is very common in these units of the upper Siegenian and the main articulate brachiopod species. The Wahnbach Formation at the famous locality Unkelmühle in the Sieg-Tal has yielded a proximal eurhenotypic fauna (Dahmer 1936*a*) with numerous specimens of *Acr. primaevus*. Some sequences in the northern Rhenish Massif have tentatively been assigned to the middle and/or upper Siegenian, but are still of uncertain age, as they have not yielded any marine fossils at all, for example the grey sandy shales, arkoses and sandstones of the Pasel Formation (Ziegler 1970) or the dark, reddish-violet and red shales and fine-grained sandstones of the Hammerlott Formation (Böger 1983) in the Ebbe Anticlinorium (Langenstrassen 2008), which are included as parts within the still cryptic 'Bunte Ebbe Gruppe' (Böger 1981, 1983), 'Ebbe-Komplex' (Timm *et al.* 1981) or 'Rimmert-Schichten' (Ziegler *et al.* 1968; Eiserhardt *et al.* 1981).

Along the southeastern margin of the Venn Massif (North Eifel), eurhenotypic brachiopod assemblages of mid-Siegenian age have been found in the Monschau Formation (Wunstorf 1932; Jansen *et al.* 2004); a locality with *Cr. crassicosta* is known from the overlying lower Rurberg Formation (Schmidt 1956, pp. 93, 105); *Hyst. hystericus*, *Cr. crassicosta* and *Rh. strigiceps* were reported from the upper Rurberg Formation (Schmidt & Schröder 1962, p. 26); and, higher in the succession, the presence of *Acr. primaevus* could be confirmed from one locality in the Heimbach Formation (Ribbert 2008, p. 292; material restudied by the present author).

The facies of quartzites and quartzitic sandstones of the Taunusquarzit Group, the middle to upper Siegenian in the southern Taunus and Hunsrück (Kutscher 1937, 1940, 1952; Jentsch & Röder 1957; Meyer 1970; Hahn 1990; Hahn & Zankl 1991; Stets & Schäfer 2002), varies from pararhenotypic with assemblages dominated by *Cr. crassicosta* to proximal eurhenotypic with, for example, abundant *Acr. primaevus* and large pelecypods (Fig. 5a).

Brachiopod biostratigraphy. The eurhenotypic sequences of the middle to upper Siegenian can largely be assigned to the *Acr. primaevus* Zone. As an approximation, the middle Siegenian corresponds with the *Mult. solitarius* Zone and the upper Siegenian with the *Hyst. hystericus* Zone. The phylogeny within the species *Acr. primaevus* still needs to be studied in greater detail, as early and late morphotypes may be more clearly distinguished and could provide a new and phylogeny-based tool to distinguish between middle and upper Siegenian strata. Mittmeyer (2008) subdivided the species into two successive subspecies (*Acr. primaevus seifensis* and *Acr. primaevus primaevus*) but still in a preliminary way. Apart from the guide fossil *Mult. solitarius*, species with biostratigraphic significance for the recognition of the middle Siegenian are, for example, *Bou. herculea*, *Fasc. sedgwicki*, '*Or.*' *ingens*, *Din. papilio* and *Mau. gosseleti*, although these may reach into the upper Siegenian, where they are rare. The upper Siegenian is characterized by *Hyst. hystericus*, *Pleb. unkelensis* and *Rh. demerathia*. Because the latter two species have also been reported from the lower part of the lower Emsian, the co-occurrence with Siegenian guide fossils should be obligatory to verify the late Siegenian age, especially when *Hyst. hystericus* is absent. After all, the higher abundance of *Trop. rhenanus* in the upper Siegenian helps to distinguish between middle and upper Siegenian.

Correlation and age. Biostratigraphic correlations with the successions in the Ardennes are possible. Based on the range of *Mult. solitarius*, the upper part of the Mirwart Formation ('Grès d'Anor') and the Villé Formation ('Grauwacke de Saint-Michel') at the southern flank of the Dinant Synclinorium are correlated with the Rhenish middle Siegenian; representatives of an early morphotype of *Acr. primaevus*, *Mau. gosseleti* and *Fasc. sedgwicki* support this assignment. According to Godefroid *et al.* (1994), *Acr. primaevus* and *Pro. personata* reach into the overlying La Roche Formation and would still prove a Siegenian age of this unit; the latter species was reported up to its upper boundary. The middle–upper Siegenian boundary may be located near the boundary between the Villé and La Roche formations as the typically middle Siegenian species *Mult. solitarius*, *Mau. gosseleti* and *Din. papilio* disappear at this level. Accordingly, the upper Siegenian could be represented by a lower part of the La Roche Formation. However, the lower Emsian species *Arduspirifer antecedens* (Frank, 1898) and *Ard. latestriatus prolatestriatus* Mittmeyer, 1973*b* have been reported from a low level within the La Roche Formation, which do not fit to the occurrences of *Pro. personata* and *Acr. primaevus* here (Godefroid *et al.* 1994); a scrutiny of these taxa may shed new light on this problem. Occurrences of *Mau. gosseleti*, an early morphotype of *Acr. primaevus* and *Pro. personata* suggest a mid-Siegenian age of the Solières Formation ('Grès et Schistes de Solières') on the northern flank of the Dinant Synclinorium (see Maillieux 1931; Godefroid & Stainier 1982; material restudied by the author).

Correlations with sections in the Armorican Massif (France), the Iberian Chains (Spain) and the Dra Plains (Morocco) can also be conducted with the help of brachiopods. Large and strongly biconvex representatives of *Rhenorensselaeria* from the Montguyon Formation of the Armorican Massif (Renaud 1942), the Santa Cruz Formation of the Iberian Chains (Carls & Valenzuela-Ríos 1998) and the Merzâ-Akhsaï Formation of the Dra Plains (Jansen 2001*a*; Schemm-Gregory 2008) are closely related to the Rhenish species *Rh. strigiceps* and point to a mid- to late Siegenian age. Additional taxa support this assignment (see discussions in Carls & Valenzuela-Ríos 1998; Jansen 2001*a*; Jansen *et al.* 2007). *Multispirifer* cf. *solitarius* from the Niguëlla Fauna in the Iberian Chains (Santa Cruz Formation), probably ancestral to *Mult. solitarius*, suggests a slightly older age than the Rhenish middle Siegenian (Carls & Valenzuela-Ríos 1998).

There are no possibilities to correlate the middle or upper Siegenian directly with the global chronostratigraphic scale, but limestones with pelagic guide fossils occur above and below the eurhenotypic, brachiopod-bearing successions of the Merzâ-Akhsaï Formation in the Dra Plains. Icriodontid conodonts, e.g. *Latericriodus steinachensis* (Al-Rawi, 1977) and ?*Caudicriodus curvicauda* (Carls & Gandl, 1969), show a mid-Pragian age (Slavík 2004; Jansen *et al.* 2007) of the basal limestone of this formation; the upper sandstones with middle or upper Siegenian brachiopod faunas are overlain by a limestone sequence at the base of the Mdâouer-el-Kbîr Formation containing pelagic microfossils of early but not earliest Emsian age in the present GSSP sense, such as the conodont *Eocostapolygnathus excavatus* (Carls & Gandl, 1969) (see Jansen *et al.* 2007). A similar succession has been described from the Iberian Chains (Carls 1987, 1988, 1999; Carls *et al.* 2008).

Palaeoenvironment, sea-level and events. The onset of marine conditions at the beginning of the mid-Siegenian is herein referred to as the 'Gensberg Event', named after a classic locality in the Siegerland. This event is documented in the Siegerland, Westerwald, Taunus, Hunsrück, North Eifel, East Eifel, Ardennes and possibly in south England. The presence of this ubiquitous transgression suggests a supraregional (eustatic?) event rather than a regionally restricted one. It is regarded as a working hypothesis for future research, although still rather speculative, that the Gensberg Event coincides with the transgressive 'Zinzilban (*kitabicus*) Event' (Yolkin *et al.* 1994) documented 35 cm below the base of the actual GSSP Emsian defined in Uzbekistan (Yolkin *et al.* 1997) or with a transgressive pulse in the middle part of the Pragian T-R cycle Ia_2 (Walliser 1998; ≈'Ia' *sensu* Johnson *et al.* 1985; Johnson & Sandberg 1989) of the global sea-level curve (Fig. 3). (On the other hand, the Zinzilban Event may even be related to the beginning of T-R cycle Ia_2 ('Ia'; Carls *et al.* 2008, p. 386).) With the Gensberg Event, brachiopod faunas that had evolved outside during the time of the 'Rhenish Gap' could immigrate into the Ardenno-Rhenish shelf region.

The sandstones and shales of the 'Siegen Normal Facies' in the Siegerland mining area represent typical and proximal eurhenotypic to pararhenotypic subfacies types that can be ascribed to a largely shallow-marine to deltaic origin (Paproth 1960; Walliser & Michels 1983; Meyer & Stets 1996; Stets & Schäfer 2002, 2009). The open-marine influence was highest in the mid-Siegenian time (Gensberg and Seifen faunas) and decreased with the beginning of the late Siegenian. The proximal eurhenotypic to pararhenotypic subfacies of quartzitic sandstones and quartzites of the Taunusquarzit Group in the Taunus and Hunsrück originated from a variety of coastal and intertidal to subtidal palaeoenvironments with partly high hydrodynamic energy (Meyer 1970; Hahn 1990; Hahn & Zankl 1991; Stets & Schäfer 2002), leading to frequent destruction of biocoenoses. The water depth increased during the deposition of this unit.

The upper middle Siegenian Seifen Fauna of the Westerwald is characterized by high biodiversity and large brachiopod shells. The rich fossil content indicates partly clear-water conditions of an open-marine shelf and very suitable living conditions for articulate brachiopods. There was probably a second transgressive pulse or just an episode with less or more irregularly distributed input of siliciclastic material (Mittmeyer 2008: 'Seifen Event').

The faunal change near the transition from the mid- to the late Siegenian, observed in the Eifel region and the Siegerland, is interpreted mainly as a result of increased siliciclastic input from land and/or falling sea-level and generally more turbid water under the influence of a prograding delta. Thick sandstone beds within the Nitztal Formation were formed in the East Eifel region with the regional 'Kürrenberg Event' (Mittmeyer 2008). The regressive phase in the Rhenish area would be consistent with the global sea-level curve, the late T-R cycle Ia_2 (Walliser 1998) in the late Pragian in the former sense (late T-R cycle 'Ia': Johnson *et al.* 1985; Johnson & Sandberg 1989; Walliser 1996). The faunal change is reflected by a shift from typical to more proximal eurhenotypic and pararhenotypic subfacies in the Siegerland, Bergisches Land, Westerwald and East Eifel region. It is characterized by a decline of claricolous brachiopods, such as '*At.*' *avirostris*, *Van. rhenana* and *Cry. minor*, and an increase of more turbidicolous ones, such as *Pleb. unkelensis*, *Hyst. hystericus*, *Rh. strigiceps* and *Cr. crassicosta*.

Comparing middle with upper Siegenian faunas, not much phylogenetic change is observable within brachiopod groups. The differences mainly result from differing palaeoecological conditions and resulting changes in the abundances of taxa. This explains why it is difficult to distinguish between the substages outside the type area of the Siegenian, even in the Taunus or the Ardennes.

The pararhenotypic subfacies of the upper Siegenian formations in the Siegerland and Bergisches Land reflects a diversity of deltaic settings with occurrences of *Cr. crassicosta*, pelecypods, fish faunas and early land plants (e.g. Walliser & Michels 1983). An episodically flooded, tidally influenced delta plain with meandering channels and tidal pools is assumed. The Pasel and Hammerlott formations, i.e. the middle Ebbe Group, in the Ebbe Anticlinorium suggest more terrestrial, limnic-fluvial and deltaic palaeoenvironments with little marine influence (Böger 1981, 1983; Timm 1981*b*; Eiserhardt *et al.* 1981) proximal to the coast of the Old Red Continent.

A more offshore, distal and subtidal palaeoenvironment is represented by the pelitic and arenaceous successions of the Siegenian part of the Wied Group ('Hunsrück Slate *sensu lato*') of the Mosel Synclinorium (Elkholy & Gad 2006). The settlement of brachiopods was only locally and temporarily possible, because the sedimentation rates were too high or the substrates too soft and probably underoxygenated. The presence of rare faunas could be related to local structural highs with more stable substrates and reduced sedimentation rates, for example represented by the enriched, distal eurhenotypic fossiliferous horizons of the Augustenthal Formation (Dahmer 1931). Here, the occurrence of numerous ripples with changing orientation indicates still relatively shallow water (Elkholy & Gad 2006).

Finally, the uppermost Siegenian Saxler Formation in the Eifel region with its rich, typical eurhenotypic faunas (Simpson 1940) marks the onset of a new transgressive phase and the spread of open-marine palaeoenvironments initiated by the 'Saxler Event' (Mittmeyer 2008), probably related to the onset of T-R cycle Ib (Johnson *et al.* 1985; Johnson & Sandberg 1989; Walliser 1996). With the rising sea-level in the latest Siegenian and earliest Emsian, the pelitic sedimentation of the Hunsrück Slate Group started in the Hunsrück and southwestern Taunus, and subsequently pelagic faunas could immigrate.

Lower Emsian brachiopod faunas

Characteristics. Highly diverse, eurhenotypic faunas from the Rhenish lower Emsian include characteristic, often brachiopod-dominated, benthic assemblages with pelecypods, trilobites, corals, ostracods, tentaculitides and crinoidal remains as additional constituents (e.g. Drevermann 1902; Fuchs 1915; Mauz 1935; Röder 1960; Fuchs 1974, 1982; Mittmeyer 2008). In spite of some Siegenian holdovers and upper Emsian successors, the lower Emsian brachiopod complex is phylogenetically rather distinct and represents a coherent unit: the

'*Arduspirifer antecedens* Fauna'. The evolutionary change within brachiopod lineages of the lower Emsian is modest, without any big disruptions, and may have taken place mainly within the area of the Ardenno-Rhenish Shelf.

Euryspirifer and *Arduspirifer* belong to the most abundant and stratigraphically important brachiopod genera. Their lower Emsian representatives evolved from Siegenian predecessors and show gradual phylogenetic change. The successive taxa represent stages within continuous developments and can be used as guide fossils (Fig. 3). The lower Emsian representatives of *Arduspirifer* split into the *latestriatus* and *antecedens* lineages (Solle 1953; Schemm-Gregory & Jansen 2005, 2006*a*, *b*), each elevated to species rank and subdivided into subspecies in the ongoing revision of the genus.

Typical and common taxa from the lower Emsian are, for example: *Leptostrophiella explanata* (Sowerby, 1842); *Pseudoleptostrophia dahmeri* (Rösler, 1954); *Crinistrophia elegans* (Drevermann, 1902); *Plicostropheodonta virgata* (Drevermann, 1902); *Iridistrophia* (*Ir.*) *maior* (Fuchs, 1915); *Platyorthis circularis circularis* (Sowerby, 1842); *Rhenoschizophoria provulvaria* (Maurer, 1886) ssp.; *Loreleiella dilatata* (Roemer, 1844); *Plebejochonetes semiradiatus* (Sowerby, 1842); *Inaequalibellirostrum inauritum* (Sandberger & Sandberger, 1856); *Straelenia dunensis* (Drevermann, 1902); *Alatiformia mediorhenana* (Fuchs, 1909); *Torosospirifer crassicosta crassicosta* (Scupin, 1900); *Ard. antecedens* (Frank, 1898); *Ard. latestriatus* (Maurer, 1886), including the subspecies *Ard. latestriatus latestriatus* (Frank, 1898) and *Ard. latestriatus prolatestriatus* Mittmeyer, 1973*b*; *Euryspirifer assimilis* (Fuchs, 1915), including the subspecies *Eur. assimilis assimilis* (Fuchs, 1915) and *Eur. assimilis latissimus* Mittmeyer, 2008; *Eur. dunensis* (Kayser, 1889); '*Athyris*' cf. *undata* (Defrance, 1828); *Meganteris ovata ovata* Maurer, 1879; and *Globithyris confluentina* (Fuchs, 1907). *Tropidoleptus rhenanus* Frech, 1897 has its onset in the Siegenian and extends up to the uppermost parts of the lower Emsian. *Leptostrophiella explanata*, *Pla. circularis*, *Rhe. provulvaria* and *Meg. ovata* first occur in the Siegenian, but their lower Emsian representatives show small differences at the subspecific level. The rhynchonellides *Lapinulus pila* (Schnur, 1851) and *Sartenaerirhynchus antiquus* (Schnur, 1853) first occur in the lower Emsian and reach into the upper Emsian.

Distribution and regional developments. The lower Emsian is well documented, with brachiopod faunas from a number of formations of the Central, West and South Eifel, the southwestern Westerwald, Middle Rhine and Lower Mosel areas, Taunus and Hunsrück. Due to strong lateral and vertical facies changes and regionally extreme thicknesses – for example up to 4000–6000 m in the Lower Mosel region (Meyer & Stets 1996) – a complex lithostratigraphy has been elaborated (e.g. Mittmeyer 2008; Requadt 2008; Meyer 2013).

The most complete and best-documented brachiopod successions are preserved in the southern Central, South and SE Eifel regions (Drevermann 1902; Mauz 1935; Simpson 1940; Röder 1960; Fuchs 1974, 1982; Mittmeyer 2008). The basal boundary of the type Emsian is located near the boundary between the Saxler and Eckfeld formations. Similar to the underlying Saxler Formation, the lowermost lower Emsian Eckfeld Formation contains a highly diverse, eurhenotypic brachiopod fauna (with *c.* 30 articulate brachiopod species; Röder 1960), and the overlying Reudelsterz Formation contains an impoverished, proximal eurhenotypic to pararhenotypic fauna (<10 articulate brachiopod species), whereas the younger, middle lower Emsian Gefell and Neichnerberg formations contain more eurhenotypic faunas again, but still with relatively low diversity (cf. Fuchs 1982). The density of fossil localities here is generally not very high. Highly abundant and diverse, typical eurhenotypic faunas of the upper lower Emsian Stadtfeld Formation (*sensu* Mittmeyer 2008; Gladbach Formation of Fuchs 1971, 1982) mark a peak in the marine development of the succession in the southern Central Eifel (Drevermann 1902; Mauz 1935; Fuchs 1982). A series of separate levels is known from the vicinity of Oberstadtfeld with a number of high-diversity assemblages (with ≈40 species of articulate brachiopods at a single locality), including, for example, abundant specimens of *Pli. virgata*, *Crin. elegans*, *Ir.* (*Ir.*) *maior*, *Lor. dilatata*, *Pleb. semiradiatus*, *Pla. circularis circularis*, *Trop. rhenanus*, *Str. dunensis*, *Eur. dunensis*, *Ard. antecedens* ssp., *Tor. crassicosta crassicosta*, '*At.*' cf. *undata*, *Cry. rhenana* and *Meg. ovata ovata*. The brachiopods are accompanied by the whole spectrum of other eurhenotypic fossils. Many fossils from older collections are labelled with the locality names 'Stadtfeld' or 'Oberstadtfeld' but come from the upper lower Emsian sandstones of the 'Humerich' (Weltersberg Member, Stadtfeld Formation), a hill 1.7 km west of Oberstadtfeld. Others are from many different forgotten localities in the vicinity and strata representing a substantial part of the lower Emsian succession. (Most of these localities are indicated on the labels with ancient local names that are hardly known and not indicated on official maps but could be reconstructed by older notes of G. Fuchs and others and also from old literature, e.g. localities 'Röder', 'Lauzert', 'Kahlenberg', 'Strich', 'Achheld', 'Unter-Hedscheid', 'Heckerwiese' and 'Vorschußberg'.)

In the southwestern Westerwald near Vallendar, similar assemblages are known from localities in the upper lower Emsian Rittersturz Formation (lower Vallendar Group; Wenndorf 1999), which yielded many well-preserved specimens of *Ard. latestriatus latestriatus* (collected by the present author). The underlying Bendorf Formation contains abundant *Eur. assimilis latissimus* and *Ard. latestriatus prolatestriatus* in outcrops in the Brexbach-Tal near Bendorf (Mittmeyer 2008), indicating the middle lower Emsian.

The uppermost part of the Rhenish lower Emsian is widely characterized by regressive trends documented from the predominantly pararhenotypic, partly proximal eurhenotypic subfacies or terrestrial facies of the Klerf and Nellenköpfchen formations (Eifel, Lower Mosel and central Middle Rhine regions: e.g. Mauz 1935; Solle 1956*a*, *b*; Wunderlich 1970; Reineck 1983; Schultka & Remy 1990; Wenndorf 1999; Stets & Schäfer 2002; Wehrmann *et al.* 2005; Franke 2006). An exception is the typical eurhenotypic fauna from the Nellenköpfchen Formation at the locality 'Kretzers Mühle' in the Westerwald (Mauz 1935; Mittmeyer 2008, p. 192), including about 30 species of articulate brachiopods. In general, the marine influence decreases from the Mosel Synclinorium in the south to the Eifel in the northwest.

The lower Emsian strata are overlain in the central Middle Rhine and lower Mosel areas by the lowermost upper Emsian Emsquarzit Formation and in the South Eifel region by the Berlé Formation; both formations represent relatively pure and very hard quartzitic sandstones or quartzites. There are first quartzitic precursor sandstone beds (= 'Vorläufer-Quarzite') in the underlying Nellenköpfchen and Klerf formations, but these are generally more impure, more micaceous and richer in plant remains; besides they are thinner bedded and may contain lower Emsian fossils or may be overlain by beds containing fossils of this age (e.g. Lippert 1939, p. 19; Solle 1972, p. 66; Meyer & Stets 1996, p. 64; Kröll 2001, pp. 79–80; Meyer 2013). The 'Emsquarzit' has also been reported from the Taunus and the Lahn and Dill regions, but the quartzitic sandstones assigned to this unit partly turned out to fall in the lower Emsian as well, for example the quartzitic sandstones near Ewersbach in the Dill Synclinorium (Thünker 1990: Höllberg Formation) and the 'Emsquarzit' near Emmershausen in the Taunus, which has yielded a single specimen of *Trop. rhenanus* (Jansen, unpublished data). As the 'Emsquarzit' occurrences in the Taunus have previously been regarded as the basal upper Emsian, separated by a hiatus from strata of mid-early Emsian (Singhofen) age and used in a discussion in favour of the presence of an elevation or island during the late early Emsian (Solle 1950*b*), new data of this kind are essential for the palaeogeographical reconstruction. The uppermost part of the lower Emsian is suggested to be lacking in parts of the Hunsrück and the Taunus; the presence of an extensive island or archipelago has been assumed (Solle 1970; Wehrmann *et al.* 2005) or at least a shallow-submarine swell, the 'Katzenelnbogen Swell' or 'Hunsrück-Taunus Swell' (cf. Meyer & Stets 1980, 1996; Weddige & Requadt 1985; Requadt 1991, 2008; Kröll 2001; Mittmeyer 2008).

In the western Taunus near the river Rhine, the Bornich Formation has yielded lowermost Emsian assemblages with *Eur. assimilis assimilis* (see Mittmeyer 1965), while the Ergeshausen Formation at the Dillenberger Mühle in the northwestern Taunus contains middle lower Emsian brachiopods with *Ard. latestriatus prolatestriatus* and *Ps. dahmeri*. In the eastern Taunus, the Spitznack and Oppershofen formations contain rich eurhenotypic faunas indicating a mid-early Emsian age. The Spitznack Formation near Wernborn, for example, has yielded multiple mass occurrences of *Ard. antecedens* (Fig. 5b), very often associated with *Pleb. semiradiatus*, *Eur. assimilis latissimus*, *Ps. dahmeri* and *Meg. ovata ovata*.

In the central and eastern parts of the Hunsrück and across the river Rhine in the southwestern Taunus, the lower Emsian succession is largely developed in the argillaceous Hunsrück Slate facies (e.g. Solle 1950*a*; Meyer & Stets 1996; Bartels *et al.* 1998, 2002), in the strict sense an essentially hercynotypic facies with ammonoids and dacryoconarides but containing in a few layers some rhenotypic brachiopods with stratigraphic significance, among these *Ard. latestriatus prolatestriatus* and *Eur. assimilis assimilis* (see Mittmeyer 2002). The Hunsrück Slate Group consists of several formations (Mittmeyer 1980; de Baets *et al.* 2013*a*), but the most typical development of this unit is represented by the Kaub Formation around the classic localities Bundenbach, Gemünden and Kaub. The Hunsrück Slate fossil lagerstätte is famous for its rich echinoderm fauna and ancient ('Cambrian-type') arthropods with preserved appendages (e.g. Bartels *et al.* 1998). A separate development is documented near Wald-Erbach in the Stromberg Unit of the southeastern Hunsrück, where the uppermost part of the lower Emsian and the transition to the upper Emsian is developed in the eurhenotypic subfacies (Wolf 1930; Mittmeyer & Geib 1967; Meyer 1970).

The lower Emsian of the northern Rhenish Massif is poorly fossiliferous. Single layers in the Schleiden Formation of the North Eifel region show a proximal eurhenotypic subfacies with rare *Eur.* ex gr. *assimilis-dunensis* and homalonitine trilobites (Ribbert 2008). The lower part of the Bensberg Formation in the Bergisches Land contains oligospecific associations with pelecypods, ostracods, eurypterides, fishes and early land plants (Hilden 2008). Due to the lack of stratigraphically significant fossils, the probable early Emsian age of the largely unfossiliferous Siesel Formation and the Rimmert Group in the Sauerland is still difficult to verify (Langenstrassen 2008).

Brachiopod biostratigraphy. The base of the lower Emsian is marked by the onset of *Acr. eckfeldensis* Mittmeyer, 2008, for example near the base of the Eckfeld Formation in the South Eifel region or the Nassau Formation in the southern Westerwald (Jentsch 1960: '*Acr. primaevus*'; Mittmeyer 2008; materials restudied by the author). The species is interpreted as a phylogenetic descendant of the Siegenian species *Acr. primaevus*; the transition probably took place during the time of the upper Saxler Formation. Additional important biostratigraphic markers for the recognition of the basal boundary interval of the classic Emsian are the first occurrences of *Ir.* (*Ir.*) *maior*, *Pla. circularis circularis*, *Tor. crassicosta crassicosta*, *Ard. latestriatus prolatestriatus* and *Eur. assimilis assimilis*. The phylogenetic transition from *Al. affinis* to *Al. mediorhenana*, proposed as a supplementary biostratigraphic marker by Mittmeyer (1973*b*, 1982*a*), is difficult to apply, because these species are only sporadically distributed.

Brachiopods are used for the biostratigraphic subdivision of the lower Emsian. The Rhenish succession has been subdivided into Ulmen, Singhofen and Vallendar groups (Solle 1950*a*; Mittmeyer & Geib 1967), which have subsequently been translated into corresponding regional substages defined by spiriferide species and subspecies (Mittmeyer 1974, 1982*a*, 2008; Schemm-Gregory & Jansen 2006*a*, *b*). (Note that the Ulmen Group had been introduced as uppermost part of the Siegenian (Solle 1950*a*) but was later revised and largely included in the lower Emsian (Mittmeyer 1973*a*, *b*, 1974); the revised concept is 'more natural' in terms of the general faunal character and closer to the tradition. This revised concept is followed in the present work.) Taxa of the genus *Arduspirifer* serve here as zonal guide fossils (Fig. 3): *Arduspirifer latestriatus prolatestriatus* indicates lower to middle parts of the lower Emsian (Ulmen–Singhofen substages), *Ard. latestriatus latestriatus* the upper part (Vallendar substage) and *Ard. antecedens* (with new subspecies still to be described) lower to upper parts. *Arduspirifer* '*arduennensis*' *initiator* Mittmeyer, 2008 from the Eckfeld Formation would be the oldest taxon of the genus to be considered here, but it is still poorly known. It is best classified as an early subspecies of *Ard. latestriatus* and could be another promising marker for the recognition of the basal lower Emsian (lower Ulmen substage). In addition, the phylogenetic succession of *Euryspirifer* taxa provides a parallel zonation; simply speaking, *Eur. assimilis assimilis* indicates the lower part, *Eur. assimilis latissimus* the middle part and *Eur. dunensis* the upper part of the lower Emsian. However, the intraspecific variability is obviously high and still hardly studied.

Apart from these spiriferides, the first occurrences of *Ps. dahmeri*, *Lap. pila*, *In. inauritum* and *Sart. antiquus* may turn out as helpful levels to subdivide the lower Emsian in its lower parts. The upper boundary of the lower Emsian is marked by the disappearance of *Ps. dahmeri*, *Rhe. provulvaria*, *Trop. rhenanus*, *Ard. latestriatus*, *Ard. antecedens*, *Eur. dunensis*, *Br. crassicosta crassicosta*, *Ir.* (*Ir.*) *maior* and *Str. dunensis*.

Correlation and age. Based on ranges of brachiopod species, the strata of the Rhenish lower Emsian can be correlated with successions in the Ardennes. Approximately in line with traditional litho- and biostratigraphic concepts (Maillieux 1936*b*, 1940; Asselberghs & Maillieux 1938; Asselberghs 1946), the boundary between the Siegenian and Emsian stages has been placed in the sequence of the southern Dinant Synclinorium within the lower part of the Pesche Formation, at the base of unit 2; it was later biostratigraphically defined by the onset of *Brachyspirifer minatus* Godefroid, 1980 (Godefroid 1980; Godefroid & Stainier 1982; Godefroid *et al.* 1994). According to brachiopod data, the position of this boundary, far above the disappearance of Rhenish Siegenian guide fossils and not far below the disappearance of *Brachyspirifer transiens transiens* Solle, 1971, is suggested to be distinctly higher than the Rhenish boundary in the sense used here, as it may approximately correlate with the top of the Ulmen Group (Godefroid 1980; Godefroid & Stainier 1982). Based on palynomorphs, the level of the Belgian boundary would also approximately coincide with the original Pragian–Zlichovian boundary in Bohemia (Steemans 1989); a similar correlation has been suggested by Carls (1987) and Carls *et al.* (2008), who evaluated the faunal succession in the Iberian Chains (Spain), where rhenotypic brachiopods co-occur with hercynotypic dacryoconarides and conodonts.

Arduspirifer latestriatus prolatestriatus, *Ard. antecedens*, *Eur. dunensis* and *Tor. crassicosta crassicosta*, index taxa from the classic Rhenish lower Emsian in the sense used here, have been reported from the Ardennes, from the Siegenian in the Belgian sense, starting from within the La Roche Formation (Godefroid & Stainier 1982, table IIb; Godefroid 1994; Godefroid *et al.* 1994, fig. 12), i.e. far below the Belgian Siegenian–Emsian boundary. Moreover, *Ard. latestriatus prolatestriatus* commonly predates *Ard. antecedens* in the Ulmen Group of the Rhenish Massif, whereas both taxa seem to first occur at the same level in the Ardennes. The different concepts of the Siegenian–Emsian boundary result from different traditions in the Ardennes and the Rhenish Massif. In the succession of the southern Dinant Synclinorium, the Siegenian–Emsian boundary level in the Rhenish sense, defined at the base of the Ulmen Group as revised by Mittmeyer (1973*a*, *b*, 1974), should be located within the La Roche Formation, between the last occurrences of *Acr. primaevus* and other Siegenian guide fossils (in the Rhenish sense) and the entry of lower Emsian guide fossils such as *Ard. latestriatus prolatestriatus*. The reported presence of the Siegenian fossil *Proschizophoria personata* throughout the La Roche Formation (Godefroid *et al.* 1994, fig. 10) and the range of *Acr. primaevus* (Godefroid *et al.* 1994, fig. 11) are problematic, as these species would overlap with *Ard. latestriatus prolatestriatus*, *Ard. antecedens* and *Eur. dunensis* (Godefroid *et al.* 1994, fig. 12). A scrutiny of the representatives of the genera *Arduspirifer*, *Acrospirifer* and *Euryspirifer* from this succession may shed new light on the correlation of the boundary in the Ardennes. In particular, it must be checked whether the basal Emsian taxa *Acr. eckfeldensis*, *Eur. assimilis assimilis* or *Ard. latestriatus initiator* are present in the sequence of the southern Dinant Synclinorium. The presence of the genus *Arduspirifer* alone does not prove an Emsian age anymore, whereas this was taken as a rule in former times; the genus appears clearly in the Siegenian of different regions in southwestern Europe (Carls 1987, p. 84, fig. 3; 1999, p. 146; 2000, p. 72), Morocco and the Rhenish Massif (Jansen 2001*a*).

In my opinion, the scope of '*Eur. dunensis*' in the sense of Godefroid (1994) is too broad. The re-examination of a number of Godefroid's specimens of *Euryspirifer* (Institut royal des Sciences naturelles de Belgique, Brussels, Belgium) by the author has shown that ('the true') *Eur. dunensis* occurs in middle parts of the Pesche Formation (unit PS 14; see Godefroid 1980, fig. 12), accompanied there by a still undescribed subspecies of *Ard. antecedens* that is regarded as being of late early Emsian (Vallendar) age. The older specimens of *Euryspirifer*, in particular those from the upper part of the La Roche Formation, which Godefroid determined as '*Eur. dunensis*', rather belong to *Eur. assimilis* ssp.

The lower Emsian in the Iberian Chains (Spain) documents an *Arduspirifer* radiation including *Ard. latestriatus prolatestriatus* and *Ard. antecedens* ssp. in unit d4a beta of the Mariposas Formation (Carls' material, restudied by the author). These taxa follow above more plesiomorphic, probably still Siegenian representatives of the same genus from the basal unit of the same formation, the d4a alpha, which is the type stratum of the Emsian zonal index conodont *Eocostapolygnathus excavatus* (Carls & Gandl, 1969). It is concluded from this succession that the level of the Siegenian–Emsian boundary in the German sense is located near the base of unit d4a beta and that the basal boundary of the GSSP Emsian is distinctly older (Carls *et al.* 2008); lower parts of the *Eoc. excavatus* Zone and the entire *Eoc. kitabicus* Zone, basal Emsian in the present GSSP sense, are probably of Siegenian age.

A correlation of upper and probably middle parts of the Rhenish lower Emsian and the Zlíchov Limestones (Zlichovian, Bohemia) is suggested by the occurrence of the strophomenide *Crinistrophia elegans* (Drevermann, 1902) (= *Crin. crinita* Havlíček, 1967) in both successions. The restudy of the type materials of *Crin. elegans* and *Crin. crinita* has confirmed Jahnke's (1971) opinion that these taxa are synonymous.

Ammonoids help in the supraregional correlation of the Rhenish lower Emsian with the pelagic biostratigraphy: *Anetoceras arduennense* (Steininger, 1853) has been found associated with *Ard. antecedens*, *Ard. latestriatus* and *Eur.* ex gr. *assimilis-dunensis* in middle lower Emsian to possibly upper lower Emsian strata in the Central Eifel region (Erben 1953, 1960; Chlupáč 1976; Mittmeyer 2008, table 3; de Baets *et al.* 2009), *Ivoites schindewolfi* de Baets *et al.*, 2013*a* in lower Emsian beds of Belgium (de Baets *et al.* 2013*b*) and representatives of *Metabacrites*, *Ivoites*, *Anetoceras*, *Erbenoceras*, *Mimosphinctes*, *Gyroceratites* and *Mimagoniatites fecundus* (Barrande, 1865) in the lower Emsian Kaub Formation of the Hunsrück Slate Group (Erben 1960; Kutscher 1969; Chlupáč 1976; de Baets *et al.* 2013*a*). These taxa range from the middle to upper Zlichovian in the sense of the Bohemian stratigraphy; *Gyroceratites*, *Mimosphinctes* and *Mim. fecundus* range into the Dalejian. In addition, the Zlichovian dacryoconaride species *Nowakia praecursor* Bouček, 1964, *Now.* aff. *zlichovensis* Bouček, 1964, *Now. hunsrueckiana* Alberti, 1983 and *Now. barrandei* Bouček & Prantl, 1959 are known from the Kaub Formation near Bundenbach (Alberti 1982*b*, 1983). The presence of *Eur. assimilis assimilis* in the Bocksberg and Obereschenbach members of the Kaub Formation (Mittmeyer 2002, p. 114) suggests an earliest Emsian (Ulmen) age of these units, but *Mim. fecundus* from the same formation (de Baets *et al.* 2013*a*) has normally been reported from strata younger than the range of *Anetoceras* as revised by de Baets *et al.* (2009), known from strata of Singhofen and possibly Vallendar age. Besides, the Hunsrück Slate Group should be overlain by the Spitznack Formation which is of mid-early Emsian age (Solle 1950*a*) and contains the stratigraphically significant subspecies *Eur. assimilis latissimus*. This contradiction of neritic–pelagic correlation remains unsolved. Tectonic disturbances, lateral facies change, resedimentation of fossils or imprecise determinations and still insufficient knowledge of the stratigraphic ranges of the fossils are possible reasons.

Palaeoenvironment, sea-level and events. The latest Siegenian 'Saxler Event' (Mittmeyer 2008) marks the onset of a transgressive period persisting into the earliest Emsian. In the western Taunus and the Hunsrück, the transition from the proximal eurhenotypic and pararhenotypic Taunusquarzit Group to the largely hercynotypic Hunsrück Slate Group falls into this phase corresponding to the Siegenian–Emsian boundary interval. Probably the same transgressive event is represented by the sharp transition from eurhenotypic sandstones of the Merzâ-Akhsaï Formation (with Siegenian brachiopods) to dark hercynotypic limestones of the basal Mdâouer-el-Kbîr Formation or Oui-n'-Mesdoûr Formation in the eastern and western Dra Plains, respectively (Morocco; Jansen *et al.* 2007) or the transition from the Santa Cruz Formation to the Mariposas Formation in the Iberian Chains (Spain; Carls *et al.* 2008), described as 'Basal Zlíchov Event' by García-Alcalde (1997). The initial pulse of T-R cycle Ib of the global sea-level curve (Johnson *et al.* 1985; Johnson & Sandberg 1989) may coincide with this transgressive event. After the correlation by Carls *et al.* (2008, fig. 1), this level would be slightly older than the important black shale and last transgressive incursion of graptolites in Bohemia (e.g. Chlupáč & Lukeš 1999; Slavík 2004; Carls *et al.* 2008: 'graptolite horizon/interval'), the '*atopus* Event' (Becker *et al.* 2012), which is commonly lumped with the slightly younger, more regional 'Basal Zlíchov Event' as originally defined by Chlupáč & Kukal (1986, 1988).

Open-shelf palaeoenvironments persisted through most of the early Emsian time in the central and southern Rhenish area. The global sea-level slowly rose, but this rise is not clearly reflected everywhere by the facies development. The Hunsrück Slate Group may represent a large part of the lower Emsian in the southwestern Rhenish Massif. Its faunal composition suggests relatively deep-water, offshore conditions as indicated by the presence of pelagic faunas. More shallow-marine conditions dominated in the eastern Taunus and the southern Eifel north of the Siegen Main Thrust, where the shelf was more elevated and more affected by coarse siliciclastic input from the land. Near the beginning of the Singhofen Group (middle lower Emsian) a faunal change is discernible with the regional 'Spitznack Event' described from the Taunus (Mittmeyer 2008); it is characterized, for example, by mass occurrences of *Lap. pila* ('*Pila*-Bank'), *Ard. antecedens* and *Ps. dahmeri*. In the southern Central Eifel region, an increasing marine influence at the transition from the Reudelsterz to the Gefell Formation is documented at this level, where mass occurrences of *Ps. dahmeri* are conspicuous. A change from largely restricted-marine to more open-marine conditions is assumed here. This was probably the time at which the ammonoids first migrated into the Eifel region and the ammonoid development in the Kaub Formation of the Hunsrück began (Erben 1960; Chlupáč 1976; de Baets *et al.* 2009, 2013*a*, *b*). Later within the early Emsian, a distinct transgressive pulse with increasing diversity and abundance of brachiopod faunas is reflected by the 'Stadtfeld Event' (Mittmeyer 2008) at the transition from the mid-early Emsian to the late early Emsian. The transgressive development was possibly enhanced by reduced sedimentation rates here (Fuchs 1982). The onset of the rich Oppershofen brachiopod fauna in the eastern Taunus (Dahmer 1939) approximately coincides with the same transgressive phase but is probably slightly older, of late Singhofen age, judging from the representatives of *Eur. assimilis* cf. *latissimus* and *Ard. antecedens* ssp. Whether the Spitznack Event corresponds to the early Emsian 'Chebbi Event' described from Morocco (Becker & Aboussalam 2011) and the Stadtfeld Event to the 'Upper Zlíchov Event' *sensu* García-Alcalde (1997), as could be suggested from their relative positions, requires confirmation by more biostratigraphic data.

During the latest early Emsian time, the open-marine development was terminated by a strong regional regression affecting large parts of the Rhenish Shelf; it was at least enhanced by very high sedimentation rates. As a result, supratidal or intertidal to shallow-subtidal palaeoenvironments prevailed, as has been concluded from sedimentary and palaeontological characters of the mainly proximal eurhenotypic to pararhenotypic or terrestrial subfacies of the Klerf and Nellenköpfchen formations in the central Rhenish Massif (Solle 1956*a*, *b*; Wunderlich 1970; Fuchs 1982; Reineck 1983; Stets & Schäfer 2002; Wehrmann *et al.* 2005). This regressive development is not reflected by the global T-R curve (Fig. 3) and is interpreted as a regional phenomenon resulting from overabundant sedimentary supply from a large delta prograding to the south (Stets & Schäfer 2002). In the northern Rhenish Massif (North Eifel, Bergisches Land, Sauerland), the lower Emsian as a whole (Schleiden Formation, Zweifall Formation in part, Bensberg and Siesel formations, Rimmert Group) is documented by poorly fossiliferous deposits of possibly limnic-fluvial, lagoonal, deltaic or restricted-marine palaeoenvironments near the coast of the Old Red Continent (Böger 1981; Breil-Schollmayer 1989;

Schultka & Remy 1990; Hilden 2008; Langenstrassen 2008; Ribbert 2008).

Upper Emsian brachiopod faunas

Characteristics. The marine faunas of the Rhenish upper Emsian are again predominantly composed of brachiopods and pelecypods, but contain the whole spectrum of other rhenotypic fossil groups as well. The event-related faunal overturn at the lower–upper Emsian transition is one of the most conspicuous ones of the Rhenish Lower Devonian (Solle 1972). It may have resulted on the one hand from the immigration of species from outside the Rhenish Shelf, or on the other hand from continued evolution of early Emsian forms within this shelf area. *Iridistrophia* (*Flabellistrophia*) *musculosa* subgen. nov. et sp. nov. (Fig. 8a, b) evolved from a species of *Ir.* (*Iridistrophia*) near this boundary. The genus *Arduspirifer* persisted from the early

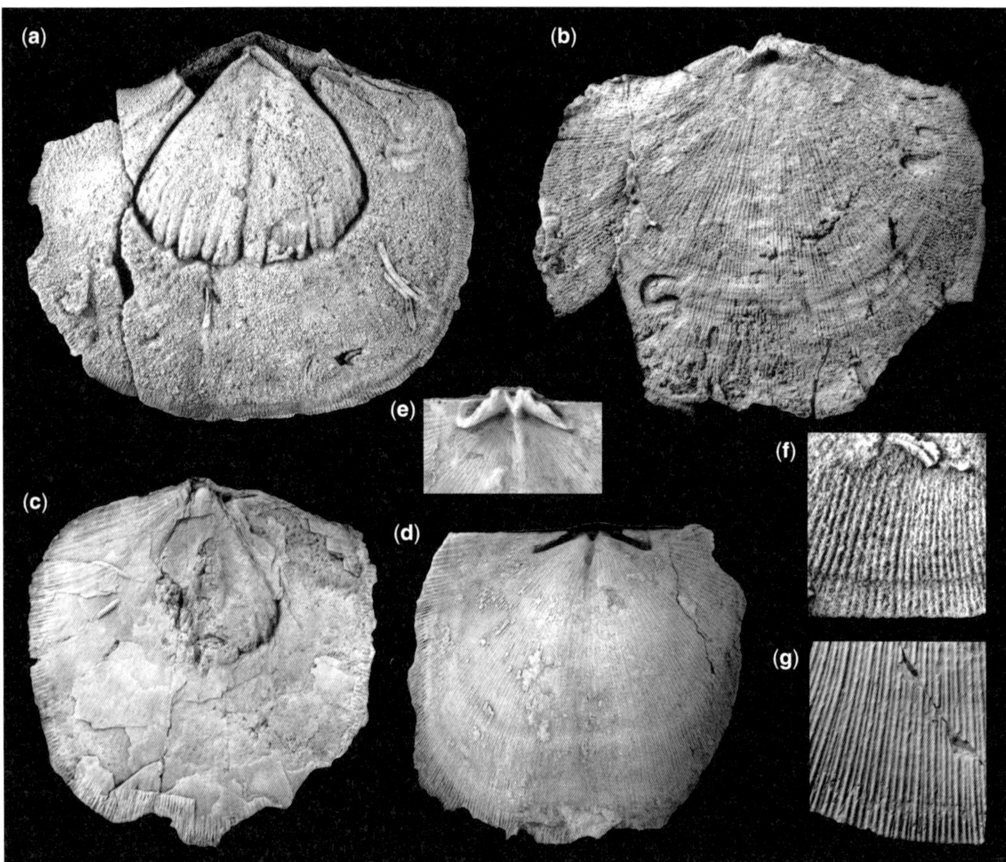

Fig. 8. (a, b) *Iridistrophia* (*Flabellistrophia*) *musculosa* subgen. nov. et sp. nov., internal mould of ventral valve (**a**) and latex cast (**b**) of corresponding external mould (×1.0), holotype SMF 65563a + b; collected by J. Hefter. Locality: Maria Roth, Kondertal, south of Winningen, Mosel River, Lower Mosel region. Stratum: Laubach Group, middle part of upper Emsian. (**c**) *Iridistrophia* (*Flab.*) *hipponyx* (Schnur, 1851), internal mould of ventral valve with remains of the shell (×1.0), SMF 66902; collected by Schmidt-Gündel & Wintgen 1999. Locality: Rohr-Nord section, 44.21–44.61 m, Rohr Syncline, North Eifel region (Brocke *et al.* 2004). Stratum: Heisdorf Formation, upper part of upper Emsian. (d, e) The same species, internal mould of dorsal valve (**d**) and latex cast of cardinalia (**e**) (×1.0), SMF 85168; collected by R. Werner 1965. Locality: Haus Mühlbach, road cut between Willwerath and Gondelsheim, Prüm Syncline, Eifel region. Stratum: Heisdorf Formation, *Alatiformia alatiformis* Zone, upper part of upper Emsian. (**f**) *Iridistrophia* (*Flab.*) *musculosa* subgen. nov. et sp. nov.; latex cast of external mould of ventral valve, detail of ornamentation (×2.0), SMF 66380.1. Locality: Miellen near Bad Ems, lower Lahn Valley. Stratum: Hohenrhein Formation, *Brachyspirifer ignoratus* Zone, lower part of upper Emsian. (**g**) *Iridistrophia* (*Flab.*) *hipponyx* (Schnur, 1851); ventral valve, detail of ornamentation (×2.0), SMF 93807a; collected by F. Drevermann. Locality: Prüm, Prüm Syncline, Eifel region. Stratum: Lauch Formation, lower part of lower Eifelian.

Emsian (the '*arduennensis* group': Solle 1953; Jansen 2001*a*; Schemm-Gregory & Jansen 2005, 2006*a*, *b*), but its early and late Emsian representatives are not further regarded as conspecific. Characteristic features of the late Emsian brachiopod evolution are the phylogenetic appearance and diversification of the genera *Paraspirifer* Wedekind, 1926 (see Solle 1971) and *Sollispirifer* Mittmeyer, 2008 (the '*mosellanus* group': Solle 1953; Schemm-Gregory & Jansen 2006*a*, *b*). The first species of *Paraspirifer* evolved from *Brachyspirifer ignoratus* (Maurer, 1883) in early late Emsian (Hohenrhein) time. Near the end of this interval, both the *Par. sandbergeri* and *Par. praecursor-cultrijugatus* branches emerged (Solle 1971); they are given subgeneric status in the present work (see the later section on systematic palaeontology): *Paraspirifer* (*Mosellospirifer*) subgen. nov. and *Par.* (*Paraspirifer*). *Paraspirifer* (*Par.*) *praecursor* Solle, 1971 gave rise to *Par.* (*Par.*) *cultrijugatus* (Roemer, 1844) and other species of this group (Solle 1971). *Sollispirifer* shows a rapid evolution from *Soll. mosellanus* (Solle, 1953) (Fig. 7f, g) to *Soll. dahmeri* (Solle, 1953), *Soll. gracilis* (Solle, 1953) and *Soll. steiningeri* (Solle, 1953); these species are important for the subdivision of the upper part of the upper Emsian (Solle's subspecies of *mosellanus* are elevated to species level in the ongoing revision of this group). The evolution within the genus *Rhenostropheodonta* Jansen, 2014*a* starts with *Rhen. rhenana* Jansen, 2014*a* (Fig. 7o, p) in the early late Emsian time, which gave rise to *Rhen. piligera* (Sandberger & Sandberger, 1856) (Fig. 7n) near the transition to the middle part of the late Emsian. The (latest early Emsian? to) earliest late Emsian species *Pachyschizophoria* sp. nov. C *sensu* Jansen, 2001*a* is regarded as the ancestor of the classic species *Pa. vulvaria* (von Schlotheim, 1820). Apart from these taxa, the following ones are very typical of upper Emsian strata: *Gibbodouvillina taeniolata* (Sandberger & Sandberger, 1856), *Platyorthis circularis transfuga* (Walther, 1903), *Chonetes sarcinulatus* (von Schlotheim, 1820), *Plebejochonetes semiradiatus* (Sowerby, 1842), *Oligoptycherhynchus daleidensis* (Roemer, 1844), *Ol. hexatomus* (Schnur, 1851), *Lapinulus pila* (Schnur, 1851), *Sartenaerirhynchus antiquus* (Schnur, 1853), *Cuninulus melanopotamicus* Sartenaer, 2005, *Subcuspidella* ex gr. *subcuspidata* (Schnur, 1851), *Al. alatiformis* (Drevermann, 1907), *Al. janseni* Gad, 2002 (Fig. 7h), *Arduspirifer arduennensis treverorum* Schemm-Gregory & Jansen, 2005 (Fig. 7i, j), *Ard. arduennensis arduennensis* (Schnur, 1853), *Euryspirifer robustiformis* Mittmeyer, 1972, *Eur. paradoxus* (von Schlotheim, 1813), *Brachyspirifer carinatus rhenanus* Solle, 1971, *Paraspirifer* (*Mosellospirifer*) *sandbergeri* Solle, 1971 (Fig. 7k), *Par.* (*Mos.*) *longimargo* Solle, 1971, *Rhenothyris compressa* (Maurer, 1886), '*Athyris*' cf. *undata* (Defrance, 1828), *Meristella follmanni* (Dahmer, 1916), *Meganteris ovata suessi* (Drevermann, 1902) and *Cryptonella macrorhyncha* (Schnur, 1853) (Fig. 7c, d).

The brachiopod evolution during the late Emsian shows many steps, extinctions and originations, serving as an excellent base for biostratigraphic subdivision. As no strong faunal turnovers have been recorded that could be ascribed to drastic extinction or emigration of substantial parts of the fauna and subsequent immigration of a new fauna, it is suggested that evolution largely took place step by step within the Ardenno-Rhenish shelf area. The term '*Euryspirifer paradoxus* Fauna' is proposed for this large late Emsian faunal complex. A more distinct, facies-related faunal change is visible in the Heisdorf Formation (uppermost Emsian) of the Eifel region, where the '*Paraspirifer cultrijugatus* Fauna' appears and persists into the lower Eifelian.

Distribution and regional developments. Brachiopod-bearing strata of the upper Emsian are particularly well exposed in the Lower Mosel and central Middle Rhine regions (e.g. Follmann 1925; Solle 1937, 1942*a*, 1976; Meyer & Stets 1996; Elkholy 1998; Kröll 2001; Gad *et al.* 2008; Mittmeyer 2008) and the southern Eifel Synclines (e.g. Lippert 1939; Struve 1961*a*, *b*; Werner 1969; Struve *et al.* 2008; Meyer 2013), but also in the Lahn and Dill regions (e.g. Solle 1942*b*, *c*; Pauly 1958; Weddige & Requadt 1985; Thünker 1990; Bender 2008; Requadt 2008), the Bergisches Land (e.g. Spriestersbach & Fuchs 1909; Spriestersbach 1925; Zygojannis 1971; Hilden 2008) and the Sauerland (Langenstrassen 1972, 2008). Due to many regional facies variations, the upper Emsian lithostratigraphy is very complex and includes numerous formations in different regions. The classic succession of the central Middle Rhine region starts with the Emsquarzit Formation ('Koblenzquarzit' of early works), which has yielded fossils at a few localities only, for example in the Rhenser Mühlental near Rhens (south of Boppard; Middle Rhine Valley). Faunal lists have been published by Viëtor (1919), Solle (1936), Mittmeyer (1972), Wenndorf (2001) and Schemm-Gregory (2004). The Emsquarzit Formation contains, apart from a diversity of pelecypods, a first typical upper Emsian brachiopod assemblage, including abundant specimens of *Rhen. rhenana*, *Ir.* (*Flab.*) *musculosa*, *Pachyschizophoria* sp. nov. C *sensu* Jansen, 2001*a*, *Ard. arduennensis treverorum* and *Br. ignoratus*. Quartzitic sandstones assigned to the 'Emsquarzit' have been reported from a very large area, but some turned out to be lower Emsian (see above). On the other hand, some classic occurrences of the 'Emsquarzit' have later been assigned to the lower Hohenrhein Formation, for example the richly fossiliferous outcrops in the Kleinbornsbach-Tal SW of the 'Kühkopf' near Koblenz (Dahmer 1948; see Solle 1972, p. 65). The Hohenrhein Formation, overlying the Emsquarzit Formation in the central Middle Rhine region, bears similar faunas, but the abundance of fossils is generally higher, as many sandstone beds are fossiliferous. Typical are mass occurrences of *Ard. arduennensis treverorum* accumulated in sandy shell beds. *Brachyspirifer ignoratus* and *Rhen. rhenana* can be very abundant as well, for example at the classic locality 'Miellen' in the Lahn Valley (Follmann 1925). *Pachyschizophoria* sp. nov. C is a typical component of the Emsquarzit Formation extending into the Hohenrhein Formation; it was phylogenetically replaced by *Pa. vulvaria*, possibly in the Hohenrhein time. The ensemble of Emsquarzit and Hohenrhein formations represents the Lahnstein Group. The following Laubach and Kondel groups, the latter including the Flaserschiefer and Kieselgallen-Schiefer formations, document a typical to distal eurhenotypic subfacies development with a series of highly diverse, brachiopod-dominated faunas (Solle 1942*a*; Mittmeyer 1982*b*, 2008; Gad *et al.* 2008). The genus *Paraspirifer* originated in the early late Emsian time, and its diversification took place in the whole middle and late parts of the late Emsian (Laubach–Kondel) interval; it is represented here by the species *Par. (Mos.) sandbergeri*, *Par. (Mos.) longimargo* and *Par. (Par.) praecursor*. In addition, very common elements of the Laubach Group are *Rhen. piligera* and *Pa. vulvaria*, whereas the Kondel Group displays large occurrences of *Soll. mosellanus*, *Soll. dahmeri*, *Al. janseni* and *Cun. melanopotamicus*. The upper Emsian succession of the Olkenbach Syncline, a southwestern partial syncline of the large Mosel Synclinorium, differs slightly in the lithological development from the succession in the central Middle Rhine area, but excellently documents the same evolution of brachiopods (Solle 1937, 1942*a*, 1976).

The uppermost Emsian Kondel Group is particularly rich in brachiopods in the Dill Synclinorium, where the Mandeln and Haigerhütte formations contain highly diverse brachiopod faunas of the distal eurhenotypic subfacies (Mandeln Formation: Dahmer 1916, 1923; Solle 1942*b*; Haigerhütte Formation: Jahnke & Michels 1982*b*; Frech 1888; Dahmer 1923; Solle 1942*b*; Carls *et al.* 1972; Jansen *et al.* 2001). As a striking peculiarity, the uppermost upper Emsian (upper Kondel) strata in the Haigerhütte section ('Papiermühle Haiger', Dill Synclinorium) have yielded many specimens of *Quadrithyris trisecta* (Kayser, 1882), an 'exotic' species of a genus common in the calcareous facies of, for example, the Barrandian area.

The marine facies of the upper Emsian in the Bergisches Land is characterized by thick siliciclastic and partly calcareous successions of the Remscheid Group bearing impoverished marine faunas with a predominance of pelecypods (Fuchs 1909; Spriestersbach 1915, 1925; Zygojannis 1971), but in places many spiriferides of the species '*Subcuspidella*' *crassifulcita* (Spriestersbach, 1915) also occur. From the evolutionary stage of this species a latest Emsian age is assumed.

In the southern Eifel Synclines, the upper Emsian reveals a separate facies starting with the Berlé Formation at the base and followed by the Wiltz, Wetteldorf and Heisdorf formations, completely documented and highly fossiliferous on the flanks of the Prüm Syncline (e.g. Werner 1969; Struve *et al.* 2008). The faunas are similarly diverse as in the Middle Rhine region but differ slightly in taxonomic composition, resulting in minor correlation problems between the two regions. A high faunal diversity has been recorded from argillaceous to sandy shales of the Wiltz Formation representing lower to middle parts of the upper Emsian, with a rich brachiopod fauna of approximately 45 species (Werner 1969; for example locality We 10, 'Ziegelei-Grube Niederprüm'), including typical eurhenotypic taxa such as *Ard. arduennensis arduennensis*, *Eur. robustiformis*, *Pa. vulvaria*, *Ol. daleidensis*, *Fascistropheodonta* sp. nov., *Gigastropheodonta* sp. and *Ir. (Flab.) musculosa*, but also abundant representatives of '*At.*' cf. *undata*, *Mer. follmanni*, *Atrypa* sp. and *Cry. macrorhyncha*. These brachiopods in combination with abundant orthoconic nautiloids indicate the distal, enriched eurhenotypic subfacies. The mass occurrence of *Arduspirifer extensus* (Solle, 1953) in association with many specimens of the 'exotic' species *Anoplia theorassensis* Maillieux, 1941 in quartzitic sandstone beds intercalated in upper parts of the Wiltz Formation is remarkable. The localities near Daleiden in the West Eifel region have yielded many brachiopods and pelecypods preserved as siliceous, undeformed, but transported and abraded internal moulds of articulated valves – this is the 'Daleiden mode of preservation' ('Daleider Versteinerungen'; Richter 1916). The overlying Wetteldorf Formation is poorer in brachiopods, but still contains *Ard. arduennensis arduennensis* and, as advanced species, *Par. (Par.) globosus* Solle, 1971, which substitutes for the species *Par. (Par.) praecursor* here, and *Soll. mosellanus*; the latter species indicates a latest Emsian (early Kondel) age. Mass occurrences of '*Subcuspidella*' *wetteldorfensis* (Richter & Richter, 1919) and '*Sub.*' *lateincisa* (Scupin, 1900) in the sandy facies development of the formation ('Wetteldorfer Sandstein', for example at the locality 'Wetteldorf quarry') point to a proximal eurhenotypic subfacies. The

Heisdorf Formation has a wider distribution in the Eifel Synclines (Krömmelbein 1955; Glinski 1961; Ochs & Wolfart 1961; Struve 1961*a*; Werner 1982; Struve *et al.* 2008; Meyer 2013). In general, the calcareous component increases, with limestones, siltstones and calcareous sandstones as common rocks, as well as pseudo-oolithic iron beds. These strata contain a different fauna – the older '*Par. cultrijugatus* Fauna'. The faunal character changes gradually, so that this fauna is not sharply distinguishable from the *Euryspirifer paradoxus* Fauna. On the other hand, it is slightly different compared to the uppermost Emsian fauna of the Middle Rhine and Lower Mosel regions. *Douvillinella filifer* (Schmidt, 1913), *Soll. steiningeri*, '*Soll.*' *schreiberi* (Happel, 1932), *Par.* (*Par.*) *cultrijugatus*, *Al. alatiformis*, *Oligoptycherhynchus wetteldorfensis* (Schmidt, 1941) and *Cimicinella cimex* (Richter & Richter, 1918) are typical species of the Heisdorf Formation, accompanied by other, typical upper Emsian species, such as *Soll. mosellanus*, *Pa. vulvaria* and *Eur. paradoxus*. The increasingly calcareous lithology in combination with a changing fossil content introduces the transition to the allorhenotypic subfacies, which is very common in the Eifelian. In the eastern Sauerland, a similar facies is represented by the *Orthocrinus* and Schmallenberg formations ('*Cultrijugatus*-Schichten'), of latest Emsian to earliest Eifelian age (Langenstrassen 1972, 2008).

In the northern Eifel Synclines, the upper Emsian succession is incomplete – Wiltz and Wetteldorf formations are largely absent, reduced in thickness or replaced by coeval strata developed in different facies (e.g. Hotz *et al.* 1955; Ochs & Wolfart 1961; Meyer 2013). The age of the supposed 'Emsquarzit' in this area is still a matter of debate (e.g. Ribbert 1993, pp. 25–27, 2008, p. 294). Brachiopods from the quartzitic 'Emsquarzit' beds of Honertseifen SW of Kronenburg (Blankenheim Syncline; Ribbert 2010, fig. 20), restudied by the author, include *Sollispirifer gracilis* (Solle, 1953) and are accordingly assigned to the uppermost Emsian. This raises the question: if the underlying strata in 'Klerf facies' have a late Emsian age as well. The age of these beds is important for the palaeogeographical reconstruction.

Finally, the enriched, distal eurhenotypic subfacies of the locality 'Kröffelbach' near Grävenwiesbach (Maurer 1886; near Kraftsolms, Solmsbach-Tal, northern Taunus) should be mentioned. With its abundance of *Ir.* (*Flab.*) *musculosa*, *Br.* cf. *ignoratus* (= '*Spirifer cultrijugatus*' in previous works), *Eur. paradoxus*, *R. compressa*, *Meg. ovata suessi* and additional species, this still poorly known fauna partly resembles the fauna from the Hohenrhein Formation of the central Middle Rhine area.

Brachiopod biostratigraphy. A number of brachiopod species have their first occurrences at the lower boundary of the upper Emsian, for example *Br. ignoratus*, *Ard. arduennensis treverorum*, *Eur. paradoxus*, *Eur. robustiformis*, *Ir.* (*Flab.*) *musculosa*, *Rhen. rhenana* and *Cry. macrorhyncha*. The *Eur. paradoxus* Zone approximately corresponds to the whole upper Emsian.

The successions of the Rhenish upper Emsian have been subdivided first into Lahnstein, Laubach and Kondel groups (Solle 1937) and later into more biostratigraphically defined, correspondent regional substages with brachiopods as the most important guide fossils (Solle 1972; Mittmeyer 1974, 1982*a*, 2008). The lower part of the upper Emsian (Lahnstein substage) is characterized by the *Ard. arduennensis treverorum* and *Br. ignoratus* zones. The ubiquitous presence of *Ard. arduennensis arduennensis* is an important character of the lower and middle parts of the upper Emsian (Lahnstein–Laubach substages), in particular in the Eifel region. Many assemblages of middle and upper parts of the upper Emsian (Laubach–Kondel substages) can be assigned to the *Par.* (*Par.*) *praecursor* Zone, whose lower boundary marks the base of the middle upper Emsian (Laubach substage) in the central Rhenish Massif. The reported occurrences of rare *Par.* (*Par.*) *praecursor* in the uppermost Hohenrhein Formation (Solle 1971; Gad 1998; Gad *et al.* 2008) are not a severe problem, because earlier transitional forms are to be expected in this continuous succession, and the biostratigraphic boundary does not have to be precisely coincident with the boundary between the Hohenrhein and Laubach formations. The lower boundary of the middle upper Emsian is best drawn at the level where fully developed *Par.* (*Par.*) *praecursor* are no longer accompanied by transitional forms of *Br. ignoratus* (cf. Solle 1972, pp. 76–77). As a new criterion, this boundary can also be recognized by the phylogenetic transition from *Rhen. rhenana* to *Rhen. piligera*. Corresponding with the evolution of the genus *Sollispirifer*, an older *Soll. mosellanus* Zone and a younger *Soll. dahmeri* Zone can be distinguished within the upper part of the upper Emsian (Kondel substage). Important innovations are the origination of *Alatiformia janseni* Gad, 2002 and *Cun. melanopotamicus* near the beginning of the Kondel time. *Douvillinella filifer* is a guide fossil of the uppermost Emsian Heisdorf interval (Werner 1969). The *Par.* (*Par.*) *cultrijugatus* Zone represents the uppermost Emsian (Kondel/Heisdorf interval) to the lower part of the lower Eifelian (Lauch interval).

Correlation and age. The upper Emsian formations of the Rhenish Massif can be correlated with the Hierges and Saint-Joseph formations and a great part of the Eau Noire Formation on the southern flank of the Dinant Synclinorium (Ardennes, Belgium), where the stratigraphic ranges of *Eur. paradoxus* and other guide fossils indicate the same age (Godefroid 1994; Godefroid *et al.* 1994; Bultynck *et al.* 2000). The precise position of the boundary level between lower and upper Emsian is difficult to ascertain in this standard succession of the Ardennes; it must be located somewhere within the very poorly fossiliferous succession from the Vireux to the Chooz Formation (Godefroid & Stainier 1988; Godefroid *et al.* 1994). The underlying Pesche Formation is still of early Emsian age (e.g. with *Eur. dunensis*), the overlying Hierges Formation of late Emsian age (e.g. with *Eur. paradoxus*).

The succession of upper Emsian Rhenish spiriferide zones is traceable in the Ardennes as well: *Ard. arduennensis arduennensis* is present in the Hierges Formation, and *Soll. mosellanus* and *Par. (Par.) praecursor* have been reported from the upper part of the Hierges Formation, the Saint-Joseph Formation and lower parts of the Eau Noire Formation (Godefroid 1977, 1980; Weddige *et al.* 1979; Godefroid *et al.* 1994; Bultynck *et al.* 2000). *Paraspirifer (Par.) cultrijugatus* is present in the Eau Noire Formation (Godefroid 1977, 1980; Bultynck *et al.* 2000). The following correlations result from the brachiopods and are also corroborated by conodont data (see below): based on the ranges of *Ard. arduennensis arduennensis* and *Soll. mosellanus* (no subspecies of *mosellanus* are indicated by the Belgian workers), the Hierges Formation corresponds approximately with the Wiltz plus Wetteldorf formations in the Eifel region and with the Hohenrhein Formation plus the Laubach Group and the basal Kondel Group in the central Middle Rhine region. The upper part of the Hierges Formation, the Saint-Joseph Formation and the basal Eau Noire Formation (?) may correlate with the lower Kondel Group, corresponding biostratigraphically to the interval between the onset of *Soll. mosellanus* and the onset of *Par. (Par.) cultrijugatus*. A precise delimitation of Lahnstein, Laubach and lower Kondel substages in the Ardennes is problematic, because *Soll. mosellanus* and *Par. (Par.) praecursor* were reported to first occur at approximately the same level within the Hierges Formation (Godefroid *et al.* 1994, figs 12 & 13; Bultynck *et al.* 2000, fig. 6). Lower and middle parts of the Eau Noire Formation correlate with the upper Kondel Group and upper parts of the Heisdorf Formation, corresponding to the interval between the reported onset of *Par. (Par.) cultrijugatus* and the onset of *Intermedites intermedius* (von Schlotheim, 1820) (see Weddige *et al.* 1979; Bultynck *et al.* 2000).

In the eastern Dra Plains (Foum Zguid Section, southern Morocco), a typical eurhenotypic succession of the upper part of the Mdâouer-el-Kbîr Formation ('Rich 3 Sandstone') contains the lower–upper Emsian boundary, recognizable by the first occurrences of *Ard. arduennensis arduennensis* and *Eur.* cf. *robustiformis* (see Jansen *et al.* 2007). At a higher level of the same succession (lower Timrhanrhart Formation), the upper Emsian ammonoid genus *Sellanarcestes* occurs in a distinct limestone unit, well before dacryoconarides, conodonts and ammonoids indicate the beginning of the Eifelian.

The base of the Rhenish upper Emsian has been correlated roughly with the base of the Dalejian Stage in Bohemia (Carls *et al.* 1972; Chlupáč 1982; Carls 1988, p. 447; Jahnke & Jansen 1998; Jansen *et al.* 2007) and the onset of the transgressive 'Daleje Event' in the sense of most authors (House 1985; Chlupáč & Kukal 1988; Walliser 1985: '*gracilis* or *cancellata* Event'; Walliser 1996: 'Mid-Emsian Event'). This correlation has mainly been made in sections with alternating rhenotypic and hercynotypic facies in western Europe and western North Africa, supported by data from the rare Rhenish localities where pelagic fossils co-occur with neritic fossils in the same beds located more or less below and above the boundary. The correlation is not necessarily precise, because the ranges of the fossils are strongly facies-controlled. There is by far no continuous record of pelagic guide fossils in the Rhenish sections. In addition, the faunal overturn in the Rhenish area is partly linked to the slight facies change from the more restricted-marine Nellenköpfchen and Klerf formations to the more open-marine Emsquarzit and Berlé formations, which may likewise be related to the regional subsidence–sedimentation history and not necessarily to the global sea-level rise of the Daleje Event.

Several workers suggested a lower position of the basal boundary of the Rhenish upper Emsian related to the base of the Dalejian based on the entry of *Nowakia cancellata* (Richter, 1854) (see Lütje 1979; Morzadec *et al.* 1981; García-Alcalde & Truyóls-Massoni 1994). In my opinion, these correlations are not convincing, because either the determinations of the brachiopods used for the recognition of the basal upper Emsian are doubtful, for example '*Schizophoria vulvaria*' in Melou (1981) or '*Ard. arduennensis arduennensis*' in Heddebaut (1981) (referred to by Morzadec *et al.* 1981), or the ranges of the taxa used are poorly known, e.g. *Tetratomia amanshauseri* (Dahmer, 1923) (see García-Alcalde & Truyóls-Massoni 1994), and this even in their Rhenish type region. One of the species sometimes discussed as guide fossil for the lower boundary of the upper Emsian, *Lapinulus pila* (Schnur, 1851) (formerly genus '*Uncinulus*'), already appears in the Rhenish Massif in the lower Emsian. In summary, the correlation is still uncertain. Furthermore, Becker (2007) and Ferrová *et al.* (2012) pointed out that a first transgressive episode of the Bohemian Daleje Shale actually started earlier than commonly suggested, with the onset of the *Nowakia elegans* Zone.

There are few possibilities of direct neritic–pelagic correlation in the Ardenno-Rhenish upper Emsian: conodont finds in the *Ard. arduennensis arduennensis* Zone of the southwestern Lahn Synclinorium enable a correlation with the *Polygnathus laticostatus* and lower *Linguipolygnathus serotinus* conodont zones (Weddige & Requadt 1985; Requadt 1990). The same age is indicated by conodonts from the lower and middle parts of the Hierges Formation, again associated with *Ard. arduennensis arduennensis*, on the southern flank of the Dinant Synclinorium in the Ardennes (Bultynck *et al.* 2000). The uppermost Emsian *Po. costatus patulus* Conodont Zone is represented by the Heisdorf Formation in the Prüm Syncline (Eifel region; Weddige 1982) and also by the Saint-Joseph Formation and lower parts of the Eau Noire Formation in the Ardennes (Bultynck *et al.* 2000). Ammonoids are very rare in the Rhenish upper Emsian: a questionable specimen of *Anarcestes plebeius* (Barrande, 1865) has been reported (Solle 1976; determined by O. H. Walliser) from the Höllenthal Formation (Laubach substage) of the Olkenbach Syncline. If the determination is correct, it would imply a late Dalejian to earliest Eifelian age. The specimen is associated with typical upper Emsian brachiopods: *Rhen. piligera*, *Gib. taeniolata*, *Ard. arduennensis* ssp., *Par.* (*Mos.*) *longimargo*, *R. compressa* and *Eur. paradoxus*. *Sellanarcestes wenkenbachi* (Kayser, 1884) is known from shales of the lower and upper parts of the Kondel Group (Solle 1942*a*, *b*, *c*, 1972; Walliser 1965; Mittmeyer 2008). The same genus is typical of the late Dalejian Třebotov Formation in Bohemia (Chlupáč & Turek 1983). It generally shows a late Emsian age and went extinct well before the Emsian–Eifelian boundary (Becker & House 1994, 2000; Klug 2002). The occurrence of the ammonoid *Anarcestes lateseptatus* (Beyrich, 1837) in the Wissenbach Formation, in one case found closely below a specimen of *Eur. paradoxus* (see Solle 1972, p. 82) and therefore most probably still in the upper Emsian, suggests a correlation with the upper part of the Třebotov Formation (Chlupáč & Turek 1983). The species, however, ranges into the basal Middle Devonian (Klug 2002). In summary, it is confirmed that the upper Emsian in the Rhenish sense correlates approximately with the Dalejian Stage in Bohemia.

Palaeoenvironment, sea-level and events. The global sea-level curve shows a transgressive pulse at the beginning and a general transgressive trend in the course of the late Emsian (late T-R cycle Ib and early Ic), interrupted by a rapid short-term regression in the middle part (time of the lower *Ling. serotinus* Zone: Johnson *et al.* 1985; Johnson & Sandberg 1989), or, according to Walliser (1996, 1998), a transgression in the early part and a regressive trend in the late part of the late Emsian. In contrast, a general transgressive trend is reflected by the facies development in the Mosel Synclinorium, where the succession from the Emsquarzit to the Hohenrhein, Laubach, Flaserschiefer and Kieselgallen-Schiefer units changes from proximal to typical and distal eurhenotypic subfacies. This is indicated by sedimentary features such as the general decrease in sand and increase of mud content of the rocks (Meyer & Stets 1996). The reconstructed palaeoenvironments changed from the intertidal to shallow and, finally, deep subtidal. The transgressive pulse at the beginning of the late Emsian is termed the 'Berlé Event' (Mittmeyer 2008) in the Rhenish area; it may approximately coincide with the Daleje Event (Carls *et al.* 1972; House 1985; Jansen *et al.* 2007) in the middle of T-R cycle Ib, which is possibly of global significance (see discussion and reservations in Becker 2007). However, the Rhenish development is influenced by the regional sedimentary history, so the onset of increasingly open-marine conditions may have resulted from reduced siliciclastic input and continued subsidence of the 'Lahn-Mosel Trough' as well. Deltaic and intertidal conditions of the Nellenköpfchen/Klerf interval, with rapid sedimentation and subsidence, changed into prevailing intertidal to shallow-subtidal conditions with erosion and resedimentation during the time of the Emsquarzit (Stets & Schäfer 2002). The environmental change enabled the immigration of brachiopods, which followed the changing distribution of suitable biotopes – a case of habitat tracking. In any case, the increased marine influence at the beginning of the late Emsian involved a faunal turnover that was possibly partly caused by the immigration of new taxa from outside the Rhenish Shelf region or from peripheral or isolated refugia. On the other hand, continued evolution from early Emsian predecessors within the shelf area is also possible. Finally, it must also be taken into consideration that frequent erosion and destruction of shells may have led to a lack of information in the basal upper Emsian, simulating a greater sharpness of the faunal shift.

The regional 'Laubach' and 'Kondel' events (Mittmeyer 2008) are documented from lithological changes and faunal turnovers near or at the bases of the corresponding lithostratigraphic groups (e.g. Meyer & Stets 1996; Kröll 2001; Mittmeyer 2008). With the Laubach Event, the genus *Paraspirifer* diversified into species of the subgenera *Par.* (*Mosellospirifer*) and *Par.* (*Paraspirifer*), and the Kondel Event approximately coincides with a further radiation of *Par.* (*Paraspirifer*) – see the phylogenetic tree by Solle (1971, diagram 1) in combination with the content of the newly defined subgenera herein. The first species of the North American subgenus *Par.* (*Laurentispirifer*) probably originated from a species of *Par.* (*Paraspirifer*) in the latest Emsian or earliest Eifelian time. It is suggested that a transgression, possibly in the course of the Choteč Event, allowed the migration of a representative of *Paraspirifer* via the North Gondwanan shelf into the Eastern Americas Realm. Another possible effect of the Kondel Event was the radiation and flourishing of the genus *Sollispirifer*, the spiriferides of the '*mosellanus* group'.

A eustatic regression at the time of the lower *Ling. serotinus* Conodont Zone, as shown by Johnson *et al.* (1985) and Johnson & Sandberg (1989), would be expected within the late Lahnstein to Laubach intervals (Weddige & Requadt 1985). It is at least not clearly reflected by the succession in the central Rhenish Massif, although Mittmeyer (2008, p. 141) mentioned 'regional sedimentation of sand and slight regression' in the Rhine and Mosel regions with the Laubach Event, also suggested by Kröll (2001) and Poschmann (2015) in their facies characterizations of the lower Laubach Formation in the Lower Mosel area. But here the sandy development actually more or less continued from the underlying Hohenrhein Formation. Apart from that, a change from siliceous to calcareous sandstones and the onset of masses of the ichnogenus *Chondrites* are typical characters of this event. The transgressive beginning of T-R cycle Ic, reported by Johnson *et al.* (1985) and Johnson & Sandberg (1989), but not shown by Walliser (1996, 1998), could then correspond to a transgressive pulse within the later Laubach to Kondel time. In fact, the development, characterized by increasingly fine-grained sedimentary rocks and distal eurhenotypic facies, suggests a deeper subtidal palaeoenvironment during this interval.

In contrast, the continuous regression in late parts of the late Emsian suggested by Walliser (1996, fig. 3; 1998) is not visible in the Rhenish successions. It is documented, for example, in the upper Dalejian of Bohemia (Třebotov Limestone: e.g. Chlupáč & Lukeš 1999) or in the upper Emsian of the western Dra Plains in southern Morocco ('Rich 4' sandstones of the upper Khebchia Formation: Jansen *et al.* 2007). In the central Rhenish Massif, this regression – if it is a global trend – may be largely masked by retarded siliciclastic input from the Old Red Continent and continued subsidence, resulting in a rising relative sea-level.

Regressive tendencies of the Kondel Event (e.g. 'Brauneisen-Sandstein' of Solle 1976) in the Olkenbach Syncline are regarded as more local effects. The presented T-R curve (Fig. 3) is an attempt to combine the most convincing elements of the different curves.

The marine, proximal eurhenotypic subfacies of the Remscheid Group (upper Emsian) in the Bergisches Land reflects siliciclastic sedimentation in a variety of restricted-marine settings in the vicinity of the Old Red Continent (Zygojannis 1971; Hilden 2008). The onset of marine conditions was governed or at least supported here by the rising global sea-level during the late Emsian, possibly correlated with the beginning of T-R cycle Ic. Brachiopod faunas are generally of low diversity, restricted to certain levels and typically dominated by the species '*Subcuspidella*' *crassifulcita*, which is not known from other parts of the Rhenish Massif; pelecypods are abundant and diverse in this facies. It is assumed that a special lagoonal facies type was developed here, where this endemic spiriferide could evolve. The differentiation of this palaeoenvironment may have been influenced by the 'keratophyric' volcanism in this area ('K 4 volcanite').

The upper Emsian succession in the southern Eifel Synclines begins with the shallow-marine quartzitic sandstones of the Berlé Formation, which resembles the Emsquarzit Formation of the central Rhenish Massif. The development above is different: The enriched, distal eurhenotypic subfacies of the Wiltz Formation with its shales and sandstones reflects deeper water and more offshore settings. However, the lack of ammonoids suggests that the palaeoenvironment was still a relatively shallow sea or a sheltered bay with restricted connection to the open sea. The regressive tendencies of the overlying Wetteldorf Formation may correlate with the above-mentioned eustatic lower *Ling. serotinus* Zone regression or the Laubach Event, but it has also been suggested that this change was caused by a rise of the 'Stavelot-Venn Island', shallowing of the neighbouring 'Wetteldorf Sea'; and increased siliciclastic input (Struve *et al.* 2008, pp. 305, 306).

In the northern Eifel synclines, parts of the upper Emsian (Wiltz and Wetteldorf formations) are either not documented or probably represented by the deltaic to terrestrial 'Klerf facies'. As indicated by the presence of the uppermost Emsian 'Emsquarzit' at Honertseifen, this shallow-marine quartzitic facies is diachronous and probably followed a northward transgression in the course of the late Emsian. The partially allorhenotypic subfacies of the high Heisdorf Formation, with its claricolous brachiopods and corals in combination with the more calcareous lithology, reflects a shallow-marine palaeoenvironment with relatively clear water after the input of siliciclastic material had diminished. Here, the *Par. cultrijugatus* Fauna flourished and persisted into the early Eifelian. The possible transgression at the beginning of T-R cycle Ic may have partly pushed back the coarse siliciclastic input and provided more open-marine conditions in parts of the Eifel depositional area, but its bathymetric differentiation (Struve *et al.* 2008) complicates the interpretation.

Lowermost Eifelian brachiopod faunas

Characteristics. Where the lower part of the lower Eifelian (lowermost Eifelian, Lauch interval) is developed in the allorhenotypic subfacies, it contains the younger '*Paraspirifer cultrijugatus* Fauna'. This fauna is typified by the brachiopod taxa *Par.* (*Par.*) *cultrijugatus* (Roemer, 1844), *Par.* (*Par.*) *frechi* Solle, 1971, *Par.* (*Par.*) *curvatissimus* Solle, 1971, *Al. alatiformis* (Drevermann, 1907), '*Sollispirifer*' *schreiberi* (Happel, 1932), *Intermedites vetustus* (Solle, 1953), *Int.* ex gr. *intermedius* (von Schlotheim, 1820), *Iridistrophia* (*Flabellistrophia*) *hipponyx* (Schnur, 1851) (covering bedding surfaces or even forming shell beds), *Rhipidomella subcordiformis* (Kayser, 1871), *Oligoptycherhynchus wetteldorfensis* (Schmidt, 1941), *Cuninulus concavus* Sartenaer, 2005, *Gypidula montana acutecostata* Spriestersbach, 1942, *Teichostrophia lepis subtilis* Struve, 1992 (Fig. 7r) and *Rhenothyris aequabilis tectiplicata* Struve, 1970*b*. In addition, leptaenine strophomenides, chonetidines, atrypides, athyridides and productellids occur. Apart from the extinction of some upper Emsian fossils near the Lower–Middle Devonian boundary and the first occurrences of the genus *Intermedites* Struve, 1995, the species *R. aequabilis* Struve, 1970*b* and a few other taxa in the lowermost part of the Eifelian, there is no radical or abrupt change in faunal composition at this boundary, but rather a stepwise succession of last and first occurrences and slight change in abundances of species during the boundary interval (Struve & Werner 1982).

Concerning the trilobites, a fauna composed of proetides (*Rhenocynproetus*), phacopides (*Pedinopariops*, *Asteropyge*) and lichides (*Ceratarges*) is present in the lowermost Eifelian. The lower part of the Lauch Formation, the calcareous Wolfenbach Member, is highly fossiliferous and rich in brachiopods, whereas the upper part, the sandy and silty marlstones of the Dorsel Member, contains a partly impoverished, partly rich brachiopod fauna (Werner 1974). The *Par. cultrijugatus* Fauna went extinct at the end of the Lauch time. It corresponds to the 'OCA Fauna' (='*orbignyanus-cultrijugatus-alatiformis* Fauna') *sensu* Struve (1982*a*), which, however, explicitely included affiliated faunas from 'many regions of the northern hemisphere' where the three index species may be replaced by presumably related species. A more critical assessment of the taxa under consideration appears to be necessary, including studies of their affinities, phylogeny and biostratigraphy. '*Uncinulus orbignyanus* (de Verneuil, 1850)' has been revised by Sartenaer (2004), who assigned 'the Rhenish *orbignyanus*' to the genus *Cuninulus* Sartenaer, 2005 and distinguished two separate species with different stratigraphic ranges, *Cun. concavus* Sartenaer, 2005 and *Cun. melanopotamicus* Sartenaer, 2005.

Distribution and regional developments. The lowermost Eifelian strata with the younger *Par. cultrijugatus* Fauna are particularly well exposed and fossiliferous in the Eifel Synclines, where this fauna occurs in the calcareous and marly, allorhenotypic successions of the Lauch Formation (e.g. Hotz *et al.* 1955; Krömmelbein 1955; Glinski 1961; Ochs & Wolfart 1961; Struve 1961*a*; Werner 1969, 1974; Struve *et al.* 2008). The Emsian–Eifelian GSSP boundary section at Wetteldorf in the Prüm Syncline contains abundant brachiopods of this unit (Struve & Werner 1982). The boundary has been placed closely below the upper boundary of the Heisdorf Formation, which is overlain by the Lauch Formation. A facially similar, but partly different succession with brachiopod faunas is exposed in the eastern Sauerland, where the stage boundary seems to be located near the boundary between the *Orthocrinus* and Schmallenberg formations (Langenstrassen 1972, 2008; Langenstrassen & Müller 1982).

In contrast, hercynotypic argillaceous shales of the largely Eifelian Wissenbach Formation follow above the Kondel Group in the Mosel, Dill and Lahn synclinoria, at least regionally starting in the uppermost Upper Emsian, where rare last representatives of rhenotypic brachiopods may demonstrate this level. Prevalent here are hercynotypic brachiopods, which are adapted to soft substrates of deep water (Schubert 1996), for example representatives of the genera *Prokopia* and *Dalejodiscus*.

Brachiopod biostratigraphy. Intermedites vetustus (or *Int. intermedius* in a broad sense due to literature data), *R. aequabilis tectiplicata* and *Tei. lepis subtilis* appear close to the base of the Eifelian and characterize the Lauch Formation, corresponding to the lower part of the *Int. vetustus* Zone. Persisting from the uppermost Emsian, *Par.* (*Par.*) *cultrijugatus*, *Cun. concavus* and *Al. alatiformis* are three guide taxa that are very typical of the Lauch Formation, and *Par.* (*Par.*) *cultrijugatus* has its main distribution here. The species have their maximum abundance in the Wolfenbach Member, the lower part of the Lauch Formation, but they persist into its upper part, the often less fossiliferous Dorsel Member, and disappear near the top of this unit. Finally, a very common species of the whole interval is *Ir.* (*Flab.*) *hipponyx*.

Correlation and age. The Lauch Formation, as a part of the basal Eifelian GSSP boundary succession, has an earliest Eifelian age. In terms of the conodont stratigraphy, the lower part of the formation is assigned to the *Polygnathus costatus partitus* Zone (Struve *et al.* 2008). According to Weddige *et al.* (1979, fig. 4) and Weddige (1988), the succeeding *Po. costatus costatus* Zone begins near the base of the Dorsel Member or within this unit. Thanks to the occurrences of conodonts, dacryoconarides and ammonoids, an almost worldwide correlation of this interval is possible. Stratigraphically important dacryoconarides, in particular the lower Eifelian subspecies *Nowakia* (*Dmitriella*) *sulcata sulcata* (Roemer, 1843) and related forms are known from the Lauch Formation (Alberti 1982*a*) and also from the hercynotypic Wissenbach Formation, several meters above last distal eurhenotypic brachiopod assemblages of the uppermost Emsian Haigerhütte Formation (Upper Kondel Group) in the Dill Synclinorium (Haigerhütte Section; Alberti 1981, 1985). The stratigraphically important ammonoid species *Pinacites jugleri* (Roemer, 1843) occurs in the lower Eifelian part of the Wissenbach Formation in the Harz Mountains (Chlupáč & Turek 1983) and again in the Haigerhütte Section in the Dill Synclinorium (Alberti 1981); it has also been reported from calcareous lenses intercalated in shales of the Fredeburg Formation in the eastern Sauerland (Langenstrassen 1972). A critical summary of its occurrences was given by Ebbighausen *et al.* (2011). These hercynotypic facies lack, however, any rhenotypic brachiopods for direct neritic–pelagic correlation. Single specimens of ammonoids from the allorhenotypic Wolfenbach Member (lower Lauch Formation) at the locality 'Held near Prüm' in the Prüm Syncline (Werner 1990) have been illustrated by Becker & House (1994). They include *Foordites platypleura* (Frech, 1889) and *Fidelites* cf. *occultus* (Barrande, 1865), which indicate the *Fo. platypleura* Zone. The same locality yielded the brachiopod species *Par.* (*Par.*) *cultrijugatus* and *Al. alatiformis*. *Foordites* cf. *platypleura* has also been reported from the hercynotypic Rupbach Formation (latest Emsian to Mid-Devonian age) of the Lahn Synclinorium (Requadt 1990), whereas *Fid. occultus* is a common species in the Choteč Limestone (Eifelian) of Bohemia, and it may also occur in the Rupbach Formation (Chlupáč & Turek 1983). In the Ardennes, a Lauch equivalent is documented with brachiopods, e.g. *Int. intermedius* (not seen, *vetustus*?) and '*Soll.*' *schreiberi*, and conodonts of the *Po. costatus partitus* Zone in the upper part of the Eau Noire Formation to the lower part of the Couvin Formation of the Couvin area (Bultynck *et al.* 2000).

Palaeoenvironment, sea-level and events. The allorhenotypic subfacies of the Lauch Formation reflects mainly shallow- and ± clear-water conditions (Faber 1980; Werner & Hahn 1988). The faunas include brachiopods regarded as claricolous, such as atrypides, reticulariid spiriferides, athyridides and pentamerides. Compared with the Early Devonian, the earliest Eifelian was a time of relatively high eustatic sea-level (middle part of T-R cycle Ic *sensu* Johnson *et al.* 1985; Johnson & Sandberg 1989; Walliser 1996), providing a shallow-marine calcareous shelf in the area of the present-day Eifel region and deeper-water, hercynotypic conditions in the central and eastern Rhenish Massif. The sedimentation area in the Eifel region was distinctly divided into moderately deep to very shallow-water or probably even subaerial settings (Struve *et al.* 2008), showing that the area was located relatively high on the shelf. The high eustatic sea-level was certainly related to the transgressive 'Basal Choteč Event' (Chlupáč & Kukal 1986, 1988; = '*jugleri* Event' *sensu* Walliser 1985), following a short time after the beginning of the Eifelian.

The genus *Intermedites* has its onset near the base of the Lauch Formation. Due to similarities to the 'group around *Rostrospirifer* and *Otospirifer*', Schemm-Gregory (2010) regarded it as a 'South Chinese immigrant'. Representatives of this genus could migrate thanks to the high sea-level near the beginning of the Eifelian, which caused a general loss of faunal endemicity during this time. This interpretation appears plausible, but Solle (1953) pleads for closer relationships between the youngest species of the Rhenish genus *Sollispirifer*, *Soll. dahmeri* (Solle, 1953), and the oldest species of *Intermedites*, *Int. vetustus* (Solle, 1953) (species in the present generic assignment), suggesting a phylogenetic transition. A detailed re-investigation of these taxa to clarify this question appears still to be desirable.

The lowermost part of the Eifelian ends with the upper boundary of the Lauch Formation and the level of the 'OCA Extinction Event'. This event marks the extinction of the '*orbignyanus-cultrijugatus-alatiformis* Fauna', which explicitly was established for a very wide palaeogeographical range and also included similar variants, subspecies and related species (see Struve 1982*a*; Struve *et al.* 1997, 2008). Because of the uncertain affiliations and stratigraphic ranges of forms replacing the three index brachiopod species in areas outside the Rhenish area, I propose the use of the local term 'Kirberg Event' for this extinction event, following the common practice of using geographical names in event nomenclature. This term has also been used as an alternative by Struve (1990, p. 261) and Struve *et al.* (2008, p. 310). The eponymous Kirberg Member is the first subunit of the overlying Nohn Formation following after the event level. Near the lithostratigraphic boundary between Lauch and

Nohn formations or slightly below it, within the *Po. costatus costatus* Zone, the brachiopod species *Par.* (*Par.*) *cultrijugatus*, *Al. alatiformis* and *Cun. concavus* disappear.

It is possible that the Kirberg Event is genetically somehow linked to the Basal Choteč Event, as has been suggested earlier (Struve *et al.* 1997), but a more detailed correlation appears necessary. The Basal Choteč Event in the Barrandian type region is characterized by an abrupt onset of darker-coloured biomicritic and biosparitic limestones and dark shales within lighter micritic limestones, and a clear faunal change, in particular in the trilobites, ammonoids and dacryoconarides (Chlupáč & Kukal 1986, 1988). Weddige (1988, p. 106) and Vodrážková *et al.* (2013) regarded the base of the Dorsel Member, the upper unit of the Lauch Formation, as the level corresponding to the onset of the Basal Choteč Event, making it distinctly older than the Kirberg Event. The event in the Eifel may even fall into the underlying Wolfenbach Member as the main transgressive phase of the event started already in the *Po. costatus partitus* Biochron (Becker & Aboussalam 2013; Vodrážková *et al.* 2013). The global event in pelagic facies obviously does not precisely correlate with the neritic event in the Eifel region, but both may represent two consecutive phases of changing palaeoenvironments. A few meters above the base of the Choteč Limestone, Vodrážková *et al.* (2013, p. 440) described a shallowing event within the *Po. costatus costatus* Zone. There is a possibly corresponding regressive, post-Choteč Event marker unit in southeastern Morocco (Klug 2002; Becker & Aboussalam 2013). The possibly coeval succession of the Dorsel Member is interpreted as regressive, and the Kirberg Event – according to Struve with hiatus character (Struve *et al.* 2008, p. 314) – could be interpreted as a sea-level lowstand. Interestingly, the genus *Paraspirifer* disappears near the upper boundary of the Lauch Formation and never reappears in the Ardenno-Rhenish area but appears with slightly modified morphology at about the same level in eastern North America in the Moorehouse Member of the Onondaga Formation (*Po. costatus costatus* Zone) and extends upwards into younger strata of the Middle Devonian (Johnson 1979; Godefroid & Fagerstrom 1983; Bartholomew & Brett 2007). This implies that the new subgenus *Par.* (*Laurentispirifer*) spread in the Appalachian Basin in the time of the lithostratigraphic member following after the suggested level of the Basal Choteč Event (Ver Straeten 2007; Brocke *et al.* 2015). As *Paraspirifer* and possible representatives of the new subgenus occur in North Africa, it is assumed that a species had migrated from the northern Gondwanan shelf in the Eastern Americas Realm during the latest Emsian or earliest Eifelian time.

The Kirberg Event introduces the interval of the 'Great Gap' *sensu lato* (Struve 1982*b*; Struve *et al.* 2008), with new brachiopod associations in the Eifelian of the Eifel region typified by reticulariid spiriferides and atrypides – these are, in stratigraphic order, the *Uexothyris* Fauna, the *Uexothyris–Gerothyris* syncharchy, and the *Gerothyris*, *Yeothyris* and *Isospinatrypa–Invertrypa* faunas (Struve 1990; Struve *et al.* 2008). Apart from *Uexothyris*, the upper part of the lower Eifelian (Nohn level) is characterized by other reticulariids, for example *Rhenothyris aequabilis tumida* Struve, 1970*b* (Fig. 7l & m).

In the central and eastern parts of the Rhenish Massif, the general transgressive trend and possibly continuing subsidence in the Early/Mid-Devonian boundary interval resulted in a change from the distal eurhenotypic subfacies to the argillaceous hercynotypic ('Wissenbach') facies – and in the extinction of the eurhenotypic brachiopod faunas. A plausible reason for this extinction is the onset of largely pelitic sedimentation with the rising sea-level and resulting soft and probably poorly oxygenated substrates on the deeper shelf not suitable for the settlement of these brachiopods.

Systematic palaeontology

This section includes only short descriptions; more details and complete synonymy lists will be published later.

Morphological abbreviations: AVIM, internal mould(s) of articulated valves; DVIM, internal mould(s) of dorsal valve(s); EM, external mould(s); IM, internal mould(s); VVEM, external mould(s) of ventral valve(s); VVIM, internal mould(s) of ventral valve(s); combinations such as VVIM + EM mean that the internal mould and the corresponding external mould (counterpart) are present.

Deposition of material: AMNH, American Museum of Natural History, New York, NY, USA; IRSNB, Institut royal des Sciences naturelles de Belgique, Brussels, Belgium; SMF, Senckenberg Research Institute and Natural History Museum, Frankfurt a. M., Germany; USNM, Smithsonian National Museum of Natural History, Washington, DC, USA; YPM, Peabody Museum of Natural History, Yale University, New Haven, CT, USA.

Phylum **Brachiopoda** Duméril, 1805
Subphylum **Rhynchonelliformea** Williams, Carlson, Brunton, Holmer & Popov, 1996
Class **Strophomenata** Williams, Carlson, Brunton, Holmer & Popov, 1996
Order **Orthotetida** Waagen, 1884
Suborder **Orthotetidina** Waagen, 1884
Superfamily **Chilidiopsoidea** Boucot, 1959

Family **Chilidiopsidae** Boucot, 1959
Subfamily **Chilidiopsinae** Boucot, 1959

Genus *Iridistrophia* Havlíček, 1965

Type species. *Orthis umbella* Barrande, 1848.

Diagnosis. Jansen (2001*a*, p. 174).

Remarks. Within the genus *Iridistrophia*, representatives with relatively short dental plates and faintly or not impressed ventral muscle fields can be distinguished from representatives with relatively long dental plates passing into well-developed, curved muscle-bounding ridges that enclose distinctly impressed ventral muscle fields (Jansen 2001*a*, pp. 175–176). Accordingly, *Iridistrophia* is subdivided into two subgenera that follow stratigraphically one after the other: *Ir.* (*Iridistrophia*) and *Ir.* (*Flabellistrophia*) subgen. nov. *Iridistrophia* (*Iridistrophia*) is most probably the phylogenetic forerunner of *Ir.* (*Flabellistrophia*), and the transition took place near the boundary between the early and late Emsian. In the central Rhenish Massif, the first fully developed representatives of *Ir.* (*Flabellistrophia*) occur in the Emsquarzit Formation, in the basal upper Emsian.

Subgenus *Iridistrophia* (*Iridistrophia*)
Havlíček, 1965
(Fig. 7q)

Type species. *Orthis umbella* Barrande, 1848.

Diagnosis. Shell medium-sized to large. Dental plates short or moderately long to long, rarely passing into distinct muscle-bounding ridges delimiting the ventral muscle field laterally; muscle-bounding ridges not continued anteriorly, often totally lacking. Ventral muscle field not or faintly impressed; diductor impressions not subdivided or more or less distinctly subdivided by weak radial ridges. Median node at the base of the bilobed cardinal process in most specimens representing isolated elevation, not joined with relatively weak dorsal median ridge.

Species included. *Orthis umbella* Barrande, 1848; *Orthis iris* Barrande, 1879; *Strophomena elongata* Barrande, 1879; *Orthothetes maior* Fuchs, 1915; *Orthothetes euzona* Fuchs, 1919; *Schellwienella praeumbracula* Kozłowski, 1929; *Orthotetina* (*Schellwienella*) *pencki* Paeckelmann & Sieverts, 1932; *Iridistrophia eodevonica* Havlíček, 1967; *Iridistrophia anorhrifensis* Jansen, 2001*a*.

Range. Lochkovian to lower Emsian.

Discussion. The presence of short dental plates could not be confirmed in the type material of the upper Silurian species '*Iridistrophia*' *chloe* Havlíček, 1992, which is housed in the Dr.-Bohuslav-Horák Museum in Rokycany, Czech Republic. The species is therefore excluded from the genus. The same applies to '*Schuchertella euzona*' described by Boucot (1960) from the 'Schistes de Mondrepuits' in the Ardennes (Boucot 1960, pp. 305–307, pl. 14, figs 1–8; IRSNB). The figured material does not belong to *Ir.* (*Ir.*) *euzona* (Fuchs, 1919).

The assignment of additional *Iridistrophia* species from the Canadian Arctic Islands (Smith 1980), Argentina (Herrera *et al.* 1998), Kazakhstan (Ushatinskaya 1975) and Australia (Farrell 1992) to the subgenus or at least the genus appears probable but should be verified by the study of original specimens.

Subgenus *Iridistrophia* (*Flabellistrophia*)
subgen. nov.
(Fig. 8a–g)

Type species. *Orthis hipponyx* Schnur, 1851.

Derivation of name. A composite of *flabellum* (Latin) = fan and the generally used suffix *-strophia*; allusion to the fan-shaped ventral muscle field.

Diagnosis. Shell large. Dental plates long and passing into distinct, anteriorly curved muscle-bounding ridges that delimit the ventral muscle field laterally and anteriorly. Ventral muscle field moderately to strongly impressed, with flabellate diductor scars subdivided by radial ridges. Median node at the base of the bilobed cardinal process usually joined with a relatively strong dorsal median ridge.

Species included. *Orthis hipponyx* Schnur, 1851; *Iridistrophia* (*Flabellistrophia*) *musculosa* subgen. et sp. nov. Questionably assigned: *Orthis undifera* Schnur, 1853; *Iridistrophia dendritica* Benedetto, 1984 (from Venezuela).

Range. Upper Emsian to Eifelian.

Comparison. *Iridistrophia* (*Flabellistrophia*) differs from the Lochkovian to lower Emsian subgenus *Ir.* (*Iridistrophia*) by generally longer dental plates and distinctly impressed ventral muscle field with diductor scars more clearly subdivided by radial ridges; the muscle field is laterally and anteriorly enclosed by curved muscle-bounding ridges. In contrast, the ventral muscle field is weakly or not impressed in *Ir.* (*Iridistrophia*), the dental plates are mostly shorter and muscle-bounding ridges are lacking or faintly developed. Juvenile or neanic specimens of *Ir.* (*Flabellistrophia*) may not show the impressed ventral muscle field and resemble representatives of the older subgenus but then often show relatively progressive muscle-bounding ridges. In *Ir.* (*Iridistrophia*), stronger muscle-bounding ridges and radial ridges within the muscle field may be simulated by deformation. The muscle fields can be better impressed in adult to gerontic specimens of phylogenetically advanced species of *Ir.* (*Iridistrophia*), but these commonly do not show muscle-bounding ridges anterior to the muscle field. The node at the base of the bilobed cardinal process is normally joined with a stronger dorsal median ridge in *Ir.* (*Flabellistrophia*), whereas the node is mostly isolated in *Ir.* (*Iridistrophia*).

The new subgenus resembles the genus *Hipparionix* Vanuxem, 1842 from the Lower Devonian of eastern North America (type species: *Hip. proximus* Vanuxem, 1842), but differs in smaller ventral muscle field and the lack of a ventral process subdividing the central apical cavity. In addition, the Rhenish subgenus shows a smaller cardinal process and a wider angle enclosed by the brachiophores. In contrast to the Rhenish subgenus, the posterolateral costellae are curved in posterior direction towards the cardinal margin in the type species of *Hipparionix*, and this is also visible on the internal moulds (studied specimens: AMNH-FI 34750-34757, 34759-34761).

Iridistrophia (*Flabellistrophia*) *musculosa* subgen. nov. et sp. nov.
(Fig. 8a, b, f)

v ? 1842 *Orthis subarachnoidea*, nob. d'Archiac & de Verneuil, Fossils Rhenish Provinces: p. 372, pl. 36, fig. 3.

v p? 2001*a* *Iridistrophia* cf. *hipponyx* (Schnur, 1851). Jansen, Brachiopoden Unter-Devon Marokko Deutschland: pp. 182, 183, pl. 16, fig. 6, pl. 17, figs 1–4.

Derivation of name. *Musculosus* (Latin) = muscular. The name draws attention to the strong development of the ventral diductor scars.

Diagnosis. Shell broadly semielliptical in outline. Ventral valve almost flat or faintly to moderately resupinate; greatest width often close to the relatively wide cardinal margin, posterolateral margins weakly to moderately curved in outline. Dorsal valve strongly convex in longitudinal section. Ornamentation uniformly costellate, showing a very regular pattern. Dental plates diverging at 70–85°; ventral muscle field large, usually reaching around half valve length, with diductor scars subdivided by strong radial ridges. Endospines often clearly developed on internal surface of ventral valve. Dorsal apical region with adductor field clearly impressed with respect to the dental sockets, umbo of internal mould often pronounced; cardinal process bilobed, with lobes steeply projecting in ventral direction.

Range. Lower to upper parts of upper Emsian.

Holotype. Internal mould of ventral valve with corresponding external mould, SMF 65563a + b, Figure 8a, b. Dimensions of the internal mould: $W = 58.2$ mm, $L = 49.4$ mm.

Type locality. Maria Roth, Konder-Tal, south of Winningen, Lower Mosel area, Rhenish Massif, Germany. Topographical mapsheet 1:25 000 Rheinland-Pfalz No. 5711 Boppard.

Type stratum. Laubach Group, middle part of upper Emsian.

Material. 1 AVIM + VVEM, 2 AVIM+ DVEM, 6 AVIM, 7 VVIM + EM, 15 VVIM, 6 VVEM, 4 DVIM + EM, 24 DVIM, 2 DVEM from different levels within the upper Emsian and from localities in the Middle Rhine, Eifel, Hunsrück, Taunus and Dill areas (inventory numbers: SMF 65322, 65324, 65335, 65566, 65821, 66041, 66176-66180, 66294, 66372, 66379-66381, 66515, 66516, 66563, 66821-66844, 66846, 66849, 66853, 66854, 66856-66859, 93811-93822).

Comparison. The new species differs from the geologically younger, uppermost Emsian to Eifelian species *Iridistrophia* (*Flabellistrophia*) *hipponyx* (Schnur, 1851) by a relatively wider cardinal margin, slightly wider shell and a larger ventral muscle field usually reaching half valve length. Ventral valves are rather flat, whereas they tend to be more resupinate in the geologically younger species. The dorsal valve is generally more convex. The ornamentation is markedly uniformly costellate in *Ir.* (*Flab.*) *musculosa*, whereas a distinct tendency to develop a more irregular pattern is visible in *Ir.* (*Flab.*) *hipponyx* (compare Fig. 8f with g). The angle enclosed by the dental plates is mostly a little wider, 70–85° in *Ir.* (*Flab.*) *musculosa* v. 50–70° in *Ir.* (*Flab.*) *hipponyx*. The cardinal process lobes are ventrally directed in *Ir.* (*Flab.*) *musculosa* and more posteroventrally in *Ir.* (*Flab.*) *hipponyx*. The dorsal apical region, including the adductor field, is generally more impressed with respect to the dental sockets in the new species, and the umbo of the dorsal internal mould is more pronounced. *Iridistrophia* (*Flab.*) *musculosa* is regarded as the phylogenetic forerunner of *Ir.* (*Flab.*) *hipponyx*; the phylogenetic transition took place during the latest Emsian (Kondel) time.

Class **Rhynchonellata** Williams, Carlson, Brunton, Holmer & Popov, 1996
Order **Rhynchonellida** Kuhn, 1949
Superfamily **Uncinuloidea** Rzhonsnitskaya, 1956
Family uncertain

Genus *Sartenaerirhynchus* gen. nov.
(Fig. 9a–i)

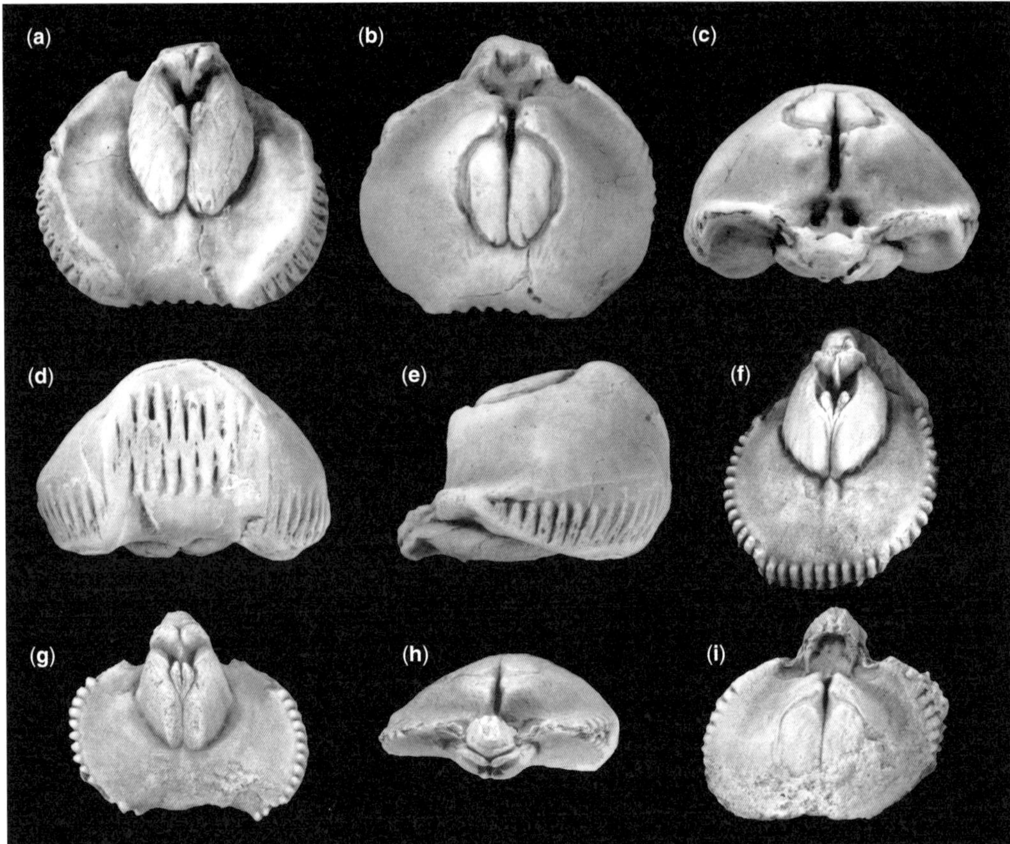

Fig. 9. (a–e) *Sartenaerirhynchus antiquus* (Schnur, 1853) gen. nov.; internal mould of articulated valves, ventral (**a**), dorsal (**b**), posterior (**c**), anterior (**d**) and lateral (**e**) views (× 1.9); Schnur collection, IPB, syntype, Institute of Palaeontology, University of Bonn. Locality: Daleiden, West Eifel region. Stratum: Wiltz Formation, *Arduspirifer arduennensis arduennensis* Zone, lower or middle part of upper Emsian. (**f**) *Sart. frontecostatus* (Drevermann, 1902) gen. nov.; internal mould of ventral valve (× 1.9), syntype SMF-Mbg. 1106. Locality: Seifen, Westerwald. Stratum: Seifen Formation, *Multispirifer solitarius* and *Acrospirifer primaevus* zones, upper part of middle Siegenian. (g–i) The same species, same locality and stratum, internal mould of articulated valves, ventral (**g**), posterior (**h**) and dorsal (**i**) views (× 1.8); syntype SMF-Mbg. 1104.

Type species. *Terebratula antiqua* Schnur, 1853.

Derivation of name. The genus is dedicated to Paul Sartenaer (1925–2015).

Diagnosis. Shell medium-sized, subpentagonal to subelliptical in outline, posteriorly acuminate, markedly dorsi-biconvex; greatest thickness of shell at the front. Anterior and anterolateral margins geniculate and with spine-like projections at the commissures; anterolateral edges of ventral valve before the geniculation slightly bent ventrally. Shell largely smooth but bearing strong, uniformly developed costae on the trails of both valves. M-shaped front lacking. Sulcus shallow, with high sulcus tongue; fold low. Dental plates not leaving incisions at internal mould (absent?), but vestigial lateral apical cavities may be present. Ventral muscle field strongly impressed, with high, bifid process originating from posterior adductor scars and directed anterodorsally; median ridge bisects the diductor scars. Cardinal process strongly developed, striate; strong dorsal median septum present; shallow septalium indicated. Dorsal adductor field slightly impressed, surrounded by thin muscle-bounding ridges.

Species included. *Terebratula antiqua* Schnur, 1853; *Uncinulus frontecostatus* Drevermann, 1902. Questionably assigned: *Rhynchonella* (*Wilsonia*) *Sancti Michaëlis* Kayser, 1889.

Range. Middle Siegenian to upper Emsian.

Discussion and comparison. The diagnosis may be supplemented in the future by additional internal characters of the dorsal valve, which can be better examined in serial sections of shells and may further elucidate the relationships of the genus. The new genus differs from the genera of the family Nucinulidae mainly by the presence of coarser costae restricted to the geniculate parts of both valves, the more angular general shape of the shell and the peculiar process within the ventral muscle field. The morphology of the ventral muscle field may even justify the erection of a new family. The Nucinulidae are represented in the Rhenish Massif by the genus *Lapinulus* Sartenaer, 2005, which has a subglobular shell, finer costellae covering the whole shell and a completely different ventral muscle field. There are some similarities to the Glossinotoechiidae, in particular the genera *Glossinotoechia* Havlíček, 1959 (type species: *Terebratula Henrici* Barrande, 1847) and *Chlupacitoechia* Havlíček, 1992 (type species: *Uncinulus* (*Glossinulus*) *chlupaci* Havlíček, 1956) (Havlíček 1961, 1992; Schumann 1965; García-Alcalde 2008). The new genus differs from both genera in largely smooth valves except for the geniculate parts, the absence of distinct dental plates and the presence of a high process within the ventral muscle field. Species of *Glossinotoechia* and *Chlupacitoechia* share the shape of the anterolateral margins of the ventral valve with the new genus: these margins are first bent in the ventral direction and then abruptly in the dorsal direction, each developing a *paries geniculatus* ('double geniculation'; García-Alcalde 2008).

The type species *Sartenaerirhynchus antiquus* (Fig. 9a–e) ranges from the lower to the upper Emsian and the second species, *Sart. frontecostatus* (Fig. 9f–i), from the middle to the upper Siegenian of the Rhenish Massif. Maillieux (1941) reported '*Uncinulus*' *frontecostatus* and '*Unc.*' *antiquus* from the Pesche Formation in the Ardennes and the latter also from the lower Emsian beds of Burg-Reuland in Luxembourg.

Order **Spiriferida** Waagen, 1883
Suborder **Delthyridina** Ivanova, 1972
Superfamily **Delthyridoidea** Phillips, 1841
Family **Hysterolitidae** Termier & Termier, 1949
Subfamily **Paraspiriferinae** Pitrat, 1965

Genus *Paraspirifer* Wedekind, 1926

Type species. *Spirifer cultrijugatus* Roemer, 1844.

Revised diagnosis. Shell large, dorsi-biconvex, compact to moderately transverse in outline, megathyrid to distinctly brachythyrid, rarely mucronate, with moderately strong, rounded, simple and bifurcating costae showing a vergence in cross section; sulcus generally deep and smooth, with high tongue and without median costa; a few lateral plications included in the depression of the sulcus; fold high and angular to rounded in cross section; micro-ornamentation fimbriate, consisting of fine concentric growth lines and single rows of micro-spines. Uncovered anterior portions of dental plates long to very short or almost absent; ventral muscle field moderately to strongly impressed. Short crural plates commonly present but hardly reaching the shell bottom; notothyrial platform absent.

Species included. Listed under the subgenera.

Subgenus *Paraspirifer* (*Paraspirifer*) Wedekind, 1926
(Fig. 10a, b, j, k)

Type species. *Spirifer cultrijugatus* Roemer, 1844.

Diagnosis. Shell more or less transverse in outline, moderately to strongly dorsi-biconvex; dorsal valve two to about four times higher than ventral valve; sulcus tongue high; fold mostly high, angular to rounded in transverse section, clearly demarcated as elevated structure against lateral fields, which are straight or concave. Costae numerous, relatively coarse, simple or bifurcating; bifurcation of costae common, scattered or rarely absent, taking place mostly late during ontogeny, in the anterior part of adult shells; total number of costae along the commissure reaches a maximum of about 20 per flank, on the internal mould normally below 20 discernible. Free portions of dental plates moderately long in ancient forms to very short or absent in younger forms; ventral muscle field strongly impressed, subrhombical to kite-shaped or subtriangular in outline, with greatest width commonly at half of its length or within the anterior half; lateral boundaries of muscle field commonly sharply bent at the greatest width; areas lateral to muscle field flattened, forming a visceral platform.

Species included. *Spirifer cultrijugatus* Roemer, 1844; *Histerolites chillonensis* Quintero & Revilla, 1966; *Paraspirifer praecursor* Solle, 1971; *Paraspirifer globosus* Solle, 1971; *Paraspirifer cultrijugatus frechi* Solle, 1971; *Paraspirifer cultrijugatus minor* Solle, 1971; *Paraspirifer curvatissimus* Solle, 1971; *Paraspirifer bucculentus* Solle, 1971; *Paraspirifer beclardi* Godefroid, 1977. The former subspecies *Par. cultrijugatus frechi* is elevated to species level, *Par. cult. minor* is regarded as semi-adult specimens of *Par. cultrijugatus*.

Range. Middle part of upper Emsian (Laubach substage) to lowermost part of Eifelian (Lauch interval).

Discussion and comparison. The subgenus *Par.* (*Paraspirifer*) includes the species of the *praecursor-cultrijugatus* group (Solle 1971); it is

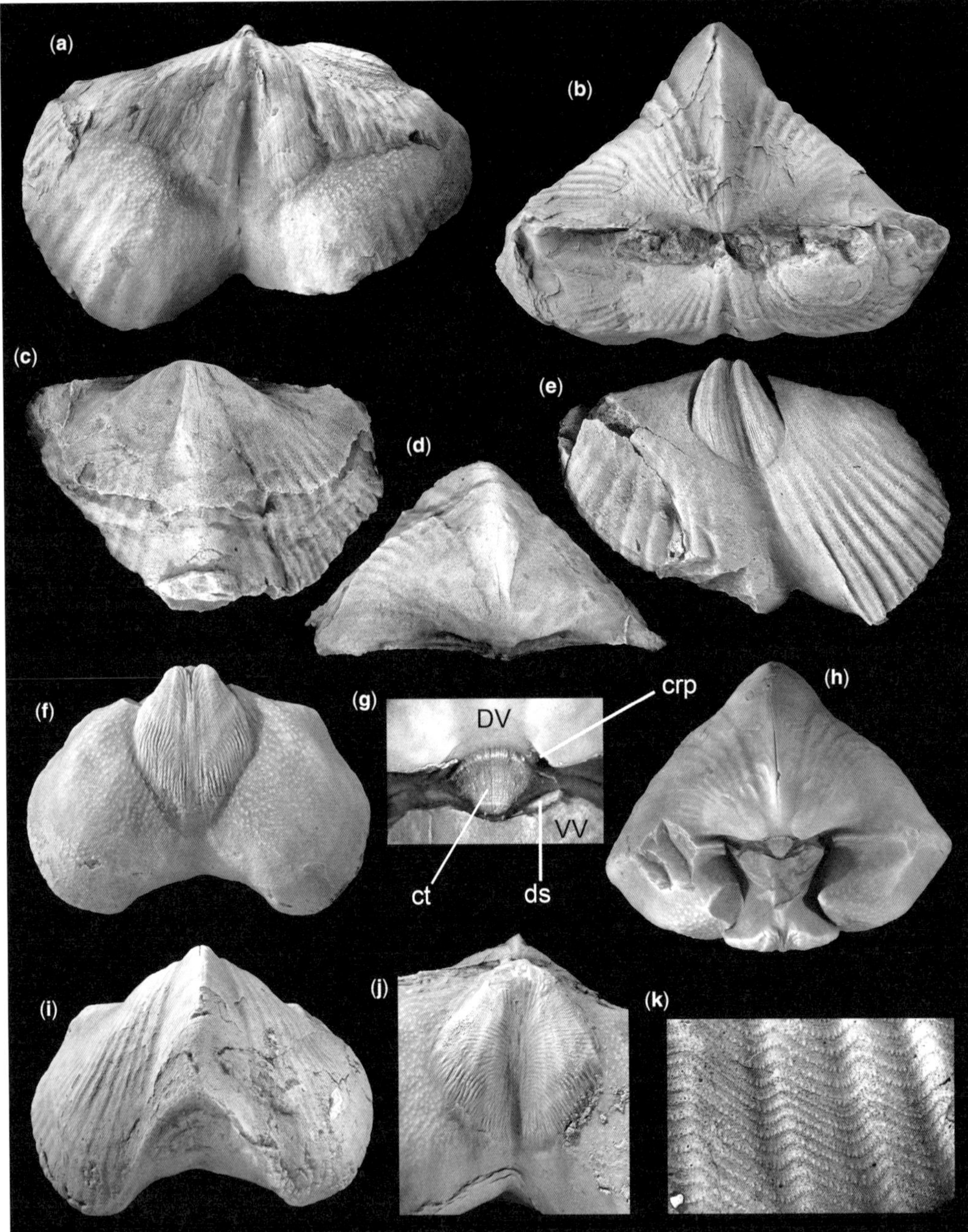

Fig. 10. (a, b) *Paraspirifer* (*Par.*) *cultrijugatus* (Roemer, 1844), articulated shell, partly exfoliated and showing internal mould, ventral (**a**) and posterior (**b**) views (× 1.0); holotype by virtue of monotypy, C. F. Roemer collection No. 12, Palaeontological Museum, University of Bonn. Locality: Eifel region. Stratum: Lauch Formation, *Par.* (*Par.*) *cultrijugatus* Zone, lower Eifelian. (c, d) *Paraspirifer* (*Mosellospirifer*) *sandbergeri* Solle, 1971, internal mould of dorsal valve, top (**c**) and posterior (**d**) views (× 1.0), SMF 93352; collected by J. Hefter 1941. Locality: Dörrbach-Tal near Koblenz. Stratum: Laubach Group, middle part of upper Emsian. (**e**) The same species, internal mould of ventral valve (× 1.0), SMF 94424a; collected by J. Hefter 1941. Locality: Karstel near Oberlahnstein. Stratum: upper Emsian beds. (f–i) *Paraspirifer* (*Laurentispirifer*) *conradi* Godefroid & Fagerstrom, 1983, internal mould of articulated valves, ventral (**f**), posterior (**h**) and dorsal (**i**) views (× 1.0), cardinalia (**g**) in posterior view

distinguished from the other subgenera by the external shape of the shells, the macro-ornamentation and the outline of the ventral muscle field (see below for details). Apart from the Ardenno-Rhenish area (e.g. Laubach Formation, Middle Rhine region; Heisdorf and Lauch formations, Eifel region), the typical subgenus occurs in deposits of Devonian Peri-Gondwanan shelf areas; it is known, for example, from the Pyrenees (e.g. Kalforro and Elorzuri formations, Basque Pyrenees), Cantabrian Mountains (Moniello Formation), Iberian Chains (Molino and Monforte formations) and the Sierra Morena (Herrera Formation, Almadén Syncline).

Subgenus *Paraspirifer* (*Mosellospirifer*) subgen. nov.
(Figs 7k & 10c–e)

Type species. *Paraspirifer sandbergeri* Solle, 1971.

Derivation of name. Named after the river Mosel (Latin: *Mosella*); the new subgenus is common in the upper Emsian of the Mosel Valley and its distributaries; combination with the genus name *Spirifer*.

Diagnosis. Shell commonly transverse in outline, subequally biconvex to slightly dorsi-biconvex; sulcus tongue low; fold moderately high and subangular to rounded in cross section, more or less clearly demarcated as elevated structure against lateral fields. Costae coarse; bifurcation frequent, taking place predominantly late during ontogeny, in anterior parts of adult shells; total number of costae along the commissure reaches a maximal number of approximately 20 per flank of internal moulds. Free portions of dental plates long to moderately long, can be short in large specimens; ventral muscle field elongate and subelliptical or suboval in outline, faintly to strongly impressed, with greatest width commonly at half of its length or in the anterior half; lateral boundaries of muscle field continuously curved convexly outward; areas lateral to ventral muscle field faintly to moderately flattened.

Species included. *Spirifer auriculatus* Sandberger & Sandberger, 1856; *Paraspirifer sandbergeri* Solle, 1971; *Paraspirifer sandbergeri longimargo* Solle, 1971 (elevated to species level; = *Paraspirifer sandbergeri brevimargo* Solle, 1971); *Paraspirifer eos* Solle, 1971; *Paraspirifer sandbergeri nepos* Solle, 1971. Questionably included: *Paraspirifer gigantea* Su, 1976; *Paraspirifer desbiensi* Bizzarro & Lespérance, 1999.

Range. Lower to upper parts of upper Emsian, rarely present in the lowermost Eifelian (Lauch interval).

Comparison and discussion. *Paraspirifer* (*Mosellospirifer*) differs from *Par.* (*Paraspirifer*) in often more abundant bifurcations of costae; in the latter subgenus, the frequency of bifurcations is generally more variable from species to species and within a species. Shells are nearly equibiconvex to weakly dorsi-biconvex in *Par.* (*Mosellospirifer*), with the dorsal valve often being only little higher than the ventral valve to about twice as high, whereas shells of the typical subgenus are commonly strongly dorsi-biconvex with dorsal valves two to four times higher than ventral valves. The most significant difference concerns the outline of the ventral muscle field: in *Par.* (*Mosellospirifer*) it is generally slender and subelliptical in outline, in contrast to the generally wider and subrhombical to kite-shaped or subtriangular muscle field of *Par.* (*Paraspirifer*). The free anterior portions of the dental plates are mostly longer in *Par.* (*Mosellospirifer*), with the exception of the ancestral species *Par.* (*Par.*) *praecursor* and *Par.* (*Par.*) *globosus* which have similarly individualized dental plates. The areas lateral to the ventral muscle field are commonly more flattened in *Par.* (*Paraspirifer*) to form individualized visceral platforms.

Paraspirifer (*Mosellospirifer*) differs from *Par.* (*Laurentispirifer*) subgen. nov. in later bifurcation of costae reaching a lower number of nearly 20 per flank along the commissure, a lower dorsal valve in relation to the ventral valve, lower sulcus tongue and often more rounded cross section of dorsal fold. In the ventral valve, the anterior free portions of the dental plates are longer, and the ventral muscle field is more slender with its greatest width located more anteriorly.

'*Paraspirifer*' *desbiensi* Bizzarro & Lespérance, 1999 from the Emsian of the Gaspé Peninsula (eastern Canada) shows many similarities to representatives of *Par.* (*Mosellospirifer*), such as the general shell form and the outline of the ventral muscle field. It differs in the presence of a flattened dorsal

Fig. 10. (*Continued*) (×3.0); paratype USNM 398830. crp, crural plate; ct, ctenophoridium; ds, dental socket; DV, dorsal valve; VV, ventral valve (structures preserved as moulds). Locality: Jefferson County, Indiana. Stratum: Jeffersonville Limestone, Eifelian. (**j**) *Paraspirifer* (*Par.*) *cultrijugatus* (Roemer, 1844), internal mould of articulated valves, ventral muscle field (×1.1), SMF 25239; collected by G. Solle 1937. Locality: Wetteldorf section, bed 191, Prüm Syncline, Eifel region. Stratum: Lauch Formation, *Par.* (*Par.*) *cultrijugatus* Zone, lower part of lower Eifelian. (**k**) *Paraspirifer* (*Par.*) *praecursor* Solle, 1971, latex cast of external mould of dorsal valve showing fimbriate micro-ornamentation (×5.0), YPM 600914; collected by G. Solle. Locality: Füllersbach-Tal. Stratum: Kieselgallen-Schiefer Formation, *Par.* (*Par.*) *praecursor* Zone, uppermost part of upper Emsian.

fold with rectangular section and fine groove, costae with rectangular profile and indistinct bifurcations, very long ventral muscle field and strong dorsal myophragm extending on entire length of valve. The species is therefore assigned with reservation to the genus.

The species '*Paraspirifer*' *gigantea* Su, 1976 from presumably Emsian beds of Inner Mongolia (north China) seems to belong to the genus *Paraspirifer* but could not be assigned with certainty to any subgenus yet. Judging from the shape of the ventral muscle field (Su 1976, pl. 107, fig. 6), affinities to *Par.* (*Mosellospirifer*) are possible.

Subgenus *Paraspirifer* (*Laurentispirifer*) subgen. nov. (Fig. 10f–i)

Type species. *Paraspirifer conradi* Godefroid & Fagerstrom, 1983.

Derivation of name. After *Laurentia* – representatives of the new subgenus are distributed on the marine shelves along this palaeocontinent; combination with the genus name *Spirifer*.

Diagnosis. Shell relatively narrow, compact to more or less transverse in outline, brachythyrid, commonly with narrow hinge line, markedly dorsibiconvex; sulcus tongue high; lateral fields of dorsal valve steep, straight or often slightly convex in transverse section; fold generally not very high, subangular to angular in cross section, moderately to poorly demarcated as elevated structure against lateral fields. Costae numerous; bifurcation of internal 4–12 costae, except the sulcus-facing, innermost costae, very common and taking place early during ontogeny, in the posterior part of adult shells; total number of costae along the commissure commonly >20 per flank, often 25–30 or higher. Free portions of dental plates generally very short or, more rarely, moderately long; ventral muscle field subelliptical to subrhombical in outline, with greatest width near half of its length or more posterior; areas lateral to muscle field more or less flattened.

Species included. *Delthyris acuminata* Conrad, 1839; *Terebratula acuminatissima* de Castelnau, 1843; *Spirifer bownockeri* Stewart, 1927; *Paraspirifer conradi* Godefroid & Fagerstrom, 1983; *Paraspirifer halli* Godefroid & Fagerstrom, 1983; *Paraspirifer clarkei* Godefroid & Fagerstrom, 1983; *Paraspirifer* sp. A Godefroid & Fagerstrom, 1983.

Range. Middle Devonian.

Discussion. Godefroid & Fagerstrom (1983) revised the classic North American taxa of *Paraspirifer*: *Par. acuminatus* (Conrad, 1839), *Par. acuminatissimus* (de Castelnau, 1843) and *Par. bownockeri* (Stewart, 1927). They introduced three new species mainly based on differences in shell form: *Par. conradi*, *Par. halli* and *Par. clarkei*. When the author examined a part of their collections (AMNH and USNM), general differences from most of the European forms became conspicuous: (1) the abundant and early bifurcation of costae resulting in a more finely costate macro-ornamentation hardly impressed on internal moulds of ventral valves; (2) the compact, predominantly rounded to subglobular shell form; (3) lateral fields of the dorsal valve often steep and slightly convex in transverse section, adjoining to an often poorly demarcated, angular fold; (4) the narrow hinge line in combination with a strongly curved lateral outline; and (5) the outline of the ventral muscle field with greatest width located more posteriorly. The ensemble of the American species of *Paraspirifer* represents a separate Middle Devonian subgenus, here named *Par.* (*Laurentispirifer*) subgen. nov. It also occurs in Venezuela, where specimens with typical morphology have been described as *Par.* aff. *Par. cultrijugatus* and *Par.* aff. *Par. acuminatus* (see Benedetto 1984, pl. 21, figs 1–8). These support the palaeobiogeographical affiliation of the respective fauna to the Eastern Americas Realm.

The known species of the new subgenus are geologically younger than the Ardenno-Rhenish forms, which disappeared with the Kirberg Event in the early Eifelian. Among the latter, *Par.* (*Par.*) *frechi* Solle, 1971 and *Par.* (*Par.*) *curvatissimus* Solle, 1971 are most similar as far as the shell form is concerned. *Paraspirifer* (*Laurentispirifer*) evolved from a species of *Par.* (*Paraspirifer*) by backward dislocation of the bifurcations and increase in number of external costae plus changes in shell shape and the ventral muscle field.

From the Rhenish Massif, an internal mould of a dorsal valve from the basal Heisdorf Formation (SMF 25329; locality Gemeinde-Steinbruch Gondelsheim, uppermost Emsian of the Eifel region; see Solle 1971, p. 139) resembles in outline and ornamentation (*c.* 25 plications per flank at the commissure) the North American species *Par.* (*Lau.*) *acuminatus* (Conrad, 1839). This specimen could be an indication of an origin of the subgenus from a Rhenish form. Mergl & Massa (2004) described and figured *Par.* aff. *cultrijugatus* of uncertain age from the central Sahara (near the Algeria–Niger frontier), which I would assign to *Par.* (*Lau.*) cf. *clarkei* Godefroid & Fagerstrom, 1983 due to the shell shape. According to Racheboeuf (1990), this area south of the Reguibat–Hoggar line represents a late Eifelian–Givetian southern Saharan biogeographical region belonging to the Eastern Americas Realm. Another specimen figured by Drot (1964:

pl. 4, fig. 5) from the Dra Plains (southern Morocco), from a bed 'immediately below the level with *Pinacites jugleri*' ('Tf 21'; Drot 1964, p. 223) and therefore presumably of earliest Eifelian age, is in shape and ornamentation transitional from *Par.* (*Par.*) *frechi* to *Par.* (*Lau.*) *acuminatus*.

Comparison. *Paraspirifer* (*Laurentispirifer*) is distinguished from the Rhenish subgenus *Par.* (*Paraspirifer*) by more frequent and earlier bifurcation of costae reaching a maximum number of commonly more than 20 or even over 30 per flank along the commissure so that the macro-ornamentation appears on the whole finer. As a result, the interior of the thickened ventral valves or their moulds hardly exhibit internal costae. In *Par.* (*Paraspirifer*) the number of costae along the commissure reaches approximately 20 per flank, which are often distinctly impressed on ventral internal moulds. The shells show in most species of *Par.* (*Laurentispirifer*) a narrower hinge line, a more compact shape, in outline more rounded cardinal extremities and more curved lateral margins; only *Par.* (*Par.*) *curvatissimus* and *Par.* (*Par.*) *bucculentus* resemble in this respect the new subgenus. In the American subgenus, the dorsal fold is often less distinctly delimited to the sides by a change in slope, sometimes resulting in a markedly triangular cross section of the dorsal valve. The lateral fields of the dorsal valve are straight or tending to be slightly convex in cross section, whereas they are slightly concave or straight in the Rhenish subgenus. The greatest width of the ventral muscle field is generally located at about its mid-length or in its posterior half in *Par.* (*Laurentispirifer*), whereas the greatest width is located in its anterior half to about its mid-length in *Par.* (*Paraspirifer*).

Summary and conclusions

The succession of Pridolian (uppermost Silurian) to lower Eifelian (lowermost Middle Devonian) rhynchonelliformean ('articulate') brachiopod faunas from the Rhenish Massif is examined on the basis of the revision of their taxa, which, however, is still continuing. This study covers the interval from the earliest Gedinnian transgression to the Kirberg Event (reintroduced term, Struve 1990; ='OCA Extinction Event', Struve 1982*a*) or the onset of the Great Gap *sensu lato* (Struve 1982*b*). The main objectives of the project 'Rhenish Lower Devonian Brachiopoda' are taxonomic revisions, the refinement of brachiopod biostratigraphy, analysis of faunal changes and reconstruction of the brachiopods' palaeoecology, palaeobiology and palaeobiogeography. In the present work, the biostratigraphic framework of the Rhenish Lower Devonian is revised, and brachiopods are used as the main guide fossils. The general faunal development is described and interpreted against the background of sedimentary sequences and the facies, discussing possible sea-level fluctuations, events and palaeoenvironmental changes.

As the spiriferide brachiopods show rapid evolution and high abundance, they are most suitable for a biostratigraphic subdivision of the Rhenish Lower Devonian. Twenty-two taxon-range zones from the Pridolian to lower Eifelian are proposed. A number of additional rhynchonelliformean brachiopod taxa are associated with these spiriferides and also used in the biostratigraphic subdivision. Correlations of the Rhenish sections with successions in the Ardennes (Belgium and northern France), the Artois region and the Armorican Massif (France), Bohemia (Czech Republic), Podolia (Ukraine), the Iberian Chains (northeastern Spain) and the Dra Plains (southern Morocco) are conducted, providing additional possibilities of neritic–pelagic correlation. It is confirmed that the lowermost part of the Gedinnian sequence in the classic sense has a Pridolian (latest Silurian) age (Godefroid & Cravatte 1999), that the basal Emsian boundary in the present GSSP sense (Yolkin *et al.* 1997) is much older than the traditional Siegenian–Emsian boundary in the Rhenish Massif (Jansen *et al.* 2007; Carls *et al.* 2008) and that the latter boundary is distinctly older than the Siegenian–Emsian boundary as defined in the Ardennes (Godefroid 1980; Godefroid & Stainier 1982).

Before brachiopods are used in biostratigraphy, the effects of the rock facies and its palaeoenvironmental significance have to be considered. Distribution and composition of the brachiopod faunas from the Rhenish Devonian are dependent on variations of the neritic–siliciclastic, rhenotypic ('Rhenish') facies, which is proposed herein to be subdivided into three main subfacies: the eurhenotypic subfacies reflects shallow-marine palaeoenvironments with turbid water and suitable living conditions mainly for diverse articulate brachiopods, pelecypods, trilobites, a few tabulate and rugose corals, and some ostracods, gastropods, bryozoans, crinoids and tentaculitides. Proximal, typical and distal variations of this subfacies show different distances from the coast and degrees of terrigenous influence. The pararhenotypic subfacies reflects restricted-marine, marginal-marine, deltaic or intertidal palaeoenvironments with changeable conditions, e.g. salinity fluctuations, suitable only for animals tolerant to these, in particular species of the terebratulide genera *Crassirensselaeria* or *Globithyris*, associated with lingulides, pelecypods, agnathans, eurypterides, ostracods and tentaculitides. The allorhenotypic subfacies reflects shallow-marine palaeoenvironments with clear water, reduced or finer siliciclastic input and a higher content of calcium carbonate in the substrates. Its faunas partially contain species of genera regarded as rhenotypic, but less typical, more clear water-preferring species occur as well, for example taxa of the reticulariid spiriferides, atrypides and pentamerides. The brachiopods are accompanied by a specific trilobite fauna and a diversity of corals.

The rhenotypic facies as a whole is typically developed from the Lochkovian to the Frasnian of Europe and North Africa, but similar facies are represented by contemporaneous sedimentary successions on all continents. The taxonomic composition of the rhenotypic faunas is very different from the composition of shallow-water benthic faunas of the lower or upper Palaeozoic. The rhenotypic faunas also strongly differ ecologically from comparable neritic faunas of the Mesozoic and the Cenozoic, which are characterized by a higher percentage of pelecypods, more motile predators and the presence of a deep infauna (Sepkoski 1990; Aberhan 1994; Bush & Bambach 2011; Allmon & Martin 2014).

The Rhenish brachiopods from the Pridolian to the lower Eifelian are documented as characteristic sets of assemblages, which follow one after the other as distinct 'faunal intervals' or 'faunas' named after characteristic spiriferide species. The changes between these faunas were mainly governed by shelf-wide or more regional environmental perturbations of different magnitudes and durations, leading to emigration or regional extinction of substantial parts of a brachiopod fauna. After these short-term events or more extended phases, a new fauna took over. Whether the faunal shifts always occurred with a severe or sudden crisis or with a more gradually changing palaeoenvironment stimulating a stepwise replacement of the fauna is still unclear. They possibly resulted from greater changes in the following factors: quantity of siliciclastic input, sedimentation rate, subsidence rate, eustatic and relative sea-level, and climate. The most distinct events defined with locality names refer to the onsets of the new faunas that may correlate with transgressions (e.g. Berlé Event) but also with regressions (e.g. Kürrenberg Event). In the discussion of the relative sea-level history of the Rhenish area, the global sea-level curves as reconstructed by Johnson *et al.* (1985), Johnson & Sandberg (1989) and Walliser (1996, 1998) are considered.

Episodes of unsuitable or hostile conditions, shelf-wide or more regional, may have lasted for several thousands or hundred thousands of years, or even longer. Evolution continued, partly in small and isolated habitats, probably within the main Ardenno-Rhenish Shelf or at its periphery – not necessarily documented in the fossil record – or in neighbouring shelf areas. Restricted panmixis may have led to rapid evolution in smaller populations. With the widespread recurrence of more suitable conditions, brachiopods immigrated again from outside or spread over the shelf from isolated habitats, and benthic biocoenoses with brachiopods could re-establish on a wider scale. This pattern is particularly obvious after the late Gedinnian to early Siegenian phase of the 'Rhenish Gap' *sensu stricto*.

Only little evolutionary change took place in the intervals of more stable palaeoenvironments and within the Ardenno-Rhenish Shelf. It was certainly fostered by events of smaller magnitude. This evolutionary change is not always easy to recognize and still requires new, detailed morphological studies of the brachiopods.

The oldest brachiopod assemblages of the middle Palaeozoic ('Variscan') succession belong to the *Quadrifarius dumontianus* Fauna, which is of earliest Gedinnian (late Pridolian) age. It occurs in rather isolated exposures in the Ebbe and Remscheid anticlinoria, the Müsen Horst and the southern Taunus. The ecological effects of the transgressive S/D-Event (Walliser 1985, 1996) or Klonk Event (Jeppsson 1998) may have caused its extinction. After a break in respective documentation, shallow-marine brachiopod facies recurs with the early Gedinnian *Howellella mercurii* Fauna, reaching its maximum development in the 'Flaserschiefer' (= Hardt Member) of the Hüinghausen Formation. The brachiopod communities appeared with the Hüinghausen Event, providing shallow-marine conditions suitable for diverse articulate brachiopods during a falling sea-level consistent with the early T-R cycle Ia_1 (Walliser 1998). The *How. mercurii* Fauna went extinct within Gedinnian time, with the onset of the 'Rhenish Gap' (*sensu stricto*) of full marine documentation, representing a continuous, strong regressive phase in the Ardenno-Rhenish area not reflected by the global sea-level curve during the mid- and late Lochkovian to early Pragian times (middle to late T-R cycle Ia_1 and early T-R cycle Ia_2) and presumably caused mainly by increasing siliciclastic input; terrestrial, limnic-fluvial, lacustrine, deltaic or intertidal conditions spread out in wide parts of the Ardenno-Rhenish region. The evolution of the Maghrebo-European brachiopod faunas during the time of the 'Rhenish Gap' took place mainly on the Peri-Gondwanan Shelf. In sections of the Armorican Massif (France), the Iberian Chains (Spain) or the Dra Plains (Morocco), the evolution of the respective faunas during this interval of about 6–8 myr length is documented. Regarding the long gap of full marine documentation in the Ardenno-Rhenish area, it is understandable that lower Gedinnian and middle Siegenian faunas are strongly dissimilar.

A first, but still weak marine influence started with the beginning of the Rhenish Siegenian, probably caused by increasing subsidence and decreasing supply of siliciclastic material or already linked to the early Pragian transgressive pulse at the beginning of T-R cycle Ia_2 (Walliser 1998; ≈'Ia' *sensu* Johnson *et al.* 1985; Johnson & Sandberg 1989); it has been described as the Hermeskeil Event (Mittmeyer 2008). As regards lithology, red rock colours widely disappear with the beginning of the lower Siegenian. The increasing marine influence led to the local establishment of conditions of the pararhenotypic subfacies, rarely proximal variations of the eurhenotypic subfacies. The conditions of the 'Rhenish Gap' *sensu stricto* ended with the Gensberg Event at the beginning of mid-Siegenian time, representing a transgression of supraregional importance. Environmental conditions of the normal marine, eurhenotypic subfacies widely spread. The mid- to late Siegenian (latest Pragian (?), early Emsian in the present GSSP sense) *Acrospirifer primaevus* Fauna appeared. It is still a working hypothesis that this development may correspond to the transgressive Zinzilban Event documented in Uzbekistan (Yolkin *et al.* 1994) or a (corresponding?) mid-Pragian (in the original sense) transgressive pulse of the global sea-level curve. In contrast, the faunal turnover near the beginning of the late Siegenian, sedimentologically expressed a little later by the voluminous sand deposition referred to as the Kürrenberg Event (Mittmeyer 2008), was governed by a regionally increased supply of siliciclastics and higher sedimentation rates, probably in connection with a eustatic sea-level fall during the late T-R cycle Ia_2. Organisms with clear-water preference diminished so that a certain combination of turbidicolous biota of the *Hysterolites hystericus* Zone became dominant.

The latest Siegenian Saxler Event (Mittmeyer 2008) marks the beginning of a transgression that may coincide with the beginning of T-R cycle Ib of the global sea-level curve (Johnson *et al.* 1985; Johnson & Sandberg 1989; Walliser 1996) and the 'Basal Zlíchov Event' *sensu* García-Alcalde (1997). The transgressive phase continued into the early Emsian and led to the facies change from the Taunusquarzit Group to the Hunsrück Slate Group in the southern Rhenish Massif. The early Emsian *Arduspirifer antecedens* Fauna shows a gradual evolution of the rhenotypic brachiopods, exemplified by the genera *Euryspirifer* and *Arduspirifer*. Their evolution most probably took place within the Ardenno-Rhenish Shelf region. After the Spitznack Event at the beginning of the mid-early Emsian (Singhofen) time had led to a slight faunal change in the brachiopods, the peak of marine influence was reached with the transgressive Stadtfeld Event (Mittmeyer 2008) near the beginning of the late early Emsian (Vallendar) time. The correlation of these events with supraregional (?) events (Chebbi and Upper Zlíchov events) is still questionable. Regressive tendencies in the latest early Emsian (Klerf/Nellenköpfchen time), mainly caused by very high sedimentation rates accommodated by concomitant strong subsidence, led to regional extinctions and migrations. After the balanced system of rapid subsidence and high sedimentation rates had ceased, the rising sea-level in connection with the Berlé Event (Mittmeyer 2008) provided a more continuous connection to the sea so that intertidal and then open-marine conditions of the eurhenotypic facies could re-establish, and the late Emsian *Euryspirifer paradoxus* Fauna developed. It is possible that the Berlé Event was linked to the global Daleje Event (House 1985; Chlupáč & Kukal 1986, 1988; Ferrová *et al.* 2012), which was merely characterized by a transgressive phase rather than an abrupt rise of the sea-level. Some of the early Emsian genera of the Rhenish Shelf survived the event with new subspecies or species that probably evolved in the same shelf area; others immigrated from areas outside. The faunal turnover at the boundary from early to late Emsian times was strong (Solle 1972), possibly a little biased by a lack of documentation. In contrast, the overall transgressive trend during the late Emsian, reflected by the successions of the central Middle Rhine and Lower Mosel regions, was accompanied by a modest, stepwise faunal change. Those faunas differ slightly from the faunas of the Eifel region. One of the most conspicuous features of the late Emsian faunas is the radiation of the genus *Paraspirifer* (see Solle 1971). In early late Emsian (Lahnstein) time, *Par.* (*Mosellospirifer*) subgen. nov. originated from *Brachyspirifer ignoratus*, with the species *Par.* (*Mos.*) *longimargo* and *Par.* (*Mos.*) *sandbergeri*. Close to the beginning of the middle part of the late Emsian (Laubach) time, *Par.* (*Paraspirifer*) *praecursor* developed as the first species of its subgenus. It is still unknown precisely when the new subgenus *Par.* (*Laurentispirifer*) originated, but it most probably evolved in latest Emsian or earliest Eifelian (Lauch) time from a species of *Par.* (*Paraspirifer*) and could have migrated via an 'Afro-Appalachian corridor' into the Eastern Americas Realm, where it flourished during the Mid-Devonian. Possible representatives of *Par.* (*Laurentispirifer*) are also present in southern Libya (Mergl & Massa 2004), Morocco (Drot 1964) and Venezuela (Benedetto 1984).

The suggested regression of the latest part of T-R cycle Ib (Johnson *et al.* 1985; Johnson & Sandberg 1989), in the time represented by the lower *Linguipolygnathus serotinus* Conodont Zone, may correlate with regressive trends of the Laubach Event (Mittmeyer 2008). The beginning of the succeeding T-R cycle Ic could then correspond with a transgressive pulse within the Laubach-Kondel interval. In the Eifel region, this change might be documented by the turnover from regressive tendencies reflected by the proximal eurhenotypic subfacies of the Wetteldorf Formation to the more open-marine conditions of the partially allorhenotypic Heisdorf Formation. The latest Emsian Kondel Event (Mittmeyer 2008) is locally accompanied by regressive effects in the central Rhenish Massif, but these are rather exceptions within an ongoing general transgressive trend. In contrast, Walliser (1996, 1998) recorded a continuous eustatic regression during late parts of late Emsian time; if this is a global development, it may have been overcompensated in the Rhenish area by reduced sedimentation rates and continued subsidence. The Kondel faunas in the Middle Rhine, Lower Mosel, Dill and Lahn regions underwent a diversification in deeper subtidal palaeoenvironments shown by the distal eurhenotypic subfacies. The strata of the Kondel Group are characterized by *Sollispirifer*, *Paraspirifer* (*Mosellospirifer*), *Par.* (*Paraspirifer*) and *Alatiformia* species and can be correlated biostratigraphically with the succession in the Eifel region by, for example, the onsets of *Soll. mosellanus* (Solle, 1953) and *Al. alatiformis* (Drevermann, 1907). While the Kondel Fauna went extinct near the Emsian–Eifelian boundary in the areas of the central and eastern Rhenish Massif, caused by the onset of deep-water, hercynotypic conditions, the allorhenotypic *Par.* Fauna of the Eifel and the Sauerland survived in shallow-water palaeoenvironments from the latest Emsian (Kondel time) into the earliest Eifelian (Lauch time). The subgenus *Paraspirifer* (*Paraspirifer*) is mainly represented here by its type species *Par.* (*Par.*) *cultrijugatus* and additional taxa of the same subgenus. Other characteristics of this interval are the high abundance of *Iridistrophia* (*Flabellistrophia*) *hipponyx* and *Cuninulus concavus*. The genus *Intermedites* appears near the lower boundary of the Eifelian Stage and may be an immigrant from outside the Rhenish area. The *Par. cultrijugatus* Fauna went extinct with the Kirberg Event within the *Polygnathus costatus costatus* Biochron at the end of the earliest Eifelian Lauch time. This event followed with some delay after the Basal Choteč Event, which is suggested to correlate approximately with the base of the Dorsel Member.

To conclude, the Pridolian to lower Eifelian sedimentary successions of the Rhenish Massif exemplify the dependence of the marine record on the interplay of past geological processes such as eustatic sea-level changes, subsidence and sedimentation. In general, the global development of the sea-level was here partly masked by regional changes in subsidence and sedimentation rates. The palaeoclimate played a role as well, as the suggested, partly tropical humid conditions resulted in strong physical erosion and fluvial transport of siliciclastic material and nutrients from the Old Red Continent into the shelf sea. The warm climate also provided favourable living conditions for brachiopods in the coastal and shallow-water palaeoenvironments of the Rhenish Sea.

I thank the following colleagues who helped during my visits, gave me access to collections and allowed the long-term loan of specimens important for this work: Sandra I. Kaiser and Georg Heumann (Steinmann-Institut, Department of Palaeontology, and Goldfuss Museum, Bonn, Germany), Fritz Geller-Grimm and Gerhard Heinrich (Museum Wiesbaden), Doris Heidelberger (Oberursel, Germany), Annelise Folie, Eric Dermience and Alain Drèze (Muséum des Sciences naturelles, Brussels, Belgium), Martina Korandová and Michal Mergl (Dr.-Bohuslav-Horák Museum, Rokycany, and University of West Bohemia, Pilsen, Czech Republic), Vojtěch Turek (National Museum, Prague, Czech Republic), Bushra Hussaini and Neil Landman (American Museum of Natural History, New York, NY, USA), Dan Levin and Mark Florence (Smithsonian National Museum of Natural History, Washington, DC, USA), and Susan Butts and Derek Briggs (Peabody Museum of Natural History, Yale University, New Haven, CT, USA). I thank Karl-Heinz Ribbert (formerly Geologischer Dienst Krefeld, Germany) for providing specimens from the North Eifel region and Silke Clasen (Magdeburg, Germany) for granting access to her diploma thesis and loaning the reference material to me.

Special thanks are due to technical assistant Erika Scheller-Wagner (Senckenberg Research Institute, Frankfurt am Main), who did a lot of the technical work such as the preparation of fossils, curatorial work and preparing marvellous latex casts. Claudia Groth (Senckenberg) prepared the excellent drawings of Figure 7. I thank the reviewers Fernando Alvarez (Department of Geology, University of Oviedo, Spain) and Adam Halamski (Institute of Paleobiology, Polish Academy of Sciences, Warszawa, Poland). Finally, Alan Lord (Senckenberg) improved the work with a number of linguistic remarks, Stan Finney (California State University, Long Beach, USA) helped with an email discussion on questions of stratigraphic nomenclature, and R. Thomas Becker (WWU Münster, Germany) conducted an intensive final editing. This is a contribution to IGCP 596 'Climate change and biodiversity patterns in the mid-Palaeozoic'.

References

Aberhan, M. 1994. Guild structure and evolution of Mesozoic benthic communities. *Palaios*, **9**, 516–545.

Alberti, G.K.B. 1962. Unterdevonische Trilobiten aus dem Frankenwald und Rheinischen Schiefergebirge (Ebbe- und Remscheider Sattel). *Geologisches Jahrbuch*, **81**, 135–156.

Alberti, G.K.B. 1981. Zur biostratigraphischen Untergliederung der Wissenbacher Schiefer (Unter- und Mittel-Devon) des östlichen Rheinischen Schiefergebirges mit Tentaculiten (Dacryoconarida). *Mitteilungen des*

Geologisch-Paläontologischen Institutes der Universität Hamburg, **50**, 77–90.

Alberti, G.K.B. 1982*a*. *Nowakia (Sulcatonowakia) sulcata sulcata* (F.A. Roemer, 1843) (Dacryoconarida) from the lowermost part of the Lauch Formation (Eifelian) of the Wetteldorf Richtschnitt. *In*: Ziegler, W. & Werner, R. (eds) On Devonian Stratigraphy and Palaeontology of the Ardenno-Rhenish Mountains and Related Devonian Matters. *Courier Forschungsinstitut Senckenberg*, **55**, 333–336.

Alberti, G.K.B. 1982*b*. Nowakiidae (Dacryoconarida) aus dem Hunsrückschiefer von Bundenbach (Rheinisches Schiefergebirge). *Senckenbergiana lethaea*, **63**, 451–463.

Alberti, G.K.B. 1983. Unterdevonische Nowakiidae (Dacryoconarida) aus dem Rheinischen Schiefergebirge, aus Oberfranken und aus N-Afrika (Algerien, Marokko). *Senckenbergiana lethaea*, **64**, 295–313.

Alberti, G.K.B. 1985. Zur biostratigraphischen Untergliederung des Greifensteiner Kalkes und der Wissenbacher Schiefer (Unter- bis Mittel-Devon, Rheinisches Schiefergebirge) mithilfe von Dacryoconarida (Tentaculiten). *Mitteilungen aus dem Geologisch-Paläontologischen Institut der Universität Hamburg*, **59**, 51–56.

Algeo, T.J. & Scheckler, S.E. 1998. Terrestrial-marine teleconnections in the Devonian: links between the evolution of land plants, weathering processes, and marine anoxic events. *Philosophical Transactions of the Royal Society of London B*, **353**, 113–130.

Allan, R.S. 1935. The fauna of the Reefton Beds, (Devonian), New Zealand; with notes on Lower Devonian qnimal communities in relation to the base of the Devonian system. *Publications of the New Zealand Geological Survey, Palaeontological Bulletin*, **14**, 1–72.

Allmon, W.D. & Martin, R.E. 2014. Seafood through time revisited: the Phanerozoic increase in marine trophic resources and its macroevolutionary consequences. *Paleobiology*, **40**, 255–286.

Al-Rawi, D. 1977. Biostratigraphische Gliederung der Tentaculiten-Schichten des Frankenwaldes mit Conodonten und Tentaculiten (Unter- und Mittel-Devon; Bayern, Deutschland). *Senckenbergiana lethaea*, **58**, 25–79.

Alvarez, F. & Racheboeuf, P.R. 1986. Sous-famille Dayiinae Waagen, 1883. *In*: Racheboeuf, P.R. (ed.) *Le Groupe de Liévin. Pridoli-Lochkovien de l'Artois (N. France). Sédimentologie – Paléontologie – Stratigraphie*. Biostratigraphie du Paléozoïque, **3**. Université de Bretagne Occidentale, Brest, 128–131.

Amler, M.R.W. 1996. Die Bivalvenfauna des Oberen Famenniums West-Europas. 2. Evolution, Paläogeographie, Paläoökologie, Systematik 2. Palaeotaxodonta und Anomalodesmata. *Geologica et Palaeontologica*, **30**, 49–117.

Anderle, H.-J. 1987. Entwicklung und Stand der Unterdevon-Stratigraphie im südlichen Taunus. *Geologisches Jahrbuch Hessen*, **115**, 81–98.

Anderle, H.-J. 2000. Gezeitensedimente in der Hermeskeil-Formation (Siegen-Stufe, Unterdevon) von Niedernhausen im Taunus (Bl. 5815 Wehen, Rheinisches Schiefergebirge). *Jahrbücher des Nassauischen Vereins für Naturkunde*, **121**, 83–94.

Anderle, H.-J. 2006. Taunus. *In*: Deutsche Stratigraphische Kommission (ed.), Heuse, T. & Leonhardt, D. (coord.) *Stratigraphie von Deutschland VII. Silur*. Schriftenreihe der Deutschen Gesellschaft für Geowissenschaften, Hannover, **46**, 45–48.

Anderle, H.-J. 2008. Südtaunus. *In*: Deutsche Stratigraphische Kommission (ed.) Weddige, K. (coord.) *Stratigraphie von Deutschland VIII. Devon*. Schriftenreihe der Deutschen Gesellschaft für Geowissenschaften, Hannover, **52**, 118–130.

d'Archiac, É.J.A. & de Verneuil, É.P. 1842. On the fossils of the older deposits in the Rhenish provinces, preceded by a general survey of the fauna of the Palaeozoic rocks, and followed by a tabular list of the organic remains of the Devonian system in Europe. *Transactions of the Geological Society of London, 2nd series*, **6**, 5–40 + 303–410, http://doi.org/10.1144/transgslb.6.2.303

Asselberghs, E. 1930. Description des faunes marines du Gedinnien de l'Ardenne. *Mémoires du Musée royal d'Histoire naturelle de Belgique*, **41**, 1–69.

Asselberghs, E. 1936. IV. Exkursion in die Siegener Schichten der Ardennen. *In*: Asselberghs, E., Henke, W., Schriel, W. & Wunstorf, W. (eds) Über eine gemeinsame Exkursion durch die Siegener Schichten des Rheinischen Schiefergebirges und der Ardennen. *Jahrbuch der Preußischen Geologischen Landesanstalt*, **56** (for 1935), 349–369.

Asselberghs, E. 1943. L'Arkose de Weismes, le Grès de Gdoumont et leur faune (Gedinnien supérieur). *Bulletin du Musée royal d'Histoire naturelle de Belgique*, **19**, 1–12.

Asselberghs, E. 1946. L'Éodévonien de l'Ardenne et des régions voisines. *Mémoires de l'Institut géologique de l'Université de Louvain*, **14**, 3–598.

Asselberghs, E. & Maillieux, E. 1938. La limite entre l'Emsien et le Siegenien sur le bord sud du bassin de Dinant. *Bulletin du Musée royal d'Histoire naturelle de Belgique*, **14**, 1–11.

Baliński, A. 2012. The brachiopod succession through the Silurian–Devonian boundary beds at Dnistrove, Podolia, Ukraine. *Acta Palaeontologica Polonica*, **57**, 897–924.

Bambach, R.K. 1999. Energetics in the global marine fauna: a connection between terrestrial diversification and change in the marine biosphere. *Geobios*, **32**, 131–144.

Barrande, J. 1847. Über die Brachiopoden der silurischen Schichten von Böhmen. *Naturwissenschaftliche Abhandlungen*, **1**, 357–475. W. Haidinger, Wien.

Barrande, J. 1848. Über die Brachiopoden der silurischen Schichten von Böhmen. *Naturwissenschaftliche Abhandlungen*, **2**, 153–256. W. Haidinger, Wien.

Barrande, J. 1865, 1867. *Système Silurien du centre de la Bohême, II. Classe des Mollusques. Ordre des Céphalopodes*. Published by the author, Prague & Paris, Pls 1–107 (1865), 1–712 (1867).

Barrande, J. 1879. *Système Silurien du centre la Bohème, V. Classe de Mollusques. Ordre des Brachiopodes*. Published by the author, Prague & Paris.

Bartels, C., Briggs, D.E.G. & Brassel, G. 1998. *The Fossils of the Hunsrück Slate. Marine Life in the Devonian*. Cambridge University Press, Cambridge.

Bartels, C., Wuttke, M. & Briggs, D.E.G. (eds) 2002. The *Nahecaris* Project. Releasing the marine life of the Devonian from the Hunsrück Slate of Bundenbach. *Metalla (Bochum)*, **9**(2), foreword, preface + 59–138.

Bartholomew, A.J. & Brett, C.E. 2007. Correlation of Middle Devonian Hamilton Group-equivalent strata in east-central North America: implications for eustasy, tectonics and faunal provinciality. *In*: Becker, R.T. & Kirchgasser, W.T. (eds) *Devonian Events and Correlations*. Geological Society, London, Special Publications, **278**, 105–131, http://doi.org/10.1144/SP278.5

Basse, M. 2002. *Eifel-Trilobiten 1. Proetida*. Goldschneck-Verlag, Korb.

Basse, M. 2003. *Eifel-Trilobiten 2. Phacopida 1*. Goldschneck-Verlag, Korb.

Basse, M. & Müller, P. 2004. *Eifel-Trilobiten III. Corynexochida, Proetida (2), Harpetida, Phacopida (2), Lichida*. Quelle & Meyer Verlag, Wiebelsheim.

Basse, M. & Müller, P. 2012. Drei Arten der Asteropyginae aus den Seifen-Schichten, Westerwald (Trilobita; Mittelsiegen-Unterstufe, Unter-Devon; Rheinisches Schiefergebirge). *Geologica et Palaeontologica*, **44**, 9–26.

Bauer, G., Fenchel, W. *et al.* 1960. Beitrag zur Geologie der Mittleren Siegener Schichten. Eine Gemeinschaftsarbeit über die Kartierungsergebnisse im Siegerland und im Wiedbezirk der Jahre 1950 bis 1956. *Abhandlungen des Hessischen Landesamtes für Bodenforschung*, **29**, 1–363.

Becker, G. & Mentzel, R. 1961. Untersuchungen im Unter-Devon des Hontheimer und Stadtkyller Sattels (Eifel). *Notizblatt des Hessischen Landesamtes für Bodenforschung*, **89**, 134–169.

Becker, R.T. 2007. Emsian substages and the Daleje Event – a consideration of conodont, dacryoconarid, ammonoid and sealevel data. *SDS Newsletter*, **22**, 29–32.

Becker, R.T. 2008. Devonian neritic-pelagic correlations – methods, case studies and problems. *In*: Königshof, P. & Linnemann, U. (eds) *Final Meeting of IGCP 497 and IGCP 499, Abstracts and Programme*. Senckenberg, Frankfurt am Main, 25–28.

Becker, R.T. & Aboussalam, Z.S. 2011. Emsian chronostratigraphy – preliminary new data and a review of the Tafilalt (SE Morocco). *SDS Newsletter*, **26**, 33–43.

Becker, R.T. & Aboussalam, Z.S. 2013. The global Chotec Event at Jebel Amelane (western Tafilalt Platform) – preliminary data. *Document de l'Institut Scientifique, Rabat*, **27**, 129–134.

Becker, R.T. & House, M.R. 1994. International Devonian goniatite zonation, Emsian to Givetian, with new records from Morocco. *Courier Forschungsinstitut Senckenberg*, **169**, 79–135.

Becker, R.T. & House, M.R. 2000. Devonian ammonoid zones and their correlation with established series and stage boundaries. *In*: Bultynck, P. (ed.) Subcommission on Devonian Stratigraphy – Fossil Groups Important for Boundary Definition. *Courier Forschungsinstitut Senckenberg*, **220**, 113–151.

Becker, R.T., Gradstein, F.M. & Hammer, O. 2012. The Devonian Period. *In*: Gradstein, F.M., Ogg, J.G., Schmitz, M.D. & Ogg, G.M. (eds) *The Geological Time Scale 2012*, Vol 2. Elsevier, Oxford, 559–601.

Béclard, F. 1887. Les fossiles coblenziens de St. Michel près de Saint Hubert. *Bulletin de la Société Belge de Géologie de Paléontologie et d'Hydrologie*, **1**, 60–97.

Bender, P. 1967. Unterdevonische Conodonten aus den Kalken von Naux (Unteres Gedinnium, Massiv von Rocroi). *Geologica et Palaeontologica*, **1**, 183–184.

Bender, P. 2008. Lahn- und Dill-Mulde. *In*: Deutsche Stratigraphische Kommission (ed.), Weddige, K. (coord.) *Stratigraphie von Deutschland VIII. Devon*. Schriftenreihe der Deutschen Gesellschaft für Geowissenschaften, **52**. Schweizerbart, Stuttgart, 221–246.

Benedetto, J.-L. 1984. *Les brachiopodes dévoniens de la Sierra de Perijá, Vénézuela. Systématique et implications paléogéographiques*. Biostratigraphie du Paléozoïque, **1**. Université de Bretagne Occidentale, Brest, 1–222.

Beyer, K. 1952. Zur Stratigraphie des obersten Gotlandiums in Mitteleuropa. *Wissenschaftliche Zeitschrift der Universität Greifswald, Jahrgang I, 1951/52. Mathematisch-naturwissenschaftliche Reihe Nr. 1, Mitteilungen aus dem Geologisch-Paläontologischen Institut der Universität Greifswald, Neue Folge*, **14**, 1–33.

Beyrich, E. 1837. *Beiträge zur Kenntnis der Versteinerungen des Rheinischen Übergangsgebirges. Erstes Heft*. F. Dümmler, Berlin.

Bignon, A., Corbacho, J. & López-Soriano, F.J. 2014. A revision of the first Asteropyginae (Trilobita, Devonian). *Geobios*, **47**, 281–291.

Bizzarro, M. & Lespérance, P.J. 1999. Systematics of some Lower and Middle Devonian spiriferid brachiopods from Gaspé with a revision of the superfamily Delthyridoidea. *Journal of Paleontology*, **73**, 1056–1077.

Blieck, A. & Jahnke, H. 1980. Pteraspiden (Vertebraten, Heterostraci) aus den Unteren Siegener Schichten und ihre stratigraphischen Konsequenzen. *Neues Jahrbuch für Geologie und Paläontologie, Abhandlungen*, **159**, 360–378.

Blieck, A., Goujet, D., Janvier, P. & Meilliez, F. 1995. Revised Upper Silurian-Lower Devonian ichthyostratigraphy of northern France and southern Belgium (Artois-Ardenne). *Bulletin du Muséum national d'Histoire naturelle, series 4*, **17**, 447–459.

Böger, H. 1981. Stratigraphische, fazielle und tektonische Zusammenhänge im Unter-Devon des Sauerlandes (Rheinisches Schiefergebirge) und der Kaledonisch-Variszische Umschwung. *Mitteilungen aus dem Geologisch-Paläontologischen Institut der Universität Hamburg*, **50**, 45–58.

Böger, H. 1983. Eine Lithostratigraphie des Unterdevons im Sauerlande und im östlichen Bergischen Lande (Rheinisches Schiefergebirge). II. Das Ebbe-Antiklinorium. *Neues Jahrbuch für Geologie und Paläontologie, Abhandlungen*, **166**, 294–326.

Borremans, G. & Bultynck, P. 1986. Conodontes du Calcaire de Naux – Gedinnien inférieur au sud immédiat du Massif de Rocroi (Ardenne française). *Aardkundige Mededelingen*, **1986**, 45–58.

Bouček, B. 1934. On some further new trilobites from the Gothlandian of Bohemia. *Bulletin international de l'Académie des Sciences de Bohême*, **1934**, 1–7.

Bouček, B. 1941. O variabilitě ramenonožců *Dayia navicula* (Sow.) a *Cyrtia exporrecta* (Wahl.) a o použití metod variační statistiky v paleontologii. *Rozpravy II. třídy České akademie věd*, **50**, 1–27.

Bouček, B. 1964. *The Tentaculites of Bohemia. Their Morphology, Taxonomy, Ecology, Phylogeny and Biostratigraphy*. Publishing House of the Czechoslovakian Academy of Sciences, Prague.

Bouček, B. & Prantl, F. 1959. Význam tentakulitů pro stratigrafii středočeského devonu. *Časopis Národního muzea, Řada přírodovědná*, **128**, 5–7.

Boucot, A.J. 1959. A new family and genus of Silurian orthotetacid brachiopods. *Journal of Paleontology*, **33**, 25–28.

Boucot, A.J. 1960. Lower Gedinnian brachiopods of Belgium. *Mémoires de l'Institut Géologique de l'Université de Louvain*, **21**, 283–324.

Boucot, A.J. 1963. The globithyrid facies of the Lower Devonian. *Senckenbergiana lethaea*, **44**, 79–84.

Boucot, A.J. 1975. *Evolution and Extinction Rate Controls*. Developments in Palaeontology and Stratigraphy, **1**. Elsevier, Amsterdam.

Boucot, A.J. 1984. Old World Realm (Rhenish–Bohemian Region) shallow-water, Early Devonian brachiopods from the Jauf Formation of Saudi Arabia. *Journal of Paleontology*, **58**, 1196–1202.

Breil-Schollmayer, A. 1989. *Zur Geologie des Siegenium und Emsium (Unterdevon) zwischen Aachen und Hellenthal (Nordeifel) unter besonderer Berücksichtigung der faziellen Entwicklung*. Dissertation an der Fakultät für Bergbau, Hüttenwesen und Geowissenschaften der Rheinisch-Westfälischen Technischen Hochschule Aachen.

Brice, D. 2000. Rhynchonellida. *In*: Bultynck, P. (ed.) Subcommission on Devonian Stratigraphy – Fossil Groups Important for Boundary Definition. *Courier Forschungsinstitut Senckenberg*, **220**, 67–70.

Brice, D., Gourvennec, R., Jahnke, H., Melou, M., Racheboeuf, P.R., Alvarez, F. & Copper, P. 1986. Brachiopodes Articulés. *In*: Racheboeuf, P.R. (ed.) *Le Groupe de Liévin. Pridoli-Lochkovien de l'Artois (N. France). Sédimentologie – Paléontologie – Stratigraphie*. Biostratigraphie du Paléozoïque, **3**. Université de Bretagne Occidentale, Brest, 97–142.

Brice, D., Carls, P., Cocks, L.R.M., Copper, P., Garcia-Alcalde, J.L., Godefroid, J. & Racheboeuf, P.R. 2000. Brachiopoda. *In*: Bultynck, P. (ed.) Subcommission on Devonian Stratigraphy – Fossil Groups Important for Boundary Definition. *Courier Forschungsinstitut Senckenberg*, **220**, 65–86.

Brocke, R., Jansen, U. et al. 2004. Fazies und Stratigraphie temporärer Schürfe (Erdgastrasse TENP 2) im Unter- und Mitteldevon der Rohr-Mulde und Sötenich-Mulde, Eifel. *Scriptum*, **11**, 7–107.

Brocke, R., Fatka, O., Lindemann, R.H., Schindler, E. & Ver Straeten, C.A. 2015. Palynology, dacryoconarids and the lower Eifelian (Middle Devonian) Basal Choteč Event: case studies from the Prague and Appalachian basins. *In*: Becker, R.T., Königshof, P. & Brett, C.E. (eds) *Devonian Climate, Sea Level and Evolutionary Events*. Geological Society, London, Special Publications, **423**. First published online September 14, 2015, http://doi.org/10.1144/SP423.8

Brunton, C.H.C., Lazarev, S.S. & Grant, R.E. 2000. Productida. *In*: Kaesler, R.L. (ed.) *Treatise on Invertebrate Palaeontology, Part H, Brachiopoda (Revised), 2–3, Linguliformea, Craniiformea, and Rhynchonelliformea*. Geological Society of America and University of Kansas, Boulder, CO, and Lawrence, KS, 350–643.

Bultynck, P. 1986. Conodontes. *In*: Racheboeuf, P.R. (ed.) *Le Groupe de Liévin. Pridoli-Lochkovien de l'Artois (N. France). Sédimentologie – Paléontologie – Stratigraphie*. Biostratigraphie du Paléozoïque, **3**. Université de Bretagne Occidentale, Brest, 201–204.

Bultynck, P., Coen-Aubert, M. & Godefroid, J. 2000. Summary of the state of correlation in the Devonian of the Ardennes (Belgium, NE France) resulting from the decisions of the SDS. *In*: Bultynck, P. (ed.) Subcommission on Devonian Stratigraphy. Recognition of Devonian Series and Stage Boundaries in Geological Areas. *Courier Forschungsinstitut Senckenberg*, **225**, 91–114.

Bush, A.M. & Bambach, R.K. 2011. Paleoecologic Megatrends in Marine Metazoa. *Annual Review of Earth and Planetary Sciences*, **39**, 241–269.

Carls, P. 1971. Stratigraphische Übereinstimmungen im höchsten Silur und tieferen Unter-Devon zwischen Keltiberien (Spanien) und Bretagne (Frankreich) und das Alter des Grès de Gdoumont (Belgien). *Neues Jahrbuch für Geologie und Paläontologie, Monatshefte*, **1971**, 195–212.

Carls, P. 1985. *Howellella* (*Hysterohowellella*) *knetschi* (Brachiopoda, Spiriferacea) aus dem tiefen Unter-Gedinnium Keltiberiens. *Senckenbergiana lethaea*, **65**, 297–326.

Carls, P. 1987. Ein Vorschlag zur biostratigraphischen Redefinition der Grenze Gedinnium/Siegenium und benachbarter Unterstufen. 1. Teil. Stratigraphische Argumente und Korrelation. *Courier Forschungsinstitut Senckenberg*, **92**, 77–121.

Carls, P. 1988. The Devonian of Celtiberia (Spain) and Devonian Paleogeography of SW Europe. *In*: McMillan, N.J., Embry, A.F. & Glass, D.J. (eds) *Devonian of the World. Vol. 1: Regional Syntheses*. Canadian Society of Petroleum Geologists, Memoirs, **14**, 421–466.

Carls, P. 1999. El Devónico de Celtiberia y sus fósiles. *In*: Gámez-Vintaned, J.A. & Liñán, E. (eds) *25 años de Paleontología Aragonesa. Homenaje al profesor Leandro Sequeiros. VI Jornados Aragonesas de Paleontología*. Institución 'Fernando el Católico' (C.S.I.C.), Zaragoza, 101–164.

Carls, P. 2000. Stratigraphical implications of some Devonian Spiriferida. *In*: Bultynck, P. (ed.) Subcommission on Devonian Stratigraphy – Fossil Groups Important for Boundary Definition. *Courier Forschungsinstitut Senckenberg*, **220**, 70–74.

Carls, P. & Gandl, J. 1969. Stratigraphie und Conodonten des Unterdevons der Östlichen Iberischen Ketten (NE-Spanien). *Neues Jahrbuch für Geologie und Paläontologie, Abhandlungen*, **132**, 155–218.

Carls, P. & Valenzuela-Ríos, J.I. 1998. The ancestry of the Rhenish Middle Siegenian brachiopod fauna in the Iberian Chains and its palaeozoogeography (Early Devonian). *Revista Española de Paleontología, número extraordinario Homenaje al Prof. Gonzalo Vidal*, 123–142.

Carls, P., Gandl, J., Groos-Uffenorde, H., Jahnke, H. & Walliser, O.H. 1972. Neue Daten zur Grenze

Unter-/Mittel-Devon. *Newsletters on Stratigraphy*, **2**, 115–147.

CARLS, P., JAHNKE, H., LUSZNAT, M. & RACHEBOEUF, P.R. 1982. On the Siegenian Stage. *In*: ZIEGLER, W. & WERNER, R. (eds) On Devonian Stratigraphy and Palaeontology of the Ardenno-Rhenish Mountains and Related Devonian Matters. *Courier Forschungsinstitut Senckenberg*, **55**, 181–198.

CARLS, P., MEYN, H. & VESPERMANN, J. 1993. Lebensraum, Entstehung und Nachfahren von *Howellella (Iberohowellella) hollmanni* n. sg., n. sp. (Spiriferacea; Lochkovium, Unter-Devon). *Senckenbergiana lethaea*, **73**, 227–267.

CARLS, P., SLAVÍK, L. & VALENZUELA-RÍOS, J.I. 2007. Revisions of conodont biostratigraphy across the Silurian–Devonian boundary. *Bulletin of Geosciences*, **82**, 145–164.

CARLS, P., SLAVÍK, L. & VALENZUELA-RÍOS, J.I. 2008. Comments on the GSSP for the basal Emsian stage boundary: the need for its redefinition. *Bulletin of Geosciences*, **83**, 383–390.

CARTER, J.L. & GOURVENNEC, R. 2006. Introduction [to 'Spiriferida']. *In*: KAESLER, R.L. (ed.) *Treatise on Invertebrate Palaeontology, Part H, Brachiopoda (Revised), 5, Rhynchonelliformea*. Geological Society of America and University of Kansas, Boulder, CO, and Lawrence, KS, 1689–1694.

DE CASTELNAU, F. 1843. *Essai sur le Système Silurien de l'Amérique septentrionale*. P. Bertrand, Paris.

CHLUPÁČ, I. 1976. The oldest goniatite faunas and their stratigraphical significance. *Lethaia*, **9**, 303–315.

CHLUPÁČ, I. 1977. Barrandian. *In*: MARTINSSON, A. (ed.) *The Silurian–Devonian Boundary. IUGS Series A, 5*. E. Schweizerbart'sche Verlagsbuchhandlung, Stuttgart, 84–95.

CHLUPÁČ, I. 1982. The Bohemian Lower Devonian stages. *In*: ZIEGLER, W. & WERNER, R. (eds) On Devonian Stratigraphy and Palaeontology of the Ardenno-Rhenish Mountains and Related Devonian Matters. *Courier Forschungsinstitut Senckenberg*, **55**, 345–400.

CHLUPÁČ, I. & HLADIL, J. 2000. The global stratotype section and point of the Silurian-Devonian boundary. *In*: BULTYNCK, P. (ed.) Subcommission on Devonian Stratigraphy. Recognition of Devonian series and stage boundaries in geological areas. *Courier Forschungsinstitut Senckenberg*, **225**, 1–7.

CHLUPÁČ, I. & KUKAL, Z. 1986. Reflection of possible global Devonian events in the Barrandian area, C.S.S.R. *In*: WALLISER, O.H. (ed.) *Global Bio-Events – A Critical Approach*. Lecture Notes in Earth Sciences, **8**. Springer, Berlin, 169–179.

CHLUPÁČ, I. & KUKAL, Z. 1988. Possible global events and the stratigraphy of the Palaeozoic of the Barrandian (Cambrian – Middle Devonian, Czechoslovakia). *Sborník geologických Věd, Journal of Geological Sciences*, **43**, 83–146.

CHLUPÁČ, I. & LUKEŠ, P. 1999. Pragian/Zlíchovian and Zlíchovian/Dalejan boundary sections in the Lower Devonian of the Barrandian area, Czech Republic. *Newsletters on Stratigraphy*, **37**, 75–100.

CHLUPÁČ, I. & TUREK, V. 1983. Devonian goniatites from the Barrandian area, Czechoslovakia. *Rozpravy ústředního ústavu geologického*, **46**, 1–159.

CHLUPÁČ, I., FEIST, R. & MORZADEC, P. 2000. Trilobites and standard Devonian boundaries. *In*: BULTYNCK, P. (ed.) Subcommission on Devonian Stratigraphy – Fossil Groups Important for Boundary Definition. *Courier Forschungsinstitut Senckenberg*, **220**, 87–98.

CLASEN, S. 1988. *Geologische Untersuchungen im Gebiet zwischen Untenrüden und Balkhausen (GK 25 Solingen 4808)*. Diplomarbeit Universität Göttingen.

CLAUSEN, C.-D. 1991. *Geologische Karte von Nordrhein-Westfalen 1:25000, Erläuterungen Blatt 4914 Kirchhundem*. Geologisches Landesamt Nordrhein-Westfalen, Krefeld.

CLAUSEN, C.-D. 1994. Das älteste Unterdevon im Müsener Horst und seine silurische Unterlage (Sauerland/ Rheinisches Schiefergebirge). *Courier Forschungsinstitut Senckenberg*, **169**, 319–327.

CONRAD, T.A. 1839. Descriptions of new species of organic remains. *Second Annual Report on the Palaeontological Department of the Survey. New York State Geological Survey*, **275**, 57–66.

COPPER, P. 1998. Evaluating the Frasnian-Famennian mass extinction: comparing brachiopod faunas. *Acta Palaeontologica Polonica*, **43**, 137–154.

CORRADINI, C. & CORRIGA, M.G. 2012. A Přídolí–Lochkovian conodont zonation in Sardinia and the Carnic Alps: implications for a global zonation scheme. *Bulletin of Geosciences*, **87**, 635–650.

CURRY, G.B. & BRUNTON, C.H.C. 2007. Stratigraphic distribution of brachiopods. *In*: SELDEN, P.A. (ed.) *Treatise on Invertebrate Palaeontology, Part H, Brachiopoda (Revised), 6: Supplement*. Geological Society of America and University of Kansas, Boulder, CO, and Lawrence, KS, H2901–H2965.

DAHMER, G. 1916. Die Fauna der obersten Koblenzschichten von Mandeln bei Dillenburg. *Jahrbuch der Königlich Preußischen Geologischen Landesanstalt*, **36** (for 1915), 174–248.

DAHMER, G. 1923. Die Fauna der obersten Koblenzschichten am Nordwestrand der Dillmulde. *Jahrbuch der Preußischen Geologischen Landesanstalt*, **42** (for 1921), 655–693.

DAHMER, G. 1931. Fauna der belgischen 'Quartzophyllades de Longlier' in Siegener Rauhflaserschichten auf Blatt Neuwied. *Jahrbuch der Preußischen Geologischen Landesanstalt*, **52**, 86–111.

DAHMER, G. 1934. Die Fauna der Seifener Schichten (Siegenstufe). *Abhandlungen der Preußischen Geologischen Landesanstalt, Neue Folge*, **147**, 1–91.

DAHMER, G. 1935. Die Fauna der Siegener Schichten in der Umgebung des Laacher Sees. *Jahrbuch der Preußischen Geologischen Landesanstalt*, **55** (for 1934), 122–141.

DAHMER, G. 1936*a*. Die Fauna der Obersten Siegener Schichten von der Unkelmühle bei Eitorf a. d. Sieg. *Abhandlungen der Preußischen Geologischen Landesanstalt, Neue Folge*, **168**, 3–36.

DAHMER, G. 1936*b*. Die Fauna der Siegener Schichten von Unkel (Bl. Königswinter). *Jahrbuch der Preußischen Geologischen Landesanstalt*, **56** (for 1935), 633–671.

DAHMER, G. 1937. Die Fauna der Siegener Schichten im Ahrgebiet. *Jahrbuch der Preußischen Geologischen Landesanstalt*, **57** (for 1936), 435–464.

DAHMER, G. 1939. Die Fauna der Unterkoblenz-Schichten (Unter-Devon) von Oppershofen (Blatt Butzbach, Hessen). *Senckenbergiana*, **21**, 119–134.

Dahmer, G. 1942. Die Fauna der 'Gedinne'-Schichten von Weismes in der Nordwest-Eifel (mit Ausschluss der Anthozoen und Trilobiten). *Senckenbergiana*, **25**, 111–156.

Dahmer, G. 1946. Gotlandium (Mittel-Ludlow) mit *Dayia navicula* im Taunus. *Senckenbergiana*, **27**, 76–84.

Dahmer, G. 1948. Die Fauna des Koblenzquarzits (Unterdevon, Oberkoblenz-Stufe) vom Kühkopf bei Koblenz. *Senckenbergiana*, **29**, 115–136.

Dahmer, G. 1951. Die Fauna der nach-ordovizischen Glieder der Verseschichten. Mit Ausschluß der Trilobiten, Crinoiden und Anthozoen. *Palaeontographica, Abt. A*, **101**, 1–152.

Dalman, J.W. 1828. *Uppställning och Beskrifning af de i Sverige funne Terebratuliter*. Kongliga Svenska vetenskaps-academiens Handlingar (for 1827). P.A. Norstedt & Söner, Stockholm, 87–155.

De Baets, K., Klug, C. & Korn, D. 2009. Anetoceratinae (Ammonoidea, Early Devonian) from the Eifel and Harz Mountains (Germany) with a revision of their genera. *Neues Jahrbuch für Geologie und Paläontologie, Abhandlungen*, **252**, 361–376.

De Baets, K., Klug, C., Korn, D., Bartels, C. & Poschmann, M. 2013*a*. Emsian Ammonoidea and the age of the Hunsrück Slate (Rhenish Mountains, Western Germany). *Palaeontographica, Abt. A*, **299**, 1–113.

De Baets, K., Goolaerts, S., Jansen, U., Rietbergen, T. & Klug, C. 2013*b*. The first record of Early Devonian ammonoids from Belgium and their stratigraphic significance. *Geologica Belgica*, **16**, 148–156.

Defrance, M. 1828. *Terebratula undata*. *In*: Cuvier, F. (ed.) *Dictionnaire des Sciences Naturelles*. F. G. Levrault, Strasbourg, and Le Normant, Paris, **53**, 155.

Drevermann, F. 1902. Die Fauna der Unterkoblenzschichten von Oberstadtfeld bei Daun in der Eifel. *Palaeontographica*, **49**, 73–120.

Drevermann, F. 1904. Die Fauna der Siegener Schichten von Seifen unweit Dierdorf (Westerwald). *Palaeontographica*, **50**, 229–287.

Drevermann, F. 1907. Paläozoische Notizen. *Bericht der Senckenbergischen Naturforschenden Gesellschaft Frankfurt a. M.*, **1907**, 125–136.

Drot, J. 1964. Rhynchonelloidea et Spiriferoidea siluro-dévoniens du Maroc pré-saharien. *Notes et Mémoires du Service géologique du Maroc*, **178**, 1–288.

Drygant, D. & Szaniawski, H. 2012. Lochkovian conodonts from Podolia, Ukraine, and their stratigraphic significance. *Acta Palaeontologica Polonica*, **57**, 833–861.

Duméril, A.M.C. 1805. *Zoologie analytique ou méthode naturelle de classification des animaux*. Allais, Paris.

Dumont, A. 1848. Mémoir sur le terrains ardennais et rhénan de l'Ardenne, du Rhin, du Brabant et du Condros. Seconde partie. Terrain rhénan. *Mémoires de l'Académie royale de Belgique, Classe des Sciences*, **22**, 165–598.

Ebbighausen, V., Becker, R.T. & Bockwinkel, J. 2011. Emsian and Eifelian ammonoids from Oufrane, eastern Dra Valley (Anti-Atlas, Morocco). *Neues Jahrbuch für Geologie und Paläontologie, Abhandlungen*, **259**, 313–379.

Eichele, O. 2014. Muscheln und Rostroconchien aus dem mittelrheinischen Unterdevon und den angrenzenden Gebieten. *Mainzer Naturwissenschaftliches Archiv, Beiheft*, **34**, 1–229.

Eisenack, A. 1934. Neue Mikrofossilien des baltischen Silurs, III und neue Mikrofossilien des böhmischen Silurs, I. *Paläontologische Zeitschrift*, **16**, 52–76.

Eiserhardt, K.-H. & Ribbert, K.-H. 2006. Nördliches Rheinisches Schiefergebirge. *In*: Deutsche Stratigraphische Kommission (ed.), Heuse, T. & Leonhardt, D. (coords) *Stratigraphie von Deutschland VII. Silur*. Schriftenreihe der Deutschen Gesellschaft für Geowissenschaften, Hannover, **46**, 38–44.

Eiserhardt, K.-H., Heyckendorf, K. & Thombansen, E. 1981. Zur Stratigraphie und Tektonik des nördlichen Ebbe-Teilsattels (Sauerland, Rheinisches Schiefergebirge). *Mitteilungen aus dem Geologisch-Paläontologischen Institut der Universität Hamburg*, **50**, 199–238.

Elkholy, H. 1998. Fazies-Untersuchungen im mittleren Ober-Ems (Laubach-Unterstufe) der Moselmulde (Unterdevon, Rheinisches Schiefergebirge). *Bonner Geowissenschaftliche Schriften*, **27**, 1–180.

Elkholy, H. & Gad, J. 2006. Die Wied-Gruppe (vormals Hunsrückschiefer): eine neue lithostratigraphische Einheit am Nordrand der Moselmulde – Untersuchungen zu ihrer faziellen und stratigraphischen Einordnung. *Mainzer geowissenschaftliche Mitteilungen*, **34**, 49–72.

Erben, H.K. 1953. Goniatitacea (Ceph.) aus dem Unterdevon und dem Unteren Mitteldevon. *Neues Jahrbuch für Geologie und Paläontologie, Abhandlungen*, **98**, 175–225.

Erben, H.K. 1960. Primitive Ammonoidea aus dem Unterdevon Fankreichs und Deutschlands. *Neues Jahrbuch für Geologie und Paläontologie, Abhandlungen*, **110**, 1–128.

Erben, H.K. 1962. Zur Analyse und Interpretation der Rheinischen und Herzynischen Magnafazies des Devons. *In*: Erben, H.K. (ed.) *Internationale Arbeitstagung über die Silur/Devon-Grenze und die Stratigraphie von Silur und Devon. Bonn-Bruxelles 1960, Symposiums Band 2*. E. Schweitzerbart'sche Verlagsbuchhandlung (Nägele und Obermiller), Stuttgart, 42–61.

Evans, K.M. 1981. A marine fauna from the Dartmouth Beds (Lower Devonian) of Cornwall. *Geological Magazine*, **118**, 517–523.

Evans, K.M. 1985. The brachiopod fauna of the Meadfoot Group (Lower Devonian) of the Torbay area, south Devon. *Geological Journal*, **20**, 81–90.

Faber, P. 1980. Fazies-Gliederung und -Entwicklung im Mittel-Devon der Eifel (Rheinisches Schiefergebirge). *Mainzer geowissenschaftliche Mitteilungen*, **8**, 83–149.

Farrell, J.R. 1992. The Garra Formation (Early Devonian: Late Lochkovian) between Cumnock and Lassas Lee, New South Wales, Australia: stratigraphic and structural setting, faunas and community sequence. *Palaeontographica, Abt. A*, **222**, 1–41.

Fenchel, W., Gies, H. & Lusznat, M. 1985. Geologische Übersicht. *In*: Fenchel, W., Gies, H. et al. (eds) *Sammelwerk Deutsche Eisenerzlagerstätten. I. Eisenerze im Grundgebirge (Varistikum). 1. Die Sideriterzgänge im Siegerland-Wied-Distrikt*. Geologisches Jahrbuch, series D, **77**. Hannover, 17–63.

Ferrová, L., Frýda, J. & Lukeš, P. 2012. High-resolution tentaculite biostratigraphy and facies development across the Early Devonian Daleje Event in the

Barrandian (Bohemia): implications for global Emsian stratigraphy. *Bulletin of Geosciences*, **87**, 587–624.

FOLLMANN, O. 1925. Die Koblenzschichten am Mittelrhein und im Moselgebiet. *Verhandlungen des Naturhistorischen Vereins der preußischen Rheinlande und Westfalens*, **78/79**, 1921/22, 1–105.

FRANK, W. 1898. Beiträge zur Geologie des südöstlichen Taunus, insbesondere der Porphyroide dieses Gebietes. *Bericht der oberhessischen Gesellschaft für Natur- und Heilkunde*, **32**, 42–76.

FRANKE, C. 2006. Die Klerf-Schichten (Unter-Devon) im Großherzogtum Luxemburg, in der Westeifel (Deutschland) und im Gebiet von Burg Reuland (Belgien): fazielle und biostratigraphische Deutungen. *In*: FRANKE, C. (ed.) *Beiträge zur Paläontologie des Unterdevons Luxemburgs, Vol. 1*. Ferrantia, **46**. Musée national d'histoire naturelle Luxembourg, Luxembourg, 42–96.

FRANKE, W. 2000. The mid-European segment of the Variscides: tectonostratigraphic units, terrane boundaries and plate tectonic evolution. *In*: FRANKE, W., HAAK, K., ONCKEN, O. & TANNER, D. (eds) *Orogenic Processes: Quantification and Modelling in the Variscan Belt*. Geological Society, London, Special Publications, **179**, 35–61, http://doi.org/10.1144/GSL.SP.2000.179.01.05

FRANKE, W., EDER, W., ENGEL, W. & LANGENSTRASSEN, F. 1978. Main Aspects of Geosynclinal Sedimentation in the Rhenohercynian Zone. *Zeitschrift der Deutschen Geologischen Gesellschaft*, **129**, 201–216.

FRECH, F. 1888. Geologie der Umgegend von Haiger bei Dillenburg (Nassau). Nebst einem palaeontologischen Anhang. *Abhandlungen zur geologischen Specialkarte von Preussen und den Thüringischen Staaten*, **8**, 1–36 [223–258].

FRECH, F. 1889. 1. Ueber das rheinische Unterdevon und die Stellung des 'Hercyn'. *Zeitschrift der Deutschen Geologischen Gesellschaft*, **1889**, 175–287.

FRECH, F. 1897. *Lethaea palaeozoica, 2. Band, 1. Lieferung. Lethaea geognostica 1*. E. Schweizerbart'sche Verlagshandlung (E. Koch), Stuttgart.

FUCHS, A. 1907. Die unterdevonischen Rensselaerien des Rheingebietes. *Jahrbuch der Königlich Preußischen Geologischen Landesanstalt und Bergakademie*, **24** (for 1903), 43–53.

FUCHS, A. 1909. Die Brachiopoden und Gastropoden der Remscheider Schichten. *In*: SPRIESTERSBACH, J. & FUCHS, A. (eds) Die Fauna der Remscheider Schichten. *Abhandlungen der Königlich Preußischen Geologischen Landesanstalt, Neue Folge*, **58**, 53–79.

FUCHS, A. 1915. Die Hunsrückschiefer und die Unterkoblenzschichten am Mittelrhein (Loreleigegend). I. Teil: Beitrag zur Kenntnis der Hunsrückschiefer- und Unterkoblenzfauna der Loreleigegend. *Abhandlungen der Königlich Preußischen Geologischen Landesanstalt, Neue Folge*, **79**, 1–79.

FUCHS, A. 1919. Beitrag zur Kenntnis der Devonfauna der Verse- und der Hobräcker Schichten des sauerländischen Faciesgebietes. *Jahrbuch der Königlich Preußischen Geologischen Landesanstalt*, **39** (for 1918), 58–95.

FUCHS, A. 1923. Über die Beziehungen des sauerländischen Faciesgebietes zur belgischen Nord- und Südfacies und ihre Bedeutung für das Alter der Verseschichten. *Jahrbuch der Preußischen Geologischen Landesanstalt*, **42** (for 1921), 839–859.

FUCHS, A. 1929. Beitrag zur Kenntnis der unteren Gedinnefauna. *Jahrbuch der Preußischen Geologischen Landesanstalt*, **50**, 194–201.

FUCHS, A. 1934. Über eine untere Gedinnefauna im Ebbesandstein des Ebbegebirges. *Zeitschrift der Deutschen Geologischen Gesellschaft*, **86**, 395–409.

FUCHS, G. 1971. Faunengemeinschaften und Fazieszonen im Unterdevon der Osteifel als Schlüssel zur Paläogeographie. *Notizblatt des Hessischen Landesamtes für Bodenforschung*, **99**, 78–105.

FUCHS, G. 1974. Das Unterdevon am Ostrand der Eifeler Nordsüd-Zone Stratigraphie, Fazies und Tektonik des Ober-Siegen und Unter-Ems im Raum zwischen der Ahr und der Linie Daun-Ulmen. *Beiträge zur naturkundlichen Forschung in Südwestdeutschland*, **2**, 3–163.

FUCHS, G. 1982. Upper Siegenian and Lower Emsian in the Eifel Hills. *In*: ZIEGLER, W. & WERNER, R. (eds) On Devonian Stratigraphy and Palaeontology of the Ardenno-Rhenish Mountains and related Devonian Matters. *Courier Forschungsinstitut Senckenberg*, **55**, 229–255.

FÜRSICH, F.T. & HURST, J.M. 1980. Euryhalinity of Palaeozoic articulate brachiopods. *Lethaia*, **13**, 303–312.

GAD, J. 1998. Paläontologische und geologische Bemerkungen über die Hohenrhein-Schichten (Rheinisches Schiefergebirge, Ober-Ems) an der Typuslokalität im unteren Lahntal. *Mainzer Geowissenschaftliche Mitteilungen*, **27**, 137–145.

GAD, J. 2002. *Alatiformia janseni* n. sp., eine neue Brachiopodenart aus dem Ober-Ems der Moselmulde (Unterdevon, Rheinisches Schiefergebirge). *Mainzer Geowissenschaftliche Mitteilungen*, **31**, 123–128.

GAD, J. 2005. Miosporen aus dem Hunsrückschiefer des Westerwaldes (Rheinisches Schiefergebirge, Unterdevon) und die stratigraphische Stellung der Mayen-Formation. *Mainzer Geowissenschaftliche Mitteilungen*, **33**, 167–218.

GAD, J., SCHÄFER, P. & WEIDENFELLER, M. 2008. *Geologische Karte von Rheinland-Pfalz 1:25000, Erläuterungen Blatt 5611 Koblenz*. Landesamt für Geologie und Bergbau Rheinland-Pfalz, Mainz.

GANDL, J. 1972. Die Acastavinae und Asteropyginae (Trilobita) Keltiberiens (NE-Spanien). *Abhandlungen der Senckenbergischen Naturforschenden Gesellschaft*, **530**, 1–184.

GARCÍA-ALCALDE, J.L. 1996. El Devónico del Dominio Astur-Leonés en la Zona Cantábrica (N de España). *Revista Española de Paleontología, número extraordinario*, 58–71.

GARCÍA-ALCALDE, J.L. 1997. North Gondwanan Emsian events. *Episodes*, **20**, 241–246.

GARCÍA-ALCALDE, J.L. 2008. Glossinotoechiidae (braquiópodos uncinuloideos) del Devónico de la Cordillera Cantábrica (N de España). *Revista Española de Paleontología*, **23**, 237–266.

GARCÍA-ALCALDE, J.L. & TRUYÓLS-MASSONI, M. 1994. Lower/Upper Emsian versus Zlichovian/Dalejan (Lower Devonian) boundary. *Newsletters on Stratigraphy*, **30**, 83–89.

GARCÍA-ALCALDE, J.L., ARBIZU, M., GARCÍA-LÓPEZ, S., LEYVA, F., MONTESINOS, R., SOTO, F. & TRUYÓLS-

MASSONI, M. 1990. Devonian stage boundaries (Lochkovian/Pragian, Pragian/Emsian, and Eifelian/Givetian) in the Cantabric region (NW Spain). *Neues Jahrbuch für Geologie und Paläontologie, Abhandlungen*, **180**, 177–207.

GLINSKI, A. 1961. Die Schichtenfolge der Rohrer Mulde (Devon der Eifel). *Senckenbergiana lethaea*, **42**, 273–289.

GODEFROID, J. 1977. Le genre *Paraspirifer* Wedekind, 1926 (Spiriferida – Brachiopode) dans l'Emsien et le Couvinien de la Belgique. *Annales de la Société géologique du Nord*, **97**, 27–44.

GODEFROID, J. 1980. Le genre *Brachyspirifer* Wedekind, R. 1926 dans le Siegenien, l'Emsien et le Couvinien du bord méridional du Synclinorium de Dinant. *Bulletin de l'Institut royal des Sciences naturelles de Belgique, Sciences de la Terre*, **52**, 1–102.

GODEFROID, J. 1982. Gedinnian lithostratigraphy and biostratigraphy of Belgium – historical subdivisions and brachiopod biostratigraphy, a synopsis. *In*: ZIEGLER, W. & WERNER, R. (eds) On Devonian Stratigraphy and Palaeontology of the Ardenno-Rhenish Mountains and Related Devonian matters. *Courier Forschungsinstitut Senckenberg*, **55**, 97–134.

GODEFROID, J. 1994. Le genre *Euryspirifer* Wedekind, 1926 (Brachiopoda, Spiriferida) dans le Dévonien inférieur de la Belgique. *Bulletin de l'Institut royal des Sciences naturelles de Belgique, Sciences de la Terre*, **64**, 57–83.

GODEFROID, J. 1995. *Dayia shirleyi* Alvarez & Racheboeuf, 1986, un brachiopode silurien dans les 'Schistes de Mondrepuis à Muno' (sud de la Belgique). *Bulletin de l'Institut royal des Sciences naturelles de Belgique, Sciences de la Terre*, **65**, 269–272.

GODEFROID, J. & CRAVATTE, T. 1999. Les brachiopodes et la limite Silurien/Dévonien à Muno (sud de la Belgique). *Bulletin de l'Institut royal des Sciences naturelles de Belgique*, **69**, 5–26.

GODEFROID, J. & FAGERSTROM, J.A. 1983. Le genre *Paraspirifer* Wedekind, R., 1926 dans le Dévonien moyen de la partie orientale de l'Amérique du Nord. *Bulletin de l'Institut royal des Sciences naturelles de Belgique, Sciences de la Terre*, **55**, 1–61.

GODEFROID, J. & STAINIER, P. 1982. Lithostratigraphy and biostratigraphy of the Belgian Siegenian on the south and south-east borders of the Dinant Synclinorium. *In*: ZIEGLER, W. & WERNER, R. (eds) On Devonian Stratigraphy and Palaeontology of the Ardenno-Rhenish Mountains and Related Devonian matters. *Courier Forschungsinstitut Senckenberg*, **55**, 139–163.

GODEFROID, J. & STAINIER, P. 1988. Les Formations de Vireux et de Chooz (Emsien Inférieur et Moyen) au bord sud du Synclinorium de Dinant entre les villages d'Olloy-surViroin (Belgique) à l'Ouest et de Chooz (France) à l'Est. *Bulletin de l'Institut royal des Sciences naturelles de Belgique*, **58**, 95–173.

GODEFROID, J., BLIECK, A. ET AL. 1994. Les formations du Dévonien inférieur du Massif de la Vesdre, da la Fenêtre du Theux et du Synclinorium de Dinant (Belgique, France). *Mémoires pour servir à l'explication des cartes géologiques et minières de la Belgique*, **38**, 1–144.

GOLDFUSS, G.A. 1829. *Petrefacta Germaniae. Erster Theil*. Arnz & Co., Düsseldorf, 77–164, pls 26–50.

GOLDRING, R. & LANGENSTRASSEN, F. 1979. Open shelf and near-shore clastic facies in the Devonian. *Special Papers in Palaeontology*, **23**, 81–97.

GOLONKA, J., ROSS, M.I. & SCOTESE, C.R. 1994. Phanerozoic paleogeographic and paleoclimatic modeling maps. *In*: EMBRY, A.F., BEAUCHAMP, B. & GLASS, D.J. (eds) *Pangea: Global Environments and Resources*. Canadian Society of Petroleum Geologists, Memoir, **17**, 1–47.

GOSSELET, J.A.A. 1880. *Esquisse géologique du Nord de la France et des contrées voisines. Premier fascicule: Terrains primaires*. Six-Horemans, Lille.

GOSSELET, J.A.A. 1888. L'Ardenne. *In*: MINISTÈRE DES TRAVEAUX PUBLIQUES (ed.) *Mémoires pour servir à l'expliclation de la carte géologique detaillée de la France*. Baudry & Cie, Paris, 1–881.

GOURVENNEC, R. 1985. Le genre *Howellella* (Brachiopoda, Spiriferida) en Europe de l'Ouest au Siluro-Dévonien. *Geobios*, **18**, 143–170.

GOURVENNEC, R. 1986. Superfamille Spiriferacea King, 1846. *In*: RACHEBOEUF, P.R. (ed.) *Le Groupe de Liévin. Pridoli-Lochkovien de l'Artois (N. France). Sédimentologie – Paléontologie – Stratigraphie*. Biostratigraphie du Paléozoïque, Université de Bretagne Occidentale, Brest, **3**, 133–136.

GOURVENNEC, R. 1989. *Brachiopodes Spiriferida du Dévonien inférieur du Massif Armoricain. Systématique, paléobiologie, évolution, biostratigraphie*. Biostratigraphie du Paléozoïque, **9**. Université de Bretagne Occidentale, Brest, 1–281.

GRIGO, M. 1993. Fazies, Inkohlungsverhältnisse und das Alter der Seifener Schichten (Rheinisches Schiefergebirge; Unter-Devon). *Neues Jahrbuch für Geologie und Paläontologie, Monatshefte*, **1993**, 497–511.

GRIGO, M. 1994. Palaeocopida und Eridostracoda (Ostracoda) aus den Seifener Schichten des Westerwaldes (Unter-Devon, Rheinisches Schiefergebirge). *Mainzer geowissenschaftliche Mitteilungen*, **23**, 81–94.

GRIGO, M., LÜTTE, B.-P. & OEKENTORP, K. 1992. Korallen (Rugosa) aus dem Unter-Devon des nördlichen Westerwaldes (Rheinisches Schiefergebirge). *Neues Jahrbuch für Geologie und Paläontologie, Monatshefte*, **1992**, 735–749.

HAHN, H.-D. 1990. *Fazies grobklastischer Gesteine des Unterdevon (Graue Phyllite bis Taunusquarzit) im Taunus (Rheinisches Schiefergebirge)*. Dissertation, Philipps-Universität Marburg/Lahn.

HAHN, H.-D. & ZANKL, H. 1991. Sedimentation in the Lower Devonian of the Taunus area (Graue Phyllite to Taunusquarzit). *Zentralblatt für Geologie und Paläontologie Teil I*, **1990**, 1509–1520.

HALAMSKI, A.T. 2008. Middle Devonian Brachiopods from the northern Part of the Holy Cross Mountains, Poland in relation to selected coeval faunas. Part One: Introduction, Lingulida, Craniida, Strophomenida, Productida, Protorthida, Orthida. *Palaeontographica, Abt. A*, **287**, 41–98.

HALAMSKI, A.T. & BALIŃSKI, A. 2013. Middle Devonian brachiopods from the southern Maïder (eastern Anti-Atlas, Morocco). *Annales Societatis Poloniae*, **83**, 243–307.

HANCE, L., DEJONGHE, L. & STEEMANS, P. 1992. Stratigraphie du Dévonien inférieur dans le Massif de la

Vesdre. *Annales de la Société Géologique de Belgique*, **115**, 119–134.
HAPPEL, L. 1932. Das Unterdevon der Prümer Mulde. *Senckenbergiana*, **14**, 331–358.
HAVLÍČEK, V. 1956. The Brachiopods of the Braník and Hlubočepy Limestones in the Immediate Vicinity of Prague. *Sborník Ústředního Ústavu Geologického*, **22**, 535–665 [in Czech].
HAVLÍČEK, V. 1959. Rhynchonellacea im böhmischen älteren Paläozoikum. (Brachiopoda). *Věstník Ústředního ústavu geologického*, **34**, 78–82.
HAVLÍČEK, V. 1961. Rhynchonelloidea des böhmischen älteren Paläozoikums. *Rozpravy Ústředního ústavu geologického*, **27**, 1–211.
HAVLÍČEK, V. 1965. Superfamily Orthotetacea (Brachiopoda) in the Bohemian and Moravian Palaeozoic. *Věstník Ústředního ústavu geologického*, **40**, 291–294.
HAVLÍČEK, V. 1967. Brachiopoda of the suborder Strophomenidina in Czechoslovakia. *Rozpravy Ústředního ústavu geologického*, **33**, 1–235.
HAVLÍČEK, V. 1992. New Silurian and Devonian Strophomenidina (Brachiopoda) in Bohemia. *Věstník Českého geologického ústavu*, **67**, 169–178.
HAVLÍČEK, V. & ŠTORCH, P. 1990. Silurian brachiopods and benthic communities in the Prague Basin (Czechoslovakia). *Rozpravy Ústředního ústavu geologického*, **48**, 1–275.
HEDDEBAUT, C. 1981. Les Brachiopodes Spiriferacea et Reticulariacea. *In*: MORZADEC, P., PARIS, F. & RACHEBOUEF, P.R. (eds) *La Tranchée de La Lézais. Emsien supérieur du Massif Armoricain, sédimentologie, paléontologie, stratigraphie*. Mémoires de la Société géologique et minéralogique de Bretagne, Rennes, **24**, 231–253.
HERRERA, Z.A., SALAS, M.J. & GIOLITTI, J.A. 1998. Chilidiopsoidea (Brachiopoda) del Devónico Inferior de la Precordillera Argentina. *Revista Española de Paleontología*, **13**, 149–166.
HILDEN, H.D. 2008. Südliches Bergisches Land und Oberbergisches Muldenvorland. *In*: DEUTSCHE STRATIGRAPHISCHE KOMMISSION (ed.), WEDDIGE, K. (coord.) *Stratigraphie von Deutschland VIII. Devon*. Schriftenreihe der Deutschen Gesellschaft für Geowissenschaften, Hannover, **46**, 392–401.
HIMMLER, K. 1976. Zur Geologie des Ebbegebirges, Sauerland. Ein neuer Fundpunkt. *Der Aufschluss*, **27**, 247–252.
HOTZ, E.E., KRÄUSEL, W. & STRUVE, W. 1955. Die Eifelmulden von Hillesheim und Ahrdorf. *In*: KRÖMMELBEIN, K., HOTZ, E.E., KRÄUSEL, W. & STRUVE, W. (eds) *Zur Geologie der Eifelkalkmulden*. Beihefte zum Geologischen Jahrbuch, **17**, 45–192.
HOUSE, M.R. 1975. Facies and time in Devonian tropical areas. *Proceedings of the Yorkshire Geological Society*, **40**, 233–288, http://doi.org/10.1144/pygs.40.2.233
HOUSE, M.R. 1985. Correlation of mid-Palaeozoic ammonoid evolutionary events with global sedimentary perturbations. *Nature*, **213**, 17–22.
HUBERT, B.L.M., ZAPALSKI, M.K., NICOLLIN, J.-P., MISTIAEN, B. & BRICE, D. 2007. Selected benthic faunas from the Devonian of the Ardennes: an estimation of palaeobiodiversity. *Acta Geologica Polonica*, **57**, 223–262.
IVANOVA, E.A. 1972. Osnovnye zakonomernosti evolyutsii spiriferid (Brachiopoda). *Palaeontologicheskiy Zhurnal*, **1972**, 28–42 [in Russian].
JAHNKE, H. 1971. Fauna und Alter der Erbslochgrauwacke (Brachiopoden und Trilobiten, Unter-Devon, Rheinisches Schiefergebirge und Harz). *Göttinger Arbeiten zur Geologie und Paläontologie*, **9**, 1–105.
JAHNKE, H. 1986. Superfamille Strophodontacea Caster, 1939. *In*: RACHEBOEUF, P.R. (ed.) *Le Groupe de Liévin. Pridoli-Lochkovien de l'Artois (N. France). Sédimentologie – Paléontologie – Stratigraphie*. Biostratigraphie du Paléozoïque, **3**. Université de Bretagne Occidentale, Brest, 107–111.
JAHNKE, H. & JANSEN, U. 1998. Subdivision of the Emsian stage – compilation of sections in W-Europe and palaeontological remarks. *Subcommission on Devonian Stratigraphy Newsletter*, **15**, 31–35.
JAHNKE, H. & MICHELS, D. 1982*a*. The Siegenian in its type region. *In*: PLODOWSKI, G., WERNER, R. & ZIEGLER, W. (eds) *Subcommission on Devonian Stratigraphy. Field Meeting on Lower and Lower Middle Devonian Stages in the Ardenno-Rhenish Type Area*. Senckenbergische Naturforschende Gesellschaft, Frankfurt am Main, 175–185.
JAHNKE, H. & MICHELS, D. 1982*b*. Upper Emsian to Middle Devonian at Haiger Hütte (Dill Syncline). Stop 32 – Section at the Haiger Hütte near Dillenburg. *In*: PLODOWSKI, G., WERNER, R. & ZIEGLER, W. (eds) *Subcommission on Devonian Stratigraphy. Field Meeting on Lower and Lower Middle Devonian Stages in the Ardenno-Rhenish Type Area*. Senckenbergische Naturforschende Gesellschaft, Frankfurt am Main, 205–212.
JANSEN, U. 1994. *Die Stropheodontacea (Brachiopoda) der Seifener Schichten (Unter-Devon, Westerwald)*. Diplomarbeit, Philipps-Universität Marburg/Lahn.
JANSEN, U. 2001*a*. Morphologie, Taxonomie und Phylogenie unter-devonischer Brachiopoden aus der Dra-Ebene (Marokko, Prä-Sahara) und dem Rheinischen Schiefergebirge (Deutschland). *Abhandlungen der Senckenbergischen Naturforschenden Gesellschaft*, **554**, 1–389.
JANSEN, U. 2001*b*. On the genus *Acrospirifer* Helmbrecht & Wedekind, 1923 (Brachiopoda, Lower Devonian). *Journal of the Czech Geological Society*, **46**, 131–141.
JANSEN, U. 2012*a*. Die Welt des frühen Devons aus der Sicht der Brachiopodenforschung. *Natur Forschung Museum*, **142**, 44–51.
JANSEN, U. 2012*b*. Revision of Rhenish Lower Devonian Brachiopoda. *In*: WITZMANN, F. & ABERHAN, M. (eds) *Centenary Meeting of the Paläontologische Gesellschaft. Programme, Abstracts and Field Guides*, 24–29 September 2012, Museum für Naturkunde Berlin. Terra Nostra, **2012/3**. Schriften der GeoUnion Alfred-Wegener-Stiftung, Berlin, 86–87.
JANSEN, U. 2012*c*. On the traditional Siegenian-Lower Emsian successions in the Rhenish Slate Mountains, in special consideration of the basal Emsian boundary and its supraregional correlation. *Subcommission on Devonian Stratigraphy Newsletter*, **27**, 21–27.
JANSEN, U. 2014*a*. Strophomenid brachiopods from the Rhenish Lower Devonian (Germany). *Bulletin of Geosciences*, **89**, 113–136.

JANSEN, U. 2014*b*. Pridolian to Early Eifelian Brachiopod Zonation of the Rhenish Massif. *In*: RÕCHA, R., PAIS, J., KULLBERG, J.C. & FINNEY, S. (eds) *STRATI 2013 First International Congress on Stratigraphy – At the Cutting Edge of Stratigraphy*. Springer, New York, 407–411.

JANSEN, U., KÖNIGSHOF, P., PLODOWSKI, G., SCHINDLER, E. & SCHINDLER, T. 2001. Pre-conference field trip: Rhein/Mosel area and Lahn/Dill Synclines, Rheinisches Schiefergebirge (13–14 May 2001). *In*: JANSEN, U., KÖNIGSHOF, P., PLODOWSKI, G. & SCHINDLER, E. (eds) *15th International Senckenberg Conference, Joint Meeting IGCP 421/SDS*, May 2001, *Field trips guidebook*. Senckenbergische Naturforschende Gesellschaft, Frankfurt am Main, 45–85.

JANSEN, U., BROCKE, R. *ET AL*. 2004. Ein Profil im Paläozoikum des südöstlichen Venn-Sattels (Ordovizium bis Unter-Devon, Rheinisches Schiefergebirge). *Scriptum*, **11**, 109–131.

JANSEN, U., LAZREQ, N., PLODOWSKI, G., SCHEMM-GREGORY, M., SCHINDLER, E. & WEDDIGE, K. 2007. Neritic-pelagic correlation in the Lower and basal Middle Devonian of the Dra Valley (Southern Anti-Atlas, Moroccan Pre-Sahara). *In*: BECKER, R.T. & KIRCHGASSER, W.T. (eds) *Devonian Events and Correlations*. Geological Society, London, Special Publications, **278**, 9–37, http://doi.org/10.1144/SP278.2

JENTSCH, S. 1960. Die Moselmulde und ihre südöstlichen Randstrukturen zwischen Lahn und Westerwald. *Notizblatt des Hessischen Landesamtes für Bodenforschung*, **88**, 190–215.

JENTSCH, S. & RÖDER, D. 1957. Zur Geologie des Taunusquarzits bei Bad Homburg. *Notizblatt des Hessischen Landesamtes für Bodenforschung*, **85**, 114–128.

JEPPSSON, L. 1998. Silurian Oceanic Events: summary of general characteristics. *In*: LANDING, E. & JOHNSON, M. (eds) *Silurian Cycles. Linkages of Dynamic Stratigraphy with Atmospheric, Oceanic, and Tectonic Changes*. New York State Museum, Albany, Bulletin, **491**, 239–257.

JOHNSON, J.G. 1979. Devonian brachiopod biostratigraphy. *Special Papers in Palaeontology*, **23**, 291–306.

JOHNSON, J.G. & SANDBERG, C.A. 1989. Devonian eustatic events in the eastern United States and their biostratigraphic responses. *In*: MCMILLAN, N.J., EMBRY, A.F. & GLASS, D.J. (eds) *Devonian of the World*. Canadian Society of Petroleum Geologists, Memoir, **14**, 171–178.

JOHNSON, J.G., KLAPPER, G. & SANDBERG, C.A. 1985. Devonian eustatic fluctuations in Euramerica. *Bulletin of the Geological Society of America*, **96**, 567–587.

JUX, U. 1971. *Rheinische Magnafazies im devonischen Weltbild*. Kölner Geographische Arbeiten. Sonderband Forschungen zur allgemeinen und regionalen Geographie, Kayser-Festschrift. Franz Steiner, Wiesbaden, 141–157.

JUX, U. 1981. Zur stratigraphischen Verbreitung bergischer Globithyriden. *Sonderveröffentlichungen des Geologischen Institutes der Universität Köln*, **41**, 93–107.

KALJO, D., MARTMA, T., GRYTSENKO, V., BRAZAUSKAS, A. & KAMINSKAS, D. 2012. Přídolí carbon isotope trend and upper Silurian to lowermost Devonian chemostratigraphy based on sections in Podolia (Ukraine) and the East Baltic area. *Estonian Journal of Earth Sciences*, **61**, 162–180.

KAUFMANN, B. 2006. Calibrating the Devonian Time Scale: a synthesis of U–Pb ID-TIMS ages and conodont stratigraphy. *Earth-Science Reviews*, **76**, 175–190.

KAYSER, E. 1871. Die Brachiopoden des Mittel- und Ober-Devon der Eifel. *Zeitschrift der Deutschen Geologischen Gesellschaft*, **23**, 491–663.

KAYSER, E. 1878. Die Fauna der ältesten Devon-Ablagerungen des Harzes. *Abhandlungen zur geologischen Specialkarte von Preussen und den Thüringischen Staaten*, **2**, I-XXIII, 1–295.

KAYSER, E. 1882. Paragraph within a report. *In*: DEUTSCHE GEOLOGISCHE GESELLSCHAFT, C. Verhandlungen der Gesellschaft, I. Protokoll der Januar-Sitzung. *Zeitschrift der Deutschen Geologischen Gesellschaft*, **34**, 198–199.

KAYSER, E. 1884. Die Orthocerasschiefer zwischen Balduinstein und Laurenburg an der Lahn. *Jahrbuch der Königlich Preußischen Geologischen Landesanstalt und Bergakademie*, **4** (for 1883), 1–56.

KAYSER, E. 1889. Die Fauna des Hauptquarzits und der Zorger Schiefer des Unterharzes. *Abhandlungen der Königlich Preußischen Geologischen Landesanstalt, Neue Folge*, **1**, 1–139.

KAYSER, E. & HOLZAPFEL, E. 1894. Ueber die stratigraphischen Beziehungen der böhmischen Stufen F,G,H Barrande's zum rheinischen Devon. *Jahrbuch der Kaiserlich Königlich Geologischen Reichsanstalt*, **44**, 479–514.

KEGEL, W. 1913. Der Taunusquarzit von Katzenelnbogen. *Abhandlungen der Königlich Preußischen Geologischen Landesanstalt, Neue Folge*, **76**, 1–162.

KEGEL, W. 1950. Sedimentation und Tektonik in der rheinischen Geosynklinale. *Zeitschrift der Deutschen Geologischen Gesellschaft*, **100**, 267–289.

KLAPPER, G. & MURPHY, M.A. 1975. Silurian-Lower Devonian Conodont Sequence in the Roberts Mountains Formation of central Nevada. *University of California Publications in Geological Sciences*, **111**, 1–62.

KLUG, C. 2002. Quantitative stratigraphy and taxonomy of late Emsian and Eifelian ammonoids of the eastern Anti-Atlas (Morocco). *Courier Forschungsinstitut Senckenberg*, **238**, 1–109.

KOCH, C. 1881. Footnote *in* KAYSER, E. Beitrag zur Kenntniss des Taunusquarzits. *Neues Jahrbuch für Mineralogie, Geologie und Paläontologie*, **1881**, 386–387.

KOCH, L., LEMKE, U. & BRAUCKMANN, C. 1990. *Vom Ordovizium bis zum Devon: Die fossile Welt des Ebbe-Gebirges*. Linnepe Verlag, Hagen.

DE KONINCK, L. 1876. Notice sur quelques fossiles recueillis par G. Dewalque dans le système Gedinnien de A. Dumont. *Annales de la Société géologique de Belgique*, **3**, 25–52.

KOSSMAT, F. 1927. Gliederung des varistischen Gebirgsbaues. *Abhandlungen des Sächsischen Geologischen Landesamts*, **1**, 1–40.

KOZŁOWSKI, R. 1929. Les brachiopodes gothlandiens de la Podolie polonaise. *Palaeontologia Polonica*, **1**, 1–254.

KRANTZ, A. 1857. Über ein neues bei Menzenberg aufgeschlossenes Petrefakten-Lager in den devonischen

Schichten. *Verhandlungen des Naturhistorischen Vereins der Preußischen Rheinlande und Westphalens*, **14**, 143–165.

Kräusel, R. & Weyland, H. 1930. Die Flora des deutschen Unterdevons. *Abhandlungen der Preußischen Geologischen Landesanstalt, Neue Folge*, **131**, 3–92.

Krebs, A. 1979. Devonian basinal facies. *Special Papers in Palaeontology*, **23**, 125–139.

Kříž, J. 1998. Silurian. *In*: Havlíček, V., *Prague Basin* [superordinate chapter]. *In*: Chlupáč, I., Havlíček, V., Kříž, J., Kukal, Z. & Štorch, P. (eds) *Palaeozoic of the Barrandian (Cambrian to Devonian)*. Czech Geological Survey, Prague, 79–101.

Kröll, R. 2001. Zur Stratigraphie, Fazies und Tektonik des Unterdevon zwischen der Untermosel und Boppard (Moselmulde, Rheinisches Schiefergebirge). *Bonner Geowissenschaftliche Schriften*, **29**, 1–261.

Krömmelbein, K. 1955. Stratigraphie und Tektonik der Salmerwald-Mulde (Devon, Eifel). *In*: Krömmelbein, K., Hotz, E.E., Kräusel, W. & Struve, W. (eds) *Zur Geologie der Eifelkalkmulden*. Beihefte zum Geologischen Jahrbuch, **17**. Amt für Bodenforschung, Hannover, 7–44.

Kuhn, O. 1949. *Lehrbuch der Paläozoologie*. E. Schweizerbart, Stuttgart.

Kutscher, F. 1937. Taunusquarzit, Throner Quarzite und Hunsrückschiefer des Hunsrücks und ihre stratigraphische Stellung. *Jahrbuch der Preußischen Geologischen Landesanstalt*, **57** (for 1936), 186–237.

Kutscher, F. 1940. Fossilvorkommen im Taunusquarzitzuge Weissfels-Hujets Sägemühle-Wehlenstein des Bl. Birkenfeld-West (Hunsrück). *Decheniana*, **99**, 105–118.

Kutscher, F. 1952. Fossilfunde im Taunusquarzit des westlichen Soonwaldes (Hunsrück). *Notizblatt des Hessischen Landesamtes für Bodenforschung*, **3**, 87–90.

Kutscher, F. 1969. Beiträge zur Sedimentation und Fossilführung des Hunsrückschiefers. 24. Die Ammonoideen-Entwicklung im Hunsrückschiefer. *Notizblatt des Hessischen Landesamtes für Bodenforschung*, **97**, 46–64.

Langenstrassen, F. 1972. Zur Fazies und Stratigraphie der Eifel-Stufe im östlichen Sauerland (Rheinisches Schiefergebirge, Bl. Schmallenberg und Girkhausen). *Göttinger Arbeiten zur Geologie und Paläontologie*, **12**, 1–106.

Langenstrassen, F. 2008. Unter- und Mittel-Devon im Sauerland. *In*: Deutsche Stratigraphische Kommission (ed.), Weddige, K. (coord.) *Stratigraphie von Deutschland VIII. Devon*. Schriftenreihe der Deutschen Gesellschaft für Geowissenschaften, Hannover, **52**, 417–438.

Langenstrassen, F. & Müller, H. 1982. The Lower/Middle Devonian boundary in the Sauerland (Latrop Anticline and Wittgenstein Syncline, eastern Rheinische Schiefergebirge). *In*: Ziegler, W. & Werner, R. (eds) On Devonian Stratigraphy and Palaeontology of the Ardenno-Rhenish Mountains and related Devonian matters. *Courier Forschungsinstitut Senckenberg*, **55**, 337–344.

Lankester, E.R. 1868–1870. *The Cephalaspidae. A Monograph of the Fishes of the Old Red Sandstone of Britain. I*. Palaeontographical Society, London.

Le Menn, J., Plusquellec, Y., Morzadec, P. & Lardeux, H. 1976. Incursion hercynienne dans les faunes rhénanes du Dévonien inférieur de la rade de Brest (Massif Armoricain). *Palaeontographica, Abt. A*, **153**, 1–61.

Leriche, M. 1912. La faune du Gedinnien inférieur de l'Ardenne. *Mémoires du Musée royal d'histoire naturelle de Belgique*, **6**, 1–58.

Leveridge, B.E. 2011. The Looe, South Devon and Tavy basins: the Devonian rifted passive margin successions. *In*: Leveridge, B.E. (ed.) The Marine Devonian of Great Britain. *Proceedings of the Geologists' Association*, **122** (special issue), 616–717.

Lippert, H.-J. 1939. Geologie der Daleider Muldengruppe. *Abhandlungen der Senckenbergischen Naturforschenden Gesellschaft*, **445**, 1–66.

Lusznat, M. 1968. *Geologische Karte von Nordrhein-Westfalen 1:25000, Erläuterungen Blatt 5113 Freudenberg*. Geologisches Landesamt Nordrhein-Westfalen, Krefeld.

Lütje, F. 1979. Biostratigraphical significance of the Devonian Dacryoconarida. *Special Papers in Palaeontology*, **23**, 281–289.

Maillieux, E. 1931. La faune des Grès et Schistes de Solières (Siegénien Moyen). *Mémoires du Musée royal d'histoire naturelle de Belgique*, **51**, 1–90.

Maillieux, E. 1936*a*. La faune et l'âge des Quartzophyllades de Longlier. *Mémoires du Musée royal d'histoire naturelle de Belgique*, **73**, 3–140.

Maillieux, E. 1936*b*. [Cited remarks by E. Maillieux which he made during a field trip]. *In*: Asselberghs, E., Henke, W., Schriel, W. & Wunstorf, W. (eds) Über eine gemeinsame Exkursion durch die Siegener Schichten des Rheinischen Schiefergebirges und der Ardennen. *Jahrbuch der Preußischen Geologischen Landesanstalt und Bergakademie*, **56** (for 1935), 357–358.

Maillieux, E. 1940. Le Siegenien de l'Ardenne et ses faunes. *Bulletin du Musée royal d'histoire naturelle de Belgique*, **16**, 1–23.

Maillieux, E. 1941. Les brachiopodes de l'Emsien de l'Ardenne. *Mémoires du Musée royal d'histoire naturelle de Belgique*, **96**, 3–74.

Maurer, F. 1879. Section within a report of the Deutsche geologische Gesellschaft. *In*: C. Verhandlungen der Gesellschaft. 3. Siebenundzwanzigste Versammlung der Deutschen geologischen Gesellschaft zu Baden. Protokoll der Sitzung vom 26. September 1879. *Zeitschrift der Deutschen geologischen Gesellschaft*, **31**, 641.

Maurer, F. 1883. Über das rheinische Unterdevon. *Zeitschrift der Deutschen Geologischen Gesellschaft*, **35**, 633–635.

Maurer, F. 1886. Die Fauna des rechtsrheinischen Unterdevon aus meiner Sammlung zum Nachweis der Gliederung. *Neues Jahrbuch für Mineralogie, Geologie und Palaeontologie*, **1882**, 3–55.

Mauz, J. 1935. Vergleichende Untersuchungen über die Unterkoblenz-Stufe bei Oberstadtfeld und Koblenz. *Abhandlungen der Senckenbergischen Naturforschenden Gesellschaft*, **429**, 1–94.

McCoy, F. 1852. Description of the British Palaeozoic Fossils. *In*: Sedgwick, A. & McCoy, F. (eds)

(1851–1855) *A Synopsis of the Classification of the British Palaeozoic Rocks, with a Detailed Systematic Description of the British Palaeozoic Fossils in the Geological Museum of the University of Cambridge*. (1), 1851, i–iv, 1–184; (2), 1852, i–viii, 185–406; (3), 1855, i–xcviii, 407–661. J. W. Parker & Son, London & Cambridge.

MEISCHNER, D. 1971. Clastic Sedimentation in the Variscan Geosyncline East of the River Rhine. *In*: MÜLLER, G. (ed.) *Sedimentology of Parts of Central Europe. Guidebook VIII*. International Sedimentological Congress, 1971, Heidelberg. Waldemar Kramer, Frankfurt am Main, 9–43.

MELOU, M. 1981. Les Brachiopodes Orthida. *In*: MORZADEC, P., PARIS, F. & RACHEBOUEF, P.R. (eds) *La Tranchée de La Lézais. Emsien supérieur du Massif Armoricain, sédimentologie, paléontologie, stratigraphie*. Mémoires de la Société géologique et minéralogique de Bretagne, Rennes, **24**, 135–148.

MERGL, M. 2003. Silicified brachiopods of the Kotýs Limestone (Lochkovian) in the Bubovice area (Barrandian, Bohemia). *Acta Musei Nationalis Pragae, Series B, Natural History*, **59**, 99–150.

MERGL, M. & MASSA, D. 2004. Devonian Brachiopods of the Tasmena Basin (Central Sahara; Algeria and North Niger). *Acta Musei Nationalis Pragae, Series B, Natural History*, **60**, 61–112.

MEYER, D.E. 1970. *Stratigraphie und Fazies des Paläozoikums im Guldenbachtal/SE-Hunsrück am Südrand des Rheinischen Schiefergebirges*. Dissertation, Universität Bonn.

MEYER, W. 1958. Geologie der Siegener Schichten zwischen Ahr und Nette. *Zeitschrift der Deutschen Geologischen Gesellschaft*, **109**, 452–462.

MEYER, W. 1965. Gliederung und Altersstellung des Unterdevons südlich der Siegener Hauptüberschiebung in der Südost-Eifel und im Westerwald (Rheinisches Schiefergebirge). *In*: SCHMIDT-THOMÉ, P. & SCHÖNENBERG, R. (eds) *Max Richter-Festschrift*. Piepersche Buchdruckerei + Verlagsanstalt, Clausthal-Zellerfeld, 35–47.

MEYER, W. 2013. *Geologie der Eifel*. 4th edn. Schweizerbart, Stuttgart.

MEYER, W. & PAHL, A. 1960. Zur Geologie der Siegener Schichten in der Osteifel und im Westerwald. *Zeitschrift der Deutschen Geologischen Gesellschaft*, **112**, 278–291.

MEYER, W. & STETS, J. 1980. Zur Paläogeographie von Unter- und Mitteldevon im westlichen und zentralen Rheinischen Schiefergebirge. *Zeitschrift der Deutschen Geologischen Gesellschaft*, **131**, 725–751.

MEYER, W. & STETS, J. 1996. *Das Rheintal zwischen Bingen und Bonn*. Sammlung geologischer Führer, **89**. Gebrüder Borntraeger, Berlin.

MITTMEYER, H.-G. 1965. Die Bornicher Schichten im Gebiet zwischen Mittelrhein und Idsteiner Senke (Taunus, Rheinisches Schiefergebirge). *Notizblatt des Hessischen Landesamtes für Bodenforschung*, **93**, 73–98.

MITTMEYER, H.-G. 1972. Delthyrididae und Spinocyrtiidae (Brachiopoda) des tiefsten Ober-Ems im Mosel-Gebiet (Ems-Quarzit, Rheinisches Schiefergebirge). *Mainzer Geowissenschaftliche Mitteilungen*, **1**, 82–121.

MITTMEYER, H.-G. 1973*a*. Die Hunsrückschiefer-Fauna des Wisper-Gebietes im Taunus. Ulmen-Gruppe, tiefes Unterems, Rheinisches Schiefergebirge. *Notizblatt des Hessischen Landesamtes für Bodenforschung*, **101**, 16–45.

MITTMEYER, H.-G. 1973*b*. Grenze Siegen/Unterems bei Bornhofen (Unter-Devon, Mittelrhein). *Mainzer geowissenschaftliche Mitteilungen*, **2**, 71–103.

MITTMEYER, H.-G. 1974. Zur Neufassung der Rheinischen Unterdevon-Stufen. *Mainzer geowissenschaftliche Mitteilungen*, **3**, 69–79.

MITTMEYER, H.-G. 1980. Zur Geologie des Hunsrückschiefers. *In*: STÜRMER, W., SCHAARSCHMIDT, F. & MITTMEYER, H.-G. (eds) *Versteinertes Leben im Röntgenlicht*. Kleine Senckenberg-Reihe, **11**. Waldemar Kramer, Frankfurt am Main, 26–33.

MITTMEYER, H.-G. 1982*a*. Rhenish Lower Devonian biostratigraphy. *In*: ZIEGLER, W. & WERNER, R. (eds) On Devonian Stratigraphy and Palaeontology of the Ardenno-Rhenish Mountains and Related Devonian Matters. *Courier Forschungsinstitut Senckenberg*, **55**, 257–269.

MITTMEYER, H.-G. 1982*b*. Lahnstein section (Middle Rhine; Emsian, Lower Devonian). *In*: PLODOWSKI, G., WERNER, R. & ZIEGLER, W. (eds) *Subcommission on Devonian Stratigraphy. Field Meeting on Lower and Lower Middle Devonian Stages in the Ardenno-Rhenish Type Area. Guidebook*. Senckenbergische Naturforschende Gesellschaft, Frankfurt am Main, 187–203.

MITTMEYER, H.-G. 1982*c*. *Geologische Karte von Rheinland-Pfalz 1:25000, Erläuterungen Blatt 5508 Kempenich*. Landesamt für Geologie und Bergbau Rheinland-Pfalz, Mainz.

MITTMEYER, H.-G. 1997. *Geologische Karte von Rheinland-Pfalz 1:25000, Erläuterungen Blatt 5608 Virneburg*. Landesamt für Geologie und Bergbau Rheinland-Pfalz, Mainz.

MITTMEYER, H.-G. 2002. Appendix 1: distribution of fossils in the middle Kaub Formation at Bundenbach; fossil content of the sandy parts of the Bocksberg Member and of the Obereschenbach Member compiled by Mittmeyer. *In*: BARTELS, C., WUTTKE, M. & BRIGGS, D.E.G. (eds) *The Nahecaris Project. Releasing the Marine Life of the Devonian from the Hunsrück Slate of Bundenbach. Metalla (Bochum)*, **9.2**. Deutsches Bergbau-Museum, Bochum, 105–122.

MITTMEYER, H.-G. 2008. Unterdevon der Mittelrheinischen und Eifeler-Typ-Gebiete (Teile von Eifel, Westerwald, Hunsrück und Taunus). *In*: DEUTSCHE STRATIGRAPHISCHE KOMMISSION (ed.), WEDDIGE, K. (coord.) *Stratigraphie von Deutschland VIII. Devon*. Schriftenreihe der Deutschen Gesellschaft für Geowissenschaften, Hannover, **52**, 139–203.

MITTMEYER, H.-G. & GEIB, K.-W. 1967. Gliederung des Unterdevons im Gebiet Warmsroth-Wald-Erbach (Stromberger Mulde). *Notizblatt des Hessischen Landesamtes für Bodenforschung*, **95**, 24–44.

MORZADEC, P. 1986. Trilobites. *In*: RACHEBOEUF, P.R. (ed.) *Le Groupe de Liévin. Pridoli-Lochkovien de l'Artois (N. France). Sédimentologie – Paléontologie – Stratigraphie*. Biostratigraphie du Paléozoïque, **3**. Université de Bretagne Occidentale, Brest, 185–200.

MORZADEC, P., PARIS, F. & RACHEBOUEF, P.R. 1981. Conclusions stratigraphiques. *In*: MORZADEC, P., PARIS, F. & RACHEBOUEF, P.R. (eds) *La Tranchée de La Lézais.*

Emsien supérieur du Massif Armoricain, sédimentologie, paléontologie, stratigraphie. Mémoires de la Société géologique et minéralogique de Bretagne, Rennes, **24**, 11–18.

MORZADEC, P., PARIS, F., PLUSQUELLEC, Y., RACHEBOUEF, P.R. & WEYANT, M. 1991. La limite Lochkovien-Praguien (Dévonien inférieur) dans le Massif Armoricain: espèces index et correlations. *Comptes Rendus de l'Académie des Sciences, Série II*, **313**, 901–908.

MÜLLER, P. 2011. Fossilien aus dem Westerwälder Devon und Karbon. *Der Aufschluss*, **62**, 219–239.

MÜLLER, P. & HAHN, G. 2010. Edrioasteroidea aus den Seifen-Schichten des Westerwaldes (Unter-Devon; Deutschland), Teil 1. *Geologica et Palaeontologica*, **43**, 83–91.

MÜLLER, R.-D. 1987. *Biostratigraphie und Fazies der Gesteine des Unteremsiums am Südostrand des Siegerländer Antiklinoriums*. Dissertation, Philipps-Universität Marburg/Lahn.

NEUMANN-MAHLKAU, P. 1970. Sedimentation und Paläogeographie zur Zeit der Gedinne-Transgression am Massiv von Stavelot-Venn. *Geologische Mitteilungen*, **9**, 311–356.

NIEHOFF, W. 1958. Die primär gerichteten Sedimentstrukturen, insbesondere die Schrägschichtung im Koblenzquarzit am Mittelrhein. *Geologische Rundschau*, **47**, 232–321.

NIKIFOROVA, O.I. 1977. Podolia. *In*: MARTINSSON, A. (ed.) *The Silurian–Devonian Boundary*. Final Report of the Committee in the Silurian–Devonian Boundary within IUGS Commission on Stratigraphy and a State of the Art Report for Project Ecostratigraphy, IUGS, Series A, no. 5E. Schweizerbart'sche Verlagsbuchhandlung, Stuttgart, 51–64.

NIKIFOROVA, O.I., MODZALEVSKAYA, T.L. & BASSETT, M.G. 1985. Review of the upper Silurian and Lower Devonian Articulate Brachiopods of Podolia. *Special Papers in Palaeontology*, **34**, 1–66.

OCHS, G. & WOLFART, R. 1961. Geologie der Blankenheimer Mulde (Devon, Eifel). *Abhandlungen der Senckenbergischen Naturforschenden Gesellschaft*, **501**, 1–100.

PAECKELMANN, W. & SIEVERTS, H. 1932. Neue Beiträge zur Kenntnis der Geologie, Palaeontologie und Petrographie der Umgegend von Konstantinopel. *Abhandlungen der Preußischen Geologischen Landesanstalt, Neue Folge*, **142**, 1–79.

PAPROTH, E. 1960. Über die Fauna der Mittleren Siegener Schichten des Siegerlandes. *Abhandlungen des Hessischen Landesamtes für Bodenforschung*, **29**, 321–339.

PARIS, F. 1986. Chitinozoaires. *In*: RACHEBOEUF, P.R. (ed.) *Le Groupe de Liévin. Pridoli-Lochkovien de l'Artois (N. France). Sédimentologie – Paléontologie – Stratigraphie*. Biostratigraphie du Paléozoïque, **3**. Université de Bretagne Occidentale, Brest, 55–62.

PARIS, F., WINCHESTER-SEETO, T., BOUMENDJEL, K. & GRAHN, Y. 2000. Toward a global biozonation of Devonian chitinozoans. *In*: BULTYNCK, P. (ed.) Subcommission on Devonian Stratigraphy – Fossil Groups Important for Boundary Definition. *Courier Forschungsinstitut Senckenberg*, **220**, 39–55.

PAULY, E. 1958. Das Devon der südwestlichen Lahnmulde und ihrer Randgebiete. *Abhandlungen des Hessischen Landesamtes für Bodenforschung*, **25**, 3–138.

PHILLIPS, J. 1841. *Figures and descriptions of the Palaeozoic fossils of Cornwall, Devon, and West Somerset. Geological Survey of Great Britain, Memoir 1*. Longman & Co., London.

PILGER, A. 1952. Zur Gliederung und Kartierung der Siegener Schichten I, II. *Geologisches Jahrbuch*, **66**, 703–721.

PILGER, A. 1954. Derzeitiger Stand der geologischen Neukartierung des Siegerlandes. *Geologisches Jahrbuch*, **69**, 27–52.

PILGER, A. & SCHMIDT, WO. 1959. Über das Vorkommen von marinen Faunen in der Unteren Siegen-Stufe des Siegerlandes. *Geologisches Jahrbuch*, **76**, 421–426.

PITRAT, C.W. 1965. Spiriferidina. *In*: BOUCOT, A.J., JOHNSON, J.G., PITRAT, C.W. & STATON, R.D. Spiriferida. *In*: MOORE, R.C. (ed.) *Treatise on Invertebrate Palaeontology, Part H, Brachiopoda, 2*. Geological Society of America and University of Kansas, Boulder, CO, and Lawrence, KS, H667–H728.

POSCHMANN, M. 2015. The corkscrew-shaped trace fossil *Helicodromites* Berger, 1957, from Rhenish Lower Devonian shallow-marine facies (Upper Emsian, SW Germany). *Paläontologische Zeitschrift*, **89**, 635–643.

POSCHMANN, M. & JANSEN, U. 2003. Lithologie und Fossilführung einiger Profile in den Siegen-Schichten des Westerwaldes (Unter-Devon, Rheinisches Schiefergebirge). *Senckenbergiana lethaea*, **83**, 157–183.

PŘIBYL, A. 1940. Die Graptolithenfauna des mittleren Ludlows von Böhmen (Oberes eß). *Věstnik geologičeskeho ústavu*, **16**, 63–74.

DE LA QUINTERO, I. & REVILLA, J. 1966. Algunas especies nuevas y otras poco conocidas. *Notas y Comunicaciones del Instituto Geológico y Minero de España*, **82**, 27–86.

QUIRING, H. 1923. Beiträge zur Geologie des Siegerlandes. III. Über Leitformen in den Siegener Schichten der Umgebung von Siegen. *Jahrbuch der Preußischen Geologischen Landesanstalt*, **43** (for 1922), 90–112.

RABIEN, A. 1956. Zur Stratigraphie und Fazies des Ober-Devons in der Waldecker Hauptmulde. *Abhandlungen des Hessischen Landesamtes für Bodenforschung*, **16**, 3–83.

RACHEBOEUF, P.R. 1981. *Chonétacés (brachiopodes) siluriens et dévoniens du sud-ouest de l'Europe (Systématique – Phylogénie – Biostratigraphie – Paléobiogéographie)*. Mémoires de la Société géologique et minéralogique de Bretagne, Rennes, **27**, 1–294.

RACHEBOEUF, P.R. 1990. Paléobiogéographie de la marge nord-gondwanienne au Dévonien inférieur et moyen: nouvelles données déduites de l'étude des Brachiopodes Chonetacés. *Comptes Rendus de l'Académie des Sciences*, **310**, 1481–1486.

RACHEBOEUF, P.R. & BABIN, C. 1986. Biostratigraphie et Corrélations. *In*: RACHEBOEUF, P.R. (ed.) *Le Groupe de Liévin. Pridoli-Lochkovien de l'Artois (N. France). Sédimentologie – Paléontologie – Stratigraphie*. Biostratigraphie du Paléozoïque, **3**. Université de Bretagne Occidentale, Brest, 31–45.

RACKI, G. & TURNAU, E. 2000. Devonian series and stage boundaries in Poland. *In*: BULTYNCK, P. (ed.) Subcommission on Devonian Stratigraphy. Recognition of Devonian Series and Stage Boundaries in Geological Areas. *Courier Forschungsinstitut Senckenberg*, **225**, 145–158.

RACKI, G., BALIŃSKI, A., WRONA, R., MAŁKOWSKI, K., DRYGANT, D. & SZANIAWSKI, H. 2012. Faunal dynamics across the Silurian–Devonian positive isotope excursions ($\delta^{13}C$, $\delta^{18}O$) in Podolia, Ukraine: comparative analysis of the Ireviken and Klonk events. *Acta Palaeontologica Polonica*, **57**, 795–832.

REINECK, H.-E. 1983. Sind die Klerfer Schichten Wattenablagerungen? *Natur und Museum*, **113**, 24–28.

RENAUD, A. 1942. *Le Dévonien du Synclinorium médian Brest-Laval*. Mémoires de la Société Géologique et Minéralogique de Bretagne, Rennes, **7**, 1–439.

REQUADT, H. 1990. *Geologische Karte von Rheinland Pfalz 1:25000, Erläuterungen Blatt 5613 Schaumburg*. Geologisches Landesamt Rheinland-Pfalz, Mainz.

REQUADT, H. 1991. Fazies und Paläogeographie des Devons in der südwestlichen Lahnmulde (Rheinisches Schiefergebirge). *Mainzer geowissenschaftliche Mitteilungen*, **20**, 229–248.

REQUADT, H. 2008. Südwestliche Lahnmulde (Rheinland-Pfalz). *In*: DEUTSCHE STRATIGRAPHISCHE KOMMISSION (ed.), WEDDIGE, K. (coord.) *Stratigraphie von Deutschland VIII. Devon*. Schriftenreihe der Deutschen Gesellschaft für Geowissenschaften, Hannover, **52**, 204–220.

RIBBERT, K.-H. 1993. *Geologische Karte von Nordrhein-Westfalen 1:25000, Erläuterungen Blatt 5504 Hellenthal*. Geologisches Landesamt Nordrhein-Westfalen, Krefeld.

RIBBERT, K.-H. 2006. Venn-Antiklinale. *In*: DEUTSCHE STRATIGRAPHISCHE KOMMISSION (ed.), HEUSE, T. & LEONHARDT, D. (coords) *Stratigraphie von Deutschland VII. Silur*. Schriftenreihe der Deutschen Gesellschaft für Geowissenschaften, Hannover, **46**, 33–37.

RIBBERT, K.-H. 2008. Unter-Devon in der Venn-Antiklinale und dem Westrand der Eifeler Kalkmuldenzone. *In*: DEUTSCHE STRATIGRAPHISCHE KOMMISSION (ed.), WEDDIGE, K. (coord.) *Stratigraphie von Deutschland VIII. Devon*. Schriftenreihe der Deutschen Gesellschaft für Geowissenschaften, Hannover, **52**, 287–296.

RIBBERT, K.-H. 2010. *Geologie im Rheinischen Schiefergebirge, Teil 1, Nordeifel*. Geologischer Dienst Nordrhein-Westfalen, Krefeld.

RICHTER, D. 1979. Die Gedinnium/Siegenium-Grenze nördlich und südlich des Hohen Venns (Nordeifel). *Zeitschrift der Deutschen Geologischen Gesellschaft*, **130**, 93–105.

RICHTER, REINH. 1854. Thüringische Tentaculiten. *Zeitschrift der Deutschen Geologischen Gesellschaft*, **6**, 275–290.

RICHTER, RUD. 1916. Die Entstehung der abgerollten 'Daleider Versteinerungen' und das Alter ihrer Mutterschichten. *Jahrbuch der Königlich Preußischen Geologischen Landesanstalt*, **37**, 247–259.

RICHTER, RUD. & RICHTER, E. 1918. Paläontologische Beobachtungen im Rheinischen Devon. I. Über einzelne Arten von *Acidaspis*, *Lichas*, *Cheirurus*, *Aristozoë*, *Prosocoelus*, *Terebratula* und *Spirophyton* aus der Eifel. *Jahrbücher des Nassauischen Vereins für Naturkunde*, **70**, 143–161.

RICHTER, RUD. & RICHTER, E. 1919. Über zwei gesteinsbildende *Spirifer*-Arten des Wetteldorfer Sandsteins. *Jahrbücher des Nassauischen Vereins für Naturkunde*, **72**, 26–38.

RICHTER, RUD. & RICHTER, E. 1937. Die Herscheider Schiefer, ein zweites Vorkommen von Ordovizium im Rheinischen Schiefergebirge, und ihre Beziehung zu den wiedergefundenen *Dayia*-Schichten. *Senckenbergiana*, **19**, 289–313.

RICHTER, RUD. & RICHTER, E. 1942. Die Trilobiten der Weismes-Schichten am Hohen Venn, mit Bemerkungen über die Malvinocaffrische Provinz. *Senckenbergiana*, **25**, 156–179.

RICHTER, RUD. & RICHTER, E. 1954. Die Trilobiten des Ebbe-Sattels und zu vergleichende Arten (Ordovizium, Gotlandium/Devon). *Abhandlungen der Senckenbergischen Naturforschenden Gesellschaft*, **488**, 1–76.

ROEMER, C.F. 1844. *Das Rheinische Ubergangsgebirge. Eine palaeontologisch-geognostische Darstellung*. Hahn'sche Hofbuchhandlung, Hannover.

ROEMER, F.A. 1843. *Die Versteinerungen des Harzgebirges*. Hahn'sche Hofbuchhandlung, Hannover.

ROSE, O. 1936. Versteinerungen im Taunusquarzit des Rheintaunus. *Jahrbuch des Nassauischen Vereins für Naturkunde*, **83**, 49–58.

RÖDER, D.H. 1960. Ulmen-Gruppe in sandiger Fazies. *Abhandlungen des Hessischen Landesamtes für Bodenforschung*, **31**, 3–65.

RÖSLER, A. 1954. Zur Fauna des rheinischen Unter-Devons. *Notizblatt des Hessischen Landesamtes für Bodenforschung*, **82**, 30–37.

RZHONSNITSKAYA, M.A. 1956. Systematization of Rhynchonellida. *In*: GUZMAN, E. & AYALA-CASTANARES, A. (eds) *Resúmenes de los Trabajos Presentados. México, 20. International Geological Congress, September 1956, (Report)*. México D.F., 125–126.

SALVADOR, A. 1994. *International Stratigraphic Guide*. 2nd edn. Geological Society of America, Boulder, CO.

SAVAGE, N.M. 1986. Classification of Paleozoic rhynchonellid brachiopods. *In*: COPPER, P. & JIN, J. (eds) *Brachiopods*. Proceedings of the Third International Brachiopod Congress, 2–5 September 1995, Sudbury/Ontario/Canada. A. A. Balkema, Rotterdam, Brookfield, 249–260.

SANDBERGER, G. & SANDBERGER, F. 1854. *Die Versteinerungen des rheinischen Schichtensystems in Nassau. Mit einer kurzgefaßten Geognosie dieses Gebietes und mit steter Berücksichtigung analoger Schichten anderer Länder* (Lfg. 7 (fascicle), Bogen 26–29 (sheets printed in the fascicle)). Kreidel & Niedner Verlagshandlung, Wiesbaden, 201–232.

SANDBERGER, G. & SANDBERGER, F. 1856. *Die Versteinerungen des rheinischen Schichtensystems in Nassau. Mit einer kurzgefaßten Geognosie dieses Gebietes und mit steter Berücksichtigung analoger Schichten anderer Länder* (Lfg. 8–9 (fascicles), I–XV, Bogen 30–71 (sheets printed in the fascicles)). Kreidel & Niedner Verlagshandlung, Wiesbaden, 233–564.

SARTENAER, P. 1998. Rhynchonellida mondeal (column B161ds97), Synclinorium de Dinant (B162ds97), Aachen-Becken, incl. Massif de la Vesdre et Fenêtre de Theux (B163ds97), Bergisches Land (B165ds97), Sauerland (B166ds97), Kellerwald (B167ds97), Thüringen (B168ds97). *In*: WEDDIGE, K. (ed.) Devonian Correlation Table. Ergänzungen 1997. *Senckenbergiana lethaea*, **77**, 314–317.

SARTENAER, P. 2004. Restatement of *Terebratula Orbignyana* De Verneuil, 1850 on the basis of the original collection. *Bulletin de l'Institut royal des Sciences naturelles de Belgique, Sciences de la Terre*, **74** (suppl), 81–88.

SARTENAER, P. 2005. New middle and late Emsian, and early Eifelian rhynchonellide brachiopod genera of the family Nucinulidae Sartenaer, 2004. *Bulletin de l'Institut royal des Sciences naturelles de Belgique, Sciences de la Terre*, **75**, 25–52.

SCHÄFER, A. & STETS, J. 1995. The Lower Devonian 'Emsquarzit' – tidal sedimentation in the Rhenish Basin (Rheinisches Schiefergebirge, Germany). *Zentralblatt für Geologie und Paläontologie, Teil I*, **1994**, 227–244.

SCHÄFER, W. 1962. *Aktuo-Paläontologie nach Studien in der Nordsee*. Verlag Waldemar Kramer, Frankfurt am Main.

SCHEMM-GREGORY, M. 2004. *Die Spiriferen-Fauna des Emsquarzits (Unter-Devon, Rheinisches Schiefergebirge). Diplomarbeit*, Philipps-Universität Marburg/Lahn.

SCHEMM-GREGORY, M. 2008. A new terebratulid brachiopod species from the Siegenian (middle Lower Devonian) of the Dra Valley, Morocco. *Palaeontology*, **51**, 793–806.

SCHEMM-GREGORY, M. 2010. *Intermedites* Struve, 1995 (Brachiopoda, Middle Devonian) – discovery of a South Chinese immigrant in Europe and North Africa. *Acta Palaeontologica Sinica*, **49**, 425–438.

SCHEMM-GREGORY, M. 2011. The howellellid branches within the delthyridoid spiriferids (Brachiopoda, Silurian to Devonian). *Memoirs of the Association of Australasian Palaeontologists*, **41**, 115–128.

SCHEMM-GREGORY, M. & JANSEN, U. 2005. *Arduspirifer arduennensis treverorum* n. ssp., eine neue Brachiopoden-Unterart aus dem tiefen Ober-Emsium des Mittelrhein-Gebiets (Unter-Devon, Rheinisches Schiefergebirge). *Mainzer geowissenschaftliche Mitteilungen*, **33**, 79–100.

SCHEMM-GREGORY, M. & JANSEN, U. 2006*a*. Annotations to the Devonian Correlation Table, B 124 di 06: Brachiopod biostratigraphy in the Lower Devonian of the Rheinisches Schiefergebirge (Germany) based on the phylogeny of *Arduspirifer* (Delthyridoidea, Brachiopoda). *Senckenbergiana lethaea*, **86**, 113–116.

SCHEMM-GREGORY, M. & JANSEN, U. 2006*b*. Phylogeny of *Arduspirifer*, Rheinisches Schiefergebirge (B124di06). *In*: WEDDIGE, K. (ed.) Devonian Correlation Table. Ergänzungen 1998. *Senckenbergiana lethaea*, **86**, 121.

SCHEMM-GREGORY, M. & JANSEN, U. 2007. A new genus of terebratulid brachiopod from the Siegenian of the Rheinisches Schiefergebirge. *Acta Palaeontologica Polonica*, **52**, 413–422.

SCHINDLER, E. 1993. Die Kellwasser-Krise (hohe Frasne-Stufe, Ober-Devon). *Göttinger Arbeiten zur Geologie und Paläontologie*, **46**, I-IV, 1–115.

SCHINDLER, T., AMLER, M.R.W. *ET AL*. 2004. Neue Erkenntnisse zur Paläontologie, Biofazies und Stratigraphie der Unterdevon-Ablagerungen (Siegen) der ICE-Neubaustrecke bei Aegidienberg (Siebengebirge, W-Deutschland). *Decheniana (Bonn)*, **157**, 135–150.

VON SCHLOTTHEIM [= SCHLOTHEIM], E.F. 1813. Beiträge zur Naturgeschichte der Versteinerungen in geognostischer Hinsicht. *In*: LEONHARD, C.C. (ed.) *Taschenbuch für die gesammte Mineralogie mit Hinsicht auf die neuesten Entdeckungen*, **7/1**, Hermann, Frankfurt am Main, 3–134.

VON SCHLOTHEIM, E.F. 1820. *Die Petrefactenkunde auf ihrem jetzigen Standpunkte durch die Beschreibung seiner Sammlung versteinerter und fossiler Überreste des Thier- und Pflanzenreiches der Vorwelt erläutert*. Beckersche Buchhandlung, Gotha.

SCHMIDT, HERM. 1926. Schwellen- und Beckenfazies im ostrheinischen Paläozoikum. *Zeitschrift der Deutschen Geologischen Gesellschaft, Monatsberichte*, **77**, 226–234.

SCHMIDT, HERM. 1962. Über Faziesbereiche im Devon Deutschlands. *In*: ERBEN, H.K. (ed.) *2. Internationale Arbeitstagung über die Silur/Devon-Grenze und die Stratigraphie von Silur und Devon, Bonn-Bruxelles 1960. Symposiums-Band*. Schweizerbart'sche Verlagsbuchhandlung, Stuttgart, 224–230.

SCHMIDT, H. 1941. Die mitteldevonischen Rhynchonelliden der Eifel. *Abhandlungen der Senckenbergischen Naturforschenden Gesellschaft*, **459**, 1–79.

SCHMIDT, W.E. 1913. *Cultrijugatus*-Zone und unteres Mitteldevon südlich der Attendorn-Elsper Doppelmulde. *Jahrbuch der Preußischen Geologischen Landesanstalt*, **33** (for 1912), 265–318.

SCHMIDT, WO. 1954. Die ersten Vertebraten-Faunen im deutschen Gedinne. *Palaeontographica, Abt. A*, **105**, 1–47.

SCHMIDT, WO. 1956. Neue Ergebnisse der Revisions-Kartierung des Hohen Venns. *Beihefte zum Geologischen Jahrbuch*, **21**, 1–146.

SCHMIDT, WO. 1958. Die ersten Agnathen und Pflanzen aus dem Taunus-Gedinnium. *Notizblatt des Hessischen Landesamtes für Bodenforschung*, **86**, 31–49.

SCHMIDT, WO. 1959. Grundlagen einer Pteraspiden-Stratigraphie im Unterdevon der Rheinischen Geosynklinale. *Fortschritte in der Geologie von Rheinland und Westfalen*, **5**, 1–82.

SCHMIDT, WO. & SCHRÖDER, E. 1962. Erläuterungen zur Geologischen Übersichtskarte der nördlichen Eifel. *In*: *Erläuterungen zur Geologischen Übersichtskarte der nördlichen Eifel, 1:100000, Hochschul-Umgebungskarte Aachen*. Geologisches Landesamt Nordrhein-Westfalen, Krefeld.

SCHNUR, J. 1851. *Die Brachiopoden aus dem Uebergangsgebirge der Eifel. Programm der vereinigten höhern Bürger- und Provinzial-Gewerbeschule zu Trier für das Schuljahr 1850–1851*. Buchdruckerei Lintz, Trier.

SCHNUR, J. 1853. Zusammenstellung und Beschreibung sämtlicher im Übergangsgebirge der Eifel vorkommenden Brachiopoden nebst Abbildungen derselben. *Palaeontographica*, **3**, 169–254.

SCHRIEL, W. & GROSS, W. 1933. Zur Stratigraphie, Tektonik und Palaeontologie des alten Unterdevons im südlichen Bergischen Lande. *Abhandlungen der Preußischen Geologischen Landesanstalt, Neue Folge*, **145**, 3–77.

SCHUBERT, M. 1996. Die dysaerobe Biofazies der Wissenbacher Schiefer. *Göttinger Arbeiten zur Geologie und Paläontologie*, **68**, 1–131.

SCHULTKA, S. & REMY, W. 1990. Ein 'Flöz'-Profil im linksrheinischen Schiefergebirge als Beispiel paralischer Verhältnisse im Ems. *Neues Jahrbuch für Geologie und Paläontologie, Abhandlungen*, **181**, 41–54.

SCHUMANN, D. 1965. Rhynchonelloidea aus dem Devon des Kantabrischen Gebirges (Nordspanien). *Neues Jahrbuch für Geologie und Paläontologie, Abhandlungen*, **123**, 41–104.

SCHWEITZER, H.J. 1983. Die Unterdevonflora des Rheinlandes. *Palaeontographica, Abt. B*, **189**, 1–138.

SCUPIN, H. 1900. Die Spiriferen Deutschlands. *Palaeontologische Abhandlungen, Neue Folge*, **4**, 207–344.

SEPKOSKI, J.J., JR 1990. Evolutionary Faunas. *In*: BRIGGS, D.E.G. & CROWTHER, P.R. (eds) *Palaeobiology – A Synthesis*. Blackwell, Oxford, 37–41.

SHIRLEY, J. 1962. Review of the correlation of the supposed Silurian strata of Artois, Westphalia, the Taunus and Polish Podolia. *In*: ERBEN, H.K. (ed.) *Symposium Silur/Devon-Grenze 1960*. E. Schweizerbart'sche Verlagsbuchhandlung (Nägele und Obermiller), Stuttgart, 234–242.

SIMPSON, S. 1940. Das Devon der Südost-Eifel zwischen Nette und Alf. *Abhandlungen der Senckenbergischen Naturforschenden Gesellschaft*, **447**, 1–81.

SLAVÍK, L. 2004. A new conodont zonation of the Pragian Stage (Lower Devonian) in the stratotype area (Barrandian, central Bohemia). *Newsletters on Stratigraphy*, **40**, 39–71.

SLAVÍK, L., CARLS, P., HLADIL, J. & KOPTÍKOVÁ, L. 2012. Subdivision of the Lochkovian Stage based on conodont faunas from the stratotype area (Prague Synform, Czech Republic). *Geological Journal*, **47**, 616–631.

SMITH, R.E. 1980. Lower Devonian Lochkovian biostratigraphy and brachiopod faunas, Canadian Arctic Islands. *Bulletin Geological Survey of Canada*, **308**, 1–155.

SOLLE, G. 1936. Revision der Fauna des Koblenzquarzits an Rhein und Mosel. *Senckenbergiana*, **18**, 154–215.

SOLLE, G. 1937. Geologie der mittleren Olkenbacher Mulde. *Abhandlungen der Senckenbergischen Naturforschenden Gesellschaft*, **436**, 1–72.

SOLLE, G. 1942*a*. Die Kondel-Gruppe (Oberkoblenz) im Südlichen Rheinischen Schiefergebirge. I-III. *Abhandlungen der Senckenbergischen Naturforschenden Gesellschaft*, **461**, 1–92.

SOLLE, G. 1942*b*. Die Kondel-Gruppe (Oberkoblenz) im Südlichen Rheinischen Schiefergebirge. IV-V. *Abhandlungen der Senckenbergischen Naturforschenden Gesellschaft*, **464**, 95–156.

SOLLE, G. 1942*c*. Die Kondel-Gruppe (Oberkoblenz) im Südlichen Rheinischen Schiefergebirge. VI-X. *Abhandlungen der Senckenbergischen Naturforschenden Gesellschaft*, **467**, 157–240.

SOLLE, G. 1950*a*. Obere Siegener Schichten, Hunsrückschiefer, tiefstes Unterkoblenz und ihre Eingliederung ins Rheinische Unterdevon. *Geologisches Jahrbuch*, **65**, 299–380.

SOLLE, G. 1950*b*. Beobachtungen und Deutungen zum Unterkoblenz in Taunus und Hunsrück. *Senckenbergiana*, **31**, 185–196.

SOLLE, G. 1953. Die Spiriferen der Gruppe *arduennensis-intermedius* im rheinischen Devon. *Abhandlungen des Hessischen Landesamtes für Bodenforschung*, **5**, 1–156.

SOLLE, G. 1956*a*. Gliederung und Aufbau der Klerfer Schichten am Nordrand der Olkenbacher Mulde (Unterdevon, Südost-Eifel). *Notizblatt des Hessischen Landesamtes für Bodenforschung*, **84**, 85–92.

SOLLE, G. 1956*b*. Die Watt-Fauna der unteren Klerfer Schichten von Greimerath (Unterdevon, Südost-Eifel). Zugleich ein Beitrag zur unterdevonischen Mollusken-Fauna. *Abhandlungen des Hessischen Landesamtes für Bodenforschung*, **17**, 1–47.

SOLLE, G. 1963. *Hysterolites hystericus* (Schlotheim) [Brachiopoda; Unterdevon], die Einstufung der oberen Graptolithen-Schiefer in Thüringen und die stratigraphische Stellung der Zone des *Monograptus hercynicus*. *Geologisches Jahrbuch*, **81**, 171–220.

SOLLE, G. 1970. Die Hunsrück-Insel im oberen Unterdevon. *Notizblatt des Hessischen Landesamtes für Bodenforschung*, **98**, 50–80.

SOLLE, G. 1971. *Brachyspirifer* und *Paraspirifer* im Rheinischen Devon. *Abhandlungen des Hessischen Landesamtes für Bodenforschung*, **59**, 1–163.

SOLLE, G. 1972. Abgrenzung und Untergliederung der Ober-Ems-Stufe, mit Bemerkungen zur Unterdevon-/Mitteldevon-Grenze. *Notizblatt des Hessischen Landesamtes für Bodenforschung*, **100**, 60–91.

SOLLE, G. 1976. Oberes Unter- und unteres Mitteldevon einer typischen Geosynklinal-Folge im südlichen Rheinischen Schiefergebirge. Die Olkenbacher Mulde. *Geologische Abhandlungen Hessen*, **74**, 1–264.

SOWERBY, J. DE C. 1839. Mollusca and Conchifers. *In*: MURCHISON, R.I. (ed.) *The Silurian System, Part 2, Organic Remains*. John Murray, London, 577–768.

SOWERBY, J. DE C. 1842. Description of Silurian fossils from the Rhenish Provinces. *In*: DE ARCHIAC, É.J.A., & DE VERNEUIL, É.P. (eds) On the Fossils of the Older Deposits in the Rhenish Provinces. *Transactions of the Geological Society, London*, **6**, 408–410, http://doi.org/10.1144/transgslb.6.2.303

SPRIESTERSBACH, J. 1915. Neue und bekannte Versteinerungen aus dem rheinischen Devon, besonders aus dem Lenneschiefer. *Abhandlungen der Königlich Preußischen Geologischen Landesanstalt, Neue Folge*, **80**, 1–80.

SPRIESTERSBACH, J. 1925. Die Oberkoblenzschichten des Bergischen Landes und Sauerlandes. *Jahrbuch der Preußischen Geologischen Landesanstalt*, **45** (for 1924), 367–450.

SPRIESTERSBACH, J. 1942. Lenneschiefer (Stratigraphie, Fazies und Fauna). *Abhandlungen des Reichsamts für Bodenforschung, Neue Folge*, **203**, 1–219.

SPRIESTERSBACH, J. & FUCHS, A. 1909. Die Fauna der Remscheider Schichten. *Abhandlungen der Königlich Preußischen Geologischen Landesanstalt, Neue Folge*, **58**, 1–81.

STEEMANS, P. 1989. Etude palynostratigraphique du Dévonien inférieur dans l'ouest de l'Europe. *Mémoires pour servir à l'explication des cartes géologiques et minières de la Belgique*, **27**, 1–453.

STEININGER, F.F. & PILLER, W.E. (eds) 1999. Empfehlungen (Richtlinien) zur Handhabung der stratigraphischen Nomenklatur. *Courier Forschungsinstitut Senckenberg*, **209**, 1–19.

STEININGER, J. 1853. *Geognostische Beschreibung der Eifel*. Lintz'sche Buchhandlung, Trier.

STETS, J. & SCHÄFER, A. 2002. Depositional environments in the Lower Devonian siliciclastics of the Rhenohercynian Basin (Rheinisches Schiefergebirge, W-Germany) – case studies and a model. *Contributions to Sedimentary Geology*, **22**, 1–78.

STETS, J. & SCHÄFER, A. 2009. The Siegenian delta: land–sea transitions at the northern margin of the Rhenohercynian Basin. *In*: KÖNIGSHOF, P. (ed.) *Devonian Change: Case Studies in Palaeogeography and Palaeoecology*. Geological Society, London, Special Publications, **314**, 37–72, http://doi.org/10.1144/SP314.3

STETS, J. & SCHÄFER, A. 2011. The Lower Devonian Rhenohercynian Rift – 20 Ma of sedimentation and tectonics (Rhenish Massif, W-Germany). *Zeitschrift der Deutschen Gesellschaft für Geowissenschaften*, **162**, 93–115.

STEWART, G.A. 1927. Fauna of the Silica Shale of Lucas County. *Geological Survey of Ohio, 4th Series*, **32**, 1–76.

STREEL, M., HIGGS, K., LOBOZIAK, S., RIEGEL, W. & STEEMANY, P. 1987. Spore Stratigraphy and correlation with faunas and floras in the type marine Devonian of the Ardenne-Rhenish regions. *Review of Palaeobotany and Palynology*, **50**, 211–229.

STRUVE, W. 1961*a*. Zur Stratigraphie der südlichen Eifler Kalkmulden (Devon: Emsium, Eifelium, Givetium). *Senckenbergiana lethaea*, **42**, 291–345.

STRUVE, W. 1961*b*. Das Eifeler Korallen-Meer. *Der Aufschluss*, Heft Eifel, 81–107.

STRUVE, W. 1963*a*. Das Korallen-Meer der Eifel vor 300 Millionen Jahren – Funde, Deutungen, Probleme. *Natur und Museum*, **93**, 237–276.

STRUVE, W. 1963*b*. Das Korallen-Meer des Eifeler Mitteldevons und seine Bewohner. *Eifeler Jahrbuch*, **1964**, 12–30, Eifelverein, Düren.

STRUVE, W. 1966. Einige Atrypinae aus dem Silurium und Devon. *Senckenbergiana lethaea*, **47**, 123–163.

STRUVE, W. 1970*a*. *Phacops*-Arten aus dem Rheinischen Devon. 1. *Senckenbergiana lethaea*, **51**, 133–189.

STRUVE, W. 1970*b*. Beiträge zur Kenntnis devonischer Brachiopoden, 16: 'Curvate Spiriferen' der Gattung *Rhenothyris* und einige andere Reticulariidae aus dem Rheinischen Devon. *Senckenbergiana lethaea*, **51**, 449–577.

STRUVE, W. 1973. Die ältesten Taunus-Fossilien. *Natur und Museum*, **103**, 349–359.

STRUVE, W. 1982*a*. The Eifelian within the Devonian frame, history, boundaries, definitions. *In*: ZIEGLER, W. & WERNER, R. (eds) On Devonian Stratigraphy and Palaeontology of the Ardenno-Rhenish Mountains and Related Devonian Matters. *Courier Forschungsinstitut Senckenberg*, **55**, 401–432.

STRUVE, W. 1982*b*. The Great Gap in the record of marine Middle Devonian. *In*: ZIEGLER, W. & WERNER, R. (eds) On Devonian Stratigraphy and Palaeontology of the Ardenno-Rhenish Mountains and Related Devonian Matters. *Courier Forschungsinstitut Senckenberg*, **55**, 433–447.

STRUVE, W. 1988. Geologic introduction. *In*: BULTYNCK, P., DREESEN, R., GROESSENS, E., STRUVE, W., WEDDIGE, K. & ZIEGLER, W. Field Trip A (22.-24. July 1988), Ardennes (Belgium) and Eifel Hills (Federal Republic of Germany). *In*: ZIEGLER, W. (ed.) 1st International Senckenberg Conference and 5th European Conodont Symposium (ECOS V), Contributions I. *Courier Forschungsinstitut Senckenberg*, **102**, 88–102.

STRUVE, W. 1990. Paläozoologie III (1986–1990). *In*: ZIEGLER, W. (ed.) Wissenschaftlicher Jahresbericht 1988/89 des Forschungsinstituts Senckenberg, Frankfurt am Main. *Courier Forschungsinstitut Senckenberg*, **127**, 251–279.

STRUVE, W. 1992. Neues zur Stratigraphie und Fauna des rhenotypen Mittel-Devon. *Senckenbergiana lethaea*, **71**, 503–624.

STRUVE, W. 1995. Die Riesen-Phacopiden aus dem Maïder, SE-marokkanische Prä-Sahara. *Senckenbergiana lethaea*, **75**, 77–129.

STRUVE, W. & WERNER, R. 1982. Brachiopods. *In*: WERNER, R. & ZIEGLER, W. Proposal of a Boundary Stratotype for the Lower/Middle Devonian Boundary (*partitus*-Boundary). *In*: ZIEGLER, W. & WERNER, R. (eds) On Devonian Stratigraphy and Palaeontology of the Ardenno-Rhenish Mountains and Related Devonian Matters. *Courier Forschungsinstitut Senckenberg*, **55**, 44–50.

STRUVE, W., PLODOWSKI, G. & WEDDIGE, K. 1997. Biostratigraphische Stufengrenzen und Events in der Prümer und Hillesheimer Mulde. 67. Jahrestagung der Paläontologischen Gesellschaft, Exkursionsführer. *Terra Nostra*, **97**, 123–167.

STRUVE, W., BASSE, M. & WEDDIGE, K. 2008. Prädevon, Ober-Emsium und Mitteldevon der Eifeler Kalkmulden-Zone. *In*: DEUTSCHE STRATIGRAPHISCHE KOMMISSION (ed.), WEDDIGE, K. (coord.) *Stratigraphie von Deutschland VIII. Devon*. Schriftenreihe der Deutschen Gesellschaft für Geowissenschaften, Hannover, **52**, 297–374.

SU, YANG-ZHENG 1976. Brachiopoda, Cambrian–Devonian. *In*: *Atlas of Palaeontology of North China, Nei Mongol (Inner Mongolia)*. Palaeozoic. Geological Publishing House, Beijing, **1**, 155–227, pls. 76–130 [in Chinese].

TERMIER, H. & TERMIER, G. 1949. Essai sur l'évolution des Spiriféridés. *Service Géologique du Maroc, Notes et Mémoirs*, **74**, 85–112.

THÜNKER, M. 1990. *Geologische Karte von Nordrhein-Westfalen 1:25000, Erläuterungen Blatt 5115 Ewersbach*. Geologisches Landesamt Nordrhein-Westfalen, Krefeld.

THÜNKER, M. 2001. *Geologische Karte von Nordrhein-Westfalen 1:25000, Erläuterungen Blatt 5114 Siegen*. Geologisches Landesamt Nordrhein-Westfalen, Krefeld.

THÜNKER, M. 2008. Unterdevon im Siegerland. *In*: DEUTSCHE STRATIGRAPHISCHE KOMMISSION (ed.), WEDDIGE, K. (coord.) *Stratigraphie von Deutschland VIII. Devon*. Schriftenreihe der Deutschen Gesellschaft für Geowissenschaften, Hannover, **52**, 252–266.

TIMM, J. 1981*a*. Zur Trilobitenstratigraphie des Silur/Devon-Grenzbereiches im Ebbe-Antiklinorium (Rheinisches Schiefergebirge). *Mitteilungen aus dem Geologisch-Paläontologischen Institut der Universität Hamburg*, **50**, 91–108.

TIMM, J. 1981*b*. Die Faziesentwicklung der ältesten Schichten des Ebbe-Antiklinoriums. *Mitteilungen aus dem Geologisch-Paläontologischen Institut der Universität Hamburg*, **50**, 147–173.

TIMM, J., DEGENS, E.T. & WIESNER, M.G. 1981. Erläuterungen zur Geologischen Karte des zentralen Ebbe-Antiklinoriums 1:25000. *Mitteilungen aus dem Geologisch-Paläontologischen Institut der Universität Hamburg*, **50**, 59–75.

USHATINSKAYA, G.T. 1975. *In*: USHATINSKAYA, G.T. & NILOVA, N.V. Brachiopody. *In*: MENNERA, V.V. (ed.) *Kharakteristika fauny pogranichnykh sloev silura i devona tsentralnogo Kazakhstana*. Materialy po Geologii Tsentralnogo Kazakhstana, **12**. Nedra, Moscow, 93–119 [in Russian].

VANDERCAMMEN, A. 1963. Spiriferidae du Dévonien de la Belgique. *Mémoires de l'Institut royal des Sciences naturelles de Belgique*, **150**, 1–179.

VANUXEM, L. 1842. *Geology of New York. Part 3, Comprising the Survey of the Third Geological District. Natural History of New York 4 (3)*. D. Appleton & Co, New York.

DE VERNEUIL, É. 1850. Tableau des fossiles du terrain dévonien du département de la Sarthe. *Bulletin de la Société géologique de France*, **7**, 778–787.

VER STRAETEN, C.A. 2007. Basinwide stratigraphic synthesys and sequence stratigraphy, upper Pragian, Emsian and Eifelian stages (Lower to Middle Devonian), Appalachian Basin. *In*: BECKER, R.T. & KIRCHGASSER, W.T. (eds) *Devonian Events and Correlations*. Geological Society, London, Special Publications, **278**, 39–81, http://doi.org/10.1144/SP278.3

VIËTOR, W. 1919. Der Koblenzquarzit, seine Fauna, Stellung und linksrheinische Verbreitung. *Jahrbuch der Königlich Preußischen Geologischen Landesanstalt*, **37** (for 1916), 317–476.

VODRÁŽKOVÁ, S., FRÝDA, J., SUTTNER, T.J., KOPTÍKOVÁ, L. & TONAROVÁ, P. 2013. Environmental changes close to the Lower-Middle Devonian boundary; the Basal Choteč Event in the Prague Basin (Czech Republic). *Facies*, **59**, 425–449.

VOGEL, K., XU, H.-K. & LANGENSTRASSEN, F. 1989. Brachiopods and their relation to facies development in the Lower and Middle Devonian of Nandan, Guangxi, South China. *In*: WALLISER, O.H. & ZIEGLER, W. (eds) Contributions to Devonian Palaeontology and Stratigraphy. Part I: Chinese-German Collaboration. Part II: Various Devonian Topics. *Courier Forschungsinstitut Senckenberg*, **110**, 17–59.

WAAGEN, W.H. 1883. Salt range fossils. I. *Productus* Limestone fossils. *Geological Survey of India, Memoirs, Palaeontologia Indica (series 13)*, **4**, 391–546.

WAAGEN, W.H. 1884. Salt range fossils. I. *Productus* Limestone fossils. *Geological Survey of India, Memoirs, Palaeontologia Indica (series 13)*, **4**, 547–610.

WALLISER, O.H. 1964. Conodonten des Silurs. *Abhandlungen des Hessischen Landesamtes für Bodenforschung*, **41**, 1–106.

WALLISER, O.H. 1965. Über *Sellanarcestes* Schindewolf 1933. *Fortschritte in der Geologie von Rheinland und Westfalen*, **9**, 87–96.

WALLISER, O.H. 1985. Natural boundaries and Commission boundaries in the Devonian. *Courier Forschungsinstitut Senckenberg*, **75**, 401–408.

WALLISER, O.H. 1996. Global events in the Devonian and Carboniferous. *In*: WALLISER, O.H. (ed.) *Global Events and Event Stratigraphy in the Phanerozoic*. Springer, Berlin, 225–250.

WALLISER, O.H. 1998. Meeresspiegel (H100di97). *In*: WEDDIGE, K. (ed.) Devon-Korrelationstabelle. Ergänzungen 1997. *Senckenbergiana lethaea*, **77**, 298.

WALLISER, O.H. & MICHELS, D. 1983. Der Ursprung des Rheinischen Schelfes im Devon. *Neues Jahrbuch für Geologie und Paläontologie, Abhandlungen*, **166**, 3–18.

WALLISER, O.H. & ZIEGLER, W. 2008. Paläogeographie und Fazies des Devons in Deutschland. *In*: DEUTSCHE STRATIGRAPHISCHE KOMMISSION (ed.), WEDDIGE, K. (coord.) *Stratigraphie von Deutschland VIII. Devon*. Schriftenreihe der Deutschen Gesellschaft für Geowissenschaften, Hannover, **52**, 17–21.

WALLISER, O.H., XU, H.-K. & YU, C.-M. 1989. Comparison of the Devonian of South China and Germany. A palaeontological cooperation programme between the P.R. China and the F.R. Germany. *In*: WALLISER, O.H. & ZIEGLER, W. (eds) Contributions to Devonian Palaeontology and Stratigraphy. Part I: Chinese-German Collaboration. Part II: Various Devonian Topics. *Courier Forschungsinstitut Senckenberg*, **110**, 5–15.

WALTHER, K. 1903. Das Unterdevon zwischen Marburg a. L. und Herborn (Nassau). *Neues Jahrbuch für Mineralogie, Geologie und Palaeontologie, Beilagen-Band*, **17**, 1–75.

WEDDIGE, K. 1982. The Wetteldorf Richtschnitt as boundary stratotype from the view point of conodont stratigraphy. *In*: ZIEGLER, W. & WERNER, R. (eds) On Devonian Stratigraphy and Palaeontology of the Ardenno-Rhenish Mountains and Related Devonian Matters. *Courier Forschungsinstitut Senckenberg*, **55**, 26–37.

WEDDIGE, K. 1988. Eifel conodonts. *In*: BULTYNCK, P., DREESEN, R., GROESSSENS, E., STRUVE, W., WEDDIGE, K. & ZIEGLER, W. Field Trip A (22–24 July 1988), Ardennes (Belgium) and Eifel Hills (Federal Republic of Germany). *In*: ZIEGLER, W. (ed.) 1st International Senckenberg Conference and 5th European Conodont Symposium (ECOS V), Contributions I. *Courier Forschungsinstitut Senckenberg*, **102**, 103–110.

WEDDIGE, K. 2003. M. y. calibration (column 0001di03). *In*: WEDDIGE, K. (ed.) Devonian Correlation Table. *Senckenbergiana lethaea*, **83**, 215.

WEDDIGE, K. (COORD.) & DEUTSCHE STRATIGRAPHISCHE KOMMISSION (ed.) 2008. *Stratigraphie von Deutschland VIII. Devon*. Schriftenreihe der Deutschen Gesellschaft für Geowissenschaften, Hannover, **52**, 1–578.

WEDDIGE, K. & REQUADT, H. 1985. Conodonten des Ober-Emsium aus dem Gebiet der Unteren Lahn (Rheinisches Schiefergebirge). *Senckenbergiana lethaea*, **66**, 347–381.

WEDDIGE, K., WERNER, R. & ZIEGLER, W. 1979. The Emsian-Eifelian Boundary. An Attempt at Correlation between the Eifel and Ardennes Regions. *Newsletters on Stratigraphy*, **8**, 159–169.

WEDDIGE, K., JANSEN, U., SCHINDLER, E., RIBBERT, K.-H., WELLER, H. & ZAGORA, K. 2005. Das Devon in der Stratigraphischen Tabelle von Deutschland 2002. *Newsletters on Stratigraphy*, **41**, 43–59.

WEDEKIND, R. 1926. Die Devonische Formation. *In*: SALOMON, W.H. (ed.) *Grundzüge der Geologie, Vol. 2, Erdgeschichte*. Schweizerbart'sche Verlagsbuchhandlung (Nägele), Stuttgart, 194–226.

WEHRMANN, A., BLIECK, A. ET AL. 2005. Paleoenvironment of an Early Devonian land–sea transition: a case study from the southern margin of the Old Red Continent (Mosel Valley, Germany). *Palaios*, **20**, 101–120.

WENNDORF, K.-W. 1990. Homalonotinae (Trilobita) aus dem Rheinischen Unter-Devon. *Palaeontographica, Abt. A*, **211**, 1–184.

WENNDORF, K.-W. 1999. Neue Fossilfunde aus dem Unterdevon von Rhein und Mosel (Geologische Karte von Rheinland Pfalz, Blatt 5611 Koblenz). Teil 1: Unterems. *Mainzer geowissenschaftliche Mitteilungen*, **28**, 63–84.

WENNDORF, K.-W. 2001. Neue Funde aus dem Unterdevon an Rhein und Mosel (Geologische Karte von Rheinland-Pfalz, Blatt 5611 Koblenz). Teil 2: tiefes Oberems (Emsquarzit). *Mainzer geowissenschaftliche Mitteilungen*, **30**, 7–42.

WENTZLAU, D. 1960. Stratinomische, stratigraphische und tektonische Untersuchungen in den Mittleren Siegener Schichten südöstlich des Siegener Schuppensattels auf den Blättern Freudenberg und Siegen. *Abhandlungen des Hessischen Landesamtes für Bodenforschung*, **29**, 157–249.

WERNER, R. 1969. Ober-Ems und tiefstes Mittel-Devon am N-Rand der Prümer Mulde (Devon, Eifel). *Senckenbergiana lethaea*, **50**, 161–237.

WERNER, R. 1974. Zur Fazies und Fauna des Dorsel-Horizontes (obere Lauch-Schichten, tiefes Mitteldevon) in den südlichen Eifeler Kalkmulden. *Mainzer geowissenschaftliche Mitteilungen*, **3**, 215–230.

WERNER, R. 1982. Wetteldorf Richtschnitt. *In*: ZIEGLER, W. & WERNER, R. (eds) On Devonian Stratigraphy and Palaeontology of the Ardenno-Rhenish Mountains and Related Devonian Matters. *Courier Forschungsinstitut Senckenberg*, **55**, 22–25.

WERNER, R. 1990. Paläozoologie IV. *In*: ZIEGLER, W. (ed.) Wissenschaftlicher Jahresbericht 1988/89 des Forschungsinstituts Senckenberg, Frankfurt am Main. *Courier Forschungsinstitut Senckenberg*, **127**, 281–284.

WERNER, R. & HAHN, H.-D. 1988. Profilaufnahme und Mikrofazies des Wetteldorfer Richtschnittes, südlich Schönecken-Wetteldorf/Eifel. *Mainzer geowissenschaftliche Mitteilungen*, **17**, 249–282.

WHITTAKER, A., COPE, J.C.W. ET AL. 1991. A guide to stratigraphical procedure. *Journal of the Geological Society, London*, **148**, 813–824, http://doi.org/10.1144/gsjgs.148.5.0813

WILLIAMS, A., CARLSON, S.J., BRUNTON, C.H.C., HOLMER, L.E. & POPOV, L.E. 1996. A supra-ordinal classification of the Brachiopoda. *Philosophical Transactions of the Royal Society of London (series B)*, **351**, 1171–1193.

WILLIAMS, A., BRUNTON, C.H.C. & WRIGHT, A.D. 2000. Orthotetida. *In*: KAESLER, R.L. (ed.) *Treatise on Invertebrate Palaeontology, Part H, Brachiopoda (Revised), 3, Linguliformea, Craniiformea, and Rhynchonelliformea (part)*. Geological Society of America and University of Kansas, Boulder, CO, and Lawrence, KS, 644–689.

WINTER, J. 1971. Brachiopoden-Morphologie und Biotop – ein Vergleich quantitativer Brachiopoden-Spektren aus Ahrdorf-Schichten (Eifelium) der Eifel. *Neues Jahrbuch für Geologie und Paläontologie, Monatshefte*, **1971**, 102–132.

WINTER, J. 1977. Fazies und Paläogeographie des Eifeler Mitteldevons mit Exkursionsbeispielen aus der Hillesheimer Mulde. *In*: MEYER, W., STOLTIDIS, I. & WINTER, J. (eds) Geologische Exkursion in den Raum Weyer – Schuld – Heyroth – Niederehe – Üxheim – Ahütte. *Decheniana*, **130**, 328–334.

WIRTH, H. 1960. Stratigraphische und fazielle Untersuchungen im Vordertaunus. *Notizblatt des Hessischen Landesamtes für Bodenforschung*, **88**, 146–166.

WOLF, M. 1930. Alter und Entstehung des Wald-Erbacher Roteisensteins. *Abhandlungen der Preußischen Geologischen Landesanstalt, N. F.*, **123**, 1–103.

WUNDERLICH, F. 1970. Genesis and environment of the 'Nellenköpfchenschichten' (Lower Emsian, Rheinian Devon) at locus typicus in comparison with modern coastal environment of the German Bay. *Journal of Sedimentary Petrology*, **40**, 102–130.

WUNSTORF, W. 1932. Die Siegener Schichten bei Monschau. *Jahrbuch der Preußischen Geologischen Landesanstalt*, **52** (for 1931), 251–256.

YOLKIN, E.A., IZOKH, N.G., SENNIKOV, N.V., YAZIKOV, A.Y. & KIM, A.I. & ERINA, M.V. 1994. Most important global sedimentological and biological events in Devonian of the South Tien Shan and in the South of West Siberia. *Stratigraphy and Geological Correlation*, **2**, 24–31 [in Russian].

YOLKIN, E.A., KIM, A.I., WEDDIGE, K., TALENT, J.A. & HOUSE, M.R. 1997. Definition of the Pragian/Emsian Stage boundary. *Episodes*, **20**, 235–240.

ZAPALSKI, M.K., HUBERT, B.L.M., NICOLLIN, J.-P., MISTIAEN, B. & BRICE, D. 2007. The palaeobiodiversity of stromatoporids, tabulates and brachiopods in the Devonian of the Ardennes – changes trough time. *Bulletin de la Société Géologique de France*, **178**, 383–390.

ZEILER, F. 1857. Versteinerungen der älteren Rheinischen Grauwacke. *Verhandlungen des Naturhistorischen Vereins der Preußischen Rheinlande und Westphalens*, **14**, 45–64.

ZIEGLER, W. 1960. Conodonten aus dem Rheinischen Unterdevon (Gedinnium) des Remscheider Sattels (Rheinisches Schiefergebirge). *Paläontologische Zeitschrift*, **34**, 169–201.

ZIEGLER, W. 1970. *Geologische Karte von Nordrhein-Westfalen 1:25000, Erläuterungen Blatt 4713 Plettenberg*, 2. Auflage. Geologisches Landesamt Nordrhein-Westfalen, Krefeld.

ZIEGLER, W. 1979. Historical subdivisions of the Devonian. *Special Papers in Palaeontology*, **23**, 23–47.

ZIEGLER, W., HILDEN, H.D. & LEUTERITZ, K. 1968. Die Neugliederung der ehemaligen Rimmert-Schichten im Ebbe-Sattel (Meßtischblatt Plettenberg). *Fortschritte in der Geologie von Rheinland und Westfalen*, **16**, 133–142.

ZYGOJANNIS, N. 1971. Die Remscheider Schichten im südlichen Bergischen Land (Rheinisches Schiefergebirge). *Sonderveröffentlichungen des Geologischen Instituts der Universität zu Köln*, **21**, 1–164.

Palynology, dacryoconarids and the lower Eifelian (Middle Devonian) Basal Choteč Event: case studies from the Prague and Appalachian basins

RAINER BROCKE[1]*, OLDRICH FATKA[2], RICHARD H. LINDEMANN[3], EBERHARD SCHINDLER[1] & CHARLES A. VER STRAETEN[4]

[1]*Senckenberg Forschungsinstitut und Naturmuseum Frankfurt, Senckenberganlage 25, 60325 Frankfurt am Main, Germany*

[2]*Department of Geology and Palaeontology, Charles University, Albertov 6, 12843 Prague 2, Czech Republic*

[3]*Skidmore College, Geoscience Department, 815 North Broadway, Saratoga Springs, NY 12866, USA*

[4]*New York State Museum, 3140 Cultural Education Center, Albany, NY 12230, USA*

**Corresponding author (e-mail: rainer.brocke@senckenberg.de)*

Abstract: During recent studies of the Basal Choteč Event (BCE) at its type locality (Na Škrábku Quarry at Choteč Village, Prague Basin of the Barrandian area, Czech Republic) and selected sections of time-equivalent strata in the Appalachian Basin (USA), palynomorphs and dacryoconarids have proven responsive to changing environmental conditions. To date, there have been no detailed reports of dacryoconarids from the Appalachian Basin (AB) and none of palynomorphs from Bohemia or elsewhere. Palynomorphs of the Barrandian area comprise a more or less monospecific assemblage of prasinophycean algae interpreted here to represent an ecological epibole. Mazuelloids and scolecodonts are also present, whereas acritarchs, spores and chitinozoans are accessory components. Prasinophytes also predominate in coeval strata of the Appalachian Basin's northern region, whereas a chitinozoan species and morphotypes possibly assignable to fungi abound in the central region. Scolecodonts and acritarchs are regionally variable throughout the interval. The former are rare in the central region of the basin but are ubiquitous and sometimes abundant in the northern region. Dacryoconarids of the Appalachian Basin are also regionally variable. The dacryoconarid fauna of the northern region, however, descended from a previous Emsian fauna that diversified during the BCE and subsequently functioned as the foundation of the upper Eifelian faunas, while dacryoconarids of the central region represent an incursion epibole of Old World forms that entered the basin at the onset of the event interval and became extinct at its close. Among the dacryoconarids there are key taxa that serve as excellent biostratigraphic markers to identify the BCE in the Appalachian Basin. In both the Prague and Appalachian basins, the BCE occurs near the maximum transgression of the Devonian Ic sequence. Additional faunal changes are found in the Appalachian Basin leading up to the main body of the event.

House (1985) first recognized the lower Eifelian (Middle Devonian) Basal Choteč Event (BCE) in West Gondwana (North African region: Vaughan & Pankhurst 2008) and the central and southern European peri-Gondwana of the so-called Old World Realm. In its type region of the Prague Basin (Czech Republic), the event interval is recorded in the lower section of the Choteč Limestone near the base of the *Polygnathus costatus* Zone (Berkyová 2009). The BCE involves the turnover of several faunal groups (Chlupáč & Kukal 1986) and a lithofacies shift to relatively deeper-water environments in temporal concert with a eustatic sea-level rise and the onset of oceanic dysoxic–anoxic conditions (Walliser 1996; House 2002). Chemostratigraphic studies have documented a moderate $\delta^{13}C$ excursion and a magnetic susceptibility spike within the BCE interval, both of which are thought to be indicative of a climate regime shift (Buggisch & Mann 2004; Koptíková 2011). Despite its evident significance in the study of Devonian bioevents and palaeobiogeography,

From: BECKER, R. T., KÖNIGSHOF, P. & BRETT, C. E. (eds) 2016. *Devonian Climate, Sea Level and Evolutionary Events*. Geological Society, London, Special Publications, **423**, 123–169.
First published online September 14, 2015, http://doi.org/10.1144/SP423.8

the faunal and lithological attributes of the BCE have received scant attention in the Appalachian Basin (AB) of the Eastern Americas Realm beyond the correlation of the event interval with the Nedrow Member of the Onondaga Formation (Ver Straeten 2007; Brett *et al.* 2009).

This paper presents hitherto unknown findings of palynomorphs and dacryoconarids (including six new species and two species in open nomenclature from the Appalachian Basin) in the BCE of the Prague and Appalachian basins, as well as detailed stratigraphic sections of the BCE interval in the latter, to facilitate comparisons of the style of the event in two separate palaeobiogeographical realms.

Research history of the Basal Choteč Event (BCE)

The term Choteč Event was introduced by House (1985), but he stressed in his 2002 overview (House 2002, p. 13) of mid-Palaeozoic events that Chlupáč & Kukal (1986, 1988) had 'given more precision' to the appellation. Even earlier, Walliser (1985) used the term *jugleri* Event for the event interval, named after the widespread goniatite *Pinacites jugleri*. However, as this goniatite species enters only at the top of the event interval (Becker & House 1994), he later supported the use of the term 'Basal Choteč Event' (Walliser 1996). The event level occurs shortly above the Emsian–Eifelian boundary (*Polygnathus patulus*/*partitus* conodont zones) and has been recognized in other areas. For example, in the Eifel Hills it is also known as the OCA (*orbignyanus–cultrijugatus–alatiformis*) Extinction Event (Struve 1982), characterized by the disappearance of the brachiopods *Uncinulus orbignyanus*, *Paraspirifer cultrijugatus* and *Alatiformia alatiformis* (e.g. Weddige 1988; Struve *et al.* 1997, 2008). As already pointed out by Walliser (1985, 1996), the Choteč Event is often connected with a lithological change from light grey limestones to dark, often dacryoconarid-rich limestones; Walliser (1985) mentioned the intercalation of dark shales. This facies change was also recorded by Chlupáč (1985) for the Barrandian area, Requadt & Weddige (1978) for the Rheinisches Schiefergebirge and Henn (1985) for northern Spain. Klug *et al.* (2000) reported it from Moroccan sections, indicating that the 'Choteč Event level' occurred within the *partitus* conodont zone. Recognition of the lithological change in the latter region had already been mentioned for the Tafilalt area in Alberti (1980), Becker & House (1994) and Kaufmann (1998). In recent years, sections in the type area of the Choteč Event have gained increasing attention (e.g. Berkyová 2009; Elrick *et al.* 2009; Berkyová & Munnecke 2010; Brocke *et al.* 2011; Koptíková 2011; Vodrážková *et al.* 2013).

Localities, material and methods

In the Prague Basin of the Barrandian area, our study is based on eight samples from the Na Škrábku Quarry near Choteč (Figs 1–4). The section at the quarry (49° 59′ 20″ N, 14° 16′ 45″ E) was first mentioned by Chlupáč (1959, p. 457); this outcrop represents the stratotype of the Choteč Formation.

This section and others in its vicinity have been studied in great detail in the context of the definition of the Emsian–Eifelian boundary and the establishment of the Choteč Event (e.g. Chlupáč 1982, 1985; Chlupáč & Kukal 1986, 1988). Several palaeontological, as well as non-biotic, aspects have been studied at this important section (e.g. Chlupáč 1959; Buggisch & Mann 2004; Koptíková 2011; Vodrážková *et al.* 2013).

In the Appalachian Basin, eastern USA, over 350 outcrops of the lower Eifelian Onondaga Formation, time-equivalent units and adjacent strata have been examined (Figs 5 & 6). Their stratigraphic trends are discussed in Ver Straeten (2007). Subsequent research by a team including four of the authors of this paper is focused on better refining Emsian and Eifelian biostratigraphy across the Appalachian Basin, utilizing various fossil organisms. Fifteen of these sites, in the northern and central parts of the basin, focus on lower Eifelian strata, including the BCE. These sites occur in the states of New York, Pennsylvania, Maryland, Virginia and West Virginia (Figs 5–7; Appendix A).

Palynological samples of the Prague and Appalachian basins were prepared by utilizing a standard treatment with HCl–HF (e.g. Traverse 2007) without any oxidation. The resulting organic residue was sieved with 10 μm nylon sieves. In some cases, samples that were rich in organic matter were also sieved in an ultrasonic bath to remove the finer organic detritus. In addition, the 'light organic fraction', which self-separates by floating away from the rest of the kerogen during the chemical preparation, was extracted for microscopic analysis. This method was tested on a few samples from the Appalachian and the Barrandian Na Škrábku section (e.g. sample 5, enriched in prasinophytes). For all samples, the studied portion of the organic residue was mounted on slides with glycerol gelatine and routinely analysed under a transmitted light microscope (Nikon Eclipse 90i). Dark-coloured to nearly opaque palynomorphs were studied with the same microscope, which was equipped with an infrared (IR) video system (for

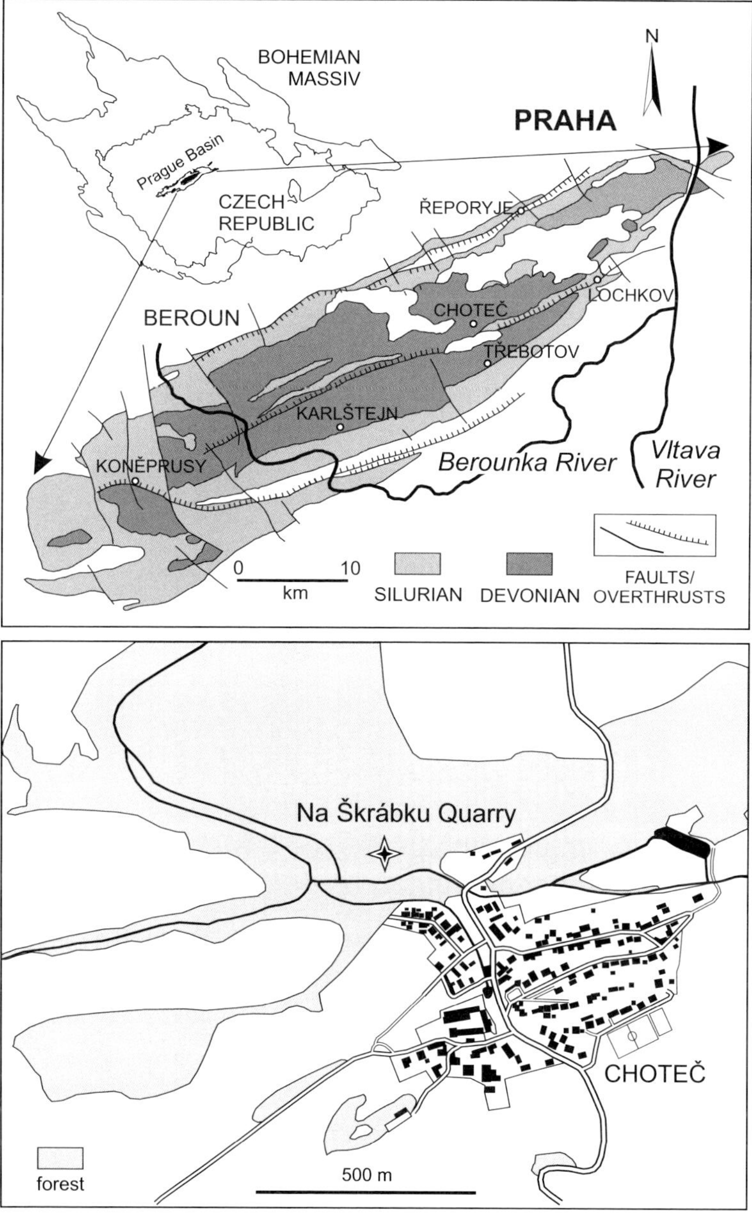

Fig. 1. Map showing the location of the Prague Basin in the Bohemian Massif, Czech Republic, and a simplified geological sketch map of the Prague Basin. The local map pinpoints the Na Škrábku Quarry near the village of Choteč, the type locality of the BCE.

specifications, see Brocke & Wilde 2001). In order to distinguish fungi from acritarchs, epifluorescence was applied.

The rock samples that were studied for dacryoconarids were collected from the Edgecliff and Nedrow members of the Onondaga Formation, and their equivalents at the localities are shown in Figure 6. They were treated by standard methods (splitting, washing, disaggregation and sieving); the better preserved shells were mounted on scanning electron microscope (SEM) stubs, sputter coated with gold, and then measured, described and documented with a SEM at magnifications of ×35 – ×3500.

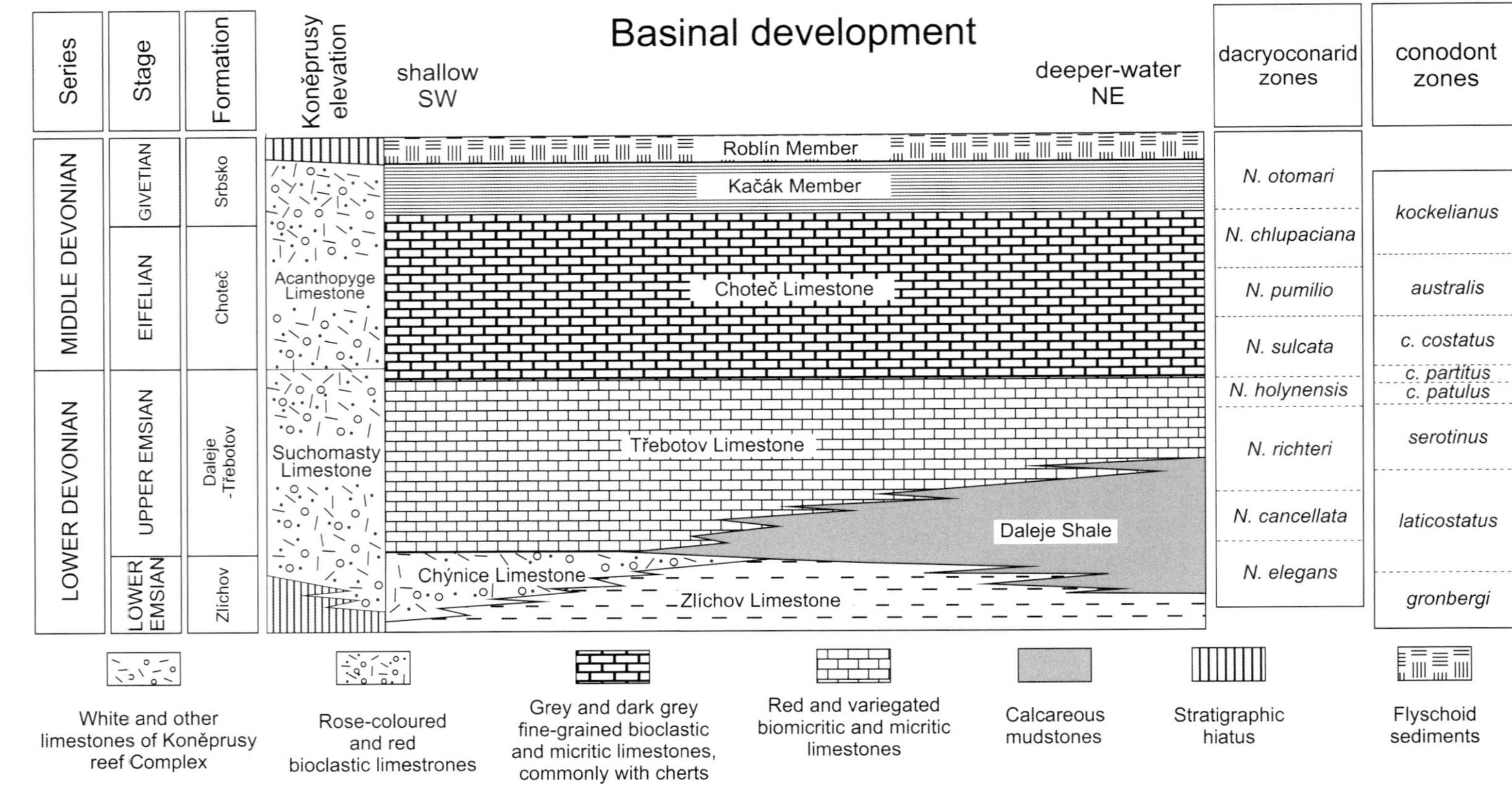

Fig. 2. Stratigraphy and facies development of the Lower–Middle Devonian of the Prague Basin; modified after Budil *et al.* (2009).

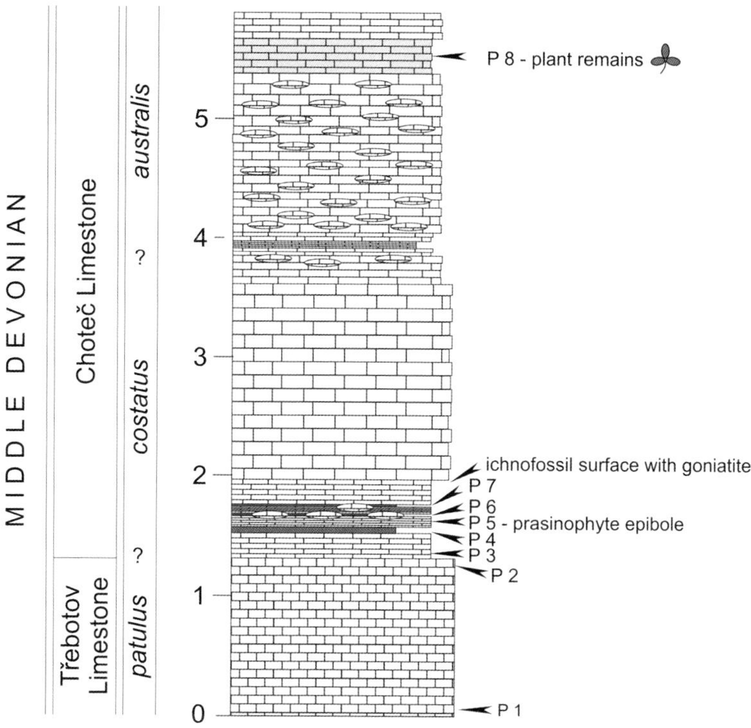

Fig. 3. Stratigraphic section in the Na Škrábku Quarry, showing the uppermost levels of the Třebotov Limestone and the lower portion of the Choteč Limestone. Numbered palynological samples ('P samples') are indicated by arrows.

Setting of the Basal Choteč Event in the Barrandian area

In the Barrandian area, the Lower–Middle Devonian boundary interval is characterized by a fully marine succession of carbonates and calcareous shales generally with common fossils (e.g. Chlupáč *et al.* 1979). The Lower–Middle Devonian boundary is drawn according to the first appearance of the conodont *Polygnathus partitus*, which falls within the *Nowakia holynensis* dacryoconarid zone. These levels are situated above the last appearance datum (LAD) of *Gyroceratites gracilis* and are characterized by a rich ammonoid fauna of the *Anarcestes lateseptatus* group. The boundary interval comprises upper levels of the Daleje–Třebotov Formation to lower levels of the Choteč Formation (Fig. 2), and is exposed at many natural outcrops and in old quarries. About 10 well-exposed boundary sequences were studied in detail (e.g. Svoboda & Prantl 1947, 1948; Chlupáč 1959; Klapper *et al.* 1978; Chlupáč *et al.* 1979, 1980; Koptíková 2011; Vodrážková *et al.* 2013).

The Daleje–Třebotov Formation includes three members: (1) the Suchomasty Limestone; (2) the Daleje Shale; and (3) the Třebotov Limestone. Interpreted to be a shallow-water deposit, the Suchomasty Limestone is developed on the SW flank of the Prague Basin. It is interpreted as a shallow-water, high-energy equivalent of the Daleje Shale and Třebotov Limestone. The Daleje Shale is characterized by greenish siliciclastic mudrocks with calcareous concretions deposited in a deeper, low-energy environment (Chlupáč *et al.* 1998). These clastic sediments are overlain by light to dark grey, fine-grained, nodular, bedded micritic–biomicritic carbonates of the Třebotov Limestone (originally defined by Svoboda & Prantl 1947; redefined by Chlupáč 1959). The fauna of the Trebetov and its preservation suggest a low-energy deeper environment. The upper part of the Třebotov Limestone is characterized by calcisiltite deposited from distal storm or turbidite currents alternating with slowly deposited and condensed hemipelagic material (Koptíková 2011).

Eustatic shallowing in the Prague Basin resulted in deposition of the upper levels of the

Fig. 4. Photographs of study intervals in the Na Škrábku Quarry. (**a**) Detail of the middle of the quarry wall; position of palynological samples indicated by arrows. (**b**) Lower surface of bed number 10; blow-up of the goniatite in the lower left. (**c**) General view of the eastern quarry wall with the boundary between the Třebotov and Choteč limestones.

Daleje–Třebotov Formation and is connected with increasing current activity associated with the immigration of new faunal elements (Chlupáč 1985).

The Choteč Formation incorporates two members: the *Acanthopyge* Limestone and the Choteč Limestone (Fig. 2). The *Acanthopyge* Limestone was deposited on the SW flank of the Prague Basin and is interpreted as a shallower-water, high-energy equivalent of the Choteč Limestone. The *Acanthopyge* Limestone is a light grey, platy, biodetrital limestone; it overlies the Suchomasty Limestone. The Choteč Limestone (defined by Svoboda & Prantl 1948; redefined by Chlupáč 1957, 1959) includes two major lithologies: grey, platy crinoidal biodetritic limestone alternating with light grey calcisiltic limestones and shaly intercalations. In more proximal depositional environments, turbidite deposits are present in the lower part of the Choteč Limestone (e.g. the Na Škrábku Quarry).

Organisms affected by the event

Diverse fossil groups have been intensively studied in Lower and Middle Devonian sequences, including the boundary interval (for a summary, see Chlupáč *et al.* 1979).

Radiolarians. Braun & Budil (1999) reported a rich radiolarian fauna from two outcrops in the uppermost levels of the Choteč Limestone (railway cut in Praha 5-Hlubočepy and the Kněží Hora Hill near Karlštejn, Fig. 1).

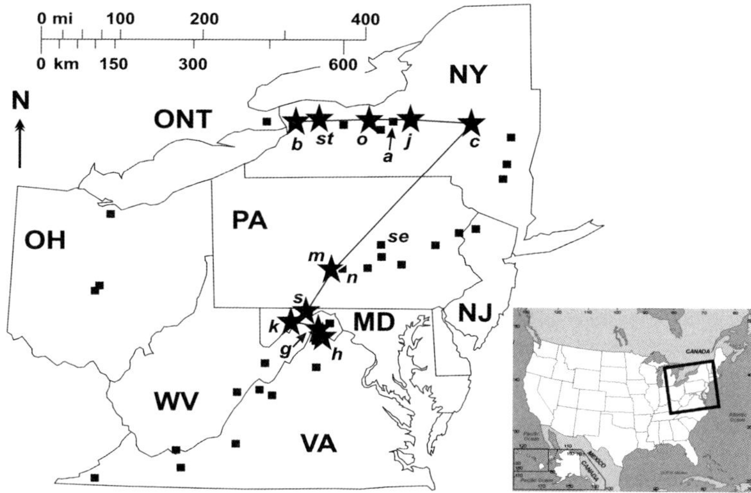

Fig. 5. Map of key outcrops and study localities, Appalachian Basin, eastern USA. Key outcrops of the lower Eifelian strata along approximately 1750 km of outcrop belt (modified after Ver Straeten 2007). Stars indicate localities from this study; other identified localities are noted in the text or in photographs. Uppercase letters denote US states and a Canadian province; lower case letters denote localities: a, Auburn; b, Buffalo; c, Cherry Valley; g, Gainesville; h, Hayfield; j, Jamesville–Nedrow; k, Keyser; m, Mapleton; MD, Maryland; o, Oak Corners; OH, Ohio; ONT, Ontario (Canada); n, Newton Hamilton; NJ, New Jersey; NY, New York; PA, Pennsylvania; s, Spring Gap; se, Selinsgrove Junction; st, Stafford; VA, Virginia; WV, West Virginia.

Foraminifers. Preliminary reports announced the occurrence of Devonian foraminifers more than 100 years ago (Schubert & Liebus 1902; Liebus & Wahner 1904; Pokorný 1958). More recently, Silurian and Devonian foraminifers were described by Holcová (2002). Detailed studies of several sections covering the Basal Choteč Event (BCE) document a distinct decrease in foraminiferal abundance (Holcová 2003, 2004*a*, *b*, *c*; Holcová & Slavík 2013).

Conodonts. Significant studies include papers by Klapper (1977), Klapper *et al.* (1978), Chlupáč *et al.* (1977), Weddige & Ziegler (1987), Galle & Hladil (1991), Zusková (1991), Hladil & Kalvoda (1993) and Klapper & Vodrážková (2013). Berkyová (2009) summarized earlier data and provided new results in a comprehensive study. The conodont zonation elaborated in the above-mentioned papers is summarized in Figure 2. The conodont zonation of Berkyová (2009) differs slightly from more recent studies on brachiopods (Mergl & Vodrážková 2012) and on environmental changes (Vodrážková *et al.* 2013); in these later studies, the base of the *Tortodus kockelianus australis* Zone is drawn at different levels.

Gastropods. Horný (1955, p. 65) announced the occurrence of 10 species of palaeozygopleurid gastropods in the Choteč Limestone; other early Devonian taxa were studied by Frýda & Bandel (1997) and Frýda *et al.* (2013). Frýda *et al.* (2008, p. 94) found out that the highly diverse palaeozygopleurid gastropods originate within the *partitus* Biozone (e.g. just below the BCE). Palaeozygopleurids are, however, completely absent above the event.

Bivalves. The absence of a modern systematic revision precludes an evaluation of this group.

Cephalopods. Chlupáč & Turek (1983) and Manda & Turek (2011) revealed considerable changes in associations of both goniatite and nautiloid cephalopods connected with the BCE. The important changeover in the goniatite fauna falls very close to the base of the *Pinacites jugleri* Zone.

Brachiopods. Brachiopods of the shallow-water facies were studied by Havlíček & Kukal (1990), who separated the *Karbous–Orbiproetus* and *Orbiproetus–Scabriscutellum* communities in the Suchomasty Limestone, and the *Karbous–Acanthopyge* Community in the *Acanthopyge* Limestone. Mergl (2001, 2008) studied linguloid, discinoid and acrotretoid brachiopods. In the first of these papers, an increase in diversity above the BCE was noted. However, more recently, Mergl

& Vodrážková (2012, p. 330) came to a quite different conclusion:

> lingulate brachiopods do not display any significant change around the Emsian/Eifelian boundary or the Basal Choteč Event (Middle Devonian, Eifelian, *Polygnathus costatus* Zone) and confirm the general uniformity of lingulate faunas in the Lower and early Middle Devonian.

Dacryoconarids. A dacryoconarid zonation for the Barrandian area was established by Bouček (1964) and new data were later provided by Lukeš (1989). This generally accepted zonation is summarized in Figure 2.

Ostracods. Přibyl & Šnajdr (1950) studied a diverse Lower–Middle Devonian ostracod fauna and reported a distinct extinction associated with the BCE at the Prastav Quarry. Šlechta (1996) revised a large number of the earlier described ostracod taxa established near the Lower–Middle Devonian boundary and also found a distinct extinction event at the Praha Barrandov locality.

Trilobites. Extensive monographs on phacopids (Chlupáč 1977), scutelluids and proetids (Šnajdr 1960, 1980), and numerous short reports on highly diverse trilobites, were summarized by Chlupáč (1983), who distinguished four major trilobite assemblages in the boundary interval:

- *Phacops–Struveaspis* Assemblage in the Třebotov Limestone (Chlupáč 1983, p. 56);
- *Koneprusites–Cyphaspides* Assemblage in the Choteč Limestone (Chlupáč 1983, p. 59) – medium diversity;
- *Acanthopyge–Phaetonellus* Assemblage in the *Acanthopyge* Limestone (Chlupáč 1983, p. 57) – high diversity;
- *Orbitoproetus–Scabriscutellum* Assemblage in the Suchomasty Limestone (Chlupáč 1983, p. 56) – high diversity (50 species).

Echinoderms. The rich crinoid fauna was studied by Bouška (1948, 1956), and more recently supplemented by Prokop (1970, 1976, 1987, 2012, 2013), Prokop & Petr (1997, 2004) and Hotchkiss *et al.* (1999, 2007).

Plant remains. Obrhel (1958, 1968) reported the occurrence of several long-ranging species from the Třebotov and Choteč limestones; Chlupáč *et al.* (1979, p. 141) mentioned, for the first time, a common occurrence of leiospheres (=prasinophytes in Vodrážková *et al.* 2013) in lower levels of the Choteč Limestone.

Organic-walled microfossils (OWM). Devonian OWM (e.g. acritarchs, prasinophytes, spores, chitinozoans, scolecodonts and fungi) from the Barrandian are known from several papers, such as Obrhel (1964), Lele (1968, 1972), Čorná (1969), McGregor (1979*a*, 1980, 1981*a*, 1983), Vavrdová (1989), Fatka (1999), Fatka & Brocke (2008) and Vavrdová & Dašková (2011). However, until recently, only a few of them dealt in particular with the BCE (Brocke *et al.* 2011, 2013; Vodrážková *et al.* 2013). Therein, the extraordinarily high number (mass occurrences) of green algae, namely prasinophytes, is emphasized. The majority of these prasinophytes belong to the genera *Tasmanites* and *Leiosphaeridia*. They are present in a dark grey limestone bed just above the base of the Choteč Limestone, which is regarded as representing the BCE level at its type locality (the Na Škrábku Quarry near Choteč village).

Abiotic effects

In an extensive study on carbon isotope stratigraphy of the Lower–Middle Devonian of southern and central Europe, Buggisch & Mann (2004, p. 528, fig. 4) described a positive excursion of $\delta^{13}C_{carb}$ from sections in the Prague Syncline.

Elrick *et al.* (2009) compared changes of the $\delta^{18}O$ from conodont apatite at two sections in the Prague Basin (the Na Škrábku Quarry and Prague-Barrandov), with one section in the western United States (the northern Antelope Range of central Nevada). They interpreted the oxygen isotopic signal to be global rather than related to local changes in seawater temperature (possibly cooling).

In a comprehensive study, Koptíková (2011) provided results of magnetic susceptibility measurements and gamma-ray logging in combination with geochemical methods. She also identified the presence of Bouma sequences in the Choteč Formation.

Setting of the Basal Choteč Event in the Appalachian Basin, eastern USA

Across the Appalachian Basin (AB), the Basal Choteč Event (BCE) falls within a carbonate–mixed carbonate and shale–bedded chert succession (Figs 5–7). The precise position of the Emsian–Eifelian stage boundary remains elusive. However, it and the BCE fall within the third-order transgression (transgressive systems tract of sequence stratigraphy) of Devonian Sequence/transgressive–regressive (T–R) Cycle Ic (of Johnson *et al.* 1985; refined regionally by Brett & Ver Straeten 1994; Ver Straeten 2007; Brett *et al.* 2011, their sequence Eif-1).

In the northern and western parts of the Appalachian Basin (e.g. New York, western New Jersey, eastern Pennsylvania, in the northern part of the

Appalachian Basin (NAB); and Ohio in the western part of the Appalachian Basin (WAB)), the Ic succession occurs within relatively shallow-ramp limestones, sometimes cherty, with minor shales in slightly deeper settings (Fig. 7c–e). In these areas, strata are assigned to the Onondaga Formation, except in central Ohio (i.e. the Columbus Formation of the WAB). Across the central part of the basin (e.g. central Pennsylvania, western Maryland, NW Virginia and NE West Virginia = CAB), the Ic succession occurs largely in deeper, more basinward facies. In the CAB, these strata are characterized by mixed argillaceous limestones and calcareous grey to dark grey shales, with minor black shales (Fig. 7a, b, f–h). These strata are assigned to the Selinsgrove Member of the Needmore Formation (Ver Straeten 2007). In the southern part of the Appalachian Basin, strata of the Selinsgrove Member (sometimes termed the calcareous shale and limestone member) transition laterally into bedded chert-dominated lithofacies, in the upper part of the Huntersville Chert Formation (Ver Straeten 2007).

Correlations of the Onondaga Limestone and its time-correlative strata between key outcrops across the northern and central Appalachian Basin are demonstrated in this study (Fig. 6). In New York's Onondaga Limestone, four member-level subdivisions are recognized: the Edgecliff, Nedrow, Moorehouse and Seneca members (Oliver 1954). These same time-correlative subdivisions are widely recognizable across the basin, although difficult to delineate in the chert-rich facies in the SW part of the basin (Ver Straeten 2007). The Moorehouse and Seneca members are separated by the Tioga B K-bentonite of Way *et al.* (1986), which has a radiometric age of 390 $\pm$ 0.5 Ma (Roden *et al.* 1990).

The sequence stratigraphy of the Onondaga Formation and equivalents, refined after the Johnson *et al.* (1985) T–R cycle model (Brett & Ver Straeten 1994; Ver Straeten 2007; Brett *et al.* 2011), is characterized by a conformable–disconformable third-order sequence boundary at the base of the succession (=base of Edgecliff Member), overlain by a transgressive systems tract (TST) extending to the top of the Nedrow Member. The position of the maximum flooding surface (i.e. the maximum landwards position of the shoreline) in moderate to deeper settings is marked by a pair of black shale beds (the lower (LBB) and upper Nedrow black beds (UBB) of Brett & Ver Straeten 1994; Ver Straeten 2007) separated by a thin limestone bed; in shallower facies, the black beds are represented by a pair of distinctly finer-grained limestone beds ($\pm$green shales). Initial aggradational sedimentation (highstand systems tract (HST)) in the lower Moorehouse Member is succeeded by shallowing patterns (falling-stage systems tract (FSST)) into the middle Moorehouse.

Following another third-order lowstand of sea level at the base of Sequence Id (generally appearing conformable), a second TST begins in the upper Moorehouse Member, which continues through the overlying Seneca Member and into the lower part of black shale-dominated strata of the overlying Marcellus strata (the Union Springs Formation of the Marcellus subgroup in New York; i.e. the lower part of the Marcellus Formation of other states except in Ohio, where the time-equivalent strata occur in the limestones of the Delaware Formation).

In the Eastern Americas Realm, the lowermost known occurrence of *Polygnathus partitus*, which marks the base of the Eifelian Stage (Ziegler & Klapper 1985), is near the base of the Nedrow Member (Klapper & Oliver 1995). The absence of this conodont from the subjacent Edgecliff Member is attributed to environmental exclusion (Klapper 1981). Currently available biostratigraphic data indicate that the base of the Eifelian Stage is at or near the base of the Edgecliff Member (Kirchgasser 2000).

The Choteč interval in the Appalachian Basin may occur through the interval from the mid-Edgecliff to the top Nedrow members and equivalents. However, the key interval of the BCE appears to be focused around the basinwide lower and upper Nedrow black beds (LBB and UBB, respectively) and their shallower-water correlatives. The base of the LBB approximates the base of the *costatus* Zone, near the maximum flooding surface of the Ic Sequence and the bathymetrically deepest facies of the Onondaga Formation (Ver Straeten 2007).

Data on palynomorphs (e.g. acritarchs, chitinozoans and spores) and dacryoconarids presented in this paper represent the first results of a team project trying to refine Emsian and Eifelian biostratigraphy in the Appalachian Basin. For both groups, no sufficient data are available yet to establish formal zonations in the Appalachian Basin. This lies beyond the scope of this paper. The details of the palynomorph and dacryoconarid records during the BCE interval are described below. Other fossil organisms (e.g. conodonts, goniatites and brachiopods) are considered when possible.

Organic-walled microfossils (OWM)

To date, no palynological data exist from this particular time interval in the Appalachian Basin. The majority of papers on Devonian acritarchs in North America address the Givetian of Ohio (e.g. Wicander & Wood 1979, 1981; Wicander 1983, 1984; Wicander & Wright 1983), Iowa (Wicander &

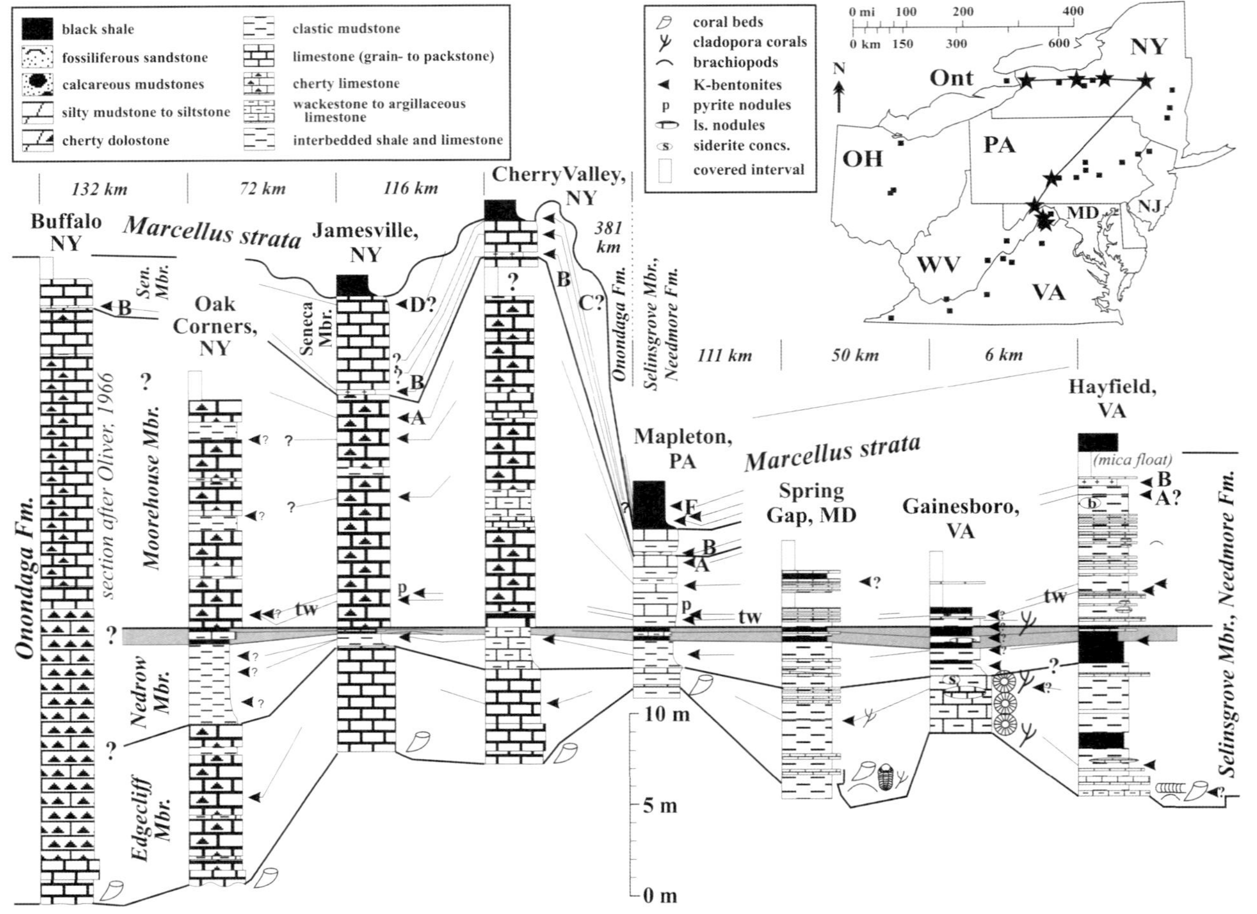

Fig. 6. Correlations of key outcrops, Onondaga Formation and correlative strata, Appalachian Basin. Study outcrops along approximately 870 km of outcrop belt, western to eastern New York to northern Virginia. Datum = top of Nedrow Member and time-equivalent strata (= top of Nedrow black beds). Lines denote chronostratigraphic correlations. Grey shaded interval denotes Nedrow black beds interval. Bold lines denote member-level divisions; thin lines denote event bed correlations; lines with arrowheads denote airfall tephra beds (K-bentonites). p, correlatable pyrite zone above twin K-bentonites; Sen, Senenca Mbr; tw, widely correlatable pair of thin K-bentonites. Capital letters A–F denote layers of the Tioga A–G K-bentonites cluster. The Jamesville section is 10 km from the Nedrow study locality. Buffalo, NY section after Oliver (1967).

Fig. 7. Photographs of the Onondaga Formation of New York and correlative strata in the study area. (**a**) Schoharie- and Onondaga-equivalent Needmore Formation (calcareous shale member and Selinsgrove Member, respectively), at Spring Gap, MD (locality 's' in Fig. 5). Chronostratigraphic correlatives of members of New York's units are denoted. Approximately 13 m of a 19 m section along the highway shown in the photograph. (**b**) Onondaga-equivalent Selinsgrove Member, Needmore Formation at Hayfield, VA (locality 'h' in Fig. 5). Chronostratigraphic correlatives of members of New York's Onondaga Formation are denoted. Note the position of the Nedrow black beds along the outcrop. Tioga B K-bentonite is covered at the top of slope, just beyond right-hand edge of the photograph. The Selinsgrove Member is approximately 18 m thick here. (**c**) Onondaga Formation in the type area, Jamesville, NY (locality 'j' in Fig. 5). Key widely correlative marker units denoted, along with member-level divisions. Note the position of the Nedrow black beds at the top of the Nedrow Member, and 'twin K-bentonites' low in the Moorehouse Member. The Onondaga Formation is 22.8 m thick here. (**d**) Lower and upper black beds of the upper Nedrow Member, along Interstate 81 at Nedrow, NY. The lower and upper Nedrow black beds are denoted by lbb and ubb, respectively. Locality is 9.5 km WSW of the Jamesville locality. (**e**) Nedrow black beds in shallower-marine limestone-dominated facies, at the abandoned Schooley Quarry, Auburn, NY (locality 'a' in Fig. 5). (**f**) Mid-Nedrow- to lower Moorehouse-equivalent strata of Selinsgrove Member, Needmore Formation, along railroad cuts at Newton Hamilton, PA (locality 'n' in Fig. 5). Note the black beds and the position of the twin K-bentonites. The black and white bar is 1.5 m. (**g**) Black beds and adjacent strata, Selinsgrove Member, near Mapleton, PA (locality 'm' in Fig. 5). Again, note the twin K-bentonites. The black beds are separated by 36 cm. (**h**) Lower part of the Selinsgrove Member, at the type section in central Pennsylvania (locality 'se' in Fig. 5). Note the position of the black beds and twin K-bentonites. The scale bar at the centre is next to 1.5 m staff. E, Edgecliff Member and correlatives; bb, Nedrow black beds; lbb, lower black bed; M, Moorehouse Member and correlatives; N, Nedrow Member and correlatives; S, Seneca Member and correlatives; tw, closely spaced 'twin' K-bentonites in lower Moorehouse and correlative strata; Ti-B, Tioga B K-bentonite; ubb, upper black bed.

Wood 1997, including chitinozoans) and Kentucky (Wood & Clendening 1985, including chitinozoans), followed by the late Devonian of Ohio (Wicander 1973*a*, *b*, *c*, 1974, 1975; Winslow 1962), Oklahoma (Wilson & Urban 1963, 1971; Wilson & Skvarla 1967), Indiana (Wicander & Loeblich 1977), Iowa (Wicander & Playford 1985, including spores) and, recently, from Illinois (Wicander & Playford 2013), and a brief report from New York (Higgs & Hughes 2012). Early Devonian (Gedinnian) records are from Oklahoma (several papers by Loeblich & Wicander: e.g. Loeblich 1970; Loeblich & Wicander 1974, 1976; Wicander 1986) and from New York State (Wicander & Schopf 1974). A comprehensive catalogue of the stratigraphic distribution of North American acritarchs was given by Wicander (1983). Further reports are from the middle–late Devonian of Canada, Alberta (Staplin 1961; Turner 1986, 1991) and Ontario from the Emsian–Eifelian interval (Legault 1973*a*, *b*; Playford 1977). Reports by Deunff from the 1950s were stratigraphically not constrained (Playford 1977).

Chitinozoans have been reported from the mid-Devonian of Michigan (Dunn & Miller 1964), Ohio (Wood 1974; Wright 1976), Indiana (Wright 1980), Illinois (Collinson & Scott 1958), Iowa (Wicander & Wood 1997) and Missouri (Urban & Kline 1970). Chitinozoans from Ontario, Canada were noted by Legault (1973*a, b*). The distribution and ranges of chitinozoans, including Devonian taxa from North America, are given by Jenkins & Legault (1979).

Devonian spores have frequently been reported from different areas in North America: the Lower Devonian of Cherry Valley, New York (Hughes & Higgs 2011), the Middle Devonian of Illinois (Peppers & Damberger 1969) and Georgia (Ravn & Benson 1988, including the Frasnian), and the Upper Devonian of Ohio (Winslow 1962), Illinois (Wicander & Playford 2013), western New York and Pennsylvania (Richardson & Ahmed 1988) and, recently, of Kentucky (Clayton *et al.* 2012; Higgs & Hughes, 2012; Rooney *et al.* 2013). In addition, the Devonian of Canada: Ontario (e.g. McGregor & Owens 1966; McGregor 1973, 1977), eastern Canada (McGregor & Camfield 1976), Canadian Arctic (McGregor 1974, 1981*a*, *b*; McGregor & Camfield 1982) and southern Saskatchewan (Playford & McGregor 1993, including acritarchs). A biostratigraphic compilation for spores of North America was published by McGregor (1979*b*). Richardson & McGregor (1986) introduced a concept of spore zonation for the Silurian and Devonian of the Old Red Continent (ORC).

Owing to the peculiarities within the Appalchian Basin, as mentioned above, a northern and a central part are discriminated.

Northern part of the Appalachian Basin (NAB)

Throughout much of New York State, the lowermost bed of the Nedrow Member is an argillaceous lime mudstone–wackestone that abruptly overlies a glauconitic and pyritic, crinoidal, poorly washed biosparite or packstone facies of the uppermost Edgecliff Member. The ichnogenus *Chondrites* is frequent in the argillaceous beds of the Nedrow Member. Bromley & Ekdale (1984), Bromley (1990) and Seilacher (2007) interpreted the *Chondrites* trace-maker to have been a chemosymbiotic organism adapted to dysaerobic environments. Schubert (1996) supports this assumption in his paper on the biofacies of the Wissenbach Schiefer of the Rhenohercynian Zone in Germany. This interpretation is also supported by Savrda & Bottjer (1986), who cited examples from the Cretaceous and the Miocene. The assignment of *Chondrites* to the group of ‘facies-breaking ichnogenera’ (by Seilacher 2007, pl. 71) leaves space for habitat interpretation and does not exclude low-oxygen conditions. The argillaceous Nedrow beds also contain a diverse, although physically diminutive, benthic macrofauna of brachiopods, bryozoans, rugose corals and trilobites: again, suggestive of a dysaerobic environment. Microfossils include ostracods, scolecodonts, *Tentaculites scalariformis*, *Homoctenus* sp., palynomorphs and the dacryoconarids described herein.

Central part of the Appalachian Basin (CAB)

At multiple localities in Virginia, Maryland and south-central Pennsylvania, the basal bed of Nedrow-equivalent strata in the Selinsgrove Member (Needmore Formation) and the Nedrow Member of New York is a fissile black shale, which abruptly overlies a medium-grey, crinoidal, poorly washed biosparite of the uppermost Edgecliff bed. Dacryoconarids within the Nedrow Member-equivalent are taxonomically distinct from those of the subjacent equivalents of the Edgecliff Member and the Schoharie Formation.

Dacryoconarids of the basal Nedrow bed and the two UBB occur in exceptionally high numbers as ecological epiboles on shale laminae surfaces. The dacryoconarid shells are frequently aligned parallel to one another to the extent that laminae surfaces may be mistaken for tectonic slickensides. Most dacryoconarids are preserved only as faint impressions of crushed and dissolved shells. Preservation is usually so poor that it is only with diligent effort that one can discern that the ‘slickensides’ are actually aligned, crushed and leached impressions of dacryoconarid shells. In instances where the shells are preserved, it is clear that the dacryoconarid-rich layers were deposited as thin coquinite layers separated by thin shale partings. Neither the coquinite layers nor the shale partings were disturbed by bioturbation and are devoid of a benthic fauna.

The Basal Choteč Event and its equivalents in other areas

As a detailed discussion of the BCE in areas other than the Prague and Appalachian basins is far beyond the scope of the present paper, only some of the more important regions are briefly mentioned here. In the German Rhenohercynian and Saxothuringian zones, the facies of contemporaneous rocks is different or difficult to recognize because they cannot be exactly correlated time-wise with the Choteč strata. However, a change from more oxygenated 'lighter'-coloured to 'darker' mostly shaly rocks of the lower Eifelian 'Wissenbach-Schiefer' facies has been recognized (e.g. Requadt & Weddige 1978; Schubert 1996). Recently, the event has been referred to in the context of carbonate facies in the Meinerzhagener Korallenkalk (Ernst *et al.* 2012). As noted before, within the Rhenohercynian (Eifel Hills) the BCE is associated with the extinction of the upper Emsian–lower Eifelian OCA brachiopod fauna (e.g. Struve 1982; Weddige 1988; Struve *et al.* 1997, 2008). Similar to the Wissenbach-Schiefer facies, reports from the Iberian Peninsula are known through Henn (1985), Montesinos (1987) and García-López & Sanz-López (2002), and from the Ossa Morena Zone by Machado *et al.* (2010). In the Moroccan Anti-Atlas, similar observations have been published (e.g. Alberti 1980; Becker & House 1994; Kaufmann 1998; Klug *et al.* 2000, 2013; Becker & Aboussalam 2013). Elrick *et al.* (2009) and Pedder (2010) have reported examples from Nevada, USA. In his recent study of rugose coral taxa and palaeobiogeography, the latter author has reported a breakdown in rugosan provincialism and the termination of the Emsian–Eifelian Great Basin coral province of Nevada within the BCE interval (Pedder 2010). The only report from South America was published by Troth *et al.* (2011), dealing with strata in Bolivia. Therein, the Choteč Event slightly post-dates the sudden incursion of goniatites in a cold-water succession.

Discussion and results

The Barrandian area

Effects on invertebrate fauna in the the Barrandian area caused by the BCE can be summarized as follows:

- Chlupáč (1983) documented a gradual change in trilobites: 90% of species and, more importantly, 50% of upper Emsian trilobite genera did not cross the level of the BCE;
- the appearance of cosmopolitan proetids, scutelluids and phacopids (Chlupáč 1983);
- the disappearance of palaeozygopleurid gastropods (Frýda *et al.* 2008);
- changes between goniatites (innovations in Pinacitidae, a decline of Mimagoniatitidae and Mimoceratidae: Chlupáč & Turek 1983), although Becker & House (1994) emphasized that *Gyroceratites* (and, subsequently, the Mimoceratidae) did not range into the Eifelian;
- in the course of the BCE, rutoceratoid generic diversity dropped dramatically; the recovery of nautiloids was slow (Manda & Turek 2011);
- dacryoconarids exhibit a modest faunal turnover (Bouček 1964), whereas Lukeš (1989) reported a stronger turnover in the Prague–Barrandov section.

Organic-walled microfossils. In total, eight palynological samples have been analysed from the type section Na Škrábku Quarry near the village of Choteč. The studied section comprises a calcareous sequence ranging from the upper Třebotov Formation to the lower Choteč Formation. There is a distinct lithological change from a light, bioclastic composition of the Třebotov Limestone to a darker bioclastic limestone alternating with a dark laminated lime–mudstone sequence of the Choteč Limestone (Figs 3 & 4). Palynological sample P 1 from the upper Třebotov Limestone does not yield any determinable figured OWM. Sample P 2 from its topmost level bears a few thin-walled prasinophytes assignable to the genus *Leiosphaeridia.* The succeeding sample P 3 from the basal Choteč Limestone is more enriched in organic matter with small *Tasmanites* (Fig. 8f), a few *Dictyotidium*; spores, such as *Acinosporites* sp., *Dibolisporites* sp., cf. *Camarozonotriletes sextantii* and small-sized trilete forms occasionally occur (Fig. 8g). Sample P 4 is situated just below the limestone bed, which is considered to represent the level of the BCE. Sample P 4 contains a higher concentration of organic matter (often dark brown–black), but only a few palynomorphs can be identified. The main components are marine palynomorphs, such as chitinozoans, scolecodonts and fragments of thin-walled prasinophytes; terrestrial spores are very rare. With the onset of the distinctive dark-grey bioclastic limestone bed (i.e. sample P 5), which is about 20 cm above the base of the Choteč Limestone (Fig. 9), the palynological assemblage changes significantly. This bed is chracterized by mass occurrences of phycomata of prasinophycean algae: large (150–600 μm) *Tasmanites* spp. (Fig. 8k) and *Leiosphaeridia* spp. are the dominant groups, forms of *Dictyotidium* spp. are frequent.

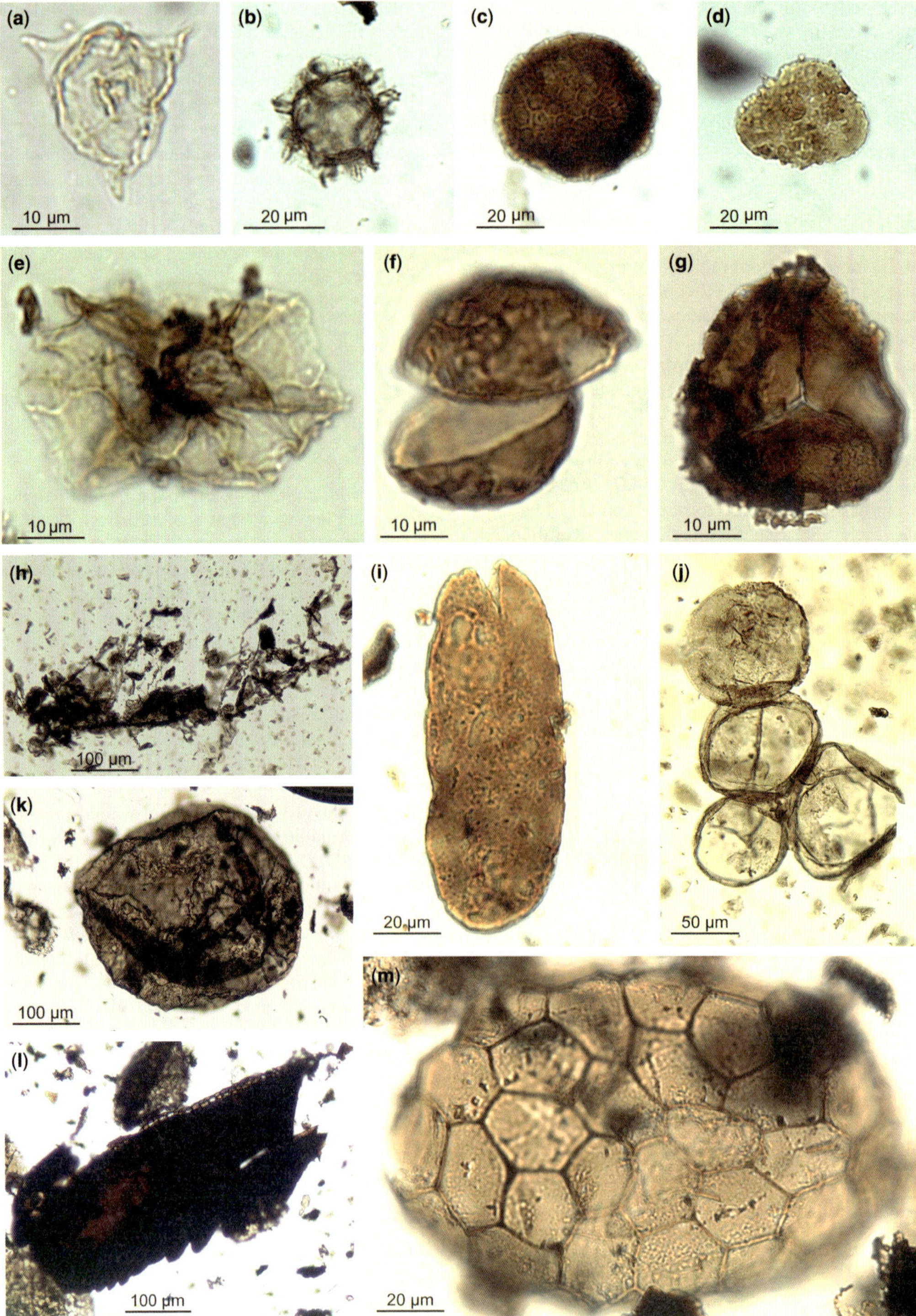

Fig. 8. Palynomorphs of the Na Škrábku Quarry, Prague Basin. The photographs were taken in transmitted light and with applied infrared video technique (IR). Indicated are localities, stratigraphic position, sample/slide numbers, England Finder (E.F.) coordinates and identification numbers (PMP) of the palynological collection at Senckenberg.

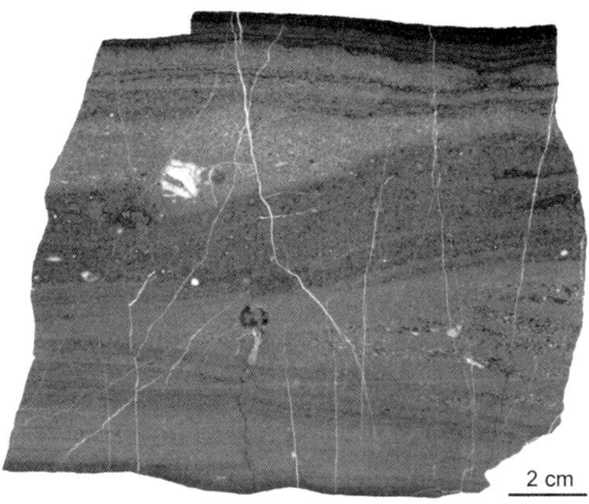

Fig. 9. Thin-section of the bed representing the palynological sample P 5. It displays the BCE in the Na Škrábku Quarry: the bed is mostly developed as a peloidal grainstone (in parts mudstone to wackestone, partly bioclastic). Repeated cross-bedding with erosional contacts indicate current activity and associated sediment transport. Prasinophytes (dark spots) occur dispersedly in the entire bed, but are enriched in the middle, in slightly coarser laminae, and also aligned in cross-stratified laminae. The photograph was taken under transmitted light microscope; collection of the Senckenberg palynological section, number PMP 235 DS.

Comparatively less frequent, but important for characterizing this assemblage, are mazuelloids that are represented by different morphotypes. Rare to very abundant are scolecodonts, small-sized acritarchs of the *Veryhachium trispinosum* group, and spores, such as *Lophotriletes devonicus* (Fig. 8d). In addition to the standard preparation method, the 'light organic fraction' of sample P 5 was studied palynologically. The microscopic analysis reveals mostly larger-sized (150–500 μm), thin-walled prasinophytes of *Leiosphaeridia* type (Fig. 8j), and agglomerations of filamentous and spherical/coccoid cyanobacteria-like organic microfossils (Fig. 8h), along with fungi-like microorganisms. The palynological assemblage of the succeeding sample P 6 differs by the appearance of moderately frequent and diverse acritarchs, such as *Cymatiosphaera* cf. *canadensis* (Fig. 8b), *C.* cf. *cornifera* (Fig. 8e), *Micrhystridium* sp., *Veryhachium trispinosum* (Fig. 8a), *Navifusa bacilla* and *Polyedryxium* sp.. Prasinophytes – mainly *Tasmanites* sp. (Fig. 8j) and *Dictyotidium* sp. (Fig. 8m) – are present, but they are smaller and not abundant; small-sized apiculate spores are rare. It is likely that this composition points to an initial recovery of the 'normal marine' phytoplankton community. However, the presence of *Cymatiosphaera* spp. may also indicate freshwater influx (see the discussion in the 'Interpretation' section). Sample P 7 shows a similar, marine-dominated composition, but the overall frequency of palynomorphs follows the decreasing trend already observed in P 6. The main species are *N. bacilla* (Fig. 8i), *Dictyotidium* cf. *variatum* (Fig. 8c) and *V. trispinosum*. Spores are represented by apiculate specimens (e.g. *Dibolisporites* sp.). Samples P 6 and P 7 are derived from the darker lime–mudstone interval; above this, the sequence changes to a medium-grey, more thick-bedded limestone succession, but this part was not sampled for the current study. Sample P 8 is from an interval of platy, greenish-grey limestones with plant fossils. However, the palynological sample yields only sparse organic matter, no palynomorphs could be identified.

The studied part of the Na Škrábku section shows palynological assemblages, which are characteristic of a relatively deeper, basinward, organic-matter-rich setting. A variable mixture of marine palynomorphs, such as acritarchs, chitinozoans, scolecodonts and prasinophytes, with less frequent spores and phytoclasts represents the 'normal marine' composition. However, in the BCE level (palynological sample P 5), a unique proliferation of prasinophytes along with a higher portion of the enigmatic mazuelloids (=muellerisphaerids) took place. Mazuelloids were found in the routine palynological residue, together with a mass occurrence of prasinophytes obtained from the residue of the conodont preparation. (e.g. Brocke *et al.* 2011). In those residues they are even more frequent and show a higher diversity of morphotypes. It is likely that the preservation potential of such mineralized shells is higher, and shows more diversification and frequency in species.

Fig. 8. (*Continued*) (**a**) *Veryhachium trispinosum* group; P 5-2, E.F. N56-2, PMP 235. (**b**) *Cymatiosphaera* cf. *C. canadensis* Deunff, 1961; P 6-1; E.F. D 54-1, PMP 236. (**c**) *Dictyotidium* cf. *D. variatum* Playford, 1977; P 5-2, E.F. S 42-4, PMP 235. (**d**) *Lophotriletes devonicus* (Naumova *ex* Chibrikova) McGregor & Camfield, 1982; P 5-2, E.F. K 52-4, PMP 235. (**e**) *Cymatiosphaera* cf. *C. cornifera* Deunff, 1955; P6-2, E.F. K 37-2, PMP 236. (**f**) *Tasmanites* sp.; P 3-2, E.F. T 36-2, PMP 233. (**g**) cf. *Camarozonotriletes sextantii* McGregor & Camfield, 1976; P 3-1, E.F. J 37-3, PMP 233. (**h**) Agglomeration of filamentous and spherical cyanobacteria-like organic microfossils (light fraction of floating kerogen); P 5-C, E.F. G 53-1, PMP 235. (**i**) *Navifusa bacilla* (Deunff, 1955) Playford, 1977; P 7-2, E.F. Y 36-4, PMP 237. (**j**) Arrangement of thin-walled prasinophytes, cf. *Leiosphaeridia* sp. (light fraction of floating kerogen); P 5-C, E.F. W 35, PMP 235. (**k**) *Tasmanites* sp. (light fraction of floating kerogen); P 5-A, E.F. Q 39, PMP 235. (**l**) Scolecodont; P 5-1, E.F. C 49, PMP 235. (**m**) *Dictyotidium* sp.; P 5-2, E.F. C 41-1, PMP 235.

Northern part of the Appalachian Basin (NAB)

Dacryoconarids. Seven dacryoconarid taxa, including six previously undescribed species, are present in the Nedrow Member of the NAB (Figs 10–12). The currently known first occurrence of *Striatostyliolina mima* n. sp. is at the base of the Edgecliff Member, and its currently known uppermost occurrence is in the mid-Givetian Ledyard Member of the Ludlowville Formation. A form of *Striatostyliolina* similar to *S. mima* is present in the upper Emsian Carlisle Center Member of the Schoharie Formation, but it is not currently certain that the two are conspecific. *Viriatellina manifesta* n. sp. may also first occur in the upper beds of the Schoharie Formation, but its currently certain first

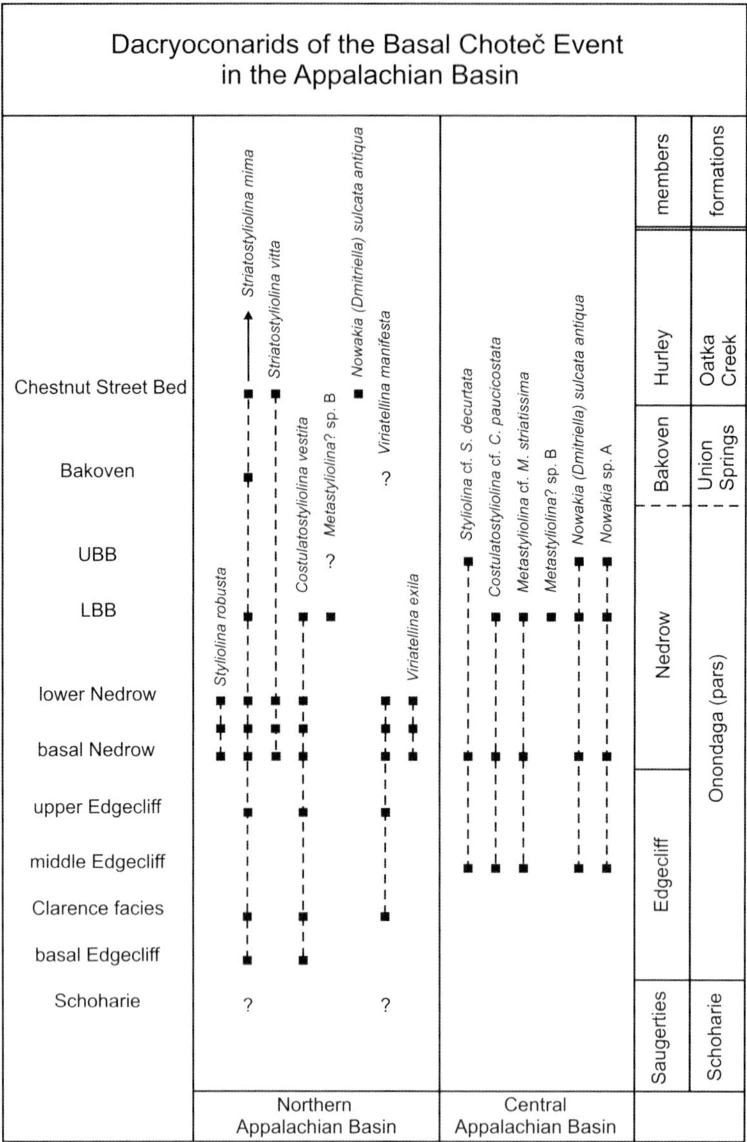

Fig. 10. Range chart of dacryoconarid taxa through the BCE interval in the northern (NAB) and central part of the Appalachian Basin (CAB). Note the very good time-restricted ranges of some taxa, making them extremely good index fossils; also note the differences in dacryoconarid taxa between the NAB and the CAB. For further explanations, see the text.

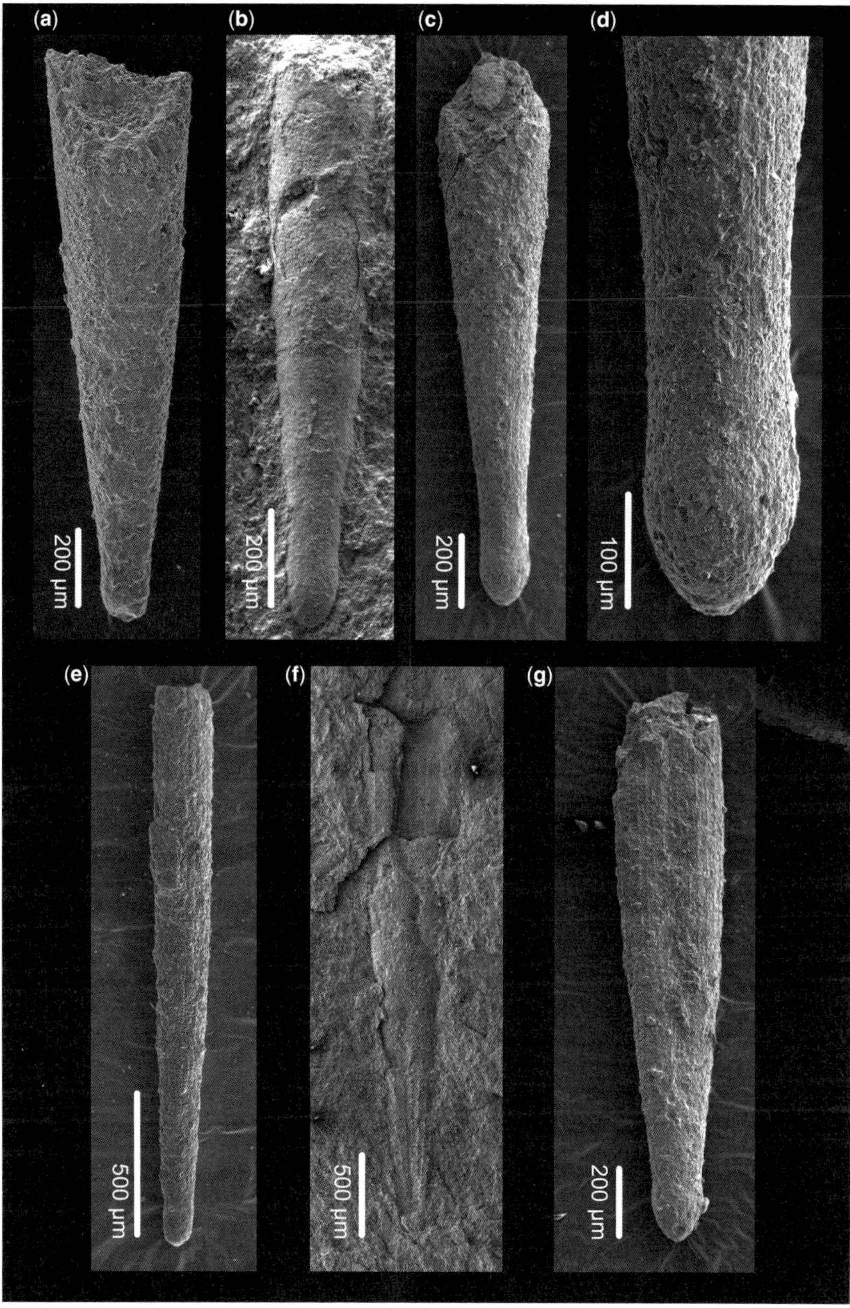

Fig. 11. Dacryoconarids of the Appalachian Basin. I. All figures are SEM images. New York State Museum (NYSM) numbers, sample localities, position in the section and E/E suite sample numbers are given where applicable. (**a**) *Styliolina robusta* n. sp.; holotype, NYSM 17229, NY Route 11, Nedrow, NY, basal Nedrow Member. (**b**) *Styliolina* cf. *S. decurtata* Bouček, 1964; NYSM 17236, Hayfield, VA, mid-Edgecliff-equivalent; E/E.07-1B-14. (**c**) & (**d**) *Striatostyliolina mima* n. sp.; holotype, NYSM 17230, US Route 20, Cherry Valley, NY, basal Nedrow Member. (**e**) *Striatostyliolina vitta* n. sp.; holotype, NYSM 17231, US Route 20, Cherry Valley, NY, 25 cm above the base of the Nedrow Member. (**f**) *Costulatostyliolina* cf. *C. paucicostata* (Bouček, 1964); NYSM 17237, Hayfield, VA, Nedrow-equivalent, LBB; E/E. 07-1B-17. (**g**) *Costulatostyliolina vestita* n. sp.; holotype, NYSM 17232, United States Route 20, Cherry Valley, NY, 25 cm above the base of the Nedrow Member.

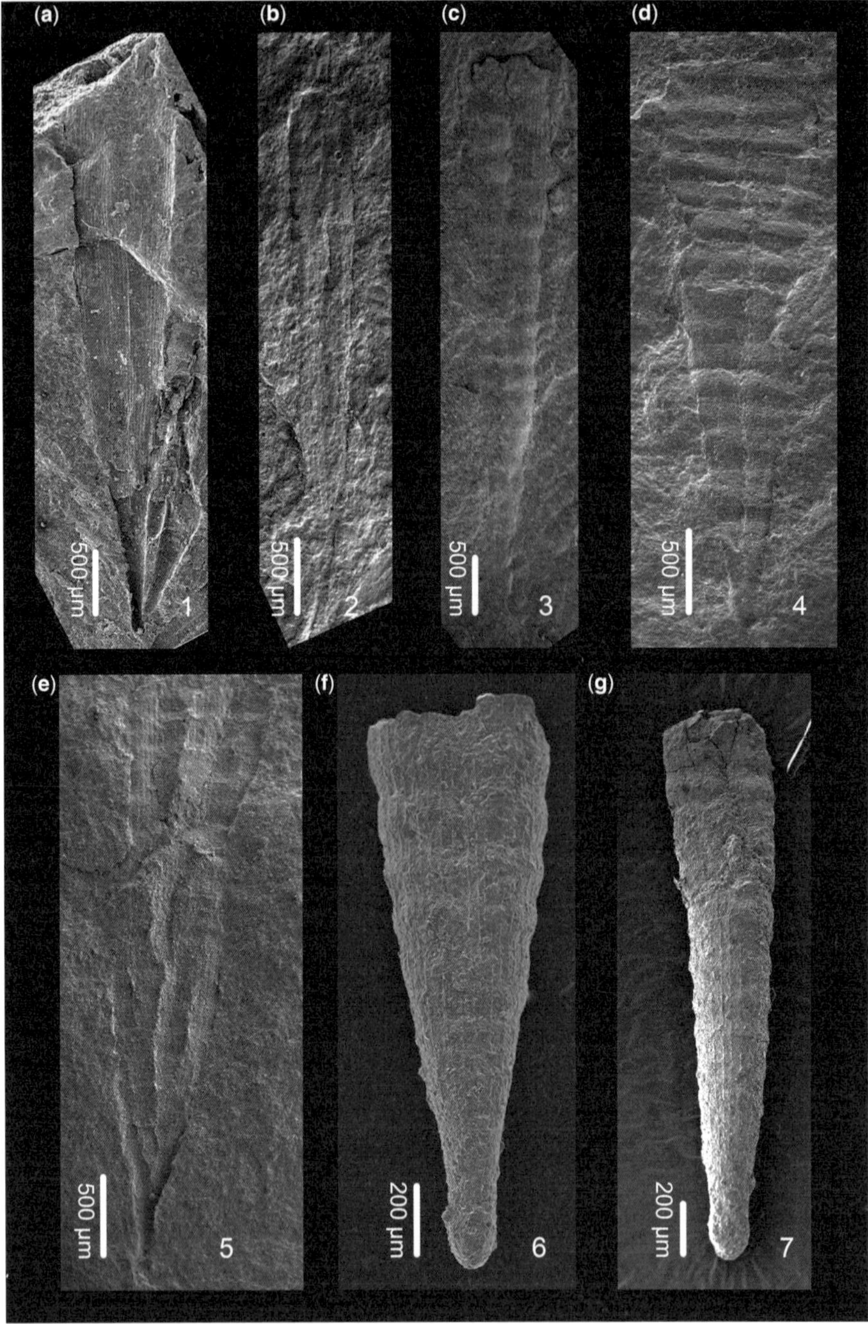

Fig. 12. Dacryoconarids of the Appalachian Basin. II. All figures are SEM images. New York State Museum (NYSM) numbers, sample localities, position in the section and E/E suite sample numbers are given where applicable. (**a**) *Metastyliolina* cf. *M. striatissima* Bouček & Prantl, 1961; NYSM 17238, Spring Gap, MD, Nedrow-equivalent, LBB; E/E. 07-5A-7. (**b**) *Metastyliolina*? sp. B; NYSM 17239, Hayfield, VA, Nedrow-equivalent, LBB; E/E 07-1B-17. (**c**) *Nowakia* (*Dmitriella*) *sulcata antiqua* Alberti, 1981; NYSM 17240, Mapleton, PA, Nedrow-equivalent, LBB; E/E 07-6-21. (**d**) & (**e**) *Nowakia* sp. A; NYSM 17235, 17241, Hayfield, VA, mid-Edgecliff-equivalent; E/E 07-1B-14. (**f**) *Viriatellina manifesta* n. sp.; NYSM 17233, Cherry Valley, NY, 25 cm above the base of the Nedrow Member. (**g**) *Viriatellina exila* n. sp.; NYSM 17234, Cherry Valley, NY, 25 cm above the base of the Nedrow Member.

occurrence is at the base of the Clarence facies of the Edgecliff Limestone. This species may also occur in the upper Eifelian, post-Onondaga Bakoven Shale and possibly higher in the Hamilton Group in conjunction with several additional forms of the genus that appear to be descendants of *V. manifesta* n. sp.. The first occurrence of *Costulatostyliolina vestita* n. sp. is also at the base of the Edgecliff Member and its uppermost occurrence is just below the two Nedrow black beds that mark the top of the Nedrow Member. *Styliolina robusta* n. sp. and *Viriatellina exila* n. sp. first occur at the base of the Nedrow and last occur immediately below the two Nedrow black beds. If this limited, short-ranged presence of the two taxa holds true in the future (i.e. between the base of the Nedrow and the two Nedrow black beds near the top of the unit), they make up ideal index fossils for the Nedrow Member below the two dark marker beds. *Striatostyliolina vitta* n. sp., which also first occurs at the base of the Nedrow, is absent from the upper Onondaga, but recurs in the Chestnut Street Beds at the base of the Oatka Creek Formation, which is in the mid-Eifelian *kockelianus* Zone and part of the Stony Hollow Event (Ver Straeten & Brett 2006; Ver Straeten 2007; Brett *et al.* 2011; DeSantis & Brett 2011).

Only two of the dacryoconarid species named above are known to occur in either of the two black beds at the top of the Nedrow in the northern AB. Although these beds are commonly devoid of dacryoconarids, *Striatostyliolina mima* n. sp., *Costulatostyliolina vestita* n. sp. and *Metastyliolina*? sp. B are occasionally present in small numbers.

Of the seven dacryoconarid taxa known to be present in the Nedrow of the northern region of the NAB during the BCE, two may predate, at least four originate early in, and five apparently became extinct late in the event interval. Most appear to be derived from similar forms that occur in the upper Emsian Schoharie Formation. Whereas four of the new species described herein are either extirpated or become extinct prior to apex of the BCE, *Viriatellina manifesta* n. sp. and *Striatostyliolina mima* n. sp. occur upsection in the Hamilton Group. One form, *Metastyliolina*? sp. B, which is restricted to one or both of the two UBB, first occurs at the base of the Nedrow in the CAB.

Other fauna. It appears that each faunal group responded to the BCE in its own particular fashion. Brachiopods seem to have been unaffected (Dutro 1981). Whereas conodonts, particularly lineages of *Polygnathus*, show no apparent deviation from the tempo of originations that had begun during late Emsian times (see Klapper 1981), ostracods underwent a profound extinction event (see Berdan 1981). Rugose coral diversity declined and provincialism began to disintegrate immediately before and during the BCE (Oliver 1977), particularly with the migration of *Synaptophyllum* into North America (Pedder 2010).

Oliver (1956) reported a subtle shift in benthic faunas during Nedrow deposition, but did not document a significant faunal turnover between the Nedrow and the upper members of the Onondaga. He did, however, document a geographical segregation of benthic faunas within the Nedrow interval between central New York and the SE part of the state. House (1981, p. 33) expanded the magnitude of this geographical segregation of faunas, reporting that his goniatite Fauna 2 in the Edgecliff and lowermost Nedrow, 'rather widely distributed in the Southern Appalachians is a fauna, not recognized in New York', and his Fauna 3, of the middle–upper Nedrow in New York, shares only one species, *Foordites* cf. *F. buttsi*, with the contemporaneous fauna of the Southern Appalachians. Dacryoconarids show this geographical segregation of AB faunas within the BCE interval more profoundly than do other faunal elements.

Organic-walled microfossils (OWM). Five sections with a focus on the Nedrow Member (Onondaga Formation) have been studied palynologically in the NAB in New York: Stafford Quarry, Oak Corners Quarry, Schooley Quarry near Auburn, Nedrow (Jamesville) and Cherry Valley. In addition, one sample from the upper Edgecliff Member collected in the Oak Corners Quarry near Phelps has been included. Altogether, 13 samples have been analysed, from which six were palynologically productive. A majority of the lower Nedrow samples are characterized by a relatively low concentration of organic matter and moderate–poor preservation of the OWM; in many cases they are highly altered. Pyrite framboids are frequent in the organic residue, partly occurring in the interior of acritarch vesicles. Most of the palynomorphs are of marine origin, whereas terrestrially derived material is limited to a few spores and phytoclasts (tracheids and woody material of unknown derivation). This type of palynofacies is typical of an overall carbonaceous composition of the sedimentary rocks, associated with a usually lower preservation potential for figured organic matter. Material from the middle and upper Nedrow Member is richer in moderately to well-preserved palynomorphs.

In the basal and lower part of the Nedrow Member (Oak Corners Quarry and Cherry Valley), chitinozoans of *Angochitina* type (Figs 13–15), cf. *Eisenackitina* sp. (mostly fragments) and a few specimens of spherical chitinozoans assigned to the genus *Hoegisphaera*, namely *Hoegisphaera* cf. *H. glabra*, are present. Acritarchs are limited in number

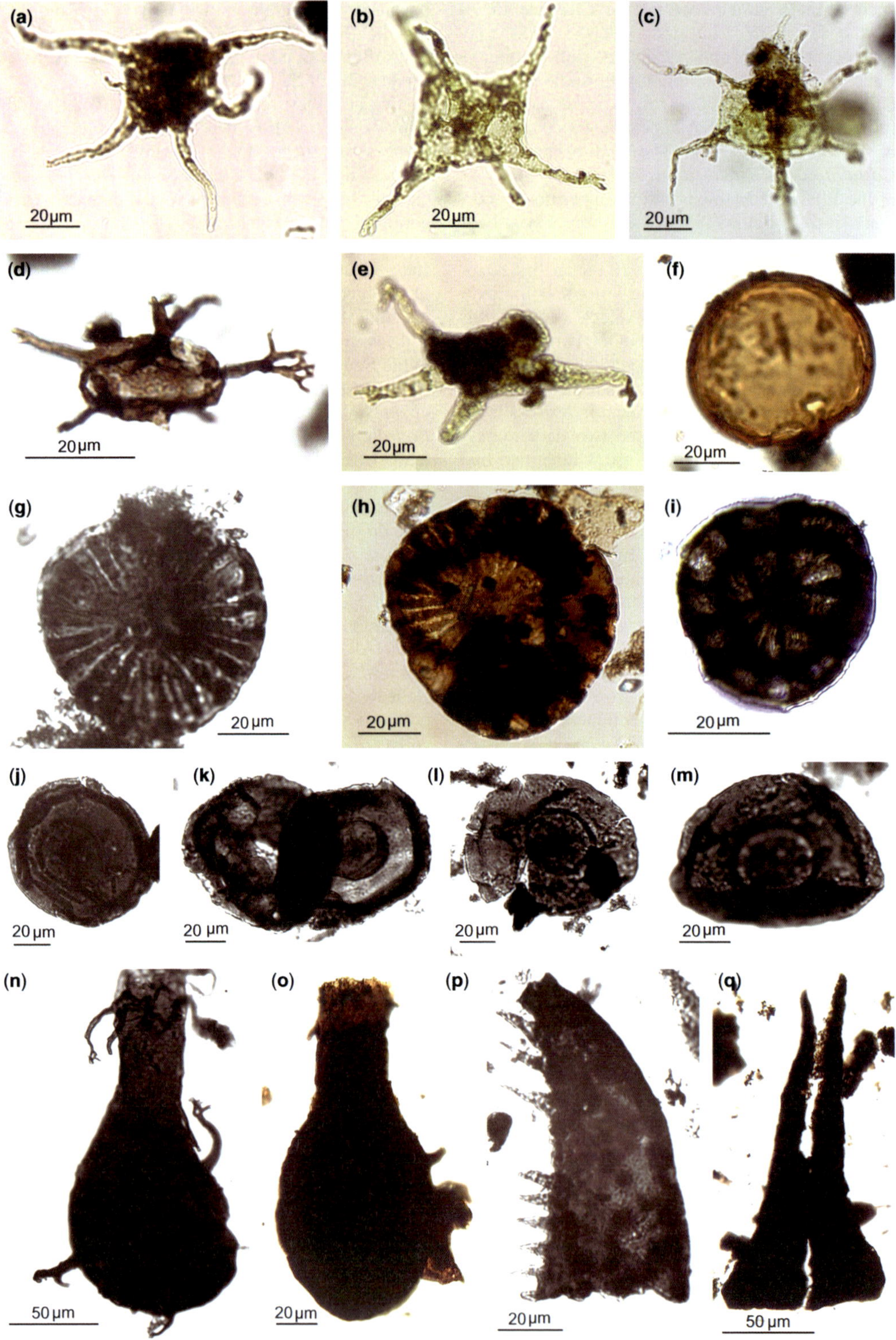
(a)
20 µm
(b)
20 µm
(c)
20 µm
(d)
20 µm
(e)
20 µm
(f)
20 µm
(g)
20 µm
(h)
20 µm
(i)
20 µm
(j)
20 µm
(k)
20 µm
(l)
20 µm
(m)
20 µm
(n)
50 µm
(o)
20 µm
(p)
20 µm
(q)
50 µm

and diversity, and are poorly preserved. Most of them are of simple morphology bearing short spines (e.g. ?*Winwaloeusia distracta* or *Gorgonisphaeridium* spp.). In addition, a few thick-walled prasinophytes of *Tasmanites* type and scolecodonts are present. The spores are sparse; specimens of ?*Retusotriletes* sp. and a few poorly preserved, sculptured trilete spores have been observed at only a few levels at Cherry Valley (sample Ned 1-2, 75 cm above the base of the Nedrow Limestone). In the same sample several chains (? conidia) of morphotypes related to the possible fungi-like *Reduviasporonites* cf. *R. stoschianus* are present (Fig. 14: for a discussion see the 'Systematic section' later in this paper).

In the middle Nedrow of the Stafford Quarry section (sample A4-1), a different picture appears regarding the preservation of the organic matter in general, and in particular of the frequency and diversity of the OWM. The maturity of the organic matter at this locality is somewhat lower compared with that from Cherry Valley or Oak Corners Quarry. It is indicated by a pale to yellow colour of acritarchs, and a light–middle brown colour of prasinophytes, spores and zooclasts. The assemblage of sample A4-1 is characterized by moderately preserved marine and terrestrial elements. Acritarchs are the prevalent components. Among them *Navifusa bacilla*, *Leiofusa estrecha*, *Multiplicisphaeridium* cf. *M. ramusculosum*, *Diexallophasis* spp., *Ozotobrachion furcillatus*, *Cymatiosphaera cornifera*, *Dictyotidium* sp., *Gorgonisphaeridium* sp., *Baltisphaeridium* cf. *B. distentum* sensu Playford, 1977, *Polyedryxium* spp. and cf. *Hapsidopalla chela* are present to abundant. Thick-walled specimens of *Tasmanites* are frequent. Spores are represented by a few specimens of *Verrucosisporites* cf. *V. polygonalis*, *Emphanisporites annulatus* (Fig. 13h), and other altered specimens of *Emphanisporites* and unclassified trilete forms. *Reduviasporonites* cf. *R stoschianus* is, beside the acritarchs, the most common taxon; it occurs as conidia-like chains. Also present are thick-walled dark hyphae (e.g. Fig. 14t). Cuticles of animal origin (zooclasts) exist sporadically.

In the upper Nedrow of Stafford Quarry (= at the position of the Nedrow black beds) palynomorphs are less numerous, but the preservation is moderate–good: acritarchs, such as *Diexallophasis simplex* (Fig. 13a), *D. remota* (Fig. 13b) and *Tasmanites* sp. (Fig. 13f), are predominant. *Navifusa bacilla*, *Polyedryxium pharaonis* (Fig. 13c), cf. *Exochoderma arca* (Fig. 13e), scolecodonts and spores occur occasionally, whereas some hyphae could represent *Reduviasporonites* (e.g. Fig. 14d, e). Some black palynomorphs are clearly filled by pyrite (no transparency when infrared microscopy is applied), others may point to reworking as they

Fig. 13. Palynomorphs of the Appalachian Basin. I. Photographs taken in transmitted light and with applied infrared video technique (IR). Indicated are localities, stratigraphic position, sample/slide numbers, England Finder (E.F.) coordinates and identification numbers (PMP) of the palynological collection at Senckenberg. (**a**) cf. *Diexallophasis simplex* Wicander & Wood, 1981, specimen with near-homomorphic unbranched, microgranulate processes; Stafford Quarry, west of Stafford, NY, shale at the Nedrow–Moorehouse contact (=the Nedrow black bed); A4-2-2, E.F. T 60-4, PMP 384. (**b**) *Diexallophasis remota* (Deunff) Playford, 1977; Stafford Quarry, NY, shale at the Nedrow–Moorehouse contact (=the Nedrow black bed); A4-2-2, E.F. T 37-1, PMP 384. (**c**) *Polyedryxium pharaonis* Deunff, 1961; Stafford Quarry, NY, shale at the Nedrow–Moorehouse contact (=the Nedrow black bed); A4-2-2, E.F. K 31-1, PMP 384. (**d**) *Multiplicisphaeridium ramusculosum* (Deflandre) Lister, 1970; Spring Gap roadcut, Maryland, close to basal Moorehouse-equivalent; E/E.11-1-13-1, E.F. H 59-2, PMP 737. (**e**) cf. *Exochoderma arca* Wicander & Wood, 1981; Stafford Quarry, NY, shale at the Nedrow–Moorehouse contact (=the Nedrow black bed); A4-2-2, E.F. J 43-2, PMP 384. (**f**) *Tasmanites* sp.; Stafford Quarry, NY, shale at the Nedrow–Moorehouse contact (=the Nedrow black bed); A4-2-2, E.F. W 44-2, PMP 384. (**g**) *Emphanisporites rotatus* McGregor emend. McGregor, 1973; Spring Gap roadcut, Maryland, basal–lower Nedrow-equivalent, IR; E/E.11-1-7A-2, E.F. Q 36-1, PMP 728. (**h**) *Emphanisporites annulatus* McGregor, 1961; Stafford Quarry, NY, middle Nedrow Member; A4-1-2, E.F. C 33-4, PMP 383. (**i**) *Emphanisporites annulatus* McGregor, 1961; Spring Gap roadcut, MD, basal–lower Nedrow-equivalent, IR; E/E.11-1-7A-2, E.F. Q 36-1, PMP 728. (**j**) *Hoegisphaera* cf. *H. glabra* Legault, 1973*a*, *b*; Keyser, WV, Middle Needmore Formation, approximate base of the Edgecliff-equivalent; E/E07.4-3; E.F. V 49-2, PMP 334. (**k**) Two specimens of *Hoegisphaera* cf. *H. glabra* Legault, 1973*a*, *b* with fragmentary rest of membrane (indicated by arrows); Hayfield, VA, approximate middle Nedrow-equivalent (below LBB), IR; E/E.11-4-12-1, E.F. F 53-1, PMP 715. (**l**) *Hoegisphaera* cf. *H. glabra* Legault, 1973*a*, *b*; Hayfield, VA, approximately middle Nedrow-equivalent (below LBB), IR; 11-4-12-1, E.F. M 60-3, PMP 715. (**m**) *Hoegisphaera* cf. *H. glabra* Legault, 1973*a*, *b*; Hayfield, VA, lower–middle Nedrow-equivalent, IR; E/E.11-4-13-2, E.F. P 43-4, PMP 716. (**n**) *Ancyrochitina* cf. *A. lezaisensis* Paris, 1981, with ring-like termination of processes (arrows); Spring Gap roadcut, MD, upper Edgecliff-equivalent, IR; E/E.11-1-5-2, E.F. K 64-4, PMP 726. (**o**) *Angochitina* sp.; Oaks Corners Quarry, NY, low in the Nedrow Member (base?); A4-9-2, E.F. X 56-2, PMP 386. (**p**) Part of the first maxillar of paulinitid or kielanoprionid scolecodont; Hayfield, VA, lower–? middle Nedrow-equivalent, IR; E/E.11-4-11-1, E.F. L 55-3, PMP 712. (**q**) Pair of carriers of ?paulinitid scolecodont; Hayfield, VA, basal Nedrow-equivalent; E/E.11-1-8-1, E.F. N 39-4, PMP 639.

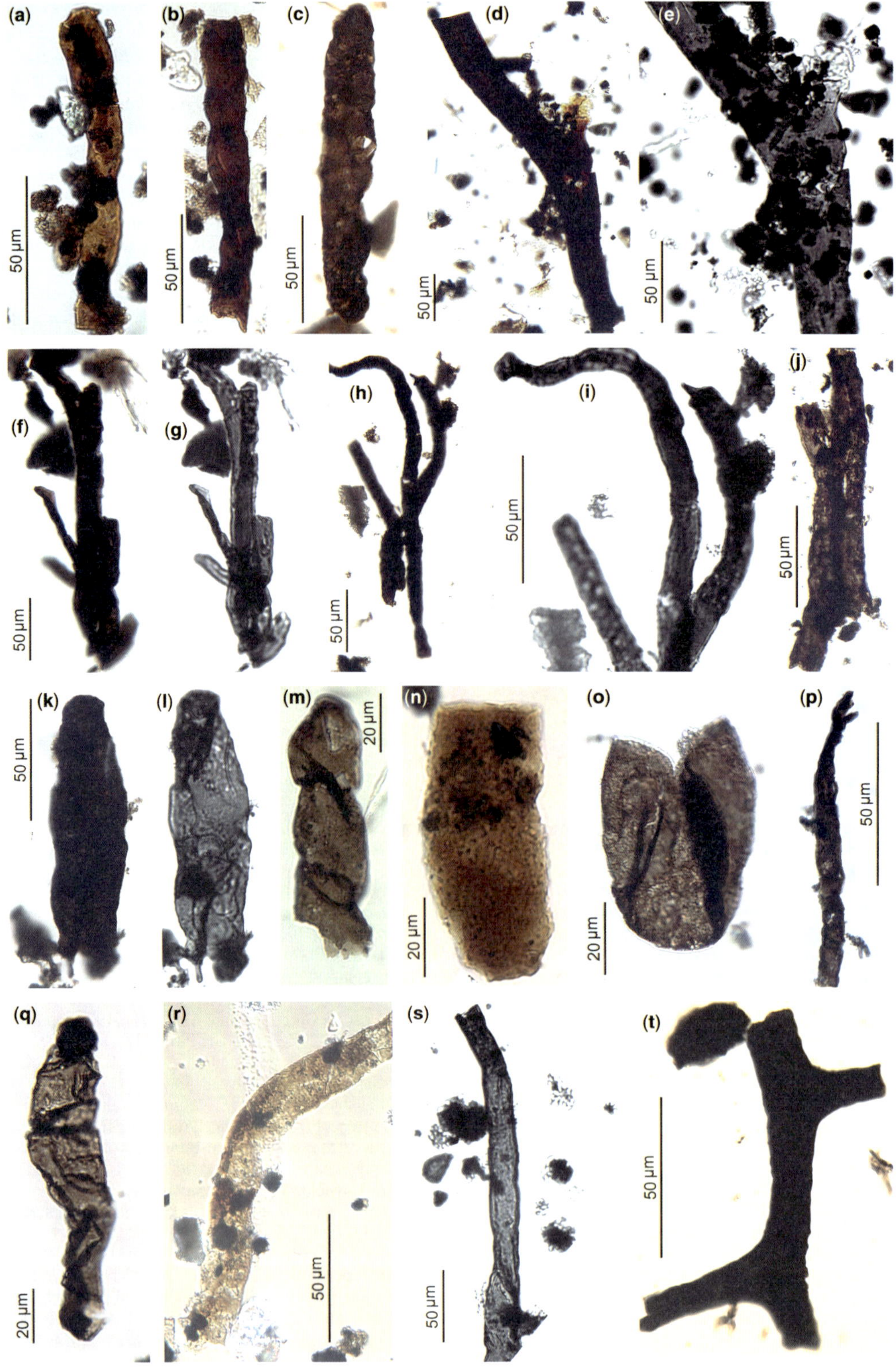
(a)
50 µm
(b)
50 µm
(c)
50 µm
(d)
50 µm
(e)
50 µm
(f)
50 µm
(g)
(h)
50 µm
(i)
50 µm
(j)
50 µm
(k)
50 µm
(l)
(m)
20 µm
(n)
20 µm
(o)
20 µm
(p)
50 µm
(q)
20 µm
(r)
50 µm
(s)
50 µm
(t)
50 µm

are darker compared to the majority of the pale to brownish palynomorphs.

The Oak Corners Quarry assemblage in the mid-Edgecliff-equivalent is even poorer in OWM, but, in contrast to the Stafford Quarry, terrestrial plant remains (tracheids) and spores are more common; cuticles of zooclasts and tubular filaments (? hyphae of cf. *R. stoschianus*: Fig. 14s) are present in low numbers. Acritarchs and prasinophytes are more or less absent.

At the Schooley Quarry, a sample from the UBB (00-03) yields mainly small-sized sculptured acritarchs, such as *Gorgonisphaeridium* sp., and other unidentified forms with short processes. Phytoclast, spores, degraded prasinophytes and zooclasts are rarely present. This sample and the one from the LBB (00-04) are rich in organic matter, but its high degree of alteration hampers the identification of the palynomorphs.

To summarize, studies of palynological assemblages from the NAB focused largely on the Nedrow Member. In the lower Nedrow, they are mainly characterized by the presence of acritarchs, whereas chitinozoans and spores are less frequent. However, the chitinozoan *Hoegisphaera* cf. *H. glabra* is also present in low numbers. *Hoegisphaera glabra* has previously been reported from higher parts of the Eifelian Columbus and Delaware limestones of Ohio (Wright 1976, 1978). Wright (1980) described a somewhat similar composition from the Middle Devonian of Indiana, but there the typical *H. glabra* occurs higher in the Hamilton Group (i.e. in the Givetian). Other reports of *H. glabra* in the Givetian are from Kentucky (Wood & Clendening 1985) and Iowa (Urban 1972; Urban & Newport 1973; Wicander & Wood 1997). Acritarchs in the middle Nedrow Member are diverse and represented by forms with longer processes (e.g. *Diexallophasis* spp., *Exochoderma arca* and *Polyedryxium pharaonis*), both indicative of open, normal marine, shelf conditions. This is also consistent with the relatively rare occurrence of spores and phytoclasts. *Tasmanites* is abundant in some levels of the middle and upper Nedrow Member; the same is true for *Reduviasporonites* cf. *R. stoschianus*. The proliferation of both taxa is in accordance with the concept of an ecological epibole, as defined by Brett & Baird (1997).

Stratigraphically important is the occurrence of *Emphanisporites annulatus*; its range is from the Emsian to the Givetian worldwide, but it is also typical of the Eifelian in North America (e.g. McGregor & Camfield 1982; Ravn & Benson 1988; Traverse & Schuyler 1994).

Central part of the Appalachian Basin (CAB)

Dacryoconarids. In the CAB, the dacryoconarid fauna of the mid-Edgecliff-equivalent strata of the Selinsgrove Member includes at least four Old World taxa (Figs 10–12) that have not previously been reported from the Eastern Americas Realm. All occur upsection in the Nedrow-equivalent to the lower of the two Nedrow black beds and two also occur in the UBB, which marks the top of the BCE interval. Only *Nowakia* (*Dmitriella*) *sulcata* cf. *N.* (*D.*) *s. antiqua* has been observed higher in the section as a member of the Stony Hollow Event fauna at the base of the Oatka Creek Formation at Cherry Valley, New York.

Although the Emsian dacryoconarid fauna of the CAB has yet to be described formally, it has been studied in sufficient detail to state with certainty that the BCE dacryoconarid fauna of the CAB was

Fig. 14. Palynomorphs of the Appalachian Basin. II. Morphotypes assigned to the species cf. *Reduviaspronites stoschianus* (Figs 1–3, 6–13, 16 & 17) and other unidentified palynomorphs mostly assigned to fungi from the upper Edgecliff Member and Nedrow Member, and equivalent strata in the Selinsgrove Member, Needmore Formation, in studied localities, Appalachian Basin, USA. Photographs were taken in transmitted light and with applied infrared video technique (IR). Indicated are locality, stratigraphic position, sample/slide numbers, England Finder (E.F.) coordinates and identification numbers (PMP) of the palynological collection at Senckenberg. (**a**) Stafford Quarry, east of Stafford, NY, middle Nedrow Member; A4-1-2, E.F. H 36-2, PMP 383. (**b**) Stafford Quarry, NY, middle Nedrow Member; A4-1-2, E.F. D 48-3, PMP 383. (**c**) Hayfield, VA, boundary interval between Edgecliff and Nedrow-equivalents; E/E.11-1-7A-2, E.F. S 32-4, PMP 729. (**d**) Stafford Quarry, NY, middle Nedrow Member; A4-1-1, E.F K 37-4, PMP 383. (**e**) Same specimen in IR showing channel-like structure in the centre of the object. (**f**) Spring Gap roadcut, MD, lower Nedrow-equivalent; E/E.11-1-7-1, E.F. S 32-4, PMP 729. (**g**) Same specimen in IR, showing internal structures such as segmentation? (**h**) Spring Gap roadcut, MD, lower Nedrow-equivalent; E/E.11-1-7A-2, E.F. Q 36-1, PMP 728. (**i**) Same specimen in IR, showing internal structures. (**j**) Spring Gap roadcut, MD, LBB; E/E.11-1-8A-1, E.F. D 61-3, PMP 731. (**k**) Spring Gap roadcut, MD, upper Edgecliff-equivalent; E/E.11-1-5-4, E.F. D H 56-2, PMP 726. (**l**) Same specimen in IR, showing vague segmentation possibly indicating cells. (**m**) Cherry Valley, NY, lower Nedrow Member; Ned. 1-2, E.F. Y 52-4. (**n**) Stafford Quarry, NY, middle Nedrow Member; A4-1-1, E.F. L 63-4. (**o**) Spring Gap roadcut, MD, upper Edgecliff-equivalent; E/E.11-1-5-2, E.F. K 64-4, PMP 726. (**p**) Spring Gap roadcut, MD, upper Edgecliff-equivalent; E/E.11-1-2-2, E.F. S 40-2, PMP 729. (**q**) Hayfield, VA, basal Nedrow-equivalent; E/E.11-1-8-1, E.F. N 39-4, PMP 639. (**r**) Photograph with interference contrast; Stafford Quarry, NY, middle Nedrow Member; A4-1-2, E.F C 44-4, PMP 383. (**s**) IR photography; Stafford Quarry, NY, middle Nedrow Member; A4-1-1, E.F. H 38-2, PMP 383. (**t**) Hayfield, VA, lower Nedrow-equivalent; E/E.11-4-11-1, E.F. L 55-3, PMP 712.

not derived from dacryoconarids that occur in either the subjacent Edgecliff-equivalent strata or below that in the upper Emsian Schoharie-equivalent strata. The Nedrow-equivalent dacryoconarid fauna consists of Old World Realm (OWR) species or their descendants, which immigrated into the CAB at the onset of the BCE and departed one way or another at the close of the event. One of these taxa immigrated to the NAB during the T–R Cycle Ic maximum flooding, which is the acme and termination of the BCE.

Organic-walled microfossils (OWM). Palynologically, two sections in the CAB – Spring Gap (Maryland) and Hayfield (Virginia) – have been analysed in great detail (Figs 15 & 16; Table 1); further samples have been studied from Gainesboro (Virginia), Keyser (West Virginia) and Mapleton (Pennsylvania).

The Spring Gap section reaches from Schoharie Formation-equivalents to equivalents of the Edgecliff, Nedrow and lower Moorhouse members, and is represented by 18 palynological samples (Fig. 15). In addition, two samples from the Nedrow black beds had previously been studied. In general, samples are characterized by a comparatively high content of organic matter with generally poor preservation and an apparent higher level of maturity compared to the NAB. Palynomorphs are mainly dark brown to opaque and often determinable only when infrared microscopy (IR) is applied. However, some levels yield material of better preservation with light–medium brown palynomorphs. Samples from the Schoharie Formation-equivalent are very poor in palynomorphs or are even barren. Only some remains of spores (retusotrilete forms), chitinozoans, *Navifusa* sp. and possible phytoclasts have been detected. The first good record comes from the transition to the Edgecliff-equivalent (sample E/E.II-1-4). This sample yields few phytoclasts, small acritarchs of the *Multiplicisphaeridium* type and a few specimens of cf. *Reduviasporonites stoschianus*. The next sample up in the section, E/E.II-1-5, is quite rich in spores (e.g. *Apiculiretusispora plicata*, *Retusotriletes rotundus*, *Emphanisporites annulatus*, *E. rotatus* and several other undetermined taxa); rarely present are phytoclasts, acritarchs, such as *Multiplicisphaeridium* sp., specimens of *Dictyotriletes* sp. and chitinozoans, such as ?*Ancyrochitina* cf. *A. lezaisensis* (Fig. 13n). Sample E/E.II-1-6 displays poorly preserved spores and a few cf. *R. stoschianus* (only assignable by IR method) and no marine palynomorphs so far. Samples E/E.II-1-7a, E/E.II-1-7b and E/E.II-1-7c in the mid-Nedrow-equivalent show the highest number of specimens and a better preservation of palynomorphs (pale to medium brown). Some of them (mainly spores) are darker, perhaps owing to reworking. Prevalent are cf. *R. stoschianus* (? conidia and hyphae) and *N. bacilla;* spores, such as *E. annulatus* (Fig. 13i), *E. rotatus* (Fig. 13g), *Retusotriletes* spp. and *Apiculiretusispora* sp., and chitinozoans (*Angochitina* spp., ?*Ancyrochitina*)

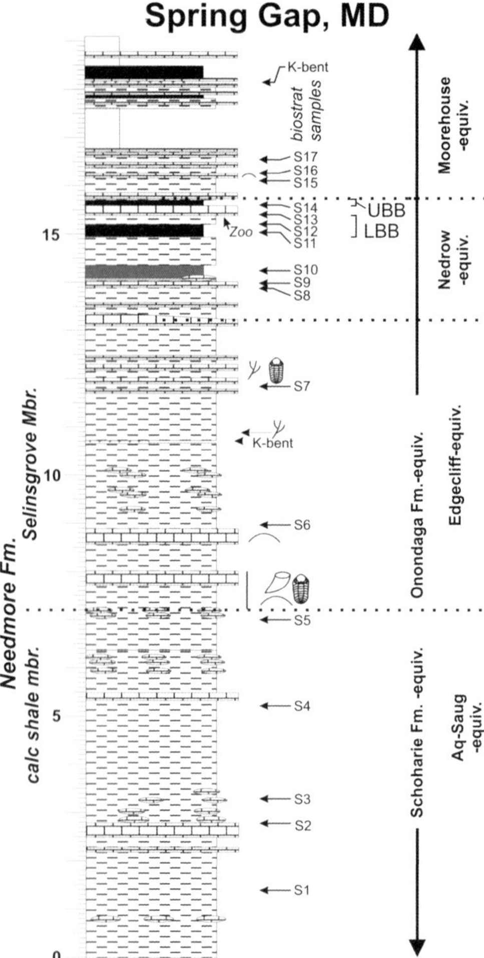

Fig. 15. Details of section at Spring Gap, MD (north side of Maryland Route 51). The upper part of the informal calcareous shale member and the Selinsgrove Member, Needmore Formation. Correlations with New York members are denoted on the far right. Note the Nedrow black beds as two black bands at and above 15 m. Arrow-heads without a line denote altered airfall volcanic tephra layers (K-bentonites). Palynology samples marked as 'E/E.11 …'; dacryoconarid samples marked as 'E/E.07 …'. The key to the symbols is given in Figure 6; for levels of samples, see Table 1. Aq-Saug, Aquetuck and Saugerties members of New York's Schoharie Formation; equiv., equivalent; LBB, lower Nedrow black beds; UBB, upper Nedrow black beds; *Zoo*, heavily bioturbated bed of *Zoophycos* traces.

are common. Acritarchs are represented by species of the genera *Diexallophasis* and *Polyedryxium*. Sample E/E.II-1-8 from the LBB is rich in organic matter, but palynomorphs are difficult to identify owing to poor preservation. Samples were additionally treated in an ultrasonic bath in order to clear them from amorphous organic matter (AOM). The most common forms are phytoclasts and cf. *R. stoschianus* (mainly hyphae up to 500 μm in length). The presence of zooclasts is likely. The UBB (E/E.II-1-9 and additional sample E/E07-5A) shows a similar picture, but spores (cf. *Grandispora*, but mostly unidentified), chitinozoan remains (cf. *Ancyrochitina*) and a few acritarchs, such as *Veryhachium polyaster*, *N. bacilla* and *Micrhystidium* spp., occur occasionally. Forms of cf. *Reduviasporonites* are common. The next prolific samples (E/E.II-1-11–E/E.II-1-13) are already located in the lower Moorehouse-equivalent. The palynospectrum is characterized by a dominance of AOM, phytoclasts, questionable spores and probable hyphae of cf. *R. stoschianus*. Marine components are represented by acritarchs, such as *Veryhachium* and *Multiplicisphaeridium ramusculosum* (Fig. 13d), chitinozoans, such as *Hoegisphaera* cf. *H. glabra*, scolecodonts in higher numbers and zooclasts.

The Hayfield roadcut section comprises upper Schoharie Formation-equivalent, Edgecliff-equivalent, Nedrow-equivalent, Moorhouse-equivalent and Seneca-equivalent strata (the last not sampled for palynomorphs). The studied interval is represented by 22 palynological samples (Fig. 16).

Samples from the upper Schoharie Formation-equivalent are poor in palynomorphs, but cf. *Reduviasporonites* and few spores have been identified. One of them recalls a specimen of *Rhabdosporites*; others belong to unidentified apiculate and zonate/pseudosaccate forms. The next prolific sample is from the basal Edgecliff-equivalent (E/EII-4-03), which bears chitinozoan fragments of possibly *Alpenachitina eisenacki* origin, scolecodonts, spores (*Acinosporites* cf. *A. lindlarensis*) and, further, unidentified opaque specimens (due to reworking?). Sample E/E.II-4-05 from a shaly interval in the Edgecliff-equivalent yields spores and *N. bacilla*; other acritarchs are rare; cf. *R. stoschianus* is sporadically present. Samples from the top of the Edgecliff-equivalent (E/E.II-4-07 and E/E.II-4-07/08) are quite rich in acritarchs (*Diexallophasis* spp., *Veryhachium* sp., cf. *Hapsidopalla chela*, *N. bacilla* and *Multiplicisphaeridium* spp.). Spores are represented by aff. *Dibolisporites* sp.; chitinozoans are preserved as fragments only. Samples E/E.II-4-09 and E/E.II-4-10 from the basal Nedrow-equivalent up to the last sample before the Nedrow black beds interval display frequent occurrences of chitinozoans (e.g. *Angochitina* sp.)

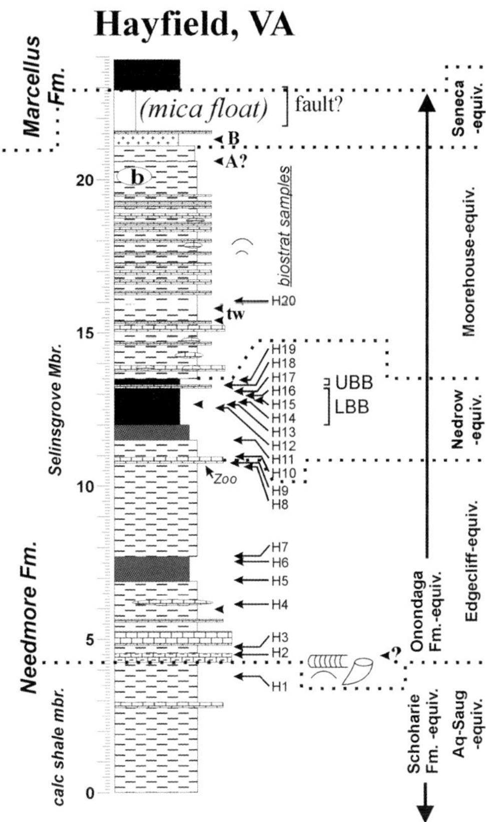

Fig. 16. Details of the section at Hayfield, VA (NE side at the intersection of US Route 50 and Virginia Route 600). The upper part of the informal calcareous shale member and the Selinsgrove Member, Needmore Formation. Correlations with New York units are denoted on the far right. The key to the symbols is given in Figure 6; for levels of samples, see Table 1. Nedrow black beds are seen in the middle of the section. An additional section below this is visible on western side of Route 600. Palynology samples marked as 'E/E.11 . . .'; dacryoconarid samples marked as 'E/E.07 . . .'. Symbols and abbreviations as in Figures 5 and 15.

and scolecodonts; some of them are comparatively large (up to 800 μm) in contrast to specimens found in older assemblages. In addition, few spores and questionable *Reduviasporonites* forms are present. Samples from the Middle Nedrow-equivalent (E/E.11-4-11–E/E.11-4-13) up to the LBB (E/E.II-4-11-15, E/E07-1b-16 and E/E.11-4-17) are distinct from most previous ones by their general high content of organic matter and significant palynological composition. Prevalent are *Hoegisphaera* cf. *H. glabra* (Fig. 13k, m) associated with aff. *Alpenachitina eisenacki* (mostly dark fragments) and large-sized scolecodonts (Fig. 13p, q).

Table 1. *Sample numbers and their levels in sections Hayfield (WV) and Spring Gap (MD) as indicated in Figures 15 and 16*

Biostratigraphy samples, Hayfield, WV	
H20	E/E.II-4-14 & E/E.07-1b-16 (12.8–12.85 m: LBB)
H19	E/E.II-4-15 (13.1–13.2 m: LBB)
H18	E/E.II-4-16 (13.3–13.35 m: UBB)
H17	E/E.II-4-17 (13.43–13.5 m: UBB)
H16	E/E.07-1b-17 (12.95–13.1 m: LBB)
H15	E/E.II-4-18 (16.05–16.15 m)
H14	E/E.II-4-13 (12.72–12.78 m: LBB)
H13	E/E.II-4-12 (12.65–12.7 m: LBB)
H12	E/E.II-4-11 (12.50–12.55 m: LBB)
H11	E/E.II-4-10 (11.50–11.62 m)
H10	E/E.II-4-09 (10.95–11.05 m: base of Nedrow-equivalent)
H9	E/E.II-4-08 (10.75–10.9 m: top Edgecliff-equivalent)
H8	E/E.II-4-07 (10.62–10.71 m)
H7	E/E.II-4-06 (7.7–7.75 m)
H6	E/E.07-1B-14 (7.55–7.6 m)
H5	E/E.II-4-05 (6.94–7.0 m)
H4	E/E.II-4-04 (6.15–6.2 m)
H3	E/E.II-4-03 (4.75–4.8 m)
H2	E/E.II-4-02 (4.5–4.6 m)
H1	E/E.II-4-01 (3.8–3.9 m)
Biostratigraphy samples, Spring Gap, MD	
S17	E/E.II-1-13 (16.5–16.6 m)
S16	E/E.II-1-4 (7.0–7.1 m)
S15	E/E.II-1-3 (3.3–3.4 m)
S14	E/E.II-1-3b (5.23–5.3 m)
S13	E/E.II-1-2 (2.8–2.9 m)
S12	E/E.II-1-1 (1.4 m)
S11	E/E.II-1-5 (8.9–9.0 m)
S10	E/E.II-1-6 (11.8–11.9 m)
S9	E/E.II-1-7a (14.0–14.1 m)
S8	E/E.II-1-7 (14.25–14.3 m)
S7	E/E.II-1-8 & E/E.07-5-7 (15.05–15.15 m)
S6	E/E.II-1-9 (15.22–15.3 m)
S5	E/E.II-1-10a (15.40–15.55 m)
S4	E/E.II-1-11 (16.1–16.15 m)
S3	E/E.II-1-10 & E/E.07-5-8 (15.6–15.7 m)
S2	E/E.II-1-12 (16.27–16.3 m)
S1	E/E.II-1-7b (13.88–13.98 m)

Spores are represented by *Apiculiretusispora* sp. and other unidentified retusotrilete forms. Hyphae of cf. *Reduviasporonites* occur in lower number whereas acritarchs are very rare; only few small specimens of *Micrhystridium* have been observed.

Upper Nedrow black bed (UBB) assemblages (E/E.II-4-16/17; E/E07-1b-19) are different from those of the LBB. Acritarchs, such as *Triangulina alargada*, *Veryhachium polyaster* and *Micrhystridium* spp., are more frequent, as is *Tasmanites* sp.; chitinozoans and scolecodonts are rare. Terrestrial material is represented by small trilete spores (e.g. cf. *Aneurospora minuta*) and phytoclasts.

Our highest sample E/E.II-4-18 from the basal Moorehouse-equivalent is less productive than those from the black beds. Few acritarchs, chitinozoans and prasinophytes have been found along with spores, whereas those species typical for the black beds (*Hoegisphaera* cf. *H. glabra* and cf. *R. stoschianus*) are absent.

From the Gainsboro section (Virginia), two samples representing the LBB and the UBB have been investigated. Sample E/E07-2-6 from the LBB yields hyphae of cf. *Reduviasporonites*. Phytoclasts are common, trilete spores are present, but are highly altered or too dark – even using the IR method – to assign them systematically.

The sample from the UBB (E-E07-2-8) displays a much better preservation of the OWM. Acritarchs are the dominant group; common are *Hapsidopalla chela* and *Micrhystridium* spp. Thin-walled prasinophytes (*Leiosphaeridia* sp.), cf. *Reduviasporonites stoschianus* and scolecodonts, as well as spores, such as *Retusotriletes* sp., cf.

Cymbosporites proteus and *Dibolisporites* sp., are frequent.

Additional samples are available from the Keyser section (West Virginia), but only a few have been studied in detail. A comprehensive analysis will be the matter of future investigations, including the application of SEM studies. In general, the overall impression of the palynological composition of the entire section is similar to that from Hayfield. However, in Keyser, morphotypes of *Hoegisphaera* occur earlier, close to the base of the Edgecliff-equivalent, just below the base of Devonian Sequence Ic (sample E/E 07.04.3: Fig. 13j). In consequence, at least the genus *Hoegisphaera* is not restricted to the middle and upper Nedrow, in particular to the black beds interval. Subsequent taxonomical studies on these forms will be directed to resolve a possible conspecific relationship to *H.* cf. *H. glabra*.

The LBB sample from the Mapleton Quarry (Pennsylvania) is very rich in AOM, but only very few figured specimens (possibly algae and spores) can be identified.

CAB samples from the boundary interval between the Edgecliff and Nedrow units up through the entire Nedrow Member-equivalents (including the black beds) show, in parts, a significantly large number of cf. *Reduviasporonites stoschianus* and *Hoegisphaera* cf. *H. glabra*, interpreted here as an ecological epibole in the respective succession. Those samples are often accompanied by large numbers of larger scolecodonts. Chitinozoans, acritarchs and spores vary in frequency and diversity in the entire Nedrow-equivalent, but, in the upper Edgecliff-equivalent, acritarchs can be abundant and diverse. The increase in chitinozoan diversity in some levels indicates a 'deeper-water', more basinal depositional setting. Stratigraphically important acritarchs are *Triangulina alargada* and *Hapsidopalla chela*, the latter is common in the Eifelian–lower Givetian of the Silica Formation of Ohio (e.g. Wicander & Wood 1981). Among the spores, *Emphanisporites annulatus* and *E. rotatus* and specimens tentatively assigned to the genera *Grandispora* and *Rhabdosporites* indicate a late Emsian–early Eifelian age.

Interpretation

This paper represents the first comparison of the Basal Choteč Event (BCE) interval between the Barrandian type area and the Appalachian Basin. As the focus clearly was set on the palynomorphs and the dacryoconarids (a comparison of these organisms has never been carried out before), other faunal elements and abiotic features have only been considered when additional information was needed (Fig. 17). Therefore, conclusions on the two groups of organisms are given in some detail; for the dacryoconarids, formal descriptions are added – eight of them represent new taxa for the AB.

Palynological studies from the Na Škrábku Quarry reveal a distinct palynological assemblage of the bed, which represents the level of the BCE. The palynological composition of this characteristic bed (sample P 5) is dominated by prasinophycean phycomata up to 600 μm in diameter. In addition, cyanobacterial agglomerations and fungi-like microfossils are present; other palynomorphs occur in very low numbers. Because this exceptional proliferation is restricted to this bed, we conclude that it represents an ecological epibole, as redefined by Brett & Baird (1997). It is widely accepted that prasinophytes are able to tolerate specific ecological conditions compared to normal marine phytoplankton, and are able to form blooms when, for example, acritarchs retreated or disappeared. When occurring in the geological record, such prasinophytes are frequently called 'disaster species', a term introduced by Tappan (1980) and subsequently applied by several authors (e.g. Riegel *et al.* 1986; Van de Schootbrugge *et al.* 2007). Causes for phytoplankton blooms are still speculative. Often such mass occurrences are explained by a temporal stratification of the water column, (e.g. by lowered salinity induced by the input of freshwater and/or restricted ocean circulation). Another possible cause is higher nutrient load (e.g. nitrogen, iron, phosphate) derived from land by fluvial or aeolian transport or as a result of marine upwelling. However, green algae, such as the prasinophytes, are able to use reduced nitrogen (i.e. ammonium) much more effectively than other algal groups, especially under conditions of black shale deposition (Prauss 2007; Riegel 2008). In this case the blooming of these opportunistic algae at the BCE might be related to enhanced nutrient supply coming from land (fluvial and/or aeolian), together with the establishment of a pycnocline induced by lower salinity from freshwater input during maxiumum transgression. In the BCE sample P 5, and particularily in the succeeding samples P 6 and P 7, the possibly freshwater tolerant genus *Cymatiosphaera* is common. Since it was also reported from the Lower Devonian freshwater ecosystem of the Rhynie chert (Dotzler *et al.* 2007), a freshwater influence seems likely. Mazuelloids, which co-occur in the same sample, are considered to reflect specific environmental conditions (e.g. nutrient-rich waters); their appearance as blooms is also associated with black shales (Kremer 2005).

In the Appalchian Basin, the interval of the BCE does not show such a distinctive bloom of prasinophytes, as is revealed in the Barrandian section. However, a higher abundance of these algae exists

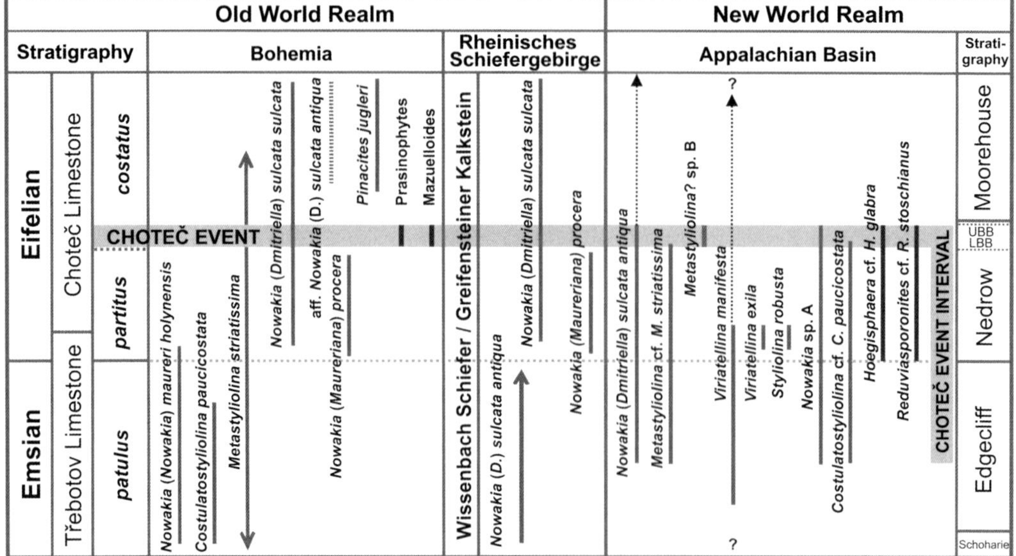

Fig. 17. Ranges of selected dacryoconarid and palynomorph taxa connected to the BCE, not to scale. In the Appalachian Basin, endemic forms exist besides the given species, see Figure 10. In the NAB, the last occurrence of *Nowakia* (*D.*) *sulcata antigua* is recorded from the Chestnut Street Bed (base of the Oatka Creek Formation), and the supposed (?) last occurrence of *Viriatellina manifesta* from the Bakoven Member. According to the dacryoconarids, an interval for the BCE ranging from the mid-Edgecliff Member to the top Nedrow Member can be identified. However, a culmination/acme in the black beds of the Nedrow Member is obvious by the restricted occurrence of *Metastyliolina*? sp. B.; cf. *Reduviasporonites stoschianus* and the chitinozoan *Hoegisphaera* sp. cf. *H. glabra* are present in the entire Nedrow Member, but seem to also prevail in the black beds. However, in the Na Škrábku section of the Barrandian area, this event is apparently restricted to a single bed (=palynological sample P 5, compare Figs 4a & 9) close to the base of the Choteč Limestone.

in the Nedrow Member of the NAB, although the relevant assemblages are not exclusively formed by prasinophytes. Here, they are accompanied by the appearance of comparatively high proportions of cf. *Reduviasporonites stoschianus* and *Hoegisphaera* cf. *H. glabra*. In the CAB, prasinophytes are present, but not abundant; dominant are cf. *R. stoschianus* and *H.* cf. *H. glabra*.

Proliferations of fungi or microfossils attributed to fungi have been frequently reported from different time periods and areas, and are usually interpreted to reflect ecological crisis – they clearly do co-occur with various bioevents. Well known is the Permian–Triassic Extinction Event, in which a fungal spike is considered to be one indicator of a disrupted ecosystem following the mass extinction (e.g. Visscher *et al.* 1996, 2011; Bercovici *et al.* 2015). Such fungi were also found in higher numbers in the Pennsylvanian of Peru (Wood & Elsik 1999) and, more recently, from the Cretaceous–Palaeogene boundary (e.g. Vajda & Bercovici 2014) and the Oligocene–Miocene boundary in Egypt (El Atfy *et al.* 2013). In the latter, the fungal peak co-occurs with freshwater algae.

Mass occurrences of chitinozoans are little known, but have been reported from the Kellwasser Crisis interval near the Frasnian–Famennian boundary in France (Paris *et al.* 1996), in which Angochitininae bloom in the respective layers of the La Serre section (Montagne Noire). *Hoegisphaera* cf. *H. glabra* does not occur in masses, but is relatively abundant in the Nedrow Member or its equivalents, more or less simultaneously with cf. *R. stoschianus*, and has not been reported in significant numbers from underlying or overlying strata of our sections. We consider that both taxa characterize the BCE and interpret their proliferation as an ecological epibole.

While in the Na Škrábku Quarry (Prague Basin), the BCE is best seen in a distinct level (sample P 5), in the Appalachian Basin, the interval ranges probably from the uppermost Edgecliff Member and its equivalents up to the top of the Nedrow Member and its equivalents. However, there the UBB obviously represents the termination of the BCE.

From a stratigrapical point of view, spores, such as *Emphanisporites annulatus* and *E. rotatus*, and acritarchs, such as *Triangulina alargada*,

Hapsidopalla chela and *Ozotobrachion furcillatus*, are common from the upper Emsian to the Givetian. Such forms and related taxa may have the potential to refine the lower Eifelian. However, more detailed taxonomy, in particular by SEM studies of the prolific palynological assemblages, is required.

Dacryoconarids are an important tool in dealing with the globally detectable BCE. They are present in the Old World Realm (OWR: e.g. Barrandian area, German Rhenohercynian, and Saxothuringian Belts, Carnic Alps, North Africa) and in the North American New World Realm (NWR: e.g. Nevada, Appalachian Basin). Dacryoconarid occurrences on continents other than Europe and North America are beyond the palaeogeographical scope of this paper. A main point of the dacryoconarid studies presented here is to show – for a number of taxa for the first time – the relationships (or discrepancies) between the 'classical' areas of the OWR and regions within the Appalachian Basin.

In the classical regions of Central Europe, numerous dacryoconarid taxa are very good indicators of both the brief time interval of the BCE and the Lower–Middle Devonian (=Emsian–Eifelian) boundary. For the history and discussion of the boundary see, for example, Ziegler & Klapper (1985), Chlupáč & Kukal (1986), Walliser (1996), Ziegler (2000) and detailed references therein. Special aspects with respect to dacryoconarids are dealt with in Bouček (1964), Alberti (1985*a*, *b*, 1993), Lukeš (1989) and Chlupáč (1985, 1998). The classical sections of Holyně (in the Barrandian area: see Lukeš 1989; Chlupáč 1985, 1998) and Haiger Hütte (in the German Rhenohercynian: see Alberti 1985*b*, 1993) yielded many of the taxa mentioned in this paper. For the regional zonation in the Carnic Alps, see Alberti (1985*a*). Main 'actors' among the dacryoconarids are *Nowakia* (*Nowakia*) *maureri* (with the subspecies *N.* (*N.*) *m. maureri* and *N.* (*N.*) *m. holynensis*), *N.* (*N.*) *holyocera*, *N.* (*Dmitriella*) *sulcata* (with the subspecies *N.* (*D.*) *s. antiqua* and *N.* (*D.*) *s. sulcata*) and *N.* (*Maureriana*) *procera*. Taxa, such as *N.* (*N.*) *m. maureri* and *N.* (*N.*) *m. holynensis*, are characteristic of the interval immediately below and right at the Emsian–Eifelian boundary. The same holds true for *N.* (*N.*) *holyocera* and *N.* (*D.*) *s. antiqua*. In the OWR, two taxa are critical markers for the onset of the Eifelian stage: *N.* (*M.*) *procera* and *N.* (*D.*) *s. sulcata*, both of which are present in the Haiger Hütte section above the boundary. This is supported by Lütke (1985) for *N.* (*M.*) *procera* from the *costatus* Zone of Nevada.

Nowakia (*M.*) *procera* is of special interest for comparison with the AB because it seems to be very closely related to a new species (*Nowakia* sp. A) from the CAB covering (most probably) a similar time interval (see Figs 10 & 17). Currently, research on dacryoconarids of the AB (and correlation with the OWR) is in progress, but some remarkable conclusions can be drawn based on the present data (for details see the 'Systematic section' later in this paper). No Eifelian taxa have previously been described formally from the AB. Ranges of the dacryoconarids from the northern and central parts of the Appalachian Basin (AB) are shown in Figure 10. There is no doubt that the AB was occupied by two discrete faunas with disparate origins and histories. The fauna of the NAB evolved from endemic Emsian ancestors and, whereas new species first and last occur within the BCE, those that survived the crisis are the founders of much of the Hamilton Group dacryoconarids. However, the BCE fauna of the CAB is an immigrant fauna of conspecific or closely related Old World taxa, which supplanted the endemic Emsian fauna and, for the most part, became extinct during the acme of the BCE. The occurrence of these taxa is an incursion epibole *sensu* Brett & Baird (1997). The taxa have immigrated within the short time interval of the BCE to the CAB, possibly migrating SW from the Old World into the Appalachian Basin embayment (Fig. 18). As far as the dacryoconarids are concerned, the event interval extends from the middle Edgecliff Member of the Onondaga Formation to the LBB and UBB of the uppermost Nedrow Member strata, and their equivalents. We regard the LBB–UBB, where most dacryoconarids became extinct, to be the acme or culmination of the BCE.

Only two dacryoconarid taxa occur in both the NAB and the CAB, and both yield peculiarities. *Metastyliolina* sp. B is present in the NAB in the LBB and probably in the UBB; in the CAB, it is present exclusively in the LBB. Regardless of the questionable specimen in the UBB of the NAB, the taxon is an excellent index fossil for the upper part of the Nedrow Member and its equivalents, and even for the very short time interval of the widespread Nedrow black beds. The second taxon shared by the northern and central areas of the Appalachian Basin is *Nowakia* (*Dmitriella*) *sulcata antiqua*. In the CAB, it covers the interval from equivalents of the middle Edgecliff Member to the UBB of the Nedrow Member. In the NAB, however, it is present only in the Chestnut Street Beds of the Cherry Valley area: that is, in much younger strata of the basal Oatka Creek Formation, probably related to the acme of the Stony Hollow Event (compareVer Straeten & Brett 2006; Ver Straeten 2007; Brett *et al.* 2011; DeSantis & Brett 2011). There is, at present, no explanation for this late appearance in the NAB; however, the range of the taxon, including a late occurrence in the upper part of the Choteč Limestone in the Barrandian area (Alberti 1993), is similar to the OWR. *Nowakia* (*D.*) *s. antiqua* has not previously been

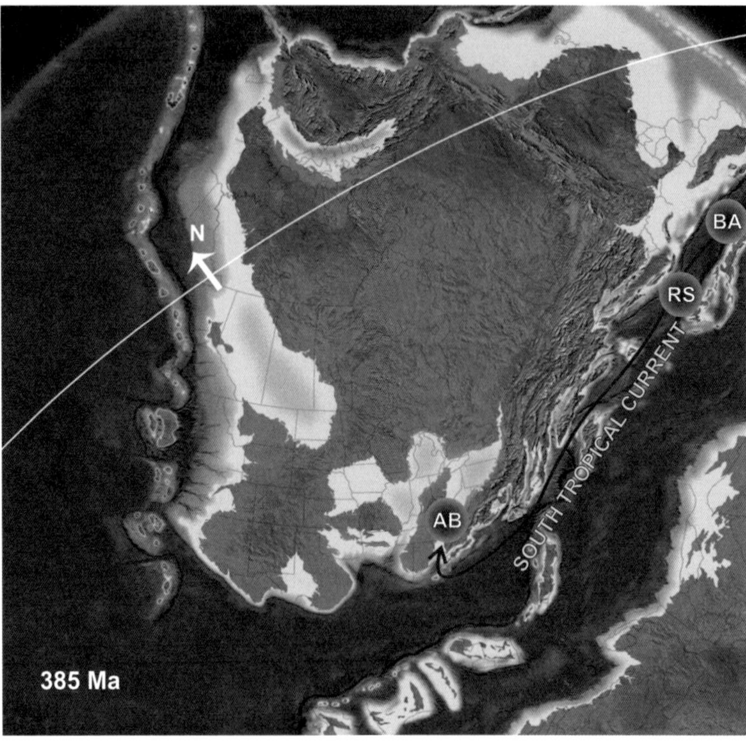

Fig. 18. Palaeogeographical reconstruction modified from Blakey (2007), with the suggested immigration path (red line) of dacryoconarid taxa from the OWR now recognized in the Appalachian Basin. BA, Barrandian area (Prague Basin, Czech Republic); RS, Rheinisches Schiefergebirge (Germany); AB, Appalachian Basin (eastern USA); south tropical current adopted from Wilde *et al.* (1991).

reported from the AB. All other taxa differ between the two subregions of the Appalachian Basin. However, they make very good index fossils and are partly conspecific with forms from the OWR (e.g. *Costulatostyliolina* cf. *paucicostata*). In the NAB, *Styliolina robusta* n. sp. and *Viriatellina exilia* n. sp. represent perfect index fossils for the lower part of the Nedrow Member. *Costulatostyliolina vestita* n. sp. and *Viriatellina manifesta* n. sp., as well as *Striatostyliolina vitta* n. sp. and *Striatostyliolina mima* n. sp., have a somewhat wider time range (for details see Figs 10 & 17).

In addition to the results derived from the two groups of fossil organisms, the presence of the two widely distributed black beds (LBB and UBB) of the upper Nedrow Member (e.g. Brett & Ver Straeten 1994; Ver Straeten & Brett 2006; Ver Straeten 2007) is a striking phenomenon. They resemble features known as 'time-specific facies' (Walliser 1984*a*, *b*, 1986; for further explanations see Brett *et al.* 2012), at least within the Appalachian basins. Recent recognition of dark shales within the Choteč Limestone of the Barrandian Holyně section (R. Brocke & O. Fatka work in progress) may even allow comparisons across much more widely separated areas.

Concluding remarks

Based on two fossil groups (palynomorphs and dacryoconarids), the Barrandian type area (Prague Basin) has been compared for the first time with the Appalachian Basin with respect to the globally recognizable BCE. In addition, previous results on sea-level fluctuations in the AB by one of us (C.A. Ver Straeten) are given. The critical event interval, through the Edgecliff and Nedrow members, represents a third-order transgressive systems tract (TST). Culmination of the transgressive phase (=maximum transgression: i.e. 'the surface of maximum flooding' in sequence stratigraphic terms) lies in the topmost Nedrow Member: that is, at the position of the widely distributed 'black beds' of the Nedrow. The transgressive succession represents the lower half of the Ic Sequence (=Ic T–R Cycle of Johnson *et al.* 1985; =Eif-1 Sequence of Brett & Ver Straeten 1994; Ver Straeten 2007; Brett *et al.* 2011) which is considered to be a third-order sequence.

The palynological studies show two distinct proliferations of OWM in the investigated areas. Palynological assemblages of the distinctive bed of the BCE level in its type locality are characterized by mass-occurrences of large prasinophytes, while those from strata below and above reveal, to a large extent, palynospectra of 'normal marine' composition. Thus, the BCE represents a bioevent recognizable by an ecological epibole of a distinct phytoplankton group. The AB palynological assemblages show abundances of the fungi-like cf. *Reduviasporonites stoschianus* and of the glabrous chitinozoan *Hoegisphaera* sp. cf. *H. glabra*. They co-occur in many levels of the Nedrow Member and its equivalents, but are absent or very rare in the strata below and above. Therefore, we interpret their occurrence in the AB to represent a somewhat similar and contemporaneous ecological epibole to the prasinophyte epibole in the Barrandian area, which corresponds to the BCE. However, whilst in the Barrandian, the event is best recognizable in a distinct bed, in the Appalachian Basin it covers an interval of several metres, terminating with the two black beds in the upper Nedrow Member. This accords with the dacryoconarid record.

Further results of the dacryoconarid studies include the recognition of an immigration event (incursion epibole) of taxa from the OWR into the Central Appalachian Basin, and a separation of dacryoconarid faunas between the northern and central parts of the Appalachian Basin. At least four Old World taxa that have not previously been reported from the Eastern Americas Realm were discovered. Only two taxa are shared by both subregions of the AB. An immigration path from the Prague Basin–Rhenohercynian Zone along the SE margin of Laurussia is suggested (Fig. 18). In addition, faunal connections with North Africa also existed in the critical time interval, as indicated by House (1973) and Oliver (1977) regarding goniatites and rugose corals, respectively. As for the dacryoconarids, at least two of the taxa detected now in the AB (*Nowakia* (*Dmitriella*) *sulcata antiqua* and *Costulatostyliolina paucicostata*) are also known from the uppermost Emsian of North Africa (Alberti 1993). Concerning biostratigraphy, some of the taxa serve as excellent index forms. Hence, correlation of the BCE interval between the studied areas is now clearly shown by their dacryoconarid successions (Figs 10 & 17).

Furthermore, the widely distributed black beds of the upper Nedrow Member (LBB and UBB) can be regarded as representing, at least regionally, a 'time-specific facies' development in the sense of Walliser (1984*a*, *b*, 1986).

Systematic section

Palynology

Fungi-like palynomorphs. In the studied upper Nedrow Member-equivalent sequences (e.g. the Nedrow black beds), palynomorphs were discovered that are morphologically close to the supposed fungal structure *Reduviasporonites stoschianus*, as described from the Pennsylvanian of Peru (Wood & Elsik 1999). Other Carboniferous records of *Reduviasporonites* (*R. chalastus*) are known from Scotland (Stephenson *et al.* 2004). The Appalachian material is, to our knowledge, the oldest record of morphotypes attributable to *Reduviasporonites*. Studied specimens are tentatively assigned to *R. stoschianus* by following the classification and morphological nomenclature of Wood & Elsik (1999). However, systematically, they are classified here as *Incertae sedis*. In doing so, we are aware of the ongoing discussion about the biological origin of the genus *Reduviasporonites*, mainly in the context of the Permian–Triassic Extinction Event. Whether those microfossils represent fungi (e.g. Visscher *et al.* 1996, 2011; Steiner *et al.* 2003), acritarchs or zygnemataceaen algae (Afonin *et al.* 2001; Foster *et al.* 2002), or whether they are even of animal origin (e.g. coelenterate polyps: Kalgutkar & Jansonius 2000), is not yet clear. Further detailed analysis, including SEM studies on the given material and comparisons with co-equal specimens from other localities, are required.

The studied material does not show fluorescence, which is typical for an algal origin. Furthermore, the morphology is variable; some forms resemble skeletal hyphae of fungi. Other specimens are uniserial tubular, often branched filaments with or without clear segmentation, which could be interpreted as conidiophores. A third morphotype shows chains of cells or dispersed cells interpreted as coinids, which is typical for *Reduviasporonites*.

In some aspects, morphotypes are similar to the so-called tubiphytes, which have been assigned to cyanobacteria (e.g. see the discussion in Riding & Guo 1992). Banded tubes, very common in palynological residues from the Silurian to Devonian, are often classified as nematoclasts (e.g. *Porcatitubulus*: Taylor & Wellman 2009; Filipak & Zaton 2011), but in many cases their origin is still uncertain. However, subspherical to elongated, relatively thin-walled forms occur, which may represent single elements of fungal chains that form conidia, but which also show morphological similarities to the acitarch *Navifusa*. Thus, a confusion of identification is possible, in particular as both 'taxa' occur in the same assemblages.

Incertae sedis

Genus *Reduviasporonites* Wilson, 1962
Type species *Reduviasporonites catenulatus* Wilson, 1962

cf. *Reduviasporonites stoschianus* (Balme, 1980) Elsik, 1999
(Fig. 14a–c, f–m, p, q)

Remarks. The Devonian specimens of cf. *Reduviasporonites stoschianus* from the Appalachian Basin show a close morphological similarity to the Pennsylvanian specimens of Peru (Wood & Elsik 1999), and in parts to those described from the Upper Permian Bellerophon Formation in Italy (Elsik 1999). Specimens of cf. *R. stoschianus* show a variety in morphology, they occur in chains of cells ('conidia'), as well as branched or unbranched hyphea-like tubes ('conidiophores') in the same samples. The 'conidiophores' are often thicker walled and provided with septa, and may show distal furcation or bulbous to truncate morphology. The inner wall is characteristically shrunken. Conidia are usually elongated rather than spherical with septa. These may also morphologically refer to *Reduviasporonites chalastus* (M. Stephenson pers. comm. 2014). Kalgutkar & Jansonius (2000) discussed the morphological variety and stratigraphic occurrence of the illustrated fossils in Elsik (1999) and Wood & Elsik (1999), and are sceptical if they are plant or fungal remains. However, since all morphotypes mentioned above are found in the same samples, we consider that they may belong to a complex of *Reduviasporonites*. If this assignment is accepted, the *Reduviasporonites* morphotypes shown in this paper are the oldest known representatives of this genus.

Chitinozoa

Numerous glabrous chitinozoan specimens found in the Appalachian assemblages recall *Hoegisphaera glabra*, which was first described from the Late Devonian (Frasnian) of Alberta, Canada (Staplin 1961), and therein designated as being restricted to this time interval (see Paris *et al.* 1999). The subsequent occurrence of comparable material is mainly from the Middle Devonian (Givetian) of North America; in consequence, these taxa with a younger record and vesicle ornamentation were considered as *Hoegisphaera* sp. cf. *H. glabra* (e.g. Legault 1973*a*, *b*). However, they were also classified as *Hoegisphaera glabra* neglecting the stratigraphic limitation (e.g. Wright 1980; Wood & Clendening 1985). Our material from the Appalachian Basin is supposedly older (? early Eifelian) and thus we follow the argument of a somewhat open designation until a more detailed analysis (i.e. SEM studies) may show their conspecific relationship.

Chitinozoa Eisenack, 1931
Order **Operculatifera** Eisenack, 1931
Family **Desmochitinidae** Eisenack, 1931, emend. Paris, 1981
Subfamily **Demochitininae** Paris, 1981
Genus *Hoegisphaera* Staplin, 1961, emend. Paris, Grahn, Nestor & Lakova, 1999
Type species: *Hoegisphaera glabra* Staplin, 1961

Hoegisphaera cf. *H. glabra*, Legault 1973*a*, *b*
(p. 91, pl. 8, figs 4–6, 8 & 10)

Description. Spherical–subspherical body shape with a circular opening bordered by a low rim, surrounding an operculum that often is dislocated. Surface smooth or slightly ornamented, a few may show degraded membranes (mat-like structure). In our material, two individuals are arranged close to each other in the equatorial plane. Legault (1973*a*, *b*) discussed the presence of membraneous sheets completely surrounding specimens. In some cases, the membraneous tissue possibly served as a connection between individuals, representing a specific mode of aggregation. It is not clear from our material whether these membraneous sheets did exist and completely enclosed the specimen or even more than one specimen. However, we cannot exclude that these membranes were lost during the diagenetic processes and/or palynological preparation.

Diameter of body: (chamber) 75–95 μm, operculum 19–42 μm.

Remarks. Specimens of *Hoegisphaera* cf. *H. glabra* found in the CAB sections (e.g. Hayfield, Spring Gap) occur in samples of coarser-grained, silty sediments, enriched in in organic matter; mostly AOM, a few spores but very rare acritarchs. They are associated with large scolecodonts, cf. *Reduviasporonites stoschianus*, and probably land-derived organic matter (OM). In sections of the NAB (e.g. the Stafford Quarry), acritarchs are more common in the contemporaneous level along with cf. *Reduviasporonites*, whereas *Hoegisphaera* sp. is less common.

*Occurrences (*Hoegisphaera glabra *and* Hoegisphaera *sp. cf.* H. glabra*).* North America: Columbus and Delaware limestones (lower and upper Eifelian, respectively), Ohio (Wright 1976, 1978); Cedar Valley Formation and Wapsipinicon Formation (Givetian), Iowa (Urban 1972; Urban & Newport 1973; Wicander & Wood 1997); North

Venon Limestone (Givetian), Indiana (Wright 1980); Boyle Dolomite (Givetian), Kentucky (Wood & Clendening 1985); Rockport Quarry Member, Hamilton Formation (Middle Devonian), Ontario (Legault 1973*a*, *b*); Duvernay Shale (Frasnian), Alberta (Staplin 1961).Amazon Basin, Brazil: *Hoegisphaera* sp. cf. *H. glabra* (early Givetian, Grahn 2011, p. 35). France: Lezais, Armorican Massif, upper Emsian (Paris 1981).

Dacryoconarids

Class **Tentaculita** Bouček, 1964
Order **Dacryoconarida** Fisher, 1962
Family **Styliolinidae** Grabau & Shimer, 1910
Genus *Styliolina* Karpinsky, 1884

Discussion. Concepts of *Styliolina* Karpinsky, 1884 and the family Styliolinidae Grabau & Shimer, 1910 are founded on the attributes of *S. nucleata* Karpinsky and *S. fissurella* (Hall), which Karpinsky (1884) regarded to be conspecific (see Fisher 1962). Lindemann & Yochelson (1994) reported that the shell walls of both species consist of a single layer of homogeneous calcite and, based largely on this attribute, proposed that the Styliolinidae be removed from the Order Dacryoconarida Fisher, 1962. Having re-examined numerous topotypes at high SEM magnifications, it is now certain that the shell wall of *S. fissurella* (Hall) is not homogeneous but is microlaminated in the manner of the shells of the dacryoconarid genera *Striatostyliolina*, *Viriatellina* and *Costulatostyliolina* (Bouček 1964; Lardeux 1969; Lindemann & Yochelson 1994). Accordingly, *Styliolina* Karpinsky is no longer regarded to be separate from the dacryoconarids.

Styliolina robusta Lindemann & Schindler n. sp.
(Fig. 11a)

Etymology. Latin, *robusta*, meaning strong, referring to the shell profile and thickness relative to that of *Styliolina fissurella* (Hall).

Holotype. The holotype NYSM 17229 (Fig. 11a) from the lowermost bed of the Nedrow Limestone in a roadcut on the east side of South Salina Street, NY, Route 11 at Nedrow, NY (N42° 57′ 40″, W76° 08′ 18″).

Material and occurrences. In excess of 50 complete individuals and many partials from the lower beds of Nedrow Member of the Onondaga Formation at the type locality and at Cherry Valley, NY.

Diagnosis. *Styliolina*, with a 25 µm-thick microlaminated shell up to 1.5 mm long and 0.53 mm wide, with a blunt-based, 0.12–0.14 mm-wide initial chamber separated by a nearly obsolete constriction from the conical proximal region, which has a growth angle of 15–18° that diminishes to 8–10° in the distal region.

Description. The shell is straight, short and wide, with relatively large growth angles in the proximal, medial and distal regions. The apex is blunt, not rounded or pointy, and devoid of an apical node.

Discussion. *Styliolina fissurella* (Hall), the only species of the genus that has been reported formally from the Devonian of the Appalachian Basin, has a shell that is only 6–10 µm thick and up to 5 mm long, with an apical growth angle of 5–7°, becoming subcylindrical in the distal region. *Styliolina robusta* n. sp. is distinct from *S. fissurella* (Hall) in each of these attributes.

Styliolina cf. *S. decurtata* Bouček, 1964
(Fig. 11b)

Figured specimen. NYSM 17236 (Fig. 11b) from the middle Edgecliff-equivalent at Hayfield, VA.

Material and occurrences. In excess of 20 complete shells and numerous partials from the middle of the Edgecliff-equivalent interval of the Selinsgrove Member of the Needmore Formation at Hayfield, VA, and from the Nedrow-equivalent upper black bed at Mapleton, PA.

Description. *Styliolina* with a 12–14 µm-thick, microlaminated shell that is up to 2.5 mm long and 0.22 mm wide, with a rounded to slightly pointy 0.8–0.95 mm-wide initial chamber separated by a long constriction from the conical proximal region, which has a growth angle of 9–12° that diminishes to subparallel in the distal region.

Discussion. *Styliolina* cf. *S. decurtata* is distinct from *S. fissurella* and *S. robusta* n. sp. in the width of its initial chamber, proximal growth angle and shell thickness. It is most similar to *S. decurtata* Bouček, 1964 from the upper Emsian–lower Eifelian Třebotov Limestone at Holyně, Czech Republic. The two differ only in the shape of the initial chamber, with that of *Styliolina* cf. *decurtata* being slightly less pointy than the specimens figured by Bouček (1964, pl. 32, figs 1 & 2).

Family **Striatostyliolinidae** Bouček, 1964
Genus *Striatostyliolina* Bouček & Prantl, 1961

Striatostyliolina mima Lindemann & Schindler n. sp.
(Fig. 11c, d)

Etymology. Latin, *mima,* actress; referring to morphological mimicry of *Styliolina fissurella* (Hall).

Holotype. NYSM 17230 (Fig. 11c, d) from the basal bed of the Nedrow Limestone in a roadcut on the north side of US Route 20 north of Cherry Valley, NY (N49° 28′ 37″, W74° 44′ 18″).

Material and occurrences. Many hundreds of complete and partial specimens. The species may first occur in the upper Emsian Schoharie and Bois Blanc formations (Oliver 1966), but that is not currently certain. Its first certain occurrence is in the basal bed of the Edgecliff Limestone. It occurs throughout the Onondaga Formation and ranges upwards into the Hamilton Group, at least as high as the mid-Givetian Ledyard Member of the Ludlowville Formation.

Diagnosis. The 20 μm-thick microlaminated shell is up to 1.3 mm long and 0.3 mm wide, with a teardrop-shaped to slightly pointy 130–140 μm-wide initial chamber separated by a long, gradual constriction from the proximal region, which has a growth angle of 10° that diminishes to subcylindrical in the distal region. Approximately 16–20 striae on a semi-circumference extend from apex to aperture.

Description. The shell is straight and narrow, with an initial chamber that varies from teardrop shaped to slightly pointy, sometimes terminating in a nearly obsolete apical node. Striae are very weak and poorly preserved. As is the case with *Styliolina*, the *Striatostyliolina mima* n. sp. shell is microlaminated, but may falsely appear to be homogeneous.

Discussion. Striatostyliolina mima n. sp. is the most abundant non-annulated dacryoconarid in Eifelian and Givetian strata of NY, and the main source of erroneous reports of *Styliolina fissurella*. It differs from species of *Striatostyliolina* described from the OWR by Bouček (1964) and Lardeux (1969) in that it is far shorter than the others, and that it has 2–4 times the number of striae. However, the striae are weakly incised into the shell surface and, even when preserved, cannot be reliably discerned with a light microscope or at low magnifications with the SEM.

Striatostyliolina vitta Lindemann & Schindler n. sp.
(Fig. 11e)

Etymology. Latin, *vitta,* a ribbon; with reference to the overall profile of the shell.

Holotype. NYSM 17231 (Fig. 11e) from the second argillaceous bed above the base of the Nedrow Member at Cherry Valley, NY.

Material and occurrences. Approximately 20 complete specimens and numerous partials from the Nedrow Member and a vast number of specimens from the basal bed of the Oatka Creek Formation. Beyond the type locality and bed, the only other currently known occurrence in the Onondaga Formation is in the basal bed of the Nedrow Member at Nedrow, NY. The currently known uppermost occurrence of *S. vitta* n. sp. is in the Chestnut Street Beds at the base of the Oatka Creek Formation at Cherry Valley, NY.

Diagnosis. The microlaminated 15 μm-thick shell is up to 2.2 mm long and 0.23 mm wide, with a round to slightly pointy 95–112 μm-wide initial chamber separated by a nearly obsolete constriction from the proximal region, which has a growth angle of 6° that diminishes to 2–3° in the distal region. The six or seven striae on a semi-circumference extend from apex to aperture.

Description. The shell is straight and exceptionally narrow. The apex of some individuals bears a nearly obsolete, 40–45 μm-wide node. Striae are 2–3 μm wide, separated by flat, 25–40 μm-wide interspaces.

Discussion. Among currently known dacryoconarid species, *Striatostyliolina vitta* n. sp. is most similar in size and shape to *Styliolina minuta* (see Bouček 1964, pl. 33).

Genus *Costulatostyliolina* Lardeux, 1969

Costulatostyliolina cf. *C. paucicostata* (Bouček, 1964)
(Fig. 11f)

Figured specimen. NYSM 17237 (Fig. 11f) from the lower of the two upper Nedrow-equivalent black beds of the Selinsgrove Member, Needmore Formation at Hayfield, VA.

Material and occurrence. In excess of 50 compressed shells and surficial impressions. At Hayfield, VA, *Costulatostyliolina* cf. *C. paucicostata* (Bouček) ranges from the middle of the Edgecliff-equivalent up to the above-stated bed. It has not been observed at any other locality.

Description. The 15 μm-thick microlaminated shell is up to 3.5 mm long and 0.4 mm wide, with a rounded, 140–160 μm-wide initial chamber separated by a nearly obsolete constriction from the proximal region, which has a growth angle of approximately 9° that diminishes to subcylindrical in the distal region. There are six–seven costae on the initial chamber and in the proximal region, and from eight to ten in the distal region of the shell.

Discussion. Lardeux (1969) designated *Costulatostyliolina paucicostata* (Bouček) as the type species of the then newly erected genus. Although Bouček (1964) described *Striatostyliolina paucicostata* as having five–seven 'ribs' (=costae) on the semicircumference, on his plate 38 (figs 1–4) he illustrated individuals with between five and 10 costae (note that the figure references in Bouček's text and figure captions incorrectly switch plates 36 and 38). The species' type unit is the upper part of the Třebotov Limestone, which places it only a little below the base of the Choteč Formation. The only discernable difference between the figured type material from the Třebotov Limestone and the specimens from the Nedrow Limestone is that the costae of the latter may be somewhat narrower and that the interspaces are somewhat wider in the distal region of the shell than those of the originals. However, this is within the range of intraspecific variability illustrated by Bouček (1964, pl. 38, figs 1–4).

Costulatostyliolina vestita Lindemann & Schindler n. sp.
(Fig. 11g)

Etymology. Latin, *vestitus*, costume; with reference to the appearance of a *Styliolina* dressed up with costae.

Holotype. NYSM 17232 (Fig. 11g) from the second argillaceous bed above the base of the Nedrow Member at Cherry Valley, NY.

Material and occurrences. Several dozen partial and complete shells. The currently known first occurrence of this species is at the base of the Edgecliff Member at Cherry Valley, NY. It also occurs in the argillaceous interval of the lower Edgecliff at Clarence, NY, the argillaceous beds of the upper Edgecliff and Nedrow members at Phelps, NY, and the basal to lower argillaceous beds of the Nedrow throughout central and east-central NY.

Diagnosis. The microlaminated 25 μm-thick shell is up to 3 mm long and 0.3 mm wide, with a round to slightly pointy, 135–145 μm-wide initial chamber separated by a long, gradual constriction from the proximal region, which has a growth angle of 9–12° that diminishes to less than 6° in the distal region. The 14–18 costae on a semi-circumference extend from apex to aperture.

Description. The shell is straight and narrow. The apex of some individuals bears a weak node that is commonly absent. Costae are weak, submicron in width, separated by flat, 10–20 μm-wide interspaces.

Discussion. The overall morphology of *Costulatostyliolina vestita* n. sp. closely resembles that of both *Styliolina fissurella* (Hall) and *Striatostyliolina mima* n. sp., but is differentiated from them by the presence of costae, which are easily corroded from the shell surface. The overall morphology of *C. vistita* n. sp. also resembles that of *C. strigata* (Hall) from the Chestnut Street Beds of the Oatka Creek Formation, which has only six–seven costae on a semi-circumference (Lindemann 2008) as opposed to 14–18.

Metastyliolina cf. *M. striatissima* Bouček & Prantl, 1961
(Fig. 12a)

Figured specimen. NYSM 17238 (Fig. 12a) from the lower of the two upper Nedrow-equivalent black beds of the Selinsgrove Member, Needmore Formation at Spring Gap, MD.

Material and occurrences. External impressions of several hundred specimens. No shells or steinkerns have been found. Beyond its occurrence at Spring Gap, MD, the taxon ranges from the middle of the Edgecliff-equivalent to the lower of the two UBB at Hayfield, VA, and Gainesboro, VA.

Description. The shell is up to 7 mm long, with a slightly pointy 80–90 μm-wide initial chamber separated by a nearly obsolete constriction from the acicular proximal region, which has a growth angle of less than 7° that diminishes to subcylindrical in the medial and distal regions. There are between eight and 10 costae on a semi-circumference in the proximal region, 18–26 in the medial region, and 32–42 in the distal region. Growth lines with irregular spacings of 75–250 μm occur in the medial and distal regions of most individuals.

Discussion. In the Prague Basin, *Metastyliolina striatissima* is the predominant dacryoconarid in the Choteč Limestone (Bouček 1964). *Metastyliolina* cf. *M. striatissima* has a comparable occurrence in the southern Appalachian Basin (AB), where its morphological attributes accord well with the species' written description and with the individuals figured by Bouček (1964, pl. 37, figs 8 & 9). The sole exception is that the AB specimens are somewhat wider in the medial and distal regions than those of the Prague Basin. This may be attributed to different degrees of secondary compaction and the consequent widening of the shell profile, which is far more pronounced in the AB. An alternative possibility, prompted by the erection of *M. striatissima grueti* by Lardeux (1969), is that either geographical or temporal subspecies exist and the AB form might be among them.

Metastyliolina? sp. B
(Fig. 12b)

Figured specimen. NYSM 17239 (Fig. 12b) from the lower of the two upper Nedrow-equivalent black beds of the Selinsgrove Member, Needmore Formation at Hayfield, VA.

Material and occurrences. Approximately 20 impressions of compressed partial specimens and four partial compressed shells. This form occurs in the lower of the two upper Nedrow-equivalent black beds at Gainesboro, VA, and Mapleton, PA, as well as in a thin bed of black shale proximal to the top of the Nedrow Limestone in the Oak Corners Quarry at Phelps, NY.

Description. The shell is up to 5 mm long with a proximal growth angle of 4–6° that diminishes to subcylindrical in the distal region. There are six–eight costae in the proximal region and 14–20 distally.

Discussion. This form is poorly known from crushed partial shells. It is questionably referred to *Metastyliolina* as opposed to *Costulatostyliolina* owing to its length, slender profile and relatively numerous costae in the distal region. Although it is poorly documented, it is included here because it occurs near the top of the Nedrow interval in both the central and northern regions of the AB.

Family **Nowakiidae** Bouček & Prantl, 1960
Subfamily **Nowakiinae** Bouček & Prantl, 1960
Genus *Nowakia* Gürich, 1896
Subgenus *Nowakia* (*Dmitriella*) Ljaschenko, 1966

Nowakia (*Dmitriella*) *sulcata antiqua*
Alberti, 1981
(Fig. 12c)

Figured specimen. NYSM 17240 (Fig. 12c) from the lower of the two upper Nedrow-equivalent black beds of the Selinsgrove Member, Needmore Formation at Mapleton, PA.

Material and occurrences. In excess of 50 unaltered shells and an equal number of partially to fully compressed specimens. The lowermost known occurrence of this taxon is at the base of the middle of the Edgecliff-equivalent at Hayfield, VA, and Gainesboro, VA. In the CAB, its zone extends to the top of the Nedrow-equivalent at Spring Gap, MD. In the NAB, this taxon is absent from the Onondaga Formation, but occurs in the Chestnut Street Beds, Hurley Member, Oatka Creek Formation at Cherry Valley, NY.

Description. The 8–10 μm-thick microlaminated shell is up to 6 mm long and 0.7 mm wide, with a 100–110 μm-wide initial chamber separated by a nearly obsolete constriction from the proximal region, which has a growth angle of 8–10°, becoming subcylindrical in the distal region. Transverse sculpture is nearly obsolete in the proximal region, progressively strengthening to narrow, blunt crested rings separated by broad, rounded swales (=sulca) with wavelengths of 200–280 μm in the medial and distal regions. There are between seven and 10 costae on the initial chamber and in the distal region, progressively increasing to 25–30 in the distal region of full-length shells.

Discussion. Nowakia (*Dmitriella*) *sulcata* has long served as an index for the Choteč Limestone (Bouček 1964) and the BCE (Chlupáč & Kukal 1986). Bouček (1964, p. 90) reported that it also occurs in the Třebotov Limestone. Although this taxon is common in Europe and North Africa, a subspecies rarely occurs in the AB (Fig. 10) (see also Brill *et al.* 2014).

Alberti (1981) first described *Nowakia* (*D.*) *s. antiqua* from the upper Emsian as the presumed ancestor of the Eifelian *N.* (*D.*) *s. sulcata*. The primary difference between the two subspecies is an evolutionary reduction in number of costae on a semi-circumference from 25 or more on *N.* (*D.*) *s. antiqua* to about half that number on *N.* (*D.*) *s. sulcata*. Alberti (1982) reported forms that are transitional between the two varieties in strata just above the base of the *partitus* Zone (i.e. the base of the Eifelian Stage) and subsequently reported on *N.* (*D.*) *s. antiqua* from the Choteč Limestone (Alberti 1993).

Nowakia sp. A
(Fig. 12d, e)

Figured specimen. NYSM 17235 (Fig. 12d, e) from the lower of the two upper Nedrow-equivalent black beds of the Selinsgrove Member, Needmore Formation at Hayfield, VA.

Material and occurrences. In excess of 100 more or less compressed specimens, most of which are external impressions of the shell. The taxon is currently known to range from the middle of the Edgecliff-equivalent at Hayfield, VA, and Gainesboro, VA, as well as the upper Nedrow-equivalent at Spring Gap, MD.

Description. The shell is up to 5.5 mm long and 0.4 mm wide, with a rounded, 130–140 μm-wide initial chamber that passes abruptly into the juvenile region, which has a growth angle of approximately 8–10° that diminishes somewhat in the distal region. Transverse sculpture consists of bluntly rounded, symmetrical ripples separated by wide swales with an average wavelength of 130–160 μm over most of the shell, diminishing in strength and wavelength behind the aperture in full-length specimens. Although there is variability between individuals, the juvenile region is typically devoid of transverse sculpture up to a shell length of 1.2–1.5 mm. There are between eight and 10 costae in the proximal region, approximately 25 in the medial region, and 30–35 in the distal region.

Discussion. The overall morphology of this form is markedly similar to that of the lower Eifelian *Nowakia* (*Maurerina*) *procera* (Maurer). This is particularly so in the long, smooth juvenile region, as well as in the apparent shape and pacing of transverse rings in the medial and distal regions. However, Alberti (1981) described *N.* (*Maurerina*) as being unique among *Nowakia* subgenera in that it is devoid of costae. Lütke (1985, pl. 3, figs 1 & 2) figured two well-preserved *N.* (*M.*) *procera* (Maurer) that are clearly devoid of costae. Thus, the form of *Nowakia* under consideration herein is precluded from the subgenus and species, and is not referred to any taxon below the generic level.

Genus *Viriatellina* Bouček, 1964

Viriatellina manifesta Lindemann & Schindler n. sp.
(Fig. 12f)

Etymology. Latin, *manifesta*, evident; referring to the obvious transverse shell sculpture of uncompressed individuals.

Holotype. NYSM 17233 (Fig. 12f) from the second argillaceous bed above the base of the Nedrow Member at Cherry Valley, NY.

Material and occurrences. Dozens of complete specimens and several hundred partials. The species is currently known from the argillaceous interval of the lower Edgecliff Limestone at Clarence, NY, and argillaceous beds of the upper Edgecliff in west-central NY, extending up to the lower 0.7 m of the Nedrow Limestone at Cherry Valley, NY. Similar forms of *Viriatellina* occur in the upper Emsian Schoharie Formation of eastern NY and the upper Eifelian Bakoven Shale of central NY, but it is not currently certain whether they are conspecific with *V. manifesta* or are steps in an evolving lineage.

Diagnosis. The microlaminated 10 μm-thick shell is up to 2.5 mm long and 0.4 mm wide, with a slightly pointy 130–150 μm-wide initial chamber separated by a weak constriction from the proximal region, which has a growth angle of 13–15° that diminishes to 8–10° in the distal region. Transverse sculpture consists of symmetrical ripples and intervening swales that begin late in the proximal region and strengthen distally. There are 16–20 costae on the semi-circumference over the full length of the shell.

Description. Costae are submicron in width, separated by flat 15–40 μm-wide interspaces in the distal region. The narrow costae are easily lost to dissolution, whereupon they appear as secondary striae, particularly in the proximal region and the initial chamber. Transverse sculpture consists of low-amplitude ripples beginning in the medial region with wavelengths of less than 0.14 mm, which progressively strengthen and lengthen to wavelengths in excess of 0.3 mm in the distal region. The ripples are flattened out during compaction of the shell, which can cause them to vanish all together. Compaction also widens the shell, producing a distinctive chevron morphology, which other dacryoconarid taxa of the Onondaga Formation do not attain.

Discussion. Only two species of this genus, *Viriatellina gracilistriata* (Hall) and *V. porteri*, have been reported formally from the Appalachian Basin (Lindemann & Yochelson 1992). Although *V. manifesta* n. sp. bears some similarity to *V. gracilistriata* (Hall), they differ markedly in that the former has 16–20 costae on a semi-circumference and the latter 36–42 in the distal region. Bouček (1964, p. 99, pl. 19) described and illustrated specimens of what he regarded to be *V. gracilistriata* (Hall) from Emsian strata of Bohemia, which Lütke (1985, p. 211) subsequently referred to *V. fortistriata* n. sp., designating one of Bouček's figured specimens as the holotype. Lardeux (1969, p. 124, text fig. 92; pl. 43, fig. 1) described and illustrated specimens of *V.* cf. *gracilistirata* (Hall) from Emsian and Eifelian strata of Brittany, which are very distinct from *V. gracilistiata* (Hall) as described by Lindemann & Yochelson (1992) and differ from *V. manifesta* n. sp. in having only 12–16 costae on a semi-circumference.

Viriatellina exila Lindemann & Schindler n. sp.
(Fig. 12g)

Etymology. Latin, *exilis*, slim; referring to slender morphology of the shell relative to *Viriatellina gracilistrata* (Hall) and *V. manifesta* n. sp.

Holotype. NYSM 17234 (Fig. 12g) from the lower 0.7 m of the Nedrow Limestone at Cherry Valley, NY.

Material and occurrences. Twenty-five specimens from the lower beds of the Nedrow at Cherry

Valley, NY. This is the only locality and interval from which this taxon is currently known for certain, although somewhat similar forms occur in the basal bed of the Nedrow at Nedrow, NY.

Diagnosis. The 25 μm-thick microlaminated shell is up to 1.4 mm long and 0.35 mm wide, with a slightly pointy, 120–130 μm-wide initial chamber separated by a nearly obsolete constriction from the proximal region, which has a growth angle of 14°–16° that diminishes to about 4° in the distal region. Transverse sculpture consists of low-amplitude ripples that begin in the proximal region, and gradually strengthen distally in amplitude and wavelength over the length of the shell. There are between eight and 12 costae on the semi-circumference over the full length of the shell.

Description. The shell is straight and slender with low-amplitude transverse ripples, the crests of which nearly mirror images of the intervening swales. The ripples begin in the proximal region with wavelengths of 120 μm, which progressively increase to 260 μm in the distal region. The low, submicron-wide costae are separated by 25–40 μm-wide interspaces. The initial chamber is slightly pointy but devoid of an apical node.

Discussion. Viriatellina exila n. sp. differs from *V. manifesta* n. sp. in being more slender in the medial and distal regions, and having fewer costae and a thicker shell wall. It is most similar to the upper Emsian *V. hercynica* of Bouček (1964, p. 95) but differs in being far shorter, having much weaker costae and lower-amplitude transverse ripples.

We would like to thank Stana Vodrážková and Petr Budil (Praha) for assistance during sampling at the Na Škrábku Quarry, and for subsequent discussions. We also thank Carl Brett (Cincinnati, OH) for constructive discussion on the Choteč Event during the SDS meeting in Morocco 2013. Petra Tonarová (Tallinn) is thanked for determining the two pictured scolecodonts. Michael Krings (München) and Michael Stephenson (Keyworth, Nottingham) contributed to the discussion on the systematics of *Reduviasporonites stoschianus*. Rachel Barrachina (Skidmore College, Saratoga Springs, NY) is thanked for help with the SEM images of the dacryoconarids. Michael Ricker and Haytham El Atfy are acknowledged for their assistance in the production of graphical charts; Jutta Oelkers-Schaefer and Gunnar Riedel (all at Senckenberg Forschungsinstitut und Naturmuseum Frankfurt) are thanked for their assistance in palynological preparation. John Marshall (Southampton) and an anonymous reviewer are thanked for constructive remarks and R. Thomas Becker (Münster) for final formal corrections. This paper is a contribution to the IGCP Project 596 'Climate change and biodiversity patterns in the Mid-Palaeozoic'.

Appendix A: Appalachian Basin study localities

- Goodrich Road roadcut, Clarence, NY: 42.984960°, −78.637273°
- Stafford Quarry, west of Stafford, NY: 42.979964°, −78.088233°
- Oaks Corners Quarry, SE of Phelps, NY: 42.932542°, −77.023682°
- Abandoned Schooley Quarry, Auburn, NY: 42.954130°, −76.562183°
- US Route 11 roadcuts, Nedrow, NY: 42.961129°, −76.138772°
- East end of Jamesville Quarry, Jamesville, NY: 42.994283°, −76.024205°
- US Route 20 roadcuts, north of Cherry Valley, NY: *c.* 42.822617°, −74.725097°
- Abandoned quarry SW of Mapleton, PA: 40.377664°, −77.953548°
- Roadcut at Spring Gap, MD: 39.564570°, −78.713113°
- Roadcut Route 8, east of Keyser, WV: 39.448554°, −78.954074°
- Abandoned quarry along Route 684, Gainesboro, VA: 39.284756°, −78.262652°
- Roadcut, NE side of intersection, US Route 50 west of Winchester, VA: 39.233224°, −78.290060°

References

Afonin, S. A., Barinova, S. S. & Krassilov, V. A. 2001. A bloom of *Tympanicysta* Balme (green algae of zygnematalean affinities) at the Permian–Triassic boundary. *Geodiversitas*, **24**, 481–487.

Alberti, G. K. B. 1980. Neue Daten zur Grenze Unter-/Mittel-Devon, vornehmlich aufgrund der Tentaculiten und Trilobiten im Tafilalt (SE-Marokko). *Neues Jahrbuch für Geologie und Paläontologie, Monatshefte*, **1980**, 581–594.

Alberti, G. K. B. 1981. *Nowakia* (*Sulcatonowakia*) *sulcata antiqua* n. ssp. aus dem oberen Teil der Heisdorf-Formation (Unter-Devon) der Hillesheimer Mulde (Eifel). *Senckenbergiana lethaea*, **61**, 445–452.

Alberti, G. K. B. 1982. *Nowakia* (*Sulcatonowakia*) *sulcata sulcata* (F. A. Roemer, 1843) (Dacryoconarida) from the lowermost part of the Lauch Formation (Eifelian) of the Wetteldorf Richschnitt. *Courier Forschungsinstitut Senckenberg*, **55**, 333–336.

Alberti, G. K. B. 1985*a*. Neue Taxa der Dacryoconarida, insbesondere der Corniculinidae n. fam., aus dem basalen Flemersbacher Tentakulitenkalk (Bayrische Faziesreihe, Unterdevon) von Oberfranken. *Mitteilungen des Geologisch-Paläontologischen Instituts der Universität Hamburg*, **59**, 39–50.

Alberti, G. K. B. 1985*b*. Zur biostratigraphischen Untergliederung des Greifensteiner Kalkes und der Wissenbacher Schiefer (Unter- bis Mittel-Devon, Rheinisches Schiefergebirge) mit Hilfe von Dacryoconariden (Tentaculiten). *Mitteilungen des Geologisch-Paläontologischen Instituts der Universität Hamburg*, **59**, 51–56.

ALBERTI, G. K. B. 1993. Dacryoconaride und homoctenide Tentaculiten des Unter- und Mittel-Devons I. *Courier Forschungsinstitut Senckenberg*, **158**, 1–229.

BALME, B. E. 1980. Palynology of Permian–Triassic boundary beds at Kap Stosch, east Greenland. *Meddelser om Grønland udgivne af kommissionen for Kidenskabelige Undersøgelser i Grønland*, **200**, 1–37.

BECKER, R. T. & ABOUSSALAM, Z. S. 2013. The global Choteč event at Jebel Amelane (Western Tafilalt Platform) – preliminary data. *In*: BECKER, R. T., EL HASSANI, A. & TAHIRI, A. (eds) *International Field Symposium 'The Devonian and Lower Carboniferous of northern Gondwana', Field Guidebook*. Document de l'Institut Scientifique, Rabat, **27**, 129–134.

BECKER, R. T. & HOUSE, M. R. 1994. International Devonian goniatite zonation, Emsian to Givetian, with new records from Morocco. *Courier Forschungsinstitut Senckenberg*, **169**, 79–135.

BERCOVICI, A., CUI, Y., FOREL, M.-B., YU, J. & VAJDA, V. 2015. Terrestrial paleoenvironment characterization across the Permian–Triassic boundary in South China. *Journal of Asian Earth Sciences*, **98**, 225–246.

BERDAN, J. M. 1981. Ostracode biostratigraphy of the lower and middle Devonian of New York. *In*: OLIVER, W. A. & KLAPPER, G. (eds) *Devonian Biostratigraphy on New York*. International Union of Geological Sciences, Subcommission on Devonian Stratigraphy, Washington, DC, 83–96.

BERKYOVÁ, S. 2009. Lower–Middle Devonian (upper Emsian–Eifelian, *serotinus–kockelianus* zones) conodont faunas from the Prague Basin, the Czech Republic. *Bulletin of Geosciences*, **84**, 667–686.

BERKYOVÁ, S. & MUNNECKE, A. 2010. 'Calcispheres' as a source of lime mud and peloids – evidence from the early Middle Devonian of the Prague Basin, the Czech Republic. *Bulletin of Geosciences*, **85**, 585–602.

BLAKEY, R. C. 2007. *Paleogeographic reconstructions of North America*. Colorado Plateau Geosystems, Inc., updated July 2011, http://jan.ucc.nau.edu/~rcb7/

BOUČEK, B. 1964. *The Tentaculites of Bohemia*. Czechoslovak Akademy of Sciences, Praha.

BOUČEK, B. & PRANTL, F. 1960. Několik nomenklatorických poznámek k nadřádu Tentaculitoidea Ljašenko. *Časopis Národního Musea Oddíl Přírodovědný*, **129**, 198–199.

BOUČEK, B. & PRANTL, F. 1961. Über einige neue Tentaculiten-Gattungen aus dem böhmischen Devon. *Věstník Ústředního ústavu geologického*, **36**, 385–388.

BOUŠKA, J. 1948. *Holynocrinus*, new crinoid genus from the Middle Devonian of Bohemia. *Journal of Paleontology*, **22**, 520–524.

BOUŠKA, J. 1956. Pisocrinidae Angelin from the Silurian and Devonian of Bohemia (Crinoidea). *Rozpravy Ústředního Ústavu Geologického*, **20**, 1–137.

BRAUN, A. & BUDIL, P. 1999. A Middle Devonian Radiolarian fauna from the Chotec limestone (Eifelian) of the Prague Basin (Barrandian, Czech Republic). *Geodiversitas*, **21**, 581–592.

BRETT, C. E. & BAIRD, G. C. 1997. Epiboles, outages and ecological evolutionary bioevents: taphonomic, ecological and biogeographic factors. *In*: BRETT, C. E. & BAIRD, G. C. (eds) *Paleontological Events, Stratigraphical, Ecological, and Evolutionary Implications*. Columbia University Press, New York, 249–284.

BRETT, C. E. & VER STRAETEN, C. A. 1994. Stratigraphy and facies relationships of the Eifelian Onondaga Limestone (Middle Devonian) in western and west central New York State. *In*: BRETT, C. E. & SCATTERDAY, J. (eds) *New York State Geological Association, 66th Annual Meeting Guidebook*. New York State Geological Association, Rochester, NY, 221–269.

BRETT, C. E., IVANY, L. C., BARTHOLOMEW, A. J., DESANTIS, M. K. & BAIRD, G. C. 2009. Devonian ecological–evolutionary subunits in the Appalachian Basin: a revision and a test of persistence and discreteness. *In*: KÖNIGSHOF, P. (ed.) *Devonian Change: Case Studies in Palaeogeography and Palaeoecology*. Geological Society, London, Special Publications, **314**, 7–36, http://doi.org/10.1144/SP314.2

BRETT, C. E., BAIRD, G. C., BARTHOLOMEW, A. J., DESANTIS, M. K. & VER STRAETEN, C. A. 2011. Sequence stratigraphy and a revised sea-level curve for the Middle Devonian of eastern North America. *Palaeogeography, Palaeoclimatology, Palaeoecology*, **304**, 21–53, http://doi.org/10.1016/j.palaeo.2010.10.009

BRETT, C. E., MCLAUGHLIN, P. I., HISTON, K., SCHINDLER, E. & FERRETTI, A. 2012. Time specific aspects of facies: examples and possible causes. *Palaeogeography, Palaeoclimatology, Palaeoecology*, **367–368**, 6–18, http://doi.org/10.1016/j.palaeo.2012.10.009

BRILL, M. E., BROWN, T. & COUNTRYMAN, M. 2014. Dacryoconarids of the Eifelian (Middle Devonian) Stony Hollow Event. *Geological Society of America, Abstracts with Programs*, **46**, 82.

BROCKE, R. & WILDE, V. 2001. Infrared video microscopy – an efficient method for the routine investigation of opaque organic-walled microfossils. *Facies*, **45**, 157–164.

BROCKE, R., BERKYOVÁ, S., FATKA, O., LINDEMANN, R. H., SCHINDLER, E. & VER STRAETEN, C. A. 2011. The early Mid-Devonian Choteč Event: do palynomorphs have the potential for long-distance correlations? *Geological Society of America, Abstracts with Programs*, **43**, 97.

BROCKE, R., FATKA, O., LINDEMANN, R. H., SCHINDLER, E. & VER STRAETEN, C. A. 2013. New biostratigraphic insights from the early Mid Devonian Choteč Event. *In*: EL HASSANI, A., BECKER, R. T. & TAHIRI, A. (eds) *International Field Symposium 'The Devonian and Lower Carboniferous of northern Gondwana', Abstracts Book*. Document de l'Institut Scientifique, Rabat, **26**, 28.

BROMLEY, R. G. 1990. *Trace Fossils: Biology and Taphonomy*. Unwin Hyman, London.

BROMLEY, R. G. & EKDALE, A. A. 1984. *Chondrites*: a trace fossil indicator of anoxia in sediments. *Science*, **224**, 872–874.

BUDIL, P., HÖRBINGER, F. & MENCL, R. 2009. Lower Devonian dalmanitid trilobites of the Prague Basin (Czech Republic). *Transactions of the Royal Society of Edinburgh, Earth and Environmental Sciences*, **99**, 61–100.

BUGGISCH, W. & MANN, U. 2004. Carbon isotope stratigraphy of Lochkovian to Eifelian limestones from the

Devonian of central and southern Europe. *International Journal of Earth Sciences*, **93**, 521–541.

Chlupáč, I. 1957. Facial development and biostratigraphy of the Lower Devonian of Central Bohemia. *Sborník Ústředního ústavu geologického, Oddíl geologický*, **20**, 277–347.

Chlupáč, I. 1959. Faciální vývoj a biostratigrafie břidlic dalejských a vápenců hlubočepských (Eifel) ve středočeském devonu. *Sborník Ústředního ústavu geologického, Oddíl geologický*, **25**, 445–511.

Chlupáč, I. 1977. *The Phacopid Trilobites of the Silurian and Devonian of Czechoslovakia*. Rozpravy Ústředního Ústavu Geologického, **43**. Vydal Ústřední ústav geologický v Academii, Prague.

Chlupáč, I. 1982. Preliminary submission for Lower Middle Devonian boundary stratotype in the Barrandian area. *Courier Forschungsinstitut Senckenberg*, **55**, 85–96.

Chlupáč, I. 1983. Trilobite assemblages in the Devonian of the Barrandian area and their relations to palaeoenvironments. *Geologica et Palaentologica*, **17**, 43–73.

Chlupáč, I. 1985. Comments of the Lower–Middle Devonian boundary. *Courier Forschungsinstitut Senckenberg*, **75**, 389–400.

Chlupáč, I. 1998. Devonian. *In*: Chlupáč, I., Havlíček, V., Kříž, J., Kukal, Z. & Štorch, P. (eds) *Palaeozoic of the Barrandian (Cambrian to Devonian)*. Czech Geological Survey, Prague, 101–133.

Chlupáč, I. & Kukal, Z. 1986. Reflection of possible global Devonian events in the Barrandian area, C.S.S.R. *Lecture Notes in Earth Sciences*, **8**, 169–179.

Chlupáč, I. & Kukal, Z. 1988. Possible global events and the stratigraphy of the Palaeozoic of the Barrandian (Cambrian–Middle Devonian, Czechoslovakia). *Sborník geologických věd, Geologie*, **43**, 83–146.

Chlupáč, I. & Turek, V. 1983. *Devonian Goniatites from the Barrandian Area, Czechoslovakia*. Rozpravy Ústředního Ústavu Geologického, **46**. Vydal Ústřední ústav geologický v Academii, Prague.

Chlupáč, I., Lukeš, P. & Zikmundová, J. 1977. *Barrandian 1977. A Field Trip Guidebook, Field Conference of the International Subcommission on Devonian Stratigraphy*. Czech Geological Survey, Prague.

Chlupáč, I., Lukeš, P. & Zikmundová, J. 1979. The Lower–Middle Devonian boundary beds in the Barrandian area, Czechoslovakia. *Geologica et Palaeontologica*, **13**, 125–156.

Chlupáč, I., Kříž, J. & Schönlaub, H. P. 1980. Field Trip E. Silurian and Devonian conodont localities of the Barrandian. *In*: Schönlaub, H. P. (ed.) *Second European Conodont Symposium (ECOS II)*. Abhandlungen der Geologischen Bundes-Anstalt, **35**, 147–180.

Chlupáč, I., Havlíček, V., Kříž, J., Kukal, Z. & Štorch, P. 1998. *Palaeozoic of the Barrandian (Cambrian to Devonian)*. Czech Geological Survey, Prague.

Clayton, G., Paterson, N. W. *et al.* 2012. Palynostratigraphy and palynofacies of Upper Devonian rocks in the Appalachian Basin, U.S.A. (abstract). *In*: *A Joint Meeting of the 45th Annual Meeting of AASP – The Palynological Society and Meeting of the CIMP – Commission Internationale de la Microflore Paléozoïque Subcommissions Program and Abstracts*, 21–25 July 2012, Lexington, KY. *AASP – The Palynological Society Newsletter*, **45** (Special Issue 1), 13–14.

Collinson, C. & Scott, A. J. 1958. *Chitinozoan Faunule of the Devonian Cesar Valley Formation*. Illinois State Geological Survey Circular, **247**.

Čorná, O. 1969. Bemerkungen zur Verbreitung palynologischer Mikrofossilien vom Präkambrium bis zum Unterkarbon. *Geologisches Zentralblatt Bratislava*, **20**, 399–416.

DeSantis, M. K. & Brett, C. E. 2011. Late Eifelian (Middle Devonian) biocrises: timing and signature of the pre-Kačák Bakoven and Stony Hollow Events in eastern North America. *Palaeogeography, Palaeoclimatology, Palaeoecology*, **304**, 113–135, http://doi.org/10.1016/j.palaeo.2010.10.013

Deunff, J. 1955. Un microplancton fossile dévonien à Hystrichosphères du continent nord-américain. *Bulletin Microscopie appliquée*, série 2, **5**, 138–149.

Deunff, J. 1961. Quelques précisions concernant les Hystrichosphères du Dévonien du Canada. *Compte rendu sommaire des séances de la Société géologique de France*, **8**, 216–218.

Dotzler, N., Taylor, T. N. & Krings, M. 2007. A prasinophycean alga of the genus *Cymatiosphaera* in the Early Devonian Rhynie chert. *Review of Palaeobotany and Palynolgy*, **147**, 106–111.

Dunn, D. L. & Miller, T. H. 1964. A distinctive chitinozoan from the Alpena Limestone (Middle Devonian) of Michigan. *Journal of Paleontology*, **38**, 725–728.

Dutro, J. T. 1981. Devonian brachiopod biostratigraphy. *In*: Oliver, W. A. & Klapper, G. (eds) *Devonian Biostratigraphy of New York*. International Union of Geological Sciences, Subcommission on Devonian Stratigraphy, Washington, DC, 67–92.

Eisenack, A. 1931. Neue Mikrofossilien des baltischen Silurs 1. *Paläontologische Zeitschrift*, **13**, 74–118.

El Atfy, H., Brocke, R. & Uhl, D. 2013. A fungal proliferation near the probable Oligocene/Miocene boundary, Nukhul Formation, Gulf of Suez, Egypt. *Journal of Micropalaeontology*, **32**, 183–195.

Elrick, M., Berkyová, S., Klapper, G., Sharp, Z., Joachimski, M. & Frýda, J. 2009. Stratigraphic and oxygen isotope evidence for My-scale glaciation driving eustasy in the Early–Middle Devonian greenhouse world. *Palaeogeography, Palaeoclimatology, Palaeoecology*, **276**, 170–181, http://doi.org/10.1016/j.palaeo.2009.03.008

Elsik,, W. C. 1999. *Reduviasporonites* Wilson, 1962: synonymy of the fungal organism involved in the Late Triassic crisis. *Palynology*, **23**, 37–41.

Ernst, A., May, A. & Marks, S. 2012. Bryozoans, corals, and microfacies of Lower Eifelian (Middle Devonian) limestones at Kierspe, Germany. *Facies*, **58**, 727–758.

Fatka, O. 1999. Organic walled microfossils of the Barrandian area: a review. *Journal of the Czech Geological Society*, **44**, 31–42.

Fatka, O. & Brocke, R. 2008. Morphological variability and method of opening of the Devonian acritarch *Navifusa bacilla* (Deunff, 1955) Playford, 1977. *Review of Palaeobotany and Palynology*, **148**, 108–123.

Filipak, P. & Zaton, M. 2011. Plant and animal cuticle remains from the Lower Devonian of southern Poland and their palaeoenvironmental significance. *Lethaia*, **44**, 397–409.

Fisher, D. W. 1962. Small conoidal shells of uncertain affinities. *In*: Moore, R. C. (ed.) *Treatise on Invertebrate Paleontology. Part W: Miscellanea*. Geological Society of America, Boulder, CO and University of Kansas Press, Lawrence, KS, 98–143.

Foster, C. B., Stephenson, M. H., Marshall, C., Logan, G. A. & Greenwood, P. F. 2002. A revision of *Reduviasporonites* Wilson, 1962: description, illustration, comparison and biological affinities. *Palynology*, **26**, 35–58.

Frýda, J. & Bandel, K. 1997. *New Early Devonian gastropods from the Plectonotus (Boucotonotus)–Palaeozygopleura community in the Prague Basin (Bohemia)*. Mitteilungen aus dem Geologisch–Paläontologischen Institut der Universität Hamburg, **80**.

Frýda, J., Ferrová, L., Berkyová, S. & Frýdová, B. 2008. A new Early Devonian palaeozygopleurid gastropod from the Prague Basin (Bohemia) with notes on the phylogeny of the Loxonematoidea. *Bulletin of Geosciences*, **83**, 93–100.

Frýda, J., Ferrová, L. & Frýdová, B. 2013. Review of palaeozygopleurid gastropods (Palaeozygopleuridae, Gastropoda) from Devonian strata of the Perunica microplate (Bohemia), with a re-evaluation of their stratigraphic distribution, notes on their ontogeny, and descriptions of new taxa. *Zootaxa*, **3669**, 469–489.

Galle, A. & Hladil, J. 1991. Lower Paleozoic corals of Bohemia and Moravia. VI. *Fossil Cnidaria including Archaeocyatha and Porifera, Münster, Germany, Excursion Guidebook*, **B3**. International Association for the Study of Fossil Cnidaria and Porifera, Westfälische Wilhelms Universität, Münster, 1–83.

Garcia-López, S. & Sanz-López, J. 2002. Devonian to Lower Carboniferous conodont biostratigraphy of the Bernesga Valley section (Cantabrian Zone, NW Spain). *Cuadernos del Museo Geominero*, **1**, 163–205.

Grabau, A. W. & Shimer, H. W. 1910. *North American Index Fossils*, Vol. II. Seiler and Co., New York.

Grahn, Y. 2011. Chapter 3: Re-examination of Silurian and Devonian Chitinozoa described and illustrated by Lange between 1949–1967. *In*: Bosetti, E. P., Grahn, Y. & Melo, J. H. G. (eds) *Essays in Honour of Frederico Waldemar Lange*. Interciencia, Rio de Janeiro, 28–115.

Gürich, G. 1896. *Das Palaeozoicum im polnischen Mittelgebirge: Russisch-Kaiserliche Mineralogische Gesellschaft*. St Petersburg, Verhandlungen, Serie 2, **32**.

Havlíček, V. & Kukal, Z. 1990. Sedimentology, benthic communities, and brachiopods in the Suchomasty (Dalejan) and *Acanthopyge* (Eifelian) Limestones of the Koněprusy area (Czechoslovakia). *Sborník geologických věd, Paleontologie*, **31**, 105–205.

Henn, A. 1985. *Biostratigraphie und Facies des hohen Unter-Devon bis tiefen Ober-Devon der Provinz Palencia, Kantabrisches Gebirge, N-Spanien*. Göttinger Arbeiten zur Geologie und Paläontologie, **26**.

Higgs, K. & Hughes, G. 2012. Palynology of some Late Givetian and Frasnian shale sequences in the Appalachian Basin of western New York State, USA (abstract). *In*: *A Joint Meeting of the 45th Annual Meeting of AASP – The Palynological Society and Meeting of the CIMP – Commission Internationale de la Microflore Paléozoïque Subcommissions Prgram and Abstracts*, 21–25 July 2012, Lexington, KY. *AASP – The Palynological Society Newsletter*, **45** (Special Issue 1), 28–29 .

Hladil, J. & Kalvoda, J. 1993. Extinction and recovery successions of the Devonian marine shoals; the Eifelian–Givetian and Frasnian-Famennian events in Moravia and Bohemia. *Bulletin of the Czech Geological Survey*, **68**, 13–23.

Holcová, K. 2002. Silurian and Devonian foraminifers and other acid-resistant microfossils from the Barrandian area. *Acta Nationalis Pragae, Series B, Natural History*, **58**, 83–140.

Holcová, K. 2003. Foraminiferal assemblages in acid residues from the 'Císařská rokle' Gorge at Srbsko (the Lower/Middle Devonian boundary interval, Barrandian area) and their paleoenvironmental significance. *Bulletin of Geosciences*, **78**, 393–403.

Holcová, K. 2004*a*. Detailed analysis of the smaller acid-resistant foraminifera from the Lower/Middle Devonian boundary beds in the Barrandian area (Czech Republic): implication for the paleoecology of the Devonian foraminifera. *Journal of Foraminiferal Research*, **34**, 214–231.

Holcová, K. 2004*b*. Foraminifers from the Lower/Middle Devonian boundary beds of the Barrandian area, Czech Republic, and their paleoecology. *Journal of Foraminiferal Research*, **34**, 214–231.

Holcová, K. 2004*c*. Silurian and Devonian foraminifera from the Barrandian area (Czech Republic). *In*: Bubík, M. & Kaminski, M. A. (eds) *Proceedings of the Sixth International Workshop on Agglutinated Foraminifera*. Grzybowski Foundation, Special Publications, **8**, 167–184.

Holcová, K. & Slavík, L. 2013. The morphogroups of small agglutinated foraminifera from the Devonian carbonate complex of the Prague Synform (Barrandian area, Czech Republic). *Palaeogeography, Palaeoclimatology, Palaeoecology*, **386**, 210–224.

Horný, R. J. 1955. Palaeozygopleuridae nov. fam. (Gastropoda) ze středočeského devonu. *Sborník Ústředního ústavu geologického, Oddíl paleontologický*, **21**, 17–159.

Hotchkiss, F. H. C., Prokop, R. J. & Petr, V. 1999. Isolated skeletal ossicles of a new brittlestar of the Family Cheiropterasteridae Spencer, 1934 (Echinodermata: Ophiuroidea) in the Lower Devonian of Bohemia (Czech Republic). *Journal of the Czech Geological Society*, **44**, 189–193.

Hotchkiss, F. H. C., Prokop, R. J. & Petr, V. 2007. Isolated ossicles of the Family Eospondylidae Spencer & Wright, 1966, in the Lower Devonian of Bohemia (Czech Republic) and correction of the systematic position of eospondylid brittlestars (Echinodermata: Ophiuroidea: Oegophiurida). *Acta Musei Nationalis Pragae, Series B: Historia Naturalis*, **63**, 3–18.

House, M. R. 1973. An analysis of Devonian Goniatite distributions. *Special Papers in Palaeontology*, **12**, 305–317.

House, M. R. 1981. Lower and Middle Devonian goniatite biostratigraphy. *In*: Oliver, W. A. & Klapper,, G. (eds) *Devonian Biostratigraphy of New York*. International Union of Geological Sciences, Subcommission on Devonian Stratigraphy, Washington, DC, 33–37.

House, M. R. 1985. Correlation of mid-Palaeozoic ammonoid evolutionary events with global sedimentary perturbations. *Nature*, **313**, 17–22.

House, M. R. 2002. Strength, timing, setting and cause of mid-Palaeozoic extinctions. *Palaeogeography, Palaeoclimatology, Palaeoecology*, **181**, 5–25.

Hughes, G. & Higgs, K. T. 2011. Correlation of spores, acritarchs and U–Pb isochrons in the Lower Devonian of Cherry Valley, New York State, USA. Palaeozoic palynology. Papers presented at the Palaeozoic palynology abstracts from the 44th AASP-TPS Meeting, 3–7 September 2011, Southampton, UK.

Jenkins, W. A. M. & Legault, J. A. 1979. Stratigraphic ranges of selected Chitinozoa. *Palynology*, **3**, 235–264.

Johnson, J. G., Klapper, G. & Sandberg, C. A. 1985. Devonian eustatic fluctuations in Euramerica. *Geological Society of America Bulletin*, **96**, 567–587.

Kalgutkar, R. M. & Jansonius, J. 2000. *Synopsis of Fossil Fungal Spores, Mycelia and Frutifications*. AASP Contribution Series, **39**.

Karpinsky, A. 1884. *Die fossilen Pteropoden am Ostabhange des Ural*. Mémoire de l'Académie des Sciences, 7th Series, **32**.

Kaufmann, B. 1998. Facies, stratigraphy and diagenesis of Middle Devonian reef- and mud-mounds in the Mader (eastern Anti-Atlas, Morocco). *Acta Geologica Polonica*, **48**, 43–106.

Kirchgasser, W. T. 2000. Correlation of stage boundaries in the Appalachian Devonian, eastern United States. *Courier Forschungsinstitut Senckenberg*, **225**, 271–284.

Klapper, G. 1977. Lower-Middle Devonian boundary conodont sequence in the Barrandian area of Czechoslovakia. *Časopis pro mineralogii a geologii*, **22**, 401–410.

Klapper, G. 1981. Review of New York Devonian conodont biostratigraphy. *In*: Oliver, W. A. & Klapper, G. (eds) *Devonian Biostratigraphy of New York*. International Union of Geological Sciences, Subcommission on Devonian Stratigraphy, Washington, DC, 57–76.

Klapper, G. & Oliver, W. A., Jr. 1995. The Detroit River Group is Middle Devonian: discussion on 'Early Devonian age of the Detroit River Group, Inferred from Arctic stromatoporoids'. *Canadian Journal of Earth Sciences*, **32**, 1070–1073.

Klapper, G. & Vodrážková, S. 2013. Ontogenetic and intraspecific variation in the late Emsian–Eifelian (Devonian) conodonts *Polygnathus serotinus* and *P. bultyncki* in the Prague Basin (Czech Republic) and Nevada (western U.S.). *Acta Geologica Polonica*, **63**, 153–174.

Klapper, G., Ziegler, W. & Mashkova, T. V. 1978. Conodonts and correlation of Lower-Middle Devonian boundary beds in the Barrandian area of Czechoslovakia. *Geologica et Palaeontologica*, **12**, 103–116.

Klug, C., Korn, D. & Reisdorf, A. 2000. Ammonoid and conodont stratigraphy of the late Emsian to early Eifelian (Devonian) at the Jebel Ouaoufilal (near Taouz, Tafilalt, Morocco). *Travaux de l'Institut Scientifique, Rabat, Série Géologique & Géographique Physical*, **20**, 45–56.

Klug, C., Korn, D., Naglik, C., Frey, L. & De Baets, K. 2013. The Lochkovian to Eifelian succession of the Amessoui Syncline (Southern Tafilalt). *In*: Becker, R. T., El Hassani, A. & Tahiri, A. (eds) *International Field Symposium 'The Devonian and Lower Carboniferous of northern Gondwana', Field Guidebook*. Document de l'Institut Scientifique, Rabat, **27**, 51–59.

Koptíková, L. 2011. Precise position of the Basal Choteč event and evolution of sedimentary environments near the Lower–Middle Devonian boundary: the magnetic susceptibility, gamma-ray spectrometric, lithological, and geochemical record of the Prague Synform (Czech Republic). *Palaeogeography, Palaeoclimatology, Palaeoecology*, **304**, 96–112, http://doi.org/10.1016/j.palaeo.2010.10.011

Kremer, B. 2005. Mazuelloids: products of post-mortem phosphatization of acanthomorphic acritarchs. *Palaios*, **20**, 27–36.

Lardeux, H. 1969. *Les Tentaculites d'Europe occidentale et d'Afrique du Nord*. Cahiers de Paléontologie, Centre National de Recherche Scientifique, Paris.

Legault, J. A. 1973*a*. Mode of aggregation of *Hoegisphaera* (Chitinozoa). *Canadian Journal of Earth Sciences*, **10**, 793–797.

Legault, J. A. 1973*b*. Chitinozoa and Acritarcha of the Hamilton Formation of southwestern Ontario. *Geological Survey of Canada Bulletin*, **221**, 1–103.

Lele, K. M. 1968. Preliminary note on the miospores and other microfossils from the Srbsko Formation (Givetian) in the Barrandian Basin (Czechoslovakia). *Abstracts of the papers presented at the Session of the International Palaeontological Union*, 20–27 August, Prague, Czechoslovakia, 16.

Lele, K. M. 1972. Observations on Middle Devonian microfossils from the Barrandian Basin, Czechoslovakia. *Review of Palaeobotany and Palynology*, **14**, 129–134.

Liebus, A. & Wahner, F. 1904. Foraminiferenfauna in den Schichten der Etage Gg3. *Sitzungsberichte des Deutschen Naturwissenschaftlich-Medizinischen Vereines für Böhmen 'Lotos' in Prag (Prag), NF24*, **52**, 11.

Lindemann, R. H. 2008. Taxonomic revision of *Styliolina fissurella strigata* (Hall): *Costulatostyliolina strigata* (Hall) from the Devonian (Eifelian) of New York. *Northeastern Geology and Environmental Sciences*, **30**, 289–294.

Lindemann, R. H. & Yochelson, E. L. 1992. *Viriatellina* (Dacryoconarida) from the Middle Devonian Ludlowville Formation at Alden, New York. *Journal of Paleontology*, **66**, 193–199.

Lindemann, R. H. & Yochelson, E. L. 1994. Redescription of *Styliolina* [INCERTAE SEDIS] – *Styliolina fissurella* (Hall) and the type species *S. nucleata* (Karpinsky). *In*: Landing, E. (ed.) *Studies in Stratigraphy and Paleontology in Honor of Donald W. Fisher*. *New York State Museum Bulletin*, **481**, 149–160.

LISTER, T. R. 1970. The acritarchs and chitinozoa from the Wenlock and Ludlow series of the Ludlow and Millichope areas, Shropshire. *Palaeontographical Society (Monograph)*, **124**, 1–100.

LJASCHENKO, G. P. 1966. Novyje rody devonskych Nowakii. *Paleontologicheskij Sbornik*, **3**, 49–53.

LOEBLICH, A. R. 1970. Morphology, ultrastructure and distribution of Paleozoic acritarchs. *In*: *Morphology, Ultrastructure and Distribution of Paleozoic Acritarchs. Proceedings of the North American Paleontological Convention, 1969, Part G*. Allen Press, Lawrence, KS, 705–788.

LOEBLICH, A. R. & WICANDER, E. R. 1974. New Early Devonian (Late Gedinnian) microphytoplankton: *demorhetium lappaceum* n.g., n.sp. from the Bois d'Arc Formation of Oklahoma, U.S.A. *Neues Jahrbuch für Geologie und Paläontologie, Monatshefte*, **12**, 707–711.

LOEBLICH, A. R. & WICANDER, E. R. 1976. Organic-walled microplankton from the Lower Devonian, Late Gedinnian Haragan and Bois d'Arc formations of Oklahoma, U.S.A., part 1. *Palaeontographica, Abteilung B*, **159**, 1–39.

LUKEŠ, P. 1989. Tentaculites from the Lower/Middle Devonian section in Prague-Barrandov. *Věstník Ústředního ústavu geologického*, **64**, 194–206.

LÜTKE, F. 1985. Devonian Tentaculites from Nevada. *Courier Forschungsinstitut Senckenberg*, **75**, 197–226.

MACHADO, G., HLADIL, J., SLAVIK, L., KOPTÍTOVA, L., MOREIRA, N., FONSECA, M. & FONSECA, P. 2010. An Emsian–Eifelian calciturbidite sequence and the possibile correlatable pattern of the basal Choteč Event in western Ossa-Morena Zone, Protugal (Odivelas Limestone). *Geologica Belgica*, **13**, 431–446.

MANDA, Š. & TUREK, V. 2011. Late Emsian Rutoceratoidea (Nautiloidea) from the Prague Basin, Czech Republic: morphology, diversity and palaeoecology. *Palaeontology*, **54**, 911–1024.

MCGREGOR, D. C. 1961. Spores with proximal radial pattern from the Devonian of Canada. *Geological Survey of Canada, Bulletin*, **76**, 1–11.

MCGREGOR, D. C. 1973. Lower and Middle Devonian spores of eastern Gaspé, Canada. I Systematics. *Palaeontographica, Abteilung B*, **142**, 1–77.

MCGREGOR, D. C. 1974. Early Devonian spores from central Ellesmere Island, Canadian Arctic. *Canadian Journal of Earth Sciences*, **11**, 70–78.

MCGREGOR, D. C. 1977. Lower and Middle Devonian spores of eastern Gaspé, Canada. II. Biostratigraphic significance. *Palaeontographica, Abteilung B*, **163**, 111–142.

MCGREGOR, D. C. 1979*a*. Devonian spores from the Barrandian region of Czechoslovakia and their significance for interfacies correlation. *Geological Survey of Canada Paper*, **79–1B**, 189–197.

MCGREGOR, D. C. 1979*b*. Devonian spores of North America. *Palynology*, **3**, 31–52.

MCGREGOR, D. C. 1980. *Palynology of samples submitted by I. Chlupáč (Czechoslovakian Geological Survey) from Devonian of the Barrandian region of Czechoslovakia*. Report F1-2-1980-DCM

MCGREGOR, D. C. 1981*a*. *Palynology of samples of Devonian rocks from the Barrandian region of Czechoslovakia, submitted by I. Chlupáč of the Czechoslovakian Geological Survey*. Report F1-7-1981-DCM

MCGREGOR, D. C. 1981*b*. Spores and the Middle–Upper Devonian boundary. *Review of Palaeobotany and Palynology*, **34**, 25–47.

MCGREGOR, D. C. 1983. *Palynology of samples of rock submitted by I. Chlupáč of the Czechoslovakian Geological Survey, from Lower Devonian strata of the Barrandian region of Czechoslovakia*. Report F1-12-1983-DCM

MCGREGOR, D. C. & CAMFIELD, M. 1976. Upper Silurian? to Middle Devonian spores of the Moose River Basin, Ontario. *Geological Survey of Canada Bulletin*, **263**, 1–63.

MCGREGOR, D. C. & CAMFIELD, M. 1982. Middle Devonian miospores from the Cape De Bray, Weatherhall, and Hecla Bay Formations of northeastern Melville Island, Canadian Arctic. *Geological Survey of Canada Bulletin*, **348**, 1–105.

MCGREGOR, D. C. & OWENS, B. 1966. Devonian spores of eastern and northern Canada. *Geological Survey of Canada Paper*, **66-30**, 1–66.

MERGL, M. 2001. Lingulate brachiopods of the Silurian and Devonian of the Barrandian. *Acta Musei nationalis Pragae, Series B – historia naturalis*, **57**, 1–49.

MERGL, M. 2008. Lingulate brachiopods from the *Acanthopyge* Limestone (Eifelian) of the Barrandian, Czech, Republic. *Bulletin of Geosciences*, **83**, 281–298, http://doi.org/10.3140/bull.geosci.2008.03.281

MERGL, M. & VODRÁŽKOVÁ, S. 2012. Emsian–Eifelian lingulate brachiopods from the Daleje–Třebotov Formation (Třebotov and Suchomasty limestones) and the Choteč Formation (Choteč and Acanthopyge limestones) from the Prague Basin; the Czech Republic. *Bulletin of Geosciences*, **87**, 315–332.

MONTESINOS, J. R. 1987. Agoniatitina y Anarcestina del Dévonico Medio de la Cordillera Cantábrica (Dominios Palentino y Asturleonés, LO de Espana. *Cuaderno Laboratoria Xeooxico de Laxe*, **12**, 99–118.

OBRHEL, J. 1958. Die Flora der Choteč Kalke und Třebotov-Kalke (Eifel) des mittelböhmischen Devons. *Sborník Ústředního ústavu geologického*, **25**, 99–107.

OBRHEL, J. 1964. Ein problematisches Mikrofossil aus dem Devon Böhmens. *Věstník Ústředního ústavu geologického*, **39**, 217–218.

OBRHEL, J. 1968. Die Silur- Devonflora des Barrandiums. *Paläontologische Abhandlungen, B*, **2**, 635–703.

OLIVER, W. A., Jr. 1954. Stratigraphy of the Onondaga Limestone (Devonian) in central New York. *Geological Society of America Bulletin*, **65**, 621–652.

OLIVER, W. A., Jr. 1956. Stratigraphy of the Onondaga Limestone (Devonian) in central New York. *Geological Society of America Bulletin*, **67**, 1441–1474.

OLIVER, W. A., Jr. 1966. The Bois Blanc and Onondaga formations in western New York and adjacent Ontario. *In*: *New York State Geological Association, Guide Book, 38th Annual Meeting*. New York State Geological Association, Rochester, NY, 32–43.

OLIVER, W. A., Jr. 1967. *Stratigraphy of the Bois Blanc Formation in New York*. United States Geological Survey, Professional Papers, **584-A**, 1–8.

Oliver, W. A., Jr. 1977. Biogeography of Late Silurian and Devonian rugose corals. *Palaeogeography, Palaeoclimatology, Palaeoecology*, **22**, 85–135.

Paris, F. 1981. *La Tranchée de la lezais emsien supérieur du Massif Armoricain. Sédimentologie, paléontologie, stratigraphie*. Mémoires de la Société géologique et minéralogique de Bretagne, **24**.

Paris, F., Girard, C., Feist, R. & Winchester-Seeto, T. 1996. Chitinozoan bio-event in the Frasnian–Famennian boundary beds at La Serre (Montagne Noire, Southern France). *Paleogeography, Palaeoclimatology, Paleoecology*, **121**, 131–145.

Paris, F., Grahn, Y., Nestor, V. & Lakova, I. 1999. A revised Chitinozoan classification. *Journal of Paleontology*, **73**, 549–570.

Pedder, A. E. H. 2010. Lower–Middle Devonian rugose coral faunas of Nevada: contribution to an understanding of the 'barren' E Zone and Choteč Event in the Great Basin. *Bulletin of Geosciences*, **85**, 1–26.

Peppers, R. A. & Damberger, H. H. 1969. Palynology and petrography of a Middle Devonian coal in Illinois. *Illinois State Geological Survey Circular*, **445**, 1–36.

Playford, G. 1977. Lower to Middle Devonian acritarchs of the Moose River Basin, Ontario. *Geological Society of Canada Bulletin*, **279**, 1–87.

Playford, G. & McGregor, D. C. 1993. Miospores and organic walled microphytoplankton of Devonian-Carboniferous boundary beds (Bakken Formation), southern Saskatchewan: a systematic and stratigraphic appraisal. *Geological Survey of Canada Bulletin*, **445**, 1–107.

Pokorný, V. 1958. Nálezy foraminifer v souvrství vápenců hlubočepských (Eifel). *Časopis pro mineralogii a geologii*, **4**, 167–169.

Prauss, M. L. 2007. Availability of reduced nitrogen chemospecies in photic zone waters as the ultimate cause for fossil prasinophyte prosperity. *Palaios*, **22**, 489–499.

Přibyl, A. & Šnajdr, M. 1950. On new Ostracoda from the Choteč Limestone-g2 (Middle Devonian) of Holyně near Praque. *Sborník Ústředního ústavu geologického Československé Republiky, Oddělení Paleontologie*, **17**, 101–119.

Prokop, R. J. 1970. Family Calceocrinidae Meek and Worthen, 1869 (Crinoidea) in the Silurian and Devonian of Bohemia. *Sborník geologických věd, Paleontologie*, **12**, 79–134.

Prokop, R. J. 1976. The genus *Edriocrinus* HALL, 1859 from the Devonian of Bohemia (Crinoidea). *Časopis pro mineralogii a geologii*, **21**, 187–191.

Prokop, R. J. 1987. The stratigraphical distribution of Devonian crinoids in the Barrandian area (Czechoslovakia). *Newsletters on Stratigraphy*, **17**, 101–107.

Prokop, R. J. 2012. A new species of the genus *Follicrinus* (Echinodermata, Crinoidea) from the Middle Devonian (Eifelian) of the Czech Republic. *Journal of the National Museum (Prague), Natural History Series*, **181**, 1–3.

Prokop, R. J. 2013. *Simakocrinus* gen. nov. (Crinoidea) from the Bohemian Early and Middle Devonian of the Barrandian area (the Czech Republic). *Acta Musei Nationalis Pragae, Series B, Historia Naturalis*, **69**, 65–68.

Prokop, R. J. & Petr, V. 1997. The genus *Pygmaeocrinus* Bouška, 1947 (Crinoidea, Inadunata) in the Devonian of the Barrandian area (Czech Republic). *Acta Musei Nationalis Pragae, Series B, Historia Naturalis*, **53**, 1–10.

Prokop, R. J. & Petr, V. 2004. Pleurocystitidae indet. (Cystoidea, Rhombifera) in the Bohemian Devonian (Czech Republic). *Journal of the National Museum, Natural History Series*, **173**, 1–5.

Ravn, R. L. & Benson, D. G. 1988. Devonian miospores and reworked acritarchs from southeastern Georgia, U.S.A. *Palynology*, **12**, 179–200.

Requadt, H. & Weddige, K. 1978. Lithostratigraphie und Conodontenfaunen der Wissenbacher Fazies und ihrer Äquivalente in der südwestlichen Lahnmulde (Rheinisches Schiefergebirge). *Mainzer Geowissenschaftliche Mitteilungen*, **7**, 183–237.

Richardson, J. B. & Ahmed, S. 1988. Miospores, zonation and correlation of Upper Devonian sequences from western New York and Pennsylvania. *In*: McMillian, N. J., Embry, A. F. & Glass, D. J. (eds) *Devonian of the World. Proceedings of the 2nd International Symposium on the Devonian System*. Canadian Society of Petroleum Geologists, Memoirs, **14**, 541–558.

Richardson, J. B. & McGregor, D. C. 1986. Silurian and Devonian spore zones of the Old Red Sandstone Continent and adjacent regions. *Geological Survey of Canada Bulletin*, **364**, 1–79.

Riding, R. & Guo, L. 1992. Affinity of Tubiphytes. *Palaeontology*, **35**, 37–49.

Riegel, W. 2008. The Late Palaeozoic Phytoplankton Blackout – artefact or evidence of global change? *Review of Palaeobotany and Palynology*, **148**, 73–90.

Riegel, W., Loh, H., Maul, B. & Prauss, M. 1986. Effects and causes in a black shale event – the Toarcian Posidonia shale of NW Germany. *Lecture Notes in Earth Science*, **8**, 267–276.

Roden, M. K.,, Parrish, R. R. & Miller, D. S. 1990. The absolute age of the Eifelian Tioga Ash Bed, Pennsylvania. *Journal of Geology*, **98**, 282–285.

Rooney, A., Clayton, G. & Goodhue, R. 2013. The dispersed spore *Retusotriletes loboziakii* sp. nov., affiliated with the enigmatic Late Devonian alga *Protosalvinia* Dawson, 1884. *Palynology*, **37**, 196–201.

Savrda, C. E. & Bottjer, D. J. 1986. Trace-fossil model for reconstruction of paleo-oxygenation in bottom waters. *Geology*, **14**, 3–6.

Schubert, M. 1996. Die dysaerobe Biofazies der Wissenbacher Schiefer (Rheinisches Schiefergebirge, Harz, Devon). *Göttinger Arbeiten zur Geologie und Paläontologie*, **68**, 1–131.

Schubert, R. J. & Liebus, A. 1902. Vorläufige Mittheilung über Foraminiferen aus dem Böhmischen Devon (Etage Gg3 Barr.). *Verhandlungen der Kaiserlich-Königlichen Geologischen Reichsanstalt*, **1902**, 66.

Seilacher, A. 2007. *Trace Fossil Analysis*. Springer, Berlin.

Šlechta, O. 1996. Ostracode faunas from the Lower/Middle Devonian section in Praha Barrandov. *Bulletin of the Czech Geological Survey*, **71**, 135–144.

Šnajdr, M. 1960. *Studie o čeledi Scutelluidae (Trilobitae)*. Rozpravy Ústředního Ústavu Geologického, **26**. Vydal Ustredni ustav geologicky v Academii, Prague.

ŠNAJDR, M. 1980. *Silurian and Devonian Proetidae (Trilobita)*. Rozpravy Ústředního Ústavu Geologického, **45**. Vydal Ustredni ustav geologicky v Academii, Prague.

STAPLIN, F. C. 1961. Reef-controlled distribution of Devonian microplankton in Alberta. *Palaeontology*, **4**, 329–424.

STEINER, M. B., ESHET, Y., RAMPINO, M. R. & SCHWINDT, D. M. 2003. Fungal abundance spike and the Permian–Triassic boundary in the Karoo Supergroup, (South Africa). *Palaeogeography, Palaeoclimatology, Palaeoecology*, **194**, 405–414.

STEPHENSON, M. H., WILLIAMS, M., LENG, M. J. & MONAGHAN, A. A. 2004. Aquatic plant microfossils of probable non-vascular origin from the Ballagan Formation (Lower Carboniferous), Midland Valley, Scotland. *Proceedings of the Yorkshire Geological Society*, **55**, 145–158, http://doi.org/10.1144/pygs.55.2.145

STRUVE, W. 1982. The Eifelian within the Devonian frame, history, boundaries, definitions. *Courier Forschungsinstitut Senckenberg*, **55**, 401–432.

STRUVE, W., PLODOWSKI, G. & WEDDIGE, K. 1997. Biostratigraphische Stufengrenzen und Events in der Prümer und Hillesheimer Mulde. *Terra Nostra*, **97/7**, 123–167.

STRUVE, W., BASSE, M. & WEDDIGE, K. 2008. Prädevon, Ober-Emsium und Mitteldevon der Eifeler Kalkmuldenzone. *Schriftenreihe der Deutschen Gesellschaft für Geowissenschaften*, **52**, 297–374.

SVOBODA, J. & PRANTL, F. 1947. O stratigrafii a tektonice staršího paleozoika v okolí Třebotova [The stratigraphy and tectonics of the Lower Palaeozoic rocks in the vicinity of Třebotov]. *Sborník Státního geologického ústavu Československé republiky*, **14**, 281–314.

SVOBODA, J. & PRANTL, F. 1948. O stratigrafiii a tektonice staršího paleozoika v okolí Chýnice. *Sborník Státního geologického ústavu Československé republiky*, **15**, 1–39.

TAPPAN, H. 1980. *The Paleobiology of Plant Protists*. W.H. Freeman & Co., San Francisco, CA.

TAYLOR, W. A. & WELLMAN, C. H. 2009. Ultrastructure of enigmatic phytoclasts (banded tubes) from the Silurian–Devonian: evidence for affinities and role in early terrestrial ecosystems. *Palaios*, **24**, 167–180.

TRAVERSE, A. 2007. *Palaeopalynology*, 2nd edn. Springer, Dordrecht.

TRAVERSE, A. & SCHUYLER, A. 1994. Palynostratigraphy of the Catskill and part of the Chemung Magnafacies, southern New York State, USA. *Courier Forschungsinstitut Senckenberg*, **169**, 261–274.

TROTH, I., MARSHALL, J. E. A., RACEY, A. & BECKER, R. T. 2011. Devonian sea-level change in Bolivia: a high palaeolatitude biostratigraphical calibration of the global sea-level curve. *Palaeogeography, Palaeoclimatology, Palaeoecology*, **304**, 3–20.

TURNER, R. E. 1986. New and revised acritarch taxa from the Upper Devonian (Frasnian) of Alberta, Canada. *Canadian Journal of Earth Sciences*, **23**, 599–607.

TURNER, R. E. 1991. New acritarch taxa from the Middle and Upper Devonian (Givetian–Frasnian) of western Canada. *Canadian Journal of Earth Sciences*, **28**, 1471–1487.

URBAN, J. B. 1972. A reexamination of Chitinozoa from the Cedar Valley Formation of Iowa with observaions on their morphology and distribution. *Bulletin of American Paleontology*, **63**, 1–43.

URBAN, J. B. & KLINE, J. K. 1970. Chitinozoa of the Cedar City Formation, Middle Devonian of Missouri. *Journal of Paleontology*, **44**, 69–76.

URBAN, J. B. & NEWPORT, R. L. 1973. Chitinozoa of the Wapsipinicon Formation (Middle Devonian) of Iowa. *Micropaleontology*, **19**, 239–246.

VAJDA, V. & BERCOVICI, A. 2014. The global vegetation pattern across the Cretaceous–Paleogene mass extinction interval; a template for other extinction events. *Global Planetary Change*, **122**, 29–49.

VAN DE SCHOOTBRUGGE, B., TREMOLADA, F. *ET AL.* 2007. End-Triassic calcification crisis and blooms or organic-walled disaster species. *Palaeogeography, Palaeoclimatology, Palaeoecology*, **244**, 126–141.

VAUGHAN, A. P. M. & PANKHURST, R. J. 2008. Tectonic overview of the West Gondwana margin. *Gondwana Research*, **13**, 150–162.

VAVRDOVÁ, M. 1989. Early Devonian palynomorphs from the Dvorce– Prokop Limestone (Barrandian region, Czechoslovakia). *Věstník Ústředního ústavu geologického*, **64**, 207–224.

VAVRDOVÁ, M. & DAŠKOVÁ, J. 2011. Middle Devonian palynomorphs from southern Moravia: an evidence of rapid change from terrestrial deltaic plain to carbonate platform conditions. *Geologica Carpathica*, **62**, 109–119.

VER STRAETEN, C. A. 2007. Basinwide stratigraphic synthesis and sequence stratigraphy, upper Pragian, Emsian and Eifelian stages (Lower to Middle Devonian), Appalachian Basin. *In*: BECKER, R. T. & KIRCHGASSER, W. T. (eds) *Devonian Events and Correlations*. Geological Society, London, Special Publications, **278**, 39–81, http://doi.org/10.1144/SP278.3

VER STRAETEN, C. A. & BRETT, C. E. 2006. Pragian to Eifelian strata (mid Lower to lower Middle Devonian), northern Appalachian Basin – a stratigraphic revision. *Northeastern Geology*, **28**, 80–95.

VISSCHER, H., BRINKHUISS, H. *ET AL.* 1996. The terminal Paleozoic fungal event: evidence of terrestrial ecosystem destabilization and collapse. *Proceedings of the National Academy of Science of the United States of America*, **93**, 2155–2158.

VISSCHER, H., SEPHTON, A. A. & LOOY, C. V. 2011. Fungal virulence at the time of the end-Permian biosphere crisis? *Geology*, **39**, 883–886.

VODRÁŽKOVÁ, S., FRÝDA, J., SUTTNER, T. J., KOPTÍKOVÁ, L. & TONAROVÁ, P. 2013. Environmental changes close to the Lower–Middle Devonian boundary; the Basal Choteč Event in the Prague Basin (Czech Republic). *Facies*, **59**, 425–449.

WALLISER, O. H. 1984*a*. Geologic processes and global events. *Terra cognita*, **4**, 17–20.

WALLISER, O. H. 1984*b*. Pleading for a Natural D/C-Boundary. *Courier Forschungsinstitut Senckenberg*, **67**, 241–246.

WALLISER, O. H. 1985. Natural boundaries and Commission boundaries in the Devonian. *Courier Forschungsinstitut Senckenberg*, **75**, 401–408.

WALLISER, O. H. 1986. *The IGCP Project 216: 'Global Biological Events in Earth History'*. Lecture Notes in the Earth Sciences, **8**. Springer, Berlin, 1–4.
WALLISER, O. H. 1996. Global events in the Devonian and Carboniferous. *In*: WALLISER, O. H. (ed.) *Global Events in the Phanerozoic*. Springer, Berlin, 225–250.
WAY, J. H., SMITH, R. C. & RODEN, M. 1986. Detailed correlations across 175 miles of the Valley and Ridge of Pennsylvania using 7 ash beds in the Tioga Zone. *In*: SEVON, W. D. (ed.) *Selected Geology of Bedford and Huntington Counties. 51st Annual Field Conference of Pennsylvania Geologists Guidebook*. Pennsylvania Bureau of Topographic and Geological Survey, Harrisburg, PA, 55–72.
WEDDIGE, K. 1988. Eifel conodonts. *Courier Forschungsinstitut Senckenberg*, **102**, 103–110.
WEDDIGE, K. & ZIEGLER, W. 1987. Lithic and faunistic ratios of conodont sample data as facial indicators. *In*: AUSTIN, R. (ed.) *Conodonts: Investigative Techniques and Applications*. British Micropalaeontological Society, London, 333–340.
WICANDER, E. R. 1973*a*. *Diversity and abundance fluctuations in a Late Devonian-Early Mississippian phytoplankton assemblagefrom northeast Ohio, U.S.A.* PhD thesis, University of California, Los Angeles, CA.
WICANDER, E. R. 1973*b*. Marine primary productivity during the Late Devonian and Early Mississippian of Ohio. *Geological Society of America Abstracts*, **5**, 121–122.
WICANDER, E. R. 1973*c*. Phytoplankton abundance and diversity during the Late Devonian and Early Mississippian of Ohio. *American Association of Petroleum Geologists Bulletin*, **54**, 812.
WICANDER, E. R. 1974. Upper Devonian–Lower Mississippian acritarchs and prasinophyceaen algae from Ohio, U.S.A. *Palaeontographica, Abteilung B*, **148**, 9–43.
WICANDER, E. R. 1975. Fluctuations in a Late Devonian–Early Mississippian phytoplankton flora of Ohio, U.S.A. *Palaeogeography, Palaeoclimatology, Palaeoecology*, **17**, 89–108.
WICANDER, E. R. 1983. *A Catalog and Biostratigraphic Distribution of North American Devonian Acritarchs*. American Association of Stratigraphic Palynologists, Contribution Series, **10**.
WICANDER, E. R. 1984. Middle Devonian acritarch biostratigraphy of North America. *Journal of Micropalaeontology*, **3**, 19–24, http://doi.org/10.1144/jm.3.2.19
WICANDER, E. R. 1986. Lower Devonian (Gedinnian) acritarchs from the Haragan Formation, Oklahoma, U.S.A. *Review of Palaeobotany and Palynology*, **47**, 327–365.
WICANDER, E. R., & LOEBLICH, A. R., Jr. 1977. Organic-walled microphytoplankton from the Upper Devonian Antrim Shale, Indiana, U.S.A. and its stratigraphic significance. *Palaeontographica, Abteilung B*, **160**, 129–165.
WICANDER, E. R. & PLAYFORD, G. 1985. Acritarchs and spores from the Upper Devonian Lime Creek Formation, Iowa, U.S.A. *Micropaleontology*, **31**, 97–138.
WICANDER, E. R. & PLAYFORD, G. 2013. Marine and terrestrial palynofloras from transitional Devonian–Mississippian strata, Illinois Basin, U.S.A. *Boletín Geológico y Minero*, **124**, 589–637.
WICANDER, E. R. & SCHOPF, J. W. 1974. Microrganisms from the Kalkberg Formation (Lower Devonian) of New York State. *Journal of Paleontology*, **40**, 74–77.
WICANDER, E. R. & WOOD, G. D. 1979. Organic-walled microphytoplankton of the Middle Devonian Silica Formation, Ohio, U.S.A. *In*: *Abstracts of the 12th Annual Meeting of the American Association of Stratigraphic Palynology*, Dallas, Texas, 31.
WICANDER, E. R. & WOOD, G. D. 1981. *Systematics and Biostratigraphy of the Organic-walled Microphytoplankton from the Middle Devonian (Givetian) Silica Formation, Ohio*. AASP Contribution Series, **8**.
WICANDER, E. R. & WOOD, G. D. 1997. The use of microphytoplankton and chitinozoans for interpreting transgressive/regressive cycles in the Rapid Member of the Cedar Valley Formation (Middle Devonian), Iowa. *Review of Palaeobotany and Palynology*, **98**, 125–152.
WICANDER, E. R. & WRIGHT, R. P. 1983. Organic-walled microphytoplankton abundance and stratigraphic distribution from the Middle Devonian Columbus and Delaware limestones of the Hamilton Quarry, Marion County, Ohio. *Ohio Academy of Science*, **83**, 2–13.
WILDE, P., BERRY, W. B. N. & QUINBY-HUNT, M. S. 1991. Silurian oceanic and atmospheric circulation and chemistry. *Special Papers in Palaeontology*, **44**, 123–143.
WILSON, L. R. 1962. A Permian fungus spore type from the Flowerpot Formation of Oklahoma. *Oklahoma Geological Notes*, **22**, 91–96.
WILSON, L. R. & SKVARLA, J. J. 1967. Electron-microscope study of the wall structure of *Quisquillites* and *Tasmanites*. *Oklahoma Geological Notes*, **27**, 54–63.
WILSON, L. R. & URBAN, J. B. 1963. An *incertae sedis* palynomorph from the Devonian of Oklahoma. *Oklahoma Geological Notes*, **23**, 16–19.
WILSON, L. R. & URBAN, J. B. 1971. Electron microscope studies of the marine palynomorph *Quisquillites*. *Micropaleontology*, **17**, 239–243.
WINSLOW, M. R. 1962. *Plant Spores and Other Microfossils from the Upper Devonian and Lower Mississippian Rocks of Ohio*. United States Geological Survey, Professional Papers, **364**.
WOOD, G. D. 1974. *Chitinozoa of Silica Formation (Middle Devonian, Ohio): Vesicle Ornamentation and Paleoecology*. Michigan State University, Publication Series, **11**, 127–162.
WOOD, G. D. & CLENDENING, J. A. 1985. Organic-walled microphytoplankton and chitinozoans from the Middle Devonian (Givetian) Boyle Dolomite of Kentucky. *Palynology*, **9**, 133–145.
WOOD, G. D. & ELSIK, W. C. 1999. Paleoecologic and stratigraphic importance of the fungus *Reduviasporonites stoschianus* from the Early–Middle Pennsylvanian of the Copacabana Formation, Peru. *Palynology*, **23**, 43–53.
WRIGHT, R. P. 1976. Occurrence, stratigraphic distribution, and abundance of Chitinozoa from the Middle Devonian Columbus limestone of Ohio. *Ohio Journal of Science*, **76**, 127–162.
WRIGHT, R. P. 1978. Biogeography of Middle Devonian chitinozoa of the midwestern United States. *Palionlogia, Numero extraordinario*, **1**, 501–505.

WRIGHT, R. P. 1980. *Middle Devonian Chitinozoa of Indiana*. Indiana Geological Survey, Special Report, **18**.

ZIEGLER, W. 2000. The Lower Eifelian Boundary. *Courier Forschungsinstitut Senckenberg*, **225**, 27–36.

ZIEGLER, W. & KLAPPER, G. 1985. Stages of the Devonian System. *Episodes*, **48**, 104–109.

ZUSKOVÁ, J. 1991. Conodont faunas from the Lower/Middle Devonian section in Praha-Barrandov. *Věstník Ústředního ústavu geologického*, **66**, 107–112.

Shallow-water facies setting around the Kačák Event: a multidisciplinary approach

P. KÖNIGSHOF[1]*, A. C. DA SILVA[2], T. J. SUTTNER[3], E. KIDO[3], J. WATERS[4], S. K. CARMICHAEL[4], U. JANSEN[1], D. PAS[2] & S. SPASSOV[5]

[1]*Senckenberg Research Institute and Natural History Museum, Senckenberganlage 25, 60325 Frankfurt, Germany*

[2]*Sedimentary Petrology, B 20, University of Liege, Sart-Tilman, B-4000 Liège, Belgium*

[3]*University of Graz, Institute of Earth Sciences, Heinrichstraße 26, A-8010 Graz, Austria*

[4]*Department of Geology, Appalachian State University, Boone, NC 28608, USA*

[5]*Geophysical Centre, Royal Meteorological Institute of Belgium, 1 rue du Centre Physique, B 5670 Dourbes (Viroinval), Belgium*

**Corresponding author (e-mail: peter.koenigshof@senckenberg.de)*

Abstract: In the Eifel area (western Rheinisches Schiefergebirge), a shallow- to deep-subtidal sequence of mixed carbonates and siltstones around the Kačák Event Interval close to the Eifelian–Givetian stage boundary was studied. An overall transgressive trend is inferred by the microfacies evolution. The stratigraphic variations of magnetic susceptibility in carbonates and in shale intervals show an overall decreasing evolution towards the top, which fits well with the transgressive trend. In addition, carbon and oxygen isotopes, and major, trace and rare earth element (REE) analysis have been used to get a better understanding of palaeoenvironmental variations in a shallow-water realm in the late Eifelian (*kockelianus* and *ensensis* conodont biozones): for example, the $\delta^{13}C$ excursion and Ce anomaly are interpreted to be the local representation of the beginning of the Kačák Event Interval, which is also consistent with the stratigraphy and microfacies analyses.

Black shales at the Eifelian–Givetian boundary sections are very common in Europe and elsewhere (summarized in House 1996), representing a global (eustatic) sea-level rise. The Kačák Event (Budil 1995; House 2002) occurs just below the Eifelian–Givetian Stage boundary at the Jebel Mech Irdane section in Morocco, the Global Stratotype Section and Point (GSSP). The Kačák Event is a typical black shale event, connected with the global rise in sea level at the base of the transgressive–regressive (T–R) Cycle Ie of Brett *et al.* (2011). The sea-level rise is connected with increasing $\delta^{13}C$ values of about 2‰ at the Eifelian–Givetian boundary (e.g. Buggisch & Mann 2004; van Geldern *et al.* 2006), and the positive excursion can be attributed to an increased burial of isotopically light organic carbon or increased riverine weathering flux (Kump *et al.* 1999; Saltzman 2002). This event is characterized by a biotic turnover that is primarily recorded in pelagic faunas, such as conodonts, cephalopods and dacryoconarids (e.g. Chlupáč & Kukal 1986; Bultynck 1987, 1989; Becker & House 2000). The dacryoconarid *Nowakia otomari* is considered to be an indicator of the Kačák Event or the *otomari* Event *sensu* Walliser (1985).

In Germany, the *otomari* Event was described in pelagic carbonates of, for instance, the Odershausen Formation in the eastern Rheinisches Schiefergebirge (eastern RSG) in the Kellerwald area (e.g. Walliser 1985; Schöne 1997) (Fig. 1). The Eifelian–Givetian boundary interval in the Eifel area (western RSG) is characterized by shallow-water successions. The *otomari* Event has been placed at the boundary between the Nims and Giesdorf members by Struve *et al.* (1997), whereas Schöne *et al.* (1998) suggested placing this event at the boundary between the Junkerberg and Freilingen formations (see Table 1). A detailed stratigraphic correlation of sections in the eastern RSG with those of the Eifel area is difficult owing to the different depositional settings (deep-water facies setting v. shallow-water facies setting) and correlation problems: for example, the lack of the dacryoconarid *Nowakia otomari* in the shallow-water setting.

The effects of anoxic events in Earth's history on shallow-water realms are still poorly documented. In this paper, we describe a shallow-water sequence around the Kačák Event Interval, with special focus on a comparison of magnetic susceptibility data, facies analysis and geochemical data in order to

From: BECKER, R. T., KÖNIGSHOF, P. & BRETT, C. E. (eds) 2016. *Devonian Climate, Sea Level and Evolutionary Events*. Geological Society, London, Special Publications, **423**, 171–199.
First published online June 10, 2015, http://doi.org/10.1144/SP423.4

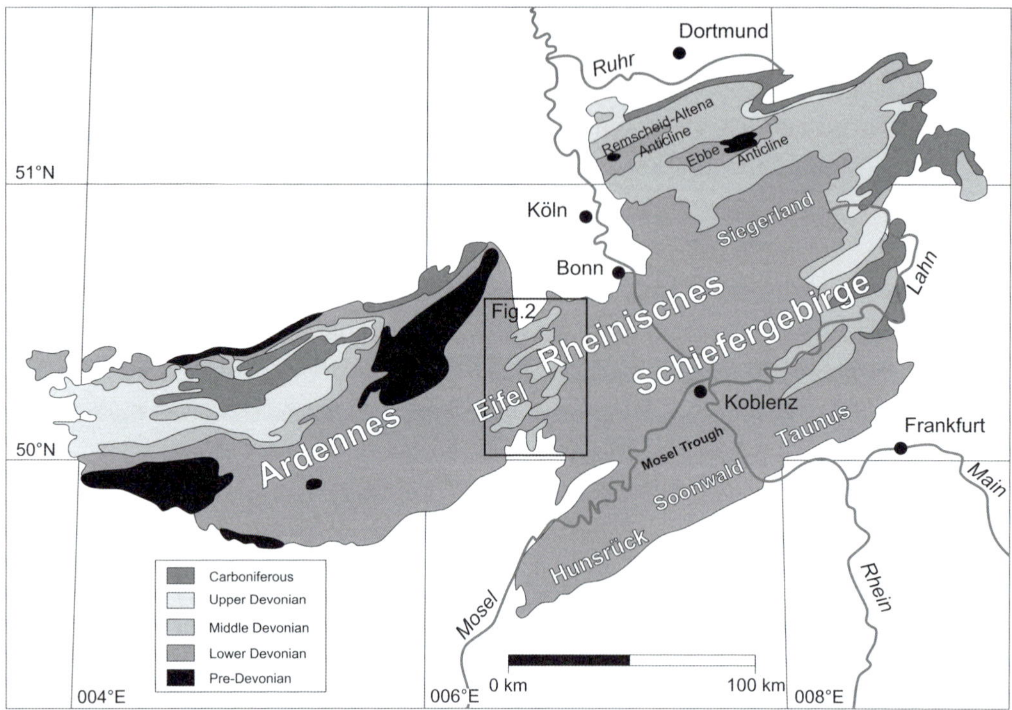

Fig. 1. Geological map of the Rheinisches Schiefergebirge (RSG) and Ardennes (slightly modified from Wehrmann *et al.* 2005). The rectangle demarcates the study area (Fig. 2).

get a better understanding of mainly climate-driven changes and sea-level fluctuations.

Geological setting

The RSG belongs to the European Variscides, which have been interpreted by a number of studies as a collage of microplates (e.g. Matte 1986; Franke *et al.* 1990; Franke & Oncken 1990, 1995; Franke 2000) successively accreted during the Early Devonian–Late Carboniferous and leading to the amalgamation of the supercontinent Pangaea (e.g. Scotese & Barret 1990; Kroner & Hahn 2003; Romer *et al.* 2003; Kroner *et al.* 2007; Linnemann *et al.* 2010; Eckelmann *et al.* 2014). It is widely accepted that the bulk of the RSG is part of the

Table 1. *Stratigraphic position of the outcrop west of Blankenheim within the Junkerberg Formation, and the transition to the Freilingen Formation*

Conodont Zones			Biostratigraphy of the Type Eifelian (*sensu* Struve 1996; R160dm96)			regional valid Member of the Blankenheim-Middle-Facies (*sensu* Ochs & Wolfart 1961 and Struve 1982)	regional geological events
			Formation	Subformation	Member		
			Ahbach Fm.				↑ *otomari* Event Interval
ensensis	Middle Devonian	Eifelian	Freilingen Fm.		Bohnert Mb.	Oberes Bankkalk Mb.	
						Brachiopoden Korallen-Mergel Mb.	*otomari* Event (after Schöne et al. 1998)
					Eilenberg Mb.	Unteres Bankkalk Mb.	
kockelianus			Junkerberg Fm.	Grauberg Sub. Fm.	Giesdorf Mb.	?	*ostiolata* Extinction Event, Great Gap (after Struve 1982)
					Nims Mb.	Blankenheim Mb.	*otomari* Event (after Struve et al. 1997)
				Heinzelt Sub. Fm	Rechert Mb.	Eisen Mb.	
					Hönselberg Mb.		
					Mussel Mb.	*Schellwienellen* Mb.	
					Klausbach Mb.	Kalksandstein Mb.	
australis			Ahrdorf Fm.	Niederehe Sub. Fm.		Schiefer Mb.	
				Betterberg Sub. Fm.	Wasen Mb.		
					Flesten Mb.	Brachiopoden Mb.	
					Köll Mb.		
					Bildstock Mb.	Bildstock Mb.	

Biostratigraphy after Struve (1996), Ochs & Wolfart (1961) and Struve (1982).
Geological events: see Schöne *et al.* (1998), Struve (1982) and Struve *et al.* (1997).

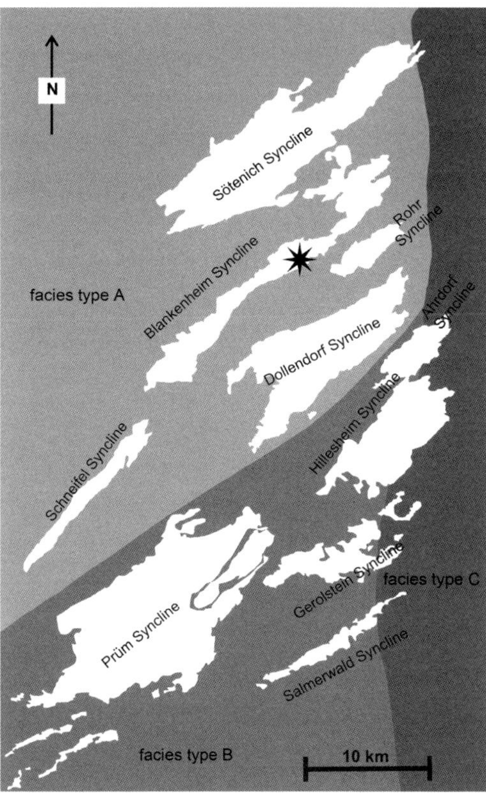

Fig. 2. Facies model of the Middle Devonian of the Eifel region (modified from Winter 1977). Facies type (**a**) facies dominated by clastic input; facies type (**b**) carbonate platforms and biostromal reefs (including the mid-Eifelian High); facies type (**c**) facies characterized by reduced clastic input and increasing carbonate sedimentation. The black star marks the investigated section near Blankenheimerdorf (50°26′29.22″N, 6°38′12.79″E).

Avalonia terrane, which was separated from Gondwana in the Early Ordovician and drifted northwards (e.g. Oncken *et al.* 2000; Tait *et al.* 2000; Torsvik & Cocks 2004; Linnemann *et al.* 2008, 2010; Romer & Hahne 2010). At about 450 Ma, Avalonia collided with Baltica in the Late Ordovician–early Silurian, which led to the closure of the Tornquist Sea. Owing to the collision of Baltica and Avalonia with Laurentia, the Iapetus Ocean was closed at around 420 Ma and Laurussia was formed (Kroner *et al.* 2007; Linnemann *et al.* 2008; Nance *et al.* 2010). The closure of the Rheic Ocean began in the late Silurian–Early Devonian and continued until the Early Carboniferous by successive closing from west to east. The ‘Amorican Terrane Assemblage’ (ATA) is believed to have separated from Gondwana in the Ordovician and followed Avalonia in a northwards direction. An island arc may have formed between Avalonia and the ATA during closure of the Rheic Ocean and accreted to Avalonia in the Early Devonian (Franke & Oncken 1995; Franke 2000). This arc is documented by magmatic rocks in the ‘Northern Phyllite Zone’ and the ‘Mid-German-Crystalline Zone’ (MGCZ) (e.g. Brinkmann 1948; Dombrowski *et al.* 1995; Reischmann *et al.* 2001). During the Early Devonian, the Rhenohercynian Basin developed as a narrow (about 250–300 km) but rather elongate (more than 2000 km) sedimentary trough south of the Old Red Continent (Stets & Schäfer 2009). Towards the south, the trough was confined by the MGCZ during the Lochkovian and Pragian. Detrital supply came from northern (Wierich 1999), as well as from southern, source areas (Hahn 1990; Hahn & Zankl 1991). Lower Devonian sequences are characterized mainly by sandstones and siltstones, whereas carbonates are less frequent. Sedimentation generally changed during the Middle Devonian. Biostromes began to flourish in the northern RSG (Pas *et al.* 2013 and references therein). In the SE RSG, biostromes are associated with volcaniclastic deposits indicating increased volcanic activity (e.g. Königshof *et al.* 1991, 2010; Nesbor *et al.* 1993; Braun *et al.* 1994; Nesbor 2004). The RSG east of the river Rhine is characterized by several autochthonous and parautochthonous (e.g. Wachendorf 1986; Meischner 1991; Schwan 1991; Bender & Königshof 1994), as well as allochthonous, units (e.g. Engel *et al.* 1983; Oczlon 1992; Franke 2000; Huckriede *et al.* 2004; Salamon & Königshof 2010; Eckelmann *et al.* 2014).

West of the river Rhine, the Eifel synclines, interpreted as part of the north–south-trending axial depression of the RSG, are the dominant structure (Fig. 2). In contrast to the entire RSG east of the river Rhine, the Eifel area shows very low to low-grade thermal alteration (e.g. Teichmüller & Teichmüller 1979; Helsen & Königshof 1994; Königshof & Werner 1994). A remagnetization event has been described in the carbonate rocks of the Devonian of the Rhenohercynian Fold-and-Thrust Belt in Belgium and NW Germany, near Cologne (Zwing *et al.* 2002, 2005, 2009; Zegers *et al.* 2003; Da Silva *et al.* 2012, 2013), but without specific studies on the Eifel area.

In the Eifel area, siliciclastics were delivered from the north during the Early Devonian and early Middle Devonian (Eifelian), but diminished during Givetian times when shallow subtropical carbonates were established over much of the region. Struve (1963) established a depositional model of the Eifel area with a north–south trending basin surrounded by landmasses, which he considered as the so-called ‘Eifel Sea Strait’. In contrast to this model, Winter (1977) defined three facies belts

(Fig. 2) in the Eifel synclines. Later, Faber (1980) modified this model, based on detailed microfacies studies, which gave evidence of small-scale cyclicity developments of a carbonate platform during the early Eifelian, and of a flat shelf lagoon during the late Eifelian and early Givetian, affecting the eastern part of the Eifel synclines. Paproth & Struve (1982) distinguished between north-, west-, and south-Eifel biofacies based on faunal differences. During the Givetian this facies differentiation broke down to some degree and mainly stromatoporoid/coral biostromes extended over the entire area. The studied section lies within the Blankenheim Syncline (Figs 2 & 3), between the villages of Blankenheim and Blankenheimerdorf, and comprises shallow-shelf mixed carbonate and siliciclastic facies of Middle Devonian age (Eifelian) accumulated on the southern margin of the former Avalonia microcontinent.

Methods

Microfossil separation and microfacies analysis

Microfacies analysis is based on facies changes observed in the field and observations from more than 80 thin sections of different formats (most of them in the format of 7.5 × 11 cm). Thin sections are stored at the Senckenberg Research Institute and Natural History Museum, Frankfurt. Fourteen rock samples, between 1 and 3 kg, were treated with formic acid diluted in water at a ratio of 1:5. Conodonts were extracted from residues by hand picking after heavy liquid separation (sodium polytungstate: density 2.79 g cm^{-3}) and are stored at the University of Graz.

Magnetic susceptibility and specific magnetic measurements

Magnetic susceptibility (χ) is considered to be a proxy parameter for detrital input, and is used for correlations of different coeval stratigraphic sections (e.g. Ellwood *et al.* 1999) and palaeoenvironmental or palaeoclimatic reconstructions (Hladil *et al.* 2009; Da Silva & Boulvain 2010; Whalen & Day 2010; Koptíková 2011; De Vleeschouwer *et al.* 2012; Da Silva *et al.* 2013). Initial magnetic susceptibility measurements on 77 samples, cleaned of surface alteration, were performed on a KLY-3S Kappabridge (AGICO, noise level 2×10^{-8} SI) at the University of Liège (Belgium). Magnetic susceptibility (MS) is expressed in $m^3\ kg^{-1}$; each data point is the average of three measurements. The sample mass was weighed with a precision of 0.01 g.

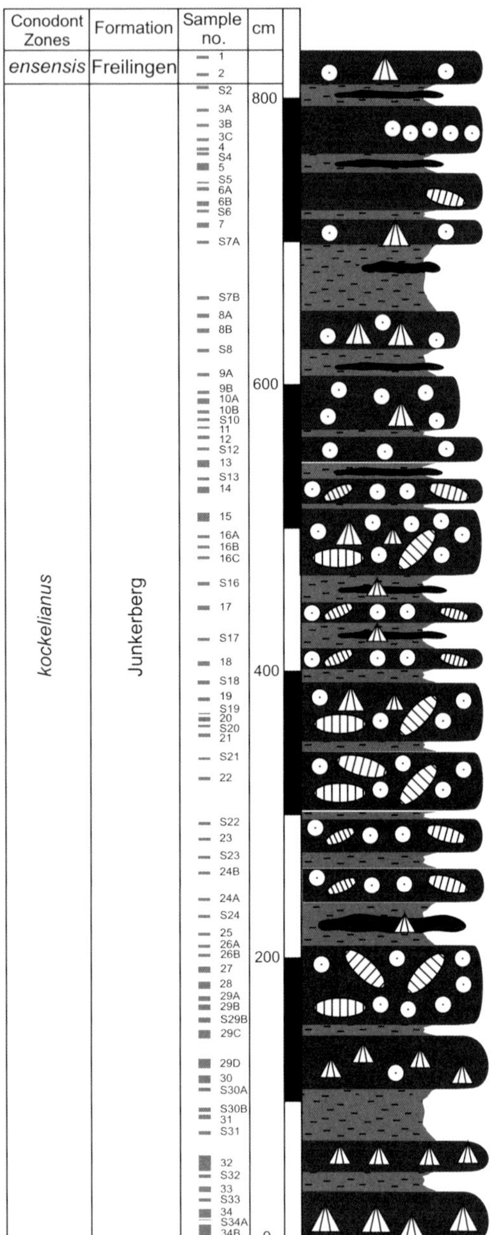

Fig. 3. Lithological column of the Blankenheim section. The section has a thickness of about 8 m, and is composed of bioclastic wackestones, floatstones and rare grainstones with intercalations of shales and siltstones.

Diagenetic processes can considerably influence the detrital susceptibility signal (primary signal); an aspect that is often underestimated (Schneider *et al.* 2004; De Vleeschouwer *et al.* 2010; Riquier *et al.*

2010; Da Silva *et al.* 2012, 2013). Because magnetic minerals ((titano)magnetite, pyrrhotite and greigite) are very sensitive to post-depositional diagenesis, it is essential to understand the nature and origin of the magnetic minerals carrying the magnetic susceptibility signal. In this respect, the nature and grain size (detrital grains are classically coarser than diagenetic grains) of the magnetic minerals should be scrutinized. Natural materials have three major magnetic behaviours: diamagnetic minerals have extremely weak negative values of MS (e.g. calcite and quartz); paramagnetic minerals have weak positive values (e.g. smectite, illite, biotite, dolomite and pyrite); and ferromagnetic (*sensu latu*) minerals have high and positive values (e.g. magnetite, pyrrhotite, hematite and goethite). In this paper, the nature and grain size of magnetic minerals is constrained through hysteresis loop measurements, acquisition of isothermal remanent magnetization (IRM), backfield curves and short-term remanence decay, which were measured with a J-coercivity rotational magnetometer at the Geophysical Centre of the Royal Meteorological Institute of Dourbes in Belgium. We selected 20 samples on a regular interval, which were cut into a cuboidal shape (approximately $0.9 \times 0.7 \times 2.5$ cm) and weighed with a precision of 0.001 g. Remanent and induced magnetizations were measured between $+500$ and -500 mT, with averaged field increments of 0.5 mT per magnetization step. The magnetization duration at each magnetization step is in the order of tenths of a second. When the backfield curve is finished, the direct current is switched off automatically and the magnetizing field decreases towards a constant residual field of about 0.4 mT within the following 0.4 s. The decay of the remaining remanence (IRM 500 mT, 0.4 s) is monitored over 100 s, and is called the short-term remanence loss.

The following parameters were derived from the hysteresis loops: saturation magnetization, M_s ($A\ m^2\ kg^{-1}$); remanent saturation magnetization, M_{rs} ($A\ m^2\ kg^{-1}$); high-field magnetic susceptibility, χ_{HF} ($m^3\ kg^{-1}$); coercive force, B_c (mT); and remanent coercive force B_{cr} (mT). M_s, M_{rs} and χ_{HF} were normalized with respect to sample mass. χ_{HF} corresponds to the high-field magnetic susceptibility values and represents the paramagnetic plus diamagnetic contributions. The ferromagnetic contribution corresponds to χ_{Ferro}, which is the total susceptibility (χ) minus the χ_{HF}. The hysteresis parameters are compared with the classic 'Day plot' of M_{rs}/M_s v. B_{cr}/B_c (Day *et al.* 1977; Dunlop 2002), allowing an assessment of the grain size of the magnetic minerals. The amount of high- and low-coercivity minerals can be roughly estimated through the high-field remanence (%), corresponding to the difference in remanence acquired at 300 and 500 mT.

Geochemical proxies

Carbon ($\delta^{13}C_{carb}$) and oxygen ($\delta^{18}O$) isotope ratios of whole-rock carbonate total organic carbon (TOC) and sulphur content were analysed for 45 rock samples collected through the section. Rock powders were produced from polished rock surfaces of hand specimens by drilling the micritic matrix under a binocular microscope. In addition, carbon isotopes were analysed on isolated prasinophytes (green algae) in 14 samples in order to compare $\delta^{13}C_{carb}$ and $\delta^{13}C_{org}$ trends. All ratios of $\delta^{13}C_{carb}$ and $\delta^{18}O$, as well as $\delta^{13}C_{org}$, are expressed in the conventional δ-notation as per mil (‰) relative to the Vienna Pee Dee Belemnite (Vienna-PDB) standard.

In this study, oxygen isotope values of conodont apatite were also measured for two samples (BL12-29c and BL12-22) to reconstruct the palaeo-sea-surface temperature. They were analysed by using a monogeneric icriodontid conodont assemblage (exclusively well-preserved platform elements). The colour alteration index (CAI) of conodont elements is very low, ranging between 1.5 and 2. The analysis of $\delta^{18}O_{apatite}$ was calibrated using a value of 21.7‰ for standard NBS120c.

Carbon and oxygen isotope analysis on whole-rock carbonate were performed using a Kiel II preparation line and Finnigan MAT Delta Plus mass spectrometer at the University of Graz (Austria). Carbon isotope analysis of organic carbon and oxygen isotope analysis on conodont apatite were performed at the GeoZentrum Nordbayern of the University of Erlangen-Nürnberg (Germany) using an elemental analyser (Carlo-Erba1110) coupled online to a ThermoFinnigan Delta Plus mass spectrometer, and a high-temperature reduction furnace (TC-EA) connected to a ThermoFinnigan Delta Plus mass spectrometer, respectively.

TOC and sulphur content were analysed on a LECO CS-300 (version 1.0, Year 1992) at the University of Graz, and are reported as C_{org} and S_{tot}. One gram of fine powder of each sample was treated three times with 2N HCl for 24 h to remove the carbonate component. Following acid treatment, the samples were rinsed to neutrality (three times with distilled water). The resulting dried powder was then analysed.

Whole-rock geochemical analyses (major, trace and rare earth element (REE)) were performed on 45 samples at Activation Laboratories in Ancaster, Ontario, Canada. Proxies for detrital input include total Al_2O_3, Rb, K_2O, TiO_2 and SiO_2. The carbonate content was measured by CaO + Loss of Ignition (LOI) and Sr. Proxies for anoxia include the chalcophile elements (S, Cu, Zn and Pb), authigenic U (where $U_{authigenic} = U_{total} - Th/3$) (Wignall & Myers 1988), Ce anomaly (Wright *et al.* 1987),

V/Cr (Pujol *et al.* 2006), $\delta^{13}C$ excursions and TOC (T. C. Bond *et al.* 2013). Primary productivity proxies include excess Ba (where Ba_{excess} = Ba − 0.15 Al_2O_3) and excess P_2O_5 (where P_{excess} = P_2O_5 − 0.15 Al_2O_3) (Pujol *et al.* 2006). Bulk $\delta^{18}O$ values are not correlated with increases in MnO concentrations, nor decreases in total Sr (ppm), nor increases in Ca/Mg ratio. When using these geochemical data in combination with microfacies observations, we therefore conclude that diagenetic alteration is not significant in any of the samples.

Results

Stratigraphy and microfacies

Microfossil biostratigraphy. The Blankenheim section exposes the Junkerberg and Freilingen formations and reaches the latest Eifelian (Table 1). The conodont assemblage is dominated by 'shallow-marine forms' of the genus *Icriodus* (Weddige 1977; Weddige & Ziegler 1979). Conodonts average approximately 10–30 elements per kg, with a higher number of specimens occurring between samples BL 12-29c and BL 12-22. Elements include icriodontids (commonly platform and rare coniform elements), one broken specimen of *Belodella* sp., two broken specimens of *Polygnathus linguiformis linguiformis* Hinde, 1879 (two different morphotypes), and two indeterminable platform fragments of *Polygnathus* sp. Icriodontid species (*Icriodus arkonensis* Stauffer, 1838, *I. regularicrescens* Bultynck, 1970 and *I. struvei* Weddige, 1977) are the most diagnostic, and constrain the section as ranging from the *Tortodus kockelianus* to the *Polygnathus ensensis* Biozone. *Icriodus regularicrescens* occurs from near the base to the top of the section. The lower boundary of the *ensensis* Biozone was tentatively set at the first occurrence of *Icriodus arkonensis* near the top of the section (sample BL 12-2) because the marker species *I. obliquimarginatus* Bischoff & Ziegler, 1957 was not found. Illustrations of representative specimens of diagnostic conodont species are provided in Figure 4.

Apart from conodonts, pyritized foraminiferid tests belonging to the genus *Pseudopalmula*, *Semitextularia*, probably *Paratikhinella*, and ostracod valves of *Polyzygia* (sample BL 12-34a) and other forms are obtained from several samples, but they are less diagnostic in terms of stratigraphy.

Macrofossil biostratigraphy. The succession from BL 12-25 to BL 12-32 contains a brachiopod fauna of biostratigraphic significance (Fig. 5):

- Sample BL 12-S 25: *Schizophoria schnuri blankenheimensis* Struve, 1965;
- Sample BL 12-28D: numerous specimens of *Iridistrophia*? cf. *undifera* (Schnur, 1853) and *Vandercammenina*? *latistriata* (Frech, 1911);
- Sample BL 12-29D: numerous *Iridistrophia*? cf. *undifera*;
- Sample BL 12-32: particularly rich in brachiopods: *Vandercammenina*? *latistriata*, *Iridistrophia*? cf. *undifera* and few poorly preserved *Athyris (Alvarezites) wolfarti* Struve, 1992?

We collected the following taxa from loose blocks in the sample BL 12: *Iridistrophia*? cf. *undifera*, *Schizophoria schnuri blankenheimensis*, *Carpinaria* vel *Subcuspidella* sp. (one specimen), *Helaspis* sp. (one specimen, probably *H. plexa* (Wolfart, 1956)) and Strophomenida gen. et sp. indet. (one specimen).

The associations suggest an assignment to the late middle Eifelian, equivalent to the Junkerberg Formation and most possibly to the Blankenheim Member ('*latistriatus*-Horizont' *sensu* Wolfart in Ochs & Wolfart 1961). The correlation of the latter with the 'Type Eifelian' succession is, however, controversial. Based on brachiopods, Struve (in Struve *et al.* 2008) assigns a 'Nims age' to the Blankenheim Member, whereas Basse & Müller (2004), arguing with the trilobite stratigraphy, regard a partial Giesdorf age as possible. The study of additional material of the spinocyrtiids from the section may provide a better-constrained age.

Microfacies analysis. The studied section is about 8 m thick and is characterized by bioclastic wackestones, floatstones and rare grainstones, with a local occurrence of laminated limestones. The limestone beds vary in thickness from centimetre- to decimetre-scale and are intercalated with shales. The section is composed of shallow-shelf mixed carbonate and siliciclastic facies. Open-marine planktonic organisms, such as dacryoconarid tentaculitids, cephalopods and/or pelagic ostracodes, are absent. Bioclasts are dominated by brachiopods, crinoids, bryozoans and subordinate numbers of trilobites, corals, algae and/or molluscs, depending on environmental changes. All bioclasts are very well-preserved due to the low diagenetic overprint (Helsen & Königshof 1994). Carbonates lack indicators of pressure solution or strong diagenetic overprint. Microfacies analysis allowed the discrimination of five microfacies, which are described from shallowest to deepest.

Microfacies 1: Bioclastic wackestone, partly strongly burrowed. In the outcrop, the grey carbonates of this facies show wavy-nodular to planar bedding. Wackestones have 20–25% bioclasts, including crinoids, brachiopod shells, bryozoans, trilobites, rare corals (rare *Thamnopora*), stromatoporoids, ostracodes (Fig. 6a–c: BL 12-23(2), BL

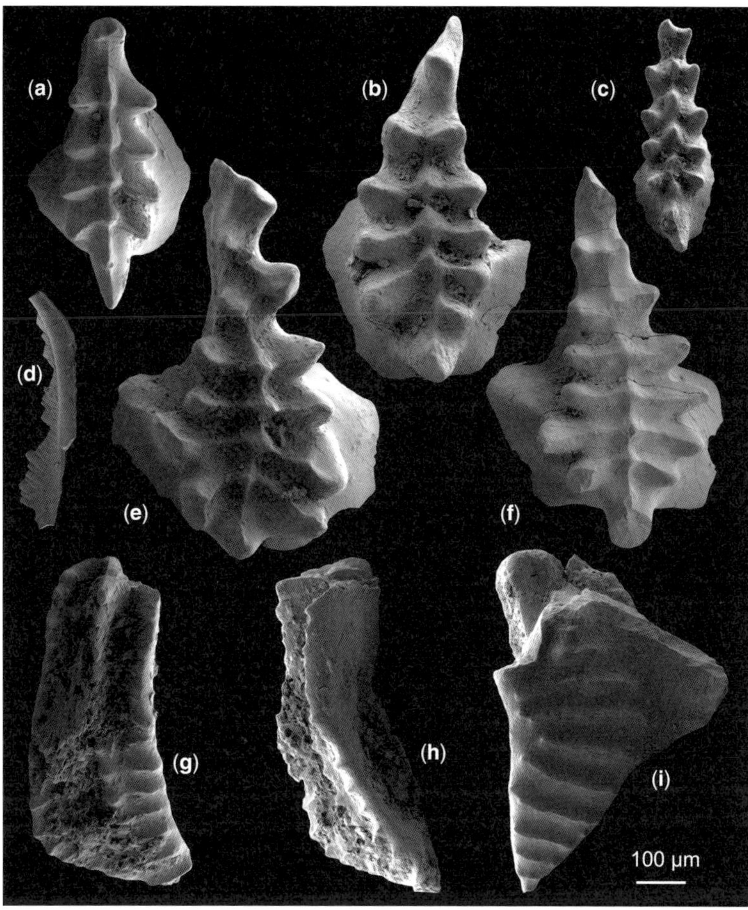

Fig. 4. (**a**) *Icriodus struvei* Weddige, 1977, I element, upper view (sample BL 12-34A). (**b**) *Icriodus werneri* Weddige, 1977, I element, upper view (sample BL 12-22). (**c**) *Icriodus regularicrescens* Bultynck, 1970, I element, upper view (sample BL 12-22). (**d**) *Belodella* sp., coniform element, upper view (sample BL 12-5). (**e**) *Icriodus retrodepressus* Bultynck, 1970, I element, upper view (sample BL 12-3). (**f**) *Icriodus arkonensis* Stauffer, 1938, I element, upper view (sample BL 12-2). (**g**) & (**h**) *Polygnathus linguiformis linguiformis* Hinde, 1879, Pa element, upper and lateral views (sample BL 12-29C). (**i**) *Polygnathus linguiformis linguiformis* Stauffer, 1879, posterior part of broken Pa element, upper view (sample BL 12-12).

12-24A and BL 12-25) and occasional gastropods. The matrix is primarily composed of a pelmicrite with a local occurrence of pelsparite. Burrowing is very common, so the texture is more or less homogenized. Minor phosphorite intraclasts show variable grain sizes (fine silt to small pebbles) and, in addition, pyrite and iron oxide crusts occur, mainly as coverings on bioclasts, such as trilobite fragments or brachiopod shells (Fig. 6d, e: BL 12-33(2) and BL 12-34A).

Interpretation: The facies points to an upper-ramp position with moderate water depth just below the fair-weather wave base. The diverse fauna includes primarily brachiopods, trilobites and bryozoans, and suggests an open-marine environment. Gastropods could be an indication of a more restricted environment in the lower part of the section but they are less common. The sediments have been reworked and were deposited close to the place of origin; the sediments are poorly sorted and bioclasts show no abrasion. The more clayey/marly sediments indicate a former soft substrate, which is strongly burrowed. More calcareous layers occur in the lower part of the sequence. These layers can be correlated with a sampled section (Ernst *et al.* 2011), which is located very close (less than 20 m) to the described section herein. These authors described a number of bryozoans with increasing calcifying taxa, such as trepostome bryozoans, encrusting worms and calcimicrobes,

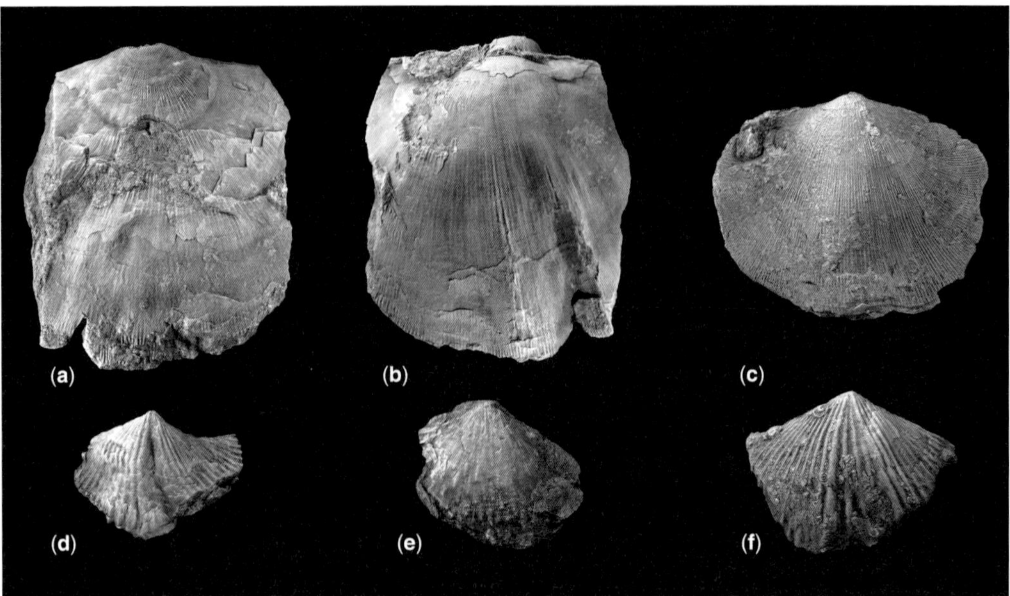

Fig. 5. Brachiopods from the section BL 12 (whitened with MgO before photographing, shown in natural size). (**a**) & (**b**) *Iridistrophia*? cf. *undifera* (Schnur, 1853), ventral (a) and dorsal (b) view of partly exfoliated articulated shell. (**c**) *Schizophoria schnuri blankenheimensis* Struve, 1965, dorsal view of articulated shell (sample BL 12-S-25). (**d**) *Carpinaria* vel *Subcuspidella* sp., ventral valve. (**e**) *Helaspis* sp., probably *H. plexa* (Wolfart, 1956), ventral valve. (**f**) *Vandercammenina*? *latistriata* (Frech, 1911), ventral valve.

particularly *Girvanella* from that horizon (Ernst *et al.* 2011, fig. 10c, d). *Girvanella* is an indicator of low sedimentation rates and may occur in a water depth of tens of metres (Flügel 2004), although Riding (1975) suggested that no confident depth limits can be set for *Girvanella* occurrences. A period with preferred calcite precipitation in a shallow-water setting within a photic zone is likely. The existence of phosphorite intraclasts and iron-oxide crusts around some bioclasts, such as trilobites, brachiopod shells (Fig. 6d, e) and corals, may suggest occasional subaerial exposition. However, Fe oxide crusts may also indicate submarine weathering of pyrite under oxic conditions (e.g. Flügel 2004), which is more likely.

Microfacies 2: Brachiopod/crinoid floatstone. The prevailing bioclasts are brachiopods (Fig. 6f: BL 12-32) and crinoids; bryozoans are less abundant. Often they occur together as a brachiopod–crinoid- or brachiopod–bryozoan-floatstone (Fig. 7a: BL 12-29C), but monospecific biota also occur, such as crinoid-, brachiopod- (Fig. 6f) or bryozoan-floatstones (Fig. 7a: BL 12-29C). The large bioclasts are embedded within a bioclastic wackestone matrix or within a lime-mudstone, as shown in Figure 6f. Large components show no abrasion and were transported a relatively short distance from the source area (many biotic components are complete) and show few effects as a result of compaction (e.g. very rare clay steams and/or stylolites). Sometimes, shells have been infested with boring organisms (Fig. 7b: BL 12-33(3)).

Interpretation: Based on the sizes of the components, the preservation, particularly the accumulation of brachiopods (which may be more or less autochthonous) and facies/sedimentological criteria, a shallow-water low-energy environment (mid-ramp setting, shallow subtidal) is inferred. The lower part of the section described herein can be correlated with another section sampled 20 m to the north (see Ernst *et al.* 2011), which also exhibited fenestrate bryozoans that are an indication of quiet water conditions and the presence of soft substrates (Ernst *et al.* 2011). Borings are common, which also suggests low sedimentation rates.

Microfacies 3: Bioclastic wackestone to grainstone. This microfacies is composed of poorly sorted wackestones to grainstones (e.g. Fig. 7c: BL 12(1)), typically with a fine-grained micritic matrix that also contains quartz grains. Burrowing is absent. This lithofacies occurs in the middle part of the sequence and in the uppermost part (sample numbers BL 12-1–BL 12-3), representing the youngest sediments of the section. Main bioclasts of up

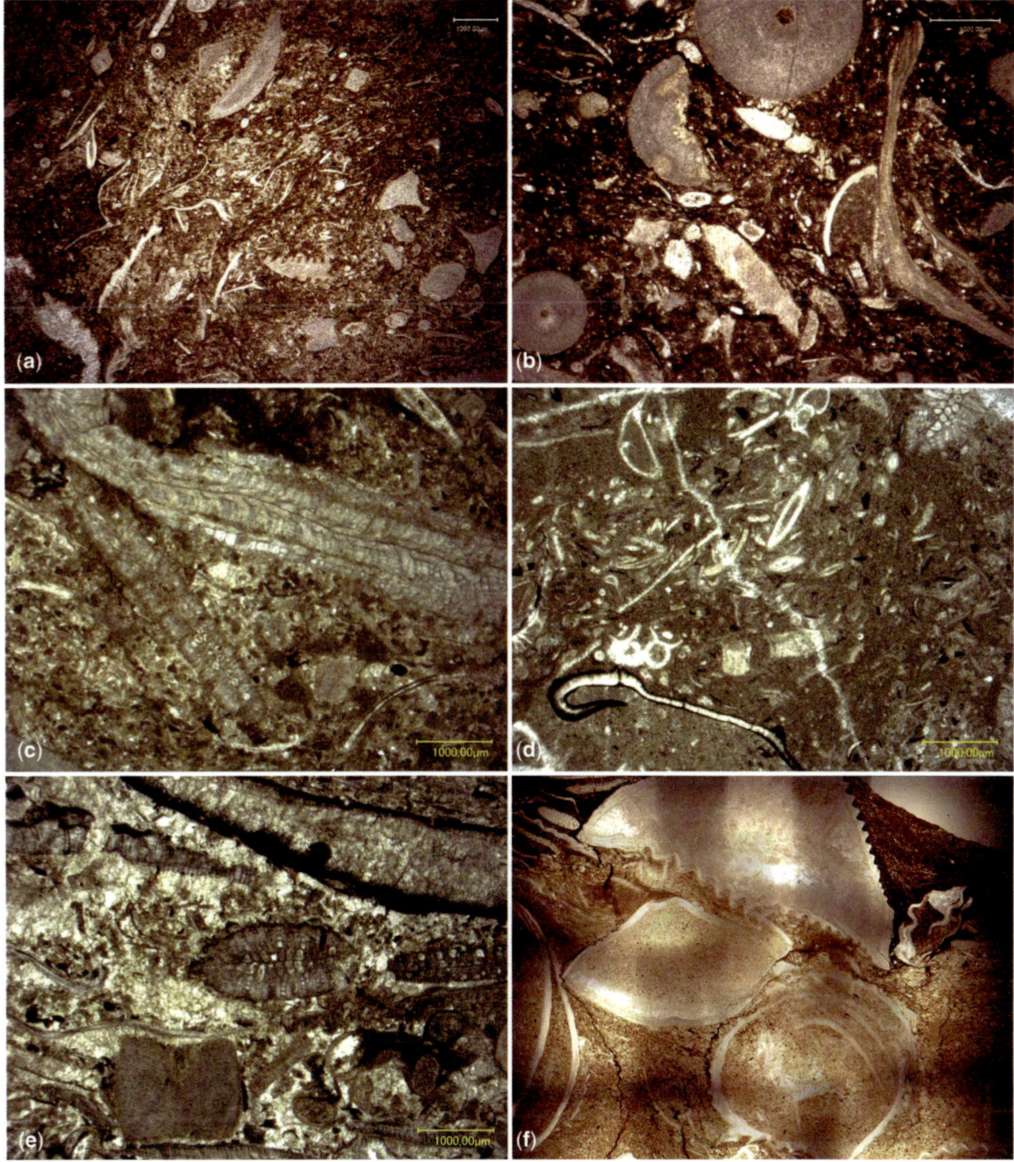

Fig. 6. (**a**)–(**e**) Allochthonous bioclastic wackestone, locally strongly burrowed due to an increasing number of endobenthic organisms. Main bioclasts are crinoids, brachiopods, bryozoans and trilobites; corals, stromatoporoids and ostracodes are less abundant. (a) Sample BL 12-23(2); (b) sample BL 12-24A; (c) sample BL 12-25; (d) & (e) occasionally small phosphorite intraclasts, pyrite crystals and iron-oxide crusts occur (d, sample BL 12-33(2); e, sample BL 12-34A). (**f**) Brachiopod–crinoid-floatstone embedded in a lime-mudstone (sample BL 12-32).

to 15% are composed of crinoids, trilobites, brachiopod shells and rare corals. Bioclasts are less abundant than in Microfacies 1 and they are much smaller. Furthermore, calcareous algae and gastropods are absent. Very small phosphorite clasts and some extraclasts occur (Fig. 7d: BL 12-3A(1)). Some layers show graded bedding.

Wackestones to grainstones, which are dominated by crinoids (Fig. 7e, f: BL 12-12(2) and BL 12-14) and common occurrences of tube worms (Fig. 8a: BL 12-20(2)), also occur, but they have less palaeoenvironmental significance as they occur in shallow-, as well as in deeper-marine settings (Flügel 2004).

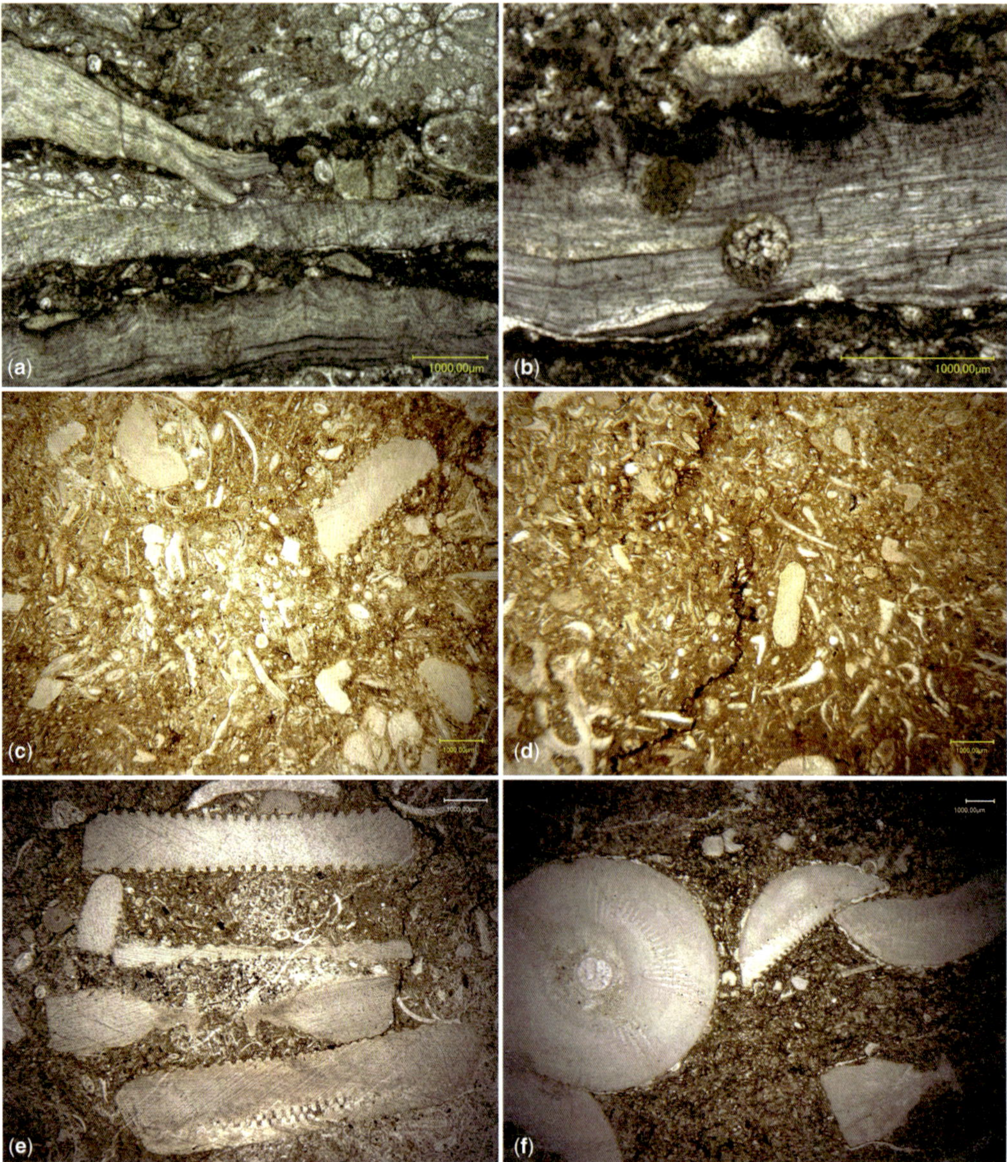

Fig. 7. (**a**) Brachiopod–bryozoan-floatstone (sample BL 12-29C); (**b**) brachiopod-floatstone; brachiopod shells are often infested by boring organisms (sample BL 12-33(3)); (**c**) & (**d**) poorly sorted bioclastic wackestone with small phosphorites and intraclasts (c, sample BL 12-1; d, sample BL 12-3A(1)); (**e**) & (**f**) crinoid-dominated wackestones (e, sample BL 12-12(2); f, sample BL 12-14).

Interpretation: The faunal association, the preservation, the poorly sorted sediment and the fine bioclastic micritic matrix suggest a generally low-energy environment, in a mid- to deep-ramp (subtidal environment) setting. The occurrence of rare grainstones is interpreted as storm layers. Sedimentation (e.g. poorly sorted, reworking) is most probably associated with a sea-level rise.

Microfacies 4: Microbioclastic wackestone. This microfacies exhibits a variable amount of biota, which ranges from below 5%, with large isolated fragments, such as ostracodes (Fig. 8b: BL 12-11(2)), up to 20% of shell hash of trilobites, brachiopods and ostracodes, among other organisms. Those layers can exhibit graded bedding. Rare bryozoans also occur (Fig. 8c: BL 12-9A). The main difference

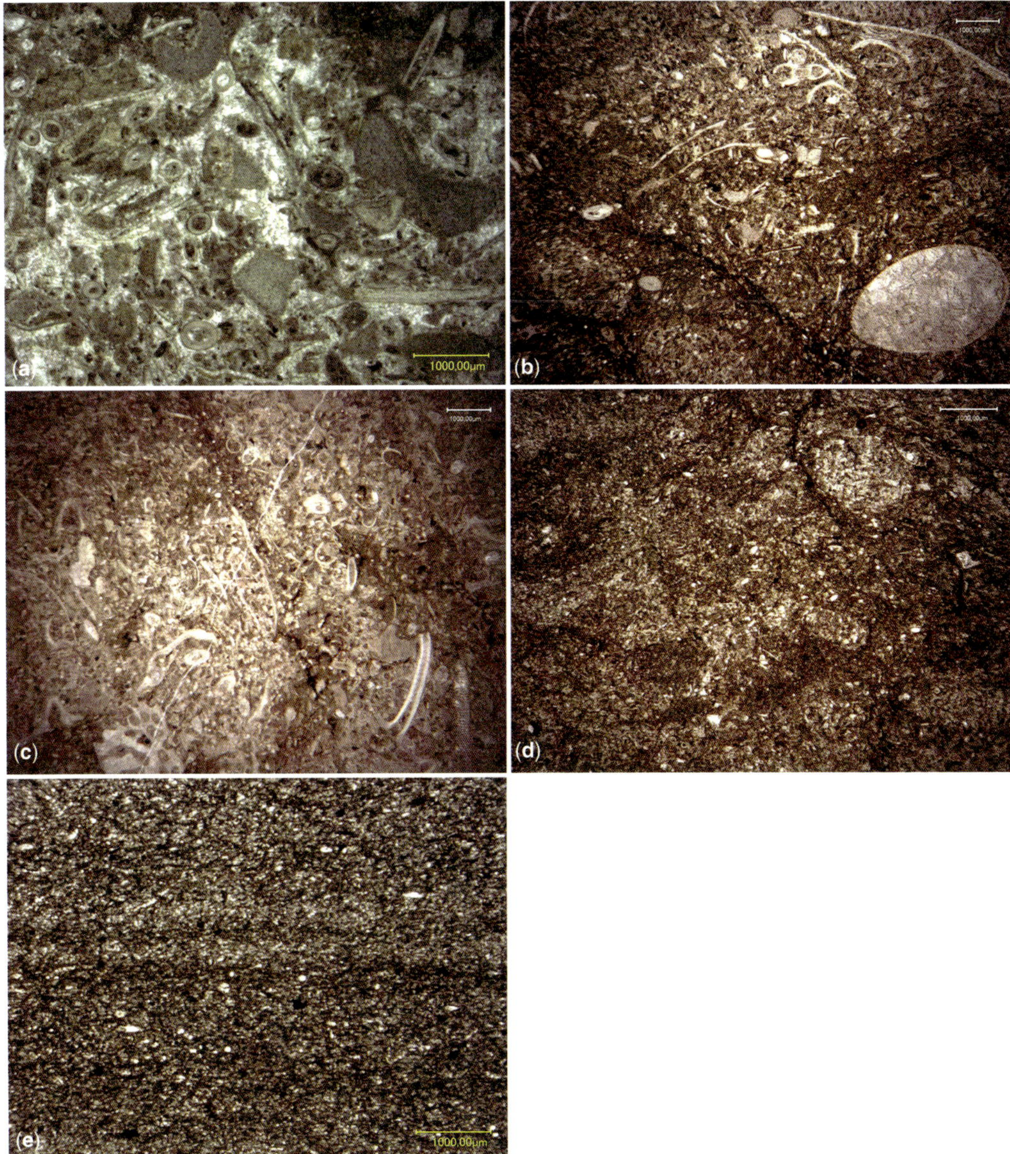

Fig. 8. (**a**) Large occurrences of tube worms in a partly sparitic wackestone (sample BL 12-20(2)); (**b**)–(**c**) microbioclastic wackestone with large fragments of ostracodes; sediment is strongly burrowed (b, sample BL 12-11(2); c, sample BL 12-9A); (**d**) matrix composed of quartz-bearing pelmicrite (sample BL 12-15(2)); (**e**) peloidal calcisiltite with rare bioclasts (sample BL 12-30).

to Microfacies 1 is the size of bioclasts (microbioclasts in contrast to large bioclasts in Microfacies 1) and the matrix. The matrix is a quartz-bearing, burrowed pelmicrite (Fig. 8d: BL 12-15(2)) containing occasionally larger crinoid and echinoderm fragments. Bryozoans are very rare. Burrowing is a common feature. A bioturbation index, in which a descriptive grade was assigned to the degree of bioturbation, was established by Taylor & Goldring (1993). The degree of bioturbation in this section can be assigned to grade 4, which is characterized by high abundance and density. This facies occurs mainly in the upper part of the section.

Interpretation: This microfacies suggests generally a low-energy, outer-ramp position. Those layers that show graded bedding are interpreted as material reworked during higher-energy events (e.g. storm deposits).

Microfacies 5: Peloidal calcisiltite. The rock is composed of densely packed micrite peloids and silt-sized terrigenous quartz grains. Bioclasts are very rare, and are composed of crinoid ossicles and brachiopod remnants. Small-scale lamination is common. This facies type occurs only in layers BL 12-30 and BL 12-31 in the lower part of the section (Fig. 8e: BL 12-30).

Interpretation: The densely packed, quartz-rich calcisiltite could be interpreted as a post-drowning phase of a carbonate platform comparable to an example from Morocco (e.g. Blomeier & Reijmer 1999). Similar finely laminated lime mudstones also occur in very shallow intertidal and supratidal environments (e.g. Gammon & James 2001), but these facies also show microbial mats, fenestral fabrics and/or desiccation features, which do not occur in Microfacies 5. Therefore, we interpreted the latter as representing a deeper, low-energy setting.

Magnetic susceptibility data

The mean initial magnetic susceptibility for the whole section is $2.25 \times 10^{-8} m^3 kg^{-1}$, which is a little bit lower but in the range of the MS marine standard $= 5.5 \times 10^{-8} m^3 kg^{-1}$ – the median value for about 11 000 lithified marine sedimentary rocks, including siltstone, limestone, marl and shale samples (Ellwood *et al.* 2011). The lowest susceptibility is $0.45 \times 10^{-8} m^3 kg^{-1}$ (no negative values, indicating that the section is not composed of very pure carbonates) and the highest value is $9.15 \times 10^{-8} m^3 kg^{-1}$.

The section is composed of an alternation of carbonates and shales, and both lithologies were sampled. It clearly appears that the shale levels have systematically higher susceptibility (mean value of $4.27 \times 10^{-8} m^3 kg^{-1}$) than the carbonates ($1.27 \times 10^{-8} m^3 kg^{-1}$) (Fig. 9).

M_s (magnetization at saturation) and χ_{Ferro} (ferro-magnetic susceptibility) are both proxies for the concentration of ferromagnetic minerals. The contribution of M_s is rather low (5.02×10^{-4} and 12.62×10^{-4} A $m^2 kg^{-1}$) and M_s data are not well correlated with χ_{in} (correlation factor $r = 0.40$:

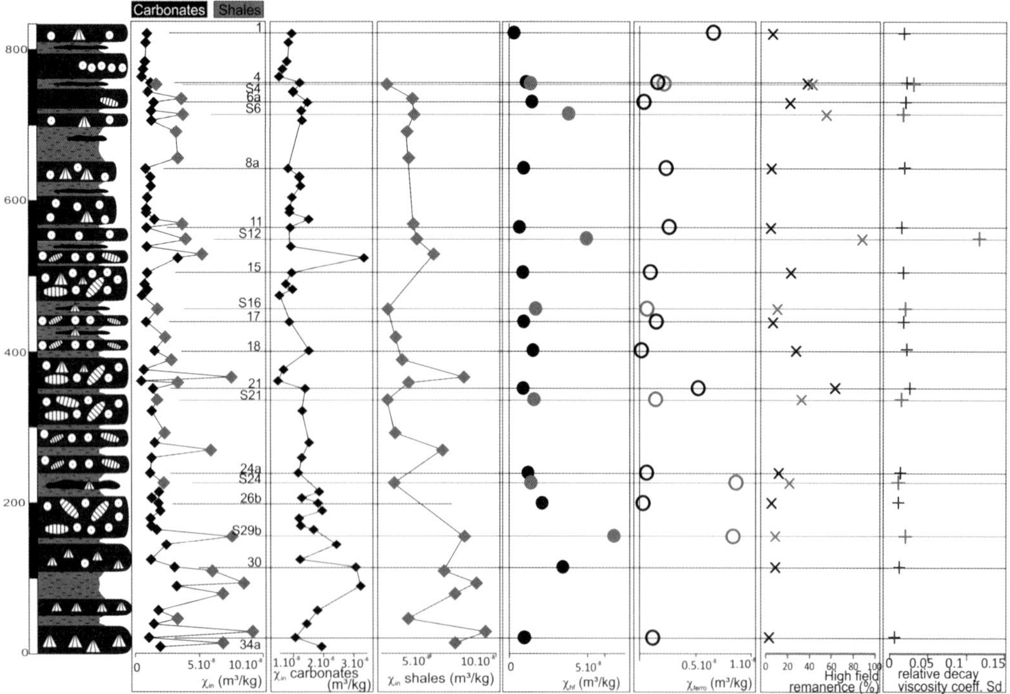

Fig. 9. Lithological column (scale bar in cm) and magnetic susceptibility evolution compared with magnetic hysteresis and backfield curve data on selected samples. χ_{in}, low-field magnetic susceptibility; M_s, magnetization at saturation; χ_{hf}, high-field magnetic susceptibility; χ_{ferro}, ferromagnetic susceptibility.

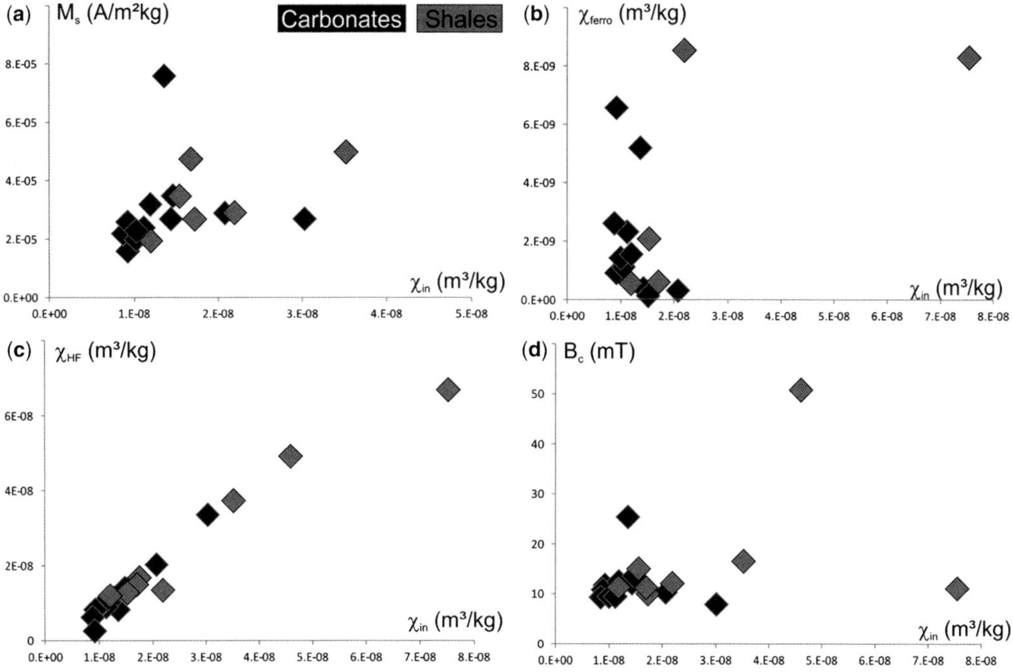

Fig. 10. Comparison between magnetic susceptibility values (χ_{in}) and hysteresis parameters from selected samples from Blankenheim section: (**a**) M_s, magnetization at saturation, $r = 0.40$; (**b**) χ_{ferro}, ferromagnetic susceptibility, $r = 0.54$; (**c**) χ_{HF}, high-field magnetic susceptibility, $r = 0.98$; (**d**) B_c, coercivity, $r = 0.40$.

Fig. 10a). χ_{Ferro} values are also very low (0.01×10^{-8} and 0.86×10^{-8} m^3 kg^{-1}) and correlate weakly with χ_{in} ($r = 0.54$: Figs 9 & 10b). These low correlations observed between χ and proxies for ferromagnetic minerals are good arguments in favour of a low influence of ferromagnetic minerals. χ_{HF} is a proxy for the sum of the paramagnetic and diamagnetic minerals. χ_{HF} values are between 0.26×10^{-8} and 6.70×10^{-8} m^3 kg^{-1} (Fig. 9). All values are positive, indicative of a relatively large amount of paramagnetic minerals. There is a very strong correlation between χ_{HF} and χ_{in} ($r = 0.98$: Figs 9 & 10c), indicating that, in this case, the magnetic signal is mostly linked to paramagnetic minerals. This fits well with the fact that the mean susceptibility values are systematically higher for clay mineral levels.

It is possible to obtain information on the nature of the ferromagnetic minerals through the back-field curve. The high-field remanence saturation represents the proportion of non-saturated (at 300 mT) magnetic minerals (high-coercivity minerals, such as hematite, as opposed to low-coercivity minerals, such as magnetite, that are easily saturated). High-field remanence is between 0 and 20% for 10 samples (small amount of high-coercivity minerals, such as hematite), between 20 and 50% for seven samples (larger amount of high-coercivity minerals), and higher than 50% for three samples (Fig. 9).

The shape of the hysteresis loops, B_{cr}/B_c and M_{rs}/M_s, provide information on the magnetic grain size, with the help of the Day plot (Day *et al.* 1977; Dunlop 2002). Main grain-size categories correspond, from the coarser to the finest, to multidomain (MD), pseudo-single domain (PSD), single domain (SD) and super paramagnetic (SP). Coarser grains are often interpreted as related to a detrital origin, while the smallest grains are interpreted as having formed during diagenesis. The samples from Blankenheim fall along the PSD and SD + MD mixing curve (Dunlop 2002), with only one sample (S12) along the SD + SP mixing curve (Fig. 11). The remanence decay is also indicative of the grain size. The relative decay viscosity coefficient (S_d, a dimensionless quantity) was calculated from the remanence decay measured over 100 s. S_d represents the slope in the IRM$_{-500\ mT}$ v. Log(time) diagram. The Blankenheim samples range between 6×10^{-3} and 30×10^{-3}, with only one sample (S12) reaching 116×10^{-3} (Fig. 9). According to Spassov & Valet (2012) S_d values between 3.9 and 6.5×10^{-3} are indicative of SD grains, while values above 100 are indicative of

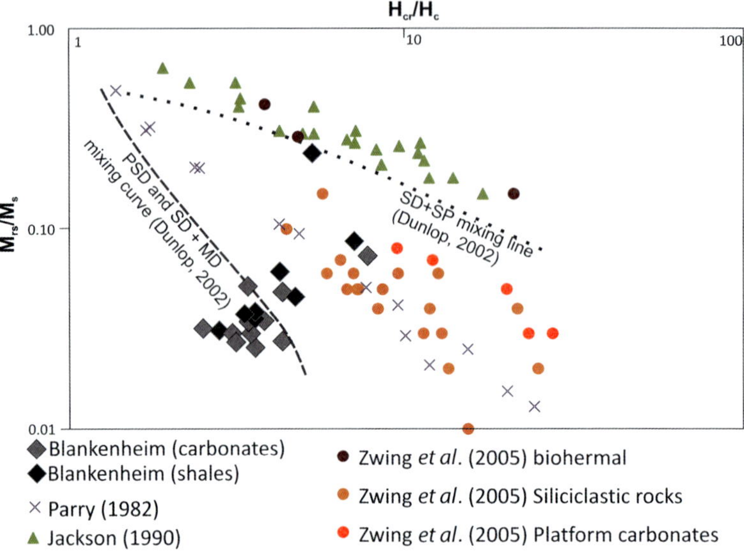

Fig. 11. Bi-logarithmic Day plot (Day *et al.* 1977) for selected samples from the Blankenheim section, compared with the mixing curves by Dunlop (2002), with classic remagnetized limestones (Jackson 1990), MD magnetite grains (Parry 1982) and with results from Zwing *et al.* (2005) on different lithologies from NE Germany (near Cologne). The majority of the Blankenheim samples fall on the left-hand side of the plot, within the PSD and SD + MD mixing curve.

SP + SD grains. This fits relatively well with the results obtained with hysteresis data, with a majority of the sample with relatively coarse grains and only one sample with very-fine SP granulometry (sample S12).

Interpretation

From the magnetic measurements, we can deduce that we have a magnetic signal, which is mostly carried by paramagnetic minerals (probably clay minerals considering the clear link between clay-dominated intervals and magnetic susceptibility highs: Fig. 12). Ferromagnetic (*sensu lato*) minerals are present in very low abundance, and they are dominated by magnetite and hematite, but these minerals have a small influence on the total bulk magnetic signal. The Day diagram (cf. Fig. 9) indicates that the ferromagnetic mineral content consists of a mixture of SD and MD, dominated by MD grains (i.e. coarse grain sizes indicating detrital origin). Only one sample shows the influence of smaller grains (BL S12). Hematite is the most abundant in this sample, suggesting that it was subjected to a stronger or different diagenetic pathway.

The Rhenohercynian zone has experienced a widespread remagnetization event, identified through detailed magnetic studies in Belgium (Molina Garza & Zijderveld 1996; Zegers *et al.* 2003; Da Silva *et al.* 2012, 2013) and in the NW RSG in Germany, near Cologne (Zwing *et al.* 2005). The latter study compared magnetic susceptibility data from biohermal carbonate rocks, platform carbonate rocks and siliciclastic rocks (Fig. 11). They identified the most important carrier of the late Palaeozoic magnetization component as magnetite. However, results were different, depending on the lithology, with MD (detrital) magnetite dominating in the siliciclastic rocks and SP (diagenetic) magnetite in the biohermal carbonates, with the platform carbonates showing intermediate hysteresis properties. Zwing *et al.* (2005) interpreted the remagnetization as a widespread event, affecting all lithologies, but that this remagnetization would be disguised by the large amount of MD magnetite in siliciclastic rocks. This remagnetization is interpreted as related to complex processes. In the Late Devonian and Early Carboniferous sedimentary rocks, the remagnetization is coeval with clay diagenesis (through the smectite to illite transition releasing iron); in the Middle Devonian strata, however, clay diagenesis and remagnetization are not coeval and pyrite oxidation processes are observed (Zwing *et al.* 2009).

In the case of the Blankenheim section, either the MD detrital magnetite 'hides' a remagnetization event, as in Zwing *et al.* (2005), or the MD magnetite has a detrital origin and no remagnetization processes occurred in this section. Helsen & Königshof (1994) have shown that the conodont alteration

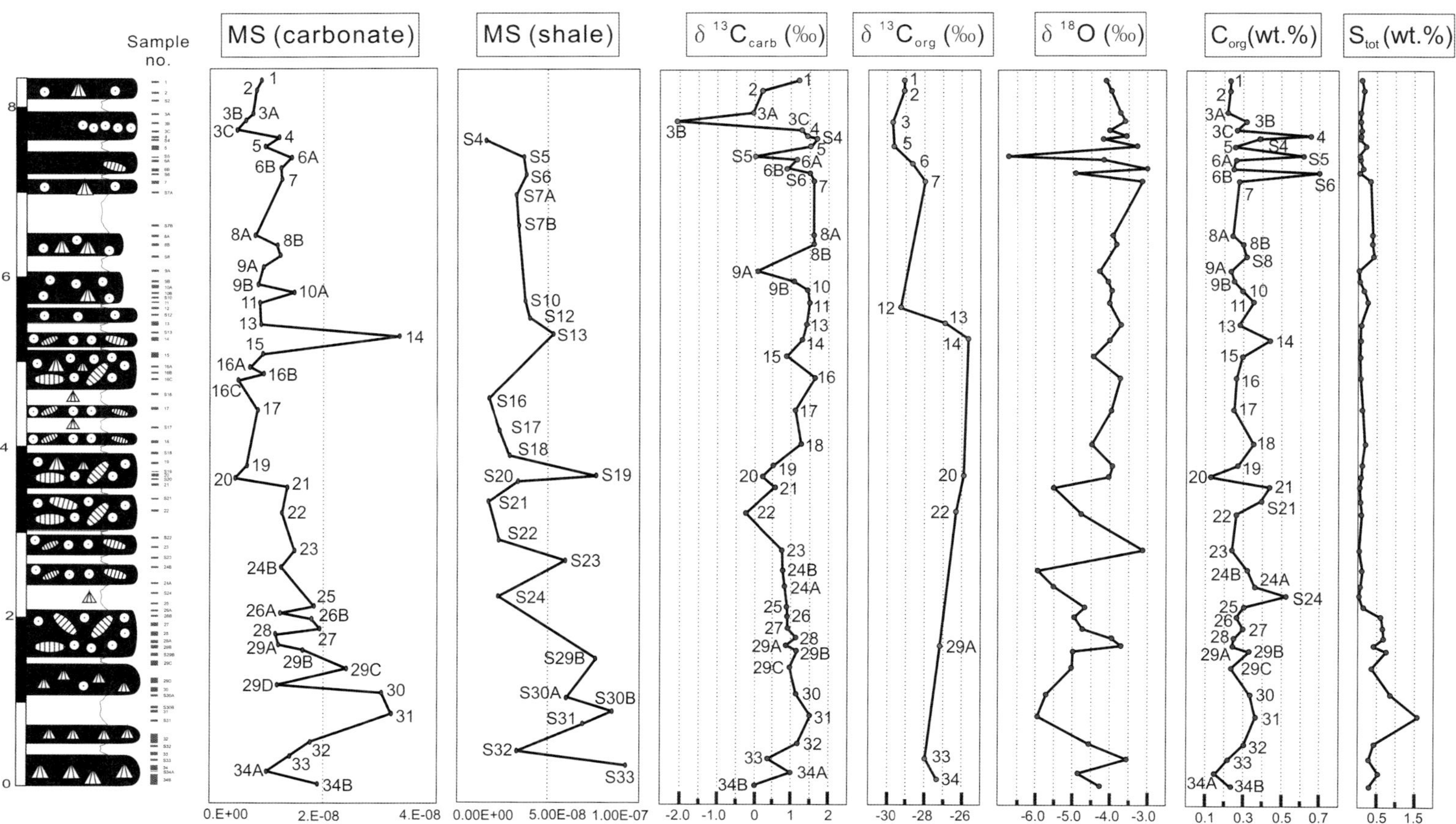

Fig. 12. Lithological column showing magnetic susceptibility evolution (in carbonates and shales), stable carbon and oxygen isotopes of bulk rocks, palynomorphs and TOC and sulphur data.

index from Lower and Middle Devonian rocks within the Eifelian area indicate little, if any, effect of heating, with alteration index ranging between 1.5 and 2.0, indicating very low temperatures of approximately 55 °C. From thin-section observations, the alteration appears to be very weak, with no indication of pressure solution and with remarkable preservation of fossils (see above). This is in accordance with the generally low S_d values, which indicate that SP grains are not very common in the samples. However, one would expect an SP presence in remagnetized samples. Low alteration and low compaction are explained by a reduced sedimentation overburden (Helsen & Königshof 1994). Figure 11 displays a Day plot, which includes the data from Jackson (1990), corresponding to classic remagnetized limestone samples (Channel & McCabe 1994), and from Parry (1982), corresponding to coarse-grain magnetite. Figure 11 also includes data from Zwing *et al.* (2005) on biohermal carbonates, platform carbonates and siliciclastic Devonian rocks, with different magnetic properties observed for the different lithologies. Our Blankenheim data are on the left-hand side of the plot, separated from the other results, and follow the PSD and SD + MD mixing curve. Thus, the data correspond to grain size probably even coarser than that observed by Zwing *et al.* (2005) on the siliciclastic rocks in the area of Cologne. The carbonate and the shale intervals in Blankenheim are carrying magnetite of the same grain size despite the lithological variations. This is in opposition to Zwing *et al.* (2005), in which the carbonates have smaller grains. In Zwing *et al.* (2005), the different grain size related to different lithologies was explained by the remagnetization event. The steady magnetite grain size in Blankenheim is in favour of a detrital origin for the magnetite grains, without any impact of a remagnetization event, although this needs to be cross-validated by further studies. Furthermore, a comparison of magnetic susceptibility results with geochemical data, and specifically with elements that are considered as proxy for detrital inputs, such as Ti, Al, Rb and Zr (Tribovillard *et al.* 2006; Calvert & Pedersen 2007), allows us to assess the influence of detrital inputs on the magnetic susceptibility signal (Riquier *et al.* 2010; Da Silva *et al.* 2012, 2013). The correlation is high (the link between magnetic susceptibility (MS) and detrital proxy elements on 45 samples: for Al_2O_3, $r = 0.84$; for TiO_2, $r = 0.83$; for Rb, $r = 0.79$; for Th, $r = 0.81$; and for Zr, $r = 0.72$), indicating a major impact of detrital inputs on the MS signal.

As mentioned earlier, the magnetic susceptibility of shale layers is systematically higher than the adjacent carbonate beds, a link not always observed. For example, Bertola *et al.* (2013) have shown that the magnetic susceptibility values were fairly similar in the carbonates and in the adjacent shale beds in two Tournaisian sections from Belgium (Bertola *et al.* 2013, fig. 10). These results on Tournaisian Belgian rocks also suggest that the input of MS-carriers stayed roughly constant during the deposition of the limestone shale alternation and that these carriers were probably ferromagnetic. In the Blankenheim section, magnetic susceptibility appears more as a proxy for shale proportion (paramagnetic minerals), which can be noted from outcrop visual observation. However, Figure 9 displays a plot with separate magnetic susceptibility curves for carbonates v. shales. Both curves show relatively similar trends, with relatively high peaks in the lower part of the section and decreasing magnetic susceptibility towards the top, which could be related to a decrease in detrital inputs, in relation with the observed transgressive trend. Ellwood *et al.* (1999) proposed that during transgression, sea level increases and the portion of landscape exposed to erosion decreases, leading to a decrease in detrital input. This link was actually observed in various carbonate platform examples (Hladil 2002; Racki *et al.* 2002; Da Silva *et al.* 2010; Whalen & Day 2010). This interpretation fits well with the observed deepening trend reflected by the facies.

Geochemical proxies

Carbon and oxygen isotopes

The carbon isotope values of bulk rock samples (Fig. 12) fall within a range of −2.1 and 1.7‰. A positive excursion of $\delta^{13}C_{carb}$ values from −0.1‰ (BL 12-34B) to 1.5‰ (BL 12-31) is observed at the base of the section. The interval between BL 12-31 and BL 12-22 is characterized by a gradual decrease in $\delta^{13}C$ from 1.5 to −0.2‰. The second positive excursion lasts from BL 12-22 until BL 12-16, peaking at 1.2 ‰. Further, two positive peaks of $\delta^{13}C_{carb}$ values are recorded within the 5–6 and 6–7 m of section separated by low values at BL 12-15 (0.9‰) and BL 12-9A (0.1‰). The values decrease between BL 12-7 and BL 12-3B, ranging from 1.6 to −2.1‰ (lowest $\delta^{13}C_{carb}$ value obtained across the entire section). Two positive peaks occur in BL 12-6A and around BL 12-S4.

Carbon isotope values of prasinophytes fluctuate between a minimum of −29.7‰ and a maximum of −25.7‰. The trend of $\delta^{13}C_{org}$ within the lowermost 1 m of the section corresponds to the trend of $\delta^{13}C_{carb}$ of the same interval. The highest $\delta^{13}C_{org}$ value is recorded in BL 12-14. A negative shift of −3.6‰ is observed from BL 12-14 to BL 12-12. Below this negative shift, the values are relatively higher than those above the shift and show an amplitude fluctuation ranging from −28 to −25.7‰ and from −29.7 to −28‰, respectively. The lowest value of $\delta^{13}C_{org}$, as well as $\delta^{13}C_{carb}$, is observed around 7.5–8 m of the section. The value increases towards the top of the section from −29.7‰ (BL 12-3) to −29.1 ‰ (BL 2-1). The increase of values in the uppermost part of the section is also observed in MS (carbonate) and $\delta^{13}C_{carb}$.

Although we provide oxygen isotopes of bulk sedimentary rocks, we are aware that $\delta^{18}O$ analyses from bulk samples do not produce reliable palaeoenvironmental results compared with data from the calcite of brachiopod shells or the PO_4 group of biogenic phosphate. In this paper, $\delta^{18}O$ values are plotted to compare variations through the section with other geochemical and geophysical proxies. $\delta^{18}O$ values of bulk-rock samples range between −6.7 and −3.1‰.

TOC and sulphur content

Total organic carbon is below 0.7% throughout the entire section (Fig. 12), with a minimum of 0.12% (BL 12-20) and a maximum of 0.69% (BL 12-S6). Sample BL 12-34B from the base of the section shows a TOC concentration of 0.24%. The content decreases to 0.16% and then increases to 0.36% approximately 1 m above the section base (BL 12-31). A second positive peak (0.51%) is observed at BL 12-S24, followed by decreasing values to approximately 3.5 m. Another positive shift to 0.43% is observed between BL 12-22 and BL 12-21. The following drop to 0.12% (BL 12-20) equates to the lowest value measured across the entire section. Two minor positive spikes are recorded at BL 12-18 (0.34%) and BL 12-14 (0.43%) between BL 12-20 and BL 12-7, followed by a low-amplitude fluctuation to approximately 7 m above the section base. Sulphur content varies between 0.02 (BL 12-S5) and 1.54% (BL 12-31). The significant positive shift near the base of the section peaks at the same level, with the first maximum in the $\delta^{13}C_{carb}$ and the TOC curves. Sulphur content then decreases to 0.36% (BL 12-29C), followed by a minor positive excursion (maximum 0.73%: BL 12-29B). Thereafter, values decrease to 0.04% (BL 12-S24), followed by an interval of low-amplitude fluctuation through to sample level BL 12-13. Above this interval, although values increase to 0.26% (BL 12-11), and to 0.33–0.41% between BL 12-S8 and BL 12-7, they consistently show less fluctuation than the section below.

Estimated palaeotemperature

In recent years, the oxygen isotope composition of conodont apatite has been used to estimate the palaeotemperature (e.g. Joachimski *et al.* 2004, 2009; Trotter *et al.* 2008). In this study, we measured exclusively icriodontid platform elements of two samples (BL 12-29c and BL 12-22). $\delta^{18}O_{apatite}$ values of both samples show a ratio of 19.2‰ (1 SD = ±0.2‰). Assuming an oxygen isotope composition of −1‰ for Middle Devonian seawater, a palaeotemperature of 29.7 °C is calculated using the temperature equation provided by Pucéat *et al.* (2010). According to Joachimski *et al.* (2009), who summarized $\delta^{18}O_{apatite}$ data of conodonts from Germany, France, the Czech Republic and the United States for the Middle Devonian using a value of 22.6‰ for NBS120c, the $\delta^{18}O_{apatite}$ values ranging from 19 to 21‰ (VSMOW) gave palaeotemperatures from 22 to 30 °C.

Major, trace and REE analysis

Principal component analysis (PCA) of normalized whole-rock values was conducted using PAST software (Hammer *et al.* 2001). The first two principal components contribute meaningful information for the interpretation of geochemical patterns throughout the section (Fig. 13). PC-1 values are interpreted as a detrital signal due to the positive and negative loadings of siliciclastic detrital indicators v. carbonate indicators (Fig. 14). Conversely, PC-2 values are consistent with the signals indicated by $\delta^{13}C$ and TOC (Fig. 15), even if they are not proxies for anoxia. Plotting individual signals against stratigraphy provides a clearer signal of detrital input (Fig. 16) and events (Figs 17 & 18).

Given the tectonic complexity of the area, we analysed samples to determine sediment provenance. Trace elements, such as Hf, Nb, Sc, Ta, Th, U, Y, Yb and Zr, are found in the titanium-bearing detrital fraction of sedimentary rocks and can be used to distinguish the maturity (Carpentier *et al.* 2013) or the tectonic environment (Wood 1980; Pearce *et al.* 1984; Bhatia & Crook 1986) of the source rock. Analysis of Th/U ratios in sediments in the Blankenheim section suggests that the sediment source is juvenile, with Th/U <3 (Fig. 19). In addition, trace-element geochemical signatures suggest that the sedimentary provenance for the detrital material is consistent with a continental island arc environment (Fig. 20). This interpretation supports earlier research that the Blankenheim section was part of the newly amalgamated Avalonian microcontinent (Franke & Oncken 1995; Franke 2000), as well as recently published data of the eastern part of the Rheinisches Schiefergebirge (e.g. Eckelmann *et al.* 2014).

Detrital signatures are seen between 61 and 148 cm (around samples BL 12-32 and BL 12-29B) above the base of the section, with two major pulses of detrital input higher up in the section at 738 and 759 cm (around samples BL 12-S5–BL 12-3C: Fig. 16). Based on increases of chalcophile elements, such as Zn, Pb, Cu and S

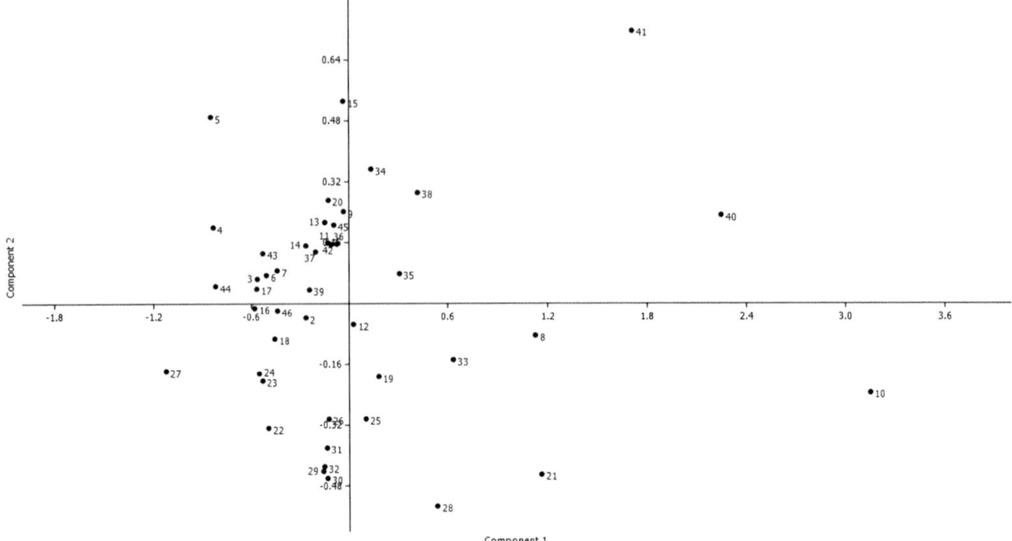

Fig. 13. Principal component analysis (PCA) of normalized whole-rock values showing PC1 v. PC2. PC1 is interpreted as a detrital signal, with PC2 showing an anoxia signal.

(Fig. 17), a signature of anoxia is suggested at 89 cm above the base of the section (samples BL 12-31 and BL 12-30), although redox proxies such as authigenic U and Ce anomalies do not support anoxia in this interval (Fig. 18). A second event at the top of the section (from 738 to 779 cm) possibly indicates anoxia through a set of staggered negative $\delta^{13}C$ excursions, a Ce anomaly and increases in TOC (Fig. 18). This upper event begins with the pulse of detrital input at 738 cm and continues through 779 cm. Although there is a spike in excess Ba, excess P and V/Cr at 365 cm (Fig. 18), we are reluctant to assign a meaning to this interval as V/Cr is a proxy for anoxia, while excess Ba and P are proxies

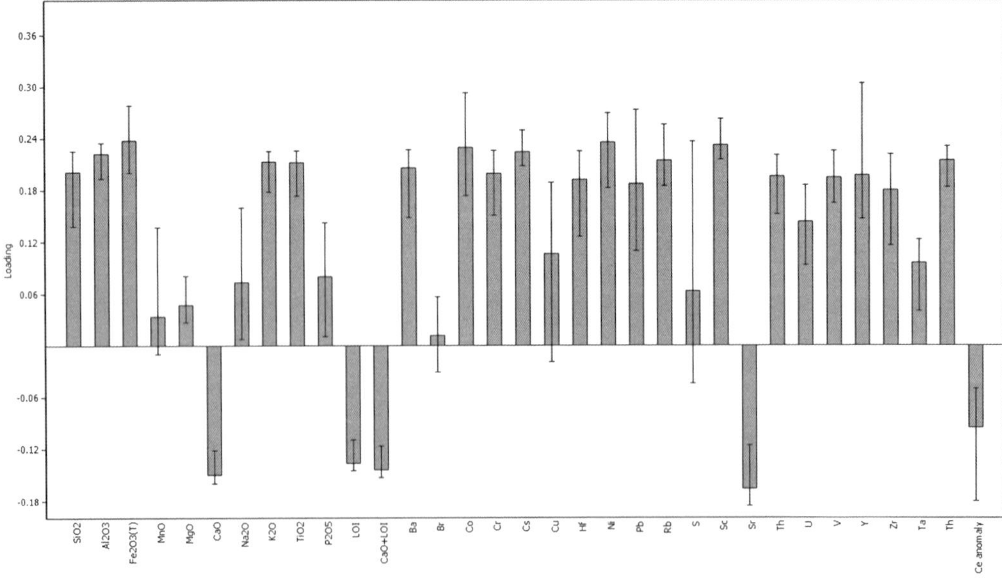

Fig. 14. PC1 loadings showing the detrital signal in contrast (positive values) to the carbonate signal (negative values).

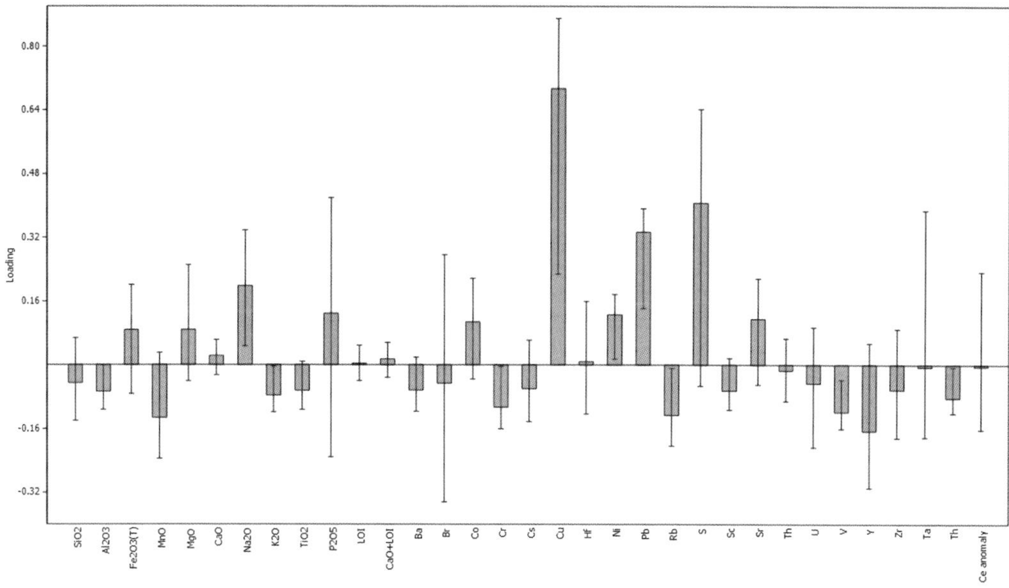

Fig. 15. PC2 loadings showing the anoxia signals, with positive loadings in chalcophile elements and P_2O_5.

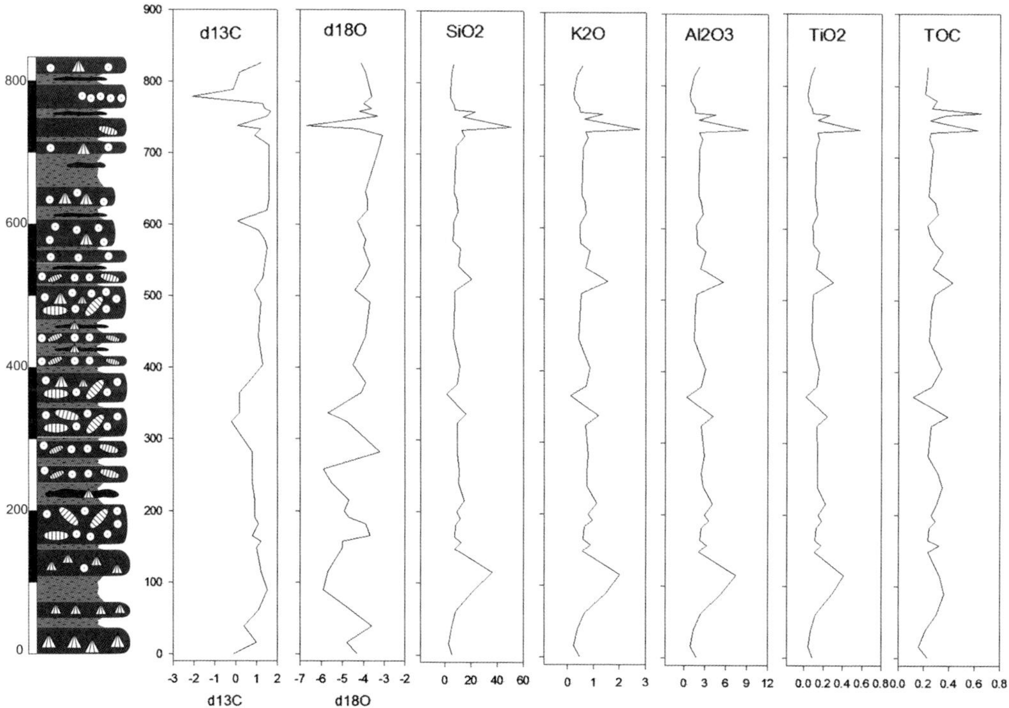

Fig. 16. Stratigraphic distribution of detrital proxies in addition to $\delta^{13}C$ and $\delta^{18}O$ signatures. There is a major excursion at 738 cm and at 759 cm, which we interpret as a major sediment influx. Note that TOC and $\delta^{18}O$ are correlated with this pulse of detrital input. An earlier flux of material can be seen at 89 cm above the base of the section.

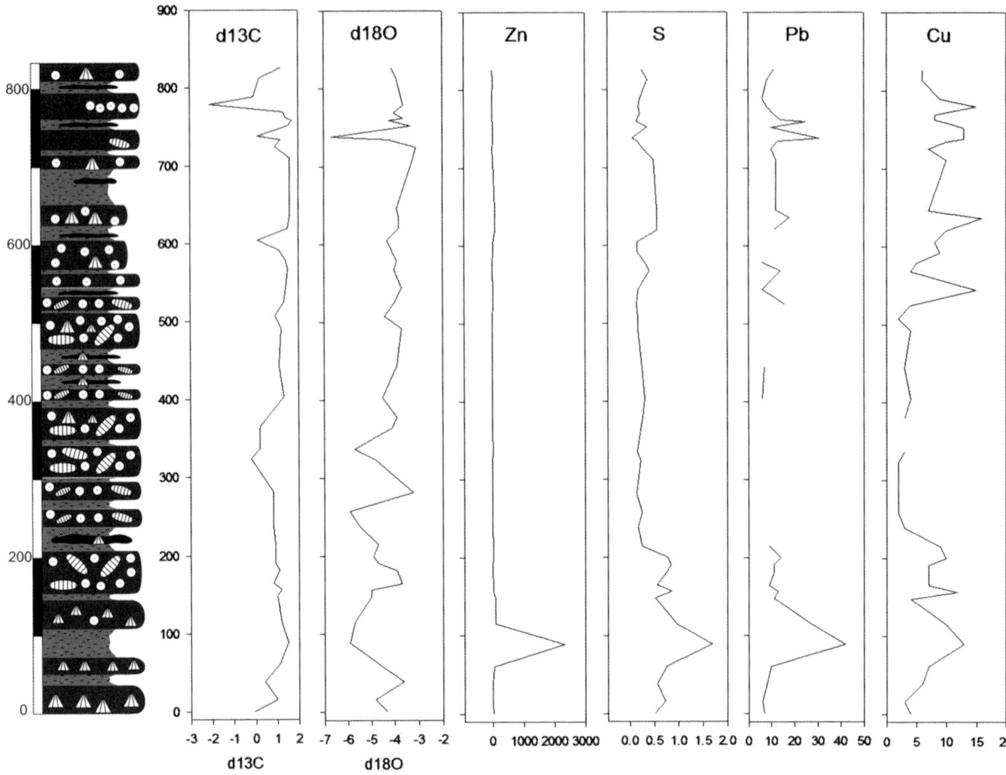

Fig. 17. Stratigraphic distribution of chalcophile element proxies in addition to $\delta^{13}C$ and $\delta^{18}O$ signatures. There is an excursion of the chalcophile elements at 89 cm above the base of the section.

for primary productivity, which are not generally seen in the same interval in known anoxia events (Dymond *et al.* 1992; Pujol *et al.* 2006), and because the use of Ba as an indicator is dependent on a variety of local water-column conditions (Von Breymann *et al.* 1992; Paytan *et al.* 2007). In addition, these signals are not replicated by the other proxies used in this study. While TOC has been used as a proxy for anoxia in many studies (Ingall *et al.* 1993; Algeo & Maynard 2004; Pujol *et al.* 2006; Marynowski & Filipiak 2007; D.P. Bond *et al.* 2013), it may not be an appropriate proxy to use in this case, as it has a positive correlation with detrital input (Fig. 15).

Interpretation

In the lower event (89 cm), elevations in chalcophile element concentrations are not correlated with redox-sensitive element anomalies in comparison to the upper potentially anoxic horizon (738–779 cm) (Fig. 17). This may possibly be due to the differences in sediment supply and flux to the environment; the lower horizon shows a gradual increase in detrital fraction elements above the increases of chalcophile elements. The reason for this discrepancy between anoxia proxies is unclear, and may represent bacterial sulphate reduction in buried sediments rather than an anoxic event at the sediment–water interface. At the top of the section, however, the sharp spike in redox-sensitive elemental anomalies and a negative excursion in $\delta^{13}C$ concurrent with, and subsequent to, two large sediment pulses may indicate anoxia at the sediment–water interface.

Discussion and conclusions

The entire section is composed of shallow subtidal to moderately deep subtidal mixed carbonates and siltstones. According to Struve (1990), the stratigraphical extent of the so-called 'Great Gap' period lasting from the lower Eifelian into the lower Givetian is characterized by sedimentary gaps and not full-marine sediments, and may also have been recognized in the Couvin area, Belgium (Bultynck & Hollevoet 1999). Phosphorite intraclasts and iron-oxide crusts around some bioclasts have been found

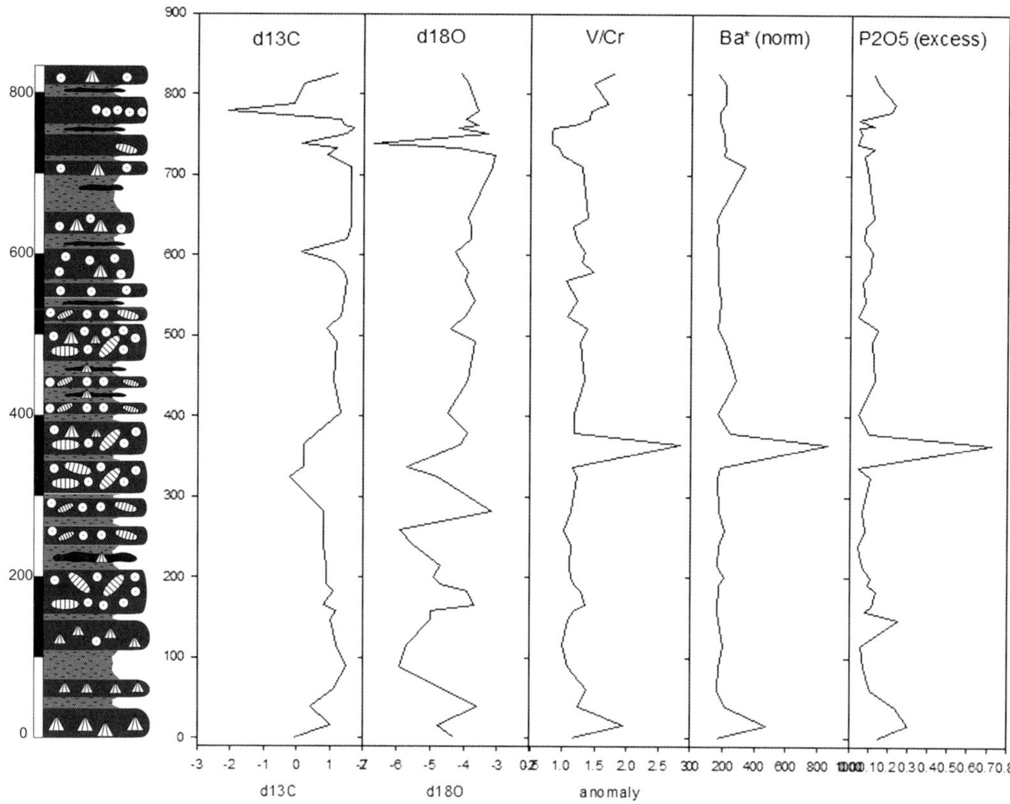

Fig. 18. Stratigraphic distribution of anoxia (V/Cr) and productivity proxies (normalized Ba, excess P_2O_5) in addition to $\delta^{13}C$ and $\delta^{18}O$ signatures. There is a major excursion at 365 cm, which we are reluctant to classify as an anoxic event or an event correlated with changes in sedimentation.

in some thin sections, which may be an indication of submarine weathering and does not necessarily mean that an area further in the north underwent subaerial exposure, as suggested by Winter (1977). Thus, we favour a shallow-marine environment for the entire section, which shows an overall slightly transgressive trend. The latter is also confirmed by magnetic susceptibility data.

According to DeSantis & Brett (2011), magnetic susceptibility data suggest a signature for a regressive phase in eastern North America for deposits of the Stony Hollow, followed by the transgressive succession of the Hurley–Cherry Valley deposits (Brett *et al.* 2011). Magnetic susceptibility data of the Blankenheim section show similar patterns in the same stratigraphic position and are also comparable to those described from the GSSP Mech Irdane, Tafilalt, Morocco (Crick *et al.* 2000). It might be possible that the lower event at the base of the Blankenheim section is based on regional variations in sediment supply and/or sea-level changes, or may be correlated with the Stony Hollow Event associated with probable warming and incursion of tropical species into the subtropical to temperate shelf region of eastern North America (DeSantis & Brett 2011).

As shown earlier, geochemical proxies of the Blankenheim section exhibit two major signals: an increase in chalcophile elements that occurs at the base of the section at 89 cm within the *kockelianus* conodont Biozone; and a second peak that occurs from 738 to 779 cm from the base of the section. Above the alluvial sediment influx at 738 cm, there is a large negative excursion in $\delta^{13}C$ and a Ce anomaly <-0.1 (Fig. 15), the latter of which has been seen in other sections with anoxia (Morad & Felitsyn 2001; Pujol *et al.* 2006; Carmichael *et al.* 2014). Although spikes in V/Cr are not readily apparent in this interval, they do show increases from a secular low (Fig. 17). Any potential authigenic U signatures are likely to have been overprinted by high levels of detrital Th in this part of the section, invalidating authigenic U as a potential anoxia tracer in this particular location.

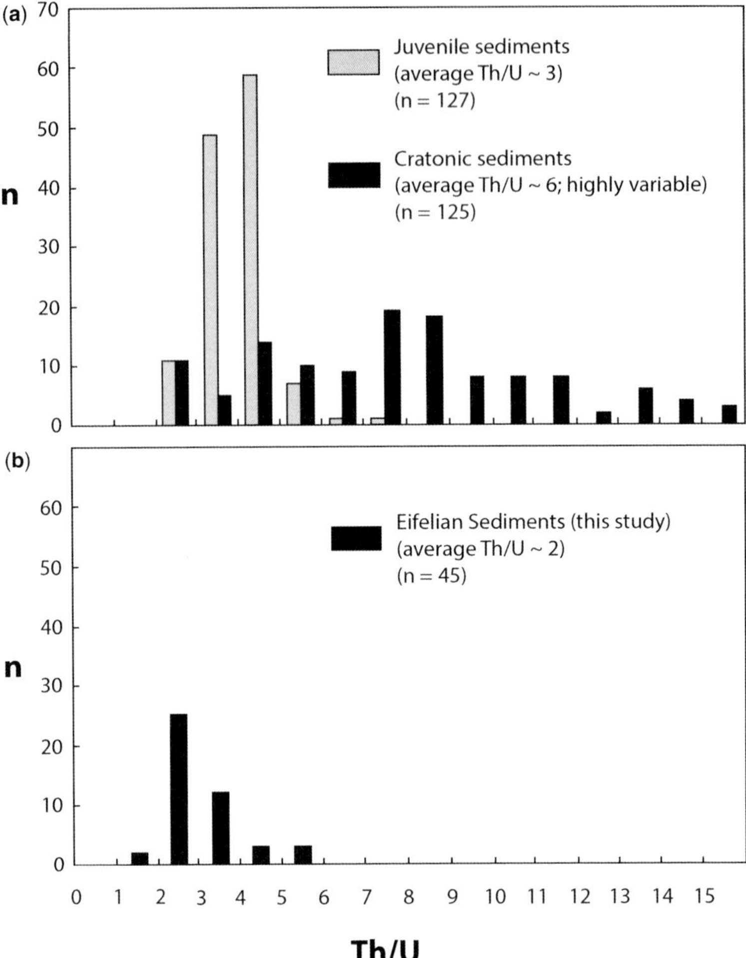

Fig. 19. (**a**) Th/U ratios in basinal sediments can be used to determine the sediment source, with Th/U values for mature, cratonic sediments with highly variable ratios (average of *c.* 6), and juvenile sediment sources with Th/U values clustered around 3 (modified from Carpentier *et al.* 2013). (**b**) The source of sediments in the Blankenheim section has a clustered Th/U value of approximately 2, suggesting a juvenile source.

While there is a phase offset between the $\delta^{13}C$ and Ce anomaly at 779 cm in comparison with the detrital pulses at 738 cm and 759 cm, this could be due to bacterial reduction of the organic matter in the sediments, which will lead to carbon isotope fractionation (Kump *et al.* 1999). This mechanism is consistent with our observation that TOC in this section is often correlated with detrital sedimentation, and the negative $\delta^{13}C$ excursion seen may be due to bacterial reduction, fractionation and mobilization of accumulated organic carbon in the detrital sediments below. While this explanation for a negative excursion is highly dependent on local conditions, negative excursions in $\delta^{13}C$ have been seen immediately prior to the Kačák Event both in Ontario, Canada (van Hengstum & Gröcke 2008) and in Morocco (Ellwood *et al.* 2003). The $\delta^{13}C$ excursion and Ce anomaly are, therefore, interpreted to be the local representation of the beginning of the 'true' Kačák Event Interval, which is also consistent with the conodont and microfacies analyses presented above. The major conclusions of our multidisciplinary approach are as follows:

- Based on micro- and macrofossils, the Blankenheim section exposes the Junkerberg and Freilingen formations, and reaches the uppermost Eifelian (*kockelianus* and *ensensis* conodont biozones).

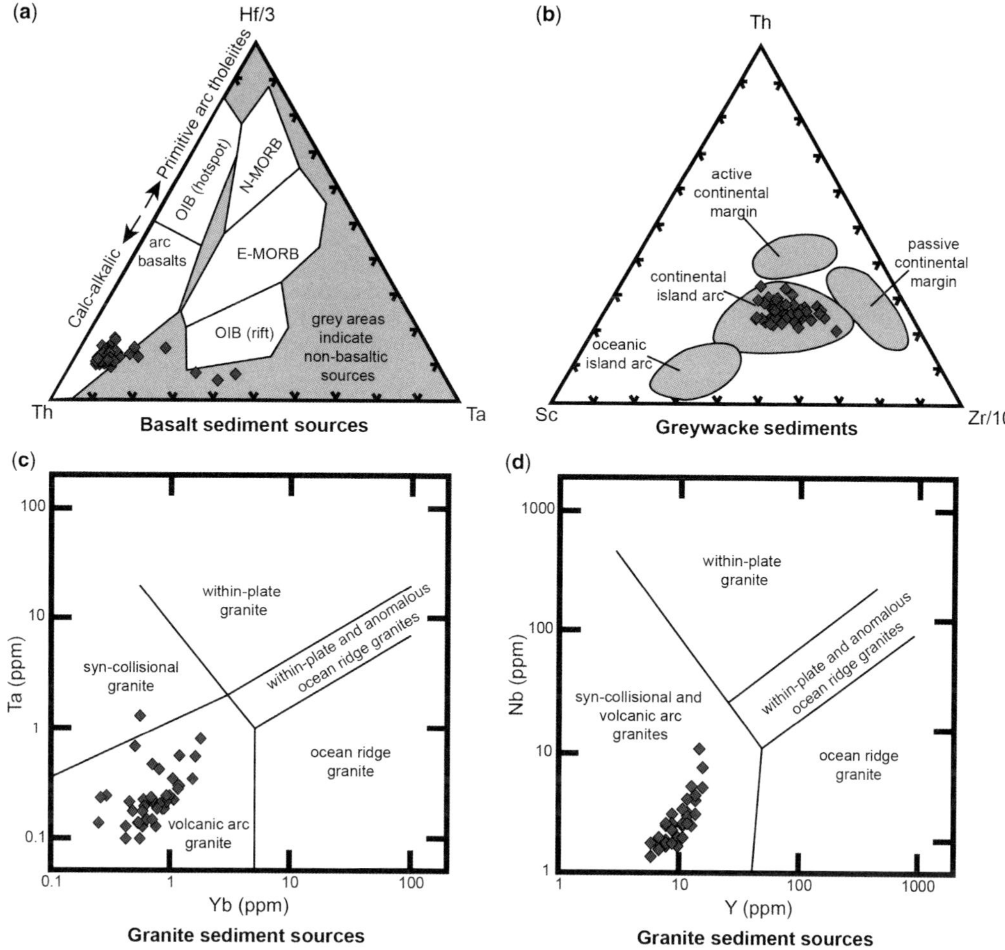

Fig. 20. Tectonomagmatic discrimination diagrams showing sediment-source signatures in detrital Ti-bearing minerals from (**a**) basalts (modified from Wood 1980), (**b**) greywacke sediments (modified from Bhatia & Crook 1986) and (**c**) & (**d**) granitic sources (modified from Pearce *et al.* 1984). Regardless of the signature suite used, the Blankenheim sediments clearly cluster as arc-volcanic signatures, consistent with deposition along the newly amalgamating Avalonia microcontinent.

- The entire section is composed of shallow- to deep-subtidal mixed carbonates and siltstones. The beginning of the Kačák Event Interval can be recognized even in the shallow-marine environment (open-marine organisms, such as tentaculites, cephalopods and/or pelagic ostracodes are absent). The overall section shows a slightly transgressive trend.
- From the magnetic measurements, we can deduce that we have a magnetic signal, which is mostly carried by paramagnetic minerals. The magnetic susceptibility of shale layers is systematically higher than the adjacent carbonate beds. Furthermore, hysteresis plots point to preserved primary coarse (MD) detrital magnetite, which is classically not the case in most of the remagnetized Rhenohercynian zone. Magnetic susceptibility (MS) shows a generally decreasing trend in the section, which could be related to a decrease in detrital input, consistent with the observed transgressive trend.
- The lower part of the sequence (89 cm above the base of the section, around samples BL 12-31 and BL 12-30) exhibits a correlation of the occurrence of chalcophile elements with the occurrence of quartz-rich calcisiltite. This event may be a result of local variations in sediment supply and/or sea-level changes or may be

associated with the pre-Kačák Stony Hollow Event. The $\delta^{13}C$ excursion and Ce anomaly in the upper part of the section (738–779 cm) are interpreted to be the local representation of the beginning of the 'real' Kačák Event Interval, which is also consistent with the conodont and microfacies analyses presented herein.

- The conodont apatite oxygen isotope record from our samples is interpreted as reflecting the palaeotemperature of the Devonian tropical and subtropical sea surface. The estimated palaeotemperature of our samples is 29.7 °C.
- In terms of plate tectonics, the analysed trace elements, such as Hf, Nb, Sc, Ta, Th, U, Y, Yb and Zr, suggest that the sediment source is juvenile and consistent with the Avalonian microcontinent. Furthermore, we can conclude that the sedimentary provenance for the detrital material is consistent with continental island-arc or ribbon continent volcanic-arc settings.
- A multidisciplinary approach has great potential for investigating palaeoenvironmental changes, particularly with respect to events in shallow-water realms.

EK and TJS are grateful for the financial support of FWF P 23775-B17. We thank Michael Joachimski (Erlangen) for measuring the oxygen isotope composition of conodont apatite. Petra Tonarova (Prague) is thanked for picking additional prasinophytes that were used for organic carbon stable isotopes analyses. Jana Anger (Senckenberg Research Institute and Natural History Museum Frankfurt) is thanked for preparing some figures. This paper is a contribution to IGCP 580 and IGCP 596.

References

ALGEO, T. J. & MAYNARD, J. B. 2004. Trace-element behavior and redox facies in core shales of Upper Pennsylvanian Kansas-type cyclothems. *Chemical Geology*, **206**, 289–318.

BASSE, M. & MÜLLER, P. 2004. *Eifel-Trilobiten. 3. Corynexochida, Proetida (2), Harpetida, Phacopida (2), Lichida*, 1st edn. Quelle & Meyer, Wiebelsheim.

BECKER, R. T. & HOUSE, M. R. 2000. Devonian ammonoid zones and their correlation with established series and stage boundaries. *Courier Forschungsinstitut Senckenberg*, **220**, 113–151.

BENDER, P. & KÖNIGSHOF, P. 1994. Regional maturation patterns of the Devonian strata in the eastern Rheinisches Schiefergebirge (Lahn-Dill area) based on conodont colour alteration (CAI). *Courier Forschungsinstitut Senckenberg*, **168**, 335–345.

BERTOLA, C., BOULVAIN, F., DA SILVA, A. C. & POTY, E. 2013. Sedimentology and magnetic susceptibility of Mississippian (Tournaisian) carbonate sections in Belgium. *Bulletin of Geosciences*, **88**, 69–82.

BHATIA, M. R. & CROOK, K. A. W. 1986. Trace element characteristics of graywackes and tectonic setting discrimination of sedimentary basins. *Contributions to Mineralogy and Petrology*, **92**, 181–193.

BISCHOFF, G. & ZIEGLER, W. 1957. Die Conodontenchronologie des Mitteldevons und des tiefsten Oberdevons. *Abhandlungen des hessischen Landesamtes für Bodenforschung*, **22**, 1–136.

BLOMEIER, D. P. G. & REIJMER, J. J. G. 1999. Drowning of a Lower Jurassic carbonate platform: Jbel Bou Dahar, High Atlas, Morocco. *Facies*, **41**, 81–110.

BOND, D. P., ZATOŃ, M., WIGNALL, P. B. & MARYNOWSKI, L. 2013. Evidence for shallow-water 'Upper Kellwasser' anoxia in the Frasnian–Famennian reefs of Alberta, Canada. *Lethaia*, **46**, 355–368.

BOND, T. C., DOHERTY, S. J. ET AL. 2013. Bounding the role of black carbon in the climate system: a scientific assessment. *Journal of Geophysical Research: Solid Earth*, **118**, 5380–5552, http://doi.org/10.1002/jgrd.50171

BRAUN, R., OETKEN, S., KÖNIGSHOF, P., KORNDER, L. & WEHRMANN, A. 1994. Development and biofacies of reef-influenced carbonates (Central Lahn Syncline, Rheinisches Schiefergebirge). *Courier Forschungsinstitut Senckenberg*, **169**, 351–386.

BRETT, C. E., BAIRD, G. C., BARTHOLOMEW, A. J., DESANTIS, K. M. & VER STRAETEN, C. A. 2011. Sequence stratigraphy and a revised sea-level curve for the Middle Devonian of eastern North America. *In*: BRETT, C. B., SCHINDLER, E. & KÖNIGSHOF, P. (eds) Sea-Level Cyclicity, Climate Change, and Bioevents in Middle Devonian Marine and Terrestrial Environments. *Palaeogeography, Palaeoclimatology, Palaeoecology*, **304**, 21–53.

BRINKMANN, R. 1948. Die Mitteldeutsche Schwelle. *Geologische Rundschau*, **36**, 56–66.

BUDIL, P. 1995. Demonstrations of the Kačák Event (Middle Devonian, uppermost Eifelian) at some Barrandian localities. *Vestnik Ceskeho Geologickeho ustavu*, **70**, 1–24.

BUGGISCH, W. & MANN, U. 2004. Carbon isotop stratigraphy of Lochkovian to Eifelian limestones from the Devonian of central and southern Europe. *International Journal of Earth Sciences, Geologische Rundschau*, **93**, 521–541.

BULTYNCK, P. 1970. Révision stratigraphique et paléontologique (brachiopodes et conodontes) de la coupe type du Couvinian. *Mémoires de l'Institute Géologique de l'Université de Louvain*, **26**, 152.

BULTYNCK, P. 1987. Pelagic and neritic condont successions from the Givetian of pre-Sahara Morocco and the Ardennes. *Bulletin de l'Institut Royal des Sciences Naturelles de Belgique, Sciences de la Terre*, **59**, 95–103.

BULTYNCK, P. 1989. Conodonts from a potential Eifelian-Givetian global boundary stratotype at Jebel Ou Driss. *Bulletin de l'Institut Royal des Sciences Naturelles de Belgique, Sciences de la Terre*, **57**, 149–181.

BULTYNCK, P. & HOLLEVOET, C. 1999. The Eifelian-Givetian boundary and Struve's Middle Devonian Great Gap in the Courvin area (Ardennes, southern Belgium). *Senckenbergiana lethaea*, **79**, 3–11.

CALVERT, S. E. & PEDERSEN, T. F. 2007. Elemental proxies for palaeoclimatic and palaeoceanographic variability in marine sediments: interpretation and application. *In*: HILLAIRE-MARCEL, C. & VERNAL, A. D. (eds)

Proxies in Late Cenozoic Paleoceanography. Developments in Marine Geology, **1**. Elsevier, Amsterdam, 567–644.

Carmichael, S. K., Waters, J. A., Suttner, T. S., Kido, E. & DeReuil, A. A. 2014. A new model for the Kellwasser Anoxia Events (Late Devonian): Shallow water anoxia in an open oceanic setting in the Central Asian Orogenic belt. *Palaeogeography, Palaeoclimatology, Palaeoecology*, **399**, 394–403.

Carpentier, M., Weis, D. & Chauvel, C. 2013. Large U loss during weathering of upper continental crust: the sedimentary record. *Chemical Geology*, **340**, 91–104.

Channel, J. E. T. & McCabe, C. 1994. Comparison of magnetic hysteresis parameters of unremagnetized and remagnetized limestones. *Journal of Geophysical Research*, **99**, 4613–4623.

Chlupáč, I. & Kukal, Z. 1986. Reflection of possible global Devonian events in the Barrandian area, C.S.S.R. *In*: Walliser, O. H. (ed.) *Global Bio-Events. A Critical Approach Proceedings of the First International Meeting of the IGCP Project 216: 'Global Biological Events in Earth History'*. Lecture Notes in Earth Sciences, **8**, 169–179.

Crick, R. E., Ellwood, B. B., El Hassani, A. & Feist, R. 2000. Proposed magnetostratigraphy susceptibility magnetostratotype for the Eifelian-Givetian GSSP (Anti-Atlas, Morocco). *Episodes*, **23**, 93–101.

Da Silva, A. C. & Boulvain, F. 2010. Magnetic susceptibility correlation of km-thick Eifelian-Frasnian sections (Ardennes and Moravia). *Geologica Belgica*, **13**, 309–318.

Da Silva, A. C., Yans, J. & Boulvain, F. 2010. Early-Middle Frasian (Early Late Devonian) sedimentology and magnetic susceptibility of the Ardennes area (Belgium): identification of severe and rapid sea-level fluctuations. *Geologica Belgica*, **13**, 319–332.

Da Silva, A. C., Dekkers, M. J., Mabille, C. & Boulvain, F. 2012. Magnetic signal and its relationship with paleoenvironments, diagenesis and remagnetization – examples from the Devonian carbonates of Belgium. *Studia Geophysica & Geodaedica*, **56**, 677–704.

Da Silva, A. C., De Vleeschouwer, D. et al. 2013. Magnetic susceptibility as a high-resolution correlation tool and as a climatic proxy in Paleozoic rocks – Merits and pitfalls: Examples from the Devonian in Belgium. *Marine and Petroleum Geology*, **46**, 173–189.

Day, R., Fuller, M. & Schmidt, V. A. 1977. Hysteresis propreties of titanomagnetites: grain-size and compositional dependence. *Physics of the Earth and Planetary Interiors*, **13**, 260–267.

DeSantis, M. K. & Brett, C. E. 2011. Late Eifelian (Middle Devonian) biocrises: Timing and signature of the pre-Kačák Bakhoven and Stony Hollow Events in eastern North America. *Palaeogeography, Palaeoclimatology, Palaeoecology*, **304**, 113–135.

De Vleeschouwer, D., Da Silva, A. C., Boulvain, F., Crucifix, M. & Claeys, P. 2012. Precessional and half-precessional climate forcing of Mid-Devonian monsoon-like dynamics. *Climate of the Past*, **7**, 1427–1455.

De Vleeschouwer, X., Petitclerc, E., Spassov, S. & Préat, A. 2010. The Givetian–Frasnian boundary at Nismes parastratotype (Belgium): the magnetic susceptibility signal controlled by ferromagnetic minerals. *In*: Da Silva, A. C. & Boulvain, F. (eds) *Magnetic Susceptibility, Correlations and Palaeozoic Environments. Geologica Belgica*, **13**, (4), 351–366.

Dombrowski, A., Henjes-Kunst, F., Höhndorf, A., Kröner, A., Okrusch, M. & Richter, P. 1995. Orthogneisses in the Spessart Crystalline Complex, Northwest Bavaria: Silurian granitoid magmatism at an active continental margin. *Geologische Rundschau*, **84**, 399–411.

Dunlop, D. J. 2002. Theory and application of the Day plot (M_{rs}/M_s v. H_{cr}/H_c) – 1. Theoretical curves and tests using titanomagnetite data. *Journal of Geophysical Research*, **107**, 1–22.

Dymond, J., Suess, E. & Lyle, M. 1992. Barium in deep-sea sediment: a geochemical proxy for paleoproductivity. *Paleoceanography*, **7**, 163–181.

Eckelmann, K., Nesbor, H.-D., Königshof, P., Linnemann, U., Hofmann, M., Lange, J.-M. & Sagawe, A. 2014. Plate interactions of Laurussia and Gondwana during the formation of Pangaea – Constraints from U–Pb LA-SF-ICP-MS detrital zircon ages of Devonian and Early Carboniferous siliciclastics of the Rhenohercynian zone, Central European Variscides. *Gondwana Research*, **25**, 1484–1500, http://doi.org/10.1016/j.gr.2013.05.018

Ellwood, B. B., Crick, R. E. & El Hassani, A. 1999. Magnetosusceptibility event and cyclostratigraphy (MSEC) method used in geological correlation of Devonian rocks from Anti-Atlas Morocco. *American Association of Petroleum Geologists Bulletin*, **83**, 1119–1134.

Ellwood, B. B., Benoist, S. L., El Hassani, A., Wheeler, C. & Crick, R. E. 2003. Impact ejecta layer from the Mid-Devonian possible connection to global mass extinctions. *Science*, **300**, 1734–1737.

Ellwood, B. B., Tomkin, J. H. et al. 2011. A climate-driven model and development of a floating point time scale for the entire Middle Devonian Givetian Stage: A test using magnetostratigraphy susceptibility as a climate proxy. *In*: Brett, C. B., Schindler, E. & Königshof, P. (eds) Sea-Level Cyclicity, Climate Change, and Bioevents in Middle Devonian Marine and Terrestrial Environments. *Palaeogeography, Palaeoclimatology, Palaeoecology*, **304**, 85–95, http://doi.org/10.1016/j.palaeo.2010.10.014

Engel, W., Franke, W., Grote, C., Weber, K., Ahrendt, H. & Eder, F. W. 1983. Nappe tectonics in the southeastern part of the Rheinisches Schiefergebirge. *In*: Martin, H. & Eder, F. W. (eds) *Intracontinental Fold Belts*. Springer, Berlin, 267–287.

Ernst, A., Königshof, P., Taylor, P. D. & Bohaty, J. 2011. Microhabitat complexity – an example from Middle Devonian bryozoans-rich sediments in the Blankenheim Syncline (northern Eifel, Rheinisches Schiefergebirge). *Palaeobiodiversity and Palaeoenvironments*, **91**, 257–284, http://doi.org/10.1007/s12549-011-0060-6

Faber, P. 1980. Fazies-Gliederung und -Entwicklung im Mittel-Devon der Eifel (Rheinisches Schiefergebirge). *Mainzer geowissenschaftliche Mitteilungen*, **8**, 83–149.

Flügel, E. 2004. *Microfacies of Carbonate Rocks*. Springer, Berlin.

Franke, W. 2000. The mid-European segment of the Variscides: tectonostratigraphic units, terrane boundaries and plate tectonic evolution. *In*: FRANKE, W., HAAK, V., ONCKEN, O. & TANNER, D. (eds) *Orogenic Processes: Quantification and Modelling in the Variscan Belt*. Geological Society, London, Special Publications, **179**, 35–61, http://doi.org/10.1144/GSL.SP.2000.179.01.05

FRANKE, W. & ONCKEN, O. 1990. Geodynamic evolution of the northcentral Variscides – a comic strip. *In*: FREEMAN, R., GIESE, P. & MUELLER, S. (eds) *The European Geotraverse*. European Science Foundation, Strasbourg, 187–194.

FRANKE, W. & ONCKEN, O. 1995. Zur prädevonischen Geschichte des Rhenohercynischen Beckens. *Nova Acta Leopoldina NF*, **71**, 53–72.

FRANKE, W., BORTFELD, R. K. *ET AL*. 1990. Crustral structur of the Rhenish Massif: results of deep seismic reflection lines DEKORP 2-North and 2-North-Q. *Geologische Rundschau*, **79**, (3), 523–566.

GAMMON, P. R. & JAMES, N. P. 2001. Palaeogeographical influence on Late Eocene biosiliceous sponge-rich sedimentation, southern Western Australia. *Sedimentology*, **48/3**, 559–584, http://doi.org/10.1046/j.1365-3091.2001.00379.x

HAHN, H. D. 1990. *Fazies grobklastischer Gesteine des Unterdevons (Graue Phyllite bis Taunusquarzit) im Taunus (Rheinisches Schiefergebirge)*. PhD thesis, University of Marburg.

HAHN, H. D. & ZANKL, H. 1991. Sedimentation in the Lower Devonian of the Taunus area (Graue Phyllite to Taunusquarzit). *Zentralblatt für Geologie und Paläontologie, Teil I*, **1990**, 1509–1520.

HAMMER, Ø., HARPER, D. A. T. & RYAN, P. D. 2001. PAST: Paleontological statistics software package for education and data analysis. *Palaeontologia electronica*, **4**, (1), art. 4.

HELSEN, S. & KÖNIGSHOF, P. 1994. Conodont thermal alteration patterns in Paleozoic rocks from Belgium, northern France and western Germany. *Geological Magazine*, **131**, 369–386.

HINDE, G. J. 1879. On conodonts from the Chazy and Cincinnati group of the Cambro-Silurian and from the Hamilton and Genesee-Shale divisions of the Devonian in Canada and the United States. *Quarterly Journal of the Geological Society of London*, **35**, 351–369.

HLADIL, J. 2002. Geophysical records of dispersed weathering products on the Frasnian carbonate platform and early Famennian ramps in Moravia, Czech Republic: proxies for eustasy and palaeoclimate. *Palaeogeography, Palaeoclimatology, Palaeoecology*, **181**, 213–250.

HLADIL, J., KOPTIKOVA, L. *ET AL*. 2009. Early Middle Frasnian platform reef strata in the Moravian Karst interpreted as recording the atmospheric dust changes: the key to understanding perturbations in the *punctata* conodont Zone. *Bulletin of Geosciences*, **84**(1), 75–106.

HOUSE, M. R. 1996. The Middle Devonian Kacak Event. *Proceedings of the Ussher Society*, **9**, 79–84.

HOUSE, M. R. 2002. Strenght, timing, setting and cause of mid-Paleozoic extinctions. *Palaeogeography, Palaeoclimatology, Palaeoecology*, **181**, 5–25.

HUCKRIEDE, H., WEMMER, W. & AHRENDT, H. 2004. Palaeogeography and tectonic structure of allochtonous units in the German part of the Rheno-Hercynian Belt (Central European Variscides). *International Journal of Earth Sciences*, **93**, 414–431.

INGALL, E. D., BUSTIN, R. M. & VAN CAPPELLEN, P. 1993. Influence of water column anoxia on the burial and preservation of carbon and phosphorus in marine shales. *Geochimica et Cosmochimica Acta*, **57**, 303–316.

JACKSON, M. 1990. Diagenetic sources of stable remanence in remagnetized paleozoic cratonic carbonates: A rock magnetic study. *Journal of Geophysical Research*, **95**, 2753–2761.

JOACHIMSKI, M. M., VAN GELDERN, R., BREISIG, S., DAY, J. & BUGGISCH, W. 2004. Oxygen isotope evolution of biogenetic calcite and apatite during the Middle and Upper Devonian. *International Journal of Earth Sciences*, **93**, 542–553.

JOACHIMSKI, M. M., BREISIG, S. *ET AL*. 2009. Devonian climate and reef evolution: insights fom oxygen isotopes in apatite. *Earth and Planetary Science Letters*, **284**, 599–609.

KOPTÍKOVÁ, L. 2011. Precise position of the Basal Choteč event and evolution of sedimentary environments near the Lower–Middle Devonian boundary: the magnetic susceptibility, gamma-ray spectrometric, lithological, and geochemical record of the Prague Synform (Czech Republic). *Palaeogeography, Palaeoclimatology, Palaeoecology*, **304**, 96–112.

KÖNIGSHOF, P. & WERNER, R. 1994. Zur Bestimmung der Versenkungstemperaturen im Devon der Eifeler Kalkmulden-Zone mit Hilfe der Conodontenfarbe. *Courier Forschungsinstitut Senckenberg*, **168**, 255–265.

KÖNIGSHOF, P., GEWEHR, B., KORNDER, L., WEHRMANN, A., BRAUN, R. & ZANKL, H. 1991. Stromatoporen-Morphotypen aus einem zentralen Riffbereich (Mitteldevon) in der südwestlichen Lahnmulde. *Geologica et Palaeontologica*, **25**, 19–35.

KÖNIGSHOF, P., NESBOR, H.-D. & FLICK, H. 2010. Volcanism and reef development in the Devonian: a case study from the Lahn syncline, Rheinisches Schiefergebirge (Germany). *Gondwana Research*, **17**, 264–280, http://doi.org/10.1016/j.gr.2009.09.006

KRONER, U. & HAHN, T. 2003. Sedimentation, Deformation and Metamorphose im Saxothuringikum während der variszischen Orogenese: die komplexe Entwicklung von Nord-Gondwana während kontinentaler Subduktion und schiefer Kollision. *In*: LINNEMANN, U. (ed.) Das Saxothuringikum – Abriss der präkambrischen und paläozoischen Geologie von Sachsen und Thüringen. *Geologica Saxonica*, **48/49**, 137–150.

KRONER, U., HAHN, T., ROMER, R. L. & LINNEMANN, U. 2007. The Variscan orogeny in the Saxo-Thuringian zone-heterogenous overprint of Cadomian/Palaeozoic peri-Gondwana crust. *In*: LINNEMANN, U., NANCE, R. D., KRAFT, P. & ZULAUF, G. (eds) *The Evolution of the Rheic Ocean: From Avalonian–Cadomian Active Margin to Alleghenian–Variscan Collision*. Geological Society of America, Special Papers, **423**, 153–172.

KUMP, I. R., ARTHUR, M. A., PATZKOWSKY, M. E., GIBBS, M. T., PINKUS, D. S. & SHEEHAN, P. M. 1999. A weathering hypothesis for glaciation at high atmosheric pCO2 during the Late Ordovician. *Palaeogeography, Palaeoclimatology, Palaeoecology*, **152**, 173–187.

LINNEMANN, U., D'LEMOS, R., DROST, K., JEFFRIES, T., GERDES, A., ROMER, R. L. & SAMSOR, S. D. 2008. Introduction (Chapter 3: The Cadomian Orogeny). *In*: MCCANN, T. (ed.) *The Geology of Central Europe*. Geological Society, London, 103–154.

LINNEMANN, U., HOFMANN, M., ROMER, R. L. & GERDES, A. 2010. Transitional stages between the Cadomian and Variscan Orogenies: Basin development and tectonomagmatic evolution of the southern margin of the Rheic Ocean in the Saxo-Thuringian Zone (North Gondwana shelf). *In*: LINNEMANN, U. & ROMER, R. L. (eds) *Pre-Mesozoic Geology of Saxo-Thuringia – From the Cadomian Active Margin to the Variscan Orogen*. Schweizerbart Science, Stuttgart, 59–98.

MARYNOWSKI, L. & FILIPIAK, P. 2007. Water column euxinia and wildfire evidence during deposition of the Upper Famennian Hangenberg event horizon from the Holy Cross Mountains (central Poland). *Geological Magazine*, **144**, 569–595.

MATTE, Ph. 1986. Tectonics and plate tectonics model for the Variscan Belt of Europe. *Tectonophysics*, **126**, 329–374.

MEISCHNER, D. 1991. Kleine Geologie des Kellerwaldes (Exkursion F). *Jahresbericht und Mitteilungen des Oberrheinischen Geologischen Vereins, NF*, **73**, 115–142.

MOLINA GARZA, R. S. & ZIJDERVELD, J. D. A. 1996. Paleomagnetism of Paleozoic strata, Brabant and Ardennes Massifs, Belgium: Implications of prefolding and postfolding Late Carboniferous secondary magnetizations for European apparent polar wander. *Journal of Geophysical Research*, **101**, 799–818.

MORAD, S. & FELITSYN, S. 2001. Identification of primary Ce-anomaly signatures in fossil biogenic apatite: implication for the Cambrian oceanic anoxia and phosphogenesis. *Sedimentary Geology*, **143**, 259–264.

NANCE, R. D., GUTIÉRREZ-ALONSO, G. *ET AL.* 2010. Evolution of the Rheic Ocean. *Gondwana Research*, **17**, 194–222.

NESBOR, H.-D. 2004. Paläozoischer Intraplattenvulkanismus im östlichen Rheinischen Schiefergebirge – Magmenentwicklung und zeitlicher Ablauf. *Geologisches Jahrbuch Hessen*, **131**, 145–182.

NESBOR, H.-D., BUGGISCH, W., FLICK, H., HORN, M. & LIPPERT, H.-J. 1993. Vulkanismus im Devon des Rhenoherzynikums. Fazielle und paläogeographische Entwicklung vulkanisch geprägter mariner Becken am Beispiel des Lahn-Dill-Gebietes. *Geologisches Jahrbuch Hessen*, **98**, 3–87.

OCHS, G. & WOLFART, R. 1961. *Geologie der Blankenheimer Mulde (Devon, Eifel)*. Abhandlungen der Senckenbergischen Naturforschenden Gesellschaft, **501**.

OCZLON, M. S. 1992. *Gondwanan and Laurussia Before and During the Variscan Orogeny in Europe and Related Areas*. Heidelberger Geowissenschaftliche Abhandlungen, **53**.

ONCKEN, O., PLESCH, A., WEBER, K., RICKEN, W. & SCHRADER, S. 2000. Passive margin detachment during arc–continent collision (Central European Variscides). *In*: FRANKE, W., HAAK, V., ONCKEN, O. & TANNER, D. (eds) *Orogenic Processes: Quantification and Modelling in the Variscan Belt*. Geological Society, London, Special Publications, **179**, 199–216, http://doi.org/10.1144/GSL.SP.2000.179.01.13

PAPROTH, E. & STRUVE, W. 1982. Bemerkungen zur Entwicklung des Givetium am Niederrhein. *Paläogeographischer Rahmen der Bohrung Schwarzbachtal 1. Senckenbergiana lethaea*, **63**, 359–376.

PARRY, L. G. 1982. Magnetization of immobilized particle dispersions with two distinct particles sizes. *Physics of the Earth and Planetary Interiors*, **28**, 230–241.

PAS, D., DA SILVA, A. C., CORNET, P., BULTYNCK, P., KÖNIGSHOF, P. & BOULVAIN, F. 2013. Sedimentary development of a continuous Middle Devonian to Mississippian section from the fore-reef fringe of the Brilon Reef Complex (Rheinisches Schiefergebirge, Germany). *Facies*, **59**, (4), 969–990, http://doi.org/10.1007/s10347-012-0351-z

PAYTAN, A., AVERYT, K., FAUL, K., GRAY, E. & THOMAS, E. 2007. Barite accumulation, ocean productivity, and Sr/Ba in barite across the Paleocene-Eocene thermal maximum. *Geology*, **35**, 1139–1142.

PEARCE, J. A., HARRIS, N. B. & TINDLE, A. G. 1984. Trace element discrimination diagrams for the tectonic interpretation of granitic rocks. *Journal of Petrology*, **25**, 956–983.

PUCÉAT, E., JOACHIMSKI, M. M. *ET AL.* 2010. Revised phosphate-water fractionation equation reassessing paleotemperatures derived from biogenic apatite. *Earth and Planetary Science Letters*, **298**, 135–142, http://doi.org/10.1016/j.epsl.2010.07.034

PUJOL, F., BERNER, Z. & STÜBEN, D. 2006. Palaeoenvironmental changes at the Frasnian/Famennian boundary in key European sections: Chemostratigraphic constraints. *Palaeogeography, Palaeoclimatology, Palaeoecology*, **240**, 120–145.

RACKI, G., RACKA, M., MATYJA, H. & DE VLEESCHOUWER, X. 2002. The Frasnian/Famennian boundary interval in the South Polish-Moravian shelf basins: integrated event-stratigraphical approach. *Palaeogeography, Palaeoclimatology, Palaeoecology*, **181**, 251–297.

REISCHMANN, T., ANTHES, G., JAECKEL, P. & ALTENBERGER, U. 2001. Age and origin of the Böllsteiner Odenwald. *Mineralogy and Petrology*, **72**, 29–44.

RIDING, R. 1975. *Girvanella* and other algae as depth indicators. *Lethaia*, **8**, 173–179, http://doi.org/10.1111/j.1502-3931.1975.tb01310.x

RIQUIER, L., AVERBUCH, O., DE VLEESCHOUWER, X. & TRIBOVILLARD, N. 2010. Diagenetic v. detrital origin of the magnetic susceptibility variations in some carbonate Frasnian-Famennian boundary sections from Northern Africa and Western Europe: implications for paleoenvironmental reconstructions. *International Journal of Earth Sciences*, **99** (Suppl. 1), S57–S73.

ROMER, R. L. & HAHNE, K. 2010. Life of the Rheic Ocean: scrolling through the shale record. *Gondwana Research*, **17**, 408–421.

ROMER, R. L., LINNEMANN, U. & GEHMLICH, M. 2003. Geochronologische und isotopengeochemische Randbedingungen für die cadomische und variszische Orogenese im Saxothuringikum. *In*: LINNEMANN, U. (ed.) Das Saxothuringikum – Abriss der präkambrischen und paläozoischen Geologie von Sachsen und Thüringen. *Geologica Saxonica*, **48/49**, 19–28.

SALAMON, M. & KÖNIGSHOF, P. 2010. Middle Devonian olistostromes in the Rheno-Hercynian (Rheinisches Schiefergebirge) – an indication of back arc rifting on a passive margin of Laurussia? *Gondwana Research*, **17**, 281–291, http://doi.org/10.1016/j.gr.2009.10.004

SALTZMAN, M. R. 2002. Carbon isotope ($\delta^{13}C$) stratigraphy across the Silurian–Devonian transition in North America: evidence for a perturbation of the global chemical cycle. *Palaeogeography, Palaeoclimatology, Palaeoecology*, **187**, 83–100.

SCHNEIDER, J., BECHSTADT, T. & MACHEL, H. G. 2004. Covariance of C- and O-isotopes with magnetic susceptibility as a result of burial diagenesis of sandstones and carbonates: an example from the Lower Devonian La Vid Group, Cantabrian Zone, NW Spain. *International Journal of Earth Science*, **93**, 990–1007.

SCHÖNE, B. R. 1997. *Der Otomari-Event und seine Auswirkungen auf die Fazies des Rhenoherzynischen Schelfs (Devon Rheinisches Schiefergebirge).* Göttinger Arbeiten zur Geologie und Paläontologie, **70**, 1–140.

SCHÖNE, B. R., BASSE, M. & MAY, A. 1998. Korrelationen des Eifelium/Givetium- Grenzbereichs im Rheinischen Schiefergebirge. *Senckenbergiana lethaea*, **77**, 233–242.

SCHWAN, W. 1991. Geologie des Acker-Bruchberg-Ilsenburg-Zuges (Oberharz) – Derzeitiger Forschungsstand und Diskussion der Probleme. *Zentralblatt für Geologie und Paläontologie, Teil 1*, **7**, 787–850.

SCOTESE, C. R. & BARRET, S. F. 1990. Gondwana's movement over the South Pole during the Palaeozoic: evidence from lithological indicators of climate. *In*: MCKERROW, W. S. & SCOTESE, C. R. (eds) *Palaeozoic Palaeogeography and Biogeography*. Geological Society, London, Memoirs, **12**, 75–85, http://doi.org/ 10.1144/GSL.MEM.1990.012.01.06

SPASSOV, S. & VALET, J. P. 2012. Detrital magnetization from redeposition experiments of different natural sediments. *Earth and Planetary Science Letters*, **351–352**, 147–157.

STAUFFER, C. R. 1938. Conodonts of the Olentangy Shale. *Journal of Paleontology*, **12**, 411–443.

STETS, J. & SCHÄFER, A. 2009. The Siegenian delta: land–sea transitions at the northern margin of the Rhenohercynian Basin. *In*: KÖNIGSHOF, P. (ed.) *Devonian Change: Case Studies in Palaeogeography and Palaeoecology*. Geological Society, London, Special Publications, **314**, 37–72, http://doi.org/10.1144/SP314.3

STRUVE, W. 1963. Das Korallen-Meer der Eifel vor 300 Millionen Jahren – Funde, Deutungen, Probleme. *Natur und Museum*, **93**, 237–276.

STRUVE, W. 1982. The Great Gap in the record of the marine Devonian. *Courier Forschungsinstitut Senckenberg*, **55**, 433–447.

STRUVE, W. 1990. Paläozoologie III (1986–1990). *In*: ZIEGLER, W. (ed.) Wissenschaftlicher Jahresbericht 1988/89 des Forschungsinstituites Senckenberg, Frankfurt am Main. *Courier Forschungsinstitut Senckenberg*, **127**, 251–279.

STRUVE, W. 1996. Trilobiten, Rheinisches Schiefergebirge, Mitteldevon. *In*: WEDDIGE, K. (ed.) Devon – Korrelationstabelle. *Senckenbergiana lethaea*, **76**, 280.

STRUVE, W., PLODOWSKI, P. & WEDDIGE, K. 1997. Biostratigraphische Stufengrenzen und Events in der Prümer und Hillesheimer Mulde. *Terra Nostra*, **97**, 123–167.

STRUVE, W., BASSE, M. & WEDDIGE, K. 2008. Prädevon, Ober-Emsium und Mitteldevon der Eifeler Kalkmulden-Zone. *In*: DEUTSCHE STRATIGRAPHISCHE KOMMISSION (eds) *Stratigraphie von Deutschland VIII. Devon*. Schriftenreihe der Deutschen Geowissenschaften, **52**, 297–374.

TAIT, J. A., SCHÄTZ, M., BACHTADSE, V. & SOFFEL, H. 2000. Palaeomagnetism and Palaeozoic paleogeography of Gondwana and European terranes. *In*: FRANKE, W., HAAK, V., ONCKEN, O. & TANNER, D. (eds) *Orogenic Processes: Quantification and Modelling in the Variscan Belt*. Geological Society, London, Special Publications, **179**, 21–34, http://doi.org/10.1144/GSL.SP.2000.179.01.04

TAYLOR, A. M. & GOLDRING, R. 1993. Description and analysis of bioturbation and ichnofabric. *Journal of the Geological Society, London*, **150**, 141–148, http://doi.org/10.1144/gsjgs.150.1.0141

TEICHMÜLLER, M. & TEICHMÜLLER, R. 1979. Ein Inkohlungsprofil entlang der linksrheinischen Geotraverse von Schleiden nach Aachen und die Inkohlung der Nord-Süd Zone der Eifel. *Fortschritte in der Geologie von Rheinland und Westfalen*, **27**, 323–355.

TORSVIK, T. H. & COCKS, L. R. M. 2004. Earth geography from 400 to 250 Ma: a palaeomagnetic, faunal and facies review. *Journal of the Geological Society, London*, **161**, 555–572, http://doi.org/10.1144/0016-764903-098

TRIBOVILLARD, N., ALGEO, T. J., LYONS, T. & RIBOULLEAU, A. 2006. Trace metals as paleoredox and paleoproductivity proxies: an update. *Chemical Geology*, **232**, 12–32.

TROTTER, J. A., WILLIAMS, I. S., BARNES, C. R., LECUYER, C. & NICOLL, R. S. 2008. Did cooling oceans trigger Ordovician biodiversification? Evidence from the conodont thermometry. *Science*, **321**, 550–554.

VAN GELDERN, R., JOACHIMSKI, M. M., DAY, J., JANSEN, U., ALVAREZ, F., YOLKIN, E. A. & MA, X.-P. 2006. Carbon, oxygen and strontium isotope records of Devonian brachiopod shell calcite. *Palaeogeography, Palaeoclimatology, Palaeoecology*, **240**, 47–67.

VAN HENGSTUM, P. J. & GRÖCKE, D. R. 2008. Stable isotope record of the Eifelian–Givetian boundary Kačák-*otomari* Event (Middle Devonian) from Hungary Hollow, Ontario, Canada. *Canadian Journal of Earth Sciences*, **45**, 353–366.

VON BREYMANN, M. T., EMEIS, K. C. & SUESS, E. 1992. Water depth and diagenetic constraints on the use of barium as a palaeoproductivity indicator. *In*: SUMMERHAYES, C. P., PRELL, W. L. & EMEIS, K. C. (eds) *Upwelling Systems: Evolution Since the Early Miocene*. Geological Society, London, Special Publications, **64**, 273–284, http://doi.org/10.1144/GSL.SP.1992.064.01.18

WACHENDORF, H. 1986. Der Harz – variszischer Bau und geodynamische Entwicklung. *Geologisches Jahrbuch*, **A91**, 3–67.

WALLISER, O. H. 1985. Natural boundaries and commission boundaries in the Devonian. *Courier Forschungsinstitut Senckenberg*, **75**, 401–408.

WEDDIGE, K. 1977. Die Conodonten der Eifel – Stufe im Typusgebiet und in benachbarten Faziesgebieten. *Senckenbergiana lethaea*, **58**, 271–419.

WEDDIGE, K. & ZIEGLER, W. 1979. Evolutionary patterns in Middle Devonian conodont genera *Polygnathus* and *Icriodus*. *Geologica et Palaeontologica*, **13**, 157–164.

WEHRMANN, A., BLIECK, A. ET AL. 2005. Palaeoenvironment and palaeoecology of intertidal deposits in a Lower Devonian siliciclastic sequence of the Mosel Region, Germany. *Palaios*, **20**, 101–120.

WHALEN, M. T. & DAY, J. 2010. Cross-basin variations in magnetic susceptibility influenced by changing sea level, paleogeography, and paleoclimate: upper Devonian, Western Canada. *Journal of Sedimentary Research*, **80**, 1109–1127.

WIERICH, F. 1999. *Orogene Prozesse im Spiegel synorogener Sedimente – Korngefügekundliche Liefergebietsanalyse siliziklastischer Sedimente im Devon des Rheinischen Schiefergebirges*. Marburger Geowissenschaften: Zeitschrift der Marburger Geowissenschaftlichen Vereinigung, **1**.

WIGNALL, P. B. & MYERS, K. J. 1988. Interpreting benthic oxygen levels in mudrocks: a new approach. *Geology*, **16**, 452–455.

WINTER, J. 1977. Excursion guide Eifel synclines. *In*: MEYER, W., STOLTIDIS, J. & WINTER, J. (eds) Geologische Exkursion in den Raum Weyer-Schuld-Heyroth-Niederehe-Üxheim-Ahütte. *Decheniana*, **130**, 322–334.

WOOD, D. A. 1980. The application of a ThHfTa diagram to problems of tectonomagmatic classification and to establishing the nature of crustal contamination of basaltic lavas of the British Tertiary Volcanic Province. *Earth and Planetary Science Letters*, **50**, 11–30.

WRIGHT, R. T., COFFIN, R. B. & LEBO, M. 1987. Dynamics of planktonic bacteria and heterotrophic microflagellates in the Parker estuary, northern Massechusetts. *Continental Shelf Research*, **7**, 1383–1397.

ZEGERS, T. E., DEKKERS, M. J. & BAILY, S. 2003. Late Carboniferous to Permian remagnetization of Devonian limestones in the Ardennes: role of temperature, fluids, and deformation. *Journal of Geophysical Research*, **108**, 5/1–5/19.

ZWING, A., BACHTADSE, V. & SOFFEL, H. C. 2002. Late Carboniferous remagnetisation of Palaeozoic rocks in the NE Rhenish Massif, Germany. *Physics and Chemistry of the Earth, Parts A/B/C*, **27**, 1179–1188.

ZWING, A., MATZKA, J., BACHTADSE, V. & SOFFEL, H. C. 2005. Rock magnetic properties of remagnetized Palaeozoic clastic and carbonate rocks from the NE Rhenish Massif, Germany. *Geophysical Journal International*, **160**, 477–486.

ZWING, A., CLAUER, N., LIEWIG, N. & BACHTADSE, V. 2009. Identification of remagnetization processes in Paleozoic sedimentary rocks of the northeast Rhenish Massif in Germany by K-Ar dating and REE tracing of authigenic illite and Fe oxides. *Journal of Geophysical Research: Solid Earth*, **114**, B06104.

Conodont biofacies of the Taghanic transgressive interval (middle Givetian): Polish record and global comparisons

K. NARKIEWICZ[1]*, M. NARKIEWICZ[1] & P. BULTYNCK[2]

[1]*Polish Geological Institute-National Research Institute, Rakowiecka 4, 00-975 Warszawa, Poland*

[2]*Department of Paleontology, Royal Belgian Institute of Natural Sciences, Brussels, Belgium*

**Corresponding author (e-mail: katarzyna.narkiewicz@pgi.gov.pl)*

Abstract: Conodont biofacies of the Lublin and Łysogóry–Radom basins in SE Poland have been analysed in five cored borehole sections in a narrow interval of the middle Givetian *Polygnathus ansatus* Zone, corresponding to the global Taghanic transgression. Assemblages exhibiting various proportions of dominant genera, *Icriodus* (I) and *Polygnathus* (P), as well as particular P species and a few accessory taxa, reflect both temporal transgression dynamics and lateral facies changes. The latter comprise transition from a brackish lagoon with intermittent open-marine influence, to a carbonate shoal and offshore marly shelf, generally characterized by P–I biofacies, but with a varying proportion of constituent genera and polygnathid species. Comparison of the Polish record with stratigraphically well-constrained, quantitative biofacies evidence worldwide allowed the construction of a 2D nearshore–offshore model for the Euramerican epicontinental faunas connected with Taghanic transgressive facies. The I/P ratio has a diagnostic value for specific subenvironments (very nearshore/shallow-water and drowned platform) but for other settings the *Polygnathus ansatus* to *Polygnathus linguiformis* ratio appears more useful. The Moroccan faunas display specific biofacies patterns tentatively explained by different climatic conditions. The conodont biofacies concept has a limited application for palaeogeographically isolated settings, including pelagic-oceanic areas of microcontinents or submarine rises. In other cases (Eastern Australia), palaeobiogeographical bias precludes direct comparisons with the Euramerican model.

Although conodont animals are still enigmatic in many respects (e.g. Blieck *et al.* 2010; Murdock *et al.* 2013), the conodont biofacies concept proved to be useful in palaeoenvironmental interpretations of Palaeozoic and Triassic deposits (Seddon & Sweet 1971; Druce 1973; Sweet 1988; Pohler & Barnes 1990). The Devonian biofacies models are among the most elaborate, with particularly advanced Late Devonian schemes (Sandberg 1976; Dreesen & Thorez 1980; Sandberg & Dreesen 1984; Sandberg *et al.* 1988, 1989, 1992; Ziegler & Sandberg 1990; Belka & Wendt 1992). Published Middle Devonian conodont biofacies interpretations are fewer and less detailed than the Late Devonian ones, often referring to wide stratigraphical intervals (Bultynck 1976; Schumacher 1976; Weddige & Ziegler 1976; Sparling 1984). Stratigraphic generalizations may lead to neglecting of evolutionary factors, for example, speciation or extinction, that provide an independent background against which ecologically controlled biofacies patterns are to be analysed (Narkiewicz & Narkiewicz 2011; see also Belka & Wendt 1992). Moreover, such a general approach may lead to ignoring the complexities of a sedimentary record that reflect environmental changes influencing a succession of conodont assemblages.

Previous studies defined Devonian conodont biofacies either qualitatively, by determining dominant forms (e.g. Seddon & Sweet 1971; Druce 1973; Chatterton 1976; Schumacher 1976), or quantitatively, by establishing numerical proportion of component form genera in studied assemblages (e.g. Sandberg 1976; Sandberg & Dreesen 1984; Belka & Wendt 1992). In the case of the Middle Devonian this approach was often based on a relative abundance of the two most common platform genera, *Icriodus* Branson & Mehl, 1938 and *Polygnathus* Hinde, 1879. Accordingly, such an icriodid to polygnathid elements ratio (I/P) was thought to reflect a range of nearshore shallow-marine to offshore deeper-marine biotopes (Weddige & Ziegler 1976; Klapper & Barrick 1978; Sparling 1984). Nevertheless, some of the authors were also able to demonstrate that certain species are indicative of particular environments (Schumacher 1976; Sparling 1983, 1984; Klapper & Lane 1985; Sandberg *et al.* 1992).

In the present paper we analyse conodont biofacies of the Taghanic transgressive interval in

From: Becker, R. T., Königshof, P. & Brett, C. E. (eds) 2016. *Devonian Climate, Sea Level and Evolutionary Events*. Geological Society, London, Special Publications, **423**, 201–222.
First published online June 10, 2015, http://doi.org/10.1144/SP423.2

the middle Givetian *Polygnathus ansatus* Zone (=former Middle *Polygnathus varcus* Zone; Ziegler *et al.* 1976). The transgression initiated one of the most prominent and long-lasting marine onlaps during the Devonian, corresponding to the IIa cycle of Johnson *et al.* (1985; see also Johnson 1970; House 1983, 1985; Johnson *et al.* 1996). Recently, there have been divergent approaches to defining the onset of the Taghanic transgression (e.g. Brett *et al.* 2011). In the present paper we adopt the traditional stratigraphical approach, in the sense of the refined IIa–Tagh depophase of Aboussalam & Becker (2011). The starting point for our considerations is the record of the Łysogóry–Radom and Lublin basins in SE Poland, whose stratigraphical framework and depositional history have been investigated recently in detail (K. Narkiewicz 2011; M. Narkiewicz 2011*a*, *b*; M. Narkiewicz *et al.* 2011). The regional Polish data and interpretations are then compared with global data available from selected basins covering a wide range of palaeogeographical and facies settings (Fig. 1).

Although the duration of the *ansatus* Zone is *c.* 1 Ma (Kaufmann 2006), the investigated transgressive interval may have corresponded to a few hundred thousand years. This is suggested by the observation that the initial Taghanic transgression is traced in the upper part of the zone, as implied by carbon isotope data of Buggisch & Joachimski (2006; see also Aboussalam 2003). The transgressive pulse induced various biotic events and processes (Klapper & Johnson 1980; House 2002; Aboussalam & Becker 2011; Brett *et al.* 2011), although it did not involve any major conodont extinctions or speciations (Aboussalam & Becker 2011). The purpose of our study is to improve our understanding of the environmental factors that controlled conodont distribution around this important stratigraphic interval. This may have further biostratigraphical and palaeoenvironmental implications for elucidating the dynamics of both eustatic and biotic events during mid-Givetian times.

Palaeogeographical and stratigraphic background of the Polish sections

The Devonian Łysogóry–Radom and Lublin basins of SE Poland formed part of a wide tropical belt of shelf sedimentation along the southern margin of the Euramerica (Laurussia) Continent (Narkiewicz 2007; Belka & Narkiewicz 2008; Fig. 1). The

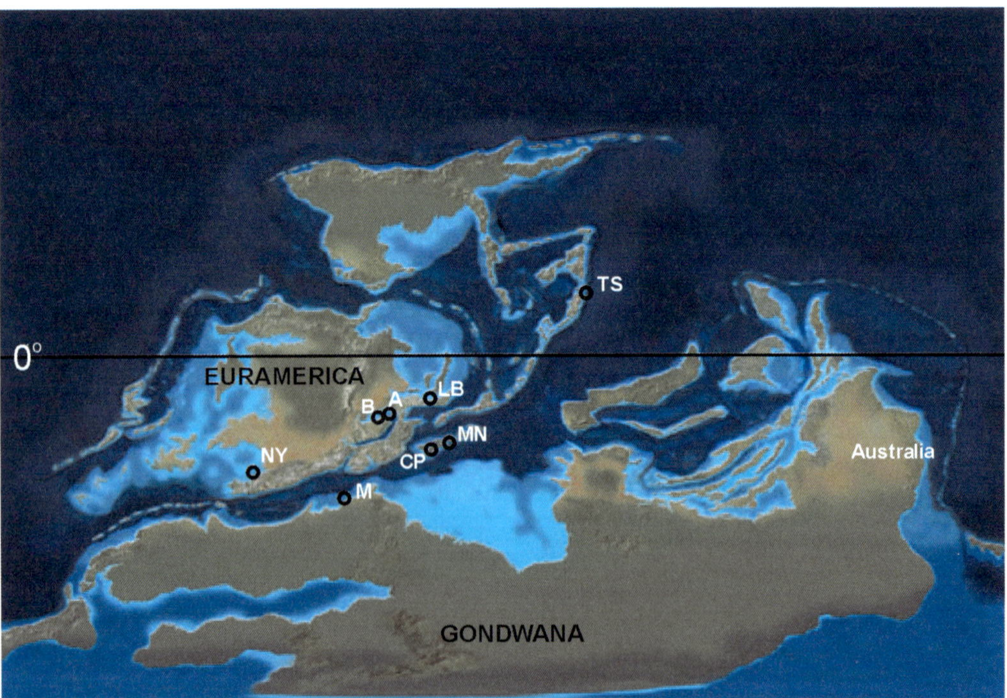

Fig. 1. Analysed localities against the Middle Devonian global palaeogeography (after Blakey 2013, http://www2.nau.edu/rcb7/370Marect.jpg). A, Ardennes; B, Boulonnais; CP, Central Pyrenees; LB, Lublin and Łysogóry-Radom basins; M, Morocco; MN, Montagne Noire; NY, New York State; TS, SE Tien-Shan.

Lublin Basin, located on a stable basement of the East European Craton, is characterized by the Middle Devonian carbonate, terrigenous and evaporitic marginal-marine to shallow open-shelf facies attaining a thickness of *c.* 200 m. The sediments pinch out and become progressively more terrigenous north- and eastwards, towards the low-relief land areas (Fig. 2) . In contrast, the sedimentation rates in the Łysogóry–Radom Basin were much higher, leading to the accumulation of up to 1400 m of strata. Open-marine calcareous–clayey sediments dominate over restricted-marine terrigenous facies.

The uppermost Lower Devonian and entire Middle Devonian of the Lublin Basin are ascribed to the Telatyń Formation (Narkiewicz 2011*b*). Five transgressive–regressive cycles, labelled T-1 to T-5, can be traced within the formation, based on core descriptions and wireline-log data (Narkiewicz 2011*b*; Fig. 3). Chronostratigraphical data and the regular facies succession suggest predominant eustatic controls on the depositional

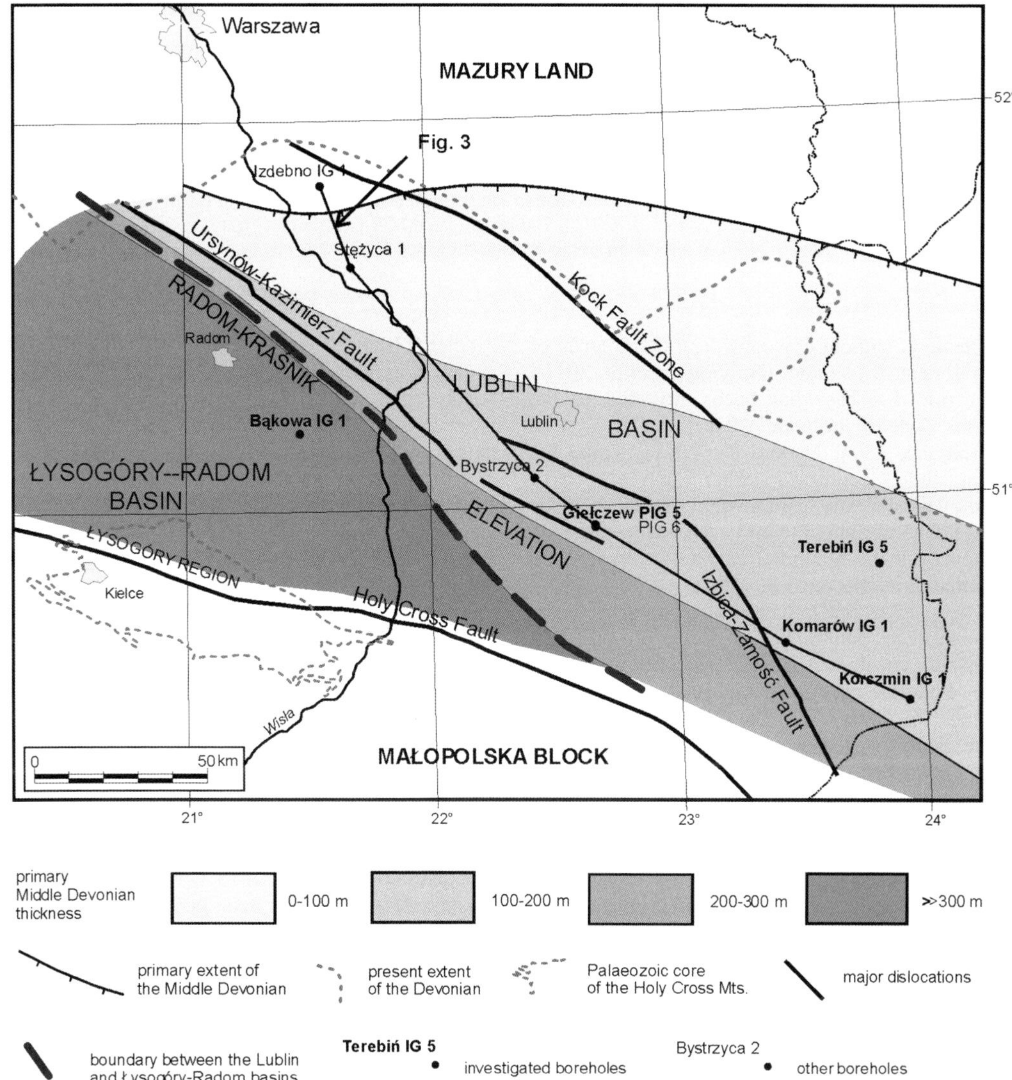

Fig. 2. Investigated borehole sections in SE Poland localized against the general tectonic framework and primary sediment thickness distribution of the Middle Devonian in the Lublin and Łysogóry–Radom basins (modified after Narkiewicz *et al.* 2011).

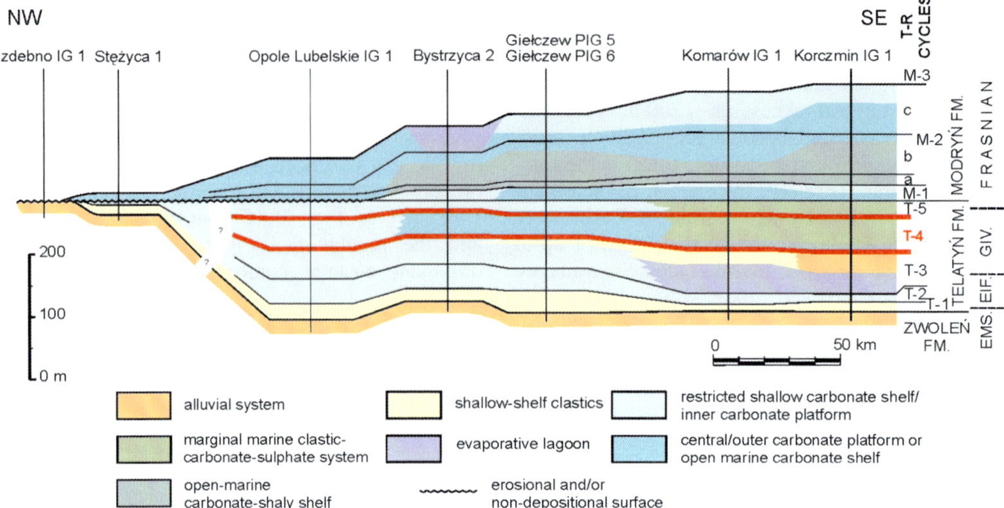

Fig. 3. Depositional architecture of the Middle Devonian to lower Frasnian strata of the Lublin Basin in a longitudinal cross-section, modified after Narkiewicz *et al.* (2011). The investigated Mid-Givetian T-4 transgressive–regressive (T–R) cycle is highlighted in red. For a location of the cross-section, see Figure 2.

architecture, without any considerable influence of synsedimentary tectonics (Narkiewicz *et al.* 2011). The cycle T-4 base has been attributed to the *ansatus* Zone of the middle Givetian (Narkiewicz & Bultynck 2007; Narkiewicz 2011). Thus, the recent studies confirmed earlier interpretation of these deposits by Narkiewicz & Narkiewicz (1998), who attributed them to the onset of the IIa eustatic cycle, i.e. the initial Taghanic transgression as defined by Johnson *et al.* (1985).

The considerable influence of regional tectonics on local facies patterns hampers the interpretation of eustatic events and cycles in the Middle Devonian of the Łysogóry–Radom Basin (Narkiewicz 2011*a*). The *ansatus* Zone has been identified here in the clayey–silty deposits of the Łaziska Member (Mbr) of the Bąkowa Formation (Narkiewicz 2011). The overlying Kunegundów Mbr is interpreted as an open calcareous–clayey shelf facies, with abundant and diverse fauna, and its base may represent the IIa transgression.

Materials

The quantitative biofacies analysis of the conodont assemblages from SE Poland was based on 407 platform and selected non-platform elements determined in 14 samples from 5 borehole sections shown in Figure 2. Representatives of four genera have been investigated: *Polygnathus*, *Icriodus*, *Belodella* Ethington, 1959 and *Neopanderodus* Ziegler & Lindström, 1971 (Table A1). The first two genera dominate, whereas the remaining two are poorly represented. Sampling and sample processing procedures, as well as taxonomic and biostratigraphical considerations, are presented by Narkiewicz & Bultynck (2007) and Narkiewicz (2011).

The investigated assemblages contain elements exhibiting various shapes and sizes (Narkiewicz & Bultynck 2007), and apparently are not hydrodynamically sorted. Therefore it is likely that they reflect primary fossil communities (cf. Broadhead *et al.* 1990; McGoff 1991). On the other hand, however, platform elements are much more abundant than ramiform and coniform ones (see Table A1) and thus proportions are inverse to those in reconstructed conodont apparatuses (Klapper & Philip 1971; Nicoll 1982, 1985; Klapper & Barrick 1983; Sandberg & Dreesen 1984; Sweet 1988; Bultynck 2003). Most probably, the non-platform elements were selectively destroyed due to both taphonomic processes (Broadhead *et al.* 1990) and sample processing. It may be argued, however, that those factors did not considerably influence the relative proportions of the platform elements that are the main subject of further considerations.

For comparative purposes we selected eight areas (located in Fig. 1) representing different widely distributed palaeogeographical and facies settings of the Taghanic transgression, and that have been adequately described in published papers. The most important criteria for our selection were: (1) good stratigraphical constraints of a transgressive event, indicating the presence of the *ansatus* Zone; (2) availability of quantitative data on conodont distribution and abundance, allowing us to define both conodont biofacies and the frequency of the key taxa; and (3) reliable palaeoenvironmental interpretations. In some analysed sections there is no apparent record of the transgressive interval within a generally deeper-water facies. These examples were nevertheless included in our considerations as they furnish valuable information on conodont biofacies variability in the critical interval. More information on particular localities and data sources is given below.

In the case of the Lower Tully Limestone of New York State in the USA (Heckel 1973) we based our comparison both on published data (Huddle 1981) and unpublished material from three samples collected by Pierre Bultynck in 1972. Two samples are from the Fabius Bed and ?Tully Valley Bed at the base of the Moravia 5b section and one sample is from the upper part of the Carpenter Falls Bed in the Bellona 9a section (Heckel 1973). Preliminary investigations included taxonomic assignments and specimen counts used for a quantitative biofacies determination.

Methodological remarks

The notion of the 'conodont biofacies' as applied here, refers to an assemblage defined based on a relative abundance of representatives of one or two dominant genera collectively, forming at least 75% of all platform elements (Sandberg *et al.* 1988). For example, the icriodid biofacies is an assemblage in which the genus *Icriodus* is represented by at least 75% of individuals in a sample, while in the polygnathid–icriodid (P–I) biofacies, elements attributed to *Polygnathus* and *Icriodus* represent together 75% or more, and the first genus dominates.

It is obvious that the higher the condont abundance in a sample, the more reliable the biofacies determination. The lower limit of the conodont frequency per sample was established by Belka & Wendt (1992) as 50 and by Bultynck *et al.* (1998) as 30 specimens. In the investigated Polish deposits the conodont frequencies are generally low, often a few to a dozen individuals or so (Narkiewicz & Bultynck 2007; Narkiewicz 2011). For purposes of the present investigations, a lower limit of 20 specimens was adopted, similar to that suggested by Klapper & Lane (1985) for Frasnian shallow-shelf facies. In a few cases, cumulative abundance from closely located samples ($\leq$0.5 m) representing the same lithofacies was accepted. The only exception was the Terebiń IG 5 section for which the samples from a 2.6 m interval were cumulated (Table A1).

Conodont biofacies patterns in SE Poland

Investigations of the Lublin Basin sections allowed the analysis both of the vertical succession of conodont biofacies and their lateral variability. All the platform genera were included and the characteristic species *Polygnathus ansatus* Ziegler, Klapper & Johnson, 1976 and *Polygnathus linguiformis* Hinde, 1879 were considered separately. Identification of the latter species is usually straightforward even for broken specimens, owing to the peculiar platform shape. *Polygnathus ansatus* is less characteristic and it cannot be totally excluded that its frequency is underestimated as some broken specimens could have been classified as *Polygnathus* sp. (Table A1). Information on the lithology and macrofauna of the sampled intervals is summarized in Table 1.

Vertical succession

In the Giełczew PIG 5 section six assemblages from the depth interval 2017.9–2002.5 m were investigated (Fig. 4). For the assemblages from 2017.6 to 2017.5 m and 2010.7 to 2010.6 m (cf. Table A1), cumulated data from two samples were used for each interval.

In the nodular wackestones forming the T-4 cycle base, a characteristic vertical succession can be observed: from the P–I biofacies with a maximum *P. ansatus* proportion, to the polygnathid (P) one (Fig. 4). *Belodella* is an accessory element whose percentage decreases from 5 to 2%. The initial small diversity of the dominant genera (two species in each) notably increases upwards (five species each) and then drops to two polygnathid and one icriodid (I) species. Among the icriodids the most abundantly represented species is *Icriodus latecarinatus* Bultynck, 1974. In the assemblage from 2010.6 to 2010.7 m (P–I biofacies) the polygnathid diversity is similar to that in the assemblage below (see Table A1), but the *P. ansatus* proportion drops by half, whereas the *P. linguiformis* percentage increases. The number of *Icriodus* species increases to three but the number of specimens decreases (Table A1).

Higher in the section (2002.5–2003.0 m) two assemblages are composed almost exclusively of icriodids whose diversity increases to four and five species, respectively. *Icriodus difficilis* Ziegler, Klapper & Johnson, 1976 dominates, attaining a proportion of 25% in each assemblage.

In the Korczmin IG 1 section three assemblages were investigated, including depth intervals of 2492.3–2492.0 m and 2489.0–2488.6 m (Fig. 4; Table A1), which comprise cumulated material

Table 1. *Description of the sampled levels from the Lublin and Łysogóry–Radom basins*

Borehole and depth (m)	Lithology	Macrofauna	Depositional system*	Conodont biofacies
Giełczew PIG 5 2002.5–2003.0	Bioturbated skeletal packstone grading upwards into cross-bedded encrinite (crinoid packstone–grainstone)	Bottom part: numerous tiny and single large crinoid fragments, branching tabulates. Upper part: almost exclusively crinoids	Shallow-water carbonate shelf	I
Giełczew PIG 5 2010.6–2010.7	Marly skeletal wackestone with larger skeletal fragments embedded in fine bioclastic matrix	Strongly reworked massive tetracorals and platy tabulates, single brachiopods and crinoids, including fragments up to 8 mm	Open-marine carbonate shelf	P–I
Giełczew PIG 5 2015.8–2017.9	Dark, grey-beige, nodular bioturbated wackestones, grading upwards into bioturbated mudstones	Single crinoid fragments and brachiopods (partly articulated)	Open-marine carbonate shelf	P P–I
Komarów IG 1 2375.6–2376.0	Dark grey, marly skeletal wackestones, partly dolomitized (fine dolosparite). Bioturbation common	Numerous fragmented brachiopods and crinoids, single branching tetracorals	Open carbonate–shaly shelf (more carbonate)	P–I
Korczmin IG 1 2485.0	Dark grey to black, marly mudstones and wackestones, homogeneous to nodular. Strong bioturbation	Crinoids, brachiopods (partly articulated), single branching tetracorals. Vertical and oblique burrows	Open carbonate–shaly shelf	P–I
Korczmin IG 1 2488.6–2489.0	Dark grey to black, marly mudstones and wackestones, homogeneous to nodular Strong bioturbation	Crinoids, brachiopods (partly articulated), single branching tetracorals. Vertical and oblique burrows	Open carbonate–shaly shelf	P
Korczmin IG 1 2492.0–2492.3	Bioturbated grey to dark grey homogeneous to nodular wackestones.	Numerous crinoid fragments (mainly small), disarticulated brachiopods	Open carbonate–shaly shelf	P–I
Terebiń IG 5 1574.6–1574.7	Grey skeletal–intraclastic wackestones–packstones. Common pyritization, including intraclasts (only rims or total)	Unsorted large brachiopods and crinoids. Distinct burrows	Brackish lagoon or estuary*	P–I
Terebiń IG 5 1576.6–1577.2	Grey skeletal–intraclastic wackestones–packstones. Common pyritization, including intraclasts (only rims or total)	Unsorted large brachiopods and crinoids. Distinct burrows	Brackish lagoon or estuary*	P–I
Bąkowa IG 1 2116.0	20 cm-thick bed of bioclastic limestone in calcareous silty shales with brachiopods	Brachiopods, gastropods, cephalopods, fine crinoid fragments. Unsorted, chaotically distributed skeletons	Clayey–silty shelf*	P–I

*Depositional systems refer to generalized intervals of a sedimentary succession; samples marked with asterisk can represent exceptional/episodic depositional conditions in a system.

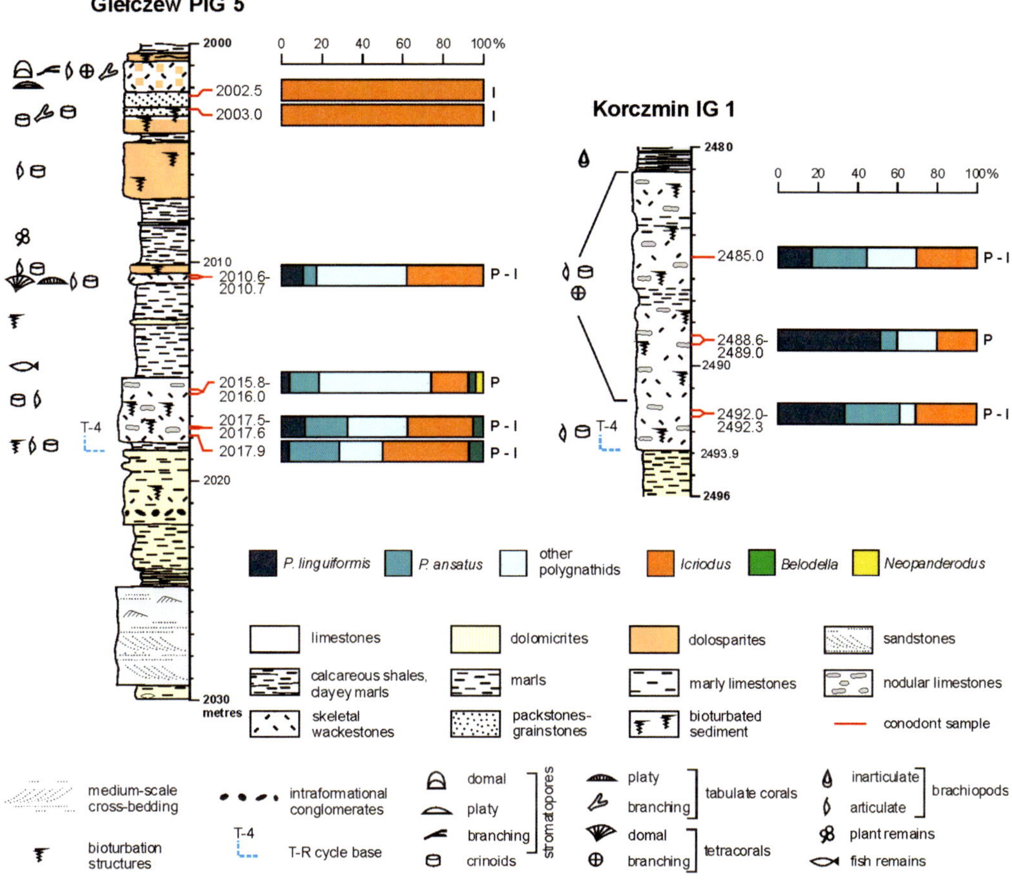

Fig. 4. Vertical conodont biofacies succession in the lower part of the T-4 cycle in the Giełczew PIG 5 and Korczmin IG 1 borehole sections (see Fig. 2 for a location). Sample depths in metres. P, polygnathid; I, icriodid.

from separate but closely located samples. Moreover, the assemblage from the depth 2485.0 m with only 19 specimens has been exceptionally taken into account as it supplements the vertical succession pattern.

The oldest assemblage is ascribed to the P–I biofacies showing a low diversity of both component genera, three and two species, respectively, and a large proportion of *P. linguiformis* and *P. ansatus*. The middle assemblage is characterized by the P biofacies, with three polygnathid species dominated by *P. linguiformis*. The diversity of slightly less abundant icriodids remains unchanged. In the upper assemblage (P–I biofacies) the number of species of both genera is similar. The percentage of still dominant *P. linguiformis* drops while that of *P. ansatus* remains the same. Among icriodids, *I. eslaensis* van Adrichem Boogaert, 1967 attains a high percentage (26%).

Lateral variability

In order to investigate the lateral variability of biofacies in the Lublin Basin we compared the conodont assemblages from the lowermost 1–4 m of the T-4 cycle in the Giełczew PIG 5, Komarów IG 1, Korczmin IG 1 and Terebiń IG 5 sections (Fig 5). Observations of the vertical variability (cf. previous section) were used to define assemblages exhibiting a consistent biofacies in the Giełczew PIG 5 section (cumulated assemblages for the depth interval 2017.5–2017.9 m) and in Korczmin IG 1 (2492.0–2492.3 m).

All investigated assemblages belong to the P–I biofacies but a closer examination of proportions of different forms reveals two extreme variants represented by the Terebiń IG 5 and Korczmin IG 1 sections. In the first one, *Polygnathus* has its smallest proportion (52%), *P. ansatus* is rare (6%) and

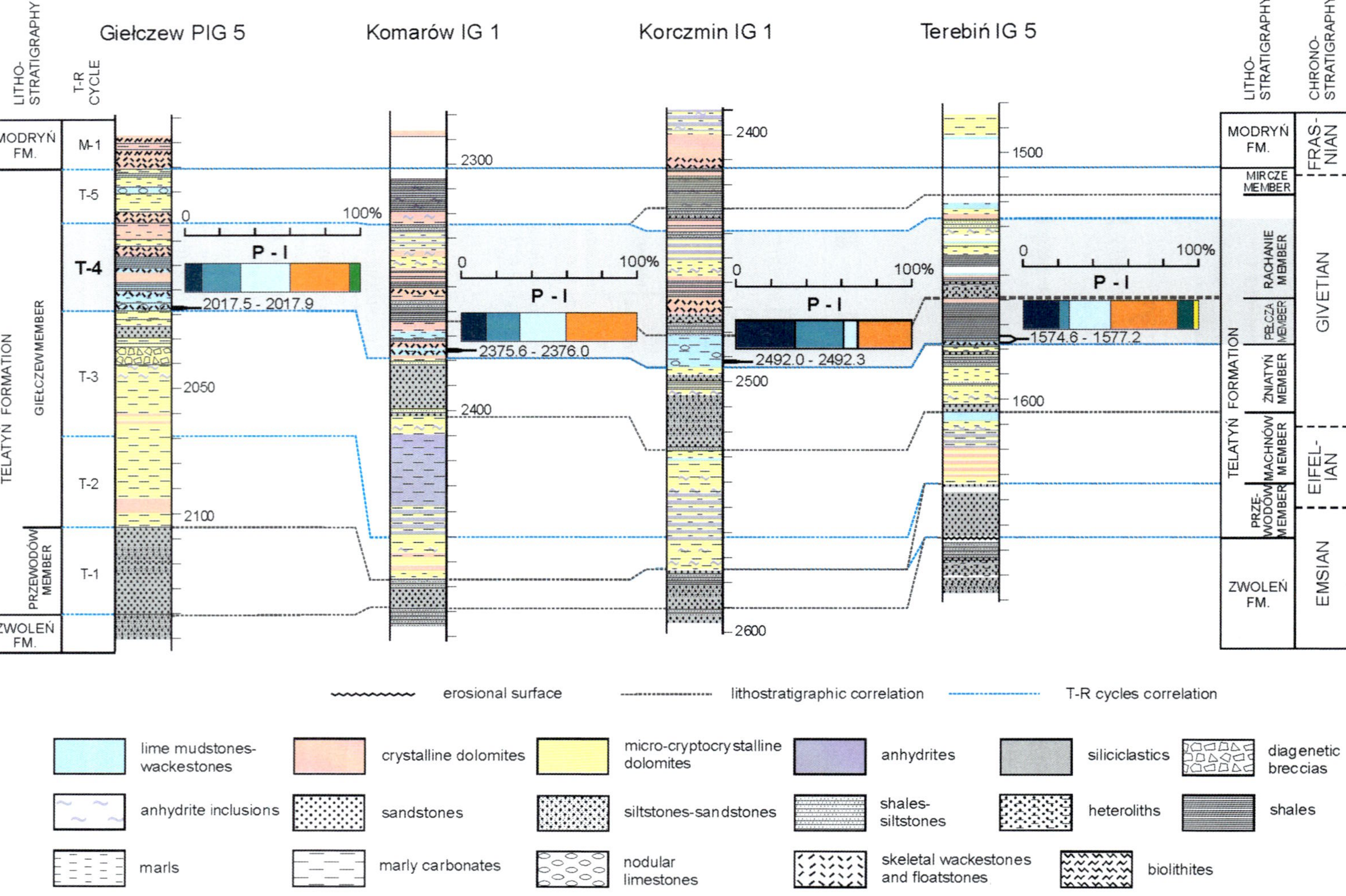

Fig. 5. Lateral variability of the conodont biofacies in the lower part of the T-4 cycle in the Lublin Basin. All depths in metres. For an explanation of the taxonomic composition of the biofacies, see Figure 4. P, polygnathid; I, icriodid.

icriodids are relatively more frequent (36%). Representatives of *Belodella* (9%) and *Neopanderodus* (3%) are accessory components but attain the highest percentages in comparison to assemblages from other sections. In the Korczmin IG 1 section the genus *Polygnathus* and its *ansatus* and *linguiformis* species attain maximum percentages (68, 29 and 23%, respectively). Notable is the total lack of *Belodella* and *Neopanderodus*. The assemblages from the sections Giełczew PIG 5 and Komarów IG 1 exhibit an intermediate composition between both extremes described above. They are characterized by a nearly identical proportion of polygnathids (61 and 62%) and similar percentages of *P. ansatus* (22 and 21%), and *P. linguiformis* (10 and 14%), respectively. In addition, icriodids occur in a similar proportion (36 and 38%). In the Giełczew PIG 5 section, *Belodella* (3%) is an accessory component.

The *ansatus* Zone assemblage from the depth of 2116 m in the Bąkowa IG 1 section located in the Łysogóry–Radom Basin was analysed for comparison with the Lublin Basin. If the overlying Kunegundów Mbr records the IIa eustatic cycle (Narkiewicz *et al.* 2011), the investigated assemblage may be slightly older than those analysed from the Lublin Basin. The polygnathid to icriodid proportion is identical as in the Komarów IG 1 section (56 and 44%, respectively), while percentages of *P. ansatus* and *P. linguiformis* (16% each) are similar. The species diversity is nevertheless larger, attaining three polygnathid and five icriodid species.

Interpretation

The successions of biofacies of the T-4 cycle in the Giełczew PIG 5 and Korczmin IG 1 sections exhibit patterns that can be related to the Taghanic transgression development (Fig. 6). Open-marine facies encroaching the areas of the former

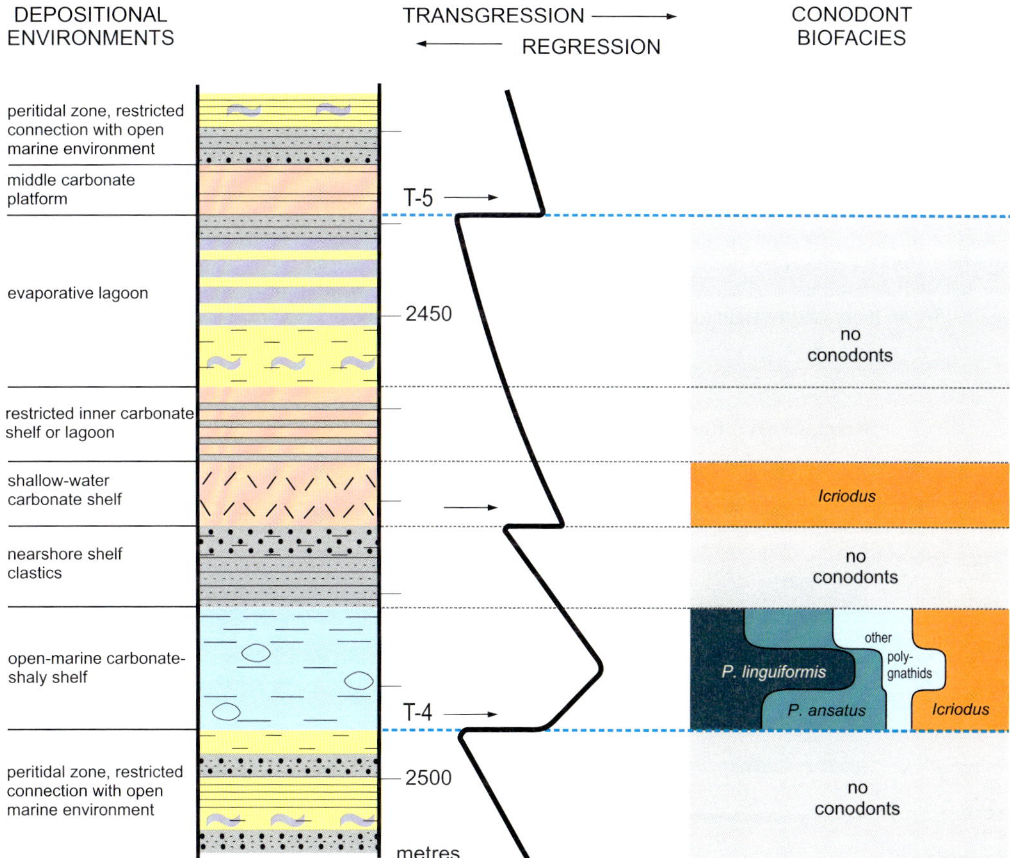

Fig. 6. Generalized vertical conodont biofacies succession of the T-4 cycle in the Lublin Basin shown against the lithology and palaeoenvironmental interpretation of the Korczmin IG 1 borehole section. Legend to lithology, see Figure 5.

marginal-marine sedimentation hosted diverse conodont assemblages. During the initial phase, icriodids were still important (P–I biofacies) while species diversity was low. The maximum percentage of *P. ansatus* indicates that the species was among the first forms occupying new niches created during the transgression.

In the successive stage, the P biofacies replaced the P–I one, which is here interpreted as the maximum flooding phase associated with maximum water depths. A characteristic peak with a relative *P. linguiformis* abundance and a drop of the *P. ansatus* percentage may be related to the most open-marine and/or deepest-water preferences of the former species as compared to the entire conodont assemblage. The subsequent recurrence of the P–I biofacies is associated with evidence of a regression, i.e. the appearance of shaly, probably intermittently estuarine or lagoonal facies, with plant and fish remains (Giełczew PIG 5), or open-marine facies but with a stronger terrigenous admixture in the carbonates (Korczmin IG 1) (Table 1). Notable are an overall drop in the *P. linguiformis* percentage and variable *P. ansatus* occurrences. It seems surprising that a few *Belodella* representatives appeared only in the early transgressive phase and during the highstand, that is, in the P biofacies. *Belodella* disappears with the first signs of the regression, which seems to depart from a simple shallow-water and/or nearshore interpretation of a Middle Devonian palaeoecology of this genus (e.g. Chatterton 1976; Sparling 1983). Perhaps the additional factor limiting belodellid habitats was a water turbidity evidenced by an increased terrigenous admixture.

An increasing regressive trend is shown by the appearance of unfossiliferous siltstones and heteroliths in the Korczmin IG 1 section, and clayey–marly shales with plant material in Giełczew PIG 5 (Fig. 4). These are succeeded by shallow-water carbonates characterized by the I biofacies, marking a subordinate transgressive pulse (Fig. 6). The same taxonomic diversity as at the onset of the transgression, and the introduction of a cosmopolitan *I. difficilis* (Table A1), suggest that the assemblage, although being generally shallow-water, was still under open-marine influences promoting faunal migration. The short-lived period of open-marine conditions was terminated by the appearance of restricted lagoonal conditions marking a prolonged regression. The following transgressive phase of the T-5 cycle (Fig. 5) reflects the successive eustatic IIb cycle of Johnson *et al.* (1985; Narkiewicz *et al.* 2011).

The T-4 transgression invaded near-coastal environments with a generally flat topography. Thus it may be assumed that its beginning was a geologically instantaneous event in the Lublin Basin, and consequently the assemblages in the basal part of the cycle (Fig. 5) reflect lateral environmental gradients in the initial transgressive phase. These assumptions are partly confirmed by largely uniform ecological conditions as evidenced by the uniform P–I biofacies (Fig. 5). The extreme biofacies variants described above represent an estuary or a brackish lagoon with intermittent open-marine influences (Terebiń IG 5) and an open carbonate–shaly shelf located in a more offshore position (Korczmin IG 1) (Fig. 7). The former environment

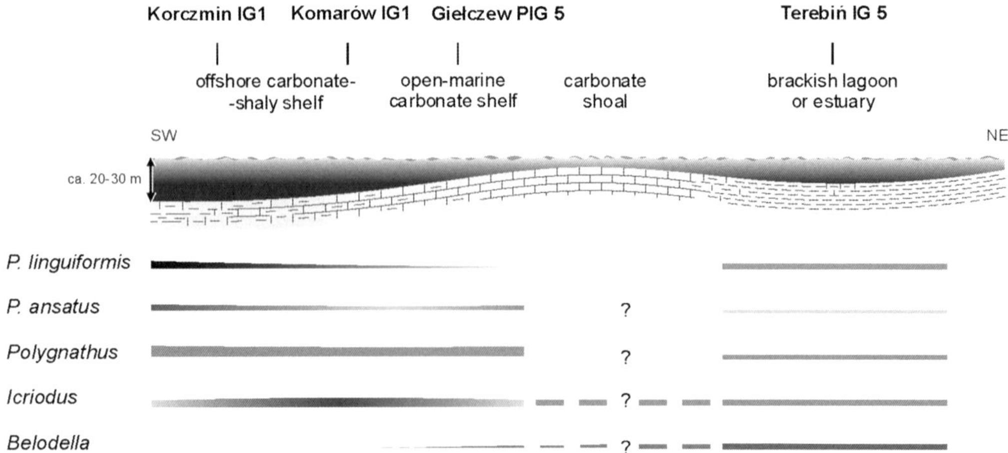

Fig. 7. Schematic distribution of conodont taxa in the framework of palaeoenvironmental interpretation of the transgressive T-4 cycle part (Taghanic transgression) in the Lublin Basin, in the offshore (SW) to nearshore (NE) cross-section.

is evidenced by prevailing grey–greenish shales with a minor carbonate content and containing plant remains and impoverished fauna, punctuated by thin burrowed wackestones with brachiopods and crinoid remains (Narkiewicz 2011*b*). On the other hand, the permanent open-marine conditions are documented in the Korczmin IG 1 section by the presence of a continuous succession of burrowed skeletal wackestones with tetracorals, brachiopods and crinoids. Giełczew PIG 5 and Komarów IG 1 can be ascribed to intermediate palaeoenvironments of an open carbonate shelf and a more proximal, open carbonate–shaly shelf, respectively. The species diversity of polygnathids and icriodids lacks any clear lateral pattern, e.g. an increase in more offshore and/or deeper-water facies.

Bathymetry is the most problematic aspect of the reconstruction shown in Figure 7. In general, the maximum palaeo-water depths presumably did not exceed a few tens of metres and this was also most probably an order of magnitude of the sea-level rise during the T-4 transgression (Narkiewicz & Narkiewicz 1998; Narkiewicz *et al.* 2011). Brackish lagoons or estuaries in the relatively nearshore Terebiń IG 5 area could have been even shallower, although their main attribute may have been freshwater influence. In Figure 7 we assume the presence of a carbonate shoal that separated the Terebiń IG 5 area from a distal part of the shelf. Such a discontinuous barrier, allowing influx of normal marine waters, could protect quiet muddy palaeoenvironments to the east. The outer slopes of the barrier were the sites of a carbonate sedimentation (Giełczew PIG 5) grading seawards into slightly deeper-water marly muds (Komarów IG 1, Korczmin IG 1).

Figure 7 shows the occurrence of particular conodont taxa, stressing their relative abundance in different depositional environments, which presumably reflects ecological preferences of the respective genera and species. It appears that the most nearshore/shallow-water element is *Belodella*, whose preferences are more complex, perhaps related also to less turbid waters. Its presence in the Terebiń IG 5 assemblages can be connected to an episode of a more open-marine carbonate deposition (cf. Table 1).

Representatives of *Icriodus* and *Polygnathus* commonly co-occur in the studied assemblages, although in variable proportions, reflecting their proximity to a basin margin. This indicates generally overlapping ecological preferences under the fairly uniform conditions of the initial transgression phase. During the course of the transgression, however, the conditions changed towards relatively deeper water, which led to a considerable elimination of icriodids during the maximum flooding phase. Subsequently, polygnathids lost their importance at the expense of icriodids with the onset of a regression. *Polygnathus linguiformis* and *P. ansatus* reach their peak abundance in the 'most offshore' Korczmin IG 1 section. Well-documented occurrences of *I. latecarinatus*, the species being remarkably abundant in the Giełczew PIG 5 section, are known only from Europe and are indicative of shallow-water facies (see Bultynck 1974, fig. 2, 1987, fig. 2). This may suggest specific environmental conditions probably due to a carbonate shoal vicinity (Fig. 7).

There is a remarkable similarity between the conodont biofacies in the Lublin Basin (particularly in Komarów IG 1) and the Łysogóry–Radom Basin. The Komarów IG 1 section is a representative of an open carbonate–shaly shelf located more distally than the clayey–silty shelf interpreted for Bąkowa IG 1 (Fig. 2; Table 1). The *ansatus* Zone assemblage from the latter section (Table A1) is related to an episode of carbonate deposition, favouring conodonts typical of offshore, less land-influenced facies. Nevertheless, the above comparison demonstrates that very similar conodont assemblages can occur in different palaeogeographical areas and may accompany different depositional systems of the *ansatus* Zone.

Review of the global record

Table 2 summarizes data on litho- and biofacies for selected localities with a well-constrained stratigraphical record of the *ansatus* Zone in different palaeogeographical and facies settings. Short comments on each locality are given below.

Boulonnais (northern France)

Available data suggest a vertical biofacies gradient: out of three analysed samples, the two lower can be assigned to the I–P biofacies (with accessory *Tortodus* Weddige, 1977), whereas the upper one exhibits predominance of icriodids over polygnathids (90% v. 10%; I biofacies) and a lack of *Tortodus*. Both *P. linguiformis* and *P. ansatus* are lacking, and the age determination of the Couderousse Mbr (Table 2) was possible owing to the co-occurrence of *Polygnathus alatus* Huddle, 1934 (first appearance in the *ansatus* Zone) and *Polygnathus pseudofoliatus* Wittekindt, 1966 (last occurrence in the same zone) (Narkiewicz & Bultynck 2010; Walliser & Bultynck 2011). The biofacies gradient corresponds to a shallowing-upward trend towards the overlying Bastien Mbr (Pelhate & Poncet 1988).

Ardennes (Belgium)

The studied material was obtained from two complementary sections of the Flohimont Mbr at

Table 2. *Summary of the data on the Taghanic transgressive level from selected localities outside Poland*

Locality *Stratigraphic unit*	Lithology and fossil assemblage	Depositional systems and/or facies	Depositional record of a transgression/deepening	Conodont abundance*	Conodont biofacies†	*P. linguiformis*‡	*P. ansatus*‡	Accessory genera	Source of information
Boulonnais (northern France) *Blacourt Fm, Couderousse Mbr*	Marly shales with nodular levels. Brachiopods, solitary tetracorals, branching tabulates, bryozoans, crinoids	Shallow-subtidal shelf facies within an overall carbonate platform succession	Distinct onlap of marly subtidal facies on a biostromal carbonate platform, then shallowing-upwards trend	2–3 (274)	I, **I–P**	0	0	*Tortodus*	Narkiewicz & Bultynck (2010)
Ardennes (Belgium) *Fromelennes Fm, Flohimont Mbr*	Marly limestones, shales and nodular shales. Abundant brachiopods (atrypids, cyrtospiriferids, stringocephalids), platy tabulates, branching stromatoporoids, gastropods, trilobites, solitary tetracorals, crinoids	Shallow-subtidal shelf temporarily developed within an overall carbonate platform succession	Change from the biostromal facies to shaly–marly facies within a 2 m-thick uppermost Mont d'Haurs Fm interval	1–3 (257)	**I**, I–P	0	**R**	*Tortodus Ozarkodina*	Bultynck (1974); Bultynck *et al.* (2001)
Manitoba (Canada) *Dawson Bay Fm, Mbr B*	Micritic–marly, partly grained and rubbly limestones with brachiopods (in places abundant) and crinoids	Shallow-subtidal, fully marine shelf	Fossiliferous marine limestones overlying barren (continental?) green shales	1–3 (242)	**I–P**, P–I, P	**0**–M	**R**–M	*Belodella*	Norris & Uyeno (1998); Day *et al.* (1996)
New York (USA) *Tully Fm, Lower Tully sensu* Heckel (1973)	Bioturbated lime mudstones to skeletal wackestones and packstones, subordinate stromatolites, ?oolites and mud-mounds. Brachiopods, sponges, solitary tetracorals, bryozoans, tentaculitoids, styliolinids, gastropods, ostracods, trilobites, echinoderms	Sediment-starved outer shelf with open-marine, partly dysoxic deposition	Submarine nondepositional/erosional surface topping fine clastics, overlain by purely open-marine carbonates	2–3 (138)	I–P, **P–I**	M–**A**	**0**–R		Heckel (1973); Huddle (1981); Baird & Brett (2003, 2008); Brett *et al.* (2011); unpublished data
Anti-Atlas (Morocco) *Tafilalt Platform*	Nodular limestones and marls, grey shales, micritic limestones; hardgrounds, detrital quartz. Styliolinids, ostracods, crinoids, agglutinated forams, cephalopods, trilobites, in places small brachiopods and solitary tetracorals, gastropods, bryozoans, calcispheres	Pelagic, shallow-water carbonate shelf with very low sedimentation rates	Discontinuity surface overlain by more marly deposits	1–3 (620)	**P**	**A**	0	*Tortodus Ozarkodina*	Aboussalam (2003); Narkiewicz & Bultynck (2010); Aboussalam & Becker (2011)
Anti-Atlas *Ma'der Basin*	Silty–shaly succession with limestone intercalations. Brachiopods, corals, trilobites, rare cephalopods	Deeper-shelf, continuous deposition	No data	2 (82)	P–I, P	**M**	0–R	*Belodella Panderodus Ozarkodina*	Narkiewicz & Bultynck (2010)
Montagne Noire (southern France) *Coumiac Fm, Lower Mbr (PB, CTS)*	Light-coloured limestones (skeletal wacke–packstones), nodular limestone intercalations, hardgrounds. Pelagic and benthic biota: styliolinids, tentaculitoids, forams, ostracods, trilobites, echinoderms, brachiopods, gastropods	PB: rhythmically bedded pelagic carbonate ramp; CTS: mostly neritic shallow ramp	PB: erosional (subaqueous) discontinuity; CTS: nodular limestone with ostracods and brachiopods overlain by styliolinid–ostracod wackestone	**2**–3 (318)	**P**	**A**	**0**–R	*Belodella* (CTS) *Tortodus* (PB, CTS)	Aboussalam (2003); Aboussalam & Becker (2011)
Central Pyrenees (Spain) *Compte Fm, Mbr A*	Nodular and well-bedded limestones, styliolinid wacke-packstones. Styliolinids, ostracods, crinoids, trilobites	Open-marine facies, ?pelagic platform	?Appearance of nodular marly limestones in purely calcareous succession	**1**–2 (83)	I–P, P–I, **P**	M–**A**	**0**–R	*Tortodus Ozarkodina Belodella*	Carls (1988); Liao & Valenzuela-Rios (2008, 2013); Liao *et al.* (2008); Gouwy *et al.* (2013)
SE Tien-Shan (Central Asia) *Kalagach Fm*	Coarsely detrital crinoid–brachiopod limestones, partly graded, with intercalations of tentaculitoid lime mudstones–wackestones	Pelagic–hemipelagic deposition, intermittent gravity-flow sedimentation	No data	2 (29)	**P**	**M**	**R**–M	*Ozarkodina Mehlina*	Bardashev & Ziegler (1985); Bardashev (1992)

*Average number of elements per sample: 1, 0–20; 2, 21–100; 3, >100; maximum number given in parentheses.
†All biofacies found.
‡Relative *P. linguiformis* and *P. ansatus* frequency: 0, none; R, rare (<10%) elements in samples; M, moderately frequent (11–50%); A, abundant (>50%).
Bold: most frequent case (if possible to indicate).

Fromelennes: a stratigraphically lower one along the D46 roadcut, and an upper section along the Houille Brook (Bultynck *et al.* 2001). Bultynck *et al.* (2001) correlated the lower and upper part of the roadcut section with the *Polygnathus rhenanus* and *ansatus* zones, respectively. The Houille Brook section was attributed to the ?*ansatus* Zone. Recent revision of the conodont material from the roadcut section revealed the presence of *P. ansatus* in sample 20 (upper part of the section; specimen figured as *P.* aff. *ansatus* in Bultynck 1987, pl. 8, fig. 9) and *P.* cf. *P. ansatus* (Fig. 8) in sample 6 from a basal part of the Flohimont Mbr (for sample location see Bultynck *et al.* 2001). The new data suggest that the transgressive basal part of the Fromelennes Formation corresponds to the base of the Taghanic Stage, as proposed earlier by Klapper & Johnson (1980, fig. 1). The roadcut section exposes well-bedded marly limestones, partly nodular, with platy and branching tabulate corals, branching stromatoporoids, brachiopods, tetracorals and crinoids. Conodonts are moderately abundant reaching tens to >100 specimens per sample. The I biofacies is typical, while the I–P biofacies is less common, with *I. latecarinatus* dominating among icriodids. The member is more shaly upwards and the macrofauna becomes less diverse, composed mostly of brachiopods and gastropods while conodonts gradually disappear.

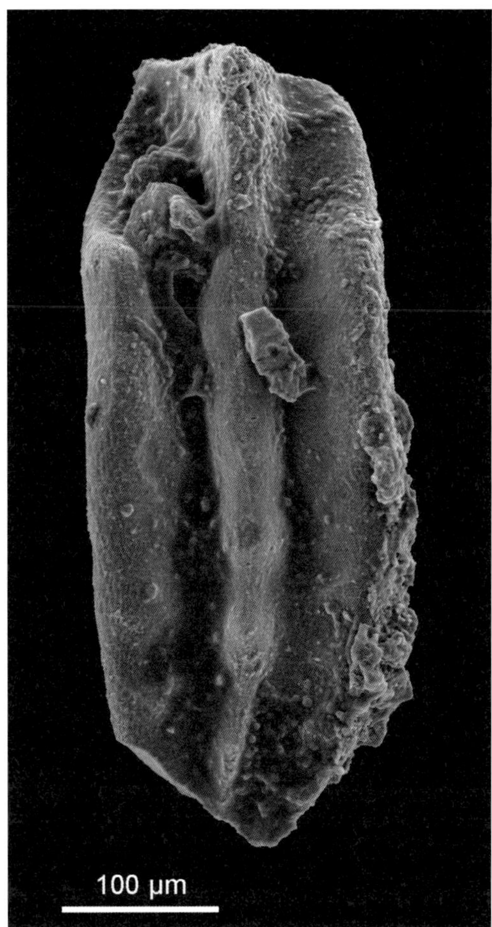

Fig. 8. *Polygnathus* cf. *P. ansatus* Ziegler, Klapper & Johnson, 1976 specimen no. MUZ PIG 1795.II.1 from Flohimont Mbr, D46 road-cut section (Ardennes), Sample 6 of Bultynck *et al.* (2001). Upper view of a juvenile specimen. Anterior trough margins are clearly seen, although the outer one is less bowed than those in *P. ansatus* representatives illustrated by Ziegler *et al.* (1976, pl. 2, figs 11–26). Other attributes, including the geniculation points located opposite each other and the anterior trough margins, which meet the free blade at the same position, conform to the species diagnosis (see Ziegler *et al.* 1976).

Manitoba (Canada)

The lowermost parts of the analysed sections of Mbr B of the Dawson Formation exhibit a maximum conodont abundance, generally more than 100 specimens per sample, probably connected with a low sedimentation rate at the beginning of the transgression. A basal I–P biofacies is characterized by relatively abundant icriodids (>50%) and less common polygnathids including *P. ansatus* (2–10%) but lacking *P. linguiformis*. More than 2 m above the base of the sections, less abundant assemblages (<50 specimens per sample) are characterized by the appearance of P–I and even P biofacies in addition to I–P. *Polygnathus ansatus* may be locally abundant (>4%, up to 41%), *P. linguiformis* appears in some samples (up to 26% in a single sample) and *Belodella* is an accessory element.

New York (USA)

There are divergent views on the stratigraphical position of the Taghanic unconformity (=Taghanic transgression level). According to the majority of previous authors it corresponds to the base of the Tully Formation (e.g. House 1983; Johnson *et al.* 1985; Aboussalam & Becker 2011), whereas Brett *et al.* (2011) place it at the base of their Middle Tully (=base of the Carpenter Falls Bed), and thus within the Lower Tully *sensu* Heckel (1973). In our biofacies analysis we considered only the Lower Tully as defined by Heckel (1973).

Averaged data for the Lower Tully, based on a collection of 1337 conodonts (Huddle 1981), exhibit the presence of P–I biofacies with polygnathids constituting 66% and icriodids 34% of all forms. Polygnathids are represented almost exclusively by *P. linguiformis*, while no data are available on the occurrence of *P. ansatus*. The assemblages from the samples collected by P. Bultynck (Table 2) shows a variable proportion of P and I elements and a dominance of *P. linguiformis* among polygnathids. The dominance is, however, much less pronounced in the uppermost sample from the Carpenter Falls Bed where notably *P. ansatus* (6%) appears (absent in the samples below).

Anti-Atlas (Morocco): Tafilalt Platform

Analysed data (Table 2) are based mostly on Narkiewicz & Bultynck (2010, Bou Tchrafine section, BT I), Aboussalam (2003, Seheb el Rassal 2) and Aboussalam & Becker (2011, Mdoura East). The sections encompass the boundary between the topmost Bou Tchrafine Group (=*Maenioceras* Marl, Aboussalam 2003) and the basal part of the overlying Achguig Gp (=lower part of the Upper *Sellagoniatites* Bed). They are stratigraphically condensed, for example, in the BT I section the *ansatus–Skeletognathus norrisi* zones interval corresponds to 1.5 m of strata. The P biofacies is here dominated by *P. linguiformis*, with accessory presence of icriodids and *Tortodus* (≤2% each). In the Seheb el Rassal 2 section, the single specimen-poor sample (D1 top, Aboussalam 2003) contains mainly *P. linguiformis* with single icriodid and *Tortodus*. Also both samples (7 and 9a) from the Mdoura East are poor (35 specimens or less), showing a prevalence of *P. linguiformis* in the P biofacies with accessory *Ozarkodina plana* (Bischoff & Ziegler, 1957).

Anti-Atlas (Morocco): Ma'der Basin

The data include samples 7 and 8 from the Bou Dîb section (Narkiewicz & Bultynck 2010), probably from the upper part of the *ansatus* Zone. Sample 7 is characterized by P biofacies with a large proportion of *P. linguiformis* (43%) and few *Ozarkodina* Branson & Mehl, 1933*a*. Sample 8 belongs to the P–I biofacies with a low percentage of *P. linguiformis* (20%), few *P. ansatus*, some *Panderodus* Ethington, 1959 and a surprisingly high proportion of *Belodella* (16%). It is difficult to decide which of the two samples is more representative, as there is a probability that Sample 8 (crinoidal limestone) may contain redeposited, more shallow-water material. It can be noted, however, that the presence of P–I biofacies in basinal facies (and absence of 'deeper-water' *Palmatolepis*-dominated assemblages) was also found in the Upper Frasnian of the Anti-Atlas by Belka & Wendt (1992).

Montagne Noire (southern France)

Investigated samples from the Taghanic Event Interval cover 32 cm in the Col de Tribes South (CTS) section and 232 cm in the Pic de Bissous (PB) (Aboussalam 2003). A low conodont frequency is typical (<50 specimens per sample) with only 2 samples (out of 11 analysed) that yielded >100 specimens per sample. In PB the transgressive level (26a/26b beds boundary) is not marked by any significant sediment and conodont biofacies change (P biofacies throughout), except for the appearance of rare *P. ansatus* and a single *Tortodus*. *Polygnathus linguiformis* remains important (62–96%) below and above the interval boundary. In CTS the conodont biofacies is similar, with dominant polygnathids, single icriodids and an accessory presence of *Tortodus*. *Polygnathus ansatus* is very rare (one specimen in a single sample). Rare *Belodella* (4%) occurs only in a single sample of grained-skeletal limestone from CTS.

Central Pyrenees (Spain)

The analysed conodont material was obtained from four sections comprising a succession of Givetian limestones, 20–30 m thick, and thus showing a moderate stratigraphic condensation (see Table 2 for references). The overall specimen frequency is poor, rarely attaining more than 20 specimens per sample, which is a particularly low number given the large size of the samples processed. In general, the assemblages are dominated by polygnathids with a highest proportion of *P. linguiformis*. The *ansatus* Zone interval does not reveal any clear depositional record of a transgressive event. In the Compte section (Liao & Valenzuela-Rios 2008) the lower part of the *ansatus* interval (well-bedded limestones) exhibits mostly P biofacies with single samples characterized by P–I and I–P biofacies. The dominance of P biofacies starts approximately parallel to an onset of nodular marly limestones, which may possibly reflect the Taghanic transgression level. There is also the suggestion of a concomitant *P. linguiformis* increase, while *Tortodus* appears higher (from bed 41a upwards).

SE Tien-Shan (Central Asia)

Notable are relatively low conodont frequencies, particularly in view of a stratigraphic condensation of the Middle Devonian, which is 34 m thick, the *Polygnathus varcus* Zone being 4.5 m thick. Polygnathids prevail, while icriodids are present in one sample (out of three). This sample, which is also relatively impoverished in *P. linguiformis*, may contain a redeposited shallow-water admixture in background pelagic sediments. *Polygnathus ansatus* is present, albeit in low percentages.

Interpretation and discussion

Construction of a global biofacies model: general problems

A common approach to a palaeoecological interpretation of conodonts is to analyse distributional patterns of particular biofacies or taxa against different sub-environments along the nearshore–offshore profile of a marine basin, thus similar to the example in Figure 7 (for published examples, see Dreesen & Thorez 1980; Sparling 1984; Sweet 1988). Is it possible to construct such a general model for the investigated *ansatus* Zone interval?

The first difficulty encountered when trying to accomplish this task is a proper correlation of compared conodont assemblages. The most general problem, that is, the placement of the studied material in a comparable time-evolutionary framework, was already mentioned in the introduction. In that respect it may be argued that, given the narrow time-interval of a single conodont zone and a lack of any important extinction–speciation events during that time, we are likely to be dealing with a coherent set of data.

The other question is that of representativeness and comparability of particular sections/basins in terms of the initiation and early stages of the global Taghanic transgression. As we have seen in the Polish Lublin Basin, there is a considerable fine-scale vertical variability in the biofacies succession, which is apparently correlated with transgression dynamics (Fig. 6). To overcome this, the lateral biofacies changes were analysed based on data representing a comparable, earliest stage of the transgression (Fig. 5). The analysed global record in some cases (Manitoba, Ardennes) does exhibit a suggestion of a vertical succession from 'shallower' icriodid-rich to 'deeper' polygnathid-rich biofacies. In most examples, however, the record appears to be either reverse, that is, shallowing-upwards (Boulonnais, ?New York) or showing hardly any vertical gradient at all. In the case of deeper-marine pelagic facies (Montagne Noire, Central Pyrenees, SE Tien-Shan) the sedimentary evidence of the transgression may be absent or its record very subtle. It may be concluded that at this stage of investigations it is impossible to achieve an adequate time-resolution to precisely correlate either the initiation of the Taghanic transgression (which may be diachronous within the *ansatus* Zone interval) or its various stages (e.g. subordinate pulses, maximum flooding phases etc.) in different basins and sections. Therefore, the conodont data analysed in this paper may be treated only as generalized information on a range of biofacies occurring in the investigated *ansatus* Zone interval.

The next difficulty in constructing a global biofacies model is the problem of potential bias introduced by provincialism and/or endemism of conodonts. Although the investigated interval marks a period of increased cosmopolitism and, at the same time, a lowered number of endemic taxa (Klapper & Johnson 1980), there is still a remarkable biogeographical conodont variability. In particular, several important icriodid taxa are endemic (most probably *I. latecarinatus*) while others (*I. difficilis*, *Icriodus brevis* Stauffer, 1940) seem to have a worldwide distribution. Moreover, although the Euramerican faunas seem to be particularly rich and diverse, other areas show impoverishment in certain taxa, even on a generic level, like the poor representation of icriodids in the Middle Devonian faunas of East Australia (Mawson & Talent 1989). The cited authors report a scarce occurrence of *P. ansatus* but its presence is doubtful as the illustrated specimens either apparently belong to *Polygnathus hemiansatus* Bultynck, 1987 (Mawson & Talent 1989, pl. 3, figs. 19–20) or their poor preservation precludes exact identification (Mawson & Talent 1989, pl. 3, fig. 21). It seems, therefore, possible that *P. ansatus* is absent from the Australian sections, probably due to palaeobiogeographical factors.

The problem of conodont palaeobiogeography is connected with a general question of environmental limits of conodont occurrences in oceanic and deeper-marine conditions (cf. Klapper & Barrick 1978). Neither of the analysed global settings represents a true oceanic or bathyal realm but three of them, Montagne Noire, Central Pyrenees and SE Tien-Shan, can be attributed to terranes or microcontinents separated by oceanic areas from Euramerica and Gondwana (Fig. 1). They represent mostly pelagic environments, probably submarine elevations without any clear connections to coastal belts of epicontinental basins. With a few exceptions (e.g. Montagne Noire, cf. Table 2) the conodont abundances per sample are notably low. This is even more striking if we take into account the stratigraphically condensed nature of the discussed strata (see above), meaning that for the same time-interval the conodont abundance in, say, the SE Tien-Shan area, was an order of magnitude lower than in the Polish or Belgian sections. It is worth noting that a similar low conodont frequency is typical for the mid-Givetian condensed limestones associated with submarine rises surrounded by deeper basins in the Rhenish area of Germany (Lottmann 1990; Aboussalam 2003). Apparently, these environments were very sparsely inhabited (or perhaps it is better to say, rarely visited) by conodonts when compared with the typical epicontinental basins analysed here.

Such rare occurrences suggest that we are dealing with environments in which the presence of conodonts was an exception rather than a rule and that were marginal to their typical habitats. The notion of a conodont biofacies in such settings (cf. Table 2) clearly has a different meaning than that for the shallow-water epicontinental shelves of Poland, Belgium or New York. For example, when saying that the P biofacies was characteristic of the *ansatus* Zone in SE Tien-Shan, we are not implying that polygnathids thrived there. Instead, we mean that those rare conodonts that for some reason entered this area were mostly polygnathids. Such observations suggest that polygnathids, being generally more deeper-water and/or offshore taxa than icriodids (e.g. Weddige & Ziegler 1976; Klapper & Barrick 1978; Sparling 1984), had a higher survival potential in oceanic conditions and were more likely candidates for occasional trans-oceanic migration than e.g. icriodids (cf. Klapper & Johnson 1980). But, at the same time, the discussed examples cannot be simply and uncritically incorporated into any global conodont biofacies model together with areas of a 'normal' conodont occurrence.

'Nearshore–offshore' biofacies model

Having the above limitations in mind it seems possible to construct a biofacies model confined to epicontinental basins of Euramerica and, partly, Gondwana continents (with the exclusion of East Australia; Mawson & Talent 1989). Figure 9a shows a composite shelf profile that is an idealized scheme combining data from the analysed localities attributable to shallow-water carbonate–terrigenous shelf to carbonate platform and outer shelf environments. Manitoba is here regarded as representative of most proximal, nearshore facies, comparable to the Terebiń IG 5 setting of the Lublin Basin (cf. Fig. 7) but perhaps more shallow-water and/or nearshore and with less terrigenous, riverine clastic input. The Boulonnais and Ardennes sections are attributed to a carbonate platform drowned by the initial Taghanic transgression, that is, onlapped by subtidal marly facies with abundant benthic shallow-water fauna. It should be stressed that the position of the platform indicated in Figure 9a

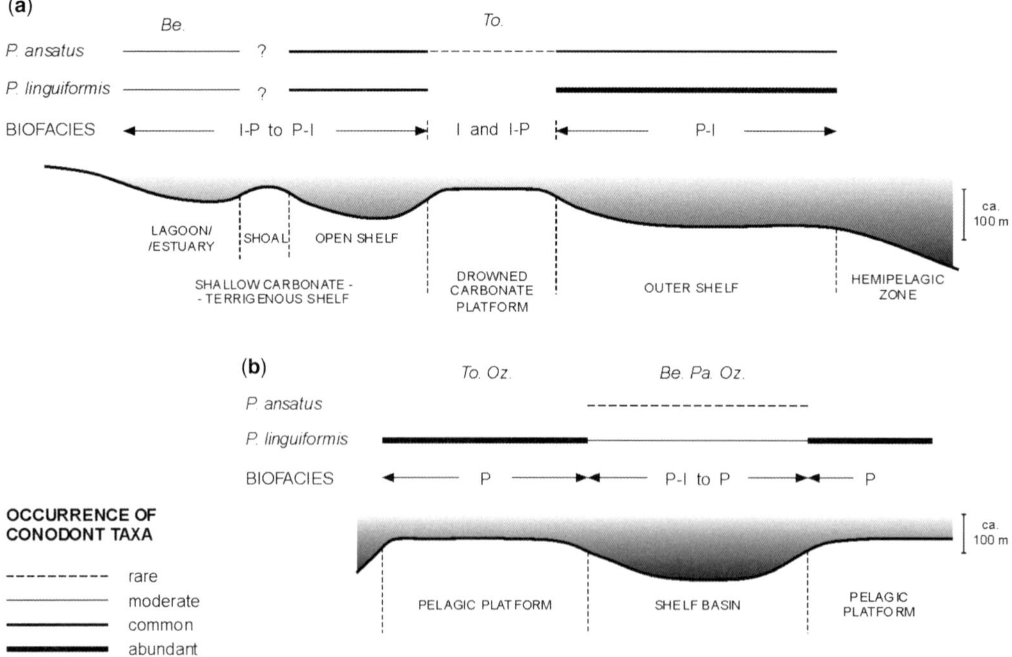

Fig. 9. Schematic biofacies models of the initial Taghanic (*ansatus* Zone) transgression. (**a**) Scheme based on various Euramerica Continent localities discussed in the text. (**b**) Scheme for the Moroccan shelf (northern Gondwana Continent). P, polygnathid; I, icriodid. Conodont generic names: *To.*, *Tortodus*; *Oz.*, *Ozarkodina*; *Be.*, *Belodella*; *Pa.*, *Panderodus*.

does not imply that it formed a barrier to the more shoreward part of the shelf, which otherwise exhibits evidence of open-marine conditions and thus good connection with the distal outer shelf. The latter, based on data from New York, is placed seawards of the carbonate platform, which, again, does not reflect the original position in the basin and is shown here only for the sake of model completeness.

As may be seen in Figure 9a, the distribution of particular biofacies does not reflect any clear onshore–offshore pattern, in particular the transition from I to P biofacies that could have been anticipated taking into account previous interpretations (Schumacher 1976; Weddige & Ziegler 1976; Klapper & Barrick 1978; Sparling 1984). Instead, the I biofacies was found rather unexpectedly in the drowned carbonate platform setting and not in the most nearshore position exemplified by the Manitoba and Lublin sections. The latter may, in fact, represent slightly offshore portions of the shelf flooded by the Taghanic transgression, and the 'pure' icriodid biofacies are yet to be found in more proximal settings. It can be also noted that the I biofacies is associated with the second, weaker pulse of the transgression in the Lublin area (Fig. 6). The interpreted facies were shallower water and, at the same time, exhibited less terrigenous admixture than those inferred for the initial transgression.

Apart from the drowned platform and probable shallow-water/nearshore settings attributable to the I biofacies, the conodont biofacies based on I/P ratio do not appear to be particularly diagnostic with respect to shelf sub-environments. The latter are characterized by mixed polygnathid–icriodid assemblages representative of P–I, partly I–P, biofacies. This more or less uniform pattern is considerably complicated when taking into account the distribution of *P. linguiformis* and *P. ansatus* (Fig. 9a). The former species exhibits a clear onshore to offshore increase in abundance and, at the same time, absence from the drowned platform facies. It attains absolute dominance in outer shelf facies of New York, both in total conodont counts and versus other polygnathids. *Polygnathus ansatus*, on the other hand, seems to reach its peak abundance in more proximal shelf settings, with rare occurrences also in a flooded carbonate platform and in outer shelf environments. Thus, it can be concluded that the *ansatus*/*linguiformis* ratio is more diagnostic in terms of the analysed biofacies models than the classic I/P ratio. The accessory genera are too rare in the investigated interval to draw any sound conclusions as to their importance as environmental indicators. Nevertheless, it can be noted that in Euramerican sections *Belodella* is confined to the most nearshore settings while *Tortodus* occurs on drowned carbonate platforms.

The Moroccan Basin is presented separately in Figure 9b as it is not easy to incorporate into the pattern described above. The Moroccan palaeogeography and facies are hardly comparable to those known from Euramerica. A shoreline position and characteristics of nearshore facies are unclear as is the notion of a 'pelagic platform' (e.g. the Tafilalt Platform). The latter is envisaged as a flat-topography shallow-water (less than 100 m deep) environment with apparently minimum subsidence and very slow, laterally discontinuous sedimentation (Wendt 1989). The biotic assemblages are diverse and contain different benthic and pelagic elements. The adjoining intra-shelf basins are characterized by higher sedimentation rates, more terrigenous input and palaeo-water depths of the order of some hundreds of metres. Given such a facies pattern one would expect a conodont biofacies distribution generally showing decreasing I/P ratio following the platform-to-basin transition. Apparently this is not the case and, in fact, an opposite trend is suggested, with the appearance of P–I biofacies in the Ma'der Basin while the Tafilalt Platform is characterized exclusively by polygnathid assemblages. This trend is further confirmed by decreased relative abundance of *P. linguiformis* in the basinal facies, with a notable appearance of rare *P. ansatus*, which is otherwise absent in the platform settings. There is also a remarkable difference in accessory genera occurring in both compared settings (Table 2; Fig. 9b). In particular, *Belodella* appears to be confined to a basinal setting in Morocco while it is present in nearshore shallow-water environments of Euramerica. It may also be noted that Late Frasnian conodont biofacies distribution does not reflect a simple platform-to-basin depositional pattern, with relative icriodid abundance being surprisingly high near the platform margins and in the basinal settings (Belka & Wendt 1992).

There is no straightforward and apparent explanation for this 'Moroccan paradox' at present. It can be noted only that the Moroccan shelf was located in higher latitudes (40 to 50°S), compared with the other basins analysed (Golonka & Gawęda 2012; Scotese 2012; Blakey 2013, http://www2.nau.edu/rcb7/370Marect.jpg). Consequently, the climate could have been colder, which imposed additional environmental constraints on the occurrence of conodont taxa. For example, some polygnathids (such as *P. linguiformis*) that were more adapted to offshore, deeper and presumably colder waters of a tropical belt, flourished in the shallower and somewhat cooler waters of the Moroccan pelagic platform in higher latitudes. A similar explanation for the distribution of the Ordovician genus *Amorphognathus* Branson & Mehl, 1933*b* and related forms is given by Sweet (1988, p. 165).

Conclusions

(1) In view of the analysed Polish and global evidence, construction of a universal conodont biofacies model for the Taghanic transgressive interval of the *ansatus* Zone is difficult, if ever possible. The main obstacles are the problem of providing a precise time-framework correlated with phases of the transgression, and the variability in palaeobiogeographical distribution of conodonts, including limits of their general global occurrence.

(2) The Polish Lublin Basin example demonstrates that biofacies successions can reflect significant short-term temporal changes of the transgression dynamics, unresolvable in terms of the current conodont zonation. These different phases should be taken into account when analysing lateral biofacies distribution, which ought to refer to a consistent, geologically instantaneous transgression phase.

(3) The quantitative biofacies studies should take into account the factor of a depositional rate of conodont-bearing sediments. The biofacies concept is to be applied with caution to stratigraphically condensed strata from the settings that appear marginal in respect to mid-Givetian conodont occurrences, e.g. palaeogeographically isolated mid-oceanic and/or deep pelagic environments. In such environments conodont assemblages are commonly poor and are dominated by *P. linguiformis* that seem to be rare immigrants from epicontinental seas.

(4) It is possible and potentially useful to construct regional biofacies models for narrow, clearly defined age intervals, like that presented in this study for the Lublin Basin in SE Poland. Such models should not only include a 3D approach to biofacies distribution against regional palaeogeography and depositional facies (Belka & Wendt 1992), but they also need to take into account the fourth dimension, that is, the factor of time (see the second point above).

(5) The more general nearshore–offshore biofacies 2D model proposed in this study is based on selective data from epicontinental shallow-marine basins of Euramerica. Whereas icriodid and I–P biofacies seem to characterize shallow, nearshore facies and drowned carbonate platforms, the I/P ratio is hardly diagnostic for other shelf settings. It is here argued that the *P. ansatus*/*P. linguiformis* ratio can be more useful than I/P in differentiating inner parts of the shelf (higher values) v. outer shelf environments (lower values).

The present work was conducted in the framework of project no. 61.2101.1302.00.0 of the Polish Geological Institute-NRI. We wish to thank Jan Turczynowicz for performing computer graphics. This is a contribution to the IGCP Project 596 'Climate change and biodiversity patterns in the Mid-Palaeozoic'. We are grateful to the reviewers and the volume editor, R. Thomas Becker, for their constructive comments and suggestions.

Appendix A: Conodont taxa in the study

Table A1. *Conodont taxa and their frequency in the analysed Polish material*

Borehole	Giełczew PIG 5						Komarów IG 1		Korczmin IG 1			Terebiń IG 5		Bąkowa IG 1
Depth (m)	2017.9	2017.5–2017.6	2015.8–2016.0	2010.6–2010.7	2003.0	2002.5	2376.0	2375.6	2492.0–2492.3	2488.6–2489.0	2485.0	1576.6–1577.2	1574.7–1574.6	2116.0
Taxa/Samples														
Polygnathus linguiformis	1	9	1	3			1	5	10	18	6	1	6	4
Polygnathus ansatus	7	16	4	2			5	4	8	4	3	2		4
Polygnathus alatus														1
Polygnathus pseudofoliatus		2												
Polygnathus xylus												1		
Polygnathus timorensis		1		2										
Polygnathus ovatinodosus		1							1		1			
Polygnathus parawebbi										1				
Polygnathus sp. indet.	6	20	15	11			4	7	4	7	3	5	2	5
Icriodus arkonensis	1				2		1		3			1		1
Icriodus obliquimarginatus				1										
Icriodus platyobliquimarginatus		1										1		
Icriodus lilliputensis		3			1	5								1
Icriodus brevis		1				1			1		1			2
Icriodus latecarinatus	3	5	2	1		2		3		2			3	
Icriodus eslaensis		2				1		2			5			1
Icriodus difficilis					5	11				1		2		1
Icriodus excavatus				1	1									
Icriodus sp. indet.	9	13	3	8	13	24	1	9	7	4	2	4	1	5
Belodella sp.	1	2	1										3	
Neopanderodus sp.													1	
Total P1 elements	28	76	25	29	22	44	12	30	34	37	19	17	16	25
Other elements	12	20	5	3	1	2	3	9	6	4	8	3	5	2

References

ABOUSSALAM, Z. S. 2003. Das 'Taghanic-Event' im höheren Mittel-Devon von West-Europa und Marokko. *Münstersche Forschungen zur Geologie und Paläontologie*, **97**, 1–332.

ABOUSSALAM, Z. S. & BECKER, R. T. 2011. The global Taghanic Biocrisis (Givetian) in the eastern Anti-Atlas, Morocco. *Palaeogeography, Palaeoclimatology, Palaeoecology*, **304**, 136–164.

BAIRD, G. C. & BRETT, C. E. 2003. Taghanic Stage shelf and off shelf deposits in New York and Pennsylvania: Faunal incursions, eustasy, and tectonics. *In*: KÖNIGSHOF, P. & SCHINDLER, E. (eds) *Mid-Palaeozoic Bio- and Geodynamics. The North Gondwana–Laurussia Interaction. Proceedings of the 15th Senckenberg Conference*. Courier Forschungsinstitut Senckenberg, **242**, 141–156.

BAIRD, G. C. & BRETT, C. E. 2008. Late Givetian Taghanic bioevents in New York: new discoveries and questions. *Bulletin of Geosciences*, **83**, 357–370.

BARDASHEV, I. A. 1992. Stratigraphy of Middle Asian Middle Devonian. *Courier Forschungsinstitut Senckenberg*, **154**, 31–83.

BARDASHEV, I. A. & ZIEGLER, W. 1985. Conodonts from a Middle Devonian section in Tadzhikistan (Kalagach Fm., Middle Asia, USRR). *Courier Forschungsinstitut Senckenberg*, **75**, 65–78.

BELKA, Z. & NARKIEWICZ, M. 2008. Devonian. *In*: MCCANN, T. (ed.) *The Geology of Central Europe: Volume 1: Precambrian and Paleozoic*. Geological Society, London, 383–410.

BELKA, Z. & WENDT, J. 1992. Conodont biofacies patterns in the Kellwasser Facies (upper Frasnian/lower Famennian) of the eastern Anti-Atlas, Morocco. *Palaeogeography, Palaeoclimatology, Palaeoecology*, **91**, 143–173.

BISCHOFF, G. & ZIEGLER, W. 1957. Die Conodontenchronologie des Mitteldevons und des tiefsten Oberdevons. *Abhandlungen des Hessischen Landesamtes für Bodenforschung*, **22**, 1–135.

BLAKEY, R. 2013. Library of Paleogeography. Colorado Plateau Goesystems, Inc., http://cpgeosystems.com/paleomaps.html

BLIECK, A., TURNER, S., BURROW, C. J., SCHULTZE, H-P., REXROAD, C. B., BULTYNCK, P. & NOWLAN, G. S. 2010. Fossils, histology, and phylogeny: why conodonts are not vertebrates. *Episodes*, **33**, 234–241.

BRANSON, E. B. & MEHL, M. G. 1933*a*. Conodonts from the Bainbridge (Silurian) of Missouri. *University of Missouri Studies*, **8**, 39–52.

BRANSON, E. B. & MEHL, M. G. 1933*b*. Conodonts from the Maquoketa-Thebes (Upper Ordovician) of Missouri. *University of Missouri Studies*, **8**, 121–131.

BRANSON, E. B. & MEHL, M. G. 1938. The conodont genus *Icriodus* and its stratigraphic distribution. *Journal of Paleontology*, **12**, 156–166.

BRETT, C. E., BAIRD, G. C. & BARTHOLOMEW, A. J. 2011. Sequence stratigraphy and a revised sea-level curve for the Middle Devonian of eastern North America. *Palaeogeography, Palaeoclimatology, Palaeoecology*, **30**, 21–53.

BROADHEAD, T. W., DRIESE, S. G. & HARVEY, J. L. 1990. Gravitational settling of conodont elements: implications for paleoecologic interpretations of conodont assemblages. *Geology*, **18**, 850–853.

BUGGISCH, W. & JOACHIMSKI, M. M. 2006. Carbon isotope stratigraphy of the Devonian of central and southern Europe. *Palaeogeography, Palaeoclimatology, Palaeoecology*, **240**, 68–88.

BULTYNCK, P. 1974. Conodontes de la Formation de Fromelennes du Givétian de l'Ardenne franco-belge. *Bulletin de l'Institut royal des Sciences naturelles de Belgique, Sciences de la Terre*, **50**, 1–30.

BULTYNCK, P. 1976. Comparative study of Middle Devonian conodonts from north Michigan (U.S.A.) and the Ardennes (Belgium–France). *Geological Association of Canada, Special Paper*, **15**, 119–141.

BULTYNCK, P. 1987. Pelagic and neritic conodont successions from the Givetian of pre-Sahara Morocco and the Ardennes. *Bulletin de l'Institut royal des Sciences naturelles de Belgique, Sciences de la Terre*, **57**, 149–181.

BULTYNCK, P. 2003. Devonian Icriodontidae: biostratigraphy, classification and remarks on paleoecology and dispersal. *Revista Española de Micropaleontologia*, **35**, 295–314.

BULTYNCK, P., HELSEN, S. & HAYDUKIEWICZ, J. 1998. Conodont succession and biofacies in upper Frasnian formations (Devonian) from the southern and central parts of the Dinant Synclinorium (Belgium) – (Timing of facies shifting and correlation with late Frasnian events). *Bulletin de l'Institut royal des Sciences naturelles de Belgique, Sciences de la Terre*, **68**, 25–75.

BULTYNCK, P., CASIER, J.-G., COEN-AUBERT, M. & GODEFROID, J. 2001. Pre-conference field trips (V1), Couvin-Philippeville-Wellin area, Ardenne (May 11–12, 2001). *In*: JANSEN, U., KÖNIGSHOF, P., PLODOWSKI, G. & SCHINDLER, E. (eds) *Field Trips Guide Book, 15th International Senckenberg Conference*. Forschungsinstitut Senckenberg, Frankfurt am Main, 1–44.

CARLS, P. 1988. The Devonian of Celtiberia (Spain) and Devonian paleogeography of SW Europe. *In*: MCMILLAN, N. J., EMBRY, A. F. & GLASS, D. J. (eds) *Devonian of the World*. Canadian Society of Petroleum Geologists, Memoir, **14**, 421–466.

CHATTERTON, B. D. E. 1976. Distribution and paleoecology of Eifelian and Early Givetian conodonts from Western and Northwestern Canada. *Geological Association of Canada, Special Paper*, **15**, 143–157.

DAY, J., UYENO, T. T., NORRIS, A. W., WITZKE, B. J. & BUNKER, B. J. 1996. Middle-Upper Devonian relative sea-level histories of central and western North American interior basins. *In*: WITZKE, B. J., LUDVIGSON, G. A. & DAY, J. (eds) *Paleozoic Sequence Stratigraphy: Views from the North America Craton*. Geological Society of America, Special Papers, **306**, 259–275.

DREESEN, R. & THOREZ, J. 1980. Sedimentary environments, conodont biofacies and paleoecology of the Belgian Famennian (Upper Devonian) – an approach. *Annales de la Société Géologique de Belgique*, **103**, 97–110.

DRUCE, E. C. 1973. Upper Paleozoic and Triassic conodont distribution and the recognition of biofacies. *In*: RHODES, F. H. T. (ed.) *Conodont Paleozoology*. Geological Society of America, Special Papers, **141**, 191–237.

ETHINGTON, R. L. 1959. Conodonts of the Ordovician Galena Formation. *Journal of Paleontology*, **33**, 257–292.

GOLONKA, J. & GAWĘDA, A. 2012. Plate tectonic evolution of the southern margin of Laurussia in the Paleozoic. *In*: SHARKOV, E. (ed.) *Tectonics – Recent Advances*. InTech, Rijeka, http://doi.org/10.5772/50009

GOUWY, S., LIAO, J.-C. & VALENZUELA-RIOS, J. I. 2013. Eifelian (Middle Devonian) to Lower Frasnian (Upper Devonian) conodont biostratigraphy in the Villech section (Spanish Central Pyrenees). *Bulletin of Geosciences*, **88**, 315–338.

HECKEL, P. H. 1973. *Nature, Origin, and Significance of the Tully Limestone*. Geological Society of America, Special Papers, **139**.

HINDE, G. J. 1879. On conodonts from the Chazy and Cincinnati group of the Cambro-Silurian, and from the Hamilton and Genesee shale divisions of the Devonian, in Canada and the United States. *Geological Society of London, Quarterly Journal*, **35**, 351–369.

HOUSE, M. R. 1983. Devonian eustatic events. *Proceedings of the Ussher Society*, **5**, 396–405.

HOUSE, M. R. 2002. Strength, timing, setting and cause of mid-Palaeozoic extinctions. *Palaeogeography, Palaeoclimatology, Palaeoecology*, **181**, 5–25.

HUDDLE, J. W. 1934. Conodonts from the New Albany Shale of Indiana. *Bulletins of American Paleontology*, **21**, 1–136.

HUDDLE, J. W. 1981. Conodonts from the Genesse Formation in Western New York. *Geological Survey Professional Paper*, **1032-B**, 1–66.

JOHNSON, J. G. 1970. Taghanic onlap and the end of North American provinciality. *Geological Society of America Bulletin*, **96**, 567–587.

JOHNSON, J. G., KLAPPER, G. & SANDBERG, C. A. 1985. Devonian eustatic fluctuations in Euramerica. *Geological Society of America Bulletin*, **81**, 2077–2106.

JOHNSON, J. G., KLAPPER, G. & ELRICK, M. 1996. Devonian transgressive–regressive cycles and biostratigraphy, northern Antelope Range, Nevada: establishment of reference horizons for global cycles. *Palaios*, **11**, 3–14.

KAUFMANN, B. 2006. Calibrating the Devonian timescale: a synthesis of U-Pb ID-TIMS ages and conodont stratigraphy. *Earth-Science Reviews*, **76**, 175–190.

KLAPPER, G. & BARRICK, J. E. 1978. Conodont ecology: pelagic v. benthic. *Lethaia*, **11**, 15–23.

KLAPPER, G. & BARRICK, J. E. 1983. Middle Devonian (Eifelian) conodonts from Spillville Formation in northern Iowa and southern Minnesota. *Journal of Paleontology*, **57**, 1212–1243.

KLAPPER, G. & JOHNSON, J. G. 1980. Endemism and dispersal of Devonian conodonts. *Journal of Paleontology*, **54**, 400–455.

KLAPPER, G. & LANE, H. R. 1985. Upper Devonian (Frasnian) conodonts of the *Polygnathus* biofacies, N.W.T., Canada. *Journal of Paleontology*, **59**, 904–951.

KLAPPER, G. & PHILIP, G. M. 1971. Devonian conodont apparatuses and their vicarious skeletal elements. *Lethaia*, **4**, 429–452.

LIAO, J.-C. & VALENZUELA-RIOS, J. I. 2008. Givetian and Early Frasnian conodonts from the Compte section (Middle–Upper Devonian, Spanish Central Pyrenees). *Geological Quarterly*, **51**, 419–442.

LIAO, J.-C. & VALENZUELA-RIOS, J. I. 2013. The Middle and Upper Devonian conodont sequence from La Guardia D'Àres sections (Spanish Central Pyrenees). *Bulletin of Geosciences*, **88**, 339–368.

LIAO, J.-C., KÖNIGSHOF, P., VALENZUELA-RIOS, J. I. & SCHINDLER, E. 2008. Depositional environment interpretation and development of the Renanué section (Upper Eifelian-Lower Frasnian; Pyrenees, N. Spain). *Bulletin of Geosciences*, **83**, 481–490.

LOTTMANN, J. 1990. Die *pumilio*-Events (Mittel Devon). *Göttinger Arbeiten zur Geologie und Paläontologie*, **44**, 1–98.

MAWSON, R. & TALENT, J. A. 1989. Late Emsian–Givetian stratigraphy and conodont biofacies – carbonate slope and offshore shoal to sheltered lagoon and nearshore carbonate ramp – Broken River, North Queensland, Australia. *Courier Forschungsinstitut Senckenberg*, **117**, 205–259.

MCGOFF, H. J. 1991. The hydrodynamics of conodont elements. *Lethaia*, **24**, 235–247.

MURDOCK, D. J. E., DONG, X-P., REPETSKI, J. E., MARONE, F., STAMPANONI, M. & DONOGHUE, P. C. J. 2013. The origin of conodonts and of vertebrate mineralized skeletons. *Nature*, **502**, 546–549, http://doi.org/10.1038/nature12645

NARKIEWICZ, K. 2011. Conodont biostratigraphy of the Middle Devonian in the Lublin area (south-eastern Poland). *Prace Państwowego Instytutu Geologicznego*, **196**, 147–192.

NARKIEWICZ, K. & BULTYNCK, P. 2007. Conodont biostratigraphy of shallow marine Givetian deposits from the Radom-Lublin area, SE Poland. *Geological Quarterly*, **51**, 419–442.

NARKIEWICZ, K. & BULTYNCK, P. 2010. The Upper Givetian (Middle Devonian) *subterminus* conodont Zone in North America, Europe and North Africa. *Journal of Paleontology*, **84**, 588–625.

NARKIEWICZ, K. & NARKIEWICZ, M. 1998. Conodont evidence for the mid-Givetian Taghanic Event in south-eastern Poland. *Palaeontologia Polonica*, **58**, 213–223.

NARKIEWICZ, K. & NARKIEWICZ, M. 2011. Conodont biofacies record of the Givetian transgressive levels in the Lublin and Łysogóry-Radom basins (SE Poland). *In*: *IGCP 596 Opening Meeting, Graz, 19–24 September 2011*, Berichte des Institutes für Erdwissenschaften, Karl-Franzens-Universitat Graz, **16**, 74–75.

NARKIEWICZ, M. 2007. Development and inversion of Devonian and Carboniferous basins in the eastern part of the Variscan foreland (Poland). *Geological Quarterly*, **51**, 231–256.

NARKIEWICZ, M. 2011*a*. Lithostratigraphy, depositional systems and transgressive-regressive cycles in the Middle Devonian to Frasnian of the Łysogóry-Radom Basin (south-eastern Poland). *Prace Państwowego Instytutu Geologicznego*, **196**, 7–52.

NARKIEWICZ, M. 2011*b*. Lithostratigraphy, depositional systems and transgressive-regressive cycles in the Devonian of the Lublin Basin (south-eastern Poland). *Prace Państwowego Instytutu Geologicznego*, **196**, 53–146.

NARKIEWICZ, M., NARKIEWICZ, K. & TURNAU, E. 2011. Devonian depositional development of the Łysogóry-Radom and Lublin basins (south-eastern Poland). *Prace Państwowego Instytutu Geologicznego*, **196**, 255–288.

NICOLL, R. S. 1982. Multielement composition of the conodont *Icriodus expansus* Branson & Mehl from the

Upper Devonian of the Canning Basin, Western Australia. *BMR Journal of Australian Geology & Geophysics*, **7**, 197–217.

NICOLL, R. S. 1985. Multielement composition of the conodont species *Polygnathus xylus xylus* Stauffer, 1940 and *Ozarkodina brevis* (Bischoff & Ziegler, 1957) from the Upper Devonian of the Canning Basin, Western Australia. *BMR Journal of Australian Geology & Geophysics*, **9**, 133–147.

NORRIS, W. A. & UYENO, T. T. 1998. Middle Devonian brachiopods, conodonts, stratigraphy, and transgressive–regressive cycles, Pine Point area, south of Great Slave Lake, district of Mackenzie, Northwest Territories. *Geological Survey of Canada*, **522**, 1–190.

PELHATE, A. & PONCET, J. 1988. Evolution sédimentaire de la Formation de Blacourt (Givétien de Ferques-Boulonnais). *Le Dévonien de Ferques-Bas Boulonnais (N-France). Biostratigraphie du Paléozoïque*, **7**, 25–36.

POHLER, S. M. L. & BARNES, C. R. 1990. Conceptual models in conodont paleoecology. *Courier Forschungsinstitut Senckenberg*, **118**, 409–440.

SANDBERG, C. A. 1976. Conodont biofacies of Late Devonian *Polygnathus styriacus* Zone in Western United States. *Geological Association of Canada, Special Paper*, **15**, 171–186.

SANDBERG, C. A. & DREESEN, R. 1984. Late Devonian icriodontid biofacies models and alternate shallow-water conodont zonation. *In*: CLARK, D. L. (ed.) *Conodont Biofacies and Provincialism*. Geological Society of America, Special Papers, **196**, 143–178.

SANDBERG, C. A., ZIEGLER, W., DREESEN, R. & BUTLER, J. L. 1988. Late Frassnian mass extinction: conodont event stratigraphy, global changes and possible causes. *Courier Forschungsinstitut Senckenberg*, **102**, 263–307.

SANDBERG, C. A., ZIEGLER, W. & BULTYNCK, P. 1989. New standard conodont zones and early *Ancyrodella* phylogeny across Middle–Upper Devonian boundary. *Courier Forschungsinstitut Senckenberg*, **110**, 195–230.

SANDBERG, C. A., ZIEGLER, W., DREESEN, R. & BUTLER, J. L. 1992. Conodont biochronology, biofacies, taxonomy, and event stratigraphy around Middle Frasnian Lion Mudmound (F2 h), Frasnes, Belgium. *Courier Forschungsinstitut Senckenberg*, **150**, 7–87.

SCHUMACHER, D. 1976. Conodont biofacies and paleoenvironments in Middle Devonian-Upper Devonian boundary beds, Central Missouri. *Geological Association of Canada, Special Paper*, **15**, 159–169.

SCOTESE, C. R. 2012. Paleomap Project, http://www.scotese.com

SEDDON, G. & SWEET, W. C. 1971. An ecologic model for conodonts. *Journal of Paleontology*, **45**, 869–880.

SPARLING, D. R. 1983. Conodont biostratigraphy and biofacies of lower Middle Devonian Limestone, north-central Ohio. *Journal of Paleontology*, **57**, 825–864.

SPARLING, D. R. 1984. Paleoecologic and paleogeographic factors in the distribution of lower Middle Devonian conodonts from north-central Ohio. *In*: CLARK, D. L. (ed.) *Conodont Biofacies and Provincialism*. Geological Society of America, Special Papers, **196**, 113–125.

STAUFFER, C. R. 1940. Conodonts from the Devonian and associated clays of Minnesota. *Journal of Paleontology*, **14**, 417–435.

SWEET, W. C. 1988. The conodonta: morphology, taxonomy, paleoecology, and evolutionary history of a long animal phylum. *Oxford Monographs on Geology and Geophysics*, **10**, 1–212.

VAN ADRICHEM BOOGAERT, H. A. 1967. Devonian and Lower Carboniferous conodonts of the Cantabrian Mountains (Spain) and their stratigraphic application. *Leidse Geologische Mededelingen*, **39**, 129–192.

WALLISER, O. & BULTYNCK, P. 2011. Extinctions, survival and innovations of conodont species during the Kačák Episode (Eifelian–Givetian) in south-eastern Morocco. *Bulletin de l'Institut royal des Sciences naturelles de Belgique, Sciences de la Terre*, **81**, 5–25.

WEDDIGE, K. 1977. Die Conodonten der Eifel-Stufe im Typusgebiet und in benachbarten Faziesgebieten. *Senckenbergiana lethaea*, **58**, 271–419.

WEDDIGE, K. & ZIEGLER, W. 1976. The significance of *Icriodus:Polygnathus* ratios in limestones from the type Eifelian, Germany. *Geological Association of Canada, Special Paper*, **15**, 187–199.

WENDT, J. 1989. Facies pattern and paleogeography of the Middle and Late Devonian in the Eastern Anti-Atlas (Morocco). *In*: MCMILLAN, N. J., EMBRY, A. F. & GLASS, D. J. (eds) *Devonian of the World*. Canadian Society of Petroleum Geologists, Memoir, Calgary, **14**, 467–480 [imprint 1988].

WITTEKINDT, H. 1966. Zur Conodonten chronologie des Mitteldevons. *Fortschritte in der Geologie von Rheinland und Westfalen*, **9**, 621–646.

ZIEGLER, W. & SANDBERG, C. A. 1990. The Late Devonian standard conodont zonation. *Courier Forschungsinstitut Senckenberg*, **121**, 1–115.

ZIEGLER, W., KLAPPER, G. & JOHNSON, J. G. 1976. Redefinition and subdivision of the *varcus*-Zone (conodonts, Middle-?Upper Devonian) in Europe and North America. *Geologica et Palaeontologica*, **10**, 109–140.

ZIEGLER, W. G. & LINDSTRÖM, M. 1971. Über *Panderodus* Ethington, 1959, und *Neopanderodus* n.g. (Conodonta) aus dem Devon. *Neues Jahrbuch für Geologie und Paläontologie Monatshefte*, **1971**(10), 628–640.

A carbonate carbon isotope record for the late Givetian (Middle Devonian) Global Taghanic Biocrisis in the type region (northern Appalachian Basin)

JAMES J. ZAMBITO IV[1]*, MICHAEL M. JOACHIMSKI[2], CARLTON E. BRETT[3], GORDON C. BAIRD[4] & Z. SARAH ABOUSSALAM[5]

[1]*Wisconsin Geological and Natural History Survey, University of Wisconsin – Extension, 3817 Mineral Point Road, Madison, WI 53705, USA*

[2]*GeoZentrum Nordbayern, Universität Erlangen-Nürnberg, Schlossgarten 5, 91054 Erlangen, Germany*

[3]*Department of Geology, University of Cincinnati, Cincinnati, OH 45221–0013, USA*

[4]*Department of Geoscience, SUNY College at Fredonia, Fredonia, NY 14063, USA*

[5]*Institute of Geology and Palaeontology, Westphalian Wilhelms University, Corrensstrasse 24, D-48149 Münster, Germany*

**Corresponding author (e-mail: jay.zambito@wgnhs.uwex.edu)*

Abstract: During the Global Taghanic Biocrisis (*c.* 385 Ma), Middle Devonian faunas worldwide underwent extinction. In the biocrisis type region, the northern Appalachian Basin, biodiversity changes occurred through three bioevents that ultimately resulted in the loss of numerous endemic taxa. Carbon isotope excursions during this biocrisis have been documented in various stratigraphic successions, but never in the type region. Herein, we reconstruct changes in $\delta^{13}C_{carb}$ from the biocrisis type region and compare these changes to local faunal transitions. An approximately 1.5‰ negative excursion corresponds to the first bioevent, a time of inferred global warming and replacement of most endemic taxa of the mid-palaeolatitude Appalachian Basin by invasive palaeoequatorial taxa. An approximately 2‰ positive excursion is associated with the second bioevent, recognized as a return of the endemic fauna and the loss of invasive taxa. This positive excursion occurs near the *Polygnathus ansatus–Ozarkodina semialternans* zonal boundary and is recognized elsewhere. Faunal cosmopolitanism associated with the third bioevent corresponds with an inflection in the carbon isotope record from negative to positive trending values, which agrees with a positive carbon record excursion seen elsewhere at the *semialternans–Schmidtognathus hermanni* zonal boundary. This new carbon isotope record provides an important reference for recognizing this biocrisis in other areas and facies.

Supplementary material: The $\delta^{13}C$ and $\delta^{18}O$ dataset collected for this study is available at http://www.geolsoc.org.uk/SUP18840.

During the Global Taghanic Biocrisis, which occurred at approximately 385 Ma (Becker *et al.* 2012), Middle Devonian faunas worldwide underwent a major extinction during a time of global warming, increased aridity, eustatic sea-level rise – known as the Taghanic Onlap – and decreased oxygenation of epicontinental seas (Johnson 1970; House 2002; Aboussalam 2003; Joachimski *et al.* 2004, 2009; van Geldern *et al.* 2006; Aboussalam & Becker 2011; Marshall *et al.* 2011; Zambito *et al.* 2012*a*, 2013; Turnau 2014). In fact, the Middle–Upper Devonian boundary was originally defined at this stratigraphic level because of the abrupt faunal changes observed in a number of taxonomic groups (Klapper *et al.* 1987; House 2002). The Taghanic Biocrisis marks the end of established Devonian faunal provinciality related to latitudinal climatic (temperature) gradients, resulting in a worldwide cosmopolitan fauna that persisted until the late Frasnian extinction (Johnson 1970; Boucot 1989; McGhee 1996 and references therein; Sandberg *et al.* 2002).

High-resolution stratigraphic studies have provided the necessary framework for detailed qualitative and quantitative palaeoecological studies of the pulsed faunal migrations, replacements, recurrences and extinctions during this biocrisis in the type region, the northern Appalachian Basin (Heckel 1973; Baird & Brett 2003, 2008; Sessa 2003; Bonelli *et al.* 2006; Zambito *et al.* 2009, 2012*a*,

From: Becker, R. T., Königshof, P. & Brett, C. E. (eds) 2016. *Devonian Climate, Sea Level and Evolutionary Events*. Geological Society, London, Special Publications, **423**, 223–233.
First published online June 26, 2015, http://doi.org/10.1144/SP423.7

b). As hypothesized by Johnson (1970) and Boucot (1989), based on biogeographical migrations, reconstructions of sea-surface temperature variations before and after the Taghanic Biocrisis interval indicate global warming coincident with the expansion of equatorial benthic faunas to higher latitudes (Boucot & Theron 2001; Aboussalam 2003; Joachimski *et al.* 2004, 2009; van Geldern *et al.* 2006; Brand *et al.* 2008; Zambito *et al.* 2013). Here, we present a carbonate $\delta^{13}C$ ($\delta^{13}C_{carb}$) record from the Taghanic Biocrisis type region and compare this isotopic record with local faunal transitions, as well as with carbon isotope record reconstruction elsewhere.

Geological and palaeoecological setting of the type region

The northern Appalachian Basin deposits of New York State and adjacent areas comprise the type region of the Taghanic Biocrisis (Fig. 1) (House 1985, 2002). In the Middle Devonian, the northern Appalachian Basin was located at approximately 30°S latitude (Fig. 1) (Cocks & Torsvik 2011). These strata were deposited in a foreland basin that formed during the Acadian Orogeny as the Laurentian and Avalonian terranes converged obliquely; the Taghanic Biocrisis is recorded in the sediments deposited as part of the 'Catskill Delta Complex' during the transition from the second to the third collisional tectophase of this orogeny (Ettensohn 2008; Ver Straeten 2010 and references therein). In the northern Appalachian Basin, the Taghanic Biocrisis is recorded primarily in the strata of the Tully Formation, but evidence for immigration and extinction related to the biocrisis is present in the uppermost beds of the underlying Hamilton Group and continues into the overlying Genesee Group (Baird & Brett 2003, 2008; reviewed in Zambito *et al.* 2012*a*). The Tully Formation in western New York State is a carbonate-dominated succession interpreted to have formed as a result of general tectonic quiescence and reduced progradation, as well as the formation of a proximal intrabasinal clastic trap during the onset of the third tectophase (Heckel 1973; Baird & Brett 2003; Baird *et al.* 2012).

Using a sequence stratigraphic approach, the Tully Formation has been divided into three informal units (lower, middle and upper), each of which have been recognized to varying extents outside of the type region and therefore are interpreted to represent eustatic sea-level changes (Baird & Brett 2003, 2008; Aboussalam & Becker 2011; Brett *et al.* 2011). Indeed, a globally recognized feature of the Taghanic Biocrisis is a eustatic sea-level rise that submerged portions of the North American Transcontinental Arch (Taghanic Onlap *sensu* Johnson 1970; Johnson *et al.* 1985). Johnson (1970) originally placed the Taghanic Onlap at the base of the Tully Formation and its correlative strata, associated with the first appearance of the marker brachiopod *Rhyssochonetes aurora*, and at the base of regional Taghanic Substage. Subsequently, its use varied from the single greatest transgression during the biocrisis (Baird & Brett 2003) to describing the overall higher sea level observed during the Taghanic Biocrisis (Aboussalam & Becker 2011). Herein, we use the term Taghanic Onlap to represent the transgression associated with the onset of upper Tully Formation deposition (the transgressive Bellona–West Brook interval in New York) and continued sea-level rise through the upper Tully (latest *Polygnathus ansatus–Ozarkodina semialternans* conodont zones) to the maximum flooding surface at or near the base of the Geneseo Formation (*Schmidtognathus hermanni* Zone). We prefer this refined definition because the base–Bellona contact is a recognizable and widespread transgressive surface that has major utility in stratigraphic correlation in the type region and beyond (Baird & Brett 2003; Brett *et al.* 2011).

Devonian faunas have been divided globally into biogeographical 'realms', the distribution of which is inferred to have been controlled by climatic gradients between the equator and poles (Johnson 1970; Koch & Boucot 1982; Boucot 1989 and references therein). At the onset of the Taghanic Biocrisis these realms in Laurentia consisted of the equatorial warm-water Old World Realm, and the higher-latitude cooler-water Eastern Americas Realm (Fig. 1a). Prior to the Taghanic Biocrisis, genera that comprised the Tully Fauna (a subset of the Old World Realm, Cordilleran Region) occurred in what is now western North America, in settings more equatorial than the Appalachian Basin. At the same time, the northern Appalachian Basin was occupied by the diverse Hamilton Fauna (a subset of the Eastern Americas Realm, Appohimchi Subprovince) for a period of approximately 4–5 myr (Brett & Baird 1995; Brett *et al.* 1996; Baird & Brett 2008). Johnson (1970) and Boucot (1989) suggested that changes in (latitudinal) climatic gradients were the primary drivers of the observed faunal transitions associated with the Taghanic Biocrisis. In particular, an episode of global warming is thought to have permitted the expansion of the ranges of certain equatorial taxa into higher latitudes. Moreover, Johnson (1970) further suggested that concurrent sea-level rise resulted in the drowning of the Trans-Continental Arch, the connecting of depositional basins and establishment of widespread cosmopolitanism (Fig. 1a). However, there was clearly also an episodic faunal influx from and

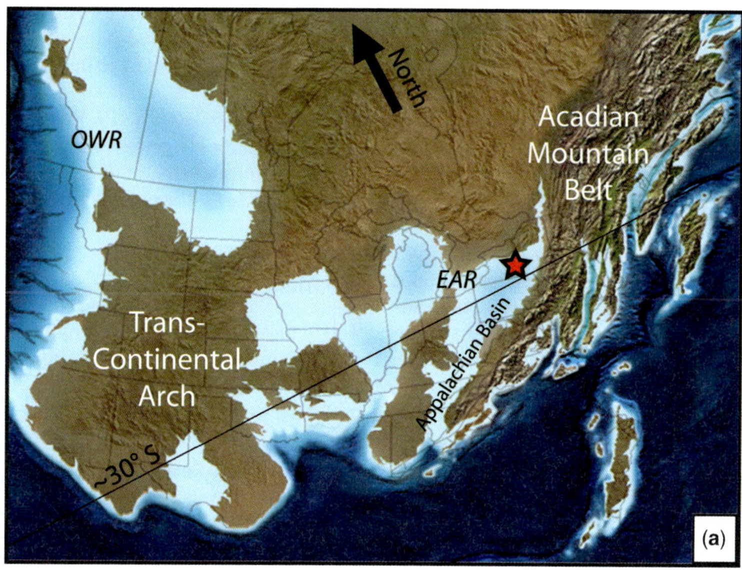

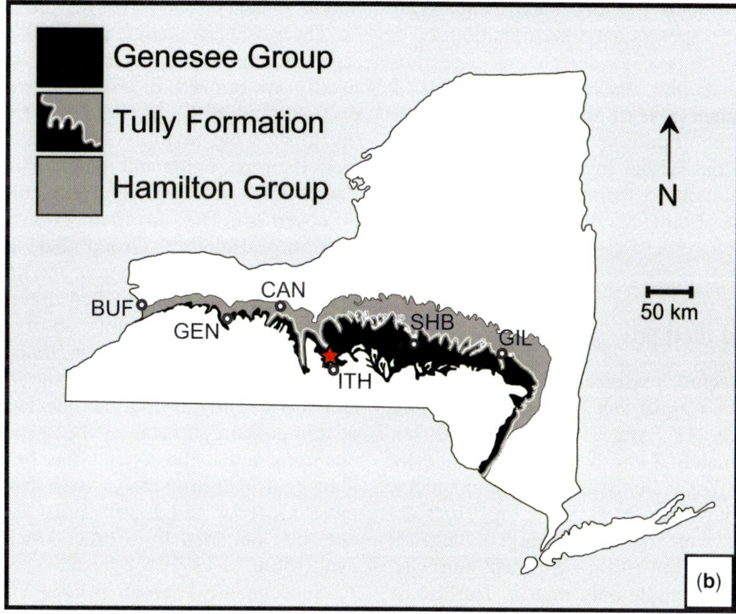

Fig. 1. (**a**) Palaeogeographical reconstruction of North America during the Taghanic Biocrisis. The Appalachian Basin strata of New York State (type section: Taughannock Falls, Trumansburg, NY, represented by a red star) that record this biocrisis were deposited approximately 30°S of the palaeo-equator. OWR and EAR denote the positions of the 'Old World Realm' and 'Eastern Americas Realm' faunas, respectively; note the position of the Trans-Continental Arch (the 'Continental Backbone' of Johnson 1970) relative to these faunas. Figure and palaeogeographical map adapted from Baird *et al.* (2012) and Blakey (2014). (**b**) Outcrop belt for the Hamilton, Tully and Genesee in the Appalachian Basin of New York State. Abbreviations for the geographical references include: BUF, Buffalo; GEN, Geneseo; CAN, Canandaigua; ITH, Ithaca; SHB, Sherburne; GIL, Gilboa. The red star denotes the Taghanic Biocrisis type section.

into the east, along the Afro-Appalachian route of faunal migration (e.g. House 1973; Oliver 1975; Rachebeuf *et al.* 2001), which was especially important for pelagic faunal groups, such as conodonts and ammonoids (e.g. Klapper & Ziegler 1967; House 1978; Aboussalam 2003).

In the Taghanic type region, at least three bioevents are recognized and have been studied in detail (Baird & Brett 2008; see the review in Zambito *et al.* 2012*a*). The polyphased nature of the Taghanic Biocrisis has also been recognized in other places, including Morocco, Montagne Noire (France) and the Rhenish Massif (Germany); however, each region has somewhat unique faunal and environmental changes (Aboussalam & Becker 2011). Initially defined by Walliser (1990, 1996) and subsequently refined by Aboussalam (2003; Aboussalam & Becker 2011), a biocrisis is composed of a series of bioevents representing sudden palaeooceanographical or biotic change, and, furthermore, biocrises should span more than one biozone and comprise a number of bioevents that themselves occur within a single biozone and/or transgressive–regressive cycle. In the type region, these bioevents include faunal migrations, replacements and extinctions through an interval of approximately 500 kyr as follows: (1) emigration and nearly complete replacement of the endemic Hamilton Fauna with the previously equatorial Tully Fauna; (2) subsequent extermination of a majority of Tully taxa and recurrence of a portion of the Hamilton Fauna; and (3) loss of much of the Hamilton Fauna (at least, locally), species turnover within some Hamilton genera, return of a few Tully taxa and further incursion of additional palaeoequatorial taxa to form the (cosmopolitan) Genesee Fauna (Baird & Brett 2008; Zambito *et al.* 2012*a*, *b*).

Materials and methods

The $\delta^{13}C_{carb}$ record presented herein is reconstructed from the Cargill Salt Company Test Core 17 from Lansing, NY (API: 31-109-13173-00-00), drilled approximately 8 km to the east of the type section for the Taghanic Biocrisis (Taughannock Falls, Trumansburg, New York State: Fig. 1b). Carbonate powders were collected using a tungsten carbide drill bit from freshly cleaned surfaces. Care was taken to sample only matrix. The use of bulk carbonate powders, drilled from fine-grained matrix, for carbon isotope analysis has become a common tool for reconstructing the carbon isotope record in the Palaeozoic, particularly in the absence or paucity of brachiopod shell calcite such as in the case of this study (Cramer & Saltzman 2005 and discussion and references therein; Buggisch & Joachmiski 2006; Weissert *et al.* 2008; Edwards & Saltzman 2014 and discussion and references therein). Carbon isotope analyses were performed at the University of Erlangen-Nürnberg, Germany. Carbonate powders were reacted with 100% phosphoric acid at 70°C using a Gasbench II connected to a ThermoFisher Five Plus mass spectrometer. All values are reported in per mil (‰) relative to V-PDB (Vienna Pee Dee Belemnite) by assigning a $\delta^{13}C$ value of +1.95‰ and a $\delta^{18}O$ value of −2.20‰ to NBS19. Precision and reproducibility of carbon isotope analyses were monitored by replicate analysis of laboratory standards and was ±0.04‰ (1 standard deviation).

Results

The dataset consists of 147 samples (Fig. 2). No relationship exists between sample lithology and $\delta^{13}C$ values, with values of both calcareous shale and limestone spanning the entirety of the $\delta^{13}C$ range observed in the studied section. Likewise, trends in the $\delta^{13}C$ profile show no relationship to lithology, as values both increase and decrease upsection within transitions from calcareous shale to limestone, and vice versa. In addition, no co-variation is observed in a cross-plot of the $\delta^{13}C$ and $\delta^{18}O$ data ($r^2 = 0.119$: Fig. 2b).

The most prominent feature of the Taghanic type region carbon isotope record is that each of the locally recognized bioevents corresponds to an excursion in $\delta^{13}C_{carb}$ (Fig. 2a). Coincident with the loss of the majority of Hamilton Fauna taxa from the type region and the incursion of the Tully Fauna (Bioevent 1), a negative excursion of −1.6‰ is observed. The negative excursion begins in uppermost Hamilton Group shale and appears to continue into the limestone of the lower Tully Formation (Carpenter Falls bed interval), across a large regional unconformity observed in the core. It should be noted, however, that three approximately 10 m-scale cycles in the upper Moscow and lower Tully formations are truncated locally at this erosion surface, so the seemingly continuous trend across the contact may be more apparent than real. Unfortunately, data from the uppermost Moscow and lower Tully formations have not been obtained from other localities to permit evaluation.

Subsequently, carbon isotope values gradually return to the apparent baseline of approximately 1.0‰ through the lower Tully. Immediately prior to the recurrence of the Hamilton Fauna (Bioevent 2), a positive shift in $\delta^{13}C_{carb}$ from 0.8 to 1.9‰ occurs within the uppermost lower Tully interval observed immediately below the unconformity-bounded Smyrna Bed, a chamosite-rich encrinite, at the base of the middle Tully. The positive $\delta^{13}C_{carb}$ trend continues, with relatively large scatter, across numerous discontinuities to a maximum value of +2.9‰ within the Bellona–West Brook interval and high $\delta^{13}C_{carb}$ values (>2.0‰) in the lower half of the Moravia Bed. Subsequently,

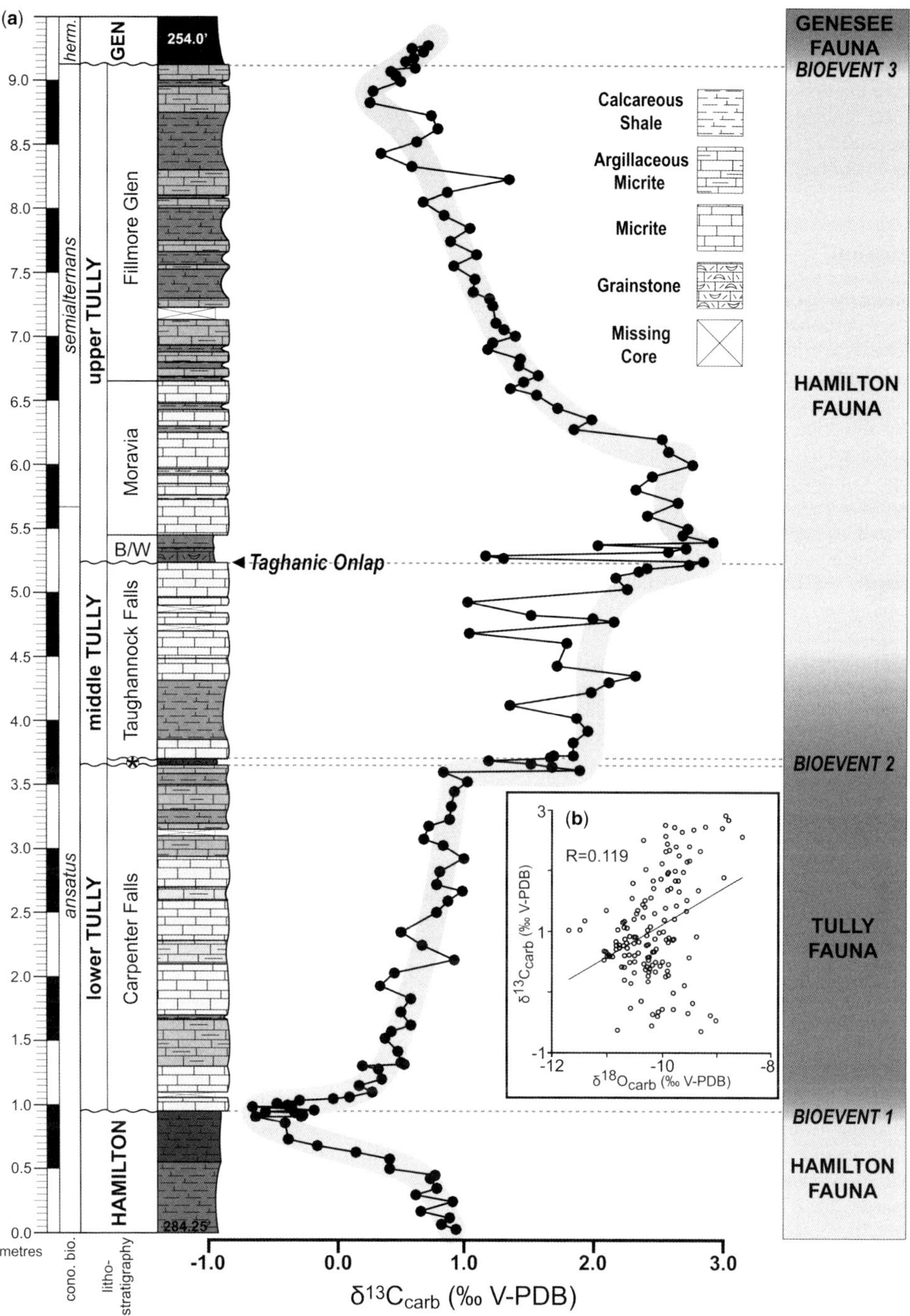

Fig. 2. (**a**) $\delta^{13}C_{carb}$ record through the Global Taghanic Biocrisis from the Cargill Salt Co. Test Core 17 (Lansing, NY). cono. bio., conodont biozonation; *Smyrna bed; B/W, Bellona/West Brook interval; *herm.*, *hermanni* zone; GEN, Genesee Group. Conodont biostratigraphy in the core is inferred from work carried out on nearby outcrops (Ziegler *et al.* 1976). Dashed lines represent the position of regional unconformities; the thick grey line highlights the generalized isotopic pattern. Footage at the top and bottom of the studied section denotes core depth. Details on the bioevents and associated faunal changes referenced are presented in Zambito *et al.* (2012*a*). (**b**) Cross-plot of the $\delta^{13}C$ and $\delta^{18}O$ data collected for this study.

carbon isotope values gradually decrease through the uppermost Moravia and Fillmore Glen intervals to values as low as 0.3‰. The onset of a second positive excursion at the top of the studied interval may be indicated by a trend to more positive values starting directly below the sharp Tully–Genesee contact.

Discussion

Whereas the oxygen isotope composition of sedimentary carbonates is easily affected by secondary diagenetic alteration, carbon isotope ratios are usually less influenced. During diagenesis, carbonates will be cemented, and metastable high-magnesium calcite and aragonitic components will be dissolved and replaced by low-magnesium calcite. The carbon isotope composition of these secondary phases is determined by the carbon isotope composition of dissolved inorganic carbon of the diagenetic solution, which in a diagenetically closed system corresponds to the carbon isotope composition of the dissolving primary carbonates. In this case, the carbon isotope composition of the sedimentary carbonate will be inherited by the diagenetically stabilized carbonate. In a diagenetically open system, isotopically light carbon from the remineralization of organic carbon or soil-derived CO_2 will result in lighter carbon isotope values of the stabilized carbonates (e.g. Weissert *et al.* 2008). Since the studied carbonates generally have a low organic carbon content (Heckel 1973), any contribution of isotopically light carbon from the remineralization of organic carbon is excluded. Further, there is no evidence of subaerial exposure and the formation of soil horizons (Brett *et al.* 2011). Consequently, we argue that the general trends in $\delta^{13}C_{carb}$ reflect primary variations. This interpretation is supported by the fact that we observe no co-variation between $\delta^{13}C_{carb}$ and $\delta^{18}O_{carb}$ (Fig. 2), and that the $\delta^{13}C_{carb}$ trends are recorded independently of developed calcareous shale and limestone lithologies.

In order to evaluate whether the trends in the $\delta^{13}C_{carb}$ record of the type region are of local or global significance, or diagenetically induced, we compared the $\delta^{13}C_{carb}$ record of the Appalachian Basin with records of coeval successions with good biostratigraphic control. Aboussalam (2003) reconstructed $\delta^{13}C_{carb}$ records for a number of European and Moroccan successions. However, a number of factors may complicate such comparisons. Foremost, in this study, strata were sampled at a much higher resolution than in previous research, and are more likely to represent the entire range of $\delta^{13}C_{carb}$ values recorded in the rocks. Secondly, although the type region is thought to be one of the most complete successions of the Taghanic interval, the rock record itself is differentially preserved at the various localities because this interval is marked by repeated, abrupt sea-level fluctuations and, therefore, numerous unconformities (Fig. 2) (Aboussalam & Becker 2011; Zambito *et al.* 2012*a*; Turnau 2014). Regardless, a comparison shows some similarities with the $\delta^{13}C_{carb}$ record for the type region and other areas (Fig. 3).

The negative carbon isotope excursion observed in the type region at the onset of the Taghanic Biocrisis is not clearly distinguishable in any other succession: although, at Pic de Bissous, a negative shift occurs within the lower Taghanic Biocrisis interval, and, at Bou Tchrafine, a shift to more negative values prior to the Taghanic Biocrisis is observed (Fig. 3). The negative excursion in New York is centred on the unconformable contact between the siliciclastic Hamilton and the carbonate Tully (Fig. 3). It is possible, although not testable with the current dataset, that the observed negative shift centred on the unconformity represents portions of two superimposed excursions that are fortuitously aligned to make an apparently continuous pattern. The trend towards lower carbon isotope ratios below the unconformity could be explained by subaerial exposure; however, similar negative $\delta^{13}C_{carb}$ values above the unconformity that gradually increase to the apparent baseline cannot be explained by this mechanism given that all strata present have a marine origin. In addition, there is no evidence for subaerial exposure at the unconformity (Brett *et al.* 2011). Alternatively, migration of diagenetic fluid along this stratigraphic contact could have created the observed isotopic profile with lightest values ($\delta^{13}C_{carb}$ values reaching a minimum of almost −1‰) along or near the plane of unconformity; higher values would be symmetrically arrayed both above and below the contact, reflecting greater distance from the diagenetic fluid conduit (unconformable surface). Indeed, the $\delta^{18}O_{carb}$ values do show approximately 2‰ scatter at or near unconformities, although a similar scatter (*c.* 1.5‰) also occurs within stratigraphic intervals where the succession is more conformable and adjacent $\delta^{13}C_{carb}$ values are relatively similar, thereby arguing against such diagenesis.

Reconstruction of a high-resolution carbon isotope profile through the onset of the Taghanic Biocrisis in more proximal settings of the type region, where the Hamilton–Tully transition is more conformable, is necessary to confirm whether this negative carbon isotope excursion represents a primary or secondary trend. However, a negative excursion at this stratigraphic level would be expected given the faunal and lithological changes associated with the Hamilton–Tully transition, interpreted to represent warming and a change in local water-mass circulation from eutrophic estuarine to oligotrophic

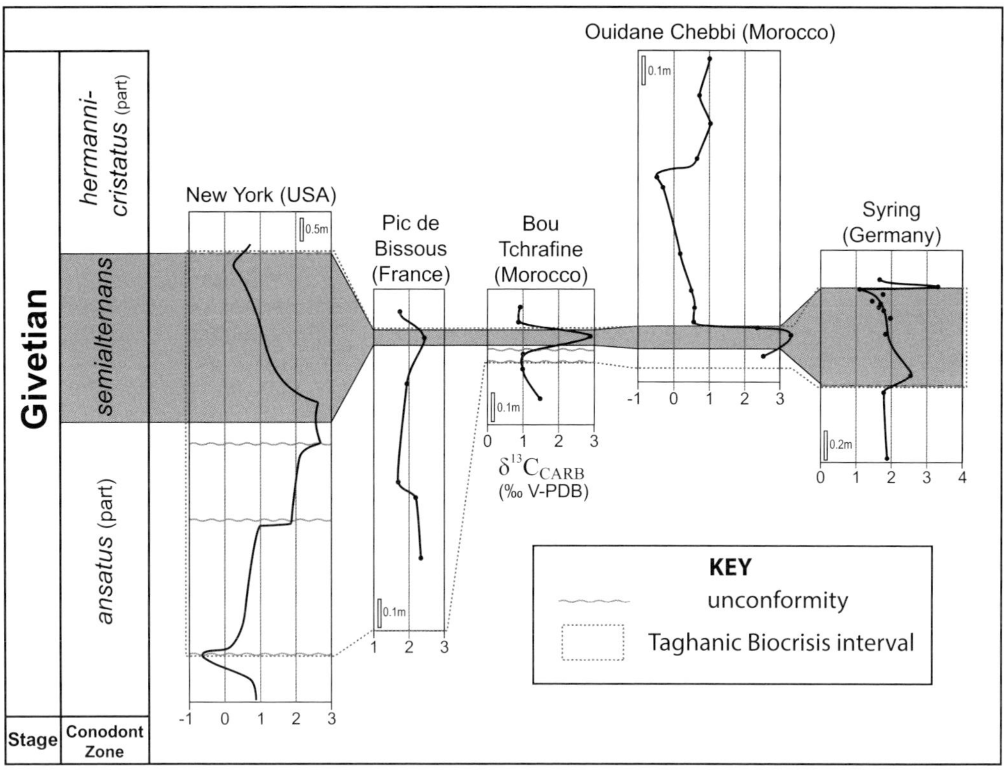

Fig. 3. Biostratigraphic correlation of Taghanic Biocrisis sections with corresponding carbonate $\delta^{13}C$ records. The $\delta^{13}C_{carb}$ record for the New York reconstruction is from this study, all other $\delta^{13}C_{carb}$ records are from Aboussalam (2003). Wavy lines represent unconformities. The dashed box represents the biocrisis interval as recognized at each section. Note that the vertical scale varies.

anti-estuarine type circulation and, therefore, more conducive to carbonate precipitation (Zambito *et al.* 2012*a*). This negative excursion, representing a relative increase in isotopically light carbon in the water column, could have resulted from a collapse in primary productivity associated with oligotrophy.

Although the absolute isotope values and stratigraphic position relative to biostratigraphic boundaries are different between successions, a positive excursion is observed in all $\delta^{13}C_{carb}$ reconstructions near the *ansatus–semialternans* zonal boundary (Fig. 3). In New York, the abrupt onset of more positive values within the uppermost *ansatus* Zone occurs slightly earlier than at other sections relative to the biostratigraphic framework. At New York and Syring (Rheinisches Schiefergebirge), where the *semialternans* Zone intervals are thickest, a similar trend in $\delta^{13}C_{carb}$ values is seen with more positive values observed in the lowermost part of the *semialternans* Zone, and values decreasing throughout this zone. Other successions that also show a positive excursion in this interval include the Ardennes (Belgium: Yans *et al.* 2007), Blauer Bruch (Rheinisches Schiefergebirge, Germany: Buggisch & Joachimski 2006, as discussed in Aboussalam & Becker 2011) and the Iowa Basin (USA.: Day *et al.* 2010; J. Day pers. comm. 2013). The positive $\delta^{13}C_{carb}$ excursion through the *ansatus–semialternans* zonal boundary is apparently a global signal and, therefore, useful for global chemostratigraphic correlation (Fig. 3) (Aboussalam 2003; Aboussalam & Becker 2011).

In regard to the biostratigraphic framework, the timing of this $\delta^{13}C_{carb}$ excursion as reconstructed in different areas seems slightly asynchronous, which can be attributed to the diachroneity in the first appearance of index taxa in different regions. Furthermore, although the *semialternans* Zone is estimated to be around 200 kyr in duration, it is bracketed by two globally recognized unconformities: the first associated with the sequence boundary associated with the Taghanic Onlap at the base of the upper Tully; and the second at the base of the Genesee Group, representing the maximum flooding surface of the Taghanic Onlap transgression (Brett *et al.* 2011 and references therein; Ellwood *et al.* 2011). The relatively early onset of a positive shift in the carbon isotope values in New York compared to other regions (Fig. 3) may, therefore, reflect that middle Tully correlatives are missing from other regions, as this sequence is not recognized in most sections outside the type region (Aboussalam & Becker 2011; Becker *et al.* 2013). Alternatively, the apparent asynchroneity of the onset may be related to differences in stratigraphic resolution at which the carbon isotope record has been reconstructed in the successions compared in Figure 3.

The high-frequency variation in the carbon isotope record through this positive excursion is a feature unique to that interval of the core, as well as the New York succession relative to elsewhere (Fig. 3). Similar to the negative excursion at the base of the Tully, the lighter carbon isotope ratios occur both above and below the unconformity separating the middle and upper Tully. Therefore, migration of diagenetic fluid along this unconformity could be responsible for the (inconsistently) lighter values within the overall positive excursion, given that an approximately 3‰ scatter in $\delta^{18}O_{carb}$ is observed in this interval. In addition, similar to the excursion at the base of the Tully, this positive $\delta^{13}C_{carb}$ excursion is associated with unconformities; although more than one positive excursion may be superimposed at these unconformities, we currently interpret this as a single, pulsed excursion because only one excursion is known from elsewhere (Fig. 3).

In New York, the positive $\delta^{13}C_{carb}$ excursion coincident with the recurrence of the Hamilton Fauna and the Taghanic Onlap is further notable in that the first abrupt positive shift occurs just below the lower–middle Tully unconformity and the second part of this positive shift occurs just below the middle–upper Tully unconformity (Fig. 2). Although there is no direct evidence for a Middle Devonian glaciation, recent studies have interpreted oxygen isotope (Elrick *et al.* 2009) and sequence stratigraphic (Brett *et al.* 2011) patterns during this time as suggestive of glacio-eustasy; the association of positive shifts in the carbon isotope ratio preceding unconformities in the New York succession may be seen in context with palaeoclimatically controlled sea-level change. This globally recognized positive carbon isotope shift is likely to be related to enhanced organic carbon burial in ocean settings that were not preserved (Aboussalam & Becker 2011). Organic carbon burial would have led to atmospheric CO_2 drawdown, which in turn could result in sea-level fall and the development of the observed unconformities if glaciers and/or ice caps existed during the Middle Devonian. At the very least, atmospheric CO_2 drawdown would lead to cooler temperatures coincident with the local extermination of the 'warm-water' Tully Fauna and the recurrence of 'cooler-water' Hamilton taxa (Zambito *et al.* 2012*a*, 2013).

The onset of a positive $\delta^{13}C_{carb}$ excursion just below the unconformable Tully–Genesee contact and corresponding to the *semialternans–hermanni* zonal boundary may be recorded in the uppermost portion of the studied succession. In comparison, a positive $\delta^{13}C_{carb}$ excursion is also observed at the base of the *hermanni* Zone at Syring, while the records from Pic de Bissous and Bou Tchrafine (Aboussalam 2003) do not show a correlative excursion at the base of this zone (Fig. 3). However, following a gradual decrease in $\delta^{13}C_{carb}$, an abrupt positive shift in $\delta^{13}C_{carb}$ has been reconstructed at Ouidane Chebbi (Fig. 3). A time-equivalent positive $\delta^{13}C_{carb}$ excursion has also been reported from the Iowa Basin (Day *et al.* 2010; J. Day pers. comm. 2013). The *hermanni* Zone is associated with widespread black shale deposition and, therefore, a positive $\delta^{13}C_{carb}$ excursion would be expected with increased organic matter burial, in particular in New York State where the Tully–Genesee transition represents a shift from predominantly carbonate to black shale. Therefore the change from a decreasing to increasing trend in $\delta^{13}C_{carb}$ may record the onset of a positive excursion, although continued sampling stratigraphically upwards into the Geneseo Shale is necessary to confirm this.

Conclusions

The Taghanic Biocrisis represents a faunal transition in response to climate change, aridity, sea-level fluctuations and dysoxic conditions. Reconstruction of a carbon isotope record in the northern Appalachian Basin, the type region for the Taghanic Biocrisis, documents changes that are coincident with, and in part may explain, faunal changes and the position of unconformities in this succession. In addition, this study provides evidence of a global positive $\delta^{13}C_{carb}$ excursion associated with the Taghanic Onlap. This excursion should prove to be a useful chemostratigraphic correlation tool for Taghanic interval strata, in particular in sections where biostratigraphic data are unavailable. Furthermore, this study highlights the importance of adequate sampling resolution and documentation of unconformities in reconstructing and comparing environmental records from different sedimentary basins.

A. Miller, T. Algeo, D. Meyer and R.T. Becker provided comments on an early draft of this manuscript. Discussions with T. Algeo, J. Day and P. McLaughlin helped to formulate the ideas presented herein. Two anonymous reviewers provided helpful comments that improved this manuscript. The New York State Geological Survey

(New York State Museum) provided access to the Cargill Lansing core.

References

Aboussalam, S. Z. 2003. Das «Taghanic-Event» im höheren Mittel-Devon von West-Europa und Marokko. *Münstersche Forschungen zur Geologie und Paläontologie*, **97**, 1–332.

Aboussalam, S. Z. & Becker, R. T. 2011. The Global Taghanic Biocrisis (Givetian) in the Eastern Anti-Atlas, Morocco. *Palaeogeography, Palaeoclimatology, Palaeoecology*, **304**, 136–164.

Baird, G. C. & Brett, C. E. 2003. Shelf and off-shelf deposits of the Tully Formation in New York and Pennsylvania: faunal incursions, eustasy and tectonics. *Courier Forschungsinstitut Senckenberg*, **242**, 141–156.

Baird, G. C. & Brett, C. E. 2008. Late Givetian Taghanic bioevents in New York State: new discoveries and questions. *Bulletin of Geosciences*, **83**, 357–370, http://doi.org/10.3140/bull.geosci.2008.04.357

Baird, G. C., Zambito, J. & Brett, C. E. 2012. Genesis of unusual lithologies associated with the Late Middle Devonian Taghanic Biocrisis in the type Taghanic succession of New York State and Pennsylvania. *Palaeogeography, Palaeoclimatology, Palaeoecology*, **367/368**, 121–136.

Becker, R. T., Gradstein, F. M. & Hammer, O. 2012. The Devonian Period. *In*: Gradstein, F. M., Ogg, J. G., Schmitz, M. D. & Ogg, G. M. (eds) *The Geological Timescale 2012*, Volume **2**. Elsevier, Amsterdam, 559–601.

Becker, R. T., Abousslam, Z. S., Baider, L., El Hassani, A. & Stichling, S. 2013. The lower and middle Devonian at El Khraouia (Southern Tifilalt). *In*: Becker, R. T., El Hassani, A. & Tahiri, A. (eds) *International Field Symposium. The Devonian and Lower Carboniferous of Northern Gondwana, Field Guidebook.* Document de l' Institut Scientifique, Rabat, Morocco, **27**, 31–40.

Blakey, R. 2014. *NAU Geology, Paleogeographic Reconstructions, 2014*, http://jan.ucc.nau.edu/~rcb7/RCB.html [last accessed 27 February 2014].

Bonelli, J. R., Jr., Bennington, J. B., Brett, C. E. & Miller, A. I. 2006. Testing for faunal stability across a regional biotic transition; quantifying stasis and variation among recurring coral-rich biofacies in the Middle Devonian Appalachian Basin. *Paleobiology*, **32**, 20–37.

Boucot, A. J. 1989 [imprint 1988]. Devonian biogeography; an update. *In*: McMillan, N. J., Embry, A. F. & Glass, D. J. (eds) *Proceedings of the 2nd International Symposium on the Devonian System, Volume III.* Canadian Society of Petroleum Geologists, Memoirs, **14**, 211–227.

Boucot, A. J. & Theron, J. N. 2001. First *Rhipidothyris* (Brachiopoda) from southern Africa: biostratigraphic, paleoecological, biogeographical significance. *Journal of the Czech Geological Society*, **46**, 155–160.

Brand, U., Azmy, K., Jiang, G. & Lee, X. 2008. Global Taghanic and Givetian seawater records: an amelioration of faunal realms, climatic conditions and high levels of atmospheric carbon dioxide. *In*: 2008 Joint Meeting of The Geological Society of America, Soil Science Society of America, American Society of Agronomy, Crop Science Society of America, Gulf Coast Association of Geological Societies with the Gulf Coast Section of SEPM, Paper 198-2. *Geological Society of America Abstracts with Programs*, **40**, (6), 265.

Brett, C. E. & Baird, G. C. 1995. Coordinated stasis and evolutionary ecology of Silurian to Middle Devonian faunas in the Appalachian Basin. *In*: Erwin, D. H. & Anstey, R. L. (eds) *New Approaches to Speciation in the Fossil Record.* Columbia University Press, New York, 285–315.

Brett, C. E., Ivany, L. C. & Schopf, K. M. 1996. Coordinated stasis; an overview. *Palaeogeography, Palaeoclimatology, Palaeoecology*, **127**, 1–20.

Brett, C. E., Baird, G. C., Bartholomew, A. J., DeSantis, M. K. & Ver Straeten, C. A. 2011. Sequence stratigraphy and a revised sea-level curve for the Middle Devonian of eastern North America. *Palaeogeography, Palaeoclimatology, Palaeoecology*, **304**, 21–53.

Buggisch, W. & Joachimski, M. M. 2006. Carbon isotope stratigraphy of the Devonian of Central and Southern Europe. *Palaeogeography, Palaeoclimatology, Palaeoecology*, **240**, 68–88.

Cocks, L. R. M. & Torsvik, T. H. 2011. The Palaeozoic geography of Laurentia and western Laurussia: a stable craton with mobile margins. *Earth-Science Reviews*, **106**, 1–51.

Cramer, B. D. & Saltzman, M. R. 2005. Sequestration of 12C in the deep ocean during the early Wenlock (Silurian) positive carbon isotope excursion. *Palaeogeography, Palaeoclimatology, Palaeoecology*, **219**, 333–349.

Day, J. E., Witzke, B. J., Bunker, B. J., Holmden, C. & Rowe, H. 2010. Epeiric $C13_{carb}$ record from the Middle and Upper Devonian Cedar Valley Group – Iowa Basin of Central North America. *Geological Society of America, Abstracts with Programs*, **42**, 514.

Edwards, C. T. & Saltzman, M. R. 2014. Carbon isotope ($\delta^{13}C_{carb}$) stratigraphy of the Lower-Middle Ordovician (Tremadocian–Darriwilian) in the Great Basin, western United States: implications for global correlation. *Palaeogeography, Palaeoclimatology, Palaeoecology*, **399**, 1–20.

Ellwood, B. B., Tomkin, J. H. et al. 2011. A climate-driven model and development of a floating point time scale for the entire Middle Devonian Givetian Stage: a test using magnetostratigraphy susceptibility as a climate proxy. *Palaeogeography, Palaeoclimatology, Palaeoecology*, **304**, 85–95.

Elrick, M., Berkyová, S., Klapper, G., Sharp, Z., Joachimski, M. & Frýda, J. 2009. Stratigraphic and oxygen isotope evidence for My-scale glaciation driving eustasy in the Early–Middle Devonian greenhouse world. *Palaeogeography, Palaeoclimatology, Palaeoecology*, **276**, 170–181.

Ettensohn, F. R. 2008. The Appalachian foreland basin in the Eastern United States. *In*: Miall, A. D. (ed.) *The Sedimentary Basins of the World.* Elsevier, Amsterdam, 105–179.

Heckel, P. H. 1973. *Nature, Origin, and Significance of the Tully Limestone; An Anomalous Unit in the Catskill*

Delta, Devonian of New York. Geological Society of America, Special Papers, **138**, 1–244.

House, M. R. 1973. An analysis of Devonian goniatite distribution. *Special Papers in Palaeontology*, **12**, 305–317.

House, M. R. 1978. Devonian ammonoids from the Appalachians and their bearing on international zonation and correlation. *Special Papers in Palaeontology*, **21**, 1–70.

House, M. R. 1985. Correlation of mid-Palaeozoic ammonoid evolutionary events with global sedimentary perturbations. *Nature*, **313**, 17–22.

House, M. R. 2002. Strength, timing, setting and cause of mid-Palaeozoic extinctions. *Palaeogeography, Palaeoclimatology, Palaeoecology*, **181**, 5–25.

Joachimski, M. M., Breisig, S., Buggisch, W., Day, J. & van Geldern, R. 2004. Oxygen isotope evolution of biogenic calcite and apatite during the Middle and Late Devonian. *International Journal of Earth Sciences*, **93**, 542–553.

Joachimski, M., Breisig, S. et al. 2009. Devonian climate and reef evolution: insights from oxygen isotopes in apatite. *Earth and Planetary Science Letters*, **284**, 599–609.

Johnson, J. G. 1970. Taghanic onlap and the end of North America Devonian provinciality. *Geological Society of America Bulletin*, **81**, 2077–2105.

Johnson, J. G., Klapper, G. & Sandberg, C. A. 1985. Devonian eustatic fluctuations in Euramerica. *Geological Society of America Bulletin*, **96**, 567–587.

Klapper, G. & Ziegler, W. 1967. Evolutionary development of the *Icriodus latericrescens* Group (Conodonta) in the Devonian of Europe and North America. *Palaeontographica*, **127**, 68–83.

Klapper, G., Feist, R. & House, M. R. 1987. Decision on the boundary stratotype for the Middle/Upper Devonian series boundary. *Episodes*, **10**, 97–101.

Koch, W. F., II. & Boucot, A. J. 1982. Temperature fluctuations in the Devonian Eastern Americas Realm. *Journal of Paleontology*, **56**, 240–243.

Marshall, J. E. A., Brown, J. F. & Astin, T. R. 2011. Recognising the Taghanic Event in the Devonian terrestrial environment and its implications for understanding land-sea interactions. *Palaeogeography, Palaeoclimatology, Palaeoecology*, **304**, 165–183.

McGhee, G. R., Jr. 1996. *The Late Devonian Mass Extinction; The Frasnian/Famennian Crisis*. Columbia University Press, New York.

Oliver, W. A., Jr. 1975. Presidential address: biogeography of Devonian rugose corals. *Journal of Paleontology*, **50**, 365–373.

Rachebeuf, P. R., Girard, C., Lethiers, F., Derycke, C., Herrera, A. & Trompette, R. 2001. Evidence for Givetian stage in the Mauritanian Adrar (West Africa): biostratigraphical data and palaeogeographical implications. *Newsletters on Stratigraphy*, **38**, 141–162.

Sandberg, C. A., Morrow, J. R. & Ziegler, W. 2002. Late Devonian sea-level changes, catastrophic events, and mass extinctions. *In*: Koeberl, C. & MacLeod, K. G. (eds) *Catastrophic Events and Mass Extinctions: Impacts and Beyond*. Geological Society of America, Special Papers, **356**, 473–487.

Sessa, J. 2003. *The dynamics of rapid, asynchronous biotic turnover in the Middle Devonian Appalachian Basin of New York*. MS thesis, University of Cincinnati, Cincinnati, OH.

Turnau, E. 2014. Floral change during the Taghanic Crisis: spore data from the Middle Devonian of northern and south-eastern Poland. *Review of Palaeobotany and Palynology*, **200**, 108–121.

van Geldern, R., Alvarez, F., Day, J., Jansen, U., Joachimski, M. M., Ma, X.-P. & Yolkin, E. A. 2006. Carbon, oxygen and strontium isotope records of Devonian brachiopod shell calcite. *In*: Buggisch, W. (ed.) Evolution of the System Earth in the Late Palaeozoic: Clues from Sedimentary Geochemistry. *Palaeogeography, Palaeoclimatology, Palaeoecology*, **240**, (1–2), 47–67.

Ver Straeten, C. A. 2010. Lessons from the foreland basin: northern Appalachian basin perspectives on the Acadian orogeny. *In*: Tollo, R. P., Bartholomew, M. J., Hibbard, J. P. & Karabinos, P. M. (eds) *From Rodinia to Pangea: The Lithotectonic Record of the Appalachian Region*. Geological Society of America, Memoirs, **206**, 251–282, http://doi.org/10.1130/2010.1206(12)

Walliser, O. H. 1990. How to define 'global bio-events'. *Lecture Notes in Earth Sciences*, **30**, 1–4.

Walliser, O. H. 1996. Global events in the Devonian and Carboniferous. *In*: Walliser, O. H. (ed.) *Global Events and Event Stratigraphy in the Phanerozoic*. Springer, Heidelberg, 225–250.

Weissert, H., Joachimski, M. M. & Sarnthein, M. 2008. Chemostratigraphy. *Newsletters on Stratigraphy*, **42**, 145–179.

Yans, J., Corfield, R. M., Racki, G. & Preat, A. 2007. Evidence for perturbation of the carbon cycle in the Middle Frasnian *punctata* Zone (Late Devonian). *Geological Magazine*, **144**, 263–270.

Zambito, J. J., IV, Baird, G. C., Brett, C. E. & Bartholomew, A. J. 2009. Depositional sequences and paleontology of the Middle–Upper Devonian transition (Genesee Group) at Ithaca, New York: a revised lithostratigraphy for the northern Appalachian Basin. *In*: Over, J. D. (ed.) Studies in Devonian Stratigraphy: Proceedings of the 2007 International Meeting of the Subcommission on Devonian Stratigraphy and IGCP499. *Palaeontographica Americana*, **63**, 49–69.

Zambito, J. J., IV, Brett, C. E. & Baird, G. C. 2012*a*. The Late Middle Devonian (Givetian) Global Taghanic Biocrisis in its type area (northern Appalachian Basin): geologically rapid faunal transitions driven by global and local environmental changes. *In*: Talent, J. A. (ed.) *Earth and Life: Global Biodiversity, Extinction Intervals and Biogeographic Perturbations through Time*. Springer, New York, 677–703.

Zambito, J. J., IV, Brett, C. E., Baird, G. C., Kolbe, S. & Miller, A. 2012*b*. New perspectives on transitions between ecological-evolutionary subunits in the 'type interval' for coordinated stasis. *Paleobiology*, **38**, 664–681.

Zambito, J. J., IV, Brett, C. E., Baird, G. C., Joachimski, M. M. & Over, D. J. 2013. The Middle Devonian Global Taghanic Biocrisis in the type region, northern Appalachian Basin, USA: new insights from paleoecology and stable isotopes. *In*: El Hassani, A., Becker, R. T. & Tahiri, A. (eds) *International Field Symposium. The Devonian and Lower Carboniferous of Northern Gondwana, Abstract Book*. Document de l' Institut Scientifique, Rabat, Morocco, **26**, 132–134.

Ziegler, W., Klapper, G. & Johnson, J. G. 1976. Redefinition and subdivision of the *varcus* Zone (conodonts, Middle-?Upper Devonian) in Europe and North America. *Geologica et Palaeontologica*, **10**, 109–140.

Kellwasser horizons, sea-level changes and brachiopod–coral crises during the late Frasnian in the Namur–Dinant Basin (southern Belgium): a synopsis

BERNARD MOTTEQUIN[1] & EDOUARD POTY[2]*

[1]*Royal Belgian Institute of Natural Sciences, Palaeontology Department, rue Vautier 29, B 1000 Brussels, Belgium*

[2]*Liège University, Animal and Human Palaeontology Unit, Allée du 6 Août, Bât. B18, Sart Tilman, B 4000 Liege 1, Belgium*

**Corresponding author (e-mail: e.poty@ulg.ac.be)*

Abstract: In Belgium, the Lower Kellwasser Event (LKW) corresponds to the relative sea-level maximum of the first ('Aisemont sequence' (AS)) of the two late Frasnian third-order sequences that are recognized here, but the Upper Kellwasser Event (UKW) may have been triggered by a series of tsunamites. The end of the middle Frasnian carbonate platform and reefs is caused by the sea-level drop and emersion of the last middle Frasnian third-order sequence ('Lion sequence') in the Lower *rhenana* Zone. The end of the 'Petit-Mont' mudmound growth during the transgressive (TST) and highstand (HST) systems tracts of the AS was caused by sea-level fall and emersion at the top of this sequence. The coral and brachiopod extinction in the Upper *rhenana* Zone, during the second late Frasnian third-order sequence ('Lambermont sequence' (LS)), is progressive and due to the widespread development of the dysoxic and anoxic facies, before the UKW. Only the LS TST has been identified. No sea-level fall has been recognized in relation to the UKW or near the Frasnian–Famennian boundary. The late Frasnian extinctions are more likely to be related to the decrease in the atmospheric oxygen rate and its impact on marine environments and, to complete, the UKW.

The late Frasnian (Late Devonian) mass extinction is traditionally considered as one of the 'Big Five' of the Phanerozoic, even if other biological events seem to have been more significant than previously thought, such as the Hangenberg biological crisis (Kaiser *et al.* 2011). One of the most striking features of the late Frasnian in basinal facies is the development of two distinct black limestone–shale horizons known in the literature as the Kellwasser horizons (e.g. Schindler 1993; Gereke 2007; Gereke & Schindler 2012), the extent of which is significant. Studies focused on basinal sections have shown that extinctions mainly recorded among invertebrate faunas were linked with these horizons, and they have subsequently been named the Kellwasser events. However, most of the research dedicated to the Frasnian–Famennian boundary in Western Europe is focused on deep-setting sections where the sedimentary record is extremely condensed, but usually well dated by conodonts and goniatites, contrary to their contemporaneous equivalents in ramp and platform settings. However, the historical type area of the Frasnian and Famennian stages, namely the Namur–Dinant Basin in southern Belgium (Fig. 1), offers a unique opportunity to detail the timing and the aftermath of the late Frasnian extinctions on shallow-water biota (brachiopods and corals).

Numerous studies have already dealt with the sedimentological, geochemical and biological changes observed in southern Belgium around the Frasnian–Famennian boundary, notably Vanguestaine *et al.* (1983), Sandberg *et al.* (1988), Streel & Vanguestaine (1989), Bultynck & Martin (1995), Casier & Devleeschouwer (1995), Claeys *et al.* (1996), Muchez *et al.* (1996), Bultynck *et al.* (1998, 2000), Streel *et al.* (2000*a*, *b*), Crick *et al.* (2002), Mottequin (2005, 2008*a*, *b*, *c*), Azmy *et al.* (2012) and Kaiho *et al.* (2013).

The aim of this paper is to document the facies changes that took place at the end of the mid-Frasnian and during the late Frasnian in southern Belgium, on the basis of sedimentary evolution, sequence stratigraphy and palaeontological data (brachiopods and corals), in order to clarify the causes of the late Frasnian extinction events. It is based mainly on the study and revision of sections in the southern margin of the Dinant Synclinorium, the Philippeville Anticlinorium (central part of the Dinant Synclinorium) and the northern area of

From: Becker, R. T., Königshof, P. & Brett, C. E. (eds) 2016. *Devonian Climate, Sea Level and Evolutionary Events*. Geological Society, London, Special Publications, **423**, 235–250.
First published online June 26, 2015, http://doi.org/10.1144/SP423.6

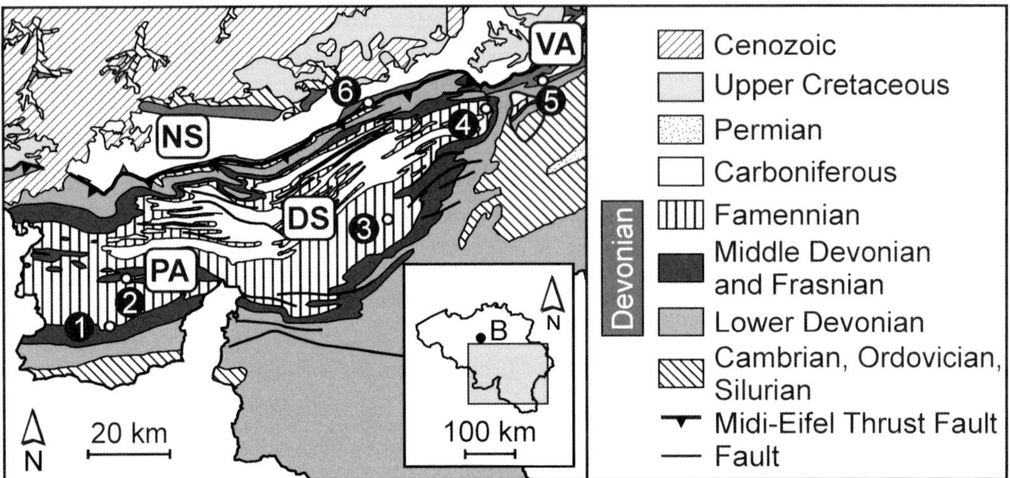

Fig. 1. Schematic geological map of southern Belgium (modified from de Béthune 1954) with the location of important sections: (1) Frasnes; (2) Neuville; (3) Deulin; (4) Hony; (5) Lambermont; (6) La Mallieue (Engis). B, Brussels; DS, Dinant Synclinorium; NS, Namur Synclinorium; PA, Philippeville Anticlinorium; VA, Vesdre area.

the Namur–Dinant Basin (northern part of the Dinant Synclinorium, the Namur Synclinorium and Vesdre area) (Fig. 1).

Geological setting

The Belgian Frasnian is well exposed in the Dinant and Namur synclinoria (see also Belanger *et al.* 2012), the Philippeville Anticlinorium, and the Vesdre area (Fig. 1). These Variscan structural elements constituted the Namur–Dinant Basin, which developed along the southern margin of Laurussia during Devonian time, and lay in the NW part of the Rheno-Hercynian Fold Belt. During the Frasnian, the facies succession reflected a ramp–platform setting with a mixed siliciclastic–carbonate sedimentation and several breaks of slope, as well as the development of carbonate build-up levels in its distal part (southern flank of the Dinant Synclinorium) (e.g. Boulvain *et al.* 2004, 2012) (Fig. 2).

During the early–mid Frasnian, a proximal northern area with limestone-dominant facies developed in opposition with a distal southern, usually deeper area, in which argillaceous facies and reefs predominated. This setting continued during the late Frasnian, but argillaceous deposits became dominant everywhere, with some occurrences of limestone levels mainly in the northern part of the basin and small reddish mudmounds in the southern area. Furthermore, the late Frasnian succession is also characterized by the widespread development of dysoxic–anoxic shaly facies (see below).

Five distinct sedimentation areas are recognized on the basis of their particular late Frasnian lithostratigraphic succession (e.g. Bultynck & Dejonghe 2002), namely the southern, central and northern parts of the Dinant Synclinorium, the Namur Synclinorium and the Vesdre area (Fig. 1).

Stratigraphy

The Frasnian lithostratigraphy of southern Belgium was recently revised by Boulvain *et al.* (1999) and summarized by Bultynck & Dejonghe (2002) (Figs 2 & 3). Biostratigraphy is based mainly on conodonts (e.g. Bultynck *et al.* 1998; Gouwy & Bultynck 2000), but they are often of limited use for accurate correlations due to the rarity of levels with rich and diverse faunas. Late Frasnian brachiopods (e.g. Sartenaer 1968*a*, 1989; Godefroid & Helsen 1998; Mottequin 2005, 2008*a*, *b*, *c*) and corals (e.g. Coen *et al.* 1977; Coen-Aubert 2000, 2012) are particularly well documented and help to clarify the dating. Bultynck & Dejonghe (2002) and Thorez *et al.* (2006) provided an overview of the Famennian lithostratigraphy of the Namur–Dinant Basin (see also Denayer *et al.* 2012). On the southern flank of the Dinant Synclinorium and in the Philippeville Anticlinorium, the lower Famennian succession corresponds to the essentially shaly Senzeille and Mariembourg formations, which were gathered within the Famenne Shale Group by Thorez & Dreesen (1986). Bultynck & Dejonghe (2002) recommended the use of the term Famenne Group to include both formations because the distinction criteria between these units were based primarily on palaeontological markers or thin oolitic ironstone levels, which are

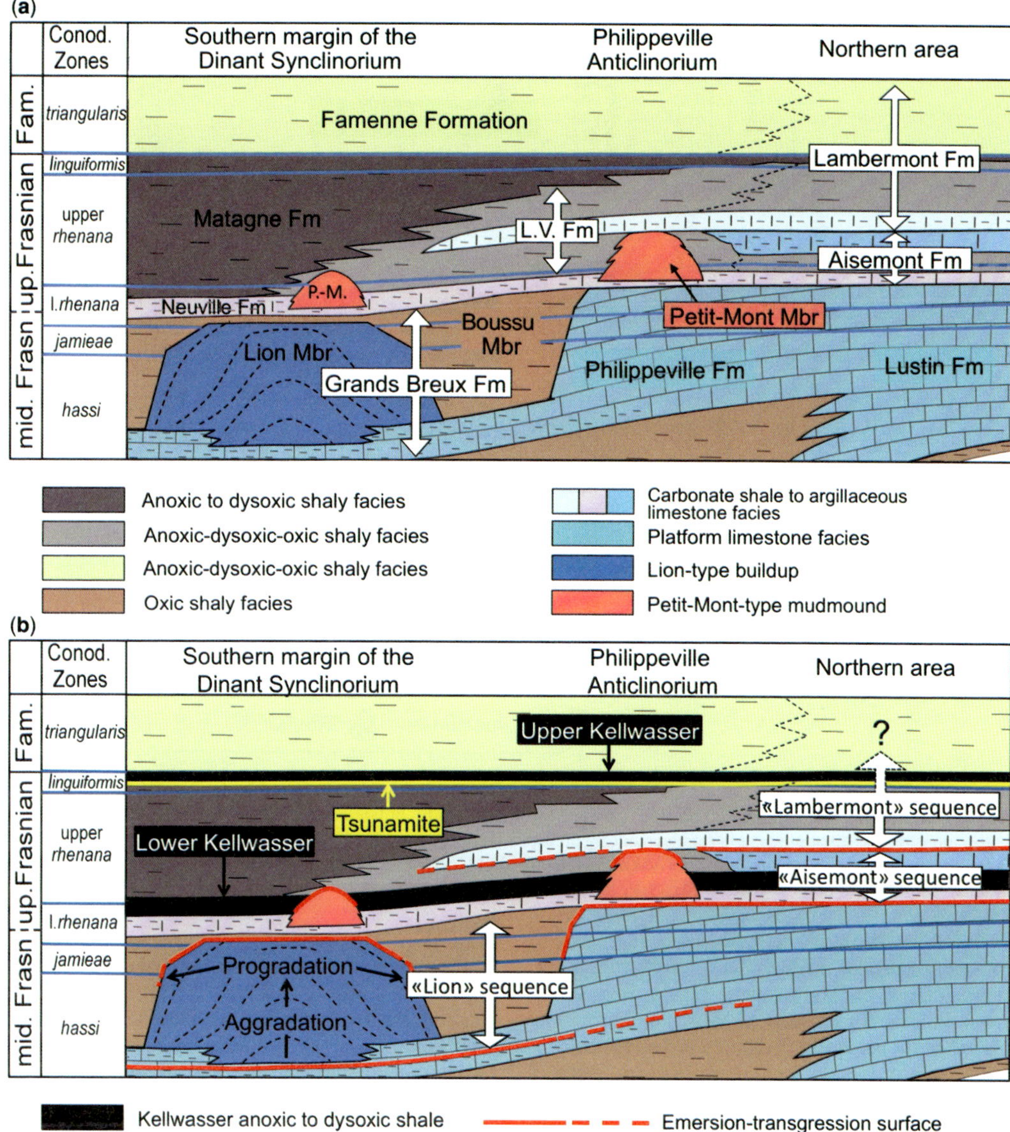

Fig. 2. General sections from the upper part of the middle Frasnian to the lowermost Famennian in southern Belgium: (**a**) bio- and lithostratigraphy; and (**b**) third-order sequences and Kellwasser horizons.

not always easy to highlight. More recently, Thorez *et al.* (2006) have maintained the distinction between the Senzeille and Mariembourg formations. However, we have retained the term Famenne Formation, which is used within the context of the modernization programme of the Geological map of Wallonia (southern Belgium), but this is not yet approved by the National Commission for Stratigraphy in Belgium (Devonian Subcommission).

The extension of the black shale facies during the late Frasnian and correlation with Kellwasser events

Fissile, dark-coloured shales reflecting dysoxic–anoxic conditions are one of the most striking characteristics of the Matagne Formation but are also recognized in parts of its contemporaneous counterparts, namely the Les Valisettes, Barvaux,

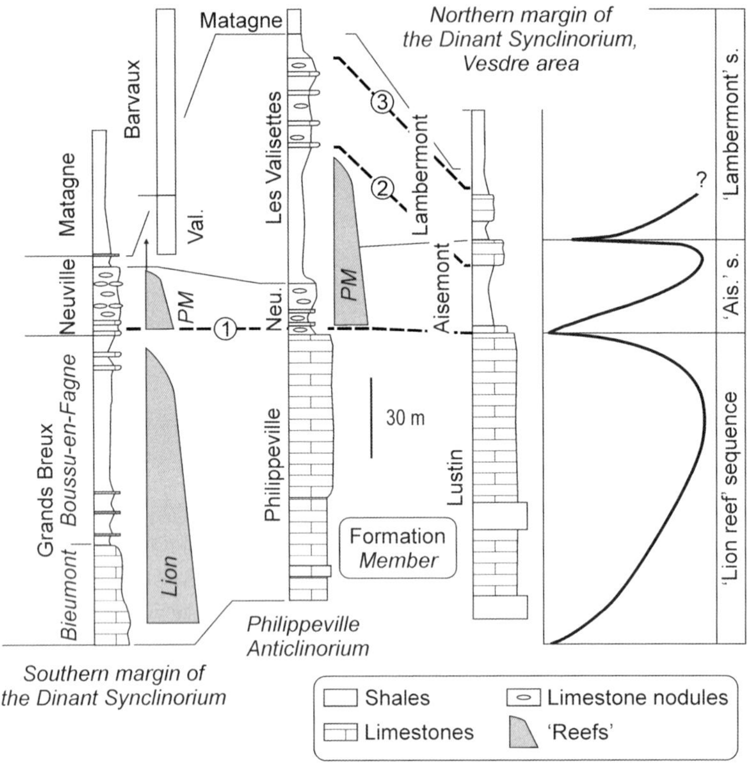

Fig. 3. Mid- to upper Frasnian lithostratigraphy of southern Belgium (modified from Boulvain *et al.* 1999) with sequence stratigraphy and significant changes observed in rugose and tabulate coral faunas: (1) extinction of thamnoporoids and disphyllids, and replacement by phillipsastreids ('fauna 1' of Coen *et al.* 1977; see also Boulvain & Coen-Aubert 1992); (2) appearance of *Iowaphyllum* ('fauna 3' of Coen *et al.* 1977); and (3) extinction of all the colonial and dissepimented solitary rugose corals. Ais., Aisemont; s., sequence.

Aisemont (middle member) and Lambermont formations (Fig. 2). They can be black, as is sometimes the case for those of the Matagne Formation, but are more generally dark-grey, greenish, brownish or purple. Besides nektonic organisms such as goniatites and cricoconarids (Matern 1931; Maillieux 1936; Gatley 1979, 1983; House & Price 1985; House & Kirchgasser 1993), the shales of the Matagne Formation include some levels rich in benthic megafaunas such as pelecypods of the subfamily Buchiolinae (Grimm 1998*a*) and brachiopods, notably the smooth and thin-shelled leiorhynchid *Ryocarhynchus tumidus* (see Sartenaer 1968*a*), which seems to have been adapted to poorly oxygenated environments (Baliński 2002), but also productids (Chonetidina) (Mottequin 2005) and lingulids (Grimm 1998*b*). Some rhynchonellids, such as leiorhynchid (e.g. *R. tumidus*) and pugnacid representatives, are generally the most common elements of Palaeozoic dysaerobic communitites (e.g. Bowen *et al.* 1974; Racki 1989; Alexander 1994; Mottequin & Legrand-Blain 2010). When the oxygenation conditions were better than those prevailing during the deposition of the Matagne shales but still low, benthic faunas were more diversified and included notably athyridid, strophomenid and spiriferid brachiopods (Mottequin 2004, 2005, 2008*a*, *b*), and small, solitary rugose corals.

The LKW and UKW events, occurring respectively within the Upper *Palmatolepis rhenana* Zone (near the base of the MN 13a Zone *sensu* Girard *et al.* 2005) and the *Palmatolepis linguiformis* Zone (MN 13b Zone), correspond to two distinct rises of anoxic waters onto the platform (Schindler 1993) that were considered by Joachimski & Buggisch (1993) as having developed during highstand periods. In the Namur–Dinant Basin, the LKW is placed in the lowermost part of the Matagne Formation on the southern border of the Dinant Synclinorium; in the Philippeville Anticlinorium, however, it is situated in the basal part of the Les Valisettes Formation (Bultynck *et al.* 1998). On the SE margin of the Dinant Synclinorium, the fissile and azoic shales of the middle part of the Les Valisettes Formation could correspond to the LKW (Mottequin 2008*b*). Poty & Chevalier (2007) correlated the middle part of the shaly member of the Aisemont Formation with the LKW (see also Bultynck *et al.* 1998; Gouwy & Bultynck 2000). In the Philippeville Anticlinorium, the Matagne Formation, the thickness of which does not exceed 10 m, corresponds to the Upper Kellwasser Event (UKW) according to Bultynck *et al.* (1998). On the northern margin of the Dinant Synclinorium (Hony section), the 2 m-thick black shale horizon, occurring at the top of the Frasnian part of the Lambermont Formation and directly overlying the tsunamite bed (see below), was correlated with the UKW by Herbosch *et al.* (1996). In the Vesdre area (Lambermont section), the black shale horizon is thicker (10 m thick), but only its 2 m-thick uppermost part, which also overlies the same tsunamite bed, is considered here as corresponding to the UKW *sensu stricto*; the underlying black shale is regarded as corresponding to a previous local input of anoxic water as it is commonly observed in the Namur–Dinant Basin.

The UKW is widespread in the Belgian Basin and developed even in places where anoxic shales were previously absent, as those of the LKW, and this particular shaly facies is developed at the Frasnian–Famennian boundary. Its base is marked everywhere by a 0.15–1 m-thick limestone bed with shaly intercalations that was interpreted by Sandberg *et al.* (1988, bed 48t in their fig. 9) as being related to a eustatic fall, but which pertain to a series of at least seven tsunamites (Figs 2 & 4) (Poty *et al.* 2014) that triggered the input of anoxic–dysoxic waters in previously oxygenated environments. The third event caused the last extinction of the Upper Frasnian but, almost everywhere in southern Belgium, most benthic groups were already decimated at that time (see below).

In the proximal areas, apart from shales corresponding to the LKW and UKW, other levels of anoxic and dysoxic shales are commonly developed; in the most distal part of the basin (southern margin of the Dinant Synclorium), however, anoxic shales are known from the base of the Lower *rhenana* Zone to the Lower *Palmatolepis triangularis* Zone (i.e. from the base of the Matagne Formation to that of the Famenne Formation), preventing the development of benthic fauna for most of the time.

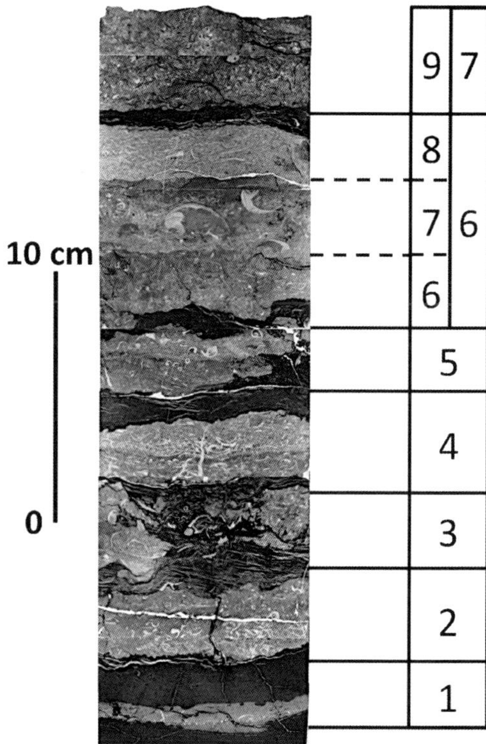

Fig. 4. View in thin section of the bed at the base of the Upper Kellwasser in the Deulin section, showing a series of limestone layers with shaly interbeds. Limestone layers correspond to shallow-water wackestone–packstone with reworked brachiopod, gastropod and ostracod shells. They are interpreted as the backwash deposit of at least seven tsunamis (limestone level 6 comprises three distinct layers that might correspond to three backcurrents without shaly intercalations).

Sequence stratigraphy

Muchez *et al.* (1996) proposed a sequence stratigraphic interpretation for the upper Frasnian and the lowermost Famennian of Belgium but this was flawed owing to incorrect conodont zonation data. A model of bathymetric evolution of the middle and upper Frasnian bioconstructions in relation to eustatic variations was proposed by Boulvain & Herbosch (1996) and Boulvain & Coen-Aubert (1997). In this model, the shallow-water limestones developed in the upper part of the Lion-type build-ups and the Petit-Mont mudmounds were considered as corresponding to a lowstand systems tract (LST), and the overlying shaly limestone and shaly facies (belonging, respectively, to the Boussu-en-Fagne Member and Les Valisettes Formation) to transgressive (TST) and highstand systems tracts (HST). This suggests to the authors that the end of the Lion and Petit-Mont build-ups, and, in more proximal areas, the end of the middle Frasnian carbonate platform, was triggered by a TST with more intense relative deepening than previously, with inputs of clays and anoxic waters, corresponding to the change of conditions between the middle and the upper Frasnian. Poty & Chevalier (2007) and Denayer & Poty (2010) argued, on one hand, that the contact between the middle Frasnian carbonate platform and the Aisemont Formation, and, on the other hand, the contacts between the Aisemont Formation and the overlying Lambermont Formation and between the Petit-Mont build-ups and the overlying Les Valisettes Formation were disconformities that correspond to emersion–transgression surfaces, and are thus third-order sequence boundaries (sequence boundary between 'Aisemont' and 'Lambermont' sequences), as previously considered by Muchez *et al.* (1996). It is considered here that the shallow-water limestone facies at the top of the Lion Member corresponds to the falling-stage systems tract (FSST) of a last middle Frasnian sequence ('Lion sequence'), and the overlying argillaceous facies of the Boussu-en-Fagne Member to the TST (LST?) of a sequence correlatable with the 'Aisemont sequence' of Poty & Chevalier (2007). The boundary between them is the same disconformity that corresponds to an erosion–transgression surface, as demonstrated by its sharp and abrasive contact (Fig. 5), rather than that described by Poty & Chevalier (2007) at the top of the Philippeville and Lustin formations.

The first upper Frasnian sequence ('Aisemont sequence' of Poty & Chevalier 2007)

In the Namur–Dinant Basin, the disconformity at the top of the mid-Frasnian carbonate platform (Lustin and Philippeville formations) to the north, and at the top of the Lion-type reefs to the south, could be correlated with the sea-level fall of the boundary between transgressive–regressive (T–R) cycles IIc and IId1 of Johnson *et al.* (1985, 1996), as discussed by Day (1998), but needs further investigation to be correlated with the cycles of Becker *et al.* (2012). This significant sea-level drop marks the first step of the late Frasnian Crisis: the end of the Lion-type reef edification and that of the mid-Frasnian carbonate platform. Such thick carbonate units did not develop in the late Frasnian of the Namur–Dinant Basin, and were replaced by Petit-Mont-type mudmounds in the distal areas and thin argillaceous limestone units, such as the upper member of the Aisemont Formation, in the proximal areas.

In the southern part of the basin, the transgression corresponding to the first upper Frasnian third-order sequence was marked by the development of argillaceous crinoidal limestones and shales rich in corals, in which thamnoporids, alveolitids and massive disphyllids (*Hexagonaria sensu lato*)

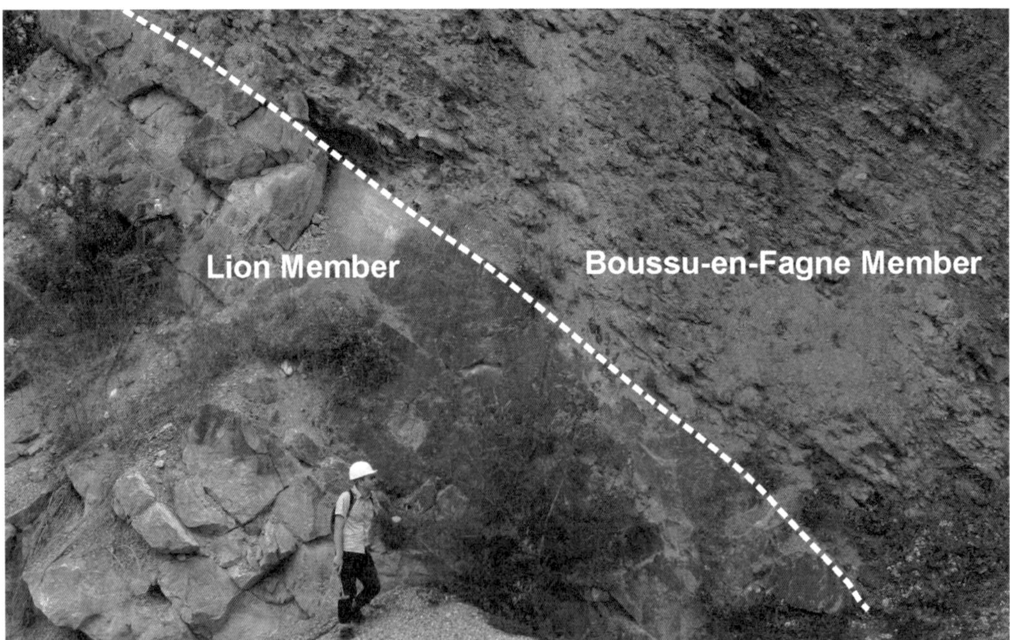

Fig. 5. Clear-cut contact between massive limestones of the Lion Member and the shales of the Boussu-en-Fagne Member (Frasnes, North quarry) corresponding to the boundary between the 'Lion' and the 'Aisemont' sequences.

are dominant. This biofacies is typical of the Boussu-en-Fagne Member (Grands Breux Formation), which not only covers the last reefs, but also includes them. The contact between the Boussu-en-Fagne Member and the Lion-type reefs is sharp, not only at their top but also laterally where there is strictly no indentation with biohermal limestones, which are very pure ($CaCO_3$ >98% according to Poty & Chevalier 2004). This observation strongly suggests that the deposition of this essential member was posterior to the growth and the end of the reefs, namely corresponding to the onset of the 'Aisemont sequence' (Fig. 6).

The Boussu-en-Fagne Member is overlain by the highly nodular shales of the Neuville Formation, which northwards, in the major part of the Philippeville Anticlinorium, directly rests on the limestones of the Philippeville Formation and marks the onset of the '*semichatovae* transgression' of Sandberg *et al.* (1989). The International Subcommission on Devonian Stratigraphy (SDS) has decided to use this level as the base for a formal upper Frasnian substage (Becker 2011). In the Neuville railway section (Fig. 7), the first 5 m of the Neuville Formation show the same rugose coral fauna dominated by the cerioid disphyllid *Hexagonaria*, as observed in the Boussu-en-Fagne Member, but the interval is devoid of thamnoporoids. Disphyllids are replaced by phillipsastreids ('Fauna 1' of Coen *et al.* 1977; see also Boulvain & Coen-Aubert 1992) after a very short interval in which both groups are found together (Coen & Coen-Aubert 1974).

Northwards, the Neuville Formation passes laterally into the lower member of the Aisemont

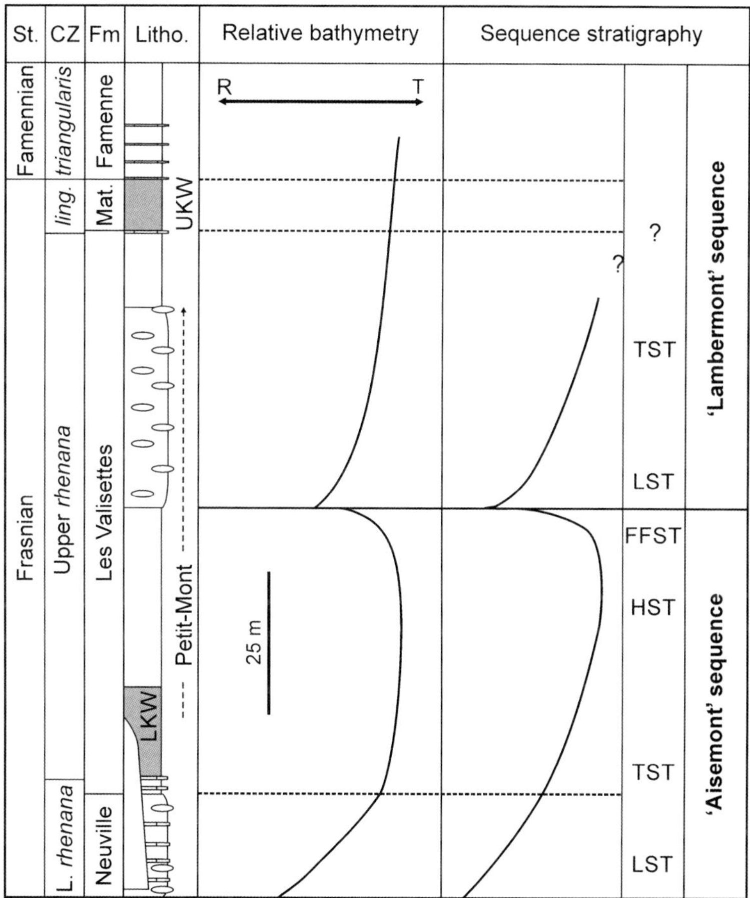

Fig. 6. Sequence stratigraphy of the upper Frasnian based on the Neuville section (Philippeville Anticlinorium). CZ, conodont zones; FFST, falling-stage systems tract; Fm, Formation; HST, highstand systems tract; Litho., lithology (see Fig. 3); LST, lowstand systems tract; St., stratigraphy; R, regression; T, transgression; TST, transgressive systems tract.

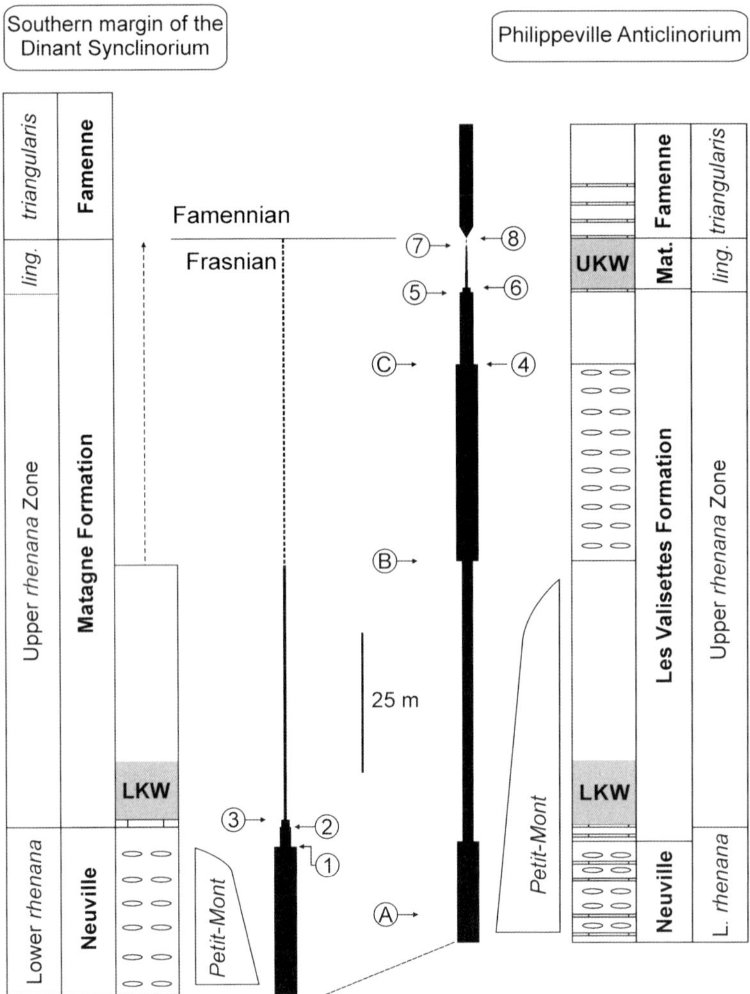

Fig. 7. Comparison between the different extinction and recovery phases observed among the brachiopods from the southern margin of the Dinant Synclinorium and the Philippeville Anticlinorium (Neuville section) (see Fig. 3 for lithology). Conodont data are from Bultynck *et al.* (1998), modified from Mottequin (2005). LKW, Lower Kellwasser; UKW, Upper Kellwasser. Brachiopods: (1) local disappearance of pentamerids; (2) local disappearance of atrypids; (3) first occurrence of *Ryocharhynchus tumidus* (rhynchonellid); (4) last occurrence of pentamerids; (5) last occurrence of atrypids; (6) first occurrence of *R. tumidus* and decimation of the last Frasnian representatives of the strophomenoids (strophomenids), productids, spiriferids, athyridids and orthids, etc.; (7) disappearance of *R. tumidus*; (8) appearance of new spiriferid, athyridid and rhynchonellid species. Corals: (**a**) extinction of thamnoporoids and disphyllids, and replacement by phillipsastreids ('fauna 1' of Coen *et al.* 1977; see also Boulvain & Coen-Aubert 1992); (**b**) appearance of *Iowaphyllum* ('fauna 3' of Coen *et al.* 1977); and (**c**) extinction of all the colonial and dissepimented solitary rugose corals.

Formation (Figs 2 & 3). On the basis of conodont, coral and brachiopod data, Poty & Chevalier (2007) have shown that the base of the TST was diachronic (see also Gouwy & Bultynck 2000; Mottequin 2005) and that there was a gap increasing northwards. Therefore, in the northern part of the Dinant Synclinorium, disphyllids are absent from the late Frasnian and levels with phillipsastreids overlie directly the mid-Frasnian limestones of the Lustin Formation (Figs 2 & 3).

The Aisemont Formation records a transgressive–regressive cycle corresponding to a third-order sequence. According to Denayer & Poty (2010), its lower member and the lower part of its middle member can be assigned to the TST. Its middle shaly member corresponds to the LKW and, instead, records a bathymetric rise, then a lowering, that can be assigned to the upper part of the TST and the lower part of the HST; the maximum flooding surface is situated in the middle of the member. The upper limestone member corresponds to the end of the HST and to the FSST (see Poty & Chevalier 2007; Denayer & Poty 2010 for more details).

The top of the Aisemont Formation is marked by a second disconformity that also corresponds to an erosional transgressive surface after emersion, and it is sharply capped by the shaly Lambermont, Famenne, Falisolle or Franc-Waret formations.

The second upper Frasnian sequence ('Lambermont sequence')

The base of the Lambermont Formation is composed of shales, which are sometimes calcareous and nodular, with brachiopods. A detailed comparison between the Hony section in the northern part of the Dinant Synclinorium, and the Lambermont section (Fig. 1) in the more proximal Vesdre area shows the presence of a gap at the base of the Lambermont Formation in the latter, which is interpreted as resulting from the retrogradation of the TST of the 'Lambermont sequence'. The recognition of this gap is based on four specific levels common to the two sections, corresponding to two brachiopod horizons (Mottequin 2005), a silty tempestite and a limestone bed with shaly intercalations, interpreted as a series of tsunamites by Poty *et al.* (2014), situated just below the UKW (Fig. 4). Therefore, again, the transgression of the second upper Frasnian third-order sequence is well marked and diachronic, as for the 'Aisemont sequence'.

While the TST of this second sequence can easily be defined, the rest of the sequence is not yet recognized because neither bathymetric trend nor emersion can really be recorded, and there is no evidence for a shallowing in the *linguiformis* Zone, as was reported by Sandberg *et al.* (2002), and Streel & Vanguestaine (1989) on the basis of miospores. For example, in the Hony section, the deposits, which are situated below and above the UKW shales, are identical and suggest the same type of environment, except for the UKW underlying tsunamite bed and the overlain tempestite bed, the latter being interpreted by the authors as corresponding to a shallowing. Therefore, the development and the stratigraphic extension of the second upper Frasnian sequence has to be defined in the Belgian succession, but it can be tentatively correlated with the second of the two cycles ('transgressions') present in the major cycle IId of Johnson *et al.* (1985) (see also Day 1998), which extends up to the Middle *triangularis* Zone. In southern Belgium, the UKW Event is not linked to this transgression and corresponds to an event that is independent of this sequence (Poty *et al.* 2014).

The development of the Petit-Mont-type mudmounds

The Petit-Mont mudmounds are considered as developing from the phillipsastreids biozone of the Neuville Formation as no disphyllids are known from the carbonate bodies (Boulvain *et al.* 1999). These mudmounds show indentations with the Neuville Formation and the base of the Les Valisettes Formation, and commonly include argillaceous limestone layers, which suggests that their growth was contemporaneous with the latter. The uppermost part of the mudmounds shows the development of shallow-water carbonates such as cryptalgal bindstones (Boulvain & Coen-Aubert 1992), which can be correlated with the FSST described in the upper part of the upper member of the Aisemont Formation by Denayer & Poty (2010). The top of these mudmounds is marked by a sharp undulating surface that was interpreted, in the Beauchâteau mudmound, as a karstic surface and a sequence boundary by Sandberg *et al.* (1992) and Muchez *et al.* (1996). This emersion is responsible for the end of the edification of the Petit-Mont type mudmounds, and the sea-level drop was correlated with debris-flow deposits occurring within the surrounding deeper facies of the Beauchâteau mudmounds (Muchez *et al.* 1996), but is not yet recorded in the more distal shaly facies of the Les Valisettes and Matagne formations.

In the Neuville section (Figs 1, 6 & 7), the lower part of the Les Valisettes Formation is made up of dark shales devoid of benthic macrofauna, except in the first metres above its base, which yielded brachiopods and small solitary rugose corals. These observations suggest an anoxic environment that can be correlated with the middle member of the Aisemont Formation (TST–lower part of HST) and, probably, with the upper member (upper part of the HST, FSST). Indeed, the upper part of the Les Valisettes Formation is marked by the development of limestone beds and calcareous shales rich in *Frechastraea* and *Iowaphyllum*, the latter being typical of the 'fauna 3' of Coen *et al.* (1977) (i.e. with the lower part of the Lambermont Formation and, therefore, with the TST of the second sequence).

Very small reddish mudmounds, reaching 1–2 m in width and 0.2–0.4 m in height, occur within the uppermost part of the Les Valisettes Formation and are the last bioconstructions known from the Belgian upper Frasnian (Fig. 8). Larger mudmounds, similar to the Petit-Mont mudmounds, may exist but, until now, this possibility was never

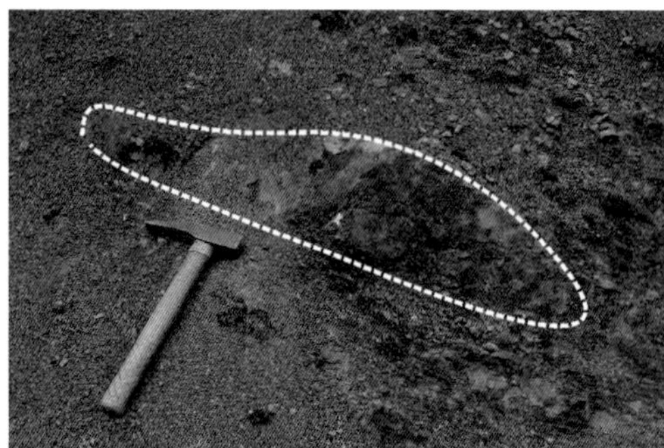

Fig. 8. Small reddish mudmound from the uppermost part of the Les Valisettes Formation (upper Frasnian), about 10 m below the tsunamite bed and the UKW: Neuville railway section (Philippeville Anticlinorium).

biostratigraphically constrained by the record of a coral fauna belonging to the 'fauna 3' or other biostratigraphical markers.

Coral extinctions in the Namur–Dinant Basin

The initial decline of the rugose corals within the Namur–Dinant Basin is recognized in the Lower *rhenana* Zone, and corresponds to the extinction of colonial disphyllids (*Disphyllum*, *Hexagonaria* and *Argutastrea*) and to their replacement by members of phillipsastreids (*Frechastraea* and *Phillipsastrea*: 'fauna 1' of Coen *et al.* 1977) (Figs 3 & 7). This coral turnover is correlated with the onset of the sea-level rise that triggered the TST of the 'Aisemont sequence', and following the fall in sea level that coincided with the top of the Lion Reef Member, and the Philippeville and Lustin formations. It is parallel to the increase in the argillaceous input, characteristic of the late Frasnian in Belgium, and possibly with a reduced oxygenation of the seawater (see the following section on 'Brachiopod extinctions in the Namur–Dinant Basin'). None the less, it was not due to the LKW *sensu stricto*, which occurred later (in the middle part of the Aisemont Formation), and induced strictly no extinction in corals and brachiopods as the corals found in the lower part of the formation are still present in the upper part, with some new species allowing the definition of a rugose coral 'fauna 2' by Coen *et al.* (1977). The arrival of the genus *Iowaphyllum* in the Les Valisettes and Lambermont formations, specifically in the lower part of the 'Lambermont' sequence, allows the definition of a 'fauna 3' (Coen *et al.* 1977). Rugose corals disappeared progressively, along with the tabulates, in the Upper *rhenana* Zone, as dysoxic and anoxic facies developed more locally, so well before the UKW (Fig. 7). The youngest Frasnian rugose corals (*Metriophyllum*) were recovered from the *linguiformis* Zone in the limestone bed at the top of the Les Valisettes Formation (Philippeville Anticlinorium: Denayer *et al.* 2012), corresponding to the 'tsunamites bed' of Poty *et al.* (2014), just below the onset of the black shales of the Matagne Formation (UKW). Until now, no rugose corals were recovered from the Belgian earliest Famennian strata (*triangularis* zones), and the first to reappear after the late Frasnian Crisis are small solitary forms occurring in the Lower *Palmatolepis crepida* Zone and belonging to the so-called *Cyathaxonia* fauna of Hill (1981) (Denayer *et al.* 2012). Accurate recent data related to tabulate corals are not available for the late Frasnian–early Famennian interval, but they also suffered in the late Frasnian Crisis and never recovered a dominant position after (Poty 1999; Poty & Chevalier 2007). *Alveolites* is very common in all the late Frasnian coral assemblages, building biostromes together with phillipsastreids in the lower part of the Aisemont Formation (Poty & Chevalier 2007), whereas thamnoporids were absent from the end of the Boussu-en-Fagne Member. *Alveolites* and uncommon *Scoliopora* disappeared with the last rugose corals and micro-mudmounds at the top of the limestone member of the Les Valisettes Formation. Latest Frasnian and early Famennian taxa belong to primitive and long-ranging genera (*Aulopora* and *Cladochonus*: Denayer *et al.* 2012). As corals, stromatoporoids disappeared at the end of the Frasnian in the Namur–Dinant Basin and only reoccurred during the Strunian (Conil 1961), along with dissepimented rugose corals, before their extinction just below the Devonian–Carboniferous boundary during the Hangenberg Crisis (Poty 1999).

Brachiopod extinctions in the Namur–Dinant Basin

Owing to their diversity and abundance in Frasnian communities, brachiopods are prime tools for evaluating the extinction events related to the late Frasnian biological crisis. Contrary to the cyrtospiriferids – probably the most emblematic Upper Devonian brachiopods, which are marked by rapid and significant diversification phases during the Frasnian and Famennian (e.g. Ma & Day 2003, 2007) – the pentamerids and the atrypids, known, respectively, from the Cambrian (Carlson & Boucot 2002, p. 923, fig. 618) and the Ordovician (Copper 2002, p. 1379, fig. 931), progressively declined and eventually became definitely extinct at the end of this stage (e.g. Copper 1998, 2002; Godefroid *in* Brice *et al.* 2000).

Brachiopods are particularly abundant in the upper Frasnian succession of the Namur–Dinant Basin (e.g. Godefroid & Helsen 1998; Mottequin 2005, 2008*a*, *b*, *c*) and, among them, cyrtospiriferids were especially common in argillaceous facies that prevailed during the late Frasnian. One of the best examples of this kind of facies are the green–purplish shales of the Barvaux Formation (Upper *rhenana* Zone), where clusters of large cyrtospiriferids occur in some levels and dominate a poorly diversified macrofauna comprising other brachiopods (mainly medium-sized strophomenids: see Mottequin 2008*c*), as well as rare tabulate and rugose corals, bivalves, gastropods and trilobites. However, cyrtospiriferids were less frequent in reefal and perireefal environments where those with high ventral area or representatives of the family Adolfiidae were preferentially developed. Cyrtospiriferids from the Barvaux Formation are generally large (up to 10 cm-wide), such as *Cyrtospirifer grabaui*: this may be related to the increase in size of the lophophore, an adaptation to an environment depleted in nutritive elements (Gourvennec 1989) and/or in oxygen. Among the Athyridida, the considerable development of the Helenathyridinae (mainly *Biernatella* and accessor *Neptunathyris*: see Mottequin 2004, 2008*b*) in the upper part of the Neuville Formation on the southern flank of the Dinant Synclinorium is noteworthy. Indeed, these small-sized and smooth athyridids, characterized by a diplospiralium, constitute a large part of the benthos in the distal part of the Namur–Dinant Basin, just before the LKW, but they became extinct by the end of the Frasnian. Furthermore, *Neptunathyris buxi* is also known from the lower part of the Matagne Formation (Upper *rhenana* Zone). Their double-spired spiralia suggests a complex lophophore with reinforced efficiency, permitting the animal to live in poorly oxygenated and/or nutrient-depleted environments (Baliński 1995). In the Neuville railway section, the top of the Neuville Formation recorded the development of a particular assemblage of thin- and smooth-shelled or poorly ornamented rhynchonellids (*Flabellulirostrum* sp. and *Navalicria compacta*) and *Biernatella abunda* that suggest a poorly oxygenated environment.

The brachiopod decline occurred in three major steps within the interval spanning the Lower *rhenana* Zone to the *linguiformis* Zone, which can be subdivided into eight phases (Fig. 7). Most brachiopod orders suffered severely and the major turnover occurred at the top of the Upper *rhenana* Zone in parallel with the severe deterioration of the oxygenation conditions preceding the UKW. Indeed, these extinction episodes were linked principally to diachronous regional facies changes related to local environments: that is, the progressive development of dysaerobic–anaerobic facies, the intensity of which depended on the proximal or distal position along the ramp. For example, atrypids disappeared at the top of the Lower *rhenana* Zone in the deeper part of the basin, just before the deposition of the dark shales of the Matagne Formation, but persisted within the Upper *rhenana* Zone in its shallow parts (e.g. Godefroid & Helsen 1998; Mottequin 2008*a*), thus below the UKW. A similar pattern is also recognized for the pentamerids, which disappeared parallel to the end of the carbonate mudmound edification (Upper *rhenana* Zone), but their extinction predates that of the atrypids (Mottequin 2008*a*). In the Namur–Dinant Basin, the decimation of the strophomenids, orthids, athyridids, spiriferids and spiriferinids (Mottequin 2005, 2008*b*, *c*) took place in two successive phases, as previously reported for the Atrypida by Godefroid & Helsen (1998).

Only an impoverished brachiopod fauna has been recorded from the *linguiformis* Zone (UKW), and is composed notably of the leiorhynchid *Ryocarhynchus tumidus* (see Sartenaer 1968*a*), adapted to poorly oxygenated environments and also known from similar facies in Poland (e.g. Baliński 2002), and also productids (Chonetidina) (Mottequin 2005). These particular poorly diversified brachiopods, the only ones able to live in such dysoxic environments, disappeared during the ultimate extinction phase at the top of the Frasnian (Fig. 7) and can be considered as disaster species. Contrary to corals, the post-extinction brachiopod recovery was rapid in the basal Famennian of southern Belgium but, despite their great abundance, the brachiopod diversity was quite low. At present, only one surviving species (i.e. the athyridid *Cleiothyridina davidsoni* (Lazarus taxon)), is definitely recognized in the *triangularis* zones (early Famennian). None the less, some species of lingulids, craniids and productids may have crossed the Frasnian–Famennian boundary, but this still needs confirmation as the material presently available is too poorly preserved or insufficient (Mottequin 2008*b*). New cosmopolitan genera appeared at the base of the Famennian (*triangularis* zones), especially among the spiriferids, athyridids (*Crinisarina*) and rhynchonellids (see, e.g., Sartenaer 1968*b*, 2001 for more details), concomitantly with new species of pre-existing orthid (*Schizophoria*, *Aulacella*) and orthotetid (*Floweria*) genera (Mottequin 2005, 2008*b*, *c*). Among the spiriferids, the early Famennian is mostly characterized by the development of the Cyrtiopsinae that progressively supplanted the Cyrtospiriferinae. The genus *Aulacella* is sometimes a common element of the epibenthos within ammonoid facies from the Famennian of Morocco and Algeria (Webster *et al.* 2005; Mottequin *et al.* 2015), but also of the Famennian

Annulata events beds, notably in Germany and Iran (Becker 1992; Becker *et al.* 2004). It is interesting to note that *Aulacella* is not recorded within the dark shales of the Matagne Formation as the environment was probably too oxygen-depleted for this genus, which includes some species adapted to deep-water and hypoxic environment.

Conclusions

The upper Frasnian of southern Belgium records two third-order sequences (T–R cycles): the 'Aisemont' and 'Lambermont' sequences. The first follows a significant drop in sea level that was responsible for stopping the development of the middle Frasnian Lion-type reefs and carbonate platform, which subsequently never recovered. It corresponds in the northern area of the Namur–Dinant Basin to the Aisemont Formation, and in the Philippeville Anticlinorium and the southern margin of the Dinant Synclinorium to the upper part of the Boussu-en-Fagne Member, the Neuville Formation and the shaly lower part of the Les Valisettes Formation. In the Aisemont Formation, the Lower Kellwasser Event, as previously stated by Poty & Chevalier (2007), can be correlated with the maximum relative sea level of the sequence corresponding to the shaly middle member: that is, the end of the TST and the lower part of the HST (Denayer & Poty 2010). Southwards, it corresponds to the lower part of the Les Valisettes Formation, but its dysoxic–anoxic facies persisted later, during the HST and FSST of the sequence (i.e. to the base of the limestone member), the latter corresponding to the onset of the second late Frasnian 'Lambermont sequence'. The extinction of middle Frasnian-type rugose corals and their replacement by upper Frasnian-type ones ('fauna 1' of Coen *et al.* 1977) was recorded in the lower part of the Neuville Formation (i.e. during the TST of the Aisemont sequence). The LKW had strictly no effect on their distribution, as stated previously by Poty & Chevalier (2007), except that corals and brachiopods are locally absent during the development of anoxic facies. This is well exemplified by brachiopods, the extinctions of which are clearly related to diachronous regional facies changes caused by transgressions. For example, atrypids and pentamerids disappeared at the top of the Lower *rhenana* Zone in the deeper part of the basin (southern margin of the Dinant Synclinorium), just before the deposition of the dark shales of the Matagne Formation, reflecting poorly oxygenated conditions, but persisted within the Upper *rhenana* Zone in its shallower parts (Philippeville Anticlinorium, northern area of the Namur–Dinant Basin). The reddish Petit-Mont-type mudmounds developed during the TST and HST of the Aisemont sequence, and their growth was stopped by the sea-level drop corresponding to the end of the sequence (FSST), as was the case for the end of Lion-type reefs. Corals are still present during the TST of the 'Lambermont sequence' ('fauna 3'), but only locally when dysoxic–anoxic facies did not develop: for example, in the limestone member of the Les Valisettes Formation, which includes the last, very small, metric, reddish mudmounds of Petit-Mont-type. Until now, it is not yet possible to recognize the HST and FSST of the 'Lambermont sequence', but it has been determined that there was no drop in sea level close to the UKW in Belgium.

The UKW rests on and seems to have been triggered by a series of tsunamites that are recorded throughout the Belgian basin (Poty *et al.* 2014). Similar storm/tsunami deposits have been also reported elsewhere in the same stratigraphic interval (Racki *et al.* 2002; Bond & Wignall 2008; Du *et al.* 2008; Carmichael *et al.* 2014). The UKW is responsible for the last Frasnian extinctions of brachiopods (e.g. cyrtospiriferids), but both last colonial (*Frechastraea*, *Phillipsastrea*, *Iowaphyllum*) and dissepimented solitary (such as *Hankaxis*) rugose corals and alveolitids tabulate corals (*Alveolites*, *Scoliopora*) were already absent.

Therefore, in the Namur–Dinant Basin, it is the increase and expansion of dysoxic–anoxic environments (and therefore the increase in argillaceous sedimentation), most probably in connection with a lowering of the atmospheric oxygen content (see discussion in McGhee 2013), that was responsible for the degradation of environments during the late Frasnian, and for most of the extinctions. The middle Frasnian carbonate platform and the Lion-type reefs had no possibility of recovering after their demise following the last middle Frasnian fall in sea level.

We are greatly indebted to Eberhard Schindler and an anonymous reviewer for their thorough reviews that improved the manuscript. This paper is a contribution to the International Geoscience Programme (IGCP) Project 596 – 'Climate Change and Biodiversity Patterns in the Mid Palaeozoic'.

References

Alexander, R. R. 1994. Distribution of pedicle boring traces and the life habits of Late Paleozoic leiorhynchid brachiopods from dysoxic habitats. *Lethaia*, **27**, 227–234.

Azmy, K., Poty, E. & Mottequin, B. 2012. Biochemostratigraphy of the upper Frasnian in the Namur–Dinant Basin, Belgium: implications for a global Frasnian–Famennian pre-event. *Palaeogeography, Palaeoclimatology, Palaeoecology*, **313–314**, 93–106.

BALIŃSKI, A. 1995. Devonian Athyridoid brachiopods with double spiralia. *Acta Palaeontologica Polonica*, **40**, 129–148.

BALIŃSKI, A. 2002. Frasnian–Famennian brachiopod extinction and recovery in southern Poland. *Acta Palaeontologica Polonica*, **47**, 289–305.

BECKER, R. T. 1992. Zur Kenntnis von Hemberg-Stufe und *Annulata*-Schiefer im Nordsauerland (Oberdevon, Rheinisches Schiefergebirge, GK 4611 Hohenlimburg). *Berliner geowissenschaftliche Abhandlungen, Reihe E*, **3**, 3–41.

BECKER, R. T. 2011. SDS Annual Report 2010 to ICS. *Subcommission on Devonian Stratigraphy, Newsletter*, **26**, 9–13.

BECKER, R. T., ASHOURI, A. R. & YAZDI, M. 2004. The Upper Devonian *Annulata* Event in the Shotori Range (eastern Iran). *Neues Jahrbuch fur Geologie und Paläontologie, Abhandlungen*, **231**, 119–143.

BECKER, R. T., GRADSTEIN, F. M. & HAMMER, O. 2012. The Devonian Period. *In*: GRADSTEIN, F. M., OGG, J. G., SCHMITZ, M. D. & OGG, G. M. (eds) *The Geological Timescale 2012, Volume 2*. Elsevier, Amsterdam, 559–601.

BELANGER, I., DELABY, S. ET AL. 2012. Redéfinition des unités structurales du front varisque utilisées dans le cadre de la nouvelle Carte géologique de Wallonie (Belgique). *Geologica Belgica*, **15**, 169–175.

BÉTHUNE, P. de. 1954. *Carte géologique de Belgique (échelle 1/500.000). Atlas de Belgique*. Académie royale de Belgique, Brussels, Planche 8.

BOND, D. P. G. & WIGNALL, P. B. 2008. The role of sea-level change and marine anoxia in the Frasnian–Famennian (Late Devonian) mass extinction. *Palaeogeography, Palaeoclimatology, Palaeoecology*, **263**, 107–118.

BOULVAIN, F. & COEN-AUBERT, M. 1992. Sédimentologie, diagenèse et stratigraphie des biohermes de marbre rouge de la partie supérieure du Frasnien belge. Compte rendu de la Session extraordinaire des Sociétés géologiques belges les 14 et 15 septembre 1990. *Bulletin de la Société belge de Géologie*, **100**, 3–55.

BOULVAIN, F. & COEN-AUBERT, M. 1997. *Le monticule frasnien de la carrière du nord à Frasnes (Belgique): sédimentologie, stratigraphie séquentielle et coraux*. Geological Survey of Belgium, Professional Papers, **285**.

BOULVAIN, F. & HERBOSCH, A. 1996. Anatomie des monticules micritiques du Frasnien belge et contexte eustatique. *Bulletin de la Société géologique de France*, **167**, 391–398.

BOULVAIN, F., BULTYNCK, P. ET AL. 1999. *Les formations du Frasnien de la Belgique*. Geological Survey of Belgium, Memoirs, **44**.

BOULVAIN, F., CORNET, P. ET AL. 2004. Reconstructing atoll-like mounds from the Frasnian of Belgium. *Facies*, **50**, 313–326.

BOULVAIN, F., COEN-AUBERT, M. ET AL. 2012. Field trip 1: Givetian and Frasnian of Southern Belgium. *Kölner Forum für Geologie und Paläontologie*, **20**, 5–49.

BOWEN, Z. P., RHOADS, D. C. & MCALESTER, A. L. 1974. Marine benthic communities in the Upper Devonian of New York. *Lethaia*, **7**, 93–120.

BRICE, D., CARLS, P., COCKS, L. R. M., COPPER, P., GARCÍA-ALCALDE, J. L., GODEFROID, J. & RACHEBOEUF, P. R. 2000. Brachiopoda. *Courier Forschungsinstitut Senckenberg*, **220**, 65–86.

BULTYNCK, P. & DEJONGHE, L. 2002. Devonian lithostratigraphic units (Belgium). *Geologica Belgica*, **4**, 39–69.

BULTYNCK, P. & MARTIN, F. 1995. Assessment of an old stratotype: the Frasnian/Famennian boundary at Senzeilles, southern Belgium. *Bulletin de l'Institut royal des Sciences naturelles de Belgique, Sciences de la Terre*, **65**, 5–34.

BULTYNCK, P., HELSEN, S. & HAYDUCKIEWICH, J. 1998. Conodont succession and biofacies in upper Frasnian formations (Devonian) from the southern and central parts of the Dinant Synclinorium (Belgium) – (Timing of facies shifting and correlation with late Frasnian events). *Bulletin de l'Institut royal des Sciences naturelles de Belgique, Sciences de la Terre*, **68**, 25–75.

BULTYNCK, P., COEN-AUBERT, M. & GODEFROID, J. 2000. Summary of the state of correlation in the Devonian of the Ardennes (Belgium-NE France) resulting from the decisions of the SDS. *Courier Forschungsinstitut Senckenberg*, **225**, 91–114.

CARLSON, S. J. & BOUCOT, A. J. 2002. Pentamerida. *In*: KAESLER, R. L. (ed.) *Treatise on Invertebrate Paleontology, Part H, Brachiopoda, 4 (Revised)*. Geological Society of America and University of Kansas, Boulder, CO and Lawrence, KS, 921–928.

CARMICHAEL, S. K., WATERS, J. A., SUTTNER, T. J., KIDO, E. & DEREUIL, A. A. 2014. A new model for the Kellwasser Anoxia Events (Late Devonian): shallow water anoxia in an open oceanic setting in the Central Asian Orogenic Belt. *Palaeogeography, Palaeoclimatology, Palaeoecology*, **399**, 394–403.

CASIER, J.-G. & DEVLEESCHOUWER, X. 1995. Arguments (Ostracodes) pour une régression culminant à proximité de la limite Frasnien–Famennien, à Sinsin. *Bulletin de l'Institut royal des Sciences naturelles de Belgique, Sciences de la Terre*, **65**, 51–58.

CLAEYS, P., KYTE, K. T., HERBOSCH, A. & CASIER, J.-G. 1996. Geochemistry of the Frasnian–Famennian boundary in Belgium: mass extinction, anoxic oceans and microtektite layer, but not much iridium? *In*: RYDER, G., FASTOVSKY, D. & GARTNER, S. (eds) *The Cretaceous–Tertiary Event and Other Catastrophes in Earth History*. Geological Society of America, Special Papers, **307**, 491–504.

COEN, M. & COEN-AUBERT, M. 1974. Conodontes et Coraux de la partie supérieure du Frasnien dans la tranchée du chemin de fer de Neuville (Massif de Philippeville, Belgique). *Bulletin de l'Institut royal des Sciences naturelles de Belgique, Sciences de la Terre*, **50**, 1–8.

COEN, M., COEN-AUBERT, M. & CORNET, P. 1977. Distribution et extension stratigraphique des récifs à '*Phillipsastrea*' dans le Frasnien de l'Ardenne. *Annales de la Société géologique du Nord*, **96**, 325–331.

COEN-AUBERT, M. 2000. Annotations to the Devonian correlation table, B142dm00–B142ds00: stratigraphic distribution of the Middle Devonian and Frasnian rugose corals from Belgium. *Senckenbergiana Lethaea*, **80**, 743–745.

Coen-Aubert, M. 2012. New species of *Frechastraea* Scrutton, 1968 at the base of the Late Frasnian in Belgium. *Geologica Belgica*, **15**, 265–272.

Conil, R. 1961. Les gîtes à stromatopores du Strunien de la Belgique. *Mémoires de l'Institut géologique de l'Université de Louvain*, **22**, 335–369.

Copper, P. 1998. Evaluating the Frasnian–Famennian mass extinction: comparing brachiopod faunas. *Acta Palaeontologica Polonica*, **43**, 137–154.

Copper, P. 2002. Atrypida. *In*: Kaesler, R. L. (ed.) *Treatise on Invertebrate Paleontology, Part H, Brachiopoda, 4 (Revised)*. Geological Society of America and University of Kansas, Boulder, CO and Lawrence, KS, 1377–1474.

Crick, R. E., Ellwood, B. B. *et al.* 2002. Magnetostratigraphy susceptibility of the Frasnian/Famennian boundary. *Palaeogeography, Palaeoclimatology, Palaeoclimatology*, **181**, 67–90.

Day, J. 1998. Distribution of latest Givetian–Frasnian Atrypida (Brachiopoda) in central and western North America. *Acta Palaeontologica Polonica*, **43**, 205–240.

Denayer, J. & Poty, E. 2010. Facies and palaeoecology of the upper member of the Aisemont Formation (Late Frasnian, S. Belgium): an unusual episode within the Late Frasnian crisis. *Geologica Belgica*, **13**, 197–212.

Denayer, J., Poty, E., Marion, J.-M. & Mottequin, B. 2012. Lower and middle Famennian (Upper Devonian) rugose corals from southern Belgium and northern France. *Geologica Belgica*, **15**, 273–284.

Du, Y., Gong, Y., Zeng, X., Huang, H., Yang, J., Zhang, Z. & Huang, Z. 2008. Devonian Frasnian–Famennian transitional event deposits of Guangxi, South China and their possible tsunami origin. *Science in China Series D, Earth Sciences*, **51**, 1570–1580.

Gatley, S. S. 1979. Upper Devonian goniatite environments in Belgium. *Proceedings of the Ussher Society*, **4**, 284–296.

Gatley, S. S. 1983. *Frasnian (Upper Devonian) goniatites from southern Belgium*. PhD thesis, Hull University.

Gereke, M. 2007. Die oberdevonische Kellwasser-Krise in der Beckenfazies von Rhenoherzynikum und Saxothuringikum (spätes Frasnium/frühestes Famennium, Deutschland). *Kölner Forum für Geologie und Paläontologie*, **17**, 1–228.

Gereke, M. & Schindler, E. 2012. 'Time-specific facies' and biological crises – The Kellwasser Event interval near the Frasnian/Famennian boundary (Late Devonian). *Palaeogeography, Palaeoclimatology, Palaeoecology*, **367–368**, 19–29.

Girard, C., Klapper, G. & Feist, R. 2005. Subdivision of the terminal Frasnian *linguiformis* conodont Zone, revision of the correlative interval of Montagne Noire Zone 13, and discussion of stratigraphically significant associated trilobites. *In*: Over, D. J., Morrow, J. R. & Wignall, P. B. (eds) *Understanding Late Devonian and Permian–Triassic Biotic and Climatic Events: Towards an Integrated Approach*. Developments in Palaeontology and Stratigraphy Series, **20**. Elsevier, Amsterdam, 181–198.

Godefroid, J. & Helsen, S. 1998.The last Frasnian Atrypida (Brachiopoda) in southern Belgium. *Acta Palaeontologica Polonica*, **43**, 241–272.

Gourvennec, R. 1989. Brachiopodes Spiriferida du Dévonien inférieur du Massif Armoricain. systématique: paléobiologie, evolution, biostratigraphie. *Biostratigraphie du Paléozoïque*, **9**, 1–281.

Gouwy, S. & Bultynck, P. 2000. Graphic correlation of Frasnian sections (Upper Devonian) in the Ardennes, Belgium. *Bulletin de l'Institut royal des Sciences naturelles de Belgique, Sciences de la Terre*, **70**, 25–52.

Grimm, M. C. 1998*a*. Frasnian inarticulate Brachiopoda of the Büdesheim Syncline (Eifel/Germany), of the Saxony Vogtland (Germany) and the Ardennes (Belgium and Northern France). *Senckenbergiana lethaea*, **77**, 73–85.

Grimm, M. C. 1998*b*. Systematik und Paläoökologie der Buchiolinae nov.subfam. (Cardiolidae, Arcoida, Lamellibranchiata, Devon). *Schweizerische Paläontologische Abhandlungen*, **118**, 1–135.

Herbosch, A., Claeys, P. & Kyte, F. T. 1996. Etude géochimique de la limite Frasnien–Famennien à Hony (Belgique). *In*: *Abstract Volume of the 17th IAS Regional African–European Meeting of Sedimentology, Sfax (Tunisia), 26–28 March*. International Association of Sedimentologists, Gent, 76.

Hill, D. 1981. Rugosa and tabulata. *In*: Teichert, C. (ed.) *Treatise of Invertebrate Palaeontology, Part F, Coelenterata, Supplement 1*. Geological Society of America and University of Kansas, Boulder, CO and Lawrence, KS.

House, M. R. & Kirchgasser, W. T. 1993. Devonian goniatite biostratigraphy and timing of facies movements in the Frasnian of eastern North America. *In*: Hailwood, E. A. & Kidd, R. B. (eds) *High Resolution Stratigraphy*. Geological Society, London, Special Publications, **70**, 267–292, http://doi.org/10.1144/GSL.SP.1993.070.01.19

House, M. R. & Price, J. D. 1985. New late Devonian genera and species of tornoceratid goniatites. *Palaeontology*, **28**, 159–188.

Joachimski, M. M. & Buggisch, W. 1993. Anoxic events in the late Frasnian – Causes of the Frasnian-Famennian crisis? *Geology*, **21**, 675–678.

Johnson, J. G., Klapper, G. & Sandberg, C. A. 1985. Devonian eustatic fluctuations in Euramerica. *Bulletin of the Geological Society of America*, **96**, 567–587.

Johnson, J. G., Klapper, G. & Elrick, M. 1996. Devonian transgressive–regressive cycles and biostratigraphy, Northern Antelope Range, Nevada: establishment of reference horizons for global cycles. *Palaios*, **11**, 3–14.

Kaiho, K., Yatsu, S., Oba, M., Gorjan, P., Casier, J.-G. & Ikeda, M. 2013. A forest fire and soil erosion event during the Late Devonian mass extinction. *Palaeogeography, Palaeoclimatology, Palaeoecology*, **392**, 272–280.

Kaiser, S. I., Becker, R. T., Steuber, T. & Aboussalam, S. Z. 2011. Climate-controlled mass extinctions, facies, and sea-level changes around the Devonian–Carboniferous boundary in the eastern Anti-Atlas (SE Morocco). *Palaeogeography, Palaeoclimatology, Palaeoecology*, **310**, 340–364.

Ma, X. P. & Day, J. 2003. Revision of selected North American and Eurasian late Devonian (Frasnian) species of *Cyrtospirifer* and *Regelia* (Brachiopoda). *Journal of Paleontology*, **77**, 267–292.

Ma, X. P. & Day, J. 2007. Morphology and revision of Late Devonian (early Famennian) *Cyrtospirifer* (Brachiopoda) and related forms from South China. *Journal of Paleontology*, **81**, 286–311.

Maillieux, E. 1936. La faune des Schistes de Matagne. *Mémoires du Musée d'Histoire naturelle de Belgique*, **77**, 1–74.

Matern, H. 1931. Die Goniatiten-Fauna der Schistes de Matagne in Belgien. *Bulletin du Musée royal d'Histoire naturelle de Belgique*, **7** (13), 1–15.

McGhee, G. R., Jr. 2013. *When the Invasion of Land Failed: The Legacy of the Devonian Extinctions (The Critical Moments and Perspectives in Earth History and Paleobiology)*. Columbia University Press, New York.

Mottequin, B. 2004. The genus *Biernatella* Baliński, 1977 (Brachiopoda) from the late Frasnian of Belgium. *Bulletin de l'Institut royal des Sciences naturelles de Belgique, Sciences de la Terre*, **74**, (Suppl.), 49–58.

Mottequin, B. 2005. *Les Brachiopodes de la transition Frasnien/Famennien dans le Bassin de Namur-Dinant (Belgique). Systématique–Paléoécologie–Biostratigraphie–Extinctions*. PhD thesis, Liège University.

Mottequin, B. 2008*a*. Late middle to late Frasnian Atrypida, Pentamerida, and Terebratulida (Brachiopoda) from southern Belgium. *Geobios*, **41**, 493–513.

Mottequin, B. 2008*b*. New observations on Upper Devonian brachiopods from the Namur–Dinant Basin (Belgium). *Geodiversitas*, **30**, 455–537.

Mottequin, B. 2008*c*. Late middle Frasnian to early Famennian (Late Devonian) strophomenid, orthotetid and athyridid brachiopods from southern Belgium. *Journal of Paleontology*, **82**, 1052–1073.

Mottequin, B. & Legrand-Blain, M. 2010. Late Tournaisian (Carboniferous) brachiopods from Mouydir (Central Sahara, Algeria). *Geological Journal*, **45**, 353–374.

Mottequin, B., Malti, F. Z., Benyoucef, M., Crônier, C., Samar, L., Randon, C. & Brice, D. 2015. Famennian rhynchonellides (Brachiopoda) from deep-water facies of the Ougarta Basin (Saoura Valley, Algeria). *Geological Magazine*, first published online 10 April 2015, http://doi.org/10.1017/S0016756814000697

Muchez, P., Boulvain, F., Dreesen, R. & Hou, H. F. 1996. Sequence stratigraphy of the Frasnian–Famennian transitional strata: a comparison between South China and southern Belgium. *Palaeogeography, Palaeoclimatology, Palaeoecology*, **123**, 289–296.

Poty, E. 1999. Famennian and Tournaisian recoveries of shallow water Rugosa following late Frasnian and late Strunian major crisis, southern Belgium and surrounding areas, Hunan (South China) and the Omolon region (NE Siberia). *Palaeogeography, Palaeoclimatology, Palaeoecology*, **154**, 11–26.

Poty, E. & Chevalier, E. 2004. *L'activité extractive en Wallonie. Situation actuelle et Perspectives*. Ministère de la Région Wallonne, Namur.

Poty, E. & Chevalier, E. 2007. Late Frasnian phillipsastreid biostromes in Belgium. *In*: Alvaro, J. J., Aretz, M., Boulvain, F., Munnecke, A., Vachard, D. & Vennin, E. (eds) *Palaeozoic Reefs and Bioaccumulations: Climatic and Evolutionary Controls*. Geological Society, London, Special Publications, **275**, 143–161, http://doi.org/10.1144/GSL.SP.2007.275.01.10

Poty, E., Denayer, J. & Mottequin, B. 2014. Tsunamis triggered the Late Frasnian Kellwasser extinction event. *In*: Cerdeño, E. (ed.) *The History of Life: A View from the Southern Hemisphere. Abstract Volume of the 4th International Palaeontological Congress, September 28–October 3, CCT-CONICET, Mendoza, Argentina*. International Palaeontological Association, Lawrence, KS, 598.

Racki, G. 1989. Articulate brachiopods and Late Paleozoic dysaerobic biofacies. *Lethaia*, **22**, 148.

Racki, G., Racka, M., Matyja, H. & Devleeschouwer, X. 2002. The Frasnian/Famennian boundary interval in the South Polish–Moravian shelf basins: integrated eventstratigraphical approach. *Palaeogeography, Palaeoclimatology, Palaeoecology*, **181**, 251–297.

Sandberg, C. A., Ziegler, W., Dreesen, R. & Butler, J. L. 1988. Late Frasnian mass extinction: conodont event stratigraphy, global changes, and possible causes. *Courier Forschungsinstitut Senckenberg*, **102**, 263–307.

Sandberg, C. A., Poole, F. G. & Johnson, J. G. 1989. Upper Devonian of western United States. *In*: McMillan, N. J., Embry, A. F. & Glass, D. J. (eds) *Devonian of the World: Proceedings of the 2nd International Symposium on the Devonian System*. Canadian Society of Petroleum Geologists, Memoirs, **14**, 183–220.

Sandberg, C. A., Ziegler, W., Dreesen, R. & Butler, J. L. 1992. Conodont biochronology, biofacies, taxonomy and event stratigraphy around Middle Frasnian Lion Mudmound (F2h), Frasnes, Belgium. *Courier Forschungsinstitut Senckenberg*, **150**, 1–87.

Sandberg, C. A., Morrow, J. R. & Ziegler, W. 2002. Late Devonian sea-level changes, catastrophic events, and mass extinction. *In*: Koeberl, C. & McLeod, K. G. (eds) *Catastrophic Events and Mass Extinctions: Impacts and Beyond*. Geological Society of America, Special Papers, **356**, 473–487.

Sartenaer, P. 1968*a*. De la validité de *Caryorhynchus* Crickmay, C. H. 1952, genre de brachiopode rhynchonellide, et sa présence dans le Frasnien supérieur d'Europe occidentale. *Bulletin de l'Institut royal des Sciences naturelles de Belgique*, **44** (34), 1–21.

Sartenaer, P. 1968*b*. De l'importance stratigraphique des rhynchonelles famenniennes situées sous la Zone à *Ptychomaletoechia omaliusi* (Gosselet, J., 1877). Sixième note: *Paromoeopygma* n. gen. *Bulletin de l'Institut royal des Sciences naturelles de Belgique*, **44** (42), 1–26.

Sartenaer, P. 1989. Deux genres Rhynchonellides nouveaux d'âge frasnien moyen et supérieur, résultant du brisement de *Calvinaria* Stainbrook, 1945. *Bulletin de l'Institut royal des Sciences naturelles de Belgique, Sciences de la Terre*, **59**, 61–77.

Sartenaer, P. 2001. Revision of the rhynchonellid brachiopod genus *Ripidiorhynchus* Sartenaer. *Geologica Belgica*, **3**, 191–213.

Schindler, E. 1993. Event-stratigraphic markers within the Kellwasser Crisis near the Frasnian/Famennian boundary (Upper Devonian) in Germany. *Palaeogeography, Palaeoclimatology, Palaeoecology*, **104**, 115–125.

STREEL, M. & VANGUESTAINE, M. 1989. Palynomorph distribution in a siliclastic layer near the Frasnian/Famennian boundary at two shelf facies localities in Belgium. *Bulletin de la Société belge de Géologie*, **98**, 109–114.

STREEL, M., CAPUTO, M. V., LOBOZIAK, S. & MELO, J. H. G. 2000*a*. Late Frasnian–Famennian climates based on palynomorph analyses and the question of the Late Devonian glaciations. *Earth-Science Reviews*, **52**, 121–173.

STREEL, M., VANGUESTAINE, M., PARDO-TRUJILLO, A. & THOMALLA, E. 2000*b*. The Frasnian-Famennian boundary sections at Hony and Sinsin (Ardenne, Belgium): new interpretation based on quantitative analysis of palynomorphs, sequence stratigraphy and climatic interpretation. *Geologica Belgica*, **3**, 271–283.

THOREZ, J. & DREESEN, J. 1986. A model of a regressive depositional system around the Old Red Continent as exemplified by a field trip in the Upper Famennian 'Psammites du Condroz' in Belgium. *Annales de la Société géologique de Belgique*, **109**, 285–323.

THOREZ, J., DREESEN, R. & STREEL, M. 2006. The Famennian stage in Belgium and neighbouring countries (Avesnois, Northern France, and Aachen area, Western Germany. *Geologica Belgica*, **9**, 27–45.

VANGUESTAINE, M., DECLAIRFAYT, T., ROUHART, A. & SMEESTERS, A. 1983. Zonation par acritarches du Frasnien supérieur-Famennien inférieur dans les bassins de Dinant, Namur, Herve et Campine (Dévonien supérieur de Belgique). *Annales de la Société géologique de Belgique*, **106**, 121–171.

WEBSTER, G. D., BECKER, R. T. & MAPLES, C. G. 2005. Biostratigraphy, paleoecology, and taxonomy of Devonian (Emsian and Famennian) crinoids from southeastern Morocco. *Journal of Paleontology*, **79**, 1052–1071.

The effect of environmental changes on the evolution and extinction of Late Devonian trilobites from the northern Canning Basin, Western Australia

KENNETH J. McNAMARA[1]* & RAIMUND FEIST[2]

[1]*Department of Earth Sciences, University of Cambridge, Downing Street, Cambridge CB2 3EQ, UK*

[2]*Institut des Sciences de l'Evolution, Laboratoire de Paléontologie, Université de Montpellier II, Place E. Bataillon, 34095 Montpellier cedex 5, France*

**Corresponding author (e-mail: kjm47@cam.ac.uk)*

Abstract: The Frasnian–Famennian Virgin Hills Formation represents fore-reef facies deposited as part of the extensive Late Devonian reef system that fringed the SW Kimberley Block in Western Australia. It contains a rich trilobite fauna dominated primarily by proetids and, to a lesser extent, harpetids, phacopids, scutelluids and odontopleurids. To date, 49 taxa have been described, 40 of these being restricted to the Frasnian. Herein five Frasnian taxa are described, three in open nomenclature, and two the new species *Telopeltis intermedia* and *Otarion fugitivum*. Evolutionary trends in the Virgin Hills trilobites are dominated by a reduction in body size and eye size and, to a lesser extent, a reduction in exoskeletal vaulting. Although recording no sedimentological signature, the fauna was strongly affected by the two globally recognized Kellwasser extinction events. The first, at the end of conodont Zone 12, affected taxa at the species and genus level. The second, within Zone 13b, had a much greater impact on the fauna, causing extinctions at the familial and ordinal levels. Evidence is presented to suggest that evolutionary trends in the trilobites during the late Frasnian reflect selection for forms adapted to low nutrient conditions. The two intensive Kellwasser extinction episodes may reflect periodic massive inputs of nutrients from the terrestrial into the shallow-marine environment.

The Frasnian–Famennian boundary marks, arguably, one of the most significant events in the history of the Trilobita. Five orders existed worldwide during the Frasnian – the Corynexochida, Odontopleurida, Harpetida, Phacopida and Proetida – and all had fossil records extending back more than 100 myr (Fortey & Owens 1997). However, only two of them, the Phacopida and Proetida, persisted into the Famennian. In this overview of a recently described Late Devonian trilobite fauna from the northern Canning Basin in Western Australia, we examine not only their patterns of evolution and extinction prior to and following the Frasnian–Famennian crisis, but also investigate why some types of trilobite might have survived while others perished.

The Late Devonian rocks of the northern Canning Basin in Western Australia at the Frasnian–Famennian transition contain by far the richest trilobite faunas of this age, with 49 species in 17 genera having been described, of which 40 species and 13 genera occur in the Frasnian, the remainder in the Famennian (Feist & Becker 1997; McNamara & Feist 2006; Feist & McNamara 2007, 2013; Feist *et al.* 2009; McNamara *et al.* 2009). A further five taxa are described herein. This is more than twice the number of species previously described from these horizons elsewhere (Feist & Schindler 1994). No genera or species are common to the Frasnian and Famennian. This trilobite fauna, which extends through the late Frasnian into the early Famennian, provides a rich source of information for assessing not only the nature of the extinction – such as whether it was one event or multiphase – but also the nature of the evolutionary changes that occurred within these lineages, before and after the extinction events. Certain parallel trends occur in a number of lineages. Their analysis, particularly in conjunction with the nature of the sediments in which they are preserved, has the potential to provide a clue to the changing environmental conditions at this time, particularly those that might have played a significant role not only in the extinction of some of the trilobites, but also of other elements of the marine invertebrate fauna.

As Feist & Schindler (1994) have pointed out, the Late Devonian was a period of diversity crisis for trilobites, particularly during the two Kellwasser events. These two environmental crises adversely affected the late Frasnian faunas of the Rheic and

From: Becker, R. T., Königshof, P. & Brett, C. E. (eds) 2016. *Devonian Climate, Sea Level and Evolutionary Events*. Geological Society, London, Special Publications, **423**, 251–271.
First published online June 10, 2015, http://dx.doi.org/10.1144/SP423.5

Prototethys oceans. In Europe and NW Africa, these crises carry a strong sedimentological signature arising from the development of anoxic conditions. It has been suggested that prolonged anoxia arose from enhanced productivity, with the principal drivers thought to have been sea-level oscillations, tectonic forcing and oceanic stratification (Sandberg *et al.* 1988; Joachimski & Buggisch 1993; Becker & House 1997; Bond *et al.* 2004; Tribovillard *et al.* 2004; Bond & Wignall 2008; Carmichael *et al.* 2014). The Lower and Upper Kellwasser events occurred at the end of conodont Zone 12 and during Zone 13b, respectively. Although they have left a strong sedimentological signature elsewhere, they have not left a clear one in the Canning Basin (Becker *et al.* 1991; George *et al.* 2014).

A stepwise diminution of diversity had been occurring in trilobite faunas since the Mid-Devonian (Taghanic Biocrisis: Aboussalam 2003) at the specific and generic levels (Lerosey-Aubril & Feist 2012). This has been ascribed to adaptation to widespread, uniform environments that developed on outer platforms at low latitudes (Feist 1991). This diversity reduction was exacerbated and accentuated by the two Kellwasser events. In this paper, we examine the extinction patterns of the trilobites to ascertain whether the Kellwasser events had the same effect on faunal diversity at the eastern end of the northern Gondwanan continent as it did in the west.

Prior to the Upper Kellwasser Event, the five trilobite orders underwent significant evolutionary changes. These particularly affected the cephalic morphology, with a number of parallel trends occurring across taxonomic boundaries. These morphological changes have the potential to provide an insight into the changing environmental conditions that existed prior to, and between, the major extinction events, and may assist in shedding light on the factors responsible for the loss of 60% of trilobite orders at the end of the Frasnian and why some orders survived while others did not.

Geological history

From the Napier Range in the NW to the Lawford Range in the SE, Late Devonian sedimentary rocks extend for about 350 km along the southern margin of the Kimberley Block in northern Western Australia (Fig. 1). Developed on the Lennard Shelf on the northern side of the Canning Basin during latest Givetian–Famennian times, these rocks represent one of the best-preserved Palaeozoic reef complexes (Playford & Lowry 1966; Playford 1980, 1981, 1984; Becker *et al.* 1991; George *et al.* 2009, 2013, 2014; Playford *et al.* 2009). The reefs were dominated by stromatoporoids in the Givetian and Frasnian, but replaced by cyanobacterial reefs in the Famennian (Playford 1980; George & Chow 2002; Wood 2004). The rich trilobite fauna is confined to the marginal slope facies of the Virgin Hills Formation. The material used in this study is derived from this formation in the southern Lawford Range, largely from McWhae Ridge (see McNamara *et al.* 2009 for details). The Virgin Hills Formation, within which the trilobites occur, consists of a condensed sequence of thinly bedded and gently dipping, very fine-grained, limestones and calcareous siltstones, many of which are hematite-rich.

Dating of the Virgin Hills Formation is based on both ammonoid and conodont biostratigraphy. The goniatite zones of the Frasnian at McWhae Ridge range from the *Playfordites tripartitus* Zone (Upper Devonian I-I2: see Becker & House 2009) to the disappearance of *Manticoceras* (Upper Devonian I-L2: Becker *et al.* 1991). The principal diagnostic Frasnian ammonoid genera of the Frasnian, *Manticoceras*, *Crickites* and *Beloceras*, became extinct during the *linguiformis* Zone at the end of the Frasnian (Becker *et al.* 1989, 1991).

Conodonts appear to offer a finer-scale zonation for the Late Devonian (Ziegler & Sandberg 1990; Klapper 2007). Ziegler & Sandberg (1990) subdivided the Frasnian into 10 conodont zones, three of which encompass the trilobite-bearing beds at McWhae Ridge: Early *Palmatolepis rhenana* Zone, Late *rhenana* Zone and *Palmatolepis linguiformis* Zone. An alternative scheme was established for the Montagne Noire sequence (Klapper 1989; Girard *et al.* 2005) and has since been adopted for the Canning Basin sequence (Klapper 2007). Klapper & Becker (1999) correlated both zonation schemes based on the German type section for the middle and upper Frasnian 'standard zones'. The three named zones of Ziegler & Sandberg (1990) equate to the upper part of Zone 12, the very latest Zone 12 and Zone 13a, and Zone 13b/c, respectively. The biostratigraphical zonation of the Famennian part of the Virgin Hills Formation is also based on conodonts.

Trilobite evolutionary trends

The oldest part of the Virgin Hills Formation that has yielded trilobites correlates with Frasnian conodont Zone 11 (Klapper 2007). Much of the described trilobite material has been collected from McWhae Ridge, in the SE part of the Lawford Range, where it ranges in age from Zone 12 to Zone 13b. The overlying Famennian extends to the *Palmatolepis rhomboidea* Zone, the earliest trilobites occurring in the Upper *Palmatolepis triangularis* Zone.

Nine species in eight genera of trilobites were collected from Zone 11 further north in the Lawford

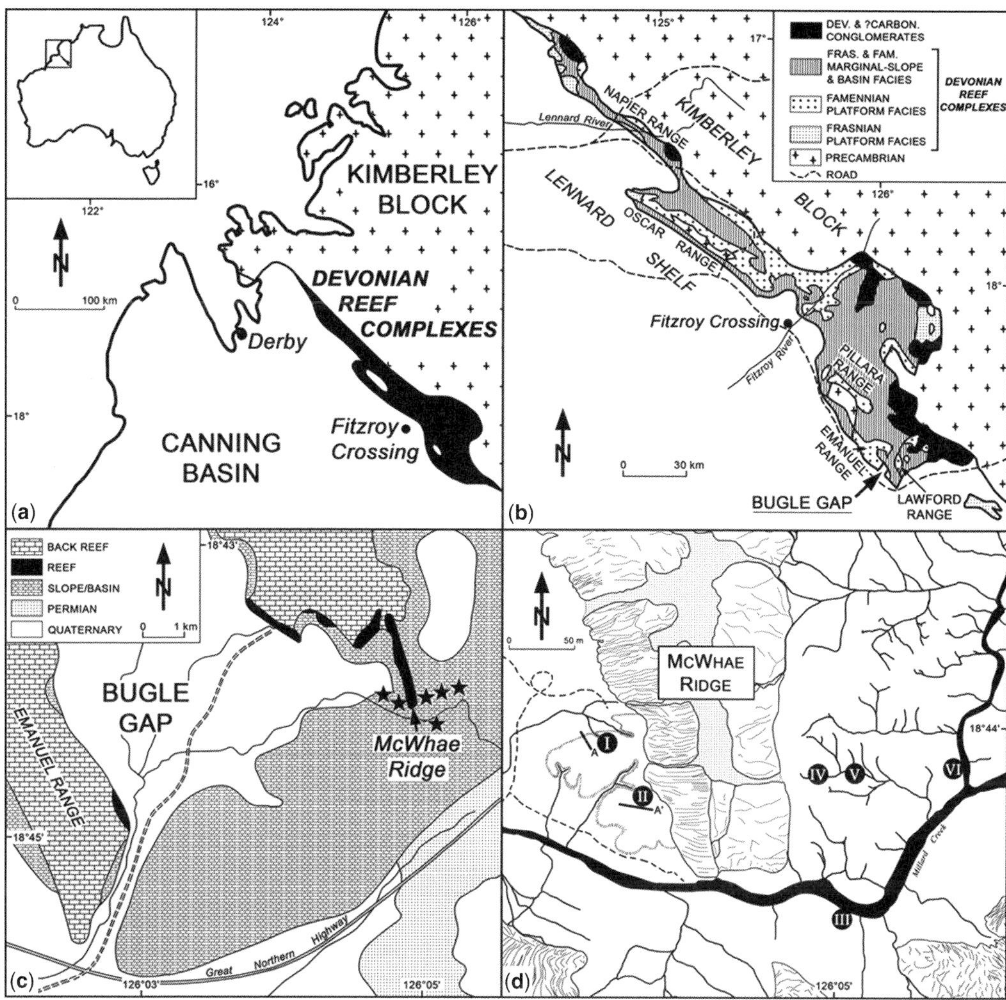

Fig. 1. Maps showing late Frasnian and early Famennian Virgin Hills Formation localities that have yielded phacopids at McWhae Ridge, South Lawford Range, WA. Localities I and II (sections A and A') on the west side of McWhae Ridge comprise predominantly Frasnian, and entirely Famennian strata, respectively. Localities III–VI on the eastern side of McWhae Ridge are all Frasnian. IV and V are the Phacopid Gully locality; VI is Calyx Corner (after Feist *et al.* 2009, fig. 1).

Range (e.g. from the 'Harpid Bed' of Windy Knolls = Section WCB 369: see Becker & House 2009) and the Horse Spring Range (see Feist & McNamara 2013 for details). Three are proetids, of which two are blind; two species each of harpetids and odontopleurids; and single species of scutelluid and a phacopid (Fig. 2). Zone 12 has the most diverse trilobite assemblage, with 17 species in 10 genera. More than half of the species are proetids, with nine forms, of which three – species of *Palpebralina*, *Rudybole* and *Pteroparia* – were blind. This zone is rich in harpetids, with five species in two genera, *Eskoharpes* and *Globoharpes*. Two species of the odontopleurid *Gondwanaspis* and the scutelluid *Telopeltis* make up the rest of the assemblage (Fig. 2).

The most depauperate assemblage occurs in the overlying Zone 13a, with just seven species in six genera. Proetids do not dominate here, having been reduced to just two oculated genera, *Palpebralina* and the tropidocoryphid *Chlupaciparia*. Odontopleurids and phacopids are each represented by two species; one of the phacopids, *Trimerocephaloides*, is blind, the other, *Acuticryphops*, is oculated but with reduced eye size. As with the other zones, corynexochids are represented by a

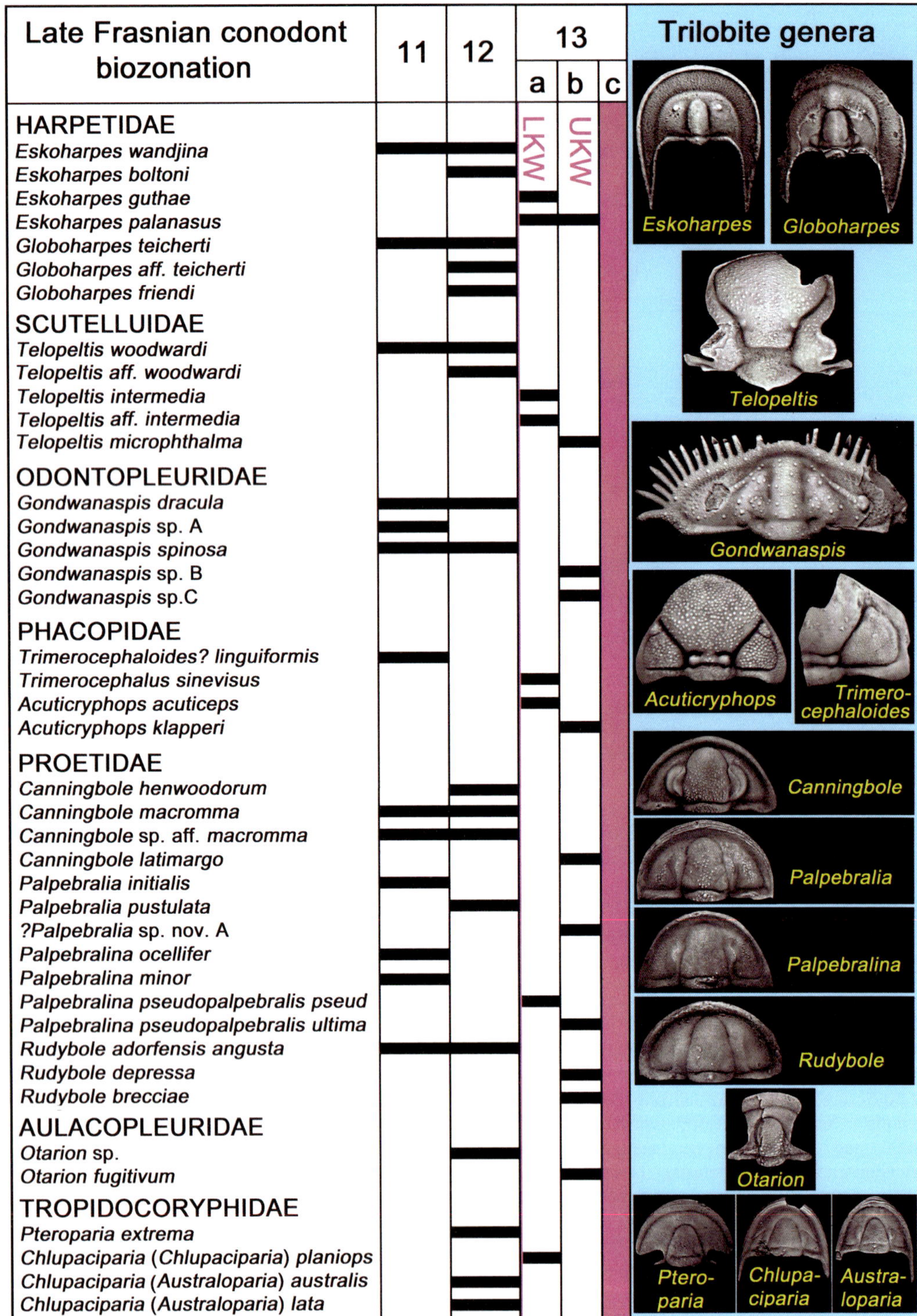

Fig. 2. Ranges of late Frasnian Virgin Hills Formation trilobites and extinctions at Lower (LKW) and Upper (UKW) Kellwasser Event horizons.

single species of the scutelluid *Telopeltis*, a genus characterized by its relatively convex exoskeleton.

The last zone of the Frasnian in which trilobites occur is Zone 13b. But like zones 11 and 12, this zone was dominated by proetids, with five species in four genera. A single phacopid, *Acuticryphops*, persisted from Zone 13a. No trilobites have been recovered from Zone 13c. Species diversity is higher in the zones immediately preceding the two Kellwasser events, demonstrating that, like other elements of the invertebrate fauna, the trilobites were severely affected by the biotic crises.

Corynexochida

This order is represented in the Virgin Hills Formation by the last known member of the order, the scutelluid *Telopeltis*. This genus is endemic to the Canning Basin, where it is one of the more common elements of the trilobite fauna, locally, in Zone 12, forming coquinas. Five forms are recognized, the oldest being *Telopeltis woodwardi* McNamara & Feist, 2006 (zones 11 and 12) and the youngest *T. microphthalma* McNamara & Feist, 2006 (Zone 13b). All species of *Telopeltis* are relatively small for scutelluids, the youngest being appreciably smaller, reaching a maximum cephalic length of only 8.5 mm, compared with 13 mm in *T. woodwardi*.

The most obvious difference between the taxa is in the degree of convexity of the exoskeleton. While *Telopeltis* is, in part, characterized by its very convex exoskeleton, the degree of convexity is variable, being appreciably less in later species (Fig. 3). Such reduction in exoskeletal convexity is not restricted to these scutelluids, occurring in other trilobite orders in the Virgin Hills Formation (see below).

The other marked evolutionary trend apparent in the cephalon involves the eye lobe, which shows an evolutionary reduction in size, being smallest in the youngest species, *T. microphthalma*, the exsagittal length being about half that of its likely older progenitor and of intermediate size in the aptly named *T. intermedia* n. sp. (Fig. 3). Moreover, the eye surface is correspondingly smaller in younger taxa, comprising irregular, curving rows that each contain up to 10 holochroal lenses in *T. microphthalma*, compared with 25 in *T. woodwardi*. Eye reduction is a common evolutionary trend in lineages in some of the other orders.

In many phylogenetically older scutelluids, early ontogenetic growth was typified by a strongly vaulted transitory pygidium that flattened during ontogeny (Chatterton 1971). The phylogenetically derived nature of *Telopeltis* suggests that possession of a strongly vaulted pygidium in adults is by paedomorphosis. However, within the *Telopeltis* clade, the reduction in pygidial convexity suggests less paedomorphic development in later species. The eye reduction is also a paedomorphic trait, the eye lobe size increasing in relative size ontogenetically in older scutelluids.

A further five paedomorphic features are present in the younger species, *T. microphthalma*, and eight in *T. woodwardi*, including glabellar furrow depth and retention of an occipital spine. Compared with most other Devonian scutelluids, this last genus is very small, suggesting that the paedomorphic features may have evolved in response to selection for earlier maturation at a smaller body size, perhaps in a stressed environment (McNamara & Feist 2008, p. 270), as we discuss further below.

Odontopleurida

Odontopleurid trilobites are present in the Frasnian part of the Virgin Hills Formation, but absent from the Famennian. The only genus is *Gondwanaspis*, which represents the last genus of the order Odontopleurida (Fig. 2). This is a geographically widespread genus, having been described from Morocco, Montagne Noire in southern France, the Harz Mountains in Germany, and Rudny Altai, Siberia, as well as from the Canning Basin in Western Australia (Feist & McNamara 2007).

Although they are rare elements of the fauna, five species of *Gondwanaspis* have been described from Frasnian zones 11, 12 and 13b in the Canning Basin (Feist & McNamara 2007). Even at the species level, forms are widely distributed: *G. dracula* having been described both from the Canning Basin and Germany. This final odontopleurid species is characterized by the possession of a wide cephalon with very low convexity, poorly defined glabellar lobes and spines or protuberances on the non-arched anterior margin of the cranidium.

Like the scutelluids, the odontopleurid species are all relatively small (see below for a more detailed discussion), reaching a maximum cephalic length of less than 5 mm. As in a number of lineages in other trilobite orders, the most significant difference between the taxa is in the size of the eye. It is smallest in the oldest species, *Gondwanaspis* sp. A in Zone 11, and largest in the youngest form, *Gondwanaspis* sp. C in Zone 13b. The stratigraphically intermediate species have intermediate-sized eyes. This trend to increase in eye size during the latter part of the Frasnian is in contrast to patterns observed in a number of other lineages documented herein.

Although the ontogenetic development of *Gondwanaspis* is unknown, a number of ontogenetic changes seemed to have been consistent across all odontopleurids. These include the appearance and increase in size of lateral glabellar lobes L1 and L2;

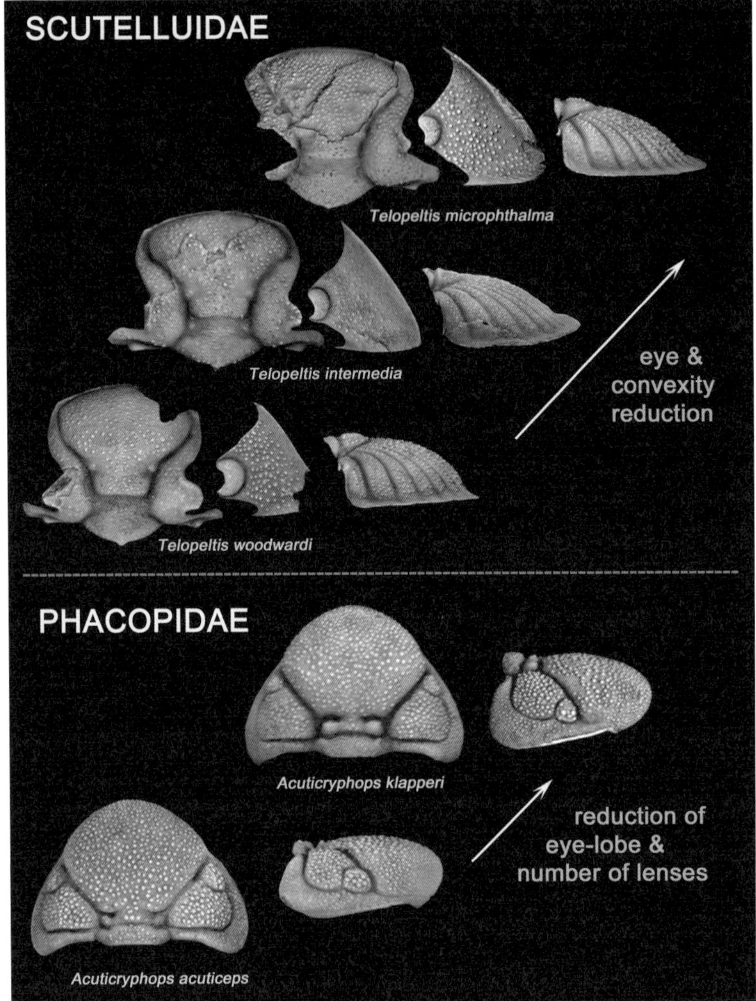

Fig. 3. Evolutionary trends in late Frasnian scutelluid and phacopid trilobites from the Virgin Hills Formation.

the posterior migration of eye lobes, which caused a decrease in the angle of convergence of the eye ridges; movement of the eye lobes closer to the glabella, resulting in a narrowing of the fixigena; an increase in tubercle concentration as, except for the occipital spine, the primary spines disappeared; anterior expansion of the glabella as the anterior margin became convex forward; and a reduction in occipital spine size (Feist & McNamara 2007).

If, as seems likely, *Gondwanaspis* evolved with a similar suite of ontogenetic changes, then many of the distinctive morphological features of the cephalon could be interpreted as paedomorphic. These include: possession of weakly developed lateral glabellar lobes; anteriorly positioned eye lobes; paucity of tubercles in many species; and transversely truncated glabella frontal lobe. However, the absence of an occipital spine is a peramorphic feature.

Phacopida

Phacopids are present in both the Frasnian and Famennian parts of the Virgin Hills Formation, although none are present in Zone 12. They occur only rarely prior to the Lower Kellwasser Event, being represented by rare specimens referred questioningly to *Trimerocephaloides*, in conodont Zone 11. Another species, *T. sinevisus* Feist, McNamara, Crônier & Lerosey-Aubril, 2009, which lacks eyes and palpebral lobes, occurs in Zone 13a. In lacking eyes, this form is reminiscent of *Trimerocephalus*, which is present in the Famennian Virgin Hills Formation. *Trimerocephaloides* is the only known blind Frasnian phacopid.

Occulated phacopids are unknown prior to the Lower Kellwasser Event at the end of Zone 12, even though Zone 12 strata contain a rich trilobite fauna. Immediately after the Lower Kellwasser Event, in conodont Zone 13a, *Acuticryphops acuticeps* (Kayser, 1889) becomes a common element of the trilobite fauna prior to the Upper Kellwasser Event (Feist *et al.* 2009). This species is also known from an equivalent horizon in the Montagne Noire in France and central Morocco. Sedimentologically, Zone 13a strata at McWhae Ridge are quite different from those of zones 12 and 13b in lacking any hematite. An evolutionary trend of paedomorphic reduction in eye lens number has been well documented in *Acuticryphops* in the French and Moroccan sequences (Feist 1995; Crônier *et al.* 2004; Feist *et al.* 2009), and also occurs in the Canning Basin, indicating that global influences were affecting the selection of forms with reduced eyes. However, whereas in the French and Moroccan sections, the reduction occurs intraspecifically, it is an interspecific event in the Canning Basin, with the smaller *A. klapperi* Feist, McNamara, Crônier & Lerosey-Aubril, 2009, which is restricted to conodont Zone 13b, possessing a reduced number of lenses in each eye and also having differences in glabellar characters (Fig. 3). *A. acuticeps* that occurs in the Virgin Hills Formation has between six and 13 lenses in each eye. In *A. klapperi*, it was reduced to between three and six lenses.

Phacopids show a higher level of biodiversity in the Famennian, although being relatively scarce elements of the fauna. Seven species in three genera, *Houseops*, *Babinops* and *Trimerocephalus*, have been described (Feist & Becker 1997; Feist *et al.* 2009). None have been found immediately following the mass-extinction event, the first occurring in the Upper *triangularis* Zone. In European sections, only blind phacopids occur in the earliest Famennian (i.e. *Nephranops* in the Middle *triangularis* Zone; compare Becker & Schreiber 1994). However, in the Canning Basin, initial recovery following the mass-extinction event in the Upper *triangularis* Zone perireefal environments was characterized by oculated forms. Presumably, these trilobites evolved from conservative ancestors with normal eyes that had survived the Kellwasser biocrises in reef-related shallow-water niches or were immigrants from outside the basin.

The discovery of these early Famennian occulated species of *Houseops* has allowed the origin of post-extinction event phacopids from shallow-water environments to be demonstrated for the first time. Unlike the trend in eye reduction seen in the late Frasnian phacopid *Acuticryphops*, the evolving *Houseops* lineage shows an increase in eye size and in lens number, from the stratigraphically oldest *H.* sp. A with 25 lenses, to *H. beckeri* Feist, McNamara, Crônier & Lerosey-Aubril, 2009 with 44 and to the youngest, *H. canningensis* Feist, McNamara, Crônier & Lerosey-Aubril, 2009, with 51 (Feist *et al.* 2009). Another evolutionary trend in species of *Houseops* is an increase in the size of the cephalon, from a maximum length of 4.1 mm in the oldest species to 9.6 mm in the youngest. This suggests that the peramorphic increase in lens numbers might have occurred by hypermorphosis.

Species of *Babinops* that occur in the Upper *crepida* and *rhomboidea* zones have even more lenses in their eyes, with 61 in *B. minor* Feist, McNamara, Crônier & Lerosey-Aubril, 2009 and 81 in *B. planiventer* Feist & Becker, 1997.

Harpetida

Two genera of Harpetida, *Eskoharpes* and *Globoharpes*, occur in the Virgin Hills Formation. They are confined to the Frasnian, with four species of *Eskoharpes* ranging through conodont zones 11–13b (McNamara *et al.* 2009). *Eskoharpes palanasus* McNamara, Feist & Ebach, 2009 from conodont Zone 13b represents the last known harpetid trilobite. The earliest, pre-Lower Kellwasser species of *Eskoharpes* (those occurring in zones 11 and 12), *E. wandjina* McNamara, Feist & Ebach, 2009 and *E. boltoni* McNamara, Feist & Ebach, 2009, share a number of features that distinguish them significantly from post-Lower Kellwasser (i.e. zones 13a and 13b) species. The older species each possess a highly vaulted cephalon with a narrow brim, and thus a strongly convex, steep genal roll. The result is that in each species the glabella is set high above the level of the low-angled, narrow brim. This vaulting is most accentuated in the oldest species, *E. wandjina*, on account of its more strongly convex glabella. Moreover, the genal areas are also more strongly swollen, resulting in the eye lobes being raised much higher above the brim than in any other species of *Eskoharpes*.

The two species that occur in zones 13a and 13b, *E. guthae* McNamara, Feist & Ebach, 2009 and *E. palanasus*, continue the evolutionary trend of reduction in cephalic vaulting, both possessing a relatively weakly convex glabella and genal areas (Fig. 4). They also evolved a relatively wider brim than the earlier species. This enhanced the longer, flatter cephalic shape, in contrast to the short, highly vaulted cephalon of the older species. The brim occupied about 25% of the cephalic length in the oldest species, *E. wandjina* (zones 11 and 12), increasing to 27–28% in *E. boltoni* (Zone 12), 31% in *E. guthae* (Zone 13a), finally up to 33–34% in the youngest species, *E. palanasus* (zones 13a and 13b).

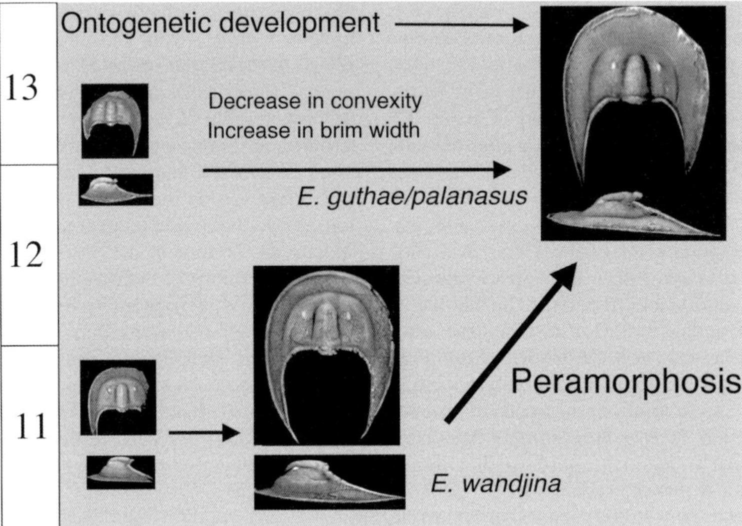

Fig. 4. Peramorphic evolution of the brim in late Frasnian *Eskoharpes* in the Virgin Hills Formation. Note also the flattening of the cephalon in descendant adults.

The reduction in cephalic convexity is a feature of a number of the late Frasnian trilobites in the Virgin Hills Formation, including the scutelluine *Telopeltis* and species of the odontopleurid *Gondwanaspis*. The functional significance of this exoskeletal flattening in a number of unrelated lineages is not clear. However, Feist & Clarkson (1989) suggested that in tropidocoryphines a similar reduction in cephalic vaulting may have been associated with the evolution to an endobenthic lifestyle. In the case of the harpetids, changes in cephalic convexity may be related to developmental constraints linking variations in glabellar and genal convexity to brim width. Jell (1978) has pointed out that Cambrian trilobites with more extensive caeca also tend to possess flatter exoskeletons; conversely, those more convex forms have less well-developed caeca. A similar pattern is evident in *Eskoharpes*, in that later, post-Lower Kellwasser forms possess flatter cephala and wider brims that bear more extensive caeca.

The increase in the width of the brim and the corresponding expansion of the caecal fields that cover them may provide an insight into some of the environmental changes that were occurring before and after the Lower Kellwasser Event. There has been much debate over the function of the pitted brim of harpetid trilobites. While most researchers in the past have argued that the pits served either a sensory function or played a role in food gathering, Přibyl & Vaněk (1986) elaborated on this concept by suggesting that the main functional significance lay with the intervening caeca, which comprise a complex system of narrow, anastomosing ridges rather than the pits. They noted the fundamental similarity between the anastomosing interpit structures in harpetids and the caeca present in many other trilobites.

Chatterton (1980) and McNamara *et al.* (2009) have considered that the functional significance of the brim, and to some extent the genal roll, had less to do with the pits, but more with the intervening anastomosing ridges between these pits, functioning as a complex series of caeca. McNamara *et al.* (2009) have pointed out how the earliest harpetids, such as *Harpides*, have clearly defined caeca, with only small depressions or, in some cases, poorly developed, minute pits between them that extended through the upper and lower lamellae. As the Harpetida evolved, so the caeca became more and more separated from each other as the pits enlarged.

It has been suggested that genal caeca may have served a respiratory function in trilobites (Jell 1978), the caeca effectively increasing the surface area of the exoskeleton. Jell (1978) suggested that the anastomosing tubes carried blood, and oxygen would have been absorbed through the cuticle. Consequently, the more well developed the circulatory caeca, so the greater the resultant surface area, and the more effective its respiratory function, especially in low-oxygenated environments.

Thus, in the harpetid fringe, the brim can be considered as an extremely large surface area for oxygen absorption due to the development of pits between the caeca (Chatterton 1980). Moreover, this would have meant that harpetids could have inhabited dysaerobic environments. This being the case, the increase in width of the brim between the pre- and post-Lower Kellwasser species of *Eskoharpes* would effectively have increased the surface area of the exoskeleton. If oxygen was absorbed through this thin, but extensive, exoskeleton, it could be argued that the increase in brim width during the late Frasnian could have been in response to a reduction in ambient oxygen levels.

Relatively complete ontogenies of the youngest species of *Eskoharpes*, *E. palanasus* (zones 13a and 13b), and of the oldest species, *E. wandjina* (zones 11 and 12), are known that allow the role of heterochrony in the evolution of *Eskoharpes* to be established. The cephalic morphological changes in *E. wandjina* during ontogeny are almost isometric. The early meraspid cephalon, like the holaspid, has a strongly vaulted glabella and genal areas, and the brim underwent little a relative increase in width during growth. The genal roll in meraspids is slightly narrower than in larger specimens, but in all growth stages was steeply inclined (see McNamara *et al.* 2009, text-fig. 2). The phylogenetically youngest species, *E. palanasus*, underwent much greater ontogenetic morphological change, particularly a widening of the brim (Fig. 4).

Consequently, meraspid cephala of *Eskoharpes palanasus* (Zone 13b) resemble the holaspids of the older species, *E. wandjina* and *E. boltoni* (Zone 12). This greater ontogenetic morphological change in the derived species indicates evolution by peramorphosis. Phylogenetically younger species in this clade are not as large as the oldest species, *E. wandjina*. Details of actual growth rates are unknown. However, peramorphosis, combined with size reduction, in the younger species suggests possible acceleration in growth rates. The principal peramorphic features are brim width, and genal roll, glabella and genal vaulting.

Proetida

The most diverse group of trilobites in the Virgin Hills Formation is the proetids. Seventeen species in six genera have been described from the Frasnian part of the formation (Feist & McNamara 2013). Only two species and two genera are known from the Famennian (Feist & Becker 1997), demonstrating that, although this order survived the Frasnian–Famennian mass-extinction event, their biodiversity was severely affected, at least in this part of Gondwana.

The morphological features that underwent most evolutionary change were predominantly the eye and the course of the facial suture (Fig. 5). All six proetid genera were affected and essentially in the same way, in that the dominant evolutionary trend was one of eye reduction. Trends of eye reduction in trilobites have frequently been documented and almost invariably they are thought to have occurred due to changes in sea level, specifically deepening, which would reduce light levels and so, it has been argued, lead to selection for reduced eyes (e.g. Richter & Richter 1926; Erben 1961; Clarkson 1967; Feist & Clarkson 1989). Given the detailed analysis of the sea-level changes that occurred during the Late Devonian in the reefal system of the Canning Basin, it is possible to test this hypothesis to assess whether there is a direct correlation between periods of eye reduction and transgressive phases. Interpreting the causative factors is discussed further below. In this section, we provide evidence for the all-pervasive nature of eye reduction in this group of trilobites. Between zones 11 and 13b, paedomorphic eye reduction or its consequences can be documented both in pteropariine and drevermanniine proetids, involving independent reduction in *Pteroparia*, *Canningbole*, *Palpebralina*, *Palpebralia* and *Rudybole*.

Tropidocoryphidae. Eye reduction has previously been documented in pteropariine tropidocoryphids in the European–North African region, leading to blindness (Feist & Schindler 1994; Feist 1995, 2002, 2003). The late Frasnian forms from the Canning Basin include *Pteroparia extrema* Feist & McNamara, 2013, from Zone 12 (Feist & McNamara 2013, fig. 5P, R–Y). This is a blind species, suggesting that eye reduction was probably a global and not just a regional phenomenon. Feist & Clarkson (1989) showed how eye reduction in Late Devonian tropidocoryphids proceeded initially along the *Longicoryphe–Erbenicoryphe* lineage, the palpebral lobe regressing, but the course of the anterior sutures remaining unchanged, and then along the *Pterocoryphe–Pteroparia* lineage. Here, the palpebral sutures straightened and shortened, while the anterior branches migrated backwards to attain a posterior position opposite the preglabellar mid-length. The extreme version of this movement is shown by the Canning Basin form, where only very small librigenae persist at the genal angles. These librigenae were not ankylosed, which was often the case in some other blind Late Devonian adult proetids (Lerosey-Aubril & Feist 2005).

However, some pteropariines retain eye lobes with lenses, the two oculated forms being *Chlupaciparia* (*Chlupaciparia*) and *Chlupaciparia* (*Australoparia*). These forms both have relatively long palpebral sutures. The eye lobe is smaller, in a more advanced position, and more globular with fewer and more prominent lenses in *C.* (*Australoparia*) than in *C.* (*Chlupaciparia*). *C.* (*Australoparia*) is known only from the Canning Basin, where it occurs in conodont Zone 12. The nominate

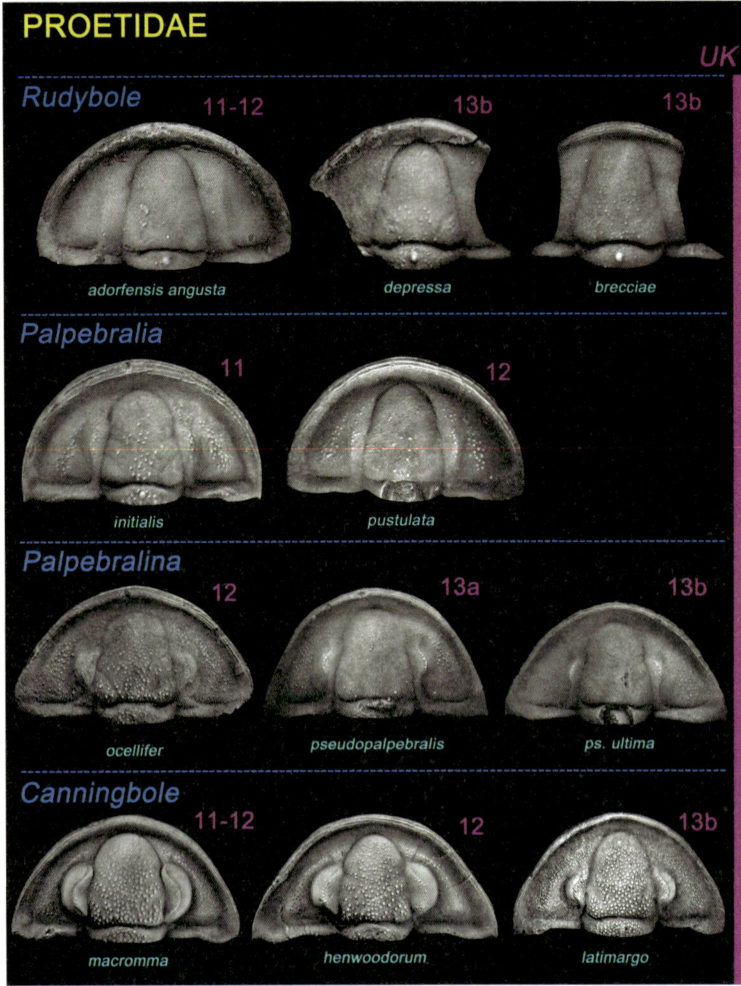

Fig. 5. Evolutionary trends of facial suture migration and eye reduction in proetid lineages in the Virgin Hills Formation through terminal Frasnian zones 11–13b, preceding the Upper Kellwasser Extinction Event (UK).

subgenus is restricted to Zone 13a in the Canning Basin, but with *Pteroparia* it continues into the latest Frasnian Zone 13b in Morocco and Germany (Feist & Schindler 1994; Feist 2002). In the younger species of *Chlupaciparia*, the eye does not get smaller but flattens. Both oculated and blind tropidocoryphids extended to the terminal Frasnian Upper Kellwasser extinction event.

Proetidae. Unlike other drevermanniine proetids, *Canningbole* possesses the typical proetid pattern of eye lobes, sigmoidal facial sutures and narrow fixigenae without eye ridges. Evolutionary changes from *C. macromma* McNamara, Feist & Ebach, 2009 (zones 11 and 12), *C. henwoodorum* McNamara, Feist & Ebach, 2009 (Zone 12) and *C. latimargo* McNamara, Feist & Ebach, 2009 (Zone 13b) mainly involve the glabellar shape, width of the anterior border and border furrow, length and size of the eye, and the degree of cephalic arching (Fig. 5). Of these, shortening and narrowing of the palpebral lobes and consequent reduction in eye size is the most significant, eye length reducing from 47% of cephalic length in the oldest taxon to 28% in the youngest (Feist & McNamara 2013). Moreover, the width of the cephalic border progressively narrows from 13% cephalic length in *C. macromma* to just 5% in *C. latimargo*, while the preglabellar field develops and widens towards the end Frasnian extinction.

The fixigenae are wider in *Palpebralina* than in *Canningbole*, and short palpebro-ocular ridges are present. Between Zone 12 and Zone 13b, the eyes become less well developed (Fig. 5). In the oldest species, *P. ocellifer* McNamara, Feist & Ebach, 2009, the eye is long, well defined and with perceptible lenses. By Zone 13a, the eye lobe of *P. pseudopalpebralis pseudopalpebralis* McNamara, Feist & Ebach, 2009 has reduced significantly to a narrow, crescentic band, lacking discernable lenses. In the youngest form, *P. pseudopalpebralis ultima* McNamara, Feist & Ebach, 2009, the eye lobe is even fainter and narrower. Extreme eye reduction is expressed in *Palpebralia* by the presence of remnant eye lobes that lack lenses and in *Rudybole*, another blind form that lacks palpebral lobes.

In other Late Devonian proetids, it has been shown that eye reduction followed a centripetal mode (Erben 1961; Lerosey-Aubril 2006). In this mode of eye reduction, the degeneration affected the more surficial elements of the eyes first (e.g. lenses, ocular field and palpebral lobe), while more internal components, especially the optic nerve, remained unaffected. Lerosey-Aubril (2006) has previously suggested that eye degeneration and blindness in Late Devonian proetoids was primarily acquired by centripetal reduction. The Canning Basin drevermanniine fauna supports this view. In *Palpebralina*, eye reduction clearly follows a centripetal mode. Eye lobes become smaller in *P. minor* McNamara, Feist & Ebach, 2009 compared with *P. ocellifer*, and bear barely perceptible lenses. This trend is continued in *P. pseudopalpebralis*, the eye lobe becoming very small and the lenses no longer visible. All species of *Palpebralina* possess faint eye ridges, indicating that the optic nerves were still normally developed in these trilobites, despite the degree of degeneration of the external components of the eye.

In the blind *Rudybole*, centripetal eye degeneration also occurs. The oldest species possess eye ridges, suggesting that the optic nerves were still present, although their functionality is questionable because the eye ridges fail to reach the facial sutures. The almost complete degeneration of the eye ridges in the youngest species, *R. brecciae* (Richter, 1913), points to an even more advanced degeneration of the internal elements of the eye complex (Fig. 5).

Trilobite extinctions

Of the five orders of trilobites that occurred in the late Frasnian, three – the Odontopleuridae, Harpetidae and Corynexochida – became extinct during the Upper Kellwasser biocrisis during Zone 13b. Unlike coeval sections in North Africa and Europe, where only the first two orders ranged into Zone 13b, all three are present at this horizon in the Canning Basin, the sole remaining corynexochid, *Telopeltis*, surviving the Lower Kellwasser biocrisis (McNamara & Feist 2006). Of the two orders that survived the end-Frasnian extinctions, the Phacopida are represented by the Phacopidae (Feist *et al.* 2009), and the Proetida by the Proetidae (Feist & McNamara 2013); the Tropidocoryphidae and Otarioninae disappeared, respectively, at the Lower and the Upper Kellwasser biocrises in the Canning Basin.

Gondwanaspis seems not to have been affected by the Lower Kellwasser biocrisis at the generic level, surviving until the Upper Kellwasser biocrisis, when it became extinct. However, of the five species present in the late Frasnian in the Canning Basin, three became extinct at the Lower Kellwasser biocrisis. Having existed since the Middle Cambrian, the extinction of *Gondwanaspis* saw the end of the Odontopleuroidea, resulting from the extreme reduction in diversity during the Late Devonian (Feist & McNamara 2007).

Telopeltis shows a pattern of extinction similar to that of *Gondwanaspis*. As the sole surviving late Frasnian scutelluid, *Telopeltis* suffered species-level extinction at the end of Zone 12, with the loss of two forms. However, two derived forms continued in Zone 13a, with smaller eyes and a less vaulted exoskeleton. The last species, and the last member of the order, became extinct within Zone 13b.

Eskoharpes and *Globoharpes* are both present in Zone 11 and both genera persist into Zone 12. *Globoharpes* was a casualty of the Lower Kellwasser biocrisis, becoming extinct at this event. In *Eskoharpes*, extinction at this level was at the species level, with the two Zone 12 taxa being replaced by a similar number in Zone 13a. A single species, *E. palanasus*, survived into Zone 13b, becoming extinct during the Upper Kellwasser biocrisis. It thus represents the last known harpetid trilobite in the Canning Basin. Other contemporaneous species at the western end of the Prototethys Ocean similarly became extinct during the Upper Kellwasser biocrisis (Feist & McNamara 2007), marking the extinction of the order Harpetida, which had existed since the late Cambrian.

The most biodiverse order of trilobites, the Proetida, most clearly shows the differing effects of the two Kellwasser biocrises. Although the Proetida survived the Frasian–Famennian mass-extinction event, this order was profoundly affected by the Upper Kellwasser Event. Like the Corynexochida, Odontopleurida and Harpetida, biodiversity of the Proetida was adversely affected at lower taxonomic levels during the Lower Kellwasser biocrisis. However, the higher biodiversity of the Proetida during the late Frasnian may have been a factor in enabling the order to survive into the Famennian (Lerosey-Aubril & Feist 2012). Although, at the species and even the genus level there are no known survivors of the order.

Within the Proetida, biodiversity is highest in the family Proetidae in zones 11 and 12, with four genera and eight species. The Lower Kellwasser biocrisis saw the same pattern as shown by the odontopleurids, scutelluids and harpetids, in that there was high extinction at the species level (none of these species surviving the event). However, all four genera survived the biocrisis, but only one, *Palpebralina*, is represented in Zone 13a, the rest reappearing in Zone 13b. Compared with zones 11 and 12, proetid biodiversity was very low in Zone 13a, with just a single species. It is, perhaps, significant that this level is thought to correspond to a regional regression (George & Chow 2002; George *et al.* 2014). Diversity was higher in the proetids and other orders during the two transgressive phases of zones 12 and 13b, with five species in four genera. All Canning Basin Frasnian proetid genera became extinct during the Upper Kellwasser biocrisis and were absent during the seven *triangularis* and *Palmatolepis crepida* zones of the early Famennian, not reappearing until the *rhomboidea* Zone (Feist & Becker 1997).

The oculated tropidocoryphid *Chlupaciparia* also survived the Lower Kellwasser biocrisis, unlike the blind *Pteroparia*. Even though they are known to extend until the Upper Kellwasser level in Morocco and Germany (Feist 1991, 2002), tropidocoryphids became extinct in the Canning Basin in Zone 13a.

Otarion, a rare aulacopleurid, persisted also until the Upper Kellwasser level. The pattern of extinctions in the Proetida confirms the important impact of the Lower Kellwasser biocrisis in causing a major erosion of trilobite diversity at the specific level that probably contributed to the extinctions that occurred at higher taxonomic ranks during the Upper Kellwasser biocrisis.

Bambach *et al.* (2004) suggested that some 'mass extinctions', including the Frasnian–Famennian, were best described as 'mass depletions', as they argued that rather than extinctions, diversity losses arose from lower origination rates. While they were looking at entire faunas, the overall high diversity of the trilobites at this time allows some assessment to be made of their suggestion. Looked at on a zone-by-zone basis, then interesting regional patterns emerge. Zone 11 saw the 'origin' of 13 species, but only five extinctions at the end of the zone. In Zone 12, 10 new species appeared, but 18 species became extinct during the Lower Kellwasser biocrisis. None of the 23 species present in these two zones survived the first late Frasnian biocrisis. Recovery saw only eight new species, all except one of which became extinct at the end of Zone 13a. The transgressive phase of Zone 13b saw the appearance of 10 further species, all of which had become extinct by Zone 13c.

Thus, leading up to the Lower Kellwasser Event there are as many species originating as becoming extinct. The same holds for the Upper Kellwasser Event. However, of the 41 species present during this last phase of the Frasnian, all had become extinct by Zone 13c, the Upper Kellwasser biocrisis. Moreover, there was the loss of three of the five orders. Even in the most 'successful' surviving order, the Proetida, two of the three families present in the late Frasnian became extinct. Thus, for the trilobites at least, the two Kellwasser biocrises were not mass depletions – they were mass extinctions. Arguably, in terms of trilobite biodiversity, these two events were the most deleterious to the evolution of the Trilobita, and, following the extinction of the Phacopida at the end of the Famennian, Trilobita biodiversity never recovered to the same extent.

Environmental factors controlling trilobite evolution

The Late Devonian in the Canning Basin was a period of dynamic fluctuations in sea level, especially during the Frasnian (Becker & House 1997). During the Frasnian, as a whole, three pulses of sea-level rise have been recognized: the first in Zone 9, the second in Zone 12 and the last during Zone 13b (Becker & House 1997; George *et al.* 2014). Only the last two of these events include trilobite-bearing beds. The two Kellwasser biocrises, which are reflected in the Canning Basin as major changes in the trilobite fauna (see below), are both associated with the initiation of major regressive episodes.

It has been well documented that while these two episodes have left a characteristic sedimentological signal elsewhere (see Carmichael *et al.* 2014 for a recent review), this is not the case in the Canning Basin (Becker *et al.* 1991; George & Chow 2002). While the Kellwasser events are marked by the occurrence of dark mudstones indicative of anoxic conditions, this is not the case with the Virgin Hills Formation, which, at least in the McWhae Ridge area, consists of red or buff-coloured limestones and calcareous siltstones. The variation arises from differences in the amount of iron present in the sediments in the form of hematite. While this has generally been ascribed to formation from detrital hematite sourced from the Kimberley Block (Playford *et al.* 2009), it is possible that the hematite might be autochthonous.

Iron precipitated by the activity of bacteria leaves distinctive sedimentological signals in the sediment (Mamet & Préat 2005, 2006; Préat *et al.* 2008), and many of these are present in the red calcareous siltstones of the Virgin Hills Formation. Sediments deposited during zones 12 and 13b at McWhae Ridge are rich in hematite. Sediments of Zone 13a, however, are buff to pale grey in colour as they lack hematite. Preliminary petrological analysis of a red calcareous siltstone from Zone 13b shows that the hematite is not dispersed evenly through the sediment, which would be expected if it were of detrital origin. Hematite is concentrated in a variety of locations, but principally in association with invertebrate skeletal material or along irregular, closely spaced laminations that may be interpreted as thin microbial mats.

Association of hematite with invertebrate fragments occurs in a number of ways, but shows patterns consistent with formation under the influence of iron microbial communities. Hematite is frequently restricted to the surface of convex-upwards bivalve shell fragments that were lying at the sediment–water interface at the time of formation of the particular microbial mat. In some cases, very small, tufted hematitic microbialites were growing off crinoid fragments (Fig. 6). The hematite also extends into a narrow zone within the stereom, but microbialite growth is away from the sediment surface. The development of these patterns of hematite growth mirror, in a more subdued manner, the growth of small hematitic microbialites in the so-called *Frutexites* horizon in the lower part of the Famennian.

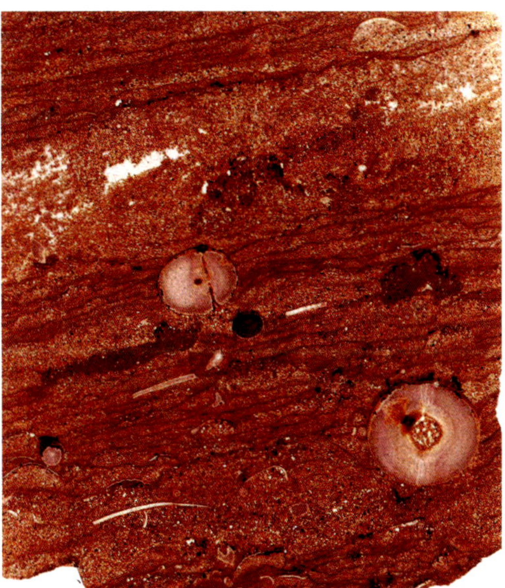

Fig. 6. Thin section of hematitic calcareous siltstone from Zone 13b at Calyx Corner, McWhae Ridge (see Fig. 1), showing the hematite concentration in multiple microbial mats, and on the tops of crinoid ossicles and shell fragments, with incipient development of tufted hematitic microbialites.

The presence of iron bacteria in the sediments at McWhae Ridge has implications for the interpretation of the causes of the evolutionary changes seen in the trilobite fauna.

Eye reduction

Arguably, the most distinctive evolutionary trend that occurs in the late Frasnian trilobites from the Canning Basin is eye reduction and/or modification of the facial sutures, especially because it occurred in a number of independent lineages. These trends have also been well recorded elsewhere but considered to be restricted to only the phacopids and proetids (Clarkson 1967; Feist & Clarkson 1989). However, as we have demonstrated here, the same phenomenon also occurs in the scutelluid *Telopeltis*. Moreover, eye reduction occurred in trilobites with both holochroal and schizochroal eyes (Clarkson *et al.* 2006). Such an evolutionary trend, occurring in three different orders and also within a number of clades within the Proetida, indicates very strong selective pressures on this trait.

Hitherto, three explanations have been proposed to explain this trend, both related to fluctuating light levels in the habitats inhabited by the trilobites. Richter & Richter (1926), Erben (1961) and Clarkson (1967) argued that eye reduction occurred in response to diminishing light levels brought on by increased water depth: in other words, during transgressive phases. Feist & Schindler (1994) argued that the trilobites were benthic and living at or below the limit of light penetration, the argument being that living in waters of little to no light penetration mitigates against producing well-developed eyes. Progressive reduction in eye size or in the number of lenses reflects, according to Clarkson *et al.* (2006), adaptation to progressive oceanic deepening. Although why, under such circumstances, the trilobites were unable to simply migrate into water of the same depth to which they were adapted as the water deepened is something of a mystery. Feist & Clarkson (1989), suggested a somewhat different causative agent. Noting that eye reduction and change in the course of the facial suture in tropidocoryphines occurred in conjunction with loss of cephalic vaulting (see Feist & Clarkson 1989, fig. 8), they suggested that, while the eye loss was a reflection of lower light levels, this was not so much due to oceanic deepening but to the adoption of an endobenthic lifestyle by the trilobites. The third explanation also involves reduced light levels. Averbuch *et al.* (2005) suggested that increasingly turbid water during periods of sea-level fall, when there was a greater influx of detrital material, could have resulted in a diminution in eye size.

The nature of the sediments in the Virgin Hills Formation enables these various hypotheses to be tested, although we should point out that studies of the sediments are, particularly with regard to the role of iron bacteria, still at a preliminary stage. However, detailed work to detect perturbations in sea levels during the late Frasnian interval in this region has been undertaken recently (see George *et al.* 2009, 2013, 2014), allowing the hypothesis of correlation between eye reduction and increasing sea level to be tested. The progressive eye reductions and modifications to the course of the facial suture lines documented between zones 11 and 13b in the Proetidae and Scutelluidae, and between zones 13a and 13b in the phacopid *Acuticryphops*, do not correlate with periods of increased palaeobathymetry. Maximum flooding occurs at the base of Zone 12, the preceding Zone 11 having been a period of rising sea level (Becker *et al.* 1993; Becker & House 1997; George *et al.* 2014, fig. 2). However, late in Zone 12 there was a marked fall in sea level, such that a major regression occurred, with another lowstand in Zone 13a. This was followed by a transgressive phase before a relative sea-level fall into Zone 13c. Thus, the two Kellwasser biocrises are both associated with major regressive episodes at their bases, yet through these episodes the overall trend is one of eye-size reduction.

If eye reduction was associated with increases in palaeobathymetry, then the opposite should also hold true, with increase in eye size occurring during periods of relative sea-level fall. However, there are indications that this may not be the case in Famennian phacopids. Although few individuals are known, there is an indication of progressive increase in eye size and number of lenses in the Famennian phacopid *Houseops* in the Virgin Hills Formation at McWhae Ridge. If so, this would suggest that there should have been ongoing sea-level fall from the Upper *triangularis* Zone to the Middle *crepida* Zone. However, the *triangularis* zones are characterized by a transgressive episode (Becker & House 1997; George *et al.* 2014). The large-eyed *Babinops* is present in the Upper *crepida* Zone, a time of high sea level. During the *rhomboidea* Zone there was a relative sea-level fall, yet the only phacopid present at this time was the blind *Trimerocephalus*. Although in European sections blind phacopids, such as *Ductina ductifrons* and *Trimerocephalus mastophthalmus*, are common during the Upper *crepida* sea-level highstand, their lack of eyes may have been in response to the evolution of an endobenthic lifestyle, and not necessarily related to palaeobathymetry (e.g. Becker & Schreiber 1994).

Thus, it is possible that factors other than sea-level change may have been responsible for the strong selection pressure on eye reduction in the late Frasnian trilobite fauna in the Virgin Hills Formation.

Body size

One very characteristic feature of the late Frasnian trilobites, not only in the Canning Basin but also elsewhere (Clarkson 1967; Clarkson *et al.* 2006), is their very small size. In fact, small body size is a feature of a number of Late Devonian marine taxa, including arthropods as a whole (see Novack-Gottshall 2008, fig. 1). For instance, in the Virgin Hills Formation, the cephala of the odontopleurid *Gondwanaspis* never exceed 4.7 mm in length, the phacopid *Acuticryphops* 6 mm, proetids 6.1 mm, scutelluids 10 mm and harpetids 15.5 mm. These are much smaller than corresponding taxa earlier in the Devonian. Moreover, even during the late Frasnian, from Zone 11 to Zone 13b, there is a reduction in maximum size in most genera. For example, *Gondwanaspis* reduced from a maximum cephalic length of 4.7 to 2.2 mm between zones 11 and 13b. During the same period *Eskoharpes* reduced from 15.5 to 8.2 mm, and *Telopeltis* from 10 to 7 mm. *Acuticryphops* reduced from 6 to 4.6 mm between zones 13a and 13b. Size decrease occurs to varying degrees within the proetids, such as from 4.5 to 2.8 mm between zones 11 and 13b in *Rudybole*.

Such phylogenetic size reduction is known to often accompany biotic crises (variously termed 'dwarfing' or 'Lilliput Effect', 'faunal stunting' or 'miniaturization': Borths & Ausich 2011). Becoming smaller can occur for a number of reasons, ranging from collapse in primary productivity to general biotic stress, causing, for instance, changes in oxygen, temperature or salinity levels. Reasons for the reduction in body size during such events include selection for earlier maturation (progenesis) under stressed conditions (see McKinney & McNamara 1991) or reduction in nutrient supply. However, it is intriguing that Canning Basin pelagic goniatites reached their maximum size in the same interval (zones 13a and 13b: Becker *et al.* 1993; Becker & House 2009).

Causes of eye and body size reduction

Experiments on a range of arthropods have shown that reduction in nutrient availability results not only in a reduction in body size of the offspring, but also a concomitant relative decrease in size of other morphological features, especially those that are 'nutrient-hungry' organs. Prime amongst these in arthropods are eyes. In living arthropods, it is known that the evolution of large eyes is costly, both in terms of their development and their operation (Niven & Laughlin 2008): for example, Laughlin *et al.* (1998) have shown that about 10% of resting metabolic rate is required just for the operation of photoreceptor cells in the blowfly, *Calliphora*. Selection pressure is therefore a fine balance between visual performance and low cost of visual production, the interplay of which can cause deviations from isometry between body size and eye size (Merry *et al.* 2011).

Merry *et al.* (2011) have demonstrated the influence of variation in nutrient supply on variation in the size of compound eyes in the Orange Sulphur butterfly, *Collas eurytheme*. Rearing some individuals in a stressed, nutrient-poor environment results in a reduced body size. Consequently, there is a covariant reduction in the size of the compound eye. Moreover, individuals reared on a low-nutrient diet experienced lower developmental rates. The Late Devonian trilobites of the Canning Basin are typified by small body size, eye reduction in many images and a frequency of paedomorphosis, possibly arising from reduced developmental rates.

Conclusions

Given that the phenotypic changes in the Canning Basin Late Devonian trilobite fauna closely mirror phenotypic changes arising from experimental reduction in nutrient input in some living arthropods, it could be argued that size and eye reduction in the trilobites might likewise have arisen from adaptations to increasingly low-nutrient conditions. As discussed above, there is some sedimentological support for oligotrophic conditions in the hematitic calcareous siltstones in which there is evidence for the activity of iron bacteria. These bacteria are known to prefer habitation of nutrient-deficient environments. Franke & Paul (1980) argued that all red beds formed in oligotrophic environments and that, especially during the Late Devonian period, organic productivity seems to have been appreciably lower than at other times. The hematitic sediments at McWhae Ridge, where the section is condensed, are suggestive of formation in relatively low-oxygen, oligotrophic conditions. Interestingly, Murphy *et al.* (2000) and Averbuch *et al.* (2005) have pointed out that coral–stromatoporoid reefs, such as are present in the Canning Basin, were also adapted to oligotrophic conditions, so it should not be assumed that low-nutrient ecosystems will be low in biodiversity – on the contrary.

The evolution of small body size and reduction in eye size across a range of taxa could, therefore, be considered to have been caused essentially by adaptation to low-nutrient conditions. For organisms adapted to such conditions today, such as those occupying the very biodiverse coral reef environments, rainforest, and the kwongan and fynbos biomes in southern Western Australia and South Africa, respectively, any relatively sudden increase in nutrient load can have a catastrophic impact on the biodiversity. Arguably, trilobite diversity during the late Frasnian times could have similarly been affected.

A number of researchers have suggested that the Late Devonian Period was a time of great changes in organic supply to the oceans, perhaps due to the rise of terrestrial flora (e.g. Algeo & Scheckler 1998; Chen *et al.* 2002). Whether this was due to an increase in nutrients causing eutrophication, as some have argued (Murphy *et al.* 2000; Averbuch *et al.* 2005), or whether the development of terrestrial soils would have had the opposite effect of trapping nutrients, such as N and P, in soils and causing nutrient starvation in the oceans (Tappan 1986) has been the subject of some debate. More recent research seems to indicate that it was more likely that the influx of nutrient-rich waters, perhaps causing eutrophication, could have had a major deleterious impact on shallow-marine invertebrate faunas (Murphy *et al.* 2000; Averbuch *et al.* 2005).

It has recently been shown that in the Late Devonian sequence in the Canning Basin there is a close relationship between total organic carbon levels and enhanced productivity, arising from dissolved nutrient influx from continental weathering as the sea levels fell (George *et al.* 2014). As we have shown, these two periods of dropping sea levels during zones 12 and 13b both correspond to the major extinction events, and are marked by corresponding spikes in total organic carbon levels.

It could therefore be argued that changing nutrient levels resulted in substantial patterns of morphological change in trilobite lineages during the late Frasnian in the Canning Basin. Initially, reduced nutrient conditions favoured the evolution of forms with small body size. A consequence of this was a diminution in eye size in various lineages of proetids, scutelluids and phacopids. These forms, adapted to the more oligotrophic conditions, would therefore have been at a severe disadvantage during periods of massive nutrient influx as the sea levels fell (Averbuch *et al.* 2005), resulting in the extinction at lower taxonomic levels at the base of the Lower Kellwasser Event. Extinctions at higher taxonomic levels during the Upper Kellwasser Event may have occurred because of even greater and more prolonged perturbations to nutrient levels. There is no sedimentological evidence that these nutrient influxes caused periods of eutrophication in the Canning Basin region. However, they may have had a direct effect on elements low in the trophic level, resulting in extinctions further up the food change. For trilobites, the consequence would have been the most severe experienced up until that time in the history of the class, with extinction of three of the five orders, and a severe diminution in the two surviving orders, the Proetida and the Phacopida.

SYSTEMATIC PALAEONTOLOGY

Material. All specimens used in this study are housed in the collections of the Western Australian Museum (WAM).

Family **Scutelluidae** Richter & Richter, 1955

Remarks. Two different taxa of *Telopeltis* McNamara & Feist, 2006 were previously described that are distinct. Besides their difference in age, they mainly differ in the size of their eye complexes and the course of the facial suture, as well as the degree of cephalic and pygidial vaulting, to such an extent that they might be considered as belonging to separate genera. New taxa displaying intermediate traits to both the older *T. woodwardi* McNamara & Feist, 2006 and the younger *T. microphthalma* McNamara & Feist, 2006 emphasize the original concept of a single genus to which all present taxa belong.

Genus *Telopeltis* McNamara & Feist, 2006

Telopeltis intermedia sp. nov.
(Fig. 3 (inverted), Fig. 7a–g)

Derivation of name. From the Latin, *intermedius* meaning 'intermediate', with reference to its intermediate position between two related taxa.

Type material. Holotype: cranidium WAM 11.269 (Fig. 7f, g) from South Lawford Range, Western Australia, Virgin Hills Formation: red, crinoid-rich calcareous siltstone, 50 cm below Upper *Beloceras* Bed, Zone 13a, western side of McWhae Ridge (Fig. 1). Paratypes: fragmentary cranidium WAM 11.270 (Fig. 7a); librigena WAM 11.271 (Fig. 7b); pygidium WAM 11.272 (Fig. 7d, e); fragment of pygidial axis WAM 11.273 (Fig. 7c), all from eastern side of McWhae Ridge, eastern Phacopid Gully, associated with *Acuticryphops acuticeps* (Kayser, 1889), Zone 13a (locality V in Fig. 1).

Additional material. One librigena and five pygidia from eastern Phacopid Gully.

Diagnosis. Cephalon with moderately down-turned profile in lateral view; small, transversely narrow palpebral lobe, a little longer than its distance from the posterior border, palpebral rim carrying row of strong nodes; short palpebro-occular ridge distally accompanied by marked palpebral furrow; marked lateral occipital lobes; pygidium widely transverse, with very small axis displaying eccentric, well-marked antero-lateral lobes; proximal end of median rib equal or slightly wider than neighbouring ribs; wide, horizontally orientated postero-lateral border. Sculpture of coarse granules on entire exoskeleton; terrace ridges restricted to anterior slope of glabella and median occipital ring.

Remarks. The new species is distinct from the ancestral *T. woodwardi* mainly in the lesser vaulting of both profiles, the anterior glabella and the post-axial pleural field in the pygidium. The anterior lateral glabellar furrows (S3), although inconspicuously impressed adaxially, become flush with the surface of the exoskeleton distally, where, in contrast to the allied species, they remain devoid of sculpture and reach the axial furrow. As the eye and the palpebral lobe is smaller compared with *T. woodwardi*, the palpebral area of the fixigenae has much increased in width, and the portion between ε and the posterior border enlarged, although the course of the posterior suture remains largely unchanged. In comparison with *T. woodwardi*, the tubercular sculpture of the new species is markedly coarser on the palpebral rim and on all parts of the pygidial axis, whereas terrace ridges diminish in density and are only perceptible, besides the median part of the occipital lobe, on the anterior-most slope of the frontal glabella.

The sculpture, outline of the preoccipital glabella, and eye lobe resemble more those of the descendant *T. microphthalma*. However, the latter is characterized by the straight, obliquely backwards course of the posterior suture and the rather wide posterior fixigenae. In comparison with *T. microphthalma*, the pygidium of the new species has a relatively smaller axis with well-defined lateral lobes that are separated from the lateral axial lobes by faint furrows, and the median pleural rib is narrower adaxially. However, the outline and vaulting are nearly identical. Pygidia and librigenae from Phacopid Gully (Zone 13a) previously assigned to *T. microphthalma* (McNamara & Feist 2006, p. 988) are now reassigned to the new species. Consequently, the range of *T. microphthalma* (McNamara & Feist 2008, fig. 1) appears to be restricted to Zone 13b.

Telopeltis aff. *intermedia* sp. nov.
(Fig. 7h, i)

Material. Cranidium WAM 11.274, from the western side of McWhae Ridge, pink calcareous siltstone, 50 cm above Upper *Beloceras* Bed, Zone 13a (Fig. 1).

Remarks. The single cranidium, occurring close to the top of Zone 13a, exhibits characteristic features,

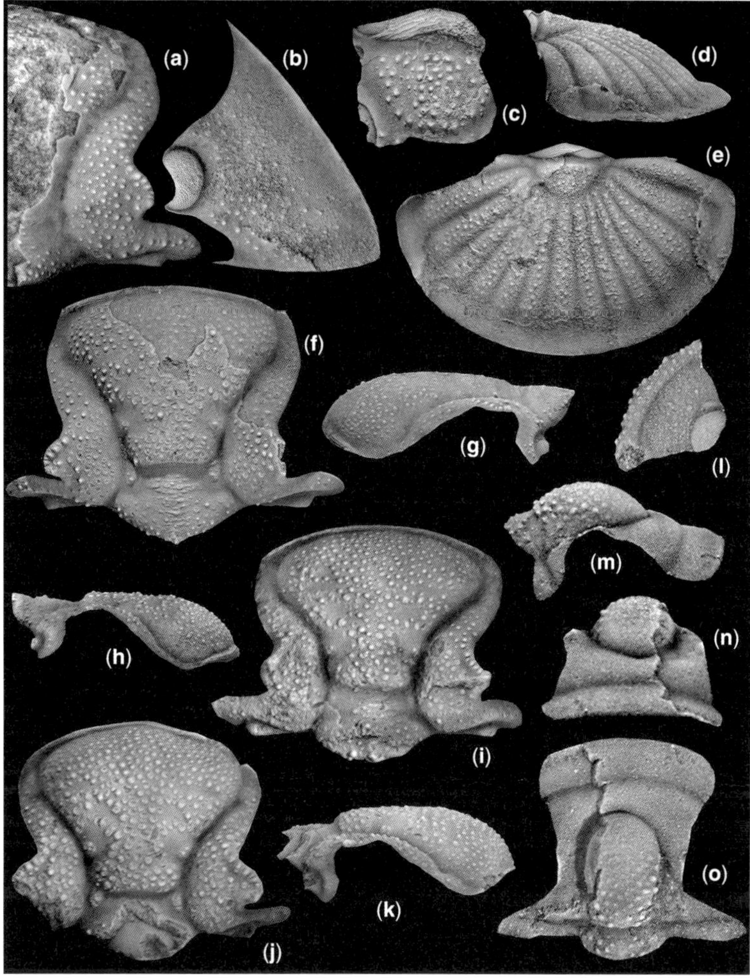

Fig. 7. (**a**)–(**g**) *Telopeltis intermedia* sp. nov.: (a) fragmentary cranidium WAM 11.270, dorsal view, ×4.4; (b) librigena WAM 11.271, dorsal view, ×3.2; (c) fragmentary pygidial axis WAM 11.273, dorsal view, ×4; (d) & (e) pygidium WAM 11.272, lateral and dorsal views, ×3.2; (f) & (g), holotype, cranidium WAM 11.269, dorsal and lateral views, ×4; (**h**) & (**i**) *Telopeltis* aff. *intermedia* sp. nov., cranidium WAM 11.274, lateral and dorsal views, ×3.7; (**j**) & (**k**) *Telopeltis* aff. *woodwardi* McNamara & Feist, 2006, cranidium WAM 11.275, dorsal and lateral views, ×3.6; (**l**) *Otarion* sp., librigena WAM 11.277, dorsal view, ×9.6. (**m**)–(**o**), *Otarion fugitivum* sp. nov., holotype, cranidium WAM 11.276, lateral, frontal and dorsal views, ×10.7.

such as the small, entire palpebral lobe, the row of coarse nodes on the palpebral rim, the palpebro-occular ridge along with a marked palpebral furrow, the antero-laterally directed and outwards-curved posterior suture, and marked lateral occipital lobes. From these features, it is clearly distinct from the slightly younger *T. microphthalma*. It shares with the latter the somewhat pointed anterior corners of the glabella, the slightly impressed S3 that do not reach the axial furrow, being separated from them by tubercular sculpture, and the absence of terrace ridges among otherwise prominent tuberculation. Particular traits of the specimen include the strong outwards curvature of the axial furrows in front of S1. More material, in particular the discovery of the pygidium, is necessary to define a possible new species.

Telopeltis aff. *woodwardi* McNamara & Feist, 2006
(Fig. 7j, k)

Material. Cranidium WAM 11.275, from the western side of McWhae Ridge, red crinoid-rich calcareous siltstone, Lower *Beloceras* Bed (= Scutelluid Bed), Zone 12 (Fig. 1).

Remarks. Among the population of *T. woodwardi* that characterizes the 'Scutelluid Bed' (= Lower *Beloceras* Bed) on the western side of McWhae Ridge, this single cranidium has a number of traits that do not correspond to the diagnosis of *T. woodwardi*. These include a much coarser granulation of the exoskeleton, rather prominent lateral occipital lobes (only a single specimen from Siphon Spring has similarly strong lobes: McNamara & Feist 2006, fig. 4), and a pronounced row of strong tubercles on the palpebral rim and palpebro-occular ridge, along with a marked palpebral furrow. These traits characterize the younger *T. intermedia*, although they are more weakly developed there. However, the strong down-curved profile of the anterior glabella, the outline of the long, antero-medially protruding frontal glabellar lobe, the marked S3 furrows that do not reach the axial furrows, the longer and wider occipital lobe, and the presence of wavy terrace ridges besides tubercles on the anterior glabella are all typical traits of *T. woodwardi*. To date, material is lacking to carry out necessary statistical analyses to appreciate whether *T.* aff. *woodwardi* constitutes a morphological variant of *T. woodwardi* or, instead, belongs to a different species that might present a phylogenetic link between *T. woodwardi* and *T. intermedia*.

Family **Aulacopleuridae** Angelin, 1854

Remarks. Devonian aulacopleurids are generally very rare in Australia, known only from *Aulacopleura* sp. and *Otarion* sp., Broken River region of NE Queensland (Feist & Talent 2000), and *Cyphaspis dabrowni* (Chatterton, 1971), Taemas area, New South Wales. In the Canning Basin region, only a single cranidium and a librigena have shown up among the otherwise very rich trilobite associations. However, the occurrence of aulacopleurids in the late and latest Frasnian emphasizes their extension up to the terminal Frasnian Kellwasser extinction in Australia, as is the case in European and North African sites (Feist & Schindler 1994; Feist 2002).

Genus *Otarion* Zenker, 1833

Otarion fugitivum sp. nov.
(Fig. 7m–o)

Derivation of name. From the Latin *fugitivus*, meaning fugitive, because of the progressive effacement of sculpture towards the anterior part of the cranidium.

Type material. Holotype: cranidium WAM 11.276, from South Lawford Range, Western Australia, Virgin Hills Formation: salmon-red crinoidal limestone with *Rudybole brecciae* (Richter, 1913), upper Phacopid Gully, east side of McWhae Ridge, Frasnian Zone 13b (locality IV in Fig. 1).

Diagnosis. Very wide (sagitally), gently vaulted anterior border and preglabellar field, posteriorly constricted, narrow glabella with uninflated front and tiny basal lateral lobes, coarse tubercles on its posterior half, attenuating anteriorly until they disappear before the front; anterior and lateral parts of cranidium smooth.

Remarks. This species was previously assigned to *Cyphaspis* ('*Cyphaspis fugitiva*' nomen nudum: McNamara & Feist 2008, fig. 1). However, the wide preglabellar field, at the expense of the markedly constricted, relatively short glabella, characterizes the genus *Otarion* rather than *Cyphaspis* (Adrain & Chatterton 1994). Despite the paucity of the material (holotype only), it exhibits unique diagnostic features that justify the assignment to a new species. Indeed, the only recognized contemporaneous aulacopleurid taxon, the European–North African *Otarion stigmatophthalmus* (Richter, 1914), is quite distinct in its cylindrical, narrow anterior border, relatively shorter preglabellar field and denser sculpture of pustules throughout the entire exoskeleton.

Otarion sp.
(Fig. 7l)

Remarks. A single incomplete aulacopleurid librigena (WAM 11.277) assigned to *Otarion* Zenker, 1833 was collected from Zone 12 at Siphon Spring, associated with *Telopeltis woodwardi* and *Canningbole henwoodorum* Feist & McNamara, 2013. It is characterized by a narrow cylindrical border covered by coarse tubercles, and a large eye accompanied by a small inflated platform in front of it. Because the posterior border and genal spine are broken off, it is not possible to decide whether the latter prolongs the course of the border roll in the same direction or if it is outwardly directed, as is typical in *Otarion*.

We gratefully acknowledge funding support from the Australian Research Council (Discovery Grant DP0664703) and the award to KJM of a Gledden Senior Visiting Fellowship at the University of Western Australia, during which time part of this research was carried out. We wish to thank Annette George for her support and encouragement and Alain Préat for helpful discussion. This is a contribution to UMR 5554, CNRS, Montpellier (ISEM 2015-037).

References

Aboussalam, Z. S. 2003. Das 'Taghanic-Event' im höheren Mittel-Devon von West-Europa und Marokko.

Münstersche Forschungen zur Geologie und Paläontologia, **97**, 1–332.

Adrain, J. M. & Chatterton, B. D. E. 1994. The aulacopleurid trilobite *Otarion*, with new species from the Silurian of northwestern Canada. *Journal of Paleontology*, **68**, 305–323.

Algeo, T. J. & Scheckler, S. E. 1998. Terrestrial-marine teleconnections in the Devonian: links between the evolution of land plants, weathering processes, and marine anoxic events. *Philosophical Transactions of the Royal Society London*, **B353**, 113–130.

Angelin, N. P. 1854. *Palaeontologica Scandinavica 1: Crustacea Formationis Transitionis. Fasciculi II.* Weigel, Lund.

Averbuch, O., Tribovillard, N., Devleeschouwer, X., Riquier, L., Mistiaen, B. & van Vliet-Lanoe, B. 2005. Mountain building-enhanced continental weathering and organic carbon burial as major causes for climatic cooling at the Frasnian–Famennian boundary (*c.* 376 Ma)? *Terra Nova*, **17**, 25–34.

Bambach, R. K., Knoll, A. H. & Wang, S. C. 2004. Origination, extinction, and mass depletions of marine diversity. *Paleobiology*, **30**, 522–542.

Becker, R. T. & House, M. R. 1997. Sea-level changes in the Upper Devonian of the Canning Basin, Western Australia. *Courier Forschungsinstitut Senckenberg*, **199**, 129–146.

Becker, R. T. & House, M. R. 2009. Devonian ammonoid biostratigraphy of the Canning Basin. *In*: Playford, P. E., Hocking, R. M. & Cockbain, A. E. (eds) *Devonian Reef Complexes of the Canning Basin, Western Australia.* Geological Survey, Western Australia, Bulletin, **145**, 415–431.

Becker, R. T. & Schreiber,, G. 1994. Zur Trilobiten-Stratigraphie im Lethmather Famennium (nördliches Rheinisches Schiefergebirge). *Berliner geowissenschaftliche Abhandlungen*, **E13**, 369–387.

Becker, R. T., Feist, R., Flajs, G., House, M. R. & Klapper, G. 1989. Frasnian–Famennian extinction events in the Devonian at Coumiac, southern France. *Comtes Rendus de l'Académie des Sciences Paris, Série II*, **309**, 259–266.

Becker, R. T., House, M. R., Kirchgasser, W. T. & Playford, P. E. 1991. Sedimentary and faunal changes across the Frasnian/Famennian boundary in the Canning Basin of Western Australia. *Historical Biology*, **5**, 183–196.

Becker, R. T., House, M. R. & Kirchgasser, W. T. 1993. Devonian goniatite biostratigraphy and timing of facies movements in the Frasnian of the Canning Basin, Western Australia. *In*: Hailwood, E. & Kidd, R. B. (eds) *High Resolution Stratigraphy*. Geological Society, London, Special Publications, **70**, 293–321, http://doi.org/10.1144/GSL.SP.1993.070.01.20

Bond, D., Wignall, P. B. & Racki, G. 2004. Extent and duration of marine anoxia during the Frasnian–Famennian (Late Devonian) mass extinction in Poland, Germany, Austria and France. *Geological Magazine*, **141**, 173–193.

Bond, D. P. & Wignall, P. B. 2008. The role of sea-level change and marine anoxia in the Frasnian–Famennian (Late Devonian) mass extinction. *Palaeogeography, Palaeoclimatology, Palaeoecology*, **263**, 107–118.

Borths, M. R. & Ausich, W. I. 2011. Ordovician–Silurian Lilliput crinoids during the end-Ordovician crisis. *Swiss Journal of Palaeontology*, **130**, 7–18.

Carmichael, S. K., Waters, J. A., Suttner, T. J., Kido, E. & DeReuil, A. A. 2014. A new model for the Kellwasser Anoxia Events (Late Devonian): shallow water anoxia in an open oceanic setting in the Central Asian Orogenic Belt. *Palaeogeography, Palaeoclimatology, Palaeoecology*, **399**, 394–403.

Chatterton, B. D. E. 1971. Taxonomy and ontogeny of Siluro-Devonian trilobites from near Yass, New South Wales. *Palaeontographica, Abt. A*, **137**, 1–108.

Chatterton, B. D. E. 1980. Ontogenetic studies of Middle Ordovician trilobites from the Esbataottine Formation; Mackenzie Mountains, Canada. *Palaeontographica, Abt. A*, **171**, 1–74.

Chen, D., Tucker, M., Shen, Y., Yans, J. & Preat, A. 2002. Carbon isotope excursions and sealevel change: implications for the Frasnian–Famennian biotic crisis. *Journal of the Geological Society, London*, **159**, 623–626, http://doi.org/10.1144/0016-764902-027

Clarkson, E., Levi-Setti, R. & Horváth, G. 2006. The eyes of trilobites: the oldest preserved visual system. *Arthropod Structure & Development*, **35**, 247–259.

Clarkson, E. N. K. 1967. Environmental significance of eye-reduction in trilobites and recent arthropods. *Marine Geology*, **5**, 367–375.

Crônier, C., Feist, R. & Auffray, J.-C. 2004. Variation in the eye of *Acuticryphops* (Phacopina, Trilobita) and its evolutionary significance: a biometric and morphometric approach. *Paleobiology*, **30**, 471–481.

Erben, H. K. 1961. Blinding and extinction of certain Proetidae (Tril.). *Journal of the Palaeontological Society of India*, **3**, 82–104.

Feist, R. 1991. The Late Devonian trilobite crises. *Historical Biology*, **5**, 197–214.

Feist, R. 1995. Effect of paedomorphosis in eye reduction on patterns of evolution and extinction in trilobites. *In*: McNamara, K. J. (ed.) *Evolutionary Change and Heterochrony*. Wiley, Chichester, 225–244.

Feist, R. 2002. Trilobites from the latest Frasnian Kellwasser Crisis in North Africa (Mrirt, central Moroccan Meseta). *Acta Palaeontologica Polonica*, **47**, 203–223.

Feist, R. 2003. Biostratigraphy of Devonian tropidocoryphid trilobites from the Montagne Noire (southern France). *Bulletin of Geosciences*, **78**, 431–446.

Feist, R. & Becker, T. 1997. Discovery of Famennian trilobites in Australia (Late Devonian, Canning Basin, NW Australia). *Geobios*, **20**, 231–242.

Feist, R. & Clarkson, E. N. K. 1989. Environmentally controlled phyletic evolution, blindness and extinction in Late Devonian tropidocoryphine trilobites. *Lethaia*, **22**, 359–373.

Feist, R. & McNamara, K. J. 2007. Biodiversity, distribution and patterns of extinction of the last odontopleuroid trilobites during the Devonian (Givetian, Frasnian). *Geological Magazine*, **144**, 777–796.

Feist, R. & McNamara, K. J. 2013. Patterns of evolution and extinction in proetid trilobites during the Late Devonian mass extinction event, Canning Basin, Western Australia. *Palaeontology*, **56**, 229–259.

FEIST, R. & SCHINDLER, E. 1994. Trilobites during the Frasnian Kellwasser Crisis in European Late Devonian cephalopod limestones. *Courier Forschungsinstitut Senckenberg*, **169**, 195–223.

FEIST, R. & TALENT, J. A. 2000. Devonian trilobites from the Broken River region of northeastern Australia. *Records of the Western Australian Museum Supplement*, **58**, 65–80.

FEIST, R., MCNAMARA, K. J., CRÔNIER, C. & LEROSEY-AUBRIL, R. 2009. Patterns of extinction and recovery of phacopid trilobites during the Frasnian–Famennian (Late Devonian) mass extinction event, Canning Basin, Western Australia. *Geological Magazine*, **146**, 12–33.

FORTEY, R. A. & OWENS, R. M. 1997. Evolutionary history. *In*: KAESLER, R. L. (ed.) *Treatise on Invertebrate Paleontology, Part O Arthropoda 1. Trilobita, revised. Volume 1*. Geological Society of America and University of Kansas Press, Lawrence, KS, 249–287.

FRANKE, W. & PAUL, J. 1980. Pelagic redbeds in the Devonian of Germany – deposition and diagenesis. *Sedimentary Geology*, **25**, 231–256.

GEORGE, A. & CHOW, N. 2002. The depositional record of the Frasnian/Famennian boundary interval in a fore-reef succession, Canning Basin, Western Australia. *Palaeogeography, Palaeoclimatology, Palaeoecology*, **181**, 347–374.

GEORGE, A., TRINAJSTIC, K. M. & CHOW, N. 2009. Frasnian reef evolution and palaeogeography, SE Lennard Shelf, Canning Basin, Australia. *In*: KÖNIGSHOF, P. (ed.) *Devonian Change: Case Studies in Palaeogeography and Palaeoecology*. Geological Society, London, Special Publications, **314**, 73–107, http://doi.org/10.1144/SP314.4

GEORGE, A., CHOW, N. & TRINAJSTIC, K. M. 2014. Oxic facies and the Late Devonian mass extinction, Canning Basin, Western Australia. *Geology*, **42**, 327–330.

GEORGE, A. D., SEYEDMEHDI, Z., & CHOW, N. 2013. Late Devonian–Early Carboniferous tectonostratigraphic framework for northern Canning Basin carbonate platform evolution. *In*: *Proceedings of the West Australian Basins Symposium 2013*. Petroleum Exploration Society of Australia, Perth, 1–16.

GIRARD, C., KLAPPER, G. & FEIST, R. 2005. Subdivision of the terminal Frasnian *linguiformis* conodont Zone, revision of the correlative interval of Montagne Noire Zone 13, and discussion of stratigraphically significant associated trilobites. *In*: OVER, J. R., MORROW, J. R. & WIGNALL, P. B. (eds) *Understanding Late Devonian and Permian–Triassic Biotic and Climatic Events: Towards an Integrated Approach*. Elsevier, Amsterdam, 181–198.

JELL, P. A. 1978. Trilobite respiration and genal caeca. *Alcheringa*, **2**, 251–260.

JOACHIMSKI, M. M. & BUGGISCH, W. 1993. Anoxic events in the late Frasnian – causes of the Frasnian–Famennian faunal crisis? *Geology*, **21**, 675–678.

KAYSER, E. 1889. Ueber einige neue oder wenig gekannte Versteinerungen des rheinischen Devons. *Zeitschrift der deutschen geologischen Gesellschaft*, **41**, 288–296.

KLAPPER, G. 1989. The Montagne Noire Frasnian (Upper Devonian) conodont succession. *In*: MCMILLAN, N. J., EMBRY, A. F. & GLASS, D. J. (eds) *Devonian of the World. Paleontology, Paleoecology, Biostratigraphy*. Canadian Society of Petroleum Geology, Memoirs, **14**, 449–468.

KLAPPER, G. 2007. Frasnian (Upper Devonian) conodont succession at Horse Spring and correlative sections, Canning Basin, Western Australia. *Journal of Paleontology*, **81**, 513–537.

KLAPPER, G. & BECKER, B. T. 1999. Comparison of Frasnian (Upper Devonian) conodont zonations. *Bolletino della Società Paleontologica Italiana*, **37**, 339–348.

LAUGHLIN, S. B., DE RUYTER VAN STEVENINCK, R. R. & ANDERSON, J. C. 1998. The metabolic cost of neural information. *Nature Neuroscience*, **1**, 36–41.

LEROSEY-AUBRIL, R. 2006. Ontogeny of *Drevermannia* and the origin of blindness in Late Devonian proetoid trilobites. *Geological Magazine*, **143**, 89–104.

LEROSEY-AUBRIL, R. & FEIST, R. 2005. Post-protaspis ontogeny of the blind cyrtosymboline *Helioproetus* (Trilobita) from the late Famennian of Thuringia, Germany. *Senckenbergiana lethaea*, **85**, 119–129.

LEROSEY-AUBRIL, R. & FEIST, R. 2012. Quantitative approach to diversity and decline in Late Palaeozoic trilobites. *In*: TALENT, J. A. (ed.) *Earth and Life*. Springer, Berlin, 535–555.

MAMET, B. & PRÉAT, A. 2005. Why is 'red marble' red? *Revista Española de Micropaleontologia*, **37**, 13–21.

MAMET, B. & PRÉAT, A. 2006. Iron-bacterial mediation in Phanerozoic red limestones: state of the art. *Sedimentary Geology*, **185**, 147–157.

MCKINNEY, M. L. & MCNAMARA, K. J. 1991. *Heterochrony: the Evolution of Ontogeny*. Plenum, New York.

MCNAMARA, K. J. & FEIST, R. 2006. New styginids from the Late Devonian of Western Australia – the last corynexochid trilobites. *Journal of Paleontology*, **80**, 981–992.

MCNAMARA, K. J. & FEIST, R. 2008. Patterns of trilobite evolution and extinction during the Frasnian/Famennian mass extinction event, Canning Basin, Western Australia. *In*: RÁBANO, I., GOZALO, R. & GARCÍA-BELLIDA, D. (eds) *Advances in Trilobite Research*. Cuadernos del Museo Geominero, **9**, 269–274.

MCNAMARA, K. J., FEIST, R. & EBACH, M. 2009. Patterns of evolution and extinction in the last harpetid trilobites during the Late Devonian (Frasnian). *Palaeontology*, **52**, 11–33.

MERRY, J. W., KEMP, D. J. & RUTOWSKI, R. L. 2011. Variation in compound eye structure: effects of diet and family. *Evolution*, **65**, 2098–2110.

MURPHY, A. E., SAGEMAN, B. B. & HOLLANDER, D. J. 2000. Eutrophication by decoupling of the marine biogeochemical cycles of C, N, and P: a mechanism for the Late Devonian mass extinction. *Geology*, **28**, 427–430.

NIVEN, J. E. & LAUGHLIN, S. B. 2008. Energy limitation as a selective pressure on the evolution of sensory systems. *Journal of Experimental Biology*, **211**, 1792–1804.

NOVACK-GOTTSHALL, P. M. 2008. Ecosystem-wide body-size trends in Cambrian-Devonian marine invertebrate lineages. *Paleobiology*, **34**, 210–228.

PLAYFORD, P. E. 1980. Devonian 'Great Barrier Reef' of the Canning Basin, Western Australia. *American*

Association of Petroleum Geologists Bulletin, **64**, 814–840.

PLAYFORD, P. E. 1981. *Devonian Reef Complexes of the Canning Basin, Western Australia. Geological Society of Australia Fifth Australian Geological Convention Field Excursion Guidebook*. Geological Society of Australia, Sydney.

PLAYFORD, P. E. 1984. Platform-margin and marginal-slope relationships in Devonian reef complexes of the Canning Basin. *In*: PURCELL, P. G. (ed.) *The Canning Basin. Proceedings of Petroleum Exploration Society of Australia/Geological Society of Australia Symposium*. Petroleum Exploration Society of Australia, Perth, 189–213.

PLAYFORD, P. E. & LOWRY, D. C. 1966. *Devonian Reef Complexes of the Canning Basin, Western Australia*. Geological Survey of Western Australia Bulletin, **118**.

PLAYFORD, P. E., HOCKING, R. M. & COCKBAIN, A. E. 2009. *Devonian Reef Complexes of the Canning Basin, Western Australia*. Geological Survey of Western Australia Bulletin, **145**.

PRÉAT, A., EL HASSANI, A. & MAMET, B. 2008. Iron bacteria in Devonian carbonates (Tafilalt, Anti-Atlas, Morocco). *Facies*, **54**, 107–120.

PŘIBYL, A. & VANĚK, J. 1986. A study of the morphology and phylogeny of the family Harpetidae Hawle and Corda, 1847 (Trilobita). *Sborník národního Muzea v Praze*, **42B**, 1–72.

RICHTER, R. 1913. Beiträge zur Kenntnis devonischer Trilobiten. Oberdevonische Trilobiten. *Abhandlungen der Senckenbergischen Naturforschenden Gesellschaft*, **31**, 341–393.

RICHTER, R. 1914. Über das Hypostom und einige Arten der Gattung *Cyphaspis*. *Centralblatt für Mineralogie*, **1914**, 306–317.

RICHTER, R. & RICHTER, E. 1926. *Die Trilobiten der Oberdevons. Beiträge zur Kenntnis devonischer Trilobiten. IV*. Abhandlungen der Preussischen Geologischen Landesanstalt Neue Folge, **99**.

RICHTER, R. & RICHTER, E. 1955. Scutelluidae n.n. (Tril.) durch 'kleine Änderung' eines Familien-Namens wegen Homonymie. *Senckenbergiana lethaea*, **36**, 291–293.

SANDBERG, C. A., ZIEGLER, W., DREESEN, R. & BUTLER, J. L. 1988. Late Frasnian mass extinction: conodont event stratigraphy, global changes, and possible causes. *Courier Forschungsinstitut Senckenberg*, **102**, 263–307.

TAPPAN, H. 1986. Phytoplankton: below the salt at the global table. *Journal of Paleontology*, **60**, 545–554.

TRIBOVILLARD, N., AVERBUCH, O., DEVLEESCHOUWER, X., RACKI, G. & RIBOULLEAU, A. 2004. Deepwater anoxia over the Frasnian–Famennian boundary (La Serre, France): a tectonically induced oceanic anoxic event? *Terra Nova*, **16**, 288–295.

WOOD, R. 2004. Palaeoecology of a post-extinction reef: Famennian (Late Devonian) of the Canning Basin, north-western Australia. *Palaeontology*, **47**, 415–445.

ZENKER, J. C. 1833. *Beiträge zur Naturgeschichte der Urwelt*. Druck und Verlag von Friedrich Mauke, Jena.

ZIEGLER, W. & SANDBERG, C. A. 1990. The Late Devonian standard conodont zonation. *Courier Forschungsinstitut Senckenberg*, **121**, 1–115.

Famennian survivor turiniid thelodonts of North and East Gondwana

VACHIK HAIRAPETIAN[1], BRETT P. A. ROELOFS[2], KATE M. TRINAJSTIC[3]* & SUSAN TURNER[3,4]

[1]*Department of Geology, Khorasgan Branch, Islamic Azad University, PO Box 81595-158, Esfahan, Iran*

[2]*Department of Geology, Curtin University, WA 6102, Australia*

[3]*WA-OIGC/Applied Chemistry, Curtin University, WA 6102, Australia*

[4]*Queensland Museum, Ancient Environments, 122 Gerler Road, Hendra, Qld 4011, Australia*

**Corresponding author (e-mail: K.Trinajstic@curtin.edu.au)*

Abstract: Microvertebrate samples from the Upper Devonian Hojedk section, southeastern Iran, and the Napier Formation, northwestern Australia, have yielded scales of agnathan thelodonts, dated as early/mid-Famennian (*crepida–marginifera/trachytera* conodont zones). These scales are referred to *Arianalepis megacostata*, a new genus and species, and *Arianalepis* sp. indet., a second indeterminate species of this new turiniid genus. Further recorded scales of *Australolepis seddoni* from the Napier Formation confirm the age range for this taxon as extending into the late Frasnian. The new remains post-date the previously youngest thelodonts from Iran and Western Australia and provide the first evidence of thelodonts surviving the Frasnian–Famennian extinction events.

Thelodonts (jawless fishes) have a long history in the fossil record, appearing at least by the Late Ordovician (Turner *et al.* 2004) and becoming extinct at some point before the end of the Devonian along with most other agnathans, with the exception being the extant cyclostomes (Märss *et al.* 2007). Despite the relative lack of taxa based on articulated remains compared to scales (*c.* 25: *c.* 140), the latter have proved useful in biostratigraphic correlation for many decades, particularly for between shallow-marine and non-marine sequences (e.g. Turner 1993, 1997; papers in Blieck & Turner 2000). Although articulated taxa are fewer (Märss *et al.* 2007), most are known from Silurian deposits and none are known from the Southern Hemisphere or in post-Emsian or younger sediments. The youngest known occurrences of thelodont taxa were previously reported from Western Australia and Iran as isolated scales in Frasnian strata (Turner & Dring 1981; Trinajstic 2000, 2001; Turner *et al.* 2002; Trinajstic & George 2009) until the discovery in the early 2000s of the first Famennian thelodont scales from Iran (Hairapetian & Turner 2003; Turner & Hairapetian 2005; Hairapetian 2008).

Here we report in detail the first Famennian thelodonts that co-occur with conodont elements and jawed vertebrate assemblages in the Hojedk section in the north of Kerman, southeastern Iran (Fig. 1), and also the recent discovery of thelodont scales from a mid-Famennian sequence in the South Oscar Range, Canning Basin, northwestern Australia (Fig. 2). The Iranian scales are placed within a new turiinid genus *Arianalepis* gen. nov., which represents the youngest known thelodont remains; the Australian material is more sparse and so its assignment is less certain. These new discoveries furthermore support the closeness of faunal relationships in the mid-Palaeozoic of northern Gondwana, especially between Iran and Australia. In addition, the finds offer an opportunity to further establish and extend biostratigraphic schemes utilizing thelodont taxa in the Devonian (Turner 1997; Long & Trinajstic 2000; Young & Turner 2000).

Geographical and geological setting

Iran

The Hojedk section (*c.* 4 km west of Haruz village and *c.* 48 km north of Kerman; 30°43′N, 57°0′E; Fig. 1) is measured on the southeastern flank of Kuh-e-Kanseh Mountain, Kerman Province, central Iran. The stratigraphy of the Hojedk Devonian strata has recently been studied in detail by Wendt *et al.* (2005), Gholamalian & Kebriaei (2008)

From: Becker, R. T., Königshof, P. & Brett, C. E. (eds) 2016. *Devonian Climate, Sea Level and Evolutionary Events*. Geological Society, London, Special Publications, **423**, 273–289.
First published online June 10, 2015, http://doi.org/10.1144/SP423.3

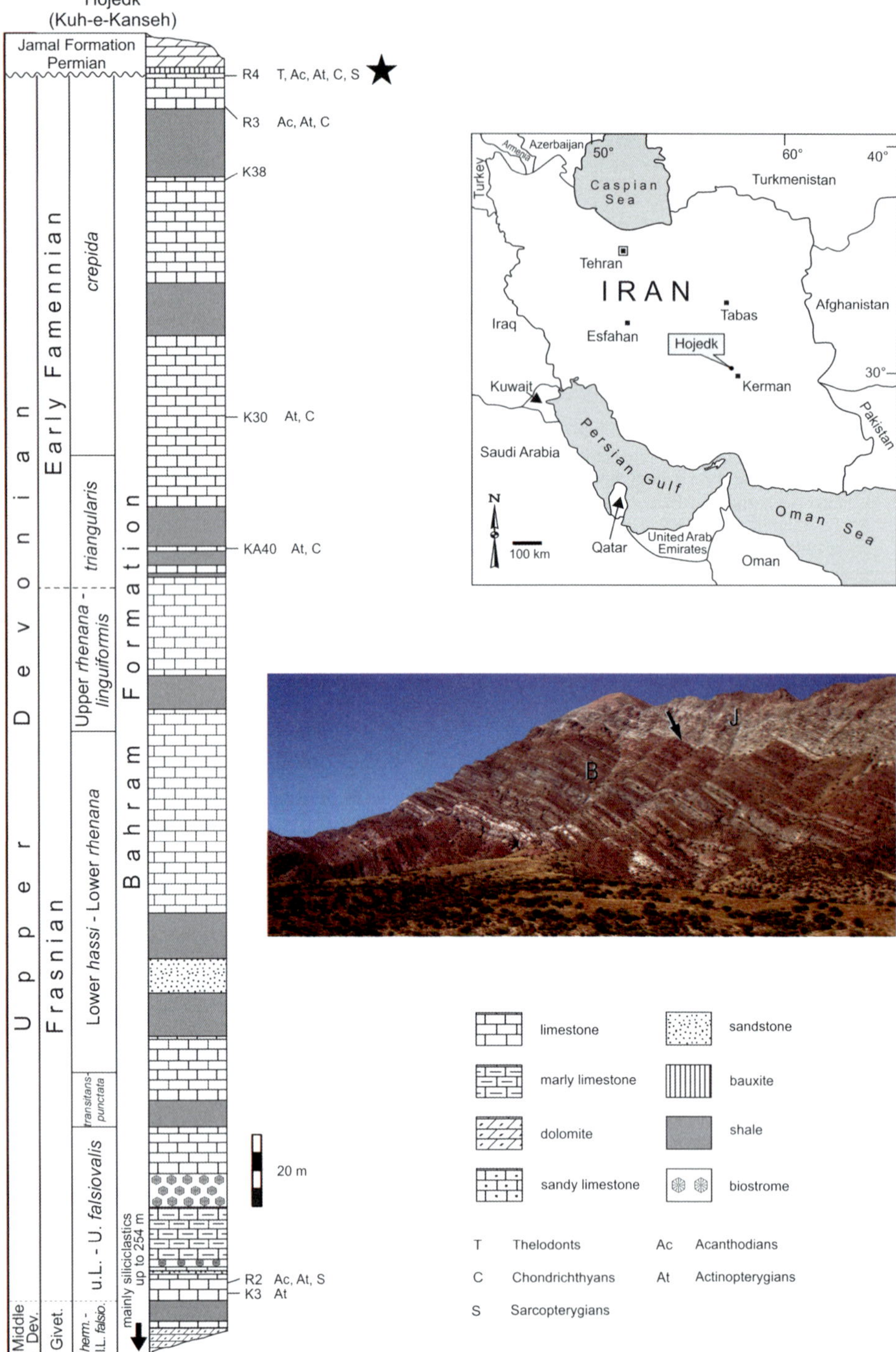

Fig. 1. Stratigraphy and location of the studied Iranian sequence. Simplified stratigraphic column of the Hojedk section in central Iran showing the main facies and fish-bearing horizons; the asterisk marks the bed with thelodont scales. Map of Iran showing the location of the section. Photograph of the Hojedk section showing the Bahram Formation (B) and the Permian Jamal dolomites (J). The distinct level of the unconformity is indicated by the arrow.

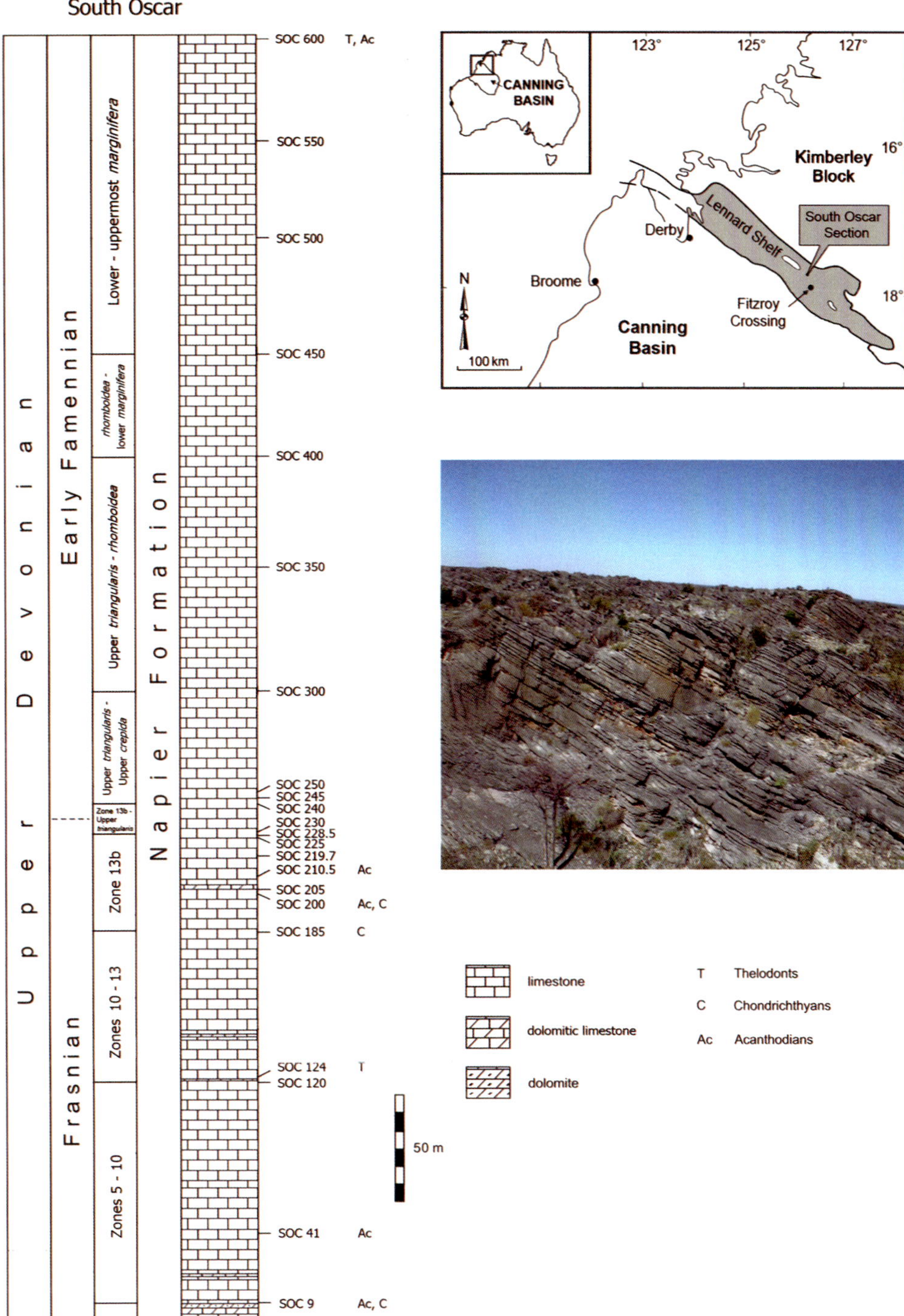

Fig. 2. Stratigraphy and location of the studied Australian sequence with simplified stratigraphic column of the South Oscar Range (SOC) section in the Canning Basin, Western Australia, showing the main facies and fish-bearing horizons (after Playton *et al.* 2013). Map of the Lennard shelf showing the location of the studied section, with inset map of Australia showing the location of the Canning Basin. Photograph of the South Oscar Range section in the Napier Formation.

and Gholamalian *et al.* (2013). The underlying Lower (?) Devonian/Eifelian Padeha Formation commences with a breccia unit followed by sandstones and dolomites. There is a disconformity at the top of the Padeha Formation, where a sandstone unit is overlain by a dolomitic horizon of the Bahram Formation containing small gastropods, large plates of the placoderm *Holonema* and plant remains; the unit has been dated as late Givetian based on conodonts (see Gholamalian & Kebriaei 2008). The limestones in the uppermost Bahram Formation are overlain with an apparent major temporal unconformity by dolomites of the Permian Jamal Formation. Conodont samples collected from the limestone beds in the upper part of the section, however, revealed an early Famennian age, from the Lower *Palmatolepis crepida* Zone (see Gholamalian & Kebriaei 2008, fig. 1, table 1, pp. 183–184).

The new thelodont material came from Sample R4, a red sandy limestone bed in the uppermost part of the section, just below the level of the Devonian/Permian unconformity (Fig. 1) (Hairapetian & Turner 2003; Turner & Hairapetian 2005; Hairapetian 2008). The bed is now dated as early Famennian based on conodonts and fish remains. Associated with the thelodont scales are conodont elements indicative of a shallow-water fauna, including *Icriodus alternatus* cf. *helmsi* Sandberg & Dreesen, 1984, *Icriodus cornutus* Sannemann, 1955, *Pelekysgnathus inclinatus* Thomas, 1949, *Polygnathus semicostatus* Branson & Mehl, 1934 and the *Polygnathus communis* group Branson & Mehl, 1934, dating of which spans the Middle–Upper *Palmatolepis crepida* Zone (Hairapetian 2008). The new thelodont taxon is associated with stratigraphical index chondrichthyan species from the Hojedk section (Fig. 1); the assemblage comprises several chondrichthyan and actinopterygian taxa, with shark teeth of phoebodonts *Phoebodus gothicus gothicus* Ginter, 1990, *Phoebodus turnerae* Ginter & Ivanov, 1992 and *Phoebodus* aff. *turnerae* Ginter & Ivanov, 1992, the protacrodonts *Protacrodus* sp. and *Deihim mansureae* Ginter, Hairapetian & Klug, 2002, the cladodontomorphs *Ctenacanthus* sp. and Elasmobranchii gen. et sp. indet., as well as sarcopterygian and actinopterygian scales, all supporting an early Famennian age (Ginter *et al.* 2002; Hairapetian 2008, fig. 4).

Australia

A section through the proximal slope facies of the Napier Formation was measured at South Oscar Range (17°55′S, 125°17′E; Fig. 2) (Playton *et al.* 2013), located along the southwestern edge of the Lennard Shelf in the Kimberley Region of NW Western Australia (Playford *et al.* 2009). The Napier Formation unconformably overlies Proterozoic basement and in outcrop ranges from early Frasnian to late Famennian. The lithology of the measured section is predominantly limestone with the basal part of the section comprising marginal and slope-derived mega breccias with minor dolomitic microbial boundstones, rudstones and breccias. The upper interval of the South Oscar Range section from approximately 185 m consists mainly of platform-derived skeletal packstone–grainstone facies as well as *in situ* stromatactoid boundstones and skeletal–peloid-coated packstones to grainstones. Overlying the mixed slope facies of the Napier Formation are the siliciclastic shelf deposits of the uppermost Devonian to Lower Carboniferous Fairfield Group (Playford & Lowry 1966). The Frasnian–Famennian boundary is placed within the upper part of the section, between 223.8 and 233.2 m, with its location based on conodont data (Hansma *et al.* 2015) in addition to the last occurrence of Frasnian stromatoporoids (Hurley 1986; Playford *et al.* 1989).

The new thelodont material comes from Sample South Oscar Range (SOC) 600 in the uppermost bed of the section and co-occurs with acanthodian and palaeoniscoid scales. This horizon has been dated as ranging from the Upper *Palmatolepis marginifera* to *Palmatolepis rugosa trachytera* conodont zones, which is supported by the overlap of *Palmatolepis gracilis gracilis* Branson & Mehl, 1934, and *Palmatolepis minuta minuta* Branson & Mehl, 1934 in Sample SOC 600. A lower *Palmatolepis rhomboidea* age is not supported because *Scaphignathus velifer* Helms, 1959 is not known below the Upper *Palmatolepis marginifera* conodont Zone within the Canning Basin and this taxon has been recovered below from Sample SOC 500 (Fig. 2). The Frasnian conodont succession at South Oscar Range is conformable with the Montagne Noire (MN) conodont zonation (Klapper 1989).

Additional thelodont scales identified here as *Australolepis seddoni* Turner & Dring, 1981 (Figs 3 & 4) have been found in the lower part of the section (SOC 124), dated as Frasnian MN Zone 10. Immediately above this horizon (MN 13b), there are shark teeth of *Stethacanthus* sp. as well as palaeoniscoid and acanthodian scales. The Frasnian part of the section has a diverse palmatolepid conodont fauna indicative of open-marine conditions (Sandberg & Ziegler 1979).

Material and methods

A buffered solution of 10% acetic acid was employed to extract the Iranian Sample R4 residue and specimens were then picked using a sieve of 0.177 mm mesh. Sixteen thelodont scales (Figs

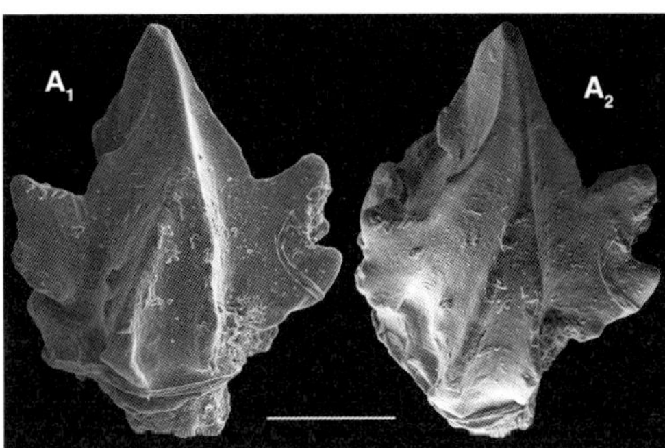

Fig 3. Scale of *Australolepis seddoni* WAM 13.10.2 from Sample SOC 124, South Oscar Range section, Canning Basin, Western Australia. A_1, dorsal view; A_2, dorsolateral view. Scale bar 0.2 mm.

5–7) have been found, most of which are broken and reddish brown in colour (Fig. 7). The scales are densely covered with adhering sedimentary quartz grains, which are not easy to remove with a needle (Fig. 7). Scanning electron micrographs were prepared in Esfahan with a Leica 360 scanning electron microscope and in the Institute of Palaeobiology, Polish Academy of Sciences (Warsaw, Poland) using a Philips XL 20. The Iranian specimens are deposited in the Department of Geology, Azad University, Esfahan (AEU).

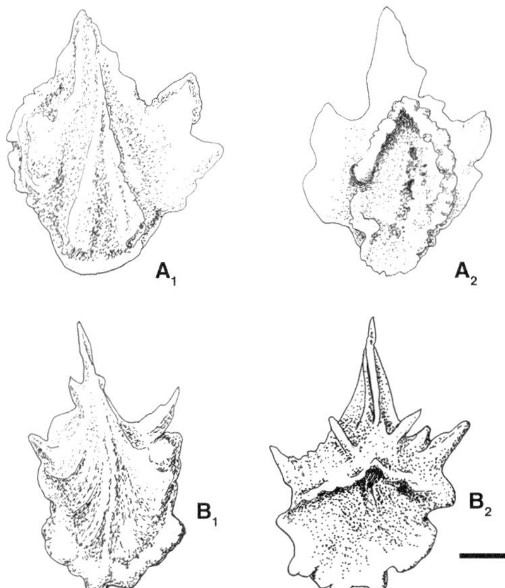

Fig. 4. Drawings of scales of *Australolepis seddoni* WAM 13.10.2 from Sample SOC 124, South Oscar Range section, Canning Basin, Western Australia. A_1, crown view; A_2, ventral view; B_1, crown view; B_2, ventral view. Scale bar 0.2 mm.

Bulk rock samples (20 kg) were taken within the measured South Oscar Range section. Acid digestion of the carbonates in 10% acetic acid and heavy-liquid separation of the residue was undertaken at Macquarie University and the fractionated residues returned for picking. One of two scales recovered from SOC 124 was photographed using a Zeiss EVO 40XVP and both scales were drawn while magnified using a binocular microscope. One scale from SOC 600 was photographed using a Leica XX2 V 7S stereomicroscope camera version 3.4.1 at the Western Australian Museum and drawn under a binocular microscope. Owing to the paucity and poor preservation of the SOC 600 scale, scanning electron microscope or histological work was not possible. Specimens are deposited in the Western Australian Museum (WAM).

The 13-fold Frasnian conodont zonation, first proposed for the MN succession (Klapper 1989) and subsequently modified to a 15-fold zonation (Girard *et al.* 2005), was used to describe the stratigraphic ranges of the conodont and vertebrate taxa recovered from the Canning Basin sections. The standard conodont zonation of Ziegler & Sandberg (1990) were used for both sections to describe Famennian age ranges and for the Frasnian in the Iranian sections, because the ranges of MN zonal markers have not yet been replicated for Iran.

SYSTEMATIC PALAEONTOLOGY

Thelodonti Jaekel, 1911

Order **Thelodontiformes** Kiaer in Kiaer & Heintz, 1932

Family **Turiniidae** Obruchev, 1964

Australolepis seddoni Turner & Dring, 1981
(Figs 3 & 4)

Material: WAM 13.10.2; one scale.

Locality and geology. South Oscar Range, Canning Basin, Western Australia (17°55′S, 125°17′E). Bed SOC 124, CZ 10, reef, margin and slope-derived skeletal limestone: Upper Devonian Napier Formation.

Stratigraphic range. Australia: Napier Formation, South Oscar Range, Canning Basin: MN Zone 10, MN Zones 6–10, Virgin Hills Formation, Horse Spring, Canning Basin; earliest Frasnian to MN Zone 10 in the Gneudna Formation type section, Carnarvon Basin, Western Australia (Turner & Dring 1981; Trinajstic 2000, 2001; Turner *et al.* 2002; Trinajstic & George 2009). In Iran the range extends from the early Frasnian *Mesotaxis falsiovalis* Zone, MN Zones 1–3, to, in the Chahriseh

Fig. 5. Scanning electron micrographs of early Famennian thelodont scales of *Arianalepis megacostata* gen. et sp. nov. from sample R4, Hojedk, Kerman Province, Iran. A, B, head scales; C, cephalopectoral scale; and D–G, trunk scale. A, AEU 720 in crown view; B, AEU 721 in lateral crown view; C, AEU 722 in lateral view; D, AEU 723 in antero-crown (D_1) and lateral (D_2) view; B, AEU 724 in lateral view; F, AEU 725 in crown view; G, AEU 726, holotype in antero-crown (G_1) and lateral (G_2) view. Scale bar 0.2 mm.

section, *Palmatolepis hassi* Zone, MN Zone 10 (Hairapetian *et al.* 2000; Yazdi & Turner 2000).

Description. The scales measure less than 1 mm in length with an elliptical base and triangular crown, terminating in a sharp apex. The anterior face of the tripartite crown has two bifurcating ribs that fuse to form a single rib that extends to the crown apex (Figs 3 & 4, drawing A_1). The anterior ribs are ornamented with small tubercles (Fig. 4, drawing A_1). A well-developed, ridged lateral lappet is present at the medial region of the scale (Figs 3 & 4, drawings A_1, B_1). The lateral extent of the lappet curves posteriorally and bifurcates to form two anteriorward-pointing projections (Fig. 4, drawing B_1). The posterior lappets are reduced compared with the medial lappets and lack the terminal projections. The mesial face adjoins the central ridge. A shallow neck separates the crown from a narrow base, which does not possess an anterior spur. The base exhibits

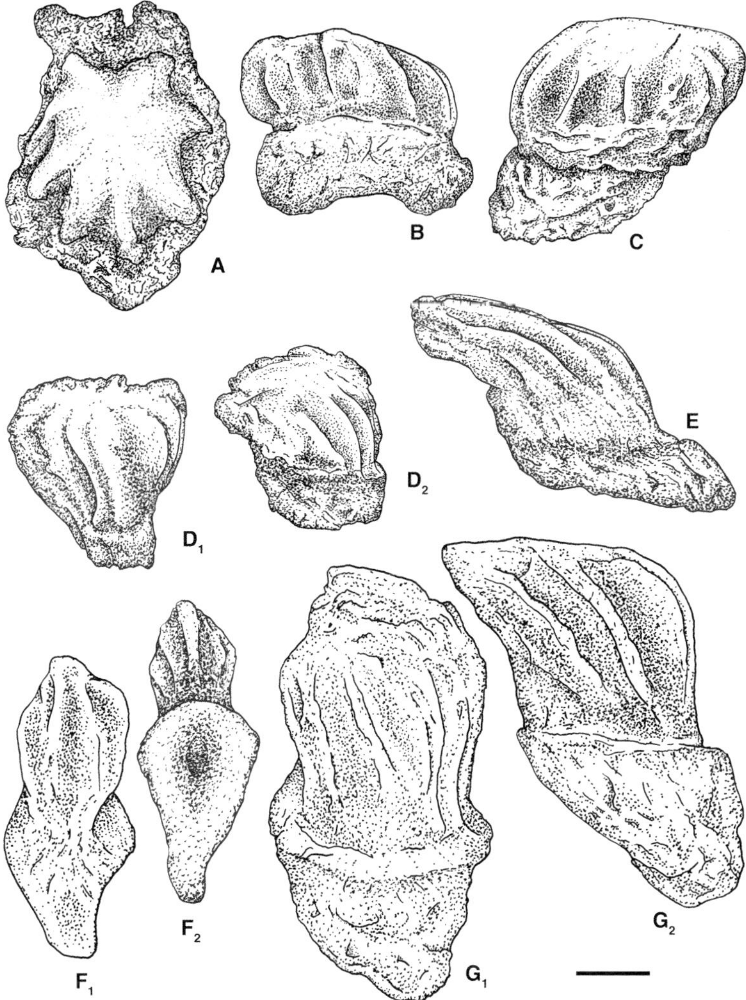

Fig. 6. Drawings of scales of *Arianalepis megacostata* gen. et sp. nov. from sample R4, Hojedk, Kerman Province, Iran. A, AEU 720 in crown view; B, AEU 721 in lateral crown view; C, AEU 722 in lateral view; D, AEU 723 in antero-crown (D_1) and lateral (D_2) view; E, AEU 724 in lateral view; F, AEU 725 in crown view (F_1) and ventral (F_2) view; G, AEU 726, holotype in antero-crown (G_1) and lateral (G_2) view. Scale bar 0.2 mm.

the lobate thickenings typical in scales of this species (Fig. 4, drawings A_2, B_2). An elongate pulp canal is visible on the underside of the scale.

Histology. As only two single scales were recovered no thin sections were made. However, the large pulp cavities are typical of turiniid histology (Märss *et al.* 2007).

Taphonomic analyses. The scale is pristine. The crowns are uniformly light brownish in colour with a white base, typical of thelodont scales, where the two hard tissues – dentine in the crown and neck, and aspidin in the base (see e.g. Märss *et al.* 2007) – are often present, with very different colours related to the varying mineralization effects based on the porosity of the tissue.

No diagenetic alteration is detectable and associated conodont remains have a conodont alteration index (CAI) of 0.5.

Remarks. These scales conform to the original diagnosis of *A. seddoni* (Turner & Dring, 1981) and are attributed to this taxon. Their small size, gracile construction, bifurcating crown ribs and smooth shallow neck, with a base wider than the crown, are all consistent with the type diagnosis. Small tubercles are present on the crown ribs, which are

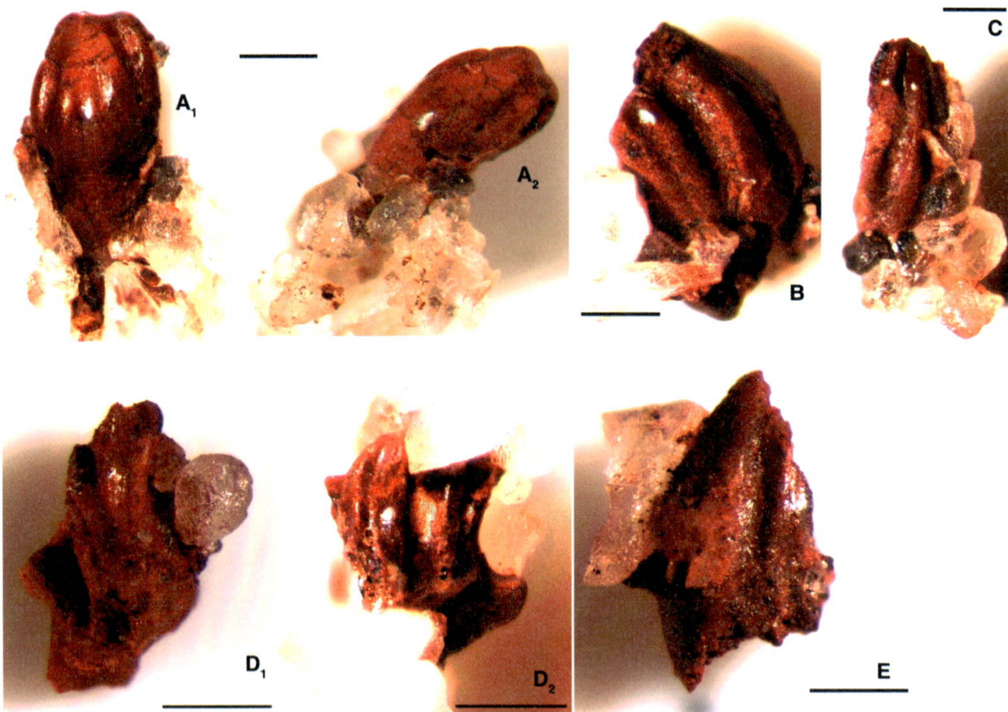

Fig. 7. Photographs of early Famennian thelodont scales from sample R4, Hojedk, Kerman Province, Iran, to show colour and taphonomic nature of the adhering quartz grains. (A) AEU 782 in crown (**A$_1$**) and lateral-crown views (**A$_2$**); (**B**) AEU 783 in lateral crown view; (**C**) AEU 783, in basal view; (D) AEU 784 in antero-crown (**D$_1$**) and lateral-crown (**D$_2$**) view; (**E**) AEU 785 in crown view. Scale bars, 0.2 mm.

also seen in scales from Horse Spring (Turner 1997; Trinajstic & George 2009). Following currently accepted concepts of squamation patterns (Märss *et al.* 2007), the scales originate from the cephalo-pectoral region of the body.

The scale conforms to those from the known uppermost range of *A. seddoni* (MN Zone 10) within Australia based on well-constrained conodont remains (Trinajstic & George 2009).

Genus *Arianalepis* gen. nov.

Etymology. From 'Ariana', an old name for the country of origin of the first Famennian thelodonts; and Greek: *lepis*, a scale.

Diagnosis. As for type and only species.

Stratigraphic range. Early Famennian, *crepida* Zone.

Type and only species. Arianalepis megacostata sp. nov.

Remarks. The Iranian and Western Australia thelodont scales are the youngest to be recorded to date; those from Iran were reported earlier and considered tentatively to be within the family Turiniidae, which is the only clade found thus far in post-Lochkovian Gondwana (Hairapetian & Turner 2003; Märss *et al.* 2007). Hairapetian & Turner (2003) and then Märss *et al.* (2007) retained them as an undetermined turiniid taxon. However, as there are significant morphological differences from earlier turiniid taxa and considering their much younger age, the new scales are referred here to a new genus (see further discussion below).

Arianalepis megacostata gen. et sp. nov.
(Figs 5–7)

Synonymy

2003 new turiniid Hairapetian & Turner: 26–27.
2005 new turiniid Turner & Hairapetian: 24.
2007 Turiniidae gen. et sp. indet. Märss *et al.*: 110.

Type specimens. Holotype: AEU 726, trunk scale (Figs 5, micrographs G_1, G_2 & 6, drawings G_1, G_2); paratypes: 15 scales (Fig. 7A–E), AEU 720–725, 782–790.

Type locality and stratigraphy. Section *c*. 4 km W of Haruz village; 30°43′N, 57°0′E, SE Kuh-e-Kanseh Mt, Hojedk, *c*. 48 km N of Kerman, central Iran. Sample R4, red sandy limestone bed, Bahram Formation, Late Devonian, early Famennian, *crepida* Zone.

Stratigraphic range. *Crepida* Zone, early Famennian.

Etymology. From Latin *mega* 'large' and *costata*, 'ribs' for the ribbed nature of the crowns.

Diagnosis. Turiniid with robust medium-sized scales with a few simple coarse ribs on crowns. Base is horizontally extended with anterior basal extensions.

Description. The scales all show a robust structure and measure less than 1 mm in length.

The head scales are rounded as in most thelodonts with robust crowns having a few, up to eight, strong ribs. There are eight ribs radiating from a high, rounded central apex to a shallow neck; wide troughs separate the ribs (Figs 5, micrographs A, B & 6, drawings A, B). Cephalopectoral scales are poorly preserved but can be seen to be more elliptical (Figs 5, micrograph C & 6, drawing C). The crowns exhibit simple coarse ribbing and well-developed bases.

Trunk scale crowns (Figs 5, micrographs D_2–G_2 & 6, drawings D_1, D_2, G_1, G_2) are characterized by simple coarse ribbing and a protruded, prominent median area, which can be flat and horizontally placed as in the holotype scale shown in Figure 5, micrograph G_1, G_2. One scale (Fig. 5, micrograph F) shows a smooth, wide median and two lateral areas at a lower level. Another scale shown in side view (Fig. 5, micrograph E) exhibits the high crown with ribs down to the shallow neck and the horizontally placed base with the scale axis at around 45°. Some indication of split ribs is seen in other scales (Fig. 7). The anterior base is developed into a single, thick basal root or spur, which extends in a lateral plane (Fig. 5, micrograph F). The holotype scale (Figs 5, micrographs G_1, G_2 & 6, drawings G_1, G_2) exhibits a relatively deeper base with an incomplete anterior anchoring root directed vertically downwards.

Unlike many turiniid taxa, especially in Gondwana (e.g. Turner 1997), there is no sign of micro-ornament on the crown surface of any scales but this is probably a factor of the removal of the external shiny layer of dentine (see taphonomic remarks below).

Histology. Owing to the relatively few scales and poorly preserved material, direct histological studies were not possible. However, all exhibit the typical single rounded pulp opening in the base found in thelodontidid scales as is usual in turiniids (cf. Märss *et al.* 2007); a quite large pulp cavity is seen in Figure 5, micrograph B, and remnants of the pulp canal in Figure 5, micrograph C. Based on comparison the crowns are formed of typical thelodont orthodentine and the base of aspidin. As noted above, the outer layer of dentine (vasodentine called 'enameloid' by some workers) is not as shiny as normal and the dentine tubules of the crown are exposed clearly in some scales (Fig. 5, micrographs G_1, G_2). The pulp cavity is relatively large and the scales are neither in their first dentine crown 'cap' stage nor senescent with large overgrown bases with closed pulp openings.

Taphonomic remarks. There are two possibilities that could explain why some of the Iranian thelodont scales are densely covered with adhering sedimentary quartz grains (e.g. Figs 5, micrographs G_1, G_2 & 7A–E). This could imply reworking and mean that the thelodont scales are older than the other remains, that is, not Famennian, and the presence in some scales of crown ribs divided into two might support this, as there is a similarity to the older Iranian turiniid *Turinia hutkensis* Blieck & Goujet, 1978. However, comparing the state of preservation of the thelodont scales with the other fish microfossils, they are identical in preservation and with the same deep red colour taken from the surrounding rock/grains. The conodont elements were also checked for evidence of reworking but all are light grey in colour, indicating an identical and low CAI, so that reworking does not seem to be a possibility.

Therefore, this leaves the alternative option that the fossils are diagenetically affected and therefore we need to look at post-depositional environmental conditions to account for these changes. The most likely explanation is that the phenomenon is a consequence of water chemistry after the scales were shed or lost from the dead animal; there is no sign of algal or fungal boring as seen in certain marginal environments (e.g. Märss *et al.* 2007). Other Gondwana samples of turiniid scales show similar adhering grains, such as those from central Australia and New Zealand, but these scales also show more extreme diagenetic effects such as silicification or loss of tissue structure (Turner 1997; Macadie 2002).

?*Arianalepis* sp. indet.
(Figs 8 & 9)

Material. WAM 13.10.1; one cephalopectoral scale.

Locality and stratigraphy. The South Oscar Range section (17°55′S, 125°17′E) lies 42 km NW of Fitzroy Crossing, South Oscar Range, Canning

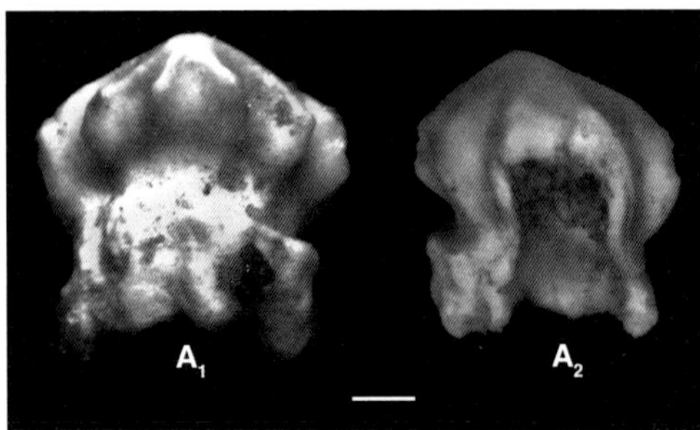

Fig. 8. Transmission microphotographs of scales of ?*Arianalepis* sp. indet. WAM 13.10.1 from sample SOC 600. A_1 and A_2 in lateral view. Scale bar 0.2 mm.

Basin, Western Australia. Sample SOC 600, white crinoidal packstone, Napier Formation, Late Devonian, Middle Famennian, *marginifera*/*trachytera* conodont zones.

Stratigraphic range. Middle Famennian, *marginifera*/*trachytera* conodont zones.

Description. The one almost complete specimen is a young, medium-sized (0.8 mm) robust scale with a small break anteriorly across the crown, neck and base (Figs $8A_2$ & 9, drawing A_{1-3}). The total range of the number of ribs cannot be determined; however, 12 broad crown ribs are reconstructed based on the spacing of the ten preserved ones. The ribs radiate from a dome-shaped apex with a centrally raised point directed posteriorly (Figs $8A_1$ & 9, drawing A_4). Two of the medial anterior ribs bifurcate halfway between the crown apex and neck–crown interface. Deep troughs separate the ribs, which diminish at a deep, narrow neck (Figs $8A_1$ & 9, drawing A_4). A series of small crenulations occurs on the terminal dorsal edge of each rib (Fig. 9, drawing A_2).

The base is approximately two-thirds the size of the crown and is separated by a high neck (Fig. 9, drawings A_2, A_3). The anterior portion of the base is broken (Figs $8A_{1-2}$ & 9, drawings A_2–A_4). The remaining preserved section indicates a roughly circular shape with a slight slope directed posteriorly. A large rounded pulp cavity and a pulp opening are present (Fig. 9, drawings A_2, A_3).

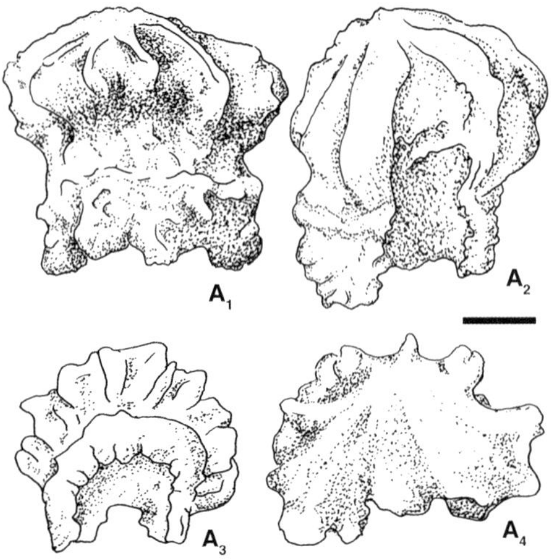

Fig. 9. Drawings of scale of ?*Arianalepis* sp. indet. WAM 13.10.1 from SOC 600. A_1 and A_2, lateral view; A_3, basal view; A_4, crown view. Scale bar, 0.2 mm.

Histology. Histology was not possible as only a single scale was recovered. However, the complete scale exhibits the typical large pulp cavity of a morphogenetically young turiniid scale where dentine has not yet grown centripetally to form a narrow pulp opening on the base.

Remarks. The scale from SOC 600 is referred tentatively to the new genus *Arianalepis* with indeterminate species, on the basis of similar morphology of a crown with few coarse ribs. Not being a body scale, however, specimen WAM 13.10.1 generally resembles all turiniid head scales and probably lacks diagnostic features. The head scale from SOC 600 has a higher number of radiating ribs: 12 compared with 8, and also seems to be more robust than the specimens from Hojedk. This might be because of normal variation, the different age of the scales, or dependent on the age of the individual animal (Märss *et al.* 2007).

Microcrenulation of scale crowns also seems to be phylogenetically significant in the Turiniidae; however, the relatively poor preservation of the Iranian scales means that micro-ornament is not preserved and, because most turiniid taxa do exhibit some ornament on the ribs, the crenulations in the Australian scale may just be due to variation within individual scales or within species.

For now we tentatively place the Australian scales within the genus *Arianalepis* gen. nov. based on the single scale.

Taphonomic analyses. The colour of the scale is uniformly white with no evidence of biogenic alteration.

Numerous, well-preserved acanthodian scales are also present within Sample SOC 600, suggesting a deeper-marine environment where thelodonts are relatively scarce (cf. Märss *et al.* 2007). The associated conodont elements are pristine with a CAI of 0.5. There is no evidence to suggest reworking in this sample as the conodont assemblage conforms to the other assemblages immediately below Sample SOC 600.

Comparison and discussion

The scales of *Arianalepis megacostata* gen. et sp. nov. from Hojedk are superficially more simple than other turiniid species from the Devonian of Gondwana and, interestingly, are most like those of the late Silurian *Turinia fuscina* Turner, 1986 of southern Australia, and mid-Late Devonian taxa, such as *T. hutkensis* Blieck & Goujet, 1978. The single complete scale from South Oscar Range shows some resemblance to the scales of *Jesslepis johnsoni* Turner, 1995 in being relatively robust, having bifurcating and deeply dissecting ribs, and a large pulp cavity. The complete scale from South Oscar Range can be distinguished based on the greater number of ribs, which do preserve micro-crenulations.

The scales from Hojedk and South Oscar Range are distinguishable from any known species. Most interesting is that these youngest thelodont scales have large robust crowns and bases unlike the slightly older taxa known from Iran and Western Australia, especially *Australolepis seddoni*, which has delicate ribbing and sculpture on the crowns. Other older Devonian taxa from the same region are *Turinia antarctica* Turner & Young, 1992 and *J. johnsoni* from the Givetian of eastern Australia and Antarctica (Turner 1995, 1997).

T. hutkensis of Iran is most similar to scales from Thailand and Antarctica (see Turner & Young 1992; Turner 1997). Some of the *Turinia* cf. *hutkensis* scales from the Chahriseh locality resemble *Turinia pagoda* Wang, Dong & Turner, 1986 and *Turinia* spp. from the western Yunnan Province and *A. seddoni* in their out-turned ridges. Scales that share features with *T. hutkensis*, such as the tight crown double-ribbing, have also been found in the central Australian basins and were earlier

Table 1. *Turiniid taxa from the Middle (Givetian) and Upper (Frasnian and Famennian) Devonian of Gondwana*

Epoch/Age	Conodont zonation		Australia: WA	Australia: NT	Australia: SA	Australia: QLD	Antarctica	Iran: SS	Iran: CEIM	China
Famennian	*Si. praesulcata*									
	Pa. gracilis expansa									
	Pa. perlobata postera									
	Pa. rugosa trachytera									
	Pa. marginifera		? *Arianalepis* sp.							
	Pa. rhomboidea									
	Pa. crepida								*Arianalepis megacostata*	
372.2	*Pa. triangularis*									
Frasnian	*Pa. linguiformis*	MN 13 c, b								
	Pa. rhenana	MN 13 a, MN 12								
		MN 11								
	Pa. jamieae	MN 10	*Australolepis seddoni*					*Australolepis seddoni*; *Turinia hutkensis*	*Australolepis seddoni*; *Turinia hutkensis*	
	Pa. rhenana	MN 9, MN 8, MN 7								
	Pa. punctata	MN 6, MN 5								
	Pa. transitans	MN 4								
382.7	*M. guanwushanensis* (= *falsiovalis*)	MN 3, MN 2, MN 1					*Turinia antarctica*			
Givetian	*norrisi*			*Turinia* sp.						
	K. disparilis									
	Sc. hermanni			?	*Turinia* sp.					
	Po. varcus			*Turinia* sp. cf. *T. pagoda*	?	*Jesslepis johnsoni*				*Turinia pagoda*
387.7	*Po. hemiansatus*									

CEIM, central–east Iran microcontinent; NT, Northern Territory; QLD, Queensland; SA, South Australia; SS, Sanadaj Sirjan; WA, Western Australia.

thought to be related to the older *Turinia australiensis* (Turner 1997: see Table 1). Could *Australolepis* be the result of paedomorphic evolution from *T. hutkensis* or *T. pagoda*? Heterochrony in the morphogenesis would leave all the scales with wide-open pulp cavities throughout life and shallow bases with only small anterior prongs, and with a base capable of only thickening sporadically, resulting in the lobate papillae that are typical of *A. seddoni* bases (cf. Fig. 4, drawing A_2). *Turinia hutkensis* and *T. antarctica* both share the fine ultrastructual lines on the crown, seen in earlier *T. australiensis* (see e.g. Turner 1997; Märss *et al.* 2007) but these are not apparent in *A. seddoni* (Fig. $4A_{1-2}$) or *A. megacostata* gen. et sp. nov. (Fig. 7, photographs A_1, D_2) although the latter does have the double ribbing seen in some scales of *T. hutkensis*.

Historically, studies on Laurasian/Laurentian thelodonts have been more numerous than those on Gondwana (Märss *et al.* 2007). This disparity began to be addressed especially during IGCP work from the 1980s onwards, when several new thelodont taxa and increased stratigraphic ranges were reported from Gondwanan faunas (e.g. Blieck *et al.* 1980; Turner *et al.* 1981, 2000; Turner 1986; Wang *et al.* 1986; Turner & Young 1992; Young & Turner 2000; Trinajstic 2001; Hairapetian 2008; Trinajstic & George 2009, table 1). Most notable was the discovery that thelodont taxa survived the Givetian–Frasnian boundary event (Turner & Dring 1981). Turiniid remains recovered from the Aztec siltstone in Antarctica (Turner & Young 1992) have shown a potential age range from the middle Givetian to the early Frasnian based on faunal association with *Antarctilamna* (Young 1993). However, the presence of the spore *Geminospora lemurata* Balme, 1960 has dated this unit as Givetian (Young 1988; Young & Turner 2000). Thelodont remains were first found in the Middle Devonian of Iran in the 1970s and other taxa in Australia, Thailand and China in conodont-dated strata (e.g. *J. johnsoni* in Queensland) (Turner & Janvier 1979; Blieck *et al.* 1980; Hamdi & Janvier 1981; Turner 1997).

The first early Frasnian thelodont scales, *A. seddoni*, were recorded from the Gneudna Formation, Carnarvon Basin, of Western Australia (Turner & Dring 1981). Spore assemblages (Balme 1988) gave a possible Givetian to Frasnian age for the lower part of the section; however, microvertebrate assemblages suggested a Frasnian age for the whole section (Trinajstic 2001). In the last decade thelodont scales identified as *T. hutkensis* and *A. seddoni* were reported from the early–middle Frasnian of Iran (e.g. Yazdi & Turner 2000; Turner *et al.* 2002) and poorly preserved scales possibly from *A. seddoni* were recovered from slightly younger Frasnian strata in the Canning Basin of Western Australia (Turner 1997, 1999). Until recently, this was the youngest record of any thelodont taxon (Turner 1997). Ranges for thelodonts have been further refined in Western Australia based on the co-occurrence of thelodonts and phoebodonts in conodont-dated sequences at Horse Springs, with the age range of *A. seddoni* extended into the upper Frasnian (Trinajstic & George 2009). Younger thelodont scales were recovered from younger early Famennian strata in the Canning Basin of Western Australia; however, as the preservation was poor these were not formally identified (Turner 1997, 1999). Subsequent studies by Hairapetian & Turner (2003) and Hairapetian (2008) reported the first Famennian scales in Iran.

Palaeobiogeography

The continuance of thelodont taxa into the Famennian may suggest a displacement post the Givetian–Frasnian extinction and the foundation of a refuge in Northern Gondwana (Kauffman & Harries 1996; Harries *et al.* 1996). As the Frasnian and Famennian turiniid fish are found in shallow to very shallow water, 'challenging' environments, for example, marginal, lagoonal and hyper-saline, were they then pre-adapted or opportunistic survivors and thus these new taxa are evidence of a survival lineage? There is no sign yet of any thelodont in the immediate earliest Famennian 'bloom' of recovery, but with evidence now in two disparate parts along the Palaeotethyan southern shore we consider *Arianalepis* gen. nov. as a Lazarus taxon.

Few agnathan fishes survived the massive Late Devonian extinctions and until recently only two thelodont taxa, *T. hutkensis* from the early–middle Frasnian of Iran (Turner *et al.* 2002) and *A. seddoni* from the early–late Frasnian (MN zones 6–10) of Western Australia and Iran (Yazdi & Turner 2000; Trinajstic 2001; Turner *et al.* 2002; Trinajstic & George 2009; Burrow *et al.* 2010) were known. However, now at least one thelodont genus with one or more species appears to have weathered the Frasnian–Famennian extinction and appears to represent a surviving Lazarus taxon (Kauffman & Harries 1996).

Australolepis seddoni is now known from the Gneudna and Virgin Hills formations of Western Australia and the Shotori Range and Chahriseh section of Iran (Turner 1997; Yazdi & Turner 2000; Turner *et al.* 2002; Trinajstic & George 2009). To date, scales have been recovered from different facies including distal and medial slope and back reef in the Canning Basin (Chow *et al.* 2013), as well as shallow carbonate ramp in Iran and the Carnarvon Basin (Hocking *et al.* 1987; Wendt *et al.* 2002; Trinajstic & George 2009). A few possible

Australolepis scales are recorded from the Late Devonian of Holy Cross Mountains (M. Ginter & S. Turner, pers. obs. 1990). The presence of thelodonts in sediments from Hull Range, Western Australia, has confirmed an early Frasnian age for these strata (Chow *et al.* 2013), where previously the date could only be constrained to the late Givetian–early Frasnian. This demonstrates again the utility of microvertebrates in general and thelodont scales in particular to help date strata where conodonts and other open-marine taxa are absent. The first appearance has been linked to conodonts by Long & Trinajstic (2000). The younger Frasnian *Australolepis* scales from the Horse Spring section (GK 364), Canning Basin, Western Australia, were also identified as *A. seddoni* (Trinajstic & George 2009). Scales from *A. seddoni* are therefore being reported from a number of sections throughout northern and East Gondwana and are proving useful for correlating early to late Frasnian sections.

Most thelodonts had died out in the Early to early Mid-Devonian and after this the only taxa are known outside of the Laurentian continental terranes (Turner 1997; Turner *et al.* 2004). Later Middle and Late Devonian thelodonts are known in East and northern Gondwana as far 'south' as Antarctica, that is, possibly up to 50–60°S in Gondwana (Turner 1997; Märss *et al.* 2007). From the records now available, it seems that the turiniids were most widespread in Gondwana in the Early to Mid-Devonian, with restriction in range by late Frasnian times to Iran and Western Australia.

The fact that at least a small population of thelodonts seems to have survived the Frasnian–Famennian Kellwasser extinction events and lived well into the Famennian is most surprising. We can only speculate on the factors that protected these turiniids from that dramatic time in vertebrate evolution (e.g. Hart 1996). However, it is clear that the area was home to several thelodont taxa from the late Silurian (Hamedi *et al.* 1997; Turner 1997; Hairapetian *et al.* 2008) onwards and their ability to adapt to high-latitude climatic zones must have been in their favour. Chen *et al.* (2002) have attributed a major transgression and eutrophic fluctuations that led to severe algal blooms especially in low-latitude continental shelves to related anoxic events. Now it is certain that least one thelodont taxon and possibly more did weather the Frasnian–Famennian extinction events with recovery and survival of the turiniids.

Blieck & Goujet (1978) and Turner (in Turner & Tarling 1982; Turner 1997) considered the relationship of Iranian thelodonts to others in Asian localities. We have seen above further evidence of the Gondwanan distribution of turiniid thelodonts and showed the links between Iran and Western Australia in the Mid- to early Late Devonian, supporting a Palaeotethyan dispersal route in shallow water between Gondwana and Euramerica

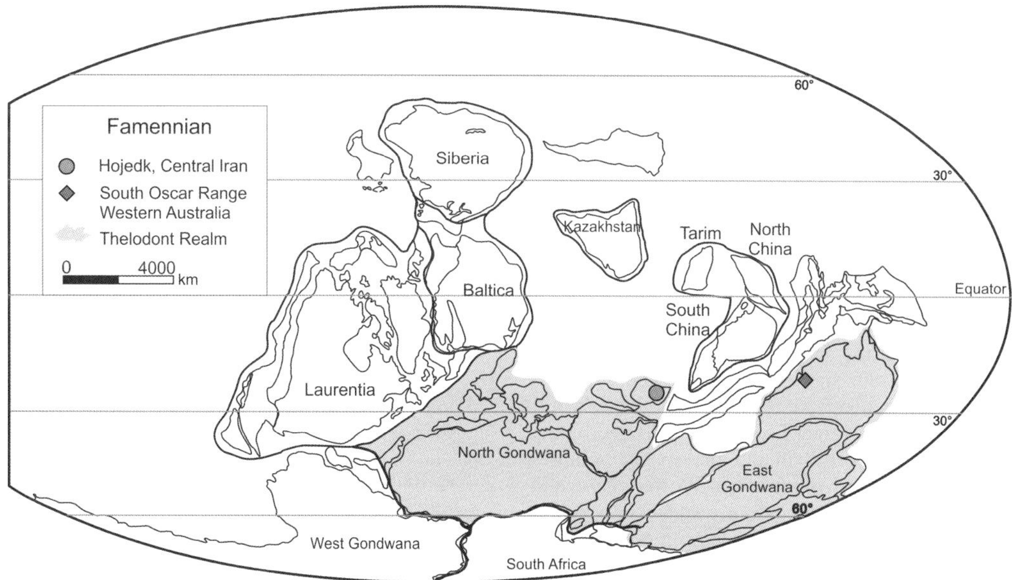

Fig. 10. Palaeogeographical map showing the position of the Famennian thelodont localities in southeastern Iran and Western Australia during the Late Devonian (Famennian). Base map after Golonka (2007) and Lebedev & Zakharenko (2010), with modifications.

(Laurentia) in the Mid–Late Devonian along the northern Gondwana shoreline. Other vertebrate taxa show a similar biogeographical pattern: Ginter *et al.* (2002) discussed the Famennian shark populations of the Palaeotethys; the youngest known ischnacanthiform *Grenfellacanthus zerinae* Long, Burrow & Ritchie, 2004 exhibits a similar pattern with a possible second species of *Grenfellacanthus* occurring in the early Famennian of Chahriseh, Iran (Long *et al.* 2004; C. Burrow, pers. comm. August 2013).

Lebedev & Zakharenko (2010) put forward a new hypothesis of vertebrate provinces for the Givetian, one of which is the Phyllolepid–Thelodont Province; several taxa of phyllolepid placoderms being endemic and earlier in East and northern Gondwana, whereas turiniid thelodonts occur there later. However, they seem to have been unaware that thelodonts did exist in western Gondwana in the Early Devonian (Turner *et al.* 2004) and also that thelodonts occur in the Broken River, North Queensland/China realm in the Mid-Devonian (Turner 1997). Nevertheless, their idea that this province might be extended into the Frasnian and beyond, because thelodonts are associated with key placoderms and chondrichthyans in northern Gondwana (Iran and Western Australia), offers a useful palaeogeographical model (Fig. 10), to which our new finds offer further support. Although they put northern Gondwana further south, the south of central Iran including Hojedk is positioned in the subtropics.

A different opinion on the palaeoposition of central Iran was presented on a recent map by Torsvik & Cocks (2013, fig. 10), which they considered at *c.* 40°S. This southern latitude still might be feasible as thelodonts are thought to have been able to live in relatively high latitudes (see Turner 1997).

Conclusions

The new genus and species of turiniid thelodont, *Arianalepis megacostata* gen. et sp. nov., found in the Upper Devonian (early Famennian) Bahram Formation of the Hojedk section of Iran, and a further uncertain species referred tentatively here to the same genus from the Upper Devonian (middle Famennian) Napier Formation of Western Australia, are younger than the previously youngest known thelodonts from Iran and Western Australia. These are the first thelodont scales known from the early–middle Famennian in both countries and worldwide, and their presence provides new data for biostratigraphic correlation between Iran and Australia. The conodont successions at South Oscar Range and in the Hojedk section constrain the dating.

Most agnathan fishes did not survive the Frasnian–Famennian event and so the presence of a survivor 'Lazarus' thelodont taxon is surprising. *Arianalepis megacostata* gen. et sp. nov. in northern Gondwana (central Iran) and the turiniid ?*Arianalepis* sp. indet. in East Gondwana (Western Australia) provide evidence of the only thelodont lineage surviving the Frasnian–Famennian extinction. These records extend the evolutionary history of the Thelodonti by some 2–6 Ma beyond the Kellwasser events into the *crepida* (Kerman) to Upper *marginifera*/*trachytera* (South Oscar Range) zones.

We emphasize here that it is now necessary to seek more diligently for further examples in the Famennian of the broad northern and eastern Gondwana area, especially in the carbonates of Western Australia.

VH thanks Prof Michal Ginter (Univ. Warsaw) for support and use of facilities in Warsaw. We thank Curtin University WA-OIGC/Applied Chemistry for use of facilities and acknowledge the use of equipment and scientific and technical assistance of the Curtin University Electron Microscope Facility, which has been partially funded by Curtin University, WA State, and Commonwealth Governments. Dr Mikael Siversson from the Western Australian Museum is thanked for photographic assistance and the use of facilities. KT and BR would like to further acknowledge funding from an Australian Research Council Grant (LP0883812), MERIWA, WA ERA, CSIRO, Buru, Chevron Australia Business Unit, Chevron Energy Technology Company, University of Greenwich, National Science Foundation and Chemostrat Inc. Field support and safety were provided by Wundargoodie Aboriginal Safaris and the Geological Survey of Western Australia. Thank you to the Mimbi Aboriginal Community, Mt Pierre Station and Fossil Downs, for field area access. BR acknowledges an Australian Postgraduate Award. This is a contribution to IGCP 596: 'Climate and biodiversity patterns in the Mid-Paleozoic (Early Devonian to Late Carboniferous)'.

References

Balme, B. E. 1960. Upper Devonian (Frasnian) spores from the Carnarvon Basin, Western Australia. *The Palaeobotanist*, **9**, 1–10.

Balme, B. E. 1988. Miospores from the Late Devonian (Early Frasnian) strata, Carnarvon Basin, Western Australia. *Palaeontographica*, **209**, 109–166.

Blieck, A. & Goujet, D. 1978. A propos de nouveau matériel de Thélodontes (Vértébrés, Agnathes) d'Iran et de Thaïlande: aperçu sur la répartition géographique et stratigraphique des Agnathes des'régions gondwaniennes' au Paléozoïque moyen. *Annales de la Société géologique du Nord*, **97**, 363–372.

Blieck, A. & Turner, S. (eds) 2000. Palaeozoic Vertebrate Biochronology and Global Marine/Non-marine Correlation. *Final report of IGCP 328 (1991–1996).* Courier Forschungsinstitut Senckenberg, **223**.

Blieck, A., Golshani, F., Goujet, D., Hamdi, A., Janvier, P., Mark-Kurik, E. & Martin, M. 1980. A new vertebrate locality in the Eifelian of the Khush-Yeilagh Formation, Eastern Alborz, Iran. *Palaeovertebrata, Montpelier*, **9**, 133–154.

Branson, E. B. & Mehl, M. G. 1934. Conodonts from the Bushberg Sandstone and equivalent formations of Missouri. *University of Missouri Studies*, **8**, 265–300.

Burrow, C. J., Turner, S. & Young, G. C. 2010. Middle Palaeozoic microvertebrate assemblages and biogeography of East Gondwana (Australasia, Antarctica). *Palaeoworld*, **19**, 37–54.

Chen, D., Tucker, M. E., Shen, Y., Yans, J. & Preat, A. 2002. Carbon isotope excursions and sea level change: implications for the Frasnian–Famennian biotic crisis. *Journal of the Geological Society, London*, **159**, 623–626, http://doi.org/10.1144/0016- 764902-027

Chow, N., George, A. D., Trinajstic, K. M. & Chen, Z. 2013. Stratal architecture and platform evolution of an early Frasnian syn-tectonic carbonate platform, Canning Basin, Australia. *Sedimentology*, **60**, 1583–1620.

Gholamalian, H. & Kebriaei, M. R. 2008. Late Devonian conodonts from the Hojedk section, Kerman Province, Southeastern Iran. *Rivista Italiana di Paleontologia e Stratigrafia*, **114**, 17–181.

Gholamalian, H., Hairapetian, V., Barfehei, N., Mangelian, S. & Faridi, P. 2013. Givetian–Frasnian boundary conodonts from Kerman province, Central Iran. *Rivista Italiana di Paleontologia e Stratigrafia*, **119**, 133–146.

Ginter, M. 1990. Late Famennian shark teeth from the Holy Cross Mountains, Central Poland. *Acta Geologica Polonica*, **40**, 69–81.

Ginter, M. & Ivanov, A. 1992. Devonian phoebodont shark teeth. *Acta Palaeontologica Polonica*, **37**, 55–75.

Ginter, M., Hairapetian, V. & Klug, C. 2002. Famennian chondrichthyans from the shelves of North Gondwana. *Acta Geologica Polonica*, **52**, 169–215.

Girard, C., Klapper, G. & Feist, R. 2005. Subdivision of the terminal Frasnian *linguiformis* conodont Zone, revision of the correlative interval of the Montagne Noire Zone 13, and discussion of the stratigraphically significant associated trilobites. *In*: Over, D. J., Morrow, J. R. & Wignall, P. B. (eds) *Understanding Late Devonian and Permian–Triassic Biotic and Climatic Events: Towards an Integrated Approach*. Developments in Paleontology and Stratigraphy Series, **20**, 181–198.

Golonka, J. 2007. Phanerozoic paleoenvironment and paleolithofacies maps. Late Palaeozoic. *Geologia*, **33**, 145–209.

Hairapetian, V. 2008. *Late Devonian–Early Carboniferous fish micro-remains from central Iran*. PhD thesis, University of Esfahan.

Hairapetian, V. & Turner, S. 2003. Upper Devonian fish microremains from eastern and southeastern Iran. *In*: Schultze, H.-P., Luksevics, E. & Unwin, D. (eds) *UNESCO-IUGS IGCP 491: The Gross Symposium 2: Advances In Palaeoichthyology*, Riga, Latvia, *September 8–14, 2003*, Abstracts, 26–27.

Hairapetian, V., Yadzi, M. & Long, J. A. 2000. Devonian vertebrate biostratigraphy of central Iran. *Records of the Western Australian Museum*, **58** (suppl.), 241–247.

Hairapetian, V., Blom, H. & Miller, C. G. 2008. Silurian thelodonts from the Niur Formation, central Iran. *Acta Palaeontologica Polonica*, **53**, 85–95.

Hamdi, B. & Janvier, P. 1981. Some conodonts and fish remains from Lower Devonian (lower part of the Khoshyeylaq Formation) north east Shahrud, Iran. *Geological Survey of Iran Reports*, **49**, 195–212.

Hamedi, M. A., Wright, A. J. *et al.* 1997. Cambrian to Silurian of east-central Iran: new biostratigraphic and biogeographic data. *Neues Jahrbuch für Geologie und Paläontologie, Monatshefte*, **7**, 412–424.

Hansma, J., Tohver, E. *et al.* 2015. Late Devonian carbonate magnetostratigraphy from the Oscar and Horse Spring Ranges, Lennard Shelf, Canning Basin, Western Australia. *Earth and Planetary Science Letters*, **409**, 232–242.

Harries, P. J., Kauffman, E. G. & Hansen, T. A. 1996. Models for biotic survival following mass extinction. *In*: Hart, M. B. (ed.) *Biotic Recovery from Mass Extinction Events*. Geological Society, London, Special Publications, **102**, 41–60, http://doi.org/10.1144/GSL.SP.1996.001.01.03

Hart, M. B. (ed.) 1996. *Biotic Recovery from Mass Extinction Events*. Geological Society, London, Special Publications, **102**, http://sp.lyellcollection.org/content/102/1.toc

Helms, J. 1959. Conodonten aus dem Saalfelder Oberdevon (Thuringen). *Geologie*, **8**, 634–677.

Hocking, R. M., Moors, H. T. & Van De Graaf, W. J. E. 1987. Geology of the Carnarvon Basin, Western Australia. *Western Australia Geological Survey Bulletin*, **133**, 1–289.

Hurley, N. F. 1986. *Geology of the Oscar Range Devonian Reef Complex, Canning Basin, Western Australia*. PhD thesis, University of Michigan.

Jaekel, O. 1911. *Die Wirbeltiere. Eine Übersicht über die fossilen und lebende Formen*. Gebrüder Borntraeger, Berlin.

Kauffman, E. G. & Harries, P. J. 1996. The importance of crisis progenitors in recovery from mass extinction. *In*: Hart, M. B. (ed.) *Biotic Recovery from Mass Extinction Events*. Geological Society, London, Special Publications, **102**, 15–39, http://doi.org/10.1144/GSL.SP.1996.001.01.02

Kiaer, J. & Heintz, A. 1932. New coelolepids from Upper Silurian on Oesel (Esthonia). *Eesti Loodusteaduse Archiiv*, **10**, 1–8.

Klapper, G. 1989. The Montagne Noire Frasnian (Upper Devonian) conodont succession. *In*: McMillan, N. J., Embry, A. F. & Glass, D. J. (eds) *Devonian of the World*. Memoir of the Canadian Society of Petroleum Geologists, **14**, 449–205.

Lebedev, O. A. & Zakharenko, G. V. 2010. Global vertebrate-based palaeozoogeographical subdivision for the Givetian–Famennian (Middle–Late Devonian): endemism–cosmopolitanism spectrum as an indicator of interprovincial faunal exchanges. *Palaeoworld*, **19**, 186–205.

Long, J. A. & Trinajstic, K. M. 2000. An overview of the Devonian microvertebrate faunas of Western

Australia. *Courier Forschungsinstitut Senckenberg*, **223**, 471–485.

LONG, J. A., BURROW, C. J. & RITCHIE, A. 2004. A new Late Devonian acanthodian fish from the Hunter Formation new Grenfell, New South Wales. *Alcheringa*, **28**, 147–156.

MACADIE, C. I. 2002. Thelodont fish and conodonts from the Early Devonian of Reefton, New Zealand. *Alcheringa*, **26**, 423–433.

MÄRSS, T., TURNER, S. & KARATAJUTE-TALIMAA, V. N. 2007. Agnatha II – Thelodonti. Volume 1B. *In*: SCHULTZE, H.-P. (ed.) *Handbook of Paleoichthyology*. Verlag Dr Friedrich Pfeil, Munich, 143.

OBRUCHEV, D. V. 1964. *Osnovy Paleontologi. v. 11. Ryby i Byescheliustnikh.* [Fundamentals of Palaeontology Fishes and Agnatha.] Nauka, Moscow [In Russian].

PLAYFORD, P. E. & LOWRY, D. C. 1966. Devonian reef complexes of the Canning Basin, Western Australia. *Geological Survey of Western Australia Bulletin*, **118**, 1–150.

PLAYFORD, P. E., HURLEY, N. F. & MIDDLETON, M. F. 1989. Reefal platform development, Devonian of the Canning Basin, Western Australia. *In*: CREVELLO, P., WILSON, J. I., SARG, J. F. & READ, J. F. (eds) *Controls on Carbonate Platform and Basin Development*. SEPM Special Publication, **44**, 187–202.

PLAYFORD, P. E., HOCKING, R. M. & COCKBAIN, A. E. 2009. Devonian reef complexes of the Canning Basin, Western Australia. *Geological Survey of Western Australia Bulletin*, **145**, 444.

PLAYTON, T. E., HOCKING, R. ET AL. 2013. Development of a regional stratigraphic framework for Upper Devonian reef complexes using integrated chronostratigraphy: Lennard Shelf, Canning Basin, Western Australia. *In*: KEEP, M. & MOSS, S. J. (eds) *The Sedimentary Basins of Western Australia IV. Proceedings of the Petroleum Exploration Society of Australia Symposium, Perth*, 1–15.

SANDBERG, C. A. & DREESEN, R. 1984. Late Devonian icriodontid biofacies models and alternate shallow water conodont zonation. *In*: CLARK, D. L. (ed.) *Conodont Biofacies and Provincialism*. Geological Society of America, Special Papers, **196**, 143–178.

SANDBERG, C. A. & ZIEGLER, W. 1979. Taxonomy and biofacies of important conodonts of Late Devonian *styriacus*-Zone, United States and Germany. *Geologica et Palaeontologica*, **13**, 173–212.

SANNEMANN, D. 1955. Oberdevonische Conodonten (to IIa). *Senckenbergiana Lethaea*, **36**, 123–156.

THOMAS, L. A. 1949. Devonian–Mississippian formations of southeast Iowa. *Geological Society of America Bulletin*, **60**, 403–438.

TORSVIK, T. H. & COCKS, L. R. M. 2013. Gondwana from top to base in space and time. *Gondwana Research*, **24**, 999–1030.

TRINAJSTIC, K. 2000. Conodonts, thelodonts and phoebodonts – together at last. *Geological Society of Australia, Abstract*, **61**, 121.

TRINAJSTIC, K. 2001. A description of additional variation seen in the scale morphology of the Frasnian thelodont *Australolepis seddoni* Turner and Dring, 1981. *Records of the Western Australian Museum*, **20**, 237–246.

TRINAJSTIC, K. & GEORGE, A. D. 2009. Microvertebrate biostratigraphy of Upper Devonian (Frasnian) carbonate rocks in Canning and Carnarvon Basins of Western Australia. *Palaeontology*, **52**, 641–659.

TURNER, S. 1986. Vertebrate fauna of the Silverband Formation, Grampians, western Victoria. *Proceedings of the Royal Society of Victoria*, **98**, 53–62.

TURNER, S. 1993. Palaeozoic microvertebrates from eastern Gondwana. *In*: LONG, J. A. (ed.) *Palaeozoic Vertebrate Biostratigraphy and Biogeography*. Belhaven Press, London, 174–207.

TURNER, S. 1995. Devonian thelodont scales (Agnatha, Thelodonti) from Queensland. *Memoirs of the Queensland Museum*, **38**, 677–685.

TURNER, S. 1997. Sequence of Devonian thelodont scale assemblages in East Gondwana. *In*: KLAPPER, G., MURPHY, M. A. & TALENT, J. A. (eds) *Paleozoic Sequence Stratigraphy, Biostratigraphy, and Biogeography: Studies in Honor of Dr J. Granville ("Jess") Johnson*. Geological Society of America, Special Papers, **321**, 295–315.

TURNER, S. 1999. Early Silurian to Early Devonian thelodont assemblages and their possible ecological significance. *In*: BOUCOT, A. J. & LAWSON, J. (eds) *Palaeocommunities: A Case Study From the Silurian and Lower Devonian*. Cambridge University Press, Cambridge, 42–78.

TURNER, S. & DRING, R. S. 1981. Late Devonian thelodonts (Agnatha) from the Gneudna Formation, Carnarvon Basin, Western Australia. *Alcheringa*, **5**, 39–48.

TURNER, S. & HAIRAPETIAN, V. 2005. Thelodonts from Gondwana. *In:* HAIRAPETIAN, V. & GINTER, M. (eds) *IGCP 491 Armenia Field Conference, Devonian Vertebrates of the Continental Margins, May 24–28*. Ichthyolith Issues, Special Publication, **8**, 24.

TURNER, S. & JANVIER, P. 1979. Middle Devonian Thelodonti (Agnatha) from the Khush–Yeilagh Formation, North-East Iran. *Geobios*, **12**, 889–892.

TURNER, S. & TARLING, D. 1982. Thelodont and other agnathan distributions as tests of Lower Palaeozoic continental reconstructions. *Palaeogeography, Palaeoclimatology, Palaeoecology*, **39**, 295–311.

TURNER, S. & YOUNG, G. C. 1992. Thelodont scales from the Middle-Late Devonian Aztec siltstone, southern Victoria Land Antarctica. *Antarctic Science*, **4**, 89–105.

TURNER, S., JONES, P. J. & DRAPER, J. J. 1981. Early Devonian thelodonts (Agnatha) from the Toko Syncline, western Queensland, and a review of other Australian discoveries. *BMR Journal of Australian Geology and Geophysics*, **6**, 51–69.

TURNER, S., BASDEN, A. & BURROW, C. J. 2000. Devonian vertebrates of Queensland. *In*: BLIECK, A. & TURNER, S. (eds) *IGCP: 328, Final Report*. Courier Forschungsinstitut Senckenberg, **223**, 487–521.

TURNER, S., BURROW, C. J., GHOLAMALIAN, H. & YAZDI, M. 2002. Late Devonian (early Frasnian) microvertebrates and conodonts from the Chahriseh area near Esfahan, Iran. *Memoirs of the Association of Australasian Palaeontologists*, **27**, 149–159.

TURNER, S., TRINAJSTIC, K., HAIRAPETIAN, V., JANVIER, P. & MACADIE, I. 2004. Thelodonts from western

Gondwana. *In*: RICHTER, M. & SMITH, M. M. (eds) *10th Early/Lower Vertebrates. IGCP 491 Symposium*, Gramado, Brazil, 45–46.

WANG, S. T., DONG, Z. & TURNER, S. 1986. Middle Devonian Turinidae (Thelodont, Agnatha) from western Yunnan, China. *Alcheringa*, **10**, 315–325.

WENDT, J., KAUFMANN, B., BELKA, Z., FARSAN, N. & KARIMI BAVANDPOUR, A. 2002. Devonian/Lower Carboniferous stratigraphy, facies patterns and palaeogeography of Iran. Part I, southeastern Iran. *Acta Geologica Polonica*, **52**, 129–168.

WENDT, J., KAUFMANN, B., BELKA, Z., FARSAN, N. & KARIMI BAVANDPOUR, A. 2005. Devonian/Lower Carboniferous stratigraphy, facies patterns and palaeogeography of Iran. Part II, northern and central Iran. *Acta Geologica Polonica*, **55**, 31–97.

YAZDI, M. & TURNER, S. 2000. Late Devonian and Carboniferous vertebrates from the Shishtu and Sardar Formations of the central Shotori Range, Iran. *Records of the Western Australian Museum*, **58** (suppl.), 223–240.

YOUNG, G. C. 1988. Antiarchs (Placoderm fishes) from the Devonian Aztec Siltstone, southern Victoria Land, Antarctica. *Palaeontographica*, **202**, 1–125.

YOUNG, G. C. 1993. The teeth of the Devonian shark *Antarctilamna*, and its relationships. *In*: *K.S.W. Campbell Symposium (A.N.U., Canberra, A.C.T., 8–10 February, 1993) Abstracts*. Association of Australasian Palaeontologists, Canberra, 28.

YOUNG, G. C. & TURNER, S. 2000. Devonian microvertebrates and marine-nonmarine correlation in East Gondwana: overview. *Courier Forschungsinstitut Senckenberg*, **223**, 453–470.

ZIEGLER, W. & SANDBERG, C. A. 1990. The Late Devonian standard conodont zonation. *Courier Forschungsinstitut Senckenberg*, **121**, 1–115.

The global *Annulata* Events: review and new data from the Rheris Basin (northern Tafilalt) of SE Morocco

S. HARTENFELS* & R. T. BECKER

Institut für Geologie und Paläontologie, WWU Münster, Corrensstrasse 24, D-48149 Münster, Germany

**Corresponding author (e-mail: shartenf@uni-muenster.de)*

Abstract: A review of the literature shows that the Famennian global *Annulata* Event(s) can be recognized as a transgressive, often hypoxic and eutrophic, interval that interrupts an overall regressive eustatic trend in more than 40 regions of North America, Europe, North Africa, Asia and Australia. According to differences in palaeogeography, sedimentology and biota, these occurrences are assigned to 10 event settings. The first detailed data on facies, ammonoid and conodont faunas are presented for the Rheris Basin of the eastern Anti-Atlas (southern Morocco) and compared with previously studied sections of the adjacent Tafilalt Platform, Tafilalt Basin and Maider Basin. The rather argillaceous succession at El Gara resembles the Tafilalt Basin (Hassi Nebech section) in its lack of black shales/limestones and similar ammonoid and conodont assemblages. However, the *Sulcoclymenia sulcata* Zone (Upper Devonian III-C2) below the *Annulata* Events contains ammonoid taxa that are unique for all of the Anti-Atlas and North Africa: *Protornoceras ornatum* Dybczynski, 1913, *Genuclymenia* aff. *angelini* (Wedekind, 1908), *Protactoclymenia* aff. *implana* (Czarnocki, 1989) and ?*Pleuroclymenia* sp. juv. The first regional record of the marker conodont *Pseudopolygnathus granulosus* Ziegler, 1962 also distinguishes the pre-event assemblage. As in many other regions, there is a major decline in ammonoids well before the Lower *Annulata* Event, which suggests an episode of extreme oligotrophy. Both *Annulata* Events at El Gara are whitish-weathered marly shales with only small specimens of *Platyclymenia* and *Prionoceras* (*sensu lato*), which are also typical for the *annulata* Zone (UD IV-A) of other Tafilalt sections, but benthonic organisms are nearly absent. This suggests local low-oxygen conditions, but only a moderate production of organic carbon, insufficient for black shale formation, unlike many German sections or in the Maider Basin (section Mrakib). The latter region represents a deeper shelf basin that had much higher productivity and a unique '*Gundolficeras–Erfoudites–Protactoclymenia–Stenoclymenia–Guerichia* biofacies' of the Lower *Annulata* Shale. The upper part of the *annulata* Zone at El Gara is characterized by *Platyclymenia* (*Platyclymenia*) *levata* n. sp. Other new taxa of the same zone in the Anti-Atlas are *Posttornoceras ascendens* n. sp. and *Stenoclymenia rectangula* n. sp. Whilst the ammonoid faunal overturn between UD III-C and UD IV-A was severe, the strong reduction in conodont diversity with the two *Annulata* Events was mostly (apart from two taxa) a palaeoecologically triggered, only episodic, feature. The comparison of the various Anti-Atlas *Annulata* Event beds and assemblages enables the distinction of event biofacies types, which reflect local differences of bathymetry, trophic conditions and seafloor ventilation.

The *Annulata* Event was named by House (1985) and, originally, it referred to a couplet of black shales in the northern Rhenish Massif (Germany) with mass occurrences of the ammonoid *Platyclymenia* (Schmidt 1924), including the name-giving species *Pl.* (*Pl.*) *annulata* (Münster, 1832). The coincident entry (not last occurrence, as stated in House 1985) of *Platyclymenia* (*Platyclymenia*) and *Prionoceras*, the oldest genus of the bradytelic superfamily Prionocerataceae, defines the base of the *Pl.* (*Pl.*) *annulata* Zone or *Prionoceras* Genozone (Fig. 1: UD IV-A) (zonal abbreviations after Becker 1993*a* and Becker & House 2000*a*). In thick successions of the Rhenish Massif (Schmidt 1924; Ziegler 1962; Becker 1992*b*; Korn 2004; Hartenfels 2011), the Holy Cross Mountains (Racka *et al.* 2010; Hartenfels 2011) or of Morocco (Becker *et al.* 2000; Webster *et al.* 2005; Hartenfels 2011; Korn *et al.* 2014: Beds N1c and N1f2), hypoxic to anoxic Lower and Upper *Annulata* Shales can be distinguished. They are separated by a better-oxygenated, bioturbated interval with one or two micrite subcycles. For these carbonate layers, Hartenfels (2011) coined the term '*Annulata* Intralimestone'. In condensed sections, there may be only one black or greenish shale, marl and/or black limestone interval (e.g. Paeckelmann 1924; Lange 1929; Ziegler 1971; Korn 2002, 2004; Hartenfels 2011; Hartenfels & Becker 2015). Apart from blooms of *Platyclymenia*, with a dominance of the smooth *Pl.*

From: Becker, R. T., Königshof, P. & Brett, C. E. (eds) 2016. *Devonian Climate, Sea Level and Evolutionary Events*. Geological Society, London, Special Publications, **423**, 291–354.
First published online August 3, 2016, http://doi.org/10.1144/SP423.14

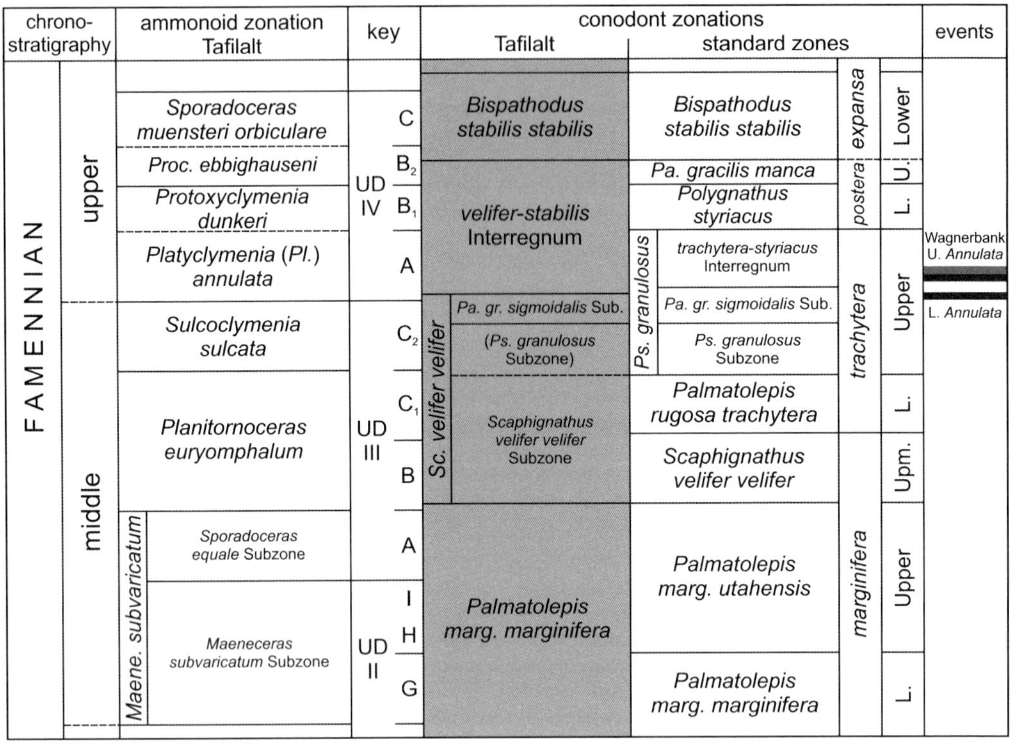

chrono-stratigraphy		ammonoid zonation Tafilalt	key		conodont zonations: Tafilalt		conodont zonations: standard zones			events
FAMENNIAN	upper	*Sporadoceras muensteri orbiculare*	UD IV	C	*Bispathodus stabilis stabilis*		*Bispathodus stabilis stabilis*	*expansa*	Lower	
		Proc. ebbighauseni		B_2	*velifer-stabilis* Interregnum		*Pa. gracilis manca*	*postera*	U.	
		Protoxyclymenia dunkeri		B_1			*Polygnathus styriacus*		L.	
		Platyclymenia (*Pl.*) *annulata*		A		*Ps. granulosus*	*trachytera-styriacus* Interregnum	*trachytera*	Upper	Wagnerbank, U. *Annulata*, L. *Annulata*
	middle	*Sulcoclymenia sulcata*	UD III	C_2	*Sc. velifer velifer*: *Pa. gr. sigmoidalis* Sub.		*Pa. gr. sigmoidalis* Sub.			
					(*Ps. granulosus* Subzone)		*Ps. granulosus* Subzone			
		Planitornoceras euryomphalum		C_1	*Scaphignathus velifer velifer* Subzone		*Palmatolepis rugosa trachytera*		L.	
				B			*Scaphignathus velifer velifer*	*marginifera*	Upm.	
		Maene. subvaricatum: *Sporadoceras equale* Subzone		A	*Palmatolepis marg. marginifera*		*Palmatolepis marg. utahensis*		Upper	
		Maeneceras subvaricatum Subzone	UD II	I, H						
				G			*Palmatolepis marg. marginifera*		L.	

Fig. 1. Correlation of ammonoid, conodont and event stratigraphy around the global *Annulata* Events in the Tafilalt.

(*Pl.*) *subnautilina* (Sandberger, 1855), mass occurrences of the bivalve *Guerichia* are typical and resemble the younger, top-Famennian Hangenberg Blackshale. Locally, *Buchiola*, the typical bivalve of the upper Frasnian Kellwasser Beds, may reappear (Becker 1992*b*). The *Annulata* Shales represent transgressive and hypoxic pulses that terminated the overall eustatic low of the main part of the middle Famennian (see the global sea-level curve of Johnson *et al.* 1985, with refinements in Becker 1993*b*). The Upper *Annulata* Event is often followed by goniatite-rich limestones, the 'Wagnerbank' of the Thuringian Bohlen section (e.g. Schindewolf 1952; Pfeiffer 1954) and its widely distributed micritic or argillaceous equivalents. It represents a regressive phase in the higher part of the *annulata* Zone (UD IV-A). Hartenfels *et al.* (2009) proposed using the base of the *annulata* Zone or Lower *Annulata* Shale for the definition of a formal upper Famennian substage.

In the conodont scale, the two *Annulata* Events fall in the upper part of the *Pseudopolygnathus granulosus* Zone (=former Upper *trachytera* Zone). In central and eastern Europe, finer subdivisions are recognized (Fig. 1). The Lower *Annulata* Shale and a Lower *Annulata* Intralimestone belong to the upper part of a regional *Palmatolepis gracilis sigmoidalis* Subzone (Hartenfels 2011). An Upper *Annulata* Intralimestone, the Upper *Annulata* Shale and the subsequent Wagnerbank interval fall in a short-ranging, but significant, *trachytera–styriacus* Interregnum. If there is only one *Annulata* Intralimestone, it can fall in either of the two intervals (Hartenfels *et al.* 2009; Hartenfels 2011). In sections that are not condensed, the *Polygnathus styriacus* Zone (=Lower *postera* Zone) begins well above the event interval and Wagnerbank or its equivalents (Hartenfels 2011; Hartenfels & Becker 2015). This corrected previous correlations based on condensed sections (e.g. Ziegler 1971; Korn & Ziegler 2002; adopted by Ziegler & Sandberg 2000; House 2002; Korn 2002; Becker *et al.* 2004).

For the miospore zonation, Racka *et al.* (2010) showed that index miospores of the East European VF and of the Western European lower VCo Zone (=Rad Interval Zone) occur just below the Lower *Annulata* Shale at Kowala. Key taxa of the next higher SP Subzone (of the VF Zone) or of the upper VCo Zone (=Cor Interval Zone) were found as minor components within the Lower *Annulata* Shale, whose palynofacies is dominated by amorphous organic matter and leiospheres. The latter pattern is locally even more pronounced in the Upper *Annulata* Shale, and is associated with a peak of euxinic conditions and with biogeochemical evidence of photic-zone anoxia. For the Ardennes, Higgs *et al.* (2013) showed that a marker shale that is tentatively correlated with the *Annulata* Event lies well above the base of the Mic Interval Zone (=upper part of GF Zone) and within a miospore-poor interval below the first record of the Rad Interval Zone (=lower VCo Zone). Both studies indicate that floral changes took place in terrestrial environments just before and within the event interval.

The *Annulata* Events are associated with significant ammonoid extinctions, short-termed spreading events and evolutionary innovations in several groups (e.g. Becker 1993*b*; Becker & Kullmann 1996; Becker *et al.* 2004; Korn 2004). Very few of the ammonoid species of the *Prolobites* Zone (UD III-C) survived into the *Annulata* Event beds. Regionally, and at the species level, this faunal overturn is dramatic and approaches 100%. But the beds just below the Lower *Annulata* Event are, in general, already very poor in ammonoids. This pattern applies to all regions that have been studied in sufficient detail (Becker & Hartenfels 2010). It documents a gradual decline well before the sudden onset of hypoxic to anoxic sedimentation, possibly due to a global trophic crisis, with extreme oligotrophy in the (sub)tropical oceans. The return of eutrophic conditions during sudden eustatic rises led to a re-population by faunal immigration and many new taxa evolved in a short period. This radiation is most pronounced in the evolution of the ammonoid families Prionoceratidae, Sporadoceratidae, Prolobitidae (Raymondiceratinae), Platyclymeniidae and Carinoclymeniidae.

So far, there are only very limited high-resolution data on carbon isotopes from the critical interval (Myrow *et al.* 2011). In the Rhenish Massif, a negative excursion below the *Annulata* Shales is in accord with extreme oligotrophy, whilst a positive spike in the Wagnerbank interval suggests eutrophic conditions and high palaeoproductivity, with an increased burial of organic carbon and improved food resources reflected in the extremely rich macrofauna contents. For the black shales, there are, so far, no isotope data, but inorganic geochemistry and C_{org} peaks at Kowala support the eutrophication model (Racka *et al.* 2010). In Thailand, the *Annulata* Event has been recognized as a major positive carbon isotope excursion of the Upper *trachytera* Zone (data in Königshof *et al.* 2012; Savage 2013).

In contrast to the ammonoids, there are no dramatic conodont extinctions or major radiations associated with the Lower and Upper *Annulata* Events. However, there was a gradual loss of seven important conodont species/subspecies: *Palmatolepis perlobata grossi* Ziegler in Kronberg *et al.* (1960) (second carbonate bed below the Lower *Annulata* Event), *Scaphignathus velifer velifer* Helms, 1959 (top Lower *Annulata* Event), *Pa. rugosa trachytera* Ziegler in Kronberg *et al.* (1960), *Pa. glabra lepta* Ziegler & Huddle, 1969 late morphotype (top Lower *Annulata* Intralimestone) and *Pa. minuta minuta* Branson & Mehl, 1934*a* (top Upper *Annulata* Intralimestone), as well as *Polygnathus duolingshanensis* Ji & Ziegler, 1993 and *Po. padovanii* Perri & Spalletta, 1990 (Wagnerbank Equivalent). Some of these extinctions can be related to stronger anoxia to euxinic conditions of the Upper *Annulata* Event Interval (Hartenfels 2011).

For other faunal groups, there are still only limited data concerning the potential evolutionary impact of the *Annulata* Events, mostly because of a lack of research. One important aspect is that shallow-water and, especially, reefal ecosystems had barely recovered from the devastating Frasnian–Famennian boundary mass extinction. Neritic carbonate platforms around the middle–upper Famennian boundary are mostly composed of crinoid shoals with a restricted benthos (e.g. the Palliser Formation of western Canada, Peterhänsli & Pratt 2008; southern Morocco, Wendt 1989). Reef complexes, such as those of the Canning Basin (Western Australia), were mostly constructed by microbes (e.g. Playford *et al.* 2009). For heterotrophic protists, the study of Rhenish sections showed a rather dramatic decline in agglutinating foraminifers with the Lower *Annulata* Event, but with variable recovery lags, and a gradual return of all taxa with the re-establishment of normal and full oxic conditions (e.g. Greifelt *et al.* 2008). The Lower *Annulata* Event also caused changes in rhynchonellid associations (e.g. extinctions of *Pugnaria* and *Leptoterorhynchus*), which can be used for correlation between SE Morocco, the Rhenish Massif and the Holy Cross Mountains (Hartenfels *et al.* 2009). Significant changes in trilobite faunas in the critical interval have been briefly discussed by Becker & Schreiber (1994) and show parallels with the ammonoid evolution. In pelagic settings, very few trilobite species survived from UD III into UD IV; they were followed shortly by the entry of at least 10 new genera (mostly proetids).

This summary shows that there is still much to be learned concerning the causation, palaeogeographically different developments and evolutionary significance of the global *Annulata* Events. In the Anti-Atlas of southern Morocco, Upper Devonian strata are superbly exposed in numerous and extensive sections, especially in the Tafilalt and Maider regions in the east. Wendt *et al.* (1984), Wendt (1989), Becker (1993*a*) and Dopieralska (2009) provided regional palaeogeographical overviews for the Famennian. A pelagic, approximately north–south-running Tafilalt Platform is bordered in the west/SW by an argillaceous Maider Basin and in the east/SE by a mixed argillaceous–carbonatic Tafilalt Basin. In the north, the carbonate platform forms a ramp and a transition towards the Rheris Basin.

Whilst sections with the *Annulata* Event(s) of the Maider Basin, northern, central and southern Tafilalt Platform, and of the Tafilalt Basin have been documented in Korn (1999), Korn *et al.* (2000), Becker *et al.* (2002) and, especially, in Hartenfels (2011), we present in this paper the first detailed account of the Rheris Basin. The faunal and sedimentological characteristics at El Gara, in comparison with the event patterns of the adjacent regions, enable us to show how the environment and biota changed within one region over the course of the global *Annulata* Event(s).

Abbreviations and repository

Conodont genera: *Al.*, *Alternognathus*; *B.*, *Bispathodus*; *Br.*, *Branmehla*; *Cae.*, *Caenodontus*; *I.*, *Icriodus*; *Pa.*, *Palmatolepis*; *Po.*, *Polygnathus*; *Proto.*, *Protognathodus*; *Ps.*, *Pseudopolygnathus*; *M.*, *Mehlina*; *Neo.*, *Neopolygnathus*; *Sc.*, *Scaphignathus*; *Si.*, *Siphonodella*.

Ammonoid genera: *Acri.*, *Acrimeroceras*; *Carino.*, *Carinoclymenia*; *Cyma.*, *Cymaclymenia*; *Enke.*, *Enkebergoceras*; *Erf.*, *Erfoudites*; *Genu.*, *Genuclymenia*; *Gund.*, *Gundolficeras*; *Kara.*, *Karaclymenia*; *Maene.*, *Maeneceras*; *Maid.*, *Maideroceras*; *Pl.*, *Platyclymenia*; *Plani.*, *Planitornoceras*; *Post.*, *Posttornoceras*; *Pr.*, *Prionoceras*; *Prae.*, *Praeglyphioceras*; *Proc.*, *Procymaclymenia*; *Prot.*, *Protoxyclymenia*; *Protacto.*, *Protactoclymenia*; *Protor.*, *Protornoceras*; *Sulco.*, *Sulcoclymenia*; *Sp.*, *Sporadoceras*; *St.*, *Stenoclymenia*; *Tri.*, *Trigonoclymenia*; *Ungu.*, *Ungusporadoceras*.

Authors of species are given when they are first quoted.
Morphological abbreviations for ammonoids: dm, diameter; uw, umbilical width; wh, whorl height; ww, whorl width; WER, whorl expansion rate; E, external lobe; A, adventitious lobe(s); L, lateral lobe(s); U, umbilical lobe; I, internal lobe.

Synonymy lists: *, introduction of a new taxon; v, material seen; e.p., *ex parte*, only a specified part of the material of a publication belongs to the discussed taxon.

All material is stored in the collection of the Geomuseum of the WWU Münster (GGM), ammonoids under B6.C-47, conodonts under B9.A-6. Mentioned GC-UG numbers refer to the collection of the Geoscience Centre of the Georg-Augusta-University Göttingen, MB.C. numbers refer to the Museum für Naturkunde, Berlin.

Global recognition of the *Annulata* Events

The *Annulata* Events and/or equivalents of the Wagnerbank can be recognized globally in a pan-tropical belt from North America to North Africa, Europe, Asia and Australia (Fig. 2). Previous compilations of distribution by Becker & Kullmann (1996), Becker *et al.* (2004) and Hartenfels (2011) are updated, revised and summarized here, roughly from west to east:

(1) **Alberta, Canadian Rocky Mountains**: Level of rare platyclymeniids in the upper part of the marginal marine (evaporitic) to neritic Costigan Member of the Palliser Formation (House & Pedder 1963; Becker & House 2000*a*); for lithology and conodont stratigraphy see Johnston & Chatterton (2001).

(2) **California**: Siliceous beds with platyclymeniids at Dugan Pond, intercalated in an up to 1100 m-thick volcanic complex of the Sierra Buttes Formation (e.g. Anderson *et al.* 1974; House 1983).

(3) **Montana**: Transgression of the Trident Member of the Three Forks Formation (House 1983). According to conodont data in Klapper (1966) and Sandberg & Klapper (1967), this pulse falls in the *granulosus* Zone (=Upper *trachytera* Zone). Ammonoid faunas of UD IV-A were described by Raymond (1909), Schindewolf (1934), House (1962) and Korn & Titus (2006).

(4) **?Nevada**: Level of clymeniids identified as '*Pleuroclymenia* aff. *ohioense* House, Gordon & Hlavin, 1986' in the upper Woodruff Formation of NE Nevada (Smith & Ketner 1975; Rolfe & Dzik 2006). They represent either ribbed and thick-whorled protactoclymeniids (as the true *Protacto. ohioense*, see the reassignment in Becker 2000) or relatives of *Pl.* (*Varioclymenia*) *americana* (Raymond, 1909) from Montana, and possibly a UD IV-A fauna.

(5) **Iowa**: Level of abundant platyclymeniids in the Maple Mill Shale of SE Iowa (Olempska & Chauffe 1999). Conodont data, summarized in Klapper *et al.* (1971), are, in general, in accord and suggest the same interval as the *Platyclymenia* fauna from Montana.

(6) **?Ohio**: Supposed *Platyclymenia–Pleuroclymenia* fauna from the basal Cleveland Shale (House 1983; House *et al.* 1986). However,

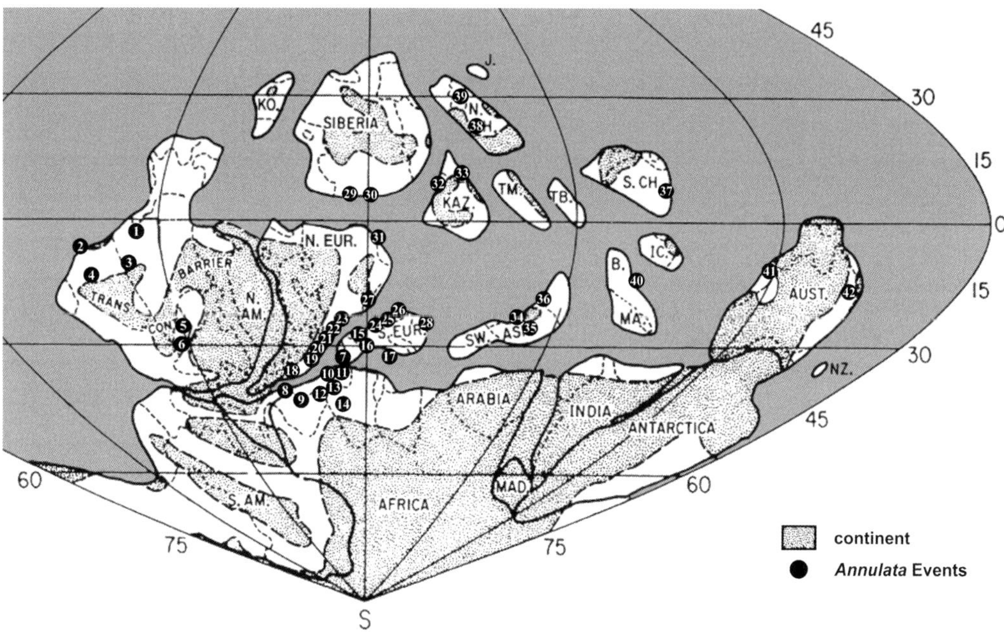

Fig. 2. Global recognition of the *Annulata* Events plotted on a plate tectonic reconstruction based on Heckel & Witzke (1979), which shows a more likely distribution of Asian plates than in other, more recent reconstructions. For details of the numbered regions (see the text).

there is no unequivocal *Platyclymenia* in the assemblage and new conodont data (Baird *et al.* 2009*a*, *b*) suggest that this transgressive pulse already falls in the *aculeatus aculeatus* Zone (=Middle *expansa* Zone).

(7) **South Portugal, northern Pyrite Belt**: Black limestones with *Platyclymenia*, conodonts and abundant guerichiids, intercalating a thick clastic, synorogenic succession near Pomarão (Pruvost 1912; van den Boogard 1963) and Mértola (Fantinet *et al.* 1976; Korn 1997).

(8) **Western Dra Valley, NW Tindouf Basin, SW Morocco**: Mixed neritic–pelagic fauna, extraordinarily rich in platyclymeniids and cyrtospriferids, within a sandy, neritic succession near the Oued Souaïssel, south of Tarfaya (Hollard 1963, 1970). In the Assa region to the east, the Famennian is first shaly and unfossiliferous, then sandy and neritic (Kaiser *et al.* 2004).

(9) **Middle Dra Valley, northern Tindouf Basin, SW Morocco**: *Platyclymenia–Prionoceras* faunas, partly from ammonoid coquina limestone, within an argillaceous succession south of Akka (UD IV-A; Hollard & Jacquemont 1956).

(10) **Western Moroccan Meseta**: Intercalation of marginal-marine arkosic sandstones and conglomerates by a deepening pulse with a few platyclymeniids in the Akrech Valley south of Rabat, **Sidi Bettache Basin** (e.g. Choubert & Faure-Muret 1961). ?Bloom of *Platyclymenia–Prionoceras–Erfoudites* faunas in red shales of the Oued Aricha, eastern Benahmed region, **western Mdakra Massif** (Termier & Termier 1951); new faunas, however, also include protoxyclymeniids, which indicate a UD IV-B age, at least for the main part of sampled beds.

(11) **Eastern part of central Moroccan Meseta**, allochthonous successions: Violet shale unit interrupting sharply a thick sequence of nodular limestone at Ziyyar, Khenifra region (Walliser *et al.* 1995; Hartenfels & Becker 2014). Further to the north, at Bou Khemis, approximately between Mrirt and Azrou, Termier *et al.* (1978) mentioned *Prionoceras*, although the higher Famennian of that region is otherwise in brachiopod facies.

(12) **Maider, SE Anti-Atlas, Morocco**: Two transgressive and hypoxic events associated with ammonoid blooms and pyritic (secondarily hematitic) red/black shale depositions, overlain by a three-fold Wagnerbank Equivalent (Becker *et al.* 1999, 2000, 2002; Korn 1999; Webster *et al.* 2005; Hartenfels 2011; Korn *et al.* 2014).

(13) **Tafilalt, SE Anti-Atlas, SE Morocco**: Marly shales and marls of the Lower and Upper *Annulata* Events with abundant ammonoids of the basal UD IV-A in the Rheris and Tafilalt basins. On the southern Tafilalt Platform (Amessoui Syncline), only the Lower *Annulata* Event is locally developed as a black limestone, which represents a shallow pelagic, highly energetic, eutrophic carbonate facies above a hiatus. On the northern Tafilalt Platform, the *Annulata* Event beds are lacking, but there are equivalents of the Wagnerbank (Becker 1992*a*; Korn 1999; Korn *et al.* 2000; Becker & House 2000*b*; Becker *et al.* 2002; Hartenfels 2011; Hartenfels *et al.* 2013).

(14) **SW Algeria, Bechar Basin**: Rich *Platyclymenia* faunas above a middle Famennian (UD III-C) ammonoid-poor interval of the northern Ben Zireg region (Massa 1965), central Saoura Valley (Graben) and Ougarta (e.g. Menchikoff 1930; Petter 1951, 1959, 1960), and the southern Gourara region (Timimoun Subbasin: Meyendorf 1939).

(15) **Cantabrian Mountains, northern Spain**: Two dark shale layers with *Platyclymenia–Prionoceras* faunas at Peña Quebrada, Gildar-Montó Unit, Palentine Domaine (Sanz-López *et al.* 1999).

(16) **Montagne Noire, southern France**: Transgressive peak within oxic pelagic ramp limestones at Col de Tribes, associated with an ammonoid bloom (Girard *et al.* 2013). Becker (1993*a*) briefly noted a correlative *Platyclymenia* onset in light-grey limestone at Coumiac.

(17) **?Sicily**: Typical conodont extinction in the very condensed Upper *trachytera* Zone and local first entry of poorly preserved ammonoids in the otherwise macrofossil-poor pelagic limestones of section Corono Mizziu I (Corradini 1998, 2002).

(18) **SW England**: Platyclymeniid bloom in the pelagic carbonate ramp of Mount Pleasant, **South Devon** (House 1963*b*; section log in House & Butcher 1973). ?Isolated fauna with *Prionoceras* and *Pleuroclymenia* in the basinal slates of the Manor Hotel Beds, NW of Dartmoor, **NW Devon** (House 1959; House & Selwood 1964; House *et al.* 1977). ?Platyclymeniid level in the basal part of the pelagic Landlake Limestone or Lower Petherwin Beds, Launceston area, **NE Cornwall** (Phillips 1841; M'Coy 1851; Selwood 1960) – now fully covered and not available for a restudy.

(19) **Ardennes, Belgium**: Up to 150 cm-thick black shale sequence ('Bocq Event') interrupting sandstones of the Montford Formation near the base of the *postera* Zone, Bocq Valley of the Dinant Syncline (Dreesen *et al.* 1989). Marker shale (Bon Mariage Shale) in the top range of *Sc. velifer velifer*, and at the boundary between the Montford and Evieux formations, Ourthe Valley (Thorez *et al.* 2006; Higgs *et al.* 2013; conodonts in Dreesen & Thorez 1994).

(20) **Germany, northern Rhenish Massif**, from west to east: 'Green *Annulata* Shale' with platyclymeniids, bivalves and trilobites at Hagen-Herbeck (Becker 1985; Becker & Schreiber 1994). Black Lower *Annulata* Shale with mass occurrences of smooth platyclymeniids and opportunistic guerichiids, and upper event interval consisting of green shales at Ziegelei Nie, north of Iserlohn-Letmathe (Koch 1984; Becker 1985, 1992*b*). Reitenberg road section and Schaumberg, both between Iserlohn-Östrich and -Grürmannsheide, with two *Annulata* Black Shales, dark-grey to olive-green weathering (Schmidt 1924; Becker 1992*b*), followed by a green shale with abundant platyclymeniids, goniatites, brachiopods and trilobites (Wagnerbank Equivalent). Two *Annulata* Black Shales with thin pyrite beds, overlain by *Platyclymenia*- and *Guerichia*-rich green, partially silty and marly, shales (argillaceous Wagnerbank Equivalents) at Oese, between Hemer and Menden (Ziegler 1962; Becker 1992*b*; Becker *et al.* 1993; Korn 2004; Hartenfels 2011). Two *Annulata* Black Shales at Oberrödinghausen railway cut and road sections, Hönne Valley (Schmidt 1924; Ziegler 1962; Korn 2004; unpublished new data). Green shale and marl intercalation within a grey to reddish, micritic limestone sequence at Beul (east of the Hönne Valley), yielding an extremely rich ammonoid fauna of UD IV-A, including *Platyclymenia* and *Prionoceras* (Denckmann 1901; Wedekind 1914; Lange 1929; Korn & Luppold 1987); the re-study of Korn (2004) proved the presence of both *Annulata* Shales. Thin black shale intercalation at Ballberg/Hövel followed by Wagnerbank equivalents (Lange 1929; Ziegler 1971; Korn & Luppold 1987; Hartenfels & Becker 2015). Thick black *Annulata* Shales overlain by dark-grey to grey, marly, fossiliferous Wagnerbank Equivalents at Effenberg (Korn & Luppold 1987; Becker 1992*b*; Korn 2004; Hartenfels 2011). Black Shale intercalations with *Platyclymenia* at localities to the south, in the Lüdenscheid Syncline (Ziegler 1970; Korn 2004). Single black *Annulata* levels at Drewer and

Eulenspiegel, Warstein region (Clausen & Leuteritz 1984; Korn *et al.* 1994; Minwegen & Herbig 1997; Korn 2002, 2004). Black shale package, intercalated by dark limestone concretions and rich ammonoid faunas of the basal UD IV-A (dominant *Platyclymenia* and *Erfoudites*) at Kattensiepen, Warstein region (Staschen 1968; Hölder 1979; Korn & Luppold 1987; Korn 2002, 2004). Transgressive *Annulata* Events in black limestone (lower event) and black shale facies (upper event) at Beringhauser Tunnel (Schülke *et al.* 2002; Schülke & Popp 2005; Hartenfels 2011). Thin, red *Annulata* Shale, yielding sparse ammonoids of the *annulata* Zone, at Enkeberg, south of the Brilon Reef Complex (Wedekind 1908; Paeckelmann 1925; Korn & Ziegler 2002; Korn 2004). **Eastern Rhenish Massif**: Rich *Platyclymenia* fauna within a red solid limestone in the Kellerwald (Schindewolf 1921; probably Wagnerbank Equivalent). **Southern Rhenish Massif**: Dark-grey, marly, 2 m-thick unit with *Platyclymenia* fauna (type locality of *Pl.* (*Pl.*) *subnautilina*) at Kirschhofen, Lahn Syncline (Sandberger 1855; Ahlburg 1918; Schindewolf 1921). Transgressive black shale intercalations with *Guerichia* east of Eibach (Rabien 1970; Buggisch *et al.* 1978; Becker 1992*b*).

(21) **Harz Mountains, Germany**: Thin, dark-grey limestone with *Buchiola* and other bivalves between light-grey limestone and faunas of UD III-C and UD IV at Altes Tal, Oberharz (Fuhrmann 1954). The Aeke Valley of the same region produced *Pl.* (*Pl.*) *annulata* from grey-bluish limestone. Born (1912) noted at the latter locality dark bluish limestone beds of the possible *annulata* Zone (UD IV-A) above lighter-grey limestone of the *delphinus* Zone (UD III-C).

(22) **Franconia and Bavarian Vogtland, SE Germany**: Intercalation of the Wagnerbank Equivalent within a homogeneous Flaserkalk sequence at Köstenhof, Franconia (=Schübelhammer, Tragelehn & Hartenfels 2002; Hartenfels 2003; Hartenfels & Tragelehn 2004). There is no evidence of both *Annulata* Events. Transgressive phase combined with the entry of a *Platyclymenia* fauna at Kirchgattendorf, Bavarian Vogtland (Schindewolf 1923).

(23) **Thuringia, Saxonian Vogtland and Wildenfelser Zwischengebirge, SE Germany**: Single *Annulata* Black Shale ('Trennschicht' in Meyer 1920) at the Bohlen, Mauxion Quarry and Gositzfelsen sections near Saalfeld a. d. Saale (Thuringia, Schwarzburg Anticline). A post-event regression is recognizable by the micritic, very fossiliferous Wagnerbank (rich in prionoceratids: Pfeiffer 1954; Helms 1959; Blumenstengel 1994; Weyer *et al.* 1996; Bartzsch *et al.* 1999, 2015; Hartenfels 2011). Two unfossiliferous black shales overlain by a *Prionoceras*-rich Wagnerbank Equivalent at the Kahlleite Quarry, Berga Anticline (Bartzsch *et al.* 1995; Gereke 2004; Hartenfels 2011). Other ammonoid-rich UD IV-A localities that probably represent the Wagnerbank are given in Müller (1956*a*). Limestone and shale with platyclymeniids and guerichiid bloom in the limestone quarry Kloschwitz, Saxonian Vogtland (Freyer 1957). Unfossiliferous, single black shale intercalation and a subsequent micritic three-fold Wagnerbank Equivalent at Grünau, Wildenfelser Zwischengebirge (Hartenfels 2011).

(24) **Carnic Alps** (Austrian–Italian border): Wagnerbank Equivalent with mass occurrence of *Prionoceras* within a homogeneous, micritic sequence that is otherwise poor in macrofossils at Pramosio Bassa (Perri *et al.* 1998; Spalletta & Perri 2001); no *Annulata* Shales. For other regions of the Carnic Alpes (Grosser Pal, Wolayer Lake), records of *Platyclymenia* faunas are listed in Flügel & Kropfitsch-Flügel (1965).

(25) **?Graz Palaeozoic, Austria**: *Platyclymenia* in the Steinberg Limestone (Heritsch 1927; Flügel & Kropfitsch-Flügel 1965); conodont data in Ebner (1980) suggest a colour change between the *trachytera* and *styriacus* zones, but there is no intercalated *Annulata* Shale.

(26) **Moravia, SE Czech Republic**: Black micritic limestone lenses with very common platyclymeniids and *Erfoudites* of the Mokra quarry and of adjacent localities near Brno (Rzehak 1910; Dvorák *et al.* 1989; Weiner 2012; Weiner & Kalvoda 2016).

(27) **Holy Cross Mountains, Poland**: Transgressive and hypoxic to euxinic *Annulata* Black Shales with mass occurrences of *Guerichia* and ammonoids of UD IV-A at Kowala (Bond & Zatoń 2003; Dzik 2006; Racka & Marynowski 2008; Racka *et al.* 2010; Hartenfels 2011). Black, organic-rich limestone with very abundant and morphologically diverse platyclymeniid fauna at Ostróvka (Czarnocki 1989; Woroncowa-Marcinowska 2006).

(28) **Balkan Terrain, western Bulgaria**: Two *Annulata* Black Shales with *Platyclymenia* fauna and guerichiids; upper part of the thick, clastic Parchar Formation, Lyubash-Golo Bardo Unit, Srednogorie Zone (Boncheva *et al.* 2011, 2015).

(29) **?Middle Urals, Bashciria, Russia**: *Platyclymenia–Protactoclymenia–Borisiclymenia* assemblage from the Terekliy River (Bogoslovskiy 1981; Korn & Klug 2002).

(30) **Eastern flank of the southern Urals, Russia**: Significant change of ammonoid faunas between UD III-C and IV-A at the Ural River, Werchneuralsk region (between levels β and γ of Perna 1914).

(31) **Mugodzhary Mountains, Aktyubinsk Oblast, Russian–Kazakh border area**: Ammonoid faunal change at Kia 1 between UD III-C and UD IV-A, separated by a unit with organic matter ('coal' in Bogoslovskiy 1969; subsequent data with less clear separation of UD III/IV assemblages in Nikolaeva & Bogoslovskiy 2005).

(32) **Karaganda Basin, Kazakhstan**: Transgressive sequence with shales and limestone with *Platyclymenia* (UD IV-A fauna) above a siliceous succession with brachiopods at the Sherubay-Nuru River, Zhanaarkin region (Bogoslovskiy 1969; Nikolaeva & Bogoslovskiy 2005). Other *Platyclymenia* faunas were mentioned by Martynova & Vorontzova (1988), but they may come from higher parts of UD IV.

(33) **Semipalatinsk region, NE Kazakhstan**: Ammonoid unit, with *Platyclymenia* and *Prionoceras* (probably at the base, but reaching at least to the top of UD IV), above a clastic and volcanic sequence, partly with brachiopods, SW of Khankeĺdiy Mountain, Chubartansk region (Bogoslovskiy 1969).

(34) **Elburz Mountains, northern Iran**: *Platyclymenia–Protactoclymenia–Erfoudites* fauna from the Zaigun Valley within the neritic Member A of the Geirud Formation (Dashtban 1995; Becker *et al.* 2004). Questionable further evidence is given by Riviere (1931, 1934) for the Roudehen Valley north of Mubarak-abad (Mobarak, Wendt *et al.* 2005) and by Becker *et al.* (2004) for a locality approximately 25 km west of Shahrud.

(35) **Central and eastern Iran**: Shotori Range: Deepening indicated by the intercalation of a single, green, marly *Annulata* Event bed, extremely rich in ammonoids (*Platyclymenia*, *Prionoceras*, *Erfoudites*), within a macrofossil-poor, micritic succession (Becker *et al.* 2004). Abadeh region: Platyclymeniid level within grey limestones of the Esteghlal Mine (Mannani & Yazdi 2015).

(36) Region of the **Khyber Pass, Pakistani–Afghan border**: Deepening episode and influx of *Platyclymenia*, the only ammonoid of the whole succession, in the upper part of the shallow-marine Ali Masjid Formation (Shah 1969, 1977).

(37) **Guangxi, South China**: Intercalated black limestones within the Tieshan Carbonate Platform, 10 km east of Guilin (Ma *et al.* 2006).

(38) **Junggar Basin, NW China**: Sudden entry of a *Platyclymenia–Prionoceras* fauna in the higher part of the thick Hongguleleng Formation of the Bulongguoer section, NW Junggar (Ma *et al.* 2011; Zong *et al.* 2014).

(39) **Great Khingan, Inner Mongolia, northern China**: Platyclymeniid–*Prionoceras* association within a condensed, pelagic carbonate platform of the Daminshan Formation (Chang 1958; Sheng 1999).

(40) **NW Thailand**: Maximum of pelagic conditions, conodont extinction and impoverishment in association with a significant positive carbon isotope excursion at Mae Sariang (Königshof *et al.* 2012; Savage 2013).

(41) **Canning Basin, Western Australia**: Hematitic, hypoxic shales of a mixed brachiopod–ammonoid biofacies with spiriferids, rhynchonellids, *Platyclymenia*, *Protactoclymenia*, *Raymondiceras* (UD IV-A) and well-preserved plant fossils in the Piker Hills Formation (Petersen 1975; Becker & House 1997, 2009). There are surprising similarities with the far distant Three Forks Shale of Montana.

(42) **New South Wales, SE Australia**: Ammonoid fauna with *Platyclymenia*, *Erfoudites* and others from the basal Mandowa Mudstone, Keepit Dam area, Tamworth Trough, intercalated within an extremely thick clastic succession (Jenkins 1968).

These occurrences can be sorted according to their different palaeogeography, sedimentology and palaeoecology into the following event settings:

(I) Condensed unit, rarely black/hypoxic, with faunal bloom interrupting a partly thick, clastic, siliceous to volcanic succession in an active, fast subsiding, orogenic basin (California, ?Nevada, Portugal, NW Devon, Bulgaria, NE Kazakhstan, Xinjiang, New South Wales).

(II) Two highly fossiliferous *Annulata* Black Shales intercalated in a mixed shaly-limy shelf basin or basin slope setting (Cantabrian Mountains, northern Rhenish Massif, Holy Cross Mountains, Maider).

(III) Single, often fossiliferous, black marl–limestone interval intercalated in pelagic oxic seamount or carbonate platform setting, partly above unconformities (northern and southern Rhenish Massif, Harz Mountains, Thuringia, Wildenfelser

Zwischengebirge, Holy Cross Mountains, Moravia, ?Moroccan Meseta, southern Tafilalt, Guangxi).

(IV) Highly fossiliferous, green shale–marl interval intercalated in a (hemi)pelagic carbonate platform or seamount setting (northern Rhenish Massif, Tafilalt Basin, eastern Iran).

(V) Short faunal bloom without hypoxic shale–marl interval in condensed and incomplete, shallow (hemi)pelagic carbonate platform settings (NE Cornwall, South Devon, Montagne Noire, ?Sicily, ?Graz Palaeozoic, Carnic Alps, Tafilalt Platform, SW Algeria, eastern Rhenish Massif, Franconia, Saxony, Mugodzhary Mountains/southern Urals, Inner Mongolia).

(VI) Inconspicuous deepening episode and carbon isotope excursion in an oxygenated pelagic carbonate platform setting (NW Thailand).

(VII) Pelagic faunal bloom in a mostly hypoxic, deep-neritic to shallow-pelagic, argillaceous shelf or foreland basin (Iowa, ?Ohio, western Dra Valley, western Moroccan Meseta).

(VIII) Hypoxic shale interval with rich, mixed neritic–pelagic faunas intercaled in a neritic or peri-reefal succession (Montana, westernmost Tindouf Basin, Western Australia).

(IX) Incursion of subordinate pelagic fauna in a neritic carbonate platform (Alberta) or shallow-marine mixed clastic–carbonatic setting (Elburz Mountains, Pakistani–Afghan border).

(X) Poorly fossiliferous, transgressive shale interval intercalated in neritic clastics, partly with minor neritic carbonates (Ardennes, western Moroccan Meseta).

So far, there has been no recognition of the *Annulata* Events in terrestrial facies, but there is a lack of investigations. The given summary supports previous interpretations (Becker 1992*b*, Hartenfels 2011) that sea-level rise and eutrophication led to enhanced productivity and blooms of plankton (mostly bacteria), nekton and specialized benthos. Local hypoxic to euxinic conditions appear to have resulted from oxygen consumption by the bacterial degradation of enhanced organic input. There is no evidence for a short-termed upslope shift of long-lasting basinal anoxia with transgression. Most basins were oxic to only weekly hypoxic in the pre-event interval.

Tafilalt conodont zonation around the *Annulata* Events

The first 'standard conodont zonation' for the Famennian was established by Ziegler (1962), and later updated, refined and revised by Ziegler (1969, 1971), Sandberg & Ziegler (1973, 1979) and Ziegler *et al.* (1974). Weddige & Ziegler (1979) proposed an autochronological concept for the branching evolution of Middle Devonian *Polygnathus* and *Icriodus*, and as the major base for conodont biozones. In a similar way, Ziegler & Sandberg (1984) attempted to base a revised, autochronological Upper Devonian 'standard conodont zonation' on the evolution of only the genus *Palmatolepis* and its ancestor *Mesotaxis*. However, in the Famennian, species of other genera are equally important, record widely correlatable speciation events within outer-shelf environments and are, therefore, biostratigraphically most useful: *Sc. velifer velifer* (*velifer* Zone, later 'hidden' under the term Uppermost *marginifera* Zone), *Ps. granulosus* (Upper *trachytera* Zone), *Po. styriacus* Ziegler in Flügel & Ziegler (1957) (Lower *postera* Zone), *B. aculeatus aculeatus* (Branson & Mehl, 1934*a*) (Middle *expansa* Zone), *B. ultimus* (Bischoff, 1957) (Upper *expansa* Zone), *Si. praesulcata* Sandberg in Sandberg *et al.* (1972) (Lower *praesulcata* Zone) and *Proto. kockeli* (Bischoff, 1957) (Upper *praesulcata* Zone). Consequently, Kaiser *et al.* (2009) and Hartenfels (2011) proposed a revised middle–uppermost Famennian zonation that is taxonomically transparent and which utilizes all conodont lineages with global distribution. This zonal concept is based on the comparison of detailed regional zonal schemes and comprises international zones, which are strictly named after their defining species (compare Spalletta *et al.* 2015). Thus, the zonation *sensu* Hartenfels (2011) represents a synthesis of new data with the former 'standard zonations' *sensu* Ziegler (1962) and Ziegler & Sandberg (1984). It includes the replacement of some palmatolepid index species with restricted geographical distribution and regionally often incomplete ranges, such as *Pa. perlobata postera* Ziegler in Kronberg *et al.* (1960) and *Pa. gracilis expansa* Sandberg & Ziegler, 1979, by more cosmopolitan or more common non-palmatolepids.

Based on Hartenfels (2011), the following regional zones and subzones are recognized in the Tafilalt (Fig. 1).

Regional Scaphignathus velifer velifer *Zone*

Lower boundary. Entry of *Sc. velifer velifer*.

Subdivision. Regional *velifer velifer* (including regionally the *granulosus* Subzone of Europe) and *gracilis sigmoidalis* subzones.

Upper boundary. Extinction of *Sc. velifer velifer*.

Common associated taxa. Within the *velifer velifer* Subzone, *Al. regularis continuus* Hartenfels, 2011,

B. stabilis zizensis Hartenfels, 2011, *Pa. gracilis semisigmoidalis* Hartenfels, 2011 (transitional subspecies between *Pa. gracilis gracilis* Branson & Mehl, 1934*a* and *Pa. gracilis sigmoidalis* Ziegler, 1962), *Po. perplexus* Thomas, 1949, *Po. subirregularis* Sandberg & Ziegler, 1979 and *Sc. velifer leptus* Ziegler & Sandberg, 1984 have their first occurrences. A regionally delayed entry within this subzone is known from *Al. regularis regularis* Ziegler & Sandberg, 1984 (elsewhere at the base of the *Sc. velifer velifer* Zone: Spalletta *et al.* 2015) and *Po. granulosus* Branson & Mehl, 1934*a* (elsewhere in the upper part of the *Palmatolepis marginifera utahensis* = Upper *marginifera* Zone). *Palmatolepis marginifera marginifera* Helms, 1959 becomes extinct. The range of *Al. beulensis* Ziegler & Sandberg, 1984 is restricted regionally to a short interval within this subzone. The entry of *Pa. gracilis sigmoidalis* defines the base of a subzone in the higher *velifer velifer* Zone. Rare *Po. homoirregularis* Ziegler, 1971 (*sensu stricto*) and *Po. margaritatus* Schäfer, 1976 first occur at the top of the *gracilis sigmoidalis* Subzone. Regionally, '*Po.*' *diversus* Helms, 1959 and *Po. nodocostatus nodocostatus* Branson & Mehl, 1934*a* have their upper range in the basal part, *Pa. glabra lepta* late morphotype and *Sc. velifer leptus* terminate within this subzone, and *Al. regularis continuus*, *Pa. minuta minuta* and *Po. granulosus* at the upper limit.

The following taxa range through the *velifer velifer* Zone: *B. stabilis vulgaris* (Dzik, 2006), *Br. ampla* (Branson & Mehl, 1934*a*), *Br. inornata* (Branson & Mehl, 1934*a*), *Caenodontus* sp. (cf. Hartenfels 2011), representatives of the '*I.*' *cornutus* Group, *M. strigosa* (Branson & Mehl, 1934*a*), *Neo. communis communis* (Branson & Mehl, 1934*b*), *Pa. gracilis gracilis*, *Pa. minuta schleizia* Helms, 1963, *Pa. perlobata helmsi* Ziegler, 1962, *Pa. perlobata maxima* Müller, 1956*b*, *Pa. perlobata schindewolfi* Müller, 1956*b* and *Po. semicostatus* Branson & Mehl, 1934*a*.

Originally, the genus *Caenodontus* was described from the Permian by Behnken (1975) and Dzik (2009), as well as from the Triassic by Kozur & Mostler (1976). Much older, Famennian, caenodontids were described by Buchroithner *et al.* (*Cae. talayotoides* Buchroithner, Flügel, Flügel & Stattegger, 1980) from Menorca, as well as by Hartenfels (2011) from the Rhenish Massif, Thuringia, Holy Cross Mountains, Maider and Tafilalt. Further Famennian records were mentioned very recently by Weiner & Kalvoda (2016) from the Moravian Karst.

Discussion. Based on the regional absence of *Pa. rugosa trachytera* and *Ps. granulosus*, Hartenfels (2011) could not recognize the international *rugosa trachytera* and *granulosus* zones in the Tafilalt, which equal the Lower and Upper *trachytera* zones of the 'standard zonation'. Thus, he introduced a regionally extended *velifer velifer* Zone. The youngest *Sc. velifer velifer* are known in the Tafilalt from the Lower *Annulata* Event Bed at Jebel Ouaoufilal Pass (southern Tafilalt), a black cephalopod limestone showing pelagic high-energy deposition. Therefore, the upper boundary of the regional *velifer velifer* Zone corresponds with the top of the Lower *Annulata* Event. The regional *velifer velifer* Zone correlates to the international *velifer velifer* and *rugosa trachytera* zones, and roughly with the lower half of the international *granulosus* Zone.

Polygnathus perplexus has been listed as an additional index species for the *granulosus* (=Upper *trachytera*) Zone in Ziegler & Sandberg (1984) and has been used as zonal marker (e.g. by Perri & Spalletta 1990). Hartenfels (2011) documented that *Po. perplexus* may enter much earlier than *Ps. granulosus* at the German Beringhauser Tunnel and Kahlleite East sections. This agrees with the range extension in Johnston & Chatterton (2001), who found the species associated with *Pa. marginifera*. A range even below *Sc. velifer* was published, without illustrations, in Belgium by Dreesen & Dusar (1974) and subsequently considered in Sandberg & Ziegler (1979).

So far, limited data (Hartenfels 2011; Hartenfels *et al.* 2013) indicate that the oldest *Pa. gracilis semisigmoidalis* can be used to trace roughly the level of the *trachytera* Zone in the Anti-Atlas. *Palmatolepis gracilis semisigmoidalis* records may explain a supposed overlap of the first *Pa. gracilis sigmoidalis* with the last *Pa. marginifera marginifera* in the Istanbul Zone of NW Turkey (Çapkinoğlu 2005), but critical specimens were not figured. Since the subspecies enters with delay in several Tafilalt sections, it is not used for a subzonal subdivision of the regional *velifer velifer* Zone.

Based on records from two sections (Jebel Erfoud and Jebel Ouaoufilal), Hartenfels (2011) established regionally the *gracilis sigmoidalis* Subzone, which is defined by the first entry of the index subspecies well below the Lower *Annulata* Event, and in association with *Sulco. sulcata* (Schindewolf, 1923) (Jebel Erfoud, Bed 22b). A corresponding *gracilis sigmoidalis* Subzone can also be separated within the *granulosus* Zone of Germany (Rhenish Massif, Saxothuringia), Poland (Holy Cross Mountains) and Austria/Italy (Carnic Alps: cf. Hartenfels 2011). The first Moroccan record of *Ps. granulosus* is restricted to three specimens from El Gara South, which co-occur with *Pa. gracilis sigmoidalis*. They provide the first Tafilalt evidence of the *granulosus* Zone, but only for its middle part (subzone): the zonal base has to be sought below.

At El Atrous, an overlap of *Pa. marginifera marginifera* and *Po. subirregularis* suggests the local recognition of the basal part of the global *rugosa trachytera* Zone (Hartenfels 2011). However, the rarity of *Po. subirregularis*, which in other sections enters irregularly, often with delay, questions the reliable biostratigraphic use of *Po. subirregularis*.

Associated with the Lower *Annulata* Event, two other rare species, *Po. homoirregularis sensu stricto* and *Po. margaritatus*, first occur. Both are known from the lower event bed at Jebel Ouaoufilal Pass (Hartenfels 2011). Shallow-water carbonates of North America (e.g. Sandberg & Poole 1977; Sandberg 1979) and Belgium (Dreesen & Thorez 1994) yielded alleged very early morphotypes of *Po. homoirregularis* that overlap with *Pa. glabra distorta* Branson & Mehl, 1934*a* and *Pa. marginifera*. This suggests a total range of the species into or below the *rugosa trachytera* Zone. However, so far, there is no detailed documentation or any illustration of these supposed oldest relatives. In this context, and due to the rarity of *Po. homoirregularis sensu stricto* and *Po. margaritatus*, the entry of both taxa is currently not a reliable biostratigraphic marker for the Lower *Annulata* Event Interval. In contrast to European successions, *Al. regularis continuus* disappears regionally in the Lower *Annulata* Event beds. *Alternognathu regularis regularis* is last encountered in the Upper *Annulata* Event Bed at El Gara South (basal part of the subsequent *velifer–stabilis* Interregnum). Therefore, the gradual loss of alternognathids seems to be a palaeoecological consequence of the *Annulata* Event in the Tafilalt.

Velifer–stabilis *Interregnum*

Lower boundary. Extinction of *Sc. velifer velifer*.

Upper boundary. Entry of *B. stabilis stabilis* Branson & Mehl, 1934*a* (=*stabilis* M2).

Common associated taxa. The entry of *B. stabilis bituberculatus* (Dzik, 2006 = *stabilis* M3) is regionally delayed (Hartenfels 2011). Elsewere (e.g. Rhenish Massif: Hartenfels 2011), *B. stabilis bituberculatus* is an alternative index species of the *gracilis manca* (=Upper *postera*) Zone. Youngest specimens are known from the basal part of the subsequent *aculeatus aculeatus* Zone. In the Tafilalt, representatives of the '*I.*' *cornutus* Group terminate within the *Annulata* Intralimestone, *Al. regularis regularis* within the Upper *Annulata* Event beds, and *B. stabilis zizensis*, *Pa. minuta schleizia*, *Pa. perlobata maxima* and *Po. subirregularis* just above the overlying Wagnerbank Equivalents. *Bispathodus stabilis vulgaris*, *Br. ampla*, *Br. inornata*, *Caenodontus* sp., *M. strigosa*, *Neo. communis communis*, *Pa. gracilis gracilis*, *Pa. gracilis semisigmoidalis*, *Pa. gracilis sigmoidalis*, *Pa. perlobata helmsi*, *Pa. perlobata schindewolfi*, *Po. homoirregularis sensu stricto*, *Po. margaritatus*, *Po. perplexus* and *Po. semicostatus* range through the zone.

Discussion. The *Annulata* Intralimestone and the Upper *Annulata* Event, as well as the Wagnerbank Equivalents, fall in the lower part of the regional *velifer–stabilis* Interregnum. Above, the absence of *Po. styriacus*, *Pa. perlobata postera*, as well as *Pa. gracilis manca* Helms, 1963, and the delayed regional entry of *B. stabilis bituberculatus* prevent the recognition of the international *styriacus* (=Lower *postera*) and *gracilis manca* zones in the Tafilalt. Since *Pa. rugosa trachytera* disappears variably slightly below, at the base of the Lower *Annulata* Event (Saxothuringia, Holy Cross Mountains), or at the top of a Lower *Annulata* Intralimestone (Rhenish Massif), the base of the Tafilalt *velifer–stabilis* Interregnum is variably coincident or slightly older than the base of the international *trachytera–styriacus* Interregnum (the upper subzone of the international *granulosus* Zone).

Rytina *et al.* (2013) documented a single specimen of *Pa. gracilis manca*, the first record for SE Morocco, associated with a questionable clydagnathid and platyclymeniids in an olistolith at Taourirt n'Khellil, in the Tinerhir region to the NW of the Tafilalt. This opened the possibility that further sampling may enable the recognition of the *gracilis manca* Zone in the eastern Anti-Atlas.

According to Sandberg & Dreesen (1984), '*I.*' *cornutus* ranges from the Middle *triangularis* Zone at least to the top of the Upper *marginifera* Zone (*marginifera utahensis* Zone). Rare possible descendants or a late morphotype were shown to occur through the *granulosus* Zone (=Upper *trachytera* Zone; see the dashed range extension in fig. 1 of Sandberg & Dreesen 1984). Unfortunately, so far, there is no illustration of these youngest relatives. Similar to older forms figured in Sandberg & Dreesen (1984), Perri & Spalletta (1990) and Sanz-López *et al.* (1999), our El Gara South samples provide two new specimens from the *Annulata* Intralimestone (basal part of the regional *velifer–stabilis* Interregnum). Their age corresponds to previous Tafilalt data in Hartenfels (2011). At Kowala (Holy Cross Mountains, Poland), specimens of the '*I.*' *cornutus* Group disappear in the first limestone nodules above the Upper *Annulata* Black Shale and at Gositzfelsen (Thuringia) in the subsequent Wagnerbank Equivalents (Hartenfels 2011). Both records mark the basal part of the European *trachytera–styriacus* Interregnum. Rytina *et al.* (2013) documented similar forms from the regional *velifer velifer* Zone at Taourirt n'Khellil.

Generally, the specimens illustrated in Sandberg & Dreesen (1984), Perri & Spalletta (1990),

Sanz-López *et al.* (1999) and Rytina *et al.* (2013), as well as our new material, differ in important aspects from the *I. cornutus* holotype *sensu* Sannemann (1955), in which the side row denticles are restricted to the anterior two-thirds of the platform. Furthermore, there are only three transversal rows of side denticles, whilst the others show seven or eight. '*Icriodus*' *cornutus* seems to comprise several taxa, which need to be studied in detail based on more material.

Tafilalt ammonoid succession around the *Annulata* Events

A regional ammonoid stratigraphy for the middle and upper Famennian of the Tafilalt has been developed by Becker (1993*a*), Korn (1999), Becker & House (2000*b*), Korn *et al.* (2000) and Becker *et al.* (2002). Additional data were published in Bockwinkel *et al.* (2002), Ebbighausen *et al.* (2002), Korn & Klug (2002), Hartenfels (2011), Becker *et al.* (2013), Hartenfels *et al.* (2013), Korn *et al.* (2013, 2014, 2015*a*, *b*), and Klein & Korn (2014). Korn *et al.* (2014) introduced a new regional prionoceratid zonation for the UD IV–UD V-A of the Tafilalt/Maider, which is discussed below. The revised regional stratigraphic chart by Becker *et al.* (2013) is further updated (Fig. 1).

Regional Maeneceras subvaricatum *Zone (UD II-G–UD III-A)*

Lower boundary. Entry of the oldest Sporadoceratidae, here reassigned to *Maene. subvaricatum* (Sobolew, 1914*a*). In previous zonal schemes, the terms *biferum* Zone (e.g. Becker 1993*a*) and *meridionale* Zone (e.g. Korn & Ziegler 2002; Becker *et al.* 2013) are synonyms.

Subdivision. There is some evidence for a succession from *Maene. subvaricatum nuntio* Becker, 1993*a* to *Maene. subvaricatum subvaricatum* and *M. latilobatum* (Schindewolf, 1923), in which the second adventitious lobe gradually increases its depth (Becker 1993*a*; Bockwinkel *et al.* 2002). This interval comprises the regional *subvaricatum* Subzone (UD II-G), which is followed by an ammonoid-poor interval (UD II-H/I). Rare occurrences of the oldest *Sporadoceras* give an upper (regional) *Sporadoceras equale* Subzone (UD III-A), which has so far only been documented in a few sections (Becker *et al.* 2002).

Upper boundary. Entry of *Plani. euryomphalum* (Wedekind, 1918).

Common associated taxa. The oxyconic and distinctive *Acri. falcisulcatum* Becker, 1993*a* occurs throughout the Tafilalt Platform (Becker 1993*a*; Ebbighausen *et al.* 2002) and also at Hassi Nebech in the Tafilalt Basin (Becker *et al.* 2002). *Acrimeroceras stella* Ebbighausen *et al.*, 2002 is locally common. Both species are restricted to the approximately lower half of the *subvaricatum* Subzone. *Armatites planidorsatus* (Münster, 1839*b*) and species of *Falcitornoceras* are found at most localities.

Discussion. A critical taxonomic re-evaluation supports the use of *M. subvaricatum* as the valid name for the oldest Moroccan sporadoceratids (see the section on 'Systematic palaeontology of ammonoids (RTB)' later in this paper). Equivalents of the German (Becker 1993*a*) and Australian (Becker & House 2009) *Posttornoceras* (UD II-H) and *Dimeroceras* zones (UD II-I) have not yet been recognized in the region.

Planitornoceras euryomphalum *Zone (UD III-B–III-C$_1$)*

Lower boundary. Entry of *Plani. euryomphalum.*

Subdivision. An upper subzone defined by the oldest clymeniids (*Pricella*) and endemic representatives of the Prolobitidae has been recognized in the southern Maider (*Afrolobites mrakibensis* Subzone, Becker *et al.* 2002), but not yet in the Tafilalt.

Upper boundary. Entry of *Sulco. sulcata.*

Common associated taxa. In sections of the Tafilalt Platform (Jebel Erfoud, Hartenfels 2011; Hamar Laghdad East, Becker *et al.* 2002), *Sp. angustisellatum* Wedekind, 1908 is an associated and partially dominant (El Khraouia, Hartenfels *et al.* 2013) marker species. This is also true for the Tafilalt Basin (Becker *et al.* 2002; Hartenfels *et al.* 2013), where ?*Erf. spiriferus* (Lange, 1929) and *Enke. varicatum* (Wedekind, 1908) co-occur.

Discussion. The zonal marker is small sized, but very distinctive, common and it is distributed more widely than previously known (Becker *et al.* 2002). This justifies a full zonal status. *Pseudoclymenia*, the index genus of UD III-B (Becker & House 2000a), is absent in Morocco.

Sulcoclymenia sulcata *Zone (UD III-C$_2$)*

Lower boundary. Entry of *Sulco. sulcata.*

Subdivision. The upper part of the *sulcata* Zone is very poor in ammonoids all over the Tafilalt.

Upper boundary. Entry of *Pl.* (*Pl.*) *annulata* subspecies, other species of the subgenus or of prionoceratids.

Common associated taxa. Sporadoceras angustisellatum is regionally widespread (Jebel Erfoud,

Hamar Laghdad, Ouidane Chebbi, Hassi Nebech and Jebel Ouaoufilal, Becker *et al.* 2002; Hartenfels 2011). New regional records of additional, rare taxa are presented here (section El Gara I).

Discussion. Previous North African *Sulco. sulcata* were known from the Tafilalt Basin and southern Maider (Becker *et al.* 2002). Its abundance at Jebel Erfoud (central Tafilalt Platform, Hartenfels 2011) and at El Gara South shows that it can define a full ammonoid zone. It provides a correlation with the highest middle Famennian of Franconia (Schindewolf 1923), Saxony (Freyer 1957) and the eastern Rhenish Massif (Korn 2004). Becker & Hartenfels (2010) showed that the rarity of ammonoids in the top part of UD III-C is a globally significant phenomenon. A pre-*Annulata* episode of extreme oligotrophy may explain the extinction of the many ammonoid species that are so characteristic of UD III-C (German *Prolobites delphinus* Zone).

Regional Platyclymenia (Platyclymenia) annulata *Zone (UD IV-A–IV-B_1)*

Lower boundary. Entry of various subspecies of *Pl.* (*Pl.*) *annulata*, other species of the subgenus or of *Prionoceras* (*sensu lato*: see Fischer & Becker 2014).

Subdivision. The Lower *Annulata* Event beds at the base of the zone from Bine Jebilet, Hassi Nebech and El Gara include *Pl.* (*Pl.*) *subnautilina subnautilina*, *Pl.* (*Pl.*) *quiringi* Müller, 1956*a*, *Pl.* (*Pl.*) *annulata* div. ssp. and *Pl.* (*Trigonoclymenia*) *protacta* (Wedekind, 1908). Among the goniatites, there are *Pr. divisum lamellosum* Korn *et al.*, 2014, *Pr. frechi* (Wedekind, 1913) n. ssp. (with more lamellose growth ornament than German typical representatives) and '*Pr.*' *vetum* Korn *et al.*, 2014 (nom. corr.; with open umbilicate early whorls).

Additional taxa of the *Annulata* Intralimestone and Upper *Annulata* Event beds are *Pl.* (*Pl.*) *annulata richteri* (Wedekind, 1914), *Pl.* (*Pl.*) *pattisoni* (Phillips, 1841), *Pl.* (*Pl.*) *semiornata* Petter, 1959, *Protactoclymenia* div. sp., *Ungu. unguiforme* Korn, Bockwinkel & Ebbighausen, 2015*a*, *b* (=*Erfoudites* n. sp. II of Bockwinkel *et al.* 2002), '*Prionoceras*' n. sp. (rare, with very wide juvenile umbilicus: Nawrath 2009) and early *Erf. rherisensis* Korn, 1999. At Bine Jebilet, a new, very compressed relative of *Pr. frechi* with up to five constrictions/whorl is common.

The Wagnerbank Equivalents and immediately overlying beds of the Tafilalt Platform are locally characterized by the entry of large-sized platyclymeniids, such as *Pl.* (*Pl.*) *ibnsinai* Korn, 1999 (possibly a junior synonym of *Clymenia nodosa*

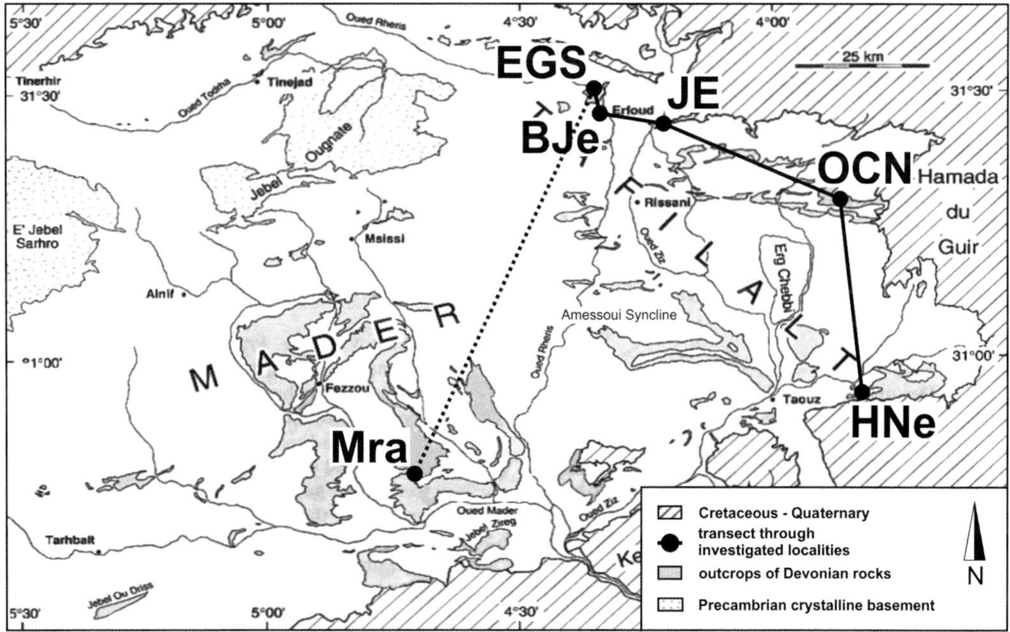

Fig. 3. Position of studied and correlated sections in the eastern Anti-Atlas. Mra, Mrakib; EGS, El Gara South; BJe, Bine Jebilet; JE, Jebel Erfoud; OCN, Ouidane Chebbi Northwest; HNe, Hassi Nebech.

Münster, 1839*a*) and *Pl.* (*Pl.*) *laxata* Czarnocki, 1989. These have been commercially exploited at Seheb-el-Rhassal and Rich Haroun, SW of Erfoud. Only at very few localities (e.g. loose at Bine Jebilet and Mfis) has *Prot. dunkeri* (Münster, 1839*a*) been found above the Wagnerbank Equivalents (Becker *et al.* 2002). Therefore, there is currently only a very poor database to separate in the Tafilalt a *dunkeri* Zone, as proposed by Korn (1999).

Upper boundary. Entry of *Proc. ebbighauseni* Klein & Korn, 2014.

Associated common taxa. Species of *Protactoclymenia*, *Pl.* (*Trigonoclymenia*) and *Erfoudites*.

Discussion. Since *Pr. divisum lamellosum* occurs frequently in the *Annulata* Event beds (e.g. at its type locality Takhbtit, rich material from Jebel Ouaoufilal Pass and Hassi Nebech), a distinction of successive *vetum* and *lamellosum* zones, as proposed by Korn *et al.* (2014), cannot be upheld. The rare Tafilalt records of *dunkeri* suggest that upper parts of the regional *annulata* Zone correlate with the lower part of the Rhenish *Prot. dunkeri* Zone (UD IV-B_1). *Protoxyclymenia* enters 10 limestone cycles above the Wagnerbank Equivalents in the basal *styriacus* Zone at Effenberg (Germany). *Franconiclymenia* appears in the very condensed Beringhauser

Fig. 4. Outcrop views of the sampled sections I (**a**) and II (**b**), just south of the El Gara settlement (background in a), ESE of Jorf, separated by a piste running to the NE. Nodular limestones form very minor, laterally traceable marker beds. LAE and UAE, Lower and Upper *Annulata* Event beds; WE, Wagnerbank Equivalents, and bed numbers.

Tunnel section in the same conodont zone (Hartenfels 2011).

Procymaclymenia ebbighauseni *Zone (UD IV-B$_2$)*

Lower boundary. Entry of *Proc. ebbighauseni*, previously included in *Proc. pudica* (Czarnocki, 1989): for example, in Becker *et al.* (2002) and Hartenfels (2011).

Upper boundary. Entry of *Sp. muensteri orbiculare* (Münster, 1832).

Common associated taxa. On the Tafilalt Platform, *Proc. ebbighauseni* is very dominant (see fauna from El Atrous North in Korn *et al.* 2000) and partly the only species found. In the Tafilalt Basin, *Erfoudites*, various *Pl.* (*Platyclymenia*) and *Protactoclymenia* may be associated (Hartenfels 2011).

Discussion. Becker *et al.* (2002) proposed a *Procymaclymenia pudica* Zone for the middle part of UD IV of the Tafilalt and Maider, which correlates with the entry of the oldest cymaclymeniids of the Rhenish Massif (e.g. *Cymaclymenia* n. sp. in Becker 1992*a*), Saxony (e.g. '*Cyma. cordata*' in Hösel 1960), Silesia (*Cymaclymenia* nov. sp. of Schindewolf 1921) and the Holy Cross Mountains (Czarnocki 1989). Korn *et al.* (2013) showed that *Proc. pudica* enters at a higher level than suggested in the Maider by loose material recorded in Becker *et al.* (2002). Subsequently, Klein & Korn (2014) assigned the Moroccan variants to a closely related new species, *Proc. ebbighauseni*. A single, oldest *in situ* specimen is at Mrakib from a limestone nodule within the upper part of Bed P1 (Webster *et al.* 2005) and the species is most common in Bed Q1, especially in a dark-grey detrital marker limestone (Bed 16d of Hartenfels 2011; Bed Q1s in Korn *et al.* 2014). A rather late entry has also been documented for the Tafilalt, especially for Bine Jebilet (26 limestone beds above the Wagnerbank Equivalents) and Hassi Nebech (Hartenfels 2011). This suggests a correlation with only the upper part of the Rhenish *dunkeri* Zone, supported by a range of *Proc. ebbighauseni* faunas into the *stabilis stabilis* Zone (=Lower *expansa* Zone) at Djebel Erfoud and Mrakib (Hartenfels 2011).

Studied sections

El Gara South

The section is situated approximately 12 km NW of Erfoud and approximately 2 km ESE of Jorf, close to the small settlement El Gara (Fig. 3). In this region, the Famennian was first recognized by Wendt *et al.* (1984). Becker (1993*a*) gave a rough cross-section for the exposed Famennian, but concentrated on the Nehdenian (UD II). Preliminary data concerning *Annulata* Events were presented by Hartenfels & Becker (2013) at the SDS-IGCP 596 Field Symposium in southern Morocco. El Gara South is exceptional since it includes taxa (ammonoids, conodonts) that have not been observed anywhere else in the Anti-Atlas.

The local landscape is flat and the outcrop conditions are generally poor. Therefore, the succession was measured bed-by-bed and sampled in two lateral trenches. The first trench (El Gara South I, GPS coordinates: 31° 28′ 23.4″ N, 004° 21′ 22.1″ W, map sheet 244 Tafilalt–Taouz) traverses both *Annulata* Events and the two-layered Wagnerbank Equivalent (Fig. 4a). In the second trench to the SE (El Gara South II, GPS coordinates: 31° 28′ 23.1″ N, 004° 21′ 21.4″ W) the overlying sequence crops out. In the field, the base of the second trench can be easily recognized by a 12 cm thick, solid, red marker limestone (Bed 19, Fig. 4b).

The studied succession ranges from UD II-G (regional *marginifera marginifera* Zone) to UD IV-A/B$_1$ (regional *velifer–stabilis* Interregnum).

El Gara South I

The succession in the first trench, El Gara South I (Fig. 4a, beds 5–12 & Figs 5–12), is as follows:

-8	Two layers of red, ferrugineous nodular limestone at the section base
-7	*c*. 90 cm greenish-brown, marly shale with small-sized goniatites
	Maene. subvaricatum subvaricatum, B6.C-47.1–B6.C-47.3, B6.C-47.4 (Fig. 6.8a, b)
-6	10–13 cm, two layers of red, ferrugineous, solid limestone with poorly preserved orthocones.
-5	*c*. 75 cm, red, nodular, poorly fossiliferous, marly shale.
-4	12–15 cm, two layers of grey, strongly nodular limestone with poorly preserved goniatites.
	Plani. euryomphalum
	Sporadoceras sp. indet.
-3a	[small oued]
	96 cm, greenish-grey, nodular, marly shale with small-sized ammonoids
	Plani. euryomphalum, B6.C-47.5 (Fig. 6.1a, b) and B6.C-47.6–B6.C-47.28
	[*Fidelites* cf. *clariondi* (Petter, 1959) juv., B6.C-47.32, Eifelian float specimen with dacryoconarids in the body chamber (Fig. 6.3a, b)]

[*Sobolewia* sp. juv., B6.C-47.31, float specimen (Fig. 6.4a, b)]
[goniatite indet., B6.C-47.30, float specimen]
orthoconic cephalopod indet., B6.C-47.29

-3b *c.* 165 cm, red, nodular, marly shale with small-sized ammonoids and other fauna
Protor. ornatum juv., B6.C-47.72 (Fig. 6.2a, b)
Sp. angustisellatum, B6.C-47.69
Maid. sapiens (Korn, 1999), B6.C-47.70 and B6.C-47.71 (Fig. 7.5a, b)
Enke. varicatum, B6.C-47.67 and B6.C-47.68 (Fig. 7.4a, b)
Sulco. sulcata, B6.C-47.33–B6.C-47.65 (C-47.37 = Fig. 7.8a, b)
Protacto. aff. *implana* Czarnocki, 1989, B6.C47.74 (Fig. 7.6a, b)
Genu. aff. *angelini* (Wedekind, 1908), 1989, B6.C-47.73 (Fig. 7.7a, b)
?*Pleuroclymenia* sp. juv., B6.C-47.66
Orthoconic cephalopods indet., B6.C-47.79-80
Buchiola sp., six specimens

-2a [small oued]
45 cm, poorly exposed, greenish-grey, nodular, marly shale

-2b 21 cm greenish-grey, nodular, marly shale with clymeniids
Protacto. aff. *subcostata* (Sobolew, 1914*b*), B6.C-47.75–B6.C-47.77 and B6.C-47.78 (Fig. 8.5a, b)
[*Pl.* (*Platyclymenia*) sp. juv., B6.C-47.81, loose from above]

-1 3–5 cm, thin, grey, nodular limestone

0 54 cm, unfossiliferous, greenish-grey, marly shale

1 3.5 cm, unfossiliferous, thin, brownish-red to slightly greenish mottled, nodular limestone (bioturbated, microsparitic mud-wackestone with ostracods, shell filaments, rare trilobites and clymeniids: Fig. 9.1); conodonts (Figs 5 & 10)
Al. regularis regularis, B9A.6-1 (Fig. 11.1a–c)
Al. regularis continuus
B. stabilis vulgaris, B9A.6-2 (Fig. 11.2)
Br. ampla, B9A.6-5 (Fig. 11.5a, b)
Br. inornata
M. strigosa
Pa. gracilis gracilis
Pa. gracilis sigmoidalis
Pa. minuta minuta, B9A.6-12 (Fig. 12.1)
Pa. minuta schleizia, B9A.6-13 (Fig. 12.2)
Pa. perlobata helmsi
Pa. perlobata schindewolfi
Ps. granulosus, B9A.6-17 (Fig. 12.6a, b) and B9A.6-18 (Fig. 12.7)
Sc. velifer velifer, B9A.6-20 (Fig. 12.9a–c)
Sc. velifer leptus, B9A.6-21 (Fig. 12.10a–c)

2a 8 cm, unfossiliferous, greenish-grey, marly shale

2b 4 cm, brownish-red, nodular limestone, partially disintegrated into isolated nodules; conodonts
Al. regularis regularis
Neo. communis communis
Pa. gracilis gracilis
Ps. granulosus, B9A.6-19 (Fig. 12.8)

3a **Lower *Annulata* Event Bed**
3 cm, brownish to whitish weathering, marly shale with small-sized ammonoids and conodonts
Pl. (*Pl.*) *subnautilina subnautilina*, B6.C-47.89–B6.C-47.96 and B6.C-47.109 (Fig. 8.3a, b)
'*Pr.*' *vetum* juv. B6.C-47.83–B6.C-47.88 (with only three concave varices/whorl until *c.* 10 mm dm, uw/dm = 0.1), B6.C-47.136 and B6.C-47.137 (largest specimen with four varices/whorl)

3b ***Annulata* Intralimestone**
3 cm thick, brownish-red nodular limestone, partially disintegrated into isolated nodules, with small-sized ammonoids and conodonts
Pl. (*Pl.*) *subnautilina subnautilina*, B6.C-47.97
Pl. (*Pl.*) *levata* n. sp., B6.C-47.98
'*Pr.*' *vetum* juv., B6.C-47.99, B6.C-47.100 (still with three constrictions/whorl)
B. stabilis vulgaris, B9A.6-3 (Fig. 11.3a, b)
'*I.*' *cornutus* Group, B9A.6-7 (Fig. 11.7a–c)
Neo. communis communis, B9A.6-8 (Fig. 11.8a, b)
Pa. gracilis gracilis
Pa. gracilis semisigmoidalis
Pa. perlobata maxima, B9A.6-15 (Fig. 12.4)
Pa. perlobata schindewolfi, B9A.6-16 (Fig. 12.5a, b)

4a **Upper *Annulata* Event Bed**
42 cm, brownish to whitish weathering, nodular, marly shale with small-sized ammonoids

	Pl. (*Pl.*) *subnautilina subnautilina*, B6.C-47.111 (Fig. 8.4a, b)
	Pl. (*Pl.*) *levata* n. sp., B6.C-47.101
	'*Pr.*' *vetum*, B6.C-47.102–B6.C-47.104, B6.C-47.110 (Fig. 6.5a, b, with a change to *c.* 90° angle of varices from *c.* 10 mm dm on)
	Al. regularis regularis
	B. stabilis vulgaris, B9A.6-4 (Fig. 11.4)
	Pa. gracilis gracilis, B9A.6-10 (Fig. 11.10)
	Pa. gracilis semisigmoidalis
	Pa. perlobata helmsi, B9A.6-14 (Fig. 12.3a, b)
	Pa. perlobata schindewolfi
4b/5a	2.5 cm, brownish to whitish weathering, marly shale with small limestone nodules at the base (4b) and greenish marly shale at the top (5a)
5b–6b	**Wagnerbank Equivalents**
	8 cm, two layers of reddish or greenish, very fossiliferous, nodular limestone (bioturbated, microsparitic cephalopod float-rudstones with goniatites and clymeniids, shell filaments, ostracods, rare trilobites, crinoid remains, echinid spines and silt; golden brown *Frutexites*-type encrustions enclose skeletal remains) (Fig. 9b), separated by 1 cm greenish, marly shale; conodonts sampled from Bed 5b (Figs 5 & 10)
	Pl. (*Pl.*) *levata* n. sp., B6.C-47.105, B6.C-47.106 (5b)
	Pl. (*Pl.*) *annulata richteri*, B6.C-47.107 (5b)
	Pl. (*Pl.*) *pattisoni* (B6.C-47.113, with ww/wh near 0.7 < 20 mm dm, as given for the holotype by Price (1982), shell smooth; 6b)
	'*Pr.*' cf. *vetum*, B6.C-47.108 (5b)
	Pr. frechi n. ssp., B6.C-47.112 (Fig. 6.7a, b; 5b)
	Guerichia sp. (5b)
	B. stabilis vulgaris (5b)
	Br. inornata (5b)
	Pa. gracilis gracilis, B9A.6-11 (Fig. 11.11; 5b)
	Pa. perlobata schindewolfi (5b)
7a–17	132.5 cm, cyclic succession of green marly shale with ammonoids and 10 intercalated layers of red, green or greenish-grey nodular limestone with a maximum thickness of 4.5 cm (Bed 9b), crossed at the top by a minor piste
	Pl. (*Pl.*) *levata* n. sp., B6.C-47.114–B6.C-47.117 (8a) and B6.C-47.119–B6.C-47.133 (8a)
	Pl. (*Pl.*) *pattisoni*, B6.C-47.118 (8a)
	Pl. (*Pl.*) *subnautilina subnautilina*, B6.C-47.146–B6.C-47.151 (16)
	Pl. (*Pl.*) *annulata richteri* B6.C-47.134 (10a)
	Pr. divisum lamellosum, B6.C-47.135, B6.C-47.136 (10a), B6.C-47.139 (12a) and B6.C-47.145 (16)
	Pr. frechi n. ssp., B.6.C-47.140 (14b) and B6.C-47.141–B6.C-47.144 (16)

El Gara South II

The succession in the first trench, El Gara South II (see Figs 4b, 8, 9 & 11), is as follows:

19	12 cm, ferruginous, solid, micritic, **red marker limestone** (bioturbated, microsparitic cephalopod float-rudstone with goniatites, clymeniids, shell filaments, ostracods, and rare trilobites) (Fig. 9c) with conodonts and corroded ammonoids at its top
	B. stabilis vulgaris
	Br. Ampla
	Br. inornata, B9A.6-6 (Fig. 11.6a, b)
	Neo. communis communis, B9A.6-9 (Fig. 11.9a, b)
	Pa. gracilis gracilis
	Pa. perlobata schindewolfi
20–25a	100 cm, cyclic sequence of green, marly shale and five intercalated layers of red, green or greenish-grey nodular limestone with a maximum thickness of 7.5 cm (Bed 23b); relatively poor in ammonoids
	Pr. takhbtitense Korn *et al.*, 2014, B6.C-47.155 (23b)
	Pr. divisum lamellosum, B6.C-47.155 (23b)
25b	18.5 cm, red, marly shale with abundant ammonoids, trilobites and bivalves
	Pl. (*Pl.*) *subnautilina subnautilina*, B6.C-47.160–B6.C-47.162

Pl. (*Pl.*) *annulata richteri*, B6.C-47.157 and B6.C-47.158
Pl. (*Pl.*) *annulata rotundata* Wedekind, 1914, B6.C-47.159 (Fig. 8.2a, b)
Pr. divisum lamellosum, B6.C-47.165 and B6.C-47.166
Pr. frechi n. ssp., B6.C-47.163 and B6.C-47.164
Ungu. unguiferum, B6.C-47.156
Guerichia sp.
proetid pygidium (on B6.C-47.160)
phacopid pygidium (on B6.C-47.163)
4 cm, greenish-grey nodular limestone with *Platyclymenia*

26a More than 60 cm (higher part covered), green and red, marly shale with limestone nodules and a rich, small-sized ammonoid and bivalve fauna at the top
Erf. rherisensis juv., B6.C-47.177 (with ww/wh = 1.07 at 8.3 mm dm)
Pl. (*Pl.*) *levata* n. sp., B6.C-47.152 (holotype, Fig. 8.8a, b), B6.C-47.176 (relatively thick specimen)
Pl. (*Pl.*) aff. *levata* n. sp., B6.C-47.181 (more strongly ribbed)
Pl. (*Pl.*) *subnautilina subnautilina*, B6.C-47.153 (Fig. 8.6, still slightly depressed at 12.5 mm wh)
Pl. (*Pl.*) *annulata richteri*, B6.C-47.179
Pl. (*Tri.*) *protacta*, B6.C-47.178
Protactoclymenia sp., B6.C-47.181 and B6.C-47.182–B6.C-47.183 (small specimen with uw/dm = 0.29 and ww/wh = 1.04 at 11.3 mm dm, differently preserved, possibly washed in)
Pr. divisum lamellosum, B6.C-47.167–B6.C-47.170
Pr. takhbtitense, B6.C-47.171 and B6.C-47.172
Pr. frechi n. ssp., B6.C-47.173 and B6.C-47.174, cf. B6.C-47.175
[*Plani. euryomphalum*, B6.C-47.184 (washed in from below)]
Guerichia sp.

Younger sediments are locally covered, but are exposed laterally, approximately 1 km to the north (Becker 1993*a*: beds O–Q with *Biloclymenia laevis* (Richter, 1848) and *Gonioclymenia* sp., UD V-A_2). Additional poorly preserved material was collected in March 1994. A low '*Cymaclymenia* ridge' yielded *Costaclymenia muensteri* (Ansted, 1838) (B6.C-47.260 and B6.C-47.261), *Erfoudites* sp. (B6.C-47.262), *Cyma. costellata* (Münster, 1832) (B6.C-47.263 and B6.C-47.264) and *Kosmoclymenia inaequistriata lamellosa* (Wedekind, 1914) (B6.C-47.265 and B6.C-47.266). This assemblage is typical for very basal parts of Famennian V (UD V-A1) and may include the Dasberg Events beds (Becker *et al.* 2002). The subsequent '*Kosmoclymenia* Ridge' produced loose *Biloclymenia* sp. (B6.C-47.267–B6.C-47.269), *Kosmo. inaequistriata inaequistriata* (B6.C-47.270), *Kosmoclymenia* sp. indet. (B6.C-47.271–B6.C-47.275), '*Mimimitoceras*' cf. *comtum* Korn, Bockwinkel & Ebbighausen, 2015*b* (B6.C-47.276–B6.C-47.278), *Cyma. costellata* (B6. C-47.279, B6.C-47.280) and *Cymaclymenia* sp. indet. (B6.C-47.281 and B6.C-47.282). This faunal is still of UD V-A age. Higher strata cannot be discerned in the wide, flat plain towards the east; there is no evidence of the widespread *Gonioclymenia* Limestone (UD V-B) of the Tafilalt Platform (Hartenfels & Becker 2012*b*).

Conodont dating. Generally, the conodont content is relatively poor at El Gara South. The two pre-event nodular layers (beds 1 and 2b) are dominated by several long-ranging taxa, such as *B. stabilis vulgaris* (=*stabilis* M1), *Br. ampla*, *Br. inornata*, *M. strigosa*, *Neo. communis communis*, *Pa. gracilis gracilis*, *Pa. perlobata helmsi* and *Pa. perlobata schindewolfi*. Based on the co-occurrence of *Pa. gracilis sigmoidalis*, *Pa. minuta minuta* and *Sc. velifer velifer*, Bed 1 falls in the regional *gracilis sigmoidalis* Subzone, the uppermost part of the extended *velifer velifer* Zone of the Tafilalt. The first Moroccan record of *Ps. granulosus* gives prospects that the international *Ps. granulosus* Zone/Subzone can be recognized in the Tafilalt in the future. The zonal/subzonal base should be sought somewhat below the current record of the index species. As in other Tafilalt successions, *Sc. velifer velifer* does not reach the Upper *Annulata* level. Both the *Annulata* Intralimestone (Bed 3b) and limestone nodules from within the Upper *Annulata* Marl (Bed 4a) provide sparse conodont faunas, which are typical for the regional *velifer–stabilis* Interregnum. There are, amongst others, *Al. regularis regularis*, *B. stabilis vulgaris*, representatives of *Neo. communis communis*, *Pa. gracilis semisigmoidalis*, *Pa. perlobata maxima* and dominant *Pa. gracilis gracilis*. Specimens of the '*I.*' *cornutus* Group show a range into the international *granulosus* Zone, which agrees with data in Sandberg & Dreesen (1984: late morphotype or specimens of a potential descendant). The lower level (Bed 5b) of the very fossiliferous Wagnerbank Equivalents yields *B. stabilis vulgaris*, *Br. inornata*, *Pa. gracilis gracilis* and *Pa. perlobata schindewolfi*. An equal association can be recognized higher, in the red marker limestone (Bed 19) at the base of El Gara South II. The absence of *B. stabilis stabilis* (=*stabilis* M2) still dates this sequence as *velifer–stabilis* Interregnum.

Ammonoid dating. The loose assemblage from Bed -7 falls in the *subvaricatum* Subzone (UD

II-G). The rather small size of the specimens and the low local diversity are distinctive and reflect the basinal setting. However, Becker (1993*a*) found associated rare *Cheiloceras* and *Falcitornoceras* in a lateral section, approximately 1 km to the north. The higher UD II and basal UD III are represented by the poorly fossiliferous higher part of beds -7 to -5.

The index species of the *euryomphalum* Zone (UD III-B/C_1) enters in Bed -4 (Fig. 6.1a, b) and ranges to Bed -3a. Based on the minor oued, this is locally the only interval where a few juvenile Middle Devonian goniatites, including *Fidelites* and *Sobolewia*, were washed in.

The index species of the *sulcata* Zone (UD III-C_2) enters abundantly in Bed -3b. Associated are the first *Maid. sapiens* of the Tafilalt (Fig. 7.5a, b), the first *Genuclymenia* of North Africa (Fig. 7.7a, b), a minute *Protor. ornatum* (Fig. 6.2a, b), the first true record of the genus for North Africa, a questionable *Pleuroclymenia* and a small *Protactoclymenia* with ventral varices (Fig. 7.6a, b), as in the Polish *Protacto. implana*. Beds -2a to 2a represent the ammonoid-poor interval at the top of UD III. The monospecific *Protacto.* aff. *subcostata* fauna from Bed -2b (Fig. 8.5a, b) is unique for the region. The highest marly shales and nodular limestones below the *Annulata* Event, beds -1 to 2b, are completely devoid of ammonoids and other macrofauna, as in all the Tafilalt sections.

The oldest platyclymeniids enter suddenly in the Lower *Annulata* Event interval (Bed 3a, basal UD IV-A). Based on their ww/wh ratios and growth lirae, the juveniles are assigned to *Pl.* (*Pl.*) *subnautilina subnautilina*. The associated, small-sized, oldest prionoceratids are identified as '*Pr.*' *vetum*. They combine an open umbilicus with only three constrictions/whorl until approximately 10 mm dm. Larger individuals develop four constrictions/whorl, as diagnostic for the species (cf. B6.C-47.110 (Fig. 6.5a, b) from the Upper *Annulata* Event Bed). This justifies the lowering of the base of the regional *vetum* Zone *sensu* Korn *et al.* (2014) to the base of UD IV-A.

Just above, the *Annulata* Intralimestone yielded the oldest *Pl.* (*Pl.*) *levata* n. sp. (B6.C-47.98). The small number of specimens collected from the Lower *Annulata* Marl prevents a reliable recognition of its lower range. Moderately rich assemblages of the regional *annulata* Zone can be found throughout the upper part of El Gara South I and II. The Wagnerbank Equivalents are characterized by three species of platyclymeniids and a new subspecies of *Pr. frechi*, which has a more lamellose growth ornament (see Fig. 6.7a) than the typical German *Pr. frechi* (see Korn *et al.* 2014) and the slightly thicker and younger '*Pr.*' *mrakibense* Korn, Bockwinkel & Ebbighausen, 2014. It is clearly thicker than the just slightly younger '*Pr.*' *lentis* Korn, Bockwinkel & Ebbighausen, 2014 from the Maider (Korn *et al.* 2014). Since it also occurs in the older *Anulata* Event beds at Bine Jebilet and Hassi Nebech (see below), it cannot be used for a subdivision of the *annulata* Zone. The first *Pr. divisum lamellosum* were locally collected from beds 10a, 12a and 16. But elsewhere it can be dominant in the Lower and Upper *Annulata* levels. With respect to the small ornament difference between the Moroccan *Pr. lamellosum* and German *Pr. divisum* (Münster, 1832), and the recognition of rare intermediates (e.g. B6.C-47.145 from Bed 16), we separate both only at the subspecies level.

Based on a simple count of shale–limestone cycles above the Wagnerbank Equivalents (beds 5b–6b), as in the German Effenberg section (Hartenfels 2011), the level of the oldest protoxyclymeniids (lower UD IV-B) can be expected near the base of Section II. Limited support is given by the local entry of *Pr. takhbtitense* in Bed 23b, which co-occurs in the Mrakib section (Bed O) with the oldest protoxyclymeniids (combined data of Webster *et al.* 2005 and Korn *et al.* 2014). However, *Pr. takhbtitense* has a lower range at its type section (black *Annulata* Limestone of UD IV-A, Lower *Annulata* Event Bed, still with *Pa. minuta minuta*: Hartenfels 2011). The rarity of protoxyclymeniids in the Tafilalt is exemplified by their absence at El Gara South in the rather rich *Platyclymenia–Prionoceras* assemblages of beds 25b and 26. Probably owing to local facies conditions, *Ungusporadoceras*, *Erfoudites* and *Pl.* (*Trigonoclymenia*) appear at El Gara with a delay and as rare taxa. Section II does not reach the level of *Proc. ebbighauseni* (UD IV-B_2), which is in accordance with the number of cycles above the *Annulata* Events in comparison with the Bine Jebilet section (see below) (Fig. 13).

Bine Jebilet

The succession (Fig. 13) is located approximately 10 km WNW of Erfoud (GPS coordinates: 31° 26′ 59.9″ N, 004° 19′ 52.7″ W, map sheet 244 Tafilalt–Taouz) and represents the transition between the Rheris Basin in the north and the condensed Tafilalt Platform in the south. First sedimentological descriptions were given by Wendt *et al.* (1984). Becker (1992*a*) studied the *Annulata* Event facies, with a focus on the ammonoid fauna (see Becker *et al.* 2002). Nawrath (2009) investigated prionoceratids of the UD IV-A, and Hartenfels *et al.* (2009) and Hartenfels (2011) provided conodont data. A detailed section log around the *Annulata* Events was given by Hartenfels (2011).

Planitornoceras euryomphalum was discovered during fieldwork in 2012 (B6.C-47.185 and B6.C-47.186). It occurs in a narrow band of greenish

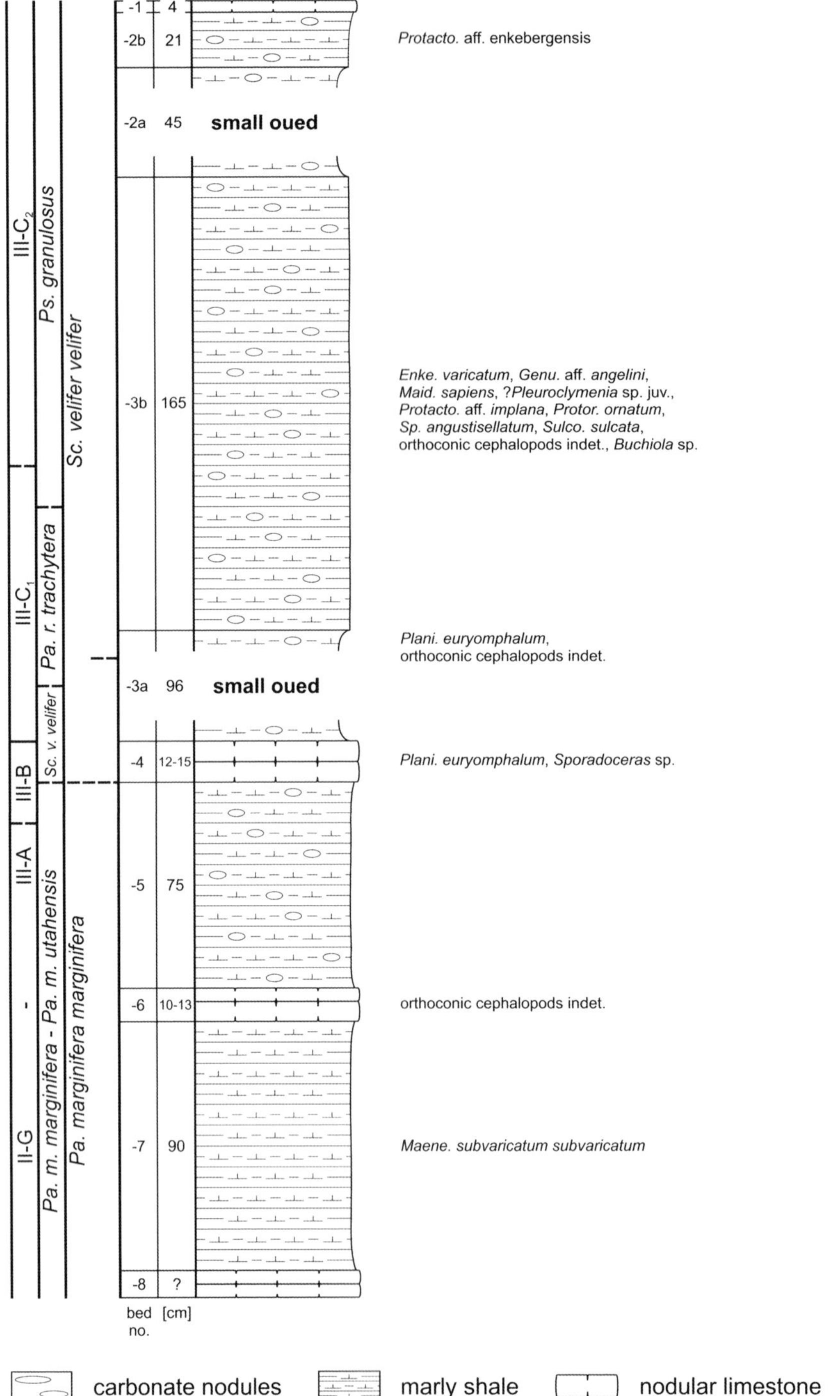

Fig. 5. Lithological log, conodont and ammonoid successions at El Gara South. LAE and UAE, Lower and Upper *Annulata* Event beds; AIL, *Annulata* Intralimestone; WE, Wagnerbank Equivalents.

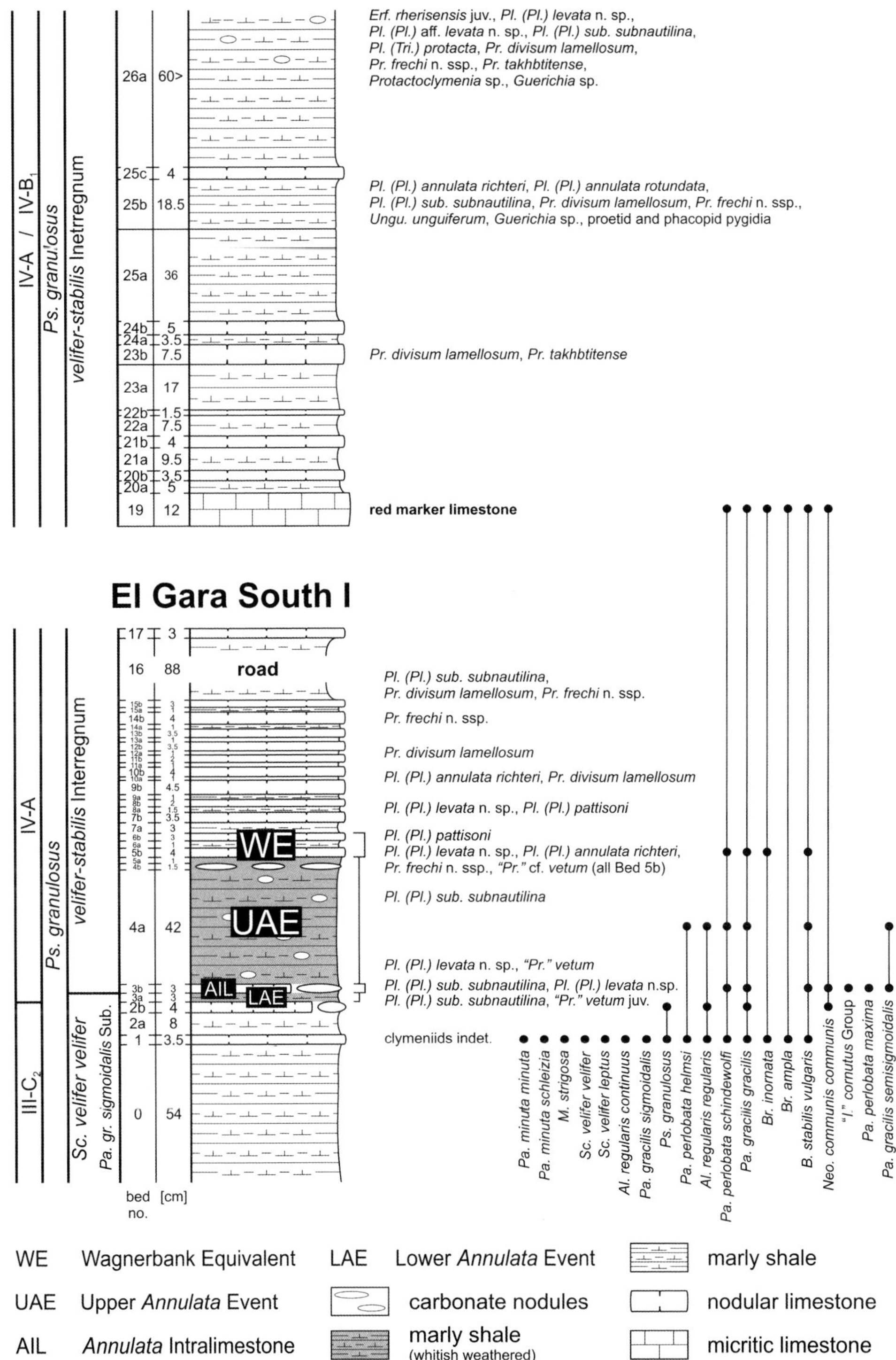

Fig. 5. *Continued.*

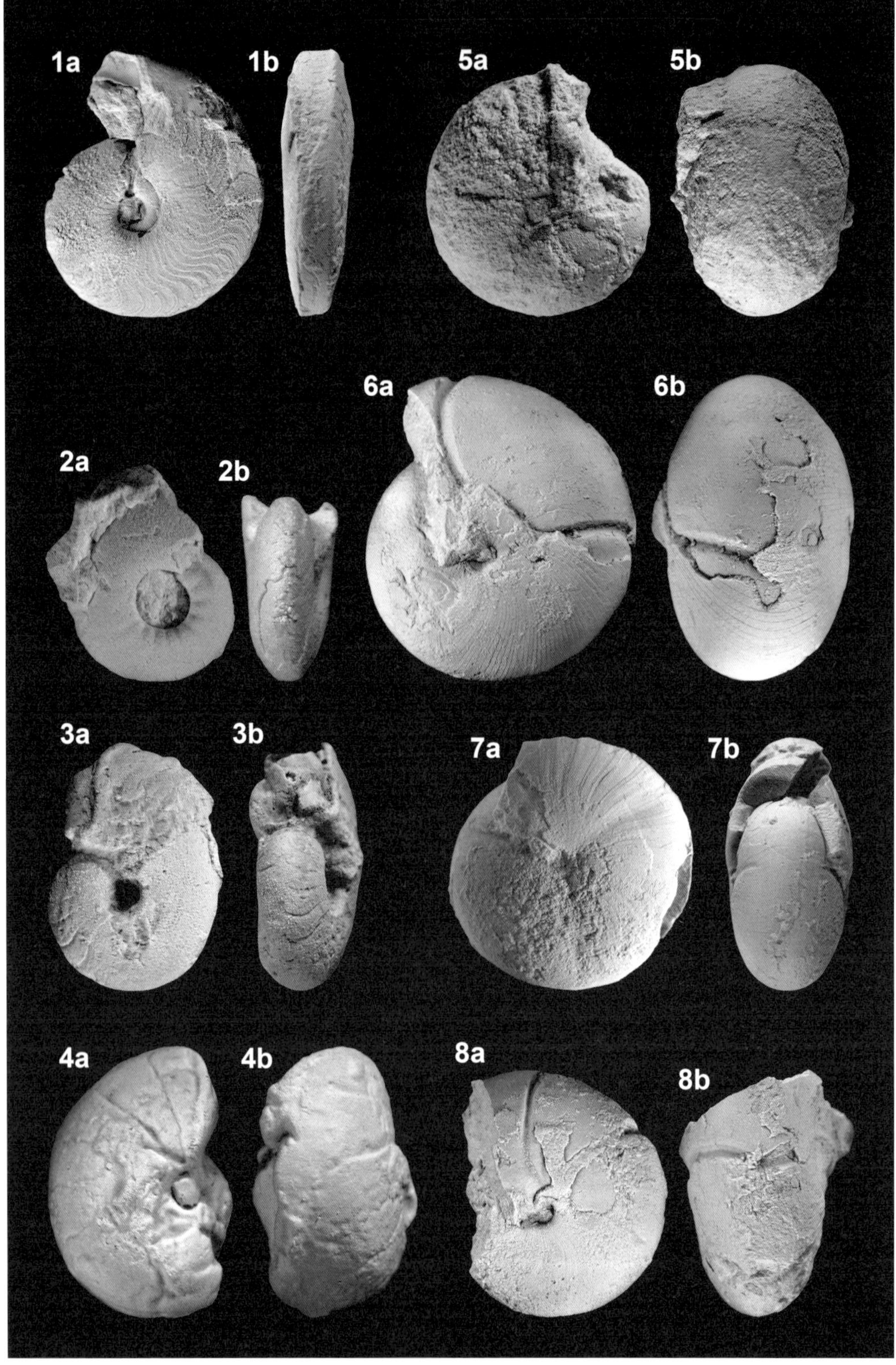
1a
1b
5a
5b
6a
6b
2a
2b
3a
3b
7a
7b
4a
4b
8a
8b

nodular limestone on the gentle back side of the local marker ridge, which contains many, often large-sized *Maeneceras* and *Acrimeroceras* (Becker 1993*a*, *subvaricatum* Subzone). Subsequently, there is an unfossiliferous plain with poor outcrop of reddish to grey nodular limestone. The two *Annulata* Events are developed as *Platyclymenia*-rich, greenish marly shales with small-sized limestone nodules (36.5 and 22 cm thick, respectively), separated by a 3 cm-thick layer of more continuous limestone nodules (greyish-green, micritic), a nodular equivalent of the *Annulata* Intralimestone. The very rich faunas are dominated by *Pl.* (*Pl.*) *subnautilina subnautilina* (including previous records of *Pl.* (*Pl.*) *pseudoflexuosa*, e.g. B6.C-47.259: Fig. 14.5a, b), but there are also some *Pl.* (*Pl.*) *quiringi* (e.g. B6.C-47.258 (Fig. 14.3a, b), clearly more evolute and with more whorls at the same size), *Pl.* (*Pl.*) *levata* n. sp. (B6.C-47.205–B6.C-47.208, but with rounded venter), *Pl.* (*Pl.*) *annulata richteri*, *Pl.* (*Tri.*) *protacta* (B6.C-47.283: Fig. 14.2a, b), *Pl.* (*Pl.*) cf. *limata* Czarnocki, 1989 (with uw/dm near 0.35), *Protacto.* aff. *lagowiense* (Sobolew, 1912) (B6.C-47.220 (Fig. 14.1a, b), B6.C-47.221 and B6.C-47.222, with subangular ventrolateral shoulders), abundant *Pr. frechi* n. ssp., a new relative of *Pr. frechi* with up to five constrictions/whorl, '*Pr.*' *vetum*, *Pr. divisum lamellosum* and *Ungu. unguiferum* (B6.C-47.187, B6.C-47.188 (Fig. 7.2, 3) and B6-C-47.190–B6.C-47.200; previously recorded as '*Erfoudites* n. sp.'). Locally very rare forms are *Pl.* (*Pl.*) aff. *quiringi* (B6.C-47.202, thicker than the typical form) and *Post. ascendens* n. sp. (B6.C-47.189: Fig. 14.4a, b). The subsequent Wagnerbank Equivalents consist of two ferruginous, reddish-brown, solid, micritic limestone beds (3.5 and 7.5 cm thick), which yield a rich ammonoid fauna, notably with *Platyclymenia* and *Prionoceras* (Hartenfels *et al.* 2009; Hartenfels 2011). Slightly higher, some fragmentary, strongly compressed, involute cyrtoclymeniids have been found. *Protoxyclymenia dunkeri*, the index of UD IV-B (B6.C-47.201: Fig. 14.8a, b), and *Pl.* (*Tri.*) *spinosa* (Münster, 1842) (B6.C-47.293: Fig. 14.7a, b) are known as loose specimens (see records in Becker *et al.* 2002). *Procymaclymenia* sp. was found 26 marly shale–limestone cycles above the Wagnerbank, near the top of the sequence (Bed 27b: B6.C-47.256 and B6.C-47.257), which becomes gradually less ammonoid-rich (Hartenfels 2011).

Jebel Erfoud

The section (Fig. 13) is exposed approximately 500 m SE of Erfoud, close to the top of the third hill along a steep ridge (GPS coordinates: 31° 25′ 49.6″ N, 004° 13′ 05.6″ W, map sheet 244 Tafilalt–Taouz). Coming from Erfoud (crossing the Oued Ziz in the direction of Merzouga), the ridge is easily marked by high telecommunication masts. Alberti (1970) called this succession 'Bordj Est' or 'Bordj d'Erfoud', Hollard (1971) called it 'Butte d'Erfoud' and Becker (1993*a*) referred to it as Jebel Erfoud. The Devonian was first described by Clariond (1934), Termier & Termier (1950) and Petter (1959, 1960). Becker (1993*a*), Korn (1999) and Becker *et al.* (2002) studied the ammonoids. In addition to the description of trilobites, Alberti (1970) provided the first conodont data. More detailed conodont investigations were conducted by Buggisch & Clausen (1972) and, more recently, by Hartenfels (2011, including a detailed bed-by-bed section log). The exposed successions belong to the strongly condensed and partly discontinuous north-central Tafilalt Platform.

At Jebel Erfoud, there is no dark *Annulata* Shale or Limestone. Above an interval with poor macrofauna (only one *Protacto.* cf. *euryomphala* (Wedekind, 1908) from Bed 30, B6.C-47.203), the Wagnerbank Equivalents consist of two, 6 and 9 cm-thick, grey, partly brownish micritic limestones (the upper one slightly marly), separated by 1.5 cm of grey marl. They yield rich and well-preserved UD IV-A ammonoid faunas (Korn 1999; Hartenfels 2011). *Procymaclymenia ebbighauseni*, the regional index of UD IV-B2, enters approximately 25 cm higher and ranges into the lower part of the *B. stabilis stabilis* Zone (Bed 39b, = Bed 27 with '*Cymaclymenia* n. sp.' of Korn 1999).

Ouidane Chebbi Northwest

The section (Fig. 13) lies in the eastern Tafilalt, approximately 45 km ESE of Erfoud, in the military

Fig. 6. Ammonoids from the middle to basal upper Famennian of El Gara South I. (**1a, b**) *Plani. euryomphalum* (Wedekind, 1918), B6.C-47.5, Bed -3a, *euryomphalum* Zone, ×2. (**2a, b**) *Protor. ornatum* Dybczynski, 1913 juv., B6.C-47.72, Bed -3b, *sulcata* Zone, showing the typical subumbilical ribbing, ×4.5. (**3a, b**) *Fidelites* cf. *clariondi* (Petter, 1959), B6.C-47.32, Bed -3a, Eifelian float specimen, ×4. (**4a, b**) *Sobolewia.* sp. juv., B6.C-47.31, Bed -3a, float specimen, ×4. (**5a, b**) '*Pr.* ' *vetum* Korn, Bockwinkel & Ebbighausen, 2014, B6.C-47. 110, Bed 4a, Upper *Annulata* Marl, lower regional *annulata* Zone, showing the transition from *c.* 120° to 90° angles between varices near 10 mm dm, ×3. (**6a, b**) *Pr. divisum lamellosum* Korn *et al.*, 2014, B6.C-47.139, Bed 12a, higher regional *annulata* Zone, ×2. (**7a, b**) *Pr. frechi* (Wedekind, 1913) n. ssp., B6.C.47.112, Bed 5b, Wagnerbank Equivalent, regional *annulata* Zone, relatively thick form with lamellose growth lines, ×2. (**8a, b**) *Maene. subvaricatum subvaricatum* (Sobolew, 1914*a*), B6.C-47.4, Bed -7, *subvaricatum* Zone, juvenile specimen, ×2.

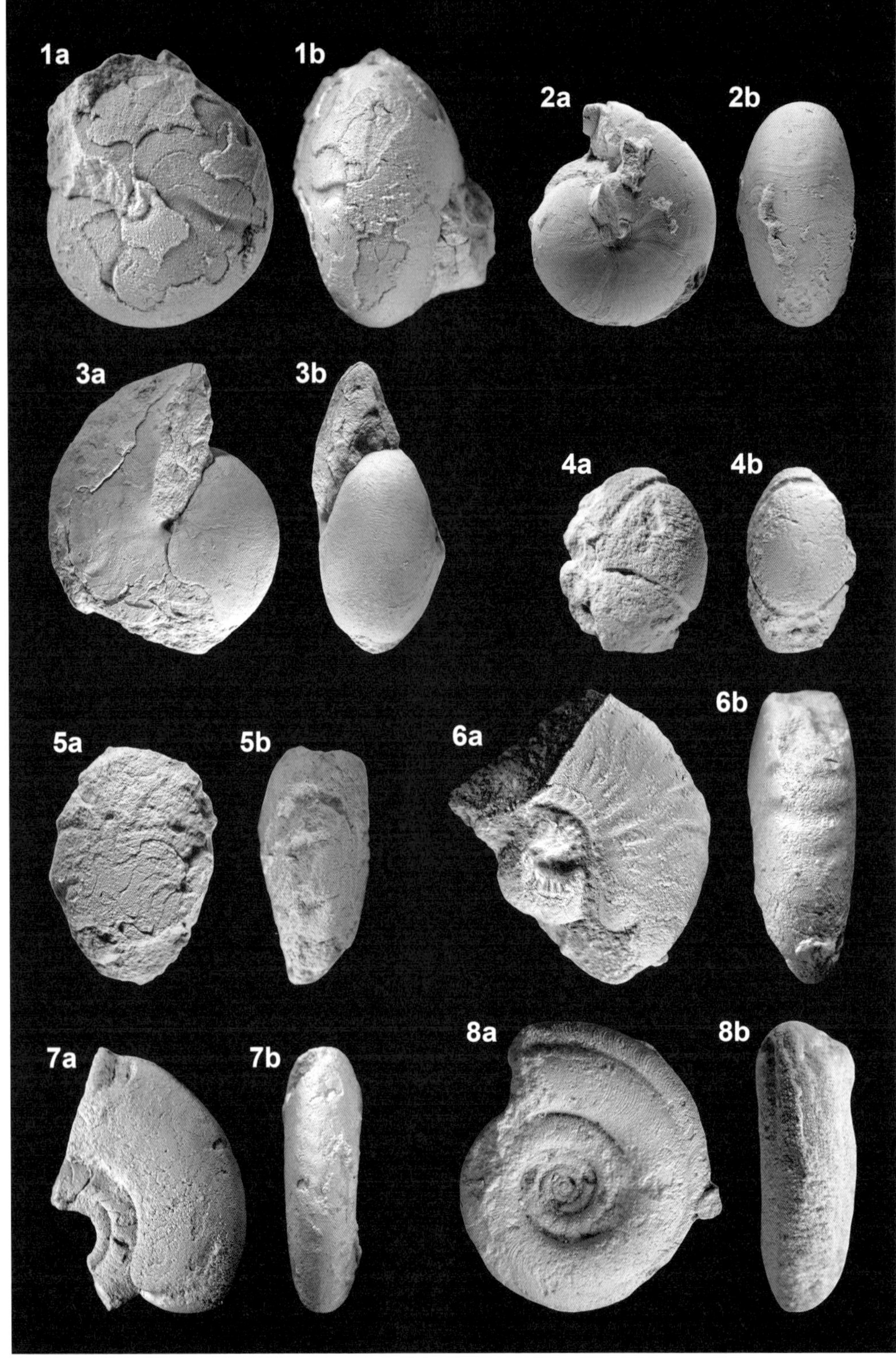
1a
1b
2a
2b
3a
3b
4a
4b
5a
5b
6a
6b
7a
7b
8a
8b

exclusion zone close to the Moroccan–Algerian border (GPS coordinates: 31° 14′ 57.4″ N, 003° 50′ 00.2″ W, map sheet 244 Tafilalt–Taouz). Devonian–Carboniferous strata are exposed in west–east-striking parallel ridges that emerge as folds under the Cretaceaous of the Hamada du Guir. Several lateral sections have previously been investigated by Klug (2002), Belka *et al.* (1999), Korn (1999), Becker *et al.* (2002), Kaiser (2005) and Kaiser *et al.* (2011). In terms of lithofacies, the studied succession is transitional between condensed sequences of the central platform and deeper settings of the adjacent Tafilalt Basin to the south (for the palaeogeography, see Wendt 1989). Hartenfels (2011) published a bed-by-bed section log, as well as conodont and ammonoid data around the *Annulata* Events and the younger Dasberg Crisis.

As at Jebel Erfoud, the *Annulata* Event beds are lacking in a minor hiatus below a Wagnerbank Equivalent, which is characterized by the sudden mass occurrence of platyclymeniids in a 6 cm-thick, ferruginous, reddish-brown, micritic limestone.

Hassi Nebech

The investigated succession (Fig. 13) crops out approximately 20 km ENE of Taouz, close to the SE foot of the Jebel Bega (GPS coordinates: 30° 55′ 40.5″ N, 003° 47′ 32.7″ W, map sheet 244 Tafilalt–Taouz). Travel directions were given in Aboussalam (2003). Becker *et al.* (2002) provided a lithological and ammonoid stratigraphic overview of the Famennian, Bockwinkel *et al.* (2002) investigated the ontogeny and variability of a rich *Maene. subvaricatum* population, Nawrath (2009) studied some prionoceratids. Hartenfels (2011) focused on the *Annulata* Events and the younger Dasberg Crisis, and provided a detailed bed-by-bed section log and conodont data.

There are lithofacies similarities with the Rheris Basin, but distinctive faunal differences. The entire *Annulata* Event Interval is sandwiched between a greenish-grey shale/marl facies with intercalated brown to brownish-red limestones and nodular limestones. The two *Annulata* Events are developed as brownish to whitish weathered (due to pyrite weathering under arid conditions), marls, yielding small-sized carbonate nodules (Hartenfels 2011). Both have abundant ammonoids and a specialized benthos of low diversity. *Platyclymenia* (*Pl.*) *subnautilina subnautilina* is dominant, *Pl.* (*Pl.*) *annulata richteri* (Hartenfels 2011, pl. 70, fig. 4) is much rarer. Among the prionoceratids, there are both ‘*Pr.*’ *vetum* and *Pr. divisum lamellosum* (Hartenfels 2011, pl. 70, fig. 9), associated with subordinate *Pr. frechi* n. ssp. and a new form with very wide juvenile umbilicus (Nawrath 2009). Accessory goniatites are *Erf. rherisensis* Korn, 1999 (N6.C-47.284-286), *Ungu. unguiferum* (B6.C-47.190 and B6.C-47.209) and *Sp. brachylobum* Frech, 1902 (B6.C-47.204). Whereas the irregular bivalve *Loxopteria* is common (Hartenfels 2011, pl. 70, figs 14a-b), unlike as at El Gara, guerichiids are locally absent. The ecology of *Loxopteria* suggests dysaerobic, not ex- or anaerobic event facies (see Nagel-Myers *et al.* 2009). It is significant, that blind, most probably endobenthonic, phacopids (see Becker & Schreiber 1994) are lacking in the event facies, but they occur in the pre- and post-event beds, which suggests reduced ventilation on the seafloor. The local absence of guerichiids suggests that the ecological conditions were only weakly eutrophic. The 16 cm-thick Lower *Annulata* Event Bed is separated from the Upper *Annulata* Event Bed (only 3.5 cm thick) by a single, 3.5 cm-thick, ferruginous, brownish to slightly red, nodular *Annulata* Intralimestone. A three-fold package of brown to red, fossiliferous (ammonoids), micritic, nodular limestones forms the Wagnerbank equivalents.

Mrakib

The section (Fig. 15) is situated in the southern Maider Basin, approximately 28 km SE of Fezzou and 2.5 km north of the Madène el Mrakib mudmound (GPS coordinates: 30° 45′ 07,8″ N, 004° 43′ 17,2″ W, map sheet 243 Todrha–Ma'der). It forms a detached ridge exposing middle Famennian–lowermost Carboniferous deposits, predominantly in fine siliciclastic facies. Becker

Fig. 7. Ammonoids from the middle to basal upper Famennian of El Gara South I (EGS) and Bine Jebilet (BJe). (**1**)–(**3**) *Ungu. unguiferum* Korn, Bockwinkel & Ebbighausen, 2015*b*: (1a, b) B6.C-47.156, EGS, Bed 25b, higher regional *annulata* Zone, rather thick paratype, ×2.5; (2a, b) B6.C-47.187, BJe, Bed 1e, Upper *Annulata* Marl, regional *annulata* Zone, showing shell surface with biconvex growth lines, ×2; (3a, b) B6.C-47.188, BJ, Bed 1e, regional *annulata* Zone, ×2. (**4a**, **b**) *Enke. varicatum* (Wedekind, 1908), B6.C-47.68, EGS, Bed -3b, *sulcata* Zone, fragmentary, ×2. (**5a**, **b**) *Maid. sapiens* (Korn, 1999), B6.C-47.71, EGS, Bed -3b, *sulcata* Zone, fragmentary specimen with typical sutures, ×2. (**6a**, **b**) *Protacto.* aff. *implana* (Czarnocki, 1989), B6.C-47.74, EGS, Bed -3b, *sulcata* Zone, fragmentary small specimen with concave dorsolateral ribs and ventral varices, ×4. (**7a**, **b**) *Genu.* aff. *angelini* (Wedekind, 1908), B6.C-47.73, EGS, Bed -3b, *sulcata* Zone, fragmentary specimen with typical septal face, ×2. (**8a**, **b**) *Sulco. sulcata* (Schindewolf, 1923), B6.C-47.37. EGS, Bed -3b, *sulcata* Zone, typical specimen with strong spiral furrows and concavo-convex growth lines, ×4.

1a
1b
2a
2b
3a
3b
4a
4b
5a
5b
6
7a
7b
8a
8b

et al. (1999, 2000, 2002), Korn (1999), Becker (2002), Hartenfels & Becker (2009) and Hartenfels (2011, cf. the section log) gave lithological overviews and provided a compilation of the ammonoid succession, which was recently updated by Korn *et al.* (2014, 2015*a*, *b*). Sartenaer (1998, 1999, 2000) described rhynchonellids, and Webster *et al.* (2005) described the crinoids in relation to a revised biostratigraphy. Unpublished studies by Nawrath (2009) and Fischer (2013) include data on prionoceratid goniatites. Corradini *et al.* (2001) sampled the section for conodonts and recognized that most carbonates (nodular limestones or solid micritic limestones) are barren. Hartenfels (2011) confirmed this negative evidence, but extracted sufficient material to follow the rough conodont zonation of the Tafilalt.

The Lower *Annulata* Shale (Bed N1c = 10a$_2$, 73 cm thick) overlies a thick succession of greenish-grey shales (upper UD III-C2) with few intercalated greyish-brown, slightly pyritic pre-event limestones, and a very restricted macrofauna consisting of rare, poorly preserved sporadoceratids and rhynchonellids. The Lower *Annulata* Event Interval contains a peculiar fauna of mostly small-sized ammonoids, gastropods and very abundant *Guerichia*. The preservation of approximately 100 available ammonoids is highly distinctive and excludes any confusion with specimens transported downslope from above. Very characteristic is the mostly incomplete preservation in dark-grey hematite of common incrustations of small-sized, altered pyrite framboids and benthic fauna, especially of *Guerichia* valves and juvenile gastropods, some of which fell into cephalopod body chambers, and the red to orange hematite–goethite weathering. The faunal composition is utterly different from all Tafilalt localities and from overlying fossiliferous greenish shales of the *annulata* Zone. The elsewhere dominant platyclymeniids and prionoceratids are locally rare (only 8 and 2% of the fauna, respectively), whilst *Gund. australe* n. sp. is dominant (37%: e.g. B6.C-47.213 and B6.C-47.214 (Fig. 16.4, 5); B6.C-47.234–B6.C-47.245), followed in abundance by small *Erf. zizensis* (25%: B6.C-47.210–B6.C-47.212 (Fig. 16.1–3); B6.C-47.246, B6.C-47.247), *Protacto.* aff. *subcostata* (20%: e.g. B6.C-47.218 (Fig. 17.1a, b) and *St. rectangula* n. sp. (8%: e.g. juvenile B6.C-47.217 (Fig. 16.8); B6.C-47.223, B6.C-47.229–B6.C-47.233 and B6.C-47.255). Among the platyclymeniids, *Pl.* (*Pl.*) *annulata richteri* (B6.C-47.287–B6.C-47.290), fragmentary *Pl.* (*Pl.*) *subnautilina subnautilina* (B6.C-47.216 (Fig. 16.7a, b); B6.C-47.248), *Pl.* (*Pl.*) *latecostata* Czarnocki, 1989 (B6.C-47.215 (Fig. 16.6a, b) first record for Gondwana, very similar to the Polish types) and fragmentary *Pl.* (*Tri.*) *protacta* (B6.C-47.249) are present. The poorly preserved *Prionoceras* are difficult to identify, but seem to include *Pr. frechi* n. ssp. (B6.C-47.250). Among the benthos, there are rare *Buchiola* and *Leptodesma* bivalves; orthocones partly belong to *Lobobactrites* cf. *paucesinuatus* Clausen, 1968 (B6.C-47.251–B6.C-47.254).

Above the poorly fossiliferous *Annulata* Intralimestone, with only one compressed protactoclymeniid (B6.C-47.227), and associated shales, the Upper *Annulata* Shale (middle Bed N1f = 11b, 56 cm thick: Fig. 14) yielded only a few squashed, red and orange weathered ammonoids that hardly can be identified (*Prionoceras* sp., *Pl.* (*Pl.*) *subnautilina* fragments). Three subsequent, very fossiliferous, greyish-brown limestones (Bed 11d = N2a, 2.5 cm; Bed 12b = N2c, 6 cm; Bed 13b = N2e, 6.5 cm) are the local Wagnerbank Equivalents. There are very rich *Prionoceras–Platyclymenia* faunas in limestone preservation, which include, as a newcomer, the large-sized *Pl.* (*Pl.*) *ibnsinai–laxata* Group. The very loose, goethitic faunas from just above are rich in various prionoceratids, especially in '*Pr.*' *lentis* Korn *et al.*, 2014, and platyclymeniids, including abundant *Pl.* (*Pl.*) *levata* n. sp. and the more involute *Pl.* (*Pl.*) *limata*. There are also loose, rare *Carino. beuelensis* (Lange, 1929) and *Kara. saharae* Becker in Becker *et al.* (2002). The base of UD IV-B1 lies within the thick interval of Bed O2a (=Bed 14e *sensu* Hartenfels 2011). This is indicated by loose, rare specimens of *Prot. dunkeri* with well-preserved ventral band (e.g. B6.C-47.224: Fig. 17.4a, b). They are unlikely to have been transported downslope from above, as the species has not been found higher (Becker *et al.* 2000, 2002; also no record in Korn *et al.* 2014). In Bed O2a, there are also goethitic

Fig. 8. Clymeniids from around the *Annulata* Events at El Gara South I. (**1a**, **b**) *Pl.* (*Pl.*) *annulata richteri* (Wedekind, 1914), B6.C-47.158, Bed 25b, higher regional *annulata* Zone, specimen with paired juvenile ribbing, ×2. (**2a**, **b**) *Pl.* (*Pl.*) *annulata rotundata* (Wedekind, 1914), B6.C-47.159, Bed 25b, typical, small, moderately evolute specimen with ribs that are getting denser and finer on median whorls, ×3. (**3**), (**4**), (**6**) *Pl.* (*Pl.*) *subnautilina subnautilina* (Sandberger, 1855): (3a, b) B6.C-47.109, Bed 3a, Lower *Annulata* Marl, basal regional *annulata* Zone, juvenile with wide whorls, ×5; (4a, b) B6.C-47.111, Bed 4a, Upper *Annulata* Marl, median-sized specimen, ×1.5; (6) B6.C-47.153 (with ww/wh = 1.04), Bed 26, upper regional *annulata* Zone, ×3. (**5a**, **b**) *Protacto.* aff. *subcostata* (Sobolew, 1914*b*), B6.C.47.78, Bed -2b, upper *sulcata* Zone, ×2. (**7**) & (**8**) *Pl.* (*Pl.*) *levata* n. sp.: (7a, b) paratype B6.C-47.121, Bed 8a, higher regional *annulata* Zone, ×3; (8a, b) holotype B6.C-47.152, Bed 26, upper regional *annulata* Zone, ×3.

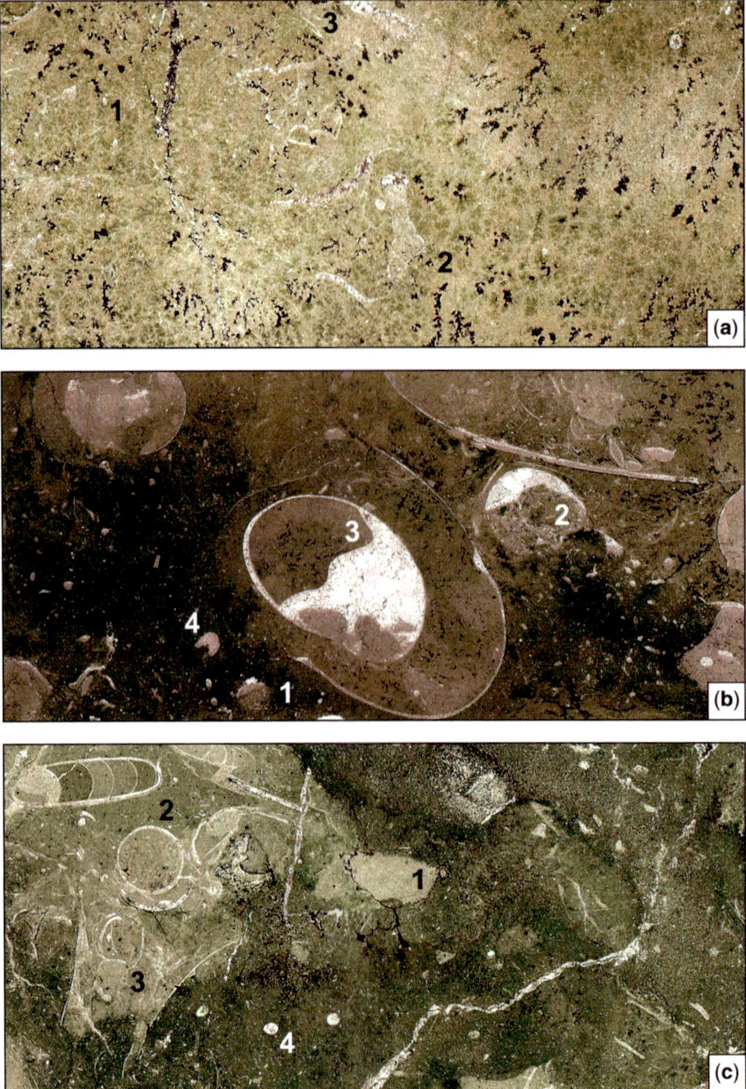

Fig. 9. Microfacies of Famennian beds at El Gara South. (**a**) Bed 1, regional *gracilis sigmoidalis* Subzone, width *c.* 29 mm: bioturbated, microsparitic mud-wackestone with ostracods (1), fragmented, thin-shelled bivalves, an evolute clymeniid with depressed whorls as in *Pl.* (*Varioclymenia*) (2), and rare trilobites (3). (**b**) Bed 5b (Wagnerbank Equivalent), regional *velifer–stabilis* Interregnum, width *c.* 33 mm; bioturbated (1), microsparitic, cephalopod float-rudstone with fine silt, platyclymeniids (2), abundant goniatites (3), shell filaments, ostracods, crinoid remains (4), rare trilobites, and echinid spines. (**c**) Bed 19 (red marker limestone), regional *velifer–stabilis* Interregnum, width *c.* 27 mm; bioturbated (1), microsparitic cephalopod float-rudstone with goniatites (2), platyclymeniids (3), shell filaments, ostracods (4), and rare trilobites.

specimens of *Prot.* cf. *wendti* (quoted as cf. *primaeva* in Becker *et al.* 2000; Webster *et al.* 2005; Hartenfels 2011) (B6.C-47.225 and B6.C-47.226: Fig. 17.3a, b). For these, a transport from above cannot be ruled out since the same form occurs also as goethitic specimens, though even more rarely, near the base of Bed Q (=Bed 16a: Webster *et al.* 2005; Hartenfels 2011). As noted by Webster *et al.* (2005), a single *Proc. ebbighauseni* (B6.C-47.228: Fig. 17.2) was found in a micritic limestone concretion high in Bed P (possibly Bed P1s of Korn *et al.* 2014). The concretion lithology is very different from the higher, coarser, dark, detrital *Procymaclymenia* Limestone (Bed 16d *sensu* Hartenfels

ammonoid zonation (key)	III-C_2			IV-A						
conodont zonation	*Ps. granulosus*									
regional conodont zonation	*Sc. velifer velifer* *Pa. gr. sigmoidalis* Sub.			*velifer-stabilis* Interregnum						
bed number	1	2b	3a	3b	4a	5a	5b	6b	19	number of platform elements/taxon
sample weight [g]	1419.5	1170.1		1868.2	686		2600.6		2541.4	
weight of residue [g]	51.4	63		179	77		686.2		791.7	
weight of dissolution [g]	1368.1	1107.1		1689.2	609		1914.4		1749.7	
number of platform elements	34	5		33	20		43		18	
platform elements/kg	25	5		20	33		23		10	
taxa/bed	15	4		7	6		4		6	
conodont-biofacies [%]										
Palmatolepis perlobata Group	11.8			21.2	10		7		16.7	
Palmatolepis minuta/gracilis Group	32.4	40		60.6	80		53.5		33.3	
Neopolygnathus		20		6.1					5.6	
Bispathodus	8.9			6.1	5		37.2		16.7	
Branmehla/Mehlina	14.7						2.3		27.8	
Alternognathus/Scaphignathus	26.5	20			5					
"*Icriodus*"				6.1						
Pseudopolygnathus	5.9	20								
Pa. minuta minuta	1									1
Pa. minuta schleizia	1									1
M. strigosa	1									1
Sc. velifer velifer	3									3
Sc. velifer leptus	2									2
Al. regularis continuus	1									1
Pa. gracilis sigmoidalis	1									1
Ps. granulosus	2	1	Lower *Annulata* Event			Upper *Annulata* Event		Wagnerbank Equivalent		3
Pa. perlobata helmsi	3	-		-	1					4
Al. regularis regularis	3	1		-	1					5
Pa. perlobata schindewolfi	1	-		6	1		3		3	14
Pa. gracilis gracilis	8	2		18	15		23		6	72
Br. inornata	2	-		-	-		1		4	7
Br. ampla	2	-		-	-		-		1	3
B. stabilis vulgaris	3	-		2	1		16		3	25
Neo. communis communis		1		2	-		-		1	4
"*I.*" *cornutus* Group				2						2
Pa. perlobata maxima				1						1
Pa. gracilis semisigmoidalis				2	1					3

El Gara South

Fig. 10. Conodont ranges, abundances, biofacies and zonation at El Gara South. The high local diversity of Bed 1 is remarkable, followed by a subsequent decline, recovery in the *Annulata* Intralimestone and a second decline, with no new taxa entering above the *Annulata* Events.

2011 = Bed Q1s of Korn *et al.* 2014). With respect to the probably derived nature of the abundant goethitic *Procymaclymenia* specimens from Beds O and P (see the discussion in Korn *et al.* 2013), the concretion specimen gives a better lower limit for the *ebbighauseni* Zone (approximately the base of Bed 15e, previously *pudica* Zone, UD IV-B2: Hartenfels & Becker 2009; Hartenfels 2011). Therefore, the false Upper *Annulata* Shale of Korn (1999), Bed P2 (=15f), represents a distinctive, transgressive and hypoxic interval low in the *ebbighauseni* Zone.

Regional comparison of sections and *Annulata* Event influences

Comparison of event litho- and biofacies

The Rheris Basin was a calm and shallow outer-shelf trough north of the Tafilalt Platform, where the settling of fine clay was only episodically reduced, leading to the deposition of thin nodular limestone interbeds. Ammonoid faunas are not pyritic/goethitic, which speaks against anoxic conditions. But the absence of benthos and the small

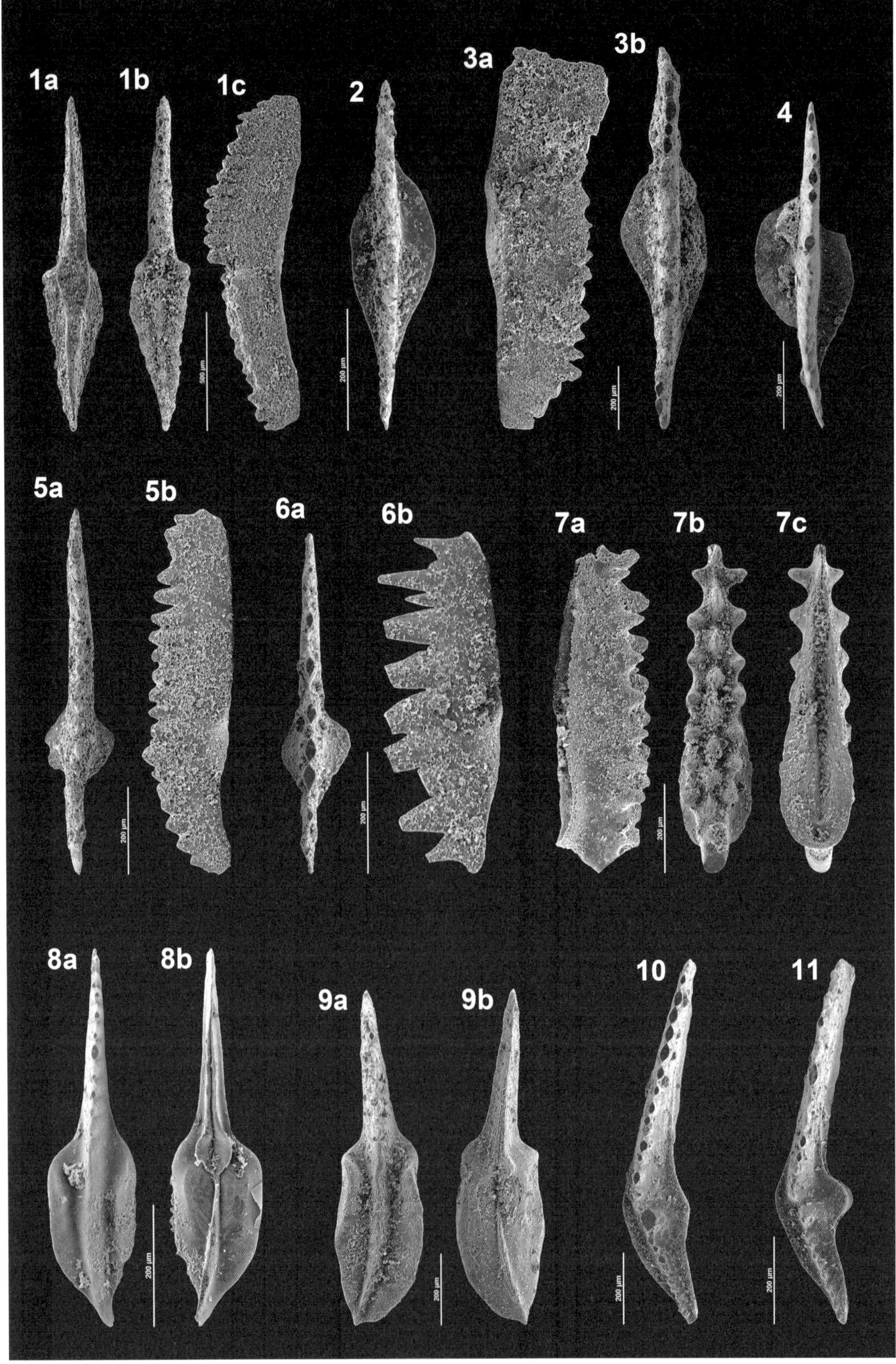
1a
1b
1c
2
3a
3b
4
5a
5b
6a
6b
7a
7b
7c
8a
8b
9a
9b
10
11
500 µm
200 µm

size of ammonoids at El Gara indicate continuously unfavourable, probably dysoxic conditions on the seafloor and a lack of food resources for mature populations. This environment culminated near the top of the *sulcata* Zone, where nearly unfossiliferous micrites (Bed 1: Fig. 9a) document strong oligotrophy. The Lower and Upper *Annulata* Events are locally developed in a brownish to whitish weathered marly shale facies, with moderately abundant small-sized ammonoids, notably *Platyclymenia* and *Prionoceras* ('low-diversity, juvenile *Platyclymenia–Prionoceras* biofacies'). These indicate improved trophic conditions, but the lack of guerichiid blooms speaks strongly against fully eutrophic conditions, as in most European event beds (Hartenfels 2011). Based on arid conditions and the extensive, deep weathering, originally present pyrite was oxidized to white sulphates. This indicates a change to reduced ventilation and anoxic conditions, at least near the seafloor. The outcrop conditions and strong weathering render the study area unsuitable for geochemical investigations, which could provide further palaeoenvironmental clues. Unique for the Tafilalt and Maider, the lower event interval is thinner than the upper (Figs 5 & 13). The *Annulata* Intralimestone represents a short episode of minor regression (reduced settling of clay) and improved oxygenation (no weathered pyrite), but ammonoids remained small. The double-layered Wagnerbank Equivalents represent an aerobic and regressive immediate post-crisis interval, characterized by the onset of a more diverse benthos, including echinids, ostracods and trilobites. The blooms of small-sized platyclymeniids and prionoceratids (Fig. 9b), in association with a low *Guerichia* content, suggest moderately eutrophic conditions, probably by bacterial nutrient-recycling from the event beds below. It is a distinctive feature of the Rheris Basin that a similar litho- and biofacies continues (until Bed 16). Above a poorly fossiliferous interval (beds 20a–23a and 24a–25a), there are two higher, moderately eutrophic marls (beds 25b and 26a, probably UD IV-B1) with *Platyclymenia–Prionoceras* faunas, in which an increasing ammonoid diversity is linked both with some *Guerichia* and oxic benthos (phacopid and proetid trilobites).

Bine Jebilet represents the transition between the Rheris Basin and the condensed Tafilalt Platform to the south. Here, the *Annulata* Events are developed in a non-pyritic, green marly shale facies, interrupting a cyclic and poorly fossiliferous succession of dominantly red, solid limestones/nodular limestones. Both event beds are characterized by ammonoid blooms, a lack of *Guerichia* and of other benthos, apart from rare *Loxopteria*, a peculiar, asymmetrical and twisted bivalve (Nagel-Myers *et al.* 2009). This indicates a sudden change to moderately eutrophic, dysoxic, but not anoxic, conditions. The ammonoid fauna, especially in the upper event interval, is larger sized than at El Gara and more diverse. The presence of additional species of *Platyclymenia* and *Prionoceras*, *Ungusporadoceras*, *Posttornoceras*, *Protactoclymenia*, and *Pleuroclymenia* is very distinctive for a 'medium diversity, normal-sized *Platyclymenia–Prionoceras* biofacies'. Congruent with El Gara South, the double-layered Wagnerbank Equivalents contain a rich, but moderately diverse, ammonoid *Prionoceras–Platyclymenia* association, with increasing amounts of the first, which continues upsection. But there is no local equivalent of the two upper fossiliferous marls of El Gara.

On the northern Tafilalt Platform, the *Annulata* Event beds are lacking, obviously in a hiatus (Hartenfels 2011) below the regressive, fossiliferous Wagnerbank Equivalent(s). The latter consist variably of two limestone cycles (Jebel Erfoud, the upper one marly) or one condensed limestone (Ouidane Chebbi Northwest). At Jebel Erfoud, the rich ammonoid fauna yield *Prionoceras* and *Platyclymenia*: at Ouidane Chebbi Northwest, however, there are only platyclymeniids.

In the Tafilalt Basin at Hassi Nebech, the *Annulata* Event sequence resembles El Gara South and Bine Jebilet, but the underlying marl–limestone alternation is richer in sporadoceratids and benthos, such as smooth rhynchonellids (*Pugnaria*), trilobites (blind *Trimerocephalus*) and bivalves

Fig. 11. Conodonts around the *Annulata* Events of El Gara South. For magnifications see scale bars. (**1a–c**) *Al. regularis regularis* Ziegler & Sandberg, 1984, B9A.6–1, Bed 1, regional *Pa. gracilis sigmoidalis* Subzone. (**2**)–(**4**) *B. stabilis vulgaris* (Dzik, 2006) = *stabilis* M2: (2) B9A.6–2, Bed 1, regional *Pa. gracilis sigmoidalis* Subzone; (3a, b) B9A.6-3, Bed 3b (*Annulata* Intralimestone), regional *velifer–stabilis* Interregnum; (4) B9A.6-4, Bed 4a (Upper *Annulata* Marl), regional *velifer–stabilis* Interregnum, broken basal cavity. (**5a**, **b**) *Br. ampla* (Branson & Mehl, 1934*a*), B9A.6-5, Bed 1, regional *Pa. gracilis sigmoidalis* Subzone. (**6a**, **b**) *Br. inornata* (Branson & Mehl, 1934*a*), B9A.6-6, Bed 19 (red marker limestone), regional *velifer–stabilis* Interregnum. (**7a–c**) '*I.*' *cornutus* Group, B9A.6-7, Bed 3b (*Annulata* Intralimestone), regional *velifer–stabilis* Interregnum. (**8**) & (**9**) *Neo. communis communis* (Branson & Mehl, 1934*b*): (8a, b) B9A.6-8, Bed 3b (*Annulata* Intralimestone), regional *velifer–stabilis* Interregnum; (9a, b) B9A.6-9, Bed 19 (red marker limestone), regional *velifer–stabilis* Interregnum. (**10**) & (**11**) *Pa. gracilis gracilis* Branson & Mehl, 1934*a*: (10) B9A.6-10, Bed 4a (Upper *Annulata* Event), regional *velifer–stabilis* Interregnum; (11) B9A.6-11, Bed 5b (Wagnerbank Equivalent), regional *velifer–stabilis* Interregnum.

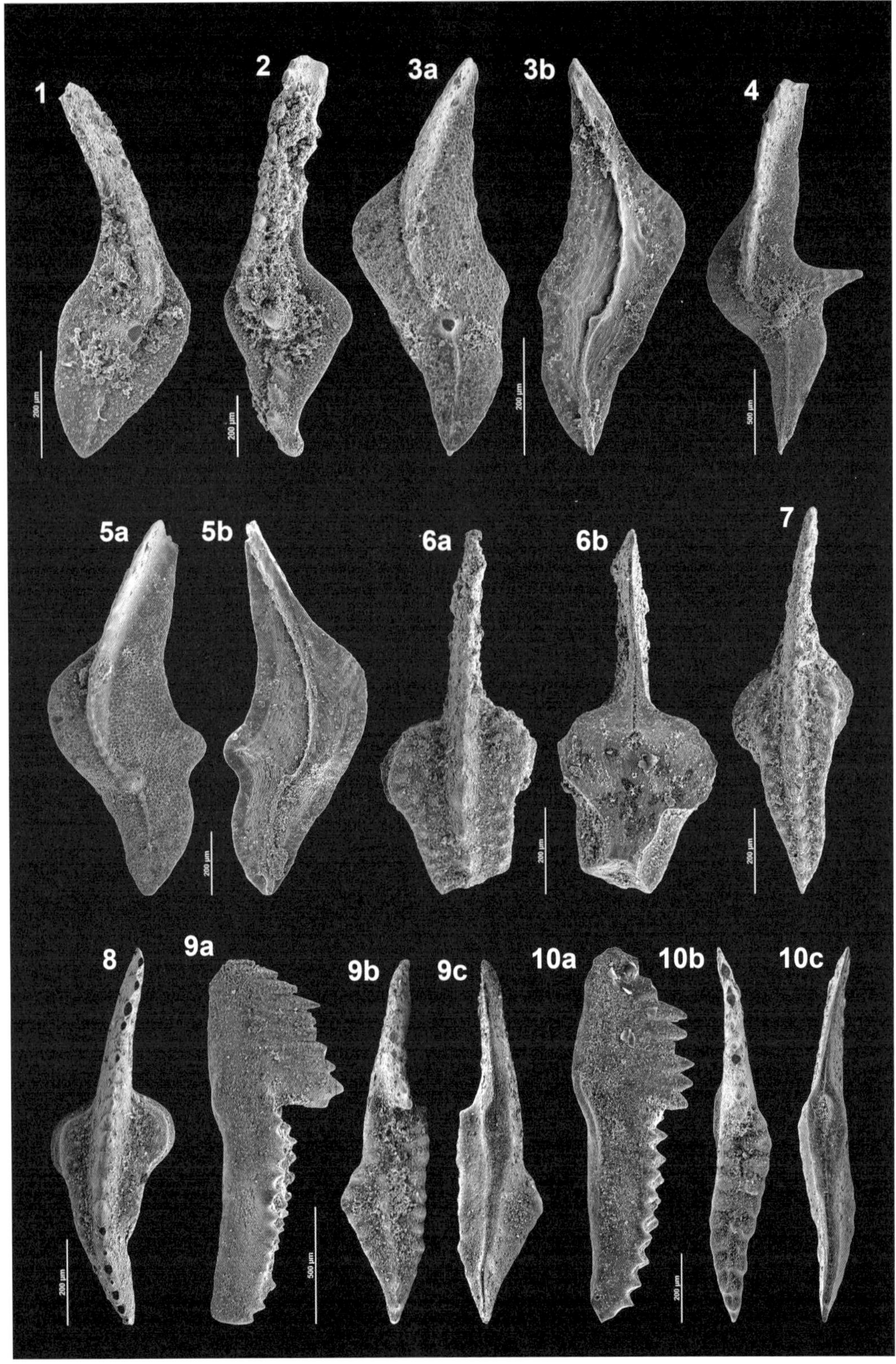
1
2
3a
3b
4
5a
5b
6a
6b
7
8
9a
9b
9c
10a
10b
10c
200 µm
500 µm

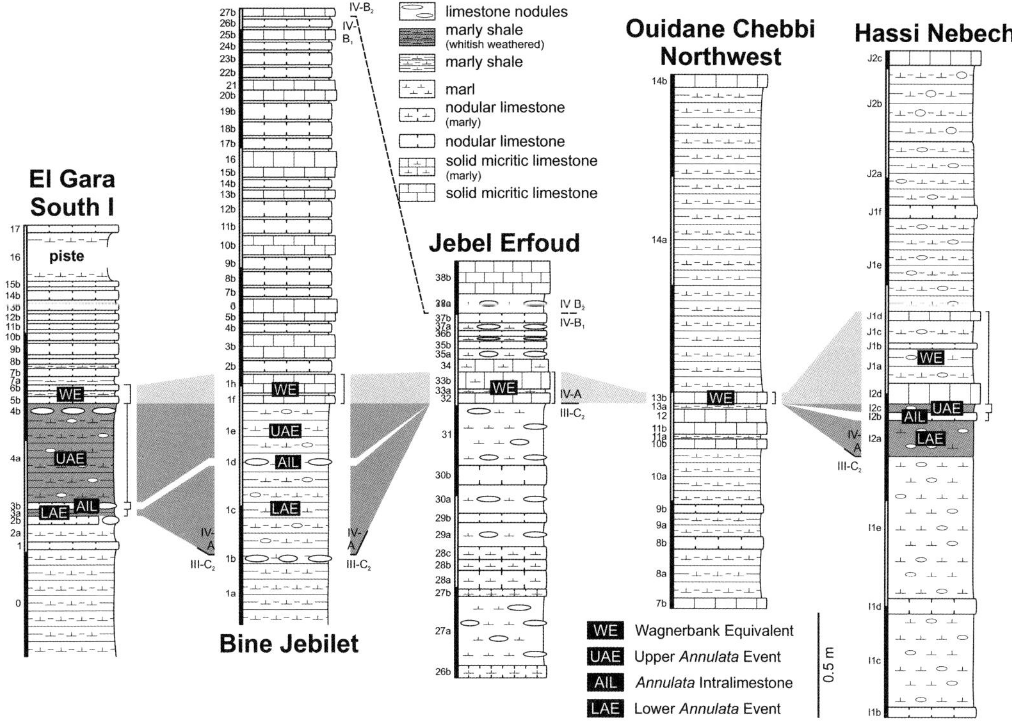

Fig. 13. Correlation of *Annulata* Event sections from the Rheris Basin (El Gara South I), northern margin of the Tafilalt Platform (Bine Jebilet), central Tafilalt Platform (Jebel Erfoud) and eastern Tafilalt Platform (Ouidane Chebbi Northwest), to the Tafilalt Basin (Hassi Nebech). The *Annulata* Event beds show reverse thicknesses within the Rheris Basin and wedge out on the elevated pelagic platform. On the eastern platform, the Wagnerbank Equivalents are condensed to a single limestone, but towards the Tafilalt Basin they become thicker, with three limestone cycles.

(*Vetupraeca*). The deeply weathered, whitish (from pyrite oxidation), marly Lower and Upper *Annulata* Event beds have abundant, normal-sized ammonoids in yellowish limestone preservation and contain common *Loxopteria*, whilst guerichiids are absent. The local event facies was dysaerobic (not ex- or anaerobic) and probably slightly better oxygenated than in the Rheris Basin. The change from the poorly fossiliferous brownish-red immediate pre-event nodular limestones to marls with mass occurrences of ammonoids suggests deepening and strongly improved trophic conditions. Both event layers yield platyclymeniids, prionoceratids and, unlike at El Gara and Bine Jebilet, *Erfoudites* ('moderately diverse, normal-sized *Platyclymenia–Prionoceras–Erfoudites–Loxopteria* biofacies'). Improved ventilation stopped the pyrite formation and induced increased bioturbation (Hartenfels 2011) of the overlying, three-fold Wagnerbank Equivalent. This level yields a rich *Platyclymenia* fauna, but, coincident with Ouidane Chebbi Northwest, there is a lack of *Prionoceras*. Upsection, faunas soon become sparse, but subsequently they increase again in abundance, which resembles El

---

Fig. 12. Conodonts around the *Annulata* Events of El Gara South. (**1**) *Pa. minuta minuta* Branson & Mehl, 1934*a*, B9A.6-12, Bed 1, regional *gracilis sigmoidalis* Subzone. (**2**) *Pa. minuta schleizia* Helms, 1963, B9A.6-13, Bed 1, regional *gracilis sigmoidalis* Subzone. (**3a, b**) *Pa. perlobata helmsi* Ziegler, 1962, B9A.6-14, Bed 4a (Upper *Annulata* Event), regional *velifer–stabilis* Interregnum. (**4**) *Pa. perlobata maxima* Müller, 1956*b*, B9A.6-15, Bed 3b (*Annulata* Intralimestone), regional *velifer–stabilis* Interregnum. (**5a, b**) *Pa. perlobata schindewolfi* Müller, 1956*b*, B9A.6-16, Bed 3b (*Annulata* Intralimestone), regional *velifer–stabilis* Interregnum. (**6**)–(**8**) *Ps. granulosus* Ziegler, 1962: (6a, b) B9A.6-17, Bed 1, regional *gracilis sigmoidalis* Subzone, broken posterior platform; (7) B9A.6-18, Bed 1, regional *gracilis sigmoidalis* Subzone; (8) B9A.6-19, Bed 2b, regional *Pa. gracilis sigmoidalis* Subzone. (**9a–c**) *Sc. velifer velifer* Helms, 1959, B9A.6-20, Bed 1, regional *gracilis sigmoidalis* Subzone. (**10a–c**) *Sc. velifer leptus* Ziegler & Sandberg, 1984, GMM B9A.6-21, Bed 1, regional *gracilis sigmoidalis* Subzone.

1a
1b
2a
2b
3a
3b
4a
4b
5a
5b
6a
6b
7a
7b
8a
8b

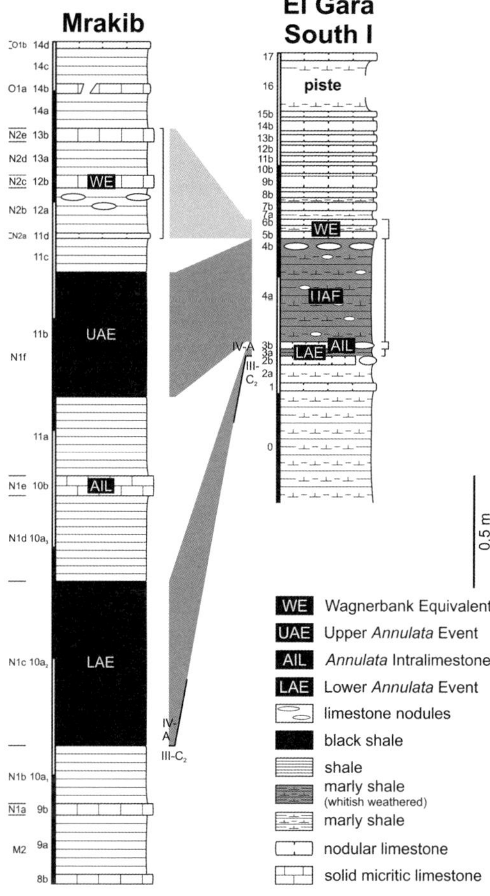

Fig. 15. Correlation of the *Annulata* Events of the Maider Basin (Mrakib) and Rheris Basin (El Gara South I), showing distinctive differences in lithology (pyritic black shales v. brownish to whitish weathered marly shales).

Gara, but with a local incoming of various protactoclymeniids in thick marl units.

In the Maider Basin at Mrakib (Fig. 15), most middle–upper Famennian beds were dysoxic and pyritic, leading to goethitic preservation of all molluscs. Above poorly fossiliferous greenish-grey shales and grey-brown micrites, the *Annulata* Events are developed as black shales that, based on pyrite alteration under arid conditions, turned into hematitic shales with an intensive red colour. The organic matter was also mostly destroyed. In comparison with the Tafilalt, the goethitic–hematitic ammonoid association of the Lower Event interval is distinctively small sized, and associated with very abundant guerichiids, some nuculids and juvenile gastropods as the only benthos. This suggests exaerobic and strongly eutrophic conditions. The 'high diversity, juvenile *Gundolficeras–Erfoudites–Protactoclymenia–Stenoclymenia–Guerichia* biofacies' is currently unique on a global scale. It does not extend above. The upper event interval represents a different 'poorly fossiliferous *Platyclymenia–Prionoceras* biofacies' without benthos that was deposited probably under strongly anoxic conditions. The subsequent three-fold Wagnerbank Equivalent correlates with successions of the Tafilalt Basin, and belongs to the better-oxygenated 'moderately diverse, normal-sized *Platyclymenia–Prionoceras–Erfoudites–Loxopteria* biofacies' known from the upper event interval of Hassi Nebech.

In summary, the diverse faunal assemblages of the Anti-Atlas *Annulata* Event beds document several distinctive and well-defined litho- and biofacies. These reflect local variations in bathymetry, trophic structures and ventilation. Regionally, there is no evidence for the widespread '*Platyclymenia–Guerichia* biofacies', characterized by mass occurrences of both taxa, of the Rhenish Massif (Becker 1992*b*; Hartenfels 2011) and Portugal (Fantinet *et al.* 1976), or of the strongly eutrophic '*Platyclymenia–Protactoclymenia–Guerichia* biofacies' of the Holy Cross Mountains (Hartenfels 2011), Moravia (Rzehak 1910; Weiner & Kalvoda 2016; additionally with *Erfoudites*) and Bulgaria (Boncheva *et al.* 2015). The *Platyclymenia–Prionoceras–Erfoudites* assemblages of the upper event interval at Hassi Nebech and of the Wagnerbank Equivalents of the Mrakib agree with the moderately deep event facies intercalating a pelagic platform of Iran (Becker *et al.* 2004). The 'low-diversity, juvenile *Platyclymenia–Prionoceras* biofacies' of El Gara is possibly also developed in the Cantabrian Mountains (Sanz-López *et al.* 1999).

In the eastern Anti-Atlas, locally variable nutrient influxes (eutrophication) caused either discrete

Fig. 14. Ammonoids from the marly *Annulata* Event beds (1–6) and overlying limestones (7–8) of Bine Jebilet; all regional *annulata* Zone. (**1a**, **b**) *Protacto.* aff. *lagowiensis* (Sobolew, 1912), B6.C-47.220, ×2. (**2a**, **b**) *Pl.* (*Tri.*) *protacta* (Wedekind, 1908), B6.C-47.283, specimen with traces of spiral ornament and alternating parabolic and non-parabolioc ribs, intermediate from *Pl.* (*Pl.*) *annulata richteri* Wedekind, 1914, ×2. (**3a, b**) *Pl.* (*Pl.*) *quiringi* Müller, 1956*a*, B6.C-47.258, ×1.5. (**4a**, **b**) *Post. ascendens* n. sp., paratype B6.C-47.189, ×2. (**5a**, **b**) *Pl.* (*Pl.*) *subnautilina subnautilina* (Sandberger, 1855), B6.C-47.259, with fewer whorls at the same size than in the directly associated *Pl.* (*Pl.*) *quiringi*, ×1.5. (**6a**, **b**) *Pl.* (*Pl.*) *annulata rotundata* Wedekind, 1908, holotype, Göttingen collection, ×1.1 (photograph by the late M.R. House). (**7a**, **b**) *Pl.* (*Tri.*) *spinosa* (Münster, 1842), B6.C-47.293, loose specimen, ×1.5. (**8a**, **b**) *Protoxy. dunkeri* (Münster, 1839*a*), B6.C-47.201, ×2.

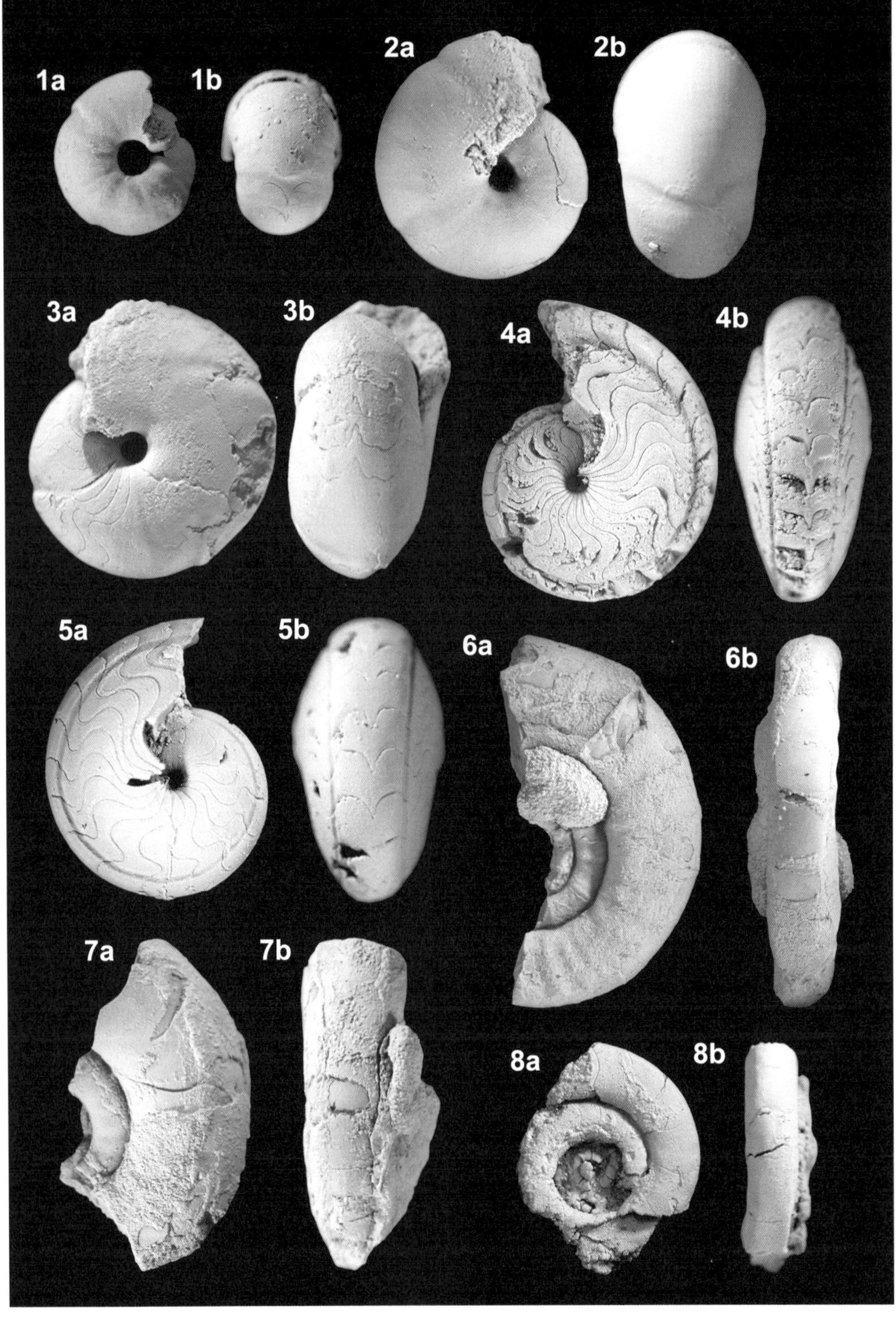
1a
1b
2a
2b
3a
3b
4a
4b
5a
5b
6a
6b
7a
7b
8a
8b

Fig. 17. Fauna from the Lower *Annulata* Marl and overlying beds at the Mrakib (Maider Basin). (**1a**, **b**) *Protacto.* aff. *subcostata* (Sobolew, 1914*b*), B6.C-47.218, Bed N1f2, basal *annulata* Zone, ×3. (**2**) *Proc. ebbighauseni* Klein & Korn, 2014, B6.C-47.228, embedded in a limestone concretion from near the top of Bed P, basal *ebbighauseni* Zone, ×1.5. (**3a**, **b**) *Protoxy.* cf. *wendti* Korn, 1999 juv., B6.C-47.225, loose from Bed O (possibly derived from above), ×3. (**4a**, **b**) *Protoxy. dunkeri* (Münster, 1839*a*), B6.C-47.224, loose from Bed O, local indicator of *dunkeri* Zone, ×2.5.

or repeatedly increased primary production during basal upper Famennian deepening intervals. By assumed analogy with the event beds of the Holy Cross Mountains (Racka *et al.* 2010), blooms of probably bacterial phytoplankton and of different opportunistic mollusks (ammonoids and specific bivalves adapted to variably eutrophic conditions) occurred, resulting in locally varying faunal associations, hypoxia and (originally) black shale deposition. It is intriguing that the strongest eutrophication signal comes from the deep Maider Basin, whilst the shallower Rheris Basin, which lay eastwards of the Ougnate Island area (palaeogeography of Dopieralska 2009), experienced only weak faunal blooms. This contradicts models for a climate and land-derived input of nutrients and trigger for Famennian black shales (e.g. Carmichael *et al.* 2016), and supports models for climate-driven episodic nutrient recycling within distal-shelf basins (e.g. Murphy *et al.* 2000; Sageman *et al.* 2003).

Comparison of conodont faunas, diversity and biofacies trends

Unlike as in ammonoids (e.g. Becker 1993*b*; Hartenfels 2011), there were no mass extinctions or

Fig. 16. Ammonoids from the Lower *Annulata* Black Shale at Mrakib (Maider Basin). (**1**)–(**3**) *Erf. zizensis* Korn, 1999 juv.: (1a, b) B6.C.210, widely open umbilicate specimen with inner flank ribbing and still without A_2-lobe, ×5; (2a, b) B6.C-47.211, specimen showing the closing umbilicus, ×5; (3a, b) B6.C-47.212, with small open umbilicus and ontogenetically just beginning A_2-lobe, ×4. (**4**)–(**5**) *Gund. australe* n. sp.: (4a, b) holotype B6.C-47.213, ×4; (5a, b) paratype B6.C-47.214, ×5. (**6a**, **b**) *Pl.* (*Pl.*) *latecostata* Czarnocki, 1989, B6.C-47.215, ×3. (**7a**, **b**) fragmentary *Pl.* (*Pl.*) *subnautilina subnautilina* (Sandberger, 1855), B6.C-47.216, ×2. (**8a**, **b**) *St. rectangula* n. sp. juv., paratype B6.C-47.217, ×5.

originations in conodonts associated with the *Annulata* Events. However, there was a gradual loss of some important conodont species/subspecies near the base of both event intervals (Figs 18–20). This difference shows a palaeoecological independence of both groups living in the same outer-shelf regions. Clearly, their evolution was not linked.

Pre-event beds. At Mrakib, Bine Jebilet and Hassi Nebech, the immediate pre-event limestones are

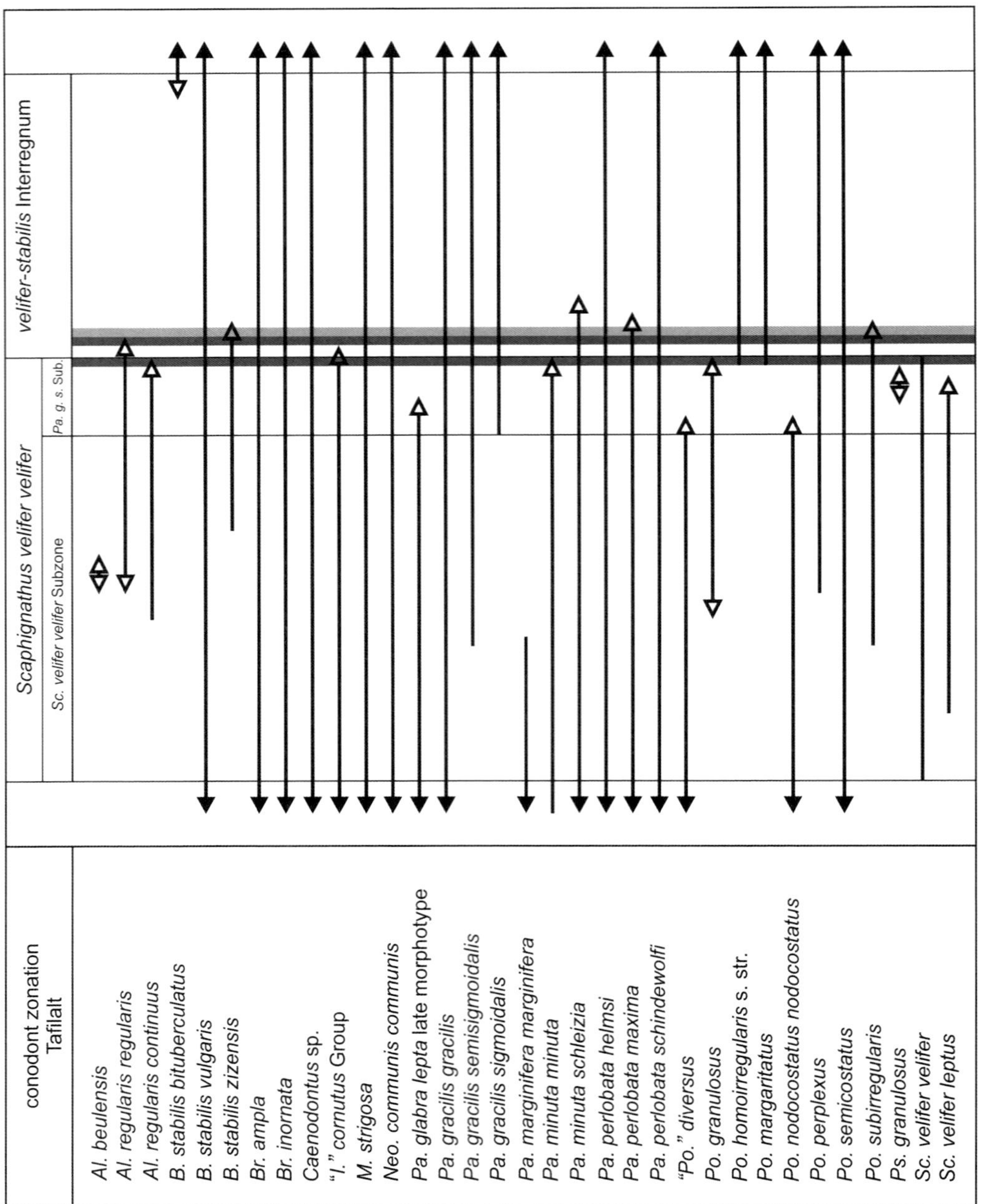

Fig. 18. Stratigraphical ranges of important conodont taxa around the Lower and Upper *Annulata* Events (dark grey) and Wagnerbank Equivalent(s) (light grey) of the Tafilalt (open arrows, longer ranges in other Moroccan regions; filled arrows, longer ranges elsewhere).

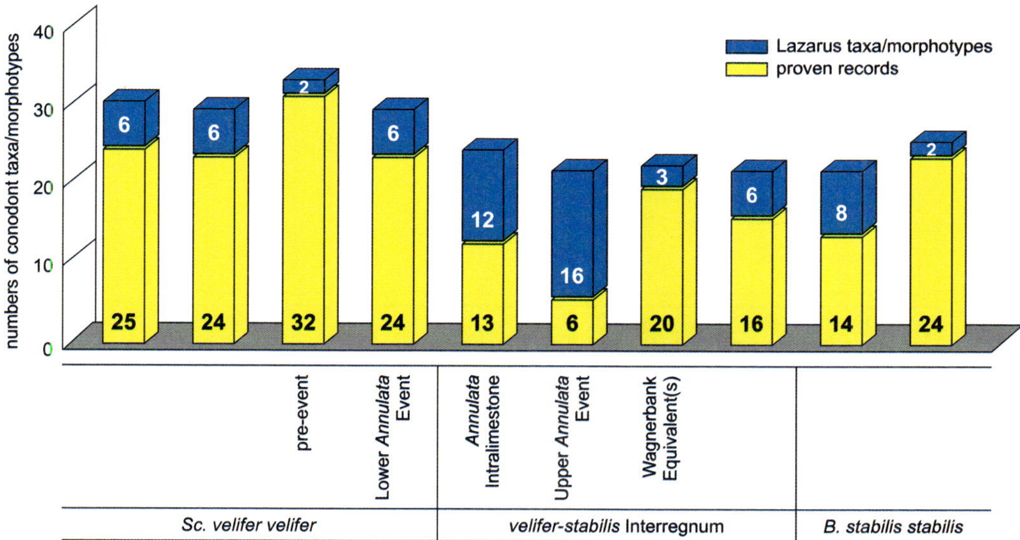

Fig. 19. Conodont diversity trends around the *Annulata* Events in the Tafilalt.

barren (cf. Corradini *et al.* 2001; Hartenfels 2011). The best conodont record is from Jebel Erfoud (Hartenfels 2011). There, dominant *Pa. gracilis gracilis*, *Br. ampla* and *B. stabilis vulgaris* indicate a palmatolepid–bispathodid biofacies *sensu* Ziegler & Weddige (1999). At Ouidane Chebbi Northwest, *Pa gracilis gracilis* dominates together with *Pa. perlobata schindewolfi*; at El Gara South, it co-occurs with additional palmatolepids, alternognathids and scaphignathids. The occurrence of *Alternognathus* and *Scaphignathus* may be used as evidence for a shallow setting, but Hartenfels (2011) documented *Sc. velifer velifer* together with *Al. regularis regularis*, *Al. regularis continuus*, *B. stabilis vulgaris* and *Br. inornata* from the basinal setting at Mrakib. Thus, *Alternognathus* and *Scaphignathus* do not indicate a shallow-water association. Generally, the pre-event limestones of the Tafilalt yield the most diverse conodont faunas of the regional *velifer velifer* Zone, with a total number of 32 taxa/morphotypes and only two additional Lazarus taxa/morphotypes.

Lower *Annulata* Event bed. In the Tafilalt, limestones of the lower event interval are only developed on the central southern platform (Amessoui Syncline: Fig. 1), where they represent a shallow, high-energetic, eutrophic, very fossiliferous (ammonoids) carbonate facies. Based on only two localities, Jebel Ouaoufilal Pass and Takhbtit West (Hartenfels 2011), the available conodont record is limited. Nevertheless, 24 different taxa/morphotypes, excluding six Lazarus taxa/morphotypes, are known so far. Again, the conodont faunas are dominated by *Pa. gracilis gracilis* and *B. stabilis vulgaris*, indicating a palmatolepid–bispathodid biofacies. At Jebel Ouaoufilal Pass, a *Br. ampla* peak is recognizable. Additional common (sub)species are *Br. inornata*, *M. strigosa*, *Neo. communis communis*, *Pa. perlobata schindewolfi* and *Po. semicostatus*. The latter is typical of inner-shelf environments (e.g. Belgium) (e.g. Dreesen & Orchard 1974), and suggests a shallow, hemipelagic setting for the black *Platyclymenia* limestone at the southern margin of the Tafilalt Platform, in agreement with the palaeogeographical reconstructions of Wendt *et al.* (1984) and Hartenfels (2011).

Annulata Intralimestone. At El Gara South, Bine Jebilet and Hassi Nebech, the dominance of *Pa. gracilis gracilis*, associated with common *Pa. perlobata schindewolfi* and *B. stabilis vulgaris*, shows that the pre-event palmatolepid–bispathodid biofacies was re-established during the regressive episode between both event intervals. In the Maider Basin, the conodont content is sparse. Alternognathids, combined with *B. stabilis vulgaris* and *Br. inornata*, indicate a different biofacies with low diversity. The current total conodont record has 13 conodont taxa/morphotypes. The temporary disappearance of 12 taxa/morphotypes marks an episodic regional diversity reduction ('Lazarus phase'), triggered by the ecological changes of the Lower *Annulata* Event.

Upper *Annulata* Event Bed. El Gara South provides, for the first time, Moroccan conodonts from the Upper *Annulata* Event interval. Again, *Pa. gracilis gracilis* is the dominant taxon. Additional palmatolepids (*Pa. gracilis semisigmoidalis*, *Pa. perlobata*

investigated localities	Mrakib								El Gara South							Bine Jebilet						Jebel Erfoud			Ouida. Ch. NW			Hassi Nebech							
	pre-event limestones		Lower *Annulata* Event	*Annulata* Intralimestone	Upper *Annulata* Event	Wagnerbank Equivalents			pre-event limestones		Lower *Annulata* Event	*Annulata* Intralimestone	Upper *Annulata* Event	Wagnerbank Equivalents		pre-event limestone	Lower *Annulata* Event	*Annulata* Intralimestone	Upper *Annulata* Event	Wagnerbank Equivalents		pre-event limestone	Wagnerbank Equivalents		pre-event limestones		Wagnerbank Equivalent	pre-event limestones		Lower *Annulata* Event	*Annulata* Intralimestone	Upper *Annulata* Event	Wagnerbank Equivalents		
bed number	6b	9b	$10a_2$	10b	11b	11d	12b	13b	1	2b	3a	3b	4a	5b	6b	1b	1c	1d	1e	1f	1h	30b	32	33b	11b	12	13b	I1b	I1d	I2a	I2b	I2c	I2d	J1b	J1d
taxa/bed	1	0	-	4	-	2	6	2	15	4	-	7	6	4	-	0	-	4	0	7	8	10	13	7	4	4	2	5	2	-	4	-	3	2	4
conodonts																																			
Al. regularis continuus				2					1																										
Al. regularis regularis				1					3	1			1																2						
B. stabilis vulgaris				1			1		3			2	1	16				7		6	5	14	71	18	3	1	2	1			1		3		5
B. stabilis zizensis																							2												
Br. inornata				2			1		2					1						3		8	9				1	1							
Br. ampla									2											1	4	19	30	6				1							
Caenodontus sp.							14	5																										1	
"*I.*" *cornutus* Group												2																							
M. strigosa									1												1	1	1	1											
Neo. communis communis							2			1		2								3	1	5	2	1	1	1									1
Pa. gracilis gracilis						16	18	1	8	2		18	15	23				7		21	44	89	149	58	15	11			1		15		35	2	7
Pa. gracilis semisigmoidalis												2	1								3		1					1							1
Pa. gracilis sigmoidalis									1															6											
Pa. minuta minuta									1													1													
Pa. minuta schleizia									1																										
Pa. perlobata helmsi									3				1								1	1	3												
Pa. perlobata maxima												1																							
Pa. perlobata schindewolfi	1					1	2		1			6	1	3				3		3	2	7	21	4	6	4					2		6		
Po. obliquicostatus																							1												
Po. perplexus																		2		2		2									1				
Po. cf. *rhabdotus*																							1												
Po. subirregularis																							3												
Ps. granulosus									2	1																									
Sc. velifer velifer									3																			2							
Sc. velifer leptus									2																										

Fig. 20. Comparison of conodont faunas around the *Annulata* Events of the correlated sections of the Maider Basin, the Rheris Basin, the Tafilalt Platform and the Tafilalt Basin.

helmsi, *Pa. perlobata schindewolfi*), *B. stabilis vulgaris* and a single *Al. regularis regularis* constitute a typical association for the palmatolepid–bispathodid biofacies. The contrast between six proven species v. 16 Lazarus taxa/morphotypes shows that the ecological conditions were still unfavourable for many (sub)species.

Post-event beds: Wagnerbank Equivalent(s). Generally, the Wagnerbank Equivalents represent a return to better-oxygenated conditions. The conodont diversity recovered from the ecological crisis, which is documented by a record of 20 different taxa/morphotypes. Comparable with the pre-event beds, the assemblages are dominated by *Pa. gracilis gracilis*, *B. stabilis vulgaris* and *Pa. perlobata schindewolfi*. At Jebel Erfoud, this association is completed by *Br. ampla*. At Mrakib, the upper two limestone layers of the three-fold Wagnerbank Equivalent yielded an abnormal, unique maximum of the normally rare and bradytelic conodont genus *Caenodontus* (Fig. 18). This represents a new conodont biofacies of a eutrophic deeper shelf basin (Hartenfels & Becker 2012*a*).

Subsequently, the proven palaeodiversity and the number of 'Lazarus taxa/morphotypes' remain more or less consistent through the higher *velifer–stabilis* Interregnum. The next radiation phase occurred within the subsequent *B. stabilis stabilis* Zone (Hartenfels 2011).

Systematic palaeontology of ammonoids (RTB)

Family **Tornoceratidae** Wedekind, 1910

Discussion. Most previous authors (e.g. Miller *et al.* 1957; Bogoslovskiy 1971; Becker 1993*a*; Korn & Klug 2002) overlooked that the term 'Tornoceratidae' was first used and explained in Wedekind (1910: p. 771), not in Arthaber (1911). This requires a correction of authorship. The subfamily name Tornoceratinae was first used in Wedekind (1918), not much later in Becker (1993*a*), as listed in Korn & Klug (2002). The latter authors provided no clear distinction of their new middle Famennian family Kirsoceratidae, which they proposed to include the genera *Pernoceras*, *Protornoceras*, *Tornia* and *Kirsoceras*. The supposedly characteristic reduction of the mid-flank A-lobe does not apply to the type genus. The Kirsoceratidae are here redefined to include only forms with a ventral saddle and a shift of the siphuncle to a subventral position.

Subfamily **Aulatornoceratidae** Becker, 1993*a* (nom. transl. Korn & Klug 2002)

Protornoceras ornatum Dybczynski, 1913 (Fig. 6.2a, b)

Description. The only available specimen is minute (B6.C-47.72, only 6.7 mm dm), but shows the convolute coiling (uw/dm = 0.3), tegoid, slightly compressed whorl form (ww/wh = 0.96), subumbilical ribbing and sutures, as in the figured Polish types, which apparently have been lost. There are six–seven fine, concavo-convex growth lirae per millimetre with a very high ventrolateral salient. The small ventral lobe is easily visible, bordered by low, relative wide outer-flank saddles.

Discussion. Dzik (2006) treated the species as a possible synonym of *Prot. polonicum* Dybczynski, 1913, but this is not supported by any hard data. All other North African species that were previously assigned to *Protornoceras* (e.g. Petter 1959; Göddertz 1987) now fall into different genera of the Aulatornoceratinae, such as *Armatites* or *Planitornoceras*, or belong to homoemorphic Givetian groups, which were first noted in House (1963*a*).

Stratigraphic range and geographical distribution. The precise age of the Polish types is unclear. No additional specimens have been found subsequently in the Holy Cross Mountains. The new Tafilalt specimen falls in the *sulcata* Zone (UD III-C2).

Subfamily **Falcitornoceratinae** Becker, 1993*a* (nom. transl. Korn & Klug 2002)

Gundolficeras australe n. sp. (Figs 16.4, 5, 21a)

Derivation of name. From the Latin *australis* = southern; due to its occurrence in southern Morocco.

Types. Holotype B6.C-47.213 (Figs 16.4, 21a), paratypes B6.C-47.214 (Fig. 16.5) and B6.C.47.234–B6.C-47.245. In addition, there are 23 poorly preserved (squashed or fragmentary) specimens, which are not designated as paratypes.

Type level and locality. Lower *Annulata* Shale (LAE) of Mrakib, southern Maider, lower *annulata* Zone (lower UD IV-A).

Diagnosis. Medium-sized species of *Gundolficeras* with tegoid, weakly compressed (from *c.* 4 mm dm on) and high whorls (WER 2.1–2.2), deeply concave, impressed umbilical wall (from 6 mm dm on), and very strong, channel-like ventrolateral furrows that disappear between 7.5 and 8 mm wh. Sutures with widely rounded inner and outer L-lobes divided by a low saddle at the umbilical seam, asymmetrical, high lateral saddle on the mid-flanks, asymmetrically rounded, moderately narrow A-lobe, moderately high, rounded ventral saddle and small, divergent E-lobe.

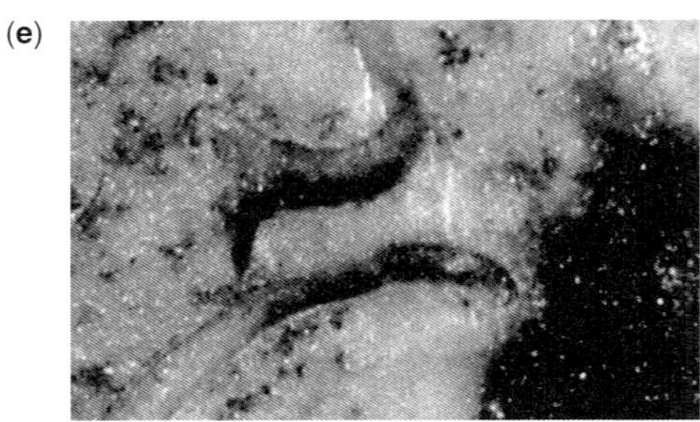

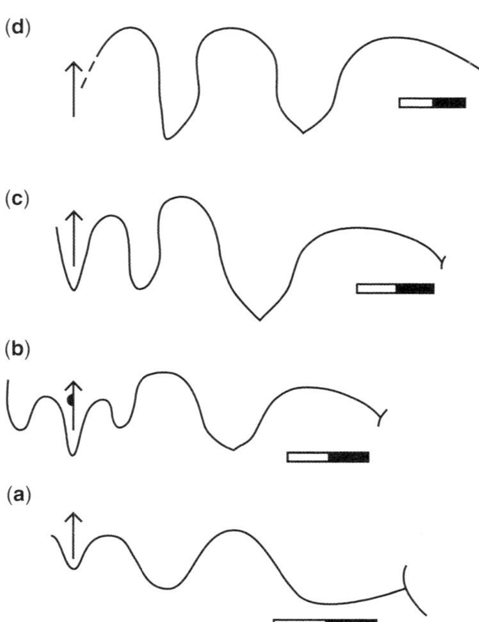

Fig. 21. Goniatites from the *Annulata* Event beds. (**a**) *Gund. australe* n. sp., holotype B6.C-47.213, suture at 5.5 mm wh. (**b**)–(**e**) *Ungu. unguiferum* Korn, Bockwinkel & Ebbighausen, 2015*b*: (b) suture of B6.C-47.190 at 6 mm wh, Lower *Annulata* Marl, Hassi Nebech; (c) suture of B6.C-47.209 at 8 mm wh, Lower *Annulata* Marl, Hassi Nebech; (d) suture of B6.C-47.196 at 11.5 mm wh, Upper *Annulata* Marl, Bine Jebilet; (e) cross-section detail showing a thin, thorn-like shell extension sealing the umbilicus in B6.C-47.193 at 13 mm wh.

Description. The holotype is the largest, but still small-sized, well-preserved specimen. The change from weakly depressed to weakly compressed whorl form occurs between 4 and 5 mm dm (paratypes B6.C-47.234 and B6.C-47.235). The ww/wh ratio decreases to values below 0.80 near 10 mm dm (holotype). All three small specimens in which WER could be measured show values of between 2.10 and 2.15 (see Table 1). All juveniles are marked by their very strong, narrow ventrolateral furrows, which run in the external lower part of the ventral saddle of the septa. However, the furrows fade very suddenly between 7.5 and 8.5 mm wh on the incomplete paratype B6.C-47.239. Consequently, all larger fragments (B6.C.47.244 and B6.C-47.245) display gently rounded outer flanks, without a trace of furrows. At *c.* 3 mm wh (6 mm dm, paratypes B6.C-47.236, B6.C-47.237), the rounded umbilical wall flattens and then turns into a deeply concave area (paratype B6.C-47.239), which was probably sealed by a thick umbilical plug. When the preservation is poor, the umbilicus falsely may appear to be open. Paratype B6.C-47.238 displays weak impressions of strongly biconvex growth ornament on the body chamber.

There is not much ontogenetic change in the sutures. The ventral saddle is low at small size (paratype B6.C-47.234) and it is rounded, with divergent flanks, at 15 mm dm. The largest fragments (paratype B6.C-47.244) are still septate at 10 mm wh, which suggests that the species reached a diameter of at least 30 mm.

Discussion. The new species is close to the contemporaneous *Gund. prescheri* Korn, 2002 from the Rhenish Massif, which, however, is thought to be smaller sized and characterized by rounded umbilical walls, at least until *c.* 5 mm wh, when the latter are already flattened in *Gund. australe* n. sp. The *prescheri* holotype is also more compressed (see Table 1), with a trochoid, not tegoid cross-section. The sutures and ornament of both species are identical. In the rotund and thicker type species of the genus, *Gund. bicaniculatum* (Petter, 1959), the ventrolateral furrows are finer than in *Gund. australe* n. sp. and disappear earlier (near 5 mm wh); the umbilicus stays open (see Becker 1995). In the also thicker and very strongly tegoid *Gund. delepinei* (Petter, 1959), the A-lobe is rounded but narrower. In *Gundolficeras* n. sp. aff. *delepinei sensu* Becker (1995) from Kazakhstan (Karaganda Basin), originally assigned by Bogoslovskiy (1971) to *Lobotornoceras bicaniculatum*, the ventrolateral furrows disappear suddenly at *c.* 4.5 mm wh, as in typical *bicaniculatum*, but the umbilicus is completely closed. The upper Famennian *Gund. escoti* (Frech, 1887) and older relatives from Germany, Poland and the Urals (see the review in Becker 1995) are characterized by asymmetric, angular A-lobes. *Gundolficeras rotersi* Korn, 2002 lacks marked furrows throughout ontogeny.

Stratigraphic range and geographical distribution. See type locality and level. Related forms, which require further study, occur in older strata of the Mrakib (*Gundolficeras* n. sp. 1 and 2 of Becker *et al.* 2000).

Table 1. *Shell parameters (in mm) of* Gund. australe *n. sp.*

No.	dm	wh	ah	ww	ww/dm	ww/wh	WER
B6.C-47.235	4	2.2	–	2.2	0.55	1.00	–
B6.C-47.234	5	3	–	2.7	0.54	0.90	–
B6.C-47.236	7.3	4.2	–	3.8	0.52	0.90	–
[holotype *prescheri*	7.8	4.9	2.6	3.5	0.45	0.71	2.25]
B6.C-47.214	8.6	5	2.7	4.4	0.51	0.88	2.12
*B6.C-47.213	11.3	6.8	3.8	5.2	0.46	0.76	2.15

*Holotype.

Family **Posttornoceratidae** Bogoslovskiy, 1962

Posttornoceras ascendens n. sp.
(Fig. 14.4a, b)

v 1995 *Posttornoceras* aff. *contiguum* Becker: 620–621, text-figs 7a, b, pl. 3, figs 1, 2
v 1997 *Posttornoceras contiguum* Becker: fig. 1a, pl. 1, fig. 11
v 2002 *Posttornoceras* aff. *contiguum* Becker: 63, figs 3C, D, 4A, B, pl. 2, figs 5–8
v 2002 *Posttornoceras* aff. *contiguum* Becker *et al.*: 170, 171
non 2006 *Posttornoceras* aff. *contiguum* Dzik: 214, figs 155G, 159

Derivation of name. Owing to the ascending height of saddles from the umbilicus to the venter.

Types. Holotype is specimen MB.C.3461 (formerly Eb-C2, Museum für Naturkunde, Berlin), the original of Becker (1995: text-fig. 7a–b, pl. 3, figs 1 & 2). MB.C.3457-59 and MB.C.3472 from Mrakib, B6.C-47.189 from Bine Jebilet (Fig. 14.4a, b) and MB.C.2175 from the Nie Brickwork Quarry (Sauerland) are paratypes.

Type locality and level. Mrakib, most probably UD IV.

Diagnosis. Species of *Posttornoceras* with extremely high (WER 2.65–2.95), compressed (ww/wh = 0.84 at 10 mm dm, 0.72 at 20 mm dm), tegoid whorls with narrowly rounded venter and shallow ventrolateral furrows until approximately 10–12 mm dm; umbilicus closed. Sutures with narrow, short E-lobe, very high, wide and asymmetrical ventral saddle, which top is somewhat flattened, very deep, asymmetrically pointed A-lobe, narrow, slightly constricted, high mid-flank saddle, subangular, asymmetrically v-shaped outer L_1-lobe, subtriangular, low inner-flank saddle, and shallow, subumbilically widened L_2-Lobe at the umbilical seam.

Discussion. Based on its very deep, pointed A-lobe and very high, wide ventral saddle, which becomes slightly higher than the mid-flank saddle towards maturity, the new species can be easily distinguished from other members of the genus. It was introduced by Becker (1995) as *Post.* aff. *contiguum* (see also the discussion in Becker 2002). In *Post. balvei* Wedekind, 1910 both flank lobes are of approximately equal depth. *Posttornoceras contiguum* (Münster, 1832) *sensu* its neotype (Becker 1993*a*, see the discussion in Becker 2002, = *Post. fallax* Korn in Korn & Ziegler 2002) is rather similar and probably ancestral; but its ventral saddle remains lower than the mid-flank saddle through ontogeny. In the Australian *Post. glenisteri* Petersen, 1975, the A-lobe is bell-shaped and the outer L_1-lobe is narrower. In *Post. posthumum* (possibly = *Wedekindoceras seidlitzi* Schindewolf, 1924; = *Post. weyeri* Korn, 1999 and *Post. contiguum sensu* Korn in Korn & Ziegler 2002) the inner-flank saddle is much higher and better rounded, and the ventral saddle lower and even more flattened at the top. In *Post. sodalis* Becker, 1995, the ventral saddle is similarly high as in *Post. ascendens* n. sp., but the inner-flank saddle is more symmetrical and the Russian species is much thicker (ww/wh *c.* 0.9 near 20 mm dm in the holotype, >0.9 in a much larger specimen from a different locality of the southern Urals, Manshia river; see Bogoslovskiy 1971). However, Nalivkina (1953) described other, more compressed specimens as *Post. contiguum*, which could be close to the new Moroccan form.

The specimen described and illustrated by Dzik (2006) as *Post.* aff. *contiguum* represents a very different new taxon with strongly pointed mid-flank saddle. It is clearly ancestral to *Discoclymenia*.

Stratigraphical range. UD III-C1 to IV-B2 (see Becker *et al.* 2002).

Geographical distribution. Southern Morocco (Tafilalt, Maider), Rhenish Massif, possibly southern Urals.

Family **Sporadoceratidae** Miller & Furnish, 1957 in Miller *et al.* (1957)
(nom. transl. Ruzhencev 1957)
Subfamily **Xenosporadoceratinae** Korn, 2002

Maeneceras subvaricatum subvaricatum
(Sobolew, 1914*a*)
(Fig. 6.8a, b)

*non 1841 *Goniatites biferus* Phillips: 120, pl. 49, fig. 230 [probably = *Erf. ungeri*]
*non 1902 *Sporadoceras subbilobatum* var. *meridionalis* Frech: 74, 81, 108, figs $35b_{1-3}$, pl. 4, figs 21a, b [probably = *Erf. ungeri*]
e.p. 1902 *Sporadoceras subbilobatum* Frech: 74, 80 [non fig. 35c]
v 1908 *Sporadoceras biferum* Wedekind: 593–594, pl. 39, figs 20 & 22, pl. 40, fig. 2
1912 *Sporadoceras biferum* Sobolew: 9
* 1914*a* ?-*Oma-dimeroceras* (*Sporadoceras*) *subvaricatum* Sobolew: 35, fig. 26, pl. 6, fig. 5a, b
* 1929 *Sporadoceras biferum* var. *sulcifera* Lange: 13, 43
1960 *Sporadoceras biferum biferum* Kullmann: 515, 516, 518, fig. 14a
1960 *Sporadoceras biferum subvaricatum* Kullmann: 515
1960 *Sporadoceras biferum sulciferum* Kullmann: 516, 518, fig. 14b, pl. 6, figs 3 & 4
v 1993*a* *Maeneceras biferum biferum* Becker: 308–310, fig. 95c, pl. 26, figs 3 & 4 [further synonymy]
2000 *Maeneceras biferum* Korn *et al.*: 70, fig. 3
2000*a* *Maeneceras subvaricatum* Becker & House: 133, 134, tab. 4
v 2000*b* *Maeneceras subvaricatum* Becker & House: 39
v 2000*c* *Maeneceras subvaricatum* Becker & House: 56
v 2002 *Maeneceras subvaricatum subvaricatum* Bockwinkel *et al.*: 285–290, text-figs 1, 2c, 4a, 6–12; pl. 1, figs 7–14; pl. 2, figs 1, 2, 9
v 2002 *Maeneceras subvaricatum subvaricatum* Becker *et al.*: 169
e.p. 2002 *Maeneceras meridionale* Korn & Ziegler: 460–461, ?text-fig. 5 [non the lectotype]
e.p. 2002 *Maeneceras meridionale* Korn & Klug: 186 [non fig. 164I]
? 2006 *Felisporadoceras subvaricatum* Dzik: 235, figs 172a, b, 181 [thick morphotype]
2009 *Maeneceras meridionale* Becker & House: 425, figs 2.2 & 2.4
non 2012 *Felisporadoceras* cf. *subvaricatum* Rakociński: 388, fig. 2 [relative of *M. rotundum*]
v 2013 *Maeneceras meridionale meridionale* Rytina *et al.*: 14
v 2013 *Maeneceras meridionale meridionale* Hartenfels *et al.*: 45, fig. 7c

Discussion. The naming of the oldest Moroccan sporadoceratid species from UD II-G has been controversial. It was first called *Sp. biferum* (Phillips, 1841) in Roch (1950) and Petter (1959), then *Maene. biferum* in Becker (1993*a*) and Korn *et al.* (2000). After it became clear that *Goniatites biferus* is based on a much younger species, which is probably a junior synonym of *Erf. ungeri* (Münster, 1832), the Polish name *Maene. subvaricatum* was applied (Becker & House 2000*b*, *c*; Bockwinkel *et al.* 2002). Inspired by the synonymies of Becker (1993*a*) and Becker & House (2000*b*), Korn (in Korn & Ziegler 2002) suggested that *Sp. subbilobatum* var. *meridionalis* Frech, 1902 has priority over *Maene. subvaricatum* and selected a lectotype from the Frech Collection at Wroclaw. This was uncritically followed by Becker & House (2009), Becker in Rytina *et al.* (2013) and Hartenfels *et al.* (2013). However, Frech (1902, p. 108) emphasized that his new 'local form' was the most common fossil in the upper Famennian (type-level = 'upper Clymenienkalk', '*Gonioclymenia* Zone', UD V) of the La Serre hill in the Montagne Noire. This was confirmed by recollecting when the higher part of Trench C was reopened (new material from UD V-A_1 with *Costaclymenia*, section log in Feist 2002). There is no evidence that the *Sp. meridionale* lectotype is from the same strata, near the base of the middle Famennian (within the Lagow Limestone, UD II-G), as the types of *Sp. subvaricatum*. If Korn's statement (in Korn & Ziegler 2002, p. 460) 'most probably [from the] early Middle Famennian' was true, the lectotype selection would have violated nomenclatorical rules (Article 75.3.6.). Frech (1902, p. 80) referred specimens from the 'middle Upper Devonian' and 'lower Clymenienkalk' of the Montagne Noire (Touriere) and other regions (Enkeberg Limestone) to '*Sp. Subbilobatum*', not to his new var. *meridionalis*. As long as the chosen lectotype is, indeed, a Frech syntype, there is no reason to doubt its upper Famennian provenience. In this context, it is very unfortunate that neither photographs nor sutures or shell dimensions of it have been published.

The illustrated suture of a La Serre specimen in Korn & Ziegler (2002, text-fig. 5F; refigured in Korn & Klug 2002, fig. 164I), which is not a syntype, differs considerably both from the sutures given by Frech (1902) for typical *Sp. meridionalis*, from all new topotypes, and from the sutures of Polish or Moroccan *subvaricatum* representatives (Sobolew 1914*a*; Bockwinkel *et al.* 2002). It is here exluded from the species. In the topotypic upper Famennian La Serre moulds, the typical spiral ornament of *Erfoudites* is not preserved. But with respect to very similar shell form (mature ww/dm near 0.5), sutures (with pointed A_1 lobe) and age, the conclusion of Bockwinkel *et al.* (2002) is maintained that *Sp. subbilobatum* var. *meridionalis* is most likely a subjective junior synonym of *Erf. ungeri*. It is not a senior synonym of the stratigraphically much older *Maene. subvaricatum*, which availability as a valid taxon name has recently been clarified (Becker & Nikolaeva 2014; Opinion 2337 2014). The *Sp. subvaricatum* original (and potential lectotype) of Sobolew (1914*a*) shows the same shell form (ww/dm ratio), adult restriction of varices and sutures as in adult Tafilalt specimens (see Bockwinkel *et al.* 2002). A Polish specimen illustrated by Dzik (2006), however, is slightly more rotund.

Since Polish topotypes are currently unavailable for study, it is currently not possible to clarify whether their early ontogeny is the same as in assumed Rhenish and Moroccan *subvaricatum* representatives. Tafilalt (from Hassi Nebech) and Maider specimens (see ?*Maene.* aff. *subvaricatum* in Becker *et al.* 2002) have rather evolute inner whorls, whilst specimens from Australia were shown by Petersen (1975, *Sp.* (*Sp.*) *biferum* of text-fig. 22F) to be strongly involute. This suggests a higher morphological and taxonomic complexity in the group than documented so far, which requires further studies.

Erfoudites Korn, 1999
(Fig. 16.1–3)

Discussion. Whilst sporadoceratids from the *Annulata* Event beds of Hassi Nebech agree in terms of sutures (with rounded A_1-lobe) and whorl expansion (>2 between 20 and 30 mm dm) with *Erf. rherisensis*, contemporaneous juveniles from the Mrakib (LAE: Fig. 16.1–3) are different. Very characteristic are a relatively wide umbilicus (uw/dm = 0.21 at 5 mm dm in B6.C-47.210: Fig. 16.1) that closes subsequently (uw/dm = 0.11 at 10.5 mm dm in B6.C-47.212: Fig. 16.3) and undulose ribbing around the umbilicus. The A_1-lobe stays narrowly rounded in the largest fragmentary specimens (B6.C-47.247, at *c.* 7 mm wh) and WER is only 1.80 at approximately 23.5 mm dm (B6.C-47.246). Therefore, they are assigned to *Erf. zizensis*, which, however, is probably a junior synonym of *Erf. moravicus* (Rzehak, 1910) and, based on a re-examination of the Göttingen lectotype, is closely related to *Erf. spirale* (Wedekind, 1918). Korn *et al.* (2015*b*) failed to discuss the relationships between these three species.

In any case, the stratigraphic range of both Tafilalt *Erfoudites* species has to be extended into the *Annulata* Event levels (basal UD IV, basal *annulata* Zone).

Subfamily **Sporadoceratinae** Miller & Furnish in Miller *et al.* (1957)

Sporadoceras muensteri muensteri
(v. Buch, 1832)

*v	1832	*Ammonites Münsteri* v. Buch: 173–174 (122–123), pl. 12, figs 4 & 5
v	1884	*Ammonites Münsteri* Beyrich: 212
v	2002	*Sporadoceras muensteri* Becker: pl. 3, figs 14 & 15 [holotype]
v e.p.	2002	*Sporadoceras muensteri* Becker *et al.*: 183, fig. 7f [only: holotype]
	2002	*Sporadoceras muensteri* Korn & Klug: 187, fig. 165
	2014	*Sporadoceras muensteri* Zong *et al.*: 187–188
non	2015*b*	*Sporadoceras muensteri* Korn *et al.*: 53–58, figs 7–12 [= *Sp. muensteri orbiculare*]

Discussion. Korn *et al.* (2015*b*) overlooked that Becker in Zong *et al.* (2014, p. 187) recognized the original of Becker (2002, pl. 3, figs 14 & 15, Münster collection at Munich AS VII 1143; suture figured in Becker *et al.* 2002, fig. 7f) as the holotype of *Sp. muensteri* (v. Buch, 1832). In accord with comments by Beyrich (1884), the original description was probably based on just one specimen of the Münster collection, the holotype by monotypy. Therefore, the Korn *et al.* (2015*b*) designation of a thicker Berlin specimen (MB.C.3503 = Becker *et al.* 2002, fig. 9d, incorrectly identified as a syntype) as lectotype is invalid. As noted by v. Buch (1832) and Zong *et al.* (2014), the type of *Sp. muensteri* is characterized by approximately equal wh and ww values at 27 mm dm, whilst in typical *Sp. orbiculare*, which Korn *et al.* (2015*b*) showed to be a rather invariable form, ww exceeds wh until approximately 40 mm wh. Therefore, both closely related forms (see comments in Becker *et al.* 2002, p. 183; Zong *et al.* 2014, p. 188) can be separated at the subspecies level. The sutures of the holotype and of the 'Korn *et al.* lectotype' are also somewhat different, especially in the shape of the A_2-lobe, but this may reflect variability.

Family **Praeglyphioceratidae** Ruzhencev, 1957

Ungusporadoceras unguiferum
Korn, Bockwinkel & Ebbighausen, 2015*b*
(Figs 7.2, 7.3, 21.2–5)

	2002	*Erfoudites* n. sp. Becker *et al.*: 170
	2011	*Erfoudites* n. sp. Becker in Hartenfels: 162, 167, figs 36 & 38
	2013	*Erfoudites* n. sp. Hartenfels *et al.*: 49, 51
*	2015*b*	*Ungusporadoceras unguiferum* Korn, Ebbighausen & Bockwinkel: 65–66, figs 19 & 20

Description. Additional material from Hassi Nebech and Bine Jebilet adds to the knowledge of the distinctive species. As noted in the original description, the fine spiral striae are best developed on the outer flank and venter (e.g. in B6.C-47.194), but there are also some near the umbilicus in B6.C-47.188 and on the inner flanks in B6.C-47.199. The crossing of undulose growth lines and spiral lines leads to a dissolution of the latter into a dense series of minute knobs in B6.C-47.199 and B6.C-47.200. An undulating, more-or-less radial wrinkle layer is preserved in B6.C-47.187 and B6.C-47.198. The umbilicus is completely closed by a thin shell flare, which produces a thorn-like extension in cross-section (Fig. 21e). Such features are known from middle–upper Famennian prolobitids (Bogoslovskiy 1969, fig. 52; Korn 2002, fig. 24), but have not yet been described from contemporaneous sporadoceratids.

Small specimens give some insights into the suture ontogeny. In B6.C-47.190 (Fig. 21b), the outer flank lobe is still small and rounded at 6 mm wh and forms a step towards the much deeper E-lobe, separated only by a low and small saddle, similar to that seen in juvenile *Praeglyphioceras* (e.g. Wedekind 1908, pl. 39, fig. 12; Nikolaeva 2011, fig. 5a). At 8 mm wh (B6.C-47.209: Fig. 21c), it is more asymmetrical and the ventral saddle remains much lower than the mid-flank saddle, as in *Prae. pseudosphaericum* (e.g. Frech 1902, fig. 35a; Wedekind 1908, pl. 39, fig. 13). At 11.5 mm wh (B6.C-47.196), the outer flank lobe becomes subangular, the ventral E-lobe is shortened, and the saddle between both is as high as the mid-flank saddle (Fig. 21d). At 12.6 mm wh (B6.C-47.198), the pointed outer flank lobe may be even deeper than the inner flank A-lobe, as it can be seen in paratype MB.C. 25512 (Korn *et al.* 2015*b*, fig. 19A).

Discussion. As noted by Korn *et al.* (2015*b*), the species differs from typical Xenosporadoceratinae by its deep and narrow supposed A_2-lobe. However, they did not note the strong similarities in shell form, strongly biconvex varices and growth lines with the somewhat enigmatic genus *Ostrovkites* Dzik, 2006, which is only known from small specimens. Owing to its ventral suture ontogeny, *Ostrovkites* was correctly transferred to the Praeglyphioceratidae by Nikolaeva (2011). The genus is not restricted to the Holy Cross Mountains, but occurs also in the Rhenish Massif (Posttornoceratidae n. sp. in Becker 1985). The early sutures of *Ungu. unguiferum* illustrated here agree with *Ostrovkites* and resemble other juvenile praeglyphioceratids. This suggests that the shell similarities are not a convergence, but an expression, of close phylogenetic relationships. Spiral ornament is very common in praeglyphioceratids, as first noted by Wedekind (1908), and subsequently described by Born (1912), Nalivkina (1953) and Nikolaeva (2011). *Ungusporadoceras* can be separated from *Ostrovkites* only by its spiral lirae, which seem to be absent in *O. numismalis* Dzik, 2006. The revised suture formula for *Ungu. unguiferum* is, as for the family, E_1E_2ALUI.

Stratigraphic range and geographical distribution. Restricted to the *annulata* Zone (UD IV-A) of the eastern Anti-Atlas (Tafilalt, Maider).

Family **Cyrtoclymeniidae** Hyatt, 1884
(nom. corr. Schindewolf 1949)

Protactoclymenia aff. *subcostata*
(Sobolew, 1914*b*)
(Figs 8.5a, b, 17.1a, b)

Description. Four fragmentary specimens from EG SI/-2b (B6.C-47.75–B6.C-47.78) belong to one species, which is characterized by a slightly subevolute (uw/dm = 0.30–0.32 at 15–23 mm dm), weakly compressed (ww/wh = 0.8–0.9) shell, with ww/dm declining from 0.42 to approximately 0.35 (at *c.* 23 mm dm in B.6-C. 47.78). The latter specimen has a whorl expansion rate of approximately 2.2. The whorl form is rounded, oval, with a slight flattening of the flanks. B.6-C.47.76 shows weakly curved, short and low ribs around the umbilicus and with a projecting, rounded ventrolateral salient. There are no clear ribs but there are remains of strongly undulose growth ornament in B.6-C.47.78 (Fig. 8.5a). Its main salient is subsymmetrical and occupies the outer half of the flank.

Similar forms occur in the Lower *Annulata* Shale of the Mrakib (e.g. B6.C-47.218: Fig. 17.1a, b). They are characterized by uw/dm ratios between 0.30 and 0.34, a gradual change from slightly depressed to slightly compressed whorl form at approximately 15 mm dm, with ww/dm falling from 0.46 to 0.38 between 12 and 16.5 mm dm, and a WER of approximately 2.3. There are specimens almost without ribbing (B6.C-47.218) and others with low, undulose ribbing around the umbilicus (B6.C-47.294).

Discussion. The protactoclymeniids of North Africa are, thus far, insufficiently studied. The El Gara and Mrakib specimens probably belong to one species with a moderate variability. *Protactoclymenia steinmanni* (Wedekind, 1908) from the *delphinus* Zone of the Rhenish Massif is based on a lectotype (erroneously re-illustrated as 'holotype' in House 1970, GC-UG 390-36), which is just slightly more evolute than the El Gara species (uw/dm = *c.* 0.35); but it lacks ribs, its regular growth lirae are prorsiradiate and its cross-section is subtrapezoidal, with flattened flanks and venter. *Protactoclymenia enkebergensis* has an uw/dm ratio of 0.31 in the lectotype (here designated, original of Wedekind 1908, fig. 1, GC-UG 390-59) and posseses very sharp, hardly curved dorsolateral ribs until approximately 20 mm dm. The type species of the genus from the *delphinus* Zone, *Protacto. pulcherrima* (Wedekind, 1908), is more involute (see lectotype GC-UG 390-37, here designated, original of Wedekind 1908, pl. 43, figs 13 & 13a). This is also true for the small-sized, contemporaneous Russian *Protacto. orientalis* (Perna, 1914) and *Protacto. pernai* (Nalivkina, 1953). The Polish *Protacto. lagowiensis* (Sobolew, 1912) from the *delphinus* Zone is just slightly more evolute (uw/dm = 0.35) and thicker (ww/wh > 0.9 at 25 mm dm). Its more pronounced ribs are weakly curved and prorsiradiate, as in relatives (aff. *lagowiense*) from the *Annulata* Event beds of Bine Jebilet (Fig. 14.1a, b). Most supposed *lagowiensis* specimens in Dzik (2006), however, belong to *Protacto. enkebergensis*.

Based on a comparison with topotypes (leg. M. Bendella), *Protacto. tenuicostata* Petter, 1960, from the UD IV of Algeria, differs in having a wider umbilicus (uw/dm > 0.35), lower ww/dm ratios through ontogeny (from 0.36 down to 0.32 between 15 and 23 mm dm), and concave ribs that reach the flank margin and with a narrower outer growth line salient. The contemporaneous *Protacto. pseudarietina* (Rzehak, 1910) from Moravia is very close to *tenuicostata*, but characterized by ventrolateral and umbilical edges. *Protactoclymenia cara* Nikolaeva & Bogoslovskiy, 2005 and *Protacto. mira* Nikolaeva & Bogoslovskiy, 2005 possess rectiradiate and biconvex growth lines, and, therefore, have to be excluded from the genus. They belong to *Cyrtoclymenia* (*sensu stricto*).

Our Tafilalt and Maider specimens seem to be closest to the poorly known *Protacto. subcostata* (Sobolew, 1914*b*). Since this comparison can only be based on the small-sized original illustration, and in the absence of any subsequent descriptions, the Moroccan specimens are currently assigned with an aff. to the Polish form.

Stratigraphic range and geographical distribution. Upper part of the *sulcata* Zone (upper part of UD III-C_2) of the northern Tafilalt and basal part of the *annulata* Zone (UD IV-A) of the Maider.

Protactoclymenia aff. *implana* (Czarnocki, 1989) (Fig. 7.6a, b)

Description. A single, corroded and incomplete clymeniid from the *sulcata* Bed at El Gara South I (B6-C-47.74) is characterized by variably spaced ventral varices, a compressed, oval cross-section, convolute coiling (uw/dm = 0.29 at *c.* 18 mm dm), ww/wh = 0.76, and concave ribs on the inner flank that are rather sharp on the inner whorls and that become more dense and less pronounced with growth. A septal face indicates a shallow, rounded flank lobe.

Discussion. Within *Protactoclymenia* there is a distinctive, small-sized, subevolute species group with ribs and lateral to ventral varices. It was partly included in *Pricella* by Nikolaeva & Bogoslovskiy (2005) because characteristic species, such as *Protacto. pinnata* (Perna, 1914), *Protacto. minuta* (Kind, 1944), *Protacto. ornata* (Kind, 1944), *Protacto. pinnatiformis* (Nalivkina, 1953) and specimens (possibly incorrectly) assigned to *Protacto. tuberculata* (Kind, 1944), tend to develop a flattened venter at maturity. A bicarinate venter and ventral lobe, as in true *Pricella*, however, never develop. The small *Clymenia pinnata* var. *rotunda* (Perna, 1914) probably represents a juvenile *Protacto. pinnata*, in which the venter is still rounded. Our specimen with rounded venter lacks the nodes of true *Protacto. tuberculata* and the ribbing is not so marked. It closely resembles the Polish *Protacto. implana*, the holotype of which has very similar uw/dm and ww/wh ratios, and similar ribbing. The varices have been illustrated in a paratype and are not restricted to the venter, but extend over the flanks. This difference results in an aff. identification for the El Gara specimen.

There is some justification to separate the *pinnata* Group at the generic or subgeneric level. In other clymeniids, however (e.g. *Flexiclymenia*, *Cymaclymenia* and *Rectoclymenia*), varices and ribbing are rather variably developed.

Stratigraphic range and geographical distribution. Upper part of the *Sulco. sulcata* Zone (upper part of UD III-C2) of the northern Tafilalt.

Family **Cymaclymeniidae** Hyatt, 1884 (nom. corr. Ruzhencev 1957)
Subfamily **Genuclymeniinae** Korn & Klug, 2002

Genuclymenia aff. *angelini* (Wedekind, 1908) (Figs 7.7a, b, 22d)

Description. The single specimen (B6-C-47.73) is fragmentary, subevolute (uw/dm = *c.* 0.39 at *c.*12.5 mm dm), rather smooth and shows a compressed (ww/wh = *c.* 0.7), oval (tegoid) cross-section. The septal face at the rear end of the body chamber includes a relatively shallow, rounded and dorsolaterally restricted A-lobe, as is typical for the genus (Fig. 22d). The growth ornament is characterized by closely spaced lirae (*c.* 6 mm) with a very high ventrolateral salient.

Discussion. Within the genus there are species groups. The type species, *Genu. frechi* Wedekind, 1908, including its synonym *Genu. frechi* var. *sulcata* Nalivkina, 1953, and the related *Genu. guembeli* Wedekind, 1908, and *Genu. discoidalis*

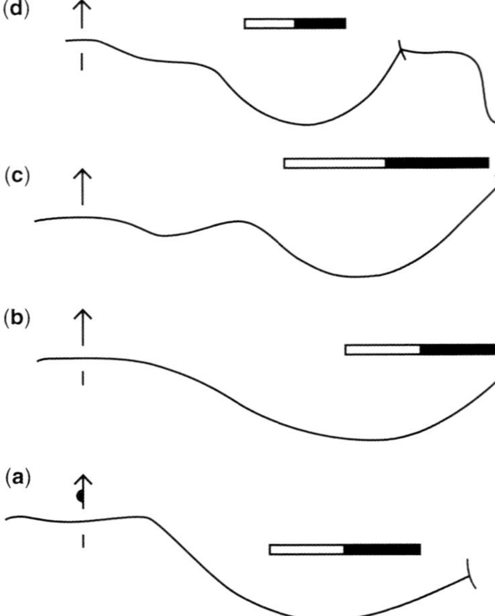

Fig. 22. Sutures of ammonoids from below and within the *Annulata* Event Interval. (**a**) *St. rectangula* n. sp., paratype B6.C-47.229, Mrakib, Lower *Annulata* Marl, at 4.5 mm wh. (**b**) *Pl.* (*Pl.*) *levata* n. sp., paratype B6.C-47.115, at 4.9 mm wh. (**c**) *Sulco. sulcata* (Schindewolf, 1923), B6.C-47.40 at 4.3 mm wh. (**d**) *Genu.* aff. *angelini* (Wedekind, 1908), B6.C-47.73, El Gara, Trench I, Bed -3b, at 6.5 mm wh.

Wedekind, 1908, are characterized by internal L-lobes, a high ventrolateral salient of the biconvex growth ornament and by furrows on the outer flank or ventrolateral edges. The latter feature is very weakly developed or lacking in the smooth *Genu. angelini* Group, which also seems to lack internal L-lobes (see Perna 1914). In the *Genu. angelini* lectotype (erroneously called holotype in House 1970, pl. 126, figs 14 & 15), whorls are thicker than in our specimen, with ww/wh = *c.* 0.95 at almost 39 mm dm. *Genuclymenia borni* Schindewolf, 1923 is characterized by a more compressed cross-section than in *Genu. angelini* but, unlike as in the El Gara form, its uw/dm ratio is only approximately 0.3. *Genuclymenia karpinskii* (Perna, 1914), *Genu. aktubensis* Nikolaeva & Bogoslovskiy, 2005 and, less typical, *Genu. polonica* Czarnocki, 1989 differ from the other species in their low ventrolateral salient of the ornament. This more-or-less smooth *aktubensis* Group is also characterized by rather deep internal L-lobes and the shallow outer-flank lobes developed by the subdivision of a wide, very shallow E-lobe.

In terms of umbilical width and cross-section, the El Gara specimen is similar to *Genu. polonica*, but it has a shallower A-lobe and no clear internal L-lobe, certainly none at the umbilical seam, as claimed for *polonica* (Czarnocki 1989, pl. 47, fig. 12c). It seems to represent a new, rather ancestral species of the *angelini* Group, which is wider umbilicate than *Genu. borni*, in which the dorsal suture is still unknown. With respect to its poor preservation, it is preliminarily identified as *Genu.* aff. *angelini*.

Based on better-preserved material, *Genuclymenia* should be divided into subgenera. *Genu. keepitensis* Jenkins, 1968 possesses a broadly convex dorsolateral saddle and umbilical L-lobe, as in the Cymaclymeniinae; therefore, it is here excluded from the genus.

Stratigraphic range and geographical distribution. Main *sulcata* Zone (lower UD III-C_2) of the northern Tafilalt.

Family **Piriclymeniidae** Korn, 1992

Sulcoclymenia sulcata (Schindewolf, 1923)
(Figs 7.8a, b, 22c)

*	1923	*Cyrtoclymenia sulcata* Schindewolf: 427–428, pl. 17, fig. 11
	1957	*Cyrtoclymenia sulcata* Freyer: 39, 61
	1960	*Platyclymenia sulcata* Petter: 31–32, pl. 2, fig. 14, pl. 3, figs 3, 3a, 5, 5b, , 6a, 9, 9a
v	2000	*Sulcoclymenia sulcata* Becker *et al.*: 79
v	2002	*Sulcoclymenia sulcata* Becker *et al.*: 167, 170, pl. 4, figs 2 & 3
	2004	*Sulcoclymenia sulcata* Korn: fig. 4

Description. The material from Bed EG SI/-3b allows some insights into the ontogeny and variability of the species. The first four whorls are depressed (B6.CV-47.45), with a decrease of ww/wh values to approximately 1.05 at 6–7 mm dm (B6-C-47.33). Median to mature whorls are compressed, with ww/wh of approximately 0.75 in the largest fragment (B6.C-47.36), which reached approximately 30 mm dm. The umbilical width decreases from around 0.5% dm of small specimens to 0.43 (B6.C-47.42) or 0.46 (B6.C-47.44) at 13–15 mm dm; there is clearly some variability. The small lectotype, here designated, illustrated by Schindewolf (1923, pl. 17, fig. 11), is a more evolute specimen. The spiral furrows are very strong (B6.C-47.43 and B6.C-47.44) or rather shallow, which partly depends on the quality of the preservation. In some specimens (B6.C-47.40 and B6.C-47.41) they are variably marked on the two sides. The strongly concavo-convex growth lirae are narrowly spaced and sometimes bundled. They form a deep and parallel-sided ventral sinus that occupies only the median part of the keel (B6.C-47.33). Episodic stronger lirae create an incipient ventral band. The largest fragment, B6.C-47.36, shows approximately seven growth lirae per millimetre and bundles of five lirae. Occasionally, there are traces of a fine, branching wrinkle layer, with approximately eight wrinkles per millimetre. Several specimens display well the typical suture, with a shallow, rounded A-lobe on the inner flank, a low and short mid-flank saddle, and a very shallow E-lobe on the outer flank (Fig. 22c). The broad ventral saddle may be higher as the flank saddle when sutures are slightly prorsiradiate.

Material from the Mrakib (Becker *et al.* 2000) does not show any difference. In some specimens (e.g. B6.C-47.291), the body chambers of the goethite moulds show delicate *ritzstreifung*, which consists of series of small pits that follow strictly the course of the growth lines. The dorsal suture consists of a deep, pointed, funnel-shaped I-lobe with a subangular, high saddle right at the umbilical seam.

Discussion. The suture formula is I:A(E–E). A wide, very shallow ventral or E-lobe is divided from approximately 6 mm dm on by a median saddle. The different suture formula, also for many other clymeniids, in Korn & Klug (2002) is changed since the main flank lobe is the probable A-lobe inherited from the tornoceratid ancestors of all clymeniids (Becker 2000).

Small size, keeled shell form and lingulate ventral sinus suggest relationships of *Sulcoclymenia* with *Genuclymenia*. This is supported by the stratigraphic succession of both genera. Since *Sulcoclymenia* clearly lacks an internal L-lobe, it should have been derived from early genuclymeniids without the latter. The septal faces of *Sulco. sulcata* and *Genu.* aff. *angelini* are very similar. Outer sutures develop in *Sulco. sulcata* in a similar way, with the subdivision of a wide juvenile E-lobe, as in *Genu. frechi* and *Genu. karpinskii* (Nikolaeva & Bogoslovskiy 2005, figs 65 & 66). *Sulcoclymenia* differs from genuclymeniids mostly in its somewhat wider umbilicus, the prominent keel and shallower inner-flank lobe (Fig. 22c). Apart from the umbilical width, there are very few similarities with the contemporaneous early Platyclymeniidae, such as the tabulate *Stenoclymenia* or larger-sized, depressed *Platyclymenia* (*Varioclymenia*). Therefore, the Piriclymeniidae are better placed in the Cyrtoclymeniaceae, not in the Clymeniaceae (=Platyclymeniaceae). As a consequence, the homoemorphy between *Ornatoclymenia*, a descendant of *Sulcoclymenia*, and *Cymaclymenia* (see Korn 1981) developed within one superfamily.

Early sutures are similar in the more strongly keeled *Glatziella letmathensis* Becker, 1997. It is now likely that this small-sized and evolute species is a younger (UD IV-B2) relative of *Sulco. sulcata*, not a member of the much younger Glatziellidae (UD VI), as originally proposed, nor a member of the Biloclymeniidae, as considered in Korn & Klug (2002). Another similar, evolute, compressed and strongly bisulcate new form occurs rarely in the South Urals (B6.C-47.292). It combines concavo-convex growth lirae (identical to *Sulcoclymenia*) with a delicate ventral band and deep, asymmetrical ('genuclymeniid') A-lobe. Both forms point to a still very incompletely known larger taxonomic complexity around *Genuclymenia–Sulcoclymenia*.

Stratigraphic range. Restricted to the approximately lower half of the *sulcata* Zone (UD III-C2).

Geographical distribution. Rhenish Massif, Franconia, Saxony (Vogtland), eastern Anti-Atlas (Tafilalt, Maider).

Family **Platyclymeniidae** Wedekind, 1914

Stenoclymenia rectangula n. sp.
(Figs 16.8a, b, 22a)

e.p.	1908	*Clymenia Sandbergeri* Wedekind: 620–621, pl. 44, fig. 9 [only]
	1910	*Clymenia Wysogorskii* Rzehak: 192–194, pl. 3, figs 4 & 5
	1914	*Platyclymenia* cf. *Wysogorskii* Wedekind: 39–40, pl. 2, fig. 7 [= Wedekind, 1908, pl. 44, fig. 9]
*e.p.	1923	*Platyclymenia prorsostriata* n. sp. Schindewolf: 453–454 [Wedekind specimen only]
?	1953	*Platyclymenia prorsostriata* Nalivkina: 122–123, pl. 5, fig. 16
?e.p.	1957	*Platyclymenia prorsostriata* Freyer: 57
	1970	*Stenoclymenia sandbergeri* House: pl. 126, figs 12 & 13 [= Wedekind 1908, pl. 44, fig. 9]
e.p.	2005	*Pricella glabra* Nikolaeva & Bogoslovskiy: 105–106, pl. 10, fig. 5 [only]
	2006	*Stenoclymenia sandbergeri* Dzik: 293, figs 215A–G & 223

Derivation of name. Due to the thin rectangular cross-section.

Type. Holotype is the specimen illustrated by Wedekind (1908, pl. 44, fig. 9), re-illustrated in Wedekind (1914), and the false lectotype of *St. sandbergeri* in House (1970).

Diagnosis. Strongly compressed (ww/wh decreasing towards 0.56–0.62 at maturity), with gently rounded flanks, ventrolateral edges and tabulate venter, widely evolute (uw/dm > 0.50); ornament with prorsiradiate, concavo-convex growth lines and without or with weak concave, undulose ribs on the outer flanks. Sutures with shallow, somewhat asymmetrical A-lobe on the flanks, subangular ventral saddle and very shallow, rounded E-lobe.

Description. The holotype has been described and illustrated by Wedekind (1908, 1914). All eight specimens from the Mrakib are incomplete and similar, especially in their thinly subtrapezoidal shell form, with maximum whorl thickness on the lower flanks. The figured juvenile (B6.C-47.217) is smooth, and has an uw/dm ratio of 0.52 and a ww/wh ratio of 0.83 at 1.8 mm wh. The ww/wh ratio declines to 0.73 at 3.3 mm wh (B6.C-47.231) and 0.65 at 5–6 mm wh (B6.C-47.230 and B6.C-47.232). It is 0.62 at 37 mm dm (10.8 mm wh) in the holotype. Weak ventrolateral ribbing is seen in B6.C-47.229, B6.C-47.231 and B6.C-47.255, but more distinctive in B6.C-47.232 and B6.C-47.233. B6.C-47.223 displays a healed shell fracture. The shallow E-lobe is visible in the smallest (B6.C-47.217) and largest fragment (B6.C-47.229: Fig. 22a). The latter is widely umbilicate (uw/dm = 0.57), which suggests mature uncoiling in the local population. The holotype has an uw/dm ratio of 0.53. Pitted flank surfaces of the Mrakib *steinkerns* can be attributed to *ritzstreifung* (impressions of ornament of the inner shell layer). In B6.C-47.232, very delicate pits are arranged in deeply rursiradiate, concave lines.

Discussion. Wedekind (1908) based his new *Clymenia Sandbergeri* from the Enkeberg (Rhenish Massif, *delphinus* Zone, do IIIβ = UD III-C) on variably smooth or markedly ribbed, strongly compressed and evolute (uw/dm > 0.50) clymeniids with distinctive ventrolateral edges, tabulate venter and prorsiradiate, concavo-convex ornament. Rzehak (1910) assigned related, smooth, tabulate specimens from the younger *annulata* Zone of Moravia to *Clymenia Wysogorskii* Frech, 1902. This inspired Wedekind (1914) to restrict *Pl. Sandbergeri* to his ribbed form and he illustrated a second specimen with sharp, concave folds, which are denser on the last half whorl. Following Rzehak's taxonomy, the smooth form was reassigned, but with reservations, to *Pl.* cf. *Wysogorskii* Frech, 1902. The latter species, however, is the much younger (uppermost Famennian; UD VI) and unrelated type species of *Trochoclymenia* (Schindewolf, 1923, 1926; Becker, 2000; Korn & Klug, 2002). Therefore, Schindewolf (1923) placed Wedekind's smooth form in his new *Pl. prorsostriata*. This taxon, however, centred on a more involute form from Gattendorf, Franconia (uw/dm = 0.45; lectotype, designated here = original of Schindewolf 1923, pl. 17, fig. 9, former Marburg collection No. 3127, recently transferred to the Senckenberg museum at Frankfurt a. M.). Subsequently, Lange (1929) assigned *Clymenia sandbergeri* to his new genus *Stenoclymenia* as it possesses a cross-section-related shallow external lobe (see also Fig. 22a). He kept Schindewolf's re-identification of Wedekind's smooth form as *Pl. prorsistriata*, although his comparison of ww/wh ratios further suggests that both are very different (0.62 in the Enkeberg form, only 0.32 in the *St. prorsostriata* lectotype). Matern (1931) called the ribbed specimen of Wedekind (1908, pl. 44, fig. 10) a 'holotype'. Although he should have used the term lectotype, this reference represents a valid type designation. This was overlooked by House (1970), who re-illustrated Wedekind's smooth specimen and designated it as a lectotype, although this clearly contradicted the intention of the original author (Wedekind 1914). Therefore, it is fortunate that House's taxonomic act was an invalid secondary type fixation.

The taxonomic separation of the ribbed true *Pl. sandbergeri*, typically with dense ribbing at maturity, and of smooth, supposed *Pl. prorsostriata* was continued by Nalivkina (1953), who described material from the *delphinus* Zone (do IIIβ = UD III-C) of the Urals. Owing to the lack of any illustration, it is not clear whether specimens from the UD III and IV of Saxony (Freyer 1957) agree all or partly either with the smooth Enkeberg or the Gattendorf form. Czarnocki (1989) illustrated from the Holy Cross Mountains a ribbed *sandbergeri* specimen, which is close to the lectotype (as selected by Matern). However, Dzik (2006) emphasized that strongly ribbed specimens are rare in Poland. He illustrated supposed *sandbergeri* variants with a smooth shell (Dzik 2006, fig. 215E–G), weak and undulose (not sharp) ribs (Dzik 2006, fig. 215D), and low, rounded ribs restricted to the outer flank (Dzik 2006, fig. 215C). He suggested a possible high intraspecific variability, but none of his specimens approaches the true/typical *St. sandbergeri* and there is no documented intergradation.

In summary, *St. sandbergeri* should be restricted to evolute specimens (uw/dm increasing from 0.47 to >0.50 at maturity) with sharp and marked ribbing that becomes dense on the last whorl. Wedekind's smooth form from Enkeberg, specimens from Moravia (Rzehak 1910), the Maider and, with some reservation, the Holy Cross Mountains (Dzik 2006) are here assigned to *St. rectangula* n. sp. It differs strictly from the extremely compressed and somewhat more involute (uw/dm = 0.45) *St. prorsostriata*. *Stenoclymenia stenomphala* Lange, 1929 is even more involute (uw/dm = 0.42), *St. valida* (Phillips, 1841) additionally characterized by dense, partly bifurcating ribs. There are some similarities with the younger, upper Famennian (UD V) *Nodosoclymenia colonia* Korn, 2004, which here is transferred to *Stenoclymenia*. It differs from *St. rectangula* n. sp. in its wider whorls (ww/wh = 0.94 at 18.4 mm dm) and slightly more nodose ventrolateral ornament, whilst its shallow E-lobe and different ornament exclude it from *Nodosoclymenia*. The poorly known *St. elliptica* Czarnocki, 1989 is possibly identical with the strongly compressed, intermediate *Pl.* (*Platyclymenia*) *pattisoni*; it is not a

Nodosoclymenia, as claimed by Dzik (2006). ?*Stenoclymenia schindewolfi* Lange, 1929 is too involute for the genus and the family. The type specimen displays a very different ornament; the species probably represents a new cyrtoclymeniid genus. The presence of parabolic ribs distinguishes *Pl. spinosa* var. *dorsocava* Schindewolf, 1923 from typical *Stenoclymenia*: as in the pair *Platyclymenia–Trigonoclymenia*, it could be separated at a subgeneric level. Nikolaeva & Bogoslovskiy (2005) described as *Pricella glabra* (Perna, 1914) at least three different species, two of which fall in *Stenoclymenia*. Specimens illustrated on their plate 10, figures 1, 3 and 4 represent a new, strongly bicarinate rather than tabulate species with undulose growth lines. The specimen of their plate 10, figure 5 is identical to the Polish specimens that are very close to *St. rectangula* n. sp., apart from a partly slightly narrower umbilicus (uw/dm = 0.47 to 0.52 at maturity).

Stratigraphic range. The holotype of the new species is from the *delphinus* Zone (UD III-C) and this is also the level of a possible Urals record (Nalivkina 1953). Moravian and our Moroccon representatives are from the *Annulata* Event Interval (lower *annulata* Zone, UD IV-A). This agrees with the range given by Freyer (1957) for his possibly conspecific '*Pl. prorsostriata*' specimens.

Geographical distribution: Rhenish Massif, Moravia, ?Saxony, Holy Cross Mountains, Urals, Maider.

Platyclymenia (*Platyclymenia*) *annulata rotundata* Wedekind, 1914
(Figs 8.2a, b, 14.6a, b)

*	1914	*Platyclymenia rotundata* Wedekind: 34, pl. 2, fig. 15
e.p.	1989	*Platyclymenia annulata annulata* Price & Korn: 279, 280–283, fig. 8a, b (only)
e.p.	2002	*Platyclymenia annulata* Korn: 589–594, figs 29A, 30, 31 & 32A–C, E, H (only)

Type. Holotype, by monotypy, the figured specimen of Wedekind (1914), here re-illustrated as Figure 14.6a, b.

Discussion. Price & Korn (1989) discussed the complex taxonomic history of *Pl. annulata* and settled its definition by the designation of a neotype. This specimen from Franconia (Mbg 3125) is remarkably evolute, with an uw/dm ratio of 0.57 at 28 mm dm. Its first whorls are smooth. Then coarse, sharp ribs appear, which become subdued and denser on the last preserved whorl. Korn (2002) illustrated a large variability of ribbing patterns in an *annulata* population from Kattensiepen (Rhenish Massif), but the uw/dm plot (Korn 2002, fig. 34) suggests that the rare very widely umbilicate specimens follow an ontogenetic path that is separate from the vast majority of measured specimens. Therefore, *Pl.* (*Pl.*) *annulata sensu stricto* is restricted here to the evolute form (with uw/dm between 0.55 and 0.60 at median stages to maturity). For the more widespread form with more or less constant uw/dm ratios between 0.42 and 0.50, the subspecies name *Pl.* (*Pl.*) *annulata rotundata* is available. The illustrated specimen from El Gara is somewhat more involute (uw/dm = 0.44) than the holotype (uw/dm = 0.49: see Fig. 14.6a).

Platyclymenia (*Pl.*) *annulata richteri*, which co-occurs at El Gara, differs in coarser ribbing, which is getting weaker, not denser, with growth. In addition, there are slight discordances of ribs and growth lines (Price & Korn 1989; Korn 2002). Its uw/dm ratios are identical with *Pl.* (*Pl.*) *annulata rotundata*. *Platyclymenia mirabilis* Wedekind, 1914, *Pl. annulata* var. *correcta* Perna, 1914 and *Pl. geminicostata* Petter, 1960 are probably junior synonyms of *annulata rotundata*. Specimens with intercalated clearly parabolic ribs are here included in *Pl.* (*Tri.*) *protacta* (Fig. 14.2a, b), but there are common intermediates between this species and *Pl.* (*Pl.*) *annulata* (cf. Korn 2002, fig. 32I–K). *Platyclymenia* (*Pl.*) *latecostata* differs from *Pl.* (*Pl.*) *annulata* in its stronger shell compression (Fig. 16.6a, b) and widely spaced ribbing of inner whorls.

Stratigraphic range. Regional *annulata* Zone (UD IV-A/B1).

Platyclymenia (*Platyclymenia*) *levata* n. sp.
(Figs 8.7, 8, 22b)

Derivation of name. From the Latin verb *levare* = to smoothen; due to the lack of ribbing of median to mature whorls.

Types. Holotype B6.C-47.152, illustrated in Figure 8.8a, b, paratypes B6.C-47.98, B6.C-47.101, B6.C-47.105, B6.C-47.106, B6.C-47.114–B6.C-47.117, B6.C-47.119–B6.C-47.121 (Fig. 8.7a, b), B6.C-47.122–B6.C-47.133 and B6.C-47.176.

Type locality and level. El Gara, Trench II, Bed 26a (upper regional *annulata* Zone).

Diagnosis. Early whorls up to 7 mm dm depressed, subsequently compressed, with ww/wh = 0.80–0.85 until approximately 30 mm dm and 0.75 at approximately 40 mm dm; moderately evolute (uw/dm = 0.40–0.48), with flattened venter bordered by subangular shoulders; ornament with fine, strongly concavo-convex, partly undulose growth lines, and variably with or without fine, sharp ribs on the inner whorls. Sutures with simple, rounded flank lobe and moderately high, flattened ventral saddle.

Table 2. *Shell parameters (in mm) in* Pl. (Pl.) levata *n. sp.*

No.	dm	wh	uw	ww	ww/dm	ww/wh	uw/dm
B6.C.47.122	7.5	2.3	3.5	2.6	0.35	1.13	0.47
B6.C-47.123	8.5	3	3.8	2.9	0.34	0.97	0.45
B6.C-47.121	11	4	4.4	3.1	0.28	0.78	0.40
B6.C-47.152	13.1	4.7	5.3	3.7	0.28	0.79	0.41 (holotype)
B6.C-47.116	15.2	5.0	6.6	4.0	0.26	0.80	0.43
B6.C-47.115	19.0	6	7.9	5.1	0.27	0.85	0.42
B6.C-47.105	*c.* 23	8.6	10.2	7	0.30	0.81	0.44
B6.C-47.114	*c.* 27.5	8.6	13	7.2	*c.* 0.27	*c.* 0.84	*c.* 0.45

Description. As in other platyclymeniids, the umbilical width is somewhat variable, with uw/dm ratios between 0.40 and 0.48 (mostly between 0.41 and 0.45). The holotype is among the less evolute variants. The cross-section is subrectangular, often with rounded ventrolateral edges, and with a flattened but gently rounded, not tabulate venter. There is also a considerable intraspecific variability concerning the absence or presence of irregularly developed juvenile ribs. The holotype has distinctive, fine, rursiradiate flank ribs up to 4.5 mm dm that become subdued on subsequent whorls. The last preserved whorl is characterized by bundled, rectiradiate, concavo-convex growth lines with higher ventrolateral than dorsolateral salient. There is no ventral band. Fine juvenile ribs may be paired (paratype B6.C-47.122, up to 4 mm dm), relatively sharp (paratype B6.C-47.123) or weak and undulose (paratype B6.C-47.116). They are lacking in paratypes B6.C-47.114, B6.C-47.115 and B6.C-47.121, and in specimens from Bine Jebilet.

Discussion. The new species belongs to the group of mostly smooth platyclymeniids around *Pl.* (*Pl.*) *subnautilina*. A similar variability of early ornament is known from Rhenish representatives (Korn 2002). The new species has a more flattened venter than *Pl.* (*Pl.*) *subnautilina* and at corresponding ontogenetic stages, when compared with the lectotype of the latter (see the discussion in Zong *et al.* 2014), it is consistently more compressed. This also applies to the junior *subnautilina* synonyms *Pl. pseudoflexuosa* Rzehak, 1910 and *Pl. ruedemanni* Wedekind, 1914. *Platyclymenia* (*Pl.*) *levata* n. sp. (Table 2) is much more compressed than the *subnautilina* population from Kattensiepen described by Korn (2002), which keeps a depressed whorl form at least up to 30 mm dm. For this form, the subspecies name *Pl.* (*Pl.*) *subnautilina beuelensis* Lange, 1929 is available, but intermediates (Korn 2002, fig. 38) preclude a separation at species level. *Platyclymenia* (*Pl.*) *glabra* Czarnocki, 1989 is a younger subjective synonym of *subnautilina beuelensis*. Even more compressed than *levata* n. sp. are *Pl.* (*Pl.*) *pattisoni* and *Pl.* (*Pl.*) *limata*. The latter species is also more involute, with uw/dm near 0.35, approaching values as in *Protactoclymenia*, and as in Anti-Atlas specimens illustrated by Korn (1999, pl. 5, fig. 7) as *Platyclymenia* sp. An examination of the Berlin type material proved that *Pl.* (*Pl.*) *quiringi* is much more evolute, with uw/dm ratios following a different, declining ontogenetic trajectory than in *subnautilina*, decreasing from 0.6 at approximately 12 mm dm (paratype MB.C.1263.2) to 0.51 at 22 mm dm (holotype MB.C.1262). A comparison of the ontogenetic changes of ww/wh values in various smooth platyclymeniids is given in Table 3. The morphometric study of Rhenish platyclymeniids by Korn (2002) shows a restricted variability of ww/wh ratios. All other named *Platyclymenia* species display distinctive ribbing, at least of median stages. As *Pl.* (*Pl.*) *pattisoni*, the new species is somewhat intermediate towards *Stenoclymenia*, but there is no shallow E-Lobe (Fig. 22b).

Table 3. *Ontogenetic changes of ww/wh ratios in smooth platyclymeniids (dm in mm)*

dm	*pattisoni*	*limata*	*levata* n. sp.	*s. subnautilina*	*s. beuelensis*
8		*c.* 1.0	1.00–1.10		
10	0.70–0.75	0.70–0.75	0.78–0.85	1.15	1.20–1.40
20	0.70	0.65–0.70	0.78–0.85	1.00–1.05	1.10–1.15
30		0.67	0.80–0.85	0.90	1.00–1.05
40	*c.* 0.55	*c.* 0.65	*c.* 0.75	*c.* 0.85	0.90–1.00

A single, more strongly ribbed specimen from Bed 26a (B6.C-47.181) shares the typical shell form of *Pl.* (*Pl.*) *levata* n. sp. and is assigned to it with reservation (aff.). It has some similarity with the inner whorls of the poorly known *Pl. recticosta* Schindewolf, 1923.

Stratigraphic range: Regional *annulata* Zone (UD IV-A/B1).

Geographical distribution: So far, only known from the eastern Anti-Atlas (El Gara and Bine Jebilet, Tafilalt, and Mrakib, Maider).

Family **Hexaclymeniidae** Lange, 1929

?*Pleuroclymenia* sp. juv.

Description: B6.C-47.66 is a juvenile (5.6 mm dm), evolute (uw/dm = 0.53) and depressed (ww/wh = *c*. 1.2), smooth clymeniid with rounded venter and rounded umbilical wall. Very peculiar are its rursiradiate, slightly biconvex growth lirae with marked ventral sinus. The flanks are occupied by a shallow, gently rounded A-lobe.

Discussion: The only known clymeniid genus of UD III-C with partly rursiradiate and slightly biconvex ornament is *Pleuroclymenia* (see the juvenile in Korn & Klug 2002, fig. 208F). However, typical species of the genus have ribbed inner whorls. It is possible that the present juvenile belongs to a new species, but its small size precludes any naming. There are possible relationships with the younger (UD IV-A) *Clymenia recticosta* Rzehak, 1910 from Moravia, the sublinear to slightly biconvex ornament of which suggests relationships with the Clymeniaceae or an iterative trend within the Platyclymeniidae.

Conclusions

- The basal upper Famennian *Annulata* Events were truly global in nature, with records from more than 40 basins/regions in the palaeo(sub)-tropical marine realms of North America, central and southern Europe, North Africa, the Urals, the Cimmerian block (Iran–Pakistan), Kazakhstan, North and South China, and Western and Eastern Australia.
- These occurrences are from a variety of sedimentary–tectonic settings: volcanic active orogenic belts (e.g. California, Bulgaria, New South Wales), siliciclastic foreland basins (eastern North America), argillaceous outer-shelf basins (Dra Valley and Maider, southern Morocco), condensed pelagic seamounts and carbonate platforms (Rhenish Massif, Montagne Noire, Tafilalt, Moroccan Meseta, Iran), shallow-water carbonate platforms (Alberta), inter-reefal basins (Canning Basin), and mixed siliciclastic–carbonatic shallow shelf ramps (Montana, Ardennes).
- This distribution is a strong argument for a global, probably climato-eustatic trigger that acted contemporaneously in many disjunct basins, independent from regional/local facies conditions, which, however, overprinted the global signal and led to very different regional sedimentary and faunal event patterns.
- In the Rheris Basin of the northern Tafilalt, where *Annulata* Event beds were studied for the first time, two moderately fossiliferous and (originally) somewhat pyritic marl intervals with a 'low-diversity, juvenile *Platyclymenia–Prionoceras* biofacies' above unfossiliferous (strongly oligotrophic) pre-event beds indicate locally only a moderate improvement of nutrient supplies, resulting in ammonoid blooms, but not in black shale deposition, and a moderate decrease in ventilation sufficient to prevent benthic life. Its combined effect caused an episodic strong reduction of conodont diversity (Lazarus phases for many taxa), but with only a few extinctions.
- Below, the upper part of the middle Famennian (UD III-C) yielded a unique small-sized ammonoid fauna with rare representatives of goniatite/clymeniid groups that were previously unknown from North Africa. This assemblage correlates with the first Anti-Atlas specimens of the important marker conodont *Ps. granulosus*.
- The comparison of El Gara with previously described *Annulata* Event sections from the adjacent Tafilalt Platform (Bine Jebilet, Jebel Erfoud, Ouidane Chebbi, Amessoui Syncline), Tafilalt Basin (Hassi Nebech) and Maider Basin (Mrakib) revealed strong differences of event lithofacies and faunas. These reflect local differences in trophic levels, bathymetry and seafloor to open-water column ventilation. Since the strongest trophic signal (*Guerichia*-juvenile ammonoid blooms in black shale) comes from the deepest offshore setting (Mrakib), it is likely that intrabasinal nutrient recycling, not an influx of land-derived nutrients, caused eutrophication.
- The *Annulata* Event macrofaunas of the eastern Anti-Atlas enable the characterization of specific biofacies (assemblage) types at each of the studied localities. Most distinctive and unique on a global scale is the 'diverse, juvenile *Gundolficeras–Erfoudites–Protactoclymenia–Stenoclymenia–Guerichia* biofacies' of the Mrakib, in which the otherwise dominant platyclymeniids and prionoceratids play a very subordinate role. The absence of the most widespread event assemblages/biofacies of Europe ('low-diversity

Platyclymenia–Guerichia and *Platyclymenia–Protactoclymenia–Guerichia* biofacies') in southern Morocco may express differences in palaeogeography or palaeolatitudes, or a mixture of both.

- The analysis of conodont biofacies across the events and between individual sections shows the dominance of typical outer-shelf palmatolepid–bispathodid biofacies, notably of the *Pa. gracilis* Group. In pre-event beds, *Scaphignathus* is not a shallow-water faunal element, as claimed in some previous conodont biofacies models, but a typical dweller of pelagic habitats. The ammonoid-rich, 'moderately diverse, normal-sized *Platyclymenia–Prionoceras–Erfoudites* biofacies' of the Wagnerbank Equivalents of the Mrakib are associated with a new *Caenodontus* conodont biofacies.
- The ammonoid faunas from around the *Annulata* Events include specimens that belong to new taxa (*Post. ascendens* n. sp., *Gund. australe* n. sp., *St. rectangula* n. sp. and *Pl.* (*Pl.*) *levata* n. sp.). There are also new regional first records and specimens, which help to solve taxonomic questions in various genera (*Maeneceras*, *Sporadoceras*, *Ungusporadoceras*, *Sulcoclymenia* and *Platyclymenia*).

We are indebted to A. El Hassani (Rabat), our principal cooperation partner in Morocco. D. Weyer (Berlin) and the late V. Ebbighausen accompanied us in the field. E. Kuropka (Münster) assisted in the conodont sample processing, G. Schreiber (Münster) produced ammonoid cross-sections, S. Helling (Münster) took some of the ammonoid photographs and T. Fährenkämper (Münster) helped with many of the illustrations. We would like to thank J. D. Over and W. T. Kirchgasser for their very helpful reviews. Our research was partly funded in the frame of the joint DFG–CNRST Maroc cooperation project Be 1367/11-1.

References

Aboussalam, Z.S. 2003. Das 'Taghanic-Event' im höheren Mittel-Devon von West-Europa und Marokko. *Münstersche Forschungen zur Geologie und Paläontologie*, **97**, 1–332.

Ahlburg, J. 1918. *Erläuterungen zur Geologischen Karte von Preußen und benachbarten deutschen Ländern*. Lieferung 208, Blatt Weilburg, 1–152.

Alberti, H. 1970. Neue Trilobiten-Faunen aus dem Ober-Devon Marokkos. *H. Martin-Festschrift, Göttinger Arbeiten zur Geologie und Paläontologie*, **5**, 15–29.

Anderson, T.B., Woodard, G.D., Strathouse, S.M. & Twichell, M.K. 1974. Geology of the late Devonian fossil locality in the Sierra Buttes Formation, Dugan Pond, Sierra City Quadrangle, California. *Geological Society of America, Abstracts with Programs*, **6**, 139.

Ansted, D.T. 1838. On a new genus of fossil multilocular shells found in the Slayte-Rocks of Cornwall. *Transactions of the Cambridge Philosophical Society*, **6**, 415–422.

Arthaber, G., von 1911. Die Trias von Albanien. *Beiträge zur Paläontologie und Geologie Österreich-Ungarns und des Orients*, **24**, 169–277.

Baird, G.C., Carr, R.K., Hannibal, J.T., Brett, C.E. & Brett, B.L. 2009*a*. Uppermost Devonian (Famennian) stratigraphy of the Cleveland area, northeastern Ohio. *In*: Brett, C.E., Bartholomew, A.J. & DeSantis, M.J. (eds) *Middle and Upper Devonian Sequences, Sea-Level, Climatic and Biotic Events in East-Central Laurentia: Kentucky, Ohio, and Michigan*. North American Paleontological Convention, Field Trip, **10**, 154–168.

Baird, G.C., Over, D.J., Sullivan, J.S., McKenzie, S.C., Schwab, J.C. & Dvorak, K.A. 2009*b*. Conodonts and the end-Devonian event stratigraphic chronology in the classic Pennsylvania 'Oil Lands' region: Latest Famennian Riceville Formation – Barea Sandstone succession. *In*: Henderson, C.M. & MacLean, C. (eds) *ICOS 2009*, Abstracts. Permophiles, **53** (Suppl. 1), 3.

Bartzsch, K., Blumenstengel, H. & Weyer, D. 1995. Ein neues Devon/Karbon-Grenzprofil am Bergaer Antiklinorium (Thüringer Schiefergebirge) – eine vorläufige Mitteilung. *Geowissenschaftliche Mitteilungen Thüringen*, **3**, 13–29.

Bartzsch, K., Blumenstengel, H. & Weyer, D. 1999. Stratigraphie des Oberdevons im Thüringischen Schiefergebirge. Teil 1: Schwarzburg-Antiklinorium. *Beiträge zur Geologie von Thüringen, Neue Folge*, **6**, 159–189.

Bartzsch, K., Gaitzsch, B., Abel, P., Hahne, K. & Weyer, D. (eds) 2015. *Exkursionsführer: Devon und Unterkarbon im Raum Saalfeld 23–25.04.25.2015*. Subkommissionen Devon und Karbon der Deutschen Stratigraphischen Kommission, Saalfeld.

Becker, R.T. 1985. Devonische Ammonoideen aus dem Raum Hohenlimburg-Letmathe (Geologisches Blatt 4611 Hohenlimburg). *Dortmunder Beiträge zur Landeskunde, naturwissenschaftliche Mitteilungen*, **19**, 19–34.

Becker, R.T. 1992*a*. Regional developments of the global hypoxic *Annulata*-Event (Middle Famennian). *In*: Walliser, O.H. (ed.) *Phanerozoic Global Bio-Events and Event-Stratigraphy*. Fifth International Conference on Global Bioevents, 16–19 February, Göttingen, Abstract Volume, 13–14.

Becker, R.T. 1992*b*. Zur Kenntnis von Hemberg-Stufe und *Annulata*-Schiefer im Nordsauerland (Oberdevon, Rheinisches Schiefergebirge, GK 4611 Hohenlimburg). *Berliner geowissenschaftliche Abhandlungen*, **E3**, 3–41.

Becker, R.T. 1993*a*. Stratigraphische Gliederung und Ammonoideen-Faunen im Nehdenium (Oberdevon II) von Europa und Nord-Afrika. *Courier Forschungsinstitut Senckenberg*, **155**, 1–405.

Becker, R.T. 1993*b*. Anoxia, eustatic changes, and Upper Devonian to lowermost Carboniferous global ammonoid diversity. *In*: House, M.R. (ed.) *The Ammonoidea: Environment, Ecology and Evolutionary Change*. Systematic Association, Special Volume, **47**, 115–163.

Becker, R.T. 1995. Taxonomy and evolution of late Famennian Tornocerataceae (Ammonoidea). *Berliner geowissenschaftliche Abhandlungen*, **E15**, 607–643.

Becker, R.T. 1997. Eine neue und älteste *Glatziella* (Clymeniida) aus dem höheren Oberdevon des Nordsauerlandes (Rheinisches Schiefergebirge). *Berliner geowissenschaftliche Abhandlungen*, **E25**, 31–41.

Becker, R.T. 2000. Taxonomy, evolutionary history and distribution of the middle to late Famennian Wocklumeriina (Ammonoidea, Clymeniida). *Mitteilungen aus dem Museum für Naturkunde in Berlin, Geowissenschaftliche Reihe*, **3**, 27–75.

Becker, R.T. 2002. *Alpinites* and other Posttornoceratidae (Goniatitida, Famennian). *Mitteilungen aus dem Museum für Naturkunde in Berlin, Geowissenschaftliche Reihe*, **5**, 51–73.

Becker, R.T. & Hartenfels, S. 2010. Der Faunenwechsel bei Ammonoideen des mittleren Famenniums (Oberdevon III/IV) – Folge einer globalen trophischen Krise in Außenschelfen? *Zitteliana, Reihe B*, **29**, 20–21.

Becker, R.T. & House, M.R. 1997. Sea-level changes in the Upper Devonian of the Canning Basin, western Australia. *Courier Forschungsinstitut Senckenberg*, **199**, 129–146.

Becker, R.T. & House, M.R. 2000*a*. Devonian ammonoid zones and their correlation with established series and stage boundaries. *Courier Forschungsinstitut Senckenberg*, **220**, 113–151.

Becker, R.T. & House, M.R. 2000*b*. The Famennian ammonoid succession at Bou Tchrafine (Anti-Atlas, southern Morocco). *Notes et Mémoires du Service Géologique du Maroc*, **399**, 37–41.

Becker, R.T. & House, M.R. 2000*c*. Devonian ammonoid succession at Jbel Amelane (Western Tafilalt, Southern Morocco). *Notes et Mémoires du Service Géologique du Maroc*, **399**, 49–56.

Becker, R.T. & House, M.R. 2009. Devonian ammonoid biostratigraphy of the Canning Basin. *In*: Playford, P.E., Hocking, R.M. & Cockbain, A.E. (eds) *Devonian Reef Complexes of the Canning Basin, Western Australia*. Geological Survey of Western Australia, Bulletin, **145**, 415–439.

Becker, R.T. & Kullmann, J. 1996. Chapter 17. Paleozoic ammonoids in space and time. *In*: Landman, N.H., Tanabe, K. & Davis, R.A. (eds) *Ammonoid Paleobiology*. Plenum Press, New York, 711–753.

Becker, R.T. & Nikolaeva, S.V. 2014. Comment on a proposal to reinstate as available the species-group names proposed for Devonian ammonoids (Mollusca, Cephalopoda) by Sobolew (1914a, 1914b). *Bulletin of Zoological Nomenclature*, **71**, 1–4.

Becker, R.T. & Schreiber, G. 1994. Zur Trilobiten-Stratigraphie im Letmather Famennium (nördliches Rheinisches Schiefergebirge). *Berliner geowissenschaftliche Abhandlungen, B. Krebs-Festschrift*, **E13**, 369–387.

Becker, R.T., Korn, D., Paproth, E. & Streel, M. 1993. Beds near the Devonian–Carboniferous boundary in the Rhenish Massif, Germany. *In*: *Guidebook, International Union of Geological Sciences*. Commission on Stratigraphy, Subcommission on Carboniferous Stratigraphy, 10–12 June, Lüttich, 1–86.

Becker, R.T., Bockwinkel, J., Ebbighausen, V. & House, M.R. 1999. Jebel Mrakib, Anti-Atlas (Morocco), a potential Upper Famennian substage boundary stratotype section. *In*: El Hassani, A. & Tahiri, A. (eds) *Excursion Guidebook, Part 1: Tafilalt and Maider (eastern Anti-Atlas)*. SDS – IGCP 421 Morocco Meeting, 23 April–1 May 1999. Institute Scientifique, Rabat, 91–107.

Becker, R.T., Bockwinkel, J., Ebbighausen, V. & House, M.R. 2000. Jebel Mrakib, Anti-Atlas (Morocco), a potential Upper Famennian substage boundary stratotype section. *Notes et Mémoires du Service Géologique du Maroc*, **399**, 75–86.

Becker, R.T., House, M.R., Bockwinkel, J., Ebbighausen, V. & Aboussalam, Z.S. 2002. Famennian ammonoid zones of the eastern Anti-Atlas (southern Morocco). *Münstersche Forschungen zur Geologie und Paläontologie*, **93**, 159–205.

Becker, R.T., Ashouri, A.R. & Yazdi, M. 2004. The Upper Devonian *Annulata* Event in the Shotori Range (eastern Iran). *Neues Jahrbiuch für Geologie und Paläontologie, Abhandlungen*, **231**, 119–143.

Becker, R.T., Aboussalam, Z.S., Hartenfels, S., El Hassani, A. & Fischer, T. 2013. The Givetian–Famennian at Oum el Jerane (Amessoui Syncline, southern Tafilalt). *In*: Becker, R.T., El Hassani, A. & Tahiri, A. (eds) *International Field Symposium 'The Devonian and Lower Carboniferous of Northern Gondwana'. Field Guidebook*. Document de l'Institut Scientifique, Rabat, **27**, 61–76.

Behnken, F.H. 1975. Leonardian and Guadalupian (Permian) conodont biostratigraphy in western and southwestern United States. *Journal of Paleontology*, **49**, 284–315.

Belka, Z., Klug, C., Kaufmann, B., Korn, D., Döring, S., Feist, R. & Wendt, J. 1999. Devonian conodont and ammonoid succession of the eastern Tafilalt (Ouidane Chebbi section), Anti-Atlas, Morocco. *Acta Geologica Polonica*, **49**, 1–23.

Beyrich, E. 1884. Erläuterungen zu den Goniatiten L. v. Buch's. *Zeitschrift der Deutschen Geologischen Gesellschaft*, **36**, 203–219.

Bischoff, G.C.O. 1957. Die Conodonten – Stratigraphie des Rhenohercynischen Unterkarbons mit Berücksichtigung der *Wocklumeria*-Stufe und der Devon/Karbon Grenze. *Abhandlungen des Hessischen Landesamtes für Bodenforschung*, **19**, 1–64.

Blumenstengel, H. 1994. Zur Bedeutung von Meeresspiegelschwankungen bei der Bildung der Oberdevonsedimente von Saalfeld, Thüringer Schiefergebirge. *Geowissenschaftliche Mitteilungen von Thüringen*, **2**, 29–44.

Bockwinkel, J., Becker, R.T. & Ebbighausen, V. 2002. Morphometry and Taxonomy of Lower Famennian Sporadoceratidae (Goniatitida) from Southern Morocco. *Abhandlungen der Geologischen Bundesanstalt*, **57**, 279–297.

Bogoslovskiy, B.I. 1962. Redkiy tip skulptury u klimeniy. *Paleontologicheskiy Zhurnal*, **1962** , 166–168.

Bogoslovskiy, B.I. 1969. Devonskie Ammonoidei. I. Agoniatity. *Trudy Paleontologicheskogo Instituta*, **124**, 1–341.

Bogoslovskiy, B.I. 1971. Devonskie Ammonoidei. II. Goniatity. *Trudy Paleontologicheskogo Instituta*, **127**, 1–228.

Bogoslovskiy, B.I. 1981. Devonskie Ammonoidei. III. Klimenii (podotriad Gonioclymeniina). *Trudy Paleontologicheskogo Instituta*, **191**, 1–123.

Boncheva, I., Sachanski, V. & Becker, R.T. 2011. Sedimentary and faunal evidence for the Late Devonian Kellwasser and *Annulata* events in the Balkan Terrane (Bulgaria). *In*: Suttner, T., Kido, E., Piller, W.E. & Königshof, P. (eds) *IGCP 596 Opening Meeting*, 19–24th September 2011, Graz, *Abstract Volume*. Berichte des Institutes für Erdwissenschaften, Karl-Franzenz-Universität Graz, **16**, 26–27.

Boncheva, I., Sachanski, V. & Becker, R.T. 2015. Sedimentary and faunal evidence for the Late Devonian Kellwasser and *Annulata* events in the Balkan Terrane (Bulgaria). *Geologica Balcanica*, **44**, 17–24.

Bond, D. & Zatoń, M. 2003. Gamma-ray spectrometry across the Upper Devonian basin succession at Kowala in the Holy Cross Mountains (Poland). *Acta Geologica Polonica*, **53**, 93–99.

Born, A. 1912. Die geologischen Verhältnisse des Oberdevons im Aeketal (Oberharz). *Neues Jahrbuch für Mineralogie, Geologie und Paläontologie, Beilage-Band*, **34**, 553–632.

Branson, E.B. & Mehl, M.G. 1934*a*. Conodonts from the Grassy Creek Shale of Missouri. *University of Missouri Studies*, **8**, 171–259.

Branson, E.B. & Mehl, M.G. 1934*b*. Conodonts from the Bushberg Sandstone and equivalent formations of Missouri. *University of Missouri Studies*, **8**, 265–300.

Buch, L., von 1832. Mittheilungen an Professor Bronn. *Jahrbuch für Mineralogie, Geognosie, Geologie und Petrefaktenkunde*, **1832**, 221–226.

Buchroithner, M.F., Flügel, E., Flügel, H.W. & Stattegger, K. 1980. Die Devongerölle des paläozoischen Flysch aus Menorca und ihre paläogeographische Verbreitung. *Neues Jahrbuch für Geologie und Paläontologie, Abhandlungen*, **159**, 172–224.

Buggisch, J. & Clausen, C.-D. 1972. Conodonten- und Goniatiten-Faunen aus dem oberen Frasnium und untern Famennium Marokkos (Tafilalt, Antiatlas). *Neues Jahrbuch für Geologie und Paläontologie, Abhandlungen*, **141**, 137–167.

Buggisch, J., Rabien, A. & Hühner, G. 1978. Biostratigraphische Parallelisierung und Faziesvergleich von oberdevonischen Becken- und Schwellen-Profilen E Dillenburg (Conodonten- und Ostracoden-Chronologie, Oberdevon I-V, Dillmulde, Rheinisches Schiefergebirge). *Geologisches Jahrbuch Hessen*, **106**, 53–115.

Çapkinoğlu, Ş. 2005. Famennian conodonts from the Ayineburnu Formation of the Istanbul Zone (NW Turkey). *Geologica Carpathica*, **56**, 113–122.

Carmichael, S.K., Waters, J.A. et al. 2016. Climate instability and tipping points in the Late Devonian: detection of the Hangenberg Event in an open oceanic island arc in the Central Asian Orogenic Belt. *Gondwana Research*, **32**, 213–231, http://doi.org/10.1016/j.gr.2015.02.009

Chang, A.C. 1958. Stratigraphy, palaeontology and palaeogeography of the ammonite fauna of the Clymenienkalk from Great Khingan with special reference to the post Devonian break (hiatus) of South China. *Acta Palaeontologica Sinica*, **6**, 71–89.

Choubert, G. & Faure-Muret, A. 1961. Note sur une flore du Carbonifère Inférieur de l'Oued Korifla (region sud de Rabat). 1. Introduction stratigraphique. *Notes du Service géologique du Maroc*, **20**, 81–89.

Clariond, L. 1934. A propos d'une coupe de la region d'Erfoud. *Compte Rendu sommaires de Société Géologique du France*, **1934**, 223–224.

Clausen, C.-D. 1968. Des Nehden in der Büdesheimer Teilmulde (Prümer Mulde/Eifel). *Fortschritte in der Geologie von Rheinland und Westfalen*, **16**, 205–232.

Clausen, C.-D. & Leuteritz, K. 1984. *Erläuterungen zu Blatt 4516 Warstein*. Geologische Karte von NRW, 1 : 25 000. Geologisches Landesamt Nordrhein-Westfalen, Krefeld, 1–155.

Corradini, C. 1998. The 'Clymeniae Limestone' in the Corona Mizziu Sections. *Rendiconti della Società Paleontologica Italiana*, **1**, 261–264.

Corradini, C. 2002. Famennian conodonts from two sections near Villasalto. *Giornale di Geologia, seria 3*, **60**, 122–135.

Corradini, C., Perri, M.C. & Spalletta, C. 2001. The Mrakib section (Morocco) as possible Upper Famennian (Devonian) stratotype section: conodont data. *SDS Newsletter*, **18**, 67–69.

Czarnocki, J. 1989. Klimenie gór Swietokrzyskich. *Prace Panstwowego Instytutu Geologicznego*, **128**, 1–91.

Dashtban, H. 1995. Upper Devonian (Famennian) goniatites from Central Alborz. *Geological Survey of Iran, GeoSciences, Scientific Quarterly Journal*, **4**, 36–43.

Denckmann, A. 1901. Ueber das Oberdevon auf Blatt Balve (Sauerland). *Jahrbuch der Königlich Preußischen geologischen Landesanstalt und Bergakademie*, **21** (for 1900), 1–19.

Dopieralska, J. 2009. Reconstructing seawater circulation on the Moroccan shelf of Gondwana during the Late Devonian: evidence from Nd isotope composition of conodonts. *Geochemistry, Geophysics, Geosystems*, **10**, Q03015, http://doi.org/10.1029/2008GC002247

Dreesen, R. & Dusar, M. 1974. Refinement of conodont biozonation in the Famennian type area. *In*: *Belgium Geological Survey, International Symposium on Belgian Micropaleontological Limits from Emsian to Viséan*, 1–10 September, Namur, Volume **13**. Geological Survey of Belgium, Brussels, 1–36.

Dreesen, R. & Orchard, M. 1974. 'Intraspecific' morphological variation within *Polygnathus semicostatus* Branson & Mehl. *In*: *Belgium Geological Survey, International Symposium on Belgian Micropaleontological Limits from Emsian to Viséan*, 1–10 September, Namur, Volume **21**. Geological Survey of Belgium, Brussels, 1–10.

Dreesen, R. & Thorez, J. 1994. Parautochthonous–allochthonous carbonates and conodont mixing in the Late Famennian (Uppermost Devonian) Condroz Sandstones of Belgium. *Courier Forschungsinstitut Senckenberg*, **188**, 159–182.

Dreesen, R., Paproth, E. & Thorez, J. 1989. Events documented in Famennian sediments (Ardenne–Rhenish Massif, Late Devonian, NW Europe). *In*: McMillan, N.J., Embry, A.F. & Glass, D.J. (eds) *Devonian of the World*. Canadian Society of Petroleum Geologists, Memoirs, **14**(II), 295–308.

Dvorák, J., Friakova, O. & Kullmann, J. 1989. Influence of volcanism on Upper Devonian black limestone and shale deposition, Czechoslovakia. *In*: McMillan, N.Y., Embry, A.F. & Glass, D.J. (eds) *Devonian of the World*. Canadian Society of Petroleum Geologists, Memoirs, **14**(III), 393–397.

DYBCZYNSKI, T. 1913. Ammonity górnego Dewonu Lielc. Wiadomosc tymczasowa. *Kosmos*, **38**, 510–525.

DZIK, J. 2006. The Famennian 'Golden Age' of conodonts and ammonoids in the Polish part of the Variscan Sea. *Palaeontologica Polonica*, **63**, 1–359.

DZIK, J. 2009. Conodont affinity of the enigmatic Carboniferous chordate *Conopiscius*. *Lethaia*, **42**, 31–38.

EBBIGHAUSEN, V., BECKER, R.T. & BOCKWINKEL, J. 2002. Morphometric Analyses and Taxonomy of oxyconic Goniatites (Paratornoceratinae n. subfam.) from the early Famennian of the Tafilalt. *Abhandlungen der Geologischen Bundesanstalt*, **57**, 167–180.

EBNER, F. 1980. Conodont localities in the surroundings of Graz/Styria. *Abhandlungen der Geologischen Bundesanstalt*, **35**, 101–127.

FANTINET, D., DREESEN, R., DUSAR, M. & TERMIER, G. 1976. Faunes famenniennes de certains horizons calcaires dans la formation quartzitophylladique aux environs de Mértola (Portugal méridional). *Portugal Servicos Geológicos Communicações*, **60**, 121–137.

FEIST, R. (ed.) 2002. *The Palaeozoic of the Montagne Noire, Southern France*. Guidebook, ECOS 8, Toulouse.

FISCHER, T. 2013. *Morphometrie, Taxonomie, Paläoökologie und Paläobiogeographie von Ammonoideen des obersten Famenniums*. Master thesis, Westfälische Wilhelms University Münster.

FISCHER, T. & BECKER, R.T. 2014. Ontogenetic morphometry, taxonomy and biogeographic aspects of Famennian (Upper Devonian) Prionoceratidae. *In*: KLUG, C. & FUCHS, D. (eds) *9th International Symposium Cephalopods – Present and Past in combination with the 5th International Symposium Coleoid Cephalopods through Time, Abstracts and Program*, Zürich, 36.

FLÜGEL, H. & KROPFITSCH-FLÜGEL, M. 1965. Ammonoidea palaeozoica. *Catalogus Fossilum Austriae*, **4**, 3–31.

FLÜGEL, H. & ZIEGLER, W. 1957. Die Gliederung des Oberdevons und Unterkarbons am Steinberg westlich von Graz mit Conodonten. *Mitteilungen des Naturwissenschaftlichen Vereins für die Steiermark*, **87**, 25–60.

FRECH, F. 1887. Ueber das Devon der Ostalpen, nebst Bemerkungen über das Silur und einem paläontologischen Anhang. *Zeitschrift der Deutschen Geologischen Gesellschaft*, **39**, 659–738.

FRECH, F. 1902. Über devonische Ammoneen. *Beiträge zur Geologie und Paläontologie Österreich-Ungarns und des Orients*, **14**, 27–112.

FREYER, G. 1957. Neue Untersuchungen im Oberdevon des Vogtlandes auf Grund des Fossilinhaltes der Kalke im Bereich der Vogtländischen Hauptmulde. *Freiberger Forschungshefte*, **C27**, 1–98.

FUHRMANN, A. 1954. Petrographie, Fauna und stratigraphische Stellung einiger Aufschlüsse im Oberharzer Oberdevon, Blatt Zellerfeld und Riefensbeek. *Geologisches Jahrbuch*, **69**, 629–652.

GEREKE, M. 2004. Das Profil Kahlleite Ost – die stratigraphische Entwicklung einer Tiefschwelle im Oberdevon des Bergaer Sattels (Thüringen). *Geologica et Palaeontologica*, **38**, 1–31.

GIRARD, C., CORNÉE, J.-J., CORRADINI, C., FRAVALO, A. & FEIST, R. 2013. Palaeoenvironmental changes at Col des Tribes (Montagne Noire, France), a reference section for the Famennian of north Gondwana-related areas. *Geological Magazine*, **151**, 864–884, http://doi.org/10.1017/S0016756813000927

GÖDDERTZ, B. 1987. Devonische Goniatiten aus SW-Algerien und ihre stratigraphische Einordnung in die Conodonten-Abfolge. *Palaeontographica, Abteilung A*, **197**, 127–220.

GREIFELT, T., BECKER, R.T. & HARTENFELS, S. 2008. Impact of the Annulata and Dasberg Events (upper Famennian) on agglutinating foraminifere assemblages in the northern Rhenish Massif (Germany). *In*: KÖNIGSHOF, P. & LINNEMANN, U. (eds) *From Gondwana and Laurussia to Pangea: Dynamics of Oceans and Supercontinents*. 20th International Senckenberg Conference and 2nd Geinitz Conference, Final Meeting of IGCP 497 and IGCP 499, Abstracts and Programme. Forschungsinstitut Senckenberg, Frankfurt am Main, 173–175.

HARTENFELS, S. 2003. *Karbonatmikrofazies und Conodontenbiofazies ausgewählter Profile im Oberdevon und Unterkarbon des Frankenwaldes und des Bayerischen Vogtlandes – Geuser, Kirchgattendorf, Köstenhof (NE-Bayern, Deutschland)*. Diplom thesis, University of Cologne.

HARTENFELS, S. 2011. Die global *Annulata*-Events und die Dasberg-Krise (Famennium, Oberdevon) in Europa und Nord-Afrika – hochauflösende Conodonten-Stratigraphie, Karbonat-Mikrofazies, Paläoökologie und Paläodiversität. *Münstersche Forschungen zur Geologie und Paläontologie*, **105**, 17–527.

HARTENFELS, S. & BECKER, R.T. 2009. Timing of the global Dasberg Crisis – Implications for Famennian eustasy and chronostratigraphy. *In*: OVER, D.J. (ed.) *Studies in Devonian Stratigraphy: Proceedings of the 2007 International Meeting of the Subcommission on Devonian Stratigraphy and IGCP 499*. Palaeontographica Americana, **63**, 71–97.

HARTENFELS, S. & BECKER, R.T. 2012*a*. Local variations of the global *Annulata* Events and Dasberg Crisis biofacies. *In*: *Proceedings of the 34th International Geological Congress 2012, Brisbane, Australia*. Abstracts-CD. Australian Geoscience Council, 3291.

HARTENFELS, S. & BECKER, R.T. 2012*b*. Conodont age and correlation oft he transgressive *Gonioclymenia* and *Kalloclymenia* Limestones (Famennian, Anti-Atlas, SE Morocco). *In*: WITZMANN, F. & ABERHAN, M. (eds) *Centenary Meeting of the Paläontologische Gesellschaft, Programme, Abstracts, and Field Guide*. Terra Nostra, Bonn, 67.

HARTENFELS, S. & BECKER, R.T. 2013. El Gara South – new data on Famennian ammonoid and conodont faunas and the Annulata Events in the Rheris Basin (northern Tafilalt, Morocco). *In*: EL HASSANI, A., BECKER, R.T. & TAHIRI, A. (eds) *International Field Symposium 'The Devonian and Lower Carboniferous of northern Gondwana'. Abstracts Book*. Document de l'Institut Scientifique, Rabat, **26**, 48–51.

HARTENFELS, S. & BECKER, R.T. 2014. New data on Upper Devonian black shale events in the eastern part of the central Moroccan Meseta. *In*: *4th International Palaeontological Congress*, Mendoza, Argentina. Abstract CD [no pagination].

HARTENFELS, S. & BECKER, R.T. 2015. Revised conodont stratigraphy of the famous Ballberg section

(Famennian, Rhenish Massive, Germany). *In*: Mottequin, B., Denayer, J., Königshof, P., Prestianni, C. & Olive, S. (eds) *IGCP 596 – SDS Symposium, Climate change and Biodiversity patterns in the Mid-Palaeozoic*. Abstracts. Strata. Travaux de Géologie sédimentaire et Paléontologie, Série 1, **16**, 66–67.

Hartenfels, S. & Tragelehn, H. 2004. Karbonatmikrofazies und Conodontenbiofazies ausgewählter Profile im Oberdevon und Unterkarbon des Frankenwaldes und des Bayerischen Vogtlandes – Geuer, Kirchgattendorf, Köstenhof (NE-Bayern, Deutschland). *In*: Reitner, J., Reich, M. & Schmidt, G. (eds) *Geobiologie. 74 Jahrestagung der Paläontologischen Gesellschaft, 2–8 Oktober 2004, Göttingen, Kurzfassungen der Vorträge und Poster*. Universitätsdrucke Göttingen, 99–101.

Hartenfels, S., Becker, R.T. & Tragelehn, H. 2009. Marker conodonts around the global *Annulata* Events and the definition of an Upper Famennian substage. *SDS Newsletter*, **24**, 40–48.

Hartenfels, S., Becker, R.T., Aboussalam, Z.S., El Hassani, A., Baider, L., Fischer, T. & Stichling, S. 2013. The Upper Devonian at El Khraouia (southern Tafilalt). *In*: Becker, R.T., El Hassani, A. & Tahiri, A. (eds) *International Field Symposium 'The Devonian and Lower Carboniferous of northern Gondwana'. Field Guidebook*. Document de l'Institut Scientifique, Rabat, **27**, 41–50.

Heckel, P.H. & Witzke, B.J. 1979. Devonian world palaeogeography determined from distribution of carbonates and related lithic palaeoclimatic indicators. *In*: House, M.R., Scrutton, C.T. & Bassett, M.G. (eds) *The Devonian System*. Special Papers in Palaeontology, **23**, 99–123.

Helms, J. 1959. Conodonten aus dem Saalfelder Oberdevon (Thüringen). *Geologie*, **8**, 634–677.

Helms, J. 1963. Zur 'Phylogenese' und Taxionomie von *Palmatolepis* (Conodontida, Oberdevon). *Geologie*, **12** , 449–485.

Heritsch, F. 1927. Eine neue Stratigraphie des Paläozoikums von Graz. *Verhandlungen der Geologischen Bundesanstalt*, **11**, 223–227.

Higgs, K.T., Prestianni, C., Streel, M. & Thorez, J. 2013. High resolution miospore stratigraphy of the Upper Famennian of eastern Belgium, and correlation with the conodont zonation. *Geologica Belgica*, **16**, 84–94.

Hölder, H. 1979. Miscellanea cephalopodica. II. Narben an Gehäusen devonischer Ammoneen. *Münstersche Forschungen zur Geologie und Paläontologie*, **29**, 47–51.

Hollard, H. 1963. Un tableau stratigraphique du Dévonien du Sud de l´Anti-Atlas. *Notes du Service Géologique du Maroc*, **23**, 105–109.

Hollard, H. 1970. Silurien–Devonien–Carbonifere. *In*: Choubert, G. & Faure-Muret, A. (eds) *Anti-Atlas Occidental et Central*. Notes et Mémoires du Service géologique, **229**, 161–188.

Hollard, H. 1971. Sur la transgression Dinantienne au Maroc Présaharien. *Compte Rendu 6è Congrès International de Stratigraphie et Géologie du Carbonifère*, **3**, 923–936.

Hollard, H. & Jacquemont, P. 1956. Le Gothlandien, le Dévonien et le Carbonifère des regions du Dra et du Zemoul (confins algéro-marocains du Sud). *Notes et Mémoires du Service géologique*, **15**, 7–33.

Hösel, G. 1960. Stratigraphische Untersuchungen im Oberdevon von Planitz bei Zwickau (Sa.). *Geologie*, **9**, 190–203.

House, M.R. 1959. Upper Devonian Ammonoids from North-West Dartmoor, Devonshire. *Proceedings of the Geologists' Association*, **70**, 315–321.

House, M.R. 1962. Observations on the ammonoid succession of the North American Devonian. *Journal of Paleontology*, **36**, 247–284.

House, M.R. 1963*a*. Evolution observed. *Discovery*, **1963**, 12–17.

House, M.R. 1963*b*. Devonian ammonoid successions and facies in Devon and Cornwall. *Quarterly Journal of the Geological Society, London*, **119**, 1–27, http://doi.org/10.1144/gsjgs.119.1.0001

House, M.R. 1970. On the origin of the clymeniid ammonoids. *Palaeontology*, **13**, 664–676.

House, M.R. 1983. Devonian eustatic events. *Proceedings of the Ussher Society*, **5**, 396–405.

House, M.R. 1985. Correlation of mid-Palaeozoic ammonoids evolutionary events with global sedimentary perturbations. *Nature*, **313**, 17–22.

House, M.R. 2002. Strength, timing, setting and cause of mid-Palaeozoic extinctions. *Palaeogeography, Palaeoclimatology, Palaeoecology*, **181**, 5–25.

House, M.R. & Butcher, N.E. 1973. Excavations in the Upper Devonian and Carboniferous rocks near Chudleigh, South Devon. *Transactions of the Royal Geological Society of Cornwall*, **20**, 199–220.

House, M.R. & Pedder, A.E.H. 1963. Devonian goniatites and stratigraphical correlations in Western Canada. *Palaeontology*, **6**, 491–539.

House, M.R. & Selwood, E.B. 1964. *Palaeozoic palaeontology in Devon and Cornwall*. Published for the 150th Anniversary of the Royal Geological Society of Cornwall. Royal Geological Society of Cornwall, Penzance, 45–86.

House, M.R., Richardson, J.B., Chaloner, W.G., Allen, J.R.L., Holland, C.H. & Westoll, T.S. 1977. *A Correlation of the Devonian Rocks in the British Isles*. Geological Society, London, Special Reports, **8**, 1–110.

House, M.R., Gordon, M., Jr. & Hlavin, W.J. 1986. Late Devonian ammonoids from Ohio and adjacent States. *Journal of Paleontology*, **60** , 126–144.

Hyatt, A. 1883/1884. Genera of fossil cephalopods. *Proceedings of the Boston Society of Natural History*, **22**, 253–338 (253–272 published in 1883, 273–338 published in 1884).

Jenkins, T.B.H. 1968. Famennian ammonoids from New South Wales. *Palaeontology*, **11**, 535–548.

Ji, Q. & Ziegler, W. 1993. The Lali Section: an excellent reference section for Upper Devonian in South China. *Courier Forschungsinstitut Senckenberg*, **157**, 1–183.

Johnson, J.G., Klapper, G. & Sandberg, C.A. 1985. Devonian eustatic fluctuations in Euramerica. *Geological Society of America Bulletin*, **96**, 567–587.

Johnston, D.I. & Chatterton, B.D.E. 2001. Upper Devonian (Famennian) conodonts from the Palliser Formation and Wabamun Group, Alberta and British Columbia, Canada. *Palaeontographica Canadiana*, **19**, 1–154.

Kaiser, S.I. 2005. *Mass extinctions, climatic and oceanographic changes at the Devonian/Carboniferous boundary*. PhD dissertation, Ruhr University Bochum.

Kaiser, S.I., Becker, R.T. *et al.* 2004. Sedimentary succession and neritic faunas around the Devonian-Carboniferous boundary at Kheneg Lakahal south of Assa (Dra Valley, SW Morocco). *In*: El Hassani, A. (ed.) *Devonian Neritic-Pelagic Correlation and Events in the Dra Valley (Western Anti-Atlas, Morocco). International Meeting on Stratigraphy, 1–10 March*. Document de l'Institut Scientifique, Rabat, **19**, 69–74.

Kaiser, S.I., Becker, R.T., Spalletta, C. & Steuber, T. 2009. High-resolution conodont stratigraphy, biofacies, and extinctions around the Hangenberg Event in pelagic successions from Austria, Italy, and France. *In*: Over, D.J. (ed.) *Studies in Devonian: Proceedings of the 2007 International Meeting of the Subcommission on Devonian Stratigraphy and IGCP 499*. Palaeontographica Americana, **63**, 97–139.

Kaiser, S.I., Becker, R.T., Steuber, T. & Aboussalam, S.Z. 2011. Climate-controlled mass extinctions, facies, and sea-level changes around the Devonian-Carboniferous boundary in the eastern Anti-Atlas (SE Morocco). *Palaeogeography, Palaeoclimatology, Palaeoecology*, **310**, 340–364.

Kind, N.V. 1944. Goniatity I klimenii zapadnogo sklona Mugodzharskikh Gor. *Uchenye zapiski Leningradskaya Universiteta, 70, Seriya geologo-pochvennykh nauk*, **11**, 137–166.

Klapper, G. 1966. *Upper Devonian and Lower Mississippian Conodont Zones in Montana, Wyoming, and South Dakota*. University of Kansas, Paleontological Contributions, **3**.

Klapper, G., Sandberg, C.A. *et al.* 1971. North American Devonian conodont biostratigraphy. *In*: Sweet, W.C. & Bergström, S.M. (eds) *Symposium on Conodont Biostratigraphy*. Geological Society of America, Memoirs, **127**, 285–316.

Klein, C. & Korn, D. 2014. A morphometric approach to conch ontogeny of *Cymaclymenia* and related genera (Ammonoidea, Late Devonian). *Fossil Record*, **17**, 1–32.

Klug, C. 2002. Quantitative stratigraphy and taxonomy of late Emsian and Eifelian ammonoids of the eastern Anti-Atlas (Morocco). *Courier Forschungsinstitut Senckenberg*, **238**, 1–109.

Koch, L. 1984. *Aus Devon, Karbon und Kreide: Die fossile Welt des Sauerlandes*. V. d. Linnepe Verlagsgesellschaft, Hagen.

Königshof, P., Savage, N.M., Lutat, P., Sardsud, A., Dopieralska, J., Belka, Z. & Racki, G. 2012. Late Devonian sedimentary record of the Paleotethys Ocean – the Mae Sariang section, northwestern Thailand. *Journal of Asian Earth Sciences*, **52**, 146–157.

Korn, D. 1981. *Cymaclymenia* – eine besonders langlebige Clymenien-Gattung (Ammonoidea, Cephalopoda). *Neues Jahrbuch für Geologie und Paläontologie, Abhandlungen*, **161**, 172–208.

Korn, D. 1992. Relationship between shell form, septal construction and suture line in clymeniid cephalopods (Ammonoidea; Upper Devonian). *Neues Jahrbuch für Geologie und Paläontologie, Abhandlungen*, **185**, 115–130.

Korn, D. 1997. The Palaeozoic ammonoids of the South Portuguese Zone. *Memórias do Instituto Geológico e Minero*, **33**, 1–132.

Korn, D. 1999. Famennian ammonoid stratigraphy of the Ma'der and Tafilalt (Eastern Anti-Atlas, Morocco). *Abhandlungen der Geologischen Bundesanstalt*, **54**, 147–179.

Korn, D. 2002. Die Ammonoideen-Fauna der *Platyclymenia annulata*-Zone vom Kattensiepen (Oberdevon, Rheinisches Schiefergebirge). *Senckenbergiana lethaea*, **82**, 557–608.

Korn, D. 2004. The mid-Famennian ammonoid succession in the Rhenish Mountains: the '*annulata* Event' reconsidered. *Geological Quarterly*, **48**, 245–252.

Korn, D. & Klug, C. 2002. Ammoneae devonicae. *In*: Riegraf, W. (ed.) *Fossilium Catalogus, I: Animalia, Pars 138*. Backhuys, Leiden, 211–237.

Korn, D. & Luppold, F.W. 1987. Nach Clymenien und Conodonten gegliederte Profile des oberen Famennium im Rheinischen Schiefergebirge. *Courier Forschungsinstitut Senckenberg*, **92**, 199–223.

Korn, D. & Titus, A. 2006. The ammonoids from the Three Fork Shale (Late Devonian) of Montana. *Fossil Record*, **9**, 198–212.

Korn, D. & Ziegler, W. 2002. The ammonoid and conodont zonation at Enkeberg (Famennian, Late Devonian; Rhenish Mountains). *Senckenbergiana lethaea*, **82**, 453–462.

Korn, D., Clausen, C.-D., Belka, Z., Leuteritz, K., Luppold, F.-W., Feist, R. & Weyer, D. 1994. Die Devon/Karbon-Grenze bei Drewer (Rheinisches Schiefergebirge). *Geologie und Paläontologie in Westfalen*, **29**, 97–147.

Korn, D., Klug, C. & Reisdorf, A. 2000. Middle Famennian ammonoid stratigraphy in the Amessoui Syncline (Late Devonian; eastern Anti-Atlas, Morocco). *Travaux de l'Institut Scientifique, Série Géologie & Géographie Physique*, **20**, 69–77.

Korn, D., Bockwinkel, J. & Ebbighausen, V. 2013. Famennian and Tournaisian strata of the Aguelmous Syncline. *In*: Becker, R.T., El Hassani, A. & Tahiri, A. (eds) *International Field Symposium 'The Devonian and Lower Carboniferous of northern Gondwana'. Field Guidebook*. Document de l'Institut Scientifique, Rabat, **27**, 121–127.

Korn, D., Bockwinkel, J. & Ebbighausen, V. 2014. Middle Famennian (Late Devonian) ammonoids from the Anti-Atlas of Morocco. 1. *Prionoceras*. *Neues Jahrbuch für Geologie und Paläontologie, Abhandlungen*, **272**, 167–204.

Korn, D., Bockwinkel, J. & Ebbighausen, V. 2015*a*. The Late Devonian ammonoid *Mimimitoceras* in the Anti-Atlas of Morocco. *Neues Jahrbuch für Geologie und Paläontologie, Abhandlungen*, **275**, 125–150.

Korn, D., Bockwinkel, J. & Ebbighausen, V. 2015*b*. Middle Famennian (Late Devonian) ammonoids from the Anti-Atlas of Morocco. 2. Sporadoceratidae. *Neues Jahrbuch für Geologie und Paläontologie, Abhandlungen*, **278**, 47–77.

Kozur, H. & Mostler, H. 1976. Neue Conodonten aus dem Jungpaläozoikum und der Trias. *Geologisch-Paläontologische Mitteilungen Innsbruck*, **6**, 1–33.

Kronberg, P., Pilger, A., Scherp, A. & Ziegler, W. 1960. Spuren altvariscischer Bewegungen im nördlichsten Teil des Rheinischen Schiefrgebirges. *Fortschritte in der Geologie von Rheinland und Westfalen*, **3**, 1–46.

Kullmann, J. 1960. Die Ammonoidea des Devon im Kantabrischen Gebirge (Nordspanien). *Abhandlungen der Akademie der Wissenschaften und Literatur, mathematisch-naturwissenschaftliche Klasse*, **1960**, 1–105.

Lange, W. 1929. Zur Kenntnis des Oberdevons am Enkeberg und bei Balve (Sauerland). *Abhandlungen der Preußischen Geologischen Landesanstalt, Neue Folge*, **119**, 1–132.

Ma, X.-P., Chen, D. & Yin, B. 2006. The Devonian of the Guilin–Xiangzhou area, south China: stratigraphy and sedimentology. *In*: *The 2nd International Palaeontological Congress, Guidebook for Field Excursion, Pre-congress Excursion A2*, 17–21 June, 2006, Beijing, China. University of Science and Technology of China Press, Hefei, 1–35.

Ma, X.-P., Zong, P. & Sun, Y.-L. 2011. The Devonian (Famennian) sequence in the western Junggar area, northern Xinjiang, China. *SDS Newsletter*, **26**, 44–49.

Mannani, M. & Yazdi, M. 2015. First report on Late Devonian *Annulata* Event in the northeast of Abadeh (Esteghlal Mine). *In*: *33rd National Geosciences Symposium*, February 2015, Iran. Geological Survey of Iran, Tehran.

Martynova, M.V. & Vorontzova, T.N. 1988. Typical sections of Devonian/Carboniferous boundary deposits in the central Kazakhtan. *In*: Golubtsov, V.K. (ed.) *The Devonian–Carboniferous Boundary at the Territory of the USSR*. Joint Stratigraphy Commission USSR, Institute of Geochemistry & Geophysics, Academy of Science of the USSR, Minsk, 181–187 [in Russian].

Massa, D. 1965. Observations sur les Séries Siluro-Dévoniennes des Confins Algéro-Marocains du Sud (1954–1955). *Notes et Mémoires, Compagne Francaise des Petroles*, **8**, 1–187.

Matern, H. 1931. Das Oberdevon der Dill-Mulde. *Abhandlungen der Preußischen Geologischen Landesanstalt, Neue Folge*, **134**, 1–139.

M'Coy, F. 1851. On some new Devonian Fossils. *Annual Magazine of Natural History*, **2**, 481–489.

Menchikoff, N. 1930. Recherches géologiques et morphologiques dans le Nord du Sahara occidental. *Revue de Géographie physique et de Géologie dynamique*, **3**, 3–147.

Meyendorf, A. 1939. La série primaire du Gourara (Sahara occidental). *Compte Rendu de l'Academie des Sciences Paris*, **209**, 228–229.

Meyer, H. 1920. *Der Bohlen bei Saalfeld in Thüringen*. Richard Clauss, Saalfeld in Thüringen.

Miller, A.K., Furnish, W.M. & Schindewolf, O.H. 1957. Paleozoic Ammonoidea. *In*: Moore, R.C. (ed.) *Treatise on Invertebrate Paleontology, Part L, Mollusca 4, Cephalopoda, Ammonoidea*. University of Kansas, Lawrence, KS, 11–79.

Minwegen, E. & Herbig, H.-G. 1997. Ein mikrobieller mudmound im oberen Famenne des nordöstlichen Sauerlandes (Rheinisches Schiefergebirge). *In*: *Sediment '97, 12. Sedimentologentreffen, 12–14. Mai 1997 am Geologischen Institut der Universität zu Köln, Kurzfassungen der Vorträge und Poster*. Terra Nostra, **1997**, 154–155.

Müller, K.J. 1956*a*. Cephalopodenfauna und Stratigraphie des Oberdevons von Schleiz und Zeulenroda in Thüringen. *Beihefte zum Geologischen Jahrbuch*, **20**, 1–93.

Müller, K.J. 1956*b*. Zur Kenntnis der Conodonten-Fauna des europäischen Devons, 1: Die Gattung *Palmatolepis*. *Abhandlungen der Senckenbergischen Naturforschenden Gesellschaft*, **494**, 1–70.

Münster, G. Graf zu. 1832. *Ueber die Planuliten und Goniatiten im Uebergangs-Kalk des Fichtelgebirges*. Birner, Bayreuth.

Münster, G. Graf zu. 1839*a*. Nachtrag zu den Clymenien des Fichtelgebirges. *Beiträge zur Petrefactenkunde*, **1**, 35–43.

Münster, G. Graf zu. 1839*b*. Nachtrag zu den Goniatiten des Fichtelgebirges. *Beiträge zur Petrefactenkunde*, **1**, 43–55.

Münster, G. Graf zu. 1842. Nachtrag zu den Versteinerungen des Übergangsgebirges mit Clymenien. *Beiträge zur Petrefactenkunde*, **6**, 112–130.

Murphy, A.E., Sageman, B.B., Ver Straeten, C.A. & Hollander, D.J. 2000. Organic carbon burial and faunal dynamics in the Appalachian Basin during the Devonian (Givetian–Famennian) greenhouse: an integrated palaeoecological and biogeochemical approach. *In*: Huber, B., MacLeod, K. & Wing, S. (eds) *Warm Climates in Earth History*. Cambridge University Press, Cambridge, 351–385.

Myrow, P.M., Strauss, J.V., Creveling, J.R., Sicard, K.R., Ripperdan, R., Sandberg, C.A. & Hartenfels, S. 2011. A carbon isotopic and sedimentological record of the latest Devonian (Famennian) from the Western U.S. and Germany. *Palaeogeography, Palaeoclimatology, Palaeoecology*, **306**, 147–159.

Nagel-Myers, J., Amler, M.R.W. & Becker, R.T. 2009. The Loxopteriinae n. subfam. (Dualinidae, Bivalvia): review of a common bivalve taxon from the Late Devonian pelacic facies. *In*: Over, D.J. (ed.) *Studies in Devonian Stratigraphy: Proceedings of the 2007 International Meeting of the Subcommission on Devonian Stratigraphy and IGCP 499*. Palaeontographica Americana, **63**, 167–191.

Nalivkina, A.K. 1953. Verknedevonskie goniatity i klimenii Mugodzhar. *Trudy Vsesoyuznogo Neftyanogo Nauchno-Issledovatel'skogo Geologo-Razvedochnogo Instituta (VNIGRI), novaya seriya*, **72**, 60–125.

Nawrath, J. 2009. *Morphometrie und Taxonomie von Prionoceraten (Ammoniten) aus dem Famennium von Franken, Marokko und des Ost-Iran*. BSc thesis, Westfälische Wilhelms University, Münster.

Nikolaeva, S.V. 2011. New Data on the Genus *Praeglyphioceras* (Praeglyphioceratidae, Ammonoidea). *Paleontological Journal*, **45**, 501–509.

Nikolaeva, S.V. & Bogoslovskiy, B.I. 2005. Devonskie Ammonoidei. IV. Klimenii (podotriad Clymeniina). *Trudy Paleontologicheskogo Instituta*, **287**, 1–220.

Olempska, E. & Chauffe, K.M. 1999. Ostracods of the Maple Mill Shale Formation (Upper Devonian) of southeastern Iowa, U.S.A. *Micropalaeontology*, **45**, 304–318.

PAECKELMANN, W. 1924. Das Devon und Carbon in der Umgebung von Balve i. Westf. *Jahrbuch der Preußischen Geologischen Landesanstalt*, **44**, 51–97.

PAECKELMANN, W. 1925. Bemerkungen über die geologischen Verhältnisse der Gegend von Brilon i. Westf. *Jahrbuch der Preußischen Geologischen Landesanstalt*, **46**, 210–230.

PERNA, A.K. 1914. Ammonei verkhnego neodevona vostochnogo sklona Urala. *Trudy Geologicheskogo Komiteta, Novaya Seriya*, **99**, 1–144.

PERRI, M.C. & SPALLETTA, C. 1990. Famennian conodonts from climenid pelagic limestone, Carnic Alps, Italy. *Palaeontographia Italica*, **77**, 55–83.

PERRI, M.C., SPALLETTA, C. & PONDRELLI, M. 1998. Late Famennian conodonts from the Pramosio Bassa section (Carnic Alps, Italy). *In*: PERRI, M.C. & SPALLETTA, C. (eds) *Southern Alps Field Trip Guidebook*. 27 June–2 July 1998. ECOS VII. Giornale di Geologia, Serie 3a, **60**, Special Issue, 228–233.

PETERHÄNSLI, A. & PRATT, B.R. 2008. The Famennian (Upper Devonian) Palliser Platform of Western Canada – architecture and depositional dynamics of a post-extinction epeiric giant. *In*: PRATT, B.R. & HOLMDEN, C. (eds) *Dynamics of Eperic Seas*. Geological Society of Canada, Special Papers, **48**, 247–281.

PETERSEN, M.S. 1975. Upper Devonian (Famennian) Ammonoids from the Canning Basin, Western Australia. *Journal of Paleontology Memoirs*, **8**, 1–55.

PETTER, G. 1951. Dévonien moyen et supérieur de la Soura et des environs d'Ougarta. *Bulletin Société Géologique de la France, 6ème Série*, **1**, 351–361.

PETTER, G. 1959. Goniatites devoniennes du Sahara. *Publications du Service de la Carte Géologique de l'Algérie, Nouvelle Série, Paléontologie*, **2**, 1–313.

PETTER, G. 1960. Clymènies du Sahara. *Publications du Service de la Carte Géologique de l'Algérie, Nouvelle Série, Paléontologie, Mémoire*, **6**, 1–58.

PFEIFFER, H. 1954. Der Bohlen bei Saalfeld/Thür. *Geologie, Beiheft*, **11**, 1–105.

PHILLIPS, J. 1841. *Figures and Descriptions of the Palaeozoic Fossils of Cornwall, Devon and West Somerset*. Longman, Brown, Green & Longmans, London.2

PLAYFORD, P.E., HOCKING, R.M. & COCKBAIN, A.E. 2009. Devonian Reef Complexes of the Canning Basin, Western Australia. *Geological Survey of Western Australia*, **145**, 1–396.

PRICE, J.D. 1982. *Some Famennian (Upper Devonian) ammonoids from north-western Europe*. PhD thesis, University of Hull.

PRICE, J.D. & KORN, D. 1989. Stratigraphically important Clymeniids (Ammonoidea) from the Famennian (Late Devonian) of the Rhenish Massif, West Germany. *Courier Forschungsinstitut Senckenberg*, **110**, 257–294.

PRUVOST, P. 1912. Sur la présence de fossiles d'âge Dévonien supérieur dans les schistes à *Néréites* de Sam-Domingos. *Communicacões da Commissão do Servico Geológico de Portugal*, **9**, 58–68.

RABIEN, A. 1970. Oberdevon. *In*: *Erläuterungen zu Blatt 5215 Dillenburg*. Geologische Karte von Hessen, 1 : 25 000. Hessisches Landesamt für Bodenforschungen, Wiesbaden, 78–83, 103–235.

RACKA, M. & MARYNOWSKI, L. 2008. Geochemical proxies of the late Famennian *Annulata* event from the Kowala quarry, Holy Cross Mountains, Poland. *In*: *Pierwszy Polski Kongres Geologiczny*, 26–28 June, Kraków, Abstracts. Polskie Towarzystwo Geologiczne, Kraków, 95.

RACKA, M., MARYNOWSKI, L., FILIPIAK, P., SOBSTEL, M., PISARZOWSKA, A. & BOND, D.P.G. 2010. Anoxic *Annulata* Events in the Late Famennian of the Holy Cross Mountains (Southern Poland): geochemical and palaeontological record. *Palaeogeography, Palaeoclimatology, Palaeoecology*, **297**, 549–575.

RAKOCIŃSKI, M. 2012. The youngest record of 'Housean pits' in Late Devonian ammonoids. *Geological Quarterly*, **56** , 387–390.

RAYMOND, P.E. 1909. The fauna of the Upper Devonian in Montana. Pt.1. The fossils of the red shales. *Annales of the Carnegie Museum*, **5**, 141–158.

RICHTER, R. 1848. *Beitrag zur Paläontologie des Thüringer Waldes. Die Grauwacke des Bohlens und des Pfaffenberges bei Saalfeld. 1. Fauna*. Arnoldi, Dresden.

RIVIERE, A. 1931. De deux espèces nouvelles dans l'Elbourz central. *Compte Rendu sommaires de Société Géologique du France*, **1931**, 73.

RIVIERE, A. 1934. Contribution à l'etude géologique de l'Elbourz (Perse). *Revue de Géographie Physique et Géologie Dynamique*, **7**, 1–185.

ROCH, E. 1950. Histoire stratigraphique du Maroc. *Division des Mines et de la Géologie, Service Géologique, Notes et Mémoires*, **80**, 1–435.

ROLFE, W.D.I. & DZIK, J. 2006. *Angustidontus*, a Late Devonian pelagic predatory crustacean. *Transactions of the Royal Society of Edinburgh, Earth Sciences*, **97**, 75–96.

RUZHENCEV, V.R. 1957. Filogeneticheskaya sistema paleozoyskikh ammonoidey. *Byulleten Moskovskogo obshchestva ispytately prirody, novaya seriya, otdel geologicheskiy*, **31**, 49–64.

RYTINA, M.-K., BECKER, R.T., ABOUSSALAM, Z.S., HARTENFELS, S., HELLING, S., STICHLING, S. & WARD, D. 2013. The allochthonous Silurian–Devonian in olistostromes at 'the southern Variscan front' (Tinerhir region, SE Morocco) – preliminary data. *In*: BECKER, R.T., EL HASSANI, A. & TAHIRI, A. (eds) *International Field Symposium 'The Devonian and Lower Carboniferous of northern Gondwana'. Field Guidebook*. Document de l'Institut Scientifique, Rabat, **27**, 11–21.

RZEHAK, A. 1910. Der Brünner Clymenienkalk. *Zeitschrift des mährischen Landesmuseums*, **10**, 149–216.

SAGEMAN, B.B., MURPHY, A.E., WERNE, J.P., VER STRAETEN, C.A., HOLLANDER, D.J. & LYONS, T.W. 2003. A tale of shales: the relative roles of production, decomposition, and dilution in the accumulation of organic-rich strata, Middle-Upper Devonian, Appalachian basin. *Chemical Geology*, **195**, 229–273.

SANDBERG, C.A. 1979. Devonian and Lower Mississippian Conodont Zonation of the Great Basin and Rocky Mountains. *Brigham Young University Geology Studies*, **26**, 87–106.

SANDBERG, C.A. & DREESEN, R. 1984. Late Devonian icriodontid biofacies models and alternate shallow-water conodont zonation. *In*: CLARK, D.L. (ed.) *Conodont Biofacies and Provincialism*. Geological Society of America, Special Papers, **196**, 143–179.

SANDBERG, C.A. & KLAPPER, G. 1967. Stratigraphy, age, and paleotectonic significance of the Cottonwood

Canyon Member of the Madison Limestone in Wyoming and Montana. *United States Geological Survey Bulletin*, **1251-B**, 1–70.

Sandberg, C.A. & Poole, F.G. 1977. Conodont biostratigraphy and depositional complexes of Upper Devonian cratonic-platform and continental-shelf rocks in the Western United States. *In*: Murphy, M.A., Berry, W.B.N. & Sandberg, C.A. (eds) *Western North America: Devonian*. University of California, Riverside Campus Museum Contribution, **4**, 144–182.

Sandberg, C.A. & Ziegler, W. 1973. Refinement of standard Upper Devonian conodont zonation based on sections in Nevada and West Germany. *Geologica et Palaeontologica*, **7**, 97–122.

Sandberg, C.A. & Ziegler, W. 1979. Taxonomy and biofacies of important conodonts of Late Devonian *styriacus*-Zone, United States and Germany. *Geologica et Palaeontologica*, **13**, 173–212.

Sandberg, C.A., Streel, M. & Scott, R.A. 1972. Comparison between conodont zonation and spore assemblages at the Devonian–Carboniferous boundary in the western and central United States and in Europe. *Septième Congrès International de Stratigraphie et de Géologie du Carbonifère, Compte Rendu*, **1**, 179–203.

Sandberger, G. 1855. *Clymenia subnautilina* (nova species), die erste und bis jetzt einzige Art aus Nassau. *Jahrbücher des Vereins für Naturkunde im Herzogthum Nassau*, **10**, 127–136.

Sannemann, D. 1955. Oberdevonische Conodonten (to IIα). *Senckenbergiana Lethaea*, **36**, 123–156.

Sanz-López, J., García-López, S., Montesinos, R.R. & Arbizu, M. 1999. Biostratigraphy and sedimentation of the Vidrieros Formation (middle Famennian–lower Tournaisian) in the Gildar-Montó unit (northwest Spain). *Bollettino della Società Paleontologica Italiana*, **37**, 393–406.

Sartenaer, P. 1998. The presence in Morocco of the late Famennian genus *Hadyrhynchia* Halíček, 1979 (rhynchonellid, brachiopod). *Bulletin de l'Institut royal de Sciences naturelles de Belgique*, **68**, 115–120.

Sartenaer, P. 1999. *Tetragonorhynchus*, new late Famennian rhynchonellid genus from Maïder, southern Morocco, and Tetragonorhynchidae n. fam. *Bulletin de l'Institut royal de Sciences naturelles de Belgique*, **69**, 67–75.

Sartenaer, P. 2000. *Phacoiderhynchus*, a new middle Famennian rhynchonellid genus from the Anti-Atlas, Morocco, and Phacoiderhynchidae n. fam. *Bulletin de l'Institut royal de Sciences naturelles de Belgique*, **70**, 75–88.

Savage, N.M. 2013. *Late Devonian Conodonts from Northwestern Thailand*. Bourland Printing, Eugene, OR.

Schäfer, W. 1976. Einige neue Conodonten aus dem höheren Oberdevon des Sauerlandes (Rheinisches Schiefergebirge). *Geologica et Palaeontologica*, **10**, 141–152.

Schindewolf, O.H. 1921. Versuch einer Paläogeographie des europäischen Oberdevonmeeres. *Zeitschrift der Deutschen Geologischen Gesellschaft*, **73**, 137–223.

Schindewolf, O.H. 1923. Beiträge zur Kenntnis des Paläozoicums in Oberfranken, Ostthüringen und dem Sächsischen Vogtlande. 1. Stratigraphie und Ammoneenfauna des Oberdevons von Hof a. S. *Neues Jahrbuch für Mineralogie, Geologie und Paläontologie, Beilagen-Band*, **49**, 250–357, 393–509.

Schindewolf, O.H. 1924. Bemerkungen zur Stratigraphie und Ammoneenfauna des Saalfelder Oberdevons. *Senckenbergiana*, **6**, 95–113.

Schindewolf, O.H. 1926. Zur Kenntnis der Devon-Karbon-Grenze in Deutschland. *Zeitschrift der Deutschen Geologischen Gesellschaft*, **78**, 88–133.

Schindewolf, O.H. 1934. Über eine Oberdevonische Ammoneen-Fauna aus den Rocky Mountains. *Neues Jahrbuch für Mineralogie, Geologie und Paläontologie, Beilagen Band (B)*, **72**, 331–350.

Schindewolf, O.H. 1949. Zur Nomenklatur der Clymenien (Cephalop., Ammon.). *Neues Jahrbuch für Mineralogie, Geologie und Paläontologie, Monatshefte (B)*, **1949**, 64–76.

Schindewolf, O.H. 1952. Über das Oberdevon und Unterkarbon von Saalfeld in Ostthüringen. *Eine Nachlese zur Stratigraphie und Ammoneen-Fauna. Senckenbergiana*, **32**, 283–306.

Schmidt, H. 1924. Zwei Cephalopoden an der Devon-Carbongrenze im Sauerland. *Jahrbuch der Preußischen Geologischen Landesanstalt*, **44**, 98–171.

Schülke, I. & Popp, A. 2005. Microfacies development, sea-level change, and conodont stratigraphy of Famennian mid- to deep platform deposits of the Beringhauser Tunnel section (Rheinisches Schiefergebirge, Germany). *Facies*, **50**, 647–664.

Schülke, I., Korn, D., Popp, A. & Ziegler, W. 2002. Potential reference section for the Early/Middle Famennian boundary at the Beringhauser Tunnel (Rheinisches Schiefergebirge, NW Germany). Document submitted to the Subcommission on Devonian Stratigraphy, Annual Meeting, Toulouse, 2002.

Selwood, E.B. 1960. Ammonoids and trilobites from the Upper Devonian and Lower Carboniferous of the Launceston area of Cornwall. *Palaeontology*, **3**, 153–185.

Shah, S.M.I. 1969. Discovery of Paleozoic rocks in the Khyber agency. *Geonews*, **1**, 31–34.

Shah, S.M.I. 1977. Stratigraphy of Pakistan. Khyber Area. *Memoirs of the Geological Survey of Pakistan*, **12**, 16–18.

Sheng, H. 1999. Late Devonian ammonoids fauna from Upper Daminshan Formation, Great Hinggan Mts., Inner Mongolia. *Professional Papers of Stratigraphy and Palaeontology, Chinese Academy of Geological Sciences*, **27**, 180–205.

Smith, J.F., Jr. & Ketner, K.B. 1975. *Stratigraphy of Paleozoic Rocks in the Carlin-Pinon Range Area, Nevada*. United States Geological Survey, Professional Papers, **867-A**.

Sobolew, D.N. 1912. O verkhnem neodevone okrestnostey Kelets. *Izvestiya Varshavskago Politekhnicheskago Instituta*, **2**, 1–14.

Sobolew, D.N. 1914*a*. Nabroski po filogenii Goniatitov. *Izvestiya Varshavskago Politekhnicheskago Instituta*, **1914**, 1–193.

Sobolew, D.N. 1914*b*. Über Clymenien und Goniatiten. *Paläontologische Zeitschrift*, **1**, 348–378.

Spalletta, C. & Perri, M.C. 2001. Subdivision and substages of the Famennian, an opinion and possible candidates for the upper part. *SDS Newsletter*, **18**, 64–66.

Spalletta, C., Perri, M.C., Corradini, C. & Over, J.D. 2015. Proposed revision of the Famennian (Upper

Devonian) standard conodont zonation. *In*: MOTTEQUIN, B., DENAYER, J., KÖNIGSHOF, P., PRESTIANNI, C. & OLIVE, S. (eds) *IGCP 596 – SDS Symposium, Climate change and Biodiversity patterns in the Mid-Palaeozoic. Abstracts. Strata*. Travaux de Géologie sédimentaire et Paléontologie, Série 1, **16**, 135–136.

STASCHEN, D. 1968. Zur Geologie des Warsteiner und Belecker Sattels (Rheinisches Schiefergebirge, Deutschland). *Münstersche Forschungen zur Geologie und Paläontologie*, **5**, 1–119.

TERMIER, H. & TERMIER, G. 1950. Paléontologie Marocaine. II. Invertebres de L'Ere primaire. Fasc. III, Mollusques. *Service Géologique, Protectorat de l'État Francais au Maroc, Notes et Mémoires*, **78**, 1–246.

TERMIER, H. & TERMIER, G. 1951. Stratigraphie et Paléontologie des Terrains Primaires de Benhamed (Chaouia sud, Maroc). *Notes et Mémoires, Direction de la Production industrielle et des Mines, Division des Mines et de la Géologie, Service Géologique*, **85**, 47–104.

TERMIER, H., TERMIER, G. & VACHARD, D. 1978. Recherches micropaléontologiques dans le Paléozoique supérieur du Maroc central. *Cahiers de Micropaléontologie*, **4**, 1–99.

THOMAS, L.A. 1949. Devonian–Mississippian formations of Southeast Iowa. *Geological Society of America Bulletin*, **60**, 403–438.

THOREZ, J., DREESEN, R. & STREEL, M. 2006. Famennian. *In*: DEJONGHE, L. (ed.) *Current Status of Chronostratigraphic Units Named from Belgium and Adjacent Areas*. Geologica Belgica, **9**, 27–45.

TRAGELEHN, H. & HARTENFELS, S. 2002. Köstenhof quarry (Frankenwald, Bavaria) – a potential reference section for the Early/Middle and the Middle/Upper Famennian boundary. Document submitted to the Subcommission on Devonian Stratigraphy, Annual Meeting, Toulouse, 2002.

VAN DEN BOOGARD, M. 1963. Conodonts of Upper Devonian and Lower Carboniferous age from southern Portugal. *Geologie en Mijnbouw*, **42**, 248–259.

WALLISER, O.H., EL HASSANI, A. & TAHIRI, A. 1995. Sur le Dévonien de la Meseta marocaine occidentale. *Courier Forschungsinstitut Senckenberg*, **188**, 21–30.

WEBSTER, G.D., BECKER, R.T. & MAPLES, C.G. 2005. Biostratigraphy, paleoecology, and taxonomy of Devonian (Emsian and Famennian) crinoids from southeastern Morocco. *Journal of Paleontology*, **79**, 1052–1071.

WEDDIGE, K. & ZIEGLER, W. 1979. Evolutionary patterns in the Middle Devonian conodont genera *Polygnathus* and *Icriodus*. *Geologica et Palaeontologica*, **13**, 157–164.

WEDEKIND, R. 1908. Die Cephalopodenfauna des höheren Oberdevon am Enkeberg. *Neues Jahrbuch für Mineralogie, Geologie und Paläontologie*, **26**, 565–633.

WEDEKIND, R. 1910. *Posttornoceras balvei* n.g. et n.sp. Ein neuer Fall von Konvergenz bei Goniatiten. *Centralblatt für Mineralogie, Geologie und Paläontologie*, **1910**, 768–771.

WEDEKIND, R. 1913. Beiträge zur Kenntnis des Oberdevon am Nordrande des Rheinischen Gebirges. 2. Zur Kenntnis der Prolobitiden. *Neues Jahrbuch für Mineralogie, Geologie und Paläontologie*, **1913**, 78–95.

WEDEKIND, R. 1914. Monographie der Clymenien des Rheinischen Gebirges. *Abhandlungen der Königlichen Gesellschaft der Wissenschaft zu Göttingen, Mathematisch-Physikalische Klasse, Neue Folge*, **10**, 1–73.

WEDEKIND, R. 1918. Die Genera der Palaeoammonoidea (Goniatiten). *Palaeontographica*, **62**, 85–184.

WEINER, T. 2012. New data on the *Annulata* Events in the Moravian Karst (Famennian, Czech Republic). *In*: WITZMANN, F. & ABERHAN, M. (eds) *Centenary Meeting of the Paläontologische Gesellschaft, Programme, Abstracts, and Field Guides*. Terra Nostra, Bonn, 192.

WEINER, T. & KALVODA, J. 2016. Biostratigraphic and sedimentary record of the *Annulata* Events in the Moravian Karst (Famennian, Czech Republic). *Facies*, **62**, Article 6, http://doi.org/10.1007/s10347-015-0456-2

WENDT, J. 1989. Facies pattern and paleogeography of the Middle and Late Devonian in the eastern Anti-Atlas (Morocco). *In*: MCMILLAN, N.J., EMBRY, A.F. & GLASS, D.J. (eds) *Devonian of the World*. Canadian Society of Petroleum Geologists, Memoirs, **14**(I), 467–480.

WENDT, J., AIGNER, T. & NEUGEBAUER, J. 1984. Cephalopod limestone deposition on a shallow pelagic ridge: the Tafilalt Platform (upper Devonian, eastern Anti-Atlas, Morocco). *Sedimentology*, **31**, 601–625.

WENDT, J., KAUFMANN, B., BELKA, Z., FARSAN, N. & BAVANDPUR, A.K. 2005. Devonian/Lower Carboniferous stratigraphy, facies patterns and palaeogeography of Iran Part II. Northern and central Iran. *Acta Geologica Polonica*, **55**, 31–97.

WEYER, D., BARTZSCH, K. & SCHNEIDER, J. 1996. *Exkursion 4: Silur, Devon und Zechstein in Ostthüringen. 66 Jahrestagung*. Paläontologische Gesellschaft, Leipzig.

WORONCOWA-MARCINOWSKA, T. 2006. Upper Devonian goniatites and co-occurring conodonts from the Holy Cross Mountains: studies of the Polish Geological Institute collections. *Annales Societatis Geologorum Poloniae*, **76**, 113–160.

ZIEGLER, W. 1962. Taxionomie und Phylogenie Oberdevonischer Conodonten und ihre stratigraphische Bedeutung. *Abhandlungen des Hessischen Landesamtes für Bodenforschung*, **38**, 1–166.

ZIEGLER, W. 1969. Eine neue Conodontenfauna aus dem höchsten Oberdevon. *Fortschritte in der Geologie von Rheinland und Westfalen*, **17**, 343–360.

ZIEGLER, W. 1970. *Erläuterungen zu Blatt 4713 Plettenberg*. Geologische Karte von NRW, 1 : 25 000. Geologisches Landesamt Nordrhein-Westfalen, Krefeld, 1–179.

ZIEGLER, W. 1971. Conodont stratigraphy of the European Devonian. *In*: SWEET, W.C. & BERGSTRÖM, S.M. (eds) *Symposium of Conodont Biostratigraphy*. Geological Society of America, Memoirs, **127**, 227–284.

ZIEGLER, W. & HUDDLE, J.W. 1969. Die *Palmatolepis glabra*-Gruppe (Conodonta) nach der Revision der Typen von Ulrich & Bassler durch J. W. Huddle. *Fortschritte in der Geologie der Rheinlande und Westfalen*, **16**, 377–386.

ZIEGLER, W. & SANDBERG, C.A. 1984. *Palmatolepis*-based revision of upper part of standard Late Devonian conodont zonation. *In*: CLARK, D.L. (ed.) *Conodont Biofacies and Provincialism*. Geological Society of America, Special Papers, **196**, 179–194.

ZIEGLER, W. & SANDBERG, C.A. 2000. Utility of Palmatolepids and Icriodontids in recognizing Upper Devonian

Series, Stage, and possible Substage boundaries. *Courier Forschungsinstitut Senckenberg*, **225**, 335–347.

Ziegler, W. & Weddige, K. 1999. Zur Biologie, Taxonomie und Chronologie der Conodonten. *Paläontologische Zeitschrift*, **73**, 1–38.

Ziegler, W., Sandberg, C.A. & Austin, R.L. 1974. Revision of *Bispathodus* group (Conodonta) in the Upper Devonian and Lower Carboniferous. *Geologica et Palaeontologica*, **8**, 97–112.

Zong, P., Becker, R.T. & Ma, X. 2014. Upper Devonian (Famennian) and Lower Carboniferous (Tournaisian) ammonoids from western Junggar, Xinjiang, northwestern China – stratigraphy, taxonomy and palaeobiogeography. *Palaeobiodiversity and Palaeoenvironments*, **95**, 159–202.

Review of chrono-, litho- and biostratigraphy across the global Hangenberg Crisis and Devonian–Carboniferous Boundary

RALPH THOMAS BECKER[1]*, SANDRA ISABELLA KAISER[2] & MARKUS ARETZ[3]

[1]*Institut für Geologie und Paläontologie, Westfälische Wilhelms-Universität, Corrensstrasse 24, D-481499 Münster, Germany*

[2]*State Museum of Natural History Stuttgart, Rosenstein 1, 70191 Stuttgart, Germany*

[3]*Géosciences Environnement Toulouse (GET), Observatoire Midi Pyrénées, Université de Toulouse, CNRS, IRD, 14 avenue E. Belin, F-31400 Toulouse, France*

**Corresponding author (e-mail: rbecker@uni-muenster.de)*

Abstract: Chrono-, litho- and biostratigraphy across the Devonian–Carboniferous transition are reviewed to provide a precise time framework for the global Hangenberg Crisis and for the current search for a revised basal Carboniferous Global Stratotype Section and Point (GSSP). The outer shelf deposits of the Rhenish Massif (Germany) form a lithological standard. Pre- (main Wocklum Limestone), lower (top Wocklum Limestone/Drewer Sandstone to Hangenberg Black Shale), middle (Hangenberg Shale/Sandstone), upper (Stockum Limestone), and post-crisis (Hangenberg Limestone) deposits are defined. Combined with the conodont, ammonoid and miospore zonations and eustatic trends, this succession can be correlated internationally. The contemporaneous successions of the Ardennes serve as a reference for shallow shelf settings. The positive and negative aspects of five options for a redefined Devonian–Carboniferous boundary level are discussed: (1) base of the black shale (main extinction level, base of *Bispathodus costatus–Protognathodus kockeli* Interregnum and LN Zone), (2) sequence boundary (widespread unconformities) or glacial and regressive peak (base of Hangenberg Sandstone), (3) base of the *kockeli* Zone and of initial postglacial transgression (base of lower Stockum Limestone), (4) entry of *Siphonodella* (*Eosiphonodella*) *sulcata* (base of upper Stockum Limestone), and (5) base of post-crisis interval (base of Hangenberg Limestone), at approximately the poorly correlated current GSSP level. Due to homonymy, *Siphonodella* (*Siphonodella*) *hassi* Ji, 1985 is renamed as *Siphonodella* (*Siphonodella*) *jii* nom. nov. Consequently, the mid-lower Tournaisian *S.* (*S.*) *hassi* Zone (previous Upper *S.* (*S.*) *duplicata* Zone) becomes the *S.* (*S.*) *jii* Zone.

Following a proposal by Paeckelmann & Schindewolf (published in 1937), the Devonian–Carboniferous Boundary (DCB) was the first chronostratigraphic level that was formally defined internationally in 1935 (Second Heerlen Congress). The boundary was defined by the entry of *Gattendorfia subinvoluta* in the Oberrödinghausen section, which is situated in the deeper-water cephalopod facies of the Rhenish Massif, Germany. However, studies on the Rhenish Stockum Limestone (e.g. Alberti *et al.* 1974) showed that the selected stratotype section contains a hiatus right at the chosen boundary level, at the base of the regional Hangenberg Limestone (HL) and base of the *Gattendorfia subinvoluta* Zone. This resulted in the search for a new boundary definition and Global Stratotype Section and Point (GSSP), with the formal implementation of an International Working Group on the Devonian–Carboniferous Boundary in 1976 (Paproth & Streel 1984*a*). After 15 years of intensive research, multiple publications, and highly controversial discussions, the new basal Carboniferous GSSP was placed within an oolithic succession (with a number of reworked conodonts), with the incoming of the supposed first *Siphonodella* (*Eosiphonodella*) *sulcata* at the base of Bed 89 in trench La Serre E′ in the Montagne Noire (southern France, Paproth *et al.* 1991). This decision was criticized by Ji *et al.* (1989) and Ziegler & Sandberg (1996), mostly because of different views concerning the taxonomic placement of intermediate specimens between *S.* (*Eo.*) *praesulcata* and *S.* (*Eo.*) *sulcata*, which had been recorded from below the GSSP bed. Subsequent work at La Serre by Kaiser (2005, 2009) confirmed that the same siphonodellid morphotype that was used to place the GSSP occurs already in Bed 84b. This new position is about 0.4 m below the original boundary and, more importantly, just above a conodont-free argillaceous interval. Therefore, the new records contradict the original suggestion (Feist & Flajs 1988) that the boundary was defined within a continuous, branching

From: Becker, R. T., Königshof, P. & Brett, C. E. (eds) 2016. *Devonian Climate, Sea Level and Evolutionary Events*. Geological Society, London, Special Publications, **423**, 355–386.
First published online January 6, 2016, http://doi.org/10.1144/SP423.10

siphonodellid chronomorphocline, the supposed *praesulcata–sulcata* transition. Since there are no criteria of bio-, chemo- or physical stratigraphy to correlate the current La Serre GSSP with precision into any other section, a second revision of the basal Carboniferous GSSP became inevitable (e.g. Kaiser & Becker 2007; Kaiser & Corradini 2008). This task is currently the focus of a new international working group (e.g. Aretz & Task Group 2011, 2013, 2014; Aretz 2014).

While the global Kellwasser Crisis at the Frasnian–Famennian boundary is widely acknowledged as one of the five major Phanerozoic mass extinction episodes, the at least equally severe global Hangenberg Crisis around the DCB has received less attention, although it wiped out complete ecosystems (e.g. all reefs and early forests), profoundly changed the lowland terrestrial ecosystems and devastated the outer shelf habitats in the marine realm. This major biotic overturn was triggered by sudden climatic changes, sea-level fluctuations, and the spread of hypoxia/anoxia in the course of a short-lived plunge from global greenhouse to icehouse conditions. A second review paper in this volume (Kaiser *et al.* 2015) focuses on the extinction patterns, abiotic events, including sequence and chemostratigraphy, and possible explanations of the Hangenberg Crisis. This review summarizes the chrono- and biostratigraphic framework of the crisis and the lithological succession in its classical German type region. It aims to serve the current search for a new basal Carboniferous GSSP and evaluates five possible levels.

Abbreviations

*ck*I, *costatus–kockeli* Interregnum; DCB, Devonian–Carboniferous boundary; HBS, Hangenberg Black Shale; HL, Hangenberg Limestone; HS, Hangenberg Shale; HSI, DCB holostratigraphical intervals of Becker (1996); HSS, Hangenberg Sandstone; LC, Lower Carboniferous/Mississippian; LST and UST, Lower and Upper Stockum Limestone; UD, Upper Devonian; WL, Wocklum Limestone.

Taxonomy: *Ac.*, *Acutimitoceras*; *Bi.*, *Bispathodus*; *Clyd.*, *Clydagnathus*; *Did.*, *Diducites*;

Eo., *Eosiphonodella*; *Pa.*, *Palmatolepis*; *Patr.*, *Patrognathus*; *Po.*, *Polygnathus*; *Post.*, *Postclymenia*; *Pr.*, *Protognathodus*; *Ps.*, *Pseudopolygnathus*; *R.*, *Retispora*; *Q.*, *Quasiendothyra*; *S.*, *Siphonodella*; *T.*, *Tournayellina*; *V.*, *Vallatisporites*.

Chronostratigraphy (Fig. 1)

The International Subcommission on Devonian Stratigraphy (SDS) voted at the end of 2003 to subdivide the Famennian formally into four substages (Becker 2005). Previously, Streel *et al.* (1998) proposed to place the base of an uppermost Famennian substage at the base of the Upper *Palmatolepis gracilis expansa* Zone *sensu* Ziegler & Sandberg (1984) to create a substage that is roughly equivalent to the Strunian of classical Belgian and French chronostratigraphy (e.g. Streel *et al.* 2006; discussions in Becker 2008). The term Strunian has been used widely/internationally for upper/uppermost Famennian shallow-water successions. However, SDS has not yet formally voted on the definition of Famennian substages. For its youngest substage there are some uncertainties concerning the taxonomy and precise range of the potentially defining marker conodont, *Bispathodus ultimus* (e.g. Kononova & Weyer 2013). The term 'uppermost Famennian' is here used in the proposed revised sense of the Belgian Strunian or of the Wocklumian (Wocklum Stufe = Famennian VI, *Wocklumeria* Stufe, Upper Devonian [UD] VI) of German regional chronostratigraphy (Becker *et al.* 2012; Fig. 1).

The selection of the basal Carboniferous GSSP at La Serre was mostly based on a morphometric study of the supposedly preserved *S.* (*Eo.*) *praesulcata–sulcata* lineage (Feist & Flajs 1988). The extensive conodont reworking within the oolithic succession was not seen as an obstacle. But the assumed entry of the first *S.* (*Eo.*) *sulcata* in Bed 89 has been questioned from the beginning (Ji *et al.* 1989; Ziegler & Sandberg 1996) and was finally disproved by Kaiser (2005, 2009). Moreover, the 'GSSP morphotype' of *S.* (*Eo.*) *sulcata* is not identical with the morphologically more advanced, lost holotype (Kaiser & Corradini 2011), which came from a much higher level (Evans *et al.* 2013). Kaiser & Corradini (2011) and Corradini *et al.* (2011) listed and reviewed the problems arising from the use of siphonodellids and/or protognathodids for the boundary definition. They concluded that the determination of the DCB position and of the index conodonts currently depends quite strongly on subjective interpretations. Siphonodellid evolution was very complex across the Hangenberg Crisis, with curved *sulcata* homeomorphs occurring in 'pre-Hangenberg beds' and several lineages ranging through the crisis interval (e.g. Corradini *et al.* 2003; Tragelehn 2010; Becker *et al.* 2013; Girard *et al.* 2013; Kalvoda *et al.* 2015; Kumpan *et al.* 2015*b*). Many sections show an event-/facies-controlled interruption of the siphonellid succession. Therefore, the old idea (Walliser 1984) to define the base of the Carboniferous by the main Hangenberg Event and mass extinction level has recently regained support. Based on the current GSSP, the Hangenberg Crisis interval includes approximately the upper third of the uppermost

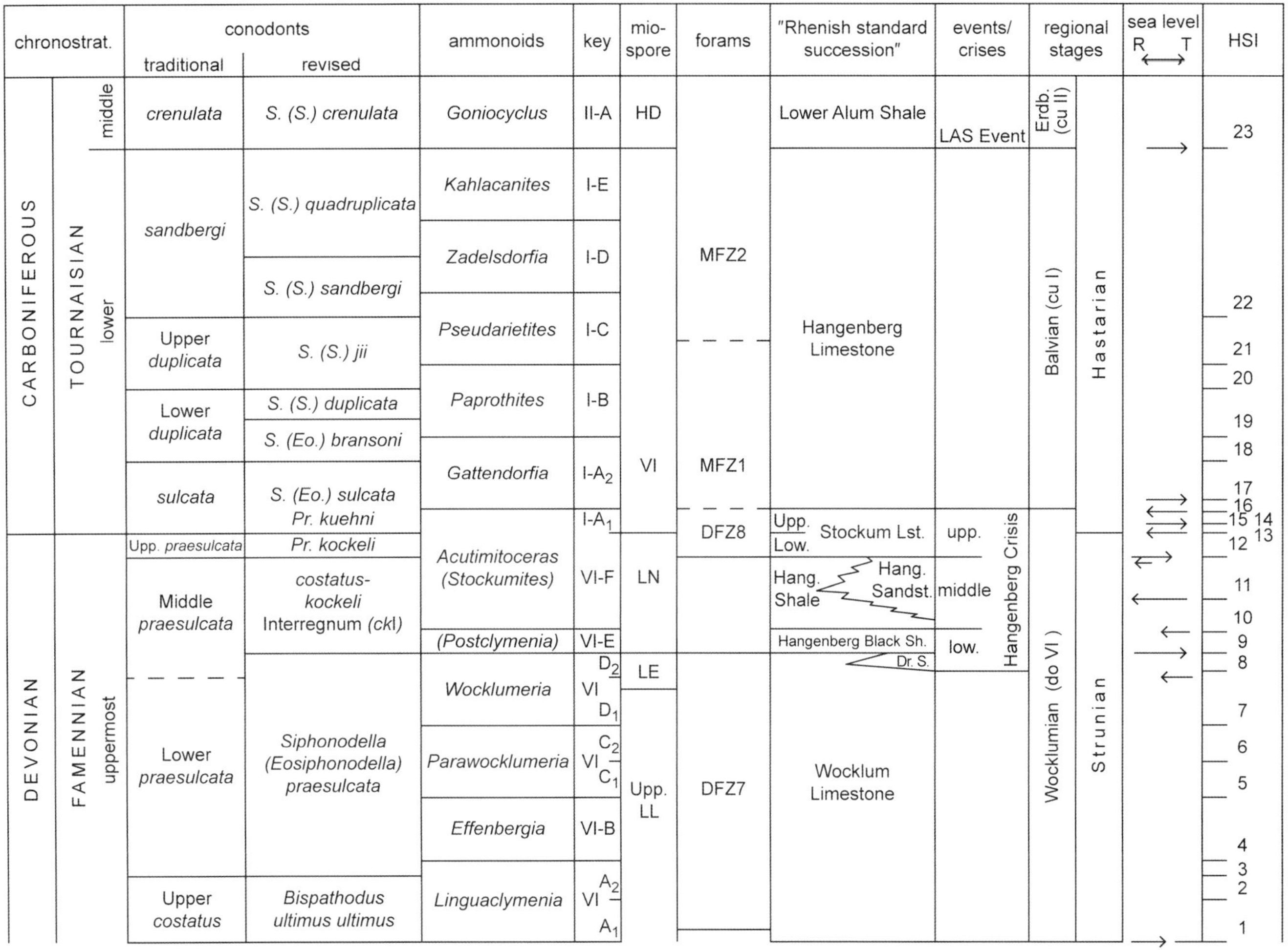

chronostrat.			conodonts traditional	conodonts revised	ammonoids	key	miospore	forams	"Rhenish standard succession"	events/crises	regional stages		HSI
CARBONIFEROUS	TOURNAISIAN	middle	*crenulata*	*S. (S.) crenulata*	*Goniocyclus*	II-A	HD		Lower Alum Shale	LAS Event	Erdb. (cu II)	Hastarian	23
		lower	*sandbergi*	*S. (S.) quadruplicata*	*Kahlacanites*	I-E	VI	MFZ2	Hangenberg Limestone		Balvian (cu I)		
				S. (S.) sandbergi	*Zadelsdorfia*	I-D							22
			Upper *duplicata*	*S. (S.) jii*	*Pseudarietites*	I-C							21, 20
			Lower *duplicata*	*S. (S.) duplicata* / *S. (Eo.) bransoni*	*Paprothites*	I-B		MFZ1					19, 18
			sulcata	*S. (Eo.) sulcata* / *Pr. kuehni*	*Gattendorfia*	I-A_2							17, 16
						I-A_1							15, 14
DEVONIAN	FAMENNIAN	uppermost	Upp. *praesulcata*	*Pr. kockeli*	*Acutimitoceras (Stockumites)*	VI-F	LN	DFZ8	Upp. / Low. Stockum Lst.	upp. (Hangenberg Crisis)	Wocklumian (do VI)	Strunian	13, 12
			Middle *praesulcata*	*costatus-kockeli* Interregnum (*ckl*)					Hang. Shale / Hang. Sandst.	middle			11, 10
					(Postclymenia)	VI-E			Hangenberg Black Sh.	low.			9
				Siphonodella (Eosiphonodella) praesulcata	*Wocklumeria*	VI D_2 / D_1	LE		Dr. S.				8, 7
			Lower *praesulcata*		*Parawocklumeria*	VI C_2 / C_1	Upp. LL	DFZ7	Wocklum Limestone				6, 5
					Effenbergia	VI-B							4
			Upper *costatus*	*Bispathodus ultimus ultimus*	*Linguaclymenia*	VI A_2 / A_1							3, 2, 1

Fig. 1. Correlation of DCB chronostratigraphy, traditional and revised conodont zones, ammonoid genozones, European miospore and foraminifere zones, the lithological 'Rhenish Standard succession', event/crisis levels, regional stages, and supposedly eustatic sea-level changes. LAS, Lower Alum Shale. The ammonoid key and the holostratigraphic intervals (HSI) follow Becker (1996).

Famennian and the basal-most Tournaisian. If the DCB is redefined in future by the main extinction, the crisis interval would begin only just below and last longer into an extended lower Tournaisian.

Two auxiliary stratotypes have been designated that are relevant for the correlation of the crisis interval across facies boundaries. Hasselbachtal in the Rhenish Massif (Becker *et al.* 1984; Becker & Paproth 1993), because of its palynomorph record and potential for terrestrial correlation (Higgs & Streel 1984), and Nanbiancun, Guangxi, South China, for its neritic setting and faunas (Yu 1988; Wang 1993); but both are not available as future new GSSP sections. The Hasselbachtal succession has recently been covered by the property owner. At Nanbiancun there are doubts concerning the precise placing of the GSSP level (Ji *et al.* 1989; Gong *et al.* 1992), and resampling of the solid limestone has become almost impossible without a powerful rock saw.

The base of the Carboniferous is simultaneously the base of the Mississippian series and of the Tournaisian stage. However, the original Tournaisian (Tn) of the Belgian type region was based on shallow-water successions that start earlier (e.g. Mortelmans 1969), while the type 'Calcaire de Tournai' is a younger unit (Hance *et al.* 2006). Traditionally, the Tournaisian has been subdivided into three substages, Tn1–Tn 3 (e.g. reviews in Mortelmans 1969; Groessens 1975). It should be noted that this traditional subdivision was abandoned in Belgium (Conil *et al.* 1977), with the introduction of the Hastarian and Ivorian considered as regional substages. However, the term 'Middle Tournaisian' is in use on a global scale, from Europe (Becker 1993*b*), to North Africa (Korn *et al.* 2002), North America (Boardman *et al.* 2013) and China (Zong *et al.* 2015). The former Tn1a or Etroeungt correlates with the widespread onset of carbonate sedimentation in the Famennian of the Ardennes. It equals roughly the proposed uppermost Famennian (Streel *et al.* 2006), while the former Tn1b correlates roughly with the current lower Tournaisian. The base of the middle Tournaisian (Tn 2) correlates with the global Lower Alum Shale Event (Becker 1993*a*, *b*; Becker *et al.* 2006). This is an important but often neglected, second-order, transgressive and anoxic extinction event that severely interrupted the post-Hangenberg Crisis biotic recovery.

Hance & Poty (2006) redefined the base of the regional Hastarian substage of Belgium to coincide with the base of the Tournaisian, although the basal bed of the name-giving Hastière Formation contains a conodont fauna of the upper crisis interval (*Protognathodus kockeli* Zone, Van Steenwinkel 1980). The redefined Hastarian is roughly identical with the revised North American Kinderhookian (Thompson 1986; Davydov *et al.* 2012). The regional Balvian stage or Lower Carboniferous I (cu I) of Germany (Schmidt 1972) equals the lower Tournaisian (Tn 1b) and, therefore, the lower Hastarian.

Lithostratigraphy (Figs 1 & 2)

The 'Rhenish Standard Succession'

The deeper-water sections of the Rhenohercynian Basin in the northern and eastern Rhenish Massif (Germany) are often used as a standard for successions spanning the DCB interval (Figs 1 & 2). Krebs (1979) summarized the general facies model for the region; Paproth *et al.* (1986) gave a palaeogeographical overview. Summaries and correlations of important Rhenish DCB sections were published by Paproth & Streel (1982), Bless *et al.* (1988, 1993), Clausen *et al.* (1989), Luppold *et al.* (1994: with a map showing the position of localities mentioned here), and Korn & Weyer (2003). Condensed, often nodular and strictly cyclic cephalopod limestones (micrite–marl couplets) dominate the Famennian strata in the shallower parts of the pelagic realm of the Rhenohercynian Basin. Deeper parts are characterized by shales that are often rich in stratigraphically important entomozoid ostracods ('Cypridinenschiefer'). They contain limestone nodules in intermediate slope settings.

The uppermost Famennian Wocklum Limestone (WL; HSI 1–7 of Becker 1996) is lithologically continuous with the underlying Dasberg Limestone, and both are mostly combined as one mapping unit. They contain a rich pelagic fauna with dominant ammonoids, orthoconic cephalopods, ostracods, conodonts, agglutinating foraminifers, some bivalves and gastropods, rhynchonellid brachiopods, mostly small-eyed to blind trilobites, and deep-water rugose corals. The WL is increasingly condensed and fossiliferous towards its top (Becker 1996). The DCB succession at its type section, at the Borkewehr near Wocklum, is illustrated in Figure 3. In some sections upper parts of the WL are missing due to shallowing upwards, extreme condensation and non-deposition (e.g. at Reitenberg Forest Quarry, Schmidt & Plessmann 1961: locality C11, the type section of several uppermost Famennian clymeniids, e.g. Korn & Price 1987; Söte 2015). On seamounts the whole WL may be lacking in a longer DCB hiatus (e.g. Beul, Paeckelmann 1938; Beringhausener Tunnel, Schülke & Popp 2005). Locally, the Drewer Sandstone, a laminated siltstone and distal turbidite, is intercalated just below the top (Korn 1991; Bless *et al.* 1993; Kumpan *et al.* 2015*a*, *b*). It marks the onset of the extended crisis interval or a crisis prelude (HSI 8 of Becker 1996).

The overlying, thick shale and silty shale package comprises the Hangenberg Shale (HS) in a wide sense. The lowermost part consists of black and dark grey shales, known as the Hangenberg Black Shale (HBS). It represents a time of restricted oxygenation and starved sedimentation (Van Steenwinkel 1993*a*, *b*; Kumpan *et al.* 2015*a*, *b*). The organic matter is amorphous and possibly of cyanobacterial origin; miospores are rare and acritarchs are almost absent (Streel 1999). The base of the HBS is often very sharp (Hasselbachtal, Oese, Oberrödinghausen, Borkewehr/Wocklum, Fig. 3) and the main extinction level for pelagic biota (HSI 9, Hangenberg Event *sensu stricto*). At Drewer the last thin *Wocklumeria*-rich nodule layers alternate above the Drewer Sandstone with moderately dark grey, pyritic/goethitic marls. This gives a more gradual transition to the HBS (Paeckelmann & Schindewolf 1937; Kumpan *et al.* 2015*a*, *b*), but the extinction was locally equally sudden. A similar pattern is known from Thuringia (Bartzsch *et al.* 1998), where the 'Rhenish Standard Succession' is fully developed. The HBS forms the main part of the lower crisis interval and contains no benthos apart from opportunistic bivalves (*Guerichia*) that are also characteristic of older Famennian black shale events (e.g. Hartenfels 2011). Their shell form and size resemble conchostracans, which explains records of contemporaneous 'conchostracan beds' in North America (see discussion in Bless *et al.* 1988; spinicaudatan level in Cole *et al.* 2015). The opportunistic bloom of *Postclymenia* is also distinctive.

The HBS is gradually succeeded by partly thick, greenish, silty shales with *Guerichia* and flattened *Ac.* (*Stockumites*) in specific layers (e.g. Paproth & Streel 1970; Korn *et al.* 1994; new specimens from Oberrödinghausen). This less hypoxic and regressive HS (Schmidt 1924) in a strict sense characterizes the onset of the middle crisis interval (HSI 10, initial Hangenberg Regression). Depending on the palaeogeographical situation, a sandstone package, the Hangenberg Sandstone (HSS), replaces the HS (thin laminated unit at Drewer, Clausen *et al.* 1989; Korn *et al.* 1994) or is developed on the top (thick turbiditic unit of Oese, Keupp & Kompa 1984; HSI 11, peak of Hangenberg Regression). Locally (near Hemer, Bomfleur 2005), the HSS cuts out the underlying HB/HBS and top of the WL (Fig. 2). This supports the transsect of Paproth (1986), who suggested that the HSS of the northern Sauerland represents a lateral extension of the erosional channel (incised valley) of the distinctive Seiler area (Gallwitz 1927). But Koch *et al.* (1970) documented that this Seiler palaeovalley was filled during three different upper/uppermost Famennian episodes with coarse siliclastics, oolites and calcareous conglomerates. A last intercalated bed of WL of their main trench was overlain by a dark, partly sandy shale and by a dark limestone, as probable HBS equivalents. This unit was followed by an oolite with reworked dark limestone clasts, which formed the base of a *c.* 190 m thick siliciclastic middle crisis interval (see Fig. 2). The poorly fossiliferous HSS becomes inconspicuous with increasing distance from the depocentre (e.g. at the Wocklum type section, Luppold *et al.* 1994; Fig. 3). Time equivalents in shaly and silty facies are indistinguishable from the HS. The whole siliciclastic interval (HBS/HSS) may be missing on seamount slope sections, such as Müssenberg (Korn 1984) or Drewer NE (Clausen *et al.* 1989), due to current induced non-deposition during sea-level fall.

The detrital facies is topped by a new development of cephalopod limestone facies, which marks the beginning of the upper crisis interval. This Stockum Limestone or the Stockum Levels are a marly–silty facies containing limestone nodules or beds. They are clearly recognizable by their *Ac.* (*Stockumites*) faunas and *Protognathodus* biofacies from the overlying HL. Based on the evolution in *Protognathodus*, Lower (HSI 12, with *Pr. kockeli*) and Upper (HSI 15, with *Pr. kuehni*) Stockum Limestones or levels can be separated (Clausen *et al.* 1989; Bless *et al.* 1993). At Drewer the HSS is followed by an upper black shale and then by a reworked unit with re-deposited pre-crisis ammonoids and conodonts at the base of the lower Stockum level (Korn 1991; Bless *et al.* 1993; Korn *et al.* 1994; Fig. 2). A second minor erosional interval can be developed locally as a thin unit of encrinite and calcareous turbidite between the lower and upper Stockum levels (Hasselbachtal, Becker *et al.* 1984; HSI 14 of Becker 1996; Fig. 2).

The post-crisis HL is less argillaceous and represents the typical cephalopod limestone facies containing *Gattendorfia* and a pelagic faunal assemblage resembling palaeoecologically (not taxonomically) the WL. Locally, a basal unconformity may be developed (Paproth & Streel 1984*b*; Bless *et al.* 1993; HSI 16). This erosional unconformity resulted in the revision of the original (at Heerlen Congress in 1935) DCB stratotype at Oberrödinghausen (Fig. 2). The HL (HSI 17–22) grades upwards into an up to 5 m thick grey shale (Korn & Weyer 2003). This poorly fossiliferous top part of the Balvian/lower Tournaisian is sharply succeeded by the black Lower Alum Shale (HSI 23). In the more recent terminology of Korn (2010), the latter unit is a member of the Oberrödinghausen and Eichenberg formations.

The main facies development of the northern Rhenish Massif can be projected into contemporaneous strata of other European basins and even to pelagic settings of other palaeocontinents, such as northern Gondwana (Kaiser *et al.* 2011) and South China (e.g. Bai & Ning 1989; Ji *et al.* 1989). Walliser (1984) coined the term "time-specific facies" and the Rhenish lithostratigraphical units assumed

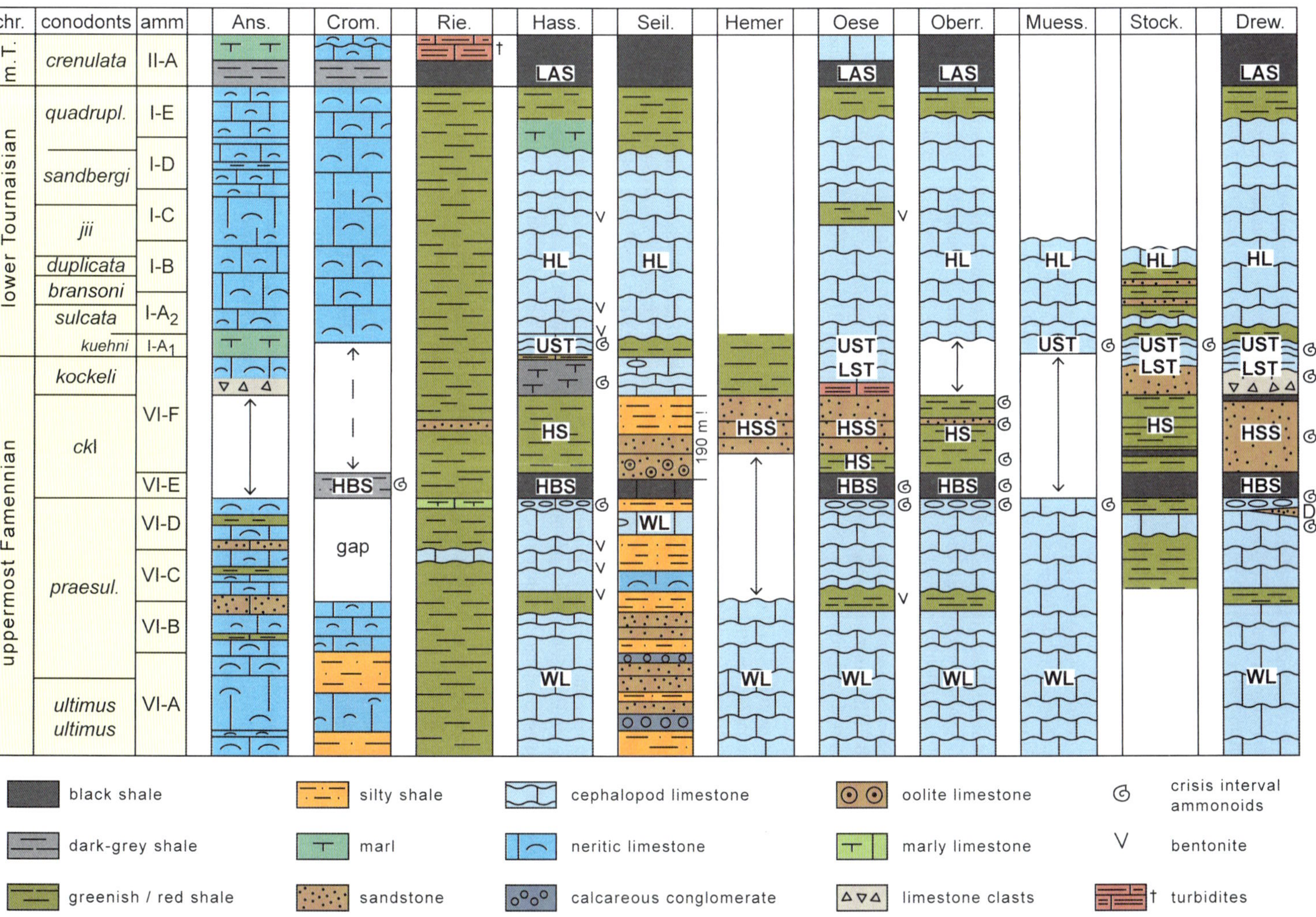
chr.
conodonts
amm
Ans.
Crom.
Rie.
Hass.
Seil.
Hemer
Oese
Oberr.
Muess.
Stock.
Drew.
m. T.
lower Tournaisian
uppermost Famennian
crenulata
quadrupl.
sandbergi
jii
duplicata
bransoni
sulcata
kuehni
kockeli
ckl
praesul.
ultimus ultimus
II-A
I-E
I-D
I-C
I-B
I-A2
I-A1
VI-F
VI-E
VI-D
VI-C
VI-B
VI-A
LAS
HL
UST
LST
HSS
HS
HBS
WL
DrS
gap
190 m !
black shale
dark-grey shale
greenish / red shale
silty shale
marl
sandstone
cephalopod limestone
neritic limestone
calcareous conglomerate
oolite limestone
marly limestone
limestone clasts
crisis interval ammonoids
bentonite
turbidites

a key role for regional and global correlation. However, the typical succession and marker beds of the crisis interval are lost in the shaly to turbiditic facies of the deeper, central and southern Rhenish Basin (e.g. Bender *et al.* 1993). In some parts of the basin, the WL is developed, but the deeper HS facies persists through all of the lower Tournaisian (e.g. Dölling & Piecha 2010).

The Rhenish neritic facies

Sections in the neritic realm are necessarily different from the 'Rhenish Standard', with transitions within the northern Rhenish Massif. In the Wuppertal area, at the Riescheid reference section (Higgs & Streel 1984, 1994; Fig. 2), the Wocklum to HL interval is replaced by a thick, poorly fossiliferous shale–calcareous siltstone succession with minor limestone interbeds ('*Cypridina* shale facies'). Within it, the crisis interval can only be identified with the help of miospores that were washed in by distal turbidites. Since the HBS is not developed, it is intriguing that the Lower Alum Shale is locally a distinctive black shale (Zimmerle *et al.* 1980) at the base of the overlying Steinberg Formation (Amler & Herbig 2006; Fig. 2).

To the NW, in the neritic facies of the western Velbert Anticline (e.g. Cromford section, Ratingen area to the west of Heiligenhaus), the WL is replaced by a cyclic, mixed carbonate–siliciclastic succession of the upper Velbert Formation (Amler & Herbig 2006; local Etroeungt beds of Paul 1939). These rocks contain a diverse fossil assemblage: bivalves, brachiopods, bryozoans, calcareous algae, microproblematica and foraminifers, including *Quasiendothyra kobeitusana* (Paul 1939; Austin *et al.* 1970*a*; Paproth & Streel 1982; Amler 1993, 1995; Weber & Wyse Jackson 2006; Ernst & Herbig 2010; Ernst *et al.* 2015). The calcareous facies/upper member of the Velbert Formation is followed by a non-anoxic equivalent of the HBS with *Postclymenia evoluta* (*Cymaclymenia euryomphala* in Paul 1939) and then by the regional Steinkothen Member of the Hastière Formation ('Ostracodenkalk' in older literature), a neritic equivalent of the HL. The overlying Pont d'Arcole Formation ('Zwischenschiefer' in older literature) represents the Lower Alum Shale event interval of the *Siphonodella* (*S.*) *crenulata* Zone.

Herbig & Mamet (2006) reconstructed for the 'Strunian' succession of the Velbert Anticline a SE-dipping carbonate ramp, ranging from turbulent, clear-water carbonates in the NW to calmer, muddy and marly sediments in the NE. The latter carries a different bryozoan fauna (Tolokonnikova *et al.* 2014) and lacks calcareous algae and foraminifers (Herbig *et al.* 2013). In the Velbert area both the pre- and post-crisis beds may turn into oolithic successions (e.g. Böger 1962; Michels 1986). Some oncolite levels (Kohleiche, Beds 05/20–29, Thomas & Zimmerle 1992) may correlate with the middle crisis interval. Paproth *et al.* (1976) showed that HS equivalents with *Post. evoluta* (quoted as *Cymaclymenia euryomphala*?) are locally intercalated (see also Herbig *et al.* 2013). The Tournaisian oolite unit has been named by Amler & Herbig (2006) as regional Laupen Member of the Hastière Formation.

The neritic succession of the Ardennes

The sections in the Namur–Dinant Basin of Belgium and northern France can be considered as some kind of reference for DCB successions in shallow-water facies (e.g. Poty *et al.* 2014). Above the predominantly siliciclastic rocks of the upper Famennian (Evieux Formation), uppermost Famennian strata, the classical Strunian facies of the Etroeungt Limestone/Formation (lateral Comblain-au-Pont and Dolhain formations of other authors), record the return of a carbonate facies. The Strunian consists of an alternation of limestone beds and shaly and marly intercalations and interbeds with cyclic stacking patterns (Van Steenwinkel 1993*a*, *b*). The amount of carbonate of these open-marine ramp deposits with storm layers increases progressively upsection. The overlying Hastière Formation consists of bedded and massif, pure limestone (Poty *et al.* 2002). The significance of reworked fauna in the lower part of the distinctive, thick, lower bed (e.g. Conil *et al.* 1986; Van Steenwinkel 1988; Casier *et al.* 2004) is debated (e.g. Poty *et al.* 2002, 2006). The base is commonly understood to represent a hiatus (e.g. Bless *et al.* 1993; Mamet & Preat 2003; Kumpan *et al.* 2014*b*;

Fig. 2. Idealized west–east facies transect of DCB beds along the northern margin of the Rhenish Massif, starting in the neritic setting of the Ardennes (Ans., Anseremme) and of the Velbert Anticline (Crom., Ratingen-Cromford), continuing through the intermediate Wuppertal region (Rie., Riescheid) and through the variably complete Sauerland sections from Hasselbachtal (Hass.) to Seiler (Seil.; Trench II), Hemer, Hemer-Oese, Oberrödinghausen Railway Cut (Oberr.), to Müssenberg (Muess.), the Stockum trenches (Stock.) and finally to Drewer (Drew.). Note the severe carbonate crisis (complete interruption of carbonate deposition) in the lower/middle crisis interval. WL, Wocklum Limestone; Dr, Drewer Sandstone; HBS, Hangenberg Black Shale; HS, Hangenberg Shale; HSS, Hangenberg Sandstone; LST, Lower Stockum Limestone; UST, Upper Stockum Limestone; HL, Hangenberg Limestone; LAS, Lower Alum Shale; m. T., middle Tournaisian; gap denotes covered outcrop. Vertical arrows show the extent of unconformities.

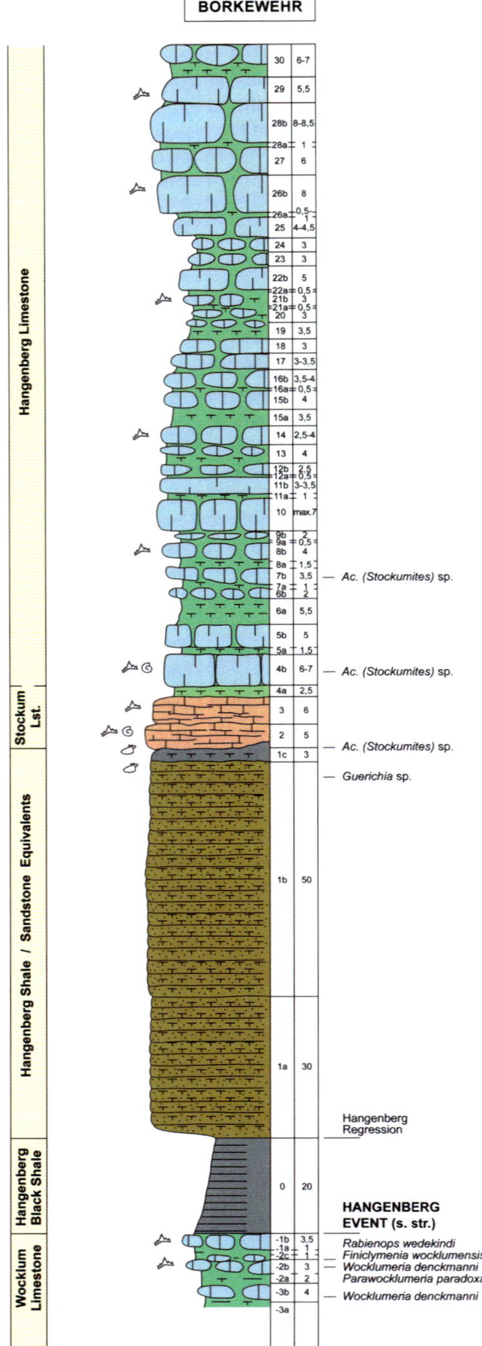

Fig. 3. The revised DCB section at Borkewehr/Wocklum, with the position of the main Hangenberg Event and showing the cyclic deposition of the WL and HL interrupted by the siliciclastic HBS and HS/HSS equivalents.

see Fig. 2), which comprises at least the time-equivalent strata of the HS and HSS. However, more recent work (Hance *et al.* 2001; Poty *et al.* 2006) indicates a shorter gap locally, and the debate concerning the completeness of the neritic sections is ongoing. For example, Mottequin & Poty (2014) noted that HBS and HSS equivalents are locally preserved between the Comblain-au-Pont and Hastière formations. Van Steenwinkel (1993*a*) placed the sequence boundary at the base of the Hastiére Formation. Kumpan *et al.* (2014*b*) re-interpreted this surface as a 'basal surface of forced regression' but above there is no other clear unconformity (sequence boundary level).

The basal unit of the Belgian Hastiére Formation (Van Steenwinkel 1980, 1984) and at least the lower part of the corresponding Avesnelles Formation in northern France (Conil *et al.* 1986) fall in the *kockeli* Zone. Eastwards, the basal unit can be followed into the basal part of the Binsfeldhammer Member of the Aachen succession (Amler & Herbig 2006). All three levels correlate with the Lower Stockum Limestone. Locally, a thin, possibly regressive, shale separates this unit from the Tournaisian (Hastarian) main part of the formation. This level has been identified by Azmy *et al.* (2009) as the Hangenberg Event (Royseux-Gare section, between beds 104 and 105, see also Austin *et al.* 1970*c*), although it more likely corresponds with the small-scale DCB regression and extinction level between the Lower and Upper Stockum limestones in the upper part of the Hangenberg Crises. A revision of the section, including conodont resampling, is ongoing.

The equivalent of the Rhenish Lower Alum Shale is found on top of the Hastière Formation, when greyish to greenish calcareous shales and siltites of the Pont d'Arcole Formation ('*peracuta* Shale') were deposited.

Biostratigraphic time-scales

Conodont zonation (Fig. 1)

The uppermost Famennian to lower Tournaisian 'standard zonation' of Ziegler & Sandberg (1984) has been updated by Ji (1985) and Kaiser *et al.* (2009). As a principle, and in order to exclude different definitions hidden under a superficially uniform nomenclature, biozones are best named after their defining index species. The uppermost Famennian has been proposed to begin with the Upper *Palmatolepis gracilis expansa* Zone (Streel *et al.* 1998) that has been renamed as the *Bispathodus ultimus* Zone by Hartenfels & Becker (2012; see equivalent regional *ultimus* Zone of the Pyrenees in Perret 1988). However, ranges of early forms/subspecies of the index taxon, such as *Bi. ultimus bartzschi* and '*Bi. ziegleri* Morphotype 1' in

Kononova & Weyer (2013), have to be excluded. This leads to a refined *ultimus ultimus* Zone. *Pseudopolygnathus marburgensis trigonicus* and *Pa. gonioclymeniae* are important alternative markers for correlation into regions that lack *Bi. ultimus ultimus*, for example North America, or when *Bi. ultimus ultimus* enters with delay (e.g. Sardinia, Mossoni *et al.* 2013). The base of the *ultimus ultimus* Zone is close to an argillaceous package in the Rhenish lower WL. Unfortunately, it is difficult to correlate into shallow-water conodont successions, such as those from the Russian Platform (Aristov 1988).

The revised subsequent *S.* (*Eo.*) *praesulcata* Zone occupies approximately the upper two-thirds of the WL. But its base is poorly defined, because the index species is often rare and affected by taxonomic problems (e.g. Tragelehn 2010; Kaiser & Corradini 2011; Kalvoda *et al.* 2015). Currently any first *Siphonodella* is taken to indicate the *praesulcata* Zone and all species of the genus with shallow, non-inverted basal cavity fall in *Siphonodella* (*Eosiphonodella*) Ji, 1985. Several ancestral or parallel lineages enter slightly earlier (Tragelehn 2010; Becker *et al.* 2012; Ji in Aretz & Task Group 2013; Kumpan *et al.* 2014*a*). These 'siphonodelloids' include *Ps. graulichi*, which is important since it allows correlation into the neritic facies of the Ardennes (Bouckaert & Dusar 1976; Bouckaert & Groessens 1976; Bouckaert *et al.* 1978) or of Kazakhstan (Martynova & Vorontzova 1983). *Polygnathus spicatus* and *Po. parapetus* are closely related (including probably *Po.* cf. *spicatus* of Austin *et al.* 1970*c*); other forms have been been wrongly assigned to *Po. symmetricus*.

The sudden and global extinction of the *Bi. costatus–ultimus* Group, of *Ps. marburgensis*, of the last *Palmatolepis*, and of other taxa, defines the base of the *costatus–kockeli* Interregnum (*ck*I) at the base of the HBS (Kaiser *et al.* 2009). This is a slightly higher level than the base of the former Middle *praesulcata* Zone, which was poorly defined by the rather diachronous disappearance of *Pa. gonioclymeniae* variably well below or near the top of the WL. Old records of *Bi. costatus* from middle/upper Tournaisian shallow-water successions of England, the Ardennes and Australia (Druce 1969, Austin *et al.* 1970*b*, *c*) have been revised to represent *Bi. aculeatus* (Ziegler *et al.* 1974). The first *Pr. meischneri* may enter as rare faunal elements just before the first siphonodellids, followed within the *praesulcata* Zone by rare and sporadic occurrences of *Pr. collinsoni* (see review in Kaiser *et al.* 2009). But the *ck*I is characterized by a pantropical spread/bloom of both early *Protognathodus* species, for example in the La Serre stratotype (Paproth *et al.* 1991), at Grüne Schneid, Austria (Kaiser 2007), at Kowala, Poland (Dzik 1997), in the Junggar Basin of NW China (Xu *et al.* 1990) and in the Gedongguan Bed of Guizhou, South China (Hou *et al.* 1985). They occur in association with long-ranging survivors, especially some polygnathids, *Neopolygnathus*, bispathodids, *Branmehla*/*Mehlina*, various pseudopolygnathids, and 'siphonodelloids'. The so far most diverse conodont faunas of the middle crisis interval are known from the Holy Cross Mountains (Dzik 1997), Graz Palaeozoic (Kaiser *et al.* 2009), southern Morocco (Becker *et al.* 2013) and Moravia (Kalvoda *et al.* 2015). Reworking of pre-Hangenberg taxa is a recurring problem. For example, several subspecies of the small-sized *Pa. gracilis* have been reported in very small numbers from the post-crisis beds of various sections (Sandberg *et al.* 1972; Luppold *et al.* 1984; Stewart & Selwood 1985; Nemirovskaya *et al.* 1993; Korn *et al.* 1994; Dzik 1997; Kalvoda *et al.* 2015). But the study of outer shelf sections with very calm and complete sedimentation showed no survival (Ji *et al.* 1987; Kaiser *et al.* 2009).

The upper crisis interval is characterized by the onset of some new taxa, especially of *Pr. kockeli*, which is also the index species of the former Upper *praesulcata* Zone and 'Lower *Protognathodus* Fauna' (Clausen *et al.* 1989). The zone occurs in the Lower Stockum Limestone and the basal unit of the Hastière Limestone (Van Steenwinkel 1980, 1988). The oldest *Po. purus* are also important at this level (Chlupác & Zikmundova 1976; Schönlaub *et al.* 1988; Kaiser 2007; Kaiser *et al.* 2009; Malec 2014). Supposedly pre-crisis specimens of *Po. purus purus* from Sardinia (Corradini *et al.* 2003; Corradini 2008; Mossoni *et al.* 2015) have never been figured. They are at odds with the rich conodont record from the uppermost Famennian of all other regions and, therefore, currently doubtful.

The debate on the morphology and precise taxonomy of *S.* (*Eo.*) *sulcata* seriously affects the recognition of the basal Carboniferous *sulcata* Zone. Despite its rarity in some regions, *Pr. kuehni* serves as an alternative index ('Upper *Protognathodus* Fauna', Alberti *et al.* 1974; Clausen *et al.* 1989; Kaiser *et al.* 2006, 2009), especially because of the crisis-/facies-controlled interruption of the *Siphonodella* record in the Stockum Limestone. The basal HL falls in the higher *sulcata* Zone (e.g. Clausen *et al.* 1989). The subsequent Tournaisian siphonodellid zonation of Sandberg *et al.* (1978) or Ji *et al.* (1987) has been refined by Ji (1985). He renamed '*S. duplicata sensu* Hass (1959)' of Sandberg *et al.* (1978) as *S.* (*S.*) *hassi* but overlooked that this name is preoccupied by *S. cooperi hassi* Thompson & Fellows, 1970 (see Becker *et al.* 2013, p. 114). Therefore, we rename this index species here as *S.* (*S.*) *jii* nom. nov., with the consequence that Ji's '*hassi* Zone', the former Upper *Siphonodella* (*S.*) *duplicata* Zone, approximately in the middle of the lower Tournaisian, has to be renamed as the *jii*

Zone. The middle Tournaisian begins with the *crenulata* Zone, which has been recognized near the base of the Rhenish Lower Alum Shale (Bless *et al.* 1993).

Q. Ji (1987) and Ji & Ziegler (1992) established a shallow-water siphonodellid zonation that has been successfully applied to the widespread neritic carbonate platforms of South China (Hunan, Q. Ji 1987; W. Ji 1987; and Coen & Groessens 1996; Guangdong, Qin *et al.* 1988; Guangxi, Qie *et al.* 2014; Hainan Island, Zhang *et al.* 2010). It has been assumed that the entry of *S.* (*Eo.*) *homosimplex* (*S. simplex* in older papers, but the holotype of that species falls in *S.* (*Eo.*) *levis*, which has priority) roughly correlates with the entry of *S. sulcata* (Q. Ji 1987; Qin *et al.* 1988). Qie *et al.* (2014), however, documented a staged entry, with rare *S. sulcata* slightly below, preceded by a thin, *Siphonodella*-free interval with the oldest *Clyd. unicornis*. The smooth shallow-water *Siphonodella* lineage provides a good marker to search for the crisis interval in underlying successions that lack index conodonts (e.g. Hance *et al.* 1994).

In European, Russian, Asian and Australian shallow-water successions, the 'standard zonation' is difficult to apply. The *Patrognathus variabilis*–*Clydagnathus plumulus* Zone of Great Britain (Austin *et al.* 1970*b*; Butler 1973: with 'siphonodelloids' identified as *Po. symmetricus*; Austin & Hill 1973), of SW Ireland (Matthews & Naylor 1973), the Russian Platform (Alekseev *et al.* 1979) and Kuznetsk Basin (Bouckaert & Boonen 1978), the corresponding *Clyd. plumulus* Zone of northern Australia (Druce 1969: with the siphonodellid '*Polygnathus* sp. B' at the base), and similar faunas from the Donetz Basin (Lipnjagow 1979), Timan (Durkina & Avchimovitch 1988), southern Siberia (Bushmina 1984), the Kashmir region (Savage 1982), northern Iran (Fallah *et al.* 2011), Xinjiang (Xiong 1991: *Clyd. plumulus* above *S.* (*Eo.*) cf. *sulcata*), and Hunan (Coen & Groessens 1996) are commonly placed at the base of the Carboniferous. A rough correlation with the *sulcata* Zone is supported by a direct association of *Clyd. plumulus* and *S.* (*Eo.*) *sulcata* in Queensland (Mory & Crane 1982) and just below *S.* (*Eo.*) *homosimplex* in Guangxi, South China (Qie *et al.* 2014). But an upper Famennian range of *Clyd. plumulus* is well established (e.g. Alekseev *et al.* 1994; Draganits *et al.* 2002; Hartenfels 2011). The species did not originate, but blossomed, in shallow settings after the Hangenberg Crisis. *Patrognathus variabilis* and *S.* (*Eo.*) *sulcata* (s.l.) enter at the same level in Wyoming (Austin *et al.* 1970*b*, p. 439). But on parts of the Russian Platform, *Patr. variabilis* may enter higher in the lower Tournaisian, above a regional *Patrognathus crassus* Zone (Alekseev *et al.* 1994; Makhlina 1996). Bushmina & Kononova (1981), by contrast, reported a direct association of *Patr. variabilis* and *I. costatus* within the Topki Formation of the Kuznetsk Basin, which suggests a regional pre-crisis range of the index species.

The rarity or absence of *Protognathodus* in these assemblages contradicts the view that it is a typical shallow-water genus. In shallow-marine facies it is difficult to recognize *kockeli* and *sulcata* Zone equivalents and to decide whether transgressions started with one or the other level. There are several examples that indicate distinctive shallow-water assemblages of the upper crisis interval. For example, Xiong & Chen (1983) recognized in Guizhou a *Patrognathus yashuiensis*–*Pr. kockeli* assemblage that was correlated with the British *variabilis*–*plumulus* Zone. In the Donetz Basin, *Patr. andersoni* co-occurs with *Clyd. plumulus* near the base of the Carboniferous, followed by other species of the genus and some *S.* (*Eo.*) *sulcata* (Lipnjagow 1979). In the Guangdong Province of China, a *Clydagnathus* fauna is sandwiched between levels with typical pre-crisis taxa and the entry of *S.* (*Eo.*) *homosimplex* (quoted as *S. simplex*, Qin *et al.* 1988). In the Jianghua County of Hunan, a fauna of low diversity with *Clyd. plumulus* occurs between the last *Icriodus* and the first *S.* (*Eo.*) *homosimplex* (*S. simplex* in W. Ji 1987). Table 1 summarizes the conodonts that died out and that survived the Hangenberg Crisis in shallow-water successions.

Ammonoid zonation (Figs 4–6)

The principles of DCB ammonoid stratigraphy were established in the German type area by the classical studies of Schindewolf (1937: uppermost Famennian, Wocklumian) and Vöhringer (1960: lower Tournaisian, Balvian). The zonal scheme has been subsequently refined by a number of publications. Differences reflect research progress and a variable preference for zonal index species, but there is little disagreement concerning the succession of marker taxa (Fig. 4). Last summaries and overviews can be found in Becker (1996), Korn (2000), Becker & House (2000) and Becker *et al.* (2002, 2015).

The base of the uppermost Famennian or Wocklumian is variably defined at the base of the WL by the incoming of (the revalidated) *Kalloclymenia subarmata*, of the rare '*Sphenoclymenia*' *brevispina*, of the oldest and distinctive *Linguaclymenia*, or the less distinctive *Muessenbiaergia laevis* (UD VI-A; Schindewolf 1937; Korn & Luppold 1987; Korn & Price 1987; Hartenfels & Becker 2012; Kononova & Weyer 2013). It is possible to establish parallel zones based on the evolution within the Prionoceratidae (see Korn 1994; Korn *et al.* 2015), Kosmoclymeniidae (see Korn & Price 1987), or on a sucession of other marker clymeniids (Fig. 4). Of special interest is the appearance of giant

Table 1. *Hangenberg Crisis victims and survivors among conodont taxa (93 species/subspecies) recorded from shallow-water successions*

Victims

Antognathus
- *volnovadensis*

Apatognathus
- *geminus*
- *libratus*
- *striatus*
- *varians cipitis*

Bispathodus
- *costatus*
- *jugosus*
- *ultimus ultimus*

Branmehla/Mehlina
- *aciedentata*
- *dentiloba*
- *fissilis*
- *mehlbrai*
- *paucidentata*
- *praelonga*

Capricornugnathus
- *capricornis*

Fungulodus/Conchodontus
- *azygoideus*
- *centronodosus*
- *chondroideus*
- *circularis*
- *conchiformis*
- *rotundatus*
- *sulcatus*
- *ziegleri*

Icriodus
- *costatus*
- *darbyensis*
- *obovatus*
- *platys*
- sp. A

Mashkovia
- *simakovi*
- *similis*

Neopolygnathus
- *'carina'* (auct.)
- *collinsoni*
- *martynovae*

'Pandorinellina'
- *"insita"*
- *nota*

Pelekysgnathus
- *bicuspidatus*
- *communis*
- *firmus*
- *inclinatus*
- *isodentatus*
- *nodosus*
- *peejavi*
- *superstes*

Polygnathus
- *changtanziensis*
- aff. *longonodosus*
- aff. *lenticularis*
- *streeli*

Pseudopolygnathus
- *conili*
- *controversus*
- cf. *micropunctatus*
- *postinodosus*
- *trigonicus*

Rhodalepis
- *polylophodontiformis*

'siphonodelloids'
- aff. *parapetus*

Tanaissognathus
- *businovensis*

Survivors

Apatognathus
- *varians ethingtoni*
- *varians varians*

Bispathodus
- *ac. aculeatus*
- *ac. anteposicornis*
- *spinulicostatus*
- *stabilis stabilis*

Branmehla/Mehlina
- *crassidentata*
- *fitzroyi*
- *inornata*
- *planiconvexa**
- *regularis*
- *strigosa*

Clydagnathus
- *cavusformis*
- *plumulus*

Neopolygnathus
- *comm. communis*
- *lectus*
- *mugodjaricus*

Omolonognathus
- *planus*

Patrognathus
- *donbassicus**
- *variabilis**

Pelekysgnathus
- *australis*

Polygnathus
- *delicatulus*
- *inornatus*
- *lenticularis*
- *lobatus*
- *longiposticus*
- *paprothae*
- *rostratus*
- *symmetricus*
- *thomasi*
- *toxophorus*
- *vogesi*
- *znepolensis**

Protognathodus
- *meischneri*

Pseudopolygnathus
- *dentilineatus*
- *nodomarginatus*
- *primus*
- *vogesi*

'siphonodelloids'
- *graulichi*
- *parapetus*
- *praesulcata* [s.l.]

In alphabetic order of genera and species.
*Denotes survival in deeper-water/uncertainty. Based on records from the Ardennes (Austin *et al.* 1970*c*; Bouckaert & Dusar 1976; Van Steenwinkel 1980, 1984; Conil *et al.* 1986), England (Austin *et al.* 1970*b*; Butler 1973), Wales (Austin & Hill 1973), Ireland (Matthews & Naylor 1973; Austin & Husri 1974), the Russian Platform (Alekseev *et al.* 1979, 1994; Aristov 1988), the Timan, northern Russia (Durkina & Avchimovitch 1988; Kuz'min 1998), the Donetz Basin, Ukraine (Lipnjagow 1979; Aisenverg *et al.* 1979), the Transcaucasus (Aristov *et al.* 1979), Azerbaijan (Grechishnikova & Levitskii 2011), the Alborz Mountains and Kerman region, Iran (Ueno *et al.* 1997; Wendt *et al.* 2002; Fallah *et al.* 2011), the Mugodzhary, southern tip of the Urals (Barskov *et al.* 1984; Gagiev *et al.* 1987), the Karaganda Basin, Kazakhstan (Martynova & Vorontzova 1988), the Kuznetsk Basin, southern Siberia (Bouckaert & Boonen 1978; Bushmina & Kononova 1981; Yolkin *et al.* 2000; Bakharev *et al.* 2011; Izokh & Andreeva 2013), Spiti and Kashmir, Himalaya, India (Savage 1982; Draganits *et al.* 2002), Xinjiang, NW China (Xiong 1991), Hunan (Q. Ji 1987; W. Ji 1987; Coen & Groessens 1996), Sichuan (Q. Ji 1987), Guangxi (Wang & Yin 1985; Wang *et al.* 1987; Shen 1994; Qie *et al.* 2014), and Guizhou (Xiong & Chen 1983), South China, Kolyma and Omolon, Russian Far East (Shilo *et al.* 1984; Gagiev & Kononova 1990; Gagiev & Bogus 1990), the Bonaparte (Druce 1969) and Canning basins (Nicoll & Druce 1979), Western Australia, Queensland and New South Wales (Mory & Crane 1982).

taxa both in the southern Urals (Bogoslovskiy 1981) and Rhenish Massif (Becker 1988; Becker *et al.* 2016). Other goniatite lineages, for example the Posttornoceratidae and Sporadoceratidae, show slow evolution in the uppermost Famennian. In the upper half of the WL or Wocklumian, the rapid evolution within the highly distinctive 'triangular clymeniids', the Wocklumeriacea, provides a very detailed zonation (e.g. Schindewolf 1937; Becker 2000). Recent work (Dzik 2006; Becker & Mapes 2010) showed that the first Wocklumeriidae, the genera *Synwocklumeria* and *Kielcensia*,

chr.	genozones	key	ev.	kosmoclymeniids	goniatites	other clymeniids/ prolecanitids
m. T.	*Goniocyclus*	II-A	LASE		*Goniocyclus ammari*	*Protocanites hollardi*
lower Tournaisian	*Kahlacanites*	I-E			*"Hasselbachia" gourara*	*Kahlacanites mariae*
	Zadelsdorfia	I-D			*Zadelsdorfia lhceni* *Paragattendorfia patens*	*Eocanites supradevonicus*
	Pseudarietites	I-C			*Pseudarietites westfalicus*	*Eocanites spiratissimus*
	Paprothites	I-B			*Paprothites dorsoplanus*	*Eocanites nodosus*
	Gattendorfia	I-A 2			*Gattendorfia subinvoluta* *Ac. (Acutimitoceras) acutum*	
	Ac. (Stockumites)	I-A 1	Hangenberg Crisis		*Ac.(Streeliceras) carinatum* *Ac. (Stockumites) prorsum* *Ac. (Stockumites) subbilobatum* Gp.	
uppermost Famennian		VI-F				*Postclymenia nigra*
	(Postclymenia)	VI-E			*"Mimimitoceras" transiens*	*Postclymenia evoluta*
	Wocklumeria	VI-D 2		*Lissoclymenia wocklumeri* *Kosmoclymenia schindewolfi*	*Mayneoceras nucleus*	*Epiwocklumeria applanata*
		VI-D 1				*Wocklumeria sphaeroides* *Wocklumeria denckmanni*
	Parawocklumeria	VI-C 2				*Parawocklumeria paradoxa*
		VI-C 1			*Effenbergia falx* *Balvia globularis*	*Parawocklumeria paprothae* *Kamptoclymenia endogona*
	Effenbergia	VI-B		*Muessenbiaergia ademmeri*	*Effenbergia lens*	*Soliclymenia paradoxa* *Glatziella helenae*
	Linguaclymenia	VI-A 2		*Muessenbiaergia bisulcata* *Muessenbiaergia parundulata*	*"Mimimitoceras" geminum*	*"Sphenoclymenia" brevispina* *Kalloclymenia subarmata*
		VI-A 1		*Muessenbiaergia sublaevis* *Linguaclymenia similis*		

Fig. 4. Correlation of DCB ammonoid genozones with the variable zones based on kosmoclymeniids, Prionocerataceae (Goniatitida) and descendent Pericyclacea, and other clymeniids.

characterize the *Parawocklumeria paradoxa* Zone (UD VI-C2). *Mayneoceras nucleus* is an important and easily recognizable alternative marker species, which ranges palaeogeographically from Europe to southern Morocco (Korn 1994; Becker *et al.* 2002), southern Algeria (Petter 1959), the Urals (Nikolaeva & Bogoslovsky 2005*a*) and South China (Ruan 1981). The last zone (UD VI-D) of the upper WL and corresponding strata is characterized by the pantropical spread of various species of *Wocklumeria* (e.g. Becker 2000; Ebbighausen & Korn 2007). Several clymeniid groups, for example *Soliclymenia*, evolute parawocklumeriids and glatziellids, most *Kalloclymenia*, the Biloclymeniidae, *Kosmoclymernia s. str.* and *Muessenbiaergia*, do not reach this level. This documents a somewhat gradual latest Famennian ammonoid extinction, which occurred prior to the Hangenberg Crisis. A *Synwocklumeria–Muessenbiaergia* assemblage from the Caucasus (Nikolaeva & Bogoslovsky 2005*b*) correlates best with the *paradoxa* Zone, not with the *Wocklumeria* level, as suggested by the authors.

The extended crisis interval begins in the Rhenish Massif (topmost WL) with a short subzone defined by the appearance of *Epiwocklumeria* (UD VI-D2; Schindewolf 1937; Becker 1996, 2000; see new specimen from Hasselbachtal, Fig. 5.1). At the same level or just slightly earlier, *Post. evoluta* enters in the Drewer Sandstone (Korn 1988). This explains a brief overlap of the latter with *Wocklumeria* in pelagic settings (e.g. Drewer) or with phacopids in neritic facies (co-occuring '*Cymaclymenia euryomphala*' and '*Phacops*' *accipitrinus* in Paul 1939). The *Epiwocklumeria applanata* Subzone was a short interval. It comprises only the last three or four thin limestone nodule layers (cycles) of the WL and probably had a duration of less than 100 ka.

The *Post. evoluta* Zone (UD VI-E; Becker 1988, 1993*b*) begins at the base of the HBS. It is not defined by the entry of the index species but by the sudden extinction of almost all ammonoids, notably of almost all clymeniids. Exceptions are some cymaclymeniids, one or two mimimitoceratids, and very rare last sporadoceratids. Table 2 summarizes the global ammonoid record of the pre-crisis (UD VI-D1) and initial crisis interval (UD VI-D2) to show the severity of extinctions at the top of the *Wocklumeria* Zone. The current record from the *applanata* Subzone is probably biased by the incomplete record of last occurrences of rare taxa (Signor–Lipps effect). This is exemplified by a new and globally youngest *Posttornoceras* from the last limestone nodule layer at Drewer (leg. Y. Gatovsky). Extinction rates increase from *c*. 75% at the family level to *c*. 87% at the species level. No genus or species is known to range through the complete crisis, and some regions suffered total extinctions (e.g. North Africa). The base of the *evoluta* Zone coincides with the conodont extinction that marks the base of the *ck*I. European and North African black shales are characterized by an opportunistic bloom and spread of *Postclymenia* (e.g. Schmidt 1924; Schindewolf 1937; Paproth & Streel 1970; Luppold *et al.* 1994; Kaiser *et al.* 2009). New HBS specimens of *Post. evoluta* (= *euryomphala*) were collected at Oese (Figs 5.2 & 5.3), Oberrödinghausen (Fig. 5.4) and also at Drewer (Becker *et al.* 2015; not listed from that level in Korn *et al.* 1994). This monospecific lower-crisis assemblage can be correlated into neritic settings of the western Rhenish Massif (Schmidt 1924; Paul 1939; Paproth *et al.* 1976; Price & Korn 1989) and beyond, into the Ardennes (Delepine 1929; Austin *et al.* 1970*a*). A corresponding youngest cymaclymeniid fauna occurs in eastern North America at the top of the Cleveland Shale of Ohio (House 1978; House *et al.* 1986) and in South China (Ma Xueping, pers. comm., May 2015). But the richest ammonoid assemblage so far of this first post-extinction phase has recently been found in Xinjiang, NW China (Zong *et al.* 2015), where it has been dated with the help of miospores and conodonts (Xu *et al.* 1990).

The initial radiation of the only surviving goniatite group (Prionoceratidae) began early in the middle crisis interval (HS) and led to the oldest, mostly badly preserved and still poorly known *Acutimitoceras* (*Stockumites*) faunas (lower UD VI-F, middle/upper *ck*I). Records are from the Rhenish Massif (Paproth & Streel 1970: Oberrödinghausen, new specimen with convex ornament from 42 cm above the HS base, Fig. 5.6; Korn *et al.* 1994: Drewer, new specimens), Thuringia (Bartzsch & Weyer 1986: Fauna 6a within the 'Hauptquarzit'), the Holy Cross Mountains, Poland (Dzik 1997: Kowala), southern Morocco (Becker in Hahn *et al.* 2012: basal Aoufital Formation of the Amessoui Syncline, southern Tafilalt), and Ohio (House *et al.* 1986: basal Bedford Shale). Many flattened Rhenish specimens show slightly biconvex growth lines and convex to slightly biconvex constrictions as in *Ac.* (*Stock.*) *procedens* (e.g. Korn *et al.* 1994: fig. 18C; new specimens, see Figs 6.3 & 6.4). Others display more strongly biconvex ornament and evolute whorls as in *Ac.* (*Stock.*) *prorsum* (Paproth & Streel 1970: pl. 24, figs 1a, b, 3; Korn *et al.* 1994: figs 18E, G–I; see also Dzik 1997: fig. 27). One new specimen from the upper part of the HS at Oberrödinghausen lacks constriction and has an almost 3 mm wide umbilicus (Fig. 6.5). Therefore, it is transitional to the younger *Ac.* (*Stockumites*) *antecedens*. The specimens from Ohio, identified in House *et al.* (1986) as *Prionoceras quadripartitum*, and as *Mimimitoceras varicosum* in House (1993), resemble *Ac.* (*Stock.*) *procedens*, especially in the discordant course of growth lines and constrictions, but have lower whorls and probably represent a different species. *Acutimitoceras* (*Stock.*) aff. *subbilobatum* from southern Morocco (Fig. 6.1), the regionally oldest post-extinction ammonoid, represents a new, compressed species with weak ribbing caused by growth-line bundling, unlike as in typical *subbilobatum*. The combined evidence from all localities suggests that both the *Ac.* (*Stock.*) *subbilobatum* (with convex ornament, early stages moderately evolute) and *Ac.* (*Stock.*) *procedens–prorsum* groups (with biconvex ornament, early stages markedly evolute) appeared just after the HBS.

Of special importance for the international correlation of the Great Basin DCB succession (western North America) are records of *Ac.* (*Stockumites*) from the lower oncolitic marker unit of the Leatham Member of the Pilot Shale (Feist & Petersen 1995). As in the Rhenish Massif (Fig. 5.5), cymaclymeniids may be associated. A new specimen of *Ac.* (*Stockumites*) with typical early whorls was collected by RTB in 2007 at southeastern Conger Mountain, Utah (Fig. 6.2, locality 9 of Morrow *et al.* 2007). These records suggest that all '*Imitoceras*' from the oncolite level and from shales just below, also from corresponding levels of the Middle Sappington Formation of Montana (Gutschick & Rodriguez 1979), belong to *Ac.* (*Stockumites*) (see previous discussion in Becker 1993*b*). This requires a significant revision of the regional stratigraphy, in contrast to the most recent study of Cole *et al.* (2015). The revised recognition of the lower/middle Hangenberg Crisis interval in the 'Conchostracan Shale' and overlying oncolithic ('algal'), sponge or brachiopod-rich limestones, siltstones and shales is completely in accord with the trilobite (*Pudoproetus*, Feist & Petersen 1995), brachiopod (*Syringothyris* faunas, e.g. Gutschick & Rodriguez 1979), and palynomorph record (upper LN Zone assemblages of Sandberg *et al.*

Fig. 5. New ammonoid records from the lower/middle Hangenberg Crisis interval of the Rhenish Massif (collection of the Geomuseum, WWU Münster, No. B6*c*. 48.1-6). (**1a**, **b**) *Epiwocklumeria applanata* Schindewolf, 1937 from the top Wocklum Limestone (Bed Ha113cN) at Hasselbachtal, lateral and ventral views, ×2. (**2**) *Post. evoluta* Schmidt, 1924 from the lower HBS of Oese showing typical sutures and traces of concavo-convex ornament, ×1.5. (**3**) *Post. evoluta* Schmidt, 1924 from the gradual HBS/HS transition at Oese, ×1.5. (**4**) *Post. evoluta* Schmidt, 1924

1972; Warren *et al.* 2014). But conodont faunas are affected by reworking of upper Famennian taxa. For example, outside the Great Basin, palmatolepids, such as *Pa. postera* and *Pa. rugosa* (see faunas in Sandberg *et al.* 1972), have never been recorded from the uppermost Famennian and have never been found in association with early protognathodids (see Kaiser *et al.* 2009) or with acutimitoceratids.

The upper crisis interval ('Lower and Upper *Protognathodus* Faunas', upper UD VI-F to Lower Carboniferous/Mississippian [LC] I-A1, 'Stockum Limestone faunas') is characterized by a rapid further radiation of acutimitoceratids and a final spread of cymaclymeniids (e.g. Barskov *et al.* 1984, Korn 1984, 1991, 1992, 1994, 2000; Luppold *et al.* 1984; Price & House 1984; Kusina 1985; House 1993). Ammonoid faunas of this interval have been reported from North America (Missouri), Ireland, the Rhenish Massif, Thuringia, Franconia, the Holy Cross Mountains, Moravia, the Carnic Alps, southern France, southern Morocco, the southern Urals (Mugodzhar) and Xinjiang (review of Becker 1993*b* and subsequent records in Korn 1994, 1999; Korn *et al.* 1994, 2004, 2007; Luppold *et al.* 1994; Becker *et al.* 2002; Korn & Weyer 2003; Korn & Feist 2007; Kaiser *et al.* 2011; Hahn *et al.* 2012; Zong *et al.* 2015). They include the oldest subglobular forms, *Ac.* (*Stockumites*) *kleinerae*, *Ac.* (*Streeliceras*), *Ac.* (*Sulcimitoceras*), and somewhat enigmatic forms, such as '*Imitoceras*' *bertchogurense* from the Mugodzhar and '*I.*' *compressum* from the Louisiana Limestone of Missouri. In ammonoid terms, the DCB is currently poorly defined by the final (cyma)clymeniid extinction. The evolutionary significance of such small-scale 'global survivor extinctions' is generally underevaluated in mass extinction scenarios.

The post-crisis lower Tournaisian and HL are characterized by the rapid diversification of Gattendorfiinae (defining the '*Gattendorfia* Stufe', base of LC I-A2) and of the oldest prolecanitids (e.g. Vöhringer 1960; Bartzsch & Weyer 1982; Price & House 1984; House 1993; Korn 1993, 2000; Ruan 1995). The fast evolution within the Pseudarietitinae provides the higher zones (LC I-B to I-D, Fig. 4), but the subfamily is currently restricted to Europe and South China (review in Becker 1993*b* and subsequent records in Dzik 1997; Korn & Weyer 2003). It may be possible to recognize the zones and subzones of Vöhringer (1960) in South China, where the best material occurs in mixed assemblages from Guizhou (Ruan 1981), while the shallower biofacies of Guangxi yielded only some isolated specimens (Yu 1988). A detailed correlation into North Africa (northern Gondwana) is more difficult and has to be based on gattendorfiids (see different views in Korn *et al.* 2007 and by Becker in Hahn *et al.* 2012). Ebbighausen *et al.* (2004), Ebbighausen & Bockwinkel (2007) and Korn *et al.* (2007) added a new *Kahlacanites* Zone at the top of the lower Tournaisian (LC I-E, Fig. 4) that is currently only known from North Africa (southern Morocco, southern Algeria). But it is of major significance for the understanding of ammonoid evolution through the global Lower Alum Shale Event, since it indicates how many lineages survived at the genus level. The subsequent middle Tournaisian is characterized by a global spread of the oldest pericyclid faunas, notably with *Goniocyclus* (e.g. Kullmann *et al.* 1990; Korn *et al.* 2002; Work 2002; Becker *et al.* 2006). In the Lower Alum Shale of the Rhenish Massif, such faunas have not been found; they are also lacking in NW China, where the *Weyerella* lineage is dominant (Ruan 1995; Zong *et al.* 2015), and in eastern Australia, which, however, yielded the same unusual prodromitid group as southern Morocco (Becker 2010).

Miospore zonation (Fig. 1)

There are many studies that describe miospore successions through the DCB (e.g. Molyneux *et al.* 1984; Coleman & Clayton 1987; Gao 1989; Higgs *et al.* 1993; Hance *et al.* 1994; Matyja & Stempien-Salek 1994; Perreira *et al.* 1996; Wagner 2001; Atta-Peters & Anan-Yorke 2003; Wicander & Playford 2013; Matyja *et al.* 2014). For early studies see the various contributions and quoted references in Streel (1969), Streel & Wagner (1970) and Playford & McGregor (1993). The principal western European miospore zonation was established by Clayton *et al.* (1977, 1978) and Higgs & Streel (1984), with updates in Maziane *et al.* (1999) and Streel (2009). Byvsheva *et al.* (1984) and Avchimovitch *et al.* (1988, 1993) provided correlations with the somewhat different zonal schemes of eastern Europe/Russia. Of special importance are palynological studies, which enable the dating and correlation of glacigenic deposits of South America (e.g. Perez-Leyton 1991; Loboziak *et al.* 1992, 1993, 1995; Vavrdová *et al.* 1991, 1996; Streel *et al.* 2000*a*, *b*, 2011; Wicander *et al.* 2011). Melo & Loboziak (2003) introduced a regional zonal scheme for the

Fig. 5. (*Continued*) from the HBS of Oberrödinghausen, ×1.5. (**5**) *Postclymenia* sp. from the HS of Oberrödinghausen, 42 cm above base, showing the typical concavo-convex ornament, ×3. (**6**) *Ac.* (*Stockumites*) sp. from the HS of Oberrödinghausen, 42 cm above base (currently the oldest specimen of the genus from the Rhenish Massif), with low aperture, some evidence of a closing umbilicus, and weakly biconvex growth lirae but without constrictions, ×2.

Table 2. *Global record of ammonoid species (sorted according to their family level systematics) from the pre-crisis lower/middle Wocklumeria Zone (UD VI-D1) and proven records from the initial crisis interval (UD VI-D2)*

Tornoceratidae,	*Balvia*	Kosmoclymeniidae
Falcitornoceratinae	*globularis**	*Lissoclymenia*
Falcitornoceras		*wocklumeri**
bilobatum Group	Sporadoceratidae	
	Sporadoceras	*Linguaclymenia*
Gundolficeras	n. sp. (from Ohio)* (+)	*similis**
delepinei	*muensteri orbiculare**	
	(= *terminus*)	Gonioclymeniidae
Posttornoceratidae	*longilobum*	*Kalloclymenia*
Posttornoceras		*pessoides*
*posthumum**	Cyrtoclymeniidae	
(= *weyeri*)	'*Cyrtoclymenia*'	*Finiclymenia*
aff. *posthumum*	*tetragona**	aff. *wocklumensis*
		*wocklumensis**
Discoclymenia	N. Gen.	
cucullata	(*Pricella* homeomorph)	Glatziellidae
	n. sp.*	*Glatziella*
Prionoceratidae,		*glaucopis**
Prionoceratinae	*Cyrtoclymenia*	
'*Mimimitoceras*'	aff. *procera**	*Postglatziella*
*geminum** (+)	*lateseptata**	*carinata*
*lentum**	(= *campanulata*)	
*liratum**	n. sp.*	Parawocklumeriidae
rotersi (auct.)* (+)		*Parawocklumeria*
	Protactoclymenia	*paprothae*
Mimimitoceras s. str.	*plicata**	*paradoxa**
alidrisii	*ventriosa*	(= *laevigata*)
*trizonatum**		
	Cymaclymeniidae	Wocklumeriidae
'*Rectimitoceras*'	*Cymaclymenia*	*Tardewocklumeria*
*quadripartitum**	*barbarae*	*perplexa?*
	carnata	
Rectimitoceras	*costellata**	*Wocklumeria*
*lineare**	*involvens** (+)	*denckmanni**
	lambidia	*oblivia**
Prionoceratidae, Balviinae	*nephroides**	*sphaeroides**
Mayneoceras	*sudetica**	
*nucleus**	*striata**(+)	*Synwocklumeria*
tetragonum		*kiensis**
	Postclymenia	*elata**
Kenseyoceras	*camerata**	
*rostratum**	(= *warsteinensis*)	*Epiwocklumeria*
(= *biforme*)	*evoluta* (+)	*applanata**

**Epiwocklumeria applanata* Subzone, UD VI-D2, *c.* the last three/four thin limestone nodule levels below the HBS and its equivalents), showing the magnitude of the main Hangenberg Extinction for all of UD VI-D (75% of subfamilies/families, 86% of genera/subgenera, 87% of species).

Mostly based on Schindewolf (1937), Korn (1981, 1988, 1993, 1994, 2000), House *et al.* (1986), Korn & Price (1987), Becker (1988, 1993*b*, 2000), Yu (1988), Ji *et al.* (1989), Bartzsch *et al.* (1998), Becker *et al.* (2002), Nikolaeva & Bogoslovsky (2005*a*), Ebbighausen & Korn (2007), Fischer (2010), Klein & Korn (2014), Korn *et al.* (2015), Zong *et al.* (2015), and new records from the Rhenish Massif and southern Morocco. Species with (+), or close relatives of them (e.g. cf. records), have been recorded as surviving into the main/higher crisis interval.

Amazon Basin. Streel (1986) and Streel *et al.* (2000*a*; 2013) commented on the age of the South American glacial sediments and summarized the correlation with the Rhenish miospore and lithostratigraphy.

Based on the Belgian reference section at Chanxhe (Streel *et al.* 2007; Maziane-Serraj *et al.* 2007), the conodont-defined (*ultimus ultimus* Zone) base of the uppermost Famennian ('Strunian') of Bed 111 falls within the higher LL Zone, just below the morphometric change from *Retispora lepidophyta lepidophyta* to *R. lepidophyta minor* (Bed 116, Maziane *et al.* 2002; Streel 2015). This enables the separation of an upper subdivision of the LL Zone (Streel 2009). In the Rhenish Massif the LL Zone is not known from the WL but from contemporaneous shales of Riescheid (Higgs & Streel 1994).

The next younger LE Zone is defined by the entry of *Indotriradites explanatus*. At Hasselbachtal in the Rhenish Massif, it has been recorded from the upper WL (Higgs & Streel 1984), from a level that falls in the lower range of *Wocklumeria* (Becker 1996, UD VI-D1) and higher *praesulcata* Zone (Kürschner *et al.* 1993). The zonal base was established in the shallow-water Comblain-au-Pont Formation of the Ardennes (Maziane *et al.* 1999), where it cannot be correlated with the conodont or ammonoid succession. However, mostly overlooked data from the southern Moroccan pelagic Tafilalt Platform (Rahmani-Antari & Lachkar 2001) confirm that the LL/LE Zone boundary can be placed slightly above the *Parawocklumeria paradoxa* Zone (UD VI-C, Jebel Erfoud, ammonoid data in Korn 1999). Therefore, the LE Zone can be used as an indicator for the proximity of the crisis interval (Streel 1999, 2015; Streel *et al.* 2000*a*). In Brazil, Melo & Loboziak (2003) recognized a corresponding Rle Zone.

The entry of *Verrucosisporites nitidus* defines the base of the LN Zone. Unfortunately, the species is not always easy to identify (Turnau *et al.* 1994) and is sometimes rare and facies sensitive (Streel *et al.* 2013). Its entry coincides with the main extinction and HBS in Germany (Higgs & Streel 1994). A much earlier record of the index species in Kentucky, from as low as the basal Cleveland Shale (Heal *et al.* 2009), has been corrected by Wicander & Playford (2013). In South American diamictite units it is not possible in certain places to separate the combined LE/LN Zones (e.g. Loboziak *et al.* 1993; Melo & Loboziak 2003).

The globally widespread *R. lepidophyta* declined gradually within the middle/upper crisis interval (middle to upper or 'atypical' LN Zone, HS/HSS; e.g. Avchimovitch *et al.* 1988; Higgs *et al.* 1993; Streel 1999, 2015). In Russia, *Tumulispora malevkensis* has been used to separate an upper division (PM Subzone, Avchimovitch 1993; Avchimovitch *et al.* 1993), which also can be traced in South China (Yang & Neves 1997: 'Pmr assemblage'). The entry of *Vallatisporites vallatus* is also a useful marker for the regressive middle crisis interval (Streel & Traverse 1978; Loboziak *et al.* 1992, 2000; Streel *et al.* 2000*b*). It has been used to define a LVa Zone in South America (Melo & Loboziak 2003), which may equal all or only the upper part of the LN Zone. *Vallatisporites vallatus* enters in Poland accordingly in the upper LN Zone (Marynowski & Filipiak 2007). It is also an important element of microfloras found just above the oncolithic unit ('algal–sponge biostrome') in the Middle Sappington Formation of Montana (Sandberg *et al.* 1972; Warren *et al.* 2014). Streel *et al.* (2013) and Streel (2015) emphasized that the RlE (LE) and LVa (LN) Zones may represent lateral facies variants of the same time interval, which has implications for the dating of glacial pulses in South America.

The final and global *R. lepidophyta* extinction defines the base of the VI Zone in the upper part of the *kockeli* Zone, between the Lower and Upper Stockum Limestones of the type area (Higgs & Streel 1984, 1994; Higgs *et al.* 1993). Other important species that disappeared are *Rugospora flexuosa*, *Diducites versabilis* and *Did. plicabilis* (e.g. Streel 2015). The zonal boundary has been used to approximate the DCB, but this correlation was based on data from the Hasselbachtal auxiliary stratotype (Becker *et al.* 1984) and Stockum (Higgs & Streel 1984), not on the La Serre GSSP level. But in the Zigan section on the western slope of the southern Urals, the extinction of *R. lepidophyta* also just predates the entry of *S. (Eo.) sulcata* (s.l., within the middle Gumerov Horizon, Pazukhin *et al.* 2009). In the Timan, the VI Zone correlates with the shallow-water *Patr. variabilis* Zone (Durkina & Avchimovitch 1988). Reworking of *R. lepidophyta* specimens in lower Tournaisian beds is a recurring problem (discussion in Playford & McGregor 1993). The entries of *Spelaeotriletes balteatus* or *Bascaudaspora mischkinensis* are used to define East European basal Carboniferous MB and PMi Subzones (Avchimovitch 1993; Byvsheva & Umnova 1993). The Lower Alum Shale Event coincides with the base of the HD Zone, defined by the entries of *Kraeuselisporites hibernicus* and *Umbonatisporites distinctus* (Higgs & Streel 1984).

Foraminifere zonation (Fig. 1)

The DCB foraminifere zonation is based on calcareous shallow-water groups, while agglutinating taxa, especially of pelagic settings, are long-ranging (e.g. Eickhoff 1973). The European–Russian zonations have been reviewed by Kulagina *et al.* (2003), Devuyst & Hance in Poty *et al.* (2006), and Kulagina (2013). They can be correlated into neritic successions of South China (e.g. Hance 1996; Hance *et al.* 2011). The uppermost Famennian is roughly characterized by the entry of *Q. kobeitusana* (Mamet *et al.* 1965; Austin *et al.* 1970*a*; Conil *et al.* 1977, 1991). The Belgian DFZ7 (formerly df3ε Zone) is defined by the presence of either *Q. kobeitusana* or *Q. konensis* (Devuyst & Hance in Poty *et al.* 2006). In the Chanxhe reference section, the base of this zone lies just above the first *Bi. ultimus ultimus* and just before the change from *R. lepidophyta lepidophyta* to *R. lepidophyta minor* (upper LL Zone, Maziane-Serraj *et al.* 2007).

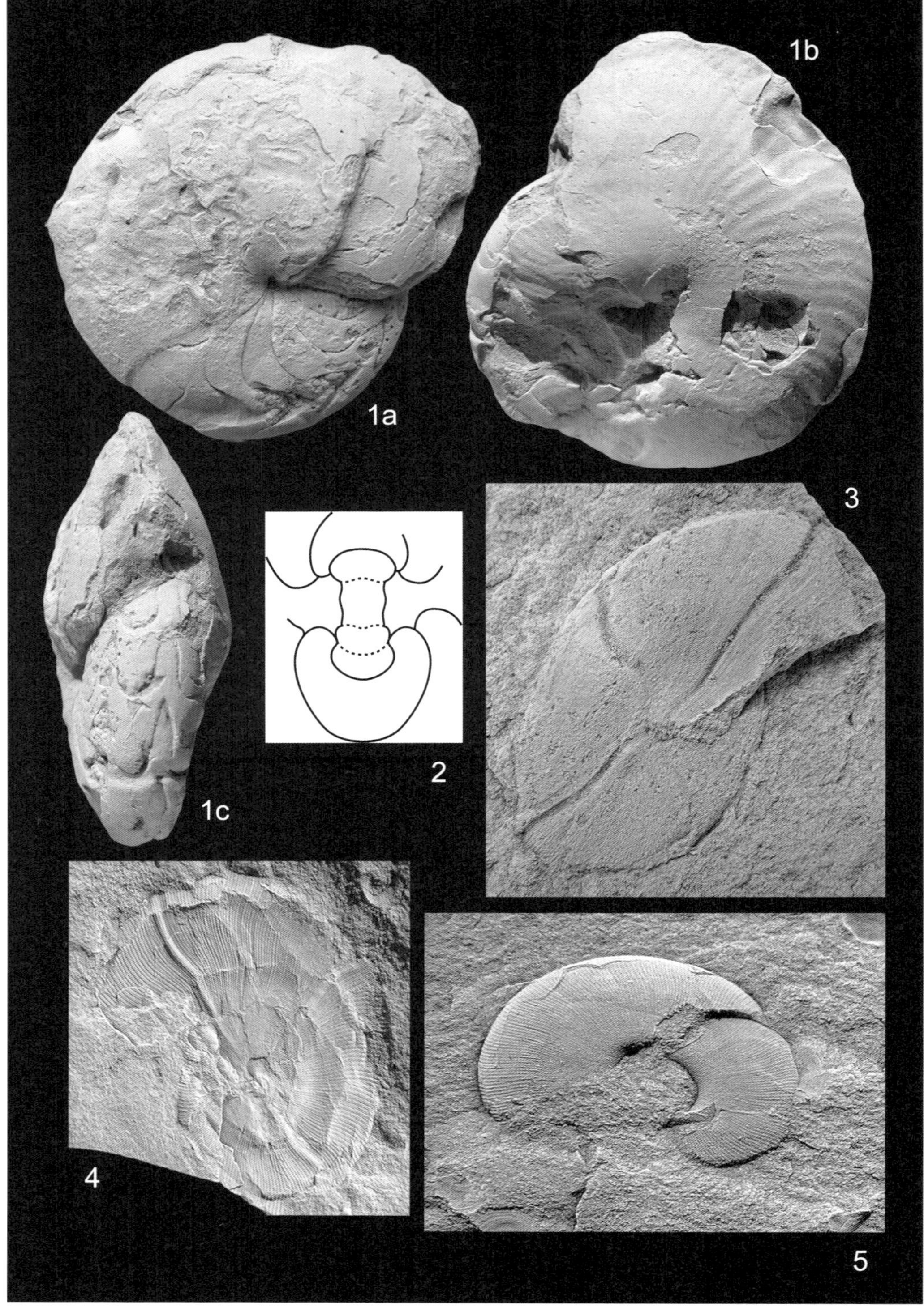
1a
1b
1c
2
3
4
5

Unfortunately, *Q. kobeitusana* has partly been used for slightly different forms, which seems to explain its entry before *Bi. ultimus ultimus* in the Belgian Anseremme section (Streel 2007).

The lower and middle crisis interval mostly lacks carbonate and is, therefore, not characterized by a specific foraminifere assemblage. The Belgian DFZ8 was mainly introduced for the problematic fauna from the Avesnelles Formation in the Avesnois, northern France. Devuyst & Hance (in Poty *et al.* 2006) considered this fauna to be Devonian, whereas Conil *et al.* (1986) had noted Carboniferous affinities and advocated correlations with the Tournaisian Hastière Formation. DFZ8 is defined by *Tournayellina pseudobeata*, which first occurs in the *kockeli* Zone of the Avesnois, northern France (Conil *et al.* 1986; Poty *et al.* 2006). But Kalvoda (in Kumpan *et al.* 2014*b*) found abundant *T. beata* and *T. pseudobeata* in the basal Avesnelles Formation and claimed a Carboniferous age of this fauna based on foraminifers typical for the MFZ2, isotope data and geophysical proxies. This new interpretation did not comment on the top-Devonian condont faunas documented by Conil *et al.* (1986). Kumpan *et al.* (2014*b*) reported *Eochernyshinella crassitheca*, an eastern European marker species of the basal Tournaisian, from the supposedly correlative (*kockeli* Zone) marker bed of the Hastière Limestone. The recently published evidence requires a lower range extension of some foraminifere taxa into the upper crisis interval unless any typical Carboniferous conodont can be recovered from the basal Avesnelles and Hastière limestones.

The post-crisis interval, for example the (main) Lower Member of the Hastière Limestone of Belgium, is characterized by simple, unilocular forms (MFZ1, Cf1α Subzone), as in the Russian *Earlandia minima* Zone (Kulagina *et al.* 2003). The same ecozone is known in South China (Hance *et al.* 2011). In the Timan this level (= regional *Bisphaera malevkensis* Zone) can be correlated both with the VI Zone and the shallow-water *Patr. variabilis* Zone (Durkina & Avchimovitch 1988). Towards the south (Moscow Syncline and Voronezh Anticline), it is correlated with the successive local *Patr. crassus* and *Patr. variabilis* Zones (Makhlina 1996). Aretz *et al.* (2014) suggested a 'Lilliput effect' (dwarfing) for this assemblage, as the consequence of the extinction of the larger-sized taxa. This crisis-influenced zone is followed higher up in the lower Tournaisian (from the middle Hastière Limestone on) by the MFZ2 (Cf1ß Subzone) with *Septabrunsiina minuta* and *T. beata*. In the Ardennes, its base seems to precede the *sandbergi* Zone (see Poty *et al.* 2006: 'upper *S.* (*S.*) *cooperi* Subzone and lower *S.* (*S.*) *obsoleta* Subzone'), which gives a rough correlation with a level within the new *jii* Zone. Devuyst & Hance (in Poty *et al.* 2006) already noted the diachronuous zonal base in the different Belgian sedimentation areas ranging from the base to the top of the middle member of the Hastière Formation. In the Urals, there is a contemporaneous *Prochernyshinella disputabilis* Zone (Kulagina *et al.* 2003; Kulagina 2013; Kalvoda *et al.* 2015), and in South China, a *Chernyshinella glomiformis–Bisphaera* Zone (e.g. Li *et al.* 2002). Kalvoda (in Kumpan *et al.* 2014*b*) reported plurilocular foraminifers already from the lower member of the Hastière Formation in the Gendron-Celles section, but maintained the use of the MFZ1 for the lower member. Kalvoda also reported more diversified faunas for the MFZ2 and even the entry of index taxa of younger zones than the MFZ2. Previous indications for this came from sections of South China (Hance *et al.* 2011).

The middle Tournaisian is characterized by the entry of several newcomers, such as *Crassiseptella*, *Palaeospiroplectammina tchernyshinensis* and *Septabrunsiina krainica* (e.g. Kalvoda in Kumpan *et al.* 2015*b*).

Options for a future basal Carboniferous GSSP

(1) *Main extinction level and global black shale event*

Advantages: It can be recognized easily by the main extinction level, especially by the extinction

Fig. 6. New ammonoid records from the middle Hangenberg Crisis interval of southern Morocco, Utah and the Rhenish Massif (collection of the Geomuseum, WWU Münster, No. B6*c*. 48.7-11). (**1a–c**) *Ac.* (*Stockumites*) aff. *subbilobatum* (Münster 1839) from a thin siderite concretion level just above the HBS equivalents at Ouaoufilal, southern Tafilalt, Morocco (the oldest representative of the genus from North Africa), two lateral and adoral views showing the distorted but compressed whorl profile, sutures and weak flank ribbing unlike typical *Ac.* (*Stock.*) *subbilobatum*, ×2. (**2**) Symmetrically reconstructed cross-section of *Ac.* (*Stockumites*) sp. from the oncolite unit of the Leatham Member (HSS time-equivalent), Pilot Shale, western Utah (Southeastern Conger Mountain, Morrow *et al.* 2007, stop 9), showing the typical, evolute early whorls of the genus, ×6. (**3**) *Ac.* (*Stockumites*) cf. *procedens* Korn, 1984 from the HSS of Drewer, with slightly biconvex growth lirae and constrictions, as typical for the species, ×2. (**4**) *Ac.* (*Stockumites*) cf. *procedens* Korn, 1984 from the HS of Oberrödinghausen, *c.* 3.8 m above base, with slightly biconvex growth lirae and constrictions, ×2. (**5**) *Ac.* (*Stockumites*) aff. *antecedens* (Vöhringer, 1960) from the HS of Oberrödinghausen, *c.* 3.8 m above base, showing biconvex growth lirae, a lack of constrictions and a rather wide umbilicus, which is only slightly enhanced by compactional fracturing ×4.

of index conodonts (*Bi. costatus–ultimus* Gp., *Ps. marburgensis* Gp., *Pa. gracilis* Gp. = *Tripodellus* in multi-element taxonomy, most *Icriodus*), ammonoids (apart from a few cymaclymeniids and mimimitoceratids), trilobites (all pelagic taxa), and of the last stromatoporoids in neritic carbonates; by sequence stratigraphy (drowning unconformity at the base of a maximum flooding interval, e.g. Van Steenwinkel 1993*a*, *b*); by magnetosusceptibility (e.g. Kumpan *et al.* 2015*a*); by element geochemistry (indicators of low oxygenation, e.g. Kumpan *et al.* 2014*a*, *b*); by gamma ray spectroscopy (Kumpan *et al.* 2014*a*, 2015); by organic geochemistry (e.g. Marynowski & Filipiak 2007); and by carbon isotope stratigraphy (see review of Kaiser *et al.* 2015: onset of the first positive spike, both in C_{org} and C_{carb}). The base of the LN Zone provides a good correlation into the terrestrial realm and into macrofossil-free high-latitude basins.

Disadvantages: Conodont reworking, partly poor distinction of LE/LN zone floras, subsequent removal by the following regression, strong condensation, and range of initial survivors into the Carboniferous (e.g. resulting consequently in 'Carboniferous clymeniids and phacopids').

(2) *Sequence boundary and regressive peak (base HSS)* (recently proposed by Kumpan *et al.* 2014*b*)

Advantages: Recognizable with the help of sequence stratigraphy (sequence boundary, e.g. Van Steenwinkel 1993*a*, *b*; Kaiser *et al.* 2011; globally widespread episode of reworking and non-deposition), palynofacies studies (e.g. Marynowski & Filipiak 2007), gamma ray spectroscopy (Kumpan *et al.* 2014*a*, *b*), and magnetic susceptibility (Kumpan *et al.* 2015*a*).

Disadvantages: No conodont, ammonoid or clear miospore boundary; possibly poorly recognizable in basinal successions with continuous shale deposition; and local controversy concerning its placing (e.g. in the Ardennes).

(3) *Base of kockeli Zone and initial postglacial transgression*

Advantages: Easily recognizable by the initial reradiation of faunas, especially with the help of index conodonts (entry of *Pr. kockeli* and of the *Po. purus* Group), ammonoids (radiation of *Acutimitoceras* faunas in association with the last clymeniids), deeper-water trilobites (entry of *Belgibole* and *Semiproetus* groups), shallow-water trilobites (e.g. *Pudoproetus* and *Brachymetopus germanicus*) and neritic foraminifers (*T. pseudobeata*); lithostratigraphy (widespread change from siliciclastics to carbonate), sequence stratigraphy (postglacial transgression, often with basal flooding unconformity), magnetic susceptibility, carbon isotope stratigraphy (second positive excursion, see Kaiser *et al.* 2015), ?oxygen isotope stratigraphy (due to the rewarming), and possibly element geochemistry (local hypoxia/anoxia).

Disadvantages: Local fossil reworking; restricted distribution of *Pr. kockeli* (Corradini *et al.* 2011); widespread facies-controlled, discontinuous conodont sampling across the base (entry of the index fossil with the first carbonate); widespread condensation in the pelagic realm; uncertainties concerning the correlation into the neritic realm (reworking in the Ardennes?); and no clear miospore boundary (difficult correlation into the terrestrial realm).

(4) *Entry of S. (Eo.) sulcata auct. (Upper Stockum level)*

Advantages: Minor evolutionary change in the *Siphonodella* lineage; arbitrarily used DCB definition of the past decades (constancy of literature), *Pr. kuehni* as alternative marker, rough correlation with the LN/VI zone boundary (correlation into the terrestrial realm), survivor extinctions (last clymeniids, phacopids, ?placoderms, ?*Quasiendothyra*), and sequence stratigraphy (minor regression or parasequence boundary).

Disadvantages: Current serious taxonomic problems (pre-crisis homeomorphs, difference between GSSP morphotype and holotype; e.g. Tragelehn 2010; Kaiser & Corradini 2011; Becker *et al.* 2013), hard to locate due to the facies-controlled interruption of the *Siphonodella* record and due to the restricted distribution and rarity of *Pr. kuehni*, no other biozone boundary, and no clear marker of physical or chemostratigraphy.

(5) *Base of post-crisis interval and main Carboniferous transgression (roughly the position of the current La Serre GSSP)*

Advantages: Main radiation of ammonoids (entry of *Gattendorfia* and *Eocanites* faunas), various trilobites and some brachiopods; base of MFZ1 (foraminifera); and sequence stratigraphy (main postglacial transgression, transgressive system tract).

Disadvantages: No clear conodont boundary; no miospore boundary (difficult correlation into the terrestrial realm); weak separation from the *kockeli* Zone transgression, especially in shallow shelf settings (difficulties of neritic conodont successions); no isotope spike (mostly only a gradual decrease of $\delta^{13}C$ values, e.g. Kaiser *et al.* 2006); and no marker of element geochemistry.

Conclusions

None of the options is perfect. But a comparison of the pro and contra arguments leads to the conclusion that (1) has (currently) the best options for global correlation.

Taxonomic notes

Siphonodella (Siphonodella) jii nom. nov.

Derivation of name: In honour of Dr Ji Qiang, for his research on DCB conodonts.

Diagnosis: See diagnosis for *S.* (*S.*) *hassi* in Ji (1985).

Holotype: Specimen illustrated by Hass (1959: pl. 49, figs 17–18).

Remarks: Sandberg *et al.* (1978) recognized that a specimen described by Hass (1959) as *S. duplicata* differs from the typical form and outlined its morphology, distribution and phylogenetic significance as '*S. duplicata sensu* Hass (1959)'. This form was named by Ji (1985) as a new species, *S.* (*S.*) *hassi*, which entry defined a *hassi* Zone, equivalent to the Upper *duplicata* Zone of Sandberg *et al.* (1978). Becker *et al.* (2013) realized that his taxon is an invalid junior homonym of *S. cooperi hassi* Thompson & Fellows, 1970, which is a stratigraphically significant subspecies at the top of the middle Tournaisian (Kinderhookian) of eastern North America (Boardman *et al.* 2013). Consequently, *S.* (*S.*) *hassi* Ji, 1985 is here renamed.

Recent fieldwork of RTB in the Rhenish Massif and in Morocco was funded by the DFG grant Be1367/11-1. Sven Hartenfels (Münster) and Harald Tragelehn (Wallenfels) collected some of the new Rhenish goniatites from the Hangenberg Crisis interval. Z. Sarah Aboussalam (Münster) took part in the Moroccan fieldwork. Traudel Fährenkämper (Münster) produced the diagrams and assisted in the compilation of the photo figures. We gratefully acknowledge the helpful reviews by D. J. Over (Geneseo) and Maurice Streel (Brussels).

References

Aisenverg, D.E., Brazhnikova, N.E. et al. 1979. The Carboniferous sequence of the Donetz Basin: a standard section for the Carboniferous system. *In*: Wagner, R.H., Higgins, A.C. & Meyen, S.V. (eds) *The Carboniferous of the U.S.S.R.* Yorkshire Geological Society Occasional Publications, **4**, 197–224.

Alberti, H., Groos-Uffenorde, H., Streel, M., Uffenorde, H. & Walliser, O.H. 1974. The stratigraphical significance of the *Protognathodus* fauna from Stockum (Devonian/Carboniferous boundary, Rhenish Schiefergebirge). *Newsletters on Stratigraphy*, **3**, 263–276.

Alekseev, A.S., Barskov, N.S. & Kononova, A.I. 1979. Conodonts of Famennian-Tournaisian boundary deposits from the Central region of the Russian Platform. *Service géologique de Belgique, Professional Paper, 1979*, 52–58.

Alekseev, A.A., Lebedev, O.A., Barskov, I.S., Barskova, M.I., Kononova, L.I. & Chizhova, V.A. 1994. On the stratigraphic position of the Famennian and Tournaisian fossil vertebrate beds in Andreyevka, Tula Region, Central Russia. *Proceedings of the Geologists' Association*, **105**, 41–52.

Amler, M. R. W. 1993. Shallow marine bivalves at the Devonian–Carboniferous boundary from the Velbert Anticline (Rhenisches Schiefergebirge). *Annales de la Société géologique de Belgique*, **115**, 405–425.

Amler, M.R. 1995. Die Bivalvenfauna des Oberen Famenniums West-Europas. 1. Einführung, Lithostratigraphie, Faunenübersicht, Systematik 1. Pteriomorphia. *Geologica et Palaeontologica*, **29**, 19–143.

Amler, M.W. & Herbig, H.-G. 2006. Ostrand der Kohlenkalk-Plattform und Übergang in das Kulm-Becken im westlichsten Deutschland zwischen Aachen und Wuppertal. *Schriftenreihe der Deutschen Gesellschaft für Geowissenschaften*, **41**, 441–477.

Aretz, M. 2014. Redefining the Devonian–Carboniferous Boundary: an overview of problems and possible solutions. *In*: Rocha, R., Pais, J., Kullberg, J.C. & Finney, S. (eds) *STRATI 2013. First International Congress on Stratigraphy, At the Cutting Edge of Stratigraphy*. Springer Geology, Zurich, 227–231.

Aretz, M. & Task Group 2011. Report of the Joint Devonian–Carboniferous Boundary GSSP Reappraisal Task Group. *Newsletter on Carboniferous Stratigraphy*, **29**, 23–26.

Aretz, M. & Task Group 2013. Report of the Joint Devonian–Carboniferous Boundary GSSP Reappraisal Task Group. *Newsletter on Carboniferous Stratigraphy*, **30**, 31–35.

Aretz, M. & Task Group 2014. Report of the Joint Devonian–Carboniferous Boundary GSSP Reappraisal Task Group. *Newsletter on Carboniferous Stratigraphy*, **31**, 26–29.

Aretz, M., Nardin, E. & Vachard, D. 2014. Diversity patterns and palaeobiogeographical relationships of latest Devonian–Lower Carboniferous foraminifers from South China: what is global, what is local? *Journal of Palaeogeography*, **3**, 35–59.

Aristov, V.A. 1988. Devonian conodonts of the Central Devonian field (Russian Platform). *Academy of Sciences of the USSR, Transactions*, **432**, 1–119 [in Russian].

Aristov, V.A., Gretchishnikova, I.A., Tchigova, V.A. & Felix, V.P. 1979. Subdivision and correlation of the Famennian and the Lower Tournaisian deposits of Transcaucasia (on brachiopods, conodonts and ostracods). *Service géologique de Belgique, Professional Paper*, **1979**, 87–95 [in Russian with extended English summary].

Atta-Peters, D. & Anan-Yorke, R. 2003. Latest Devonian and Early Carboniferous pteridophytic spores from the Sekondi Group of Ghana. *Revista Española de Micropaleontologia*, **35**, 9–27.

Austin, R.L. & Hill, P.J. 1973. A Lower Avonian (K Zone) Conodont Fauna from near Tintern, Monmouthshire, Wales. *Geologica et Palaeontologica*, **7**, 234–234.

Austin, R.L. & Husri, S. 1974. Dinantian conodont faunas of County Clare, County Limerick and County Leitrim. *International Symposium on Belgian

Micropalaeontological Limits, From Emsian to Visean, 1–10 September 1974, Namur, Publication, **3**, 18–69.

AUSTIN, R.L., CONIL, R. ET AL. 1970*a*. Transitional Devonian/Carboniferous sequences between Hook Head (Ireland) and Bohlen (D.D.R.). *In*: STREEL, M. & WAGNER, R.H. (eds) *Colloque sur la Stratigraphie du Carbonifère*. Les Congres et Colloques de l'Université de Liege, **55**, 172–178.

AUSTIN, R., DRUCE, E.C., RHODES, F. H. T. & WILLIAMS, J.A. 1970*b*. The value of conodonts in the recognition of the Devonian–Carboniferous boundary, with particular reference to Great Britain. *Compte Rendu Sixième Congrès International de Stratigraphie et de Géologie du Carbonifère*, 11–16 September 1967, Sheffield, **II**, 431–445.

AUSTIN, R., CONIL, R., RHODES, F. & STREEL, M. 1970*c*. Conodontes, spores et foraminifères du Tournaisian Inférieur dans la Vallée du Hoyoux. *Annales de la Société géologique de Belgique*, **93**, 305–315.

AVCHIMOVITCH, V.I. 1993. Zonation and spore complexes of the Devonian and Carboniferous boundary deposits of the Pripyat Depression (Byelorussia). *Annales de la Société géologique de Belgique*, **115**, 425–451.

AVCHIMOVITCH, V.I., BYVSHEVA, T.V., HIGGS, K., STREEL, M. & UMNOVA, V.T. 1988. Miospore systematics and stratigraphic correlation of Devonian–Carboniferous Boundary deposits in the European part of the USSR and Western Europe. *Courier Forschungsinstitut Senckenberg*, **100**, 169–191.

AVCHIMOVITCH, V.I., TURNAU, E. & CLAYTON, G. 1993. Correlation of uppermost Devonian and Lower Carboniferous miospore zonations in Byelorussia, Poland and Western Europe. *Annales de la Société géologique de Belgique*, **115**, 453–458.

AZMY, K., POTY, E. & BRAND, U. 2009. High-resolution isotope stratigraphy of the Devonian–Carboniferous boundary in the Namur–Dinant Basin, Belgium. *Sedimentary Geology*, **216**, 117–124.

BAI, S.-L. & NING, Z.-S. 1989. Faunal change and events across the Devonian–Carboniferous boundary of Huangmao section, Guangxi, South China. *Canadian Society of Petroleum Geologists, Memoir*, **14**, 147–157.

BAKHAREV, N.K., IZOKH, N.G., OBUT, O.T. & TALENT, J.A. (eds) 2011. Middle-Upper Devonian and Lower Carboniferous biostratigraphy of Kuznetsk Basin. *Field Excursion Guidebook, International Conference 'Biostratigraphy, Paleogeography and Events in Devonian and Lower Carboniferous (SDS/IGCP 596 Joint Field Meeting)'*, 20 July–10 August 2011, Novosibirsk, Novosibirsk Publishing House, SB RAS, 1–97.

BARSKOV, I.S., SIMAKOV, K.V. ET AL. 1984. Devonian–Carboniferous transitional deposits of the Berchogur section, Mugodzary, USSR. *Courier Forschungsinstitut Senckenberg*, **67**, 207–230.

BARTZSCH, K. & WEYER, D. 1982. Zur Stratigraphie des Untertournai (*Gattendorfia-Stufe*) von Saalfeld im Thüringischen Schiefergebirge. *Abhandlungen und Berichte für Naturkunde und Vorgeschichte*, **12**, 3–54.

BARTZSCH, K. & WEYER, D. 1986. Biostratigraphie der Devon/Karbon-Grenze im Bohlen-Profil bei Saalfeld (Thüringen, DDR). *Zeitschrift für Geologische Wissenschaften*, **14**, 147–152.

BARTZSCH, K., HAHNE, K. & WEYER, D. 1998. Der Hangenberg-Event (Devon/Karbon-Grenze) im Bohlen-Profil von Saalfeld (Thüringisches Schiefergebirge). *Abhandlungen und Berichte für Naturkunde*, **20**, 37–58.

BECKER, R.T. 1988. Ammonoids from the Devonian-Carboniferous Boundary in the Hasselbach Valley (northern Rhenish Slate Mountains). *Courier Forschungsinstitut Senckenberg*, **100**, 193–213.

BECKER, R.T. 1993*a*. Anoxia, eustatic changes, and Upper Devonian to lowermost Carboniferous global ammonoid diversity. *In*: HOUSE, M.R. (ed.) *The Ammonoidea, Environment, Ecology, and Evolutionary Change*. Systematic Association, Special Volume, **47**, 115–164.

BECKER, R.T. 1993*b*. Analysis of ammonoid palaeobiogeography in relation to the global Hangenberg (terminal Devonian) and Lower Alum Shale (Middle Tournaisian) events. *Annales de la Société géologique de Belgique*, **115**, 459–473.

BECKER, R.T. 1996. New faunal records and holostratigraphic correlation of the Hasselbachtal D/C-boundary auxiliary stratotype (Germany). *Annales de la Société géologique de Belgique*, **117**, 19–45.

BECKER, R.T. 2000. Taxonomy, evolutionary history, and distribution of the middle to late Famennian Wocklumeriina (Ammonoidea, Clymeniida). *Mitteilungen aus dem Museum für Naturkunde in Berlin, Geowissenschaftliche Reihe*, **3**, 27–75.

BECKER, R.T. 2005. Minutes of the SDS Business Meeting, Rabat, 1st of March 2004. *SDS Newsletter*, **21**, 3–8.

BECKER, R.T. (ed.) 2008. Subcommission on Devonian stratigraphy. *SDS Newsletter*, **23**, 1–27.

BECKER, R.T. 2010. The chance of evolutionary success – the case of the oldest (Tournaisian) ceratitoid ammonoids (Prodromitidae, Tornoceratina). *8th International Symposium 'Cephalopods – Present and Past'*, 30 August–3 September 2010, Dijon, Abstracts Volume, 25–26.

BECKER, R.T. & HOUSE, M.R. 2000. Devonian ammonoid zones and their correlation with established series and stage boundaries. *Courier Forschungsinstitut Senckenberg*, **220**, 113–151.

BECKER, R.T. & MAPES, R.H. 2010. Uppermost Devonian ammonoids from Oklahoma and their palaeobiogeographic significance. *Acta Geologica Polonica*, **60**, 139–163.

BECKER, R.T. & PAPROTH, E. 1993. Auxiliary stratotype sectionsv for the global stratotype section and point (GSSP) for the Devonian–Carboniferous boundary: Hasselbachtal. *Annales de la Société géologique de Belgique*, **115**, 703–706.

BECKER, R.T., BLESS, M. J. M. ET AL. 1984. Hasselbachtal, the section best displaying the Devonian-Carboniferous boundary beds in the Rhenish Massif (Rheinisches Schiefergebirge). *Courier Forschungsinstitut Senckenberg*, **67**, 181–191.

BECKER, R.T., HOUSE, M.R., BOCKWINKEL, J., EBBIGHAUSEN, V. & ABOUSSALAM, Z.S. 2002. Famennian ammonoid zones of the eastern Anti-Atlas (southern Morocco). *Münstersche Forschungen zur Geologie und Paläontologie*, **93**, 159–2005.

BECKER, R.T., KAISER, S.I. & ABOUSSALAM, Z.S. 2006. The Lower Alum Shale Event (Middle Tournaisian) in Morocco – facies and faunal changes. *In*: ARETZ, M. & HERBIG, H.-G. (eds) *Carboniferous Conference Cologne, From Platform to Basin*, 4–10 September 2006, Program and Abstracts. Kölner Forum für Geologie und Paläontologie, **15**, 7–8.

BECKER, R.T., GRADSTEIN, F.M. & HAMMER, O. 2012. The Devonian period. *In*: GRADSTEIN, F.M., OGG, J.G., SCHMITZ, M. & OGG, G. (eds) *The Geologic Time Scale 2012*. Elsevier, Amsterdam, **2**, 559–601.

BECKER, R.T., HARTENFELS, S., ABOUSSALAM, Z.S., TRAGELEHN, H., BRICE, D. & EL HASSANI, A. 2013. The Devonian–Carboniferous boundary at Lalla Mimouna (northern Maider) – a progress report. *In*: BECKER, R.T., EL HASSANI, A. & TAHIRI, A. (eds) *International Field Symposium 'The Devonian and Lower Carboniferous of northern Gondwana'*, 22–29 March 2013, Field Guidebook. Documents de l'Institut Scientifique, Rabat, **27**, 109–120.

BECKER, R.T., HARTENFELS, S. & WEYER, D. 2015 (in press). The Famennian to Lower Viséan at Drewer (northern Rhenish Massive). *Münstersche Forschungen zur Geologie und Paläontologie*, (pre) **108**, 122–140.

BENDER, P., HERBIG, H.-G., GURSKY, H.-J. & AMLER, M.W. 1993. Beckensedimente im Oberdevon und Unterkarbon des östlichen Rheinischen Schiefergebirges – Fazies, Paläogeographie und Meeresspiegelschwankungen. *Geologica et Palaeontologica*, **27**, 332–356.

BLESS, M. J. M., SIMAKOV, K.V. & STREEL, M. 1988. Advantages and disadvantages of a conodont-based or event-stratigraphic Devonian–Carboniferous boundary. *Courier Forschungsinstitut Senckenberg*, **100**, 3–14.

BLESS, M. J. M., BECKER, R.T., HIGGS, K., PAPROTH, E. & STREEL, M. 1993. Eustatic cycles around the Devonian–Carboniferous boundary and the sedimentary and fossil record in Sauerland (Federal Republic of Germany). *Annales de la Société géologique de Belgique*, **115**, 689–702.

BOARDMAN, D.R., THOMPSON, T.L., GODWIN, C., MAZULLO, S.J., WILHITE, B.W. & MORRIS, B.T. 2013. High-resolution conodont zonation for Kinderhookian (Middle Tournaisian) and Osagean (Upper Tournaisian–Lower Visean) Strata of the western edge of the Ozark Plateau, North America. *Shale Shaker*, **2013**, 98–151.

BÖGER, H. 1962. Zur Stratigraphie des Unterkarbons im Velberter Sattel. *Decheniana*, **114**, 133–170.

BOGOSLOVSKIY, B.I. 1981. Devonskie ammonoidei. III. Klimenii (Podotryad Gonioclymeniina). *Trudy Paleontologicheskogo Instituta, Akademiya Nauk SSSR*, **191**, 1–123.

BOMFLEUR, B. 2005. *Geologische Neukartierung des Raumes Hemer im Bereich der Deutschen Grundkarten 4228–4230 (Landhausen, Urbecke, Hüingsen-West), Märkischer Kreis*. BSc thesis, WWU Münster.

BOUCKAERT, J. & BOONEN, P. 1978. A conodont fauna of the Dinantian from the Kuznetsk Basin (southern Siberia). *Annales de la Société géologique de Belgique*, **100**, 157–165.

BOUCKAERT, J. & DUSAR, M. 1976. Description du sondage de Tohogne. *Service Géologique de Belgique, Professional Paper*, **1976**, 1–10.

BOUCKAERT, J. & GROESSENS, E. 1976. *Polygnathus paprothae, Pseudopolygnathus conili, Pseudopolygnathus graulichi:* espèces nouvelles à la limite Dévonien–Carbonifère. *Annales de la Société géologique de Belgique*, **99**, 587–599.

BOUCKAERT, J., CONIL, R., DUSAR, M. & STREEL, M. 1978. Stratigraphic interpretation of the Tohogne borehole (Province of Luxembourg). Devonian–Carboniferous transition. *Annales de la Société géologique de Belgique*, **100**, 87–101.

BUSHMINA, L.S. 1984. Mikrofauna i biostratigrafija nizhnego karbona (jug Zapodnoj Sibiri). *Trudy Instituta geologii i geofisiki*, **599**, 1–127.

BUSHMINA, L.S. & KONONOVA, L.I. 1981. Microfauna and biostratigraphy of the Devonian-Carboniferous beds (of the south of western Siberia). *Academy of Sciences, Siberian Branch, Institute of Geology and Geophysics, Transaction*, **459**, 1–124 [in Russian].

BUTLER, M. 1973. Lower Carboniferous conodont faunas from the eastern Mendips, England. *Palaeontology*, **16**, 477–517.

BYVSHEVA, T.V. & UMNOVA, N.I. 1993. Palynological characteristics of the lower part of the Carboniferous of the central region of the Russian Platform. *Annales de la Société géologique de Belgique*, **115**, 519–529.

BYVSHEVA, T.V., HIGGS, K. & STREEL, M. 1984. Spore correlations between the Rhenish Slate Mountains and the Russian Platform near the Devonian–Carboniferous boundary. *Courier Forschungsinstitut Senckenberg*, **67**, 37–45.

CASIER, J.-G., MAMET, B., PRÉAT, A. & SANDBERG, C.A. 2004. Sedimentology, conodonts and ostracods of the Devonian – Carboniferous strata of the Anseremme railway bridge section, Dinant Basin, Belgium. *Bulletin de l'Institut Royal des Sciences Naturelles de Belgique, Sciences de la Terre*, **74**, 45–68.

CHLUPÁC, I. & ZIKMUNDOVA, J. 1976. The Devonian and Lower Carboniferous in the Nepasice bore in East Bohemia. *Vestnik Ústredního ústavu geolického*, **51**, 269–278.

CLAUSEN, C.-D., LEUTERITZ, K., ZIEGLER, W. & KORN, D. 1989. Ausgewählte Profile an der Devon/Karbon-Grenze im Sauerland (Rheinisches Schiefergebirge). *Fortschritte in der Geologie von Rheinland und Westfalen*, **35**, 161–226.

CLAYTON, G., COQUEL, R., DOUBINGER, J., GUEINN, K.J., LOBOZIAK, S., OWENS, B. & STREEL, M. 1977. Carboniferous miospores of Western Europe: illustration and zonation. *Mededelingen Rijks Geologische Dienst*, **29**, 1–71.

CLAYTON, G., HIGGS, K., KEEGAN, J.B. & SEVASTOPULO, G.D. 1978. Correlation of the palynological zonation of the Dinantian rocks of the British Isles. *Palinologia*, **1**, 137–147.

COEN, M. & GROESSENS, E. 1996. Conodonts from the Devonian–Carboniferous transition beds of central Hunan, South China. *Mémoires de l'Institut Géologique de l'Université de Louvain*, **36**, 21–28.

COLE, D., MYROW, P.M., FIKE, D.A., HAKIM, A. & GEHRELS, G.E. 2015. Uppermost Devonian (Famennian) to Lower Mississippian events of the western

U.S.: stratigraphy, sedimentology, chemostratigraphy, and detrital zircon geochronology. *Palaeogeography, Palaeoclimatology, Palaeoecology*, **427**, 1–19.

COLEMAN, U. & CLAYTON, G. 1987. Palynostratigraphy and palynofacies of the uppermost Devonian and Lower Mississippian of eastern Kentucky (U.S.A.), and correlation with Western Europe. *Courier Forschungsinstitut Senckenberg*, **98**, 75–93.

CONIL, R., GROESSENS, E. & PIRLET, H. 1977. Nouvelle charte stratigraphique du Dinantien type de la Belgique. *Annales de la Société géologique du Nord*, **96**, 363–371.

CONIL, R., DREESEN, R., LENTZ, M.-A., LYS, M. & PLODOWSKI, G. 1986. The Devono-Carboniferous transition in the Franco-Belgian Basin with reference to foraminifera and brachiopods. *Annales de la Société géologique de Belgique*, **109**, 19–26.

CONIL, R., GROESSENS, E., LALOUX, M., POTY, E. & TOURNEUR, F. 1991. Carboniferous guide to foraminifera, corals and conodonts in the Franco-Belgian and Campine Basins: their potential for widespread correlation. *Courier Forschungsinstitut Senckenberg*, **130**, 15–30.

CORRADINI, C. 2008. Revision of Famennian–Tournaisian (Late Devonian–Early Carboniferous) conodont biostratigraphy of Sardinia, Italy. *Revue de micropaléontologie*, **51**, 123–132.

CORRADINI, C., BARCA, S. & SPALLETTA, C. 2003. Late Devonian–Early Carboniferous conodonts from the 'Clymeniae limestones' of SE Sardinia (Italy). *Courier Forschungsinstitut Senckenberg*, **245**, 227–253.

CORRADINI, C., KAISER, S.I., PERRI, M.C. & SPALLETTA, C. 2011. Conodont genus *Protognathodus* and its potential as a tool for defining the Devonian/Carboniferous boundary. *Rivista Italiana di Paleontologia e Stratigrafia*, **117**, 15–28.

DAVYDOV, V.I., KORN, D. & SCHMITZ, M.D. 2012. The Carboniferous period. *In*: GRADSTEIN, F.M., OGG, J.G., SCHMITZ, M. & OGG, G. (eds) *The Geologic Time Scale 2012*. Elsevier, Amsterdam, **2**, 603–651.

DELEPINE, G. 1929. Sur la présence de *Cymaclymenia camerata* Schind. dans la zone d'Etroeungt à Sémeries (nord de la France). *Annales de Société géologique du Nord*, **54**, 99–103.

DÖLLING, B. & PIECHA, M. 2010. Die Kernbohrung Anröchte 4415/1001 – neue Erkenntnisse über das Paläozoikum im Lippstädter Gewölbe. *Schriftenreihe der Deutschen Gesellschaft für Geowissenschaften*, **73**, 45–52.

DRAGANITS, E., MAWSON, R., TALENT, J.A. & KRYSTYN, L. 2002. Lithostratigraphy, conodont biostratigraphy and depositional environments of the Middle Devonian (Givetian) to Early Carboniferous (Tournaisian) Lipak Formation in the Pin Valley of Spiti (NW India). *Rivista Italiana di Paleontologia e Stratigrafia*, **108**, 7–35.

DRUCE, E.C. 1969. Devonian and Carboniferous Conodonts from the Bonaparte Gulf Basin, Northern Australia, and their use in international correlation. *Bureau of Mineral Resources, Geology and Geophysics, Bulletin*, **98**, 1–243.

DURKINA, A.V. & AVCHIMOVITCH, V.I. 1988. The reference sections of the Devonian–Carboniferous boundary deposits in the Timan–Pechora province. *In*: *The Devonian-Carboniferous Boundary at the Territory of the USSR*. Nauka I Technica, Minsk, 87–101.

DZIK, J. 1997. Emergence and succession of Carboniferous conodont and ammonoid communities in the Polish part of the Variscan Sea. *Acta Palaeontologica Polonica*, **42**, 57–164.

DZIK, J. 2006. The Famennian 'Golden Age' of conodonts and ammonoids in the Polish part of the Variscan Sea. *Palaeontologia Polonica*, **63**, 1–360.

EBBIGHAUSEN, V. & BOCKWINKEL, J. 2007. Tournaisian (Early Carboniferous/Mississippian) ammonoids from the Ma'der Basin (Anti-Atlas, Morocco). *Fossil Record*, **10**, 125–163.

EBBIGHAUSEN, V. & KORN, D. 2007. Conch geometry and ontogenetic trajectories in the triangularly coiled Late Devonian ammonoid *Wocklumeria* and related genera. *Neues Jahrbuch für Geologie und Paläontologie, Abhandlungen*, **244**, 9–41.

EBBIGHAUSEN, V., BOCKWINKEL, J., KORN, D. & WEYER, D. 2004. Early Tournaisian ammonoids from Timimoun (Gourara, Algeria). *Mitteilungen aus dem Museum für Naturkunde in Berlin, Geowissenschaftliche Reihe*, **7**, 133–152.

EICKHOFF, G. 1973. Das hohe Oberdevon und tiefe Unterkarbon im Bahneinschnitt Oberrödinghausen bei Menden (Rheinisches Schiefergebirge). *Compte Rendu Septième Congrés International de Stratigraphie et de Géologie du Carbonifère*, 23–28 August 1971, Krefeld, **2**, 417–434.

ERNST, A. & HERBIG, H.-G. 2010. Stenolaemate bryozoans from the latest Devonian (Uppermost Famennian) of Western Germany. *Geologica Belgica*, **13**, 173–182.

ERNST, A., TOLOKONNIKOVA, Z. & HERBIG, H.-G. 2015. Uppermost Famennian bryozoans from Ratingen (Velbert Anticline, Rhenish Massif/Germany) – taxonomy, facies dependencies and palaeobiogeographic implications. *Geologica Belgica*, **18**, 37–47.

EVANS, S.D., OVER, D.J., DAY, J.E. & HASENMUELLER, N.R. 2013. The Devonian/Carboniferous boundary and the holotype of *Siphonodella sulcata* (Huddle, 1934) in the upper New Albany Shale, Illinois Basin, southern Indiana. *In*: EL HASSANI, A., BECKER, R.T. & TAHIRI, A. (eds) *International Field Symposium 'The Devonian and Lower Carboniferous of northern Gondwana'*, 22–29 March 2013, Abstracts Book. Documents de l'Institut Scientifique, Rabat, **26**, 42–43.

FALLAH, A., HAMEDI, B. & MOSADDEGH, H. 2011. Carboniferous conodont biostratigraphy in Kiyasar region and introduction of 7 biozones comparable to World Standard Conodont Zonation. *Geosciences, Scientific Quarterly Journal*, **20**, 117–122, 193 [in Farsi with English summary].

FEIST, R. & FLAJS, G. 1988. Index conodonts, trilobites and environment of the Devonian–Carboniferous Boundary beds at La Serre (Montagne Noire, France). *Courier Forschungsinstitut Senckenberg*, **100**, 53–107.

FEIST, R. & PETERSEN, M.S. 1995. Origin and spread of *Pudoproetus*, a survivor of the Late Devonian trilobite crisis. *Journal of Paleontology*, **69**, 99–109.

FISCHER, T. 2010. *Paläopathologie und Paläoökologie von Ammonoideen des Obersten Famenniums (UD VI) aus Südmarokko (Tafilalt, Maider)*. BSc thesis, WWU Münster.

GAGIEV, M.X. & BOGUS, O.I. 1990. Oporniy razrez famensko-turneiskikh otlozhenii ugo-zapadnoi chasti prikolymskogo podniatia (Severo-Vostok SSSR). *In*: ERINA, E.A. (ed.) *Novoe v paleontologii i biostratigrafii paleozoia aziatskoi chasti SSSR*. Trudy Instituta Geologii i Geofiziki, souza SSR, **770**, 119–132.

GAGIEV, M.H. & KONONOVA, L.I. 1990. The Upper Devonian and Lower Carboniferous sequences in the Kamenka River section (Kolyma River Basin, The Soviet North-East). Stratigraphic description. Conodonta. *Courier Forschungsinstitut Senckenberg*, **118**, 81–103.

GAGIEV, M.K., KONONOVA, L.I. & PAZUKHIN, V.N. 1987. Konodontiy. *In*: MASLOV, V.A. (ed.) *Fauna i biostratigrafia pogranichnikh otlozhenii devona i karbona Berchogura (Mugodzhariy)*. Akademia Nauk SSSR, Bashkirskiy Filial, Institut Geologii, 1–121.

GALLWITZ, H. 1927. Stratigraphische und tektonische Untersuchungen an der Devon-Carbongrenze des Sauerlandes. *Jahrbuch der Preußischen Geologischen Landesanstalt*, **48**, 487–527.

GAO, L. 1989. Palynostratigraphy at the Devonian-Carboniferous boundary in the Himalayan region, Xizang (Tibet). *Canadian Society of Petroleum Geologists, Memoir*, **14**, 159–170.

GIRARD, C., CORNÉE, J.-J., CORRADINI, C., FRAVALO, A. & FEIST, R. 2013. Palaeoenvironmental changes at Col de Tribes (Montagne Noire, France), a reference section for the Famennian of north Gondwana-related areas. *Geological Magazine*, **151**, 864–884.

GONG, X., HUANG, H., ZHANG, M. & HUANG, Q. 1992. *The Stratigraphic Classification and Correlation of Carbonate Rocks of Upper Devonian and Lower Carboniferous in Guilin Karst Region*. Guangxi Science and Technology Publishing House.

GRECHISHNIKOVA, I.A. & LEVITSKII, E.S. 2011. The Famennian–Lower Carboniferous reference section Geran-Kalasi (Nakhichevan Autonomous Region, Azerbaijan). *Stratigraphy and Geological Correlations*, **19**, 21–43.

GROESSENS, E. 1975. Distribution de conodontes dans le Dinantien de la Belgique. *International Symposium on Belgian Micropaleontological Limits, From Emsian to Viséan, 1974*, Namur, **17**, 1–193.

GUTSCHICK, R.C. & RODRIGUEZ, J. 1979. Biostratigraphy of the Pilot Shale (Devonian–Mississippian) and contemporaneous strata in Utah, Nevada, and Montana. *Brigham Young University, Geological Studies*, **26**, 37–63.

HAHN, G., MÜLLER, P. & BECKER, R.T. 2012. Unterkarbonische Trilobiten aus dem Anti-Atlas (S-Marokko). *Geologica et Palaeontologica*, **44**, 37–74.

HANCE, L. 1996. Foraminiferal biostratigraphy of the Devonian–Carboniferous boundary and Tournaisian strata in Central Hunan Province, South China. *Mémoires de l'Institut Géologique de l'Université de Louvain*, **36**, 29–53.

HANCE, L. & POTY, E. 2006. Hastarian. *Geologica Belgica*, **9**, 111–116.

HANCE, L., MUCHEZ, P. ET AL. 1994. Biostratigraphy and sequence stratigraphy at the Devonian–Carboniferous transition in southern China (Hunan Province). Comparison with southern Belgium. *Annales de la Société géologique de Belgique*, **116**, 359–378.

HANCE, L., POTY, E. & DEVUYST, F.-X. 2001. Stratigraphie séquentielle du Dinantien type (Belgique) et corrélation avec le Nord de la France (Boulonnais, Avesnois). *Bulletin de la Société Géologique de France*, **172**, 411–426.

HANCE, L., POTY, E. & DEVUYST, F.-X. 2006. Tournaisian. *Geologica Belgica*, **9**, 47–53.

HANCE, L., HOU, H. & VACHARD, D. 2011. *Upper Famennian to Visean Foraminifers and Some Carbonate Microproblematica from South China*. Geological Publishing House, Beijing.

HARTENFELS, S. 2011. Die globalen *Annulata*-Events und die Dasberg-Krise (Famennium, Oberdevon) in Europa und Nord-Afrika – hochauflösende Conodonten-Stratigraphie, Karbonat-Mikrofazies, Paläoökologie und Paläodiversität. *Münstersche Forschungen zur Geologie und Paläontologie*, **105**, 17–527.

HARTENFELS, S. & BECKER, R.T. 2012. Conodont age and correlation of the transgressive *Gonioclymenia* and *Kalloclymenia* Limestones (Famennian, Anti-Atlas, SE Morocco). *Terra Nostra*, **2012**, 67.

HASS, W.H. 1959. Conodonts from the Chappel Limestone of Texas. *United States Geological Survey, Professional Paper*, **294-J**, 365–399.

HEAL, S., PATERSON, N., EBLE, C., MASON, C.E., GODHUE, R., LARSSON, N. & CLAYTON, G. 2009. Preliminary palynological results from the Late Devonian–Early Mississippian of Morehead and adjacent areas. *In*: ETTENSOHN, F.R., LIERMAN, R.T., MASON, C.E. & CLAYTON, G. (eds) *Changing Physical and Biotic Conditions on Eastern Laurussia: Evidence from Late Devonian to Middle Mississippian Basinal and Deltaic Sediments of Northeastern Kentucky, U.S.A.* North American Paleontological Convention-2009, Cincinnati, Field Trip No. 2, 65–69, 80.

HERBIG, H.-G. & MAMET, B. 2006. A muddy to clear-water ramp, latest Devonian, Velbert Anticline (Rheinisches Schiefergebirge, Germany). *Geologica et Palaeontologica*, **40**, 1–25.

HERBIG, H.-G., LOBOVA, D. & SEEKAMP, V. 2013. Sea-level history during the birth of a foreland basin : the Famennian-Visean of 'Velbert 4', westernmost Rhenish Massif, Germany. *In*: ROCHA, R., PAIS, J., KULLBERG, J.C. & FINNEY, S. (eds) *STRATI 2013, First International Congress on Stratigraphy, At the Cutting Edge of Stratigraphy*. Springer Geology, Zurich, 397–402.

HIGGS, K. & STREEL, M. 1984. Spore stratigraphy at the Devonian–Carboniferous boundary in the northern 'Rheinisches Schiefergebirge', Germany. *Courier Forschungsinstitut Senckenberg*, **67**, 157–179.

HIGGS, K.T. & STREEL, M. 1994. Palynological age for the lower part of the Hangenberg Shales in Sauerland, Germany. *Annales de la Société géologique de Belgique*, **116**, 243–247.

HIGGS, K.T., STREEL, M., KORN, D. & PAPROTH, E. 1993. Palynological data from the Devonian–Carboniferous boundary beds in the new Stockum trench II and the Hasselbachtal borehole, northern Rhenish Massif, Germany. *Annales de la Société géologique de Belgique*, **115**, 551–557.

HOU, H., JI, Q. ET AL. 1985. *Muhua Sections of Devonian–Carboniferous Boundary Beds*. Geological Publishing House, Beijing.

HOUSE, M.R. 1978. Devonian ammonoids from the Appalachians and their bearing on international zonation and correlation. *Special Papers in Palaeontology*, **21**, 1–70.

HOUSE, M.R. 1993. Earliest Carboniferous goniatite recovery after the Hangenberg Event. *Annales de la Société géologique de Belgique*, **115**, 559–579.

HOUSE, M.R., GORDON, M., JR & HLAVIN, W.J. 1986. Late Devonian ammonoids from Ohio and adjacent states. *Journal of Paleontology*, **60**, 126–144.

IZOKH, N.G. & ANDREEVA, E.S. 2013. Uppermost Famennian conodonts from Kuznetsk basin (south of west Siberia). *In*: EL HASSANI, A., BECKER, R.T. & TAHIRI, A. (eds) *International Field Symposium 'The Devonian and Lower Carboniferous of Northern Gondwana'*, Abstract Book. Document de l'Institut Scientifique, Rabat, **26**, 65.

JI, Q. 1985. Study on the phylogeny, taxonomy, zonation and biofacies of *Siphonodella* (conodonta). *Institute of Geology, Bulletin*, **11**, 51–75.

JI, Q. 1987. The Devonian–Carboniferous boundary in shallow-water facies areas of China as based on conodonts. *Acta Geologica Sinica*, **61**, 11–12.

JI, Q. & ZIEGLER, W. 1992. Phylogeny, speciation and zonation of *Siphonodella* of shallow water facies (conodonta, early Carboniferous). *Courier Forschungsinstitut Senckenberg*, **154**, 223–251.

JI, Q., ZHANG, Z.-H., CHEN, X.-Z. & WANG, G.-B. 1987. Studies of the Devonian–Carboniferous boundary in Zhaisha, Luzhei of Guangxi. *Journal of Stratigraphy*, **11**, 213–217 [in Chinese].

JI, Q., WANG, Z. ET AL. 1989. *The Dapoushang Section, An Excellent Section for the Devonian-Carboniferous Boundary Stratotype in China*. Science Press, Beijing.

JI, W. 1987. Early Carboniferous conodonts from Jianghua County of Hunan Province, and their stratigraphic value – with a discussion on the mid-Aikuanian Event. *Bulletin of the Institute of Geology, Chinese Academy of Geological Sciences*, **16**, 115–141 [in Chinese with English summary].

KAISER, S.I. 2005. *Mass extinctions, climatic and oceanographic changes at the Devonian–Carboniferous boundary*. PhD thesis, Fakultät für Geowissenschaften, Ruhr-Universität Bochum, http://www-brs.ub.ruhr-uni-bochum.de/netahtml/HSS/Diss/KaiserSandraIsa bella/diss.pdf

KAISER, S.I. 2007. Conodontenstratigraphie und Geochemie ($\delta^{13}C_{carb}$, $\delta^{13}C_{org}$, $\delta^{18}O_{phosph}$) aus dem Devon-Karbon Grenzbereich der Karnischen Alpen. *Jahrbuch der geologischen Bundesanstalt*, **146**, 301–314.

KAISER, S.I. 2009. The Devonian/Carboniferous stratotype section La Serre (Montagne Noire) revisited. *Newsletters on Stratigraphy*, **43**, 195–205.

KAISER, S.I. & BECKER, R.T. 2007. The required revision of the Devonian-Carboniferous boundary. *In*: WANG, Y., ZHANG, H. & WANG, X. (eds) *XVI International Congress on the Carboniferous and Permian, Abstracts. Journal of Stratigraphy*, **31**(suppl. 1), 95.

KAISER, S.I. & CORRADINI, C. 2008. Should the Devonian/Carboniferous boundary be redefined? *SDS Newsletter*, **23**, 55–56.

KAISER, S.I. & CORRADINI, C. 2011. The early siphonodellids (Conodonta, Late Devonian–Early Carboniferous): overview and taxonomic state. *Neues Jahrbuch für Geologie und Paläontologie, Abhandlungen*, **261**, 19–35.

KAISER, S.I., STEUBER, T., BECKER, R.T. & JOACHIMSKI, M.M. 2006. Geochemical evidence for major environmental change at the Devonian–Carboniferous boundary in the Carnic Alps and the Rhenish Massif. *Palaeogeography, Palaeoclimatology, Palaeoecology*, **240**, 146–160.

KAISER, S.I., BECKER, R.T., SPALLETTA, C. & STEUBER, T. 2009. High-resolution conodont stratigraphy, biofacies, and extinctions around the Hangenberg Event in pelagic successions from Austria, Italy, and France. *Palaeontologica Americana*, **63**, 97–139.

KAISER, S.I., BECKER, R.T., STEUBER, T. & ABOUSSALAM, Z.S. 2011. Climate-controlled mass extinctions, facies, and sea-level changes around the Devonian-Carboniferous boundary in the eastern Anti-Atlas (SE Morocco). *Palaeogeography, Palaeoclimatology, Palaeoecology*, **310**, 340–364.

KAISER, S.I., ARETZ, M. & BECKER, R.T. 2015. The global Hangenberg Crisis (Devonian–Carboniferous transition): review of a first-order mass extinction. *In*: BECKER, R.T., KÖNIGSHOF, P. & BRETT, C.E. (eds) *Devonian Climate, Sea Level and Evolutionary Events*. Geological Society, London, Special Publications, **423**. First published online November 11, 2015, http://doi.org/10.1144/SP423.9

KALVODA, J., KUMPAN, T. & BÁBEK, O. 2015. Upper Famennian and Lower Tournaisian sections of the Moravian Karst (Moravio-Silesian Zone, Czech Republic): a proposed key area for correlation of the conodont and foraminiferal zonations. *Geological Journal*, **50**, 17–38. First published online July 25, 2013, http://doi.org/10.1002/gj.2523

KEUPP, H. & KOMPA, R. 1984. Mikrofazielle und sedimentologische Untersuchungen an Devon/Karbon Profilen am Nordrand des rechtsrheinischen Schiefergebirges. *In*: PAPROTH, E. & STREEL, M. (eds) *The Devonian–Carboniferous Boundary*. Courier Forschungsinstitut Senckenberg, **67**, 139–142.

KLEIN, C. & KORN, D. 2014. A morphometric approach to conch ontogeny of *Cymaclymenia* and related genera (Ammonoidea, Late Devonian). *Fossil Record*, **17**, 1–32.

KOCH, M., LEUTERITZ, K. & ZIEGLER, W. 1970. Alter, Fazies und Paläogeographie der Oberdevon/Unterkarbon-Schichtenfolge an der Seiler bei Iserlohn. *Fortschritte in der Geologie vom Rheinland und Westfalen*, **17**, 679–732.

KONONOVA, L.I. & WEYER, D. 2013. Upper Famennian conodonts from the Breternitz Member (Upper Clymeniid Beds) of the Saalfeld region, Thuringia (Germany). *Freiberger Forschungshefte*, **C545**(psf 21), 15–97.

KORN, D. 1981. *Cymaclymenia* – eine besonders langlebige Clymenien-Gattung (Ammonoidea, Cephalopoda). *Neues Jahrbuch für Geologie und Paläontologie, Abhandlungen*, **161**, 172–208.

KORN, D. 1984. Die Goniatiten der Stockumer Imitoceras-Kalklinsen (Ammonoidea; Devon/Karbon-Grenze). *Courier Forschungsinstitut Senckenberg*, **67**, 71–89.

KORN, D. 1988. On the stratigraphical occurrence of *Cymaclymenia evoluta* (H. Schmidt, 1924) at the type locality. *Courier Forschungsinstitut Senckenberg*, **100**, 215–216.

Korn, D. 1991. Three-dimensionally preserved clymeniids from the Hangenberg Black Shale of Drewer (Cephalopoda, Ammonoidea; Devonian–Carboniferous boundary, Rhenish Massif). *Neues Jahrbuch für Geologie und Paläontologie, Monatshefte*, **1991**, 553–563.

Korn, D. 1992. Ammonoideen aus dem Oberdevon und Unterkarbon von Aprath, Schurf Steinbergerbach und Straßeneinschnitt Kohleiche. *In*: Thomas, E. (ed.) *Oberdevon und Unterkarbon von Aprath im Bergischen Land (Nördliches Rheinisches Schiefergebirge). Ein Symposium zum Neubau der Bundesstraße 224.* Sven von Loga, Cologne, 169–182.

Korn, D. 1993. The ammonoid faunal change near the Devonian–Carboniferous boundary. *Annales de la Société géologique de Belgique*, **115**, 581–593.

Korn, D. 1994. Devonische und karbonische Prionoceraten (Cephalopoda, Ammonoidea) aus dem Rheinischen Schiefergebirge. *Geologie und Paläontologie in Westfalen*, **30**, 1–85.

Korn, D. 1999. Famennian ammonoid stratigraphy of the Ma'der and Tafilalt (eastern Anti-Atlas, Morocco). *Abhandlungen der Geologischen Bundesanstalt*, **54**, 147–179.

Korn, D. 2000. Morphospace occupation of ammonoids over the Devonian–Carboniferous boundary. *Paläontologische Zeitschrift*, **74**, 247–257.

Korn, D. 2010. Lithostratigraphy and biostratigraphy of the Kulm succession in the Rhenish Mountains. *Zeitschrift der deutschen Gesellschaft für Geowissenschaften*, **161**, 431–453.

Korn, D. & Feist, R. 2007. Early Carboniferous ammonoid faunas and stratigraphy of the Montagne Noire (France). *Fossil Record*, **10**, 99–124.

Korn, D. & Luppold, F.W. 1987. Nach Clymenien und Conodonten gegliederte Profile des oberen Famennium im Rheinischen Schiefergebirge. *Courier Forschungsinstitut Senckenberg*, **92**, 199–223.

Korn, D. & Price, J. 1987. Taxonomy and phylogeny of the Kosmoclymeniinae subfam. nov. (Cephalopoda, Ammonoidea, Clymeniida). *Courier Forschungsinstitut Senckenberg*, **92**, 5–75.

Korn, D. & Weyer, D. 2003. High resolution stratigraphy of the Devonian–Carboniferous transitional beds in the Rhenish Mountains. *Mitteilungen aus dem Museum für Naturkunde in Berlin, Geowissenschaftliche Reihe*, **6**, 79–124.

Korn, D., Clausen, C.-D., Belka, Z., Leuteritz, K., Luppold, F.W., Feist, R. & Weyer, D. 1994. Die Devon/Karbon-Grenze bei Drewer (Rheinisches Schiefergebirge). *Geologie und Paläontologie Westfalens*, **29**, 97–147.

Korn, D., Ebbighausen, V. & Bockwinkel, J. 2002. Palaeogeographical meaning of a Middle Tournaisian ammonoid fauna from Morocco. *Geologica et Palaeontologica*, **36**, 79–86.

Korn, D., Belka, Z., Fröhlich, S., Rücklin, M. & Wendt, J. 2004. The youngest African clymeniids (Ammonoidea, Late Devonian) – failed survivors of the Hangenberg Event. *Lethaia*, **37**, 307–315.

Korn, D., Bockwinkel, J. & Ebbighausen, V. 2007. Tournaisian and Viséan ammonoid stratigraphy in North Africa. *Neues Jahrbuch für Geologie und Paläontologie, Abhandlungen*, **243**, 127–148.

Korn, D., Bockwinkel, J. & Ebbighausen, V. 2015. The Late Devonian ammonoid *Mimimitoceras* in the Anti-Atlas of Morocco. *Neues Jahrbuch für Geologie und Paläontologie, Abhandlungen*, **275** , 125–150.

Krebs, W. 1979. Devonian basinal facies. *In*: House, M.R., Scrutton, C.T. & Bassett, M.G. (eds) *The Devonian System*. Special Papers in Palaeontology, **23**, 125–139.

Kürschner, W., Becker, R.T., Buhl, D. & Veizer, J. 1993. Strontium isotopes in conodonts: Devonian–Carboniferous transition, the northern Rhenish Slate Mountains, Germany. *Annales de la Société géologique de Belgique*, **115**, 595–621.

Kulagina, E.I. 2013. Taxonomic diversity of foraminifers of the Devonian–Carboniferous boundary interval in the South Urals. *Bulletin of Geosciences*, **88**, 265–282.

Kulagina, E.I., Gibshman, N.B. & Pazukhin, V.N. 2003. Foraminiferal zonal standard for the Lower Carboniferous of Russia and its correlation with the conodont zonation. *Rivista Italiana di Paleontologia e Stratigrafia*, **109**, 173–185.

Kullmann, J., Korn, D. & Weyer, D. 1990. Ammonoid zonation of the Lower Carboniferous subsystem. *Courier Forschungsinstitut Senckenberg*, **130**, 127–131.

Kumpan, T., Babek, O., Kalvoda, J., Frydas, J. & Grygar, T.M. 2014*a*. A high-resolution, multiproxy stratigraphic analysis of the Devonian–Carboniferous boundary sections in the Moravian Karst (Czech Republic) and a correlation with the Carnic Alps (Austria). *Geological Magazine*, **151**, 201–215. First published online May 2, 2013, http://doi.org/10.1017/S0016756812001057

Kumpan, T., Bábek, O., Kalvoda, J., Grygar, T.M. & Frýda, J. 2014*b*. Sea-level and environmental changes around the Devonian–Carboniferous boundary in the Namur–Dinant Basin (S Belgium, NE France): a multi-proxy stratigraphic analysis of carbonate ramp archives and its use in regional and interregional correlations. *Sedimentary Geology*, **311**, 43–59.

Kumpan, T., Bábek, O., Kalvoda, J., Matys Grygar, T., Frýda, J., Becker, R.T. & Hartenfels, S. 2015*a*. Petrophysical and geochemical signature of the Hangenberg events: an integrated stratigraphy of the Devonian–Carboniferous boundary interval in the Northern Rhenish Massif (Avalonia, Germany). *Bulletin of Geosciences*, **90**, 667–694.

Kumpan, T., Bábek, O., Kalvoda, J., Grygar, T.M. & Frýda, J. 2015*b*. High-resolution stratigraphy of the Devonian-Carboniferous boundary in Europe: multidisciplinary approach. *In*: Mottequin, B., Denayer, J., Königshof, P., Prestianni, C. & Olive, S. (eds) *IGCP 596 – SDS Symposium, Climate Change and Biodiversity Patterns in the Mid-Palaeozoic*, 20–22 September 2015, Brussels, Belgium. Strata, série 1, **16**, 77–78.

Kusina, L.F. 1985. K revizii roda *Imitoceras* (Ammonoidea). *Paleontologicheskiy Zhurnal*, **1985**, 35–48, pl. 3.

Kuz'min, A.V. 1998. Conodonts from the shallow water Devonian–Carboniferous Boundary deposits of the Middle Pechora uplift. *Paleontologicheskiy*

Zhurnal, **1998**, 63–66 [in Russian with short English summary].

Li, Z.-L., Xu, J.-B., Li, Y.-H., Liu, R.-T., Cai, N., Liao, K.-L. & Shi, X.-X. 2002. Classification of chronostratigraphic and the biota from the end of Devonian Period to early Carboniferous Epoch of northern Guangxi. *Guangxi Geology*, **15**, 13–20.

Lipnjagow, O.M. 1979. The conodonts of C[t]1a and C[t]1b of the Donetz Basin. *Service géologique de Belgique, Professional Paper*, **1979**, 42–49.

Loboziak, S., Streel, M., Caputo, M.V. & De Melo, J. H. G. 1992. Middle Devonian to Lower Carboniferous miospore stratigraphy in the central Parnaíba Basin (Brazil). *Annales de la Société géologique de Belgique*, **115**, 215–226.

Loboziak, S., Streel, M., Caputo, M.V. & Melo, J. H. G. 1993. Middle Devonian to Lower Carboniferous miospores from selected boreholes in Amazonas and Parnaíba Basins (Brazil), additional data, synthesis, and correlation. *Documents des Laboratoires de Géologie de la Faculté des Sciences de Lyon*, **125**, 277–289.

Loboziak, S., de Melo, J. H. G., Steemans, P. & Rodrigues Barrilari, I.M. 1995. Miospore evidence for pre-Emsien and latest Famennian sedimentation in the Devonian of the Paraná Basin, South Brazil. *Anais da Academia Brasileira de Ciências*, **67**, 391–392.

Loboziak, S., Caputo, M.V. & Melo, J. H. G. 2000. Middle Devonian–Tournaisian miospore biostratigraphy in the southwestern outcrop belt of the Parnaíba Basin, North-Central Brazil. *Revue de Micropaléontologie*, **43**, 301–318.

Luppold, F.W., Hahn, G. & Korn, D. 1984. Trilobiten-, Ammonoideen- und Conodonten-Stratigraphie des Devon/Karbon-Grenzprofiles auf dem Müssenberg (Rheinisches Schiefergebirge). *Courier Forschungsinstitut Senckenberg*, **67**, 91–121.

Luppold, F.-W., Clausen, C.-D., Korn, D. & Stoppel, D. 1994. Devon/Karbon-Grenzprofile im Bereich von Remscheid-Altenaer Sattel, Warsteiner Sattel, Briloner Sattel und Attendorn-Elsper Doppelmulde (Rheinisches Schiefergebirge). *Geologie und Paläontologie in Westfalen*, **29**, 7–69.

Makhlina, M.K. 1996. Cyclic stratigraphy, facies and fauna of the Lower Carboniferous (Dinantian) of the Moscow Syneclise and Voronezh Anteclise. *In*: Strogen, P., Sommerville, I.D. & Jones, G.L. (eds) *Recent Advances in Lower Carboniferous Geology*. Geological Society, London, Special Publications, **107**, 359–364, http://doi.org/10.1144/GSL.SP.1996.107.01.25

Malec, J. 2014. The Devonian/Carboniferous boundary in the Holy Cross Mountains (Poland). *Geological Quarterly*, **58**, 217–234.

Mamet, B. & Preat, A. 2003. Sur les difficules d'interprétation des hiatus stratigraphiques (exemple tire de la transition Dévono-Carbonifère, Bassin de Dinant). *Geologica Belgica*, **6**, 49–65.

Mamet, B., Mortelsman, G. & Sartenaer, P. 1965. Réflexions á propos du Calcaire d'Etroeungt. *Société belge de Géologie, Paléontologie et Hydrologie, Bulletin*, **64**, 41–51.

Martynova, M.V. & Vorontzova, T.N. 1988. Typical sections of the Devonian/Carboniferous boundary deposits in Central Kazakhstan. *In*: *The Devonian–Carboniferous boundary at the territory of the USSR*. Nauka I TechnicA, Minsk, 181–187 [in Russian].

Marynowski, L. & Filipiak, P. 2007. Water column euxinia and wildfire evidence during deposition of the Upper Famennian Hangenberg event horizon from the Holy Cross Mountains (central Poland). *Geological Magazine*, **144**, 569–595.

Matthews, S.C. & Naylor, D. 1973. Lower Carboniferous conodont faunas from south-west Ireland. *Palaeontology*, **16**, 335–380.

Matyja, H. & Stempien-Salek, M. 1994. Devonian/Carboniferous boundary and the associated phenomena in western Pomerania (NW Poland). *Annales de la Société géologique de Belgique*, **116**, 249–263.

Matyja, H., Sobien, K., Marynowski, L., Stempien-Salek, M. & Malkowski, K. 2014. The expression of the Hangenberg Event (latest Devonian) in a relatively shallow-marine succession (Pommeranian Basin, Poland): the results of a multi-proxy investigation. *Geological Magazine*, **152**, 400–428, http://doi.org/10.1017/S001675681400034X

Maziane, N., Higgs, K. & Streel, M. 1999. Revision of the late Famennian miospore zonation scheme in eastern Belgium. *Journal of Micropaleontology*, **18**, 17–25, http://doi.org/10.1144/jm.18.1.17

Maziane, N., Higgs, K. & Streel, M. 2002. Biometry and paleoenvironment of *Retispora lepidophyta* (Kedo) Playford 1976 and associated miospores in the latest Famennian nearshore marine facies, eastern Belgium. *Review of Palaeobotany and Palynology*, **118**, 211–226.

Maziane-Serraj, N., Hartkopf-Fröder, C., Streel, M. & Thorez, J. 2007. Palynomorph distribution and bathymetry in the Chanxhe section (eastern Belgium), reference for the neritic late to latest Famennian transition (Late Devonian). *Geologica Belgica*, **10**, 170–175.

Melo, J. H. G. & Loboziak, S. 2003. Devonian–Early Carboniferous miospore biostratigraphy of the Amazon Basin, Northern Brazil. *Review of Palaeobotany and Palynology*, **124**, 131–202.

Michels, D. 1986. Ökologie und Fazies des jüngsten Ober-Devon von Velbert (Rheinisches Schiefergebirge). *Göttinger Arbeiten zur Geologie und Paläontologie*, **29**, 1–86.

Molyneux, S., Manger, W.L. & Owens, B. 1984. Preliminary account of Late Devonian palynomorph assemblages from the Bedford Shale and Berea Sandstone Formations of central Ohio, U.S.A. *Journal of Micropaleontology*, **3**, 41–51, http://doi.org/10.1144/jm.3.2.41

Morrow, J.R., Sandberg, C.A., Warme, J.E., Murphy, M.A. & Over, D.J. 2007. *Devonian Shelf-to-Slope Facies and Events, Central Great Basin, Nevada and Utah, U.S.A.* Subcomission on Devonian Stratigraphy, SDS 2007 Field Trip Guidebook, 1–93.

Mortelmans, G. 1969. L'Étage Tournaisien dans sa Localité-Type. *Compte Rendu Sixième Congrés International de Stratigraphie et de Géologie du Carbonifère*, 11–16 September 1967, Sheffield, **1**, 19–43.

Mory, A.J. & Crane, D.T. 1982. Early Carboniferous *Siphonodella* (Conodonta) faunas from eastern Australia. *Alcheringa*, **6**, 275–303.

MOSSONI, A., CORRADINI, C. & SPALLETTA, C. 2013. Famennian–Tournaisian conodonts from the Monte Taccu section (Sardinia, Italy). *In*: ALBANESI, G.L. & ORTEGA, G. (eds) *Conodonts from the Andes, 3rd International Conodont Symposium*. Asociación Paleontológica Argentinia, Publicación Especial, **13**, 85–90.

MOSSONI, A., CARTA, N., CORRADINI, C. & SPALLETTA, C. 2015. Conodonts across the Devonian/Carboniferous boundary in SE Sardinia (Italy). *Bulletin of Geosciences*, **90**, 371–388.

MOTTEQUIN, B. & POTY, E. 2014. The uppermost Famennian Hangenberg Event in the Namur-Dinant Basin (southern Belgium). *In*: KIDO, E., WATERS, J.A., ARIUNCHIMEG, Y., GONCHIGDORI, S., DA SILVA, A.C., WHALEN, M., STUTTNER, T.J. & KÖNIGSHOF, P. (eds) *IGCP 596 & IGCP 580 Joint Meeting and Field Workshop, International Symposium in Mongolia*, 5–18 August 2014, Ulaanbaatar, Mongolia, Abstract Volume. Berichte des Institutes für Erdwissenschaften, Karl-Franzenz-Universität Graz, **19**, 36–37.

MÜNSTER, G. (GRAF ZU) 1839. Nachtrag zu den Goniatiten des Fichtelgebirges. *Beträge zur Petrefactenkunde*, **1**, 43–55.

NEMIROVSKAYA, T.I., CHERMNYKH, V.A., KONONOVA, L.I. & PAZUKHIN, V.N. 1993. Conodonts of the Devonian–Carboniferous boundary section, Kozhim, Polar Urals, Russia. *Annales de la Société géologique de Belgique*, **115**, 629–647.

NICOLL, R.S. & DRUCE, E.C. 1979. Conodonts from the Fairfield Group, Canning Basin, Western Australia. *Bureau of Mineral Resources, Geology and Geophysics, Bulletin*, **190**, 1–134.

NIKOLAEVA, S.V. & BOGOSLOVSKY, B.I. 2005*a*. Late Famennian Ammonoids from the Upper Part of the Kiya Formation of the South Urals. *Paleontological Journal*, **39**(suppl. 5), S527–S537.

NIKOLAEVA, S.V. & BOGOSLOVSKY, B.I. 2005*b*. On the occurrence of Famennian ammonoids in the Northern Caucasus. *Paleontologicheskii Zhurnal*, **2005**, 10–19 [English translation by MAIK 'Nauka/ Interperiodica'].

PAECKELMANN, W. 1938. Erläuterungen zu Blatt Balve. Geologische Karte von Preussen und benachbarten deutschen Ländern, Nr. 2655 (neue Nr. 4613), 1–70.

PAECKELMANN, W. & SCHINDEWOLF, O.H. 1937. Die Devon-Karbon-Grenze. *Compte Rendu 2e Congrés Carbonifère, 1935*, Heerlen, **2**, 703–714.

PAPROTH, E. 1986. An introduction to a field trip to the Late Devonian outcrops in the Northern Rheinisches Schiefergebirge (Federal Republic of Germany). *Annales de la Société géologique de Belgique*, **109**, 275–284.

PAPROTH, E. & STREEL, M. 1970. Corrélations biostratigraphiques près de la limite Dévonien/Carbonifère entre les facies littoraux ardenais et les facies bathyaux rhénans. *In*: STREEL, M. & WAGNER, R.H. (eds) *Colloque sur la Stratigraphie du Carbonifère*. Les Congres et Colloques de l'Université de Liege, **55**, 365–398.

PAPROTH, E. & STREEL, M. 1982. *Devonian–Carboniferous transitional beds of the northern 'Rheinisches Schiefergebirge'*. Guidebook, IUGS Working Group on the Devonian/Carboniferous Boundary, Liège, 1–63.

PAPROTH, E. & STREEL, M. 1984*a*. The IUGS Devonian–Carboniferous Working Group: a report on activities, 1978–1984. *Courier Forschungsinstitut Senckenberg*, **67**, 5–9.

PAPROTH, E. & STREEL, M. 1984*b*. Precision and practicability: on the definition of the DCB. *In*: PAPROTH, E. & STREEL, M. (eds) *The Devonian–Carboniferous Boundary*. Courier Forschungsinstitut Senckenberg, **67**, 255–258.

PAPROTH, E., STOPPEL, D. & CONIL, R. 1976. Revision micropaleontologique des sites dinantiens de Zippenhaus et de Cromford (Allemagne). *Bulletin de la Société belge de Géologie, de Paléontologie et d'Hydrologie*, **82**, 51–139 [imprint 1973].

PAPROTH, E., DREESEN, R. & THOREZ, J. 1986. Famennian paleogeography and event stratigraphy of northwestern Europe. *Annales de la Société géologique de Belgique*, **109**, 175–186.

PAPROTH, E., FEIST, R. & FLAJS, G. 1991. Decision on the Devonian–Carboniferous boundary stratotype. *Episodes*, **14**, 331–336.

PAUL, H. 1939. Die Etroeungt-Schichten des Betrgischen Landes. *Jahrbuch der preußischen geologischen Landesanstalt*, **59**, 647–726.

PAZUKHIN, V.N., KULAGINA, E.I. & SEDAEVA, K.M. 2009. The Devonian/Carboniferous boundary on the western slope of the South Urals. *In*: *Proceedings of the International Field Meeting 'The historical type sections, proposed and potential GSSP of the Carboniferous in Russia', Carboniferous Type Sections in Russia and Potential Global Stratotypes: Southern Urals Session*, 13–18 August 2009, Ufa-Sibai. DesignPolygraph Service Ltd., Ufa, 22–33 [in Russian with English abstract].

PEREZ-LEYTON, M. 1991. Miospores du Dévonien moyen et supérieur de la coupe de Bermejo-La Angostura (Sud-Est de la Bolivie). *Annales de la Société géologique de Belgique*, **113**, 373–389.

PERREIRA, Z., CLAYTON, G. & OLIVEIRA, J.T. 1996. Palynostratigraphy of the Devonian–Carboniferous boundary in southwest Portugal. *Annales de la Société géologique de Belgique*, **117**, 189–199.

PERRET, M.-F. 1988. Le passage du Dévonien au Carbonifère dans les Pyrenées, Zonation par conodontes. *Courier Forschungsinstitut Senckenberg*, **100**, 39–52.

PETTER, G. 1959. *Goniatites dévoniennes du Sahara*. Service de la Carte géologique de l'Algérie, série A, **3166**.

PLAYFORD, G. & MCGREGOR, D.C. 1993. Miospores and organic-walled microphytoplankton of Devonian–Carboniferous boundary beds (Bakken Formation), southern Saskatchewan: a systematic and stratigraphic appraisal. *Geological Survey of Canada, Bulletin*, **445**, 1–107.

POTY, E., HANCE, L., LEES, A. & HENNEBERT, M. 2002. Dinantian lithostratigraphic units (Belgium). *In*: BULTYNCK, P. & DEJONGHE, L. (eds) *Guide to a Revised Stratigraphic Scale for Belgium*. Geologica Belgica, **4**, 69–93.

POTY, E., DEVUYST, F.-X. & HANCE, L. 2006. Upper Devonian and Mississippian foraminiferal and rugose coral zonation of Belgium and Northern France: a tool for Eurasian correlations. *Geological Magazine*, **143**, 829–857.

Poty, E., Aretz, M. & Hance, L. 2014. Belgian substages as a basis for an international chronostratigraphic division of the Tournaisian and Viséan. *Geological Magazine*, **151**, 229–243.

Price, J.D. & House, M.R. 1984. Ammonoids near the Devonian–Carboniferous boundary. *Courier Forschungsinstitut Senckenberg*, **67**, 15–22.

Price, J.D. & Korn, D. 1989. Stratigraphically important Clymeniids (Ammonoidea) from the Famennian (Late Devonian) of the Rhenish Massif, West Germany. *Courier Forschungsinstitut Senckenberg*, **110**, 257–294.

Qie, W., Zhang, X., Du, Y., Yang, B., Ji, W. & Luo, G. 2014. Conodont biostratigraphy of Tournaisian shallow-water carbonates in central Guangxi, South China. *Geobios*, **47**, 389–401, http://doi.org/10.1016/j.geobios.2014.09.005

Qin, G., Zhao, R. & Ji, Q. 1988. Late Devonian and Early Carboniferous conodonts from northern Guangdong and their stratigraphic significance. *Acta Micropalaeontologica Sinica*, **5**, 57–71.

Rahmani-Antari, K. & Lachkar, G. 2001. Contribution à l'ètude biostratigraphique du Dévonien et du Carbonifère de la plate-forme marocaine. Datation et corrélations. *Revue de Micropaléontologie*, **44**, 159–183.

Ruan, Y. 1981. *Devonian and Earliest Carboniferous Ammonoids from Guangxi and Guizhou.* Memoirs of Nanjing Institute of Geology and Palaeontology, Academia Sinica, **15** [in Chinese with English summary].

Ruan, Y. 1995. Review of the Carboniferous ammonoid zones in China. *Palaeontologia Cathayana*, **6**, 345–364.

Sandberg, C.A., Streel, M. & Scott, R.A. 1972. Comparison between conodont zonation and spore assemblages at the Devonian–Carboniferous boundary in the western and central United States and in Europe. *Compte Rendu Septième Congrés International de Stratigraphie et de Géologie du Carbonifère*, 23–28 August 1971, Krefeld, **1**, 179–203.

Sandberg, C.A., Ziegler, W., Leuteritz, K. & Brill, S.M. 1978. Phylogeny, speciation, and zonation of *Siphonodella* (Conodonta, Upper Devonian and Lower Carboniferous). *Newsletters on Stratigraphy*, **7**, 102–120.

Savage, N.M. 1982. Early Lower Carboniferous conodonts from the *Syringothyris* Limestone of Kashmir. *Journal of the Geological Society of India*, **23**, 101–111.

Schindewolf, O.H. 1937. Zur Stratigraphie und Paläontologie der Wocklumer Schichten. *Abhandlungen der Preußischen Geologischen Landesanstalt, n. F.*, **178**, 1–132.

Schmidt, H. 1924. Zwei Cephalopodenfaunen an der Devon-Carbongrenze im Sauerland. *Jahrbuch der Preußischen geologischen Landesanstalt*, **44**, 98–171.

Schmidt, H. 1972. Balvium: Neuer Name für Unterkarbon I. *Compte Rendu Septième Congrès International de Stratigraphique et de Géologie du Carbonifère*, 23–28 August 1971, Krefeld, **1**, 205–207.

Schmidt, H. & Plessmann, W. 1961. *Sauerland.* Sammlung Geologischer Führer, **39**, Bornträger, Berlin.

Schönlaub, H.P., Feist, R. & Korn, D. 1988. The Devonian–Carboniferous boundary at the section 'Grüne Schneid' (Carnic Alps, Austria): a preliminary report. *Courier Forschungsinstitut Senckenberg*, **100**, 149–167.

Schülke, I. & Popp, A. 2005. Microfacies development, sea-level change, and conodont stratigraphy of Famennian mid- to deep platform deposits of the Beringhauser Tunnel section (Rheinisches Schiefergebirge, Germany). *Facies*, **50**, 647–664.

Shen, J.-W. 1994. New conodont data from benthic D/C boundary beds in Guilin. *Acta Micropalaeontologica Sinica*, **11**, 503–514 [in Chinese with English summary].

Shilo, N.A., Bouckaert, J. *et al.* 1984. Sedimentological and paleontological Atlas of the late Famennian and Tournaisian deposits in the Omolon region (NE-USSR). *Annales de la Société géologique de Belgique*, **107**, 137–247.

Söte, T. 2015. *Stratigraphie und Fazies an der Devon/Karbon-Grenze im Forststeinbruch Reigern (Hachen, Nordsauerland).* BSc thesis, Westfälische Wilhelms University, Münster.

Stewart, I.J. & Selwood, E.B. 1985. Devonian–Carboniferous biostratigraphy and conodont localities of the Launceston area and St Melion outlier. *In*: Austin, R. & Armstrong, H.A. (eds) *Fourth European Conodont Symposium (ECOS IV), Nottingham 1985, Field Excursion Guidebook A*, University of Southampton, 63–90.

Streel, M. 1969. Corrélations palynologiques entre les sediments de transition Dévonien/Dinantien dans les bassins Ardenno-Rhénans. *Compte Rendu Sixième Congrés International de Stratigraphie et de Géologie du Carbonifère*, 11–16 September 1967, Sheffield, **1**, 3–18.

Streel, M. 1986. Biostratigraphie par spores du Dévonien Ardenno-Rhénan. *Annales de la Société Géologique du Nord*, **105**, 85–95.

Streel, M. 1999. Quantitative palynology of Famennian events in the Ardenne–Rhine regions. *Abhandlungen der Geologischen Bundesanstalt*, **54**, 201–212.

Streel, M. 2007. Comment on 'Comment on Proposed Uppermost Famennian substage' by TM Charles A. Sandberg. *SDS Newsletter*, **22**, 49–51.

Streel, M. 2009. Upper Devonian miospore and conodont zone correlation in Western Europe. *In*: Königshof, P. (ed.) *Devonian Change: Case Studies in Palaeogeography and Palaeoecology*. Geological Society, London, Special Publications, **314**, 163–176, http://doi.org/10.1144/SP314.9

Streel, M. 2015. Palynomorphs (miospores, acritarchs, prasinophytes) before and during the Hangenberg crisis. *In*: Mottequin, B., Denayer, J., Königshof, P., Prestianni, C. & Olive, S. (eds) *IGCP 596 – SDS Symposium, Climate Change and Biodiversity Patterns in the Mid-Palaeozoic*, 20–22 September 2015, Brussels, Belgium, Strata, série 1, **16**, 140–143.

Streel, M. & Traverse, A. 1978. Spores from the Devonian/Mississippian transition near the Horseshoe Curve section, Altoona, Pennsylvania, U.S.A. *Review of Palaeobotany and Palynology*, **26**, 21–39.

Streel, M. & Wagner, R.H. (eds) 1970. *Colloque sur la Stratigraphie du Carbonifère*. Les Congres et Colloques de l'Universite de Liege, 55.

STREEL, M., BRICE, D. *ET AL.* 1998. Proposal for a Strunian Substage and a subdivision of the Famennian Stage into four Substages. *SDS Newsletter*, **15**, 47–52.

STREEL, M., CAPUTO, M.V., LOBOZIAK, S. & MELO, J. H. G. 2000*a*. Late Frasnian–Famennian climates based on palynomorph analyses and the question of the Late Devonian glaciations. *Earth Science Reviews*, **52**, 121–173.

STREEL, M., CAPUTO, M.V., LOBOZIAK, S., MELO, J. H. G. & THOREZ, J. 2000*b*. Palynology and sedimentology of laminites and tillites from the latest Famennian of the Parnaíba Basin, Brazil. *Geologica Belgica*, **3**, 87–96.

STREEL, M., BRICE, D. & MISTIAEN, B. 2006. Strunian. *Geologica Belgica*, **9**, 105–109.

STREEL, M., MAZIANE-SERRAJ, N., MARSHALL, J.E. & THOREZ, J. 2007. A reference section for neritic facies at the transition late to latest Famennian. *SDS Newsletter*, **22**, 34–40.

STREEL, M., CAPUTO, M.V. & MELO, J. H. G. 2011. What do latest Famennian and Viséan diamictites from Western Gondwana tell us? *In*: SUTTNER, T.J., KIDO, E., PILLER, W.E. & KÖNIGSHOF, P. (eds) *IGCP 596 Opening Meeting*, 19–24 September 2011, Graz. Berichte des Institutes für Erdwissenschaften der Karl-Franzens-Universität Graz, **16**, 83–84.

STREEL, M., CAPUTO, M.V., MELO, J. H. G. & PEREZ-LEYTON, M. 2013. What do latest Famennian and Viséan diamictites from Western Gondwana tell us? *Palaeobiodiversity and Palaeonevironments*, **93**, 299–316.

THOMAS, E. & ZIMMERLE, W. 1992. Geologie der Baustelle B224n bei Aprath, vom Tunnel 'Im großen Busch' bis 'Straßeneinschnitt Kohleiche'. *In*: THOMAS, E. (ed.) *Oberdevon und Unterkarbon von Aprath im Bergischen Land (Nördliches Rheinisches Schiefergebirge). Ein Symposium zum Neubau der Bundesstraße 224*. Verlag Sven von Loga, Köln, 8–100.

THOMPSON, T.L. 1986. Paleozoic Succession in Missouri, part 4, Mississippian System. *Missouri Department of Natural Resources, Division of Geology and Land Survey, Report of Investigations*, **70**, 1–189.

THOMPSON, T.L. & FELLOWS, L.D. 1970. Stratigraphy and Conodont Biostratigraphy of Kinderhookian and Osagean (lower Mississippian) Rocks of Southwestern Missouri and Adjacent States. *Missouri Geological Survey and Water Resources, Report of Investigations*, **45**, 1–263.

TOLOKONNIKOVA, Z., ERNST, A. & HERBIG, H.-G. 2014. Famennian (Upper Devonian) bryozoans from borehole Velbert 4, Rhenish Massif (Germany). *Neues Jahrbuch für Geologie und Paläontologie, Abhandlungen*, **271**, 25–44.

TRAGELEHN, H. 2010. Short note on the origin of the conodont genus *Siphonodella* in the uppermost Famennian. *SDS Newsletter*, **25**, 41–43.

TURNAU, E., AVCHIMOVITCH, V.I., BYVSHEVA, T.V., CLAYTON, G., HIGGS, K.T. & OWENS, B. 1994. Taxonomy and stratigraphical distribution of *Verrucosporites nitidus* Playford, 1964 and related species. *Review of Palaeobotany and Palynology*, **81**, 289–295.

UENO, K., WATANABE, D., IGO, H., KAKUWA, Y. & MATSUMOTO, R. 1997. Early Carboniferous foraminifers from the Mobarak Formation in Shahmirzad, northeastern Alborz Mountains, northern Iran. *In*: ROSS, C.A., ROSS, J. R. P. & BRENCKLE, P.L. (eds) *Late Paleozoic Foraminifera; Their Biostratigraphy, Evolution, and Paleoecology*. Cushman Foundation for Foraminiferal Research, Special Publications, **36**, 149–152.

VAN STEENWINKEL, M. 1980. Sedimentation and conodont stratigraphy of the Hastière Limestone, lowermost Dinantian, Anseremme, Belgium. *Mededelingen Rijks geologische Dienst*, **32**, 30–33.

VAN STEENWINKEL, M. 1984. The Devonian–Carboniferous boundary in the vicinity of Dinant, Belgium. *Courier Forschungsinstitut Senckenberg*, **67**, 57–69.

VAN STEENWINKEL, M. 1988. *The sedimentary history of the Dinant platform during the Devonian/Carboniferous transition*. PhD thesis, Catholic University Leuven, Belgium.

VAN STEENWINKEL, M. 1993*a*. The Devonian–Carboniferous boundary in southern Belgium: biostratigraphic identification criteria of sequence boundaries. *Special Publications of the International Association of Sedimentology*, **18**, 237–246.

VAN STEENWINKEL, M. 1993*b*. The Devonian–Carboniferous boundary: comparison between the Dinant synclinorium and the northern border of the Rhenish Slate Mountains, a sequence-stratigraphic view. *Annales de la Société géologique de Belgique*, **115**, 665–681.

VAVRDOVÁ, M., ISAACSON, P.E., DIAZ, E. & BEK, J. 1991. Palinologia del límite Devónico-Carbonifera entorno al lago Titikaka, Bolivia: resultados preliminares. *Revista Técnica de YPFB*, **12**, 303–313.

VAVRDOVÁ, M., BEK, J., DUFKA, P. & ISAACSON, P.E. 1996. Palynology of the Devonian (Lochkovian to Tournaisian) sequence, Madre de Dios Basin, northern Bolivia. *Vestník Ceského Geologického Ustavu*, **71**, 333–349.

VÖHRINGER, E. 1960. Die Goniatiten der unterkarbonischen *Gattendorfia*-Stufe im Hönnetal. *Fortschritte in der Geologie von Rheinland und Westfalen*, **3**, 107–196.

WAGNER, P. 2001. Palynology at the Devonian/Carboniferous Boundary: discussion of the sedimentological environment at sampled sections in Portugal, Ireland, U.S.A., and Canada. *Zentralblatt für Geologie und Paläontologie*, **1**, 185–197.

WALLISER, O.H. 1984. Pleading for a natural D/C-boundary. *Courier Forschungsinstitut Senckenberg*, **67**, 241–246.

WANG, C. 1993. Auxiliary stratotype sections for the global stratotype section and point (GSSP) for the Devonian-Carboniferous boundary: Nanbiancun. *Annales de la Société géologique de Belgique*, **115**, 707–708.

WANG, C.-Y. & YIN, B.-A. 1985. An important Devonian–Carboniferous boundary stratotype in neritic facies of Yishan County, Guangxi. *Acta Micropalaeontologica Sinica*, **2**, 28–48.

WANG, C.-Y., YIN, B.-A. *ET AL.* 1987. Devonian–Carboniferous boundary sections in Yishan area, Guangxi. *In*: WANG, C.-Y. (ed.) *Carboniferous Boundaries in China*. Science Press, Beijing, 22–53.

WARREN, A., DI PASQUO, M.M., GRADER, G.W., ISAACSON, P.E. & RODRIGUEZ, A.P. 2014. Latest Famennian Middle Sappington Shale: *lepidophyta–Verrucosisporites nitidus* (LN) Zone at the Logan Gulch type section, Montana, USA. *GSA Annual Meeting, Abstracts with Programs*, **46**, 163.

Weber, H.-M. & Wyse Jackson, P.N. 2006. Bryozoen. *Schriftenreihe der Deutschen Gesellschaft für Geowissenschaften*, **41**, 101–105.

Wendt, J., Kaufmann, B., Belka, Z., Farsan, N. & Bavandpur, A.K. 2002. Devonian/Lower Carboniferous stratigraphy, facies patterns and palaeogeography of Iran. Part I. Southeastern Iran. *Acta Geologica Polonica*, **52**, 129–168.

Wicander, R. & Playford, G. 2013. Marine and terrestrial palynofloras from transitional Devonian–Mississippian strata, Illinois Basin, U.S.A. *Boletin Geológico y Minero*, **124**, 589–637.

Wicander, R., Clayton, G., Marshall, J. E. A., Troth, I. & Racey, A. 2011. Was the latest Devonian glaciation a multiple event? New palynological evidence from Bolivia. *Palaeogeography, Palaeoclimatology, Palaeoecology*, **305**, 75–83.

Work, D.M. 2002. The lower Mississippian (Kinderhookian) ammonoid *Goniocyclus* from the Hannibal Shale, Missouri. *Journal of Paleontology*, **76**, 187–189.

Xiong, J.-F. 1991. Discovery of conodonts of Yan Guan period and boundary between Devonian and Carboniferous in Bachu, Xinjiang. *Xinjiang Petroleum Geology*, **12**, 118–127.

Xiong, J.-F. & Chen, L.-Z. 1983. The Carboniferous conodonts of Guizhou. *Papers of Stratigraphy & Paleontology of Guizhou*, **1**, 33–53 [in Chinese with English summary].

Xu, H.-K., Cai, C.-Y., Liao, W.-H. & Lu, L.-C. 1990. Hongguleleng Formation in Western Junggar and the boundary between Devonian and Carboniferous. *Journal of Stratigraphy*, **14**, 292–301 [in Chinese with English abstract].

Yang, W. & Neves, R. 1997. Palynological study of the Devonian–Carboniferous boundary in the vicinity of the International Auxiliary Stratotype Section, Guilin, China. *Proceedings of the 30th International Geological Congress*, **12**, 95–107.

Yolkin, E.A., Gratsianova, R.T. *et al.* 2000. Devonian standard boundaries within the shelf belt of the Siberian Old Continent (southern part of western Siberia, Mongolia, Russian Far East) and in the South Tien Shan. *Courier Forschungsinstitut Senckenberg*, **225**, 303–318.

Yu, C. (ed.) 1988. *Devonian–Carboniferous Boundary in Nanbiancun, Guilin, China. Aspects and Records*. Science Press, Beijing.

Zhang, R., Wang, C., Yao, H., Niu, Z., Wang, J. & Wu, J. 2010. Late Devonian–Early Carboniferous conodonts from the Hainan Island. *Acta Micropalaeontologica Sinica*, **2010**, 45–59.

Ziegler, W. & Sandberg, C.A. 1984. *Palmatolepis*-based revision of upper part of standard Late Devonian conodont zonation. *In*: Clark, D.L. (ed.) *Conodont Biofacies and Provincialism*. Geological Society of America, Special Papers, **196**, 179–194, http://doi.org/10.1130/SPE196-p179

Ziegler, W. & Sandberg, C.A. 1996. Reflexions on Frasnian and Famennian Stage boundary decisions as a guide to future deliberations. *Newsletters on Stratigraphy*, **33**, 157–180.

Ziegler, W.S., Sandberg, C.A. & Austin, R.L. 1974. Revision of *Bispathodus* group (Conodonta) in the Upper Devonian and Lower Carboniferous. *Geologica et Palaeontologica*, **8**, 97–112.

Zimmerle, W., Gaida, K.-H., Gedenk, R., Koch, R. & Paproth, E. 1980. Sedimentological, mineralogical, and organic-geochemical analyses of Upper Devonian and Lower Carboniferous strata of Riescheid, Federal Republic of Germany. *Mededelingen Rijks Geologische Dienst*, **32**, 34–43.

Zong, P., Becker, R.T. & Ma, X.-P. 2015. Upper Devonian (Famennian) and Lower Carboniferous (Tournaisian) ammonoids from western Junggar, Xinjiang, northwestern China – stratigraphy, taxonomy and palaeobiogeography. *Palaeodiversity and Palaeoenvironments*, **95**, 159–202. First published online September 20, 2014, http://doi.org/10.1007/s12549-014-0171-y

The global Hangenberg Crisis (Devonian–Carboniferous transition): review of a first-order mass extinction

SANDRA ISABELLA KAISER[1]*, MARKUS ARETZ[2] & RALPH THOMAS BECKER[3]

[1]*State Museum of Natural History Stuttgart, Rosenstein 1, 70191 Stuttgart, Germany*

[2]*Géosciences Environnement Toulouse (GET), Observatoire Midi Pyrénées, Université de Toulouse, CNRS, IRD, 14 avenue E. Belin, F-31400 Toulouse, France*

[3]*Institut für Geologie und Paläontologie, Westfälische Wilhelms-Universität, Corrensstrasse 24, D-48149 Münster, Germany*

**Corresponding author (e-mail: dr.sandra.kaiser@gmail.com)*

Abstract: The global Hangenberg Crisis near the Devonian–Carboniferous boundary (DCB) represents a mass extinction that is of the same scale as the so-called 'Big Five' first-order Phanerozoic events. It played an important role in the evolution of many faunal groups and destroyed complete ecosystems but affected marine and terrestrial environments at slightly different times within a short time span of *c.* 100–300 kyr. The lower crisis interval in the uppermost Famennian started as a prelude with a minor eustatic sea-level fall, followed rather abruptly by pantropically widespread black shale deposition (Hangenberg Black Shale and equivalents). This transgressive and hypoxic/anoxic phase coincided with a global carbonate crisis and perturbation of the global carbon cycle as evidenced by a distinctive positive carbon isotope excursion, probably as a consequence of climate/salinity-driven oceanic overturns and outer-shelf eutrophication. It is the main extinction level for marine biota, especially for ammonoids, trilobites, conodonts, stromatoporoids, corals, some sharks, and deeper-water ostracodes, but probably also for placoderms, chitinozoans and early tetrapods. Extinction rates were lower for brachiopods, neritic ostracodes, bryozoans and echinoderms. Extinction patterns were similar in widely separate basins of the western and eastern Prototethys, while a contemporaneous marine macrofauna record from high latitudes is missing altogether. The middle crisis interval is characterized by a gradual but major eustatic sea-level fall, probably in the scale of more than 100 m, that caused the progradation of shallow-water siliciclastics (Hangenberg Sandstone and equivalents) and produced widespread unconformities due to reworking and non-deposition. The glacio-eustatic origin of this global regression is proven by miospore correlation with widespread diamictites of South America and South and North Africa, and by the evidence for significant tropical mountain glaciers in eastern North America. This isolated and short-lived plunge from global greenhouse into icehouse conditions may follow the significant drawdown of atmospheric CO_2 levels due to the prior massive burial of organic carbon during the global deposition of black shales. Increased carbon recycling by intensified terrestrial erosion in combination with the arrested burial of carbonates may have led to a gradual rise of CO_2 levels, re-warming, and a parallel increase in the influx of land-derived nutrients. The upper crisis interval in the uppermost Famennian is characterized by initial post-glacial transgression and a second global carbon isotope spike, as well as by opportunistic faunal blooms and the early re-radiation of several fossil groups. Minor reworking events and unconformities give evidence for continuing smaller-scale oscillations of sea-level and palaeoclimate. These may explain the terrestrial floral change near the Famennian–Tournaisian boundary and contemporaneous, evolutionarily highly significant extinctions of survivors of the main crisis. Still poorly understood small-scale events wiped out the last clymeniid ammonoids, phacopid trilobites, placoderms and some widespread brachiopod and foraminiferan groups. The post-crisis interval in the lower Tournaisian is marked by continuing eustatic rise (e.g. flooding of the Old Red Continent), and significant radiations in a renewed greenhouse time. But the recovery had not yet reached the pre-crisis level when it was suddenly interrupted by the global, second-order Lower Alum Shale Event at the base of the middle Tournaisian.

Middle to Late Palaeozoic times were characterized by a complex succession of global bioevents of variable magnitude (e.g. House 1985; Walliser 1996). The factors controlling these events are still under debate, and among others the evolution of land plants, volcanism, impacts, salinity changes, global carbonate crisis by oceanic acidification, perturbation of the carbon cycle, anoxia, rapid sea-level changes, icehouse and super-greenhouse conditions have been evoked (e.g. Caputo 1985; Algeo &

From: BECKER, R. T., KÖNIGSHOF, P. & BRETT, C. E. (eds) 2016. *Devonian Climate, Sea Level and Evolutionary Events*. Geological Society, London, Special Publications, **423**, 387–437.
First published online November 11, 2015, http://doi.org/10.1144/SP423.9

Scheckler 1998; Racki 2005). Within the Middle–Upper Palaeozoic succession a major biocrisis occurred close to the Devonian–Carboniferous boundary (DCB). Based on German lithological marker units it is known as global Hangenberg Biocrisis or Hangenberg Event and represents one of the major extinction events of the Phanerozoic. It has a magnitude and evolutionary significance comparable with the first-order mass extinction at the Frasnian–Famennian boundary (Sepkoski 1996). The two extinction intervals are separated by *c.* 13 myr (Becker *et al.* 2012) and have to be distinguished as separate biosphere turnovers with different and distinctive environmental and faunal changes (e.g. Sallan & Coates 2010). According to Sepkoski (1996) the generic extinction rate at the DCB exceeds 45%, and the family extinction rate is *c.* 20%. But these estimates are poorly constrained for many fossil groups. Often they are not based on high-resolution data that separate the significant second/third-order extinctions within the Famennian that took part in the course of the global Condroz, *Annulata* and Dasberg Events (e.g. Becker 1993*a*; Walliser 1996; House 2002; Hartenfels & Becker 2009). Hiatuses, marked sudden lithological changes, including the deposition of black shales and sandstones, the breakdown of carbonate deposition, and the appearance of diamictites and dropstones in different palaeogeographical settings, give evidence of major environmental changes at the DCB. Recent studies focused on interdisciplinary methods, e.g. combining biostratigraphy, sedimentology and chemostratigraphy. They support sudden climate and sea-level changes, as well as perturbations of the global carbon cycle at the end of the Famennian (e.g. Cramer *et al.* 2006; Kaiser *et al.* 2006, 2008, 2011; Marynowski *et al.* 2012; Kumpan *et al.* 2013, 2015).

The study of global extinction events of the Phanerozoic, accompanied by environmental changes, is fundamental for our understanding of the dynamics and stability of climate and marine ecosystems in Earth history. This review of the global Hangenberg Crisis at the DCB provides an overview of a number of previously reported new data and concepts and gives insights into the timing, causes and consequences of this somewhat neglected sixth first-order mass extinction. A review of DCB chrono- and biostratigraphy (Becker *et al.* this volume, in press) enabled us to critically evaluate almost 100 different successions in 37 countries. The classic pelagic successions of the Rhenish Massif (Germany) and the neritic successions of the Ardennes Shelf (Belgium, northern France) are used to correlate the different successions of North and South America, Europe, Africa, Asia and Australia. There are no data for Antarctica. Since a discussion of all published DCB sections would require an extensive monograph, details are presented in a condensed table format. In this way, this review offers an extensive DCB/Hangenberg Crisis bibliography that can be used for future, more specific research.

Abbreviations

*ck*I = *Bispathodus costatus–Protognathodus kockeli* Interregnum, DCB = Devonian–Carboniferous boundary, HBS = Hangenberg Black Shale, HS = Hangenberg Shale, HSS = Hangenberg Sandstone, OAE = oceanic anoxic events, UD = Upper Devonian.

Taxonomy: Ac. = *Acutimitoceras*, *Bi.* = *Bispathodus*, *Clyd.* = *Clydagnathus*, *Eo.* = *Eosiphonodella*, *Pr.* = *Protognathodus*, *Ps.* = *Pseudopolygnathus*, *R.* = *Retispora*, *Q.* = *Quasiendothyra*, *S.* = *Siphonodella*, *V.* = *Vallatisporites*. For the explanation of the spore zones and the ammonoid zonal key (UD VI to LC II) see Becker *et al.* (this volume, in press).

Stratigraphic and geochronological frame

The Hangenberg Crisis has to be set into the regional/global, and the traditional as well as current chronostratigraphic timescales (see review in this volume by Becker *et al.*). Global correlation of DCB sections is currently achieved with the help of bio-, chemo- and sequence stratigraphy. The biostratigraphic framework (Fig. 1) is based on conodonts, ammonoids, miospores and foraminifers, depending on the studied facies realm. Generally it is very detailed, especially when integrated schemes of the main fossil groups are used. Many DCB biozones had durations of only 100–300 kyr (see interpolations in Becker *et al.* 2012). Based on geochronological dating of Polish ash layers (Myrow *et al.* 2014) the main event or black shale interval lasted between less than 50 and 190 kyr. This supports a very sudden extinction at the base and estimates that the whole (extended) crisis interval represents only between 100 kyr (Sandberg & Ziegler 1996) and several hundred kyr (Becker *et al.* 2012; De Vleeschouwer *et al.* 2013).

As outlined in previous publications (e.g. Kaiser *et al.* 2011; Becker *et al.* this volume, in review) the lithological pelagic succession of the Rhenish Massif (Fig. 2) and its correlative neritic beds of the Ardennes (see Becker *et al.* this volume, in press) can serve as a standard for global correlation. Together with detailed biostratigraphic data, it enables the international correlation of Hangenberg Crisis intervals and unconformities summarized in Tables 1–5. An initial shallowing at the top of the *Siphonodella* (*Eosiphonodella*) *praesulcata*

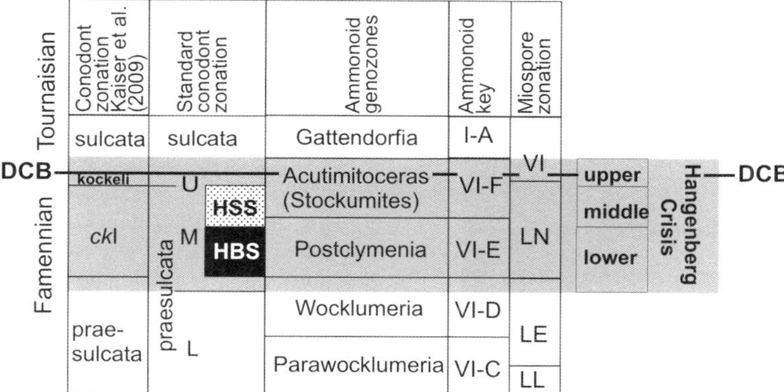

Fig. 1. Biostratigraphy around the DCB in the northern Rhenish Massif. Ammonoid zonal keys modified after Becker & House (2000); miospore zonation after Higgs & Streel (1994); *ck*I, *costatus–kockeli* Interregnum after Kaiser *et al.* (2009); HSS, Hangenberg Sandstone; HBS, Hangenberg Black Shale. For the foraminifera zonation at the DCB see Kalvoda (2002) and Poty *et al.* (2006).

Zone (top UD VI-D, LE miospore zone) marks the beginning of the crisis interval (Drewer Sandstone level, Fig. 2). It is followed by a widespread and sudden transgressive, hypoxic/anoxic event (Hangenberg Black Shale (HBS) and equivalents), the main lower crisis interval (UD VI-E, lower *Bispathodus costatus–Protognathodus kockeli* Interregnum = *ck*I, basal LN Zone). The subsequent regressive Hangenberg Shale (HS) and Hangenberg Sandstone (HSS) and their equivalents form the

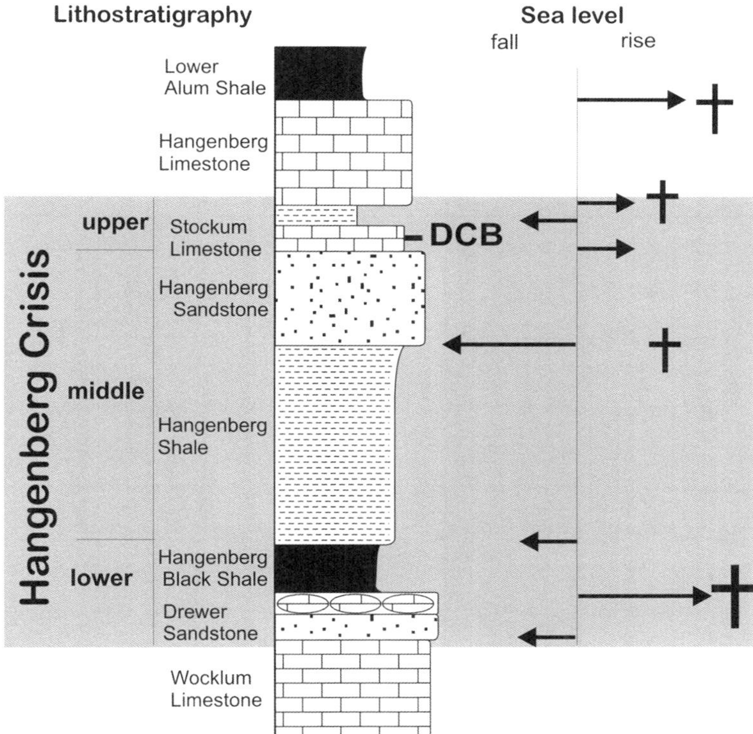

Fig. 2. Extinction episodes, sedimentology, and sea-level changes at the DCB. Lithological section scheme after the 'Rhenish standard succession' (see Becker *et al.* this volume, in press). Crosses denote extinction episodes.

middle crisis interval (middle/upper *ck*I and LN Zone). The widespread return of carbonates (Stockum Limestone and equivalents) is characteristic of the upper crisis interval (*kockeli* Zone to lower *Siphonodella* (*Eosiphonodella*) *sulcata* Zone with *Pr. kuehni*, top LN to basal VI zones).

The Hangenberg Biocrisis

The strong decline of biodiversity at the end of the Devonian has been a subject of a number studies that were summarized, e.g. by Walliser (1984, 1996), Bless *et al.* (1993), Coen *et al.* (1996), Webb (2002), and Kaiser (2005). In the 'Rhenish standard succession', the Hangenberg Crisis comprises the interval from the top of the Wocklum Limestone and *Siphonodella praesulcata* Zone (*sensu* Kaiser *et al.* 2009 = old basal Middle *praesulcata* Zone; UD VI-D2, *c.* base of LE Zone) to the base of the Hangenberg Limestone *sensu stricto* (*c.* middle *sulcata* Zone, base of *Gattendorfia* or *Acutimitoceras* (*Acutimitoceras*) *acutum* Zone, LC I-A2, lower VI Zone). It was not a single catastrophic event but a multiphased crisis, because extinctions among different fossil groups partly took place at different times (Figs 2, 3 & 4). The main extinction of the lower crisis interval occurred during the deposition of the HBS and its equivalents. This was followed by minor extinctions in the middle/upper crisis interval and in the pelagic realm, by a final minor faunal overturn (upper crisis interval) in the basal Tournaisian (Walliser 1984; Becker 1996). Affected were numerous taxa of terrestrial, shallow and deep open-marine ecosystems as shown in Figures 2 and 3 and described below. However, detailed compilations are still lacking for many fossil groups, especially taxon ranges plotted against refined biozonations. For non-ammonoid cephalopods, gastropods, non-stromatoporoid sponges, and arthropods, other than trilobites and ostracodes, the database is currently too crude to evaluate possible extinction patterns. Recent reviews (e.g. Waters *et al.* 2014) suggest that Carboniferous-type echinoderms actually originated in the Famennian. Furthermore, since there are no DCB macrofossil assemblages from any (sub)polar region, it is not possible to establish direct influence of glacial advances on faunas. In general, palaeolatitudinal influences on extinction patterns still have to be worked out for all fossil groups.

Conodonts

Early syntheses of the composition of conodont faunas across the DCB were published by Austin

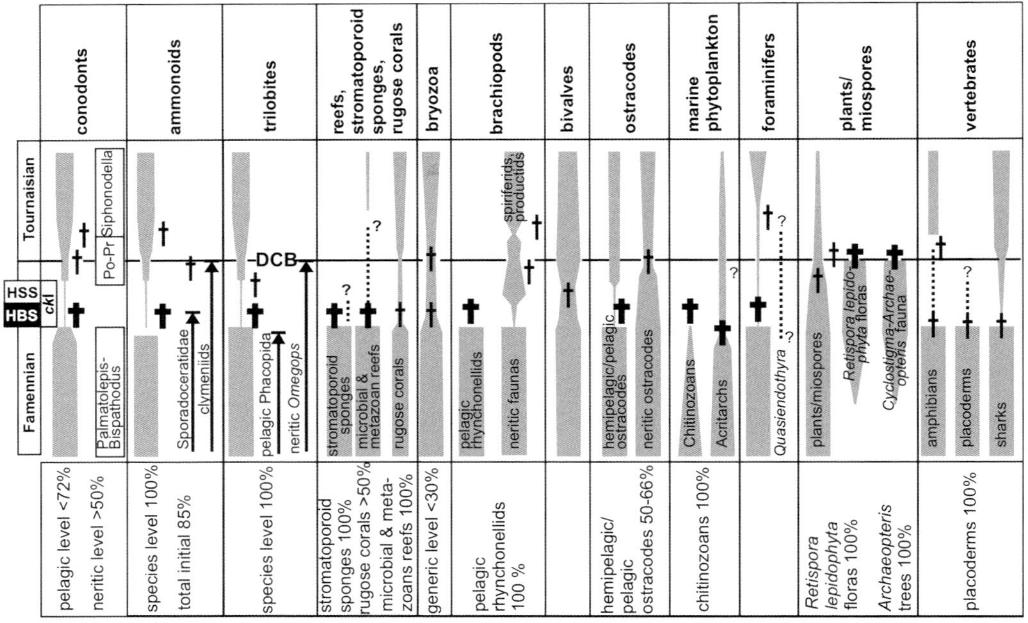

Fig. 3. Overview of fossil groups affected by the Hangenberg Crisis (light grey). Grey bars denote radiations, extinctions, and diversity changes. Crosses denote extinctions during the Hangenberg Biocrisis. The width of the bars shows the estimated relative abundance of taxa (not to scale). Note that the state of knowledge is different for the particular taxa.

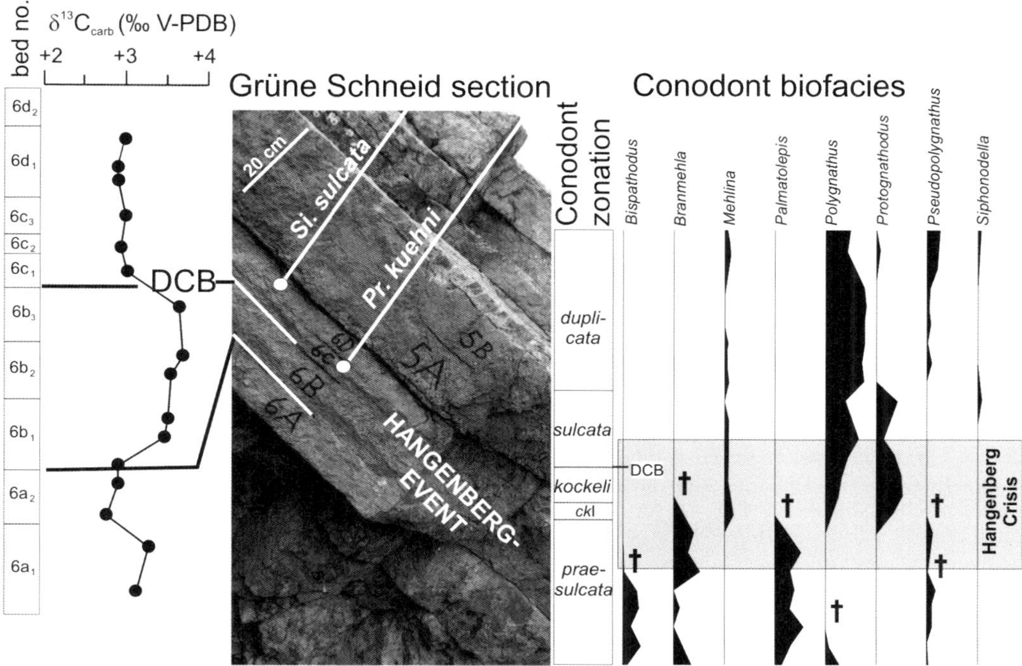

Fig. 4. Highly condensed DCB successions in continuous limestone successions at Grüne Schneid (Carnic Alps, Austria), major conodont biofacies change among different genera, and carbon isotope values. Crosses denote conodont extinctions during the main Hangenberg Event, time equivalent to a positive carbon isotope excursion and the deposition of strongly condensed, calcareous HBS equivalents. Data modified from Kaiser (2007). *ck*I after Kaiser *et al.* (2009).

et al. (1970*b*) and Dreesen *et al.* (1986). Kaiser *et al.* (2009) provided more specific reviews of DCB conodont extinctions, mainly based on pelagic faunas. The main extinction among conodonts occurred during the global deposition of the HBS. Mossoni *et al.* (2015) observed a decrease of conodont abundance in the immediate pre-crisis beds (*praesulcata* Zone) of Sardinia. Abundant and widespread uppermost Famennian taxa, such as the last palmatolepids (*Palmatolepis gracilis* Group = *Tripodellus* of apparatus taxonomy), several polygnathids, the *Pseudopolygnathus marburgensis* Group, several branmehlids, and the *Bi. costatus-ultimus* Group completely disappeared with the onset of global anoxia, although this extinction event can be obscured by subsequent reworking. The pre-crisis palmatolepid–bispathodid biofacies shifted into a crisis interval polygnathid–protognathodid biofacies, which in post-crisis time was replaced by the siphonodellid biofacies (Fig. 4, Perri & Spalletta 1998; Kaiser 2007; Kaiser *et al.* 2008). Since *Protognathodus* is missing or very rare in most neritic successions, its bloom within the crisis interval of deeper-water sections is a signal of opportunistic palaeoecology and not a sea-level indicator (see Kaiser 2005; Corradini *et al.* 2011; Mossoni *et al.* 2015). Total pelagic extinction rates are near 40% of species, with variable higher local values of 55–72% (Kaiser *et al.* 2009). There were no differences between tropical successions of the western and eastern Prototethys (Kaiser *et al.* 2009). The most complete transition through the crisis has been recorded from the Russian Far East (Gagiev & Kononova 1990), where mixed neritic–pelagic faunas require more detailed studies.

Extinction patterns are generally complex in neritic settings, with significant differences between local faunas of Europe, Russia, China, and Australia. Globally, more than 80 conodont species/subspecies have been reported from uppermost Famennian shallow-water settings, not counting the enigmatic *Fungulodus*/*Conchodontus* species, which are conodonts (Donoghue & Chauffe 1998). Based on a new database that covers 29 successions from Europe, Asia, and Australia, less than 50% of the shallow-water taxa survived, and many survivors disappeared locally. Among the victims are endemic genera, such as *Tanaissognathus*, *Mashkovia*, *Capricornugnathus* and *Antognathus*, and many endemic species of *Pelekysgnathus* and

Apatognathus. Of special importance is the final extinction of the Devonian marker genus *Icriodus*, e.g. on the Russian Platform (Aristov 1988), in the Mugodzhary of the South Urals (Maslov 1987) and in South China (W. Ji 1987). However, there are reports that the genus survived in the Russian Far East (Shilo *et al.* 1984), with one species (*Icriodus obstinatus*) supposedly restricted to the higher part of the lower Tournaisian. The main survivors in the shallow realm were species of *Polygnathus*, *Pseudopolygnathus*, *Clydagnathus*, *Bispathodus*, *Branmehla*, *Mehlina* and 'siphonodelloids'. These are the genera that occur widely both in neritic and pelagic settings. Therefore, either a wide dispersal or a broad palaeoecological spectrum enabled survival.

The basal Tournaisian diversity is only slightly (*c.* 20%) lower than that of pre-crisis time, because re-radiation was rapid in time equivalents of the *kockeli* and *sulcata* zones (Fig. 4). An intriguing detail is the abundant occurrence of the genus *Cryptotaxis* in the upper crisis interval (Louisiana Limestone) of Illinois/Missouri (Chauffe & Nichols 1995). This genus is absent from tropical Famennian faunas but rather dominant in the rare and restricted Frasnian/lower Famennian conodont faunas of the high latitudes, such as Brazil (Hünicken *et al.* 1989; Cardoso *et al.* 2015) and Bolivia (Over *et al.* 2009). Its oldest low-latitude occurrence is from uppermost Famennian beds within the Woodford Shale of Oklahoma, together with the youngest *Palmatolepis* (Over 1992), and from a lag sandstone just below the Louisiana Limestone (Sandberg *et al.* 1972). The unique distribution patterns suggest a cooling-controlled and short-term palaeogeographical expansion of the genus in the wider crisis interval.

Ammonoids

The first-order DCB ammonoid extinctions and radiations have been addressed by Price & House (1984), Korn (1986, 1993, 2000), Becker (1993*a*, *b*), Becker & Korn (1997), Kullmann (1994, 2000) and Sprey (2002). As noted by Becker (1993*a*) and Korn (2000), there are small-scale but distinctive pre-Hangenberg extinction episodes in the upper Wocklumian (within and at the end of UD VI-C), which caused the loss of several clymeniid groups, such as the evolute triangularly coiled members of the Parawocklumeriidae, evolute glatziellids, various kosmoclymeniids and the prionoceratid *Effenbergia*. Involute glatziellids, the last *Kalloclymenia*, and the last Biloclymeniidae range to the regressive base of the crisis interval. But an almost complete extinction took place during the initial anoxic interval, the main Hangenberg Event, with an extinction rate of about 85% (Becker 1993*a*, *b*; House 1996; for details, see Becker *et al.* this volume, in press). The morphospace occupation changed considerably, especially with a marked loss of longidomic, demersal and widely evolute megaplanktonic groups (Korn 2000; Sprey 2002). At the species level, no taxon is known to have passed through the complete crisis interval. Some cymaclymeniids (few species of *Cymaclymenia* and *Postclymenia*) survived the initial event phase and then spread pantropically but finally became extinct near the end of the crisis interval (e.g. Korn *et al.* 2004). This is a typical example for a survivor extinction, which was very important, since it prevented a post-crisis recovery of 'Carboniferous clymeniids'. The Sporadoceratidae survived for a very brief time into the lower crisis interval, but this is based on very sparse records (isolated single specimens) from Ohio (House *et al.* 1986) and Xinjiang (Zong *et al.* 2014). The main surviving lineage was the Prionoceratidae, represented by two groups of mimimitoceratids, one each with evolute and involute early stages (Fischer & Becker 2014; Zong *et al.* 2014). The initial recovery in the middle/upper crisis interval was characterized by a change to forms with serpenticonic early whorls that characterize the widespread *Acutimitoceras* faunas, especially of the Stockum levels. These declined considerably in parallel with the final clymeniid extinction, which underlines the significance of the small-scale but global extinction at the top of the crisis interval.

The richest DCB ammonoid faunas are known from Germany, Poland, Morocco and South China. The Russian post-crisis lower Tournaisian record is poor. The available faunas do not indicate any significant extinction differences between areas of the widely separate western and eastern Prototethys realms. A better initial survival in Xinjiang (Zong *et al.* 2014) may reflect the absence of anoxia of the lower crisis interval in that region, but this requires more detailed studies.

The lower Tournaisian re-radiation led to a strong increase of diversity (e.g. Becker 1993*a*, *b*) and disparity (Korn 2000; Sprey 2002), but the pre-crisis level was not yet reached when the anoxic Lower Alum Shale Event in the middle Tournaisian caused the next sudden global extinction (Kullmann 1994). However, faunas from around the lower/middle Tournaisian boundary of Morocco (Korn *et al.* 2002, 2007; Bockwinkel & Ebbighausen 2006; Becker *et al.* 2006) suggest that this was mostly a significant species-level extinction, whilst many genera survived.

Trilobites

Trilobite extinction and survival patterns associated with the Hangenberg Crisis have been reviewed by

Brauckmann & Brauckmann (1986), Hahn (1990), Brauckmann *et al.* (1993), Hahn *et al.* (1994) and Chlupáč *et al.* (2000). No species is known to have passed through the complete crisis interval, and there are only a few surviving genera within the Proetida, represented by the Brachymetopidae (a neritic family), Proetidae (Drevermanniinae) and Phillipsiidae (with two surviving subfamilies, the Archegoninae = Cyrtosymbolinae, but alternatively placed in the Proetidae, and Weaniinae). Therefore, the re-radiation must have come from survivors within those groups, which evolved within still unknown refugia areas. A re-dispersal of descendent species began in the upper (*kockeli* Zone) and/or post-crisis interval. There is an open debate whether some forms, such as *Pudoproetus*, represent Elvis Taxa homoeomorphic with Middle Givetian genera (in this case of the Proetinae), or whether they indicate a survival in refugia areas for more than 15 Ma (see Feist & Petersen 1995). The second interpretation (e.g. Yuan & Xiang 1998; Hahn *et al.* 2012) implies an enigmatic prevention of recovery and spread in practically all of the Upper Devonian (UD) and an even more enigmatic palaeoecological factor that enabled a sudden spread to North America, North Africa, Europe, Central Asia and South China at the end of the crisis interval. A similar discussion is required to clarify whether there was true survival of proetid genera or whether Carboniferous taxa represent iterative homoemorphs, e.g. in the case of the Famennian *Drevermannia* (*Drevermannia*) and the Carboniferous *Dr.* (*Pseudodrevermannia*) and *Dr.* (*Paradrevermannia*) (see Gandl *et al.* 2015).

The main trilobite extinction was equally severe in western (Europe–North Africa) and eastern (South China) Prototethys regions (see Yuan & Xiang 1998). In the pelagic realm the last Phacopida, represented by the blind *Dianops*, the small-eyed *Weyerites*, and the genus *Rabienops* ('*Phacops' granulatus* Group) with median-sized eyes, died out at the onset of the HBS. The shallow-water genus *Omegops*, however, obviously survived in Xinjiang briefly into the lower/middle crisis interval, based on co-occurrences with miospores of the LN Zone and the youngest *Cymaclymenia* faunas (Zong *et al.* 2012, 2014). Based on geochemistry, Carmichael *et al.* (2015) placed the HBS level in the Bulongguor type section of the Junggar Basin much below the phacopid extinction. This questions the assumption that *Omegops* specimens from the *kockeli* Zone of the Ardennes (Conil *et al.* 1986) are reworked. The globally youngest phacopid is a probably reworked specimen from the *sulcata* Zone of the DCB stratotype (Flajs & Feist 1988). It seems likely that the final demise of the order was a case of 'survivor extinction', as in the case of the contemporaneous clymeniids. So far there is no explanation as to why the widespread phacopids died out in the neritic realm whilst the associated Brachymetopidae managed to survive in widely separate regions of the Rhenish Massif (e.g. Michels 1986) and South China (Yuan & Xiang 1998).

The post-crisis recovery was very fast in both magnafacies and led to the appearance of many new proetid subfamilies and genera (e.g. Brauckmann *et al.* 1993).

Reefs, stromatoporoid sponges and corals

The primary Devonian reef-builders, such as stromatoporoid sponges and tabulate corals, suffered severely during the global Kellwasser Crisis around the Frasnian–Famennian boundary, and their Famennian diversity was significantly reduced (Scrutton 1997; Webb 2002). Lower to uppermost Famennian reefs were mostly built by calcimicrobes (Dreesen *et al.* 1985; Aretz & Chevalier 2007). However, in the Canning Basin (Wood 2007) microbial–sponge reefs are already known in the lower Fammenian (*Palmatolepis triangularis* Zone). The uppermost Famennian saw a slight global recovery of metazoan biostromes, e.g. of the Ardennes–western Rhenish Massif and in South China. Webb (2002) published an overview and distribution map of Famennian microbial and metazoan reefs. But no reef complex at all, including microbial build-ups, survived the Hangenberg Crisis. On the global scale only a single lower Tournaisian microbial reef has been described from the basal Gudman Formation of Queensland, eastern Australia (Webb 1998, 2005). However, it is not reliably dated; published conodonts from the base of the formation (Mory & Crane 1982; Webb 2005) indicate the pre-crisis *praesulcata* Zone. The dating of microbial boundstones across the DCB of the northern Urals (Antoshkina 1998) is equally arbitrary. At least the main Lower Carboniferous reefal recovery started only after the Lower Alum Shale Event, with a few microbial–metazoan reefs in Eastern Australia (Aretz & Webb 2007) and the famous Waulsortian Mounds (e.g. Lees & Miller 1995; Aretz & Chevalier 2007) in the upper Tournaisian (Ivorian).

With a considerable delay after the severe Frasnian–Famennian mass extinction, stromatoporoid sponges re-diversified in the 'Strunian' biostromes (e.g. Stearn 1987; Stock 2005; Poty 2007) but the various European, Russian and southern Chinese assemblages were very different (Bogoyavlenskaya 1982; Mistiaen *et al.* 1998; Mistiaen & Weyer 1999). The group became totally extinct during the Hangenberg Crisis, probably during the initial phase/main extinction (Weber 2000; Poty 2007). However, unconformities (potentially incomplete top of pre-crisis levels) and fossil-poor neritic sediments hampered a precise dating of the extinction level (e.g. Casier *et al.* 2005; Azmy *et al.* 2009). Mistiaen (1996) documented that stromatoporoids range to the very top of the Menggongao Formation, the base of the crisis interval, in central Hunan. In the northern Caucasus, stromatoporoid sponges disappear simultaneously with *Quasiendothyra kobeitusana* faunas (Puporev & Chegodaev 1982), as is the case for the Ardennes Shelf (Weber 2000; Poty 2007). Cockbain (1989) discussed briefly the supposed Viséan to Permian recurrences. These are mostly based on forms that are not stromatoporoid sponges or refer to specimens of doubtful provenance. In any case, the various homoeomorphic Mesozoic sponges with a calcareous basal skeleton should not be called stromatoporoid sponges.

Only a few of the deep-water rugose corals survived the onset of anoxia during the HBS (summary of German records in Poty 1986). They represent long-ranging, different families of several suborders. Homoemorphy and simple morphologies affect the recognition of extinction and survival patterns. The drastic effect of the Hangenberg Crisis on shallow-water rugose corals is reflected in major taxonomic differences between pre- and post-Famennian taxa, e.g. on the Ardennes Shelf (Poty 1986; Denayer *et al.* 2011). Only very few survivors are known, a pattern seen on the widely separate shelf platforms of Europe, South China and the Russian Far East (Poty 1999). It should be emphasized that the scarcity of colonial rugose corals in the pre- and post-crises intervals is globally documented. Characteristic for the Strunian rugose corals is the presence of homoeomorphs of Viséan taxa (e.g. the Strunian '*Palaeosmilia*', '*Clisiophyllum*', '*Dibunophyllum*', the latter now named *Bounophyllum* Chwieduk, 2005). Fan *et al.* (2003) summarized the records of 'Strunian' to Tournaisian corals of the Sichuan–Qingling Mts in their review of Chinese Palaeozoic corals. A first re-radiation began in the lower Tournaisian, but more diversified faunas appeared only after the Lower Alum Shale Event in the middle Tournaisian (Poty 1999; Denayer *et al.* 2011).

Tabulate corals are known in many DCB sections, but detailed information is rather limited; most often syringoporid corals are mentioned. Records for the Ardennes Shelf (Tourneur *et al.* 1989) and South China (Tourneur in Hance *et al.* 1994; Mistiaen in Milhau *et al.* 1997) are insufficient to outline general extinction and survival patterns.

Bryozoa

There are only a few publications that deal with the influence of the Hangenberg Crisis on bryozoan diversity and evolution. A compilation of ranges through stages by Horowitz & Pachut (1993) showed that more than 70% of the Famennian genera survived into the Lower Carboniferous. Accordingly, Gutak *et al.* (2008) did not observe a major impact of the crisis on assemblages from southern Siberia, although there is a facies-controlled regional record gap for the lower Tournaisian. In recent years more and more data for the composition of Famennian and Mississippian bryozoan assemblages have become available (e.g. Ernst & Herbig 2010; Tolokonnikova & Ernst 2010; Ernst 2013; Tolokonnikova *et al.* 2014*a*, *b*, 2015; Ernst *et al.* 2015). Although the quantification and the spatial distribution of these data are limited, they seem to confirm some of the older assumptions. The end-Fammenian was a time of a major diversification among bryozoans at the species and genus level, and a significant number of these new taxa crossed the DCB and flourished in Mississippian times. The Hangenberg Crisis apparently did not have a major impact on the group, although Ernst (2013) noted increased extinction rates at the DCB. The recent work shows that only a small fraction of the bryozoan assemblages of the Famennian and Tournaisian are properly studied. Thus new data could drastically change our understanding of the influence of the Hangenberg Crisis on this fossil group.

Brachiopods

Brachiopod ranges across the Hangenberg Crisis have been compiled by Legrand-Blain & Martinez Chacon (1988), Legrand-Blain (1991, 1995), Poletaev & Lazarev (1995), Nicollin & Brice (2004), Brice *et al.* (2005, 2007) and, most recently, by Mottequin *et al.* (2014). Whilst the few deeper-water uppermost Famennian rhynchonellids (e.g. Sartenaer 1997; Halamski & Balinski 2009) and chonetids (Afanaseva 2002) did not survive the HBS, at least at the species level (see the new post-crisis *Rozmanaria* of Bartzsch *et al.* 2015), small-sized orthids, such as *Aulacella*, and inarticulate taxa that are often tolerant to oxygen deficiency, did. It is much more difficult to recognize the Hangenberg Crisis based on brachiopods in shallow-water siliciclastic or carbonate shelf settings (e.g. Zong *et al.* 2012). Many spiriferids and productids that are widespread in the Tournaisian have pre-crisis Famennian roots. Locally there are strong faunal overturns that reflect ecological changes across the DCB, e.g. in the neritic succession of the Rhenish Velbert Anticline (summary of ranges in Legrand-Blain 1995). During the regressive middle crisis interval a range of survivors spread with the neritic, siliciclastic wedges of the HSS and its equivalents (e.g. in southern Morocco; Becker *et al.* 2013*a*). But several uppermost Famennian genera, such as *Hadyrhyncha*, *Sphenospira*, *Araratella* and *Rigauxia*, disappeared with the upper/post-crisis transgression (Mottequin *et al.* 2014). Therefore, survivor extinctions, as known from the ammonoids and miospores, can also be found among shallow-water brachiopods.

Bivalves

Amler (1993) commented on bivalve faunas across the DCB and emphasized the transitional nature of rich assemblages that characterize the uppermost Famennian neritic facies (e.g. Amler *et al.* 1990; Amler 1995; see also Mergl *et al.* 2001). Although there were local extinctions and changes in assemblages triggered by sea-level and substrate fluctuation, the overall extinction rate is low. In contrast to the pattern in ostracodes and brachiopods, this also applies to the pelagic settings. These were characterized by a low number of small-sized Palaeotaxodonta and Pteriomorpha that were tolerant to low-oxygen conditions. In the upper crisis interval a minor bloom of a few larger-sized forms has been noted at some localities (Schmidt 1924; Becker 1996). The similarly large-sized genus *Posidonia*, which is so extremely abundant and globally widespread in Upper Viséan black shales, had pre-crisis ancestors in southern Morocco (new records). The genus must have survived in unknown refugia regions and became a Lazarus Taxon.

Plants/miospores

The uppermost Famennian was characterized globally by a surprisingly similar vegetation composed of the widespread *Retispora lepidophyta* palynomorph assemblages of coastal swamp environments (e.g. Streel *et al.* 2000*a*). *Cyclostigma–Archaeopteris* forests spread at the same time and created the first significant coal measures, notably on Bear Island (Kaiser 1970). Subsequently, land plants (Fairon-Demaret 1986, 1996; Jarvis 1990; Algeo & Scheckler 1998; Decombeix *et al.* 2011) and miospores (*Lepidophyta* Flora; e.g. Streel *et al.* 2000*a*; Streel & Marshall 2006) suffered badly from the Hangenberg Crisis. But this terrestrial ecosystem turnover, especially the global extinction of *Archaeopteris* trees and of the *R. lepidophyta* floras, occurred somewhat later than the main marine extinction phase, near the end of the extended crisis interval (LN/VI Zone boundary). At that time marine invertebrates had already begun to re-radiate, but the terrestrial crisis correlates approximately with the level of survivor extinctions in ammonoids, trilobites and brachiopods. Edwards *et al.* (2000) noted that the *Rhacophyton* floral complex and the widespread lycopsid *Leptophloeum* also did not reach the Carboniferous. Marshall *et al.* (2013) found it surprising that trees coped with the arid and cool glaciation times in the palaeotropical settings of Greenland, only to disappear for all of the lower Tournaisian when it became warm and wet again. Decombeix *et al.* (2011) also placed the main radiation of Lower Carboniferous trees in the middle/upper Tournaisian, after the Lower Alum Shale Event.

The widely quoted 'Algeo *et al.* model' (Algeo *et al.* 1995), which suggests terrestrial–marine teleconnections between the spread of land plants, soil erosion/nutrient fluxes and marine black shale events and extinctions, has no factual basis, at least around the DCB. There are no miospore spikes or macroflora blooms that indicate a significantly increased vegetation cover precisely at the time of the HBS. Some authors seem desperate to find any palaeobotanical evidence. For example, Carmichael *et al.* (2015) used a Frasnian lycopsid flora of Xinjiang (Zhulumute Formation) to postulate the presence of large forests in the uppermost Famennian (Heishantou Formation) of the Central Asian oceanic arc system that regionally could have triggered marine nutrient loading.

Marine phytoplankton

It has long been known that acritarchs showed a significant decline towards the end-Devonian (e.g. Maziane & Vanguestaine 1996). Le Hérissé *et al.* (2000) noted that the most important episode of acritarch extinction occurred near the end of the Famennian but regretted the lack of any detailed studies. The lower Tournaisian saw no recovery or new characteristics. Strother (2008) explained the decline in acritarchs at the end of the Devonian by lower pCO_2 and higher organic/inorganic nutrient conditions. It is interesting to note that Mullins & Servais (2008) were unable to separate several occurrences of Famennian and lower Tournaisian acritarchs, underlining the difficulties of precisely dating appearances and extinctions in these forms. However, their data confirm the steady decline of acritarch diversity from the Devonian into the Carboniferous (Strother 2008).

Chitinozoans declined gradually through the Famennian and finally become extinct during the Hangenberg Crisis, with the last rare records from pre-crisis beds of North Africa (Paris *et al.* 2000).

Foraminifers

Calcareous foraminifers suffered badly, since their habitat, which was characterized by extremely low sedimentation rates, vanished when carbonate production and deposition ceased (Hance 1996; Kalvoda 2002; Hance *et al.* 2011; Kalvoda *et al.* 2015). The well-diversified assemblages of the Famennian were replaced in the basal Tournaisian by impoverished assemblages (e.g. Belgian and China MFZ1, *Earlandia minima* Zone of the Urals), which mainly contain small and simple forms. Aretz *et al.* (2014) suggested a 'Lilliput effect' (dwarfing) as a consequence of the extinction of the larger-sized taxa. The presence of *Quasiendothyra* in the basal Tournaisian is still a matter of debate. Often considered to have become extinct in the topmost Devonian (e.g. Conil *et al.* 1991; Herbig 2006; Hance *et al.* 2011; Aretz *et al.* 2014), *Quasiendothyra* species are considered to have survived regionally and briefly into the post-crisis Tournaisian (e.g. Mamet 1985; Kulagina 2013; Kalvoda *et al.* 2015). The lower Tournaisian recovery was slow, which gives an episode of 'evolutionary standstill', with few originations and extinctions, in the immediate post-crisis interval, a time with a low level of endemism (Aretz *et al.* 2014).

Ostracodes

Changes of ostracod faunas across the DCB have been reviewed by Tschigova (1970), Bless *et al.* (1986) and Becker & Blumenstengel (1995). The main HBS extinction episode is well marked in the planktonic entomozoids by the onset of the *Maternella* (*Maternella*) *hemisphaerica–Richterina* (*Richterina*) *latior* Interregnum. The initial re-radiation started in the Stockum Limestone (Groos-Uffenorde & Rabien 2014). In general, pelagic assemblages suffered more strongly than those of inner shelf platforms. Approximately 50% of pelagic/hemipelagic ostracod species disappeared (Walliser 1996). In Thuringia and the Montagne Noire, the regional rates are higher, at 66% (Blumenstengel 1993; Casier *et al.* 2002: sample 37/38 boundary).

By contrast, many neritic taxa of the 'Eifelian ecotype' survived. Casier *et al.* (2003, 2004, 2005) documented changes across the Strunian–Hastarian boundary of the Ardennes, where the ecological changes across the DCB resulted in local faunal turnovers involving 30–50% of the species. Tschigova (1970) noted genus-level extinctions in five different ostracod families, resulting in the final disappearance of 2 of 18 families. Unfortunately, there is no update of this old summary. Wang (2004) documented the significance of the Hangenberg Crisis for the final extinction of the long-ranging (Ordovician–Devonian) and often large-sized Leperditicopida, which had recovered in the upper/uppermost Famennian from the global Kellwasser Crisis.

Vertebrates

Vertebrates suffered badly from the Hangenberg Crisis and *c.* 50% of diversity was lost. Sallan & Coates (2010, p. 10131) claimed 'The Hangenberg Event represents a previously unrecognized bottleneck in the evolutionary history of vertebrates as a whole and a historical contingency that shaped the roots of modern biodiversity'. However, it has long been known that armoured fish (placoderms) died out at the end of the Devonian (e.g. Lelievre & Goujet 1986; Long 1995; Janvier 1996). Uppermost Famennian aquatic tetrapods, the Ichthyostegalia, and their more advanced Lower Carboniferous descendents are separated from each other by 'Romer's Gap', which has recently been assumed to be based on collection failure (Smithson *et al.* 2012).

There is still some uncertainty concerning the precise age of the last placoderms. On the Russian Platform they occur in the Khovanshchina Formation together with the youngest Devonian tetrapod (*Tulerpeton*) and a pre-crisis conodont fauna (Alekseev *et al.* 1994). However, there is an enigmatic assemblage from the basal Köprülü Shales of SE Turkey, where placoderms (*Groenlandaspis*) are associated with Carboniferous-type fishes, such as the sarcopterygian *Strepsodus* (Janvier *et al.* 1984). This 'Zap Fauna' comes from the base of a transgressive black shale above marginal marine dolomites and sandstones that yielded LE Zone miospores (Higgs *et al.* 2002). Therefore, it may represent a rare lower crisis interval fauna, in which a last placoderm had survived briefly. This hypothesis should be tested by more detailed work. In any case it is remarkable that the Hangenberg Crisis was so far reaching that it wiped out a highly diverse clade that had an ecological range from the freshwater settings of the Old Red Continent to inner and outer shelves of Europe, northern Gondwana, Asia and Australia, and to hypoxic basins of the Appalachian foreland. It should be of special interest to date precisely the last remnants of the Devonian marine top predators, the titanichthyids.

By contrast, the extinction rate in sharks was relatively low, although some of the dominant and widespread outer-shelf genera, such as *Phoebodus*, disappeared with the HBS (e.g. Ivanov 1996; Ginter & Ivanov 2000). Actinopterygians, chondrichthyans and tetrapods strongly radiated in the Lower Carboniferous but only slowly in the lower Tournaisian interval between the Hangenberg Crisis and Lower Alum Shale Event. Sharks of this time consist mostly of survivors in the strata of the Urals (Ivanov 1996), whilst there are more newcomers on the East European Platform (Lebedev 1996).

Lithology, hiatuses and sea-level changes

Stratigraphical gaps and facies changes related to major eustatic sea-level and climate changes characterize the Hangenberg Crisis (Fig. 5). Since regressions/eustatic falls caused strong erosion on the

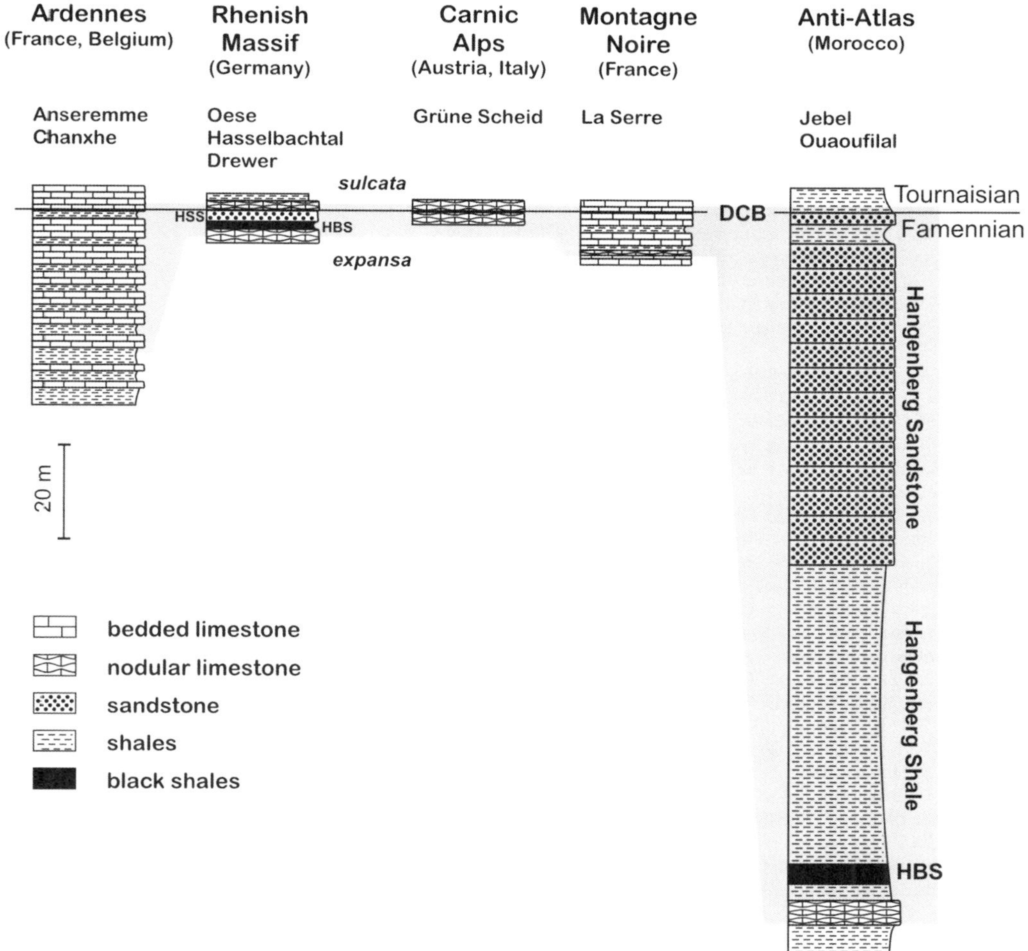

Fig. 5. Correlation of idealized DCB successions (selected reference sections) from different palaeogeographical settings of Europe and North Africa, indicating strong differences in thickness and facies during the Hangenberg Crisis (grey).

shallow shelves, these changes are best preserved in the pelagic records. Principles of DCB eustasy and sequence stratigraphy were first outlined by Van Steenwinkel (1993*a*, *b*) and Bless *et al.* (1993).

A drop in sea level at the end of the *praesulcata* Zone (*sensu* Kaiser *et al.* 2009) is indicated in the Rhenish Massif by shallowing upwards at the top of the Wocklum Limestone (Becker 1996; Streel 1999), by clastic intercalations (Fig. 2; Drewer Sandstone; Becker 1993*a*; Bless *et al.* 1993; Korn *et al.* 1994) and by an increase of ooids in shallow siliciclastic setting (Michels 1986). Elsewhere, this lowstand level marks a time of non-deposition and is characterized by reduced mud accumulation due to increasing bottom currents, erosion and sediment bypass. Comparable examples are the Gattendorf section of Franconia (Schindewolf 1923; Korn 1993) and the Dzikowiec (Ebersdorf) section of Silesia (southern Poland; Schindewolf 1937; Dzik 1997). The same trend resulted in the Anti-Atlas, Morocco, in increasing condensation, the recurrence of shallow-water faunal elements (e.g. large-eyed phacopids) with an episode of improved seafloor oxygenation, or the complete absence of sediments of the *Wocklumeria* Genozone (VI-D; Korn 1999; Kaiser *et al.* 2011). This non-deposition episode of pelagic platform settings continues into western Algeria (Weyant 1988). Widely distant basins of other continents show extreme condensation of the uppermost Famennian (e.g. Carnic Alps, Montagne Noire; Figs 4 & 5). But the eustatic signals may be regionally overprinted, e.g. in the

Table 1. *Overview of reported equivalents of the Rhenish Hangenberg Black Shale (HBS, including non-organic rich transgressive shales and dark limestones, lower crisis interval) and selected references*

1. ?Alberta: black shale between Exshaw (with *S. praesulcata*) and Banff formations at Fiddle River and Nordegg (Savoy *et al.* 1999)
2. SE Nevada–NE Utah–west Montana: 'Conchostracan Shale' (Sandberg *et al.* 1972; Gutschick & Rodriguez 1979); 'spinicaudatan' level in Cole *et al.* (2015)
3. North Dakota: ?'Conchostracan bed' at the top of the Lower Bakken Formation (Thrasher 1987)
4. Ohio, Kentucky: fossiliferous dark shale at the top of the Cleveland Shale with *Cymaclymenia* and LN Zone miospores, locally with erratic boulders (House 1978; House *et al.* 1986; Pashin & Ettensohn 1992; Ettensohn *et al.* 2007, 2009; Baird *et al.* 2009)
5. ?Cornwall, SW England: the oldest black slates of the Yeolmbridge Formation above the Strayerpark Slate with pre-crisis conodonts (Stewart 1981)
6. ?Southern margin of Brabant Massif, Belgium: subsurface transgressive siltstones with LN Zone (Loboziak *et al.* 1994)
7. Ardennes, Belgium: Pont de Scay section, black shale at top of Comblain-au-Pont Formation (Mottequin & Poty 2014)
8. Thuringia, Germany: Schwarzburg Anticline, Saalfeld region, HBS, sandwiched between hematite (originally pyrite) layers (Bartzsch & Weyer 1986; Bartzsch *et al.* 1998, 1999, 2015); Bergaer Anticline, Kahlleite, 'Alaunschiefer' = HBS at top of Kapfenberg Member of the Göschitz Formation (Bartzsch *et al.* 1995, 2001; Gereke 2004)
9. Moravia, Czech Republic: Lesni Lom Quarry, laminate unit with characteristic, positive carbonate isotope excursion (Kumpan *et al.* 2013; Kalvoda *et al.* 2015)
10. Graz Palaeozoic, Austria: very thin shale at the base of the *ck*I at Trolp Quarry (Kaiser *et al.* 2009)
11. Carnic Alps, Austria: black shale at Kronhofgraben (Schönlaub 1969; Kaiser 2007; Kaiser *et al.* 2008), thin dark limestone at Grüne Schneid (Schönlaub *et al.* 1988; Kaiser *et al.* 2006)
12. Carnic Alps, Italy: black shale at Plan di Zermula (Perri & Spalletta 2000*a*, *b*; Kaiser *et al.* 2008)
13. Western Armorican Massif, France: Brest region, Kermerrien Formation, black shales with laminated sandstones and olistolites of LN Zone (Rolet *et al.* 1986)
14. Montagne Noire, southern France: dark shale unit at La Serre (Flajs & Feist 1988), Puech de la Suque (Lethiers & Feist 1991; Kaiser *et al.* 2009), and Col des Tribes (Girard *et al.* 2013)
15. South Portuguese Zone, Portugal: Iberian Pyrite Belt, black slate with major sulphide-ore bodies (González *et al.* 2006; Sáez *et al.* 2008)
16. Sardinia: Bruncu Bullai section (Mossoni *et al.* 2015)
17. Moroccan Meseta: Oulmès region, black shale of Upper Member of Bou Gzem Formation at Ain Jemaa (Kaiser *et al.* 2007)
18. Tafilalt and Maider, eastern Anti-Atlas, Morocco: black shale, regionally weathered to white or red sulphate- and/or hematite-rich shale, at base of Fezzou and Aoufital Formations (Korn 1999; Becker *et al.* 2000, 2002, 2013*a*; Kaiser 2005; Kaiser *et al.* 2011)
19. Holy Cross Mountains, Poland: thin black shale at Kowala (Olempska 1997; Trela & Malec 2007; Marynowski & Filipiak 2007; Malec 2014; Myrow *et al.* 2014) and in the Bolechowice IG1 borehole (Filipiak 2004)
20. Silesia, southern Poland: Dzikowiec section, thin shale between middle and upper Wapnica Formation (Mistiaen & Weyer 1999)
21. Polar Urals, Russia: Kozhim section, black shale in the upper part of Zigansky Horizon (Bed 57, Nemirovskaya *et al.* 1993; Sobolev *et al.* 2000)
22. ?Nakhichevan Autonomous region, Armenia/Azerbaijan, Caucasus: Geran-Kalasi reference section, dark-grey shale of Unit 14, 'basal Tournaisian' (Grechishnikova & Levitskii 2011)
23. Alborz Mountains, northern Iran: calcareous interval of lower LN Zone within higher Geirud Formation (Ghavidel-Syooki 1994)
24. Shotori Range, eastern Iran: thin black shale at base of 'Mush Horizon' of basal Shishtu-2 Formation (Bahrami *et al.* 2011)
25. ?Himalaya regions, NW India: Spiti, black shale unit above a last fauna with *Icriodus* (Draganits *et al.* 1999)
26. Xinjiang, NW China: Emuha section, transgressive level with *Pr. collinsoni* Fauna, cymaclymeniids, and LN Zone miospores (Xu *et al.* 1990; Zong *et al.* 2014); the geochemically proven anoxic interval, without black shale, of the Boulonguor section (Carmichael *et al.* 2015) is not biostratigraphically dated and possibly an older level, but it overlies a regressive interval
27. Guangxi Province, South China: Huangmao section, lower, black Changshun Shale (Bai *et al.* 1987, 1994; Bai & Ning 1989); Lali section, black shales of lower Tangkou Member of Wangyou Formation (Su *et al.* 1988; Ji & Ziegler 1993); Huilong section, thin dark shale at base of Yaoyunling Formation (Jin *et al.* 2007); ?Zaisha section, thin shale with LN Zone (Ji *et al.* 1987); ?Haiyang section, thin, laminated shale unit between pre- and post-crisis beds (Ji & Ziegler 1992); Banchen section, Qinzhou County, basal chert-shale facies, black Changshun Shale (Bai *et al.* 1994)

(*Continued*)

Table 1. *Continued*

28. Guizhou Province, South China: Muhua section, lower Gedongguan Bed (Hou *et al.* 1985); Huishui section, black marl (lower Unit 12) above last limestone with stromatoporoids (Unit 11, near top of Lower Member of the Kolaoho Formation, Wu *et al.* 1987); Dayin section, black shale at DCB (Yuan & Xiang 1998)
29. ?Yunnan, South China: Shidian area, Daizhaimen section, thin black shale of extremely condensed and incomplete DCB interval (Carls & Gong 1992)
30. Jiangsu Province, South China: black, calcareous shale between siltstones with Famennian macroplants and VI Zone miospores (Bai *et al.* 1994)
31. NW Thailand: repeated claims of the Hangenberg Event in the Mae Sariang section are contradicted by typical pre-crisis conodont faunas (Savage 2013) that range to the top of the documented succession
32. Vietnam, Cat Ba Island: black shale within Pho Han Formation (Bed 116, Komatsu *et al.* 2014)
33. Russian Far East, pre-Kolyma Anticline: thin shale unit between pre-crisis and *sulcata* Zone conodont faunas (Gagiev & Bogus 1990)

See also Figures 6 and 7.

Great Basin of the western United States (Gutschick & Rodriguez 1979), in eastern Iran (Bahrami *et al.* 2011) or in Yunnan (Carls & Gong 1992).

With an almost pantropical distribution (Fig. 6), the Rhenish HBS and its equivalents (Figs 2 & 5; Table 1) follow at the base of the *ck*I and LN Zone. Laminated, pyrite-rich and hypoxic to euxinic black shales (Fig. 7) replaced the pre-crisis pelagic cephalopod limestones in middle and deeper shelf positions (Becker 1993*a*, *b*; Korn *et al.* 1994; Dzik 1997; Marynowski & Filipiak 2007).

This major, brief transgressive episode corresponds to a sudden eustatic rise, followed by maximum flooding and basin starvation (TST; Becker 1993*a*, *b*; Bless *et al.* 1993; Van Steenwinkel 1993*a*, *b*: Walliser 1996; Wagner 2001; Kaiser *et al.* 2011). The onset of the HBS correlates with the main extinction phase of the Hangenberg Crisis (Figs 2 & 3).

The beginning of sea-level fall and regression in the higher *ck*I and LN Zone is marked in the Rhenish Massif by the silty, green-grey Rhenish HS (Figs 2 & 5), a highstand deposit (Van Steenwinkel 1993*a*, *b*; HST or Forced Regression System Tract). Regressive shaly sediments deposited during this time interval are known from many different mid- to low-latitude settings (Fig. 8; Table 2), but regionally this initial eustatic sea-level drop may have caused the onset of non-deposition.

The Hangenberg Sandstone and its equivalent clastic deposits of the upper *ck*I and upper LN Zone (Figs 2, 5, 7d & 9c; Table 3) represent the lowstand

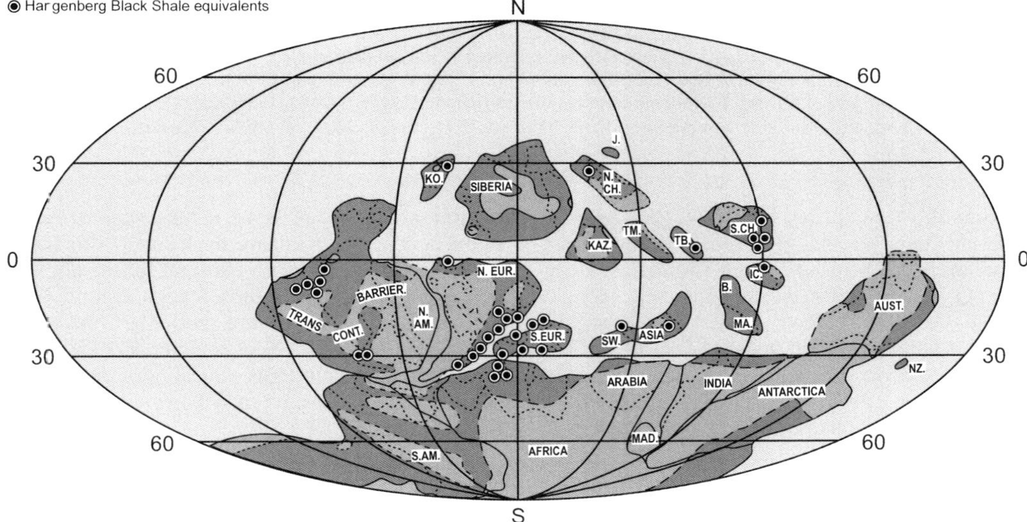

Fig. 6. Global distribution of the HBS and its equivalents plotted on a plate tectonic reconstruction that assumes a narrow western and a free eastern Prototethys (see Becker *et al.* 2012; for details of records see Table 1). Palaeogeographical map modified after Heckel & Witzke (1979).

Fig. 7. Field images of DCB outcrops. (**a**) HBS and overlying, locally thin HSS intercalated between the uppermost Famennian Wocklum Limestone and lower Tournaisian Hangenberg Limestone (both in nodular, cyclic cephalopod limestone facies). Drewer, Rhenish Massif, western Germany; photo R. T. Becker. (**b**) HBS intercalated in uppermost Famennian and lower Tournaisian cephalopod limestones. Kronhofgraben, Carnic Alps, Austria; photo S. I. Kaiser. (**c**) HBS equivalent at M'Karig, easternmost Tafilalt, Anti-Atlas, Morocco; photo S. I. Kaiser. The originally black, pyrite-rich shales are secondarily weathered to white and red, hematite- and sulphate-rich shales. (**d**) HBS and HS equivalents (Upper Member of local Bou Gzem Formation) and HSS equivalent (Táarraft Formation) at Ain Jemaa, Oulmes region, Moroccan Meseta. Conodonts from clymeniid-rich, underlying nodular limestones (Middle Member of Bou Gzem Formation) indicate an uppermost Famennian age (Kaiser *et al.* 2007); photo S. I. Kaiser.

deposits above a sequence boundary. The regionally different coarse clastics represent slope and basin-floor fans (e.g. Rhenish HSS and Thuringian 'Hangender Quarzit') or incised valley fills (Seiler conglomerate, Germany, Paproth 1986; Van Steenwinkel 1993*b*; southern Tafilalt, Kaiser *et al.* 2011). This level is found worldwide (Fig. 8; Sandberg *et al.* 1988; Becker 1996). It is time-equivalent to the (main) glaciation pulse at the end of the Famennian (see the section 'Causes of the Hangenberg Crisis – Sea-level changes and the end-Devonian glaciation'), evident by hiatuses in neritic successions, by widespread intercalations of nearshore sediments into deeper-water pelagic sediments (e.g. brachiopod sandstones or oolites/oncolites) or by an extremely condensed limestone facies of pelagic platforms (Figs 4, 5 & 9a, b). In the nearshore facies of the Ardennes, centimetre-thick siliciclastic interbeds, such as the horizon between limestone beds 103 and 104 in the Royseux Station section or in the Chanxhe 3 and Modave sections (Conil *et al.* 1986; Dreesen *et al.* 1993), could be an expression of a HSS equivalent, thus indicating a more complete succession, as often acknowledged in this facies realm.

The globally recognized unconformities (Fig. 10; Table 4) probably correlate with the sequence boundary and the subsequent episode of non-deposition. But fossil-poor or reworked sediments from different regions separating the Famennian and Tournaisian have hampered the precise dating and correlation of event beds. This resulted in

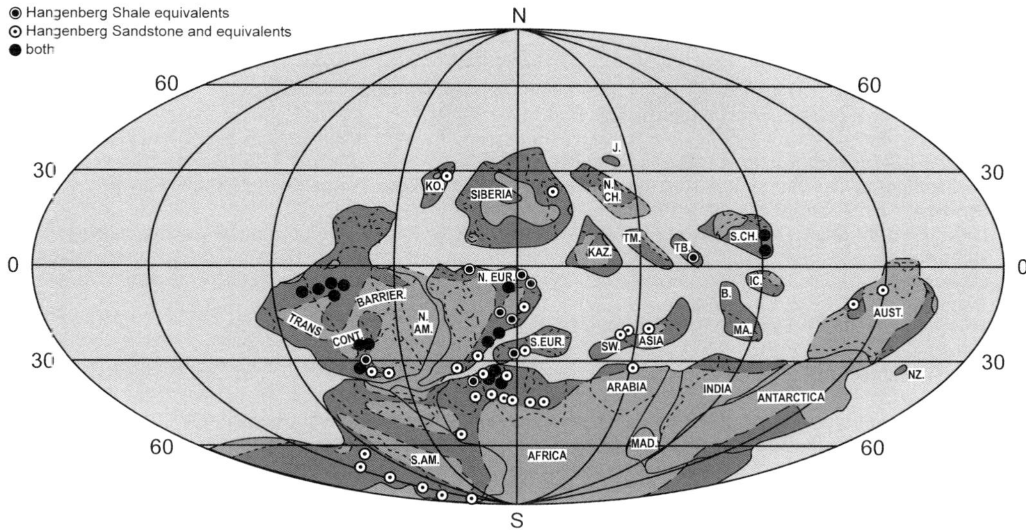

Fig. 8. Global distribution of the HS, HSS or of both. For record details see Tables 2 and 3. Palaeogeographical map modified after Heckel & Witzke (1979).

different interpretations (or even misinterpretations) of geochemical proxies, the fossil and the sedimentary record, and the exact time of end-Devonian glaciation episodes (see the section 'Causes of the Hangenberg Crisis – Sea-level changes and the end-Devonian glaciation'; Flajs & Feist 1988; Brand *et al.* 2004; Azmy *et al.* 2009; Kaiser 2009; Myrow *et al.* 2011; Wicander *et al.* 2011).

A return to pelagic limestone deposition occurred in the terminal Devonian (Upper *praesulcata* Zone = *kockeli* transgression, TST) of western Europe (Fig. 2, equivalents of the lower Stockum Limestone, Bless *et al.* 1993), North America (e.g. Louisana Limestone, Cramer *et al.* 2006, 2008), the Urals, South China and other regions (Fig. 11; complete compilation in Table 5). In the neritic realm the retreat of the shorelines resulted in the re-onset of sedimentation, reworking and redeposition of topmost Devonian material (e.g. conglomerate at the base of the Hastière Fm; Van Steenwinkel 1993*b*), and the return of carbonate facies (e.g. Hastière and Avesnelles formations of the Ardennes). Neritic conodont assemblages at the base of transgression in numerous other basins cannot be correlated clearly either with the *kockeli* or the next-higher *sulcata* Zone. But the 'Rhenish Standard Succession' shows that minor erosional events, probably caused by minor sea-level falls (parasequences), occurred at the base, within (lower/upper Stockum level boundary), and at the end of the upper crisis interval (base of Hangenberg Limestone; Bless *et al.* 1993). Their recognition in other successions is often hampered by strong condensation and gaps of longer duration, although they have been pivotal to understanding the survivor extinctions and the discrepancy between the main marine and terrestrial extinctions.

The subsequent main basal Tournaisian (Mississippian) transgression reflects a major eustatic rise. The continental plains near the former Famennian shorelines were flooded, as seen in the change from the youngest Old Red Sandstone deposits (LN miospore zone) to marine deposits in the VI miospore Zone on the British Isles (Austin & Hill 1973; Clayton *et al.* 1986; McNestry 1988). Sedimentation restarted above unconformities in eastern North America (e.g. Coleman & Clayton 1987), on the European Brabant Massif (Conil *et al.* 1993), on the Russian Platform (Alekseev *et al.* 1979, 1994; Simakov 1994), in North Africa (Conrad *et al.* 1986; Kaiser *et al.* 2011), in the Turkish Taurides (Hartkopf-Fröder in Plodowski & Salanci 1990) and in SE Asia (Zhang 1987). In South America and South Africa, the diamictite facies gradually gave way to dark, organic-rich mudstones that suggest expanding open-marine conditions (Marshall *et al.* 2002). Lower Tournaisian ammonoids of Chile (House 1996) prove a significant warming of the high latitudes.

Following a phase of weakly fluctuating global sea-level in the main part of the lower Tournaisian there was a gradual rise at its end (*Siphonodella* (*Siphonodella*) *quadruplicata* Zone, LC I-E, poorly fossiliferous, shaly, upper part of Hangenberg Limestone and its equivalents). The next global anoxic and transgressive event, the next time of

Table 2. *Overview of reported equivalents of the (regressive) Rhenish Hangenberg Shale (HS) and selected references*

1. Confusion Range, Utah, to W Montana: thin green shale above 'Conchostracan Shale' with possible *Acutimitoceras* (*Stockumites*) (Gutschick & Rodriguez 1979)
2. North Dakota: silty Unit 1 of Middle Bakken Formation with lower *Syringothyris* Fauna and possible *Acutimitoceras* (*Stockumites*) (Thrasher 1987)
3. Saskatchewan, Canada: Unit A (offshore facies) of Middle Bakken Formation, LN Zone (Playford & McGregor 1993; Smith & Bustin 1998)
4. Ohio: Bedford Shale (Molyneux *et al.* 1984; Coleman & Clayton 1987; Gutschick & Sandberg 1991)
5. Illinois: upper Saverton Shale of Pike County with LN Zone (Wicander & Playford 2013), corresponding siltstone and lag sandstone with reworked conodonts of upper Saverton Shale, Calhoun County (Sandberg *et al.* 1972; Collinson *et al.* 1979)
6. Pennsylvania: basal shale of Middle Pocono Formation with *Vallatisporites vallatus* (Streel & Traverse 1978)
7. East Greenland: lower Obrutschew Bjerg Formation with LN Zone (Marshall *et al.* 2002; Marshall & Astin 2009)
8. Franconia, Germany: Kirchgattendorf, thin shale unit below Stockum level (Korn 1993)
9. Thuringia, Germany: Saalfeld region, 'Schieferfuss' = basal part of Obernitz Member of Gleitsch Formation (Bartzsch *et al.* 2015); Bergaer Anticline, Kahlleite, Schleiz, Hangenbergschiefer = Rödersdorf Member of the Göschitz Formation (Weyer 1977; Bartzsch *et al.* 1995, 2001; Gereke 2004)
10. Pyrenees, France: thin shale units below Stockum levels at Milles (Ariége) and Saubette (Haute Pyrénées, Perret 1988)
11. Western Meseta, Morocco: Oulmès region, green shale of Upper Member of Bou Gzem Formation at Ain Jemaa (Kaiser *et al.* 2007)
12. Eastern Meseta, Morocco: Doukkala Basin, upper shale unit with LN Zone (Rahmani-Antari & Lachkar 2001)
13. Tafilalt and Maider, eastern Anti-Atlas, Morocco: green shales of lower Fezzou and Aoufital formations (Becker *et al.* 2002, 2013*a*; Kaiser *et al.* 2011)
14. Western Dra Valley, western Anti-Atlas, Morocco: silty and unfossiliferous Kheneg Lakahal Member of Tazout Formation (Kaiser *et al.* 2004; Becker in Hahn *et al.* 2012)
15. Pommerania, Poland: LN Zone part of subsurface Sapolno Calcareous Shale Formation (Matyja *et al.* 2014)
16. Holy Cross Mountains, Poland: sandy shale with LN Zone and *Ac.* (*Stockumites*) at Kowala (Filipiak 2004; Marynowski *et al.* 2012; Malec 2014; Myrow *et al.* 2014)
17. Udmurtia, Russian Platform: shale with *Verrucosisporites nitidus* (LN Zone = ml_0 level) of the lower Malevka Formation (Byvsheva *et al.* 1984)
18. Northern Urals, Russia: supposedly regressive DCB shale unit of Podcherem section 15 (Zhuravlev & Tolmacheva 1995)
19. Western slope, southern Urals, Russia: argillaceous lower Gumerov Horizon with *Tumulispora malevkensis* and *Retispora lepidophyta* (PM Zone, Pazukhin *et al.* 2009)
20. Mugodzhar, southern Urals: shaly Member 2 of Dganganin Formation, with LN Zone, Berchogur (Barskov *et al.* 1984)
21. Kurdistan, Iraq: northern thrust zone, shale unit with LN Zone miospores in the middle of the Ora Formation (Naqishbandi *et al.* 2010)
22. Tibet: upper part of shaly, silty Zhangdon Formation with LN Zone (Gao 1989; Fan *et al.* 2003)
23. Guangxi, South China: Huangmao section: upper, green Changshun Shale (Bai *et al.* 1987; Bai & Ning 1989); Lali section, upper Tangkou Member of Wangyou Formation (Su *et al.* 1988; Ji & Ziegler 1993)
24. Central Hunan: Malanbian section, 2.4 m shale below Hangenberg Sandstone equivalent at top of Menggongao Formation (Muchez 1996)

See also Figure 8.

maximum flooding, is marked by the Lower Alum Shale (Fig. 2) at the base of the middle Tournaisian (Johnson *et al.* 1985; Becker 1993*a*, *b*; Siegmund *et al.* 2002). It is associated with widespread black shales, black limestones or cherty sediments in many regions (e.g. Schönlaub *et al.* 1988; Bai & Ning 1989; Korn *et al.* 1994; Zhuravlev 1998; Kaiser *et al.* 2011; Mossoni *et al.* 2015). In the sequence stratigraphic model for third-order sequences in the neritic facies of Hance *et al.* (2001), this maximum flooding corresponds to the maximum flooding surface of sequence 2.

Causes of the Hangenberg Crisis

Scenarios claimed to explain the DCB environmental changes range from enhanced magmatic activity, rapid eustatic fluctuations, tectonics, asteroid impacts, severe climatic oscillations, anoxia, the

Fig. 9. Field images of DCB outcrops. (**a**) Condensed and continuous DCB limestone successions at La Serre stratotype, Montagne Noire, southern France, showing the position of the current Global Stratotype Section and Point and the position of the oldest *S.* (*Eo.*) *sulcata* based on data by Kaiser (2009); photo S. I. Kaiser. (**b**) Condensed DCB successions in pelagic limestone facies, with extremely thin representation of the crisis interval, at Dapoushang, Guizhou, South China; photo Ji Qiang & Wang Chen-Yuan. (**c**) HSS at Bou Tlidat, Maider, Anti-Atlas, SE Morocco. The several hundred metres-thick successions consist of turbiditic and non-turbiditic sandstones, partly with oscillation or current ripples, which indicate shallowing upwards; photo S. I. Kaiser. (**d**) Granite dropstone, first discovered by M. J. Robinson in 2006 (see Lierman & Mason 2007; Ettensohn *et al.* 2007 and Lierman *et al.* 2009), Logan Hollow Branch, Rowan County, Kentucky; uppermost Famennian, LN biozone. The dropstone lies at the top of the Cleveland Shale Member of the Ohio Shale, just below the contact with the overlying Bedford Shale.

occurrence of wildfires, shifts in atmospheric composition (rise in O_2 and fall in CO_2), to the expansion of terrestrial plants and global spread of swamp vegetation (e.g. Wang *et al.* 1993; Algeo & Scheckler 1998; Caplan & Bustin 1999; Streel *et al.* 2000*a*; Filipiak & Racki 2010; Kumpan *et al.* 2014). New high-resolution and multidisciplinary studies, including biostratigraphy, sedimentology and geochemistry of pelagic and shallow-water deposits (Kumpan *et al.* 2013, 2014, 2015; Qie *et al.* 2015), have provided new steps towards a better understanding of the Hangenberg Crisis.

Impact evidence

Geochemical evidence for a DCB impact event has been published by Bai *et al.* (1987, 1994) and Bai & Ning (1989), who observed iridium and nickel spikes in HBS equivalents of South China (Changshun Shale), microtectites, and element ratios of the black shale that supposedly match a meteoritic rather than a volcanic source. These results have not been validated for any other region. However, the major, 120 km diameter, Woodleigh impact structure of Western Australia, which was adjacent at the time, has been dated by Glikson *et al.* (2005) as 359 ± 4 Ma. The core age correlates exactly with the Hangenberg Crisis, but there are no corresponding crater-fill sediments or fallout ejecta beds. An even larger, up to 200 km diameter crater has been identified in South Australia (Glikson *et al.* 2013), but its age is even more unclear, although a possible 360 Ma age has been reported in public media (e.g. *Die Welt*, *ABC Science*, *The*

Table 3. *Overview on reported equivalents of the regressive Rhenish Hangenberg Sandstone (HSS; corresponding sandstones, conglomerates, diamictite levels, oolites or microbial biostromes) and selected references*

1. SE Nevada–NE Utah–west Montana, Great Basin, western North America: oncolite marker unit with *Pr. meischneri*, *Pr. collinsoni*, *S. praesulcata*, reworked pre-crisis conodonts, *Ac.* (*Stockumites*), *Syringothyris* fauna, and overlying siltstones and shales of upper LN Zone (with *Vallatisporites vallatus*; Units 2–6 of Middle Pilot Shale, Leatham Formation, and Sappington Member, respectively (Sandberg *et al.* 1972; Gutschick & Rodriguez 1979; Warren *et al.* 2014)
2. North Dakota: sandy, poorly fossiliferous Unit 2 of Middle Bakken Formation (Thrasher 1987)
3. Saskatchewan, Canada: main part of Middle Bakken Formation with LN Zone (Playford & McGregor 1993; Smith & Bustin 1998; Wagner 2001)
4. Ohio, Kentucky: Berea Sandstone (Coleman & Clayton 1987; Gutschick & Sandberg 1991; Pashin & Ettensohn 1992)
5. Pennsylvania–Maryland–West Virginia–Virginia: diamictitic unit of lower Spechty Kopf and Rockwell formations, correlative Cussewango Sandstone and Cloyd Conglomerate (Brezinski *et al.* 2008, 2010; Baird *et al.* 2009)
6. Nova Scotia, eastern Canada: basal Horton Group (Martel *et al.* 1993)
7. Bolivia: diamictitic Itacua Formation with LN Zone (Wicander *et al.* 2011; Streel *et al.* 2013), all or main part (with LN Zone) of diamictitic Toregua Formation (Vavrdová *et al.* 1996), diamictitic lowermost Cumana Formation (Vavrdová *et al.* 1991), diamictites of Saipura Formation (Perez-Leyton 1991)
8. Amazon Basin, Brazil: diamictitic upper Curiri Formation with LN Zone (Melo & Loboziak 2003)
9. Parnaíba Basin, Brazil: diamictitic upper Cabeças Formation with LN Zone (Streel 1986; Loboziak *et al.* 1992; Streel *et al.* 2000*a*, *b*)
10. Paraná Basin, Brazil: diamictictic interval of LN Zone within Itararé Group (Loboziak *et al.* 1995)
11. Southern Ireland: top of the Irish Old Head Sandstone (Clayton *et al.* 1986)
12. Ardennes, Belgium–Germany: Vesdre–Aachen region, 2 m thick sandstone–siltstone unit (Mottequin & Poty 2014)
13. Seiler area, Germany: *c.* 190 m thick upper oolite and siliciclastic unit (Koch *et al.* 1970)
14. Thuringia, Germany: Schwarzburg Anticline, Saalfeld region, basinal facies, 'Hangender Quarzit' = main Obernitz Member of the Gleitsch Formation, including pyritic shale interbeds (Bartzsch & Weyer 1986; Bartzsch *et al.* 1999, 2015)
15. Montagne Noire, southern France: lower oolite unit of La Serre stratotype (Flajs & Feist 1988)
16. Cantabrian Mountains, northern Spain: Bernesga Valley, northern Léon, main part of Ermita Formation (García-López & Sanz-López 2002)
17. Western Meseta, Morocco: Oulmès region, quartzitic Táaraft Formation at Ain Jemaa, Upper Moulay Hassane Formation of El Hammam Zone, Jebel Akala Quartzites of the Sidi Bettache Basin and Ben Slimane region, similar quartzites forming the top Chabet el Baya Formation of the SE Mdakra Massif, quartzites at the top Foum-el-Mejez Formation, Rehamna (Kaiser *et al.* 2007; new unpublished data)
18. Tafilalt and Maider, eastern Anti-Atlas, Morocco: main sandstones of Fezzou and Aoufital formations (Becker *et al.* 2002, 2013*a*; Kaiser *et al.* 2011)
19. Iguidi Sub-basin, Tindouf Basin, SW Algeria: oolithic prodeltaic succession (Guerrak & Chauvel 1985)
20. Saoura Valley, southern Algeria: Marhouma and Ouarourout Sandstones (Petter 1960)
21. Ahnet and Mouydir Basins, north of Hoggar, southern Algeria: Lower Khenig Sandstone (Conrad *et al.* 1986; Wendt *et al.* 2006)
22. Illizi Basin: borehole GDT1 in the SE, sandstone unit at the top of the Illerene Formation with *V. nitidus* (LN Zone, Abdesselam-Rouighi & Coquel 1997)
23. Rhadames Basin, western Libya: uppermost Famennian Tahara Formation with diamictites and *Retispora lepidophyta* (Streel *et al.* 2000*a*, *b*)
24. Western Ghana: LN Zone sandstone within the Takoradi Shale Formation (Atta-Peters & Anan-Yorke 2003)
25. South Africa: Peerdepoort Member and associated diamictites, Witpoort Formation (Almond *et al.* 2002)
26. Pripyat Depression, Byelorussia: shale–sandstone alternation of LE/LN (LE_1-PLE) zones (Avchimovitch *et al.* 1988, 1993)
27. Udmurtia, eastern Russian Platform: sandstones of Malevka Suite (Byvsheva *et al.* 1984)
28. Northern Caucasus: sandstone interval above the last *Palmatolepis* faunas (Puporev & Chegodaev 1982)
29. Transcaucasus, southern Russia: sandstone unit (Karaulov & Gretschischnikova 1997)
30. Alborz Mountains, northern Iran: sandstone unit of upper LN Zone in higher Geirud Formation (Ghavidel-Syooki 1994)
31. Kuznetsk Basin, southern Siberia, Russia: sandstones of upper Abyshevo Formation (Karaulov & Gretschischnikova 1997)

(*Continued*)

Table 3. *Continued*

32. Central Hunan, South China: siliciclastics of the topmost Menggongao Formation (Hance *et al.* 1994; Muchez 1996; Tan *et al.* 1996)
33. Guangxi, South China: Guping section, sandstone shale alternation of Member 1 of Luzhai Formation, higher LN Zone with *Tumulispora malevkensis* (Yang & Neves 1997)
34. Kolyma, Russian Far East: Kamenka River section, thin sandstone unit (Gagiev & Kononova 1990; Gagiev 1997)
35. Canning Basin, NW Australia: lower Yellow Drum Sandstone (Nicoll & Druce 1979)
36. Bonaparte Basin, northern Australia: between Ningbing and Burt Range Limestones (Druce 1969)

See also Figures 6, 7d and 9.

Malaysian Times, *Australasian Science*, February/March 2013).

The current data leave open the possibility that significant Australian impacts caused the southern Chinese geochemical signatures and contributed to the sudden palaeoclimatic perturbations near the DCB. But such interpretations are currently merely a hypothesis to be followed by future studies.

Sea-level changes and the end-Devonian glaciation

The glaciation episode at the end of the Devonian abruptly terminated a period of more than 80 myr of greenhouse conditions since the Hirnantian Glaciation at the end of the Ordovician (Simon *et al.* 2007). Although there have been periods of warmer and cooler phases, e.g. in the Middle Devonian (Joachimski *et al.* 2009), and several authors (e.g. Elrick *et al.* 2009) claim Middle or pre-Hangenberg UD glacio-eustatics to explain large-scale, sudden, global sea-level fluctuations, these climate and sea-level changes are not associated with glacial deposits. The DCB glacial phase was not the onset of the long-lived icehouse conditions that predominated in the Upper Carboniferous and Permian. The Tournaisian and Viséan were mostly greenhouse times, interrupted only briefly by glaciation pulses near the middle/upper Tournaisian boundary, within the upper Viséan and Serpukhovian (Caputo *et al.* 2008; Meor *et al.* 2014). Terminal Famennian glacial sediments of the combined LE/LN zones include polymict striated and faceted clasts, dropstones and glacial pavements (Figs 9d & 12). They occur widely in South America (e.g. Loboziak *et al.* 1993, 1995; Isaacson *et al.* 1999; Melo *et al.* 1999; Dino 2000; Padilha de Quadros 2000; Streel

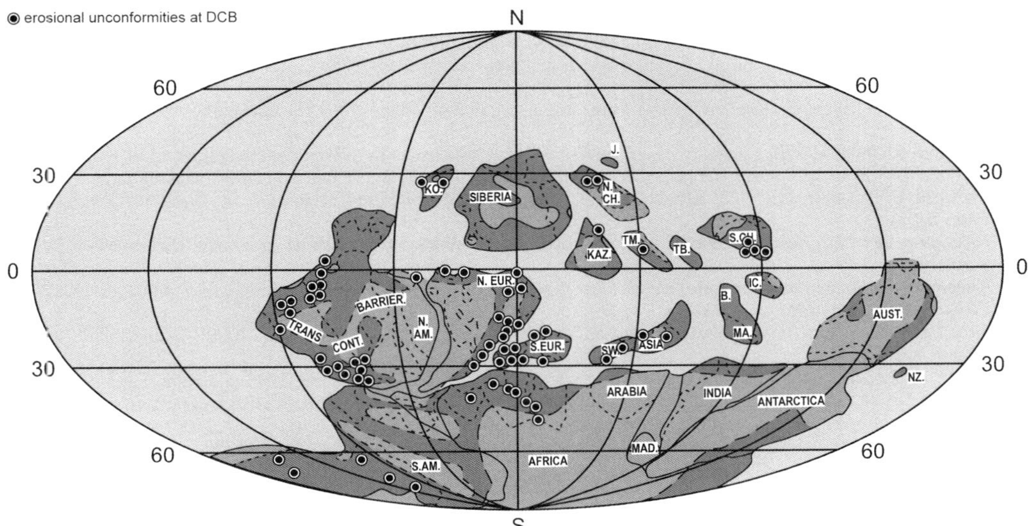

Fig. 10. Global distribution of unconformities caused by the glacioeustatic DCB regression. For record details see Table 4. Palaeogeographical map modified after Heckel & Witzke (1979).

Table 4. *Overview on reported erosional unconformities at the DCB and selected references*

1. Quesnel Terrane, British Columbia: within Tk'emlups Formation, between Harper Mountain Pebble Beds with upper Famennian conodonts and mudstones with lower Tournaisian conodonts (Beatty 2002, 2003)
2. Alberta, NW Canada: within Exshaw Formation of the type region (e.g. Jura Creek, Crowsnest Pass), between pre-crisis Black Shale Member and lower Tournaisian Siltstone Member, including the goniatite bed (Macqueen & Sandberg 1970; Richards & Higgins 1988; Johnston *et al.* 2010); below typical Banff Formation at Fiddle River and Nordegg (Savoy *et al.* 1999)
3. Wyoming and South Dakota: within Englewood Formation, within the Cottonwood Canyon Member or below the Madison Limestone (Klapper & Furnish 1962; Klapper 1966; Sandberg & Klapper 1967; Sando & Sandberg 1987)
4. North-central Utah: within Fitchville Formation (Sandberg & Poole 1977; Gutschick & Rodriguez 1979; Clark *et al.* 2014)
5. Woodruff Basin, Nevada: below Chainman Shale (Sandberg & Poole 1977; Sandberg *et al.* 2003)
6. SE Nevada–NE Utah–Montana, Great Basin: lag sandstone/top of Unit 3, below 'Conchostracan Shale' (lower crisis interval) of Middle Pilot Shale, Leatham Formation and Sappington Member (Gutschick & Rodriguez 1979); between top Sappington Member (top middle crisis interval) and middle Tournaisian Lodgepole Formation of Montana (Sandberg *et al.* 1972)
7. NW Arizona to SE Nevada: between upper Famennian Crystall Pass Member of Sultan Limestone and lower Tournaisian lower Whitmore Wash Member of Redwall Limestone (Ritter 1991)
8. New Mexico–Arizona: below Redwall Limestone, Escabrosa Limestone and Keating Formation (Armstrong *et al.* 1980; Moore 1988)
9. Colorado: palaeokarst and breccia unit within Coffe Pot Member of Dyer Formation (Myrow *et al.* 2011; Wistort *et al.* 2014)
10. Mississippi Valley, Missouri/Illinois: within upper Saverton Shale or below Louisiana Limestone, *kockeli* Zone (Sandberg *et al.* 1972)
11. Kentucky–Ohio: minor unconformities at the base of Bedford Shale (base of middle crisis interval) and at the base of the Berea Sandstone (peak regression, Baird *et al.* 2009)
12. Pennsylvania–Maryland–West Virginia–SW Virginia: below Cussewango Sandstone, an equivalent of the lower Berea Sandstone and Spechty Kopf diamictites, below Cloyd Conglomerate (peak regression; Baird *et al.* 2009; Brezinski *et al.* 2010)
13. Tennessee–Alabama: between Chattanooga Shale and Maury Formation (Hass 1956; Over 2007)
14. Oklahoma–Arkansas: between Chattanooga Shale and St. Joe Limestone (Kelly *et al.* 1997)
15. Oklahoma: within black Woodford Shale (Over 1992)
16. Central Texas: Houy Formation (Cloud *et al.* 1957)
17. Bolivia, Altiplano: below diamictitic Cumaná Formation (Diaz-Martinez & Isaacson 1994)
18. Amazon Basin, Brazil: between lower and diamictitic upper Curiri Formation (Melo & Loboziak 2003)
19. Parnaíba Basin, Brazil: below diamictitic upper Cabeças Formation (Loboziak *et al.* 1992; Streel *et al.* 2000*a*, *b*)
20. Argentinia, Precordillera. Extensive gap, with the absence of 'Strunian' palynomorphs in reworked floras (López-Gamundi & Rosello 1993; Amenabár *et al.* 2009)
21. Northern Chile, Andes region: below Middle Zorritas Formation with VI Zone miospores and lower Tournaisian goniatites (House 1996; Rubinstein *et al.* 1996)
22. Spitsbergen: level of Svalbardian deformation (Piepjohn *et al.* 2000)
23. Ardennes, Belgium: below basal Hastière Limestone (Van Steenwinkel 1988; Casier *et al.* 2002, 2004; Kumpan *et al.* 2014)
24. Niederrhein subsurface, Germany: below lower dolomite, an equivalent of the Hastière Limestone (Bless *et al.* 1988)
25. Northern Rhenish Massif, Germany: large gaps or missing siliclastics in the middle crisis interval of seamount sections (e.g. Reigern Forest Quarry, Beul, Enkeberg, Trockenbrück, Beringhauser Tunnel, Müssenberg, Drewer NE; Paeckelmann 1938; Kronberg *et al.* 1960; Luppold *et al.* 1984; Clausen *et al.* 1989; Schülke & Popp 2005)
26. Franconia, Germany: Kirchgattendorf, missing upper part of Wocklum Limestone equivalent (Schindewolf 1923; Korn 1993)
27. Thuringia, Germany: Bergaer Anticline, Schleiz region, between HBS and Hangenberg Limestone equivalents (Weyer 1977; Bartzsch *et al.* 2001)
28. Graz Palaeozoic, Austria: extremely condensed/incomplete *ck*I at Trolp (Kaiser *et al.* 2009); Weihermühle section (Ebner 1980*b*)
29. Carnic Alps, Austria/Italy: within extremely condensed succession, between *ck*I (HBS equivalent) and *kockeli* Zone, at Grüne Schneid (Schönlaub *et al.* 1988; Kaiser *et al.* 2006, 2009), between HBS and post-crisis Hangenberg Limestone equivalent at Kronhofgraben (Schönlaub *et al.* 1992; Kaiser *et al.* 2006) and Plan di Zermula (Perri & Spalletta 2000*a*, *b*; Kaiser *et al.* 2006); below *kockeli* Zone of Plöcken area (Gedik 1974)
30. Western Armorican Massif, France: Brest region, between Kermerrien and Kertanguy formations (Rolet *et al.* 1986)
31. Montagne Noire: between HBS and lower Stockum level (*kockeli* Zone) at Puech de la Suque (Lethiers & Feist 1991; Kaiser *et al.* 2006, 2009) and above HBS at Col des Tribes (Girard *et al.* 2013)

(Continued)

Table 4. *Continued*

32. Pyrenees, southern France: Arize (Ariége) Massif, between HS and lower Stockum level (*kockeli* Zone) at Milles (Cygan & Perret 2002; Kaiser *et al.* 2006, 2009); Atlantic Pyrenees, gap within condensed limestones in Garcet and Moustardé sections and more extensive gap in the Pont d'Urdos section (Perret 1988; Perret & Majesté-Menjoulas 2002*a*, *b*)
33. Cantabrian Mountains, northern Spain: Palentine Domain, Gildar-Montó Unit: within top Vidrieros Formation, boundary of *praesulcata* and *kockeli* Zone (no lower/middle crisis interval, Sanz-López *et al.* 1999)
34. Sardinia: Monte Taccu, *ckI* to *sulcata* Zone (Corradini *et al.* 2003; Corradini 2008; Mossoni *et al.* 2013); Bruncu Bullai section, between HBS equivalent and *sulcata* Zone (Mossoni *et al.* 2015)
35. SW Spain, southern Central Iberian Zone: between extremely condensed uppermost Famennian (*Bispathodus ultimus ultimus* Zone) and upper Tournaisian strata (García-López *et al.* 1999)
36. Tafilalt Platform: between UD V-B to UD VI-C and HBS or younger strata (Becker *et al.* 2002; Kaiser *et al.* 2011), between Givetian and upper Tournaisian in parts of the southern Tafilalt (new data)
37. Taoudeni Basin, northern Mali (Legrand-Blain 1985)
38. Bechar Basin, Western Algeria: absence of UD VI (Weyant 1988)
39. Grand Erg Occidental (Mac Mahon Basin), western Algeria: within 'Série argileuse', between upper Famennian (LL Zone) and upper Tournaisian miospore levels (Lanzoni & Magloire 1969; Streel 1986; Coquel & Abdesselam-Rouighi 2000)
40. Illizi Basin, eastern Algeria: locally between upper Famennian Illerène Sandstone and upper Tournaisian Hassi Issendjel Formation (Conrad *et al.* 1986; Streel 1986)
41. Rhadames Basin, western Libya: between uppermost Famennian Tahara Formation with diamictites and *Retispora lepidophyta* and upper Tournaisian shales of the Mrar Formation (Massa & Moreau-Benoit 1985; Conrad *et al.* 1986; Streel 1986)
42. Djado Sub-Basin, SW Libya to northern Niger: extensive gap below upper Tournaisian Mrar Formation (Mergl *et al.* 2001)
43. Pomerania, Poland: subsurface (Matyja & Stempien-Salek 1994)
44. Holy Cross Mountains, Poland: Ostrowka seamount (Szulczewski 1978; Szulczewski *et al.* 1996; Malec 2014)
45. Silesia, SE Poland: Dzikowiec section, absence of UD VI-D and of the middle/upper crisis interval (Schindewolf 1937; Dzik 1997; Mistiaen & Weyer 1999); Gologowy section, gap between 'Mid-to late *Palmatolepis expansa*' and 'Late' *Siphonodella duplicata-Siphonodella sandbergi* Zones' (Haydukiewicz 1981)
46. East European shelf, Russian Platform: between Khovan and Malevka horizons (Alekseev *et al.* 1979; Byvsheva & Umnova 1993) or between Khovanshchina and Kupavna formations (Tula region, Alekseev *et al.* 1994); Moscow Syncline and Voronezh Anticline, gap of main part of Gumerovo Horizon (Makhlina 1996)
47. Timan, northern Russia: at top of limestones with *Quasiendothyra kobeitusana* foraminifera and LE Zone miospores, below shales of the VI Zone (no LN Zone, Durkina & Avchimovitch 1988)
48. ?Polar Urals, Russia: Kozhim section, sharp contact of HBS equivalent and base of Humerovsky Horizon with mixed conodont fauna (Bed 57/58 boundary, Nemirovskaya *et al.* 1993; Sobolev *et al.* 2000)
49. Western slope, southern Ural: strongly condensed upper Gumerovsky Horizon at Sikaza section, with apparently mixed pre/post-crisis conodont fauna (Kononova 1979; Kochetkova *et al.* 1985; Kulagina *et al.* 2003; Artyushkova *et al.* 2011)
50. Central and eastern Taurides, Turkey: extensive gap below higher Tournaisian/Viséan (Göncüoglu *et al.* 2007)
51. Alborz, northern Iran: boundary of Geirud and Mobarak formations (Fallah *et al.* 2011)
52. Shotori Range, eastern Iran: extreme condensation at top of 'Cephalopod Bed' and below *sulcata* Zone of 'Mush Horizon' in the basal Shishtu-2 Formation (Yazdi 1999; Bahrami *et al.* 2011)
53. Zeravshan Range, eastern Uzbekistan: Kulé section, within Novchomok Formation, at reworking unit Bed 3 (Erina in Yolkin *et al.* 2008; new data)
54. Xinjiang, NW China: Hebukehe area, local erosional contact between Hongguleleng and Heishantou formations (Zong *et al.* 2014)
55. Tarim Basin, NW China (Zhou & Chen 1992)
56. Southern Mongolia: below lower Tournaisian basal Arynshand Formation (Wang & Minjin 2004)
57. Guangxi, South China: Etoucun section, at sharp contact between Etoucun and Yaoyunling Formations (Shen 1994; Jin *et al.* 2007); Huilong section: iron crust at top of Etoucun Formation (new record; section description see Jin *et al.* 2007); ?Huangmao section, boundary of Changshun Shale and Wangyou Formation (no *kockeli* Zone, Bai *et al.* 1987; Bai & Ning 1989); Longkou section: between limestones with the last *Palmatolepis* (Bed 10) and the first *S.* (*Eo.*) *sulcata* (Bed 11, Yu 1988); Haiyang section, below *sulcata* Zone (Ji & Ziegler 1992)
58. Guizhou, South China: Dapoushang section, top of Daihua Formation, followed by very thin tuff bed (Ji *et al.* 1989; Liu *et al.* 2012)
59. Yunnan, South China: Shidian area, Daizhaimen section, gaps below and above questionable HBS (Carls & Gong 1992); laterally much more extensive gap (most of the Famennian, Li & Duan 1993)
60. Sichuan, China: between Changtanzi Formation with pre-crisis conodonts and Heiyanwo Formation with *S.* (*S.*) *duplicata* in its lower part (Q. Ji 1987)
61. Omolon, Russian Far East (Gagiev 1997)
62. Kolyma, Russian Far East: pre-Kolyma Anticline, absence of DCB regressive level and (true) *kockeli* Zone (Gagiev & Bogus 1990; the alleged Upper *S. praesulcata* level includes pre-crisis conodonts and lacks *Pr. kockeli*)

See also Figure 10.

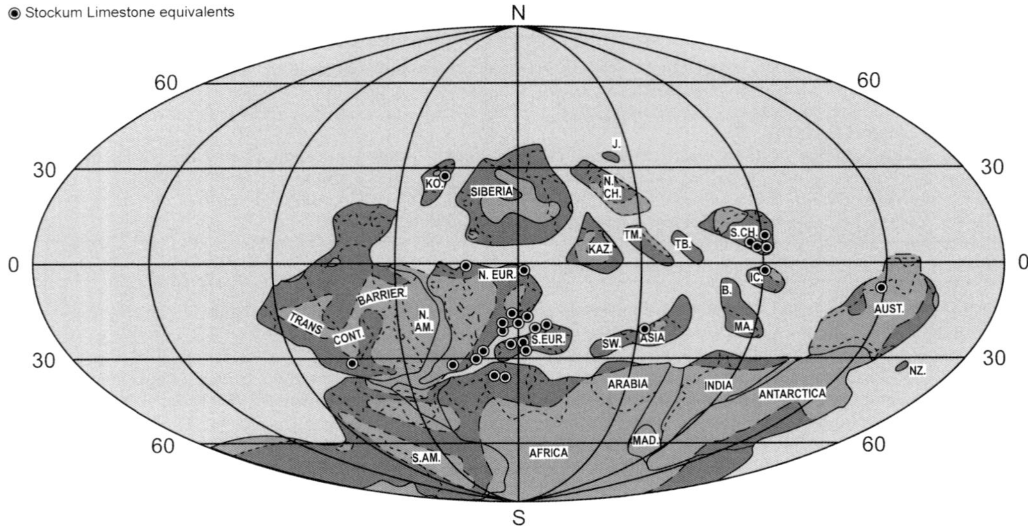

Fig. 11. Global distribution of Stockum Limestones-equivalents (upper crisis interval). For record details see Table 5. Palaeogeographical map modified after Heckel & Witzke (1979).

et al. 2000*b*) and in South Africa (Streel & Theron 1999; Almond *et al.* 2002). Possibly contemporaneous glacial sediments from central Africa are not reliably dated (Lang *et al.* 1991; Streel *et al.* 2000*a*). But there are diamictites in the uppermost Famennian Tahara Formation of western Libya (Streel *et al.* 2000*a*). In eastern North America there are contemporaneous large dropstones imbedded in the Ohio Shale of Kentucky (LN Zone; Ettensohn *et al.* 2007) and the massive diamictites of the Spechty Kopf (Fig. 12) and Rockwell formations of Pennsylvania (Brezinski *et al.* 2008, 2009, 2010). These prove significant mountain glaciers on top of the tropical but very high Appalachian palaeo-ranges. Palynological analyses in Brazil and Greenland indicate cold-humid conditions in South America during the LN Zone (Streel *et al.* 2000*b*) but cool-arid conditions on the Old Red Continent (Marshall *et al.* 2002). The major progradation of a large delta complex in North Africa has been attributed to increased humidity in northern Gondwana (Kaiser *et al.* 2011).

The complexity of the boundary interval is considered in recent studies of terrestrial sediments of Greenland and South America, indicating that the DCB glaciation was a multiphase event with several glacial/interglacial phases (Marshall 2010; Wicander *et al.* 2011). A first glaciation episode of Bolivia was considered to have started questionably as early as in the upper Famennian VCo Zone, with subsequent alternating glacial and interglacial episodes in the LL, LE and LN Zones (Wicander *et al.* 2011). This interval would correlate with the time span from the *Palmatolepis perlobata postera*/*Palmatolepis gracilis expansa* to the *praesulcata* Zone and *ck*I, an interval with enhanced carbon burial and several small-scale bio- and lithoevents, such as the global Dasberg Event and Epinette and Etreoungt Events (Kaiser *et al.* 2008; Hartenfels & Becker 2009). However, Streel (1986, 2000), Streel & Marshall (2006) and Streel *et al.* (2013) doubted the stratigraphical evidence for glacial deposits that are older than the LE/LN Zones previously suggested by Isaacson *et al.* (2008). With respect to the marine setting of the South American diamictites, it has to be stressed that most records of glacigenic sediments are from the late to final glaciation phase, when ice sheets had expanded on to the sea and collapsed, releasing their sediment load (Marshall *et al.* 2002).

Eustatically induced sea-level changes on the shelves of different continents across the DCB give further indirect evidence of major climate changes. The major sea-level fall (Johnson *et al.* 1985; Sandberg *et al.* 1988, 2002; Bless *et al.* 1993) and the widespread deposition of regressive sediments (HS, HSS; Tables 2 & 3) in the tropical realm is time-equivalent to the high-latitude glaciation pulse, which proves its glacio-eustatic nature. Recent studies in the Anti-Atlas of southern Morocco suggest that it had a scale of 100 m or more, a typical amplitude of glacio-eustatics. Cyclic siliciclastic sedimentation suggests there were smaller-scale, high-frequency, glacially induced sea-level oscillations within the glacial interval (Kaiser *et al.* 2011), which is a well known

Table 5. *Overview on reported equivalents of the Stockum Limestones (*kockeli *to* kuehni*/lower* sulcata *zones, upper crisis interval) and selected references*

1. Mississipi Valley, Missouri/Illinois: Lousiana Limestone with *Pr. kockeli* (Sandberg *et al.* 1972; Chauffe & Nichols 1995)
2. Ireland, Munster Basin: *Ac.* (*Stockumites*) level within basal VI Zone, basal Castle Slate Member of Kinsale Formation (Clayton *et al.* 1974; Matthews 1983)
3. Ardennes, Belgium: lower massive unit of Hastière Limestone with *Pr. kockeli* and possibly the limestone just above with *Bi. sculderus* and *Ps. expansus* (Van Steenwinkel 1980, 1988)
4. Ardennes, Avesnois, northern France: lower Avesnelles Limestone, Tn 1bα, partly with *Pr. kockeli* (Austin *et al.* 1970*a*; Conil *et al.* 1986)
5. Franconia, Germany: Kirchgattendorf, *Ac.* (*Stockumites*) level (Schindewolf 1923; Korn 1993)
6. Thuringia, Germany: Schwarzburg Anticline, Saalfeld region, 'Stockum faunas 6a/6b', base of Pfaffenberg Member of Gleitsch Formation (Bartzsch & Weyer 1986; Bartzsch *et al.* 2015); Bergaer Anticline, Kahlleite, lower part of 'Hangenbergkalk' with *Protognathodus* faunas (Bartzsch *et al.* 1995)
7. Graz Palaeozoic, Austria: Trolp Quarry with successive condensed *Pr. kockeli* and *kuehni* levels (Ebner 1980*a*; Kaiser *et al.* 2009)
8. Carnic Alps, Austria: thin limestone of *kockeli* Zone near the Plöcken Pass (Gedik 1974); thin limestones of successive *kockeli* and *kuehni* zones (Schönlaub *et al.* 1988); thin limestones of Grüne Schneid with successive *kockeli* and *kuehni* levels (Schönlaub *et al.* 1988, 1992; Kaiser 2007)
9. Pyrenees, southern France: Milles, Arize (Ariége) Massif, successive, condensed *kockeli* and *kuehni* levels (Perret 1988; Cygan & Perret 2002; Kaiser *et al.* 2009); Saubette ('Haute Pyrénées', Perret 1988); Moustardé (Atlantic Pyrenees, Perret 1988)
10. Montagne Noire, southern France: Puech de la Suque, *kockeli* Zone (Boyer *et al.* 1968; Lethiers & Feist 1991; Korn & Feist 2007, Kaiser *et al.* 2009)
11. Cantabrian Mountains, northern Spain: Palentine, Gildar–Montó Unit, within top Vidrieros Formation (van Adrichem Boogaert 1967; Sanz-López *et al.* 1999)
12. Maider and Tafilalt, Anti-Atlas, southern Morocco: thin siliciclastic units with *Ac.* (*Stockumites*) faunas at the top of the Fezzou and Aoufital Formation (Bou Tlidat and Mkarig, Kaiser *et al.* 2011); thin limestone with *Ac.* (*Stockumites*), *Postclymenia*, and *Pr. kockeli* at Lalla Mimouna South (Korn *et al.* 2004, 2007; Becker *et al.* 2013*a*)
13. Holy Cross Mountains, Poland: ?limestone–shale unit of (upper) LN Zone in 'Zareby IG' borehole (Filipiak 2004); Kowala, Unit C with *Protognathodus* and, in the higher part, with *Polygnathus purus* (Malec 2014; Myrow *et al.* 2014)
14. Eastern Bohemia, Czechia: Stockum level of Nepasize borehole (Chlupáč & Zikmundova 1976)
15. Moravia, Czechia: Lesni Lom Quarry, successive thin *kockeli* and *kuehnei* levels (Kalvoda & Kukal 1987; Kalvoda *et al.* 2015; Kumpan *et al.* 2013)
16. Northern Urals, Russia: limestone unit of *praesulcata* Zone above HS equivalent in Podcherem section 15 (Zhuravlev & Tolmacheva 1995)
17. Mugodzhar, southern Urals: Member 3 of Dganganin Formation with *Ac.* (*Stockumites*) faunas (Barskov *et al.* 1984)
18. Alborz Mountains, Iran: probably the basal Mobarak Formation, limestones below the entry of *S.* (*Eo.*) *sulcata* (Habibi *et al.* 2008)
19. Vietnam, Cat Ba Island: dark limestone with red algae and lower part of dark limestone with *S.* (*Eo.*) *sulcata* (Beds 117–120, Komatsu *et al.* 2014)
20. Guangxi Province, South China: Lali section, limestones of upper Tangkou Member (Bed 104), at level of VI Zone (Ji & Ziegler 1993); ?Long'an section, basal Long'an Formation with *Protognathodus–Clydagnathus* fauna (Qie *et al.* 2015); Yishan, thin *kockeli* Zone within neritic Rongxian Formation (Wang *et al.* 1987); Nanbiancun, Beds 52/lower 53 (Yu 1988; widely overlooked update in Gong *et al.* 1991)
21. Guizhou Province, South China: Wangyou section, thin limestones of Gedongguan Bed with successive *kockeli* and *kuehni* faunas (Wu *et al.* 1987); thin limestones of Gedongguan Bed, Limushan, Muhua, Gedongguan and Dapoushang sections (Hou *et al.* 1985; Ji *et al.* 1989; Liu *et al.* 2012)
22. Central Hunan, South China: ?conodont-poor basal Malanbian Formation below the onset of *Siphonodella* faunas (Coen & Groessens 1996)
23. Guangdong, China: level of *Clydagnathus gilwernensis* Assemblage high in the lower part of the Menggongao Formation (Qin *et al.* 1988)
24. Kolyma, Russian Far East: Kamenka River section, successive levels with *Pr. kockeli* and *Pr. kuehni* (Gagiev & Kononova 1990)
25. ?Bonaparte Basin, northern Australia: basal Burt Range Limestone, *Clyd. plumulus* Zone (Druce 1969)

See also Figure 11.

Fig. 12. Field images of DCB outcrops. (**a–d**) Glacigenic sediments, Spechty Kopf Formation, eastern Pennsylvania, Appalachian Mountains, North America; uppermost Famennian, LN Biozone. Photos: (a) by S. I. Kaiser and (b–d) by M. Caputo.

icehouse pattern, e.g. from Quaternary glaciations (see the section 'Developments in time – the glaciation and regression').

It is possible that the minor sea-level fall at the top of the *praesulcata* Zone (*sensu* Kaiser *et al.* 2009) and within the LE Zone – the level of the Drewer Sandstone shown in Figure 2 – reflects a first and short-term glacial advance just before the onset of the HBS equivalents elsewhere (Streel 1999; Streel *et al.* 2000*a*). It precedes the main Hangenberg Extinction but may have triggered a first decline in some fossil groups (e.g. *Palmatolepis gonioclymeniae* among conodonts, the last glatziellids among ammonoids).

Anoxia and the global carbon cycle

Facies changes (e.g. from carbonate to siliciclastics), sedimentary gaps, highly condensed successions, reworking, diagenesis, weathering, taxonomic problems and the absence of index fossils previously hampered the search for characteristic isotope records across the Hangenberg Crisis (e.g. Xu *et al.* 1986; Schönlaub *et al.* 1992; Azmy *et al.* 2009). To be effective, chemostratigraphy requires the application of interdisciplinary methods accompanied with the highest time resolution of sampling (see discussion in Kaiser 2009; Kumpan *et al.* 2014, 2015; Carmichael *et al.* 2015). Changes in the global carbon cycle and hypoxia/anoxia, up to photic-zone euxinia, have been revealed by several geochemical and carbon isotope studies around the DCB in Europe, North America and Asia (Brand *et al.* 2004; Kaiser 2005; Buggisch & Joachimski 2006; Kaiser *et al.* 2006, 2008; Marynowski & Filipiak 2007; Trela & Malec 2007; Cramer *et al.* 2008; Clark *et al.* 2009; Matyja *et al.* 2010, 2014; Day *et al.* 2011; Königshof *et al.* 2012; Marynowski *et al.* 2012; Kumpan *et al.* 2013, 2014; Cole *et al.* 2015; Qie *et al.* 2015; see also the summary in Saltzman & Thomas 2012). The studies of Kaiser (2005), Kaiser *et al.* (2006, 2008), Kumpan *et al.* (2013) and Day *et al.* (2011) indicate two distinct positive isotope excursions, one in the *ck*I (HBS level) and one in the *kockeli* Zone (Figs 4, 13 & 14). These reflect enhanced burial of organic carbon-rich

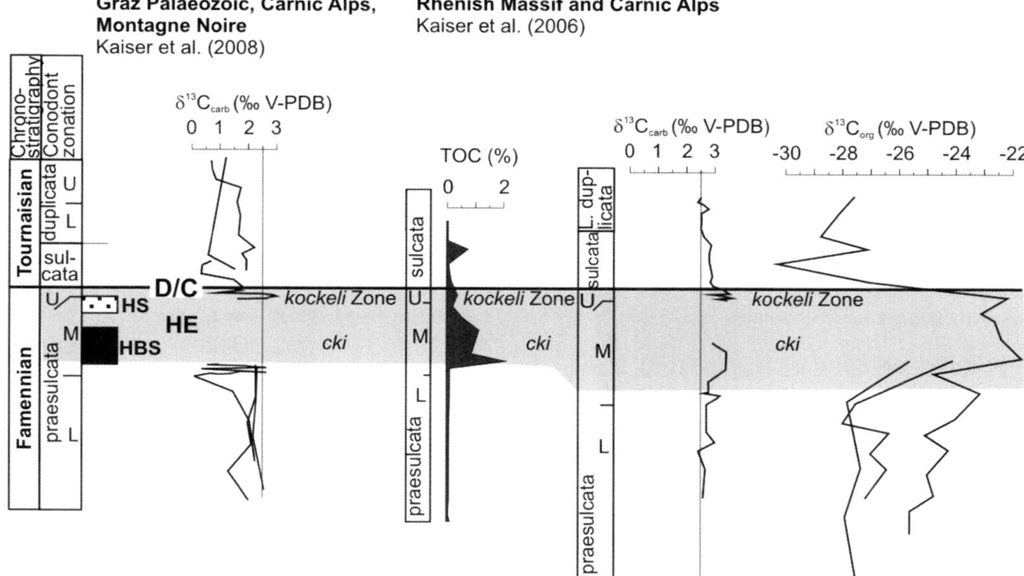

Fig. 13. Carbon isotopes ($\delta^{13}C_{carb,org}$) and TOC (total organic carbon) from DCB sections in Europe. Two distinct isotope excursions were found in the *ck*I and *kockeli* Zone during the Hangenberg Crisis.

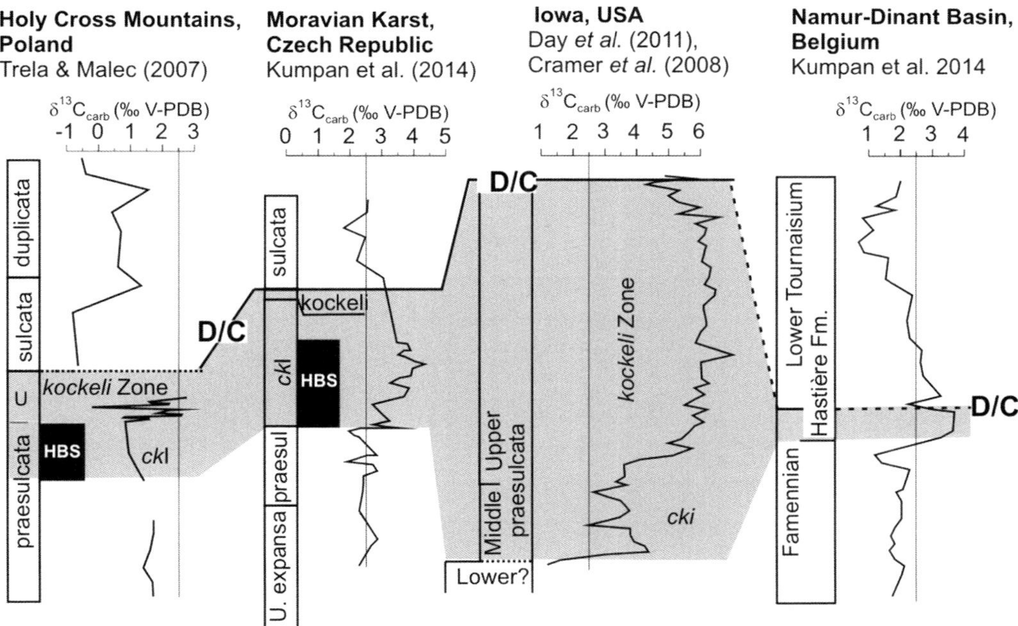

Fig. 14. Carbon isotopes ($\delta^{13}C_{carb}$) from DCB sections in Europe and North America. Two distinct isotope excursions were found in the *ck*I and/or *kockeli* Zone during the Hangenberg Crisis.

sediments during at least two different phases. Accordingly, a positive carbon isotope excursion of up to +4‰ $\delta^{13}C_{carb}$ and −21‰ $\delta^{13}C_{org}$ was measured in carbonates and sedimentary organic matter of the HBS and equivalents (*ck*I) of the Carnic Alps, Rhenish Massif and Moravian Karst (Kaiser 2005; Kaiser *et al.* 2006; Kumpan *et al.* 2013). The isotope excursion is accompanied by a high content of sedimentary total organic carbon (Kaiser *et al.* 2006; Fig. 13). A smaller spike was encountered in Hunan (less than +3‰ $\delta^{13}C_{carb}$, Bai *et al.* 1994) but is biostratigraphically less well constrained. A small to large negative isotopic excursion in HBS equivalents of South China (Guizhou: Muhua, Bai *et al.* 1994; Dapoushang, Ji *et al.* 1989; Guangxi: Huangmao, Bai *et al.* 1994) are best explained by diagenetic alteration (organic carbon oxidation and re-mobilization during calcite recrystallization, e.g. Qie *et al.* 2015).

A positive excursion of up to +6‰ $\delta^{13}C_{carb}$ and −22‰ $\delta^{13}C_{org}$ of the *kockeli* Zone (Figs 13 & 14) was found in limestones and brachiopods of the Holy Cross Mountains (Trela & Malec 2007; Malec 2014), Carnic Alps, Rhenish Massif, Graz Palaeozoic (Kaiser *et al.* 2006, 2008), Namur-Dinant Basin (Kumpan *et al.* 2014), Louisiana Limestone of Illinois and Missouri (Cramer *et al.* 2008; Clark *et al.* 2009; Day *et al.* 2011), and in several regions in China (Fig. 15; Qie *et al.* 2015). The *kockeli* excursion is time-equivalent to the transgression and the resumption of carbonate sedimentation (e.g. Stockum Limestone, Louisana Limestone, basal Hastière Formation) immediately after the glacial episode. It coincides with the initial radiation of some fossil groups (Fig. 3), such as conodonts, ammonoids, ostracodes and calcareous foraminifers, during re-warming (see the section 'Developments in time – the glaciation and regression').

Unfortunately, the exact timing of positive excursions is sometimes obscured by sedimentary perturbations associated with the eustatic fluctuations and by imprecise biostratigraphic dating (Fig. 15). Accordingly, brachiopods and ooids from the La Serre DCB stratotype section in the Montagne Noire produced positive carbon isotope values (Brand *et al.* 2004; Buggisch & Joachimski 2006), but the local strong reworking compromises the precise age of the measured fossils and sediment particles (see Flajs & Feist 1988; Casier *et al.* 2002; Brand *et al.* 2004; Kaiser *et al.* 2006; Kaiser 2009). A positive carbon isotope excursion in carbonates assigned to the Hangenberg Event interval was reported from Nevada, Utah (Great Basin; Saltzman 2005) and Colorado (Myrow *et al.* 2011, 2014). The peak values of more than 5‰ $\delta^{13}C_{carb}$ of the Great Basin are obviously not younger than the *expansa* Zone (Saltzman 2005) and more likely correlate with a minor but distinctive positive shift in the Middle/Upper *expansa* Zone during the Etroeungt and Epinette Events, as first described in Europe by Kaiser *et al.* (2008) and, more recently, by Kumpan *et al.* (2014). The distinctive isotope excursion in the Coffee Pot Member of the Dyer Formation, Colorado, first lacked a precise biostratigraphic dating (Myrow *et al.* 2011). New conodont data (Wistort *et al.* 2014) confirm that the position of peak values above a brecciated palaeokarst

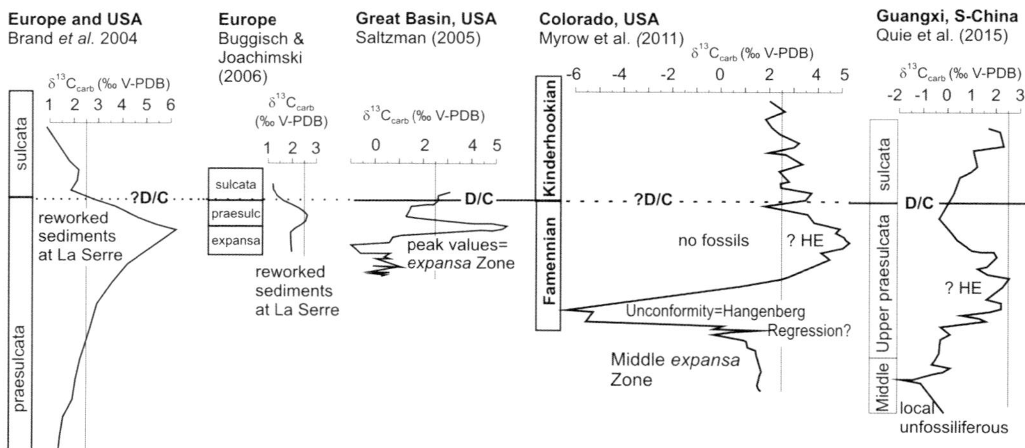

Fig. 15. Carbon isotopes ($\delta^{13}C_{carb}$) from DCB sections in Europe, North America and China. Peak values from Brand *et al.* (2004) and Buggisch & Joachimski (2006) originated from the North American Louisana Limestone and from oolithic limestones at the La Serre stratotype (France).

level could represent the (upper) Hangenberg Crisis interval.

The reconstruction of carbon isotope curves from carbonate is seriously hampered by the global carbonate crisis, with either gaps or predominant siliciclastics in the middle crisis interval. Most recently, Cole *et al.* (2015) published moderately high values, between +2 and +3‰, for the oncolitic marker unit within the Middle Sappington Formation of Montana and the Middle Pilot Shale of Utah. In the corresponding unit of the Middle Leatham Formation of Utah, values up to +4‰ were measured. Based on traditional regional conodont dating, Cole *et al.* (2015) assigned these units to pre-Hangenberg levels, although they overlie the regional HBS equivalent ('Conchostracan Shale') and although the ammonoid (widespread *Acutimitoceras* (*Stockumites*)), trilobite (*Pudoproetus*) and brachiopod (*Syringothyris*) faunas and recent miospore studies (Warren *et al.* 2014: upper LN Zone) demonstrate a post-extinction, middle/upper crisis interval age. The conodont faunas include unusual admixtures of species, which suggests reworking, but there are no index species of the *kockeli* Zone. The revised age of the oncolithic unit suggests that moderately positive values may regionally characterize the middle crisis interval. Only the subsequent Upper Sappington Member of Montana shows a prolonged moderately positive plateau, whilst there is no evidence for a higher positive excursion in the poorly dated Upper Pilot Shale and Upper Leatham Formation of Utah (Cole *et al.* 2015).

It is a general feature of carbon isotope stratigraphy that peak values differ between different regions (for a discussion, see Kumpan *et al.* 2014; Myrow *et al.* 2014; Qie *et al.* 2015), and therefore this feature cannot be used for interpretations concerning the exact timing and correlation of carbon isotope peaks. In South China, a positive carbon isotope peak measured in carbonates from several regions was assigned to the Upper *praesulcata* Zone (=*kockeli* Zone) by Qie *et al.* (2015; Fig. 15). In the absence of zonally diagnostic conodonts in the studied shallow-water sections, the positive excursions have been used as the correlation tool there. However, the risk of circular reasoning must be considered when inferring conodont ages from isotope data instead of using conodonts to date isotopic spikes.

Developments in time

The correlation of biostratigraphic, faunal, sedimentological and geochemical records demonstrates that the global Hangenberg Crisis was caused by a complex pattern of palaeoenvironmental changes. Four different phases of extinctions and abiotic changes can be recognized and have to be explained in any reasonable scenario: (1) the regressive minor prelude at the base of the lower crisis interval (Drewer Sandstone level in the Rhenish Massif or gaps and unconformities, e.g. in Morocco, Kaiser *et al.* 2011); (2) the transgressive main phase of hypoxia/anoxia (HBS level and equivalents), with the main marine extinction and increased carbon burial of the lower crisis interval; (3) the glaciation and regression of the middle crisis interval (HS and HSS level and equivalents); and (4) the post-glacial transgression (Stockum limestone level) with the second maximum of carbon burial, fluctuating (unstable) sea-level, the first faunal recovery, and survivor and terrestrial extinctions.

It is also important that the Hangenberg Crisis is only the second peak (subsequent to the global Kellwasser Crisis) of a complex succession of Devonian to Tournaisian global events, with many similarities between the two first-order crisis and second- to third-order smaller-scale extinctions and pantropical black shale episodes (e.g. House 1985, 2002; Becker 1993*a*; Walliser 1996). Any explanation and any realistic scenario requires abiotic developments that happened more frequently but culminated for some reason near the DCB. The better-preserved oceanic anoxic events (OAEs) of the Cretaceous greenhouse time can serve as guides to explain their Devonian–Lower Carboniferous counterparts, although the plate tectonic configuration was different in the mid-Palaeozoic.

The regressive prelude

The regressive trend of the initial crisis interval may reflect a first but still minor glacial pulse in the LE Zone (e.g. Streel 1999), but it has not been detected in reconstructed seawater temperatures based on oxygen isotope values of conodont phosphate from low latitudes. Only the Grüne Schneid section of the Carnic Alps indicates falling seawater temperature in the *praesulcata* Zone, but below the initial crisis interval (Kaiser *et al.* 2006). The prelude interval is characterized, at least in Europe, by warming (Kaiser *et al.* 2006; Kaiser 2007: oxygen isotopes from conodont phosphate of Grüne Schneid; De Vleeschouwer *et al.* 2013: Kowala calcite). However, the database is still small and not conclusive in other sections (Kaiser *et al.* 2006). It would be important to establish more detailed palaeotemperature profiles along a latitudinal gradient. The cosmopolitan Lepidophyta Floras, which had not yet started to decline, suggest a very equable global climate at that time (Streel & Marshall 2006). This pattern would be at odds with a first glaciation in South America and simultaneous warming in Europe. Streel (1999) explained the Rhenish Drewer

Sandstone by increased rainfall after the dry climate of most of the uppermost Famennian. The latter post-dates only slightly the first lake-forming humid event (LL-LE transition) of East Greenland, which was part of the Old Red Continent (Streel & Marshall 2006). Monsoonal activity may have fluctuated along a latitudinal gradient within the tropics/subtropics and independent from the high-latitude climate.

If the first glaciation started before the HBS, its triggering mechanism and gradual onset after more than 80 myr of global greenhouse climate and after the Middle Devonian extinction of the cold-water Malvinokaffric faunas by global warming (e.g. Troth *et al.* 2011) would be rather enigmatic. Explanations for short-lived glaciations that interrupted greenhouse times, such as a gamma-ray burst (Melott *et al.* 2004) or massive erosion of basaltic volcanites (Young *et al.* 2009), which have been invoked to explain the end-Ordovician glaciation, are rather speculative. The second hypothesis does not work for the Devonian–Carboniferous transition with respect to the very different strontium isotope trends as a record of weathering patterns (see strontium isotope curve of Veizer *et al.* 1997).

Carmichael *et al.* (2015) claim that the anoxic interval of the Hangenberg Crisis can be recognized in the northern Junggar Basin of Xinjiang, as a part of the Central Asian oceanic arc system, with the help of geochemistry but not by any black shale development. The identified level lies within the Heishantou Formation above a significant regression. If biostratigraphic data can confirm the correlation, then the regressive prelude would be very pronounced in parts of that region or it was regionally enhanced by tectonics.

The black shale and main extinction event

The *ck*I excursion of carbon isotopes indicates a global change in the isotopic composition of marine-dissolved inorganic carbon and atmospheric CO_2, which resulted from the massive burial of organic matter by the widespread deposition of the HBS and its equivalents in low-latitude shelf basins (Kaiser *et al.* 2006; Kumpan *et al.* 2013, 2015). The proximal cause for high organic carbon burial rates must have been enhanced bioproductivity during warming seawater temperatures, as indicated by oxygen isotope analyses of conodont phosphate (Kaiser *et al.* 2006; see Kuypers *et al.* 2002 for a Cretaceous OAE example). The ultimate causes of globally widespread black shales are discussed controversially, but a suddenly increased availability of nutrients is essential for blooms of organic carbon-forming primary producers. Most of the Rhenish HBS organic matter studied by Marynowski & Filipiak (2007) is amorphous and may have a cyanobacterial origin, as in Cretaceous OAE deposits (e.g. Kuypers *et al.* 2004; Karakitsios *et al.* 2007; van Bentum *et al.* 2012). Cyanobacterial blooms reflect a peculiar style of nutrient use and had the potential to alter the open-marine ecosystem structure considerably, with possibly severe consequences for plankton consumers and the higher food web. Even more significant is the geochemical evidence of green sulphur bacteria in HBS equivalents of the Holy Cross Mountains (Marynowski & Filipiak 2007) because they prove that anoxia reached regionally the photic zone. However, in Rhenish sections the HBS was not fully anoxic (Kumpan *et al.* 2015).

Tropical outer-shelf eutrophication may have been caused by climate- and salinity-driven upwelling events that inverted the oceanic stratification contemporaneously but in many different shelf basins. This is the model of climate-controlled anoxic overturns in the oceans. It could occur much more easily than today, because the vertical temperature gradients were much lower in greenhouse times. Rather high and increasing temperatures of bottom water have been proven for Cretaceous OAEs (Huber *et al.* 2002; Gustafson *et al.* 2003). The sinking of warm and salty shelf water, formed during surface heating and increased evaporation, provides the driving mechanism for deeper-water inversion (see Friedrich *et al.* 2008). Murphy *et al.* (2000) and Sageman *et al.* (2003) showed how high productivity levels of Devonian black shales could have been maintained for a significant time by efficient oceanic nutrient recycling, independent from continental influxes.

Alternatively, enhanced coastal (Perkins *et al.* 2008; see also Piper & Calvert 2009) and equatorial (Caplan & Bustin 2001) upwelling or high primary productivity due to an elevated external nutrient input and fluxes of terrestrial runoff were proposed to result in Upper Devonian black shale formation (Algeo & Scheckler 1998; Rimmer *et al.* 2004; 'top-down' model of Carmichael *et al.* 2015). These authors postulated also that the spreading of land plants during warm–humid climates at the end of the Devonian, before the short latest Devonian glacial interval, triggered increasing continental weathering and a higher nutrient flux. However, the spread of land plants with deep root complexes may have kept nutrients in the soil rather than released them (see Boucot & Gray 2001).

Furthermore, there is no quantitative evidence in the spore and macrofossil record for a sudden and significant increase of land vegetation precisely at HBS time. The miospore content of the HBS is very sparse in the Rhenish Massif, both in terms of abundance and diversity (Higgs & Streel 1994). In contrast to many other Famennian black shales, we are not aware of any trunks and other large plant remains in the HBS, despite their much improved preservation potential in dysoxic to anoxic facies. Based on analyses of the Huron Shale Member of the Ohio Shale (upper Famennian) and the Sunbury Shale (middle Tournaisian), multiple factors, such as increasing productivity and high nutrient supply related to enhanced terrestrial weathering, were proposed to have influenced the regional accumulation of organic matter (Rimmer *et al.* 2004), but this study provides no data for the locally very thin HBS equivalents of the basal Bedford Shale. A significant role for upwelling has been shown for other Famennian black shales of North America (Smith & Bustin 1998). For example the Exshaw and Cleveland shales were formed in upwelling regimes (Robl & Barron 1988; Caplan & Bustin 2001), although in the second case at a time of high influx of terrigenous organic matter (Rimmer *et al.* 2004).

It is easily possible that nutrients from different sources led regionally to eutrophication. However, it is important to note that the HBS occurs mostly in offshore carbonate platform settings, far away from rivers that would have imported diluted nutrients and terrestrial organic matter. The Appalachian foreland with its significant black shales west of an erosive mountain belt is not the right model for most of the HBS settings. It is intriguing that it is difficult to spot HBS equivalents in eastern North America. In the nearshore neritic environments of Europe, North Africa and South China, where land-derived nutrients should have arrived first and more constantly, there is little evidence of plankton blooms or black shales (as would be predicted in nutrient-loaded, coastal dead zones, McGlathery *et al.* 2007). Since HBS equivalents occur even on isolated pelagic platforms, far away from land that could have provided an increased nutrient influx (e.g. within the western Prototethys: Carnic Alps, Sardinia; eastern Prototethys: Vietnam), the climatically driven oceanic overturn model offers a better HBS explanation than increased erosion and a dominant terrestrial nutrient source, the 'top-down' model of nutrient flux by Carmichael *et al.* (2015). It requires vertical nutrient transport but does not imply a rise and spread of oxygen minimum zones with transgression, the typical 'bottom-up' model criticized by Carmichael *et al.* (2015). In the Rhenish type region the pre-crisis Famennian basinal facies ('Cypridinenschiefer') is a red (oxic) or green (oxic to slightly dysoxic) shale and, as in many other regions, there is no evidence for a black shale facies moving up the shelf slope.

An intensified release of atmospheric CO_2 due to enhanced seafloor spreading and associated volcanism has been suggested to result in climate warming and sea-level rises in the Upper Devonian. In combination with episodes of tectonic uplift this may have caused enhanced weathering fluxes (Van Geldern *et al.* 2006). However, there is no evidence for major uplift events at the time of the Hangenberg Crisis, certainly not in the Rhenish type region of the HBS. Evidence for close links between significant volcanism and intrabasinal tectonic movements, which resulted in the subsequent formation of massive sulphide deposits within equivalents of the HBS, were provided by González *et al.* (2006) for SW Spain. Hao (2001) showed that the DCB beds of the Tarim Basin of NW China consist of alternating basaltic tuffs and evaporites. The latter highlight the terminal Devonian arid conditions near the equator. In the Holy Cross Mountains, the HBS equivalent is sandwiched between thin ash layers (e.g. Myrow *et al.* 2014). However, there was no major flood basalt province at the time (see latest review by Bond & Wignall 2014), and a volcanic trigger of end-Devonian climatic warming needs a much better documentation. If massive volcanic degassing occurred in the giant Panthalassia ocean, the evidence may have been completely destroyed by subsequent subduction. Adams *et al.* (2010) proposed that the Cretaceous OAE2 was triggered similarly by volcanism. An associated considerable sulphate release had the potential to increase significantly oceanic nutrient recycling (carbon remineralization) and primary productivity but would also change the oceanic pH values. Their model provides a link (cascade) between climatic warming, black shale formation, isotope spikes and the cessation of carbonate deposition due to seawater acidification.

The transgressive nature of the HBS is undoubted and in agreement with palynofacies data (Higgs & Streel 1994; Streel 1999; Marynowski & Filipiak 2007) and with the interpretation of the Cretaceous OAEs (e.g. Grötsch *et al.* 1998). In southern Morocco the HBS may overlie unconformities (Kaiser *et al.* 2011). In the Rhenish Massif and Holy Cross Mountains it is characterized by decreasing terrestrial influence. A still underexplored aspect is the possibility that the HBS and other Famennian anoxic events resulted at least partly from peak interferences (insolation nodes) of Milankovitch cycles, as has been proposed for the OAE2 (Mitchell *et al.* 2008). The uppermost Famennian and lower Tournaisian pelagic seamount facies of Germany (both in the Rhenish Massif and in Thuringia) and the Holy Cross Mountains is strongly cyclical (e.g. Bartzsch & Weyer 1982; Korn & Weyer 2003; De Vleeschouwer *et al.* 2013; Kononova & Weyer 2013). Interruptions by some marker shales may represent minor cycle nodes before and after the HBS. Detailed geochemistry also revealed the cyclical nature of uppermost

Famennian and lower Tournaisian strata of the Ardennes (Kumpan *et al.* 2014). De Vleeschouwer *et al.* (2013) suggested that it was especially the intensity of eccentricity cycles that determined high-latitude warming, which could explain both the initial warming and the re-warming. If volcanic degassing and strong eccentricity coincided, this could explain why the HBS was more severe than the previous (Dasberg Crisis) and later (Lower Alum Shale Event) black shale events.

The main extinction agent for pelagic biota was the sudden oxygen deficiency reaching the upper water column, probably in combination with changes of ocean chemistry (changes of salinity and temperature gradients, seawater acidification) and the marine food web. Details still have to be worked out. The carbonate crisis is evident in most successions (see Becker *et al.* 2013*b*; Kumpan *et al.* 2013). HBS equivalent limestones are extremely condensed and restricted to very few localities (Kaiser *et al.* 2006, 2009). The DCB carbonate crisis forms a strong contrast to the Kellwasser Crisis, where black limestones were deposited widely across the Frasnian–Famennian boundary.

The glaciation and regression

High organic carbon burial rates during the HBS deposition in the middle crisis interval may have resulted in a significant lowering of the atmospheric pCO_2 and climatic cooling, subsequent worldwide regression, and finally a potentially biologically triggered glaciation on Gondwana (Kaiser *et al.* 2006). For OAE2 at the Cenomanian–Turonian boundary, Kuypers *et al.* (1998) suggested a 50–90% decrease in atmospheric CO_2 levels related to massive organic carbon burial. HBS and OAE2 include comparable amounts of black shales and the impressive Cretaceous CO_2 values can probably be transferred to an end-Devonian model. But in the case of the DCB, organism blooms obviously not only caused a subsequent climatic cooling but even a major glaciation. Streel (1999) stressed that the development of wet conditions in high latitudes may have been more significant than a temperature drop to facilitate the growth of large ice sheets. It is difficult to construct a link between organic burial and CO_2 levels on one side and the latitudinal distribution of rain/snowfall on the other side, but a poleward export of humidity must have taken place during the transition from the HBS to the subsequent glacial phase.

Adams *et al.* (2010) suggested that the stabilization of volcanogenic sulphate levels by pyrite formation and burial would decrease the nutrient recycling, which provides an autocyclic mechanism to end eutrophication cells, high organic productivity and black shale deposition. Regionally this happened very fast, as exemplified in the Holy Cross Mountains by the sudden change from HBS equivalents with overwhelmingly marine, mostly amorphous organic matter to an overlying marl with exclusively terrigenous organic particles (Marynowski & Filipiak 2007).

The globally widespread unconformities and erosion episodes were caused by the glacio-eustatic sea-level fall in the scale of up to 100 m or more (Kaiser *et al.* 2011). There is still a dearth of palaeotemperature data for the middle crisis interval. The minimum temperatures found by Brand *et al.* (2004) in brachiopods from the upper part of Bed 82 (Middle Siliciclastic–Calcareous Unit) at La Serre are not well constrained biostratigraphically but suggest that the cooling peaked near the end of the *ck*I, followed by a very fast subsequent rebound (conodont phosphate data from Grüne Schneid, Kaiser *et al.* 2006). In the Kowala succession of the Holy Cross Mountains, which is characterized by very low thermal alteration, isotope values of calcite suggest the lowest seawater temperatures in the middle of the *ck*I, just above the HBS (De Vleeschouwer *et al.* 2013; middle of Unit B, Malec 2014). A much better database covering sections from different palaeolatitudes and of varying palaeobathymetry is required to understand the DCB palaeotemperature trends.

As discussed earlier, a contribution of major Australian impact events to sudden cooling cannot be ruled out, but this is currently hypothetical. Studies of Upper Eocene impacts (Vonhof *et al.* 2000), showed that impact cooling is a very short-lived phenomenon and it requires feedback mechanisms to have longer-lasting effects in the scale of 100 kyr and more.

The post-glacial transgression, organic burial and late extinctions

The transgression in the *kockeli* Zone (Upper *praesulcata* Zone) can be related to the final meltdown of the large Gondwana ice sheets that had reached the sea and of low-latitude mountain glaciers during re-warming. On the previously arid Old Red Continent, close to the palaeoequator, a strengthening of the monsoon caused major lakes to fill quickly (Marshall *et al.* 2002). But the re-warming needs to be explained. During the preceding peak of low sea-level, large lowlands (former shelf areas) became exposed, and erosion of organic matter (including HBS reworking) and carbonate platforms took place. In other areas large delta systems prograded and eroded into underlying marine carbonates (Kaiser *et al.* 2011). The consequent release of CO_2 into the atmosphere may have ended the geologically short icehouse episode autocyclically, especially because carbonate deposition (burial) was arrested at the same time and low shelf productivity after the nutrient recycling slowed with pyrite burial (model of Adams *et al.* 2010). The re-warming in the *kockeli* Zone is indicated by still-restricted data of $\delta^{18}O_{phosph}$ with a suggested temperature increase of 4°C (Kaiser *et al.* 2006, 2008) and by the complete disappearance of glacigenic sediments. The melting of ice-sheets in southern and western Gondwana resulted in a fast-rising sea-level and coastal erosion. Eventually, the combination of increased terrestrial nutrient influx due to strong erosion and the re-warming led to the re-onset of carbonate production and high productivity. On platforms far away from land masses nutrient recycling may have restarted or new upwelling zones may have formed due to new changes in the vertical temperature gradients. This complex scenario produced $\delta^{13}C$ values similar to those of the preceding main Hangenberg extinction event in the *ck*I. But the second spike in the *kockeli* Zone coincided only locally with the deposition of thin black shales (e.g. Hasselbachtal, Becker *et al.* 1984; Drewer, Korn *et al.* 1994; southern Morocco, Kaiser *et al.* 2011) and the isotopic excursion may be stretched over thicker successions. This difference indicates that the trigger mechanism of both isotope excursions was not identical. In any case, the isotopic excursion in the *kockeli* Zone can be referred to the post-glacial eustatic sea-level rise (Kaiser *et al.* 2011; Qie *et al.* 2015). Unfortunately, available strontium isotope data, which could reveal changes of continental erosion rates, are still too episodic and combine mixed data from widely different basins (Brand *et al.* 2004: brachiopod shells) or they reveal a strong diagenetic overprint (Kürschner *et al.* 1993: conodont phosphate).

The 'Rhenish Standard Succession', with its distinctive minor unconformities, suggests that the sea-level oscillated in the upper crisis interval, which probably reflects climatic fluctuations (Bless *et al.* 1993; Streel 1999). This interval continued into the basal-most Carboniferous as currently defined. A regression right at the DCB (Fig. 2) seems to have been of special importance, since it corresponds to the significant survivor extinctions in ammonoids, trilobites, brachiopods and foraminifers (end-*kockeli* Zone). The terrestrial extinction was roughly contemporaneous. But so far there is no explanation why the rather minor DCB regression could kill both shallow (e.g. large-eyed phacopids) and deeper (e.g. cymaclymeniids) marine taxa that persisted through the much more profound climatic and oceanographic perturbations of the preceding main crisis. Similarly, Marshall *et al.* (2013) remarked on the disappearance of trees on the Old Red Continent even as the climate began to be warm and wet again. Much more future attention should be given to the final episode of the Hangenberg Crisis, because it shaped post-crisis ecosystems as importantly as the main extinction level by wiping out the last representatives of higher clades, which otherwise could have recovered.

Conclusions

(1) The global Hangenberg Crisis was a first-order mass extinction in the scale of the 'Big Five' extinctions. It was at least as severe as the global Kellwasser Crisis at the Frasnian–Famennian boundary.

(2) It was a prolonged biocrisis that lasted several 100 kyr and affected numerous fossil groups of marine and terrestrial realms, partly at different times. All ecosystems were affected, and many long-ranging groups became totally extinct.

(3) Two ecosystems vanished completely and did not recover before the middle/upper Tournaisian: the marine reefs and the *Archaeopteris* forests on land.

(4) Several long-ranging, higher-level animal clades died out completely: the stromatoporoid sponges, the phacopid trilobites, chitinozoans, placoderms and Ichthyostegalia. Among the ammonoids all post-Devonian families go back to just one survivor group. The term 'Romer's Gap' highlights the still significant record gap for tetrapods across the DCB.

(5) Conodonts, sharks, proetid trilobites, ostracodes, brachiopods, corals, acritarchs and foraminifers suffered at the lower taxonomic level.

(6) Pelagic extinction patterns are similar in widely separate basins of the western and eastern Prototethys, but survival was higher for several shallow-water groups (e.g. bivalves, brachiopods, bryozoa).

(7) In many other groups the database is still very crude and there is no review of taxon ranges at the required fine biostratigraphical scale.

(8) The crisis can be subdivided into clearly defined lower, middle and upper parts. As a prelude the lower crisis interval started at the top of the *praesulcata* Zone (upper UD VI-D, LE Zone) with a minor sea-level fall, which may reflect an initial, still minor glacial phase.

(9) During climatic warming and transgression, in many low latitude outer-shelf settings, contemporaneous eutrophication caused blooms of primary producers (e.g. cyanobacteria, green sulphur bacteria), the deposition of black muds (HBS, lower *ck*I, UD VI-E,

lower LN Zone), a massive burial of organic carbon, a positive carbon isotope excursion, and the onset of hypoxia/anoxia and local euxinia reaching the photic zone, caused by the bacterial degradation of organic matter, which means strong oxygen consumption.

(10) The black shale event was probably a consequence of climate/salinity-driven synchronous overturns and sustained nutrient recycling in open-marine settings that mostly were far away from land and erosive mountain ranges. There is no evidence for oxygen minimum zones migrating upslope with transgression.

(11) In the HBS interval there is no palaeobotanical evidence for a suddenly increased vegetation cover or increased soil erosion that could have caused a sudden, significant discharge of land-derived nutrients.

(12) The global carbonate crisis and the marine mass extinction coincided with the anoxic event but certainly involved not only oxygen deficiency but also other palaeoceanographic factors that delimit living conditions of biota, such as ocean acidity, temperature and salinity changes, gradients of these in the water column, rapid drowning of shallow habitats, seasonality, and changes of the food web structure.

(13) An ultimate volcanogenic trigger of the warming, associated with a significant outgassing of carbon and sulphur dioxide, can be postulated, but there is no preserved record of a major DCB volcanic province. If it was positioned in the giant Panthalassia Ocean the evidence may have been lost. Interference ('nodes') of Milankovitch cycles was possibly a different/additional but decisive trigger for climate warming in the lower crisis interval.

(14) Alleged geochemical evidence for impact signatures in southern Chinese sections as well as the possible DCB age of major impact craters in Australia require further research.

(15) The middle crisis interval (upper *ck*I, LN Zone, lower UD VI-F) is defined by the formation of major ice sheets in South America and South Africa and of mountain glaciers in low latitudes (Appalachians: Spechty Kopf; Hoggar: Tahara Formation).

(16) The sudden glaciation after more than 80 myr of largely global greenhouse climate was probably triggered by the massive drawdown of atmospheric CO_2 (probably $>50\%$) due to the HBS organic carbon burial. It led to a major glacio-eustatic sea-level fall, probably on the scale of up to 100 m or more, the widespread progradation of shallow-water siliciclastics, and common unconformities due to reworking and non-deposition (end-Devonian sequence boundary).

(17) The upper crisis interval (*kockeli* Zone) is characterized by initial post-glacial transgression, the widespread re-onset of carbonate deposition, a second carbon isotope spike, opportunistic blooms (e.g. *Protognathodus* conodont biofacies) and the early re-radiation of several marine fossil groups.

(18) The re-warming may have been an autocyclic response to the increased recycling of organic carbon and carbonate weathering on the large exposed shelf areas while organic productivity was low and when almost no CO_2 was stored as carbonate.

(19) Both the increased availability of land-derived and recycled nutrients and new upwelling cells can explain the second isotope excursion, which was linked in low latitudes with only minor local black shale deposition.

(20) Minor reworking events and unconformities suggest continuing oscillations of global sea-level and palaeoclimate in the upper crisis interval (Stockum levels; *Pr. kockeli* to *Pr. kuehni* interval). They probably correlate with heretofore poorly understood final extinctions of the last clymeniid ammonoids, phacopid trilobites, placoderms and some widespread brachiopod and foraminifera groups (*Quasiendothyra*).

(21) Correlation of the marine survivor extinctions with the terrestrial floral change (LN/VI Zone boundary) was hardly a coincidence, but the link between the two developments is not yet clear. Near the equator the fate of arid climate-adapted vegetation was sealed by strong monsoonal activity and drowning.

(22) The post-crisis lower Tournaisian is marked by continuing eustatic rise, which caused for example the flooding of the Old Red Continent and NW Gondwana and expanding shelf seas in the high latitudes of South America and South Africa.

(23) The significant radiation in a renewed greenhouse time had not yet reached the pre-crisis diversity level when it was interrupted by the global, second-order, transgressive and anoxic Lower Alum Shale Event at the base of the middle Tournaisian. The Lower Alum Shale Event bears important similarities with the HBS and most likely reflects a repetition of at least some of its triggering mechanisms.

Multidisciplinary and high-resolution approaches are essential to achieve further progress in the understanding of the 'Sixth Phanerozoic Mass Extinction'. The idea of biotically triggered glaciations deserves considerably more research effort. Major open questions are: the currently very low resolution of data for several major fossil groups; refined dating of last representatives of major clades (e.g. last placoderms, therein of the last titanichthyids); neglected possible impact signatures; evidence for a volcanic origin of the initial climatic warming; the HBS nutrient cycle; an autecological understanding of survival and extinctions; a better, more complete and more detailed calculation of diversity fluctuations (local, regional and global curves; data along palaeolatitudinal gradients); more geochemical data that provide ideas on palaeotemperatures and palaeosalinity (regional v. global patterns, for benthic and surface dwellers); the causes and timing of the survivor extinctions; the palaeoecology of the terrestrial environmental change; and the precise timing of the recovery (e.g. precise age of the first Carboniferous reefs). The data available so far permit the formulation of various hypotheses and our new space–time model for abiotic change and biotic responses and interactions at the DCB. But there are still more open questions than solid knowledge.

The manuscript benefits from a number of discussions with colleagues working at the DCB. We acknowledge the useful comments, suggestions and corrections by G. Baird, L. C. Sallan, and C. E. Brett. A special thanks is given to E. Poty (Liège) for his support and steady interest in the subject. RTB conducted recent DCB research in Morocco and the Rhenish Massif in the frame of DFG Project Be 1367/11-1.

References

Abdesselam-Rouighi, F.-F. & Coquel, R. 1997. Palynologie du Dévonien terminal-Carbonifère inferieur dans le sud-east du Bassin d'Illizi (Sahara Algerien). Position des premieres lycospores dans la serie stratigraphique. *Annales de la Société Géologique du Nord*, **5**, 47–57.

Adams, D. D., Hurtgen, M. T. & Sageman, B. B. 2010. Volcanic triggering of a biogeochemical cascade during Oceanic Anoxic Event 2. *Nature Geoscience*, **3**, 201–204.

Afanaseva, G. A. 2002. Brakhiopody otryada Chonetida iz basseynovykh fatsiy pogranichnykh otlozheniy devona í karbona Tyuringskikh i Reynskikh Slantsevykh Gor (Germaniya). *Paleontologicheskiy Zhurnal*, **2002**, 57–62.

Alekseev, A. A., Lebedev, O. A., Barskov, I. S., Barskova, M. I., Kononova, L. I. & Chzhova, V. A. 1994. On the stratigraphic position of the Famennian and Tournaisian fossil vertebrate beds in Andreyevka, Tula region, Central Russia. *Proceedings of the Geologist's Association*, **105**, 41–52.

Alekseev, A. S., Barskov, N. S. & Kononova, A. I. 1979. Conodonts of Famennian-Tournaisian boundary deposits from the central region of the Russian Platform. *Service géologique de Belgique, Professional Paper*, **1979**, 52–58.

Algeo, T. J. & Scheckler, S. E. 1998. Terrestrial-marine teleconnections in the Devonian: links between the evolution of land plants, weathering processes, and marine anoxic events. *Philosophical Transactions of the Royal Society of London, (B): Biological Sciences*, **353**, 113–130.

Algeo, T. J., Berner, R. A., Maynard, J. B. & Scheckler, S. E. 1995. Late Devonian oceanic anoxic events and biotic crises: 'rooted' in the evolution of vascular plants. *GSA Today*, **5**, 63–66.

Almond, J., Marshall, J. & Evans, F. 2002. Latest Devonian and earliest Carboniferous glacial events in South Africa. *In*: *16th International Sedimentological Congress*. Rands Afrikaans University, Johannesburg, Abstract Volume, 11–12.

Amenabár, C. R., di Pasquo, M. & Azcuy, C. L. 2009. Palynofloras of the Chigua (Devonian) and Malimán (Mississppian) formations from the Precordillera Argentina: age, correlation and discussion of D/C boundary. *Revista España de Micropaleontologia*, **41**, 217–239.

Amler, M. R. W. 1993. Shallow marine bivalves at the Devonian–Carboniferous boundary from the Velbert Anticline (Rhenisches Schiefergebirge). *Annales de la Société de géologique Belgique*, **115**, 405–425.

Amler, M. R. W. 1995. Die Bivalvenfauna des Oberen Famenniums West-Europas. 1. Einführung, Lithostratigraphie, Faunenübersicht, Systematik 1. Pteriomorphia. *Geologica et Palaeontologica*, **29**, 19–143.

Amler, M. R. W., Thomas, E., Weber, K. M. & Stephan, W. 1990. Bivalven des höchsten Oberdevons im Bergischen Land (Strunium; nördliches Rheinisches Schiefergebirge). *Geologica et Palaeontologica*, **24**, 41–63.

Antoshkina, A. I. 1998. Organic buildups and reefs on the Palaeozoic carbonate platform margin, Pechora Urals, Russia. *Sedimentary Geology*, **118**, 187–211.

Aretz, M. & Chevalier, E. 2007. After the collapse of stromatoporid sponge–coral reefs – The Famennian and Dinantian reefs of Belgium: much more than Waulsortian mounds. *In*: Álvaro, J.-J., Aretz, M., Boulvain, F., Munnecke, A., Vachard, D. & Vennin, E. (eds) *Palaeozoic Reefs and Bioaccumulations: Climatic and Evolutionary Controls*. Geological Society, London, Special Publications, **275**, 163–188, http://doi.org/10.1144/GSL.SP.2007.275.01.11

Aretz, M. & Webb, G. E. 2007. Western European and eastern Australian Mississippian shallow-water reefs: a comparison. *In*: Wong, T. E. (ed.) *Proceedings of the XVth International Congress on Carboniferous and Permian Stratigraphy*, 10–16 August 2003, Utrecht, Netherlands. Royal Netherlands Academy of Arts and Sciences, Amsterdam, 433–441.

Aretz, M., Nardin, E. & Vachard, D. 2014. Diversity patterns and palaeobiogeographical relationships of latest Devonian-Lower Carboniferous foraminifers from South China: what is global, what is local? *Journal of Palaeogeography*, **3**, 35–59.

Aristov, V. A. 1988. Devonian conodonts of the Central Devonian field (Russian Platform). *Academy of Sciences of the USSR, Transactions*, **432**, 1–119 [in Russian].

Armstrong, A. K., Mamet, B. L. & Repetski, J. E. 1980. The Mississippian System of New Mexico and southern Arizona. *In*: Fouch, T. D. & Magatham, E. R. (eds) *Palaeozoic Paleogeography of West-Central United States*. Society for Sedimentary Geology, Rocky Mountain section, Denver, CO, 82–99.

Artyushkova, O. V., Maslov, V. A., Pazukhin, V. N., Kulagina, E. I., Tagarieva, R. C., Mizenz, L. I. & Mizenz, A. G. 2011. Devonian and Lower Carboniferous type sections of the western South Urals. *In*: *International Conference 'Biostratigraphy, Palaeogeography and Events in Devonian and Lower Carboniferous', in Memory of Evgeny A. Yolkin, Pre-Conference Field Excursion Guidebook*. Institute of Geology, Ufa Scientific Center, 20 July–10 August 2011, Ufa, Novosibirsk, Russia, 1–62.

Atta-Peters, D. & Anan-Yorke, R. 2003. Latest Devonian and Early Carboniferous pteridophytic spores from the Sekondi Group of Ghana. *Revista Española de Micropaleontologia*, **35**, 9–27.

Austin, R. L. & Hill, P. J. 1973. A Lower Avonian (K Zone) Conodont Fauna from near Tintern, Monmouthshire, Wales. *Geologica et Palaeontologica*, **7**, 234–234.

Austin, R. L., Conil, R. *et al.* 1970*a*. Transitional Devonian/Carboniferous Sequences between Hook Head (Ireland) and Bohlen (D.D.R.). *In*: Streel, M. & Wagner, R. H. (eds) *Colloque sur la Stratigraphie du Carbonifère*. Les Congres et Colloques de ĺUniversité de Liege, **55**, 172–178.

Austin, R. L., Druce, E. C., Rhodes, F. H. T. & Williams, J. A. 1970*b*. The value of conodonts in the recognition of the Devonian–Carboniferous boundary, with particular reference to Great Britain. *In*: *Compte Rendu Sixième Congrès International de Stratigraphie et de Géologie du Carbonifère*, 11–16 September 1967, Sheffield, **II**, 431–445.

Avchimovitch, V. I., Turnau, E. & Clayton, G. 1993. Correlation of uppermost Devonian and Lower Carboniferous miospore zonations in Byelorussia, Poland and western Europe. *Annales de la Société géologique de Belgique*, **115**, 453–458.

Avchimovitch, V. I., Byvcheva, T. V., Higgs, K., Streel, M. & Umnova, V. T. 1988. Miospore systematics and stratigraphic correlation of Devonian–Carboniferous Boundary deposits in the European part of the USSR and western Europe. *Courier Forschungs-Institut Senckenberg*, **100**, 169–191.

Azmy, K., Poty, E. & Brand, U. 2009. High-resolution isotope stratigraphy of the Devonian–Carboniferous boundary in the Namur–Dinant Basin, Belgium. *Sedimentary Geology*, **216**, 117–124.

Bahrami, A., Corradini, C. & Yazdi, M. 2011. Upper Devonian-Lower Carboniferous conodont biostratigraphy in the Shotori Range, Tabas area, Central-East Iran Microplate. *Bolletino della Società Paleontologica Italiana*, **50**, 35–53.

Bai, S.-L. & Ning, Z.-S. 1989. *Faunal Change and Events across the Devonian–Carboniferous Boundary of Huangmao Section, Guangxi, South China*. Canadian Society of Petroleum Geologists, Calgary, Alberta, Memoir, **14**, 147–157.

Bai, S.-L., Ning, Z.-S. & Orth, C. J. 1987. Zonation and geochemical anomaly of the Devonian/Carboniferous boundary beds of Huangmao, Guangxi. *Acta Scientarum Naturalium Universitatis Pekinensis*, **1987**, 105–111 [in Chinese with English summary].

Bai, S.-L., Bai, Z. Q., Ma, X. P., Wang, D. R. & Sun, Y. L. 1994. *Devonian Events and Biostratigraphy of South China*. Peking University Press, Beijing.

Baird, G. C., Carr, R. K., Hannibal, J. T., Brett, C. E. & Brett, B. L. 2009. Uppermost Devonian (Famennian) stratigraphy of the Cleveland area, northeastern Ohio. *In*: Brett, C. E., Bartholomew, A. J. & DeSantis, M. J. (eds) *Middle and Upper Devonian Sequences, Sea-Level, Climatic and Biotic Events in East-Central Laurentia: Kentucky, Ohio, and Michigan*. Field Trip Guidebook for North American Paleontological Convention – 2009.

Barskov, I. S., Simakov, K. V. *et al.* 1984. Devonian–Carboniferous transitional deposits of the Berchogur section, Mugodzary, USSR. *Courier Forschungsinstitut Senckenberg*, **67**, 207–230.

Bartzsch, K. & Weyer, D. 1982 (for 1981). Zur Stratigraphie des Untertournai (Gattendorfia-Stufe) von Saalfeld im Thüringischen Schiefergebirge. *Abhandlungen, Berichte, Naturkunde und Vorgeschichte*, Magdeburg, **12**, 3–53.

Bartzsch, K. & Weyer, D. 1986. Biostratigraphie der Devon/Karbon-Grenze im Bohlen-Profil bei Saalfeld (Thüringen, DDR). *Zeitschrift für Geologische Wissenschaften*, **14**, 147–152.

Bartzsch, K., Blumenstengel, H. & Weyer, D. 1995. Ein neues Devon/Karbon-Grenzprofil am Bergaer Antiklinorium (Thüringer Schiefergebirge) – eine vorläufige Mitteilung. *Geowissenschaftliche Mitteilungen von Thüringen*, **3**, 13–29.

Bartzsch, K., Hahne, K. & Weyer, D. 1998. Der Hangenberg-Event (Devon/Karbon-Grenze) im Bohlen-Profil von Saalfeld (Thüringisches Schiefergebirge). *Abhandlungen und Berichte für Naturkunde*, **20**, 37–58.

Bartzsch, K., Blumenstengel, H. & Weyer, D. 1999. Stratigraphie des Oberdevons im Thüringischen Schiefergebirge. Teil 1: Schwarzburg-Antiklinorium. *Beiträge zur Geologie Thüringens, N.F.*, **6**, 159–189.

Bartzsch, K., Blumenstengel, H. & Weyer, D. 2001. Stratigraphie des Oberdevons im Thüringischen Schiefergebirge. Teil 2: Berga-Antiklinorium. *Beiträge zur Geologie Thüringens, N.F.*, **8**, 303–327.

Bartzsch, K., Gaitzsch, B., Abel, P., Hahne, K. & Weyer, D. 2015. Oberdevon und Unterkarbon im Raum Saalfeld. *In*: *Exkursionsführer, Subkommissionen Devon und Karbon der Deutschen Stratigraphischen Kommission*, 25 April 2015, Saalfeld, 23.04. bis 1–48.

Beatty, T. W. 2002. New geological and paleontological data from the Harper Ranch Group, Kamloops, British Columbia. *Geological Survey of Canada, Current Research*, **2002**, 1–9.

Beatty, T. W. 2003. *Stratigraphy of the Harper Ranch Group and Tectonic History of the Quesnel Terrane in the Area of Kamloops, British Columbia*. MSc thesis, Department of Earth Sciences, Simon Fraser University.

Becker, G. & Blumenstengel, H. 1995. The importance of the Hangenberg event on ostracod distribution at the DCB in the Thuringian and Rhein. Schiefergebirge. *In*: Riha, J. (ed.) *Ostracoda and Biostratigraphy*. Balkema, Rotterdam, 67–78.

Becker, R. T. 1993*a*. Anoxia, eustatic changes, and Upper Devonian to Lowermost Carboniferous global ammonoid diversity. *In*: House, M. R. (ed.) *The Ammonoidea, Environment, Ecology, and Evolutionary Change*. Systematics Association, Special Volume, **47**. Clarendon Press, Oxford, 115–164.

Becker, R. T. 1993*b*. Analysis of ammonoid palaeobiogeography in relation to the global Hangenberg (terminal Devonian) and Lower Alum Shale (Middle Tournaisian) events. *Annales de la Société géologique de Belgique*, **115**, 459–473.

Becker, R. T. 1996. New faunal records and holostratigraphic correlation of the Hasselbachtal D/C-boundary auxiliary stratotype (Germany). *Annales de la Société géologique de Belgique*, **117**, 19–45.

Becker, R. T. & House, M. R. 2000. Devonian ammonoid zones and their correlation with established series and stage boundaries. *Courier Forschungsinstitut Senckenberg*, **220**, 113–151.

Becker, R. T. & Korn, D. 1997. Ammonoid extinctions and radiations around the D/C boundary. *In*: Cejchan, P. & Hladil, J. (eds) *UNESCO-IGCP Project #335 'Biotic Recovery from Mass Extinctions', Final Conference 'Recoveries ´97'*, Eurocongress Centre, Prague, Czech Republic, Abstracts Book, 13–14.

Becker, R. T., Bless, M. J. M. *et al.* 1984. Hasselbachtal, the section best displaying the Devonian-Carboniferous boundary beds in the Rhenish Massif (Rheinisches Schiefergebirge). *Courier Forschungsinstitut Senckenberg*, **67**, 181–191.

Becker, R. T., Bockwinkel, J., Ebbighausen, V. & House, M. R. 2000. Jebel Mrakib, Anti-Atlas (Morocco), a potential Upper Famennian substage boundary stratotype section. *Notes et Mémoires du Service Géologique du Maroc*, **399**, 75–86.

Becker, R. T., House, M. R., Bockwinkel, J., Ebbighausen, V. & Aboussalam, Z. S. 2002. Famennian ammonoid zones of the eastern Anti-Atlas (southern Morocco). *Münstersche Forschungen zur Geologie und Paläontologie*, **93**, 159–2005.

Becker, R. T., Kaiser, S. I. & Aboussalam, Z. S. 2006. The Lower Alum Shale Event (Middle Tournaisian) in Morocco – facies and faunal changes. *In*: Aretz, M. & Herbig, H.-G. (eds) *Carboniferous Conference Cologne, From Platform to Basin,* September 4–10, 2006. Kölner Forum für Geologie und Paläontologie, Program and Abstracts, **15**, 7–8.

Becker, R. T., Gradstein, F. M. & Hammer, O. 2012. The Devonian period. *In*: Gradstein, F. M., Ogg, J. G., Schmitz, M. & Ogg, G. (eds) *The Geologic Time Scale 2012*. Elsevier, Amsterdam, **2**, 559–601.

Becker, R. T., Hartenfels, S., Aboussalam, Z. S., Tragelehn, H., Brice, D. & El Hassani, A. 2013*a*. The Devonian–Carboniferous boundary at Lalla Mimouna (northern Maider) – a progress report. *In*: Becker, R. T., El Hassani, A. & Tahiri, A. (eds) *International Field Symposium 'The Devonian and Lower Carboniferous of northern Gondwana', 22nd to 29th March 2013, Field Guidebook*. Documents de ĺInstitut Scientifique, Rabat, **27**, 109–120.

Becker, R. T., Aboussalam, Z. S., Hartenfels, S., El Hassani, A. & Baidder, L. 2013*b*. The global carbonate crisis at the Devonian–Carboniferous transition in Morocco. *In*: Reitner, J., Yang, Q., Wang, Y. & Reich, M. (eds) *Palaeobiology and Geobiology of Fossil Lagerstätten through Earth History*. A Joint Conference of the 'Paläontologische Gesellschaft' and the 'Palaeontological Society of China', Göttingen, Germany, September 23–27, 2013. Universitätsverlag Göttingen, Göttingen, Abstract Volume, **19**.

Becker, R. T., Kaiser, S. & Aretz, M. In press. Review of chrono-, litho- and biostratigraphy around the global Hangenberg Crisis and Devonian–Carboniferous boundary. *In*: Becker, R. T., Königshof, P. & Brett, C. E. (eds) *Devonian Climate, Sea Level and Evolutionary Events*. Geological Society, London, Special Publications, **423**, http://doi.org/10.1144/SP423.10

Bless, M. J. M., Crasquin, S., Groos-Uffenorde, H. & Lethiers, F. 1986. Late Devonian to Dinantian ostracodes (comments on taxonomy, stratigraphy, and paleoecology). *Annales de la Société géologique de Belgique*, **109**, 1–8.

Bless, M. J. M., Simakov, K. V. & Streel, M. 1988. Advantages and disadvantages of a conodont-based or event-stratigraphic Devonian-Carboniferous boundary. *Courier Forschungsinstitut Senckenberg*, **100**, 3–14.

Bless, M. J. M., Becker, R. T., Higgs, K., Paproth, E. & Streel, M. 1993. Eustatic cycles around the Devonian–Carboniferous boundary and the sedimentary and fossil record in Sauerland (Federal Republic of Germany). *Annales de la Société géologique de Belgique*, **115**, 689–702.

Blumenstengel, H. 1993. Ostracodes from the Devonian-Carboniferous boundary beds in Thuringia (Germany). *Annales de la Société géologique de Belgique*, **115**, 483–489.

Bockwinkel, J. & Ebbighausen, V. 2006. A new ammonoid fauna from the *Gattendorfia-Eocanites* Genozone of the Anti-Atlas (Early Carboniferous; Morocco). *Fossil Record*, **9**, 87–129.

Bogoyavlenskaya, O. V. 1982. Stromatoporaty poznego devona-rannego karbona. *Paleontologicheskiy Zhurnal*, **1982**, 33–38.

Bond, D. P. & Wignall, P. B. 2014. Large igneous provinces and mass extinctions: an update. *In*: Keller, G. & Kerr, A. C. (eds) *Volcanism, Impacts, and Mass Extinctions: Causes and Effects*. Geological Society of America, Special Papers, **505**, 29–55, http://doi.org/10.1130/2014.2505(02)

Boucot, A. J. & Gray, J. 2001. A critique of Phanerozoic climate models involving changes in the CO_2 content of the atmosphere. *Earth Science Reviews*, **56**, 1–159.

Boyer, F., Krylatov, F., LeFévre, J. & Stoppel, D. 1968. Le Dévonien Supérieur et la limite Dévono-Carbonifère en Montagne Noire (France), lithostratigraphie-biostratigraphie (conodontes). *Bulletin du Centre de Recherches Pau*, **2**, 1–212.

BRAND, U., LEGRAND-BLAIN, M. & STREEL, M. 2004. Biochemostratigraphy of the DCB global stratotype section and point, Griotte Formation, La Serre, Montagne Noire, France. *Palaeogeography, Palaeoclimatology, Palaeoecology*, **205**, 337–357.

BRAUCKMANN, C. & BRAUCKMANN, B. 1986. Famennian trilobites: an outline of their stratigraphical importance. *Annales de la Société géologique de Belgique*, **109**, 9–17.

BRAUCKMANN, C., CHLUPÁČ, I. & FEIST, R. 1993. Trilobites at the Devonian–Carboniferous boundary. *Annales de la Société geologique de Belgique*, **115**, 507–518.

BREZINSKI, D. K., CECIL, C. B., SKEMA, V. W. & STAMM, R. 2008. Late Devonian glacial deposits from the eastern United States signal an end of the mid-Paleozoic warm period. *Palaeogeography, Palaeoclimatology, Palaeoecology*, **268**, 143–151.

BREZINSKI, D. K., CECIL, C. B., SKEMA, V. W. & KERTIS, C. A. 2009. Evidence for long-term climate change in Upper Devonian strata of the central Appalachians. *Palaeogeography, Palaeoclimatology, Palaeoecology*, **284**, 315–325.

BREZINSKI, D. K., CECIL, C. B. & SKEMA, V. W. 2010. Late Devonian glacigenic and associated facies from the central Appalachian Basin, eastern United States. *Geological Society of America Bulletin*, **122**, 265–281.

BRICE, D., LEGRAND-BLAIN, M. & NICOLLIN, J.-P. 2005. New data on Late Devonian and Early Carboniferous brachiopods from NW Sahara: Morocco, Algeria. *Annales de la Société géologique du Nord*, **12**, 1–45.

BRICE, D., LEGRAND-BLAIN, M. & NICOLLIN, J.-P. 2007. Brachiopod faunal changes across the Devonian–Carboniferous boundary in NW Sahara (Morocco, Algeria). *In*: BECKER, R. T. & KIRCHGASSER, W. T. (eds) *Devonian Events and Correlations*. Geological Society, London, Special Publications, **278**, 261–271, http://doi.org/10.1144/SP278.12

BUGGISCH, W. & JOACHIMSKI, M. M. 2006. Carbon isotope stratigraphy of the Devonian of Central and Southern Europe. *Palaeogeography, Palaeoclimatology, Palaeoecology*, **240**, 68–88.

BYVSHEVA, T. V. & UMNOVA, N. I. 1993. Palynological characteristics of the lower part of the Carboniferous of the central region of the Russian Platform. *Annales de la Société géologique de Belgique*, **155**, 519–529.

BYVSHEVA, T. V., HIGGS, K. & STREEL, M. 1984. Spore correlations between the Rhenish Slate Mountains and the Russian Platform near the Devonian–Carboniferous boundary. *Courier Forschungsinstitut Senckenberg*, **67**, 37–45.

CAPLAN, M. L. & BUSTIN, R. M. 1999. Devonian–Carboniferous Hangenberg mass extinction event, widespread organic-rich mudrock and anoxia: causes and consequences. *Palaeogeography, Palaeoclimatology, Palaeoecology*, **148**, 187–207.

CAPLAN, M. L. & BUSTIN, R. M. 2001. Palaeoenvironmental and palaeoceanographic controls on black, laminated mudrock deposition: example from Devonian–Carboniferous strata, Alberta, Canada. *Sedimentary Geology*, **145**, 45–72.

CAPUTO, M. V. 1985. Late Devonian glaciation in South America. *Palaeogeography, Palaeoclimatology, Palaeoecology*, **205**, 337–357.

CAPUTO, M. V., MELO, J. H. G., STREEL, M. & ISBELL, J. L. 2008. Late Devonian and Early Carboniferous glacial records of South America. *In*: FIELDING, C. R., FRANK, T. D. & ISBELL, J. L. (eds) *Resolving the Late Paleozoic Ice Age in Time and Space*. Geological Society of America Special Papers, **441**, 161–173.

CARDOSO, C. N., SANZ-LÓPEZ, J., BLANCO-FERRERA, S., LEMOS, V. B. & SCOMAZZON, A. K. 2015. Frasnian conodonts at high palaeolatitude (Amazonas Basin, north Brazil). *Palaeogeography, Palaeoclimatology, Palaeoecology*, **418**, 57–64.

CARLS, P. & GONG, D. 1992. Devonian and Early Carboniferous Conodonts from Shidian (Western Yunnan, China). *Courier Forschungsinstitut Senckenberg*, **154**, 179–221.

CARMICHAEL, S. K., WATERS, J. A. *ET AL.* 2015. Climate instability and tipping points in the Late Devonian: detection of the Hangenberg Event in an open oceanic island arc in the Central Asian Orogenic Belt. *Gondwana Research*. First published online 18 March 2015. http://doi.org/10.1016/j.gr.2015.02.009

CASIER, J.-G., LETHIERS, F. & PRÉAT, A. 2002. Ostracods and sedimentology of the Devonian–Carboniferous stratotype section (La Serre, Montagne Noire, France). *Bulletin de lInstitut royal des Sciences naturelles de Belgique, Sciences de la Terre*, **72**, 43–68.

CASIER, J.-G., LEBON, A., MAMET, B. & PREAT, A. 2003. Ostracods and lithofacies close to the Devonian-Carboniferous boundary in the Chanxhe and Rivage sections, northeastern part of the Dinant Basin, Belgium. *Bulletin de lInstitut royal des Sciences naturelles de Belgique, Sciences de la Terre*, **73**, 83–107.

CASIER, J.-G., MAMET, B., PRÉAT, A. & SANDBERG, C. A. 2004. Sedimentology, conodonts and ostracods of the Devonian–Carboniferous strata of the Anseremme railway bridge section, Dinant Basin, Belgium. *Bulletin de lInstitut royal des Sciences naturelles de Belgique, Sciences de la Terre*, **74**, 45–68.

CASIER, J.-G., LEBON, A., MAMET, B. & PRÉAT, A. 2005. Ostracods and lithofacies close to the Devonian–Carboniferous boundary in the Chanxhe and Rivage sections, northeastern part of the Dinant Basin, Belgium. *Bulletin de lInstitut royal des Sciences naturelles de Belgique, Sciences de la Terre*, **75**, 95–126.

CHAUFFE, K. M. & NICHOLS, P. A. 1995. Multielement conodont species from the Louisiana Limestone (Upper Devonian) of west-central Illinois and northeastern Missouri, U.S.A. *Micropaleontology*, **41**, 171–186.

CHLUPÁČ, I. & ZIKMUNDOVA, J. 1976. The Devonian and Lower Carboniferous in the Nepasice bore in East Bohemia. *Vestnik Ústredního ústavu geolického*, **51**, 269–278.

CHLUPÁČ, I., FEIST, R. & MORZADEC, P. 2000. Trilobites and standard Devonian stage boundaries. *Courier Forschungsinstitut Senckenberg*, **220**, 87–98.

CHWIEDUK, E. 2005. Late Devonian and Early Carboniferous Rugosa from Western Pomerania, Northern Poland. *Acta Geologica Polonica*, **55**, 393–443.

CLARK, D. L., DERENTHAL, D., KOWALLISS, B. J. & RITTER, S. M. 2014. The major Pre-Mississippian unconformity in Rock Canyon, Central Wasatch Range, Utah. *Geology of the Intermountain West*, **1**, 1–5.

Clark, S., Day, J., Ellwood, B., Harry, R. & Tomkin, J. 2009. Astronomical tuning of integrated Upper Famennian–Early Carboniferous faunal, carbon isotope and high resolution magnetic susceptibility records: Western Illinois Basin. *SDS Newsletter*, **24**, 27–35.

Clausen, C.-D., Leuteritz, K., Ziegler, W. & Korn, D. 1989. Ausgewählte Profile an der Devon/Karbon-Grenze im Sauerland (Rheinisches Schiefergebirge). *Fortschritte in der Geologie von Rheinland und Westfalen*, **35**, 161–226.

Clayton, G., Higgs, K., Gueinn, J. J. & Van Gelder, A. 1974. Palynological correlations in the Cork Beds (Upper Devonian–?Upper Carboniferous) of southern Ireland. *Proceedings of the Royal Irish Academy*, **74**, 145–155.

Clayton, G., Graham, J. R., Higgs, K., Sevastopulo, G. D. & Welsh, A. 1986. Late Devonian and Early Carboniferous palaeogeography of southern Ireland and southwest Britain. *Annales de la Société géologique de Belgique*, **109**, 103–111.

Cloud, P. E., Barnes, V. E., Jr. & Hass, W. H. 1957. Devonian–Mississippian transition in Central Texas. *Geological Society of America Bulletin*, **68**, 807–816.

Cockbain, A. E. 1989. Distribution of Frasnian and Famennian stromatoporoids. *In*: Jell, P. A. & Pickett, J. W. (eds) *Fossil Cnidaria 5, Proceedings of the Fifth International Symposium on Fossil Cnidaria.* Memoir of the Association of Australasian Palaeontologists, **8**, 339–345.

Coen, M. & Groessens, E. 1996. Conodonts from the Devonian–Carboniferous transition beds of central Hunan, South China. *Mémoires de ĺInstitut Géologique de ĺUniversité de Louvain*, **36**, 21–28.

Coen, M., Hance, L. & Hou, H. F. (eds) 1996. Papers on the Devonian–Carboniferous transition beds of central Hunan, South China. *Memoires de ĺInstitut Géologique de ĺUniversité de Louvain*, **36**, 3–13.

Cole, D., Myrow, P. M., Fike, D. A., Hakim, A. & Gehrels, G. E. 2015. Uppermost Devonian (Famennian) to Lower Mississippian events of the western U.S.: stratigraphy, sedimentology, chemostratigraphy, and detrital zircon geochronology. *Palaeogeography, Palaeoclimatology, Palaeoecology*, **427**, 1–19.

Coleman, U. & Clayton, G. 1987. Palynostratigraphy and palynofacies of the uppermost Devonian and Lower Mississippian of eastern Kentucky (U.S.A.), and correlation with western Europe. *Courier Forschungsinstitut Senckenberg*, **98**, 75–93.

Collinson, C., Norby, R. D., Baxter, J. W. & Thompson, T. L. 1979. Stratigraphy of the Mississippian stratotype: Upper Mississippi Valley, U.S.A. *In*: *Ninth International Congress of Carboniferous Stratigraphy and Geology*. Field Trip 8, Illinois State Geological Survey.

Conil, R., Dreesen, R., Lentz, M.-A., Lys, M. & Plodowski, G. 1986. The Devono-Carboniferous transition in the Franco-Belgian Basin with reference to Foraminifera and brachiopods. *Annales de la Société géologique de Belgique*, **109**, 19–26.

Conil, R., Groessens, E., Laloux, M., Poty, E. & Tourneur, F. 1991. Carboniferous guide foraminifera, corals and conodonts in the Franco-Belgian and Campine Basins: their potential for widespread correlation. *Courier Forschungsinstitut Senckenberg*, **130**, 15–30.

Conil, R., Dreesen, R., Dusar, M., Gilissen, E., Poty, E., Streel, M. & Thorez, J. 1993. The Early Carboniferous Transgression on the Brabant Massif (W. Flanders, Belgium). *In*: Streel, M. (ed.) *Early Carboniferous Stratigraphy, Liège 1993*. Services associés de paléontologie de ĺULg, Liège, Meeting Program and Abstracts, **18**.

Conrad, J., Massa, D. & Weyant, M. 1986. Late Devonian regression and Early Carboniferous transgression on the northern African platform. *Annales de la Société géologique de Belgique*, **109**, 113–122.

Coquel, R. & Abdesselam-Rouighi, F. 2000. Révision palynostratigraphique du dévonien terminal-carbonifère inférieur dans le Grand Erg occidental (bassin de Béchar) Sahara Algérien. *Revue de Micropaléontologie*, **43**, 353–364.

Corradini, C. 2008. Revision of Famennian–Tournaisian (Late Devonian–Early Carboniferous) conodont biostratigraphy of Sardinia, Italy. *Revue de Micropaléontologie*, **51**, 123–132.

Corradini, C., Barca, S. & Spalletta, C. 2003. Late Devonian–Early Carboniferous conodonts from the 'Clymeniae limestones' of SE Sardinia (Italy). *Courier Forschungsinstitut Senckenberg*, **245**, 227–253.

Corradini, C., Kaiser, S. I., Perri, M. C. & Spalletta, C. 2011. Conodont genus *Protognathodus* and its potential as a tool for defining the Devonian/Carboniferous boundary. *Rivista Italiana di Paleontologia e Stratigrafia*, **117**, 15–28.

Corradini, C., Spalletta, C., Kaiser, S. I. & Matyja, H. 2013. Overview of conodonts across the Devonian/Carboniferous boundary. *In*: Albanesi, G. L. & Ortega, G. (eds) *Conodonts from the Andes, 3rd International Conodont Symposium*. Asociación Paleontológica Argentina, Publicatión Especial. Buenos Aires, Argentina, **13**, 13–16.

Cramer, B. D., Saltzman, M. R. & Kleffner, M. A. 2006. Spatial and temporal variability in organic carbon burial during global positive carbon isotope excursions: new insight from high resolution carbon isotope stratigraphy from the type area of the Niagaran Provincial Series. *Stratigraphy*, **2**, 327–340.

Cramer, B. D., Saltzman, M. R., Day, J. E. & Witzke, B. J. 2008. Record of the Late Devonian Hangenberg global positive carbon-isotope excursion in an epeiric sea setting: carbonate production, organic-carbon burial and paleoceanography during the Late Famennian. *In*: Holmden, H. & Pratt, B. R. (eds) *Dynamics of Epeiric Seas: Sedimentological, Paleontological and Geochemical Perspectives*. Geological Association of Canada, Special Paper, **48**, 103–118.

Cygan, C. & Perret, M.-F. 2002. Conodonts from the Upper Devonian–Lower Carboniferous succession of Milles (Arize Massif). *In*: *ECOS VIII France-Spain 2002.* Pyrenees Field Trip Guide Book, Université Paul Sabatier, Toulouse, 56–62.

Day, J., Witzke, B. & Rowe, H. 2011. Development of an epeirical subtropical paleoclimate record from western Euramerica: late Frasnian–earliest Tournaisian stable carbon isotope record from the Yellow Spring–New Albany Groups of the northwestern Illinois Basin.

In: *Geological Society of America Annual Meeting, Abstracts with Programs*, **43**, 151.

Decombeix, A.-L., Meyer-Berthaud, B. & Galtier, J. 2011. Transitional changes in arborescent lignophytes at the Devonian-Carboniferous boundary. *Journal of the Geological Society, London*, **168**, 547–557, http://doi.org/10.1144/0016-76492010-074

Denayer, J., Poty, E. & Aretz, M. 2011. Uppermost Devonian and Dinantian rugose corals from Southern Belgium and surrounding areas. *Kölner Forum für Geologie und Paläontologie*, **20**, 151–201.

De Vleeschouwer, D., Rakocinski, M., Racki, G., Bond, D. P. G., Sobien, K. & Clayes, P. 2013. The astronomical rhythm of Late Devonian climate change (Kowala section, Holy Cross Mountains, Poland). *Earth and Planetary Science Letters*, **365**, 25–37.

Diaz-Martinez, E. & Isaacson, P. E. 1994. Late Devonian glacially-influenced marine sedimentation in western Gondwana: the Cumana Formation, Altiplano, Bolivia. *In*: Embry, A. F., Beauchamp, B. & Glass, D. J. (eds) *Pangea: Global Environments and Reseources*. Canadian Society of Petroleum Geologists, Memoir, **17**, 511–522.

Dino, R. 2000. Palynostratigraphy of the Silurian and Devonian sequence of the Paraná Basin, Brazil. *In*: Rodrigues, M. A. C. & Pereira, E. (eds) *Ordovician–Devonian Palynostratigraphy in Western Gondwana: Update, Problems and Perspectives*. UERJ, Rio de Janeiro, 27–61.

Donoghue, P. C. J. & Chauffe, K. M. 1998. *Conchodontus*, *Mitrellataxis* and *Fungulodus*: conodonts, fish or both? *Lethaia*, **31**, 283–292.

Draganits, E., Mawson, R., Talent, J. A. & Krystyn, L. 1999. Lithostratigraphy, conodont biostratigraphy and depositional environments of the Middle Devonian (Givetian) to Early Carboniferous (Tournaisian) Lipak Formation in the Pin Valley of Spiti (NW India). *Rivista Italiana di Paleontologia e Stratigrafia*, **108**, 7–35.

Dreesen, R., Bless, M. J., Conil, R., Flajs, G. & Laschet, C. 1985. Depositional environment, paleoecology and diagenetic history of the 'marbre rouge à crinoides de Baelen' (late Upper Devonian, Verviers Synclinorium, eastern Belgium). *Annales de la Société géologique de Belgique*, **108**, 311–359.

Dreesen, R., Sandberg, C. A. & Ziegler, W. 1986. Review of Late Devonian and Early Carboniferous conodont biostratigraphy and biofacies models as applied to the Ardenne Shelf. *Annales de la Société géologique de Belgique*, **109**, 27–42.

Dreesen, R., Poty, E., Streel, M. & Thorez, J. 1993. *Late Famennian to Namurian in the Eastern Ardenne, Belgium: Guidebook*. International Union of Geological Sciences, Subcommission on Carboniferous Stratigraphy, Liège.

Druce, E. C. 1969. *Devonian and Carboniferous Conodonts from the Bonaparte Gulf Basin, Northern Australia, and Their Use in International Correlation*. Bureau of Mineral Resources, Geology and Geophysics, Canberra, Australia, Bulletin, **98**, 1–243.

Durkina, A. V. & Avchimovitch, V. I. 1988. The reference sections of the Devonian-Carboniferous boundary deposits in the Timan–Pechora province. *In*: Golubtsov, V. K. et al. (eds) *The Devonian–Carboniferous Boundary at the Territory of the USSR*. Nauka I Technica, Minsk, 87–101.

Dzik, J. 1997. Emergence and succession of Carboniferous conodont and ammonoid communities in the Polish part of the Variscan sea. *Acta Palaeontologica Polonica*, **42**, 57–164.

Ebner, F. 1980*a*. Conodont localities in the surrounding of Graz/Styria. *Abhandlungen der Geologischen Bundesanstalt*, **35**, 101–127.

Ebner, F. 1980*b*. Steinbergkalke und Sanzenkogel-Schichten im Kalvarienbergzug W von Gratwein. *Mitteilungen der naturwissenschaftlichen Vereinigung Steiermark*, **115**, 53–61.

Edwards, D., Fairon-Demaret, M. & Berry, C. M. 2000. Plant megafossils in Devonian stratigraphy: a progress report. *Courier Forschungsinstitut Senckenberg*, **220**, 35–37.

Elrick, M., Berkyova, S., Klapper, G., Sharp, Z., Joachimski, M. & Fryda, J. 2009. Stratigraphic and oxygen isotope evidence for My-scale glaciation driving eustasy in the Early–Middle Devonian greenhouse world. *Palaeogeography, Palaeoclimatology, Palaeoecology*, **276**, 170–181.

Ernst, A. 2013. *Diversity Dynamics and Evolutionary Patterns of the Palaeozoic Stenolaemate Bryozoa*. Habilitationschrift (unpublished), Christian-Albrechts-Universität zu Kiel.

Ernst, A. & Herbig, H.-G. 2010. Stenolaemate bryozoans from the Latest Devonian (uppermost Famennian) of Western Germany. *Geologica Belgica*, **13**, 173–182.

Ernst, A., Tolokonnikova, Z. & Herbig, H.-G. 2015. Uppermost Famennian bryozoans from Ratingen (Velbert Anticline, Rhenish Massif/Germany) – Taxonomy, facies dependencies and palaeobiogeographic implications. *Geologica Belgica*, **18**, 37–47.

Ettensohn, F. R., Lierman, T. R. & Mason, C. E. 2007. Dropstones, glaciation and black shales: New information on black-shale origins fort he upper Ohio Shale in northeastern Kentucky. *In*: *American Association of Petroleum Geologists, Eastern Section Meeting, Abstracts with Programs*, 33–34.

Ettensohn, F. R., Lierman, R. T., Mason, C. E. & Clayton, G. 2009. Changing physical and biotic conditions on eastern Laurussia: evidence from Late Devonian to Middle Mississippian basinal and deltaic sediments of northeastern Kentucky, USA. North American Paleontological Convention, Field Trip 2, 20 June 2009.

Fairon-Demaret, M. 1986. Some uppermost Devonian megafloras: a stratigraphical review. *Annales de la Société géologique de Belgique*, **109**, 43–48.

Fairon-Demaret, M. 1996. The plant remains from the Late Famennian of Belgium: a review. *The Palaeobotanist*, **45**, 201–208.

Fallah, A., Hamedi, B. & Mosaddegh, H. 2011. Carboniferous conodont biostratigraphy in Kiyasar region and introduction of 7 biozones comparable to world standard conodont zonation. *Geosciences, Scientific Quarterly Journal*, **20**, 117–122, 193 [in Farsi with English summary].

Fan, Y., Yu, X. et al. 2003. *The Late Palaeozoic Rugose Corals of Xizang (Tibet) and Adjacent Regions and Their Palaeobiogeography*. Hunan Science and

Technology Press, National Natural Science Foundation China, Series Geosciences, 1–679.

FEIST, R. & PETERSEN, M. S. 1995. Origin and spread of *Pudoproetus*, a survivor of the Late Devonian trilobite crisis. *Journal of Paleontology*, **69**, 99–109.

FILIPIAK, P. 2004. Miospore stratigraphy of Upper Famennian and Lower Carboniferous deposits of the Holy Cross Mountains (central Poland). *Review of Palaeobotany and Palynology*, **128**, 291–322.

FILIPIAK, P. & RACKI, G. 2010. Proliferation of abnormal palynoflora during the end-Devonian biotic crisis. *Geological Quaterly*, **54**, 1–14.

FISCHER, T. & BECKER, R. T. 2014. Ontogenetic morphometry, taxonomy and biogeographic aspects of Famennian (Upper Devonian) Prionoceratidae. *In*: KLUG, C. & FUCHS, D. (eds) *9th International Cephalopod Symposium, Cephalopods – Present and Past, in Combination with the 5th International Symposium, Coleoid Cephalopods through Time*. Abstracts and Program, **36**, 4–14 September, University of Zürich, Switzerland.

FLAJS, G. & FEIST, R. 1988. Index conodonts, trilobites and environment of the Devonian–Carboniferous boundary beds at La Serre (Montagne Noire, France). *In*: FLAJS, G., FEIST, R. & ZIEGLER, W. (eds) *Devonian–Carboniferous Boundary – Results of Recent Studies*. Courier Forschungsinstitut Senckenberg, **100**, 53–107.

FRIEDRICH, O., ERBACHER, J., MORIYA, K., WILSON, P. A. & KUHNERT, H. 2008. Warm saline intermediate waters in the Cretaceous tropical Atlantic Ocean. *Nature Geoscience*, **1**, 453–457.

GAGIEV, M. H. 1997. Sedimentary evolution and sea-level fluctuations in the Devonian of North-east Asia. *Courier Forschungsinstitut Senckenberg*, **199**, 75–82.

GAGIEV, M. H. & BOGUS, O. J. 1990. Oporiyi razrez famensko-turnejskikh otloszhenii ugo-zapadnoi chasti prikolymeskogo podniatia. *Trudy Instituta Geologii i Geofiziki*, **770**, 119–132.

GAGIEV, M. H. & KONONOVA, L. I. 1990. The Upper Devonian and Lower Carboniferous Sequences in the Kamenka River section (Kolyma River Basin, the Soviet North-East). *Stratigraphic description. Conodonta. Courier Forschungsinstitut Senckenberg*, **118**, 81–03.

GANDL, J., FERRER, E., MAGRANS, J. & SANZ LOPÉZ, J. 2015. Trilobiten aus dem Unter-Karbon des Katalonischen Küstengebirges (NE-Spanien). *Abhandlungen der Senckenbergischen Gesellschaft für Naturforschung*, **571**, 1–83.

GAO, L. 1989. *Palynostratigraphy at the Devonian-Carboniferous Boundary in the Himalayan Region, Xizang (Tibet)*. Canadian Society of Petroleum Geologists, Memoir, **14**, 159–170.

GARCÍA-LÓPEZ, S. & SANZ-LÓPEZ, J. 2002. Devonian to Lower Carboniferous conodont biostratigraphy of the Bernesga Valley section (Cantabrian zone, NW Spain). *Cuadernos del Museo Geominero*, **1**, 163–205.

GARCÍA-LÓPEZ, S., SANZ LÓPEZ, J. & PARDO ALONSO, M. V. 1999. Conodontos (bioestratigrafía, biofacies y paleotemperaturas) de los sinclinales de Almadén y Guadalmez (Deonico-Carbonifero Inderior), Zona Centroíberica meridional, España. *Revista Española de Paleontología*, no extr. Homenaja al Prof. J. Truyols, 161–172.

GEDIK, I. 1974. Conodonten aus dem Unterkarbon der Karnischen Alpen. *Abhandlungen der Geologischen Bundesanstalt*, **31**, 1–43.

GEREKE, M. 2004. Das Profil Kahlleite Ost – die stratigraphische Entwicklung einer Tiefschwelle im Oberdevon des Bergaer Sattels (Thueringen). *Geologica et Palaeontologica*, **38**, 1–31.

GHAVIDEL-SYOOKI, M. 1994. Upper Devonian acritarchs and miospores from the Geirud Formation in Central Alborz Range, Northern Iran. *Journal of Science, Islamic Republic Iran*, **5**, 103–122.

GINTER, M. & IVANOV, A. 2000. Stratigraphic distribution of chondrichthyans in the Devonian on the East European Platform margin. *Courier Forschungsinstitut Senckenberg*, **223**, 325–339.

GIRARD, C., CORNÉE, J.-J., CORRADINI, C., FRAVALO, A. & FEIST, R. 2013. Palaeoenvironmental changes at Col de Tribes (Montagne Noire, France), a reference section for the Famennian of north Gondwana-related areas. *Geological Magazine*, **151**, 864–884.

GLIKSON, A. Y., MORY, A. J., IASKY, R. P., PIRAJNO, F., GOLDING, S. D. & UYSAL, I. T. 2005. Woodleigh, Southern Carnarvon Basin, Western Australia: history of discovery, Late Devonian age, and geophysical and morphometric evidence for a 120 km-diameter impact structure. *Australian Journal of Earth Sciences*, **52**, 545–553.

GLIKSON, A. Y., UYSAL, I. T., GERALD, J. D. & SAYGIN, E. 2013. Geophysical anomalies and quartz microstructures, Eastern Warburton Basin, North-east Souh Australia: tectonic or impact metamorphic origin. *Tectonophysics*, **589**, 57–76.

GÖNCÜOGLU, M. C., CARPINOGLU, S. *ET AL*. 2007. The Mississippian in the Central and Eastern Taurides (Turkey): constraints on the tectonic setting of the Tauride–Anatolide Platform. *Geologica Carpathica*, **58**, 427–442.

GONG, X., HUANG, H., ZHANG, M. & HUANG, Q. 1991. *The Stratigraphic Classification and Correlation of Carbonate Rocks of Upper Devonian and Lower Carboniferous in Guilin Karst Region*. Guangxi Science and Technology Publishing House.

GONZÁLEZ, F., MORENO, C. & SANTOS, A. 2006. The massive sulphide event in the Iberian Pyrite Belt: confirmatory evidence from the Sotiel-Coronada Mine. *Geological Magazine*, **143**, 821–827.

GRECHISHNIKOVA, L. A. & LEVITSKII, E. S. 2011. The Famennian–Lower Carboniferous Reference Section Geran-Kalasi (Nakhichevan Autonomous region, Azerbaijan). *Stratigraphy and Geological Correlation*, **19**, 21–43.

GROOS-UFFENORDE, H. & RABIEN, A. 2014. Zur Verbreitung pelagischer Ostracoden im Devon Deutschlands. *Geologisches Jahrbuch Hessen*, **138**, 37–47.

GRÖTSCH, J., BILLING, I. & VAHRENKAMP, V. 1998. Carbon-isotope stratigraphy in shallow-water carbonates: implications for Cretaceous black-shale deposition. *Sedimentology*, **45**, 623–634.

GUERRAK, S. & CHAUVEL, J. J. 1985. Les minéralisations ferrifères du Sahara Algerien: le gisement de fer oolithique de Mecheri Abdelazis (basin de Tindouf). *Mineralia Deposita*, **20**, 249–259.

GUSTAFSON, M., HOLBOURN, A. & KUHNT, W. 2003. Changes in Northeast Atlantic temperature and carbon flux during the Cenomanian/Turonian paleoceanographic event: the Goban Spur stable isotope record. *Palaeogeography, Palaeoclimatology, Palaeoecology*, **201**, 51–66.

Gutak, J. M., Tolokonnikova, Z. A. & Ruban, D. A. 2008. Bryozoan diversity in southern Siberia at the Devonian–Carboniferous transition: new data confirm a resistivity to two mass extinctions. *Palaeogeography, Palaeoclimatology, Palaeoecology*, **264**, 93–99.

Gutschick, R. C. & Rodriguez, J. 1979. Biostratigraphy of the Pilot Shale (Devonian–Mississippian) and contemporaneous strata in Utah, Nevada, and Montana. *Brigham Young University Geology Studies*, **26**, 37–63.

Gutschick, R. C. & Sandberg, C. A. 1991. Upper Devonian biostratigraphy of Michigan Basin. *In*: Catacosinos, P. A. & Daniels, P. A., Jr. (eds) *Early Sedimentary Evolution of the Michigan Basin*. Geological Society of America Special Papers, **256**, 155–179.

Habibi, T., Corradini, C. & Yazdi, M. 2008. Conodont biostratigraphy of the Upper Devonian–Lower Carboniferous Shahmirzad section, central Alborz, Iran. *Geobios*, **41**, 763–777.

Hahn, G. 1990. Palaeogeographic distribution and biostratigraphic significance of Lower Carboniferous trilobites: a review. *Courier Forschungsinstitut Senckenberg*, **130**, 199–205.

Hahn, G., Hahn, R. & Brauckmann, C. 1994. Trilobiten mit '*Drevermannia*-Habitus' im Unter-Karbon. *Courier Forschungsinstitut Senckenberg*, **169**, 155–193.

Hahn, G., Müller, P. & Becker, R. T. 2012. Unterkarbonische Trilobiten aus dem Anti-Atlas (S-Marokko). *Geologica et Palaeontologica*, **44**, 37–74.

Halamski, A. T. & Balinski, A. 2009. Latest Famennian brachiopods from Kowala, Holy Cross Mountains, Poland. *Acta Paleontologica Polonica*, **54**, 289–306.

Hance, L. 1996. Foraminiferal biostratigraphy of the Devonian–Carboniferous boundary and Tournaisian strata in Central Hunan Province, South China. *Mémoires de lInstitut Géologique de lUniversité de Louvain*, **36**, 29–53.

Hance, L., Muchez, P. *et al.* 1994. Biostratigraphy and sequence stratigraphy at the Devonian–Carboniferous transition in southern China (Hunan Province). *Comparison with southern Belgium. Annales de la Société géologique de Belgique*, **116**, 359–378.

Hance, L., Poty, E. & Devuyst, F.-X. 2001. Stratigraphie séquentielle du Dinantien type (Belgique) et corrélation avec le Nord de la France (Boulonnais, Avesnois). *Bulletin de la Société Géologique de France*, **172**, 411–426.

Hance, L., Hou, H. & Vachard, D. 2011. *Upper Famennian to Visean Foraminifers and Some Carbonate Microproblematica from South China*. Geological Publishing House, Beijing.

Hao, W.-C. 2001. The Devonian–Carboniferous boundary and events at Bachu, Xinjiang, Northwestern China. *International Geology Review*, **43**, 276–284.

Hartenfels, S. & Becker, R. T. 2009. Timing of the global Dasberg Crisis – implications for Famennian eustasy and chronostratigraphy. *Palaeontographica Americana*, **63**, 69–95.

Hass, W. H. 1956. *Age and Correlation of the Chattanooga Shale and the Maury Formation*. US Geological Survey, Reston, VA, Professional Paper, **286**.

Haydukiewicz, J. 1981. Pelagicne utwory turneju w poludniowo-zachodniej czesci Gor Bardzkich. *Geologica Sudetica*, **16**, 219–226.

Heckel, P. H. & Witzke, B. J. 1979. Devonian world paleogeography determined from distribution of carbonates and related lithic palaeoclimatic indicators. *In*: House, M. R., Scrutton, C. T. & Bassett, M. G. (eds) *The Devonian System*. Special Papers in Paleontology, **23**, 99–123.

Herbig, H.-G. 2006. Kalkschalige Kleinforaminiferen. *In*: Deutsche Stratigraphische Kommission (eds) *Stratigraphie von Deutschland VI. Unterkarbon (Mississippium)*. Schriftenreihe der deutschen Gesellschaft für Geowissenschaften, **41**, 250–270.

Higgs, K. T. & Streel, M. 1994. Palynological age for the lower part of the Hangenberg Shales in Sauerland, Germany. *Annales de la Société géologique de Belgique*, **116**, 243–247.

Higgs, K. T., Finucane, D. & Tunbridge, I. P. 2002. Late Devonian and early Carboniferous microfloras from the Hakkan Province of southeastern Turkey. *Review of Palaeobotany and Palynology*, **118**, 141–156.

Horowitz, A. S. & Pachut, J. F. 1993. Specific, generic, and familial diversity of Devonian bryozoans. *Journal of Paleontology*, **67**, 42–52.

Hou, H., Ji, Q. *et al.* 1985. *Muhua Sections of Devonian–Carboniferous Boundary Beds*. Geological Publishing House, Beijing.

House, M. R. 1978. *Devonian Ammonoids from the Appalachians and Their Bearing on International Zonation and Correlation*. Special Papers in Palaeontology, **21**.

House, M. R. 1985. Correlation of mid-Palaeozoic ammonoid evolutionary events with global sedimentary perturbations. *Nature*, **313**, 17–22.

House, M. R. 1996. An *Eocanites* fauna from the Early Carboniferous of Chile and its palaeogeographic implications. *Annales de la Société géologique de Belgique*, **117**, 95–105.

House, M. R. 2002. Strength, timing, setting and cause of mid-Palaeozoic extinctions. *Palaeogeography, Palaeoclimatology, Palaeoecology*, **181**, 5–25.

House, M. R., Gordon, M. J. & Hlavin, W. J. 1986. Late Devonian ammonoids from Ohio and adjacent states. *Journal of Paleontology*, **60**, 126–144.

Huber, B. T., Norris, R. D. & MacLeod, K. G. 2002. Deep-sea paleotemperature record of extreme warmth during the Cretaceous. *Geology*, **30**, 123–126.

Hünicken, M. A., De Melo, J. H. G. & Lemos, V. B. 1989. *Devonian Conodonts from the Upper Amazonas Basin; Northwestern Brazil*. Canadian Society of Petroleum Geologists, Memoir, **14**, 479–483.

Isaacson, P. E., Hladil, J., Shen, J.-W., Kalvoda, J. & Grader, G. 1999. Late Devonian (Famennian) glaciation in South America and marine offlap on other continents. *Abhandlungen der Geologischen Bundesanstalt*, **54**, 239–257.

Isaacson, P. E., Díaz-Martínez, E., Grader, G. W., Kalvoda, J., Babek, O. & Devuyst, F. X. 2008. Late Devonian–earliest Mississippian glaciation in Gondwanaland and its biogeographic consequences.

Palaeogeography, Palaeoclimatology, Palaeoecology, **268**, 126–142.

Ivanov, A. 1996. The Early Carboniferous chondrichthyans of the South Urals, Russia. *In*: Strogen, P., Sommerville, I. D. & Jones, G. L. (eds) *Recent Advances in Lower Carboniferous Geology*. Geological Society, London, Special Publications, **107**, 417–425, http://doi.org/10.1144/GSL.SP.1996.107.01.29

Janvier, P. 1996. *Early Vertebrates*. Clarendon, Oxford.

Janvier, P., Lethiers, F., Monod, O. & Balkas, Ö. 1984. Discovery of a vertebrate fauna at the Devonian-Carboniferous boundary in SE Turkey (Hakkari Province). *Journal of Petroleum Geology*, **7**, 147–168.

Jarvis, E. 1990. New palynological data on the age of the Kiltorcan Flora of Co. Kilkenny, Ireland. *Journal of Micropaleontology*, **9**, 87–94, http://doi.org/10.1144/jm.9.1.87

Ji, Q. 1987. The Devonian–Carboniferous boundary in shallow-water facies areas of China as based on conodonts. *Acta Geologica Sinica*, **61**, 11–22.

Ji, Q. & Ziegler, W. 1992. Introduction to some Late Devonian sequences in the Guilin area of Guangxi, South China. *Courier Forschungsinstitut Senckenberg*, **154**, 149–177.

Ji, Q. & Ziegler, W. 1993. The Lali section: an excellent reference section for Upper Devonian in South China. *Courier Forschungsinstitut Senckenberg*, **157**, 1–183.

Ji, Q., Zhang, Z.-H., Chen, X.-Z. & Wang, G.-B. 1987. Studies of the Devonian–Carboniferous boundary in Zhaisha, Luzhei of Guangxi. *Journal of Stratigraphy*, **11**, 213–217 [in Chinese].

Ji, Q., Wang, Z. et al. 1989. *The Dapoushang Section, An Excellent Section for the Devonian–Carboniferous Boundary Stratotype in China*. Science Press, Beijing.

Ji, W. 1987. Early Carboniferous conodonts from Jianghua County of Hunan Province, and their stratigraphic value – with a discussion on the mid-Aikuanian Event. *Bulletin of the Institute of Geology, Chinese Academy of Geological Sciences*, **16**, 115–141 [in Chinese with English summary].

Jin, X., Devuyst, F.-X., Hance, L., Poty, E., Aretz, M., Yin, B. & Hou,& H. 2007. Stratigraphy and lithofacies of the Tournaisian and Visean in the Guilin–Liuzhou area, Guangxi, South China. *XVI International Congress on the Carboniferous and Permian*, 21–24 June 2007, Nanjing, China, Post-Congress Excursion C, 25–28 June 2007, Guide Book for Field Excursion, 1–41.

Joachimski, M. M., Breisig, S. et al. 2009. Devonian climate and reef evolution: Insights from oxygen isotopes in apatite. *Earth and Planetary Science Letters*, **284**, 599–609.

Johnson, J. G., Klapper, G. & Sandberg, C. A. 1985. Devonian eustatic fluctuations in Euramerica. *Geological Society of America Bulletin*, **96**, 567–587.

Johnston, D. I., Henderson, C. M. & Schmidt, M. J. 2010. Upper Devonian to Lower Mississippian conodont biostratigraphy of uppermost Wabamun Group and Palliser Formation to lowermost Banff and Lodgepole formations, southern Alberta and southeastern British Columbia, Canada: implications for correlations and sequence stratigraphy. *Bulletin of Canadian Petroleum Geology*, **58**, 295–341.

Kaiser, H. 1970. Die '*Hymenozonotriletes lepidophytus*-Zone' auf der Bäreninsel. *In*: Streel, M. & Wagner, R. H. (eds) *Colloque sur la Stratigraphie du Carbonifère*. Université de Liege, Les Congres et Colloques de lUniversité de Liege, **55**, 285–287.

Kaiser, S. I. 2005. *Mass Extinctions, Climatic and -Oceanographic Changes at the Devonian–Carboniferous Boundary*. PhD thesis, Fakultät für Geowissenschaften, Ruhr-Universität Bochum. http://www-brs.ub.ruhr-uni-bochum.de/netahtml/HSS/Diss/KaiserSandraIsabella/diss.pdf

Kaiser, S. I. 2007. Conodontenstratigraphie und Geochemie ($\delta^{13}C_{carb}$,$\delta^{13}C_{org}$,$\delta^{18}O_{phosph}$) aus dem Devon-Karbon Grenzbereich der Karnischen Alpen. *Jahrbuch der geologischen Bundesanstalt*, **146**, 301–314.

Kaiser, S. I. 2009. The Devonian/Carboniferous stratotype section La Serre (Montagne Noire) revisited. *Newsletters on Stratigraphy*, **43**, 195–205.

Kaiser, S. I., Becker, R. T. et al. 2004. Sedimentary succession and neritic faunas around the Devonian–Carboniferous boundary at Kheneg Lakahal south of Assa (Dra Valley, SW Morocco). *Documents de lInstitut Scientifique Rabat*, **19**, 93–100.

Kaiser, S. I., Steuber, T., Becker, R. T. & Joachimski, M. M. 2006. Geochemical evidence for major environmental change at the Devonian–Carboniferous boundary in the Carnic Alps and the Rhenish Massif. *Palaeogeography, Palaeoclimatology, Palaeoecology*, **240**, 146–160.

Kaiser, S. I., Becker, R. T. & El Hassani, A. 2007. Middle to Late Famennian successions at Ain Jemaa (Moroccan Meseta) – implications for regional correlation, event stratigraphy and synsedimentary tectonics of NW Gondwana. *In*: Becker, R. T. & Kirchgasser, W. T. (eds) *Devonian Events and Correlations*. Geological Society, London, Special Publications, **278**, 237–260, http://doi.org/10.1144/SP278.11

Kaiser, S. I., Steuber, T. & Becker, R. T. 2008. Environmental change during the Late Famennian and Early Tournaisian (Late Devonian–Early Carboniferous) – implications from stable isotopes and conodont biofacies in southern Europe. *In*: Aretz, M., Herbig, H.-G. & Somerville, I. D. (eds) *Carboniferous Platforms and Basins*. *Geological Journal*, **43**, 241–260.

Kaiser, S. I., Becker, R. T., Spalletta, C. & Steuber, T. 2009. High-resolution conodont stratigraphy, biofacies, and extinctions around the Hangenberg Event in pelagic successions from Austria, Italy, and France. *Palaeontolographica Americana*, **63**, 97–139.

Kaiser, S. I., Becker, R. T., Steuber, T. & Aboussalam, Z. S. 2011. Climate-controlled mass extinctions, facies, and sea-level changes around the Devonian-Carboniferous boundary in the eastern Anti-Atlas (SE Morocco). *Palaeogeography, Palaeoclimatology, Palaeoecology*, **310**, 340–364.

Kalvoda, J. 2002. Late Devonian–Early Carboniferous foraminiferal fauna: zonations, evolutionary events, paleobiogeography and tectonic implications. *Folia Facultatis scientiarium naturalium Universitatis Masarykianae Brunensis, Geologia*, **39**, 1–213.

Kalvoda, J. & Kukal, Z. 1987. Devonian–Carboniferous boundary in the Moravian Karst at Lesní Lom Quarry, Brno-Ĺsen, Czechoslovakia. *Courier Forschungsinstitut Senckenberg*, **98**, 95–117.

KALVODA, J., KUMPAN, T. & BÁBEK, O. 2015. Upper Famennian and Lower Tournaisian sections of the Moravian Karst (Moravio-Silesian Zone, Czech Republic): a proposed key area for correlation of the conodont and foraminiferal zonations. *Geological Journal*, **50**, 17–38.

KARAKITSIOS, V., TSIKOS, H., VAN BREUGEL, Y., KOLETTI, L., SINNGHE DAMSTÉ, J. S. & JENKYNS, H. C. 2007. First evidence for the Cenomanian–Turonian oceanic anoxic event (OAE2, 'Bonarelli' event) from the Ionian Zone, western continental Greece. *International Journal of Earth Science*, **96**, 343–352.

KARAULOV, V. B. & GRETSCHISCHNIKOVA, I. A. 1997. Devonian eustatic fluctuations in North Eurasia. *Courier Forschungsinstitut Senckenberg*, **199**, 13–23.

KELLY, B. K., MANGER, W. L. & KLAPPER, G. 1997. Conodonts from the Chattanooga Shale and the Devonian–Mississippian boundary, Southern Ozark region. *Geological Society of America, Abstracts & Program*, **29**, 264.

KLAPPER, G. 1966. Upper Devonian and Lower Mississippian conodont zones in Montana, Wyoming, and South Dakota. *University of Kansas Paleontological Contributions*, **1966**, 1–43.

KLAPPER, G. & FURNISH, W. M. 1962. Devonian–Mississippian Englewood Formation in Black Hills, South Dakota. *American Association of Petroleum Geologists Bulletin*, **46**, 2071–2078.

KOCH, M., LEUTERITZ, K. & ZIEGLER, W. 1970. Alter, Fazies und Paläogeographie der Oberdevon/Unterkarbon-Schichtenfolge an der Seiler bei Iserlohn. *Fortschritte in der Geologie vom Rheinland und Westfalen*, **17**, 679–732.

KOCHETKOVA, N. M., PAZUKHIN, V. N., REITLINGER, E. A. & SINICHYIA, Z. A. 1985. *Opornye razrezy pogranechnykh otloszhenii Devona u Karbona zapadnogo sklona Yazhnogo Urala*. Akademii Nauk, Magadan.

KÖNIGSHOF, P., SAVAGE, N. M., LUTAT, P., SARDSUD, A., DOPIERALSKA, J., BELKA, Z. & RACKI, G. 2012. Late Devonian sedimentary record of the Paleotethys Ocean – The Mae Sariang section, northwestern Thailand. *Journal of Asian Earth Sciences*, **52**, 146–157.

KOMATSU, T., KATO, S. *ET AL.* 2014. Devonian–Carboniferous transition containing a Hangenberg Black Shale equivalent in the Pho Han Formation on Cat Ba Island, northeastern Vietnam. *Palaeogeography, Palaeoclimatology, Palaeoecology*, **404**, 30–43.

KONONOVA, L. I. 1979. Upper Frasnian, Famennian and Tournaisian conodonts of the Sikaza River section (southern Urals). *Service de géologique de Belgique, Professional Papers*, **161**, 74–86.

KONONOVA, L. I. & WEYER, D. 2013. Upper Famennian conodonts from the Breternitz Member (Upper Clymeniid Beds) of the Saalfeld region, Thuringia (Germany). *Freiberger Forschungshefte*, **C545**, 15–97.

KORN, D. 1986. Ammonoid evolution in late Famennian and early Tournaisian. *Annales de la Société Géologique de Belgique*, Liège, **109**, 49–54.

KORN, D. 1993. The ammonoid faunal change near the Devonian–Carboniferous boundary. *Annales de la Société géologique de Belgique*, **115**, 581–593.

KORN, D. 1999. Famennian ammonoid stratigraphy of the Máder and Tafilalt (eastern Anti-Atlas, Morocco). *Abhandlungen der Geologischen Bundesanstalt*, **54**, 147–179.

KORN, D. 2000. Morphospace occupation of ammonoids over the Devonian–Carboniferous boundary. *Paläontologische Zeitschrift*, **74**, 247–257.

KORN, D. & FEIST, R. 2007. Early Carboniferous ammonoid faunas and stratigraphy of the Montagne Noire (France). *Fossil Record*, **10**, 99–124.

KORN, D. & WEYER, D. 2003. High resolution stratigraphy of the Devonian–Carboniferous transitional beds in the Rhenish Mountains. *Mitteilungen aus dem Museum für Naturkunde in Berlin, Geowissenschaftliche Reihe*, **6**, 79–124.

KORN, D., CLAUSEN, C.-D., BELKA, Z., LEUTERITZ, K., LUPPOLD, F. W., FEIST, R. & WEYER, D. 1994. Die Devon/Karbon-Grenze bei Drewer (Rheinisches Schiefergebirge). *Geologie und Paläontologie Westfalens*, **29**, 97–147.

KORN, D., EBBIGHAUSEN, V. & BOCKWINKEL, J. 2002. Palaeogeographical meaning of a Middle Tournaisian ammonoid fauna from Morocco. *Geologica et Palaeontologica*, **36**, 79–86.

KORN, D., BELKA, Z., FRÖHLICH, S., RÜCKLIN, M. & WENDT, J. 2004. The youngest African clymeniids (Ammonoidea, Late Devonian) – failed survivors of the Hangenberg Event. *Lethaia*, **37**, 307–315.

KORN, D., BOCKWINKEL, J. & EBBIGHAUSEN, V. 2007. Tournaisian and Viséan ammonoid stratigraphy in North Africa. *Neues Jahrbuch für Geologie und Paläontologie, Abhandlungen*, **243**, 127–148.

KRONBERG, P., PILGER, A., SCHERP, A. & ZIEGLER, W. 1960. Spuren altvariscischer Bewegungen im nordöstlichen Teil des Rheinischen Schiefergebirges. *Fortschritte in der Geologie von Rheinland und Westfalen*, **3**, 1–46.

KULAGINA, E. I. 2013. Taxonomic diversity of foraminifers of the Devonian–Carboniferous boundary interval in the South Urals. *Bulletin of Geosciences*, **88**, 265–282.

KULAGINA, E. I., GIBSHMAN, N. B. & PAZUKHIN, V. N. 2003. Foraminiferal zonal standard for the Lower Carboniferous of Russia and its correlation with the conodont zonation. *Rivista Italiana di Paleontologia e Stratigrafia*, **109**, 173–185.

KULLMANN, J. 1994. Diversity fluctuations in ammonoid evolution from Devonian to mid-Carboniferous. *Courier Forschungsinstitut Senckenberg*, **169**, 137–141.

KULLMANN, J. 2000. Ammonoid turnover at the Devonian–Carboniferous boundary. *Revue de Paléobiologie, special volume*, **8**, 169–180.

KUMPAN, T., BÁBEK, O., KALVODA, J., FRÝDA, J. & GRYGAR, T. M. 2013. A high-resolution, multiproxy stratigraphic analysis of the Devonian–Carboniferous boundary sections in the Moravian Karst (Czech Republic) and a correlation with the Carnic Alps (Austria). *Geological Magazine*, **151**, 201–215, http://doi.org/10.1017/S0016756812001057

KUMPAN, T., BÁBEK, O., KALVODA, J., GRYGAR, T. M. & FRÝDA, J. 2014. Sea-level and environmental changes around the Devonian-Carboniferous boundary in the Namur-Dinant Basin (S Belgium, NE France): a multiproxy stratigraphic analysis of carbonate ramp archives and its use in regional and interregional correlations. *Sedimentary Geology*, **311**, 43–59.

Kumpan, T., Bábek, O., Kalvoda, J., Matys Grygar, T., Frýda, J., Becker, R. T. & Hartenfels, S. 2015. Petrophysical and geochemical signature of the Hangenberg Events: an integrated stratigraphy of the Devonian-Carboniferous boundary interval in the Northern Rhenish Massif (Avalonia, Germany). *Bulletin of Geosciences*, **90**, 667–694, http://doi.org/10.3140/bull.geosci.1547

Kürschner, W., Becker, R. T., Buhl, D. & Veizer, J. 1993. Strontium isotopes in conodonts: Devonian–Carboniferous transition, the northern Rhenish Slate Mountains, Germany. *In*: Streel, M., Sevastopulo, G. & Paproth, E. (eds) Devonian–Carboniferous boundary. *Annales Société géologique de Belgique*, **115**, 595–621.

Kuypers, M. M. M., Schouten, S. & Sinninghe Damsté, J. S. 1998. The Cenomanian/Turonian oceanic anoxic event: response of the atmospheric CO_2 level. *Mineralogical Magazine*, **62A**, 836–837.

Kuypers, M. M. M., Pancost, R. D., Nijenhuis, I. A. & Sinninghe Damsté, J. S. 2002. Enhanced productivity led to increased organic carbon burial in the euxinic North Atlantic basin during the late Cenomanian oceanic anoxic event. *Paleoceanography*, **17**, 3-1–3-13.

Kuypers, M. M. M., van Breugel, Y., Schouten, S., Erba, E. & Sinninghe Damsté, J. S. 2004. N_2-fixing cyanonacteria supplied nutrient N for Cretaceous oceanic anoxic events. *Geology*, **32**, 853–856.

Lang, J., Yahaya, M., El Hamet, M. O., Besombes, J. C. & Cazoulat, M. 1991. Depôts glaciaires du Carbonifère inférieur à ĺOuest de l̀Air (Niger). *Geologische Rundschau*, **80**, 611–622.

Lanzoni, E. & Magloire, L. 1969. Associations palynologiques et leurs applications stratigraphiques dans le Dévonien supérieur et Carbonifère inférieur du Grand Erg occidental (Sahara Algérien). *Revue de Institute Francaise de Pétrole*, **24**, 441–469.

Le Hérissé, A., Servais, T. & Wicander, R. 2000. Devonian acritarchs and related forms. *Courier Forschungsinstitut Senckenberg*, **220**, 195–205.

Lebedev, O. A. 1996. Fish assemblages in the Tournaisian-Viséan environments of the East European Platform. *In*: Strogen, P., Sommerville, I. D. & Jones, G. L. (eds) *Recent Advances in Lower Carboniferous Geology*. Geological Society, London, Special Publications, **107**, 387–415, http://doi.org/10.1144/GSL.SP.1996.107.01.28

Lees, A. & Miller, J. 1995. Waulsortian Banks. *In*: Monty, C. L. V., Bosence, D. W. J., Bridges, P. H. & Pratt, B. R. (eds) *Carbonate Mud-mounds, Their Origin and Evolution*. International Association of Sedimentologists, Special Publications, **23**. Blackwell Science, Oxford, 191–271.

Legrand-Blain, M. 1985. Taoudeni Basin. *In*: Wagner, R. H., Winkler Prins, C. F. & Granados, L. F. (eds) *The Carboniferous of the World. II. Australia, Indian Subcontinent, South Africa, South America & North Africa*. IUGS Publication, **20**. Madrid, Spain, 323–325.

Legrand-Blain, M. 1991. Brachiopods as potential boundary-defining organisms in the Lower Carboniferous of western Europe: recent data and productid distribution. *Courier Forschungsinstitut Senckenberg*, **130**, 157–171.

Legrand-Blain, M. 1995. Relations entre les domains d'Europe occidentale, d́Europe meridionale (Montagne Noire) et d́Afrique du Nord a la limite Dévonien–Carbonifère: les donnees des brachiopods. *Bulletin de la Société belge de Géologie*, **103**, 77–97.

Legrand-Blain, M. & Martinez Chacon, M.-L. 1988. Brachiopods at the Devonian–Carboniferous boundary, La Serre (Montagne Noire; Hérault, France): preliminary report. *Courier Forschungsinstitut Senckenberg*, **100**, 119–127.

Lelievre, H. & Goujet, D. 1986. Biostratigraphic significance of some uppermost Devonian placoderms. *Annales de la Société géologique de Belgique*, **109**, 55–59.

Lethiers, F. & Feist, R. 1991. Ostracodes, stratigraphie et bathymétrie du passage Dévonien–Carbonifère au Viséen Inférieur en Montagne Noire (France). *Geobios*, **24**, 71–104.

Li, R.-J. & Duan, L.-L. 1993. Early Carboniferous conodont sequence in the Baoshan-Shidian area, Yunnan and its stratigraphic significance. *Acta Micropalaeontologica Sinica*, **1993**, 37–52.

Lierman, R. T. & Mason, C. E. 2007. Upper Devonian glaciation in the Ohio Shale of east-central Kentucky. *Geological Society of America Annual Meeting, Abstracts with Programs*, **39**, 70.

Lierman, R. T., Mason, C. E., Ettensohn, F. R. & Clayton, G. 2009. Stop 3: granitic dropstone embedded in the uppermost Cleveland Shale Member of Ohio Shale. *In*: Brett, C. E., Bartholomew, A. J. & DeSantis, M. K. (eds) *Middle and Upper Devonian Sequences, Sea Level, Climatic and Biotic Events in East-Central Laurentia: Kentucky, Ohio, and Michigan*. North American Paleontological Convention 2009, Field Trip 10, 27 June–3 July 2009, 1–186.

Liu, Y.-Q., Ji, Q., Kuang, H.-W., Jiang, X.-J., Xu, H. & Peng, N. 2012. U–Pb zircon age, sedimentary facies, and sequence stratigraphy of the Devonian–Carboniferous boundary, Dapoushang section, Guizhou, China. *Palaeoworld*, **21**, 100–107.

Loboziak, S., Streel, M., Caputo, M. V. & De Melo, J. H. G. 1992. Middle Devonian to Lower Carboniferous miospore stratigraphy in the central Parnaíba Basin (Brazil). *Annales de la Société géologique de Belgique*, **115**, 215–226.

Loboziak, S., Streel, M., Caputo, M. V. & Melo, J. H. G. 1993. Middle Devonian to Lower Carboniferous miospores from selected boreholes in Amazonas and Parnaíba Basins (Brazil), additional data, synthesis, and correlation. *Documents des Laboratoires de Géologie de la Faculté des Sciences de Lyon*, **125**, 277–289.

Loboziak, S., Streel, M., Dusar, M., Boulvain, F. & De Geyter, G. 1994. Late Devonian-Early Carboniferous miospores from the Menen Borehole, Namur Synclinorium, Belgium. *Review of Palaeobotany and Palynology*, **80**, 55–63.

Loboziak, S., de Melo, J. H. G., Steemans, P. & Rodrigues Barrilari, I. M. 1995. Miospore evidence for pre-Emsien and latest Famennian sedimentation in the Devonian of the Paraná Basin, South Brazil.

Anais da Academia Brasileira de Ciências, **67**, 391–392.

Long, J. A. 1995. *The Rise of Fishes: 500 Million Years of Evolution*. Johns Hopkins University Press, Baltimore, MD.

López-Gamundi, O. R. & Rosello, E. A. 1993. Devonian–Carboniferous unconformity in Argentina and its relation to the Eo-Hercynian orogeny in southern South America. *Geologische Rundschau*, **82**, 136–147.

Luppold, F. W., Hahn, G. & Korn, D. 1984. Trilobiten-, Ammonoideen- und Conodonten-Stratigraphie des Devon/Karbon-Grenzprofiles auf dem Müssenberg (Rheinisches Schiefergebirge). *Courier Forschungsinstitut Senckenberg*, **67**, 91–121.

MacQueen, R. W. & Sandberg, C. A. 1970. Stratigraphy, age, and interregional correlation of the Exshaw Formation, Alberta, Rocky Mountains. *Bulletin of Canadian Petroleum Geology*, **18**, 32–66.

Makhlina, M. K. 1996. Cyclic stratigraphy, facies and fauna of the Lower Carboniferous (Dinantian) of the Moscow Syneclise and Voronezh Anteclise. *In*: Strogen, P., Sommerville, I. D. & Jones, G. L. (eds) *Recent Advances in Lower Carboniferous Geology*. Geological Society, London, Special Publications, **107**, 359–364, http://doi.org/10.1144/GSL.SP.1996.107.01.25

Malec, J. 2014. The Devonian/Carboniferous boundary in the Holy Cross Mountains (Poland). *Geological Quarterly*, **58**, 217–234.

Mamet, B. 1985. On the presence of Quasiendothyridae in Arctic Alaska. *In*: *Compte Rendu 10th International Congress of Carboniferous Stratigraphy and Geology, 1983*, Madrid, **4**, 144–145.

Marshall, J. E. A. 2010. The Late Devonian and Early Carboniferous terrestrial climatic record. *In*: *Programme & Abstracts, International Palaeontological Congress*, 28 June–3 July 2010, London, 263.

Marshall, J. E. A. & Astin, T. 2009. Why the terrestrial upper Famennian is important to the SDS. *Palaeontographica Americana*, **63**, 217–218.

Marshall, J. E. A., Astin, T. R., Evans, F. & Almond, J. 2002. The palaeoclimatic significance of the Devonian–Carboniferous boundary. *In*: *Geology of the Devonian System, Proceedings of the International Symposium*, 9–12 July 2000, Syktyvkar, Komi Republic, 23–25.

Marshall, J. E. A., Lakin, J. A. & Finney, S. M. 2013. Terrestrial climate and ecosystem change from the Devonian–Carboniferous boundary to the earliest Viséan interval in East Greenland. *In*: El Hassani, A., Becker, R. T. & Tahiri, A. (eds) *International Field Symposium 'The Devonian and Lower Carboniferous of Northern Gondwana'*, 22–29 March 2013. Documents de ĺInstitut Scientifique, Rabat, Abstracts Book, **26**, 81–82.

Martel, A. T., McGregor, D. C. & Utting, J. 1993. Stratigraphic significance of Upper Devonian and Lower Carboniferous miospores from the type area of the Horton Group, Nova Scotia. *Canadian Journal of Earth Sciences*, **30**, 1091–1098.

Marynowski, L. & Filipiak, P. 2007. Water column euxinia and wildfire evidence during deposition of the Upper Famennian Hangenberg event horizon from the Holy Cross Mountains (central Poland). *Geological Magazine*, **144**, 569–595.

Marynowski, L., Záton, M., Rakocinski, M., Filipiak, P., Kurkiewicz, S. & Pearce, T. J. 2012. Deciphering the upper Famennian Hangenberg Black Shale depositional environments based on multi-proxy record. *Palaeogeography, Palaeoclimatology, Palaeoecology*, **346–347**, 66–86.

Maslov, V. A. (ed.) 1987. *Fauna I biostratigrafia pogranichnikh otloszhenii devona I karbona Berchogura (Mugodzhary)*. Akademia Nauk SSSR, Bashkirskii Filiali, Institut geologii.

Massa, D. & Moreau-Benoit, A. 1985. Apport de nouvelles donnes palynologiques a la biostratigraphie et a la paléogeographie du Dévonien de Libye (sud du Bassin de Rhadamès). *Sciences Géologiques, Bulletin*, **38**, 5–18.

Matthews, S. C. 1983. An occurrence of Lower Carboniferous (Gattendorfia-Stufe) ammonoids in South West Ireland. *Neues Jahrbuch für Geologie und Paläontologie, Monatshefte*, **1983**, 293–299.

Matyja, H. & Stempien-Salek, M. 1994. Devonian/Carboniferous boundary and the associated phenomena in western Pomerania (NW Poland). *Annales de la Société de géologique de Belgique*, **116**, 249–263.

Matyja, H., Malkowski, K., Sobien, K. & Stempien-Salek, M. 2010. Devonian–Carboniferous boundary in Poland: conodont and miospore successions and event stratigraphy. *In*: *Programme & Abstracts, International Palaeontological Congress*, 28 June–3 July 2010, London, 268.

Matyja, H., Sobien, K., Marynowski, L., Stempien-Salek, M. & Malkowski, K. 2014. The expression of the Hangenberg Event (latest Devonian) in a relatively shallow-marine succession (Pomeranian Basin, Poland): the results of a multi-proxy investigation. *Geological Magazine*, **153**, 429–443, http://doi.org/10.1017/S001675681400034X

Maziane, N. & Vanguestaine, M. 1996. Acritarchs from the Uppermost Famennian at Chanxhe and Tohogne (eastern Belgium). *Acta Universitatis Carolinae Geologica*, **40**, 527–530.

McGlathery, K. J., Sundback, K. & Anderson, I. C. 2007. Eutrophication in shallow coastal bays and lagoons: the role of plants in the coastal filter. *Marine Ecology Progress Series*, **348**, 1–18.

McNestry, A. 1988. The palynostratigraphy of two uppermost Devonian–Lower Carboniferous borehole sections in South Wales. *Review of Palaeobotany and Palynology*, **56**, 69–87.

Melo, J. H. G. & Loboziak, S. 2003. Devonian–Early Carboniferous miospore biostratigraphy of the Amazon Basin, Northern Brazil. *Review of Palaeobotany and Palynology*, **124**, 131–202.

Melo, J. H. G., Barrilari, I. M. R., Quadros, L. P., Loboziak, S. & Matsuda, N. S. 1999. *Miospore-based Correlation of the Late Devonian Curuá Group*. Petrobas, Amazon Basin, Brazil.

Melott, A. L., Lieberman, B. S. *et al.* 2004. Did a gamma-ray burst initiate the late Ordovician mass extinction? *International Journal of Astrobiology*, **3**, 55–61.

MEOR, H. A. H., AUNG, A.-K., BECKER, R. T., ABDUL RAHMAN, N. A., FATT NG, T., GHANI, A. A. & SHUIB, M. K. 2014. Stratigraphy and palaeoenvironmental evolution of the mid- to upper Palaeozoic succession in Northwest Peninsular Malaysia. *Journal of Asian Earth Sciences*, **83**, 60–79.

MERGL, M., MASSA, D. & PLAUCHUT, B. 2001. Devonian and Carboniferous brachiopods and bivalves of the Djado Sub-Basin (North Niger, SW Libya). *Journal of the Czech Geological Survey*, **46**, 169–188.

MICHELS, D. 1986. Ökologie und Fazies des jüngsten Ober-Devon von Velbert (Rheinisches Schiefergebirge). *Göttinger Arbeiten für Geologie und Paläontologie*, **29**, 1–86.

MILHAU, B., MISTIAEN, B. ET AL. 1997. Comparative faunal content of Strunian (Devonian) between Etaoucun (Guilin, Guangxi, South China) and the stratotype area (Etroeungt, Avesnois, north of France). *Proceedings of the 30th International Geological Congress, Beijing*, **12**, 79–94.

MISTIAEN, B. 1996. Stromatoporoids from the Late Devonian (Strunian) Menggongao Formation, China. *Mémoires de lInstitut Géologique de lUniversité de Louvain*, **36**, 141–152.

MISTIAEN, B. & WEYER, D. 1999. Late Devonian stromatoporoid from the Sudetes Mountains (Poland) and endemicity of the Upper Famennian and Uppermost Famennian (='Strunian') stromatoporoid fauna in western Europe. *Senckenbergiana lethaeae*, **79**, 51–61.

MISTIAEN, B., MILHAU, B., KHATIR, A., HOU, H., VACHARD, D. & WU, X. 1998. Famennian terminal (Strunien) d'Etroeungt (Avesnois, nord de la France) et d'Etaoucun (Guangxi, Chine du sud). Incidences paléogéographiques des donnees relatives aux stromatopores et ostracodes. *Annales de la Société Géologique du Nord*, **6**, 97–104.

MITCHELL, R. N., BICE, D. M., MONTANARI, A., CLEAVELAND, L. C., CHRISTIANSON, K. T., COCCIONI, R. & HINNOV, L. A. 2008. Oceanic anoxic cycles? Orbital prelude to the Bonarelli Level (OAE 2). *Earth and Planetary Science Letters*, **267**, 1–16.

MOLYNEUX, S., MANGER, W. L. & OWENS, B. 1984. Preliminary account of Late Devonian palynomorph assemblages from the Bedford Shale and Berea Sandstone Formations of central Ohio, USA. *Journal of Micropaleontology*, **3**, 41–51, http://doi.org/10.1144/jm.3.2.41

MOORE, D. 1988. Upper Devonian–Lower Mississippian conodont biostratigraphy and depositional patterns, southwestern New Mexico and southeastern Arizona. *New Mexico Geology*, **10**, 25–32.

MORY, A. J. & CRANE, D. T. 1982. Early Carboniferous *Siphonodella* (Conodonta) faunas from eastern Australia. *Alcheringa*, **6**, 275–303.

MOSSONI, A., CORRADINI, C. & SPALLETTA, C. 2013. Famennian–Tournaisian conodonts from the Monte Taccu section (Sardinia, Italy). *In*: ALBANESI, G. L. & ORTEGA, G. (eds) *Conodonts from the Andes, 3rd International Conodont Symposium*. Asociación Paleontológica Argentinia, Publicación Especial, **13**. Mendoza, Argentina, 85–90.

MOSSONI, A., CARTA, N., CORRADINI, C. & SPALLETTA, C. 2015. Conodonts across the Devonian/Carboniferous boundary in SE Sardinia (Italy). *Bulletin of Geosciences*, **90**, 371–388.

MOTTEQUIN, B. & POTY, E. 2014. The uppermost Famennian Hangenberg Event in the Namur–Dinant Basin (southern Belgium). *In*: KIDO, E., WATERS, J. A. ET AL. (eds) *IGCP 596 & IGCP 580 Joint Meeting and Field Workshop, International Symposium in Mongolia, Ulaanbaatar, Mongolia, 5–18th August 2014*. Berichte des Institutes für Erdwissenschaften, Karl-Franzenz-Universität Graz, Abstract Volume, **19**, 36–37.

MOTTEQUIN, B., BRICE, D. & LEGRAND-BLAIN, M. 2014. Biostratigraphic significance of brachiopods near the Devonian–Carboniferous boundary. *Geological Magazine*, **151**, 216–218.

MUCHEZ, P. 1996. Sea-level variations at the Devonian–Carboniferous transition in South China. *Mémoires de lInstitut géologique de lUniversité Louvain*, **36**, 193–202.

MULLINS, G. L. & SERVAIS, T. 2008. The diversity of the Carboniferous phytoplankton. *Review of Palaeobotany and Palynology*, **149**, 29–49.

MURPHY, A. E., SAGEMANN, B. B., HOLLANDER, D. J., LYONS, T. W. & BRETT, C. 2000. Black shale deposition and faunal overturn in the Devonian Appalachian Basin: clastic starvation, seasonal watercolumn mixing, and efficient biolimiting nutrient recycling. *Paleoceanography*, **15**, 280–291.

MYROW, P. M., STRAUSS, J. V., CREVELING, J. R., SICARD, K. R., RIPPERDAN, R., SANDBERG, C. A. & HARTENFELS, S. 2011. A carbon isotopic and sedimentological record of the latest Devonian (Famennian) from the Western U.S. and Germany. *Palaeogeography, Palaeoclimatology, Palaeoecology*, **306**, 147–159.

MYROW, P. M., RAMEZANI, J., HANSON, A. E., BOWRING, S. A., RACKI, G. & RAKOCINSKI, M. 2014. High-precision U-Pb age and duration of the latest Devonian (Famennian) Hangenberg event, and its implications. *Terra Nova*, **26**, 222–229.

NAQISHBANDI, S. R., SHERWANI, G. H. & REDHA, D. N. 2010. Palynological study of Ora and the Upper Part of Kaista Formation in Zakho area, Iraqi Kurdistan region. *Journal of Kirkuk University, Scientific Studies*, **5**, 50–74.

NEMIROVSKAYA, T. I., CHERMNYKH, V. A., KONONOVA, L. I. & PAZUKHIN, V. N. 1993. Conodonts of the Devonian–Carboniferous boundary section, Kozhim, Polar Urals, Russia. *Annales de la Société de géologique Belgique*, **115**, 629–647.

NICOLL, R. S. & DRUCE, E. C. 1979. Conodonts from the Fairfield Group, Canning Basin, Western Australia. *BMR Bulletin*, **190**, 1–134.

NICOLLIN, J.-P. & BRICE, D. 2004. Biostratigraphical value of some Strunian (Devonian, uppermost Famennian) Productidina, Rhynchonellida, Spiriferida, Spiriferina brachiopods. *Geobios*, **57**, 437–453.

OLEMPSKA, E. 1997. Changes in benthic ostracod assemblages across the Devonian–Carboniferous boundary in the Holy Cross Mountains, Poland. *Acta Palaeontologica Polonica*, **42**, 291–332.

OVER, D. J. 1992. Conodonts and the Devonian–Carboniferous boundary in the upper Woodford Shale, Arbuckle Mountains, South-Central Oklahoma. *Journal of Paleontology*, **66**, 293–311.

Over, D. J. 2007. Conodont biostratigraphy of the Chattanooga Shale, Middle and Upper Devonian, southern Appalachian Basin, eastern United States. *Journal of Paleontology*, **81**, 1194–1217.

Over, D. J., De La Rue, S., Isaacson, P. & Ellwood, B. 2009. Upper Devonian conodonts from black shales of the high latitude Tomachi Formation, Madre de Dios Basin, northern Bolivia. *Palaeontographica Americana*, **62**, 89–99.

Padilha de Quadros, L. 2000. Silurian–Devonian acritarch assemblages from Paraná Basin: an update and correlation with northern Brazilian basins. *In*: Rodrigues, M. A. C. & Pereira, E. (eds) *Ordovician–Devonian Palynostratigraphy in Western Gondwana: Update, Problems and Perspectives*. UERJ, Rio de Janeiro, 105–145.

Paeckelmann, W. 1938. Erläuterungen zu Blatt Balve. *Geologische Karte von Preussen und benachbarten deutschen Ländern*, **2655** (neue Nr. 4613), 1–70.

Paproth, E. 1986. An introduction to a field trip to the late Devonian outcrops in the northern Rheinisches Schiefergebirge (Federal Republic of Germany). *Annales de la Société geologique de Belgique*, **109**, 275–284.

Paris, F., Winchester-Seto, T., Boumendjel, K. & Grahn, Y. 2000. Toward a global biozonation of Devonian chitinozoans. *Courier Forschungsinstitut Senckenberg*, **220**, 39–55.

Pashin, J. C. & Ettensohn, F. R. 1992. Paleoecology and sedimentology of the dysaerobic Bedford fauna (Late Devonain), Ohio and Kentucky (USA). *Palaeogeography, Palaeoclimatology, Palaeoecology*, **91**, 21–34.

Pazukhin, V. N., Kulagina, E. I. & Sedaeva, K. M. 2009. The Devonian/Carboniferous boundary on the western slope of the South Urals. *In*: *Proceedings of the International Field Meeting 'The historical type sections, proposed and potential GSSP of the Carboniferous in Russia', Carboniferous Type Sections in Russia and Potential Global Stratotypes: Southern Urals Session*, August 13–18, 2009, Ufa-Sibai. DesignPolygraph Service Ufa, 22–33 [in Russian with English abstract].

Perez-Leyton, M. 1991. Miospores du Dévonien moyen et supérieur de la coupe de Bermejo-La Angostura (Sud-Est de la Bolivie). *Annales de la Société géologique de Belgique*, **113**, 373–389.

Perkins, R. B., Piper, D. Z. & Mason, C. E. 2008. Trace-element budgets in the Ohio/Sunbury shales of Kentucky: constraints on ocean circulation and primary productivity in the Devonian–Mississippian Appalachian Basin. *Palaeogeography, Palaeoclimatology, Palaeoecology*, **265**, 14–29.

Perret, M.-F. 1988. Le passage du Dévonien au Carbonifère dans les Pyrenées, Zonation par conodontes. *Courier Forschungsinstitut Senckenberg*, **100**, 39–52.

Perret, M.-F. & Majesté-Menjoulas, C. 2002*a*. *Conodonts from the Upper Devonian–Mississippian Succession of Garcet (P.-A.). ECOS VIII France-Spain 2002*. Pyrenees Field Trip Guide Book, Université Paul Sabatier, Toulouse, 71–75.

Perret, M.-F. & Majesté-Menjoulas, C. 2002*b*. *Conodonts from the Outcrop Along the Devonian-Carboniferous Section (Urdos Bridge Section). ECOS VIII France-Spain 2002*. Pyrenees Field Trip Guide Book, Université Paul Sabatier, Toulouse, 87–88.

Perri, M. C. & Spalletta, C. 1998. Late Famennian conodonts of the Malpasso section (Carnic Alps, Italy). *In*: Perri, M. C. & Spalletta, C. (eds) *Seventh International Conodont Symposium held in Europe*. Southern Alps Field Trip Guide Book, June 27–July 2, 1998. Giornale di Geologia, Serie 3a, Special Issue, **60**, 220–227.

Perri, M. C. & Spalletta, C. 2000*a*. Late Devonian–Early Carboniferous transgressions and regressions in the Carnic Alps (Italy). *Records of the Western Australian Museum, Supplements*, **58**, 305–319.

Perri, M. C. & Spalletta, C. 2000*b*. Hangenberg Event al limite Devoniano/Carbonifero al Monte Zermula, Alpi Carniche, Italia. *Giornale di Geologia, Serie 3a*, **62**, 31–40.

Petter, G. 1960. Clymènies du Sahara. *Publications du Service de la Carte Géologique de l'Algérie, nouvelle série, Paléontologie, Memoire*, **6**, 1–58.

Piepjohn, K., Brinkmann, L., Grewing, A. & Kerp, H. 2000. New data on the age of the uppermost ORS and the lowermost post-ORS strata in Dickson Land (Spitsbergen) and implications for the age of the Svalbardian deformation. *In*: Friend, P. F. & Williams, B. P. J. (eds) *New Perspectives on the Old Red Sandstone*. Geological Society, London, Special Publications, **180**, 603–609, http://doi.org/10.1144/GSL.SP.2000.180.01.32

Piper, D. Z. & Calvert, S. E. 2009. A marine biogeochemical perspective on black shale deposition. *Earth Science Review*, **95**, 63–96.

Playford, G. & McGregor, D. C. 1993. Miospores and organic-walled microphytoplankton of Devonian-Carboniferous boundary beds (Bakken Formation), southern Saskatchewan: a systematic and stratigraphic appraisal. *Geological Survey of Canada, Bulletin*, **445**, 1–107.

Plodowski, G. & Salanci, A. 1990. Devon/Karbon-Grenze in Anatolien. *Courier Forschungsinstitut Senckenberg*, **127**, 238–249.

Poletaev, V. I. & Lazarev, S. S. 1995. General stratigraphic scale and brachiopod evolution in the Late Devonian and Carboniferous suequatorial belt. *Bulletin de la Société belge de Géologie*, **103**, 99–107.

Poty, E. 1986. Late Devonian to early Tournaisian Rugose corals. *Annales de la Société géologique de Belgique*, **109**, 65–74.

Poty, E. 1999. Famennian and Tournaisian recoveries of shallow water Rugosa following late Frasnian and late Strunian major crisis, southern Belgium and surrounding area, Hunan (South China) and the Omolon region (NE Siberia). *Palaeogeography, Palaeoclimatology, Palaeoecology*, **154**, 11–26.

Poty, E. 2007. Latest Famennian (Strunian) stromatoporoid biostromes. *In*: Alvaro, J. J., Vennin, E., Munnecke, A., Boulvain, F., Vachard, D. & Aretz, M. (eds) *Catalogue of Facies from Palaeozoic Reefs and Bioaccumulations*. Muséum national d'histoire naturelle, Paris, Mémoires, **195**, 221–223.

Poty, E., Devuyst, F.-X. & Hance, L. 2006. Upper Devonian and Mississippian foraminiferal and rugose coral zonation of Belgium and Northern France: a tool for Eurasian correlations. *Geological Magazine*, **143**, 829–857.

PRICE, J. D. & HOUSE, M. R. 1984. Ammonoids near the Devonian–Carboniferous boundary. *Courier Forschungsinstitut Senckenberg*, **67**, 15–22.

PUPOREV, Y. & CHEGODAEV, L. D. 1982. Biostratigraphy of the Devonian and Carboniferous boundary deposits. Devonian and Carboniferous boundary in the northern Caucasus. *In*: *Tikhookeanskaya Nauchnaya Assotsiatsiya, Severo-Vostochniy Kompteksny Nauchno-Issledovatlskiy Institut, Dapnevost ochniyi nauchniy Tsentrali.* Akademia Nauk SSSR, Magadan, 3–8 [in Russian].

QIE, W., LIU, J. ET AL. 2015. Local overprints on the global carbonate $\delta^{13}C$ signal in Devonian–Carboniferous boundary successions of South China. *Palaeogeography, Palaeoclimatology, Palaeoecology*, **418**, 290–303.

QIN, G., ZHAO, R. & JI, Q. 1988. Late Devonian and Early Carboniferous conodonts from northern Guangdong and their stratigraphic significance. *Acta Micropalaeontologica Sinica*, **5**, 57–71.

RACKI, G. 2005. Toward understanding Late Devonian global events: few answers, many questions. *In*: OVER, D. J., MORROW, J. R. & WIGNALL, P. B. (eds) *Understanding Late Devonian and Permian–Triassic Biotic and Climatic Events:Towards an Integrated Approach.* Developments in Palaeontology & Stratigraphy, **20**. Elsevier, Amsterdam, 5–36.

RAHMANI-ANTARI, K. & LACHKAR, G. 2001. Contribution à Íètude biostratigraphique du Dévonien et du Carbonifère de la plate-forme marocaine. Datation et corrélations. *Revue de Micropaléontologie*, **44**, 159–183.

RICHARDS, B. C. & HIGGINS, A. C. 1988. Devonian–Carboniferous boundary beds of the Palliser and Exshaw Formations at Jura Creek, Rocky Mountains, Southwestern Alberta. *In*: MCMILLAN, N. J., EMBRY, A. F. & GLASS, D. J. (eds) *Devonian of the World.* Canadian Society of Petroleum Geologists, Memoir, **14**, 399–412.

RIMMER, S. M., THOMPSON, J. A., GOODNIGHT, S. A. & ROBL, T. L. 2004. Multiple controls on the preservation of organic matter in Devonian–Mississippian marine black shales: geochemical and petrographic evidence. *Palaeogeography, Palaeoclimatology, Palaeoecology*, **215**, 125–154.

RITTER, S. M. 1991. Conodont-based revision of Upper Devonian–Lower Pennsylvanian stratigraphy in the Lake Mead region of Northwestern Arizona and Southeastern Nevada. *Brigham Young University Geology Studies*, **37**, 125–138.

ROBL, T. L. & BARRON, L. S. 1988. The geochemistry of Black Shales in Central Kentucky and its relationships to inter-basinal correlation and depositional environments. *In*: MCMILLAN, N. J., EMBRY, A. F. & GLASS, D. J. (eds) *Devonian of the World.* Canadian Society of Petroleum Geologists, Memoir, **14**, 377–396.

ROLET, J., PLUSQUELLEC, Y., BABIN, C. & DEUNFF, J. 1986. Famennian regression and Strunian grabens in the Armorican Massif. A key area: western Brittany. *Annales de la Société géologique de Belgique*, **109**, 197–203.

RUBINSTEIN, C., NIEMEYER, H. & URZÚA, F. 1996. Primeros resultados palinológicos en la Formaion Zorritas, Dévonico-Carboniféro de Sierra de Almeida, Región de Antofagasta, Chile. *Revista Geológica de Chile*, **23**, 81–95.

SÁEZ, R., MORENO, C. & GONZÁLEZ, F. 2008. Synchronous deposition of massive sulphide deposits in the Iberian Pyrite Belt: new data from Las Herrerías and La Torerera ore-bodies. *Compte Rendu Geosciences*, **340**, 829–839.

SAGEMAN, B. B., MURPHY, A. E., WERNE, J. P., VER STRAETEN, C. A., HOLLANDER, D. J. & LYONS, T. W. 2003. A tale of shales: the relative roles of production, decomposition, and dilution in the accumulation of organic-rich strata, Middle-Upper Devonian, Appalachian basin. *Chemical Geology*, **195**, 229–273.

SALLAN, L. C. & COATES, M. I. 2010. End-Devonian extinction and a bottleneck in the early evolution of modern jawed vertebrates. *Proceedings of the National Academy of Sciences*, 107, 10 131–10 135.

SALTZMAN, M. R. 2005. Phosphorus, nitrogen, and the redox evolution of the Paleozoic oceans. *Geology*, **33**, 573–576.

SALTZMAN, M. R. & THOMAS, E. 2012. Carbon isotope stratigraphy. *In*: GRADSTEIN, F. M., OGG, J. G., SCHMITZ, M. & OGG, G. (eds) *The Geologic Time Scale 2012*, Vol. **1**. Elsevier, Amsterdam, 207–232.

SANDBERG, C. A. & KLAPPER, G. 1967. Stratigraphy, age, and paleotectonic significance of the Cottonwood Canyon Member of the Madison Limestone in Wyoming and Montana. *US Geological Survey Bulletin*, **1251-B**, 1–70.

SANDBERG, C. A. & POOLE, F. G. 1977. *Conodont Biostratigraphy and Depositional Complexes of Upper Devonian Cratonic-Platform and Continental-Shelf Rocks in the Western United States.* California University, Riverside, CA, Campus Museum Contributions, **4**, 144–182.

SANDBERG, C. A. & ZIEGLER, W. 1996. Devonian conodont biochronology in geologic time calibration. *Senckenbergiana lethaeae*, **76**, 259–265.

SANDBERG, C. A., STREEL, M. & SCOTT, R. A. 1972. Comparison between conodont zonation and spore assemblages at the Devonian–Carboniferous boundary in the western and central United States and in Europe. *In*: *Compte Rendu, 7th International Congress of Carboniferous Stratigraphy and Geology*, 23–28 August, Krefeld, Germany, **1**, 179–203.

SANDBERG, C. A., POOLE, F. G. & JOHNSON, J. G. 1988. Upper Devonian of Western United States. *In*: MCMILLAN, N. J., EMBRY, A. F. & GLASS, D. J. (eds) *Devonian of the World.* Canadian Society of Petroleum Geologists, Memoir, **14**, 183–220.

SANDBERG, C. A., MORROW, J. R. & ZIEGLER, W. 2002. Late Devonian sea-level changes, catastrophic events, mass extinctions. *In*: KOEBERL, C. & MACLEOD, K. G. (eds) *Catastrophic Events and Mass Extinctions: Impacts and Beyond.* Geological Society of America Special Papers, **356**, 473–487.

SANDBERG, C. A., MORROW, J. R., POOLE, F. G. & ZIEGLER, W. 2003. Middle Devonian to Early Carboniferous event stratigraphy of Devils Gate and Northern Antelope Range sections, Nevada, U.S.A. *Courier Forschungsinstitut Senckenberg*, **242**, 187–207.

SANDO, W. J. & SANDBERG, C. A. 1987. New interpretations of Paleozoic stratigraphy and history in the northern Laramie Range and vicinity, Southeast Wyoming. *US Geological Survey, Professional Paper*, **1450**, 1–39.

SANZ-LÓPEZ, J., GARCÍA-LÓPEZ, S., MONTESINOS, J. R. & ARBIZU, M. 1999. Biostratigraphy and sedimentation of the Vidrieros Formation (middle Famennian–lower Tournaisian) in the Gildar-Montó unit (northwest Spain). *Bolletino della Società Paleontologica Italiana*, **37**, 393–406.

SARTENAER, P. 1997. *Novaplatirostrum*, late Famennian rhynchonellid genus from Sauerland and Thuringia. *Bulletin de lInstitut royal des Sciences naturelles de Belgique, Sciences de la Terre*, **67**, 25–37.

SAVAGE N. M. 2013. *Late Devonian conodonts from Northwestern Thailand*. Bourland Printing, Trinity Press.

SAVOY, L. E., HARRIS, A. G. & MOUNTJOY, E. W. 1999. Extension of lithofacies and conodont biofacies models of Late Devonian to Early Carboniferous carbonate ramp and black shale systems, southern Canadian Rocky Mountains. *Canadian Journal of Earth Sciences*, **36**, 1281–1298.

SCHINDEWOLF, O. H. 1923. Beiträge zur Kenntnis des Palaeozoikums in Oberfranken, Ostthüringen und dem Sächsischen Vogtland. *I. Stratigraphie und Ammoneenfauna des Oberdefons von Hof a. d. Saale. Neues Jahrbuch für Mineralogie. Geologie und Paläontologie, Beilage-Band*, **49**, 250–357, 393–509.

SCHINDEWOLF, O. H. 1937. Zur Stratigraphie und Paläontologie der Wocklumer Schichten. *Abhandlungen der Preußischen Geologischen Landesanstalt, n. F.*, **178**, 1–132.

SCHMIDT, H. 1924. Zwei Cephalopodenfaunen an der Devon-Carbongrenze im Sauerland. *Jahrbuch der-Preußischen Geologischen Landesanstalt*, **44**, 98–171.

SCHÖNLAUB, H. P. 1969. Conodonten aus dem Oberdevon und Unterkarbon des Kronhofgrabens (Karnische Alpen, Österreich). *Jahrbuch der Geologischen Bundesanstalt*, **112**, 321–354.

SCHÖNLAUB, H. P., FEIST, R. & KORN, D. 1988. The Devonian–Carboniferous boundary at the section 'Grüne Schneid' (Carnic Alps, Austria): a preliminary report. *Courier Forschungsinstitut Senckenberg*, **100**, 149–167.

SCHÖNLAUB, H. P., ATTREP, M. *ET AL.* 1992. The Devonian/Carboniferous boundary in the Carnic Alps (Austria) – a multidisciplinary approach. *Jahrbuch der Geologischen Bundesanstalt*, **135**, 57–98.

SCHÜLKE, I. & POPP, A. 2005. Microfacies development, sea-level change, and conodont stratigraphy of Famennian mid- to deep platform deposits of the Beringhauser Tunnel section (Rheinisches Schiefergebirge, Germany). *Facies*, **50**, 647–664.

SCRUTTON, C. T. 1997. The Palaeozoic corals, I: origins and relationships. *Proceedings of the Yorkshire Geological Society*, **51**, 177–208, http://doi.org/10.1144/pygs.51.3.177

SEPKOSKI, J. J., JR. 1996. Patterns of Phanerozoic extinction a perspective from global data bases. *In*: WALLISER, O. H. (ed.) *Global Events and Event Stratigraphy in the Phanerozoic*. Springer, Berlin, 35–51.

SHEN, J.-W. 1994. New conodont data from benthic D/C boundary beds in Guilin. *Acta Micropalaeontologica Sinica*, **11**, 503–514.

SHILO, N. A., BOUCKAERT, J. *ET AL.* 1984. Sedimentological and paleontological atlas of the late Famennian and Tournaisian deposits in the Omolon region (NE–SSSR). *Annales de la Société géologique de Belgique*, **107**, 137–247.

SIEGMUND, H., TRAPPE, J. & OSCHMANN, W. 2002. Sequence stratigraphic and genetic aspects of the Tournaisian 'Liegender Alaunschiefer' and adjacent beds. *International Journal of Earth Sciences*, **91**, 934–949.

SIMAKOV, K. V. 1994. The Devonian-Carboniferous boundary in Russian Eurasia. *In*: THURSTON, D. K. & FUJITA, K. (eds) *Proceedings, International Conference on Arctic Margins, Anchorage*, September 1992, Alaska. OCS Study, Mineral Management Service, US Department of the Interior, Anchorage, 3–9.

SIMON, L., GODDERIS, Y., BUGGISCH, W., STRAUSS, H. & JOACHIMSKI, M. M. 2007. Modeling the carbon and sulfur isotope compositions of marine sediments: climate evolution during the Devonian. *Chemical Geology*, **246**, 19–38.

SMITH, G. M. & BUSTIN, R. M. 1998. Production and preservation of organic matter during deposition of the Bakken Formation (Late Devonian and Early Mississippian), Williston Basin. *Palaeogeography, Palaeoclimatology, Palaeoecology*, **142**, 185–200.

SMITHSON, T. R., WOOD, S. P., MARSHALL, J. E. A. & CLACK, J. A. 2012. Earliest Carboniferous tetrapod and arthropod faunas from Scotland populate Romer's Gap. *Proceedings of the National Academy of Sciences*, **109**, 4532–4537.

SOBOLEV, D. B., ZHURAVLEV, A. V. & TSYGANKO, V. S. 2000. Stop 8. Upper Devonian–Lower Carboniferous succession on the Kozhym River. *Ichthyolith Issues, Special Publications, Supplements*, **6**, 101–111.

SPREY, A. 2002. Morphometrie und Paläoökologie von Ammonoideen vor, während und nach globalen Faunenkrisen. *Münstersche Forschungen zur Geologie und Paläontologie*, **95**, 1–158.

STEARN, C. W. 1987. Effect of the Frasnian/Famennian extinction event on stromatoporoids. *Geology*, **15**, 677–680.

STEWART, I. J. 1981. Late Devonian and Lower Carboniferous conodonts from north Cornwall and their stratigraphical significance. *Proceedings of the Ussher Society*, **5**, 179–185.

STOCK, C. W. 2005. Devonian stromatoporoid originations, extinctions, and palaeobiogeography: how they relate to the Frasnian–Famennian extinction. *In*: OVER, D. J., MORROW, J. R. & WIGNALL, P. B. (eds) *Understanding Late Devonian and Permian-Triassic Biotic and Climatic Events: Towards an Integrated Approach*. Developments in Palaeontology & Stratigraphy, **20**. Elsevier, Amsterdam, 71–92.

STREEL, M. 1986. Miospore contribution to the upper Famennian–Strunian event stratigraphy. *Annales de la Société géologique de Belgique*, **109**, 75–92.

STREEL, M. 1999. Quantitative palynology of Famennian Events in the Ardenne–Rhine regions. *Abhandlungen der Geologischen Bundesanstalt*, **54**, 201–212.

STREEL, M. 2000. Global Famennian climates based on palynomorph quantitative analysis. *In*: RODRIGUES, M. A. C. & PEREIRA, E. (eds) *Ordovician–Devonian Palynostratigraphy in Western Gondwana: Update,*

Problems and Perspectives. UERJ, Rio de Janeiro, 77–103.

Streel, M. & Marshall, J. E. A. 2006. Devonian–Carboniferous boundary global correlations and their paleogeographic implications for the assembly of Pangaea. *In*: Wong, T. E. (ed.) *Proceedings of the XVth International Congress on Carboniferous and Permian Stratigraphy*, 10–16 August 2003, Utrecht, the Netherlands. Royal Netherlands Academy of Arts and Sciences, 481–496.

Streel, M. & Theron, J. N. 1999. The Devonian–Carboniferous boundary in South Africa and the age of the earliest episode of the Dwyka glaciation: new palynological result. *Episodes*, **22**, 41–44.

Streel, M. & Traverse, A. 1978. Spores from the Devonian/Mississippian transition near the Horseshoe Curve section, Altoona, Pennsylvania, U.S.A. *Review of Palaeobotany and Palynology*, **26**, 21–39.

Streel, M., Caputo, M. V., Loboziak, S. & Melo, J. H. G. 2000*a*. Late Frasnian–Famennian climates based on palynomorph analyses and the question of the Late Devonian glaciations. *Earth Science Reviews*, **52**, 121–173.

Streel, M., Caputo, M. V., Loboziak, S., Melo, J. H. G. & Thorez, J. 2000*b*. Palynology and sedimentology of laminites and tillites from the latest Famennian of the Parnaíba Basin, Brazil. *Geologica Belgica*, **3**, 87–96.

Streel, M., Caputo, M. V., Melo, J. H. G. & Perez-Leyton, M. 2013. What do latest Famennian and Viséan diamictites from Western Gondwana tell us? *Palaeobiodiversity and Palaeonevironments*, **93**, 299–316.

Strother, P. K. 2008. A speculative review of factors controlling the evolution of phytoplankton during Paleozoic time. *Revue de Micropaléontologie*, **51**, 9–21.

Su, Y., Wei, R. & Kuang, G. 1988. Conodonts from Devonian–Carboniferous boundary beds of Duolingshan, Yishan County of Guangxi, and their stratigraphic significance. *Acta Micropaleontologica Sinica*, **5**, 183–194.

Szulczewski, M. 1978. The nature of unconformities in the Upper Devonian–Lower Carboniferous condensed sequence in the Holy Cross Mts. *Acta Geologica Polonica*, **28**, 283–298.

Szulczewski, M., Belka, Z. & Skompski, S. 1996. The drowning of a carbonate platform: an example from the Devonian–Carboniferous of the southwestern Holy Cross Mountains, Poland. *Sedimentary Geology*, **106**, 21–49.

Tan, Z., Dong, Z., Coen, M., Hance, L., Hou, H. & Muchez, P. 1996. Famennian and Tournaisian lithostratigraphy of Central Hunan, South China. *Memoires de ĺInstitut Géologique de ĺUniversité de Louvain*, **36**, 11–19.

Thrasher, E. 1987. Macrofossils and stratigraphic subdivisions of the Bakken Formation (Devonian–Mississippian), Williston Basin, North Dakota. *In*: Carlson, C. G. & Christoper, J. E. (eds) *Fifth International Williston Basin Symposium*, Saskatchewan Geological Society, Regina, Special Publications, 53–67.

Tolokonnikova, Z. & Ernst, A. 2010. Palaeobiogeography of Famennian (Late Devonian) bryozoans. *Palaeogeography, Palaeoclimatology, Palaeoecology*, **298**, 360–369.

Tolokonnikova, Z., Ernst, A. & Herbig, H.-G. 2014*a*. Famennian (Upper Devonian) bryozoans from borehole Velbert 4, Rhenish Slate Massif (Germany). *Neues Jahrbuch fur Geologie und Paläontologie, Abhandlungen*, **273**, 25–44.

Tolokonnikova, Z., Ernst, A. & Wyse-Jackson, P. N. 2014*b*. Palaeobiogeography and diversification of Tournaisian–Viséan bryozoans (lower–middle Mississippian, Carboniferous) from Eurasia. *Palaeogeography, Palaeoclimatology, Palaeoecology*, **414**, 200–211.

Tolokonnikova, Z., Ernst, A., Poty, E. & Mottequin, B. 2015. Middle and uppermost Famennian (Upper Devonian) bryozoans from southern Belgium. *Bulletin of Geosciences*, **90**, 33–49.

Tourneur, F., Conil, R. & Poty, E. 1989. Données préliminaires sur les Tabulés et les Chaetetides du Dinantien de la Belgique. *Bulletin de la Société belge de Géologie*, **98**, 401–442.

Trela, W. & Malec, J. 2007. Zapis δ^{13}C w osadach pogranicza dewonu i karbonu w poludniowej czesci Gor Swietokrzyskich. *Przeglad Geologiczny*, **55**, 411–415.

Troth, I., Marshall, J. E. A., Racey, A. & Becker, R. T. 2011. Devonian sea-level change in Bolivia: a high palaeolatitude biostratigraphical calibration of the global sea-level curve. *Palaeogeography, Palaeoclimatology, Palaeoecology*, **304**, 3–20.

Tschigova, V. A. 1970. Correlation of Devonian and Carboniferous boundary beds in eastern and western Europe according to data resulting from the study of Ostracoda. *Compte Rendu Sixième Congrès International de Stratigraphie et de Géologie du Carbonifère*, 11–16 September 1967, Sheffield, **II**, 547–555.

Van Adrichem Boogaert, H. A. 1967. Devonian and Lower Carboniferous conodonts of the Cantabrian Mountains (Spain) and their stratigraphic application. *Leidse Geologische Mededelingen*, **39**, 129–192.

Van Bentum, E. C., Reichart, G.-J. & Sinninghe Damsté, J. S. 2012. Organic matter provenance, palaeoproductivity and bottom water anoxia during the Cenomanian/Turonian oceanic anoxic event in the Newfoundland Basin (northern proto North Atlantic Ocean). *Organic Geochemistry*, **50**, 11–18.

Van Geldern, R., Joachimski, M. M., Day, J., Jansen, U., Alvarez, F., Yolkin, E. A. & Ma, X. P. 2006. Carbon, oxygen and strontium isotope records of Devonian brachiopod shell calcite. *Palaeogeography, Palaeoclimatology, Palaeoecology*, **240**, 47–67.

Van Steenwinkel, M. 1980. Sedimentation and conodont stratigraphy of the Hastière Limestone, Lowermost Dinantian, Anseremme, Belgium. *Mededelingen Rijks geologische Dienst*, **32**, 30–33.

Van Steenwinkel, M. 1988. *The Sedimentary History of the Dinant Platform During the Devonian/Carboniferous Transition*. PhD thesis, Catholic University Leuven, Belgium.

Van Steenwinkel, M. 1993*a*. The Devonian–Carboniferous boundary in southern Belgium: biostratigraphic identification criteria of sequence boundaries. *Special Publications of the International Association of Sedimentology*, **18**, 237–246.

VAN STEENWINKEL, M. 1993*b*. The Devonian–Carboniferous boundary: comparison between the Dinant synclinorium and the northern border of the Rhenish Slate Mountains, a sequence-stratigraphic view. *Annales de la Société géologique de Belgique*, **115**, 665–681.

VAVRDOVÁ, M., ISAACSON, P. E., DIAZ, E. & BEK, J. 1991. Palinologia del límite Devónico-Carbonifera entorno al lago Titikaka, Bolivia: resultados preliminares. *Revista Técnica de YPFB*, **12**, 303–313.

VAVRDOVÁ, M., BEK, J., DUFKA, P. & ISAACSON, P. E. 1996. Palynology of the Devonian (Lochkovian to Tournaisian) sequence, Madre de Dios Basin, northern Bolivia. *Vestník Ceského Geologického Ustavu*, **71**, 333–349.

VEIZER, J., BUHL, D. *ET AL.* 1997. Strontium isotope stratigraphy: potential resolution and event correlation. *Palaeogeography, Palaeoclimatology, Palaeoecology*, **132**, 65–77.

VONHOF, H. B., SMIT, J., BRINKHUIS, H., MONTANARI, A. & NEDERBRAGT, A. J. 2000. Global cooling accelerated by early late Eocene impacts. *Geology*, **28**, 687–690.

WAGNER, P. 2001. Palynology at the Devonian/Carboniferous boundary: discussion of the sedimentological environment at sampled sections in Portugal, Ireland, U.S.A., and Canada. *Zentralblatt für Geologie und Paläontologie*, **1**, 185–197.

WALLISER, O. H. 1984. Pleading for a natural D/C-boundary. *Courier Forschungsinstitut Senckenberg*, **67**, 241–246.

WALLISER, O. H. 1996. Global events in the Devonian and Carboniferous. *In*: WALLISER, O. H. (ed.) *Global Events and Event Stratigraphy in the Phanerozoic*. Springer, Berlin, 225–250.

WANG, C. & MINJIN, C. 2004. Early Carboniferous age for Arynshand Formation, Mongolian South Gobi, based on conodonts. *Alcheringa*, **28**, 433–440.

WANG, C., YIN, B. *ET AL.* 1987. Devonian–Carboniferous Boundary sections in Yishan area, Guangxi. *In*: WANG, C. (ed.) *Carboniferous Boundaries in China*. Contribution to the 11th International Congress of Carboniferous Stratigraphy and Geology, 1987, Beijing, China. Science Press, Beijing, 22–43.

WANG, K., ATTREP, M., JR. & ORTH, C. J. 1993. Global iridium anomaly, mass extinction, and redox change at the Devonian-Carboniferous boundary. *Geology*, **19**, 776–779.

WANG, S. 2004. Mass extinction of Late Devonian Leperditicopids (Ostracoda). *In*: RONG, J. & FANG, Z. (eds) *Mass Extinction and Recovery. Evidences from the Palaeozoic and Triassic of South China*. University of Science and Technology of China Press, Hefei, Anhui Province, China, 357–366, 1054 [in Chinese with English summary].

WARREN, A., DI PASQUO, M. M., GRADER, G. W., ISAACSON, P. E. & RODRIGUEZ, A. P. 2014. Latest Famennian Middle Sappington Shale: *Lepidophyta-Verrucosisporites nitidus* (LN) Zone at the Logan Gulch type section, Montana, USA. *Geological Society of America Annual Meeting, Abstracts with Programs*, **46**, 163.

WATERS, J., SUTTNER, T. J., KIDO, E. & CARMICHAEL, S. K. 2014. The 'Age of Crinoids' began in the Late Devonian. *Geological Society of America Annual Meeting, Abstracts with Programs*, **46**, 80.

WEBB, G. E. 1998. Earliest known Carboniferous shallow-water reefs, Gudman Formation (TN1b), Queensland, Australia: implications for Late Devonian reef collapse and recovery. *Geology*, **26**, 951–994.

WEBB, G. E. 2002. Latest Devonian and Early Carboniferous reefs: depressed reef building after the Middle Paleozoic collapse. *In*: KIESSLING, W., FLÜGEL, E. & GOLONKA, J. (eds) *Phanerozoic Reef Patterns*. SEPM Special Publications, **72**, 239–269.

WEBB, G. E. 2005. Quantitative analysis and paleoecology of earliest Mississippian microbial reefs, Gudman Formation, Queensland, Australia: not just post-disaster phenomena. *Journal of Sedimentary Research*, **75**, 877–896.

WEBER, H. M. 2000. *Die karbonatischen Flachwasserschelfe im europäischen Oberfamennium (Strunium) – Fazies, Mikrobiota und Stromatoporen-Faunen*. PhD thesis, Mathematisch-Naturwissenschaftliche Fakultät der Universität zu Köln.

WENDT, J., KAUFMANN, B., BELKA, Z., KLUG, C. & LUBESEDER, S. 2006. Sedimentary evolution of a Palaeozoic basin and ridge system: the Middle and Upper Devonian of the Ahnet and Mouydir (Algerian Sahara). *Geological Magazine*, **143**, 269–299.

WEYANT, M. 1988. Relationships between Devonian and Carboniferous strata near the northern confines of the Bechar basin, Algeria. *Courier Forschungsinstitut Senckenberg*, **100**, 235–245.

WEYER, D. 1977. Ammonoidea aus dem Untertournai von Schleiz (Ostthüringisches Schiefergebirge). *Zeitschrift für geologische Wissenschaften*, **5**, 167–185.

WICANDER, R. & PLAYFORD, G. 2013. Marine and terrestrial palynofloras from transitional Devonian-Mississippian strata, Illinois Basin, U.S.A. *Boletin Geológico y Minero*, **124**, 589–637.

WICANDER, R., CLAYTON, G., MARSHALL, J. E. A., TROTH, I. & RACEY, A. 2011. Was the latest Devonian glaciation a multiple event? New palynological evidence from Bolivia. *Palaeogeography, Palaeoclimatology, Palaeoecology*, **305**, 75–83.

WISTORT, Z. P., OVER, D. J., HAGADORN, J. W., SOAR, L. K. & BULLECKS, J. 2014. Conodonts from the Upper Devonian–Lower Carboniferous Coffee Pot Member of the Dyer Formation Formation, Chaffee Group, Colorado. *Geological Society of America Annual Meeting, Abstracts with Programs*, **46**, 756.

WOOD, R. 2007. Famennian microbial-sponge reefs, Canning Basin, Australia. *In*: VENNIN, E., ARETZ, M., BOULVAIN, F. & MUNNECKE, A. (eds) *Facies from Palaeozoic Reefs and Bioaccumulations*. Muséum national d'histoire naturelle, Paris, Mémoires, **195**, 217–223.

WU, W., ZHANG, L., ZHAO, X., JIN, Y. & LIAO, Z. 1987. Carboniferous stratigraphy in China. *In*: ZHANG, L. (ed.) *Contribution to the 11th International Congress of Carboniferous Stratigraphy and Geology, 1987, Beijing, China*. Science Press, Beijing, 1–160.

XU, D.-Y., YAN, Z., ZHANG, Q.-W., SHEN, Z.-D., SUN, Y.-Y. & YE, L.-F. 1986. Significance of a δ13C anomaly near the Devonian/Carboniferous boundary at the Muhua section, South China. *Nature*, **321**, 854–855.

XU, H.-K., CAI, C.-Y., LIAO, W.-H. & LU, L.-C. 1990. Hongguleleng Formation in Western Junggar and the boundary between Devonian and Carboniferous. *Journal of Stratigraphy*, **14**, 292–301 [in Chinese with English abstract].

YANG, W. & NEVES, R. 1997. Palynological study of the Devonian–Carboniferous boundary in the vicinity of the International Auxiliary Stratotype section, Guilin, China. *In*: *Proceedings of the 30th International Geological Congress*, 21–27 August, Bejing, China, **12**, 95–107.

YAZDI, M. 1999. Late Devonian–Carboniferous conodonts from Eastern Iran. *Rivista Italiana di Paleontologia e Stratigrafia*, **105**, 167–200.

YOLKIN, E. A., KIM, A. I. & TALENT, J. A. (eds) 2008. Devonian sequences of the Kitab Reserve Area. *In*: *International Conference 'Global Alignments of Lower Devonian Carbonate and Clastic Sequences' (SDS/IGCP 499 Project Joint Field Meeting), Field Excursion Guidebook*, 25 August–3 September 2008, Kitab State Geological Reserve, Uzbekistan. SB RAS, Novosibirsk.

YOUNG, S. A., SALTZMAN, M. R., FOLAND, K. A., LINDER, J. S. & KUMP, L. R. 2009. A major drop in seawater $^{87}Sr/^{86}Sr$ during the Middle Ordovician (Darriwilian): links to volcanism and climate? *Geology*, **37**, 951–954.

YU, C. (ed.) 1988. *Devonian–Carboniferous Boundary in Nanbiancun, Guilin, China. Aspects and Records*. Science Press, Beijing.

YUAN, J. & XIANG, L. 1998. *Trilobite Fauna at the Devonian–Carboniferous Boundary in South China (S-Guizhou and N-Guangxi)*. National Museum of Natural Science, Taiwan, Special Publication, **6**, 1–281.

ZHANG, L. 1987. Carboniferous Stratigraphy in China. *In*: *Contributions to the 11th International Congress of Carboniferous Stratigraphy and Geology*. Science Press, Beijing.

ZHOU, Z. & CHEN, P. 1992. *Biostratigraphy and Geological Evolution of Tarim*. Science Press, Beijing.

ZHURAVLEV, A. V. 1998. The mid-Tournaisian event in the Northern Urals and conodont dynamics. *Proceedings of the Geological Association*, **109**, 161–168.

ZHURAVLEV, A. V. & TOLMACHEVA, T. J. 1995. Ecological recovery of conodont communities after the Cambrian/Ordovician and Devonian/Carboniferous eustatic events. *Courier Forschungsinstitut Senckenberg*, **182**, 313–323.

ZONG, P., MA, X.-P. & SUN, Y.-L. 2012. Productide, athyridide and terebratulide brachiopods across the Devonian-Carboniferous boundary in western Junggar, northwestern China. *Acta Palaeontologica Sinica*, **51**, 416–435.

ZONG, P., BECKER, R. T. & MA, X.-P. 2014. Upper Devonian (Famennian) and Lower Carboniferous (Tournaisian) ammonoids from western Junggar, Xinjiang, northwestern China – stratigraphy, taxonomy and palaeobiogeography. *Palaeodiversity and Palaeoenvironments*, **95**, 159–202, http://doi.org/10.1007/s12549-014-0171-y

Greenhouse to icehouse: a biostratigraphic review of latest Devonian–Mississippian glaciations and their global effects

J. A. LAKIN[1]*, J. E. A. MARSHALL[1], I. TROTH[2] & I. C. HARDING[1]

[1]*Ocean and Earth Science, University of Southampton, National Oceanography Centre, European Way, Southampton SO14 3ZH, UK*

[2]*BG Group plc, 100 Thames Valley Park Drive, Reading RG6 1PT, UK*

**Corresponding author (e-mail: jalakin@gmail.com)*

Abstract: The latest Devonian–Mississippian interval records the long-term transition from Devonian greenhouse conditions into the Late Palaeozoic Ice Age (LPIA). This transition was punctuated by three short glaciation events in the latest Famennian, mid-Tournaisian and Visean stages, respectively. Primary evidence for glaciation is based on diamictite deposits and striated pavements in South America, Appalachia and Africa. The aim of this review is to assess the primary biostratigraphic and sedimentological data constraining diamictite deposits through this transition. These data are then compared to the wider record of eustasy, mass extinction and isotope stratigraphy in the lower palaeolatitudes. Precise age determinations are vital to integrate high- and low-palaeolatitude datasets, and to understand the glacial control on wider global changes. Palynological techniques currently provide the best biostratigraphic tool to date these glacial deposits and to correlate the effects of glaciation globally. This review highlights a high degree of uncertainty in the known history of early LPIA glaciation as much of the primary stratigraphic data are limited and/or unpublished. Future high-resolution stratigraphic studies are needed to constrain the history of glaciation both spatially and temporally through the latest Devonian and Mississippian.

The latest Devonian–Mississippian time period saw the onset of the Late Palaeozoic Ice Age (LPIA), marked by three short 'precursor' glaciations in the latest Famennian, mid-Tournaisian and Visean, respectively (Caputo *et al.* 2008; Montañez & Poulsen 2013). In contrast, the earlier Devonian Period had globally warm climates, high sea levels and widespread reef complexes at median palaeolatitudes (Copper 2002; Joachimski *et al.* 2009). The latest Devonian and Mississippian therefore record a transitionary period between these two first-order climate modes, which was characterized by at least three distinct glacial episodes and long-term declining atmospheric CO_2 concentrations (Berner 2006). Within this transitionary period are global oceanic anoxic events associated with mass extinction, eustatic changes and isotopic excursions that approximately coincide with the glaciations in the latest Famennian and mid-Tournaisian (Streel 1986; Caplan & Bustin 1999; House 2002; Saltzman 2002; Kaiser *et al.* 2007, 2008, 2011, 2015; Sallan & Coates 2010; McGhee 2013; Yao *et al.* 2015).

Evidence for LPIA glaciation in Gondwana is based on diamictite deposits and striated pavements in both South America and Africa (Caputo *et al.* 2008; Isaacson *et al.* 2008). Recent work describing latest Devonian diamictites in Appalachia also provides evidence for an additional glacial centre in the low palaeolatitudes of Euramerica (Brezinski *et al.* 2008, 2010). The precise age of these glacial deposits is crucial in understanding the relationship between LPIA glaciation and its wider effects on eustasy, environmental change and mass extinction. There is uncertainty, however, over the exact timing of these glacial events. Several authors argue for an expanded range of glaciation that includes the Mid Devonian (Elrick *et al.* 2009) and Frasnian–Famennian (Isaacson *et al.* 1999; Streel *et al.* 2000*a*; McGhee 2013) based on indirect evidence from isotopes, palaeontology and palynology. Furthermore, the glaciation near the Devonian–Carboniferous (D–C) boundary has been regarded as both discrete (i.e. an individual precursor event) (Caputo *et al.* 2008) or intermittent throughout the entire Famennian–earliest Mississippian (Isaacson *et al.* 1999, 2008; Sandberg *et al.* 2002).

Two recent reviews of latest Devonian and Mississippian glaciation are provided by Caputo *et al.* (2008) and Isaacson *et al.* (2008). For reviews of the wider LPIA, see Fielding *et al.* (2008), Montañez & Poulsen (2013) and Limarino *et al.* (2014). This review does not reiterate these previous contributions but rather approaches the glacial histories of the latest Devonian and Mississippian from a biostratigraphic standpoint. The aims are to: (1) discuss published biostratigraphic schemes,

From: Becker, R. T., Königshof, P. & Brett, C. E. (eds) 2016. *Devonian Climate, Sea Level and Evolutionary Events*. Geological Society, London, Special Publications, **423**, 439–464.
First published online April 15, 2016, http://doi.org/10.1144/SP423.12

their limitations and correlation into South America; (2) assess the primary biostratigraphic and sedimentological evidence for glacial histories; and (3) to integrate these histories into a global context of mass extinction and environmental change. This will facilitate future research by providing an assessment of the timing of glaciations from the latest Devonian and Mississippian interval.

The review is divided into 'near-field' and 'far-field' phenomena. The former refers to those areas with direct evidence for glaciation (i.e. diamictites and/or striated pavements). 'Far field' refers to those areas with only proxy evidence for glaciation, which includes sequence-stratigraphic and palaeoenvironmental observations.

Background

Most of the continents at this time were part of either Euramerica or Gondwana (Fig. 1a). Gondwana was situated in the southern palaeolatitudes, and consisted of South America, Africa, the Arabian Plate, India, Australia and Antarctica.

The Panthalassic Ocean bordered western Gondwana along an active margin (Sempere 1995). To the north of Gondwana was the closing Variscan Sea. The subsequent collision of Gondwana with Euramerica during this closure was a key part in the assembly of Pangaea in the Late Palaeozoic. Putative ice centres have been suggested for the Brazilian Shield, the Guiana Shield, the Puna Arch and the Arequipa Massif in central South America, and along an orogenic highland belt on the southern active margin of Euramerica (Fig. 1a, b) (Díaz Martínez & Isaacson 1994; Isaacson *et al.* 2008; Brezinski *et al.* 2008, 2010). In addition, there are poorly constrained ice centres postulated in central Africa (Fig. 1c) (Isaacson *et al.* 2008).

The Late Devonian is an important interval in the evolution and diversification of land plants. This includes an increase in arborescence (i.e. trees became taller and with deeper root systems) and the evolution of seeds, which allowed vegetation to colonize dryer upland areas (Algeo & Scheckler 1998). The resulting increase in deep-soil formation and continental weathering may have caused the postulated reduction in atmospheric CO_2: thus, possibly setting the stage for glaciation in the latest Devonian and Carboniferous (Algeo *et al.* 1995; Algeo & Scheckler 1998; Berner 2006).

Published biostratigraphic schemes

In Europe and North America, Late Devonian and Mississippian successions are primarily dated using conodonts and goniatites (Becker *et al.* 2012; Davydov *et al.* 2012). These schemes are supplemented by miospore biostratigraphy, which is very well established in Western Europe (Clayton *et al.* 1977; Streel *et al.* 1987; Higgs *et al.* 1988; Maziane *et al.* 1999). The published global and regional biostratigraphic schemes discussed in this section are summarized in Figure 2.

The D–C boundary Global Stratotype Section & Point (GSSP) is at La Serre, France, and is based on the supposed evolutionary lineage of the conodonts *Siphonodella praesulcata* to *S. sulcata*. The first occurrence of the latter was used to define the base of the *sulcata* Zone and of the Carboniferous Period (Paproth & Streel 1984; Paproth *et al.* 1991). However, the practical use of the siphonodellids has been questioned by Kaiser (2009), based on the occurrence of *S. sulcata* stratigraphically below the level that defines the GSSP. In addition, the GSSP is picked within a reworked oolitic sequence that hinders its correlation potential. This issue is compounded by difficulties in taxonomic discrimination between *S. praesulcata* and *S. sulcata*, which may represent two morphological end members of a single species (Kaiser & Corradini 2011; Corradini *et al.* 2013). These findings reduce confidence in both the current placement of the D–C boundary GSSP and its inter-regional correlation. This is significant as the *praesulcata–sulcata* interval contains the first of the precursor glaciations in the latest Famennian and also the end-Devonian mass extinction.

The correlation of global schemes into South America

The spore record. The correlation of global biostratigraphic schemes into Gondwanan South America is of significant importance as this region has more reported occurrences of glacial diamictites than any other. Unfortunately, the dating of these sediments is problematic as conodonts and goniatites are extraordinarily rare in South America. In addition, the macro-fauna available for regional biostratigraphy, such as brachiopods, are largely endemic (Isaacson 1977). Hence, palynological analyses have the greatest biostratigraphic potential owing to the cosmopolitan nature of some miospore species that can be used for extra-Gondwanan correlation (Playford 1990) (Table 1). Significant research has improved Palaeozoic South American miospore taxonomy and established a miospore biostratigraphic scheme for the Amazon Basin in Brazil (Loboziak *et al.* 1986, 1999, 2000*a*, *b*, 2005; Melo & Loboziak 2000, 2003; Loboziak & Melo 2002; Playford & Melo 2009, 2010, 2012; Melo & Playford 2012). The Amazon Basin scheme has also been calibrated with established miospore biostratigraphy from Western Europe (i.e. Clayton *et al.* 1977; Streel *et al.* 1987). The reference sections for the Amazon Basin

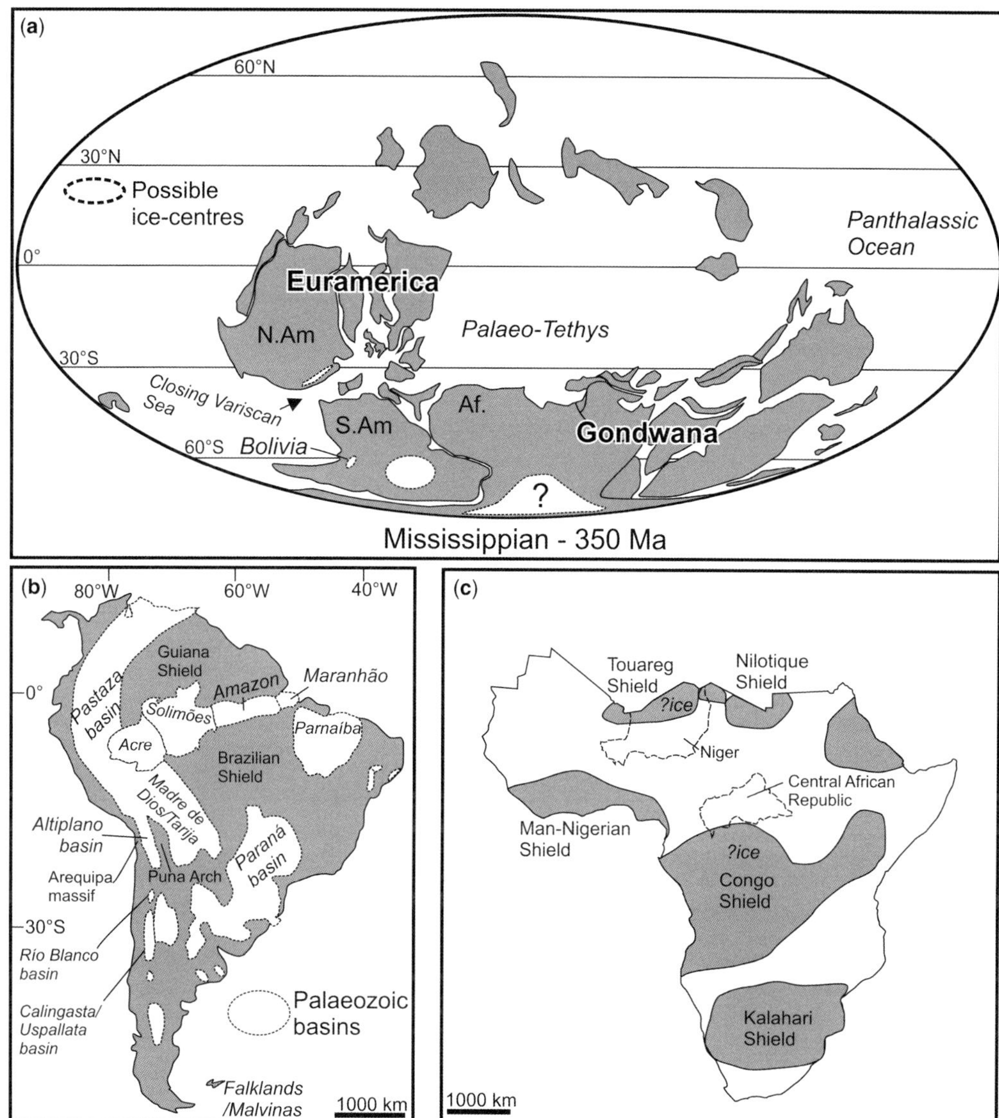

Fig. 1. (**a**) Palaeotectonic reconstruction for the Early Carboniferous (350 Ma) with putative ice centres overlain, redrawn from Domeier & Torsvik (2014). N. Am, North America; S. Am, South America; Af., Africa. (**b**) Palaeozoic basin map of South America, redrawn from Caputo *et al.* (2008). (**c**) Location of the Central African Republic and Niger in sub-Saharan Africa, with shield regions overlain, redrawn from Isaacson *et al.* (2008).

scheme unfortunately contain large stratigraphic gaps, which may therefore limit its usefulness outside the basin (Melo & Loboziak 2003; Melo & Playford 2012). Research into Palaeozoic palynostratigraphy in South America is ongoing, which will improve future correlations between Gondwana and Euramerica (e.g. see Troth *et al.* 2011; di Pasquo 2015; di Pasquo *et al.* 2015; Marshall 2016).

In the latest Famennian, the index miospore *Retispora lepidophyta* has a near-global distribution in a wide variety of settings. In Western Europe, the first occurrences of *Knoxisporites literatus*, *Indotriradites explanatus* (Luber) Playford, 1990 and *Verrucosisporites nitidus* subdivide the total range of *R. lepidophyta* into the LL (*lepidophyta–literatus*), LE (*lepidophyta–explanatus*)

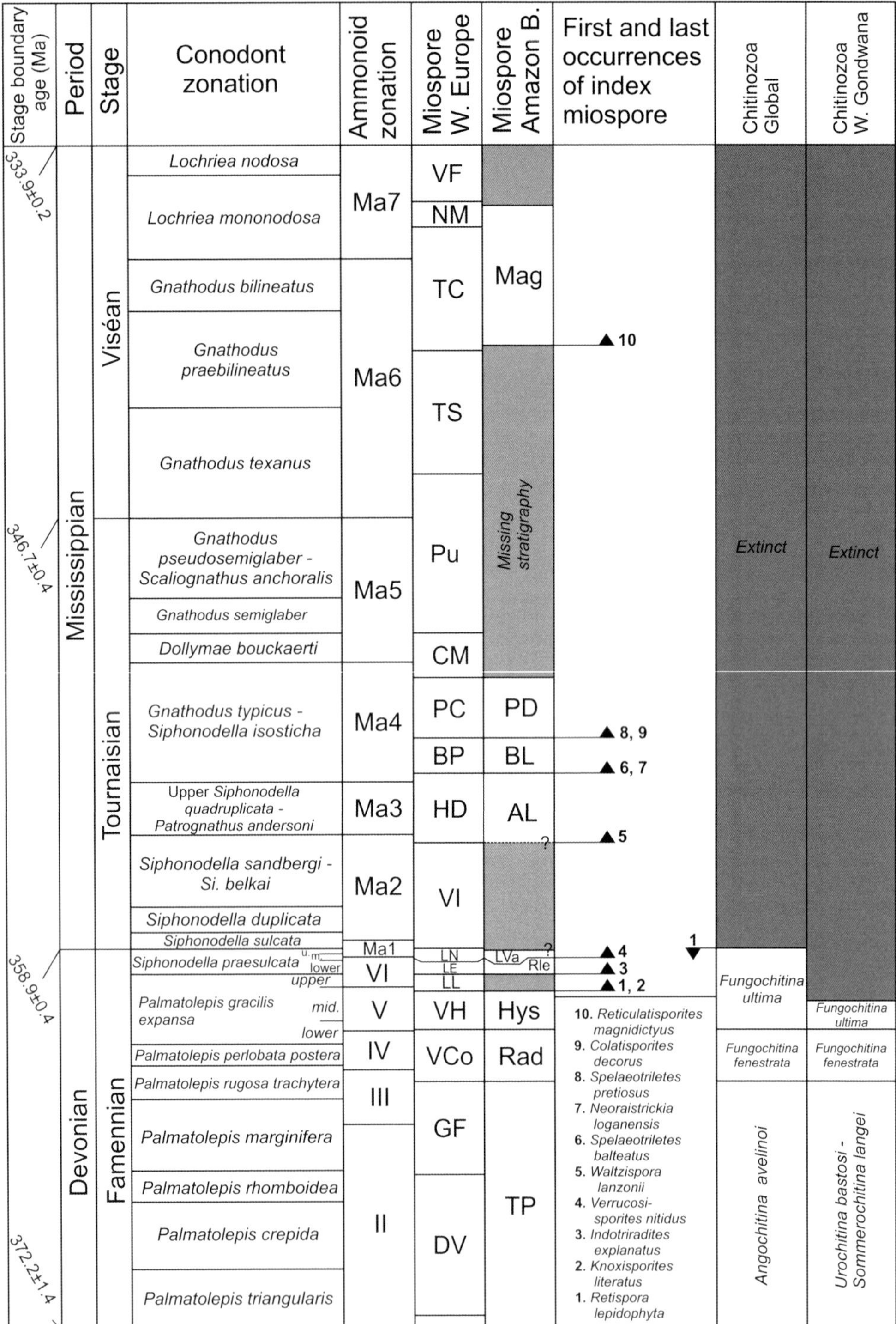

Fig. 2. Published biostratigraphic schemes for the Famennian, Tournaisian and Visean stages. Chronostratigraphic age, Stage/Period boundaries, and standard conodont and ammonoid schemes from Becker *et al.* (2012) and Davydov *et al.* (2012) (with corrections from Becker pers. comm.).

Table 1. *List of selected miospore, acritarch and prasinophyte taxa useful for stratigraphic correlation between Euramerica and Gondwana*

Miospore species
Retispora lepidophyta (Kedo) Playford, 1976
Knoxisporites literatus (Waltz) Playford, 1962
Indotriradites explanatus (Luber) Playford, 1990
Verrucosisporites nitidus Playford, 1964
Vallatisporites vallatus Hacquebard, 1957
Waltzispora lanzonii Daemon, 1974
Neoraistrickia loganensis (Winslow) Coleman & Clayton, 1987
Spelaeotriletes balteatus (Playford) Higgs, 1996
Spelaeotriletes pretiosus (Playford) Utting, 1987
Raistrickia clavata (Hacquebard) Playford, 1964
Reticulatisporites magnidictyus Playford & Helby, 1968
Rugospora australiensis (Playford and Helby) Jones & Truswell, 1992
Verrucosisporites quasigobbettii Jones & Truswell, 1992

Acritarch and prasinophyte species
Chomotriletes vedugensis Naumova, 1953
Gorgonisphaeridium ohioense (Winslow) Wicander, 1974
Maranhites mosesii (Brito) González, 2009
Stellinium micropolygonale (Stockmans and Williére) Playford, 1977
Umbellisphaeridium saharicum Jardiné *et al.*, 1972
Umbellisphaeridium deflandrei (Moreau-Benoit) Jardiné *et al.*, 1972
Horologinella quadrispina Jardiné *et al.*, 1972
(?)*Schizocystia bicornuta* Jardiné *et al.*, 1974

and LN (*lepidophyta–nitidus*) biozones, respectively (Streel *et al.* 1987). However, *K. literatus* and *V. nitidus* are rarely observed in the Amazon Basin miospore assemblages, and so the LE–LN zones are often undifferentiated with the LL Zone either missing or not identified (Melo & Loboziak 2003; Melo & Playford 2012). Therefore, Melo & Loboziak (2003) proposed a lower Rle (*lepidophyta*) Zone and an upper LVa (*lepidophyta–vallatus*) Zone in the Amazon Basin, based on the first occurrences of *R. lepidophyta* and *Vallatisporites vallatus*. The inception of *V. vallatus* is synchronous with that of *V. nitidus* in Western Europe: therefore, the Rle and LVa zones were argued to be South American equivalents of the LE and LN zones, respectively.

The extinction of *R. lepidophyta* occurs 14 cm below the D–C boundary as defined in the GSSP paratype section at Hasselbachtal, Germany (Higgs & Streel 1993). This extinction may not be entirely synchronous with the problematic boundary GSSP, but it is considered a useful global marker for the D–C boundary (Davydov *et al.* 2012). In the Amazon Basin, the last occurrence of *R. lepidophyta* and the first occurrence of *Waltzispora lanzonii* define the LVa–AL (*Radiizonates arcuatus–lanzonii*) zonal boundary. However, this boundary is an erosional surface and does not represent a continuously preserved D–C boundary sequence (Fig. 2) (Melo & Playford 2012).

In the Mississippian, there are several key miospores that exist in both Gondwana and Euramerica, the first occurrences of which can therefore be used for correlation. These include *Neoraistrickia loganensis* for the early-mid-Tournaisian and several species of *Spelaeotriletes*, such as *S. balteatus* and *S. pretiosus* (Playford *et al.* 2001; Brittain & Higgs 2007; Playford & Melo 2009).

Acritarch, prasinophyte and chitinozoan records

Acritarchs and prasinophytes represent the preserved cysts of marine phytoplankton. Their use in Devonian biostratigraphy is more limited than that of spores, but many Late Devonian species are known from both Euramerica and Gondwana (Le Hérissé *et al.* 2000; Molyneux *et al.* 2013). Such cosmopolitan taxa include *Chomotriletes vedugensis*, *Gorgonisphaeridium ohioense* and *Stellinium micropolygonale* (Molyneux *et al.* 2013) (Table 1). There is a distinct Famennian assemblage in Gondwana characterized by *Umbellisphaeridium saharicum*, *U. deflandrei*, *Horologinella quadrispina*, (?)*Schizocystia bicornuta* and *Maranhites mosesii* (Vavrdová & Isaacson 1999; Le Hérissé *et al.* 2000; Molyneux *et al.* 2013). *U. saharicum*, in particular, is associated with *R. lepidophyta* in South America (Daemon & Contreiras 1971; Vavrdová & Isaacson 1999; Wicander *et al.* 2011).

Acritarchs are of limited biostratigraphic value in the Mississippian as their abundance and diversity became increasingly diminished across the D–C boundary and into the Carboniferous (Mullins & Servais 2008). Marine phytoplankton did not recover until the Mesozoic. This period of low phytoplankton diversity has been termed the Late Palaeozoic Phytoplankton Blackout (Riegel 2008). However, a lack of detailed D–C reference sections for which there are comprehensive palynological records means that the decline in phytoplankton diversity is poorly constrained (Le Hérissé *et al.* 2000; Mullins & Servais 2008).

A global chitinozoan biostratigraphic scheme has been proposed for the Devonian, which is integrated with a regional scheme for Western Gondwana (Paris *et al.* 2000; Grahn 2005). Similarly to other fossil groups, however, there is a significant degree of endemism in South America (Troth 2006). Chitinozoa are useful for the wider Devonian Period, but suffer total extinction near the D–C boundary. In South America, this extinction is reported in the VH Miospore Zone (Grahn 2005; Grahn *et al.* 2006).

Near-field latest Famennian glaciation

The review by Caputo (1985) first argued for the glacial character of Famennian diamictites in Brazilian basins, which had until that point remained controversial. Since then, expressions of latest Famennian glaciations have been found to have a much wider geographical extent (Fig. 3). The primary evidence is based on diamictite deposits and striated pavements in Brazil, Peru and Bolivia of central South America (Caputo 1985; Caputo *et al.* 2008; Isaacson *et al.* 2008). In Brazil, this evidence is known from the Solimões, Amazon/Maranhão, Parnaíba and Paraná basins that border the Brazilian Shield (Caputo *et al.* 2008). In Bolivia and Peru, there are diamictites in the Altiplano Basin, and in the Sub-Andean Madre de Dios and Tarija basins (Díaz Martínez & Isaacson 1994; Isaacson *et al.* 1999). Brezinski *et al.* (2008, 2010) have described diamictites from Appalachia, on what was then the southern active margin of Euramerica. These are interpreted to have been deposited in a terrestrial foreland basin under subglacial, englacial and supraglacial settings (Brezinski *et al.* 2010). Evidence for glaciation in Appalachia broadens the geographical extent of latest Famennian glaciation into the lower palaeolatitudes.

Stratigraphic evidence from Brazil

The Maranhão and Amazon basins. The Maranhão and Amazon basins contain the Curiri Formation, which outcrops along two 500 km-long belts on the northern and southern basin margins (Caputo *et al.* 2008). An erosive unconformity separates the upper Curiri Formation from the diamictite-free lower Curiri Formation (Melo & Loboziak 2003). In subsurface wells, the upper Curiri Formation is an approximately 50 m-thick diamictite unit with subordinate siltstone beds (Cunha *et al.* 2007). Maximum glacial advance is represented by lobate diamictite deposits (4000 km^2) that contain 'floating' heterogeneous clasts and which overlie offshore to shoreface black shales and siltstones (Carozzi 1979). The diamictites are associated with incised subglacial channels and slump structures. The depositional model of Carozzi (1979) suggests glaciomarine conditions with ice-rafted debris and grounded ice sheets. Retreat occurs in the very latest Devonian. The upper Curiri Formation grades, both laterally and vertically, into the non-glacial Oriximiná Formation.

The Curiri Formation is associated with the enigmatic *Protosalvinia* fossil, of probably land plant origin, and the ichnofossil *Spirophyton,* a common feature of the uppermost Devonian in northern Brazilian basins (Carozzi 1979; Quijada *et al.* 2015). *Protosalvinia* is Famennian in age but probably pre-*R. lepidophyta* (Phillips *et al.* 1972; Loboziak *et al.* 1997; Over *et al.* 2009). The diamictites of the upper Curiri Formation contain *R. lepidophyta* and are assigned an undifferentiated LE–LN age (Loboziak *et al.* 1997). Melo & Loboziak (2003) indicated an equivalent Rle–LVa age for the upper Curiri Formation. However, there is no lithostratigraphic section that shows the vertical distribution of diamictites compared with the palynostratigraphy of Melo & Loboziak (2003).

Isaacson *et al.* (2008) reported a VCo, VH and LE–LN age range for glaciation in the Amazon Basin. This is probably based on Cunha *et al.* (1994), who showed the Curiri Formation as belonging to Zones 'VII–VIII' of Daemon & Contreiras (1971). These zones correspond roughly to the VCo–LN spore zones of Europe (Melo & Loboziak 2003). However, diamictites only occur within the upper unit of the Curiri Formation and not in the lower unit, so this extended age range down into the VCo Zone is likely to be an overestimate.

The Solimões Basin. The Jaraqui Diamictite Member (Jandiatuba Formation) is restricted to the subsurface in the Solimões Basin (Filho *et al.* 2007; Caputo *et al.* 2008; Isaacson *et al.* 2008). The unit is up to 50 m thick, and has a distinctive wireline response characterized by high gamma values and sharp changes in porosity, density and resistivity curves (Eiras *et al.* 1994). Eiras *et al.* (1994) labelled the member as glaciomarine and of Famennian–Tournaisian in age, while Filho *et al.* (2007) showed that it is restricted to the Famennian. The chronostratigraphic chart in Eiras *et al.* (1994) shows the Jaraqui Diamictite Member as belonging to the '*lepidophytus–spelaeotriletes*' Zone. The name '*lepidophytus*' refers to *R. lepidophyta* and so a latest Famennian age (LL–LE–LN) is likely.

Much of the primary sedimentological and biostratigraphic evidence for the glaciomarine deposits in this basin appears to derive from internal and unpublished company reports (e.g. Quadros 1988; Caputo & Silva 1990; Eiras *et al.* 1994). With the exception of Eiras *et al.* (1994), detailed information for diamictites in the Solimões Basin is not easily available. Although Caputo *et al.* (2008) have provided a brief lithological description of the diamictites and claimed that they are coincident with the upper Curiri Formation in the Amazon Basin.

The Parnaíba Basin. The Parnaíba Basin contains the Cabeças Formation, which can be observed both at outcrop and in the subsurface (Caputo *et al.* 2008). This formation consists of fine- to

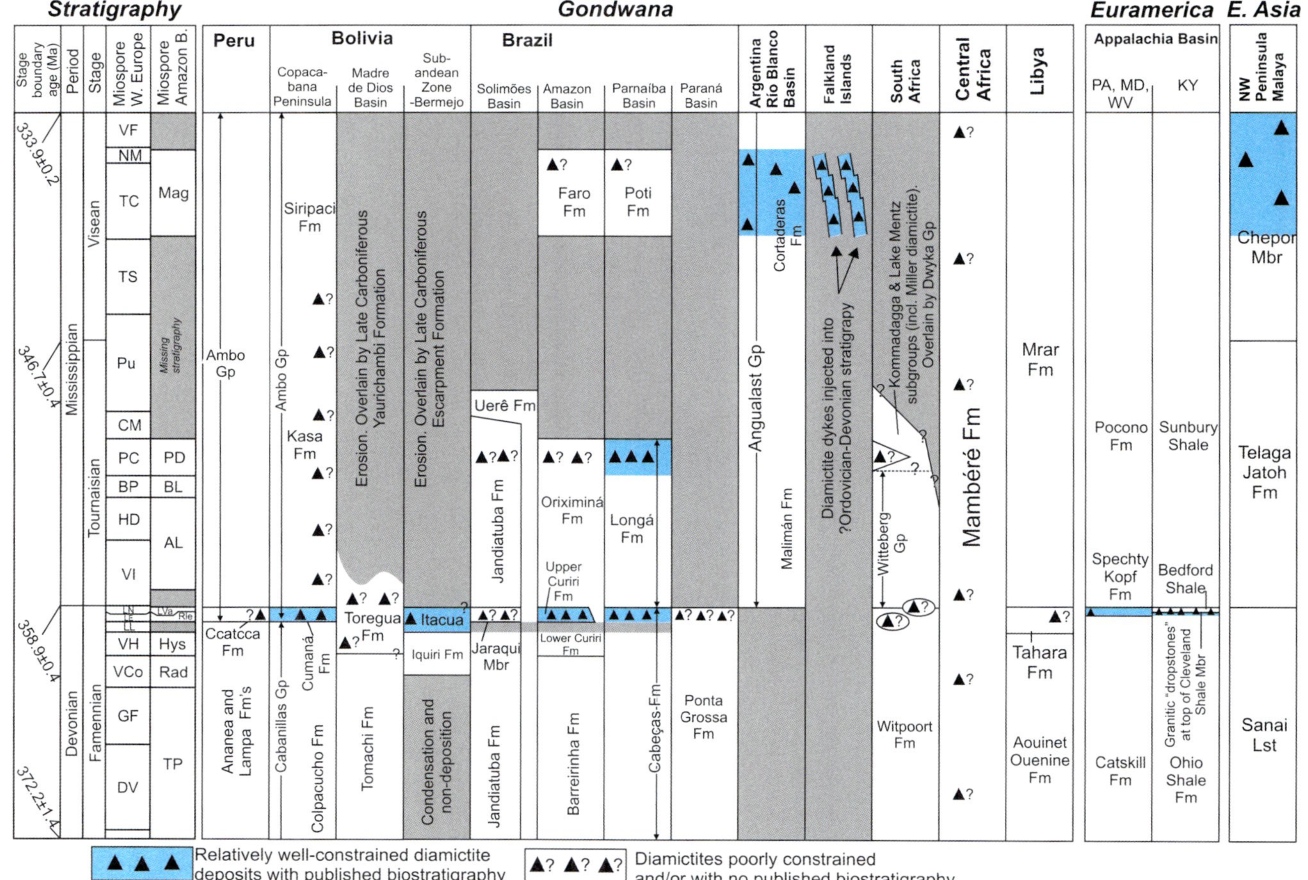

Fig. 3. A summarized overview of published diamictite occurrences and their stratigraphic position. Peru stratigraphy is referenced from Díaz-Martínez (2004); Bolivia from Díaz Martínez & Isaacson (1994), Isaacson *et al.* (1995), Troth (2006) and Wicander *et al.* (2011); Brazil from Cunha *et al.* (2007), Filho *et al.* (2007), Milani *et al.* (2007), Vaz *et al.* (2007) Caputo *et al.* (2008) and Melo & Playford (2012); Argentina from Perez Loinaze *et al.* (2010) and Limarino *et al.* (2014); the Falkland Islands/Las Malvina from Hyam *et al.* (1997); South Africa from Theron (1993), Evans (1999) and Streel & Theron (1999); Central Africa from Isaacson *et al.* (2008); Libya stratigraphy based on Klett (2000), with diamictites postulated in the Tahara Formation by Streel *et al.* (2000*a*); the USA Appalachian Basin from Brezinski *et al.* (2010) and Ettensohn *et al.* (2009) (state abbreviations: PA, Pennsylvania; MD, Maryland; WV, West Virginia; KY, Kentucky); and Malaysia from Meor *et al.* (2014). Fm, Formation; Gp, Group; Mbr, Member; Lst, limestone.

medium-grained, cross-bedded sandstones with interbedded siltstones and shales, which are interpreted as representing fluvial, deltaic–shelfal depositional environments (Caputo 1985; Góes & Feijó 1994; Vaz *et al.* 2007; Caputo *et al.* 2008). Diamictites occur with greater frequency in the upper part of the Cabeças Formation (Vaz *et al.* 2007). They contain striated clasts, and rest upon an unconformity surface and striated pavement (Caputo *et al.* 2008). The orientation of the striations suggests a glacial source area on the Brazilian Shield (Caputo *et al.* 2008). Equivalent deep-water strata are represented by varve-like rhythmites that contain marine acritarchs (Streel *et al.* 2000*b*).

The diamictites in the Parnaíba Basin were broadly inferred to be synchronous with those from the Amazon Basin by Carozzi (1980) based on their association with a *Protosalvinia–Spirophyton* assemblage, similar to the Amazon Basin. Rhythmites in the deeper part of the basin were designated a LN age based on a palynological analysis of individual millimetre-scale rhythmites (Streel *et al.* 2000*b*). In the subsurface, the Cabeças Formation was attributed an undifferentiated LE–LN age by Loboziak *et al.* (1992). A later investigation of sections in the Tocantins River Valley produced a LN Zone age for a single sample from the Cabeças Formation (Loboziak *et al.* 2000*a*).

The palynological investigations in the Parnaíba Basin are essentially based on single-spot samples. With the exception of Streel *et al.* (2000*b*), it is unknown whether these spot samples specifically date the diamictite facies or sediments in the associated stratigraphic sequences. There is no available palynostratigraphy to constrain the age of the incision surface at the base of the diamictites or the timing of the retreat, although final retreat is stated to have occurred immediately below the D–C boundary (Caputo *et al.* 2008).

The Paraná Basin. Milani *et al.* (2007) informally described a 1.5 m-thick diamictite interval as the 'Ortigueira Diamictite' in the uppermost part of the Late Devonian Ponta Grossa Formation. The unit is unconformably overlain by diamictites of the Late Carboniferous–Early Permian Itararé Group (Milani *et al.* 2007). It was extremely difficult to differentiate these latest Devonian diamictites from the overlying Itararé Group. It was only when they were analysed palynologically that a latest Famennian age was constrained (Loboziak *et al.* 1995*b*). The diamictites contain the following miospores *R. lepidophyta, V.* cf. *vallatus* and *Vallatisporites hystricosus*, and were interpreted to be of LN age (Loboziak *et al.* 1995*b*). Loboziak *et al.* (1995*b*) is, however, a preliminary report and so does not contain any illustrations/descriptions of the palynological material or any measured sections. Caputo *et al.* (2008) described the diamictites as grey, muddy with clasts of varying size. No further descriptions of latest Famennian diamictites in the Paraná Basin have been published and, without detailed sedimentary logs and facies descriptions, the depositional environment remains unknown.

Stratigraphic evidence from Bolivia

The Bolivian Altiplano. The latest Famennian glaciation in the Bolivian Altiplano is represented by the Cumaná Formation exposed near Lake Titicaca (Fig. 4a, b). The entire unit contains striated and polished exotic clasts (Díaz Martínez & Isaacson 1994). There are three generalized lithofacies: the lowermost lithofacies comprises outer-shelf laminated shales with ice-rafted dropstones (Díaz Martínez & Isaacson 1994; Díaz-Martínez *et al.* 1999); the second lithofacies consists of massive muddy to sandy diamictites with large blocks, interpreted as subaqueous debris flows; the uppermost lithofacies is the most proximal, and consists of interbedded cross-stratified sandstones and diamictites deposited within periglacial subaqueous outwash fans. This vertical association of lithofacies was interpreted as a single glacial advance into a glaciomarine environment (Díaz Martínez & Isaacson 1994).

Palynological samples from the Cumaná Formation and associated stratigraphy were collected from Villa Molino, Hinchaka and Isle del Sol by Díaz-Martínez *et al.* (1999). The samples contain *R. lepidophyta* and *I. explanatus*. The absence of *V. nitidus* means that the LN Zone cannot be recognized. The diamictites were also associated with a relatively diverse marine acritarch and prasinophyte assemblage, which includes *U. saharicum, Maranhites mosesii, Pterospermella* spp. and *Exochoderma irregulare*. The D–C boundary was recognized at Villa Molino between samples 9a and 9b (Fig. 4c). Both *R. lepidophyta* and Devonian phytoplankton were reported to disappear above the boundary (Díaz-Martínez *et al.* 1999). Sample 6 at Hinchaka was collected above the diamictites (Fig. 4c), and contained only a single acritarch species and no *R. lepidophyta*, which suggests that the D–C boundary and phytoplankton extinctions occur within or immediately above the diamictite.

Sub-Andean Zone – the Madre de Dios and Tarija basins. Diamictites in the central Cordillera and Sub-Andean Zones are assigned to the Itacua Formation. In the eastern Cordillera, diamictites are known in the Saipuru Formation and have been attributed to the *R. lepidophyta* Zone (Suárez-Soruco & López-Pugliessi 1983; Caputo *et al.* 2008).

Towards the SW, in the Madre de Dios Basin, diamictites form part of the Toregua Formation (Isaacson *et al.* 1995). The Toregua Formation was

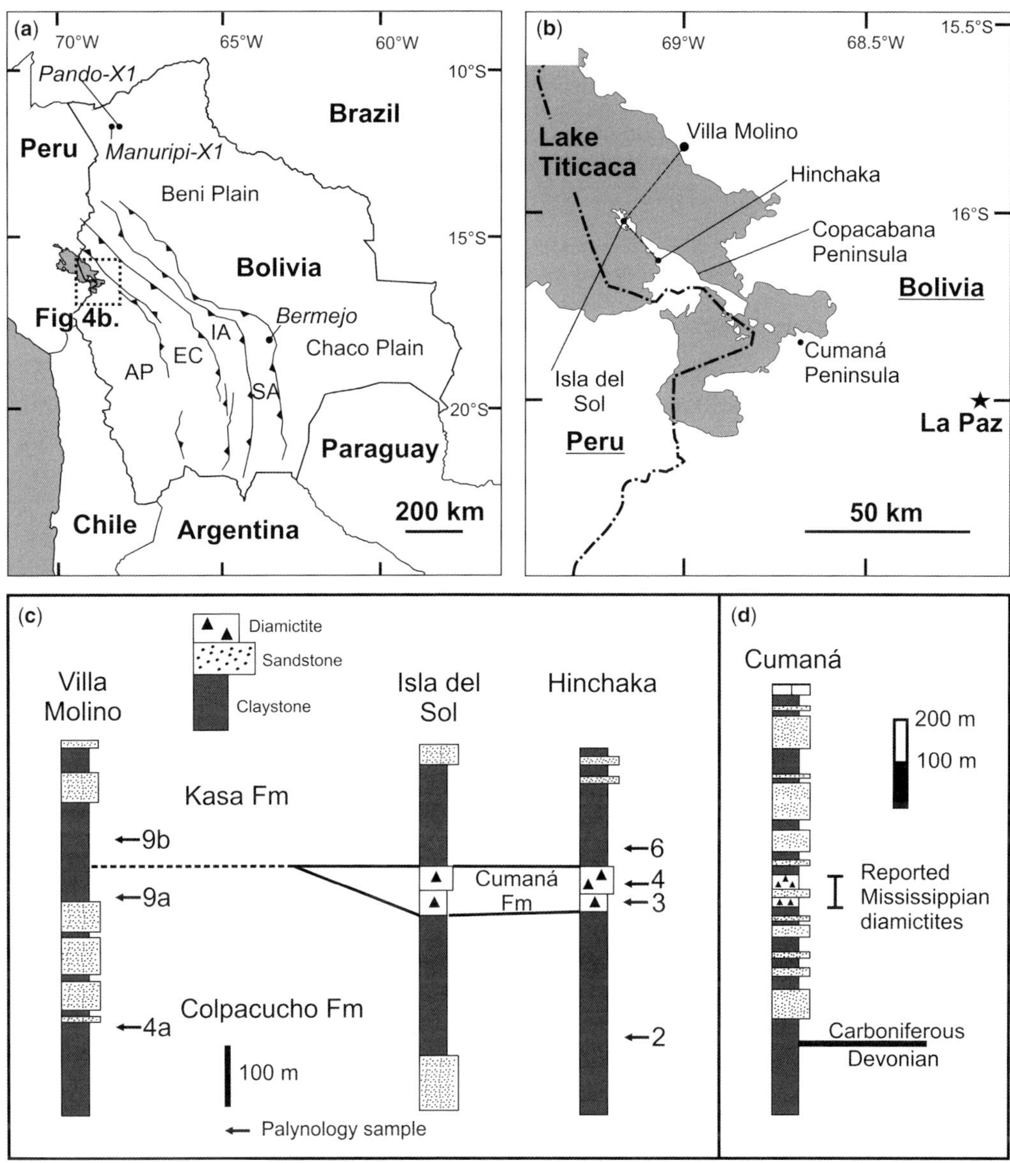

Fig. 4. Stratigraphic evidence for latest Famennian glaciation in the Bolivian Altiplano. (**a**) Tectonic map of Bolivia redrawn from Sempere (1995) and Barnes *et al.* (2012). AP, Altiplano; EC, Eastern Cordillera; IA, Inter-Andean Zone; SA, Sub-Andean Zone. The locations of the Manuripi X-1 and Pando X-1 onshore wells are shown, as is the Bermejo section of Wicander *et al.* (2011) (see Fig. 5). (**b**) Map of the Bolivian Altiplano near Lake Titicaca, with the location of key sections shown. (**c**) The Villa Molino, Hinchaka and Isle del Sol sections, which have been sampled for palynology. Redrawn from Díaz-Martínez *et al.* (1999). (**d**) Stratigraphic log from the Cumaná Peninsula that shows the stratigraphic position of the Mississippian diamictites. Redrawn from Díaz Martínez (1996).

reported to straddle the D–C boundary, as based on the first downhole occurrence of *R. lepidophyta* and *U. saharicum* in the 'Pando X-1' and 'Manuripi X-1' wells (Isaacson *et al.* 1995) (see Fig. 4a for their location).

Sub-Andean Zone – the Bermejo section. The most detailed available biostratigraphic evidence regarding latest Devonian glaciation in the Sub-Andean Zone is from Wicander *et al.* (2011), presenting a sedimentary log through the 18 m-thick diamictite

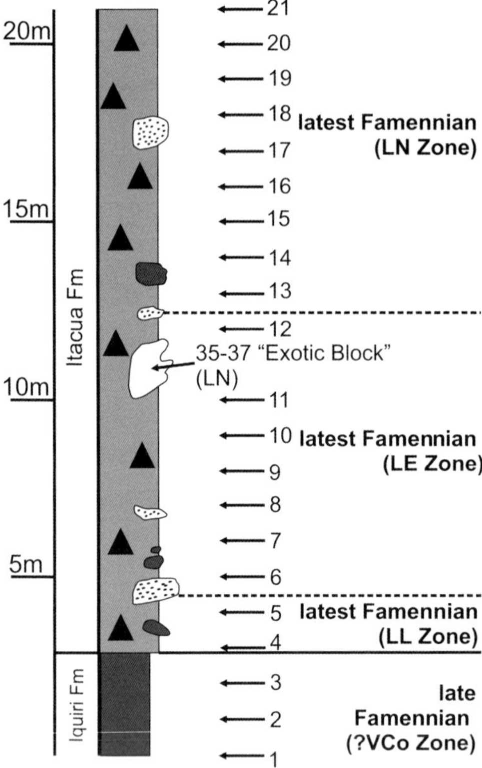

Fig. 5. Measured section of the latest Famennian and glaciomarine Itacua Formation at Bermejo. See Figure 4a for the geographical location of this section and Figure 4c for the lithology symbol key. Numbered arrows represent collected samples. Redrawn from Wicander *et al.* (2011).

unit representing the Itacua Formation at Bermejo (Fig. 5; see Fig. 4a for the geographical location). The unit contains sandstone lenses, exotic clasts, sandstone lithic boulders and overlies a sheared basal contact. It was interpreted as a glaciomarine environment based on the presence of marine acritarchs. The palynological analysis suggests a sequence of glacial events through the LL, LE and LN zones based on the stepwise occurrence of *R. lepidophyta*, *I. explanatus* and *V. nitidus*. This is significant as it potentially extends the onset of glaciation into the older LL Zone stratigraphy, which is typically missing in Brazil.

There is a degree of palynological reworking observed in the Bermejo section, which may affect any biostratigraphic interpretations. Wicander *et al.* (2011) recognized that several of the acritarch and prasinophyte species in the diamictites Itacua Formation were probably reworked from the Middle Devonian. In addition, Late Devonian miospore species have been reworked into the Mississippian Saipuri Formation sitting directly above the Itacua diamictites (Perez-Leyton 1991; Streel *et al.* 2013). The latest Devonian miospores in the Itacua Formation, however, are likely not to have been reworked considering that no exclusively Carboniferous species were identified by Wicander *et al.* (2011).

Before the publication of the data from Bermejo, the Bolivian diamictites had been interpreted as representing a single glacial advance (e.g. Díaz Martínez & Isaacson 1994). Wicander *et al.* (2011), in contrast, suggested that a series of glacial events spanned the entire LL–LE–LN range of *R. lepidophyta*. This range defines the 'Strunian' interval in Belgium, which has an estimated duration of 1–3 myr, suggesting that the glaciations were of a similar duration (Trapp *et al.* 2004; Streel *et al.* 2006). A sandstone lithic boulder clast (i.e. a non-diamictite lithology) in the Itacua Formation at Bermejo was palynologically dated as being of LN Zone age ('exotic block' in Fig. 5). The surrounding country rock, however, was interpreted as the older LE Zone. This suggests that the boulder was transported by ice in a frozen state and redeposited into older unconsolidated sediment (Wicander *et al.* 2011). The 1–3 myr duration and evidence of ice reworking makes it unlikely this is a single glacial cycle. Rather, the succession was interpreted as a composite of several deglaciation events shedding sediment into a glaciomarine environment (Wicander *et al.* 2011).

Streel *et al.* (2013) challenged the interpretation of Wicander *et al.* (2011) and argued that the entire Itacua Formation is more likely to have been deposited over a single 100 kyr glacial event within part of the LE–LN zones. They base their challenge on three lines of reasoning: (1) that the LE–LN zones are not easily differentiated in Brazil owing to the rarity of *V. nitidus* specimens. The absence, therefore, of *V. nitidus* in the country rock surrounding the exotic boulder would not necessarily be indicative of the LE Zone; (2) that the taxonomic assignment of their figured specimen of *V. nitidus* is questionable; and (3) that the record at Bermejo appears to contradict the far-field eustatic and oxygen isotope response observed in Europe, which indicates a cold regressive interval within a much narrower age range in the LN Zone only (Walliser 1984; Kaiser *et al.* 2007, 2008, 2015; Streel *et al.* 2013).

The palynological sampling at Bermejo was systematic over a continuous section and the published range chart in Wicander *et al.* (2011) only shows *V. nitidus* occurrences in the upper part of the Itacua Formation. The occurrence of *V. nitidus* was not sporadic, but was, instead, stratigraphically confined, which is why the upper part of the Itacua Formation was interpreted as being of LN Zone age.

The observation of rare *V. nitidus* in Brazilian spore assemblages may therefore not be comparable to contemporaneous assemblages in Bolivia. In addition, using an indirect far-field record to reinterpret primary near-field biostratigraphic data may be unwise and could lead to circular reasoning. Whether glaciation in the latest Devonian consists of a single short event or a composite of many is a contentious issue and requires further investigation. Furthermore, stacked glacial cycles may be difficult to recognize, as the younger glacial cycles would remove evidence of earlier glaciation during erosive ice advances.

Stratigraphic evidence from Peru

Diamictites in SE Peru have been reported in various conference abstracts and termed the Ccatcca Formation (Carlotto *et al.* 2004; Cerpa *et al.* 2004; Díaz-Martínez 2004). This formation has been correlated to the Cumaná Formation in Bolivia based on its lithological similarity and is stated to be of late Famennian age (Díaz-Martínez 2004; Isaacson *et al.* 2008). The Ccatcca Formation is approximately 100 m thick and has been split into three distinct units (Isaacson *et al.* 2008). The lowest consists of laminated shales containing dropstones, overlain by massive diamictites with evidence of gravitational reworking. The uppermost unit consists of sandstones with 'disharmonic' folding and hummocky–swaley cross-stratification (Isaacson *et al.* 2008). This vertical association is similar to the Cumaná Formation on the Bolivian Altiplano 400 km to the SE.

Stratigraphic evidence from Libya

Latest Famennian glaciation is postulated to have existed in south Libya by Streel *et al.* (2000*a*). This is based on an unpublished study of the A.1-NC 58 wildcat well in Ghadames Basin. It was noted that this well contained conglomerates, sandstones and diamictites in the Tahara Formation that were interpreted to be of glacial origin (Streel *et al.* 2000*a*). The Tahara Formation is latest Famennian age and contains *R. lepidophyta* (Coquel & Moreau-Benoit 1986).

Stratigraphic evidence from South Africa

Putative pre-Dwyka Group glacial events have been recognized in the Devonian–Early Carboniferous Witteburg Group by Almond *et al.* (2002) in South Africa. The stratigraphic lowest of these is within the Perdepoort Member of Famennian age. The Perdepoort Member is predominantly a texturally mature orthoquartzite, which contains lenticular to tabulate diamictite units. These diamictites are clast-poor, sandy, weakly bedded and have a lateral extent of hundreds of kilometres. They were interpreted as slump deposits formed during rapid deglaciation events by Almond *et al.* (2002). A glacial interpretation is supported by the reports of clasts within the diamictite-bearing striated surfaces (Almond *et al.* 2002).

Stratigraphic evidence from North America

Recent evidence has strongly suggested an additional glacial centre in the Appalachian Basin, then situated in the temperate latitudes (Cecil *et al.* 2004; Brezinski *et al.* 2008, 2010). This is based on mudstone and diamictite sequences in the Spechty Kopf and Rockwell formations exposed along a 400 km outcrop belt (Brezinski *et al.* 2008, 2010; Brezinski & Cecil 2015). These were deposited within a glacio-lacustrine to proglacial terrestrial environment, and consist of a single preserved ice advance and retreat trend (Brezinski *et al.* 2010). Evidence for glaciation also extends into Kentucky, where a 3 tonne granitic boulder clast (i.e. a dropstone) can be observed in the black mudstones of the Cleveland Member of the offshore-marine Ohio Shale Formation (Lierman & Mason 2007; Ettensohn 2008; Ettensohn *et al.* 2008, 2009). Ice probably nucleated on highland areas generated by the Acadian Orogeny and expanded towards the NW into the central Appalachian Basin (Lierman 2007).

The diamictites are reported to contain *R. lepidophyta* and correlate to the *praesulcata* Condont Zone (Brezinski *et al.* 2008; Ettensohn 2008). Woodrow & Richardson (2006) reported that the diamictite sequences of the Spechty Kopf Formation represent a minor portion of the LE Zone. In contrast, the exotic granitic boulders in the Ohio Shale Formation are apparently situated within country rock dated as being of LN age. The latter LN age determinations of the Ohio Shale were cited as unpublished written communications (Ettensohn *et al.* 2009, p. 31; Brezinski *et al.* 2010, p. 276). An LE–LN zonal range is not unreasonable, but insufficient biostratigraphic data are currently available, and it is not known whether the palynological analyses were based on spot samples or a more systematic study. Furthermore, Rooney *et al.* (2015) noted that detailed biostratigraphic resolution of the Cleveland Member is difficult owing to the uniform palynological assemblage that is dominated by amorphous organic matter.

Near-field mid-Tournaisian glaciation

The mid-Tournaisian glacial event has a lesser geographical extent compared to that of the latest Famennian, and is only known from Brazil (Fig. 3) and possibly South Africa. Detailed evidence for the mid-Tournaisian glaciation event is poorly constrained, with limited published data (Caputo *et al.* 2008).

Brazilian basins

The Amazon Basin. In the Amazon Basin, the Tournaisian section of the Oriximiná Formation is reported to contain diamictites restricted to the subsurface (Caputo *et al.* 2008). They consist of claystones with mixed sand, gravel and pebble clasts. However, there is no reference to diamictite facies in published well sections that penetrate the Oriximiná Formation (e.g. Cunha *et al.* 2007; Melo & Playford 2012). The stratigraphic distribution, depositional environment and age determination of diamictites in the Oriximiná Formation are therefore unknown.

The Solimões Basin. In the subsurface of the Solimões Basin, Tournaisian diamictites are reported in the upper part of the Jaraqui Member (Caputo *et al.* 2008). However, Filho *et al.* (2007) showed the Jaraqui Member to be restricted to the Famennian only. The diamictites are reported to have been correlated to the BP–PC zones of Western Europe by Caputo *et al.* (2008): however, two of the cited references for this age determination are unpublished reports (Loboziak *et al.* 1994*a*, *b*). The third cited reference, Loboziak *et al.* (1995*a*), does not mention Tournaisian diamictites, and provides no palynological descriptions or illustrations. Caputo *et al.* (2008), however, did provide a relatively detailed lithological description of the diamictites: although, no original sources were cited. As such, interpretations of the depositional environment and age of diamictites in the Solimões Basin cannot be validated.

The Parnaíba Basin. The Parnaíba Basin provides the most detailed account of mid-Tournaisian diamictites, which are constrained to the upper part of the Longá Formation (Playford *et al.* 2012). The diamictites are 63 m thick in the UN-24 well in the northern part of the basin, and are interbedded with thin shales and sandstones (Playford *et al.* 2012). Diamictites are poorly sorted, mudstone-dominated and with randomly orientated pebble clasts. They are interpreted as ice rainout deposits affected by gravitational reworking and ice-keel scouring (Lobato 2010, reported in Playford *et al.* 2012). The diamictites contain a well-preserved palynological assemblage that was correlated to the PC Zone (Playford *et al.* 2012). This was based on the occurrences of *Colatisporites decorus*, *N. loganensis*, *S. balteatus*, *S. pretiosus* and *Raistrickia clavata*, amongst others.

South Africa

The LPIA in South Africa is represented by diamictites of the Dwyka Group, which unconformably overly Late Devonian–Mississippian pre-glacial sediments. This basal contact contains evidence of glacially influenced soft-sediment deformation in the uppermost pre-glacial Waaipoort Formation, indicating the latter was still unconsolidated sediment by the time of the first glacial advance in South Africa (Streel & Theron 1999). The biostratigraphic dating of the Devonian and Mississippian sequences has historically been difficult due to contradictory ages suggested by miospore, floral and vertebrate remains, and the high degree of thermal maturity that degrades and darkens palynological material (Theron 1993). It was Streel & Theron (1999) who described identifiable palynological material for the first time in the Waaipoort Formation: miospores including *S. balteatus* and *S. pretiosus* were interpreted to be no older than mid-Tournaisian. As these sediments contain evidence of glacially influenced soft-sediment deformation, these palynological data suggest a mid-Tournaisian phase of glaciation in South Africa immediately after the deposition of the uppermost Waaipoort Formation.

The Miller Diamictite in the Kommadagga Subgroup is laterally discontinuous but sits stratigraphically above the Waaitpoort Formation and below the Dwyka Group (Swart 1982; Theron 1993; Evans 1999). The units association with the underlying Waaipoort Formation could suggest a glacial origin for this unit. The Miller diamictite is up to 6 m in thickness and consists of fine- to medium-grained muddy–sandy matrix with clasts up to 2 cm in diameter (Swart 1982). The depositional environment of the Miller Diamictite was described as 'problematic' by Swart (1982), who interpreted it as a debris-flow deposit within a prograding delta on the basis of there being no conclusive evidence of a glacial origin. Further work is needed to better constrain the depositional environment and age of the Miller Diamictite. Despite being succeeded by the overlying Dwyka Group, of definitive glacial origin, these units are separated by a significant unconformity (Swart 1982; Isbell *et al.* 2008), and so any interpretations derived from the Dwyka Group should not be applied to the underlying sequence.

Near-field Visean glaciation

The known evidence for Visean glaciation comes from the Brazilian basins, Argentina, the Falkland Islands and Malaysia (Fig. 3) (Hyam *et al.* 1997; Caputo *et al.* 2008; Limarino *et al.* 2014; Meor *et al.* 2014).

Brazil

The Faro Formation, in the Amazon Basin, consists of approximately 400 m of sandstones and shales interpreted as representing fluvial–deltaic to storm-influenced shelfal conditions (Cunha *et al.* 1994, 2007). Diamictites are reported from the subsurface of 'possible glacial derivation' by Caputo *et al.* (2008). These are potentially correlative to similar deposits in the Jandiatuba Formation of the Solimões Basin.

In the Parnaíba Basin, Caputo *et al.* (2008) described sandstones and diamictites in the Poti Formation, which contains clasts of varying size dispersed in a muddy matrix. These are reported to be exposed at outcrop. The sedimentological information reported in Caputo *et al.* (2008) comes from an unpublished PhD thesis (Andrade 1972). Vaz *et al.* (2007) described the Poti Formation in the subsurface as tidal to deltaic in nature, consisting of a lower unit of sandstones overlain by shales and coals, and made no mention of diamictites.

The Faro and Poti formations have been constrained to the Mag Zone in the midde–late Visean, defined by the range of the miospore *Reticulatisporites magnidictyus* (Melo & Loboziak 2000, 2003; Melo & Playford 2012). Streel *et al.* (2013) provided a list of miospore species present in the Poti Formation, with photomicrograph plates included. No diamictite facies are shown in published well or outcrop data that include either the Faro or Poti formations (Melo & Loboziak 2000; Cunha *et al.* 2007; Vaz *et al.* 2007; Melo & Playford 2012). Detailed sedimentological and biostratigraphic data regarding Visean diamictites in Brazilian basins are therefore extremely limited and difficult to validate.

Stratigraphic evidence from Argentina and the Falkland Islands/Islas Malvinas

The evidence for Visean glaciation in the Río Blanco Basin of Argentina is very well constrained, and is based on integrated sedimentological, palynological and U–Pb ages (Gulbranson *et al.* 2010; Perez Loinaze *et al.* 2010), and represents the initiation of the Visean–Serpukhovian Glacial Stage in the LPIA (Isbell *et al.* 2003; Limarino *et al.* 2014). Diamictites belong to the Cortaderas Formation, and consist of dropstones in shale facies and massive–stratified diamictites interpreted as an ice-distal glaciomarine environment (Perez Loinaze *et al.* 2010). There is a rich and well-preserved miospore assemblage that contains *R. magnidictyus*, *Rugospora australiensis* and *Verrucosisporites quasigobbettii*. These are of biozonal significance and have first occurrences in the mid–late Visean. The spore assemblage was determined by Perez Loinaze *et al.* (2010) as belonging to the MQ (*Reticulatisporites magnidictyus–Verrucosisporites quasigobbettii*) Biozone of western Argentina, correlated to the Mag Zone in the Amazon Basin (Melo & Loboziak 2003; Perez Loinaze 2007; Melo & Playford 2012). It is interesting that evidence of glaciation in the Mag Zone of the Visean in South America contrasts with early Visean warm-water faunal elements in northern Argentina, characterized by the goniatite *Michiganites scalabrinii* (House 1996).

In the Falkland Islands, there are sub-vertical diamictite dykes hosted within (?)Ordovician–Devonian strata (Hyam *et al.* 1997). These were interpreted as the downwards injection of diamicton into a sedimentary host rock during an episode of glaciation. The diamictites were analysed palynologically, and were found to contain *R. magnidictyus* and *V. quasigobbettii* (Hyam *et al.* 1997). This indicates a late Visean age with a palynoflora correlative to that of Argentina.

Bolivia: Mississippian glaciation(s)?

Diamictite deposits reported in the Tournaisian–Visean Kasa Formation in the Bolivian Altiplano are interpreted as sediment gravity flows (Fig. 4d) (Oviedo Gomez 1965; Díaz Martínez 1991, 1996; Díaz-Martínez *et al.* 1993). These are thought to be the distal expression of proglacial outbursts that would have occurred in more proximal coastal or braided alluvial settings (Díaz Martínez 1993; Díaz-Martínez *et al.* 1993; Díaz Martínez & Isaacson 1994; Isaacson *et al.* 2008). These deposits are not known to contain independent ice indicators, such as faceted clasts or striated pavements, nor are they constrained biostratigraphically. Therefore, to prove whether these deposits represent additional Mississippian glaciations in the Bolivian Altiplano requires additional study (Caputo *et al.* 2008).

Diamictites of glaciomarine origin are reported in the Mississippian Kaka Formation of the Bolivian Sub-Andean Belt (Suárez-Soruco 2000; Caputo *et al.* 2008). These deposits are not associated with any published sedimentological or biostratigraphic data, but are said to be correlative to the Kasa Formation in the Bolivian Altiplano (Caputo *et al.* 2008).

An early Visean age has been assigned to diamictites of the Itacua Formation at Balapuca in southern Bolivia (di Pasquo 2007*a*, *b*). At Bermejo, some 500 km north of Balapuca, the lowermost 18 m of the Itacua Formation have been dated as latest Famennian (Wicander *et al.* 2011). The Sub-Andean mid-Palaeozoic record contains several hiatuses and it is likely that older glacial episodes have been directly, and unconformably, overlain by younger ones in the preserved stratigraphic record (Streel *et al.* 2013).

Stratigraphic evidence in East Asia

The Chepor Member in the Kubang Pasu Formation of NW Malaysia contains a diamictite facies (pebble clasts in muddy sandstones) interpreted as suspension fall-out deposits in a glaciomarine environment (Meor *et al.* 2014). The Chepor Member is fossiliferous, with diverse communities of brachiopods, trilobites, gastropods, tabulate corals and bivalves (Meor *et al.* 2014). A Visean–late Visean age was interpreted by Meor *et al.* (2014) primarily based on the bivalve *Posidonia* sp., the trilobite assemblage and the goniatite genus *Praedaraelites*.

Near-field glaciation of uncertain age in Central Africa

The Mambéré Formation crops out in the SW of the Central African Republic and is interpreted as lacustrine to glaciolacustrine (Censier *et al.* 1985). Based on palaeomagnetic arguments, an age within the Mid Devonian–Mississippian interval was estimated. This fits generally into the presence of Late Devonian–Mississippian glaciations, but provides no further detail and should be regarded as questionably dated. There are additional reported occurrences of pre-Visean diamictites in Niger (Lang *et al.* 1991; Isaacson *et al.* 2008).

Far-field evidence

The Hangenberg Crisis and end-Devonian mass extinction

The Hangenberg Crisis in the latest Devonian and earliest Tournaisian was a protracted biotic event associated with mass extinction, eustatic changes and positive carbon isotope excursions (Walliser 1984; Brand *et al.* 2004; Kaiser *et al.* 2007, 2008, 2011, 2015). The associated mass extinction is considered a 'top-6' event in terms of ecological severity (McGhee *et al.* 2012, 2013). Its effects were wide ranging, affecting the marine benthic realm (ostracods, stromatoporoids, and trilobites) and the pelagic realm (conodonts, ammonoids and forams), and led to the extinction of terrestrial vertebrates and land plants (Becker 1993; Caplan & Bustin 1999; Hallam & Wignall 1999; Streel *et al.* 2000*a*; Clack 2007; Sallan & Coates 2010; Kaiser *et al.* 2015).

The Hangenberg Crisis *sensu stricto* is defined in reference sections exposed in the Rhenish Massif of Western Europe (Fig. 6), which begin with the transgressive Hangenberg Black Shale (HBS) in the latest Famennian mid *praesulcata* Zone. The HBS contains the main extinction pulse, and is correlated with geographically widespread black shale deposition, marine anoxia, and positive isotope excursions in both organic and inorganic carbon (Fig. 7) (Kaiser *et al.* 2007). Positive carbon isotope excursions associated with the Hangenberg Crisis are widely believed to have been caused by a global increase in the burial of organic carbon (Kaiser *et al.* 2008, 2015).

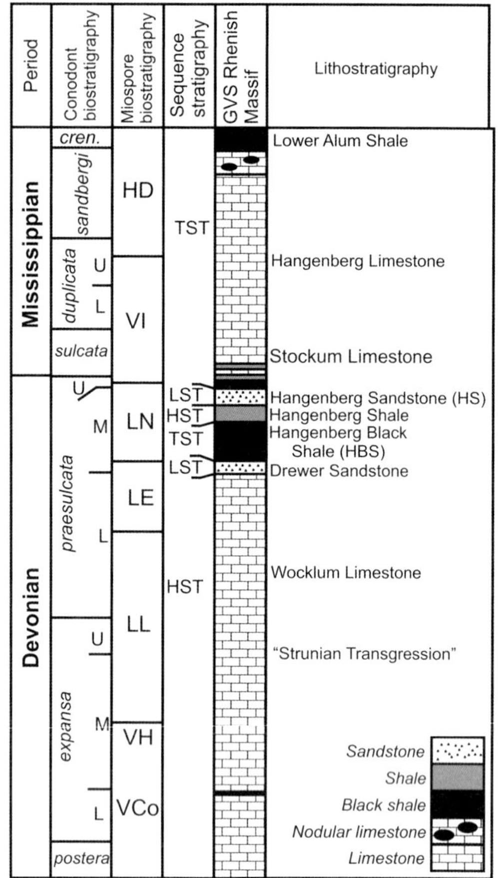

Fig. 6. Lithostratigraphy and biostratigraphy of the Hangenberg Crisis *sensu stricto* in the Rhenish Massif, Germany. Redrawn from Kaiser *et al.* (2011). LST, lowstand systems tract; TST, transgressive systems tract; HST, highstand systems tract.

The HBS is overlain by a regressive unit known as the Hangenberg Sandstone (HS) in the Rhenish Massif. Elsewhere in the Rhenish Massif, some 100 m of incision is represented by the Seiler Channel, which is infilled by the Seiler Conglomerate and Hangenberg Shale. The incision is interpreted as being due to sea-level drawdown constrained to the LN Zone (Bless *et al.* 1992; Higgs & Streel 1993). Regression across Western and Eastern Europe has been correlated with the HS using gamma-ray, geochemical and sedimentological

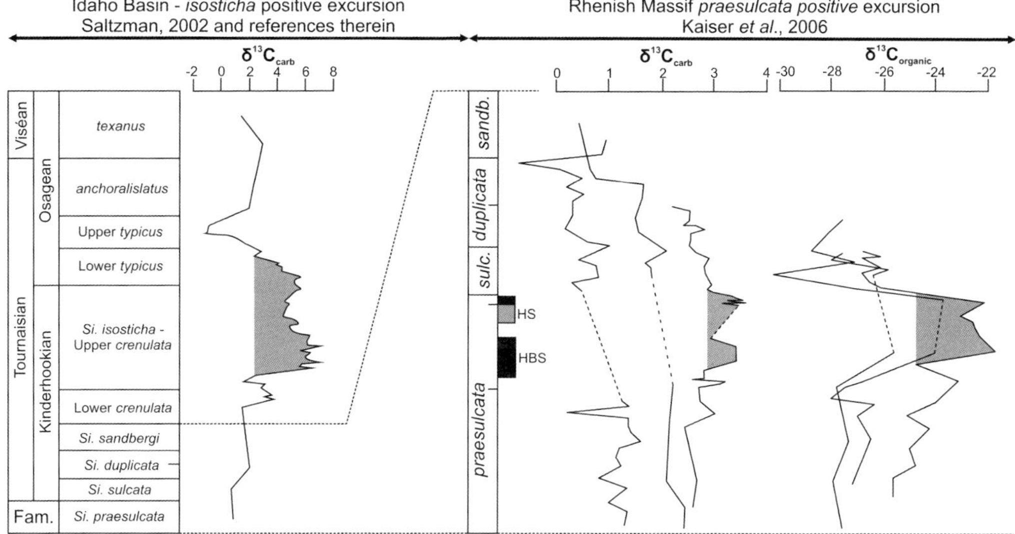

Fig. 7. Positive carbon isotope excursions in the latest Famennian *praesulcata* conodont Zone and the mid-Tournaisian *isostichia* conodont Zone from reference sections in the Rhenish Massif and the Idaho Basin. HBS, Hangenberg Black Shale *sensu stricto*; HS, Hangenberg Sandstone *sensu stricto*. Isotope curves have been redrawn from Saltzman (2002) and Kaiser *et al.* (2006).

proxies (Kumpan *et al.* 2013, 2014). Incision is also observed near the D–C boundary throughout North America (fig. 15 in Brezinski *et al.* 2010). In northern Gondwana, there is a significant pulse of siliciclastic sediment and 100 m of relative sea-level fall immediately below the D–C boundary in the Moroccan Anti-Atlas, which was correlated with the HS in the Rhenish Massif (Kaiser *et al.* 2011). In the Central Asian Orogenic Belt, there is evidence of an increased detrital supply just below the D–C boundary, which extends evidence for a eustatic drawdown into an open-ocean island arc environment (Carmichael *et al.* 2016). Regression on a global scale, which is represented by the HS in the Rhenish Massif, has been inferred to be synchronous with the main pulse of glaciation at high palaeolatitudes (Kaiser *et al.* 2011).

Above the HS, there is marine transgression containing a secondary extinction pulse at the D–C boundary, which affected the miospores, acritarchs/prasinophytes and clymeniid ammonoids (Kaiser *et al.* 2011, 2015). There is also a secondary isotope excursion constrained to this level (Kaiser *et al.* 2015).

Late Devonian biotic events, which include the Hangenberg Crisis, have been shown to group into long-period (2.4 myr) eccentricity cycles (de Vleeschouwer *et al.* 2013). Within this longer-term cyclicity are shorter periodicities of 100 kyr in duration. The transgressive–regressive couplet of the Hangenberg Crisis may have occurred over one such short 100 kyr eccentricity cycle. This would be consistent with the suggestion by Streel *et al.* (2013) of a single 100 kyr glacial cycle in Gondwana, and potentially the single advance of glaciation both in Appalachia and the Bolivian Altiplano (Díaz Martínez & Isaacson 1994; Brezinski *et al.* 2008, 2010). However, more precise information is needed as to the nature and timing of the precursor glaciations before an orbital control can be identified.

Lower Alum Shale and Siphonodella isostichia *isotope excursions*

In Germany, there is a transgressive black shale unit known as the 'Liegender Alaunschiefer' (Lower Alum Shale), interpreted as the result of eutrophication, anoxia and high organic productivity (Becker 1993; Siegmund *et al.* 2002). The Lower Alum Shale is overlain by prograding highstand carbonates and an erosional sequence boundary in the Velbert area (fig. 2 in Siegmund *et al.* 2002). The stratigraphic extent of the Liegender Alaunschiefer ranges through the mid–late Tournaisian (the Tn2–Tn3 zones), which overlaps with Tournaisian glaciation in South America during the PC–PD zones (Siegmund *et al.* 2002; Playford *et al.* 2012). Transgression and anoxic conditions during the Lower Alum Shale have been correlated with the Gondwanan record in the Anti-Atlas Mountains, Morocco, and dated to the base of the mid-

Tournaisian (Kaiser *et al.* 2011, 2013). Weathered black shales, interpreted as representing anoxic conditions, are overlain by regressive sandstones at the El Atrous section, a lithological signature similar to that of the Hangenberg Crisis lower down in the same section (Kaiser 2005; Kaiser *et al.* 2011, 2013, 2015). The Lower Alum Shale could represent a mid-Tournaisian 'Hangenberg Crisis equivalent', in which glaciation is associated with a transgressive–regressive couplet and marine anoxia in the far-field records. Further study is needed to integrate these near-field and far-field records in the mid-Tournaisian.

In North America, there is a $\leq$7‰ positive carbon isotope $\delta^{13}C_{carbonate}$ excursion within the mid-Tournaisian *Siphonodella isosticha* conodont Zone of the Kinderhookian regional stage (Fig. 7). This is roughly coincident with both the Lower Alum Shale and mid-Tournaisian glaciation in Gondwana (Saltzman 2002; Playford *et al.* 2012). Positive carbon isotope excursions can be correlated with the Russian Platform (Mii *et al.* 1999) and South China (Qie *et al.* 2011, 2015; Yao *et al.* 2015). The cause of these excursions is interpreted as the global-scale burial of organic carbon (Yao *et al.* 2015). Despite the apparent global scale of the event, there is a wide variation in the magnitudes and absolute values of $\delta^{13}C_{carbonate}$ excursions, which Yao *et al.* (2015) attributed to spatial differences in marine nutrient concentrations.

Mid-Tournaisian $\delta^{13}C_{carbonate}$ excursions in South China are accompanied by a positive shift in $\delta^{15}N$ of 1.5–4.2‰, which was interpreted to reflect enhanced water-column denitrification (Yao *et al.* 2015). Significantly, this positive shift in $\delta^{15}N$ did not return to pre-excursion values, but rather remained relatively positive into the lower part of the *G. typicus* conodont Zone. This positive shift in the mid-Tournaisian is inferred to be the initiation of elevated global $\delta^{15}N$ values throughout the entire LPIA (Algeo *et al.* 2014; Yao *et al.* 2015) caused by changes in oceanic circulation and a lower eustatic sea level during icehouse climate modes, which favoured enhanced water-column denitrification in continental-margin oxygen minimum zones. These perturbations in carbon–nitrogen isotopes are synchronous with sustained decreases in oxygen isotopic data, suggesting long-term global cooling from the mid-Tournaisian onwards (Buggisch *et al.* 2008; Yao *et al.* 2015). Yao *et al.* (2015) have interpreted these isotopic trends as reflecting the mid-Tournaisian onset of sustained continental glaciation during the LPIA.

There is additional evidence of Tournaisian regression and palaeovalley incision throughout North America at the Kinderhookian–Osagean regional stage boundary (Kammer & Matchen 2008). This implies that regression and incision in North America may immediately post-date positive carbon isotope excursions reported from the mid- to late Kinderhookian (Saltzman 2002).

Visean

Smith & Read (2000) interpreted a Visean onset of the LPIA within the *G. bilineatus* conodont Zone, based on a sequence stratigraphic interpretation of the Illinois Basin. This is correlative to the Mag Zone in the Amazon Basin (Fig. 2). They describe five sequences that have extensive deep palaeovalley incision at their sequence boundaries. This suggests that the Visean glaciation event may have consisted of multiple glacial cycles of advance and retreat, resulting in multiple stacked sedimentary cycles in the Illinois Basin.

Discussion

Uncertainties in the near-field record

This review has highlighted a relative paucity of detailed analyses of LPIA glacial deposits in the public domain. Brazil is particularly problematic, not least because many reported diamictite occurrences are from proprietary subsurface well data (Caputo *et al.* 2008). In Figure 8, only those reports of glaciation that are either stratigraphically and/or biostratigraphically constrained with original and fully published data are shown. This demonstrates that only a minority of studies provide integrated sedimentological, biostratigraphic and/or chronostratigraphic data through the diamictite sequences (e.g. Hyam *et al.* 1997; Díaz-Martínez *et al.* 1999; di Pasquo 2007*a*, *b*; Gulbranson *et al.* 2010; Perez Loinaze *et al.* 2010; Wicander *et al.* 2011; Playford *et al.* 2012). The LPIA can be shown to be strongly diachronous across certain regions and between individual basins (Montañez & Poulsen 2013; Limarino *et al.* 2014). This diachroneity, combined with the limited amount of published data, means that it is difficult to confidently assess the precise timing of glacial events in the latest Devonian–early Mississippian interval.

Where latest Famennian diamictites have been palynologically analysed, they are typically associated with *R. lepidophyta* and *U. saharicum*. As such, there is no direct near-field evidence that support an extended duration of glaciation in the Middle Devonian (Elrick *et al.* 2009), at the Frasnian–Famennian boundary (Streel *et al.* 2000*a*) nor through the entire Famennian–earliest Tournaisian (Isaacson *et al.* 1999, 2008; Sandberg *et al.* 2002). Diamictites are typically associated with sheared or erosional basal contacts. There is little biostratigraphic constraint on the timing of erosion and ice

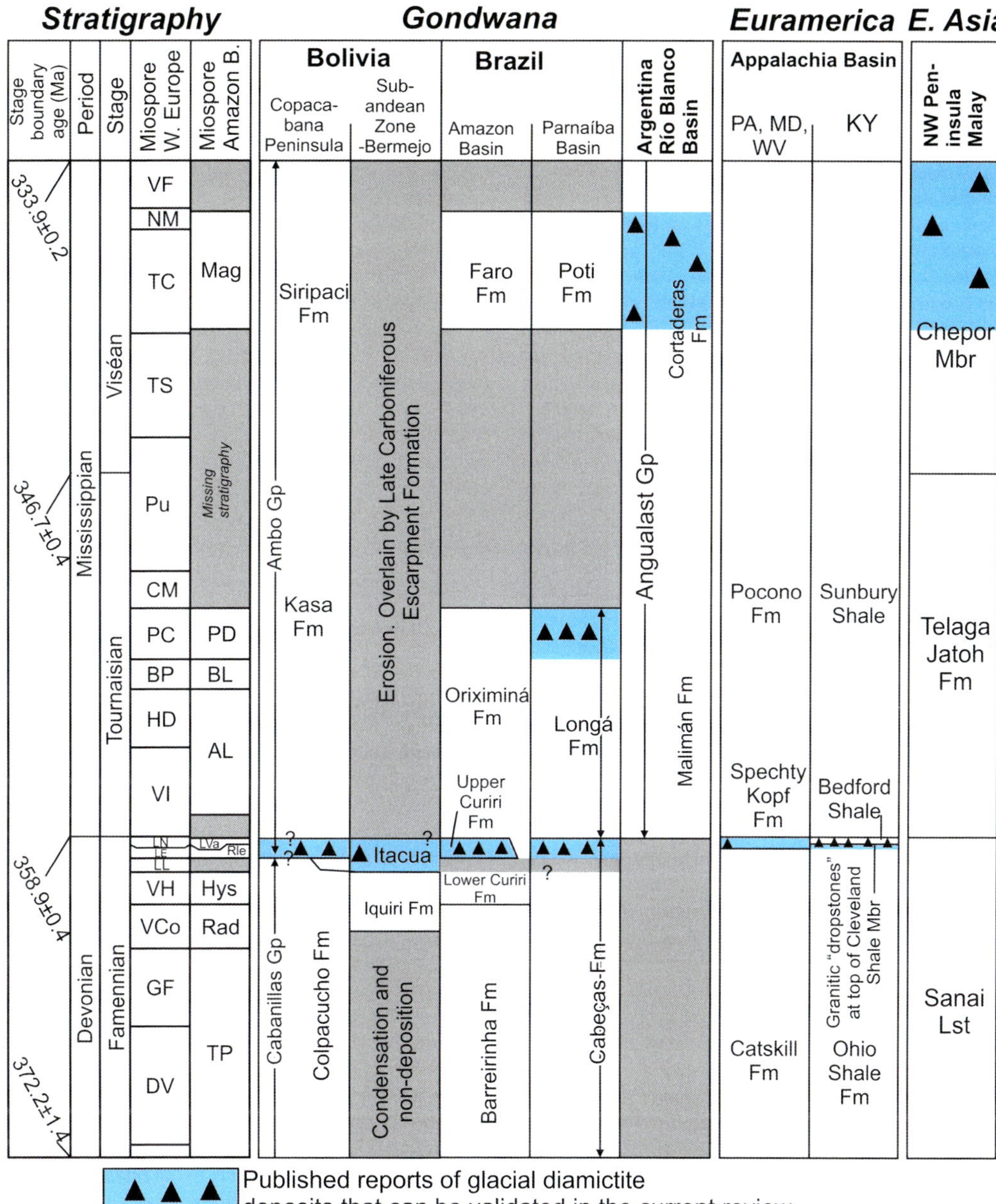

Fig. 8. Summary of published diamictite occurrences that are well constrained and are associated with detailed published sedimentological, stratigraphic and/or biostratigraphic data. PA, Pennsylvania; MD, Maryland; WV, West Virginia; KY, Kentucky. Fm, Formation; Gp, Group; Mbr, Member; Lst, limestone.

retreat, which would give a maximum estimate of glacial duration.

The validation of mid-Tournaisian and Visean glaciations in Brazil and Bolivia is difficult. Mid-Tournaisian glaciation is supported by only one detailed published study (Playford *et al.* 2012). Visean glaciation in Brazil is not, as yet, supported by any published measured sections or palynological descriptions. The reported diamictites in the Mississippian Kasa Formation in Bolivia are interesting but need to be accurately dated. Further work constraining Mississippian diamictite occurrences would test the hypothesis of Yao *et al.* (2015) of established, permanent continental glaciation from the mid-Tournaisian onwards.

Far-field integration

Palynological methods have shown great potential for the dating and global correlation of uppermost Devonian–lower Carboniferous sediments, with assemblages characterized by the miospore *R. lepidophyta* and the acritarch *U. saharicum*. However, the distribution, evolution and extinction of palynological groups over the latest Famennian glaciation require further study (di Pasquo & Azcuy 1997; Le Hérissé *et al.* 2000; di Pasquo 2007*c*; Mullins & Servais 2008). It is not currently known whether palynological extinctions are synchronous with the main glacial pulse or with the initial post-glacial transgression. Owing to problems with the D–C boundaries as defined in both Europe and the Amazon Basin (i.e. the problems with the GSSP in the former and erosion in the latter), the timing of these extinctions could provide an additional D–C boundary proxy. By defining the relationship between glaciation and palynological extinctions, it will then be possible to determine how these compare to wider palaeoclimatic and glacio-eustatic changes.

An additional correlation tool would be to recognize the wider climatic, oceanographic, biotic and isotopic events to which the glaciations were likely to have been related. The latest Famennian and mid-Tournaisian glacial events, in particular, are associated with positive carbon and nitrogen isotope excursions, indicating that glaciation is reflecting and/or driving wider changes in the global Earth system (Saltzman 2002; Kaiser *et al.* 2007, 2008).

There are several potential triggering mechanisms for these events (see Kaiser *et al.* 2015): one being the long-term decline in atmospheric CO_2, possibly related to the expansion of terrestrial vegetation (Algeo *et al.* 1995; Berner 2006). These large-scale processes may have controlled the long-term greenhouse to icehouse transition, but do not adequately explain discrete glaciation events or positive carbon isotope excursions, which occurred on much shorter timescales. The trigger for global organic carbon burial at the Hangenberg Crisis is debatable, but appears to have been geographically wide ranging, evidenced by anoxic conditions both in continental basins and open-ocean environments (Kaiser *et al.* 2007, 2011; Carmichael *et al.* 2016). The known geographical extent of anoxia and isotopic excursions in the mid-Tournaisian is increasing, with evidence in both Euramerica and Gondwana (Siegmund *et al.* 2002; Kaiser 2005; Kaiser *et al.* 2011, 2013; Le Yao *et al.* 2015). The coincidence of global environmental perturbations with evidence for glaciation at the Hangenberg Crisis and mid-Tournaisian Lower Alum Shale suggests similar global triggering mechanisms between these two events (see Kaiser *et al.* 2015).

Any glacial–interglacial cycles within each glacial episode would be orbitally controlled and their correlation could provide additional stratigraphic constraint, as based on the concepts of sequence stratigraphy and cyclostratigraphy. It remains necessary to determine whether latest Famennian glaciation was associated with a single glacial event lasting 100 kyr or multiple stacked glacial advances of an estimated 1–3 myr in duration (Wicander *et al.* 2011; Streel *et al.* 2013). The mid-Tournaisian event appears to be correlated with one regressive event, whereas the Visean sequences are associated with multiple stacked sequence boundaries, which suggests that they may be a composite of many glacial–interglacial cycles. Future research will need to be driven by high-resolution multi-proxy stratigraphic studies in order to answer these questions.

Conclusions

- This review highlights uncertainty in the known stratigraphic and geographical extent of glaciation in the latest Devonian and Mississippian owing to a relative lack of detailed published glacial sections in which sedimentological and biostratigraphic data are integrated.
- This paucity of data leads to uncertainty between direct near-field records and the wider far-field effect on glacioeustasy, mass extinction and carbon isotope stratigraphy.
- The latest Famennian–Mississippian represents a long-term greenhouse to icehouse transition, probably controlled by the long-term drawdown of atmospheric CO_2.
- Superimposed on this transition are short and discrete glacial events that can be correlated to regressive intervals, mass extinctions, positive carbon isotope excursions and global marine anoxia. These near-globally recognized events are the Hangenberg Crisis in the latest Devonian and the Lower Alum Shale Event in the mid-Tournaisian.
- Palynological analyses can be employed constructively to date near-field deposits, and to correlate the far-field effects globally as a result of cosmopolitan plant miospores recognized in Gondwana and Euramerica. The most widely recognized miospore during the latest Famennian glaciation event is *R. lepidophyta*, which in South America is typically associated with the marine acritarch *U. saharicum*.
- Future work is needed to further constrain near-field glacial records. This requires multi-proxy

and high-resolution stratigraphic studies in which sedimentology, palynology and geochemical techniques are synergistically combined.

The preparation of this manuscript was undertaken during the PhD studies of the corresponding author, which was financially supported by the Engineering and Physical Sciences Research Council (EPSRC) and the Graduate School of the National Oceanography Centre, Southampton. R. Thomas Becker, Maurice Streel and Mercedes di Pasquo are gratefully acknowledged for their comments, corrections and additional references, which made for a much improved manuscript.

References

ALGEO, T.J. & SCHECKLER, S.E. 1998. Terrestrial–marine teleconnections in the Devonian: links between the evolution of land plants, weathering processes, and marine anoxic events. *Philosophical Transactions of the Royal Society B*, **353**, 113–130, http://doi.org/10.1098/rstb.1998.0195

ALGEO, T.J., BERNER, R.A., MAYNARD, J.B. & SCHECKLER, S.S. 1995. Late Devonian oceanic anoxic events and biotic crises: 'rooted' in the evolution of vascular land plants? *GSA Today*, **5**, 64–66.

ALGEO, T.J., MEYERS, P.A., ROBINSON, R.S., ROWE, H. & JIANG, G.Q. 2014. Icehouse–greenhouse variations in marine denitrification. *Biogeosciences*, **11**, 1273–1295.

ALMOND, J., MARSHALL, J. & EVANS, F. 2002. Latest Devonian and earliest Carboniferous glacial events in South Africa. *In*: *Abstract Volume of the 16th IAS International Sedimentological Congress, Rand Afrikaans University, Johannesburg, South Africa*. International Association of Sedimentologists, Gent, 11–12.

ANDRADE, S.M. 1972. *Geologia do Sudeste de Itacajá, Bacia do Parnaíba (Estado de Goiás)*. PhD dissertation, Universidade de São Paulo.

BARNES, J.B., EHLERS, T.A., INSEI, N., MCQUARRIE, N. & POULSEN, C.J. 2012. Linking orography, climate, and exhumation across the central Andes. *Geology*, **40**, 1135–1138, http://doi.org/10.1130/G33229.1

BECKER, R.T. 1993. Analysis of ammonoid palaeobiogeography in relation to the global Hangenberg (terminal Devonian) and Lower Alum Shale (middle Tournasian) events. *Annales de la Société Géologique de Belgique*, **115**, 459–473.

BECKER, R.T., GRADSTEIN, F.M. & HAMMER, O. 2012. The Devonian Period. *In*: GRADSTEIN, F.M., OGG, J.G., SCHMITZ, M.D. & OGG, G.M. (eds) *The Geological Time Scale 2012*. Elsevier, Amsterdam, 559–602.

BERNER, R.A. 2006. GEOCARBSULF: a combined model for Phanerozoic atmospheric O_2 and CO_2. *Geochimica et Cosmochimica Acta*, **70**, 5653–5664, http://doi.org/10.1016/j.gca.2005.11.032

BLESS, M.J.M., BECKER, R.T., HIGGS, K.T., PAPROTH, E. & STREEL, M. 1992. Eustatic cycles around the Devonian–Carboniferous boundary and the sedimentary and fossil record in Sauerland (Federal Republic of Germany). *Annales de la Societe Geologique de Belgique*, **115**, 689–702.

BRAND, U., LEGRAND-BLAIN, M. & STREEL, M. 2004. Biochemostratigraphy of the Devonian–Carboniferous boundary global stratotype section and point, Griotte Formation, La Serre, Montagne Noire, France. *Palaeogeography, Palaeoclimatology, Palaeoecology*, **205**, 337–357, http://doi.org/10.1016/j.palaeo.2003.12.015

BREZINSKI, D.K. & CECIL, C.B. 2015. Late Devonian climatic change and resultant glacigenic facies of western Maryland. *In*: BREZINSKI, D.K., HALKA, J.P. & ORTT, R.A., JR, (eds) *Tripping from the Fall Line: Field Excursions for the GSA Annual Meeting, Baltimore, 2015*. Geological Society of America, Field Guides, **40**, 85–108, http://doi.org/10.1130/2015.0040(05);

BREZINSKI, D.K., CECIL, C.B., SKEMA, V.W. & STAMM, R. 2008. Late Devonian glacial deposits from the eastern United States signal and end of the mid-Paleozoic warm period. *Palaeogeography, Palaeoclimatology, Palaeoecology*, **268**, 143–151, http://doi.org/10.1016/j.palaeo.2008.03.042

BREZINSKI, D.K., CECIL, C.B. & SKEMA, V.W. 2010. Late Devonian glacigenic and associated facies from the central Appalachian Basin, eastern United States. *Geological Society of America Bulletin*, **122**, 265–281, http://doi.org/10.1130/B26556.1

BRITTAIN, J.M. & HIGGS, K.T. 2007. The early Carboniferous *Spelaeotriletes balteatus – S. pretiosus* Miospore Complex: defining the base of the *Spelaeotriletes pretiosus – Raistrickia clavata* (PC) Miospore Zone. *Comunicacoes Geologicas*, **94**, 109–123.

BUGGISCH, W., JOACHIMSKI, M.M., SEVASTOPULO, G. & MORROW, J.R. 2008. Mississippian $\delta^{13}C_{carb}$ and conodont apatite $\delta^{18}O$ records – their relation to the Late Palaeozoic Glaciation. *Palaeogeography, Palaeoclimatology, Palaeoecology*, **268**, 273–292, http://doi.org/10.1016/j.palaeo.2008.03.043

CAPLAN, M.L. & BUSTIN, R.M. 1999. Devonian–Carboniferous Hangenberg mass extinction event, widespread organic-rich mudrock and anoxia: causes and consequences. *Palaeogeography, Palaeoclimatology, Palaeoecology*, **148**, 187–207, http://doi.org/10.1016/S0031-0182(98)00218-1

CAPUTO, M.V. 1985. Late Devonian glaciations in South America. *Palaeogeography, Palaeoclimatology, Palaeoecology*, **51**, 291–317, http://doi.org/10.1016/0031-0182(85)90090-2

CAPUTO, M.V. & SILVA, O.B. 1990. Sedimentacao e tectonico da Bacia do Solimoes. *In*: RAJA GABAGLIA, G.P. & SILVA, O.B. (eds) *Origem e evolução de bacias sedimentares*. CENPES, Rio de Janeiro, 169–192.

CAPUTO, M.V., MELO, J.H.G., STREEL, M. & ISBELL, J.L. 2008. Late Devonian and Early Carboniferous glacial records of South America. *In*: FIELDING, C.R., FRANK, T.D. & ISBELL, J.L. (eds) *Resolving the Late Paleozoic Ice Age in Time and Space*. Geological Society of America, Special Papers, **441**, 161–173.

CARLOTTO, V., DÍAZ-MARTÍNEZ, E., CERPA, L., ARISPE, O. & CÁRDENAS, J. 2004. Late Devonian glaciation in the northern Central Andes: new evidence from southeast Peru. *In*: *32nd International Geological Congress, Florence*, 20–28 August 2004, Abstracts Volume: Part 2, abstract 205-12, 96.

CARMICHAEL, S.K., WATERS, J.A. *ET AL*. 2016. Climate instability and tipping points in the Late Devonian: detection of the Hangenberg Event in an open oceanic island arc in the Central Asian Orogenic Belt. *Gondwana Research*, **32**, 213–231. http://doi.org/10.1016/j.gr.2015.02.009

Carozzi, A.V. 1979. Petroleum geology in the Paleozoic clastics of the middle Amazon Basin, Brazil. *Journal of Petroleum Geology*, **2**, 55–74.

Carozzi, A.V. 1980. Tectonic control and petroleum geology of the Paleozoic clastics of the Maranhao Basin, Brazil. *Journal of Petroleum Geology*, **2**, 239–410.

Cecil, C.B., Brezinski, D.K. & Dulong, F. 2004. The Paleozoic record of changes in global climate and sea level: Central Appalachian Basin. *In*: Southworth, S. & Burton, W. (eds) *Geology of the National Capital Region – Field Trip Guidebook*. United States Geological Survey, Circular, **1264**, 77–135.

Censier, C., Henry, B., Horen, H. & Lang, J. 1985. Âde dévono-carbonifère de la formation glaciaire de la Mambéré (Republique Centrafricaine): arguments paléomagnetiques et géologiques. *Bulletin de la Société Géologique de France*, **166**, 23–35.

Cerpa, L., Carlotto, V., Arispe, O., Díaz-Martínez, E., Cárdenas, J., Valderrama, P. & Bermúdez, O. 2004. Formación Ccatca (Devónico superior): sedimentación glaciomarina en la Cordillera Oriental de la región de Cusco. *XII Congreso Peruano de Geología, Resúmenes Extendidos. Sociedad Geologíca del Perú, Publicación Especial*, **6**, 424–427.

Clack, J.A. 2007. Devonian climate change, breathing, and the origin of the tetrapod stem group. *Integrative and Comparative Biology*, **47**, 510–523, http://doi.org/10.1093/icb/icm055

Clayton, G., Coquel, R., Doubinger, J., Gueinn, K.J., Loboziak, S., Owens, B. & Streel, M. 1977. Carboniferous miospores from Western Europe: illustrations and zonation. *Mededelingen Rijks Geologische Dienst*, **29**, 1–71.

Coleman, U. & Clayton, G. 1987. Palynostratigraphy and palynofacies of the uppermost Devonian and Lower Mississippian of eastern Kentucky (U.S.A.), and correlation with western Europe. *Courier Forschungsinstitut Senckenberg*, **98**, 75–93.

Copper, P. 2002. Reef development at the Frasnian/Famennian mass extinction boundary. *Palaeogeography, Palaeoclimatology, Palaeoecology*, **181**, 27–65, http://doi.org/10.1016/S0031-0182(01)00472-2

Corradini, C., Spalletti, C., Kaiser, S.I. & Matyja, H. 2013. Overview of conodonts across the Devonian/Carboniferous boundary. *In*: Albanesi, G.L. & Ortega, G. (eds) *Conodonts from the Andes*. Proceedings of the 3rd International Conodont Symposium & Regional Field Meeting of the IGCP Project 591. Asociación Paleontológica Argentina, Buenos Aires, 13–16.

Coquel, R. & Moreau-Benoit, A. 1986. Les spores des Series struniennes et tournaisiennes de Libye occidentale. *Revue de Micropaléontologie*, **29**, 17–43.

Cunha, P.R.C., Gonzaga, F.G., Coutinho, L.F.C. & JFeijo, F.J. 1994. Bacio do Amazonas. *Boletim de Geosciencias da Petrobras*, **8**, 47–55.

Cunha, P.R.C., Melo, J.H.G. & Silva, O.B. 2007. Bacia do Amazonas. *Boletim de Geociências da Petrobras*, **15**, 227–251.

Daemon, R.F. 1974. Palinomorfos-guias do Devoniano Superior e Carbonífero Inferior das bacias do Amazonas e Parnaíba. *Anais da Academia Brasileira de Ciências*, **46**, 549–587.

Daemon, R.F. & Contreiras, C.J.A. 1971. Zoneamento palinológico da Bacia do Amazonas. *Anais do XXV Congresso Brasileiro de Geologia, São Paulo*, **3**, 79–88.

Davydov, V.I., Korn, D. & Schmitz, M.D. 2012. The Carboniferous Period. *In*: Gradstein, F.M., Ogg, G.M., Schmitz, M.D. & Ogg, G.M. (eds) *The Geological Timescale 2012*. Elsevier, Amsterdam, 603–652.

de Vleeschouwer, D.D., Rakocinski, M., Racki, G., Bond, D.P.G., Sobien, K. & Claeys, P. 2013. The astronomical rhythm of Late-Devonian climate change (Kowala section, Holy Cross Mountains, Poland). *Earth and Planetary Science Letters*, **365**, 25–37, http://doi.org/10.1016/j.epsl.2013.01.016

di Pasquo, M. 2007*a*. Asociaciones palinológicas en las formaciones Los Monos (Devóico) e Itacua (Carbonifero Inferior) en Balapuca (Cuencia Tarija), sur de Bolivia. Parte 1. Formación Los Monos. *Revista Geológica de Chile*, **34**, 97–137.

di Pasquo, M. 2007*b*. Asociaciones palinológicas en las formaciones Los Monos (Devóico) e Itacua (Carbonifero Inferior) en Balapuca (Cuencia Tarija), sur de Bolivia. Parte 2. Asociaciones de la Formación Itacua e interpretació estratigráfica y cronología de las formaciones Los Monos e Itacua. *Revista Geológica de Chile*, **34**, 163–198.

di Pasquo, M. 2007*c*. State of the art of the Devonian palynological records in northern Argentina, southern Bolivia and northwestern Paraguay. *In*: Aceñolaza, G.F., Vergel, M., Peralta, S. & Herbst, R. (eds) *Field Meeting of the IGCP 499-UNESCO 'Devonian Land–Sea Interaction: Evolution of Ecosystems and Climate (DEVEC), 2007*, San Juan, 70–73.

di Pasquo, M. 2015. First record of *Lagenicula mixta* (Winslow) Wellman *et al.* in Bolivia: biostratigraphic and palaeogeographic significance. *In*: *Libro de Résumenes XVI Simposio Argentino de Paleobotánica y Palinología*, 26–29 May 2015, La Plata, 56.

di Pasquo, M. & Azcuy, C.L. 1997. Palinomorfos retrabajados en el Carbonífero Tardío de la Cuenca Tarija (Argentina) y su aplicación a la datación en eventos diastróficos. *Geociências*, **2**, 28–42.

di Pasquo, M., Wood, G.M., Isaacson, P. & Grader, G. 2015. Palynost ratigraphic reevaluation of the Manuripi-X1 (1541–1150 m interval), Madre de Dios basin, northern Bolivia: recycled Devonian species and their implication for the timing and duration of Gondwanan glaciation. *In*: *Libro de Résumenes XVI Simposio Argentino de Paleobotánica y Palinología*, 26–29 May 2015, La Plata, 30.

Díaz Martínez, E. 1991. Litoestratigrafia del Carbonifero del Altiplano de Bolivia. *Revista Tecnica de YPFB*, **12**, 295–302.

Díaz Martínez, E. 1993. The Carboniferous sequence of the northern Altiplano of Bolivia: from glacial-marine to carbonate deposition. *Comptes Rendus XII ICC-P*, **2**, 203–222.

Díaz Martínez, E. 1996. Sintesis estratigrafica y geodynamica del Carbonifero de Bolivia. *In*: *Memorias del*

XII Congreso Geologico de Bolivia, 10–13 October 1996, Tarija, Bolivia, 355–367.

Díaz-Martínez, E. 2004. La glaciacion del Devonico superior en Sudamerica: estado del conocimiento y perspectivas. *In*: *XII Congreso Peruano de Geologia. Resumenes Extendidos*. Sociedad Geologica del Peru, Lima, 440–443.

Díaz Martínez, E. & Isaacson, P.E. 1994. Late Devonian glacially-influenced marine sedimentation in western Gondwana: the Cumana Formation, Altiplano, Bolivia. *In*: Embry, A.F., Beauchamp, B. & Glass, D.J. (eds) *Pangea: Global Environmens and Resources*. Canadian Society of Petroleum Geologists, Memoirs, **17**, 511–522.

Díaz-Martínez, E., Isaacson, P.E. & Sablock, P.E. 1993. Late Paleozoic latitudinal shift of Gondwana: stratigraphic/sedimentologic and biogeographic evidence from Bolivia. *Documents des Laboratoires Géologiques de Lyon*, **25**, 511–522.

Díaz-Martínez, E., Vavrdová, M., Bek, J. & Isaacson, P. 1999. Late Devonian (Famennian) Glaciation in Western Gondwana: evidence from the Central Andes. *Abhandlungen der Geologischen Bundesanstalt*, **54**, 213–237.

Domeier, M. & Torsvik, T.H. 2014. Plate tectonics in the late Paleozoic. *Geoscience Frontiers*, **5**, 303–350.

Eiras, J.F., Becker, C.R. *et al.* 1994. Bacia do Solimoes. *Boletim de Geociências da Petrobras*, **8**, 17–45.

Elrick, M., Berkyova, S., Klapper, G., Sharp, Z., Joachimski, M. & Fryda, J. 2009. Stratigraphic and oxygen isotope evidence for My-scale glaciation driving eustacy in the Early–Middle Devonian greenhouse world. *Palaeogeography, Palaeoclimatology, Palaeoecology*, **276**, 170–181, http://doi.org/10.1016/j.palaeo.2009.03.008

Ettensohn, F.R. 2008. Dropstone, glaciation and black shales: new inferences on black shale origins from the upper Ohio Shale in northeastern Kentucky. Paper presented at the PAPG Radisson Hotel, Greentree, PA, 17 April 2008.

Ettensohn, F.R., Lierman, R.T., Mason, C.E., Dennis, A.J. & Anderson, E.D. 2008. Kentucky dropstones and the case for Late Devonian Alpine Glaciation in the Appalachian Basin (USA) with implications for Appalachian tectonics and black-shale sedimentation. *Geological Society of America Abstracts with Programs*, **40**, 395.

Ettensohn, F.R., Lierman, R.T. & Mason, C.E. 2009. Upper Devonian–Lower Mississippian clastic rocks in northeastern Kentucky: evidence for Acadian alpine glaciation and models for source-rock and reservoir-rock development in the eastern United States. *In*: *American Institute of Professional Geologists – Kentucky Section Spring Field Trip*, 18 April 2009. American Institute of Professional Geologists – Kentucky Section, Lexington, KY, 1–63.

Evans, F.J. 1999. Palaeobiology of Early Carboniferous lacustrine biota of the Waaipoort Formation (Witteberg Group), South Africa. *Palaeontologia Africana*, **35**, 1–6.

Fielding, C.R., Frank, T.D. & Isbell, J.L. 2008. The late Paleozoic ice age – A review of current understanding and synthesis of global climate patterns. *In*: Fielding, C.R., Frank, T.D. & Isbell, J.L. (eds) *Resolving the Late Paleozoic Ice Age in Time and Space*. Geological Society of America, Special Papers, **441**, 343–354.

Filho, J.R.W., Eiras, J.F. & Vaz, P.T. 2007. Bacia do Solimões. *Boletim de Geociências da Petrobras*, **15**, 217–225.

Góes, A.M.O. & Feijó, F.J. 1994. Bacia do Parnaiba. *Boletim de Geociências da Petrobras*, **8**, 57–67.

González, F. 2009. Reappraisal of the organic-walled microphytoplankton genus *Maranhites*: morphology, excystment, and speciation. *Review of Palaeobotany and Palynology*, **154**, 6–21, http://doi.org/10.1016/j.revpalbo.2008.11.004

Grahn, Y. 2005. Devonian chitinozoan biozones of Western Gondwana. *Acta Geologica Polonica*, **55**, 211–277.

Grahn, Y., Melo, J.H.G. & Loboziak, S. 2006. Integrated Middle and Late Devonian miospore and chitinozoan zonation of the Parnaíba Basin, Brazil: an update. *Revista Brasileira de Paleontologia*, **9**, 283–294.

Gulbranson, E.L., Montañez, I.P., Schmitz, M.D., Limarino, C.O., Isbell, J.L., Marenssi, S. & Crowley, J.L. 2010. High-precision U–Pb calibration of Carboniferous glaciation and climate history, Paganzo Group, NW Argentina. *Geological Society of America Bulletin*, **122**, 1480–1498.

Hacquebard, P.A. 1957. Plant spores in coal from the Horton group (Mississippian) of Nova Scotia. *Micropaleontology*, **3**, 301–324.

Hallam, A. & Wignall, P.B. 1999. Mass extinctions and sea-level changes. *Earth-Science Reviews*, **48**, 217–250, http://doi.org/10.1016/S0012-8252(99)00055-0

Higgs, K.T. 1996. Taxonomic and systematic study of some Tournaisian (Hastrian) spores from Belgium. *Review of Palaeobotany and Palynology*, **93**, 269–297.

Higgs, K.T. & Streel, M. 1993. Palynological age for the lower part of the Hangenberg Shales in Sauerland, Germany. *Annales de la Societe Geologique de Belgique*, **116**, 243–247.

Higgs, K.T., Clayton, G. & Keegan, J.B. 1988. *Stratigraphy and Systematic Palynology of the Tournaisian Rocks of Ireland*. Geological Survey of Ireland, Special Papers, **7**.

House, M.R. 1996. An *Eocanites* fauna from the early Carboniferous of Chile and its palaeogeographic implications. *Annales de la Société Géologique de Belgique*, **117**, 95–105.

House, M.R. 2002. Strength, timing, setting and cause of mid-Palaeozoic extinctions. *Palaeogeography, Palaeoclimatology, Palaeoecology*, **181**, 5–25, http://doi.org/10.1016/S0031-0182(01)00471-0

Hyam, D.M., Marshall, J.E.A. & Sanderson, D.J. 1997. Carboniferous diamictite dykes in the Falklands Islands. *Journal of African Earth Sciences*, **25**, 505–517, http://doi.org/10.1016/S0899-5362(97)00122-X

Isaacson, P. 1977. Devonian stratigraphy and brachiopod paleontology of Bolivia. *Palaeontographica, Abt. A*, **155**, 133–192.

Isaacson, P.E., Palmer, B.A., Mamet, B.L., Cooke, J.C. & Sanders, D.E. 1995. Devonian–Carboniferous stratigraphy in the Madre de Dios Basin, Bolivia:

Pando X-1 and Manuripi X-1 wells. *In*: Tankard, A.J., Suarez, R. & Welsink, H.J. (eds) *Petroleum Basins of South America*. American Association of Petroleum Geologists Memoirs, **62**, 501–511.

Isaacson, P.E., Hladil, J., Jian-Wei, S., Kalvoda, J. & Grader, G. 1999. Late Devonian (Famennian) glaciation in South America and marine offlap on other continents. *Abhandlungen der Geologischen Bundesanstalt*, **54**, 239–257.

Isaacson, P., Díaz Martínez, E., Grader, G., Kalvoda, J., Babek, O. & Devuyst, F.X. 2008. Late Devonian–earliest Mississippian glaciation in Gondwanaland and its biogeographic consequences. *Palaeogeography, Palaeoclimatology, Palaeoecology*, **268**, 126–142, http://doi.org/10.1016/j.palaeo.2008.03.047

Isbell, J.L., Cole, D. & Catuneanu, O. 2008. Carboniferous–Permian glaciation in the main Karoo Basin, South Africa: stratigraphy, depositional controls, and glacial dynamics. *In*: Fielding, C.R., Frank, T.D. & Isbell, J.L. (eds) *Resolving the Late Paleozoic Ice Age in Time and Space*. Geological Society of America, Special Papers, **441**, 71–82.

Isbell, J.L., Miller, M.F., Wolfe, K.L. & Lenaker, P.A. 2003. Timing of late Paleozoic glaciation in Gondwana: was glaciation responsible for the development of northern hemisphere cyclothems? *In*: Chan, M.A. & Archer, A.W. (eds) *Extreme Depositional Environments: Mega End Members in Geologic Time*. Geological Society of America, Special Papers, **370**, 5–24.

Jardiné, S., Combaz, A., Magloire, L., Peniguel, G. & Vachey, G. 1972. Acritarches du Silurien terminal et du Dévonien du Sahara Algérien. *In*: *Comptes Rendus, 7th Congrés International de Stratigraphie et de Géologie du Carbonifére, Krefield*, August 1971, Volume **1**, 295–311, pls 1–3.

Jardiné, S., Combaz, A., Magloire, L., Peniguel, G. & Vachey, G. 1974. Distribution stratigraphique des acritarches dans le Paleozoïque du Sahara Algérien. *Review of Palaeobotany and Palynology*, **18**, 99–129, http://doi.org/10.1016/0034-6667(74)90012-8

Joachimski, M., Breisig, S. *et al.* 2009. Devonian climate and reef evolution; Insights from oxygen isotopes in apatite. *Earth and Planetary Science Letters*, **284**, 599–609, http://doi.org/10.1016/j.epsl.2009.05.028

Jones, M.J. & Truswell, E.M. 1992. Late Carboniferous and Early Permian palynostratigraphy of the Joe Joe Group, southern Galilee Basin, Queensland, and implications for Gondwana stratigraphy. *Bureau of Mineral Resources, Geology and Geophysics*, **13**, 143–185.

Kaiser, S.I. 2005. *Mass Extinctions, Climatic and Oceanographic Changes at the Devonian/Carboniferous Boundary*. PhD thesis, Ruhr-Universität Bochum.

Kaiser, S.I. 2009. The Devonian/Carboniferous boundary stratotype section (La Serre, France) revisited. *Newsletters on Stratigraphy*, **43**, 195–205, http://doi.org/10.1127/0078-0421/2009/0043-0195

Kaiser, S.I. & Corradini, C. 2011. The early siphonodellids (Conodonta, Late Devonian–Early Carboniferous): overview and taxanomic state. *Neues Jahrbuch für Geologie und Paläontologie Abhandlungen*, **261**, 19–35.

Kaiser, S.I., Steuber, T., Becker, R.T. & Joachimski, M.M. 2006. Geochemical evidence for major environmental change at the Devonian–Carboniferous boundary in the Carnic Alps and the Rhenish Massif. *Palaeogeography, Palaeoclimatology, Palaeoecology*, **240**, 146–160, http://doi.org/10.1016/j.palaeo.2006.03.048

Kaiser, S.I., Becker, R.T. & Hassani, A.E. 2007. Middle to Late Famennian successions at Ain Jemaa (Moroccan Meseta) – implications for regional correlation, event stratigraphy and synsedimentary tectonics of NW Gondwana. *In*: Becker, R.T. & Kirchgasser, W.T. (eds) *Devonian Events and Correlation*. Geological Society, London, Special Publications, **278**, 237–260, http://doi.org/10.1144/SP278.11

Kaiser, S.I., Steuber, T. & Becker, R.T. 2008. Environmental change during the Late Famennian and Early Tournasian (Late Devonian–Early Carboniferous): implications from stable isotopes and conodont biofacies in Southern Europe. *Geological Journal*, **43**, 241–260, http://doi.org/10.1002/gj.1111

Kaiser, S.I., Becker, R.T., Steuber, T. & Aboussalam, S.Z. 2011. Climate controlled mass extinctions, facies, and sea-level changes around the Devonian–Carboniferous boundary in the eastern Anti-Atlas (SE Morocco). *Palaeogeography, Palaeoclimatology, Palaeoecology*, **310**, 340–364, http://doi.org/10.1016/j.palaeo.2011.07.026

Kaiser, S.I., Becker, R.T., Hartenfels, S. & Aboussalam, Z.S. 2013. Middle Famennian to middle Tournaisian stratigraphy at El Atrous (Amessoui syncline, southern Tafilalt). *Documents de l'Institut Scientifique, Rabat*, **27**, 77–87.

Kaiser, S.I., Aretz, M. & Becker, R.T. 2015. The global Hangenberg Crisis (Devonian–Carboniferous transition): review of a first-order mass extinction. *In*: Becker, R.T., Königshof, P. & Brett, C.E. (eds) *Devonian Climate, Sea Level and Evolutionary Events*. Geological Society, London, Special Publications, **423**. First published online November 11, 2015, http://doi.org/10.1144/SP423.9

Kammer, T. & Matchen, D.L. 2008. Evidence for eustasy at the Kinderhookian–Osagean (Mississippian) boundary in the United States: response to Tournaisian glaciation. *In*: Fielding, C.R., Frank, T.D. & Isbell, J.L. (eds) *Resolving the Late Paleozoic Ice Age in Time and Space*. Geological Society of America, Special Papers, **441**, 261–274.

Klett, T.R. 2000. *Total Petroleum Systems of the Trias/Ghadames Province, Algeria, Tunisia, and Libya – The Tanezzuft–Oued Mya, Tanezzuft–Melrhir, and Tanezzuft–Ghadames*. United States Geological Survey, Bulletin, **2202-C**.

Kumpan, T., Babek, O., Kalvoda, J., Fryda, J. & Grygar, T.M. 2013. A high-resolution, multiproxy stratigraphic analysis of the Devonian–Carboniferous boundary sections in the Moravian Karst (Czech Republic) and a correlation with the Carnic Alps (Austria). *Geological Magazine*, **151**, 201–215, http://doi.org/10.1017/S0016756812001057

Kumpan, T., Bábek, O., Kalvoda, J., Grygar, T.M. & Frýda, J. 2014. Sea-level and environmental changes around the Devonian–Carboniferous boundary in the Namur–Dinant Basin (S Belgium, NE France): amultiproxy stratigraphic analysis of carbonate ramp archives and its use in regional and interregional correlations.

Sedimentary Geology, **311**, 43–59, http://doi.org/10.1016/j.sedgeo.2014.06.007

Lang, J., Yahava, M., Humet, O.E., Besombres, J.C. & Cazoulate, M. 1991. Depôts glaciares du Carbonifere inferieur a l'ouest de l'Air (Niger). *Geologische Rundschau*, **80**, 611–622.

Le Hérissé, A., Servais, T. & Wicander, R. 2000. Devonian acritarchs and related forms. *Courier Forschungsinstitut Senckenberg*, **220**, 195–205.

Lierman, R.T. 2007. Upper Devonian glaciation in the Ohio Shale of east-central Kentucky. *Geological Society of America Abstracts with Programs*, **39**, 70.

Lierman, R.T. & Mason, C.E. 2007. Upper Devonian glaciation in the Ohio Shale of east-central Kentucky. *Geological Society of America Abstracts with Programs*, **39**, 70.

Limarino, C.O., Cesari, S.N., Spalletti, L.A., Taboada, A.C., Isbell, J.L., Geuna, S. & Gulbranson, E.L. 2014. A paleoclimatic review of southern South America during the late Paleozoic: a record from icehouse to extreme greenhouse conditions. *Gondwana Research*, **25**, 1396–1421, http://doi.org/10.1016/j.gr.2012.12.022

Lobato, G. 2010. *Estratigrafia do intervalo neodevoniano/mississipiano da Bacia do Parnaíba em testemunhos de sondagem*. MSc dissertation, Universidade Federal do Rio de Janeiro.

Loboziak, S. & Melo, J.H.G. 2002. Devonian miospore successions of Western Gondwana: update and correlation with Southern Euramerican miospore zones. *Review of Palaeobotany and Palynology*, **121**, 133–148, http://doi.org/10.1016/S0034-6667(01)00098-7

Loboziak, S., Clayton, G. & Owens, B. 1986. *Aratrisporites saharaensis* sp. nov. A characteristic Lower Carboniferous miospore species of North Africa. *Geobios*, **19**, 497–503.

Loboziak, S., Streel, M., Caputo, M.V. & Melo, J.H.G. 1992. Middle Devonian to lower Carboniferous miospore stratigraphy in the central Parnaiba Basin (Brazil). *Annales de la Societe Geologique de Belgique*, **115**, 215–226.

Loboziak, S., Melo, J.H.G., Barrilari, I.M.R. & Streel, M. 1994*a*. *Devonian–Dinantian miospore biostratigraphy of the Solimões and Parnaíba Basins (with considerations on the Devonian of the Paraná Basin)*. Petrobras Internal Report. Petrobras/Cenpes, Rio de Janeiro (2 vols).

Loboziak, S., Melo, J.H.G., Quadros, L.P., Daemon, R.F. & Barrilari, I.M.R. 1994*b*. Biocronoestratigrafi a dos palinomorfos do Devoniano Médio–Carbonífero Inferior das bacias do Solimões e Parnaíba: Estado da arte. *In*: *Petrobras, Sintex II – 2 Seminário de Interpretação Exploratória: Rio de Janeiro, Atas*, 51–56.

Loboziak, S., Melo, J.H.G., Quadros, L.P., Barrilari, I.M.R. & Streel, M. 1995*a*. Biocronostratigrafia de miosporos do Devoniano medoi-Carbonifero inferior das bacias do Solimões e Parnaíba (estado da arte). *Anais da Academia Brasileira de Ciências*, **67**, 394–395.

Loboziak, S., Melo, J.H.G., Steemans, P. & Barrilari, I.M.R. 1995*b*. Miospore evidence for pre-Emsian and latest Famennian sedimentation in the Devonian of the Paraná Basin, south Brazil. *Anais da Academia Brasileira de Ciências*, **67**, 391–392.

Loboziak, S., Melo, J.H.G., Quadros, L.P. & Streel, M. 1997. Palynological evaluation of the Famennian *Protosalvinia* (*Foerstia*) Zone in the Amazon Basin, northern Brazil: a preliminary study. *Review of Palaeobotany and Palynology*, **96**, 31–45.

Loboziak, S., Melo, J.H.G., Playford, G. & Streel, M. 1999. The *Indotriradites dolianitii* Morphon, a distinctive group of miospore species from the Lower Carboniferous of Gondwana. *Review of Palaeobotany and Palynology*, **107**, 17–22, http://doi.org/10.1016/S0034-6667(99)00016-0

Loboziak, S., Caputo, M.V. & Melo, J.H.G. 2000*a*. Middle Devonian-Tournaisian miospore biostratigraphy in the southwestern outcrop belt of the Parnaiba basin, north-central Brazil. *Revue de Micropaleontologie*, **43**, 301–318, http://doi.org/10.1016/S0035-1598(00)90154-5

Loboziak, S., Playford, G. & Melo, J.H.G. 2000*b*. *Radiizonates arcuatus*, a distinctive new miospore species from the Lower Carboniferous of Western Gondwana. *Review of Palaeobotany and Palynology*, **109**, 271–277, http://doi.org/10.1016/S0034-6667(99)00059-7

Loboziak, S., Melo, J.H.G. & Streel, M. 2005. Devonian palynostratigraphy in western Gondwana. *In*: Koutsoukos, E.A.M. (ed.) *Applied Stratigraphy*. Topics in Geobiology, **23**. Springer, Dordrecht, 73–99.

Marshall, J.E.A. 2016. Palynological calibration of Devonian events at near-polar palaeolatitudes in the Falkland Islands, South America. *In*: Becker, R.T., Königshof, P. & Brett, C.E. (eds) *Devonian Climate, Sea Level and Evolutionary Events*. Geological Society, London, Special Publications, **423**. First published online June 7, 2016, http://doi.org/10.1144/SP423.13

Maziane, N., Higgs, K.T. & Streel, M. 1999. Revision of the late Famennian miospore zonation scheme in eastern Belgium. *Journal of Micropalaeontology*, **18**, 17–25, http://doi.org/10.1144/jm.18.1.17

McGhee, G.R. 2013. *When the Invasion of Land Failed: The Legacy of Devonian Extinctions*. Columbia University Press, New York.

McGhee, G.R., Sheehan, M.E., Bottjer, D.J. & Droser, M.L. 2012. Ecological ranking of Phanerozoic biodiversity crises: the Serpukhovian (early Carboniferous) crisis had a greater ecological impact than the end-Ordovician. *Geology*, **40**, 147–150, http://doi.org/10.1130/G32679.1

McGhee, G.R., Clapham, M.E., Sheehan, M.E., Bottjer, D.J. & Droser, M.L. 2013. A new ecological-severity ranking of major Phanerozoic biodiversity crises. *Palaeogeography, Palaeoclimatology, Palaeoecology*, **15**, 260–270, http://doi.org/10.1016/j.palaeo.2012.12.019

Melo, J.H.G. & Loboziak, S. 2000. Visean miospore biostratigraphy and correlation of the Poti Formation (Parnaba Basin, norther Brazil). *Review of Palaeobotany and Palynology*, **112**, 147–165, http://doi.org/10.1016/S0034-6667(00)00043-9

Melo, J.H.G. & Loboziak, S. 2003. Devonian–Early Carboniferous miospore biostratigraphy of the Amazon Basin, Northern Brazil. *Review of Palaeobotany and*

Palynology, **124**, 131–202, http://doi.org/10.1016/S0034-6667(02)00184-7

Melo, J.H.G. & Playford, G. 2012. Miospore palynology and biostratigraphy of Mississippian strata of the Amazona Basin, northern Brazil. Part Two. *AASP Contribution Series*, **47**, 93–201.

Meor, H.A.H., Aye-Ko, A., Becker, R.T., Noor, A.A.R., Tham, F.N., Azman, A.G. & Mustaffa, K.S. 2014. Stratigraphy and palaeoenvironmental evolution of the mid- to upper Palaeozoic succession in Northwest Peninsula Malaysia. *Journal of Asian Earth Sciences*, **83**, 60–79, http://doi.org/10.1016/j.jseaes.2014.01.016

Mii, H.-S., Grossman, E.L. & Yancey, T.E. 1999. Carboniferous isotope stratigraphies of North America: implications for Carboniferous paleoceanography and Mississippian glaciation. *Geological Society of America Bulletin*, **111**, 960–973, http://doi.org/10.1130/0016-7606(1999)111<0960:CISONA>2.3.CO;2

Milani, E.J., Melo, J.H.G., Souza, P.A., Fernandez, L.A. & França, A.B. 2007. Bacia do Paraná. *Boletim de Geociências da Petrobras*, **15**, 265–287.

Molyneux, S.G., Delabroye, A., Wicander, R. & Servais, T. 2013. Biogeography of early to mid Palaeozoic (Cambrian–Devonian) marine phytoplankton. *In*: Harper, D.A.T. & Servais, T. (eds) *Early Palaeozoic Biogeography and Palaeogeography*. Geological Society, London, Memoirs, **38**, 365–397, http://doi.org/10.1144/M38.23

Montañez, I.P. & Poulsen, C.J. 2013. The Late Paleozoic Ice Age: an evolving paradigm. *Annual Review of Earth and Planetary Sciences*, **41**, 629–656, http://doi.org/10.1146/annurev.earth.031 208.100118

Mullins, G.F. & Servais, T. 2008. The diversity of the Carboniferous phytoplankton. *Review of Palaeobotany and Palynology*, **149**, 29–49.

Naumova, S.N. 1953. Upper Devonian spore–pollen assemblage of the Russian Platform and their stratigraphic significance. *Akademiya Nauk SSSR Trudy Geologicheskogo Instituta*, **143**, 1–204 [in Russian].

Over, D.J., Lazar, R., Baird, G.C., Schieber, J. & Ettensohn, F.R. 2009. *Protosalvinia* Dawson and Associated Conodonts of the Upper Trachytera Zone, Famennian, Upper Devonian, in the Eastern United States. *Journal of Paleontology*, **83**, 70–79, http://doi.org/10.1666/08-058R.1

Oviedo Gomez, C. 1965. Estratigrafía de la Península de Copacabana Lago Titicaca – Departamento de La Paz. *Boletín del Instituto Boliviano del Petróleo*, **5**, 5–15.

Paproth, E. & Streel, M. (eds). 1984. The Devonian–Carboniferous Boundary. *Courier Forschungsinstitut Senckenberg*, **67**, 1–258.

Paproth, E., Feist, R. & Flajs, G. 1991. Decision on the Devonian–Carboniferous boundary stratotype. *Episodes*, **14**, 331–336.

Paris, F., Winchester-Seeto, T., Boumendjel, K. & Grahn, Y. 2000. Toward a global biozonation of Devonian chitinozoans. *Courier Forschungsinstitut Senckenberg*, **220**, 39–55.

Perez-Leyton, M. 1991. Miospores du Devonien Moyen et Superieur de la coupe de Bermejo-La Angostura (sud-est de la Bolivie). *Annales de la Société Géologique de Belgique*, **113**, 373–389.

Perez Loinaze, V.S. 2007. A Mississippian palynological biozone for Southern Gondwana. *Palynology*, **31**, 101–118.

Perez Loinaze, V.S., Limarino, C.O. & Cesari, S. 2010. Glacial events in Carboniferous sequences from Paganzo and Río Blanco Basins (Northwest Argentina): palynology and depositional setting. *Geologica Acta*, **8**, 399–418.

Phillips, T.L., Niklas, K.J. & Andrews, H.N. 1972. Morphology and vertical distribution of protosalvinia (Foerstia) from the new Albany shale (Upper Devonian). *Review of Palaeobotany and Palynology*, **14**, 171–196, http://doi.org/10.1016/0034-6667(72)90018-8

Playford, G. 1962. Lower Carboniferous microfloras of Spitsbergen, Part B. *Palaeontology*, **5**, 619–678.

Playford, G. 1964. Miospores from the Mississippian Horton Group, eastern Canada. *Geological Society of Canada Bulletin*, **107**, 1–47.

Playford, G. 1976. Plant microfossils from the Upper Devonian and Lower Carboniferous of the Canning Basin, Western Australia. *Palaeontographica, Abteilung B*, **158**, 1–71.

Playford, G. 1977. Lower to Middle Devonian acritarchs of the Moose River Basin, Ontario. *Energy, Mines and Resources Canada*, **279**, 1–96.

Playford, G. 1990. Australian Lower Carboniferous miospores relevant to extra-Gondwanan correlations: an evaluation. *In*: Brenckle, P.L. & Manger, W.L. (eds) *Intercontinental Correlation and Division of the Carboniferous System. Courier Forschungsinstitut Senckenberg*, **130**, 85–125.

Playford, G. & Helby, R. 1968. Spores from a Carboniferous section in the Hunter Valley, New South Wales. *Journal of the Geological Society of Australia*, **15**, 103–119, http://doi.org/10.1080/00167616808728684

Playford, G. & Melo, J.H.G. 2009. The Mississippian miospore *Neoraistrickia loganensis* (Winslow, 1962) Coleman and Clayton, 1987: morphological variation and stratigraphic and palaeogeographic distribution. *Revista Española de Micropaleontología*, **41**, 241–254.

Playford, G. & Melo, J.H.G. 2010. Morphological variation and distribution of the Tournaisian (Early Mississippian) miospore *Waltzispora lanzonii* Daemon 1974. *Neues Jahrbuch für Geologie und Paläontologie, Abhandlungen*, **256**, 183–193, http://doi.org/10.1127/0077-7749/2010/0043

Playford, G. & Melo, J.H.G. 2012. Miospore palynology and biostratigraphy of Mississippian strata of the Amazonas Basin, northern Brazil. Part One. *AASP Contribution Series*, **47**, 3–89.

Playford, G., Dino, R. & Marques-Toigo, M. 2001. The upper Paleozoic miospore genus *Spelaeotriletes* Neves and Owens, 1966, and constituent Gondwanan species. *Journal of South American Earth Sciences*, **14**, 593–608.

Playford, G., Borghi, L., Lobato, G. & Melo, J.H.G. 2012. Palynological dating and correlation of Early Mississippian (Tournaisian) diamictite sections, Parnaíba Basin, northeastern Brazil. *Revista Española de Micropaleontología*, **44**, 1–22.

Qie, W., Zhang, X., Du, Y. & Zhang, Y. 2011. Lower Carboniferous carbon isotope stratigraphy in South

China: implications for the Late Paleozoic glaciation. *Science China Earth Sciences*, **54**, 84–92.

Qie, W., Liu, J. *et al.* 2015. Local overprints on the global carbonate δ^{13}C signal in Devonian–Carboniferous boundary successions of South China. *Palaeogeography, Palaeoclimatology, Palaeoecology*, **418**, 290–303, http://doi.org/10.1016/j.palaeo.2014.11.022

Quadros, L.P. 1988. Zonamento bioestratigrafico do Paleozoico Inferior e Medio (Secao Marinha) da Bacia do Solimoes. *Boletim das Geociências, Petrobras*, **2**, 95–109.

Quijada, M., Riboulleau, A., Strother, P., Taylor, W., Mezzetti, A. & Versteegh, G.J.M. 2015. *Protosalvinia* revisited, new evidence for a land plant affinity. *Review of Palaeobotany and Palynology*. First published online 9 November 2015, http://doi.org/10.1016/j.revpalbo.2015.10.008

Riegel, W. 2008. The Late Palaeozoic phytoplankton blackout – artefact or evidence of global change? *Review of Palaeobotany and Palynology*, **148**, 73–90, http://doi.org/10.1016/j.revpalbo.2006.12.006

Rooney, A., Goodhue, R. & Clayton, G. 2015. Stable nitrogen isotope analysis of the Upper Devonian palynomorph *Tasmanites*. *Palaeogeography, Palaeoclimatology, Palaeoecology*, **429**, 13–21, http://doi.org/10.1016/j.palaeo.2015.04.010

Sallan, L.C. & Coates, M.I. 2010. End-Devonian extinction and a bottleneck in the early evolution of modern jawed vertebrates. *Proceedings of the National Academy of Sciences of the United States of America*, **107**, 10,131–10,135, http://doi.org/10.1073/pnas.0914000107

Saltzman, M.R. 2002. Carbon and oxygen isotope stratigraphy of the Lower Mississippian (Kinderhookian–lower Osagean), western United States: implications for seawater chemistry and glaciation. *Geological Society of America Bulletin*, **114**, 96–108.

Sandberg, C.A., Morrow, J.R. & Ziegler, W. 2002. Late Devonian sea-level changes, catastrophic events, and mass extinctions. *In*: Koeberl, C. & MacLeod, K.G. (eds) *Catastrophic Events and Mass Extinctions: Impacts and Beyond*. Geological Society of America, Special Papers, **356**, 473–487.

Sempere, T. 1995. Phanerozoic evolution of Bolivia and adjacent regions. *In*: Tankard, A.J., Suarez, R. & Welsink, H.J. (eds) *Petroleum basins of South America*. American Association of Petroleum Geologists, Memoirs, **62**, 207–230.

Siegmund, H., Trappe, J. & Oschmann, W. 2002. Sequence stratigraphic and genetic aspects of the Tournasian "Liegender Alaunschiefer" and adjacent beds. *International Journal of Earth Sciences*, **91**, 934–949, http://doi.org/10.1007/s00531-001-0252-9

Smith, L.B.J. & Read, J.F. 2000. Rapid onset of late Paleozoic glaciation on Gondwana: evidence from Upper Mississippian strata of the Midcontinent, United States. *Geology*, **28**, 279–282.

Streel, M. 1986. Miospore contribution to the Upper Famennian–Strunian event stratigraphy. *In*: Bless, M.J.M. & Streel, M. (eds) Late Devonian Events Around The Old Red Continent. *Annales de la Société Géologique de Belgique*, **109**, 75–92.

Streel, M. & Theron, J.H. 1999. The Devonian–Carboniferous boundary in South Africa and the age of the earliest episode of the Dwyka glaciation: new palynological result. *Episodes*, **22**, 41–44.

Streel, M., Higgs, K.T., Loboziak, S., Riegel, W. & Steemans, P. 1987. Spore stratigraphy and correlation with faunas and floras in the type marine Devonian of the Ardenne-Rhenish regions. *Review of Palaeobotany and Palynology*, **50**, 211–229, http://doi.org/10.1016/0034-6667(87)90001-7

Streel, M., Caputo, M.V., Loboziak, S. & Melo, J.H.G. 2000*a*. Late Frasnian–Famennian climates based on palynomorph analyses and the question of the Late Devonian glaciations. *Earth-Science Reviews*, **52**, 121–173, http://doi.org/10.1016/S0012-8252(00)00026-X

Streel, M., Caputo, M.V., Loboziak, S., Melo, J.H.G. & Thorez, J. 2000*b*. Palynology and sedimentology of laminites and tillites from the latest Famennian of the Parnaiba basin, Brazil. *Geologica Belgica*, **3**, 87–96.

Streel, M., Brice, D. & Mistiaen, B. 2006. Strunian. *Geologica Belgica*, **9**, 105–109.

Streel, M., Caputo, M.V., Melo, J.H.G. & Perez-Leyton, M. 2013. What do latest Famennian and Mississippian miospores from South American diamictites tell us? *Palaeobiodiversity and Palaeoenvironments*, **93**, 299–316, http://doi.org/10.1007/s12549-012-0109-1

Suárez-Soruco, R. 2000. Sierras Subandinas/Subandean Ranges. *In*: Suárez-Soruco, R. (ed.) Compendio de Geologia de Bolivia. *Revista Tecnica de Yacimientos Petroliferos Fiscales Bolivianos*, **18**(1–2), 77–100.

Suárez-Soruco, R. & López-Pugliessi, M. 1983. Formació Saipuru, nuevo nombre formacional para representar a los sedimentos superiores del Ciclo Cordillerano (Devónico Superior-Carbónico Inferior), Bolivia. *Revista Técnica de Yacimientos Petrolíferos Fiscales de Bolivia, La Paz*, **9**, 189–202.

Swart, R. 1982. *The Stratigraphy and Structure of the Kommadagga Subgroup and Contiguous Rocks*. MSc thesis, Rhodes University, Grahamstown.

Theron, J.H. 1993. The Devonian–Carboniferous boundary in South Africa. *Annales de la Societe Geologique de Belgique*, **116**, 291–300.

Trapp, E., Kaufman, B., Mezger, K., Korn, D. & Weyer, D. 2004. Numerical calibration of the Devonian–Carboniferous boundary: two new U–Pb isotope dilution–thermal ionization mass spectrometry single-zircon ages from Hasselbachtal. *Geology*, **32**, 857–860, http://doi.org/10.1130/G20644.1

Troth, I. 2006. *The Applications of Palynostratigraphy to the Devonian of Bolivia*. PhD thesis, University of Southampton.

Troth, I., Marshall, J.E.A., Racey, A. & Becker, R.T. 2011. Devonian sea-level change in Bolivia: a high palaeolatitude biostratigraphical calibration of the global sea-level curve. *Palaeogeography, Palaeoclimatology, Palaeoecology*, **304**, 3–20, http:// doi.org/10.1016/j.palaeo.2010.10.008

Utting, J. 1987. Palynology of the Lower Carboniferous Windsor group and Windsor-Canso Boundary beds of Nova Scotia, and their equivalents in Quebec, New Brunswick and Newfoundland. *Geological Survey of Canada Bulletin*, **374**, 1–93.

Vavrdová, M. & Isaacson, P.E. 1999. Late Famennian phytogeographic provincialism: evidence for a limited separation of Gondwana and Laurentia. *Abhandlungen der Geologischen Bundesanstalt*, **54**, 543–463.

Vaz, P.T., Mata Rezende, N.G.A., Filho, J.R.W. & Travassos, W.A.S. 2007. Bacia do Parnaíba. *Boletim de Geociências da Petrobras*, **15**, 253–263.

Walliser, O.H. 1984. Pleading for a natural D/C-boundary. *In*: Paproth, E. & Streel, M. (eds) The Devonian–Carboniferous Boundary. *Courier Forschungsinstitut Senckenberg*, **67**, 241–246.

Wicander, R. 1974. Upper Devonian–Lower Mississippian acritarchs and prasinophycean algae from Ohio, U.S.A. *Palaeontographica, Abteilung B*, **148**, 9–43.

Wicander, R., Clayton, G., Marshall, J.E.A., Troth, I. & Racey, A. 2011. Was the latest Devonian glaciation a multiple event? New palynological evidence from Bolivia. *Palaeogeography, Palaeoclimatology, Palaeoecology*, **305**, 75–83, http://doi.org/10.1016/j.palaeo.2011.02.016

Woodrow, D.L. & Richardson, J.B. 2006. What WAS Going on in the Late Devonian. AAPG Search and Discovery article 90059. *In*: *Official Program and Abstracts, Eastern Section, American Association of Petroleum Geologists 35th Annual Meeting*, 8–11 Ocober 2006, Buffalo, NY, http://www.searchanddiscovery.com/abstracts/html/2006/eastern/abstracts/woodrow.htm

Yao, L., Qie, W. *et al.* 2015. The TICE event: perturbation of carbon–nitrogen cycles during the mid-Tournaisian (Early Carboniferous) greenhouse–icehouse transition. *Chemical Geology*, **401**, 1–14.

Index

Page numbers in *italic* denote Figures. Page numbers in **bold** denote Tables.